AF506277

DUKE
MATHEMATICAL
JOURNAL

A special volume in honor of

Yuri Ivanovich Manin

Editorial Committee for the special volume

I. R. Shafarevitch   Phillip A. Griffiths

Nicholas Katz   Barry Mazur

Miles Reid, Chairman

Morris Weisfeld
Managing Editor

Volume 54
1987

Duke University Press

CONTENTS

Volume 54, 1987

DUKE MATHEMATICAL JOURNAL

A special volume in honor of

Yuri Ivanovich Manin

YURI IVANOVICH MANIN

Yuri Ivanovich Manin was born on February 16, 1937, in the city of Simferopol—the capital of Crimea. His father died in the war; his mother was a specialist in literature. Manin's mathematical capabilities became apparent early in his life: when he was still a schoolboy he wrote a paper in which he slightly improved an estimate by I. M. Vinogradov on the number of integer points on a sphere. In 1953 he entered Moscow University. This was a rare period in the history of Mekh-Mat (the mechanics-mathematics faculty), which has played a role in the mathematical life of the USSR roughly equivalent to that of the Ecole Normale in France. Anosov and Golod were in the same year as Manin, and one course below him were Arnol'd, Kirillov, Novikov and Tyurin. Such a blossoming of talent has not been repeated since. Manin immediately immersed himself in the atmosphere of intense creative work which is characteristic of Mekh-Mat. By his second year he was already an active participant in the seminar led by A. O. Gel'fond and I. R. Shafarevich. The goal of the seminar was to acquaint specialists in analytic number theory with the work of H. Hasse and A. Weil on the analogues of the Riemann hypotheses for algebraic varieties over a finite field. These results gave excellent estimates of trigonometric sums which were necessary to number theorists, to whom algebraic methods were unfamiliar and difficult to comprehend. The first published work of Manin appeared at this time: an elementary proof of Hasse's theorem—an analogue of the Riemann hypothesis for the ζ-function of an elliptic curve over a finite field.

In the same period Manin formed the interests which would define his work over the next fifteen years: algebraic number theory, encompassing algebraic geometry to an ever-growing extent. He carried out this work, after finishing his graduate studies, at the Steklov Mathematical Institute, which has played as great a role in the lives of many mathematicians as Mekh-Mat. If one looks at a list of the mathematicians who have worked there, one meets the names of most of the famous mathematicians in the USSR—in spite of the fact that only 60–70 mathematicians worked at the Institute at any one time. The main landmarks in this direction of Manin's work are: the proof of the functional analog of the Mordell Conjecture and the theory of Gauss-Manin connections (so named by Grothendieck), the arithmetic of cubic surfaces and varieties, the solution (along with his student V. A. Iskovskikh) of the Luroth problem for threefolds, and the creation of the p-adic theory of automorphic functions (see his reports at the International Congresses of Mathematicians at Moscow, Nice, Helsinki).

Beginning in the 1970's, Manin's interests shifted to the realm of mathematical physics, which was quite natural in view of the remarkable links which have been discovered between it and algebraic geometry. Here are some themes which characterize the variety of ideas and results which arise here: the integration of the Benney equation, the construction of instantons (in the famous four-author work "ADNM"), the study of homogeneous spaces in supergeometry. These ideas have also found application in the area of Manin's previous interests. For example, the study of the Polyakov measure of a (super)string led to a beautiful formula for the values of a Selberg zeta function of an algebraic curve at integer points. A large cycle of works has given rise to the intriguing conception of "new dimensions in mathematics."

Work with his students plays a large role in Manin's life. He is one of the most popular professors at Mekh-Mat. Each direction of his work attracts a new group of adherents, who continue to work in this area after Manin's own interests have moved onward.

Mathematics is not Manin's only interest. He has been interested in literature and linguistics for a long time, and not at all in the manner of a dilletante. In journals infinitely far removed from mathematics one may find Yu. I. Manin's review of a book (by V. V. Ivanov) or his article on a literary topic (see Revue des Études Slaves, Vol. 55, No. 3, Paris, 1983).

I. R. Shafarevich

CYCLES, CURVES AND VECTOR BUNDLES ON AN ALGEBRAIC SURFACE

A. N. TYURIN

To Yu. I. Manin on his 50th birthday

Introduction. Let X be a smooth algebraic surface. ξ an effective 0-cycle on X and $\mathscr{D}$ a divisor class. If the ideal sheaf J_ξ satisfies $h^1(J_\xi(\mathscr{D})) \neq 0$, then Serre duality and the Ext construction give a torsion-free sheaf $E(\xi, \mathscr{D})$ on X. Geometrical properties of ξ and $\mathscr{D}$ can be described in terms of those of $E(\xi, \mathscr{D})$ and vice versa. For example, if $\deg \xi = 1$ (or 2) and $h^1(\mathcal{O}_X(\mathscr{D})) = 0$, then $h^1(J_x(\mathscr{D})) > 0$ (or $h^1(J_{x,y}(\mathscr{D})) > 0$) means that x is a fixed point of the complete linear system $|\mathscr{D}|$ (respectively that the points x and y are not distinguished by the curves of $|\mathscr{D}|$), and so on. Using simple properties of $E(x, \mathscr{D})$, $E(x + y, \mathscr{D})$, one can get, for example, all the results on complete linear systems on a $K3$ surface, and can prove following I. Reider [8] that $|2K_x|$ has no fixed points on an algebraic surface of general type X with $K_X^2 \geqslant 5$, and so on. Conversely, by the description of fixed points of the complete linear system $|3K_X|$ on the Severi-Dolgachev surface $X = F_{2,3}$, S. Donaldson determined the structure of the moduli space of stable rank 2 vector bundles on $F_{2,3}$ with $c_1 = 0$, $c_2 = 1$.

In our earlier paper [10] we used this method to describe the variety of special cycles on a polarized $K3$ surface. In the present paper we apply this construction for the description of

(i) the structure of the moduli space of simple vector bundles on a $K3$ surface (§4, Ch. II);

(ii) the specific character of Brill-Noether theory for smooth curves on a $K3$ surface (R. Lazarsfeld's construction) (§5, Ch. II);

(iii) the special case of the infinitesimal Mumford-Harris conjecture (§6, Ch. II).

In Chapter I we collect together the simplest properties of the construction $(\xi, \mathscr{D}) \mapsto E(\xi, \mathscr{D})$ on a regular algebraic surface X.

We use the standard notation of [2] and [3].

Chapter I. The construction

§1. Cycles and divisor classes. On a smooth algebraic surface X which is regular $(h^1(\mathcal{O}_X) = 0)$, a 0-dimensional subscheme $\xi \subset X$ of length d is called a *cycle* of degree d on X. Each cycle is defined by the ideal sheaf $J_\xi \subset \mathcal{O}_X$ or by the

Received July 15, 1986.

structure sheaf $\mathcal{O}_\xi$ with the standard exact sequence

$$0 \to J_\xi \overset{i}{\to} \mathcal{O}_X \overset{\text{res}}{\longrightarrow} \mathcal{O}_\xi \to 0. \tag{1.1}$$

A cycle ξ_1 is called a *subcycle* of ξ if $J_\xi \subset J_{\xi_1}$.

The Hilbert scheme $\widetilde{S^d X}$ of all cycles of degree d on X is a smooth $2d$-dimensional variety (see, for example, [5]); the map $\xi \mapsto \operatorname{Supp} \xi$ from $\widetilde{S^d X}$ onto the dth symmetric product $S^d X$ is a birational morphism.

Let $\mathcal{D} \in \operatorname{Pic} X$ be a divisor class on X. Tensoring (1.1) by $\mathcal{O}_X(\mathcal{D})$ gives

$$0 \to J_\xi(\mathcal{D}) \overset{i}{\to} \mathcal{O}_X(\mathcal{D}) \overset{\text{res}}{\longrightarrow} \mathcal{O}_\xi(\mathcal{D}) \to 0, \tag{1.2}$$

and taking the cohomology exact sequence we get

$$\left.\begin{array}{l} 0 \to H^0\big(J_\xi(\mathcal{D})\big) \to H^0(\mathcal{O}_X(\mathcal{D})) \to \mathbf{C}^d \to H^1\big(J_\xi(\mathcal{D})\big) \to H^1(\mathcal{O}_X(\mathcal{D})) \to 0 \\ 0 \to H^2\big(J_\xi(\mathcal{D})\big) \to H^2(\mathcal{O}_X(\mathcal{D})) \to 0. \end{array}\right\} \tag{1.3}$$

Definition 1.1.   (1) The nonnegative integer

$$\delta(\mathcal{D}, \xi) = h^1\big(J_\xi(\mathcal{D})\big) - h^1(\mathcal{O}_X(\mathcal{D})) \tag{1.4}$$

is called the *index of $\mathcal{D}$-speciality* of ξ;

(2) if $\delta(\mathcal{D}, \xi) > 0$, then ξ is called a *$\mathcal{D}$-special* cycle. The exact sequence (1.3) gives

$$h^0\big(J_\xi(\mathcal{D})\big) = h^0(\mathcal{O}_X(\mathcal{D})) - d + \delta(\mathcal{D}, \xi). \tag{1.5}$$

The index of speciality for fixed $\mathcal{D} \in \operatorname{Pic} X$ gives the filtration $\widetilde{S^d X} \supset \widetilde{S^d_{\mathcal{D},1} X} \supset \cdots \supset \widetilde{S^d_{\mathcal{D},k} X} \supset \ldots$, where

$$\widetilde{S^d_{\mathcal{D},k} X} = \big\{ \xi \in S^d X \,|\, \delta(\mathcal{D}, \xi) \geqslant k \big\}; \tag{1.6}$$

and for fixed ξ it gives the function $\delta(\mathcal{D}, \xi)$ on $\operatorname{Pic} X$. By (1.5)

$$\delta(\mathcal{D}, \xi) \leqslant d, \tag{1.7}$$

and if $\delta(\mathcal{D}, \xi) \neq d$, then $|\mathcal{D}| \neq \varnothing$.

Let $\xi_1 \subset \xi$ be a subcycle. Then we have

$$\left.\begin{array}{l} 0 \to J_\xi(\mathcal{D}) \overset{i}{\to} J_{\xi_1}(\mathcal{D}) \to S \to 0, \\[4pt] \text{with } \operatorname{Supp} S \subset \operatorname{Supp} \xi \text{ and } h^0(S) = \deg \xi - \deg \xi_1 = d - d_1; \end{array}\right\} \tag{1.8}$$

and the cohomology exact sequence of (1.8) is

$$0 \to H^0\big(J_\xi(\mathcal{D})\big) \to H^0\big(J_{\xi_1}(\mathcal{D})\big) \to \mathbf{C}^{\,d-d_1}$$

$$\xrightarrow{\;j\;} H^1\big(J_\xi(\mathcal{D})\big) \xrightarrow{H^1(i)} H^1\big(J_{\xi_1}(\mathcal{D})\big) \to 0. \tag{1.9}$$

From this

$$\delta(\mathcal{D}, \xi) \geqslant \delta(\mathcal{D}, \xi_1).$$

Definition 1.2. A cycle ξ is *$\mathcal{D}$-stable* if

$$\delta(\mathcal{D}, \xi_1) < \delta(\mathcal{D}, \xi)$$

for each subcycle $\xi_1 \subset \xi$. Let $\xi \in \widetilde{S^d X}$ be some cycle and $h^1(J_\xi(\mathcal{D})) \neq 0$. Then by Serre duality (see [3], p. 109)

$$\mathrm{Ext}^1_{\mathcal{O}_X}\big(J_\xi(\mathcal{D}), K_X\big) = \big[\mathrm{Ext}^1_{\mathcal{O}_X}\big(\mathcal{O}_X, J_\xi(\mathcal{D})\big)\big]^* = H^1\big(J_\xi(\mathcal{D})\big)^*. \tag{1.10}$$

Thus

$$\mathrm{Ext}^1_{\mathcal{O}_X}\big(J_\xi(\mathcal{D}), H^1\big(J_\xi(\mathcal{D})\big) \otimes K_X\big) = \mathrm{Hom}\big(H^1\big(J_\xi(\mathcal{D})\big), H^1\big(J_\xi(\mathcal{D})\big)\big). \tag{1.11}$$

Consider the extension

$$0 \to H^1\big(J_\xi(\mathcal{D})\big) \otimes K_X \xrightarrow{\;i\;} E(\xi, \mathcal{D}) \to J_\xi(\mathcal{D}) \to 0 \tag{1.12}$$

with cocycle

$$\mathrm{id} \in \mathrm{Hom}\big(H^1\big(J_\xi(\mathcal{D})\big), H^1\big(J_\xi(\mathcal{D})\big)\big).$$

We need two general assertions of the theory of extensions:

PROPOSITION 1.1.   *Let X be a smooth variety and*

$$0 \to L_0 \to E \to L_1 \to 0$$

an extension with cocycle $h \in \mathrm{Ext}^1_{\mathcal{O}_X}(L_1, L_0)$. Let L_0 be a locally free sheaf and K_X the canonical class of X. Applying the functor $\mathrm{Hom}(L_0,\ ) \otimes K_X$ to the extension exact sequence gives the exact sequence

$$0 \to \mathrm{End}\, L_0 \otimes K_X \to \mathrm{Hom}(L_0, E) \otimes K_X \to \mathrm{Hom}(L_0, L_1) \otimes K_X \to 0.$$

Consider the coboundary homomorphism

$$H^{n-1}(\mathrm{Hom}(L_0, L_1) \otimes K_X) \overset{\delta}{\to} H^n(\mathrm{End}\, L_0 \otimes K_X)$$

and its Serre dual homomorphism

$$H^0(\mathrm{End}\, L_0) \overset{\delta^*}{\to} \mathrm{Ext}^1_{\mathcal{O}_X}(L_1, L_0).$$

Then $h = -\delta^*(\mathrm{id})$ where $\mathrm{id} \in H^0(\mathrm{End}\, L_0)$ is the identity homomorphism.

PROPOSITION 1.2. *Let X be a smooth variety and L_0, L_1, L_2 sheaves on X; suppose that for $i = 1, 2$ extensions*

$$0 \to L_0 \to E_i \to L_i \to 0$$

are given with cocycles $e_i \in \mathrm{Ext}^1_{\mathcal{O}_X}(L_i, L_0)$. Then for a homomorphism

$$\varphi\colon L_1 \to L_2,$$

there exists a commutative diagram

$$
\begin{array}{ccccccccc}
0 & \to & L_0 & \to & E_1 & \to & L_1 & \to & 0 \\
 & & \downarrow \mathrm{id} & & \downarrow \Phi & & \downarrow \varphi & & \\
0 & \to & L_0 & \to & E_2 & \to & L_2 & \to & 0
\end{array}
$$

if and only if the induced homomorphism

$$\tilde{\varphi}\colon \mathrm{Ext}^1_{\mathcal{O}_X}(L_2, L_0) \to \mathrm{Ext}^1_{\mathcal{O}_X}(L_1, L_0)$$

satisfies the equality

$$\tilde{\varphi}(e_2) = e_1.$$

Both propositions can be proved easily using general properties of extensions (see, for example, §1 of [1]). These assertions provide the simplest properties of $E(\xi, \mathcal{D})$ (1.12). First of all,

$$\left. \begin{array}{l} \mathrm{rk}\, E(\xi, \mathcal{D}) = h^1\big(J_\xi(\mathcal{D})\big) + 1, \; c_1\big(E(\xi, \mathcal{D}) \otimes K_X^*\big) = \mathcal{D} - K_X \\ \text{and } c_2\big(E(\xi, \mathcal{D}) \otimes K_X^*\big) = \deg \xi = d. \end{array} \right\} \quad (1.13)$$

LEMMA 1.1. (1) $h^0(E(\xi, \mathcal{D})) = h^1(J_\xi(\mathcal{D})) \cdot p_g + h^0(J_\xi(\mathcal{D}))$;
(2) $h^1(E(\xi, \mathcal{D})) = 0$;
(3) $h^2(E(\xi, \mathcal{D})) = h^2(\mathcal{O}_X(\mathcal{D}))$;
(4) *if* $h^0(J_\xi(\mathcal{D} - K_X)) = 0$, *then*

$$\mathrm{Ext}^0_{\mathcal{O}_X}(E(\xi, \mathcal{D}), E(\xi, \mathcal{D})) = 1 + h^1\big(J_\xi(\mathcal{D})\big) \cdot h^2(\mathcal{O}_X(\mathcal{D})).$$

Proof. The cohomology exact sequence of (1.12) gives equation (1) and the exact sequence

$$0 \to H^1(E(\xi, \mathscr{D})) \to H^1(J_\xi(\mathscr{D})) \xrightarrow{\delta} H^1(J_\xi(\mathscr{D})) \otimes H^2(K_X)$$

$$\underset{\mathbb{C}}{\|}$$

$$\to H^2(E(\xi, \mathscr{D})) \to H^2(J_\xi(\mathscr{D})) \to 0.$$

By Proposition 1.1, δ is an isomorphism and equation (2) follows. By (1.3) one gas (3).

If $h^0(J_\xi(\mathscr{D} - K_X)) = 0$ then $H^0(E(\xi, \mathscr{D}) \otimes K_X^*) = H^1(J_\xi(\mathscr{D}))$ and $i \otimes K_X^*$ (1.12) is the canonical homomorphism given by global sections. Thus all the endomorphisms of $E(\xi, \mathscr{D})$ preserve the exact sequence (1.12) and we get

$$0 \to H^0\big(\mathrm{Hom}\big(E(\xi, \mathscr{D}), H^1(J_\xi(\mathscr{D})) \otimes K_X\big) \to \mathrm{Ext}^0_{\mathscr{O}_X}(E(\xi, \mathscr{D}), E(\xi, \mathscr{D}))$$

$$\xrightarrow{j} \mathrm{Ext}^0_{\mathscr{O}_X}(J_\xi(\mathscr{D}), J_\xi(\mathscr{D})). \tag{1.14}$$

But $\mathrm{Ext}^0_{\mathscr{O}_X}(J_\xi, J_\xi) \subset \mathrm{Ext}^0_{\mathscr{O}_X}(J_\xi, \mathscr{O}_X)$ and by Serre duality

$$\mathrm{Ext}^0_{\mathscr{O}_X}(J_\xi, \mathscr{O}_X)^* = \mathrm{Ext}^2_{\mathscr{O}_X}(\mathscr{O}_X, J_\xi \otimes K_X) = H^2(J_\xi \otimes K_X) \overset{(1.3)}{=} H^2(K_X).$$

$$\underset{\mathbb{C}}{\|}$$

Thus $\mathrm{Ext}^0_{\mathscr{O}_X}(J_\xi(\mathscr{D}), J_\xi(\mathscr{D})) = \mathrm{Ext}^0_{\mathscr{O}_X}(\mathscr{O}_X, \mathscr{O}_X) = \mathbb{C}$ and j in (1.14) is an epimorphism. This gives (4). □

LEMMA 1.2.　*If $\xi \in \widetilde{S^d X}$ is $\mathscr{D}$-stable, then $E(\xi, \mathscr{D})$ is locally free, that is, a vector bundle.*

Proof. $E(\xi, \mathscr{D})$ is torsion-free and its bidual $E(\xi, \mathscr{D})^{**}$ is reflexive and hence locally free. The canonical homomorphism $E(\xi, \mathscr{D}) \to E(\xi, \mathscr{D})^{**}$ gives

$$0 \to E(\xi, \mathscr{D}) \xrightarrow{can} E(\xi, \mathscr{D})^{**} \to S \to 0, \tag{1.15}$$

where $\dim \mathrm{Supp}\, S = 0$.

In the commutative diagram

$$
\begin{array}{ccccccccc}
0 & \to & H^1(J_\xi(\mathscr{D})) \otimes K_X & \to & E(\xi, \mathscr{D}) & \to & J_\xi(\mathscr{D}) & \to & 0 \\
 & & \downarrow \mathrm{id} & & \downarrow can & & \downarrow \varphi & & \\
0 & \to & H^1(J_\xi(\mathscr{D})) \otimes K_X & \to & E(\xi, \mathscr{D})^{**} & \to & \mathscr{L} & \to & 0,
\end{array}
$$

the sheaf $\mathscr{L}$ has rank 1 and is locally free outside a finite set of points of $\mathrm{Supp}\, \xi$. $\mathrm{Supp}\, \mathrm{Tors}\, \mathscr{L} \subset \mathrm{Supp}\, \xi$ and hence $\mathrm{Ext}^1_{\mathscr{O}_X}(K_X, \mathrm{Tors}\, \mathscr{L}) = 0$ and $\mathrm{Tors}\, \mathscr{L} \neq 0$ would imply that $\mathrm{Tors}\, E(\xi, \mathscr{D})^{**} \neq 0$. Thus $\mathrm{Tors}\, \mathscr{L} = 0$ and $\mathscr{L} = J_{\xi_1}(\mathscr{D}_1)$ with $\mathscr{D}_1 \in \mathrm{Pic}\, X$ (see for example [3], p. 102, Example 0.4). The exact sequence (1.15) gives

$c_1(E(\xi, \mathscr{D})) = c_1(E(\xi, \mathscr{D})^{**})$. Thus $\mathscr{D}_1 = \mathscr{D}$ and $\mathscr{L} = J_{\xi_1}(\mathscr{D})$, where ξ_1 is a subcycle of ξ:

$$
\begin{array}{ccccccc}
0 & \to & H^1(J_\xi(\mathscr{D})) \otimes K_X & \to & E(\xi, \mathscr{D}) & \to & J_\xi(\mathscr{D}) & \to 0 \\
& & \| & & \downarrow & & \downarrow \varphi & \\
0 & \to & H^1(J_\xi(\mathscr{D})) \otimes K_X & \to & E(\xi, \mathscr{D})^{**} & \to & J_{\xi_1}(\mathscr{D}) & \to 0.
\end{array}
\qquad (1.16)
$$

The last extension is defined by a cocycle

$$
e_2 \in \mathrm{Ext}^1_{\mathscr{O}_X}\Big(J_{\xi_1}(\mathscr{D}), H^1\big(J_\xi(\mathscr{D})\big) \otimes K_X \Big) \overset{(1.10)}{=} \mathrm{Hom}\big(H^1\big(J_{\xi_1}(\mathscr{D})\big), H^1\big(J_\xi(\mathscr{D})\big)\big).
$$

$$(1.17)$$

The homomorphism φ of (1.16) defines homomorphisms $H^1(\varphi)$: $H^1(J_\xi(\mathscr{D})) \to H^1(J_{\xi_1}(\mathscr{D}))$ and

$$
\tilde{\varphi}: \mathrm{Hom}(H^1(J_{\xi_1}(\mathscr{D})), H^1(J_\xi(\mathscr{D}))) \to \mathrm{Hom}(H^1(J_\xi(\mathscr{D})), H^1(J_\xi(\mathscr{D})))
$$
$$
\|
\qquad\qquad\qquad\qquad\qquad\qquad\qquad\qquad \|
$$
$$
\mathrm{Ext}^1_{\mathscr{O}_X}(J_{\xi_1}(\mathscr{D}), H^1(J_\xi(\mathscr{D})) \otimes K_X) \to \mathrm{Ext}^1_{\mathscr{O}_X}(J_\xi(\mathscr{D}), \quad H^1(J_\xi(\mathscr{D})) \otimes K_X).
$$

In that case by Proposition 1.2 one as $\tilde{\varphi}(e_2) = \mathrm{id}$ (see (1.17) and (1.3)) and the composite

$$
H^1\big(J_\xi(\mathscr{D})\big) \xrightarrow{H^1(\varphi)} H^1\big(J_{\xi_1}(\mathscr{D})\big) \overset{e_2}{\to} H^1\big(J_\xi(\mathscr{D})\big)
$$

is an isomorphism. Thus

$$
h^1\big(J_{\xi_1}(\mathscr{D})\big) \geqslant h^1\big(J_\xi(\mathscr{D})\big)
$$

and

$$
\delta(\mathscr{D}, \xi_1) = \delta(\mathscr{D}, \xi).
$$

If ξ is $\mathscr{D}$-stable, then $\xi_1 = \xi$ and *can* is an isomorphism. $\qquad\qquad\square$

For each subcycle $\xi_1 \subset \xi$ one has the diagram:

$$
\begin{array}{ccccccc}
& & 0 & & 0 & & \\
& & \downarrow & & \downarrow & & \\
0 \to & \mathrm{im}\, j \otimes K_X & & \to \mathrm{im}\, j \otimes K_X & \to & 0 & \\
& & \downarrow & & \downarrow & & \downarrow \\
0 \to & H^1(J_\xi(\mathscr{D})) \otimes K_X & \to & E(\xi, \mathscr{D}) & \to & J_\xi(\mathscr{D}) & \to 0 \\
& & \downarrow & & \downarrow \Phi & & \downarrow \\
0 \to & H^1(J_{\xi_1}(\mathscr{D})) \otimes K_X & \to & E(\xi_1, \mathscr{D}) & \to & J_{\xi_1}(\mathscr{D}) & \to 0 \\
& & \downarrow & & \downarrow & & \downarrow \\
& & 0 & \to & S & \to S & \to 0 \\
& & & & \downarrow & & \downarrow \\
& & & & 0 & & 0
\end{array}
\qquad (1.18)
$$

where the right-hand column is (1.18) and the left-hand column is the end of (1.9) tensored with K_X.

Definition 1.3. A subcycle $\xi_1 \subset \xi$ for which $\delta(\mathscr{D}, \xi_1) = \delta(\mathscr{D}, \xi)$ is called a *minimal subcycle* if ξ_1 is $\mathscr{D}$-stable. Such a cycle is denoted by $\xi_{\mathscr{D}\text{-min}} \subset \xi$.

LEMMA 1.3.   *For minimal* $\xi_{\mathscr{D}\text{-min}} \subset \xi$ *we have*
(1) $E(\xi_{\mathscr{D}\text{-min}}, \mathscr{D}) = E(\xi, \mathscr{D})^{**}$;
(2) *The exact sequence* (1.15) *is the middle column of* (1.18).

Proof.   By Lemma 1.2, $E(\xi_{\mathscr{D}\text{-min}}, \mathscr{D})$ is locally free and the homomorphism $E(\xi, \mathscr{D}) \overset{\Phi}{\to} E(\xi_{\mathscr{D}\text{-min}}, \mathscr{D})$ of (1.18) is the composite

$$E(\xi, \mathscr{D}) \overset{\text{can}}{\text{---}} E(\xi, \mathscr{D}_\Phi)^{**} \overset{\Phi'}{\to} E(\xi_{\mathscr{D}\text{-min}}, \mathscr{D}).$$

Further, $\mathrm{rk}\, E(\xi, \mathscr{D}) = \mathrm{rk}\, E(\xi_{\mathscr{D}\text{-min}}, \mathscr{D})$, $c_1(E(\xi, \mathscr{D})^{**}) = c_1(E(\xi_{\mathscr{D}\text{-min}}, \mathscr{D}))$ and Φ' is an isomorphism at a general point $x \in X$. Thus $\Lambda^{\max}\Phi'$ is an isomorphism, and so is Φ'. This gives (1) and (2). $\square$

COROLLARY 1.   *If* ξ *is not* $\mathscr{D}$-*stable, then* $E(\xi, \mathscr{D})$ *is not locally free. Indeed, in this case* $E(\xi, \mathscr{D}) \ne E(\xi, \mathscr{D})^{**} = E(\xi_{\mathscr{D}\text{-min}}, \mathscr{D})$.

COROLLARY 2.   *The subcycle* $\xi_{\mathscr{D}\text{-min}} \subset \xi$ *is unique. Hence the variety* $NS^d_{\mathscr{D}, \delta} \subset S^d_{\mathscr{D}, \delta} X$ *of non* $\mathscr{D}$-*stable cycles is stratified by the function* $\deg \xi_{\mathscr{D}\text{ min}}$

$$NS^d_{\mathscr{D}, \delta} = \bigcup_{i=1}^{d} NS^{d, i}_{\mathscr{D}, \delta}, \quad \text{where } NS^{d, i}_{\mathscr{D}, \delta} = \left\{ \xi \in NS^d_{\mathscr{D}, \delta} \,|\, \deg \xi_{\mathscr{D}\text{-min}} = i \right\} \quad (1.19)$$

and each component is birationally equivalent to the product

$$NS^{d, i}_{\mathscr{D}, \delta} \overset{\text{bir}}{=} \widetilde{S^i_{\mathscr{D}, \delta} X} \times S^{d-i} X \tag{1.20}$$

where $\delta = \delta(\mathscr{D}, \xi)$.

Suppose that the class $\mathscr{D} - K_X \in \mathrm{Pic}\, X$ contains a smooth curve C (that is, $\mathscr{D} = K_X + C$ in $\mathrm{Pic}\, X$), and that the cycle ξ is contained in C. Then ξ defines the invertible sheaf $\mathscr{O}_C(\xi)$ in $\mathrm{Pic}\, C$.

LEMMA 1.4.   *In this case one has*
(1) $\delta(C, \xi) = d + h^0(\mathscr{O}_C(C - \xi)) - h^0(\mathscr{O}_C(C))$;
(2) $h^1(J_\xi(\mathscr{D})) = \delta(\mathscr{D}, \xi)$;
(3) $\delta(\mathscr{D}, \xi) = h^0(\mathscr{O}_C(\xi)) - 1 = \dim |\xi|$.
(4) *The complete linear series* $|\xi|$ *on* C *is fixed point free if and only if the cycle* ξ *is* $\mathscr{D}$-*stable.*

Proof.   First of all, $h^1(\mathscr{O}_X(C) \otimes K_X) = 0$ gives 2). From (1.3), $h^2(\mathscr{O}_X(C + K_X)) = h^0(J_\xi(C + K_X)) = 0$. Let $J_\xi^C = \mathscr{O}_C(-\xi)$ be the ideal sheaf of $\xi \subset C$.

Then we have the exact sequence

$$0 \to J_C \to J_\xi \to J_\xi^C \to 0,$$
$$\| \qquad\qquad \|$$
$$\mathcal{O}_X(-C) \quad \mathcal{O}_C(-\xi) \tag{1.21}$$

and tensoring by $\mathcal{O}_X(C)$, we get

$$0 \to \mathcal{O}_X \to J_\xi(C) \to \mathcal{O}_C(C - \xi) \to 0.$$

The cohomology exact sequence gives

$$h^0\big(J_\xi(C)\big) = h^0\big(\mathcal{O}_C(C - \xi)\big) + 1.$$

Substituting this into (1.5) proves (1). Tensoring the exact sequence (1.21) by $\mathcal{O}_X(C + K_X)$ gives

$$0 \to \mathcal{O}_X(K_X) \to J_\xi(\mathcal{D}) \to \mathcal{O}_C(K_C - \xi) \to 0, \tag{1.22}$$

where K_C is the canonical class of C. From this

$$0 \to H^1\big(J_\xi(\mathcal{D})\big) \to H^1\big(\mathcal{O}_C(K_C - \xi)\big) \to H^2(K_X) \to 0,$$

and one obtains (3).

If the complete linear series $|\xi|$ on C has fixed points, that is $|\xi| = |\xi_1| + \xi_F$, then $h^0(\mathcal{O}_C(\xi)) = h^0(\mathcal{O}_C(\xi_1))$ and by (2) and (3), $\delta(\mathcal{D}, \xi) = \delta(\mathcal{D}, \xi_1)$. Thus ξ is not $\mathcal{D}$-stable, $\xi_{\mathcal{D}\text{-min}} = \xi_1$ and one has (4). $\qquad\square$

§2. *K*-block and *K*-fitting.   Let X be a smooth regular surface.

Definition 2.1.   A torsion-free sheaf E is *regular*, if

$$h^1(E) = 0, \quad h^2(E) = h^2\big(\det E \otimes K_X^{-\text{rk } E+1}\big).$$

Definition 2.2.   Let E be a torsion-free sheaf and V a vector space with $\text{rk } V = \text{rk } E - 1$. A homomorphism $S \bmod GL(V)$

$$V \otimes L_X \overset{S}{\to} E \tag{2.1}$$

is called a *K-block* of E, if there is a point $x \in X$ for which $S_{|x}$ is a monomorphism.

Each vector $v \in V$ defines a homomorphism $v: K_X \to E$. Let us consider the set

$$\Delta(S) = \big\{ v \in V, x \in X \,|\, v_{|x} = 0 \big\} \subset \mathbb{P}V \times X. \tag{2.2}$$

Definition 2.3.　A K-block S (2.1) is called *regular*, if dim $\Delta(S) = 0$.
The projections to X and to $\mathbb{P}V$ define two cycles

$$\xi(S) = pr_X(\Delta(S)) \quad \text{and} \quad \zeta(S) = pr_{\mathbb{P}V}(\Delta(S)). \tag{2.3}$$

Definition 2.4.　The cycle $\xi(S)$ is called the *degeneracy cycle* of the K-block S.

LEMMA 2.1.　*The homomorphism S of a regular K-block* (2.1) *extends to the exact sequence*

$$0 \to V \otimes K_X \overset{S}{\to} E \to J_{\xi(S)}(\mathscr{D}) \to 0, \tag{2.4}$$

where $\mathscr{D} = c_1(E) - \operatorname{rk} V \cdot K_X$, $\xi(S)$ *is the degeneracy cycle of the K-block S* (2.3) *and*

$$\deg \xi(S) = c_2(E \otimes K_X^*).$$

Proof.　By definition, $\operatorname{rk} \operatorname{coker} S = 1$. If $\operatorname{Tors} \operatorname{coker} S \neq 0$ then E contains a torsion-free subsheaf E' fitting into an exact sequence

$$0 \to V \otimes K_X \to E' \to \operatorname{Tors} \operatorname{coker} S \to 0.$$

But in this case $\operatorname{Supp} \operatorname{Tors} \operatorname{coker} S - 1$ has codimension 1, and dim $\Delta(S) = 1$, which contradicts the regularity of S. Thus $\operatorname{coker} S$ is a rank 1 torsion-free sheaf, so that $\operatorname{coker} S = J_\xi(\mathscr{D})$, where ξ is a cycle on X (see the proof of Lemma 1.2). For each point $x \in \operatorname{Supp} \xi$ there is a unique line $\mathbb{C} \cdot v \subset V$ for which $v(x) = 0$. This implies that $\xi = \xi(S)$ as subschemes of X.

Example.　The standard exact sequence (1.12) of $E(\xi, \mathscr{D})$ defines the regular K-block i of $E(\xi, \mathscr{D})$.

LEMMA 2.2.　*Let E be a regular torsion-free sheaf and S a regular K-block of E. Then $E = E(\xi(S), \mathscr{D})$, where $\mathscr{D} = c_1(E) - (\operatorname{rk} E - 1) \cdot K_X$.*

Proof.　The end of the cohomology exact sequence of (2.4) is

$$0 \to H^1\big(J_{\xi(S)}(\mathscr{D})\big) \to$$

$$\overset{\delta}{\to} V \otimes H^2(K_X) \to H^2(E) \to H^2\big(J_{\xi(S)}(\mathscr{D})\big) \overset{(1.3)}{=} H^2(\mathscr{O}_X(\mathscr{D})).$$
$$\shortparallel$$
$$\mathbb{C}$$

Under the isomorphism $\delta\colon H^1(J_{\xi(S)}(\mathscr{D})) \to V$, (2.4) becomes the exact sequence (1.12) and Proposition 1.1 shows that the extensions coincide.　　□

LEMMA 2.3.　*If S and S' are K-blocks of a regular torsion-free sheaf E and $\xi(S) = \xi(S') = \xi$, then there exists $g \in \operatorname{Aut} E$ such that $S' = g(S)$.*

Proof. By Lemma 2.2 one has the diagram

$$
\begin{array}{ccccccccc}
0 & \to & V \otimes K_X & \overset{S}{\to} & E & \to & J_\xi(\mathscr{D}) & \to & 0 \\
 & & j \otimes \mathrm{id} \downarrow & & \downarrow g & & \downarrow \mathrm{id} & & \\
0 & \to & V' \otimes K_X & \overset{S'}{\to} & E & \to & J_\xi(\mathscr{D}) & \to & 0
\end{array}
$$

where the isomorphism f of V and V' is such that $\delta'f\delta^{-1} = \mathrm{id}$ where δ: $H^1(J_\xi(\mathscr{D})) \to V$ is the cocycle of the first extension and δ': $H^1(J_\xi(\mathscr{D})) \to V'$ is that of the second (see the proof of Lemma 2.2). Then by Proposition 1.2 we get an isomorphism $g \in \mathrm{Aut}\ E$, and $g(S) = S'$. $\square$

Definition 2.5. Let E be a torsion-free sheaf and W a vector space with rk W = rk E. A homomorphism $\varphi \bmod GL(V)$

$$
W \otimes K_X \overset{\varphi}{\to} E \tag{2.5}
$$

is called a *K-fitting* of E if there is a point $x \in X$ such that $\varphi_{|x}$ is an isomorphism.

Suppose that $c_1(E) = \mathrm{rk}\ E \cdot K_X + C$. The homomorphism $\wedge^{\max}\varphi$ defines a section

$$
s_\varphi \colon \mathscr{O}_X \to \mathscr{O}_X(C),
$$

and the zero-scheme of s_φ is a curve C belonging to the complete linear system $|c_1(E) - \mathrm{rk}\ E \cdot K_X|$ on X.

Definition 2.6. (1) The curve $C = (s_\varphi)_0$ is called the *degeneracy curve* of the *K*-fitting φ (2.5);
   (2) A *K*-fitting φ (2.5) is called *regular* if
      (a) its degeneracy curve C is smooth,
and   (b) coker $\varphi \in \mathrm{Pic}\ C$.

LEMMA 2.4. *If a torsion-free sheaf E has a regular K-fitting then E is locally free.*

Proof. The homomorphism φ (2.5) of the *K*-fitting can be extended to an exact sequence

$$
0 \to W \otimes K_X \overset{\varphi}{\to} E \to \mathscr{L} \to 0 \tag{2.6}
$$

where Supp $\mathscr{L} = C$ and $\mathscr{L} \in \mathrm{Pic}\ C$.
Applying the functor $\mathscr{H}\!om(\ , \mathscr{O}_X)$ to this exact sequence gives

$$
0 \to E^* \to W^* \otimes K_X^* \overset{\delta}{\to} \mathscr{E}\!xt^1_{\mathscr{O}_X}(\mathscr{L}, \mathscr{O}_X) \to 0. \tag{2.7}
$$

Since $\mathscr{L} \in \text{Pic } C$,

$$\mathscr{E}xt^1_{\mathcal{O}_X}(\mathscr{L}, \mathcal{O}_X) = \mathscr{L}^* \otimes \mathscr{E}xt^1_{\mathcal{O}_X}(\mathcal{O}_C, \mathcal{O}_X) = \mathscr{L}^* \otimes \mathcal{O}_C(C). \qquad (2.8)$$
$$\|$$
$$\mathcal{O}_C(C)$$

Thus, if $\mathscr{L} = \mathcal{O}_C(K_C - \xi)$, where ξ is a cycle on C, then δ of (2.7) is an epimorphism if and only if

$$W^* \otimes \mathcal{O}_C \xrightarrow{\delta \otimes K_{X|C}} \mathcal{O}_C(C + K_X) \otimes \mathscr{L}^* = \mathcal{O}_C(\xi) \qquad (2.9)$$

is an epimorphism.

Applying the function $\mathscr{H}om(\ , \mathcal{O}_X)$ to (2.7) we have

$$0 \to W \otimes K_X \xrightarrow{\varphi} E^{**} \to \mathscr{L} \to 0,$$

and an isomorphism $E = E^{**}$ by equations (2.8). $\qquad \square$

Hence a regular K-fitting of a vector bundle E defines the linear series (2.9) on C without fixed points. (But, a priori, non-complete).

Definition 2.7.   A regular fitting φ (2.5) is *complete* if there is an isomorphism $W^* = H^0(\mathcal{O}_C(\xi))$ such that the map $\delta \otimes K_{X|C}$ of (2.9) is the canonical morphism given by global sections.

LEMMA 2.5.   *A regular K-fitting φ (2.5) of a vector bundle E is complete if and only if E is regular.*

Proof.   First of all, by regularity one has $h^2(E) = h^2(\mathcal{O}_X(C + K_X)) = 0$ and for δ of (2.7), $H^0(\ker \delta \otimes K_X) = H^0(E^* \otimes K_X)$. Thus $H^0(\ker \delta \otimes K_X) = 0$ and the cohomology exact sequence of (2.7) tensored with K_X is

$$0 \to W^* \to H^0(\mathcal{O}_C(\xi)) \to H^1(E^* \otimes K_X) \to 0.$$

Hence the completeness of the linear series (2.9) is equivalent to

$$h^1(E^* \otimes K_X) = h^1(E) = h^2(E) = 0. \qquad \square$$

COROLLARY.   *The class $(E, \varphi) \bmod \text{Aut } E$, where E is a regular bundle and φ a regular K-fitting of E, is equivalent to the class $(C, |\xi|) \bmod \text{Aut } C$, where C is the degeneracy curve of φ and $|\xi|$ is a complete linear series without fixed points (2.9).*

Indeed, if $|\xi|$ is fixed point free, then the canonical homomorphism

$$H^0(\mathcal{O}_C(\xi)) = H^1(\mathcal{O}_C(K_C - \xi)) \otimes \mathcal{O}_C \xrightarrow{can} \mathcal{O}_C(\xi) \to 0$$

                                    A. N. TYURIN

is an epimorphism, which can be extended to an $\mathcal{O}_X$-epimorphism

$$H^1(\mathcal{O}_C(K_C - \xi)) \otimes \mathcal{O}_X \to \mathcal{O}_C(\xi) \to 0.$$

Tensoring this with $K_{\tilde{X}}^*$ gives the exact sequence

$$0 \to E^* \xrightarrow{\varphi^*} H^1(\mathcal{O}_C(K_C - \xi))^* \otimes K_{\tilde{X}}^* \xrightarrow{\text{can} \otimes K_{\tilde{X}}^*} \mathcal{O}_C(\xi - K_{X|C}) \to 0. \quad (2.10)$$

Applying the functor $\mathrm{Hom}(\ , \mathcal{O}_X)$ to (2.10) one has the exact sequence of a regular K-fitting

$$0 \to H^1(\mathcal{O}_C(K_C - \xi)) \otimes K_X \xrightarrow{\varphi} E = E(|\xi|, C) \to \mathcal{O}_C(K_C - \xi) \to 0 \quad (2.11)$$

of the regular vector bundle $E(|\xi|, C)$.

If $g \in \mathrm{Aut}\, C$, then

$$E(|\xi|, C) = E(g^*(|\xi|), C). \quad (2.12)$$

Definition 2.8. The class $(|\xi|, C) \bmod \mathrm{Aut}\, C$ is called the *geometrical equivalent* of the class $(E, \varphi) \bmod \mathrm{Aut}\, E$.

Each complete linear series $|\xi|$ on a smooth curve C defines a residual complete linear series $|\xi'| = |K_C - \xi|$. If the residual series $|\xi'|$ is also fixed point free, then one has the "residual" vector bundle $E(|\xi'|, C)$.

LEMMA 2.6. (1) *The vector bundles $E(|\xi|, C)$ and $E(|\xi'|, C)$ are related by the exact sequence*

$$0 \to E^*(|\xi'|, C) \otimes K_X \xrightarrow{\text{can}^*} H^0(E(|\xi|, C) \otimes K_{\tilde{K}}^*) \otimes \mathcal{O}_X \xrightarrow{\text{can}} E(|\xi|, C) \otimes K_{\tilde{X}}^* \to 0$$

$$(2.13)$$

(2) *The isomorphism $H^0(E(|\xi|, C) \otimes K_{\tilde{X}}^*)^* = H^0(E(|\xi'|, C) \otimes K_{\tilde{X}}^*)$ is defined by applying * to the exact sequence*

$$0 \to H^1(\mathcal{O}_C(K_C - \xi)) \to H^0(E(|\xi|, C) \otimes K_{\tilde{X}}^*) \to H^0(\mathcal{O}_C(C - \xi)) \to 0$$

and by Serre duality on C:

$$
\begin{array}{ccccccc}
0 \to & H^0(\mathcal{O}_C(\xi'))^* & \to & H^0(E(|\xi|, C) \otimes K_{\tilde{X}}^*)^* & \to & H^1(\mathcal{O}_C(K_C - \xi)) & \to 0 \\
& \| & & \| & & \| & \\
0 \to & H^1(\mathcal{O}_C(K_C - \xi')) & \to & H^0(E(|\xi'|, C) \otimes K_{\tilde{X}}^*) & \to & H^0(\mathcal{O}_C(\xi)) & \to 0.
\end{array}
$$

Proof.   In the diagram

$$
\begin{array}{ccc}
& 0 & 0 \\
& \downarrow & \downarrow \\
0 \qquad\qquad \to H^1(\mathcal{O}_C(K_C - \xi)) \otimes \mathcal{O}_X \xrightarrow{\varphi \otimes K_X^*} E(|\xi|, C) \otimes K_X^* \to \mathcal{O}_C(C - \xi) \to 0 \\
\downarrow \qquad\qquad \downarrow \qquad\qquad \downarrow \\
0 \to \ker can \qquad \to H^0(E(|\xi|, C) \otimes K_X^*) \otimes \mathcal{O}_X \xrightarrow{can} E(|\xi|, C) \otimes K_X^* \quad \to 0 \\
\downarrow \qquad\qquad \downarrow \qquad\qquad \downarrow \\
0 \to H^0(\mathcal{O}_C(\xi')) \otimes \mathcal{O}_X \to H^0(\mathcal{O}_C(\xi')) \otimes \mathcal{O}_X \to \qquad 0 \\
\downarrow \qquad\qquad \downarrow \\
\mathcal{O}_C(C - \xi) \qquad 0 \\
\downarrow \\
0
\end{array}
$$

$$(2.14)$$

the last term of the left-hand column is equal to the last term of the first row by the snake lemma. Thus by definition $\ker can = E^*(|\xi'|, C) \otimes K_X$. $\qquad\square$

Now each section $s \in H^0(\mathcal{O}_C(\xi))$ defines a hyperplane $V_s \subset H^1(\mathcal{O}_C(K_C - \xi))$ by

$$
V_s = s^\perp = \left\{ h \in H^1(\mathcal{O}_C(K_C - \xi)) | \langle h, s \rangle = 0 \right\}. \qquad (2.15)
$$

LEMMA 2.7.   (1) *The restriction to $V_s \otimes K_X$, of φ of (2.11) is a regular K-block of $E(|\xi|, C)$*

$$
\varphi_{V_s \otimes K_X} : V_s \otimes K_X \to E(|\xi|, C).
$$

(2) *The degeneracy cycle of this K-block is given by*

$$
\xi(\varphi_{|V_s \otimes K_X}) = (s)_0.
$$

Proof.   Both assertions come from the diagram

$$
\begin{array}{ccc}
& & 0 \\
& & \downarrow \\
0 & 0 & K_X = J_C(C + K_X) \\
\downarrow & \downarrow & \downarrow \\
0 \to V_s \otimes K_X \xrightarrow{\varphi_{|V_s \otimes K_X}} E(|\xi|, C) \to J_\xi(C + K_X) \to 0 \\
\downarrow & \downarrow & \downarrow \\
0 \to H^1(\mathcal{O}_C(K_C - \xi)) \otimes K_X \xrightarrow{\varphi} E(|\xi|, C) \to \mathcal{O}_C(-\xi) \otimes \mathcal{O}_C(C + K_X) \to 0 \\
\downarrow & \downarrow & \downarrow \\
K_X & 0 & 0 \\
\downarrow \\
0
\end{array}
$$

where the right-hand column is the exact sequence (1.22) and ξ is the degeneracy cycle of the K-block $\varphi_{|V_s \otimes K_X}$. $\qquad\qquad\square$

COROLLARY. *For each* $\xi \in |\xi|$,

$$E(\xi, C + K_X) = E(|\xi|, C)$$

(*by* (1.12) *and* (2.11)).

§3. The K-block and K-fitting maps. Let E be a vector bundle on X with rk $E = r + 1$, and $G(r, H^0(E \otimes K_X^*))$ the Grassmanian of r-dimensional subspaces of $H^0(E \otimes H_X^*)$.

The variety of regular K-blocks of E

$$B(E) = \left\{ V \subset H^0(E \otimes K_X^*) \middle| \begin{array}{c} V \otimes K_X \xrightarrow{\;can \otimes K_{X|V}\;} E \\[4pt] \text{is a regular } K\text{-block} \end{array} \right\} \qquad (3.1)$$

is a Zariski open set of $G(r, H^0(E \otimes K_X^*))$ (possibly empty).

One defines the K-block map

$$\varphi_E^B: B(E) \to \widetilde{S^d X}, \quad \text{where} \quad d = c_2(E \otimes K_X^*) \quad \text{by} \quad \varphi_E^B(S) = \xi(S). \quad (3.2)$$

The image of this map

$$\mathscr{S}(E) = \varphi_E^B(B(E)) \subset \widetilde{S^d X} \qquad (3.3)$$

is an irreducible unirational variety.

LEMMA 3.1. *Let E be a regular vector bundle with* rk $E = r + 1$ *and* $\mathscr{D} = c_1(E) - rK_X$, *and* $S \in B(E)$ *a regular K-block of E. Then* $\delta(\mathscr{D}, \xi(S)) = r - h^1(\mathcal{O}_X(\mathscr{D}))$.

Proof. The end of the cohomology exact sequence of (2.4) gives our assertion.

COROLLARY. *Let E be a regular vector bundle with* rk $E = r + 1$ *and* $\mathscr{D} = c_1(E) - rK_X$; *then*
(1) $\mathscr{S}(E) \subset \widetilde{SS_{\mathscr{D}, r - h^1(\mathcal{O}_X(\mathscr{D}))}^d} X$
where $SS_{\mathscr{D}, k}^d \subset \widetilde{S^d X}$ *is the subvariety of $\mathscr{D}$-stable cycles;*
(2) $\mathscr{S}(E) = B(E)/\text{Aut } E$.

Lemma 2.3 gives these assertions.

LEMMA 3.2. *For a regular K-block S of a regular vector bundle E with* rk $E = r + 1$ *and* $\mathscr{D} = c_1(E) - rK_X$, *let*

$$\text{Aut}_S E = \{ g \in \text{Aut } E | g(S) = S \}.$$

Then

$$\dim \operatorname{Aut}_S E = 1 + r \cdot h^2(\mathcal{O}_X(\mathcal{D})).$$

Proof.   The elements of $\operatorname{Aut}_S E$ preserve the exact sequence (2.4). Thus one has

$$1 \to H^0(\operatorname{Hom}(E, V \otimes K_X)) \to \operatorname{Aut}_S E \to \operatorname{Aut} J_\xi(\mathcal{D}) \to 1. \qquad (3.4)$$
$$\qquad\qquad \| \qquad\qquad\qquad\qquad\qquad\qquad\qquad \|$$
$$\qquad V \otimes H^2(E) \qquad\qquad\qquad\qquad\qquad \mathbf{C}^*$$

The regularity of E gives our assertion. $\square$

COROLLARY.   *If* $\mathcal{S}(E) \neq \varnothing$ *then*

$$\dim \mathcal{S}(E) = r\Big[h^0\big(J_{\xi(S)}(\mathcal{D} - K_X)\big) + h^2(\mathcal{O}_X(\mathcal{D}))\Big] - (\dim \operatorname{Aut} E - 1)$$

where S is any K-block of E.

Indeed, the cohomology exact sequence of (2.4) tensored with K_X^* gives the dimension of $B(E)$. Assertion (2) of Corollary of Lemma 3.1 and Lemma 3.2 give our equation.

Now suppose that the class $\mathcal{D} - K_X = c_1(E) - \operatorname{rk} E \cdot K_X$ contains a smooth curve C and that the cycle $\xi(S) \subset C$. Then

$$\dim \mathcal{S}(E) = r\big(h^0(\mathcal{O}_C(C - \xi)) + 1\big) - (\dim \operatorname{Aut} E - 1). \qquad (3.5)$$
$$\qquad\qquad\qquad \|$$
$$\qquad h^0(\mathcal{O}_C(\xi'))$$

The complete linear system $|C|$ on X contains a Zariski open set

$$|C|_0 = \{C' \in |C| \,|\, C' \text{ is smooth}\}. \qquad (3.6)$$

Definition 3.1.   A pair $\tilde{X} = (X, C)$, where X is a smooth surface and $|C|_0 \neq \varnothing$ is called a *q–p* (*quasipolarized*) surface.

A *q–p* surface $\tilde{X}$ defines the family of smooth curves

$$\begin{array}{c} \mathscr{X}(\tilde{X}) \subset X \times |C|_0 \\ \downarrow \pi \\ |C|_0 \end{array} \qquad (3.7)$$

where $\mathscr{X}(\tilde{X}) = \{(x, C) \,|\, x \in C\}$, and the family

$$\begin{array}{c} \operatorname{Pic}_d(\tilde{X}) \\ \downarrow \pi \quad \text{with } \pi^{-1}(C) = \operatorname{Pic}_d C \\ |C|_0 \end{array} \qquad (3.8)$$

of relative Picard varieties of degree d.

For each $C \in |C|_0$ (see [2], p. 153) let

$$W_d^r(C) = \left\{ \mathscr{L} \in Pic_d C \,|\, h^0(\mathscr{L}) \geqslant r + 1 \right\}. \tag{3.9}$$

For the family (3.7), let

$$\mathscr{W}_d^r(\tilde{X}) = \left\{ (\mathscr{L}, C) \,|\, \mathscr{L} \in W_d^r(C) \right\}. \tag{3.10}$$

This is a subvariety of $\mathrm{Pic}_d(\tilde{X})$ (3.8):

$$
\begin{array}{ccc}
\mathscr{W}_d^r(\tilde{X}) & \subset & \mathrm{Pic}_d(\tilde{X}) \\
\pi \searrow & & \swarrow \\
& |C|_0. &
\end{array}
\tag{3.11}
$$

Let E be a regular vector bundle on X with rk $E = r + 1$ and $c_1(E) - $ rk $E \cdot K_X = C$, and $G(r + 1, H^0(E \otimes K_X^*))$ the Grassmanian of $(r + 1)$-dimensional subspaces of $H^0(E \otimes K_X^*)$. The variety of regular K-fittings

$$F(E) = \left\{ W \subset H^0(E \otimes K_X^*) \,\middle|\, \begin{array}{c} W \otimes K_X \xrightarrow{\ can \,\otimes\, K_{X|W}\ } E \\ \text{is a regular } K\text{-fitting} \end{array} \right\} \tag{3.12}$$

is a Zariski-open subset of $G(r + 1, H^0(E \otimes K_X^*))$ (possibly empty). One defines the K-fitting map

$$\varphi_E^F\colon F(E) \to \mathscr{W}_\alpha^r(\tilde{X}) \quad \text{by} \quad \varphi_E^F(\varphi) = (|\xi|, C). \tag{3.13}$$

Its image

$$\mathscr{W}(E) = F(E)/\mathrm{Aut}\, E = \varphi_E^F(F(E)) \subset \mathscr{W}_d^r(\tilde{X}) \tag{3.14}$$

is an irreducible unirational variety. By Definition 2.8,

$$\mathscr{W}(E) = F(E)/\mathrm{Aut}\, E. \tag{3.15}$$

The image of the composite

$$F(E) \xrightarrow{\varphi_E^F} \mathscr{W}(E) \xrightarrow{\pi} |C|_0, \tag{3.16}$$

where π is the projection (3.11), contains all degeneracy curves of all regular K-fittings of E;

$$\mathscr{D}(E) = \left(\varphi_E^F \circ \pi \right)(F(E)). \tag{3.17}$$

The differential of the composite (3.16) at a point $W \in F(E)$ has a simple interpretation: the regular K-fitting W of a regular sheaf E defines an isomorphism $E = E(|\xi|, C)$. The tangent space

$$TF(E)_W = TG\big(r + 1, H^0(F \otimes K_X^*)\big)_W = W^* \otimes H^0(E \otimes K_X^*)/W.$$

But tensoring the exact sequence (2.11) for $E(|\xi|, C)$ by K_X^* gives

$$W^* = H^0(\mathcal{O}_C(\xi)), \; H^0(E \otimes K_x^*)/W = H^0(\mathcal{O}_C(C - \xi)).$$

From this

$$TF(E)_W = H^0(\mathcal{O}_C(\xi)) \otimes H^0(\mathcal{O}_C(C - \xi)). \tag{3.18}$$

But the tangent space of $|C|_0$ at the point C is

$$T|C|_C = H^0(\mathcal{O}_C(C)) \tag{3.19}$$

(by the regularity of X).

LEMMA 3.4.   *The differential of the composite* (3.16)

$$d\big(\varphi_E^F \circ \pi\big) \colon H^0(\mathcal{O}_C(\xi)) \otimes H^0(\mathcal{O}_C(C - \xi)) \to H^0(\mathcal{O}_C(C))$$

is the cup-product homomorphism.

Proof.   It is sufficient to check that for fixed $s \in H^0(\mathcal{O}_C(\xi))$ with $(s)_0 = \xi$, the projectivization of the restriction

$$d\big(\varphi_E^F \circ \pi\big) \colon s \otimes H^0(\mathcal{O}_C(C - \xi)) \to H^0(\mathcal{O}_C(C))$$

is the embedding $(|C^2 - \xi| + \xi) \subset |C^2|$, where C^2 denotes the self-intersection class $C_{|C}$.

By Lemma 2.7 the section $s \in H^0(\mathcal{O}_C(\xi))$ defines a regular K-block $V_s \subset W$ (2.16). The variety of regular K-fittings containing the K-block V_s with $(s)_0 = \xi$,

$$F(E)_\xi = \big\{ W \in F(E) | W \supset V_s \big\} \tag{3.20}$$

is a Zariski open set of the projective space $\mathbb{P}H^0(J_\xi(C))$

$$F(E)_\xi = \mathbb{P}H^0\big(J_\xi(C)\big)_0. \tag{3.21}$$

Since a K-fitting $W' \supset V_s$ is defined by $s' \in H^0(J_\xi(C))$, which can be lifted to the homomorphism $\bar{s}\colon K_X \to E(\xi, C + K_X) = E(|\xi|, C)$,

$$0 \to V_s \otimes \mathcal{O}_X \xrightarrow{\;S \otimes K_X^*\;} \left[E(\xi, C + K_X) = E(|\xi|, C) \right] \otimes K_X^* \to J_\xi(C) \to 0$$

$$\bar{s} \otimes K_X^* \searrow \quad \nearrow s',$$

$$\mathcal{O}_X$$

$$(3.22)$$

we have $W' = V_s \oplus (\bar{s} \otimes K_X^*)$.

The restriction of $\varphi_E^F \circ \pi$ (3.16) to $F(E)_\xi$ is the projectivization of the embedding $H^0(J_\xi(C)) \hookrightarrow H^0(\mathcal{O}_X(C))$,

$$F(E)_\xi = |C - \xi|_0 \subset \mathcal{D}(E) \subset |C|_0. \qquad (3.23)$$

The restriction of this embedding to the degeneracy curve C of the K-fitting W gives our interpretation of the differential.

Now, the varieties $\mathcal{S}(E)$ and $\mathcal{W}(E)$ are related by the flag relation: let

$$\mathcal{P}_d^r(\tilde{X}) = \left\{ \xi \subset C \mid (|\xi|, C) \in \mathcal{W}_d^r(\tilde{X}) \right\};$$

$$\mathcal{P}(E) = \left\{ V \subset W \mid V \in B(E), W \in F(E) \right\}; \qquad (3.24)$$

$$S_r^d(C) = \left\{ \xi \subset C \mid \dim |\xi| \geqslant r \right\}.$$

Then the natural projections define the diagram

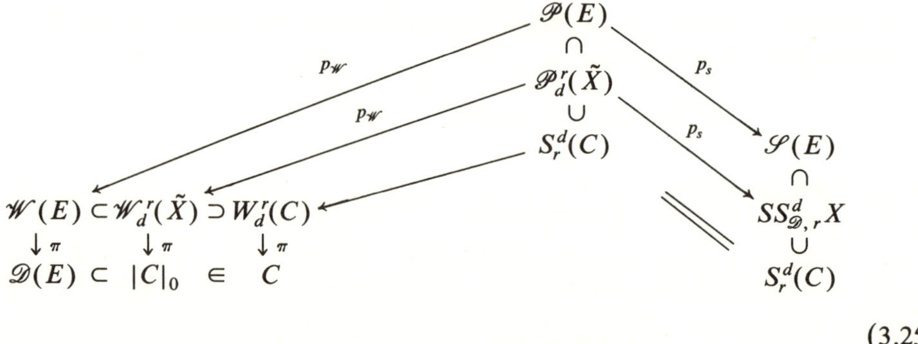

$$(3.25)$$

The projection $p_{\mathcal{W}}$ defines a $\mathbb{P}^r$-bundle structure. But

$$p_s^{-1}(\xi) = \mathbb{P} H^0\big(J_\xi(C)\big)_0$$

and

$$\dim p_s^{-1}(\xi) = \dim |C| - d + \delta(C, \zeta) = h^0(\mathcal{O}_C(C - \xi)). \qquad (3.26)$$

By (3.23), the composite

$$(p_{\mathscr{W}} \circ \pi): p_s^{-1}(\xi) \to |C|_0 \tag{3.27}$$

is the embedding $|C - \xi| \subset |C|$. $\qquad\square$

Chapter II. Applications

§4. Moduli spaces of simple sheaves and symmetric products of a $K3$-surface

Definition 4.1. (1) The $\mathbb{Z}$-module

$$\tilde{H}(X) = H^0(X, \mathbb{Z}) \oplus H^2(X, \mathbb{Z}) \oplus H^4(X, \mathbb{Z})$$

$$\begin{array}{ccccc}
 & \| & & \| & \\
\cup & \mathbb{Z} & \cup & \mathbb{Z} & \\
 & \cup & & \cup & \\
v & = \quad (r & , \quad \mathscr{D} & , \quad s &)
\end{array}$$

with the integral bilinear form:

$$\langle v, v' \rangle = \mathscr{D} \cdot \mathscr{D}' - (r \cdot s' + r' \cdot s)$$

for $v = (r, \mathscr{D}, s)$, $v' = (r', \mathscr{D}', s')$ is called the *Mukai lattice* of X.
(2) The sublattice of the Mukai lattice

$$\tilde{H}_a(X) = H^0(X, \mathbb{Z}) \oplus \operatorname{Pic} X \oplus H^4(X, \mathbb{Z})$$

is called the *algebraic Mukai lattice*.
(3) For each sheaf E on X with Chern classes $c_1(E) = \mathscr{D}$, $c_2(E) \in H^4(X, \mathbb{Z})$ $= \mathbb{Z}$, the vector

$$v(E) = \left(\operatorname{rk} E, \mathscr{D}, \tfrac{1}{2}\mathscr{D}^2 - c_2 + \operatorname{rk} E\right) \in \tilde{H}_a(X) \tag{4.1}$$

is called the *Mukai vector* of E.
By the Riemann-Roch theorem and Serre duality one gets

$$v^2(E) = \langle v(E), v(E) \rangle = -\chi(E, E)$$

$$= \dim \operatorname{Ext}^1_{\mathscr{O}_X}(E, E) - 2 \dim \operatorname{Ext}^0_{\mathscr{O}_X}(E, E),$$

$$\chi(E) = h^0(E) - h^1(E) + h^2(E) - r + s. \tag{4.2}$$

The Picard lattice $\operatorname{Pic} X$ acts on $\tilde{H}_a(X)$: for $\mathscr{D} \in \operatorname{Pic} X$,

$$T_{\mathscr{D}}(r, C, s) = \left(r, C + r\mathscr{D}, s + \frac{r}{2} \cdot \mathscr{D}^2 + \mathscr{D} \cdot C\right), \tag{4.3}$$

and one has the involution

$$v^*(r, \mathscr{D}, s) = (r, -\mathscr{D}, s).$$

For the Mukai vector of a sheaf E,

$$T_{\mathscr{D}}(v(E)) = v(E \otimes \mathcal{O}_X(\mathscr{D})) \quad \text{and} \quad v^*(E) = v(E^*). \tag{4.4}$$

LEMMA 4.1.   (1) $v(J_\xi(\mathscr{D})) = (1, \mathscr{D}, g_{\mathscr{D}} - d)$, where $\mathscr{D} \in \operatorname{Pic} X$,

$$g_{\mathscr{D}} = \frac{\mathscr{D}^2}{2} + 1 \quad \text{and} \quad d = \deg \xi; \tag{4.5}$$

(2) $v(E(\xi, \mathscr{D})) = (r + 1, \mathscr{D}, g_{\mathscr{D}} - d + r)$;
(3) $v(E(|\xi|, C)) = (r + 1, C, g_C - d + r) = (h^0(\mathcal{O}_C(\xi)), C, h^0(\mathcal{O}_C(K_C - \xi)))$;
(4) $v^2(E(\xi, \mathscr{D})) + 2 = 2\rho(r, d, g_{\mathscr{D}})$, where

$$\rho(r, d, g_{\mathscr{D}}) = g_{\mathscr{D}} - (r + 1)(g_{\mathscr{D}} - d + r) \tag{4.6}$$

is the Brill-Noether number (see [2], p. 159).

(5) $\dim B(E(\xi, \mathscr{D})) = \begin{cases} d - \rho - r \cdot h^2(\mathcal{O}_X\mathscr{D})) & \text{if } g_{\mathscr{D}} - d + r \geqslant 1, \\ 0 & \text{if } g_{\mathscr{D}} - d + r \leqslant 0. \end{cases}$

(6) $\dim F(E(|\xi|, C)) = g_C - \rho$.
(7) $\chi(E(|\xi|, C)) = g_C - \operatorname{Cliff}|\xi|$,
where $\operatorname{Cliff}|\xi| = d - 2r$ is the Clifford number of $|\xi|$.

All the equations can be checked by direct calculation.
Let

$$\tilde{H}_a^+(X) = \{(r, \mathscr{D}, s) \in \tilde{H}_a(X) | r \geqslant 2\}. \tag{4.7}$$

PROPOSITION 4.1 (Schwarzenberger [4].   *For each $v \in \tilde{H}_a^+(X)$ there exists a vector bundle E on X such that $v = v(E)$ (4.1).*

Definition 4.2.   A sheaf E on X is *simple* if $\operatorname{Ext}^0_{\mathcal{O}_X}(E, E) = \mathbb{C}$.
For each $v \in \tilde{H}_a(X)$ let $\operatorname{Sp}(v)$ be the moduli space of simple torsion-free sheaves with Mukai vector v.

PROPOSITION 4.2 (Mukai [3]).   *If $\operatorname{Sp}(v) \neq \varnothing$, then*
(1) $\operatorname{Sp}(v)$ *is smooth.*
(2) *For $E \in \operatorname{Sp}(v)$, the tangent space*

$$T\operatorname{Sp}(v)_E = \operatorname{Ext}^1_{\mathcal{O}}(E, E). \tag{4.8}$$

(3) $\dim \operatorname{Sp}(v) = v^2(E) + 2 = 2\rho(v)$, *where*

$$\rho(v(r + 1, \mathscr{D}, s)) = g_{\mathscr{D}} - (r + 1) \cdot s. \tag{4.9}$$

One has the operations

(i) $$\mathrm{Sp}(v) \leftrightarrow \mathrm{Sp}(v^*) \quad \text{by } E \longleftrightarrow E^*$$

and

(ii) $$\mathrm{Sp}(v) \leftrightarrow \mathrm{Sp}(T_{\mathscr{D}}(v)) \quad \text{by } E \longleftrightarrow E \otimes \mathcal{O}_X(\mathscr{D})$$

(see (4.3)).

If E is regular and $H^0(E) \otimes \mathcal{O}_X \xrightarrow{can} E$ is an epimorphism, then we can define $E' = (\ker can)^*$ by

$$0 \to (E')^* \to H^0(E) \otimes \mathcal{O}_X \xrightarrow{can} E \to 0 \qquad (4.10)$$

LEMMA 4.2.   (1) $E' = (\ker\ can)^*$ *is regular*;
(2) h^0 $(\mathrm{End}\ E) = h^0(\mathrm{End}\ E')$.

Proof.   The cohomology exact sequence of (4.10) and Serre duality give the equations for h^i. Since $H^1(E^*) = 0$ it follows that $H^0(E)^* = H^0(E')$ and *can* for E' is an epimorphism.

Consider the product morphism

$$\mu : H^0(E) \otimes H^0(E') \to H^0(E \otimes E').$$

The first parts of the cohomology exact sequences of (4.10) tensored with E' and its dual give us $H^0(\mathrm{End}\ E) = \ker \mu = H^0(\mathrm{End}\ E')$.   □

COROLLARY.   *The operation* $E \longleftrightarrow E'$ *defines a birational map*

$$\mathrm{Sp}(r + 1, \mathscr{D}, s) \overset{bir}{\longleftrightarrow} \mathrm{Sp}(s, \mathscr{D}, r + 1) \qquad (4.11)$$

which is called the reflection.

This reflection (4.11) is one step of the "Euclidean algorithm" by which A. N. Rudakov [9] obtained the refined description of exceptional bundles on the projective plane.

Let $C_+ \subset \mathrm{Pic}\ X$ be the cone of effective divisors, $C_- = -C_+$ the cone of anti-effective divisors and $\overline{C}_- = \mathrm{Pic}\ X - C_-$.

THEOREM 4.1.   (1) *For each* $\mathscr{D} \in \overline{C}_-$ *and each* $d > h^0(\mathcal{O}_X(\mathscr{D}))$ *the morphism*

$$\alpha : \widetilde{S^d X} \to \mathrm{Sp}(d - g_{\mathscr{D}}, \mathscr{D}, -1) \qquad (4.12)$$

given by $\alpha(\xi) = E(\xi, \mathscr{D})$ (1.12) *is birational.*

Proof. By (1.5), for general $\xi \in \widetilde{S^d X}$ we have $h^1(J_\xi(\mathscr{D})) = d - g_{\mathscr{D}} + 1$. By (4) of Lemma 1.1, for general ξ the sheaf $E(\xi, \mathscr{D})$ is locally free and simple. Now if $h^0(J_\xi(\mathscr{D})) = 0$ then

$$J_\xi(\mathscr{D}) = E(\xi, \mathscr{D})/can\left(H^0(E, \mathscr{D})\right) \otimes \mathscr{O}_X) \tag{4.13}$$

and α (4.12) is invertible. By Proposition 4.2, (3) $\dim \mathrm{Sp}(d - g_{\mathscr{D}}, \mathscr{D}, -1) = 2d = \dim \widetilde{S^d X}$. $\square$

COROLLARY 1.   *If $X \subset \mathbb{P}^g = |C|^*$ is a polarization of X then for any $k > 0$, $r > 0$ we have*

$$\mathrm{Sp}(r, kC, -1) \overset{\text{bir}}{=} S^{k^2}(g - 1) + r_X.$$

COROLLARY 2.   *If R is a smooth rational curve on X, then for any $k \geqslant 1$,*

$$\mathrm{Sp}(k^2, kR, -1) \overset{\text{bir}}{=} X.$$

COROLLARY 3.   *For any $\mathscr{D} \in \mathrm{Pic}\, X$ and any $r \geqslant \dim|\mathscr{D}|$, there is a component of $\mathrm{Sp}(r + 1, \mathscr{D}, -1)$ which is birational to a symmetric product of X.*

Indeed, by Theorem 4.1 and the operation $*$ we get the assertion. Consider the subset $\Delta \subset \tilde{H}_d^+(X)$ given by

$$\Delta = \left\{(r + 1, \mathscr{D}, -1) \in \tilde{H}_a^+(X) | r > \dim|\mathscr{D}| - g_{\mathscr{D}}\right\} \tag{4.14}$$

and $\Omega(X) \subset \tilde{H}_a^+(X)$ which is the union of the images of Δ under all operations (4.9) and (4.11). Then we have the following result.

COROLLARY 4.   *For $v \in \Omega(x)$ the variety $\mathrm{Sp}(v) \neq \varnothing$ and has a component which is birational to a symmetric product of X.*

Example.   Let $\mathscr{D}_1, \mathscr{D}_2 \in \mathrm{Pic}\, X$ with $\mathscr{D}_1\mathscr{D}_2 = 3$ and $(\mathscr{D}_1 - \mathscr{D}_2)^2 \neq 0$. Then $(2, \mathscr{D}_1 + \mathscr{D}_2, 2) \in \Omega(X)$.
Indeed,

$$T_{(-\mathscr{D}_2)}(2, \mathscr{D}_1 + \mathscr{D}_2, 2) = (2, \mathscr{D}_1 - \mathscr{D}_2, -1).$$

I hope that by applying all operations of the "Euclidean algorithm" (see the remark below (4.11)), one can prove that $\mathrm{Sp}(r, C, s)$ is birational to a symmetric product of X if and only if $\mathrm{Sp}(r, C, s) \neq \varnothing$ and there exists $\mathscr{D} \in \mathrm{Pic}\, X$ such that the greatest common divisor of $r, C \cdot \mathscr{D}, s$ is equal to 1.

The tangent bundle TX of a $K3$ surface X is simple and $v(TX) = (2, 0, -22)$. Thus by [3], TX is a nonsingular point of an irreducible component

$$\tau(X) \subset \mathrm{Sp}(2, 0, -22) \tag{4.15}$$

of the moduli space. By Proposition 4.2, (3), $\tau(X)$ is smooth and $\dim \tau(X) = 90$.

CONJECTURE. *The symplectic varieties $\tau(X)$ and $\widetilde{S^{45}X}$ are isogenous.*

Remark. The variety $\tau(X)$ is defined for any regular surface X for which $h^0(S^2 \Omega X) = 0$, and

$$\dim \tau(X) = 5(9p_a - K_X^2) = 5(\chi - 3p_a) = 5(h^{1,1} - p_a). \tag{4.16}$$

Do there exist regular surfaces for which the construction of Chapter I gives the birational type of $\tau(X)$?

§5. Brill-Noether theory for smooth curves on a $K3$ surface. The results of this section are contained in Lazarsfeld's paper [6].[1]

Definition 5.1. A smooth curve C (with $g_C \geqslant 5$) is called a Brill-Noether curve if

$$\dim W_d^r(C) = \rho(r, d, g_C) \quad \text{whenever } \rho < g_C,$$

where $W_d^r(C)$ is the variety (3.9), $\rho(r, d, g_c)$ is the Brill-Noether number (4.6); and $W_d^r(C) = \varnothing$ if $\rho < 0$.

LEMMA 5.1. *Let X be a smooth regular surface with $p_g \neq 0$ and $|C|$ a complete linear system on X without reducible curves. Then any torsion-free sheaf E on X with $h^2(E) = 0$, $c_1(E) = \mathrm{rk}\, E \cdot K_X + C$ and $F(E) \neq \varnothing$ is simple.*

Proof. First of all, for any K-fitting $\varphi \colon W \otimes K_X \to E$ the degeneracy set is a curve; if there is an exact sequence

$$0 \to W \otimes K_X \xrightarrow{\varphi} E \to S \to 0$$

with $\dim \mathrm{Supp}\, E = 0$ then the end of its cohomology exact sequence provides $0 \to W \otimes H^2(K_X) \to H^2(E) \to 0$, that is $h^2(E) \neq 0$. Now each K-fitting $\varphi \colon W \otimes K_X \to E$ defines a K-fitting $\varphi_1 \colon W_1 \otimes K_X \to E_1$ of any quotient sheaf $E \to E_1 \to 0$. If $E = E_1 \oplus E_2$ then the K-fittings $\varphi_i \colon W_i \otimes K_X \to E_i$ show that $c_1(E_i) = (\mathrm{rk}\, E_i) K_X + C_i$ and $C = C_1 + C_2$. Thus E is indecomposable; then following [1] we can define

$$N = \{ a \in H^0(\mathrm{End}\, E) \mid a^r = 0 \},$$

---

[1] The author is grateful to M. Reid for some extremely helpful discussions of this preprint.

and $\qquad$ (5.1)

$$\operatorname{Aut}_N E = \{\lambda \cdot \operatorname{id} + a | \lambda \in \mathbb{C}^*, a \in N\}.$$

The unipotent group $\operatorname{Aut}_N E$ acts on $\mathbb{P}H^0(E \otimes K_X^*)$, has the fixed point s and preserves the exact sequence

$$0 \to K_X \otimes \mathcal{O}_X(C_1) \to E \to E_1 \to 0, \qquad (5.2)$$

where $C_1 \subset (s)_0$ is a zero curve of s. In this case we have

$$1 \leftarrow \operatorname{Aut} E_1 \leftarrow \operatorname{Aut} E \leftarrow \operatorname{Ext}^0_{\mathcal{O}_X}(E, K_X \otimes \mathcal{O}_X(C_1)) \leftarrow 1. \qquad (5.3)$$

The end of the cohomology exact sequence of (5.2) gives $h^2(E_1) = h^2(E) = 0$, and E_1 has a K-fitting with degeneracy curve C_2. Then $C = C_1 + C_2$. If $C_1 = \varnothing$ and $\operatorname{Aut} E_1 \neq \mathbb{C}^*$ then by induction on $\operatorname{rk} E$ we have $C_2 = C_2' + C_2''$. If $\operatorname{Aut} E_1 = \mathbb{C}^*$, then $\operatorname{Ext}^0_{\mathcal{O}_X}(E, K_X) \neq 0$ and by Serre duality $h^2(E) \neq 0$. $\qquad \square$

THEOREM 5.1. *Let $\tilde{X}$ be a $K3$ surface quasipolarized by a complete linear system $|C|$ without reducible curves. Then*
(1) *for any smooth curve $C \in |C|_0$, if $\rho(r, d, g_C) < 0$ then $W_d^r(C) = \varnothing$;*
(2) *a general curve $C \in |C|_0$ is a Brill-Noether curve.*

Proof. For $C \in |C|_0$ and $\xi \in |\xi| \in W_d^r(C)$, by Lemmas 2.3 and 5.1 the block map (3.2)

$$\varphi^B_{E(|\xi|, C)} \colon F(E(|\xi|, C)) \to \widetilde{S^d X}$$

is an embedding. But $B(E(\xi, C))$ is a rational $(d - \rho)$-dimensional subvariety (see Lemma 4.1, (5)) of the smooth symplectic $2d$-dimensional variety $\widetilde{S^d X}$. Thus $d - \rho \leqslant d = \frac{1}{2} \dim \widetilde{S^d X}$ and we obtain assertion 1).

For any smooth curve C, if $\rho(r, d, g_C) > 0$ then $\dim W_d^r(C) \geqslant \rho$ (see [2], p. 181). Hence it is sufficient to prove that

$$\dim\left(\mathscr{W}_d^r(\tilde{X}) - \mathscr{W}_d^{r+1}(\tilde{X})\right) \leqslant \rho + g. \qquad (5.4)$$

By the diagram (3.25) and (3.26),

$$\dim p_{\mathscr{W}}^{-1}\left\{ p_s\left(\mathscr{W}_d^r(\tilde{X}) - \mathscr{W}_d^{r+1}(\tilde{X})\right)\right\} = \dim\left(\mathscr{W}_d^r(\tilde{X}) - \mathscr{W}_d^{r+1}(\tilde{X})\right) + d - g_C,$$

$$p_{\mathscr{W}}^{-1}\left\{ p_s\left(\mathscr{W}_d^r(\tilde{X}) - \mathscr{W}_d^{r+1}(\tilde{X})\right)\right\} \subset SS_{C,r}^d X - SS_{C,r+1}^d X,$$

where $SS_{C,r}^d X$ is the variety of C-stable cycles of (1.6). For (5.4) it is sufficient to prove

$$\dim\left(SS_{C,r}^d X - SS_{C,r+1}^d X\right) \leqslant \rho + d. \qquad (5.5)$$

But the fibre of the map

$$\alpha \colon \left(SS^d_{C,r}X - SS^d_{C,r+1}X \right) \to \mathrm{Sp}(r+1, C, g_C - d + r)$$

given by $\alpha(\xi) = E(\xi, C)$ is $\alpha^{-1}(E(\xi, C)) = B(E(\xi, C))$. Hence by Lemma 4.1,

$$\dim\left(SS^d_{C,r}X - SS^d_{C,r+1} \right)$$

$$\leqslant \underbrace{\dim \mathrm{Sp}(r+1, C, g_C - d + r)}_{2\rho} + \underbrace{\dim\ B(E(\xi, C)).}_{d-\rho} \qquad \square$$

§6. The infinitesimal Harris-Mumford conjecture.   Let $|C|$ be a complete linear system on a $K3$ surface X with $|C|_0 \neq \varnothing$. Consider the fibration (3.11)

$$\left(\mathscr{W}^r_d(\tilde{X}) - \mathscr{W}^{r+1}_d(\tilde{X}) \right) \xrightarrow{\pi} |C|_0. \tag{6.1}$$

Definition 6.1.   We say that a linear series $|\xi| \in W^r_d(C)$ is
(1) *non-singular* if $|\xi|$ is a non-singular point of $W^r_d(C) - W^{r+1}_d(C)$;
(2) *infinitesimally rigid* if the formal tangent space $TW^r_d(C)_{|\xi|} = 0$ (see [2], p. 189);
(3) *$\tilde{X}$-general* if the image of the component of $\mathscr{W}^r_d(\tilde{X}) - \mathscr{W}^{r+1}_d(\tilde{X})$ containing $(|\xi|, C)$ under π (6.1) is a Zariski open subset of $|C|$.

THEOREM 6.1.   *Any infinitesimally rigid $|\xi|$ on $C \in |C|_0$ is $\tilde{X}$-general.*

Proof.   By the theory of infinitesimal variations of linear systems on C (see [2], p. 189),

$$\dim TW^r_d(C)_{|\xi|} = \rho(r, d, g_C) + \dim \ker \mu_0, \tag{6.2}$$

where

$$\mu_0 \colon H^0\big(\mathcal{O}_C(\xi)\big) \otimes H^0\big(\mathcal{O}_C(K_C - \xi)\big) \to H^0\big(\mathcal{O}_C(K_C)\big) \tag{6.3}$$

is the Petri map induced by cup-product.
Let $|\xi|$ be an infinitesimally rigid linear system on $C \in |C|_0$. Then for the vector bundle $E(|\xi|, C)$ (2.11), consider the map (3.16) and its image $\mathscr{D}(E(|\xi|, C))$. By Lemma 3.4 the differential of this map at $(|\xi|, C)$ is the map (6.3). But rk $H^0\big(\mathcal{O}_C(\xi)\big) \otimes H^0\big(\mathcal{O}_C(K_C - \xi)\big) = g - \rho$ and $\dim \ker \mu_0 = \rho$. Thus the differential of (3.16) at $(|\xi|, C)$ is an epimorphism and $\mathscr{D}(E(|\xi|, C))$ is a Zariski open subset of $|C|$. $\square$

LEMMA 6.1.   *If $|\xi|$ is infinitesimally rigid on $C \in |C|_0$ then $E(|\xi|, C)$ is an infinitesimally rigid vector bundle, that is $h^1(\mathrm{End}\ E(|\xi|, C)) = 0$.*

26 A. N. TYURIN

Proof. By (3.15),

$$\dim \mathscr{W}(E) = \dim F(E) - \dim \operatorname{Aut} E + \dim \operatorname{Aut}_{\mathscr{W}} E,$$

$$g \overset{\parallel}{=} \rho \qquad \tfrac{1}{2}h^1(\operatorname{End} E) - \rho - 1 \qquad \overset{\parallel}{1}$$

and (6.4)

$$\dim \mathscr{W}(E(|\xi|, C)) = g_C - \tfrac{1}{2}h^1(\operatorname{End} E(|\xi|, C)).$$

Since $\dim \mathscr{D}(E(|\xi|, C)) \leqslant \dim \mathscr{W}(E(|\xi|, C))$, it follows that if $\dim \mathscr{D}(E(|\xi|, C)) = g$ then $h^1(\operatorname{End} E(|\xi|, C)) = 0$. $\square$

The variety $W_d^r(C)$ defines the family of vector bundles:

$$\{E(|\zeta|, C)\}, \ |\zeta| \in W_d^r(C). \tag{6.5}$$

If $|\xi|$ is non-singular, then the Kodaira-Spencer homomorphism

$$k_{|\xi|} \colon TW_d^r(C)_{|\xi|} \to H^1(\operatorname{End} E(|\xi|, C)) \tag{6.6}$$

exists.

The following assertion can be called the infinitesimal Harris-Mumford conjecture.

CONJECTURE. *If the Kodaira-Spencer homomorphism* (6.6) *is a monomorphism, then* $|\xi|$ *is* $\tilde{X}$-*general.*

By Lemma 4.1 (7) the Clifford number is the Euler characteristic of a sheaf. Therefore one may hope to get the proof of the Mumford-Harris conjecture from the infinitesimal one.

REFERENCES

1. M. F. ATIYAH, *Complex analytic connections in fibre bundles*, Trans. Amer. Math. Soc. **85** NI (1957), 181–207.
2. E. ARBARELLO, M. CORNALBA, P. A. GRIFFITHS AND J. HARRIS, *Geometry of Algebraic Curves*, Vol. I, Springer-Verlag, 1985.
3. S. MUKAI, *Symplectic structure of the moduli space of sheaves on an abelian or K3 surface*, Invent. Math. **77** (1984), 101–116.
4. R. L. E. SCHWARZENBERGER, *Vector bundles on algebraic surfaces*, Proc. London Math. Soc. **11** (1961), 601–622.
5. J. FOGARTY, *Algebraic families on an algebraic surface*, Amer. J. Math. **90** (1968), 511–521.
6. R. LAZARSFELD, *Brill-Noether-Petri without degenerations*, J. Diff. Geom. **23** (1986), 299–307.
7. M. GREEN AND R. LAZARSFELD, *Special divisors on curves on a K3 surface*, UCLA preprint (1986).
8. I. REIDER, *Vector bundles of rank 2 and linear systems on algebraic surfaces*, Univ. of Oklahoma preprint (1986).
9. A. L. GORODENKSEV AND A. N. RUDAKOV, *Exceptional vector bundles on projective space*, Duke Math. J. this issue.
10. A. N. TYURIN, *Special zero-cycles on polarised K3 surfaces*, Izvestiya ANSSR Ser. Mat. **51** N1 (1987).

STEKLOV MATHEMATICAL INSTITUTE, MOSCOW USSR

ON THE HOMOLOGY OF POISSON ALGEBRA OF ODD VARIABLES

O. OGIEVETSKY

To my teacher Yu. I. Manin on the occasion of his 50th birthday

Traditional algebraic K-theory is based on the group $GL(\infty)$. In the case of supergroups there are two analogs of this group; $GL(\infty|\infty)$ and $Q(\infty)$. Hence one can define two superanalogs of K-groups. They seem to complement each other; namely they seem to be the parts of different parity of bigger "super K-groups" (see, e.g. [5]). But in the supercase we have another remarkable group closely related to GL and Q. It is the group of canonical transformations in mechanics with odd variables. This group can be regarded as the classical (as opposed to quantum) version of GL and Q. Namely, this group has a nontrivial deformation; if the number of odd variables is even, it deforms into the group of GL-type and if the number is odd, it deforms into the group of Q-type. (Recall that in the case of anticommuting variables symplectic form is symmetric in the usual sense, so mechanics can be defined for any number of variables, not necessarily even, as in the case of commuting variables.) So it seems tempting to regard "the classical limit" of K-theory. This limit is very interesting. The computation of the classical additive $K_2^+(\Lambda)$ (it is sketched in §1) shows that it contains $\Omega^1(\Lambda)$. (Here Λ is a ring under investigation and $\Omega^1(\Lambda)$ is the space of differential one-forms of Λ.) This result should be compared with the "quantum" one: the usual $K_2^+(\Lambda)$ contains $\Omega^1(\Lambda)/d\Lambda$ rather than $\Omega^1(\Lambda)$ [2], [3]. It suggests the idea that the role of the classical additive Milnor group $K_i^+(\Lambda)$ is played by $\Omega^{i-1}(\Lambda)$. We show in §2 that the primitive part of the i-th stable homology group does really contain $\Omega^{i-1}(\Lambda)$ as the direct summand. This constitutes the main result of the paper. In §2.5 we make some notes about the first unstable homology group and the restriction of related cocycles to the algebra of linear canonical transformations.

Next, in §3 we investigate the stability questions. The naive embedding of the group of canonical transformations of N variables to the one of $N + 1$ variables (the embedded transformation acts only on the first N variables) does not suit us. One of the reasons is that the quantum deformations of these two groups are quite different. Another reason is that this embedding induces the zero map on the greater part of homology groups and so it cannot be used to define stable homology groups. Thus we have to look for another embedding. And really it

Received October 9, 1986.

does exist. It maps the group of canonical transformations of N variables right in the group of canonical transformations of $N + 2$ variables.

In §4 we discuss the quantization procedure. We find the point where the need for factorization over the exact forms arises.

In a later publication we hope to discuss more closely the algebraic structure of homology groups and to investigate cocycles like those arising in §4 and their generalizations, which give non-Milnor part of K^+-groups.

It is a pleasure to express my deep gratitude to A. Beilinson, D. Leites and V. Schechtman for many valuable conversations.

§0. Notations. In this section we introduce some notations and definitions which are used throughout the paper.

$\mathscr{K}$ is any field of zero characteristic; $\Lambda = \Lambda_0 \oplus \Lambda_1$ is a supercommutative algebra over $\mathscr{K}$. We use the letters $u, v, \ldots$ for even elements of Λ, the letters $\xi, \eta, \ldots$ for odd ones. Generic elements of Λ are denoted by $X, Y, \ldots$; $p(X) = 0, 1$ is the parity of X. $\Omega^i\Lambda$ is the space of differential i-forms, i.e., the combinations of elements $dX_1 \ldots dX_i \cdot X_{i+1}$ satisfying the properties: (a) $dX\,dY = (-1)^{(p(X)+1)(p(Y)+1)}\,dY\,dX$; (b) $dX \cdot Y = (-1)^{(p(X)+1)p(Y)}Y\,dX$; (c) $d(XY) = dX \cdot Y + (-1)^{p(X)}X\,dY$.

We shall frequently use ordered sets $M = \{m_1, \ldots, m_k\}$, where $1 \leqslant m_i \leqslant N$, $m_i \neq m_j$ for $i \neq j$. The number of elements in M is denoted by M. The ordered set $\{m_k, \ldots, m_1\}$ (with reversed order) is denoted by M^+. If M, L are two ordered sets, then define $ML = (M \cup L)\setminus(M \cap L)$, where the order is chosen in such a way that each element of $L\setminus M$ comes after all elements from $M\setminus L$. We have $(ML)^+ = L^+M^+$.

Consider the space of even functions of odd variables $\theta_1, \ldots, \theta_N$ with values in Λ, i.e., the space of linear combinations of elements $X\theta_{m_1} \ldots \theta_{m_k}$, where $p(X) \equiv k$ mod 2. The operation

$$[f, g] = \sum_{i=1}^{N} \frac{\partial f}{\partial \theta_i} \frac{\partial g}{\partial \theta_i} \tag{1}$$

makes this space the Lie algebra which is denoted by $\mathrm{Po}(N; \Lambda)$. It has a center which consists of elements $u \cdot 1$; the factor of $\mathrm{Po}(N; \Lambda)$ by this center is denoted by $\mathscr{H}(N; \Lambda)$. If a coefficient u in $u \cdot 1$ takes values only if $\Lambda_1^2 \subset \Lambda_0$ then we come to a subalgebra in $\mathrm{Po}(N; \Lambda)$ which we denote by $\widetilde{\mathrm{Po}}(N; \Lambda)$. It is this algebra that we study in the paper. The formula $f \mapsto \theta_i\,\partial f/\partial \theta_j - \theta_j\,\partial f/\partial \theta_i$ defines an automorphism of $\mathrm{Po}(N; \Lambda)$ which is denoted by T_{ij}.

Define the integral on the space of functions by the formula

$$\int X\theta^M = 0 \quad \text{if } \#M < N$$

$$\int X\theta_1 \ldots \theta_N = X.$$

We will study homology groups of $\widetilde{\mathrm{Po}}(N; \Lambda)$. If α is some cycle, then denote its image in the homology group by $[\alpha]$. If β, γ are two chains, then the notation $\beta \sim \gamma$ means that β is homological to γ.

If W is a $\mathbb{Z}_2$-graded space then $(W)_i$, $i = 0, 1$ are its graded components.

Finally, $(2n)!! = 2 \cdot 4 \cdot \cdots \cdot 2n$, $(2n + 1)!! = 1 \cdot 3 \cdot 5 \cdot 2/\cdot(2n + 1)$; $0!! = 1$.

§1. "Classical" K_2^+. The computation of $H_2(\mathscr{H}(N; \Lambda))$ where Λ is a supercommutative ring, is based on the explicit construction of the universal extension of $\mathscr{H}(N; \Lambda)$. (Recall that the second homology group is the center of such extension.) This central extension is analogous to the Steinberg algebra for $\mathfrak{gl}(\infty)$ (for the multiplicative version see [4]). We omit all the details. They can be found in [6].

1.1.   Consider the ordered sets $I = \{i_1, \ldots, i_k\}$ as in §0. If σ is a permutation of the set $\{1, \ldots, k\}$ then denote the set $\{i_{\sigma(1)}, \ldots, i_{\sigma(k)}\}$ by I^σ.

Definition.   Let $\mathscr{H} \mathrm{St}(N; \Lambda)$ (Steinberg algebra for $\mathscr{H}(N; \Lambda)$) be the Lie algebra generated by elements I_X, where $X \in \Lambda$, $p(X) \equiv \#I \mod 2$, and relations

$$I_X^\sigma = \mathrm{sgn}\, \sigma \cdot I_X$$

$$I_X + I_Y = I_{X+Y}$$

$$[(iI)_X, (Jj)_Y] = (JI)_{XY} \quad \text{if } I \cap J = \varnothing, I \cup J \neq \varnothing, i \notin I \cup J$$

$$[I_X, J_Y] = 0 \quad \text{if } I \neq J \text{ as unordered sets but } \#(I \cap J) \geqslant 2.$$

$$[I_X, J_Y] = 0 \quad \text{if } I \cap J = \varnothing, I \cup J \neq \{1, \ldots, N\}.$$

1.2.   The following formula defines a mapping from the algebra $\mathscr{H} \mathrm{St}(N; \Lambda)$ onto $\mathscr{H}(N; \Lambda)$:

$$I_X \mapsto X\theta_{i_1} \ldots \theta_{i_k}$$

where $I = \{i_1, \ldots, i_k\}$.

THEOREM (Cf. [6]).   *This mapping gives the universal central extension of* $\mathscr{H}(N; \Lambda)$.

1.3.   Let $N \geqslant 5$. One can show that the center of $\mathscr{H} \mathrm{St}(N)$ is generated by elements of three types:

$$A_i(\xi, \eta) = \left[i_\xi, i_\eta\right]; \qquad 1 \leqslant i \leqslant N; \xi, \eta \in \Lambda_1,$$

$$B_{ij}(u, v) = \left[ij_u, ij_v\right]; \qquad 1 \leqslant i < j \leqslant N; u, v \in \Lambda_0,$$

$$C_{K, L}(X, Y) = [K_X, L_Y], \quad \text{where } K \cap L = \varnothing, K \cup L = \{1, \ldots, N\}.$$

Then one can check that the elements $A_i(\xi, \eta)$ do not depend on i, $B_{ij}(u, v)$ do not depend on i, j and depend only on the images of u, v in Λ_{red}, where $\Lambda_{\mathrm{red}} = \Lambda/(\Lambda_1)$ is the factor of Λ by the ideal (Λ_1) generated by all odd elements. Finally, $C_{K,L}(X, Y)$ depend only on $a = \#K$ and the sign ε determined by the equality $\theta_{k_1} \ldots \theta_{k_a} \theta_{l_1} \ldots \theta_{l_{N-a}} = \varepsilon \theta_1 \ldots \theta_N$. If $\varepsilon = 1$ then denote the corresponding $C_{K,L}$ by C_a. It turns out that all the elements C_α can be expressed in terms of C_2 and C_1. Now let T be the wedge square of the module Λ_1 over Λ_0; that is, elements in T are of the form $\xi \wedge \eta$ with $\xi \wedge \eta = -\eta \wedge \xi$, $\xi \wedge u\eta = u\xi \wedge \eta$. One can show that the following formulas

$$\xi \wedge \eta \mapsto [i\xi, i\eta]$$

$\bar{u}\, d\bar{v} \mapsto [ij_u, ij_v]$ where u, v are arbitrary preimages of $\bar{u}, \bar{v} \in \Lambda_{\mathrm{red}}$

$$\begin{cases} du \cdot v \mapsto C_2(u, v) \\ d\xi \cdot \eta \mapsto -C_1(\xi, \eta) - \dfrac{1}{N-4} C_2(\xi\eta, 1) \end{cases} \quad \text{for } N \text{ even}$$

$$\begin{cases} du \cdot \xi \mapsto C_2(u, \xi) \\ d\xi \cdot u \mapsto C_1(\xi, u) - \dfrac{1}{N-3} C_1(\xi u, 1) \end{cases} \quad \text{for } N \text{ odd}$$

give the well-defined mappings

$$T \to H_2(\mathscr{H}\, \mathrm{St}(N; \Lambda), \mathscr{K})$$

$$\Omega^1\Lambda_{\mathrm{red}}/d\Lambda_{\mathrm{red}} \to H_2(\mathscr{H}\, \mathrm{St}(N; \Lambda), \mathscr{K})$$

$$(\Omega^1\Lambda)_{(N+1)\bmod 2} \to H_2(\mathscr{H}\, \mathrm{St}(N; \Lambda), \mathscr{K}).$$

1.4.   Let $f, g \in \mathscr{H}(N; \Lambda)$. Define the mappings

$$\alpha_A: \mathscr{H}(N; \Lambda) \wedge \mathscr{H}(N; \Lambda) \to T$$

$$\alpha_B: \mathscr{H}(N; \Lambda) \wedge \mathscr{H}(N; \Lambda) \to \Omega^1\Lambda_{\mathrm{red}}/d\Lambda_{\mathrm{red}}$$

$$\alpha_C: \mathscr{H}(N; \Lambda) \wedge \mathscr{H}(N; \Lambda) \to (\Omega^1\Lambda)_{(N+1)\bmod 2}$$

(where $\wedge$ means the wedge product over the basic field) by the formulas

$$\alpha_A(f, g) = \sum_i \frac{\partial}{\partial \theta_i}(\Pi_{-1}f) \wedge \frac{\partial}{\partial \theta_i}(\Pi_{-1}g)$$

where Π_{-1} is the projector on the eigenspace of E corresponding to eigenvalue

-1. Note that $\Pi_{-1} = 1/\Gamma(-E)$ where $\Gamma(z)$ is Euler gamma-function;

$$\alpha_B(f, g) = \sum_{i, j} \left[\frac{\partial}{\partial \theta_i} \frac{\partial}{\partial \theta_j} d\Pi_0 \bar{f} \cdot \frac{\partial}{\partial \theta_i} \frac{\partial}{\partial \theta_j} \Pi_0 \bar{g} \right]$$

where $\bar{f}, \bar{g}$ are the images of $f, g \in \Lambda$ in Λ_{red}, $[d\bar{u} \cdot \bar{v}]$ is the image of the form $d\bar{u} \cdot v$ in $\Omega^1 \Lambda_{\mathrm{red}}/d\Lambda_{\mathrm{red}}$, Π_0 is the projector on the eigenspace of E corresponding to eigenvalue 0,

$$\Pi_0 = \frac{1}{\Gamma(1 - E)} - \frac{1}{\Gamma(-E)};$$

$$\alpha_c(f, g) = \int df \cdot Eg - dg \cdot Ef.$$

LEMMA. $\alpha_A, \alpha_B, \alpha_C$ *are the cocycles of* $\mathscr{H}(N; \Lambda)$

1.5.   Combining §1.3 and §1.4 we obtain the following theorem.

THEOREM (Cf. [6]).

$$H_2(\mathscr{H}(N; \Lambda), \mathscr{K}) = (\Omega^1 \Lambda)_{(N+1) \bmod 2} \oplus \Omega^1 \Lambda_{\mathrm{red}}/d\Lambda_{\mathrm{red}} \oplus T.$$

§2. Higher homology groups.   In §2.1–2.3 we show that $H_i(\widetilde{\mathrm{Po}}(N; \Lambda), \mathscr{K})$ contains $\Omega^{i-1}(\Lambda)$ as the direct summand when $i < N/2$. §2.4 contains some results about cycles which will be useful in §4. In addition, in §2.5 we discuss the first unstable homology group.

2.1.   First, let us construct a mapping from $H_i(\widetilde{\mathrm{Po}}(N; \Lambda), \mathscr{K})$ to $\Omega^{i-1}(\Lambda)$. It is given by a cocycle. Let

$$\varphi_i(A_1, A_2, \ldots, A_i) = \frac{1}{(N - 2i)!!} \int \sum_{j=1}^{i} (-1)^{i+j} dA_1 \ldots EA_j \ldots dA_i \quad (2)$$

where E is Euler operator (see §0). (The factor $1/(N - 2i)!!$ is needed to ensure agreement with the stability map.)

PROPOSITION. φ_i *is a cocycle.*

Proof.   Consider the summands in $\delta\varphi_i(A_1, \ldots, A_{i+1})$ which contain $dA_1, \ldots, dA_{i-1}$. Since $\delta\varphi_i$ is antisymmetric it is sufficient to prove that the sum of such summands equals zero. Denote this sum by P and let $\Omega = dA_1 \ldots dA_{i-1}$. Then

$$P = \int \sum_{j=1}^{i-1} (-1)^j \left[dA_j, A_{i+1} \right] dA_1 \ldots \widehat{dA_j} \ldots dA_{i-1} \cdot EA_i$$

$$+ \sum_{j=1}^{i-1} (-1)^{j-1} \left[dA_j, A_i \right] dA_1 \ldots \widehat{dA_j} \ldots dA_{i-1} \cdot EA_{i+1}$$

$$- E\left(\left[A_i, A_{i+1} \right] \right) \cdot \Omega$$

32   O. OGIEVETSKY

where $\widehat{dA_j}$ means that the factor dA_j is deleted. Let $EA_j = (a_j - 2)A_j$. Integrating the first and the third terms by parts we obtain

$$P = \int A_{i+1}\left\{ \sum_{j,k}(-1)^{j+1}\frac{\partial}{\partial\theta_k}(dA_j)\frac{\partial}{\partial\theta_k}\left(dA_1\ldots\widehat{dA_j}\ldots dA_{i-1}\cdot A_i\right)(a_i - 2)\right.$$

$$+\sum_{j,k}(-1)^{j+1}\frac{\partial}{\partial\theta_k}(dA_j)\frac{\partial}{\partial\theta_k}(A_i)\,dA_1\ldots\widehat{dA_j}\ldots dA_{i-1}(a_{i+1}-2)$$

$$\left. -(a_i + a_{i+1} - 4)\sum_k\frac{\partial}{\partial\theta_k}(A_i)\frac{\partial}{\partial\theta_k}(\Omega)\right\}.$$

It is easy to see that the factor in the curly brackets equals zero. ∎

2.2.   Now we give the explicit expressions for cycles. Let

$$C_i(u) = u\theta_i\theta_{i+1},$$

$$C_i(\xi, \eta) = \tfrac{1}{2}\left(\xi\theta_i \wedge \eta\theta_{i+1}\theta_{i+2}\theta_{i+3} + \eta\theta_i \wedge \xi\theta_{i+1}\theta_{i+2}\theta_{i+3}\right).$$

PROPOSITION.   *Let $i < N/2$. Then the following formulas define a mapping ρ_i:* $\Omega^{i-1}(\Lambda) \to H_i(\widetilde{\mathrm{Po}}(N; \Lambda), \mathcal{X})$

$$du_1\ldots du_t\,d\xi_1\,d\eta_1\ldots d\xi_k\,d\eta_k\cdot X$$

$$\mapsto \left[\mu C_1(u_1) \wedge \cdots \wedge C_{2t-1}(u_t) \wedge C_{2t+2}(\xi_1, \eta_2) \wedge \cdots \wedge\right.$$

$$\left. C_{2t+4k-3}(\xi_k, \eta_k) \wedge X\theta_{2t+4k+1}\ldots\theta_N\right], \qquad p(X) \equiv N\bmod 2 \quad (3)$$

$$du_1\ldots du_t\,d\xi_1\,d\eta_1\ldots d\xi_{k-1}\,d\eta_{k-1}\,d\xi_k$$

$$\cdot X \mapsto \left[-\mu C_1(u_1) \wedge \cdots \wedge C_{2t-1}(u_t)\right.$$

$$\wedge C_{2t+1}(\xi_1, \eta_1) \wedge \cdots \wedge C_{2t+4k-7}(\xi_{k-1}, \eta_{k-1}) \wedge \tfrac{1}{2}(\xi_k\theta_{2t+4k-3}$$

$$\left. \wedge X\theta_{2t+4k-2}\ldots\theta_N + \xi_k\theta_{2t+4k-3}\theta_{2t+4k-2}\theta_{2t+4k-1} \wedge X\theta_{2t+4k}\ldots\theta_N)\right],$$

$$p(X) \equiv (N + 1)\bmod 2 \quad (4)$$

where $\mu = (N - 2i - 2)!!$ with $i - 1 = t + 2k$ in the first formula and $i - 1 = (t + 2k - 1)$ in the second one; [ ] means the image in the homology group of a cycle standing in the brackets.

Proof. First, we have to check the (anti)symmetry of constructed expressions in du, $d\xi$, $d\eta$. We shall widely use the following remark:

Remark. The automorphisms T_{ij} (see §0) are inner and so they act trivially on homology groups (it is obviously so in the case when $\mathbb{C} \subset \mathcal{K}$; other cases can be reduced to this one by Lefschetz principle).

After this the only nontrivial thing remaining is the symmetry of $C_1(\xi_1, \eta_1) \wedge C_5(\xi_2, \eta_2)$ in ξ_1, ξ_2. Using the automorphisms T_{ij}, we see that

$$\xi_1\theta_1 \wedge \eta_1\theta_2\theta_3\theta_4 \wedge \xi_2\theta_5 \wedge \eta_2\theta_6\theta_{\theta 8} \sim \xi_2\theta_1 \wedge \eta_1\theta_2\theta_3\theta_4 \wedge \xi_1\theta_5 \wedge \eta_2\theta_6\theta_7\theta_8,$$

$$\xi_1\theta_1\theta_2\theta_2 \wedge \eta_1\theta_4 \wedge \xi_2\theta_5\theta_6\theta_7 \wedge \eta_2\theta_8 \sim \xi_2\theta_1\theta_2\theta_3\eta_1\theta_4 \wedge \xi_1\theta_5\theta_6\theta_7 \wedge \eta_2\theta_8.$$

Therefore, if

$$A = C_1(\xi_1, \eta_1) \wedge C_5(\xi_2, \eta_2) - C_1(\xi_2, \eta_1) \wedge C_5(\xi_1, \eta_2)$$

then

$$A \sim \eta_1\theta_2\theta_3\theta_4 \wedge \eta_2\theta_5 \wedge \left(-\xi_1\theta_1 \wedge \xi_2\theta_6\theta_7\theta_8 - \xi_2\theta_2\theta_3\theta_4 \wedge \xi_1\theta_5\right)$$

$$+ \eta_1\theta_1 \wedge \eta_2\theta_6\theta_7\theta_8(\xi_1\theta_2\theta_2\theta_4 \wedge \xi_2\theta_5 - \xi_2\theta_2\theta_3\theta_4 \wedge \xi_1\theta_5).$$

Using the identity

$$0 \sim \delta(\theta_1\theta_2\theta_3\theta_4 \wedge \xi\theta_1 \wedge \eta\theta_1)$$

$$= \xi\theta_2\theta_3\theta_4 \wedge \eta\theta_1 = \xi\theta_1 \wedge \eta\theta_2\theta_3\theta_4 + \xi\eta \wedge \theta_1\theta_2\theta_3\theta_4 \tag{5}$$

we can write

$$A \sim \eta_1\theta_2\theta_3\theta_4 \wedge \eta_2\theta_5 \wedge \xi_1\xi_2 \wedge \theta_1\theta_6\theta_7\theta_8 + \eta_1\theta_1 \wedge \eta_2\theta_6\theta_7\theta_8 \wedge \xi_1\xi_2 \wedge \theta_2\theta_3\theta_4\theta_5.$$

Applying the automorphism $T_{15} \circ T_{26} \circ T_{37} \circ T_{48}$ to the second summand, we obtain

$$A \sim \xi_1\xi_2 \wedge \theta_1\theta_6\theta_7\theta_8 \wedge \left(\eta_1\theta_2\theta_3\theta_4 \wedge \eta_2\theta_5 + \eta_1\theta_5 \wedge \eta_2\theta_2\theta_3\theta_4\right).$$

Again using the identity (5), we obtain

$$A \sim -\xi_1\xi_2 \wedge \eta_1\eta_2 \wedge \theta_1\theta_6\theta_7\theta_8 \wedge \theta_2\theta_3\theta_4\theta_5.$$

Applying $T_{12} \circ T_{63} \circ T_{74} \circ T_{85}$ to $\theta_1\theta_6\theta_7\theta_8 \wedge \theta_2\theta_3\theta_4\theta_5$ we see that

$$\theta_1\theta_6\theta_7\theta_8 \wedge \theta_2\theta_3\theta_4\theta_5 \sim \theta_2\theta_3\theta_4\theta_5 \wedge \theta_1\theta_6\theta_7\theta_8 = -\theta_1\theta_6\theta_7\theta_8 \wedge \theta_2\theta_3\theta_4\theta_5.$$

Hence, $A \sim 0$ as claimed.

It remains to show that the cycles above satisfy the Leibnitz rule up to a boundary. It can be done by examining several cases depending on the type ((3) or (4)) of the form. The details are purely technical and we omit them. ∎

2.3. THEOREM. *Let $i < N/2$. Then $(\Omega^{i-1}(\Lambda))_{i+1+N \bmod 2}$ is a direct summand of $H_i(\widetilde{Po})(N; \Lambda), \mathscr{K})$.*

Proof. The cocycle φ_i constructed in §2.1 gives the mapping φ_i: $H_i(\widetilde{Po}(N; \Lambda), \mathscr{K}) \to \Omega^{i-1}(\Lambda)$. This mapping is obviously subjective. Next, it is easy to see that the mapping ρ_i from §2.2 is the right inverse to φ_i up to a sign. This proves the theorem. ∎

2.4.   One can try to generalize the cycles from §2.2 in the following way. Let M_t, $t = 1, \ldots, T$ be ordered sets (as in §0) such that $\theta^{M_1} \ldots \theta^{M_t} = \theta_1 \ldots \theta_N$. Then it is easy to see that any chain of the form $\alpha = X_1 \theta^{M_1} \wedge \cdots \wedge X_T \theta^{M_T}$ (where $p(X_t) \equiv \#M_t \bmod 2$) is a cycle. But such cycles do not give any new elements in the homology group. In this part we explain this fact:

Fix a positive integer $m, m \leqslant N$. Let $\alpha = X_4 \theta^{M_4} \wedge \cdots \wedge X_T \theta^{M_T}$ where $\theta^{M_4} \ldots \theta^{M_T} = \theta_m \theta_{m+1} \ldots \theta_N$. For a positive integer $i < m$ set

$$\mathscr{D}_i(X, Y) = X\theta_1 \ldots \theta_i \wedge Y\theta_{i+1} \ldots \theta_m \wedge \alpha \tag{6}$$

where $p(X) \equiv i \bmod 2$, $p(Y) \equiv (m - i) \bmod 2$. Let $\theta_1 \theta^{M_1} \theta^{M_2} \theta^{M_3} = \theta_1 \ldots \theta_m$, $\#M_1 = i$, $\#M_2 = j$, $\#M_3 = m - i - j - 1$. Taking the boundary of the chain $X\theta_1 \theta^{M_1} \wedge Y\theta_1 \theta^{M_2} \wedge Z\theta_1 \theta^{M_3} \wedge \alpha$ and using the remark from the proof of the Proposition in §2.2, we obtain the following set of equations for $\mathscr{D}_i(X, Y)$:

$$\mathscr{D}_{i+j}(XY, Z) = \mathscr{D}_{j+1}(Y, ZX) + (-1)^{(j+1)(m+1)} \mathscr{D}_{i+1}(X, YZ). \tag{7}$$

In addition, let $\theta^I \theta^J = \theta_1 \ldots \theta_m$, $\#I = i$, $\#J = m - i$. Writing $X\theta^I \wedge Y\theta^J = -Y\theta^J \wedge X\theta^I$ and rearranging θ's with the help of the automorphisms T_{jk}, we obtain the following symmetry property of $\mathscr{D}_i(X, Y)$:

$$\mathscr{D}_i(X, Y) = (-1)^{i(m-i)} \mathscr{D}_{m-i}(Y, X). \tag{8}$$

The set of equations (7), (8) is not hard to solve.

PROPOSITION.   *If m is even, then*

$$\mathscr{D}_{2i}(u, v) = \mathscr{D}_2(u, v) - \frac{2(i - 1)}{m - 4} \mathscr{D}_2(uv, 1)$$

$$\mathscr{D}_{2i+1}(\xi, \eta) = \mathscr{D}_1(\xi, \eta) + \frac{2i}{m - 4} \mathscr{D}_2(\xi\eta, 1). \tag{9}$$

If m is odd, then

$$\mathscr{D}_{2i}(u, \xi) = \mathscr{D}_2(u, \xi) - \frac{2(i-1)}{m-3}\mathscr{D}_1(u\xi, 1)$$

$$\mathscr{D}_{2i+1}(\xi, u) = -\mathscr{D}_2(u, \xi) + \frac{m-3-2i}{m-3}\mathscr{D}_1(\mu\xi, 1). \qquad (10)$$

Proof. We shall give the proof only in the case of even m. The considerations in the case of odd m are quite analogous. Let e_i be the difference between $\mathscr{D}_i$ and its expression as given in the Proposition. Then the elements e_{2i} satisfy the following equations:

$$e_2(u, v) = 0$$

$$e_{2i+2j-2}(uv, w) = e_{2i}(u, vw) + e_{2_j}(v, uw)$$

$$e_{2i}(u, v) = -e_{m-2i}(v, u).$$

Substituting $j = 1$ in the second equation and using the first one we obtain

$$e_{2i}(uv, w) = e_{2i}(u, vw).$$

Hence $e_{2i}(u, v) = e_{2i}(1, uv)$. Let $f_{2i}(u) = e_{2i}(1, u)$. Then

$$f_{2i+2j-2}(u) = f_{2i}(u) + f_{2j}(u)$$

and we conclude that $f_{2i}(u) = (i-1)f_4(u)$. In addition $f_{2i}(u) = -f_{m-2i}(u)$ by (8); therefore, $(i-1)f_4 = -(m-i-1)f_4$. It follows that $f_4 = 0$ and therefore $f_{2i} = 0$.

Next, it is easy to see that $e_1(\xi, \eta) = 0$ and $e_{2i+1}(u\xi, \eta) = e_{2j+1}(\xi, u\eta)$. So $e_{2i+1}(\xi, \eta) = 0$ and this concludes the proof. ∎

Properly applying this proposition to first and second, second and third, ... wedge factors in the cycle $X_1\theta^{M_1} \wedge \cdots \wedge X_T\theta^{M_T}$ (where $\theta^{M_1} \ldots \theta^{M_T} = \theta_1 \ldots \theta_N$) one can easily express it in terms of the standard cycles introduced in §2.2.

2.5. Homology groups H_i with $i \geqslant N/2$ are also very interesting. In this part we discuss the case when $\Lambda = \Lambda_0$ is a commutative ring, $N = 2n$ and $i = n$. First of all let us look at n-cycles of the form $u_1\theta_1\theta_2 \wedge \cdots \wedge u_n\theta_{2n-1}\theta_{2n}$. The cocycle φ_n applied to these cycles obviously gives zero. But now we can define the cocycle with values in $\Omega^{n-1}/d\Omega^{n-2}$.

$$\chi^n(f_1, \ldots, f_n) = \int [df_1 \ldots df_{n-1} \cdot fn] \qquad (11)$$

where $[\omega]$ means the image of the form $\omega \in \Omega^{n-1}$ in $\Omega^{n-1}/d\Omega^{n-2}$.

Next, computing the boundary of a chain

$$u_1\theta_1\theta_2 \wedge \cdots \wedge u_{n-2}\theta_{2n-5}\theta_{2n-4} \wedge u_{n-1}\theta_{2n-3}\theta_{2n} \wedge \theta_{2n-2}\theta_{2n} \wedge \theta_{2n-1}\theta_{2n}$$

and applying the remark from the proof of Lemma in §2.2, one can check that the cycle $u_1\theta_1\theta_2 \wedge \cdots \wedge u_{n-1}\theta_{2n-3}\theta_{2n-2} \wedge \theta_{2n-1}\theta_{2n}$ is homological to zero. So the formula

$$du_1 \ldots du_{n-1} \cdot u_n \mapsto [u_1\theta_1\theta_2 \wedge \cdots \wedge u_n\theta_{2n-1}\theta_{2n}]$$

defines a mapping from $\Omega^{n-1}/d\Omega^{n-2}$ to H_n. Combining this with the cocycle χ_n we conclude that in contrast with $H_{i,\ i<n}$, the homology group H_n contains $\Omega^{n-1}/d\Omega^{n-2}$ as the direct summand.

Note. Elements $u\theta_i\theta_j$ from the algebra $\mathfrak{so}(2n; \Lambda)$, serves as the generator L_{ij} of rotations in the plane (ij). The restriction of χ_n gives the following cocycle of $\mathfrak{so}(2n; \Lambda)$ with values in $\Omega^{n-1}/d\Omega^{n-2}$:

$$\chi_n\left(u_1 L_{i_1 j_1}, \ldots, u_n L_{i_n j_n}\right) = [du_1 \ldots du_{n-1} \cdot u_n]\varepsilon_{i_1 j_1 \ldots i_n j_n}$$

where $\varepsilon_{i_1 j_1 \ldots i_n j_n}$ is the absolutely antisymmetric symbol.

The analogous cocycle for the algebra $\mathfrak{so}(2n + 1; \Lambda)$ takes values in nontrivial representation. Namely, let $\{e_i\}$ be the basis of $(2n + 1)$-dimensional representation V of $\mathfrak{so}(2n + 1)$. Then the following formula defines the n-cocycle of $\mathfrak{so}(2n + 1; \Lambda)$ with coefficients in $V \otimes \Omega^n$:

$$\psi_n\left(u_1 L_{i_1 j_1}, \ldots, u_n L_{i_n j_n}\right) = du_1 \ldots du_n \sum_k \varepsilon_{i_1 j_1 \ldots i_n j_n k} \cdot e_k.$$

§3. Stability

3.1.   Now we turn to a stability map. As we remarked in the introduction, the simple-minded embedding $\text{Po}(N) \to \text{Po}(N + 1)$ (under which functions of N variables map to functions which do not depend on the $(N + 1)$th variable) does not give us what we want. In fact, one can easily show that this embedding induces the zero mapping on the parts of homology groups which we are studying now. Next, this embedding does not agree with quantization, since $\text{Po}(2n)$ and $\text{Po}(2n + 1)$ deform into algebras of quite different types (see §4.1). So we need another embedding, under which the algebra $\text{Po}(N)$ maps right into $\text{Po}(N + 2)$ and this embedding cannot be factorized through $\text{Po}(N + 1)$. Such embedding can be constructed with the help of the Euler operator E. Let $\theta_{N+1}, \theta_{N+2}$ be two new odd variables. For $f \in \text{Po}(N)$ define

$$\varepsilon(f) = f + E(f) \cdot \theta_{N+1}\theta_{N+2} \in \text{Po}(N + 2). \tag{12}$$

LEMMA. *ε is an embedding.*

Proof. Straightforward. ∎

Definition. Let $\mathrm{Po}_0 = \varprojlim_{\varepsilon} \mathrm{Po}(2n)$, $\mathrm{Po}_1 = \varprojlim_{\varepsilon} \mathrm{Po}(2n+1)$.

3.2. All the constructions from §2 agree with ε.

LEMMA. *Let $i < N/2$. Then the following diagrams are commutative:*

$$(1) \qquad H_i(\widetilde{\mathrm{Po}}(N; \Lambda), \mathscr{K}) \xrightarrow{\varepsilon} H_i(\widetilde{\mathrm{Po}}(N+2; \Lambda), \mathscr{K})$$

$$\searrow \varphi_i \qquad \varphi_i \swarrow$$

$$\Omega^{i-1}(\Lambda)$$

where φ_i is the cocycle defined by (2);

$$(2) \qquad H_i(\widetilde{\mathrm{Po}}(N; \Lambda), \mathscr{K}) \xrightarrow{\varepsilon} H_i(\widetilde{\mathrm{Po}}(N+2; \Lambda), \mathscr{K})$$

$$\searrow \rho_i \qquad \rho_i \swarrow$$

$$\Omega^{i-1}(\Lambda)$$

where ρ_i is the mapping from the proposition in §2.2.

Proof. (1) It easily follows from the identity

$$\int EX_1 \cdot X_2 \cdots \cdots X_m + X_1 \cdot EX_2 \cdots \cdots X_m + \cdots + X_1 \cdots \cdots X_{m-1}$$

$$\cdot EX_m = (N - 2m) \int X_1 \ldots X_m.$$

(2) Let α be some cycle of $\widetilde{\mathrm{Po}}(N; \Lambda)$. Set $t = \theta_{N+1}\theta_{N+2}$. We can consider $\varepsilon(\alpha)$ as the cycle of $\widetilde{\mathrm{Po}}(N; \Lambda[t])$. Now let α be the image of a differential form ω. Obviously, $\varepsilon(C_i(u)) = C_i(u)$. Next,

$$\varepsilon(2c_i(\xi, \eta)) = 2C_i(\xi, \eta) - \xi t\theta_i \wedge \eta\theta^I + \xi\theta_i \wedge \eta t\theta^I - \eta t\theta_i \wedge \xi\theta^I$$

$$+ \eta\theta_i \wedge \xi t\theta^I - \xi t\theta_i \wedge \eta t\theta^I - \eta t\theta_i \wedge \xi t\theta^I$$

where $I = \{i + 1, i + 2, i + 3\}$. Using the equations (7) one can check that the sum of all terms in the right hand side, except $2C_i(\xi, \eta)$, is homological to zero.

38   O. OGIEVETSKY

So $\varepsilon(C_i(\xi, \eta)) \sim C_i(\xi, \eta)$. Now it is easy to compare $\varepsilon(\alpha)$ with the image of ω directly in $H_*(\widetilde{\mathrm{Po}}(N + 2; \Lambda), \mathcal{X})$ and to conclude the proof. We leave it to the courageous reader. ∎

3.3.   Using this lemma we come to the following variant of Theorem 2.3:

THEOREM. $H_i(\widetilde{\mathrm{Po}_\mu}(\Lambda), \mathcal{X})$ *contains* $(\Omega^{i-1}(\Lambda))_{(\mu+i-2) \bmod 2}$ *as the direct summand* (*here* $\mu = 0, 1$).

§4. Quantization.   The algebra $\mathrm{Po}(N)$ has a nontrivial deformation to the corresponding "quantum" algebra. We want to investigate what happens when we try to quantize all the previous constructions.

4.1.   We recall some concepts in a convenient form. Let f, g be functions of odd variables $\theta_1, \ldots, \theta_N$. Set $\partial_{I_0} f = \partial/\partial\theta_{i_1} \ldots \partial/\partial\theta_{i_k} f$, where $I_0 = \{i_1, \ldots, i_k\}$ is an order set. The following expression defines a new function, which is denoted $f * g$:

$$f * g + \sum_{\text{unordered } I} \hbar^{\#I} \partial_{I_0} f \cdot \partial_{I_0^+} g \tag{13}$$

where I_0 is the set I supplied with any order (it is easy to see that the corresponding expression does not depend on the choice of order on I).

Fact.   The operation $*$ is associative, $(f * g) * t = f * (g * t)$

$$\left(= \sum_{I, K, L} \hbar^{\#I + \#K + \#L} \partial_{K_0 I_0} f \cdot \partial_{I_0^+ L_0} g \cdot \partial_{L_0^+ K_0^+} t \right.$$

with the above notations). (The formula (13) gives the composition law for Weyl symbols (see [1])).

Note.   This formula has a remarkable property: it does not contain denominators; so, in contrast with the even quantum mechanics, the odd one can be defined by the formula (13) is any characteristic.

Now let $s = \hbar^2$. Then the deformation of $\mathrm{Po}(N)$ is given by the formula (with the above notations):

$$[f, g] = \sum_{I: \, \#I \text{ is odd}} s^{\frac{1}{2}(\#I - 1)} \partial_{I_0} f \cdot \partial_{I_0} + g.$$

For a while let $\mathbb{C}$ be the basic field. If $N = 2n$ then dividing the variables $\theta_1, \ldots, \theta_{2n}$ into pairs $(\theta_1, \theta_2), (\theta_3, \theta_4), \ldots$ and going to the basis $\theta_{2i+1} \pm \sqrt{-1}\,\theta_{2i}$, we see by standard argument that the deformed algebra is the full algebra of differential operators on the space of functions of variables $\theta_{2i+1} + \sqrt{-1}\,\theta_{2i}$. In the case of odd variables the full algebra of differential operators is the full algebra of linear operators on the space of functions of these variables (it is easily seen by counting the dimensions of both algebras). So we see that $\mathrm{Po}(2n)$ deforms into $\mathfrak{gl}(2^{n-1}|2^{n-1})$. Now let $N = 2n + 1$. One can add a new variable, θ_{2n+2}. Then the deformation of $\mathrm{Po}(2n + 1)$ becomes a subalgebra in $\mathfrak{gl}(2^n|2^n)$, consisting of the matrices which commute ith θ_{2n+2}. Since $\theta_{2n+2}^2 = 1$ one can find a basis in which the operator θ_{2n+2} has the form $\begin{pmatrix} 0 & \square \\ \square & 0 \end{pmatrix}$. Matrices commuting with this matrix have the form $\begin{pmatrix} A & B \\ -B & A \end{pmatrix}$. This subalgebra is $Q(2^n)$. Therefore, $\mathrm{Po}(2n + 1)$ deforms into $Q(2^n)$.

Note. One can use Weyl symbols to define the deformation of the universal central extension of $\mathrm{Po}(N)$.

4.2. Let us try to construct cycles analogous to those of §2.2. For simplicity the discussion is restricted to the case $N = 2n$, $\Lambda = \Lambda_0$. As in §2.4, if $\theta^{M_1} \ldots \theta^{M_T} = \theta_2 \ldots \theta_N$, then $\beta = X_1 \theta^{M_1} \wedge \cdots \wedge X_T \theta^{M_T}$ is a cycle. Again fix m, let $\alpha = X_4 \theta^{M_4} \wedge \cdots \wedge X_T \theta^{M_T}$ and define $\mathcal{D}_i(X, Y) = X\theta_1 \ldots \theta_i \wedge Y\theta_{i+1} \ldots \theta_m \wedge \alpha$. Let $\theta^{M_0}\theta^{M_1}\theta^{M_2}\theta^{M_3} = \theta_1 \ldots \theta_m$, $\#M_0 = 2l + 1$, $\#M_1 = i$, $\#M_2 = j$, $\#M_3 = m - i - j - 2l - 1$. Taking the boundary of the chain $X\theta^{M_0}\theta^{M_1} \wedge Y\theta^{M_0}\theta^{M_2} \wedge Z\theta^{M_0}\theta^{M_3} \wedge \alpha$, we obtain the following set of equations for $\mathcal{D}_i(X, Y)$:

$$\mathcal{D}_{i+j}(uv, w) = \mathcal{D}_{j+p}(v, uw) + \mathcal{D}_{i+p}(u, vw). \tag{14}$$

(Here i, j, p are odd positive integers). As in the classical case, $\mathcal{D}_i(X, Y)$ satisfy the symmetry properties (8).

We see that the set of equations (14) is stronger than (7), since now p is an arbitrary odd positive integer, not necessarily 1, as in (7). Proposition from §2.3 gives the solution to the set of equations (7), (8). Substituting it in (14) we obtain

$$(p - 1)\mathcal{D}_2(u, 1) = 0.$$

So at the quantum level we have $\mathcal{D}_2(u, 1) + 0$. This means that any cycle of the form

$$u_1\theta_1\theta_2 \wedge u_2\theta_3\theta_4 \wedge \cdots \wedge u_i\theta_{2i-1}\theta_{2i} \wedge \theta_{2i+1} \ldots \theta_N$$

is homologous to zero. But the exact differentials map precisely into cycles of this form (see §2.2). Thus, the exact differentials turn out to lie in the kernel of the quantum version of the mapping

$$\Omega^i \to H_{i+1}.$$

Another explanation of this phenomenon can be given using perturbation theory. We shall briefly illustrate it by the example of the second homology group. Let $\lambda(f, g) = \Sigma_{I, \#I=3} \partial_{I_0} f \cdot \partial_{I_0^+} g$ (with the notations of §3.1) be the cocycle defining the infinitesimal deformation. Let $\varphi_2(f, g) = \int df \cdot Eg - dg \cdot Ef$ be the cocycle with values in Ω^1. Let us try to find the infinitesimal deformation

$\varphi_2 + s\chi$ (with $s^2 = 0$) of φ_2. Then χ (as the cochain of the classical algebra) must obey the following equations

$$\delta\chi(f, g, t) = \tau(f, g, t)$$

where

$$\tau(f, g, t) = -\varphi_2(\lambda(f, g), t) + \varphi_2(\lambda(f, t), g - \varphi_2(\lambda(g, t), f).$$

One can compute:

$$\tau(f, g, t) = \tfrac{2}{3}dp(f, g, t)$$

where

$$p(f, g, t) = \int \sum_{i, j, k} \left(\frac{\partial}{\partial \theta_k} \frac{\partial}{\partial \theta_j} f \right) \left(\frac{\partial}{\partial \theta_j} \frac{\partial}{\partial \theta_i} g \right) \left(\frac{\partial}{\partial \theta_i} \frac{\partial}{\partial \theta_k} t \right).$$

Note. Of course, we could say in advance that τ would be a full derivative because we know that the cocycle with values in $\Omega^1/d\Lambda$ *can* be extended to the quantum algebra.

Now it is easy to see that the cocycle p is nontrivial. (Indeed, define the cocycle $p_0(f, g)$ with values in adjoint representation by the formula $\int p_0(f, g)t = \int p(f, g, t)$. Then one can check p_0 is proportional to λ so the triviality of p would imply the triviality of λ and conversely. But we know that λ is nontrivial.) Thus, we conclude that φ_2 can be extended to the quantum algebra only if we factorize over exact differentials.

REFERENCES

1. F. A. BEREZIN AND M. S. MARINOV, *Particle spin dynamics as the Grassmann variant of classical mechanics*, Annals of Phys., **104** (1977), 336–362.
2. B. L. FEIGIN AND B. L. TSYGAN, *Additive K-theory and crystaline cohomologies*, Funct. Anal. and Appl. **19** (1985), 52–62.
3. J. L. LODAY AND D. QUILLEN, *Cyclic homology and the Lie algebra homology of matrices*, Comment. Math. Helvetici **59** (1984), 565–591.
4. J. MILNOR, *Introduction to algebraic K-theory*, Princeton Univ. Press, 1971.
5. O. OGIEVETSKY, *K_2-functor of supercommutative rings*, Lebedev Inst. Preprint.
6. ______, *The second homology group of Poisson algebra of odd variables over the supercommutative ring*, Lebedev Inst. preprint.

P. N. LEBEDEV PHYSICAL INSTITUTE, DEPARTMENT OF THEORETICAL PHYSICS, ACADEMY OF SCIENCES OF THE USSR

ON THE MONODROMY GROUPS ATTACHED TO CERTAIN FAMILIES OF EXPONENTIAL SUMS

Dedicated to Y. I. Manin on his fiftieth birthday

NICHOLAS M. KATZ

Introduction. The main theme of this paper is that very innocuous looking one-parameter families of exponential sums over finite fields can have quite strong variation as the parameter moves. Even when the parameter variety is the affine line $\mathbf{A}^1$ or the multiplicative group $\mathbf{G}_m$ over a finite field, the algebraic group G_{geom} which controls the variation is often "as large as possible". These results are the finite-field analogue of the fact that in characteristic zero, very simple differential equations on $\mathbf{A}^1$ and on $\mathbf{G}_m$ can have very large differential galois groups.

In Part 1 we develop some general results concerning irreducible lisse sheaves on open curves in characteristic $p > 0$, in part modeled on [Ka-2] and [Ka-Pi]. In Parts 2 and 3 we calculate G_{geom} for the one-parameter families of exponential sums in characteristic p which correspond to the Kloosterman and Airy differential equations respectively of any rank $n \geqslant 2$. It is very striking that in both of these cases, our results on G_{geom} are in perfect analogy with the results of [Ka-2] and [Ka-Pi] on the differential galois group G_{gal} of the corresponding differential equation, as soon as $p > 2n + 1$. Indeed, one can speculate that in some future "motivic grand unification", the two sorts of results will both be "realizations" of a single motivic result. For the moment, we must be content to offer the results themselves as indirect evidence for the existence of such a unification.

Part 1. Lisse sheaves on open curves: general results. Throughout this paper, we fix a prime number p, an algebraically closed field k of characteristic p, a prime number $\ell \neq p$, and an algebraic closure $\overline{\mathbf{Q}}_\ell$ of $\mathbf{Q}_\ell$. Let U be a smooth connected affine curve over k, and $\mathscr{F}$ a lisse $\overline{\mathbf{Q}}_\ell$-sheaf on U of rank $n \geqslant 1$. We denote by π_1 the fundamental group $\pi_1(U, \overline{\eta})$ of U with base point a geometric generic point $\overline{\eta}$ of U, by ρ the n-dimensional $\overline{\mathbf{Q}}_\ell$-representation of π_1 on $\mathscr{F}_{\overline{\eta}}$ which $\mathscr{F}$ "is", by G_{geom} the Zariski closure of $\rho(\pi_1)$ in $\mathrm{GL}(n, \overline{\mathbf{Q}}_\ell)$, and by $(G_{\text{geom}})^0$ the identity component of G_{geom}. We say that $\mathscr{F}$ is irreducible if ρ is irreducible as a representation of π_1, or equivalently if G_{geom} acts irreducibly in its given n-dimensional representation. We say that $\mathscr{F}$, or ρ, is Lie-irreducible if the restriction of ρ to $(G_{\text{geom}})^0$ is irreducible, or equivalently if the restriction of ρ

Received October 11, 1986.

to every open subgroup of π_1 is irreducible. (More generally, a continuous finite-dimensional representation ρ of a topological group is called Lie-irreducible if its restriction to every open subgroup of finite index is irreducible: clearly ρ is Lie-irreducible if and only if its restriction to some, or to every, open subgroup of finite index is Lie-irreducible.)

PROPOSITION 1. *Suppose that $\mathscr{F}$ is irreducible. Then either ρ is induced from a representation of a proper open subgroup of π_1, or ρ is Lie-irreducible, or there exists an integer $d \geqslant 2$ dividing n and a factorization of ρ as a tensor product $\tau \otimes \omega$, where τ is a Lie-irreducible representation of π_1 of dimension n/d, and where ω is an irreducible representation of π_1 of dimension d which factors through a finite quotient of π_1. In this last case, the pair (τ, ω) is unique up to replacing it by $(\tau \otimes \chi, \omega \otimes \chi^{-1})$ with χ a character of π_1 of finite order.*

Proof. Because $(G_{\mathrm{geom}})^0$ is a normal subgroup of G_{geom} of finite index, either ρ as a G_{geom}-representation is induced from a proper subgroup H which contains $(G_{\mathrm{geom}})^0$, or its restriction to $(G_{\mathrm{geom}})^0$ is isotypical. If ρ is induced from an H, then ρ as π_1-representation is induced from the open subgroup $\rho^{-1}(H)$, and there is nothing to prove. If ρ is not induced, then its restriction to $(G_{\mathrm{geom}})^0$ is isotypical, say isomorphic to d copies of an irreducible representation τ_0 of $(G_{\mathrm{geom}})^0$. If $d = 1$, this exactly means that ρ is Lie-irreducible.

So suppose $d \geqslant 2$. Choose **any** open normal subgroup K of π_1 which is sufficiently small that $\rho(K)$ is contained in $(G_{\mathrm{geom}})^0$. Then $\rho(K)$ is Zariski dense in $(G_{\mathrm{geom}})^0$, so τ_0 is K-irreducible, and $\rho|K \approx d\tau_0$. Therefore the isomorphism class of τ_0 as K-representation is invariant by π_1-conjugation. Therefore τ_0 extends to a projective representation, say τ_1, or π_1. Because we are working on a smooth affine curve U over an algebraically closed field, we know that $H^2(\pi_1, \_\_)$ vanishes, so there is no obstruction to lifting the irreducible projective representation τ_1 to a (necessarily irreducible) linear representation τ_2 of π_1. Notice that on K, τ_2 is projectively equal to τ_0, so of the form $\tau_0 \otimes \chi$ for some character χ of K. Consider the d-dimensional representation ω_2 of π_1 defined as $\mathrm{Hom}_K(\tau_2, \rho)$. Its restriction to K is scalar, namely χ^{-1}, and χ^{-d} is the restriction to K of the character $\det(\omega_2)$. Again by the vanishing of H^2, any $\overline{\mathbf{Q}}_\ell^\times$-valued character of π_1 has a dth root. Twisting τ_2 by a dth root of $\det(\omega_2)$, we obtain a representation τ of π_1 whose restriction to K differs from τ_0 by a character of order d. Shrinking K, we may assume that τ actually coincides with τ_0 on K. Because $\tau_0|K$ is Lie-irreducible, it follows immediately that τ is Lie-irreducible. If we define ω to be $\mathrm{Hom}_K(\tau, \rho)$, we have the asserted factorization of ρ as $\tau \otimes \omega$, with ω factoring through a finite quotient of π_1, and with τ Lie-irreducible. That ω is itself irreducible results from the fact that $\tau \otimes \omega = \rho$ is irreducible.

To see the uniqueness of such a factorization of ρ, let $\tau' \otimes \omega'$ be another, and choose an open normal subgroup K of π_1 on which both ω and ω' are trivial. Then $\rho|K$ is equal both to $\dim(\omega)$ copies of $\tau|K$, and to $\dim(\omega')$ copies of $\tau'|K$, with both of these irreducible. Therefore ω and ω' have the same dimension, and $\tau|K \approx \tau'|K$. Therefore the space $\mathrm{Hom}_K(\tau, \tau')$ is a one-dimensional representation χ of π_1 which is trivial on K, so of finite order, and by construction we have $\tau \otimes \chi \approx \tau'$. Writing ω as $\mathrm{Hom}_K(\tau, \rho)$, and ω' as $\mathrm{Hom}_K(\tau', \rho)$, we find $\omega \approx \omega' \otimes \chi$, as required.   QED

Remark 2.   Conversely, if τ is any Lie-irreducible representation of π_1, and ω any irreducible representation of π_1 which factors through a finite quotient, then the tensor product $\tau \otimes \omega$ is irreducible. For let Λ be any subrepresentation. Choose an open normal subgroup K on which ω is trivial. Then $\Lambda | K$ is a sum of copies of $\tau | K$, so we have $\Lambda \approx \tau \otimes \mathrm{Hom}_K(\tau, \Lambda)$ as π_1-representation. But the second factor $\mathrm{Hom}_K(\tau, \Lambda)$ is a subrepresentation of $\mathrm{Hom}_K(\tau, \tau \otimes \omega) \approx \omega$, so is either 0 or ω itself.

COROLLARY 3 (of Proposition 1).   *Suppose that $\mathscr{F}$ is irreducible, that its determinant is of finite order, and that its rank n is a prime number. Then either $\mathscr{F}$ is induced from a character of a subgroup of π_1 of index n, or it is Lie-irreducible, or it factors through a finite quotient of π_1.*

LEMMA 4.   *A Lie-irreducible $\mathscr{F}$ whose determinant is of finite order has $(G_{\mathrm{geom}})^0$ semisimple. If in addition (1): the determinant of $\mathscr{F}$ has finite order prime to p, (2): the rank n of $\mathscr{F}$ is prime to p, (3): the rank r of $(G_{\mathrm{geom}})^0$ is $< p$ (e.g., if $n < p$, or if $n < 2p$ and $(G_{\mathrm{geom}})^0$ lies inside SO or Sp), then the index of $(G_{\mathrm{geom}})^0$ in G_{geom} is prime to p.*

Proof.   Once $\det(\rho)$ is of finte order, $(G_{\mathrm{geom}})^0$ lies in $\mathrm{SL}(n, \overline{\mathbf{Q}}_\ell)$, and as it acts irreducibly, it must be semisimple. Suppose now that in addition the conditions (1), (2), and (3) all hold. Consider the action of π_1 on $(G_{\mathrm{geom}})^0$ by conjugation. Let us denote by K the subgroup of π_1 which acts by inner automorphism. The quotient π_1/K injects into the automorphism group of the Dynkin diagram of $(G_{\mathrm{geom}})^0$, which is in turn a subgroup of the symmetric group on $r = \mathrm{rank}((G_{\mathrm{geom}})^0)$ letters. But as $(G_{\mathrm{geom}})^0$ has rank $r < p$, π_1/K is prime-to-p. Because $(G_{\mathrm{geom}})^0$ acts irreducibly, $\rho(K)$ lands in $\mathbf{G}_m \cdot (G_{\mathrm{geom}})^0$. Taking determinants, and using (1) and (2), we see that a subgroup K^0 of K of index prime to p lands in $(G_{\mathrm{geom}})^0$. The index of $(G_{\mathrm{geom}})^0$ in G_{geom} divides that of K^0 in π_1, so is prime to p.   QED

To go further, we need to use the Feit-Thompson sharpening [Fe-Th] of Jordan's theorem. Jordan's theorem (cf. [Cu-Re], 36.13) is the fact that there exists a function $c(n)$ of n alone such that if Γ is a finite subgroup of $\mathrm{GL}(n, K)$, K any field of characteristic zero, then there exists a normal abelian subgroup Γ_0 of Γ such that the quotient group Γ/Γ_0 has order $\leqslant c(n)$. In particular, if p is a prime number which is $> c(n)$, then any p-Sylow subgroup Γ_p of Γ must lie in Γ_0, and it must be a p-Sylow subgroup of the abelian group Γ_0. Therefore Γ_p is both unique, hence normal in Γ, and it is abelian. The Feit-Thompson sharpening is the fact that if $p > 2n + 1$ (rather than $p > c(n)$), any p-Sylow subgroup of Γ is both normal and abelian.

PROPOSITION 5. *Suppose that U is the affine line $\mathbf{A}^1$, and that $p > 2n + 1$. Then any irreducible $\mathcal{F}$ of rank n is Lie-irreducible. If in addition $\det(\mathcal{F})$ is trivial, then G_{geom} is connected; in general, $\det(\mathcal{F})$ has finite p-power order, say order q, and G_{geom} is the product of $(G_{\mathrm{geom}})^0$ with the finite subgroup μ_q of the scalars. The group $(G_{\mathrm{geom}})^0$ is semisimple.*

Proof. The basic fact we must exploit is that for $\mathbf{A}^1$, the fundamental group π_1 has no nonzero continuous homomorphism to a finite group of order prime to p. Suppose first that ρ is induced from a proper open subgroup K of π_1. Then the index d of K is a divisor of n, and so $d \leqslant n$, whence $d < p$. But K contains an open subgroup K_0 which is normal in π_1 and of index dividing $d!$. As $d!$ is prime to p, we must have $K_0 = \pi_1$, contradiction. So this case cannot arise. Therefore (by Proposition 1) if ρ is not Lie-irreducible, it must be of the form $\tau \otimes \omega$, with ω an irreducible finite-image representation of π_1 of dimension $d \geqslant 2$, d some divisor of n. Let us denote by Γ the image of ω in $\mathrm{GL}(d, \overline{\mathbf{Q}}_\ell)$. Since $p > 2n + 1 \geqslant 2d + 1$, Feit-Thompson tells us that "the" p-Sylow subgroup Γ_p of Γ is both normal and abelian. But the quotient Γ/Γ_p is a prime-to-p quotient group of π_1, so it must be trivial. Therefore $\Gamma = \Gamma_p$ is itself abelian: as ω is irreducible, we must have $d = 1$, again a contradiction. Therefore ρ is Lie-irreducible.

Now consider the character $\chi = \det(\rho)$ of π_1. It actually takes values in the units $\mathbf{U}_\lambda$ of the integer ring of a finite extension E_λ of $\mathbf{Q}_\ell$, and this group is an extension of a finite group by a pro-ℓ group (the units in $1 + (2\ell)$). Because we are on $\mathbf{A}^1$, any pro-ℓ valued character of π_1 is trivial, so the vanishing of H^2 and the cohomology sequence shows that χ has finite order. Once the order of χ is finite, we write χ as a product of characters of finite prime-power order, of which only the p-power character can possibly be nontrivial on $\mathbf{A}^1$. Therefore $\det(\rho)$ is of finite p-power order.

Because $\det(\rho)$ has finite p-power order, say q, and n is prime to p, $\det(\rho)$ has a unique nth root ω. Twisting ρ by ω^{-1}, say $\Lambda = \rho \otimes \omega^{-1}$, we have $\rho = \omega \otimes \Lambda$, and $\det(\Lambda)$ is trivial. Let us denote by G (resp. G_1) the group G_{geom} for Λ (resp. ρ). Lemma 4 shows that G^0 is semisimple and G/G^0 is prime to p. As G/G^0 is also a quotient of π_1, and we are on $\mathbf{A}^1$, we have $G = G^0$. Because $\rho = \omega \otimes \Lambda$, G_1 lies inside the product $\mu_q \cdot G$. Because ω has finite order these two groups have the same identity component. So $G \subset G_1 \subset \mu_q \cdot G$, and taking determinants shows $G_1 = \mu_q \cdot G$.   QED

We now consider the case when U is $\mathbf{G}_m$. For every integer $d \geqslant 1$ which is prime to p, we denote by $[d] \colon \mathbf{G}_m \to \mathbf{G}_m$ the dth power map, i.e., the d-fold Kummer covering of $\mathbf{G}_m$ by itself. We say that an $\mathcal{F}$ on $\mathbf{G}_m$ is Kummer-induced if it is of the form $[d]_* \mathcal{G}$ for some lisse $\mathcal{G}$ on $\mathbf{G}_m$, and some $d \geqslant 2$ prime to p.

PROPOSITION 6. *Suppose that U is $\mathbf{G}_m$, that $p > 2n + 1$, and that $\mathcal{F}$ is irreducible of rank n. Then $\mathcal{F}$ is either Kummer-induced or Lie-irreducible. If $\mathcal{F}$ is not Kummer-induced, and if its determinant is of finite order prime to p, then*

$(G_{\text{geom}})^0$ *is semisimple, and the group of components* $G_{\text{geom}}/(G_{\text{geom}})^0$ *is cyclic of finite order prime to* p.

Proof. The fact we must exploit is that on $\mathbf{G}_m$, there is a unique prime-to-p quotient group of π_1 of every finite order $d \geqslant 1$ prime to p, namely the cyclic quotient corresponding to the d-fold Kummer covering. Suppose first that $\mathscr{F}$ is induced, from a proper subgroup K of π_1. Then the index d of K divides n, so $d < p$, and hence K contains an open subgroup K_0 which is normal in π_1 of index dividing $d!$, so prime to p. Therefore the quotient π_1/K_0 is cyclic, and hence K is itself normal in π_1 of index d prime to p, so by unicity K is the subgroup corresponding to the d-fold Kummer covering. Thus our $\mathscr{F}$ is Kummer-induced if it is induced.

Suppose now that $\mathscr{F}$ is neither induced nor Lie-irreducible. Then by Proposition 1 , the corresponding representation ρ must be of the form $\tau \otimes \omega$, with ω an irreducible finite-image representation of π_1 of dimension $d \geqslant 2$, d some divisor of n. Let us denote by Γ the image of ω in $\mathrm{GL}(d, \overline{\mathbf{Q}}_\ell)$. Since $p > 2n + 1 \geqslant 2d + 1$, Feit-Thompson tells us that "the" p-Sylow subgroup Γ_p of Γ is both normal and abelian. But the quotient Γ/Γ_p is a prime-to-p quotient group of π_1, so it is cyclic prime-to-p. Therefore Γ is an extension of a prime-to-p cyclic group by a finite abelian p-group, so by Schur-Zassenhaus Γ is the semidirect product of Γ_p with a cyclic prime-to-p group. But any irreducible representation ω of such a Γ is either a character or is induced from a character of a (necessarily normal) subgroup of index prime to p. (Proof by induction on the index of Γ_p in Γ: as Γ_p is normal in Γ, either ω is induced from an irreducible representation of a proper subgroup Γ^0 of Γ which contains Γ_p, and the induction hypothesis applies, or the restriction of ω to the normal abelian Γ_p is isotypical, so d copies of a character χ of Γ_p which is invariant by Γ-conjugation. Using the semi-direct product structure, we see that χ extends to a character, say χ, of Γ. Twisting ω by the inverse of χ, we reduce to the case when the irreducible representation ω of Γ factors through the abelian quotient Γ/Γ_p, which is possible only if $d = 1$.) Therefore as $d \geqslant 2$, ω itself is Kummer-induced, and hence by the projection formula so is $\tau \otimes \omega$, contradiction.

Suppose now that $\mathscr{F}$ is not Kummer-induced, i.e., that it is Lie-irreducible, and that its determinant is of finite order prime to p. Then by Lemma 4, $(G_{\text{geom}})^0$ is semisimple and the group of components $G_{\text{geom}}/(G_{\text{geom}})^0$ is a prime-to-p quotient of π_1, necessarily cyclic because we are on $\mathbf{G}_m$.   QED

The real problem in applications is recognizing whether or not a given $\mathscr{F}$ is Lie-irreducible in the first place; it is here that working on $\mathbf{A}^1$ or $\mathbf{G}_m$ seems to be essential. Overlooking this problem for the moment, we have the following general result. In it, we return to our open curve U. We denote by C its complete nonsingular model.

THEOREM 7.   *Suppose that* $\mathscr{F}$ *is Lie-irreducible of rank n on U, that* $p > 2n + 1$, *that* $\det(\mathscr{F})$ *is of finite order prime to* p, *and that there exists a point* ∞ *in* $C - U$

at which all the breaks of $\mathcal{F}$ as I_∞-representation have exact denominator n. Then $(G_{\text{geom}})^0$ is one of the following subgroups of $\mathrm{GL}(n)$:

$\mathrm{SL}(n)$, *if $n = 2$ or if n is odd*
$\mathrm{SL}(n)$ *or* $\mathrm{Sp}(n)$ *or* $\mathrm{SO}(n)$ *if n is even and* ≥ 4.

Moreover, in the case $(G_{\text{geom}})^0$ is $\mathrm{SO}(n)$, G_{geom} is not contained in $\mathbf{G}_m \cdot \mathrm{SO}(n)$.

Proof. We will show that Theorems 1 and 2 of [Ka-Pi] apply to the subgroup G of G_{geom} generated by $(G_{\text{geom}})^0$ and by $\rho(I_\infty)$, and to its given n-dimensional representation $\rho|G$. Let us denote by P_∞ the wild inertia subgroup of I_∞, and by γ an element of I_∞ which generates I_∞/P_∞. The fact that ρ has all its ∞-breaks with exact denominator n, with n prime to p and $\dim(\rho) = n$, insures that the restriction of ρ to P_∞ is the direct sum of n distinct characters χ_i of P_∞, and that the conjugation-induced action of γ cyclically permutes these n characters (cf. [Ka-1], 1.14).

Let us admit temporarily that $\rho(P_\infty)$ lies in a unique maximal torus T of $(G_{\text{geom}})^0$. Because T centralizes $\rho(P_\infty)$, T must respect each of the n one-dimensional χ_i-eigenspaces L_i of $\rho(P_\infty)$, and the action of T on each L_i is by a character χ_i, whose restrictions to $\rho(P_\infty)$ is χ_i. The element $B = \rho(\gamma)$ of G normalizes both $\rho(P_\infty)$ and $(G_{\text{geom}})^0$, so by the unicity of T, it must normalize T as well. Because B cyclically permutes the n one-dimensional χ_i-eigenspaces L_i of $\rho(P_\infty)$, it cyclically permutes the n distinct (because their restrictions to P_∞ are distinct) characters χ_i of T. Because $\rho(P_\infty)$ lies in $(G_{\text{geom}})^0 = G^0$, the quotient G/G^0 is a finite tame quotient of I_∞, so is cyclic (and prime-to-p, but this is irrelevant for now). Finally, the given representation of G is not the tensor product of two strictly lower-dimensional representations of G, because the break hypothesis insures that already as a representation of $\rho(I_\infty)$ it is not such a tensor product (cf. [Ka-2], 2.5.9.3 for the completely analogous argument in the differential galois context). Thus all the hypotheses of Theorems 1 and 2 of [Ka-Pi] are verified, whence the asserted list of possibilities for $(G_{\text{geom}})^0$.

It remains to explain why $\rho(P_\infty)$ lies in a unique maximal torus T of $(G_{\text{geom}})^0$. It lies in $(G_{\text{geom}})^0$ because, by Lemma 4, the index of $(G_{\text{geom}})^0$ in G_{geom} is prime to p, while $\rho(P_\infty)$ is a finite abelian p-group. Because $\rho(P_\infty)$ acts with n distinct characters, its centralizer Z_{GL} in the ambient $\mathrm{GL}(n)$ is the entire group D of diagonal (with respect to a $\rho(P_\infty)$-eigenbasis) matrices in $\mathrm{GL}(n)$. Therefore its centralizer Z in $(G_{\text{geom}})^0$ lies in the torus D, and hence Z^0 is itself a torus. But any torus T of $(G_{\text{geom}})^0$ containing $\rho(P_\infty)$ trivially lies in Z^0, so that, once we prove that there exists a torus T of $(G_{\text{geom}})^0$ containing $\rho(P_\infty)$, we will conclude that Z^0 is the unique maximal one. This is a special case of the following general fact, applied to $\Gamma = \rho(P_\infty)$ and to $G = (G_{\text{geom}})^0$, for both the statement and proof of which I am indebted to J. Bernstein.

PROPOSITION 8. *Let G be a connected reductive group of rank r over an algebraically closed field of characteristic zero, and Γ a finite abelian subgroup of G whose order is divisible only by primes $p > r + 1$. Then Γ lies in a torus of G.*

Proof. The proof is by induction on $\dim(G)$. Write G as the product of its connected center by its derived subgroup; this gives a central extension

$$0 \to G^{\mathrm{der}} \cap Z(G)^0 \to G^{\mathrm{der}} \times Z(G)^0 \to G \to 0$$

whose kernel, being a central subgroup of a connected semisimple group G^{der} of rank $\leqslant r$, is (say by classification!) of order prime to that of Γ. Therefore Γ lifts uniquely to an abelian subgroup, still denoted Γ, of $G^{\mathrm{der}} \times Z(G)^0$, and this lifted Γ is a subgroup of the product of its projections $\mathrm{pr}_1(\Gamma) \times \mathrm{pr}_2(\Gamma)$. Thus it suffices to show that $\mathrm{pr}_1(\Gamma)$ lies in a torus of G^{der}. If G was not semisimple to begin with, this holds by induction.

If G is semisimple, let $G^{\#}$ denote the universal covering group of G, which is again a central extension of G with kernel of order prime to that of Γ. Thus Γ lifts uniquely to an abelian subgroup, still denoted Γ, of $G^{\#}$. Because the order of Γ is prime to that of $Z(G^{\#})$, either Γ is trivial and there is nothing to prove, or the centralizer $Z(\gamma)$ of any nontrivial γ in Γ is a proper closed subgroup. By a result of Steinberg (cf. [St], Corollary 8.5) the centralizer of a semisimple element in a simply connected semisimple group is itself a connected reductive group. Because Γ lies in this centralizer, we are through by induction. This concludes the proof of Proposition 8, and with it the proof of Theorem 7.   QED

Combining Theorem 7 and Propositions 5 and 6, we obtain the following two results.

THEOREM 9. *Suppose that $p > 2n + 1$. Let $\mathscr{F}$ be irreducible of rank n on $\mathbf{A}^1$, and suppose that at ∞ all the breaks of $\mathscr{F}$ as I_∞-representation have exact denominator n. Then $(G_{\mathrm{geom}})^0$ is one of the following subgroups of $\mathrm{GL}(n)$:*

$\mathrm{SL}(n)$, *if n is odd*
$\mathrm{SL}(n)$ *or* $\mathrm{Sp}(n)$ *if n is even.*

Moreover, G_{geom} is the product of $(G_{\mathrm{geom}})^0$ with the finite p-power subgroup μ_q of the scalars which is the image of its determinant.

Proof. Just combine Proposition 5 with Theorem 7, and notice that on $\mathbf{A}^1$ the $\mathrm{SO}(n)$ case is ruled out by the "moreover" in Theorem 7.   QED

THEOREM 10. *Suppose that $p > 2n + 1$. Let $\mathscr{F}$ be irreducible of rank n on $\mathbf{G}_m$, and not Kummer-induced. Suppose that at ∞ all the breaks of $\mathscr{F}$ as I_∞-representation have exact denominator n. Then $(G_{\mathrm{geom}})^0$ is one of the following subgroups of $\mathrm{GL}(n)$:*

$\mathrm{SL}(n)$, *if $n = 2$ or if n is odd*
$\mathrm{SL}(n)$ *or* $\mathrm{Sp}(n)$ *or* $\mathrm{SO}(n)$ *if n is even and* $\geqslant 4$.

Moreover, in the case $(G_{\mathrm{geom}})^0$ is $\mathrm{SO}(n)$, G_{geom} is not contained in $\mathbf{G}_m \cdot \mathrm{SO}(n)$.

Proof. This is just Proposition 6 combined with Theorem 7.   QED

Part 2. Application to Kloosterman sheaves (cf. [Ka-1]). Let $\mathbf{F}_q$ be a finite subfield of k, Ψ a nontrivial $(\overline{\mathbf{Q}}_\ell)^\times$-valued additive character of $\mathbf{F}_q$, $n \geqslant 1$ an integer, and $\chi_1, \ldots, \chi_n$ a set of n not necessarily distinct, possibly trivial $(\overline{\mathbf{Q}}_\ell)^\times$-valued multiplicative characters of $(\mathbf{F}_q)^\times$. We denote by $\mathrm{Kl}_\Psi(\chi_1, \ldots, \chi_n)$, or simply by Kl, the corresponding Kloosterman sheaf on $\mathbf{G}_m$ (i.e., the sheaf denoted $\mathrm{Kl}_\Psi(\chi_1, \ldots, \chi_n; 1, \ldots, 1)$ with all $b_i = 1$ in [Ka-1], 4.1.1). One knows that Kl is lisse and irreducible of rank n on $\mathbf{G}_m$, that on $\mathbf{G}_m/\mathbf{F}_q$ it is pure of weight $n - 1$, that all its breaks at ∞ are $1/n$, and that its local monodromy at zero is tame, a successive extension of the χ_i. One also knows that if $n \geqslant 2$, the determinant of Kl is the tame character $\Pi\chi_i$. We will sometimes refer to $(\chi_1, \ldots, \chi_n)$ as the set of exponents of Kl at zero, to emphasize the analogy with the Kloosterman differential equation (cf. [Ka-Pi]). The results of this section are in perfect analogy with, and inspired by, the results of [Ka-Pi] on the differential galois groups of Kloosterman equations.

In discussing Kloosterman sheaves, it is useful to recall that by means of the Lang isogenies relative to the various finite subfields $\mathbf{F}_q$ of k, i.e., the Kummer coverings of degrees $q - 1$, the prime-to-p completion of π_1 is the inverse limit, via the norm maps, of all the $(\mathbf{F}_q)^\times$. This allows us to view multiplicative characters of variable finite subfields of k, extended to larger ones by composition with the relative norm, as being precisely the characters of π_1 of finite, prime-to-p order.

LEMMA 11 (R. Pink).   *If a Kloosterman sheaf is induced, then it is Kummer-induced.*

Proof. Let us denote by ρ the corresponding representation of π_1 on the n-dimensional $\overline{\mathbf{Q}}_\ell$-space $V = \mathrm{Kl}_{\overline{\eta}}$. Because ρ is irreducible, it is induced if and only if there exists a decomposition of V as the direct sum of $d \geqslant 2$ subspaces W_i which are transitively permuted among each other by the action of π_1, and in this case the representation is induced from the stabilizer K_i of any chosen W_i acting on W_i. Notice that the W_i all have the same dimension, say r, and that $rd = n$. The permutation representation of π_1 on the set of the W_i is a homomorphism β of π_1 to the symmetric group S_d which factors through ρ, hence is tame at zero and has all breaks $\leqslant 1/n$ at ∞. But S_d embeds into $\mathrm{GL}(d - 1)$ by the augmentation representation, so the composite is a $d - 1$ dimensional representation β of π_1 which is tame at zero and whose Swan conductor at ∞ is $\leqslant (1/n)(d - 1) < 1$. Therefore β is tame at both zero and ∞, and hence its image in S_d must be a cyclic group of order prime to p. But its image is a transitive subgroup, so being cyclic it must be generated by a single d-cycle. Therefore all the W_i have the same stabilizer, namely $\ker(\beta)$, which is a normal subgroup of π_1 of index d. Because the image of β is prime-to-p, it follows that d is prime to p, whence ρ is induced from the d-fold Kummer covering.   QED

Let us say that a collection of $n \geqslant 2$ multiplicative characters $(\chi_1, \ldots, \chi_n)$ is Kummer-induced if there exists a prime-to-p divisor $d \geqslant 2$ of n, and a set of

$r = n/d$ multiplicative characters $\tau_1, \ldots, \tau_r$ of some finite extension of $\mathbf{F}_q$ containing the d th roots of unity, such that the $n = d \times r$ characters χ_i (when composed with the relative norm), give exactly all (counted with multiplicities) the d th roots of the r characters τ_i.

LEMMA 12.   *A Kloosterman sheaf* $\mathrm{Kl}_\Psi(\chi_1, \ldots, \chi_n)$ *is induced if and only if the set* $(\chi_1, \ldots, \chi_n)$ *of its exponents at zero is Kummer-induced.*

Proof.   By Pink's lemma above, if Kl is induced it is Kummer-induced, say $[d]_*(\mathscr{G})$ for some lisse $\mathscr{G}$ on $\mathbf{G}_m$, with $d \geqslant 2$ a prime-to-p divisor of n. Therefore $\mathscr{G}$ is lisse of rank $r = n/d$, tame and quasiunipotent at zero, and all its breaks at ∞ are $1/r$. By ([Ka-1], 8.7.1), $\mathscr{G}$ is itself a Kloosterman sheaf $\mathrm{Kl}_\psi(\tau_1, \ldots, \tau_r)$ for some ψ, possibly different from Ψ, and some τ_i. Applying the direct image formula of ([Ka-1], 5.6.2) to compute $[d]_*(\mathrm{Kl}_\psi(\tau_1, \ldots, \tau_r))$, we find that $\mathrm{Kl}_\Psi(\chi_1, \ldots, \chi_n)$ is a multiplicative translate of Kl_Ψ (all d th roots of the τ_i). Looking at its local monodromy at zero, we see that the χ_i are precisely all the d th roots of the τ_i (counted with multiplicities).

Conversely, if $(\chi_1, \ldots, \chi_n)$ is Kummer-induced, then ([Ka-1], 5.6.2) expresses $\mathrm{Kl}_\Psi(\chi_1, \ldots, \chi_n)$ as the Kummer-induction of a Kloosterman sheaf of rank r.

QED

THEOREM 13.   *Suppose that* $p > 2n + 1$, *that* $n \geqslant 2$, *and that* $\mathrm{Kl}_\Psi(\chi_1, \ldots, \chi_n)$ *is not Kummer-induced. Then* $(G_{\mathrm{geom}})^0$ *is one of the following subgroups of* $\mathrm{GL}(n)$:

$\mathrm{SL}(n)$, *if* $n = 2$ *or if* n *is odd*
$\mathrm{SL}(n)$ *or* $\mathrm{Sp}(n)$ *or* $\mathrm{SO}(n)$ *if* n *is even and* $\geqslant 4$.

Moreover, in the case $(G_{\mathrm{geom}})^0$ *is* $\mathrm{SO}(n)$, G_{geom} *is not contained in* $\mathbf{G}_m \cdot \mathrm{SO}(n)$.

Proof.   This is just a special case of Theorem 10, in view of the above discussion of Kloosterman sheaves.   QED

Algorithm 14.   Hypotheses as in Theorem 13 above, suppose that n is even and $n \geqslant 4$. Here is the algorithm to decide which of the three possibilities SL, Sp, or SO one has for $(G_{\mathrm{geom}})^0$. One looks for a multiplicative character τ (of some finite overfield of $\mathbf{F}_q$) such that the set of characters ω_i defined as $\omega_i = \chi_i/\tau$ is stable under inversion. Notice that if such a τ exists, then $\tau^{2n} = (\Pi \chi_i)^2$, so there are at most $2n$ candidates to examine. If no such τ exists, then we are in the SL case. If such a τ exists, then $(\Pi \omega_i)^2$ is trivial. If $\Pi \omega_i$ is itself trivial, then we are in the Sp case, and if not then we are in the SO case.

To see that this algorithm is correct, we argue as follows. Suppose first that such a τ exists. Replacing the χ_i by the ω_i amounts to twisting Kl by τ^{-1}, and has no effect on $(G_{\mathrm{geom}})^0$. This reduces us to the case where our Kl is self-dual (cf. [Ka-1], 4.1.4, 4.1.7). Because Kl is irreducible, the autoduality is unique up to a scalar, so is either alternating or symmetric. In the first case, G_{geom} lies in Sp; since Sp lies in SL, the determinant $\Pi \omega_i$ is trivial. In the second case, G_{geom} lies in the orthogonal group O, but by Theorem 13 it does not lie in SO, which shows that in this case the determinant $\Pi\omega_i$ is nontrivial.

Conversely, suppose that $(G_{\text{geom}})^0$ is either Sp or SO. Then G_{geom} lies in the normalizer, inside the ambient GL, of either Sp or SO, so G_{geom} itself lies in either $\mathbf{G}_m \cdot$ Sp or in $\mathbf{G}_m \cdot$ O. In either case, forming the square of the $\mathbf{G}_m$-factor is a well-defined character χ of finite order of π_1 which factors through ρ, so is tame both at zero (because ρ is), and at ∞ (because its break at ∞ is both an integer and is $\leqslant 1/n$, the largest break of ρ at ∞). Therefore χ is a prime-to-p character of π_1, and as such it has a prime-to-p square root, say Λ. Twisting our Kl by Λ^{-1}, we do not change $(G_{\text{geom}})^0$, and we turn our Kl into a self-dual one (i.e. into one whose G_{geom} is contained in either Sp or O) whose exponents at zero, the χ_1/Λ, are stable by inversion. Thus we may take τ to be Λ, and we are in the situation of the last paragraph. Hence the algorithm is correct in all cases.

Remark 15. The attentive reader may be disturbed by the fact that the above algorithm seems to rule out the SO case in characteristic $p = 2$. He may be even more disturbed by the apparent incompatibility of Theorem 13 with the fact, proven in ([Ka-1], 1.1.1) that for $p = 2$, n **odd** and all χ_i trivial, one has $(G_{\text{geom}})^0 = \text{SO}(n)$ for $n \neq 7$, and G_2 for $n = 7$. The resolution of this perplexity comes in recalling that, independently of our use of the Feit-Thompson sharpening of Jordan's theorem, which required $p > 2n + 1$ but which was not used at all in [Ka-1], we use the hypothesis $p > \text{rank}((G_{\text{geom}})^0)$ in Lemma 4 to prove that the index of $(G_{\text{geom}})^0$ in G_{geom} is prime to p, and the hypothesis that $p > 1 + \text{rank}((G_{\text{geom}})^0)$ in proving, via Proposition 8, that $\rho(P_\infty)$ lies in a torus T of $(G_{\text{geom}})^0$.

We now turn to the equidistribution consequences of Theorem 13. Recall from ([Ka-1], 4.1.1) that when we view $\text{Kl}_\Psi(\chi_1, \ldots, \chi_n)$ on $\mathbf{G}_m/\mathbf{F}_q$, then for $\mathbf{E}$ any finite extension of $\mathbf{F}_q$, and t any $\mathbf{E}$-valued point of $\mathbf{G}_m$, the trace of the corresponding Frobenius $\text{Frob}_{t,\mathbf{E}}$ is the Kloosterman sum

$$(-1)^{n-1} \sum_{\substack{x_1 x_2 \ldots x_n = t \\ \text{all } x_i \text{ in } \mathbf{E}}} \Psi(x_1 + \cdots + x_n)\chi_1(x_1)\chi_2(x_2)\ldots\chi_n(x_n),$$

where ψ (resp. each χ_i) is extended to $\mathbf{E}$ via the relative trace (resp. norm).

COROLLARY 16 (Sato-Tate law for Kloosterman sums). *Hypotheses and notations as in Theorem* 13, *suppose that all the* χ_i *are characters of* $(\mathbf{F}_q)^\times$. *There exists a constant* α *in* $(\overline{\mathbf{Q}}_\ell)^\times$, *of the form*

$$\alpha = (\text{a root of unity}) \times q^{-(n-1)/2},$$

such that after tensoring Kl *with the geometrically constant rank one sheaf on* $\mathbf{G}_m/\mathbf{F}_q$ *given by* $\alpha^{\deg}$, *the resulting lisse sheaf* $\text{Kl}(\alpha)$ *on* $\mathbf{G}_m/\mathbf{F}_q$ *has all of its Frobenii in* G_{geom}, *and is pure of weight zero. For any embedding of* $\overline{\mathbf{Q}}_\ell$ *into* $\mathbf{C}$, *and any choice of a maximal compact subgroup* K *of the Lie group of* $\mathbf{C}$-*valued points of* G_{geom}, *the conjugacy class of the semisimple part of each Frobenius* $\text{Frob}_{t,\mathbf{E}}$ *of*

Kl(α) *meets K in a single conjugacy class, denoted $\theta(t, \mathbf{E})$, of K. The conjugacy classes $\theta(t, \mathbf{E})$ are equidistributed in the space $K^{\#}$ of conjugacy classes of K with respect to normalized Haar measure, in any of the three senses of equidistribution of* ([Ka-1], 3.5).

Proof. We first show that there exists a constant α in $(\overline{\mathbf{Q}}_{\ell})^{\times}$ such that the Frobenii of Kl(α) all land in G_{geom}. This results from the explicit determination of G_{geom}, for in all cases considered its normalizer in the ambient GL(n) is equal to $\mathbf{G}_m \cdot G_{\text{geom}}$. The intersection of $\mathbf{G}_m$ with G_{geom} is a finite subgroup μ_N, so "formation of the Nth power of the $\mathbf{G}_m$-factor" is a well-defined character of the π_1 of $\mathbf{G}_m/\mathbf{F}_q$ which is geometrically constant, so of the form $\beta^{\deg}$ for some β in $(\overline{\mathbf{Q}}_{\ell})^{\times}$. We may take for α the inverse of any Nth root of β.

Because G_{geom} has a determinant of finite order, it follows that Kl(α) is pure of weight zero (since it is pure of some weight, and its determinant is of finite order). Because the determinant of Kl as a character of the arithmetic π_1 is of the form $(\Pi\chi_i)A^{\deg}$, where A is $\pm q^{n(n-1)/2}$ (cf. [Ka-1], 7.4.1.3), any α such that Kl(α) has a determinant of finite order is necessarily of the asserted form. The equidistribution now follows immediately from Deligne's Weil II, as explained at length in [Ka-1], Chapter 3). QED

Part 3. Elementary Fourier transforms on $\mathbf{A}^1$. We begin this section by explaining heuristically the classical forebearer of the finite-field situation we will study. Fix an integer $n \geqslant 2$, and a polynomial $f(x) = \Sigma a_i x^i$ in C[x] of degree n. We denote by ∂_x the derivation d/dx. The exponential $e^{f(x)}$ is annihilated by the first order operator $\partial_x - f'(x)$, so its Fourier transform

$$FT(e^f)(t) = \int e^{f(x) + tx}\, dx$$

is annihilated by the $n - 1$st order operator $f'(\partial_t) + t$, i.e., by (minus) the formal Fourier transform of $\partial_x - f'(x)$. If $n \geqslant 3$, the Wronskian of $f'(\partial_t) + t$ is annihilated by the first order operator $na_n\partial_t + (n - I)a_{n-1}$, so for $n \geqslant 3$ it is equal to a constant times e^{-bt}, for $b = (n - 1)a_{n-1}/na_n$.

In ([Ka-2], 4.2.7, 4.2.8, 4.2.10) we computed the differential galois group G_{gal} of the $n - 1$st order operator $f'(\partial_t) + t$, by exploiting the fact that all of its slopes at ∞ are equal to $n/(n - 1)$. The result is that for $n \geqslant 3$, G_{gal} is one of the following four subgroups of GL($n - 1$):

SL($n - 1$) or $\mathbf{G}_m \cdot$ SL($n - 1$), if either $n - 1$ is odd, or if $n - 1$ is even and there exist no constants c, d such that the polynomial $f(x + c) + d$ is an odd function of x,

Sp($n - 1$) or $\mathbf{G}_m \cdot$ Sp($n - 1$), if $n - 1$ is even and there exist constants c, d such that the polynomial $f(x + c) + d$ is an odd function of x.

The homotheties $\mathbf{G}_m$ are absent if and only if the coefficient a_{n-1} of $f(x) = \Sigma a_i x^i$ vanishes.

52   NICHOLAS M. KATZ

We now turn to the situation with which we shall be concerned for the remainder of this section. Let $\mathbf{F}$ be a finite subfield of k, Ψ a nontrivial $(\overline{\mathbf{Q}}_\ell)^\times$-valued additive character of $\mathbf{F}$, $n \geq 2$ an integer, and $f(x) = \Sigma a_i x^i$ a polynomial of degree n in $k[x]$. The lisse rank one sheaf $\mathscr{L}_{\Psi(f)} = f^*(\mathscr{L}_\Psi)$ on $\mathbf{A}^1$ is the analogue of e^f (cf. [Ka-1], 4.3, 8.2). We denote by $\mathscr{F}$ (or $\mathscr{F}_f$) the naive Fourier transform $\mathrm{NFT}_\Psi \mathscr{L}_{\Psi(f)}$ of $\mathscr{L}_{\Psi(f)}$ (cf. [Ka-1], 8.4 and 8.5). The ℓ-adic analogue of the fact that $\mathrm{FT}(e^f)(t) = \int e^{f(x)+tx}\,dx$ satisfies a linear differential equation of order $n - 1$ with all ∞-slopes $n/(n-1)$, whose Wronskian is e^{-bt} for $n \geq 3$, is the following theorem.

THEOREM 17. *Notations as above, suppose in addition that $n \geq 2$ is prime to p. Then*

(1) *$\mathscr{F}$ is lisse on $\mathbf{A}^1$ of rank $n - 1$, and all of its breaks at ∞ are equal to $n/(n-1)$.*
(2) *If $f(x)$ has coefficients in $\mathbf{F}$, then $\mathscr{F}$ lives naturally on $\mathbf{A}^1$ over $\mathbf{F}$, and there it is pure of weight one. For any finite overfield $\mathbf{E}$ of $\mathbf{F}$, and any t in $\mathbf{E} = \mathbf{A}^1(\mathbf{E})$, the trace of Frobenius at t is given by*

$$\mathrm{trace}(\mathrm{Frob}_{t,\mathbf{E}} | \mathscr{F}_t) = - \sum_{x \in \mathbf{E}} \Psi_\mathbf{E}(f(x) + tx),$$

where $\Psi_\mathbf{E}$ denotes the additive character $\Psi(\mathrm{Trace}_{\mathbf{E}/\mathbf{F}}(-))$ of $\mathbf{E}$.
(3) *If $n \geq 3$, $\det(\mathscr{F}) \otimes \mathscr{L}_{\Psi(bt)}$ is geometrically constant for $b = (n-1)a_{n-1}/na_n$.*

Proof. (1) Because n is prime to p, the (unique) break of $\mathscr{L}_{\Psi(f)}$ at ∞ is n. Because $n \geq 2$, and $\mathscr{L}_{\Psi(f)}$ is lisse of rank one, it follows that $\mathscr{L}_{\Psi(f)}$ is an irreducible Fourier sheaf (by [K-1], 8.3.1(1), 8.4). Therefore (cf. [Ka-1], 8.2.5, 8.4.1, and 8.5.8(1)) $\mathscr{F}$ is an irreducible Fourier sheaf, which is lisse on $\mathbf{A}^1$, and all of its ∞-breaks are > 1 (because its NFT is the lisse sheaf $\mathscr{L}_{\Psi(f)}$). From the Euler-Poincare formula ([Ka-1], 8.5.3) one sees that if two lisse sheaves on $\mathbf{A}^1$ are inverse NFT's of each other, then the sum of their ranks is the Swan conductor of each at ∞; this shows that $\mathscr{F}$ has rank $n - 1$, and Swan conductor n at ∞. Since all the $n - 1$ ∞-breaks of $\mathscr{F}$ are > 1, and their sum is n, they must each be equal to $n/(n-1)$. [For if we write each as $1 + \lambda_i$, then $\Sigma \lambda_i = 1$, each λ_i is > 0 and rational, and the multiplicity of each of the distinct numbers which occur among the λ_i is multiple of its exact denominator, so already the partial sum of only those λ_i equal to a given one contributes at least 1 to the sum, whence all the λ_i must be equal.] This concludes the proof of (1).

Suppose now that f has coefficients in $\mathbf{F}$. Then (2) follows from the basic facts about Fourier transform and the Lefschetz trace formula (cf. [Ka-1], 2.3, 8.2.5(3), 8.2.4).

Suppose now that $n \geq 3$. We must prove that $\det(\mathscr{F}) \otimes \mathscr{L}_{\Psi(bt)}$ is geometrically constant. Consider the universal family of monic polynomials of degree n over an $\mathbf{F}$-scheme. The parameter space S of this family is the $\mathbf{G}_m \times \mathbf{A}^n$ over $\mathbf{F}$

with coordinates the coefficients $A_n, \ldots, A_0$ of $f_{\text{univ}}(x) = \Sigma A_i x^i$. From Deligne's semicontinuity theorem (cf. [La-1]) one sees that over S we have a lisse $\mathscr{F}_{\text{univ}}$ which is lisse of rank $n - 1$ and whose trace of Frobenius at the E-valued point corresponding to a given polynomial $f(x)$ with coefficients in $\mathbf{E}$ is the exponential sum $-\Sigma \Psi_{\mathbf{E}}(f(x))$ over x in $\mathbf{E}$. Our $\mathscr{F}$ is just the pullback to $\mathbf{A}^1$ of $\mathscr{F}_{\text{univ}}$ by the map

$$\varphi: \mathbf{A}^1 \to S, \qquad t \to (a_n, \ldots, a_2, t + a_1, a_0).$$

The idea is to compute the traces of Frobenius on $\det(\mathscr{F}_{\text{univ}})$ by the following trick of Hasse-Davenport (cf. [Ha-Da], Section 3, 11, pages 162–165). Given an E-valued point of S, corresponding to an f, consider the L-function of $\mathbf{A}^1$ over $\mathbf{E}$ with coefficients in $\mathscr{L}_{\Psi(f)}$. The cohomological expression of this L-function is

$$L(T) = \det\left(1 - T\text{Frob}_{\mathbf{E}}|H^1\left(\mathbf{A}^1, \mathscr{L}_{\Psi(f)}\right)\right),$$

which exhibits it as a polynomial of degree $n - 1$ in T, and shows that

$$\det\left(\text{Frob}_{f,\mathbf{E}}|\mathscr{F}_{\text{univ}}\right) = (-1)^{n-1} \times (\text{the coef. of } T^{n-1} \text{ in } L(T)).$$

The naive additive expression of this same L-function as a sum over all effective divisors of $\mathbf{A}^1$ over $\mathbf{E}$ (i.e., over all monic polynomials in $\mathbf{F}[x]$) shows that for any $d \geqslant 1$, the coefficient of T^d in $L(T)$ is

$$\sum_{\substack{\text{monic } h \text{ of deg } d \text{ over } \mathbf{E}}} \Psi_E\left(\sum_{\substack{\text{zeroes } \alpha \text{ of } h, \text{ with mult.}}} f(\alpha)\right).$$

Notice that the innermost sum is actually E-rational. Indeed, if we denote by $N_i(h)$ the Newton symmetric functions of the roots of h, then the innermost sum is just $\Sigma a_i N_i(h)$, where $f(x) = \Sigma a_i x^i$.

Now we specialize to the case $d = n - 1$. The monic h's of degree $n - 1$ over $\mathbf{E}$ are given by their coefficients $s_1, \ldots, s_{n-1}$, i.e., by the elementary symmetric functions of their roots. The N_i are universal polynomials in the s_j, so we obtain the formula

$$\text{Frob}_{f,\mathbf{E}}|\det(\mathscr{F}_{\text{univ}}) = (-1)^{n-1} \sum \Psi_{\mathbf{E}}\left(\sum a_i N_i(s_1, \ldots, s_{n-1})\right),$$

the outer sum over all $(s_1, \ldots, s_{n-1})$ in $\mathbf{E}^{n-1}$. One easily sees by isobaricity that among $N_0, \ldots, N_n$, only the last two N_{n-1} and N_n can involve s_{n-1}, and one readily calculates that, for $n - 1 > 1$, the involvement is

$$N_{n-1}(s_1, \ldots, s_{n-1}) = (-1)^n (n - 1) s_{n-1} + \text{a poly. in } s_1, \ldots, s_{n-2},$$

$$N_n(s_1, \ldots, s_{n-1}) = (-1)^n (n) s_1 s_{n-1} + \text{a poly. in } s_1, \ldots, s_{n-2}.$$

Substituting, we find that

$$(-1)^{n-1}\mathrm{Frob}_{f,\mathbf{E}}|\det(\mathscr{F}_{\mathrm{univ}}) = \sum \Psi_{\mathbf{E}}\Big((-1)^n s_{n-1}\big[na_n s_1 + (n-1)a_{n-1}\big] + a_1 s_1$$

$$+ (n-1)a_0 + \sum_{i \geqslant 2} a_i P_i(s_1, \ldots, s_{n-2})\Big).$$

Summing last over the variable s_{n-1}, we see that only the terms with

$$s_1 = -(n-1)a_{n-1}/na_n$$

survive, and that

$$(-1)^{n-1}\mathrm{Frob}_{f,\mathbf{E}}|\det(\mathscr{F}_{\mathrm{univ}}) = q\Psi_{\mathbf{E}}(-a_1(n-1)a_{n-1}/na_n)$$

$$\times \sum \Psi_{\mathbf{E}}\Big((n-1)a_0 + \sum_{i \geqslant 2} a_i P_i(s_1, \ldots, s_{n-2})\Big),$$

the outermost sum over all $(s_1, \ldots, s_{n-2})$ in $\mathbf{E}^{n-2}$ for which

$$s_1 = -(n-1)a_{n-1}/na_n.$$

We now exploit the fact that the second factor above is independent of a_1. To do this, let us denote by Z the closed subscheme of S defined by the condition $A_1 = 0$, by $i: Z \to S$ the inclusion, and by $\pi: S \to Z$ the linear projection $(A_n, \ldots, A_1, A_0) \to (A_n, \ldots, A_2, 0, A_0)$. The independence of a_1 of the second factor above means that, by Chebataroff, we have an isomorphism of lisse rank one sheaves on S

$$\det(\mathscr{F}_{\mathrm{univ}}) \otimes \mathscr{L}_{\Psi}(A_1(n-1)A_{n-1}/nA_n) \approx \pi^*i^*(\det(\mathscr{F}_{\mathrm{univ}})).$$

In the Fourier transform situation we are studying, our $\det(\mathscr{F})$ is obtained from $\det(\mathscr{F}_{\mathrm{univ}})$ by pulling back by the map

$$\varphi: \mathbf{A}^1 \to S, \qquad t \to (a_n, \ldots, a_2, t + a_1, a_0).$$

The composite map $i\pi\varphi$ is the map $t \to (a_n, \ldots, a_2, 0, a_0)$, which is geometrically constant. Therefore $\det(\mathscr{F}) \otimes \mathscr{L}_{\Psi(b(t+a_1))}$ is geometrically constant, and hence so is $\det(\mathscr{F}) \otimes \mathscr{L}_{\Psi(bt)}$ itself.   QED

COROLLARY 18. *Hypotheses and notations as in Theorem 17 above, suppose that $n - 1 \equiv 0 \bmod p$. Then G_{geom} lies in $\mathrm{SL}(n)$.*

Proof. Indeed, we have $b = 0$.   QED

THEOREM 19. *Notations as in Theorem* 17 *above, suppose further that $n \geqslant 3$ and that $p > 2n + 1$. Then G_{Geom} is one of the following subgroups of* $\mathrm{GL}(n - 1)$:

$\mathrm{SL}(n - 1)$ *or* $\mu_p \cdot \mathrm{SL}(n - 1)$, *if either $n - 1$ is odd, or if $n - 1$ is even and there exist no constants c, d in k such that the polynomial $f(x + c) + d$ is an odd function of x,*

$\mathrm{Sp}(n - 1)$ *or* $\mu_p \cdot \mathrm{Sp}(n - 1)$, *if $n - 1$ is even and there exist constants c, d such that the polynomial $f(x + c) + d$ is an odd function of x.*

The homotheties μ_p are absent if and only if the coefficient a_{n-1} of $f(x) = \Sigma a_i x^i$ vanishes.

Proof.   This is almost entirely a formal consequence of Theorems 9 and 17(1). The only point that needs to be explained is how, when $n - 1$ is even, one distinguishes the SL case from the Sp case. By an additive translation in the x variable, we may suppose that $a_{n-1} = 0$, and hence that G_{geom} is equal to either $\mathrm{SL}(n - 1)$ or to $\mathrm{Sp}(n - 1)$. The keypoint is that up to an additive inversion, formation of the Verdier dual D commutes with Fourier transform (cf. [La-2], 1.3.2.2):

$$D\big(\mathrm{FT}_\Psi(K)\big) = \mathrm{FT}_\Psi\big(D([x \to -x]^*(K))\big).$$

In our down-to-earth situation, taking for K the single sheaf $\mathscr{L}_{\Psi(f)}$, we have

$$D\big([x \to -x]^*(\mathscr{L}_{\Psi(f)})\big) = \mathscr{L}_{\Psi(g)},$$

where $g(x) = -f(-x)$. The above duality formula then says that $D(\mathscr{F}_f) = \mathscr{F}_g$. By Fourier inversion, then, $\mathscr{F}_f$ is self-dual if and only if $\mathscr{L}_{\Psi(f)}$ is isomorphic to $\mathscr{L}_{\Psi(g)}$, i.e., if and only if the sheaf $\mathscr{L}_{\Psi(f)} \otimes (\mathscr{L}_{\Psi(g)})^{-1} = \mathscr{L}_{\Psi(f-g)}$ is constant. This is trivially the case if $f - g$ is constant. If $f - g$ is not a constant, then as $f - g$ has degree $< p$, the Swan conductor of $\mathscr{L}_{\Psi(f-g)}$ at ∞ is $\deg(f - g)$. Therefore, $\mathscr{F}_f$ is self-dual if and only if $f - g$ is a constant.   QED

COROLLARY 20 (Sato-Tate law for one-variable polynomial sums).   *Hypotheses and notations as in Theorem* 17 *above, suppose further that the polynomial f has coefficients in $\mathscr{F}_q$. Then there exists a constant α in $(\overline{\mathbf{Q}}_\ell)^\times$, of the form*

$$\alpha = (a\ root\ of\ unity) \times q^{-1/2},$$

such that after tensoring $\mathscr{F}$ with the geometrically constant rank one sheaf on $\mathbf{A}^1/\mathbf{F}_q$ given by $\alpha^{\deg}$, the resulting lisse sheaf $\mathscr{F}(\alpha)$ on $\mathbf{A}^1/\mathbf{F}_q$ has all of its Frobenii in G_{geom}, and is pure of weight zero. For any embedding of $\overline{\mathbf{Q}}_\ell$ into $\mathbf{C}$, and any choice of a maximal compact subgroup K of the Lie group of $\mathbf{C}$-valued points of G_{geom}, the conjugacy class of the semisimple part of each Frobenius $\mathrm{Frob}_{t,\mathbf{E}}$ of $\mathscr{F}(\alpha)$ meets K

in a single conjugacy class, denoted $\theta(t, \mathbf{E})$, of K. The conjugacy classes $\theta(t, \mathbf{E})$ are equidistributed in the space $K^{\#}$ of conjugacy classes of K with respect to normalized Haar measure, in any of the three senses of equidistribution of ([Ka-1], 3.5).

Proof. The proof of Corollary 16 works mutatis mutandis except for the evaluation of α up to roots of unity. To evaluate α, it suffices to show that in the universal case, when we write $\det(\mathscr{F}_{\mathrm{univ}})$ in the form (char. of finite order) $\times A^{\deg}$, for some A (this is always possible, cf. [De-1], 1.3.4(1)), we have $A = $ (a root of unity) $\times q^{(n-1)/2}$. To show this, we first may make a finite extension of $\mathbf{F}_q$, and reduce to the case when all the nth roots of unity lie in $\mathbf{F}_q$. We may take $A = \det(\mathrm{Frob}\,|\,H_c^1(\mathbf{A}^1, \mathscr{L}_{\Psi(f)}))$ for a **single** choice of $f(x)$ of degree n over $\mathbf{F}_q$. We take $f(x) = x^n$; an elementary calculation then gives $A = \Pi(-g(\Psi, \chi))$, the product over all the nontrivial characters χ of order dividing n.   QED

Remark 21.   In the case $n = 3$, $f(x) = x^3$, the question of how the sums $\Sigma_x \Psi(f(x) + tx)$ are distributed was first raised by Birch's note [Bi], and first worked out by Livne ([Li]), whose work was the starting point of my own interest in such sums.

REFERENCES

[Bi]   B. Birch, *How the number of points of an elliptic curve over a fixed prime field varies*, J. London Math. Soc. **43** (1968), 57–60.

[Cu-Re]   C. W. Curtis and I. Reiner, *Representation Theory of Finite Groups and Associative Algebras*, Interscience Publ., New York and London, 1962.

[De]   P. Deligne, *La conjecture de Weil* II, Pub. Math. I.H.E.S. **52** (1981), 137–252.

[Fe-Th]   W. Feit and J. Thompson, *On groups which have a faithful representation of degree less than* $(p-1)/2$, Pacific J. Math. **4** (1961), 1257–1262.

[Ha-Da]   H. Hasse and H. Davenport, *Die Nullstellen der Kongruenzzetafunktionen in gewissen zyklischen Fällen*, J. Reine Angew. Math. **172**, (1934), 151–182.

[Ka-1]   N. Katz, *Gauss Sums, Kloosterman Sums, and Monodromy Groups*, Annals of Math. Study **113**, to appear.

[Ka-2]   ______, *On the calculation of some differential galois groups,*, Inv. Math. **87** (1987), 13–61.

[Ka-Pi]   ______ and R. Pink, *A note on pseudo-CM representations and differential galois groups*, Duke Mathematical J., this issue.

[La-1]   G. Laumon, Semicontinuité du conducteur de Swan (d'apres Deligne), in *Characteristique d'Euler-Poincaré*, Seminaire de l'ENS 1978/79, Asterisque 82-83, 1981, 173–219.

[La-2]   ______, *Transformation de Fourier, constantes d'équations fonctionnelles et conjectures de Weil*, Pub. Math. I.H.E.S., to appear.

[Li]   R. Livne, *Applications of Hodge theory to Birch's conjectures concerning the average distribution of cubic exponential sums*, to appear.

[St]   R. Steinberg, *Endomorphisms of Linear Algebraic Groups*. Memoirs of the Amer. Math. Soc. **80**, 1968.

Department of Mathematics, Princeton University, Princeton, New Jersey 08544

A NOTE ON PSEUDO-CM REPRESENTATIONS AND DIFFERENTIAL GALOIS GROUPS

Dedicated to Y. I. Manin on his fiftieth birthday

NICHOLAS M. KATZ AND RICHARD PINK

Introduction. This note, a sequel to [Ka-1], falls into two parts. In the first, we give a criterion for a connected semisimple algebraic subgroup of $GL(n)$ to be one of the following subgroups:

$SL(n)$, if $n \geqslant 2$
$Sp(n)$ or $SO(n)$, if n is even and $\geqslant 4$.

The criterion is based on the classification of what we call "pseudo-CM representations", a natural generalization of the notion of "CM representation" introduced in [Ka-1]. The second part applies these results to determine the differential galois groups of some concrete differential equations, including the general Kloosterman equation on $\mathbf{G}_m$.

Part 1. Throughout this section, k is an algebraically closed field of characteristic zero, n is an integer $\geqslant 2$, V is an n-dimensional vector space over k, G is a Zariski closed subgroup of $GL(V)$, and G^0 is the identity component of G.

THEOREM 1.   *Suppose that*

(1) G^0 *lies in* $SL(V)$.
(2) *As G^0-representation, V is irreducible.*
(3) *There exists an element g in G, and a connected torus T in G^0 such that*
  (a) *As T-representation, V is the direct sum of n distinct characters $c_1, \ldots, c_n$.*
  (b) *T is $\mathrm{Ad}(g)$-stable, i.e., $gTg^{-1} = T$.*
  (c) *The automorphism $\mathrm{Ad}(g)$ of T cyclically permutes the n characters $c_1, \ldots, c_n$.*

Then there exist

an integer $r \geqslant 1$
a factorization of n as $n = n_1 \ldots n_r$, with all $n_i \geqslant 2$ and the n_i pairwise relatively prime.
algebraic groups $G_1, \ldots, G_r$, with each G_i equal to one of the groups
  $SL(n_i)$ *is odd or* $= 2$
  $SL(n_i)$ *or* $Sp(n_i)$ *or* $SO(n_i)$ *if n_i is even* $\geqslant 4$.

Received October 11, 1986. Second author supported by deutscher Akademischer Austauschdienst (DADD) during the academic year 1985–86.

such that (G^0, V) is isomorphic to $(\Pi(G_i), \otimes(\text{std}(n_i)))$, where $\text{std}(n_i)$ denotes the standard n_i-dimensional representation of G_i.

THEOREM 2. *Hypotheses as in Theorem 1, suppose in addition that*

(4) *The group of components G/G^0 is cyclic.*
(5) *As representation of G, V is not isomorphic to the tensor product of two strictly lower-dimensional representations of G.*

Then $r = 1$, i.e., G^0 is either

$\text{SL}(n)$ if $n = 2$ or n odd
$\text{SL}(n)$ or $\text{SO}(n)$ or $\text{Sp}(n)$ if n is even and $\geqslant 4$.

Moreover, in the case $G^0 = \text{SO}(n)$, G is not contained in $\mathbf{G}_m \cdot \text{SO}(n)$.

Proof. We first explain how Theorem 2 follows from Theorem 1. Consider the action of G on G^0 by conjugation. This action must respect the individual factors G_i of G^0, simply because between the various groups G_i in question, there are no nontrivial homomorphisms. By the unicity of the decomposition of an irreducible representation of a product group into a tensor product of irreducible representations of the factors, each representation $\text{std}(n_i)$ of G_i, viewed as a representation of G^0, must remain isomorphic to itself under G-conjugation. Because G/G^0 is cyclic, any irreducible representation of G^0 whose isomorphism class is G-invariant extends to an irreducible representation of G. In particular, the representations $\text{std}(n_i)$ of G^0 each extend to representations $\mathbf{std}(\mathbf{n_i})$ of G itself. Consider the representation $\otimes(\mathbf{std}(\mathbf{n_i}))$ of G. It coincides on the normal subgroup G^0 with the G^0-irreducible representation V of G, so it is G-irreducible, and as a G-representation it must be the tensor product of V with a character c of G/G^0 (namely the one-dimensional space of G^0-homomorphisms from V to $\otimes(\mathbf{std}(\mathbf{n_i}))$). Twisting a single one of the $\text{std}(n_i)$ by the inverse of this character, we find that V is itself a tensor product of strictly lower-dimensional representations of G unless $r = 1$.

If $G^0 = \text{SO}(n)$, then G cannot lie in $\mathbf{G}_m \cdot \text{SO}(n)$, for then the element g in hypothesis (3) of Theorem 1 would induce an inner automorphism of G^0, so could be rechosen to lie in G^0, in which case we would find that the standard representation of $\text{SO}(n)$ is CM, and this is known (cf. [Ka-1], 3.2.7) not to be the case. This completes the deduction of Theorem 2 from Theorem 1.

We now turn to the proof of Theorem 1. Because G^0 is a connected irreducible subgroup of $\text{SL}(n)$, it is automatically semisimple (cf. [Ka-2], 11.5.3.2). We reduce to the case when the torus T is a maximal torus of G^0. Let $\mathbf{T}$ be any maximal torus of G^0 which contains the given torus T. Then we have obvious inclusions

$$\mathbf{T} \subset Z_G(T) \subset Z_{GL(V)}(T).$$

The hypothesis (3a) of Theorem 1 shows that $Z_{GL(V)}(T)$ is precisely the group of all diagonal (with respect to a T-eigenbasis of V) matrices in $\mathrm{GL}(V)$. Therefore $(Z_G(T))^0$ is itself a torus, so by maximality of $\mathbf{T}$ it must be equal to $\mathbf{T}$ itself. In particular there is a unique maximal torus $\mathbf{T}$ of G^0 which contains T. By unicity, $\mathbf{T}$ must be $\mathrm{Ad}(g)$-stable if T is. Because T acts on V as the direct sum of n distinct characters c_i, $\mathbf{T}$ respects each c_i-eigenspace. The characters $\mathbf{c_i}$ of $\mathbf{T}$ on these eigenspaces must be n distinct characters of $\mathbf{T}$ which are cyclically permuted by $\mathrm{Ad}(g)$, simply because the $\mathbf{c_i}$ are determined by their restrictions c_i to T, and this holds for the c_i. This completes the reduction to the case when T is maximal.

We now state and prove a classification theorem for pseudo-CM representations of semisimple Lie algebras, which, when applied to $\mathrm{Lie}(G^0)$ and its given embedding into $\mathrm{End}(V)$, leads immediately to Theorem 1 exactly as in Chapter 3 of [Ka-1].

Let $\mathfrak{G}$ be a nonzero semisimple Lie algebra over k, $\mathfrak{T}$ a cartan subalgebra, and β a Lie algebra automorphism of $\mathfrak{G}$ which maps $\mathfrak{T}$ to itself. We say that a finite-dimensional representation V of $\mathfrak{G}$ is pseudo-CM, or pseudo-CM with respect to β and $\mathfrak{T}$, if the following conditions hold:

(1) V is faithful and irreducible.
(2) As $\mathfrak{T}$-representation, V is the direct sum of $n = \dim(V)$ distinct weights $w_1, \ldots, w_n$ of $\mathfrak{T}$.
(3) V is $\mathfrak{G}$-isomorphic to its β-transform, i.e., there exists an element B in $\mathrm{GL}(V)$ which, acting by conjugation on $\mathrm{End}(V)$, normalizes (the images of) $\mathfrak{G}$ and $\mathfrak{T}$, and induces β on $\mathfrak{G}$.
(4) The element B, acting on V, cyclically permutes the n distinct one-dimensional weight spaces of $\mathfrak{T}$.

(In the application, the element B in $\mathrm{GL}(V)$ will be the element g occurring in hypothesis (3) of Theorem 1, and β will be $\mathrm{Ad}(g)$.)

THEOREM 3. *Notations as above, let V be a pseudo-CM representation of a semisimple $\mathfrak{G}$. Then there exist*

an integer $r \geqslant 1$,
a factorization of n as $n = n_1 \ldots n_r$ with all $n_i \geqslant 2$ and the n_i pairwise relatively prime,
Lie algebras $\mathfrak{G}_1, \ldots, \mathfrak{G}_r$, with each $\mathfrak{G}_i$ equal to one of
   $\mathfrak{SL}(n_i)$ *if n_i is odd or $= 2$*
   $\mathfrak{SL}(n_i)$ *or $\mathfrak{Sp}(n_i)$ or $\mathfrak{SO}(n_i)$ if n_i is even $\geqslant 4$,*

such that $(\mathfrak{G}, V)$ is isomorphic to $(\Pi(\mathfrak{G}_i), \otimes(\mathrm{std}(n_i)))$, where $\mathrm{std}(n_i)$ denotes the standard n_i-dimensional representation of $\mathfrak{G}_i$.

Proof. First write $\mathfrak{G}$ as a product of simple Lie algebras $\mathfrak{G}_i$, and then factor V as $\otimes(V_i)$, where V_i is an irreducible representation of $\mathfrak{G}$ which factors through $\mathfrak{G}_i$. Because the set of simple factors of $\mathfrak{G}$ and the corresponding tensor factors of V are unique, the β-invariance of V shows that the β-transform of a given V_i

must be one of the V_j. Group together in clumps the factors $(\mathfrak{G}_i, V_i)$ according to β-orbit. This gives us a new, "coarser" factorization of $(\mathfrak{G}, V)$ as a product of situations $(\mathfrak{G}_i, V_i)$ each of which is β-stable by construction. Let us denote by β_i the automorphism of $\mathfrak{G}_i$ induced by β, so that β is the product of the β_i, and choose elements B_i (unique up to scalars) in $\mathrm{GL}(V_i)$ inducing β_i, such that B is their tensor product. The cartan subalgebra $\mathfrak{T}$ is the product of its projections $\mathfrak{T}_i$ onto the factors $\mathfrak{G}_i$, and each of these $\mathfrak{T}_i$ is necessarily normalized by its B_i. It is immediate (compare [Ka-1], 3.1.6) that V is a pseudo-CM representation of $\mathfrak{G}$ with respect to β and $\mathfrak{T}$ if and only if both the following two conditions hold:

(a) the numbers $n_i = \dim(V_i)$ are pairwise relatively prime,
(b) for each i, V_i is a pseudo-CM representation of $\mathfrak{G}_i$ w.r.t. β_i and $\mathfrak{T}_i$.

This reduces us to showing that if $(\mathfrak{G}, V)$ is isotypical, i.e., a product of several copies of a single situation (simple Lie algebra $\mathfrak{G}_0$, irreducible representation V_0) and if V is a pseudo-CM representation of $\mathfrak{G}$ of dimension $n \geq 2$ with respect to an automorphism β which cyclically permutes the factors, then the only possibilities for $\mathfrak{G}$ are

$\mathfrak{S}\mathfrak{L}(n)$ if $n = 2$ or if n odd
$\mathfrak{S}\mathfrak{L}(n)$ or $\mathfrak{S}\mathfrak{O}(n)$ or $\mathfrak{S}\mathfrak{p}(n)$ if n even ≥ 4,

with $V = \mathrm{std}(n)$. To do this, we first show that ither $\mathfrak{G}$ is itself simple, or $\mathfrak{G}$ is $\mathfrak{S}\mathfrak{O}(4)$, with $V = \mathrm{std}(4)$. Indeed, if there are $k \geq 2$ factors cyclically permuted by β, and if we use β to identify them successively, we may suppose the automorphism B of the kth tensor power $(V_0)^{\otimes k}$ of V_0 to be of the form

$$v_1 \otimes v_2 \otimes \cdots \otimes v_k \mapsto v_2 \otimes v_3 \otimes \cdots \otimes v_k \otimes A(v_1),$$

for some A in $\mathrm{GL}(V_0)$ which normalizes both $\mathfrak{G}_0$ and $\mathfrak{T}_0$. The kth iterate of B is the automorphism $A \otimes \cdots \otimes A$, under which any weight of the form $(w, w, \ldots, w)$ of $\mathfrak{T} = \mathfrak{T}_0 \times \cdots \times \mathfrak{T}_0$ has an orbit of cardinality at most $\dim(V_0)$. Therefore the B-orbit of such a weight has cardinality at most $k \cdot \dim(V_0)$. By hypothesis, there is a single B-orbit, and its cardinality is $\dim(V)$. Thus we obtain the inequalities

$$k \cdot \dim(V_0) \geq \dim(V) = \left(\dim(V_0)\right)^k, \quad \text{with } k \geq 2 \text{ and } \dim(V_0) \geq 2,$$

which are only satisfied for $k = 2 = \dim(V_0)$, in which case $\mathfrak{G}_0$ must be $\mathfrak{S}\mathfrak{L}(2)$, $\mathfrak{G}$ is $\mathfrak{S}\mathfrak{L}(2) \times \mathfrak{S}\mathfrak{L}(2) = \mathfrak{S}\mathfrak{O}(4)$, and V is $\mathrm{std}(2) \otimes \mathrm{std}(2) = \mathrm{std}(4)$. We leave to the reader the verification that the standard representation of $\mathfrak{S}\mathfrak{O}(2k)$ is in fact a pseudo-CM representation for any $k \geq 2$.

It remains only to treat the case in which $\mathfrak{G}$ is simple. If β is inner, then V is a CM-representation, and by ([Ka-1], 3.2.7) we know that $\mathfrak{G}$ is either $\mathfrak{S}\mathfrak{L}(n)$ or (if n is even) $\mathfrak{S}\mathfrak{p}(n)$, with $V = \mathrm{std}(n)$.

Suppose now that β is not inner. Because the weights of V are transitively permuted by an automorphism β of $(\mathfrak{G}, \mathfrak{T})$, they all have the same length with respect to any W-invariant scalar product on the $\mathbf{Q}$-span of the roots of $(\mathfrak{G}, \mathfrak{T})$. Therefore V is a minuscule representation of $\mathfrak{G}$, so in terms of any choice of base of the root system of $(\mathfrak{G}, \mathfrak{T})$, V is a fundamental representation ω_α. The β-invariance of V (together with Bourbaki Lie Chapter 8, Section 5, No. 2, corollary of Proposition 4 applied to β) shows that α is fixed under the nontrivial automorphism of the diagram induced by β. A glance at the tables (Bourbaki Lie Chapter 8, Section 7, No. 3, and Chapter 6, Planches) shows that the only such minuscule representations are the standard representation of $\mathfrak{SO}(2k)$ for $k \geqslant 3$, and the jth exterior power of the standard representation of $\mathfrak{SL}(2j)$ for $j \geqslant 3$. (The case $j = 2$ coincides with the case $k = 3$.) It remains only to check that the second case is not a pseudo-CM representation of $\mathfrak{SL}(2j)$ for $j \geqslant 3$.

For this, we argue as follows. The square ∂ of β is inner, so corresponds to an element of the symmetric group on $2j$ letters, while the weights of the V in question correspond to the subsets of these $2j$ letters having exactly j elements. Because β is supported to permute the weights transitively, ∂ can have at most two orbits acting on the weights. Consider the decomposition of ∂ as a disjoint product of cycles, of lengths $d_1, \ldots, d_r$, with each $d_i \geqslant 1$, $d_1 + \cdots + d_r = 2j$. Let S_i be the "support" of the ith cycle in ∂, so that the S_i are just the orbits of ∂ acting on the set $(1, 2, \ldots, 2j)$. For any subset K consisting of j elements, let us define

$$e_i = e_i(K) = \operatorname{card}(K \text{ intersects } S_i).$$

The e_i are clearly constant on ∂-orbits, and, equally clearly, any set of integers $(e_1, \ldots, e_r)$ satisfying

$$(*) \qquad 0 \leqslant e_i \leqslant d_i \quad \text{for } i = 1, \ldots, r, \text{ and } e_1 + \cdots + e_r = j,$$

actually occurs for some K with j elements. Because ∂ has at most two orbits acting on the set of K's, there can be at most two solutions $(e_1, \ldots, e_r)$ of the above equations $(*)$. This implies that either $r = 1$, or that $r = 2$ and $d_1 = 1$, $d_2 = 2j - 1$. In either case, ∂ has order at most $2j$, so $\dim(V) \leqslant 4j$. But binomial coefficients increase towards the middle, so we have

$$4j \geqslant \dim(V) = \dim\big(\Lambda^j(\operatorname{std}(2j))\big) \geqslant \dim\big(\Lambda^2(\operatorname{std}(2j))\big) = j(2j - 1)$$

a contradiction since $j \geqslant 3$. This completes the proof of Theorem 3, and concludes Part 1.   QED

Part 2.   We continue to work over an algebraically closed field k of characteristic zero. Let U be a smooth connected affine curve over k, X its complete nonsingular model, $\infty \in X - U$ a "point at infinity", ω a fiber-functor on

D.E.(X/k), $n \geqslant 2$ an integer, $\mathbf{V}$ a D.E. on X/k of rank n, $V = \omega(\mathbf{V})$ its fiber at ω, $G_{\mathrm{gal}} \subset \mathrm{GL}(V)$ its differential galois group with respect to ω, and $(G_{\mathrm{gal}})^0$ the identity component of G_{gal}.

THEOREM 4. *Notations as above, suppose that*

(1) $\mathrm{Det}(\mathbf{V})$ *is of finite order, i.e.,* $(G_{\mathrm{gal}})^0$ *lies in* $\mathrm{SL}(n) = \mathrm{SL}(V)$.
(2) *As representation of* $(G_{\mathrm{gal}})^0$, V *is irreducible.*
(3) *At* ∞, *all the slopes of* $\mathbf{V}$ *have exact denominator* n.

Then $(G_{\mathrm{gal}})^0$ *is either*

$\mathrm{SL}(n)$ *if* $n = 2$ *or* n *odd*
$\mathrm{SL}(n)$ *or* $\mathrm{SO}(n)$ *or* $\mathrm{Sp}(n)$ *if* n *is even and* $\geqslant 4$.

Moreover, in the case $(G_{\mathrm{gal}})^0 = \mathrm{SO}(n)$, G_{gal} *is not contained in* $\mathbf{G}_m \cdot \mathrm{SO}(n)$.

Proof. We apply Theorem 2 not to G_{gal} but rather to the subgroup G of G_{gal} generated by $(G_{\mathrm{gal}})^0$ and by the image $\rho(I_\infty)$ of I_∞ in G_{gal}. By ([Ka-1], 2.6.3), the image of I_∞ in the finite group $G_{\mathrm{gal}}/(G_{\mathrm{gal}})^0$ is cyclic, so $G^0 = (G_{\mathrm{gal}})^0$, and $G/(G_{\mathrm{gal}})^0$ is cyclic. To apply Theorem 2 we take for the torus T the image of $(I_\infty)^{(0+)}$, and we take for $g \in G$ the image of any element in I_∞ which generates its unique cyclic quotient of order n. By ([Ka-1], 2.5.9.3 and 2.6.6), all the conditions (1) through (5) of Theorems 1 and 2 are satisfied.   QED

Remark 4.1.   In the proof of Theorem 4 above, we could also have taken for T the smaller but "more explicit" torus which is the image in G_{gal} of $(I_\infty)^{(r/n)}$, where r/n is the unique slope at ∞ (cf. [Ka-1], 2.6.6).

In order for the above theorem to be useful, we need a criterion to decide the G^0-irreducibility of an irreducible representation of G, when G/G^0 is cyclic.

LEMMA 5. *Let V be a finite-dimensional k-vector space, $G \subset \mathrm{GL}(V)$ a Zariski closed subgroup of $\mathrm{GL}(V)$, and $H \subset G$ a Zariski closed normal subgroup of G such that the quotient G/H is a finite cyclic group. Suppose that V is G-irreducible. Then either V is H-irreducible, or V as G-representation is induced from a proper subgroup K of G which contains H.*

Proof. Because H is a normal subgroup of G of finite index and V is G-irreducible, either V is induced as in the assertion of the lemma or the restriction of V to H is H-isotypical, say $V = rW$ as H-representation, where W is an irreducible representation of H. By Jordan-Hölder theory, the isomorphism class of W must be invariant by G-conjugation. Because G/H is cyclic, W extends to a representation $\mathbf{W}$ of G. We have a natural map of G-representations

$$\mathbf{W} \otimes \mathrm{Hom}_H(\mathbf{W}, V) \to V,$$

which, being an isomorphism of H-representations, must be a G-isomorphism as well. Because V is G-irreducible, its two tensor factors must themselves be

G-irreducible. Thus $\mathrm{Hom}_H(W, V)$ is G-irreducible. As it is an r-dimensional representation which factors through the cyclic quotient G/H, it can only be G-irreducible if $r = 1$, i.e., if V is itself H-irreducible.   QED

We now turn to some concrete applications of the theory developed so far. We work on the curve $U = \mathbf{G}_m$. For every integer $d \geq 1$, we denote by

$$[d]: \mathbf{G}_m \to \mathbf{G}_m$$

the dth power mapping. We say that a D.E. $\mathbf{V}$ on $\mathbf{G}_m$ is "Kummer-induced" if it is of the form $[d]_*(\mathbf{W})$ for some $d \geq 2$ and some D.E. $\mathbf{W}$ on $\mathbf{G}_m$ (cf. [Ka-1], 1.4.6 and 1.4.7 for the compatibility with the representation-theoretic sense of "induced"). Note that if $\mathbf{V}$ is Kummer-induced, then its rank n must be divisible by d, and the quotient n/d is the rank of $\mathbf{W}$.

COROLLARY 6.   *Notations as above, let $\mathbf{V}$ be an irreducible D.E. on $\mathbf{G}_m$. Then either $\mathbf{V}$ is Kummer-induced, or V is $(G_{\mathrm{gal}})^0$-irreducible.*

Proof.   By ([Ka-1], 1.2.5 and 1.4.4), the finite quotient $G_{\mathrm{gal}}/(G_{\mathrm{gal}})^0$, for any D.E. on $\mathbf{G}_m$, is finite cyclic, being a finite quotient of $\pi_1(\mathbf{G}_m)$ corresponding to some Kummer covering of $\mathbf{G}_m$ by itself. The result now follows from Lemma 5, in virtue of ([Ka-1], 1.4.6).   QED

If $\mathbf{V}$ is regular singular at zero, there is a simple criterion, in terms of its exponents at zero, to insure that it is not Kummer-induced. Let $a_1, \ldots, a_n$ be these exponents, viewed as lying in the additive quotient group $k/\mathbf{Z}$. If in fact $\mathbf{V} = [d]_*(\mathbf{W})$, then $\mathbf{W}$ is itself regular singular at zero, and if $b_1, \ldots, b_{n/d}$ are its exponents at zero, then the n exponents a_i of $\mathbf{V}$ at zero are simply the n quantities

$$(b_i + j)/d \bmod \mathbf{Z}, \quad \text{for } i = 1, \ldots, n/d \text{ and } j = 1, \ldots, d.$$

Let us say that a set of n not-necessarily-distinct elements of $k/\mathbf{Z}$ is Kummer-induced if there exist a divisor $d \geq 2$ of n and n/d elements b_i of $k/\mathbf{Z}$ such that the a_i are the n quantities displayed above. Thus we have:

CRITERION 7.   *Let $\mathbf{V}$ be a D.E. on $\mathbf{G}_m$ which is regular singular at zero. If the exponents of $\mathbf{V}$ at zero are not Kummer-induced, then $\mathbf{V}$ is not Kummer-induced.*

Combining the above results, we find

THEOREM 8.   *Let $\mathbf{V}$ be a D.E. on $\mathbf{G}_m$ of rank $n \geq 2$. Suppose that*

(1) *Det($\mathbf{V}$) is of finite order.*
(2) *At ∞, all the slopes of $\mathbf{V}$ have exact denominator n.*
(3) *At zero, $\mathbf{V}$ is regular singular and its exponents are not Kummer-induced.*

Then $(G_{gal})^0$ is either

$SL(n)$ *if $n = 2$ or n odd*
$SL(n)$ *or $SO(n)$ or $Sp(n)$ if n is even and $\geqslant 4$.*

Moreover, in the case $(G_{gal})^0 = SO(n)$, G_{gal} is not contained in $\mathbf{G}_m \cdot SO(n)$.

Example 9.   Let $n \geqslant 2$, $a_1, \ldots, a_n$ a set of n elements in k whose image in $k/\mathbf{Z}$ is not Kummer-induced, and whose sum lies in $\mathbf{Q}$. Denote by D the invariant derivation zd/dz on $\mathbf{G}_m$. Let $m \geqslant 1$ be an integer prime to n, and $Q_m(z)$ a polynomial of degree m with $Q_m(0) = 0$. Then the D.E. on $\mathbf{G}_m$ corresponding to the n'th order operator

$$\Pi(D - a_i) - Q_m(z)$$

satisfies all the hypotheses of Theorem 8 (its exponents at zero are the $a_i \bmod \mathbf{Z}$, its determinant is the first order operator $D - \Sigma a_i$, and its slopes at ∞ are all equal to m/n), and hence its conclusion as well. In the particular case $m = 1$, we find the general Kloosterman equation, of which a very special case was treated in [Ka-1].

Algorithm 10.   Here is a simple algorithm in terms of exponents to determine, when n is even and $\geqslant 4$ and the exponents are not Kummer-induced, which of the three possibilities SL, Sp, or SO for $(G_{gal})^0$ one has in the Kloosterman case $m = 1$. One first looks for a number c in k such that for the quantities b_i defined as $a_i - c$, the two sets of exponents $(b_1, \ldots, b_n)$ and $(-b_1, \ldots, -b_n)$ coincide $(\bmod \mathbf{Z})$. If no such c exists, then we are in the SL case. If such a c exists, then Σb_i is either an integer or a half-integer, and we are in the Sp or SO cases accordingly.

To see that this algorithm is correct, suppose first that such a quantity c exists. Then c is necessarily a rational number (because Σa_i is assumed rational), so the first-order operator $D - c$ is of finite order. Therefore tensoring our given D.E. by $D - c$, which exactly has the effect of replacing the a_i by the b_i, does not change the identity component of G_{gal}, and reduces us to the case where our Kloosterman equation is self-dual (cf. [Ka-1], 4.5.2 and 1.5.3). If the autoduality, unique up to a scalar, is alternating, then G_{gal} lies in the corresponding symplectic group Sp; since Sp lies in SL, the determinant $D - \Sigma b_i$ is trivial, which means exactly that Σb_i is an integer. If the autoduality is symmetric, then G_{gal} lies in the corresponding orthogonal group O, but by Theorem 8 it does not lie in SO; thus the determinant has order two but is nontrivial, which means exactly that Σb_i is a half-integer but not an integer.

Conversely, suppose that $(G_{gal})^0$ is equal to Sp or to SO. Then G_{gal} lies in the normalizer, inside GL, of either Sp or SO, so G_{gal} is itself contained in either $\mathbf{G}_m \cdot Sp$ or $\mathbf{G}_m \cdot O$. In either case, forming the square of the $\mathbf{G}_m$-factor is a well-defined character χ of G_{gal}, which corresponds to a rank one D.E. on $\mathbf{G}_m$ which is regular singular at both zero (because the original Kloosterman D.E. is) and at ∞ (because its slope at ∞ is an integer which is $\leqslant 1/n$, the largest [and only] slope at ∞ of the original Kloosterman D.E., cf. [Ka-1], 2.5.8). Such a rank one D.E. must be of the form $D - d$, for some element d in k. Now define c to be $-d/2$. The $D - c$ is an inverse square root of $D - d$, so tensoring with $D - c$ turns our original Kloosterman D.E. into a self-dual one (i.e., into one whose G_{gal} is contained in either Sp or 0), whose exponents at zero, the b_i, are consequently stable mod $\mathbf{Z}$ by negation. This means exactly that we are in the situation of the previous paragraph. Hence the algorithm is correct in all cases.

References

[Bbk-1] N. Bourbaki, *Groupes et algebres de Lie*, Chapitres 4, 5, et 6, Paris, Masson, 1981.

[Bbk-2] ______, *Groupes et Algebres de Lie*, Chapitres 7 et 8, Paris, Diffusion CCLS, 1975.

[Ka-1] N. Katz, *On the calculation of some differential galois groups*, Inv. Math., **87** (1987) 13–61.

[Ka-2] ______, *Gauss Sums, Kloosterman Sums, and Monodromy Groups*, Annals of Math. Study **113**, to appear.

Katz: Department of Mathematics, Princeton University, Princeton, New Jersey 08544

Pink: Mathematisches Institut der Universität Bonn, Beringstrasse 1, D-5300 Bonn 1, Federal Republic of Germany

Vol. 54, No. 1   DUKE MATHEMATICAL JOURNAL ©   1987

REPRESENTATION THEORY AND THE CUSPIDAL GROUP OF $X(p)$

BENEDICT H. GROSS

To Yu. I. Manin

Let p be a prime number, and let $X(p)$ denote the modular curve which classifies elliptic curves with a level p structure. Then $X(p)$ is a normal covering of the moduli $X(1)$ of elliptic curves, with Galois group isomorphic to $\mathrm{GL}_2(\mathbb{Z}/p\mathbb{Z})/\langle \pm 1 \rangle$. Hecke used the complex representation theory of the geometric Galois group $\mathrm{SL}_2(\mathbb{Z}/p\mathbb{Z})/\langle \pm 1 \rangle$ to study certain holomorphic differentials on $X(p)$ [1; §7, 8]. In this note we will use the representation theory of $\mathrm{GL}_2(\mathbb{Z}/p\mathbb{Z})$ in characteristic p to study the cuspidal subgroup C in the Jacobian.

A general argument due to Manin and Drinfeld shows that C is a finite group [3, Cor. 3.6], and Kubert and Lang have obtained a formula for its order [2, pg. 118]. Here we will determine the structure of $C \otimes \mathbb{Z}_p$ as a module over the Galois group of $X(p)$. In particular, using the non semi-simplicity of certain modular representations of $\mathrm{GL}_2(\mathbb{Z}/p\mathbb{Z})$, we will show that for $p \geqslant 5$ the quotient C/pC has dimension $\geqslant (p - 5)(p - 1)/4$ over $\mathbb{Z}/p\mathbb{Z}$, with equality holding if and only if the prime p is regular.

§1. The canonical model of $X(p)$.

In this section, we recall some basic facts about the curve $X(p)$. The proofs may be found in the books of Kubert-Lang [2] and Shimura [5, Ch. 6].

The moduli $X(1)$ of elliptic curves is defined over $\mathbf{Q}$ and has function field $k = \mathbf{Q}(j)$, where j is Dedekind's modular function. Let E be an elliptic curve over k with invariant $j(E) = j$. Let L be the normal extension of k which is generated by the co-ordinates of the nontrivial p-division points on E over $\bar{k}$. The Galois group of L over k acts $\mathbb{Z}/p\mathbb{Z}$-linearly on E_p, and this representation gives an isomorphism: $\mathrm{Gal}(L/k) \simeq \mathrm{Aut}(E_p)$. If we fix an isomorphism $E_p \simeq (\mathbb{Z}/p\mathbb{Z})^2$ by choosing a basis of E_p over $\mathbb{Z}/p\mathbb{Z}$, we obtain an isomorphism $\mathrm{Gal}(L/k) \simeq \mathrm{GL}_2(\mathbb{Z}/p\mathbb{Z})$.

The group $\mathrm{Aut}(E) = \langle \pm 1 \rangle$ embeds as a subgroup of $\mathrm{Aut}(E_p)$, so acts on the field L. We let K be the fixed field; then K is generated over k be the x-co-ordinates of the nontrivial p-division points: $x_a = x_{-a}$ for $a \in A = E_p - \{0\}/\langle \pm 1 \rangle$. The field K is normal over k and, unlike L, is independent of the curve E chosen with invariant j. We have a canonical isomorphism $G =$

Received November 29, 1986.

$\mathrm{Gal}(K/k) \simeq \mathrm{Aut}(E_p)/\langle \pm 1 \rangle$; once a basis for E_p is chosen we have an isomorphism $G \simeq \mathrm{GL}_2(\mathbb{Z}/p\mathbb{Z})/\langle \pm 1 \rangle$.

The Weil pairing gives a nondegenerate alternating bilinear form on E_p with values in μ_p, and the symplectic automorphisms of E_p contain $\mathrm{Aut}(E)$. Hence the subfield k' of L fixed by the symplectic automorphisms, which is the constant field extension $k(\mu_p)$, is contained in K. We have the field diagram

$$
G \simeq \mathrm{GL}_2(\mathbb{Z}/p\mathbb{Z})/\langle \pm 1 \rangle
\left\{
\begin{array}{l}
 \\
 \\
\left(\begin{array}{l} K = k(x_a)_{a \in A} \\ \\ k' = k(\mu_p) \\ \\ k = \mathbb{Q}(j) \end{array} \right.
\end{array}
\right.
$$

$$
\begin{array}{c}
L = k(E_p) \\
\langle \pm 1 \rangle \mid \\
K = k(x_a)_{a \in A} \\
\mid \\
k' = k(\mu_p) \\
\mid \\
k = \mathbb{Q}(j)
\end{array}
\left.\begin{array}{c} \\ \\ \\ \\ \end{array}\right\} \mathrm{SL}_2(\mathbb{Z}/p\mathbb{Z})/\langle \pm 1 \rangle
$$

We define $X(p)$ as the complete, nonsingular, connected curve over $\mathbb{Q}(\mu_p)$ with function field K. The curve $X(p)$ is geometrically connected, as $\mathbb{Q}(\mu_p)$ is equal to the algebraic closure of $\mathbb{Q}$ in K. The group $G = \mathrm{Gal}(K/k)$ acts on $X(p)$, although only the subgroup isomorphic to $\mathrm{SL}_2(\mathbb{Z}/p\mathbb{Z})/\langle \pm 1 \rangle$ acts geometrically. We remark that K is classically described as the subfield of modular functions for the group $\Gamma(p) = \ker(\mathrm{SL}_2(\mathbb{Z}) \to \mathrm{SL}_2(\mathbb{Z}/p\mathbb{Z}))$ whose Fourier expansions at the cusps have coefficients in the subfield $\mathbb{Q}(\mu_p)$ of $\mathbb{C}$.

§2. The cuspidal group and central decomposition. We shall henceforth assume that $p \geqslant 5$, for simplicity. The curve $X(p)$ has $(p^2 - 1)/2$ cusps, which are all rational over $\mathbb{Q}(\mu_p)$ and correspond to degenerations in the moduli problem. These correspond to the places of K which lie above the place $j = \infty$ of $k = \mathbb{Q}(j)$ and are permuted transitively by the Galois group $G = \mathrm{Gal}(K/k)$. Let D be the free abelian group of divisors supported on the cusps of $X(p)$, and let D^0 be the subgroup of divisors of degree zero in D.

Let M be the modular units in K^*; this is the subgroup of functions whose divisors are supported on the cusps of $X(p)$. Let N be the quotient of M by the subgroup $\mathbb{Q}(\mu_p)^*$ of constant functions. The divisor map gives an exact sequence of G-modules

$$
0 \to N \xrightarrow[\mathrm{div}]{} D^0 \to C \to 0
$$

where C is the cuspidal subgroup in the Jacobian. Applying the snake lemma to the endomorphism multiplication by p of this exact sequence, we obtain the basic exact sequence of $\mathbb{Z}/p\mathbb{Z}[G]$-modules which we will study:

$$
(2.1) \qquad 0 \to C[p] \to N/pN \xrightarrow[\mathrm{div}]{} D^0/pD^0 \to C/pC \to 0.
$$

The center of $\mathrm{Gal}(L/k)$ acts via the character χ on the irreducible representation E_p over $\mathbb{Z}/p\mathbb{Z}$. The characters of the center $Z \simeq (\mathbb{Z}/p\mathbb{Z})^*/\langle \pm 1 \rangle$ of $G = \mathrm{Gal}(K/k)$ in characteristic p are all of the form χ^{2n} with $0 \leqslant 2n < p - 1$. If W is a finite dimensional $\mathbb{Z}/p\mathbb{Z}[G]$-module and $W(2n)$ is the sub-module on which Z acts by the character χ^{2n}, we obtain a direct sum decomposition $W \overset{\circ}{=} \oplus_{2n=0}^{p-3} W(2n)$ as the order of Z is prime to p. This decomposition is functorial and exact in the category of finite dimensional $\mathbb{Z}/p\mathbb{Z}[G]$-modules.

Our aim in the next four sections is to show that for $2 < 2n < p - 1$ the two G-modules $N/pN(2n)$ and $D^0/pD^0(2n)$ are *not* isomorphic. Hence the map $\mathrm{div}_{2n}\colon N/pN(2n) \to D^0/pD^0(2n)$ cannot be an isomorphism and the groups $C[p](2n)$ and $C/pC(2n)$ are nonzero by (2.1). In fact, we shall show that the kernel of div_{2n} contains, and the cokernel of div_{2n} maps surjectively to, the irreducible $\mathbb{Z}/p\mathbb{Z}[G]$ module $\mathrm{Sym}^{2m}(E_p^*)$, where $E_p^* = \mathrm{Hom}(E_p, \mathbb{Z}/p\mathbb{Z})$ is the dual of E_p, $2m + 2n = p - 1$, and Sym^{2m} denotes the $2m$th symmetric power representation.

The image of the resulting injection

$$(2.2) \qquad \bigoplus_{2m=2}^{p-5} \mathrm{Sym}^{2m}(E_p^*) \hookrightarrow N/pN$$

is represented by the Weierstrass units of Kubert and Lang [2, pg. 50ff]. The surjection

$$(2.3) \qquad C/pC \to \bigoplus_{2m=2}^{p-5} \mathrm{Sym}^{2m}(E_p^*)$$

defines what Kubert and Lang call the special quotient of the cuspidal group [2, pg. 133ff]. For us, the existence of these units and divisor classes on $X(p)$ is forced by the nonsemisimplicity of the principal series representations of $\mathrm{GL}_2(\mathbb{Z}/p\mathbb{Z})$ in characteristic p.

§3. Siegel units.

The Siegel units are a subgroup of specific modular units in the function field K of the curve $X(p)$, on which the action of the group $G = \mathrm{Gal}(K/k)$ is known. The fact that their image in N has index prime to p allows us to determine the structure of the $\mathbb{Z}/p\mathbb{Z}[G]$-module $N/pN(2n)$. In this section we will recall some of the properties of Siegel units, referring the reader to the book of Kubert and Lang [2] for the proofs, and will identify the central component $N/pN(2n)$ as a representation in the principal series of $G \simeq \mathrm{GL}_2(\mathbb{Z}/p\mathbb{Z})/\langle \pm 1 \rangle$.

The Fricke family of modular functions $\{ g_a^{12p} \}$ on $\Gamma(p)$ studied in [2, Ch. 2, §1] correspond to a single modular function F on $X_1(p)$ over $\mathbb{Q}$. The function F can of course be given algebraically on pairs $(\mathscr{E}, P)$ of an elliptic curve $\mathscr{E}$ with a nontrivial p-torsion point P. We will not do this here, but will simply give an analytic formula for F on Tate curves. If $\mathscr{E}$ is the Tate curve $\mathbf{G}_m/q^{\mathbf{Z}}$ with a p-torsion point t satisfying $t^p = q^\alpha$ (normalized so that $0 \leqslant \alpha < p$), we have

$$(3.1) \qquad F\left(\mathbf{G}_m/q^{\mathbf{Z}}, t\right) = t^{6\alpha - 6p} q^p \theta(t)^{12p},$$

where $\theta(x) = (1 - x)\prod_{n \geqslant 1}(1 - q^n x)(1 - q^n x^{-1})$ is the Tate theta function. Finally, we note that $F(\mathscr{E}, P) = F(\mathscr{E}, -P)$.

If we evaluate F on the pair (E, a), where E is our curve over $k = \mathbf{Q}(j)$ with invariant j and $a \in A = E_p - (0)/\langle \pm 1 \rangle$, we obtain a modular unit $u_a = F(E, a)$ in the function field $k(x_a) \subseteq K$. The group $G = \mathrm{Gal}(K/k)$ acts transitively on the units u_a by permutation: $(u_a)^\sigma = u_{\sigma a}$. The stabilizer of u_a in this action is the subgroup $H_a \subseteq G$ which fixes the extension $k(x_a)$. If we choose a basis $\langle e_1, e_2 \rangle$ for E_p over $\mathbf{Z}/p\mathbf{Z}$ with e_1 in the class of a, then H_a is identified with the subgroup $\begin{pmatrix} \pm 1 & * \\ 0 & * \end{pmatrix}/\langle \pm 1 \rangle$ in $\mathrm{GL}_2(\mathbf{Z}/p\mathbf{Z})/\langle \pm 1 \rangle$.

For each $a \in A$, let v_a be a pth root of u_a in an algebraic closure of K. The Kummer extension $K(\mu_{p^2}, v_a)_{a \in A}$ has degree p^4 over K, and is the function field of the modular curve $X(p^2)$ over $\mathbf{Q}(\mu_{p^2})$. Let V be the subgroup of its multiplicative group generated by μ_p and the units v_a. Then V/μ_p is freely generated by the $\frac{1}{2}(p^2 - 1)$ units v_a, and the Galois group acts on this module through its quotient G: $\sigma(v_a) \equiv v_{\sigma a} \pmod{\mu_p}$. Consequently, the representation of G on V/μ_p is isomorphic to the induced representation $\mathrm{Ind}_{H_a}^G \mathbf{Z}$, where H_a is the stability subgroup described in the previous paragraph.

Let $\mathbf{a} = \Sigma m(a)a$ be an element in the free abelian group on A, and put $v_\mathbf{a} = \prod v_a^{m(a)}$. Then $v_\mathbf{a}$ lies in the subfield K if and only if $\mathbf{a}$ satisfies the quadratic relations of Kubert and Lang [2, pg. 76]: for any quadratic form λ on E_p with values in $\mathbf{Z}/p\mathbf{Z}$ we must have $\Sigma m(a)\lambda(a) = 0$ in $\mathbf{Z}/p\mathbf{Z}$. Let U be the group of M generated by μ_p and the units $v_\mathbf{a}$ which lie in K^*; this is the subgroup of Siegel units we will study. By the above observation, we have an exact sequence of G-modules;

$$0 \to U/\mu_p \to V/\mu_p \xrightarrow[\alpha]{} \mathrm{Sym}^2\left(E_p\right) \to 0.$$

Here $\mathrm{Sym}^2(E_p)$ is viewed as the dual of the space $\mathrm{Sym}^2(E_p^*)$ of quadratic forms λ on E_p, and α maps the element $v_\mathbf{a}$ to the linear functional $\lambda \mapsto \Sigma m(a)\lambda(a)$. Since the central character of $\mathrm{Sym}^2(E_p)$ is equal to χ^2, which is also the central character of the 1-dimensional representation μ_p, we find that

$$(3.2) \qquad U/pU(2n) \simeq \mathrm{Ind}_{H_a}^G 1(2n)$$

for $2n \neq 2$, where 1 denotes the trivial representation of H_a on $\mathbf{Z}/p\mathbf{Z}$.

Finally, the map $U \to N$ has kernel generated by μ_p and p^{12} and cokernel of order prime to p, by a fundamental result of Kubert and Lang [2, Thm. 1.3, pg.

83]. Hence

$$(3.3) \qquad N/pN(2n) \simeq \operatorname{Ind}_{H_a}^{G} 1(2n)$$

for $2n \neq 0, 2$.

It remains to determine the central components of the induced representation $\operatorname{Ind}_{H_a}^{G} 1$ over $\mathbb{Z}/p\mathbb{Z}$. Let $Z \simeq (\mathbb{Z}/p\mathbb{Z})^*/\langle \pm 1 \rangle$ be the center of G, and B be the subgroup $B = H_a \times Z$. If we choose a basis $\langle e_1, e_2 \rangle$ with e_1 in the class of a, then B is the standard Borel subgroup $\left(\begin{smallmatrix} * & * \\ 0 & * \end{smallmatrix} \right)/\langle \pm 1 \rangle$ of $\operatorname{GL}_2(\mathbb{Z}/p\mathbb{Z})/\langle \pm 1 \rangle$.

By the transitivity of induction we find that $\operatorname{Ind}_{H_a}^{G} 1 \simeq \operatorname{Ind}_B^G(\operatorname{Ind}_{H_a}^{B} 1)$. In fact, induction from B to G commutes with the decomposition via central characters, so

$$(3.4) \qquad \operatorname{Ind}_{H_a}^{G} 1(2n) \simeq \operatorname{Ind}_B^G\!\left(\operatorname{Ind}_{H_a}^{B} 1(2n) \right)$$

for all n. But $B = H_a \times Z$, so $\operatorname{Ind}_{H_a}^{B} 1 \simeq 1_{H_a} \otimes R_Z$ where R is the regular representation of Z [4, pg. 37]. The regular representation of Z decomposes over $\mathbb{Z}/p\mathbb{Z}$ as a direct sum $\oplus_{2n=0}^{p-3} \chi^{2n}$, so we find

$$(3.5) \qquad \operatorname{Ind}_{H_a}^{B} 1(2n) \simeq \left(\chi^{2n}, 1 \right)$$

where $(\chi^{2n}, 1)$ is the character of B which maps the matrix $\left(\begin{smallmatrix} a & b \\ 0 & d \end{smallmatrix} \right)/\langle \pm 1 \rangle$ to a^{2n}. Combining formulas (3.3)–(3.5), we have proven the following.

PROPOSITION 3.6. *If* $2 < 2n < p - 1$ *then we have an isomorphism of* $\mathbb{Z}/p\mathbb{Z}[G]$ *modules*:

$$N/pN(2n) \simeq \operatorname{Ind}_B^G\!\left(\chi^{2n}, 1 \right)$$

where $(\chi^{2n}, 1)$ *is the character* $\left(\begin{smallmatrix} a & b \\ 0 & d \end{smallmatrix} \right)/\langle \pm 1 \rangle \mapsto a^{2n}$ *of* B.

§4. Cuspidal divisors. The cusps w of $X(p)$ correspond to the places of K which divide the place $j = \infty$ of k. They are permuted transitively by G and D is the free abelian group they generate. Hence $D \simeq \operatorname{Ind}_{H_w}^{G} \mathbb{Z}$, where H_w is the subgroup fixing any cusp w. This gives an isomorphism of $\mathbb{Z}/p\mathbb{Z}[G]$-modules

$$(4.1) \qquad D^0/pD^0(2n) \simeq \operatorname{Ind}_{H_w}^{G} 1(2n)$$

for $0 < 2n < p - 1$.

It remains to identify the subgroup H_w. We will show that there is a bijection of G-sets between the cusps of $X(p)$ and the set of quadratic functions λ: $A \to \mathbb{Z}/p\mathbb{Z}$ which are the squares of nontrivial linear forms on E_p. If we choose a basis $\langle e_1, e_2 \rangle$ for E_p over $\mathbb{Z}/p\mathbb{Z}$ and let ℓ be the linear form $\alpha e_1 + \beta e_2 \mapsto \beta$, then the subgroup H_λ of $\operatorname{GL}_2(\mathbb{Z}/p\mathbb{Z})/\langle \pm 1 \rangle$ which stabilizes the quadratic function $\lambda = \ell^2$ is equal to $\left(\begin{smallmatrix} * & * \\ 0 & \pm 1 \end{smallmatrix}\right)/\langle \pm 1 \rangle$. Since this may be identified with H_w in (4.1), an argument similar to the one used at the end of §3 gives the following result.

PROPOSITION 4.2. *If* $0 < 2n < p - 1$ *then we have an isomorphism of* $\mathbb{Z}/p\mathbb{Z}[G]$ *modules*:

$$D^0/pD^0(2n) \simeq \mathrm{Ind}_B^G(1, \chi^{2n})$$

where $(1, \chi^{2n})$ *is the character* $\left(\begin{smallmatrix} a & b \\ 0 & d \end{smallmatrix}\right)/\langle \pm 1 \rangle \mapsto d^{2n}$ *of* B.

It remains to establish our bijection of G-sets. Over the completion $k_\infty = \mathbb{Q}((1/j)) = \mathbb{Q}((q))$ we may fix a rigid analytic isomorphism $E \simeq \mathbb{G}_m/q^{\mathbb{Z}}$, where q is obtained by inverting the usual expansion $j = 1/q + 744 + 196884q + \cdots$. A place w' of $L = k(E_p)$ which divides $j = \infty$ gives an embedding of L into the field $k_\infty(\mu_p)(q^{1/p})$ generated by the p-torsion on the Tate curve. This embedding gives us the nondegenerate linear form $\ell\colon E_p \to \mathbb{Z}/p\mathbb{Z}$ defined by

$$(4.2) \qquad \ell(v) = \mathrm{ord}_\pi(t_v) \quad (\mathrm{mod}\ p)$$

where $\pi = q^{1/p}$ is the uniformizing parameter of the completion and t_v is the coordinate of v in the rigid analytic model. It is easily checked that ℓ determines w' and that $\lambda = \ell^2\colon A \to \mathbb{Z}/p\mathbb{Z}$ determines the place w_λ that w' divides in K. The G-action on λ is by the usual action on linear forms: $\lambda^\sigma(a^\sigma) = \lambda(a)$.

Although we will never need the orders of the Siegel units u_a at the cusps w_λ of $X(p)$, we wish to note the simple formula; which follows from (3.1):

$$(4.3) \qquad \mathrm{ord}_{w_\lambda}(u_a) \equiv 6\lambda(a) \quad (\mathrm{mod}\ p).$$

This shows that the quadratic relations are a necessary condition for the product $v_{\mathbf{a}} = \prod v_a^{m(a)} = (\prod u_a^{m(a)})^{1/p}$ to lie in K^*.

§5. The principal series in characteristic p. This section is purely representation-theoretic, and has nothing to do with the modular curve $X(p)$. As such, we will temporarily change our notation. Let G be the finite group $\mathrm{GL}_2(\mathbb{Z}/p\mathbb{Z})$ and let V denote the standard 2-dimensional representation of G. Let χ be the central character of V; then $\chi\left(\begin{smallmatrix} a & 0 \\ 0 & a \end{smallmatrix}\right) = a$ for all $a \in (\mathbb{Z}/p\mathbb{Z})^*$. We will assume that $p > 2$ so that $\chi \neq 1$.

Let B denote the standard Borel subgroup $\left\{\left(\begin{smallmatrix} a & b \\ 0 & d \end{smallmatrix}\right) : ad \neq 0\right\}$ and let

$$U = \left\{\left(\begin{smallmatrix} 1 & b \\ 0 & 1 \end{smallmatrix}\right) : b \in \mathbb{Z}/p\mathbb{Z}\right\}$$

denote the unipotent subgroup of B. Fix an integer k with $0 < k < p - 1$ and let

$j = p - 1 - k$. Then χ^k is a nontrivial character of $(\mathbb{Z}/p\mathbb{Z})^*$ with inverse χ^j. We let $(1, \chi^k)$ denote the character of B taking the matrix $\begin{pmatrix} a & b \\ 0 & d \end{pmatrix}$ to $\chi^k(d) = d^k$.

PROPOSITION 5.1. *The composition series for the induced representation* $\mathrm{Ind}_B^G(1, \chi^k)$ *is given by the exact sequence of* $\mathbb{Z}/p\mathbb{Z}[G]$*-modules*:

$$0 \to \mathrm{Sym}^k(V) \to \mathrm{Ind}_B^G(1, \chi^k) \to \mathrm{Sym}^j(V^*) \to 0.$$

This sequence is not split.

Proof. The induced representation $\mathrm{Ind}_B^G(1, \chi^k)$ occurs in the space of functions

$$\left\{ f: G \to \mathbb{Z}/p\mathbb{Z} : f\left[\begin{pmatrix} a & b \\ 0 & d \end{pmatrix} g\right] = d^k f(g) \right\}$$

of dimension $p + 1$, on which $\alpha \in G$ acts by right translations $f^\alpha(g) = f(g\alpha)$ [4, pg. 31]. The subspace of functions fixed by U contains the function $f_1\begin{pmatrix} a & b \\ c & d \end{pmatrix} = c^k$, which is supported on $G - B$, as well as a function f_2 supported on B and defined by $f_2\begin{pmatrix} a & b \\ 0 & d \end{pmatrix} = d^k$ there. These functions are clearly linearly independent. The Bruhat decomposition: $G = B \,\dot\cup\, BwB = B \,\dot\cup\, BwU$ with $w = \begin{pmatrix} 0 & 1 \\ 1 & 0 \end{pmatrix}$ shows that they span the U-invariants in the function space.

The subgroup B, which normalizes U, acts on its fixed space via characters of the torus B/U. In this case, B acts on f_1 via the character $\chi_1\begin{pmatrix} a & b \\ 0 & d \end{pmatrix} = a^k$ and on f_2 via the character $\chi_2\begin{pmatrix} a & b \\ 0 & d \end{pmatrix} = d^k$. These characters χ_i describe the highest weight vectors of the possible irreducible summands of the induced representation [6, Ch. 13]. The irreducible representation of G with central character χ^k and highest weight χ_1 is $\mathrm{Sym}^k(V)$; the representation with central character χ^k and highest weight χ_2 is $\mathrm{Sym}^j(V^*)$.

Since the G-translates of f_2 span the induced representation (as G acts transitively on the $p + 1$ cosets of B), only the character χ_1 corresponds to an irreducible summand. This gives the submodule $\mathrm{Sym}^k(V)$ spanned by the G-translates of f_1: $f\begin{pmatrix} a & b \\ c & d \end{pmatrix} = F_k(c, d)$ where F_k is a homogeneous polynomial of degree k. Since f_2 gives a highest weight vector in the quotient, which has dimension $j + 1 = p + 1 - (k + 1)$ equal to the dimension of $\mathrm{Sym}^j(V^*)$, the quotient is irreducible and the proposition is proved.

For any character α of $(\mathbb{Z}/p\mathbb{Z})^*$ the map $D_\alpha f(g) = f(g) \cdot \alpha(\det g)$ induces a G-isomorphism

$$(5.2) \qquad\qquad \mathrm{Ind}_B^G(\alpha, \chi^k\alpha) \simeq \mathrm{Ind}_B^G(1, \chi^k) \otimes \alpha(\det).$$

Since tensor product with $\alpha(\det)$ is exact, this allows us to find the composition factors of the principal series representation $\mathrm{Ind}_B^G(\alpha, \beta)$ when $\alpha \neq \beta$ from

Proposition 5.1. In particular, we obtain a nonsplit exact sequence

$$(5.3) \qquad 0 \to \mathrm{Sym}^j(V^*) \to \mathrm{Ind}_B^G(\chi^k, 1) \to \mathrm{Sym}^k(V) \to 0$$

using the isomorphisms: $\mathrm{Sym}^j(V) \otimes \chi^k(\det) \simeq \mathrm{Sym}^j(V^*)$ and $\mathrm{Sym}^k(V^*) \otimes \chi^k(\det) \simeq \mathrm{Sym}^k(V)$.

We remark that $\mathrm{Ind}_B^G(1,1) = \mathrm{Ind}_B^G 1$ is the direct sum of the trivial representation and the irreducible representation $\mathrm{Sym}^{p-1}(V) \simeq \mathrm{Sym}^{p-1}(V^*)$. This completes our discussion of the principal series.

§6. The divisor homomorphism. We now return to our original notation. Combining our determination of the modules $N/pN(2n)$ and $D^0/pD^0(2n)$ for $2 < 2n < p - 1$ in Propositions 3.6 and 4.2 with the composition series for the induced representations determined in §5, we obtain the following.

THEOREM 6.1. *If $2 < 2n < p - 1$ and $2n + 2m = p - 1$, we have nonsplit exact sequences of $\mathbb{Z}/p\mathbb{Z}[G]$-modules*

$$\begin{cases} 0 \to \mathrm{Sym}^{2m}(E_p^*) \to N/pN(2n) \to \mathrm{Sym}^{2n}(E_p) \to 0 \\ 0 \to \mathrm{Sym}^{2n}(E_p) \to D^0/pD^0(2n) \to \mathrm{Sym}^{2m}(E_p^*) \to 0. \end{cases}$$

Since $\mathrm{Sym}^{2n}(E_p)$ is not isomorphic to $\mathrm{Sym}^{2m}(E_p^*)$ (even when $2n = 2m$), we see that the G-modules $N/pN(2n)$ and $D^0/pD^0(2n)$ are *not* isomorphic. In fact, the G-homomorphism

$$\mathrm{div}_{2n} \colon N/pN(2n) \to D^0/pD^0(2n)$$

has kernel containing $\mathrm{Sym}^{2m}(E_p^*)$ and cokernel mapping surjectivity to $\mathrm{Sym}^{2m}(E_p^*)$. To determine the exact kernel and cokernel, we observe that by Theorem 6.1, the divisor map induces a G-homomorphism

$$\underline{\mathrm{div}}_{2n} : \mathrm{Sym}^{2n}(E_p) \to \mathrm{Sym}^{2n}(E_p),$$

which is either zero or an isomorphism, by Schur's lemma. An easy calculation shows that $\underline{\mathrm{div}}_{2n} = 0$ if and only if the numerator of the Bernoulli number B_{2m+2} is divisible by p. This, together with the fact that div_0 and div_2 are always isomorphisms, gives the statement on the dimension of C/pC made in the introduction.

A bit more analysis gives the structure of $C \otimes \mathbb{Z}_p$ as a module over the Galois group. Applying the Teichmüller lifting to the central character χ of $\mathrm{Gal}(L/k)$ acting on E_p, we get a character ω with values in $\mathbb{Z}_p^*$. We can therefore decompose $C \otimes \mathbb{Z}_p$ as a direct sum of eigencomponents $C \otimes \mathbb{Z}_p(2n)$ on which the center Z of G acts via the character ω^{2n}. Likewise the induced representations $\mathrm{Ind}_B^G(\omega^{2n}, 1)$ and $\mathrm{Ind}_B^G(1, \omega^{2n})$ may be defined on rank $p + 1$ modules over $\mathbb{Z}_p$.

Assume that $0 < 2n < p - 1$, so ω^{2n} is nontrivial, and define $2m$ by $2n + 2m = p - 1$. The generator $w = \begin{pmatrix} 0 & 1 \\ 1 & 0 \end{pmatrix}$ of the Weyl group gives an intertwining operator

$$\phi_w \colon \mathrm{Ind}_B^G(\omega^{2n}, 1) \to \mathrm{Ind}_B^G(1, \omega^{2n})$$

which is defined on functions by $\phi_w f(g) = \Sigma_U f(wug)$. One can show that ϕ_w is an injection of $\mathbb{Z}_p[G]$-modules, with cokernel isomorphic to $\mathrm{Sym}^{2m}(E_p^*)$. Define the Bernoulli number

$$(6.2) \qquad B_{2,\,\omega^{2m}} = \frac{1}{p} \sum_{a=1}^{p-1} \omega^{2m}(a) a^2$$

in $\mathbb{Q}_p$. When $2n \neq 2$, $B_{2,\,\omega^{2m}}$ is a p-adic integer, and our main result is the following.

THEOREM 6.3.   1) $C \otimes \mathbb{Z}_p(2n) = 0$ for $2n = 0, 2$.
2) *For* $2 < 2n < p - 1$ *and* $2n + 2m = p - 1$ *we have an isomorphism of* $\mathbb{Z}_p[G]$-modules

$$C \otimes \mathbb{Z}_p(2n) \simeq \mathrm{coker}\big(B_{2,\,\omega^{2m}}\!\cdot\!\phi_w \colon \mathrm{Ind}(\omega^{2n}, 1) \to \mathrm{Ind}(1, \omega^{2n})\big).$$

We will not prove Theorem 6.3 here, although it follows quite easily from our previous results. We remark that (6.3) is compatible with our previous remarks on the map $\underline{\mathrm{div}}_{2n}$, using Kummer's congruence:

$$\frac{1}{2} B_{2,\,\omega^{2m}} \equiv \frac{1}{2 + 2m} B_{2m+2} \pmod{p}.$$

REFERENCES

1. E. HECKE, *Grundlagen einer Theorie der Integralgruppen und der Integralperioden bei den Normalteilern der Modulgruppe.* Math. Ann. **116** (1939), 469–510 (= Math. Werke **38**, 731–772).
2. D. KUBERT AND S. LANG, *Modular Units.* Grundlehren **224**, Springer-Verlag, 1981.
3. Y. MANIN, *Parabolic points and zeta-functions of modular curves.* Izv. Akad. Nauk. SSSR, Vol. **6**, No. **1** (1972) AMS Translations, 19–64.
4. J.-P. SERRE, *Linear Representations of Finite Groups.* Graduate Texts **42**, Springer-Verlag, 1977.
5. G. SHIMURA, *Introduction to the Arithmetic Theory of Automorphic Functions.* Princeton University Press, 1971.
6. R. STEINBERG, *Lectures on Chevalley Groups.* Yale University Press, 1967.

DEPARTMENT OF MATHEMATICS, HARVARD UNIVERSITY, CAMBRIDGE, MASSACHUSETTS 02138.

A FUNCTIONAL EQUATION OF THE NON-ARCHIMEDIAN RANKIN CONVOLUTION

Dedicated to Yuri Ivanovich Manin on the occasion of his 50th birthday

A. A. PANCHISHKIN

§0. Introduction. Let p be a prime number and S a finite set of prime numbers containing p. The aim of this paper is to establish a functional equation satisfied by the S-adic L-functions, which are obtained by the non-Archimedian interpolation of the special values of the Rankin convolution of two elliptic cusp forms of different weight. Let N be an arbitrary positive integer. Let f be a cusp form of weight $k \geqslant 2$ for the congruence subgroup $\Gamma_0(N)$ with a character ψ modulo N, which is, in addition, a primitive form of conductor $C(f)$, i.e., normalized new form of the exact level $C(f)$ dividing N. Let g be another primitive form of conductor $C(g)$ and weight $l < k$ for $\Gamma_0(N)$ with a character ω. Write $e(z) = \exp(2\pi i z)$. Suppose that Fourier expansions of f and g are given by

$$(0.1) \qquad f = \sum_{n=1}^{\infty} a(n)e(nz), \qquad g = \sum_{n=1}^{\infty} b(n)e(nz).$$

The Rankin convolution of f and g is denoted by

$$(0.2) \qquad \mathscr{D}_N(s, f, g) = L_N(2s + 2 - k - l, \omega\psi)L(s, f, g),$$

where $L(s, f, g) = \sum_{n=1}^{\infty} a(n)b(n)n^{-s}$ and $L_N(2s + 2 - k - l, \omega\psi)$ is the Dirichlet L series of $\omega\psi$ with the Euler factors at the primes dividing N removed from its Euler product. It is known (from Rankin [14] and Selberg [16]) that $\mathscr{D}_N(s, f, g)$ has a holomorphic continuation over the whole complex plane and it satisfies a functional equation, which in the simplest case of $N = 1$ has the form

$$(0.3) \qquad \Psi(s, f, g) = (-1)^k \Psi(k + l - 1 - s, f, g),$$

where

$$(0.4) \qquad \Psi(s, f, g) = \gamma(s)\mathscr{D}_N(s, f, g)$$

with the Γ-factor $\gamma(s) = (2\pi)^{-2s}\Gamma(s)\Gamma(s + 1 - l)$.

Received August 15, 1986.

78   A. A. PANCHISHKIN

On the other hand, it is known (again essentially from Rankin [15], but see also
[17], [18] and [9]) that the number

$$(0.5) \qquad \frac{\Psi(l + r, f, g)}{\pi^{1-l}\langle f, f\rangle_{C(f)}}$$

is algebraic for all integers r with $0 \leqslant r \leqslant k - l - 1$. Here

$$\langle f, f\rangle_{C(f)} = \int_{H/\Gamma_0(C(f))} |f(z)|^2 y^{k-2} \, dx \, dy,$$

where $z = x + iy$, $H/\Gamma_0(C(f))$ denotes a fundamental domain for H modulo
$\Gamma_0(C(f))$.

For a positive integer A denote by $S(A)$ the set of all prime numbers, dividing
A. Let χ be a Dirichlet character of conductor $C(\chi)$ such that $S(\chi) \subset S$, where
$S(\chi)$ denotes $S(C(\chi))$. Denote by $\mathbb{Z}_S$ the S-adic completion of $\mathbb{Z}$ and let $\mathbb{Z}_S^\times$ be
the group of units of the compact ring $\mathbb{Z}_S^\times$. Then χ induces a character of finite
order of the group $\mathbb{Z}_S^\times$, which is denoted by the same letter χ, $\chi: \mathbb{Z}_S^\times \to \overline{\mathbb{Q}}^\times$. Let
$\mathbb{C}_p = \overline{\mathbb{Q}}_p$ denote the Tate field (the completion of an algebraic closure of $\mathbb{Q}_p$)
with the valuation $|\cdot|_p$, normalized by $|p|_p = p^{-1}$. We fix an embedding i_p:
$\overline{\mathbb{Q}} \hookrightarrow \mathbb{C}_p$ (the symbol i_p will be often omitted from our formulae). Let $g(\chi) =$
$\sum_{n=1}^\infty \chi(n)b(n)e(nz)$ denote the twist of g. Starting with (0.5) we establish S-adic
interpolation of algebraic numbers of the form

$$i_p\left(\frac{\Psi(l + r, f, g(\chi))}{\pi^{1-l}\langle f, f\rangle_{C(f)}}\right) \quad \text{with } 0 \leqslant r \leqslant k - l - 1.$$

The construction is based entirely on the method of Rankin and its p-adic
version given recently by H. Hida [2]. The resulting S-adic $\mathbb{C}_p$-valued L-functions
are essentially non-Archimedean Mellin transforms of certain bounded measures,
which generalize those in [2]. We prove that these functions satisfy a functional
equation. Yu. I. Manin suggested in [7], [8] that non-Archimedean L-functions
can be attached to various objects such as automorphic representations, algebraic
varieties over number fields, and Galois representations, so that the functions and
relations between them should provide as with new significant number-theoretic
facts. It is a pleasure for the author to consider this work as a step in this
direction.

Notation.   Let H be the upper half complex plane. The group $\mathrm{GL}_2^+(\mathbb{R})$ of real
2×2 matrices with positive determinant acts on H via linear fractional transfor-
mations. If $\gamma = \left(\begin{smallmatrix} a & b \\ c & d \end{smallmatrix}\right) \in \mathrm{GL}_2^+(\mathbb{R})$ and $f(z)$ is any function on H, we define, for
each $k \in \mathbb{Z}$,

$$(f|_k\gamma)(Z) = (\det(\gamma))^{k/2} f(\gamma(z))(cz + d)^{-k}.$$

For each positive integer N we write $\mathcal{M}_k(\Gamma_1(N))$ for the space of holomorphic modular forms of weight k for $\Gamma_1(N)$, and $\mathcal{S}_k(\Gamma_1(N))$ for the subspace of cusp forms. For each character ψ modulo N we write

$$\mathcal{M}_k(\Gamma_0(N), \psi) = \left\{ f \in \mathcal{M}_k(\Gamma_1(N)) \middle| f|_k \begin{pmatrix} a & b \\ c & d \end{pmatrix} \right.$$

$$\left. = \psi(d) f \text{ for all } \begin{pmatrix} a & b \\ c & d \end{pmatrix} \in \Gamma_0(N) \right\}$$

and put

$$\mathcal{S}_k(\Gamma_0(N), \psi) = \mathcal{S}_k(\Gamma_1(N)) \cap \mathcal{M}_k(\Gamma_0(N), \psi).$$

For $f_1 \in \mathcal{S}_k(\Gamma_0(N), \psi)$ with $k \geqslant 1$ the Petersson inner product of f_1 with $f_2 \in \mathcal{S}_k(\Gamma_0(N), \psi)$ is defined by

$$\langle f_2, f_1 \rangle_N = \int_{H/\Gamma_0(N)} \overline{f_2(z)}\, f_1(z) y^{k-2} \, dx \, dy.$$

§1. Statement of main results

1.1. Let us describe the p-adic analytic group X_S, on which our S-adic L-functions are defined [21]. There is the isomorphism

$$\mathbb{Z}_S^\times \cong \bigotimes_{q \in S} \mathbb{Z}_q^\times$$

and let x_p denote the homomorphism $x_p \colon \mathbb{Z}_S^\times \to \mathbb{C}_p^\times$ obtained by composition of the projection $\mathbb{Z}_S^\times \to \mathbb{Z}_p^\times$ and the embedding $\mathbb{Z}_p^\times \hookrightarrow \mathbb{C}_p^\times$. Write

$$X(\cdot) = \operatorname{Hom}_{contin}\left(\cdot, \mathbb{C}_p^\times\right)$$

for the group of p-adic characters of a topological group and define $X_S = X(\mathbb{Z}_S^\times)$, then $x_p \in X_S$. Put $\varepsilon = 1$ or 2 according as $p > 2$ or $p = 2$, and

$$U = \left\{ x \in \mathbb{Z}_p^\times \,\middle|\, x \equiv 1 (\mathrm{mod}\ p^\varepsilon) \right\},$$

then there is the decomposition

$$X_S = X\!\left((\mathbb{Z}/p^\varepsilon\mathbb{Z}) \times \prod_{p \neq q \in S} \mathbb{Z}_q^\times\right) \times X(U).$$

The analytic $\mathbb{C}_p$-structure on $X(U)$ is defined by the isomorphism

$$\vartheta \colon X(U) \overset{\sim}{\to} T = \left\{ x \in \mathbb{C}_p^\times \,\middle|\, |x - 1|_p < 1 \right\} \quad \text{with } \vartheta(x) = x(1 + p^\varepsilon).$$

We identify the torsion subgroup $X_S^{\text{tors}} \subset X_S$ with the set of all primitive Dirichlet characters χ such that $S(\chi) \subset S$, by the projection $\mathbb{Z}_S^{\times} \to (\mathbb{Z}/C(\chi)\mathbb{Z})^{\times}$ and the embedding i_p. An analytic function $F: T \to \mathbb{C}_p$ is by definition the sum of a series $\sum_{i=0}^{\infty} a_i (u - 1)^i$ with $a_i \in \mathbb{C}_p$, which is convergent for all $u \in T$. The notion of an analytic function is then extended to X_S. It is known that a bounded $\mathbb{C}_p$-analytic function F on X_S is uniquely determined by its values $F(\chi)$ with χ in the discrete subgroup $X_S^{\text{tors}} \subset X_S$ [21], [6]. If μ is a bounded $\mathbb{C}_p$-valued measure on $\mathbb{Z}_S$ then its Mellin transform $L_\mu(x) = \int_{\mathbb{Z}_S^{\times}} x \, d\mu$ $(x \in X_S)$ is a bounded $\mathbb{C}_p$-analytic function [6], [10], [21].

1.2. We assume now that the following condition on f and g are satisfied:

$$(1.2) \qquad \left| i_p(a(q)) \right|_p = 1 \quad \text{for all } q \in S,$$

$$(1.3) \qquad S \cap S(C(f)) = S \cap S(C(g)) = \varnothing,$$

$$(1.4) \qquad (C(f), C(g)) = 1, \quad \text{i.e. } S(C(f)) \cap S(C(g)) = \varnothing,$$

$$(1.5) \qquad a(C(f)) \neq 0, \qquad b(C(g)) \neq 0.$$

We put $N = C(f)C(g)$ and denote by $\alpha(q)$ a root of the Hecke polynomial $X^2 - a(q)X + \psi(q)q^{k-1}$ such that for $q \in S$ $|i_p(\alpha(q))|_p = 1$, and let $\alpha'(q)$ denote the other root. Then we have that the numbers

$$\hat{\alpha}(q) = \psi^{-1}(q)\alpha(q), \qquad \hat{\alpha}'(q) = \psi^{-1}(q)\alpha'(q)$$

coincide with roots of the complex-conjugate polynomial $X^2 - \overline{a(q)}X + \psi^{-1}(q)q^{k-1}$, because the Hecke operator $T(q)$ is a ψ-hermitian operator on $\mathscr{S}_k(\Gamma_0(N), \psi)$. Similarly, if

$$X^2 - b(q)X + \omega(q)q^{l-1} = (X - \beta(q))(X - \beta'(q)),$$

then

$$\hat{\beta}(q) = \omega^{-1}(q)\beta(q) \quad \text{and} \quad \hat{\beta}'(q) = \omega^{-1}(q)\beta'(q)$$

coincide with the roots of the polynomial $X^2 - \overline{b(q)}\, X + \omega^{-1}(q)q^{l-1}$. Let us extend the definition of $\alpha(n)$, $\alpha'(n)$, $\beta(n)$, $\beta'(n)$ to all positive integers n by multiplicativity, and put

$$f^{\rho} = \sum_{n=1}^{\infty} \overline{a(n)}\, e(nz), \qquad g^{\rho} = \sum_{n=1}^{\infty} \overline{b(n)}\, e(nz).$$

1.3. THEOREM. *Under the assumptions* (1.2)–(1.5) *there exists a bounded analytic function*

$$\Psi: X_S \to \mathbb{C}_p, \qquad \Psi(x) = \Psi_S(x, f, g),$$

which is uniquely determined by the following condition: for each $\chi \in X_S^{\mathrm{tors}}$ of conductor M, and r with $0 \leqslant r \leqslant k - l - 1$ the value $\Psi(\chi x_p^r)$ is given by the image under i_p of the following algebraic number

$$(1.6) \qquad (-1)^r \omega(M) \frac{G(\chi)^2 M^{l+2r-1}}{N^{2l+2r}\alpha(M)^2} \frac{\Psi(l+r, f, g^\rho(\chi^{-1}))}{\pi^{1-l}\langle f, f\rangle_{C(f)}} A(r, \chi)$$

where

$$(1.7) \quad A(r, \chi) = \prod_{q \in S \setminus S(\chi)} \left(1 - \chi(q)\alpha^{-1}(q)\beta(q)q^r\right)$$

$$\times \left(1 - \chi(q)\alpha^{-1}(q)\beta'(q)q^r\right)\left(1 - \chi^{-1}(q)\alpha'(q)\hat{\beta}(q)q^{-l-r}\right)$$

$$\times \left(1 - \chi^{-1}(q)\alpha'(q)\hat{\beta}'(q)q^{-l-r}\right),$$

$G(\chi) = \sum_{a=1}^{A-1}\chi(a)e(a/A)$ *is the Gauss sum for* χ.

1.4. For a positive integer A, put $\tau(A) = \begin{pmatrix} 0 & -1 \\ A & 0 \end{pmatrix}$ then it follows from the theory of primitive forms [5] that

$$f|_k\tau(C(f)) = W(f)f, \qquad g|_l\tau(C(g)) = W(g)g,$$

with some $W(f), W(g) \in \mathbb{C}$ such that $|W(f)| = |W(g)| = 1$, $W(f^\rho) = \overline{W(f)}$, $W(g^\rho) = \overline{W(g)}$. If χ is a Dirichlet character of conductor A coprime to $C(g)$ then the cusp form $g(\chi) \in \mathscr{S}_l(\Gamma_0(C(g)A^2, \omega\chi^2))$ is primitive of conductor $C(g)A^2$, so that

$$W(g(\chi)) = \omega(A)\chi(C(g))\frac{G(\chi)^2}{A}W(g).$$

It is known (from Li [5], p. 146) that the following functional equation holds:

$$(1.8) \qquad \Psi(s, f, g^\rho(\overline{\chi})) = B_\chi(s)\Psi(k + l - 1 - s, f^\rho, g(\chi))$$

with

$$B_\chi(s) = (NA^2)^{k+l-2s}(-1)^k\omega^{-1}(C(f))\psi(C(g))$$

$$\times\chi^{-2}(A)\psi\omega^{-1}(A)\frac{G(\chi^{-1})^4}{A^2}W(f)^2\overline{W(g)}^2.$$

1.5. THEOREM. *Under the assumptions as above the functional equation*

$$(1.9) \qquad \Psi(x) = C_{f,g} \frac{N^{k-l-1}}{x(N)^2} \hat{\Psi}\left(x_p^{k-l-1} x^{-1}\right)$$

holds for all $x \in X_S$, *where*

$$(1.10) \qquad C_{f,g} = i_p\left((-1)^{l+1} \omega^{-1}(C(f)) \psi(C(g)) W(f)^2 W(g^\rho)^2\right),$$

and $\hat{\Psi}(x) = \Psi_S(x, f^\rho, g^\rho)$ *is defined by Theorem* 1.3 *with* (f, g, α) *replaced by* $(f^\rho, g^\rho, \hat{\alpha})$.

1.6. We now summarize the rest of the paper. First, we introduce S-adic modular forms and S-ordinary forms in §2, and describe a linear function, generalizing the Petersson inner product by a fixed S-ordinary form. Then we discuss in §3 the holomorphic projection operator and real analytic Eisenstein series, which are slightly more general than those in [2]. This permits us to treat all integers s in the critical strip $l \leqslant s \leqslant k - 1$ of (0.2) equally. Main S-adic construction is given in §4. The detailed calculation of some non-Archimedian integrals from §4 provides us in §5 with both the formula (1.6) and the functional equation (1.9).

In his earlier note [13] the author constructed certain p-adic distributions attached to (0.2), but that attempt of p-adic construction was less successful. Unfortunately, the statement about denominators ([13], p. 230) is not quite correct. However, similar construction worked well in case of symmetric squares of the Hecke series [1] (see also [11], [12]).

We refer the reader to [6], [10], [21] for general facts about bounded measures and their Mellin transforms.

§2. *S*-adic modular forms

2.1. Let N be an arbitrary positive integer and ψ be a character modulo N. Let A denote either a finite extension K of $\mathbb{Q}_p$ or the ring $\mathbb{O}_k$ of the p-adic integers in K. Then we write

$$(2.1) \qquad \mathscr{M}_k(\Gamma_1(N); A) \quad \text{and} \quad \mathscr{M}_k(\Gamma_0(N), \psi; A)$$

for the A-submodules of $A[[q]]$ which are defined in [2] by q-expansions of the corresponding true modular forms with algebraic coefficients and by extension of scalars to A. For every

$$f = \sum_{n=0}^{\infty} a(n, f) q^n \in \mathscr{M}_k(\Gamma_1(N); A)$$

a p-adic norm $|f|_p$ of f is defined by

$$|f|_p = \sup_n |a(n, f)|_p.$$

Put

$$\mathcal{M}_k(N, S; A) = \bigcup_M \mathcal{M}_k(\Gamma_1(NM); A)$$

$$\mathcal{M}_k(N, \psi, S; A) = \bigcup_M \mathcal{M}_k(\Gamma_0(NM), \psi; A),$$

where M runs over positive integers with $S(M) \subset S$.

Let $\overline{\mathcal{M}}_k(N, S; A)$ (resp. $\overline{\mathcal{M}}_k(N, \psi, S; A)$) be the completion of $\mathcal{M}_k(N, S; A)$ (resp. $\mathcal{M}_k(N, \psi, S; A)$) for the norm $|\cdot|_p$. Any element of $\overline{\mathcal{M}}_k(N, S; K)$ will be called an S-adic modular form. Next, define the Hecke algebras

$$(2.2) \qquad \mathcal{H}_k(N, S, \mathcal{O}_K) = \varprojlim \mathcal{H}_k(\Gamma_1(NM); \mathcal{O}_K),$$

$$(2.3) \qquad \mathcal{H}_k(N, \psi, S; \mathcal{O}_K) = \varprojlim \mathcal{H}_k(\Gamma_0(NM), \psi; \mathcal{O}_K),$$

where $\varprojlim$ is taken for M with $S(M) \subset M$ and

$$(2.4) \qquad \mathcal{H}_k(\Gamma_1(NM); \mathcal{O}_K), \ \mathcal{H}_k(\Gamma_0(NM), \psi; \mathcal{O}_K)$$

are defined in [2]. The compact algebras (2.2), (2.3) are topologically generated by the Hecke operators, which act on q-expansions via standard formulae.

2.2. Let us now define the ordinary part $\overline{\mathcal{M}}_k^0(N, \psi, S; A)$ of $\overline{\mathcal{M}}_k(N, \psi, S; A)$ as an A-submodule given by the image of certain idempotent $e = \varprojlim e_M \in \mathcal{H}_k(N, \psi, S; \mathcal{O}_K)$, where e_M is defined in a similar way as in [2]: we put $M_0 = \prod_{q \in S} q$, then by definition $e_M = \lim_{r \to \infty} T(M_0)^{p^r u}$ with $T(M_0) \in \mathcal{H}_k(\Gamma_0(NM), \psi; \mathcal{O}_K)$ for a positive integer u prime to p, which is chosen so that e_M is an idempotent. We show that

$$(2.5) \qquad \overline{\mathcal{M}}_k^0(N, \psi, S; A) \text{ is contained in } \mathcal{M}_k(\Gamma_0(NM_0), \psi; A)$$

by the argument similar to that in [2]. A true eigenform which belong to $\overline{\mathcal{M}}_k^0(N, \psi, S; K)$ (for some sufficiently large K) will be called an S-ordinary modular form.

2.3. For $f = \sum_{n=1}^{\infty} a(n) q^n$ as in the Introduction with $(N, M_0) = 1$ we put $C_0 = M_0 C(f)$ and define $f_0 = \sum_{n=1}^{\infty} a(n, f, f_0) q^n$ by

$$(2.6) \qquad f_0 = \sum_{d \mid M_0} \mu(d) \alpha'(d) f | V(d)$$

where μ is the Moebius function, $f|V(d) = \sum_{n=1}^{\infty} a(n)q^{dn}$. The definition (2.6) is equivalent to the identity

$$\sum_{n=1}^{\infty} a(n, f_0)n^{-s} = \sum_{n=1}^{\infty} a(n)n^{-s} \prod_{q \in S} (1 - \alpha'(q)q^{-s}).$$

Then we have $f_0|e = f_0$, i.e., f_0 is S-ordinary and it follows from the theory of new forms that f_0 is the unique eigenform in $\mathcal{M}_k(\Gamma_0(C_0), \psi; K)$ with the property: $a(q, f_0) = \alpha(q)$ for $q \in S$ and $a(n, f_0) = a(n)$ for n coprime to M_0. As a consequence, the natural ring homomorphism of $\mathcal{H}_k(\Gamma_0(C_0), \psi, K)$ onto K, defined by $T(n) \mapsto a(n, f_0)$, is split, so that there is the algebra direct sum decomposition

$$\mathcal{H}_k(\Gamma_0(C_0), \psi; K) \cong \mathcal{A} \oplus K.$$

The idempotent corresponding to the direct summand K is denoted by 1_f.

Let us define a bounded p-adic linear form $l_f = l_{f, \alpha}$,

$$l_f : \overline{\mathcal{M}}_k(C(f), \psi, S; K) \to K$$

by

$$(2.7) \qquad l_f(F) = a(1, F|el_f),$$

where $a(1, F|el_f)$ is the first q-expansion coefficient of $F|el_f$.

2.4. PROPOSITION. *Assume that the finite algebraic number field K_0 contains all the Fourier coefficients of the form f_0. Then, for every M such that $S(M) \subset S$ and every $F \in \mathcal{M}_k(\Gamma_0(C_0 M), \psi)$ with Fourier coefficients in K_0, we have that $l_f(F) \in i_p(K_0)$ and*

$$(2.8) \qquad l_f(F) = \frac{M^{k-1}}{\alpha(M)} \frac{\langle h_M, F \rangle_{C_0 M}}{\langle h, f_0 \rangle_{C_0}},$$

where $C_0 = C(f)M_0$,

$$(2.9) \qquad h = f_0^\rho|_k \begin{pmatrix} 0 & -1 \\ C_0 & 0 \end{pmatrix}, \qquad h_M(z) = h(Mz).$$

This is essentially Proposition 4.5 of [2]. The linear function l_f gives a generalization of the Petersson inner product by f to the S-adic case.

§3. Real analytic Eisenstein series

3.1. Let $\mathcal{M}_{r, k}(\Gamma_0(N), \psi)$ denote the complex vector space of real analytic modular forms of type r and weight k for $\Gamma_0(N)$ with a Dirichlet character ψ modulo N (see [19]). We recall that if $F \in \mathcal{M}_{r, k}(\Gamma_0(N), \psi)$ then F has the form

$$F = \sum_{j=0}^{r} \mathrm{Im}(z)^{-j} g_j(z) \text{ with } g_j(z) = \sum_{n=0}^{\infty} b(j, n)e(nz)$$

holomorphic on H, and

$$F|_k\gamma = \psi(d)F \quad \text{for all } \gamma = \begin{pmatrix} a & b \\ c & d \end{pmatrix} \in \Gamma_0(N).$$

The corresponding vector space for the group $\Gamma_1(N)$ is denoted by $M_{r,k}(\Gamma_1(N))$ and we put

$$(3.1) \qquad \mathscr{M}_{r,k}(N, S) = \bigcup_M \mathscr{M}_{r,k}(\Gamma_1(NM)),$$

where M runs over positive integers with $S(M) \subset S$.

3.2. Define Eisenstein series by

$$(3.2) \quad E_{m,N}(z, s; a, b) = \sum_{0 \neq (c,d) \equiv (a,b) \,(\mathrm{mod}\, N)} (cz + d)^{-m}|cz + d|^{-2s}$$

for $a, b \in \mathbb{Z}/N\mathbb{Z}$. The series (3.2) have analytic continuation as functions of s and satisfy a functional equation [4]. Let r and m be integers such that $r \geqslant 0$, $r + m > 0$. We introduce the series

$$(3.3)$$

$$E_{r,m,M}(a) = \frac{\Gamma(m+r)(NM)^{m-1}}{(-4\pi y)^r(-2\pi i)^m} \sum_{b \,(\mathrm{mod}\, NM)} e\left(-\frac{ab}{NM}\right) E_{m+2r, NM}(Z, -r; 0, b).$$

Then for $NM > 1$ and $a \not\equiv 0 \,(\mathrm{mod}\, NM)$ we have that $E_{r,m,M}(a) \in \mathscr{M}_{r,m+2r}(N, S)$ and its Fourier expansion is explicitly given by

$$(3.4) \quad E_{r,m,M}(a) = \varepsilon_{r,m,M}(a) + (4\pi y)^{-r} \sum_{n=1}^{\infty} \left(\sum_{\substack{d|n \\ d \equiv a \,(\mathrm{mod}\, NM)}} \mathrm{sgn}(d)d^{m-1} \right)$$

$$\times W(4\pi ny, m + r, -r)e(nz)$$

where $W(y, \alpha, \beta)$ denotes the Whittaker function [19] and

$$\varepsilon_{r,m,M}(a) = (-4\pi y)^{-r}\left[\frac{\Gamma(s+r)}{\Gamma(s)}\zeta(1 - s; a, NM)\right]\Bigg|_{s=m}$$

with $\zeta(s; a, NM) = \sum_{0 < n \equiv a \,(\mathrm{mod}\, NM)} n^{-s}$ being the partial Riemann zeta function (note that $\varepsilon_{r,m,M}(a) = 0$ for $m < 0$).

3.3.   Next we define a compact ring Y by

$$Y = \varprojlim Y_M, \qquad Y_M = \mathbb{Z}/NM\mathbb{Z}.$$

It is seen from the formula (3.4) that the series (3.3) define a distribution on Y with values in $\mathcal{M}_{r,\,m+2r}(N, S)$

3.4.   *Remark.*   In case $m > 0$ the series (3.3) can be obtained by applying the operator $\delta_m^{(r)}$ of Shimura [17] to a holomorphic Eisenstein series as in [2] but we do not use this fact.

§4. The S-adic Rankin formula

4.1.   Let f and g be as in Introduction and §1. The classical Rankin formula reads as follows (see [14], [17]):

$$(4.1) \quad 2(4\pi)^{-s}\Gamma(s)\mathscr{D}_N(s, f, g) = \left\langle f^\rho, gE_{k-l}(s + 1 - k)\mathrm{Im}(z)^{s+1-k}\right\rangle_N$$

with the Eisenstein series

$$E_{k-l}(s + 1 - k) = \sum_{0 \neq (m, n)} \psi\omega(n)(mNz + n)^{-(k-l)}|mNz + n|^{-2s}.$$

In the S-adic case we consider the compact ring $Y = \varprojlim_M \mathbb{Z}/NM\mathbb{Z}$ $(S(M) \subset S)$ and put, for $a \in \mathbb{Z}/NM\mathbb{Z}$, and for each $b > 1$ prime to NM_0,

$$(4.2a) \qquad E^b_{r,\,m,\,M}(a) = E_{r,\,m,\,M}(a) - b^m E_{r,\,m,\,M}(b^{-1}a),$$

where $b^{-1} \in Y_M$ and $E_{r,\,m,\,M}(a)$ is defined by (3.3). Then (4.2a) is a distribution on Y with values in $\mathcal{M}_{r,\,m+2r}(N, S)$. Put, for $a \in Y_M$,

$$(4.2b) \qquad g_M(a) = \sum_{0 < n \equiv a \,(\mathrm{mod}\, M)} b(n)q^n$$

so that (4.2b) is a distribution on Y with values in $\mathscr{S}_l(N, S) = \mathscr{S}_{0,\,l}(N, S)$. Consider, for each $y \in Y_M$, the following multiplicative convolution of (4.2a), (4.2b)

$$\Phi_M(y, r) = \sum_{a \in Y_M^\times} \psi\bar{\omega}(a)H\big[g_M(a^2 y)E^b_{r,\,k-l-2r,\,M}(a)\big],$$

where H denotes the holomorphic projection operator [20], so that $\Phi_M(y, r)$ takes values in $\mathscr{S}_k(\Gamma_0(NM^2), \psi; K)$ and the coefficients of its q-expansion are p-adically bounded. Therefore, for a suitable finite extension K of $\mathbb{Q}_p$ we get a bounded measure with values in $\overline{\mathcal{M}}_k(N, S; K)$. Next, we define its ordinary part

$\Phi_M^0(y, r) = e\Phi_M(y, r)$ (see (2.4)), which takes values in $\mathcal{M}_k(\Gamma_0(NM_0), \psi; K)$ by (2.5). Now we apply the linear form l from §2 and define a bounded K-valued measure ϑ^b on Y as in [2]:

$$(4.3) \qquad \int_Y \vartheta(y)\, d\vartheta^b(y) = l_f\left[\mathrm{Tr}\left(\int_Y \vartheta(y)\, d\Phi^0(y, 0)\right)\right]$$

for any locally constant function $\vartheta\colon Y \to \overline{\mathbf{Q}}$, where Tr denotes the trace operator of $\mathcal{M}_k(\Gamma_0(NM_0), \psi; K)$ onto $\mathcal{M}_k(\Gamma_0(C_0), \psi; K)$, which is a bounded linear operator, defined by a system of representatives for $\Gamma_0(NM_0) \setminus \Gamma_0(C_0)$. Let $\vartheta\colon Y \to \overline{\mathbf{Q}}$ be a spherical function on Y with a character χ of finite order of the group $Y^\times$, so that

$$\vartheta(zy) = \chi(z)\vartheta(y) \qquad (z \in Y^\times,\ y \in Y).$$

We then define the twist of g by

$$g(\vartheta) = \sum_{n=0}^{\infty} \vartheta(n)b(n)e(nz)$$

then $g(\vartheta) \in \mathcal{M}_l(\Gamma_0(N^2M^1), \xi)$ for a sufficiently large M' such that $S(M') \subset S$ and $\xi = \chi^2\omega$.

4.2. PROPOSITION. *The notation and assumptions being as above, for $0 \leqslant r \leqslant k - l - 1$, the value of the S-adic integral*

$$\int_Y \vartheta(y)y_p^r\, d\vartheta^b(y)$$

is given by the image under i_p of the following algebraic number

$$(4.4) \qquad t\big(1 - b^m\psi\bar{\xi}(b)\big)\alpha(M^1)^{-1}$$

$$\times \frac{\mathcal{D}_{NM}\big(l + r, f_0|_k\gamma,\ g(\vartheta)|_l\tau(M^1N^2)\big)}{\pi^{l+2r-1}\langle h, f_0\rangle_{C_0}}$$

where $m = k - l - 2r$, $y_p\colon Y \to \mathbf{Z}_p$ is the canonical projection,

$$t = (\sqrt{-1})^{2k-m}2^{1-2k-m}\Gamma(1 + r)\Gamma(r + 1)N^{-m}M_0^{1-k/2}M'^{(l+2r)/2},$$

f_0 is defined by (2.6),

$$h = f_0^\rho|\tau(C_0), \qquad \gamma = \begin{pmatrix} M^1N^2/C(f) & 0 \\ 0 & 1 \end{pmatrix}, \qquad C_0 = C(f)M_0.$$

The essential part of the proof is to obtain the following p-adic equality

$$(4.5) \qquad \int_Y \vartheta(y)\, d\Phi^0(y, r) = (-1)^r \int_Y \vartheta(y) y_p^r\, d\Phi^0(y, 0).$$

For $0 \leqslant r \leqslant (k - l)/2 - 1$ it can be proved in a very similar way as in [2]. In the general case we use the integral formula for the projection operator H [20] and the Fourier expansion (3.4), so that the assumption $m > 0$ in [2] is not actually necessary.

§5. The S-adic Rankin convolution

5.1.   Let μ^b denote the bounded $\mathbb{C}_p$-valued measure on $\mathbb{Z}_S$ defined by ϑ^b and the canonical projection $y_S\colon Y \to \mathbb{Z}_S$. Consider its Mellin transform $L_{\mu^b}\colon X_S \to \mathbb{C}_p$, which is a bounded analytic function on X_S. For a Dirichlet character $\chi \in X_S^{\mathrm{tors}}$ of conductor M $(S(M) \subset S)$ and r with $0 \leqslant r \leqslant k - l - 1$ we evaluate $L_{\mu^b}(\chi x_p^r)$ by Proposition 4.2 with $M' = M^2 M_0$, $\vartheta(y) = (\chi \circ y_S)(y)$. The explicit calculation of Euler factors in the right hand side of (4.4) shows that

$$i_p^{-1}\!\left(L_{\mu^b}(\chi x_p^r)\right) = \frac{(-1)^r \omega(M) G(\chi^2) M^{l+2r-1}}{N^{2l+2r}\alpha(M)^2}\, \frac{\Psi\!\left(l + r, f, g^\rho(\chi^{-1})\right)}{\pi^{1-l}\langle f, f\rangle_{C(f)}}$$

$$\times C(f, g) A(r, \chi)\!\left(1 - b^{k-l-2r}\psi\chi^{-2}\omega^{-1}(b)\right),$$

where $A(r, \chi)$ is given by (1.7),

$$C(f, g) = (\sqrt{-1})^{k+l} 2^{1+k+l}\alpha(M_0) M_0^{1-k/2} N^{k+l}$$

$$\times \frac{a(C(f))\overline{b(C(g))}}{C(f)^{k/2} C(g)^{l/2}}\, \frac{\langle f, f\rangle_{C(f)}}{\langle h, f_0\rangle_{C_0}}.$$

5.2.   We define the S-adic Rankin convolution by the following equality

$$(5.2) \qquad \Psi(x) = \Psi_S(x, f, g)$$

$$= i_p(C(f, g))^{-1}\!\left(1 - b^{k-l}\psi\omega^{-1}(b)x^{-2}(b)\right)^{-1} L_{\mu^b}(x)$$

as an analytic function $\Psi\colon X_S \to \mathbb{C}_p$. It is seen from the definition that the function (5.2) is uniquely determined by its values at $x = \chi x_p^r$ ($\chi \in X_S^{\mathrm{tors}}$, $0 \leqslant r \leqslant k - l - 1$), which are given by (5.1). Hence, $\Psi(x)$ does not depend on the choice of b. The functional equation is then deduced from (1.8) with $s = l + r$ by comparing the values of the functions $\Psi(x)$ and $\hat{\Psi}(x_p^{k-l-1}x^{-1})$ at the points $x = \chi x_p^r$. Finally, the definition (5.2) implies that $\psi(x)$ is bounded, if we take into account the functional equation (1.9).

REFERENCES

1. B. Arnaud, *Interpolation p-adique d'un produit de Rankin*, Comptes Rendus Acad. Sc. Paris, ser. I, **299** (1984), 527–530.

2. H. Hida, *A p-adic measure attached to the zeta functins associated with two elliptic modular forms. I*, Invent. Math. **79** (1985), 159–195.

3. H. Jacquet, *Automorphic functions on* GL(2). *Part II*, Lecture Notes in Math. **278**, Springer Verlag, Berlin and New York, 1972.

4. N. M. Katz, *p-adic interpolation of real analytic Eisenstein series*, Ann. Math. **104** (1976), 459–571.

5. W. Ch.-W. Li, *L-series of Rankin type and their functional equations*, Math. Annalen **244** (1979), 135–166.

6. Yu. I. Manin, *Periods of cusp forms and p-adic Hecke series* Mat. Sbornik **92** (134) (1973), 378–401 (in Russian).

7. ______, *Non-archimedian integration and Jacquet-Langlands p-adic L-functions*, Uspekhi Mat. Nauk 31: 1 (1976), 5–54 (in Russian).

8. ______, *Modular forms and number theory*, Proc. Int. Congr. Math. Helsinki, 16–23 August 1978. Helsinki, 1980, pp. 177–186.

9. ______, and A. A. Panchishkin, *Convolutions of Hecke series and their values at integers*, Mat. Sbornik **104** (1977), 617–651 (in Russian).

10. B. Mazur and H. P. F. Swinnerton-Dyer, *Arithmetic of Weil curves*, Invent. Math. **25** (1974), 1–61.

11. A. A. Panchishkin, *Symmetric squares of Hecke series and their values at integers*, Mat. Sbornik **108** (1979), 393–417 (in Russian).

12. ______, *Complex valued measures associated with Euler products*, Trudy seminara imeni I. G. Petrovskogo, **7** (1981), 239–244 (in Russian).

13. ______, *Le prolongement p-adique analytique des fonctions L de Rankin I., II*, Comptes Rendus Acad. Sc. Paris, ser. I, **295** (1982), 51–53; 227–230.

14. R. A. Rankin, *Contribution to the theory of Ramanujan's function (n) and similar arithmetical functions, I, II*, Proc. Camb. Phil. Soc. **35** (1939), 351–372.

15. ______, *The scalar product of modular forms.*, Proc. London Math. Soc. **2** (3) (1952), 198–217.

16. A. Selberg, *Bemerkungen über eine Dirichletsche Reihe, die mit der Theorie der Modulformen nahe verbunden ist*, Arch. Math. Naturwid. **43** (1940), 47–50.

17. G. Shimura, *The special values of the zeta-functions associated with cusp forms*, Comm. in Pure Appl. Math. Math. **29** (1976), 783–804.

18. ______, *On the periods of modular forms*, Math. Annalen **229** (1977), 211–221.

19. J. Sturm, *Special values of zeta-functions and Eisenstein series of half integral weight*, Amer. J. Math. **102** (1980), 219–240.

20. ______, *The critical values of zeta-functions associated to the symplectic group*, Duke Math. J. **48** (1981), 327–350.

21. M. M. Vishik, *Non-archimedian measures attached to Dirichlet series*, Mat. Sbornik **99** (1976), 248–260 (in Russian).

22. André Weil, *Über die Bestimmung Dirichletscher Reihen durch Funktionalgleichungen*, Math. Annalen **168** (1967), 149–156.

Moscow State University, Department of Mathematics, Moscow 119 899 USSR

STABLE BUNDLES AND INTEGRABLE SYSTEMS

NIGEL HITCHIN

§1. Introduction.   The moduli spaces of stable vector bundles over a Riemann surface are algebraic varieties of a very special nature. They have been studied for the past twenty years from the point of view of algebraic geometry, number theory and the Yang-Mills equations. We adopt here another viewpoint, considering the *symplectic geometry* of their cotangent bundles. These turn out to be algebraically completely integrable Hamiltonian systems in a very natural way. For rank 2 bundles of odd degree this appeared as a byproduct of an investigation [5] into certain solutions of the self-dual Yang-Mills equations, a subject on which Yuri Manin has had a profound influence.

The cotangent bundle T^*N of an n-dimensional complex manifold N is a completely integrable Hamiltonian system if there exist n functionally independent, Poisson-commuting holomorphic functions on T^*N. These functions for the case where N is the moduli space of stable G-bundles on a compact Riemann surface M, and G a complex semisimple Lie group are easy to describe. The tangent space of the moduli space at a point is identified with the sheaf cohomology group $H^1(M; \mathfrak{g})$ where $\mathfrak{g}$ is a holomorphic bundle of Lie algebras. By Serre duality the cotangent space is $H^0(M; \mathfrak{g} \otimes K)$. An invariant polynomial of degree d on the Lie algebra then gives rise to a map from this cotangent space to the space $H^0(M; K^d)$ of differentials of degree d on M. Taking a basis for the ring of invariant polynomials yields a map to the vector space $W = \bigoplus_{i=1}^{k} H^0(M; K^{d_i})$ where d_i are the degrees of the basic invariant polynomials. Somewhat miraculously, the dimension of this vector space is always equal to the dimension of the moduli space N, thus providing the n functions.

The Hamiltonian vector fields corresponding to these functions give n commuting vector fields along the fibres of the map to W. The system is called *algebraically* completely integrable if the generic fibre is an open set in an abelian variety and the vector fields are linear. For the system above, at least for the case where G is a classical group, this also turns out to be true, the abelian variety being either a Jacobian or a Prym variety of a curve covering M. The construction of this curve, and its corresponding Jacobian, parallels the solution of differential equations of "spinning top" type involving isospectral deformations of a matrix of polynomials in one variable. A point of the cotangent bundle of the moduli space consists of a stable vector bundle V (with G-structure) and a holomorphic section $\Phi \in H^0(M; \mathfrak{g} \otimes K)$, which gives a holomorphic map Φ: $V \to V \otimes K$. We form the curve of eigenvalues S defined by the equation

Received November 29, 1986.

$\det(\lambda - \Phi) = 0$ in the total space of the cotangent bundle of M. Over S there is a natural eigenspace line bundle L which defines a point in the Jacobian of S. Fixing the values of the invariant polynomials fixes the coefficients of this equation and hence also the curve. Thus the only further variation possible is to change the line bundle. For $G = \mathrm{GL}(m, \mathbb{C})$, the line bundle can take all possible values in the Jacobian. For $G = \mathrm{Sp}(m, \mathbb{C})$ or $\mathrm{SO}(2m + 1, \mathbb{C})$ the bundle is restricted to lie in the Prym variety of S with respect to an involution with fixed points. For $G = \mathrm{SO}(2m, \mathbb{C})$, the curve S is singular but the line bundle is defined on its desingularization, which has an involution without fixed points. It is the Prym variety of this curve which is a generic fibre in this case.

Having established the existence of the integrable system we apply it in the realm of algebraic geometry to compute some sheaf cohomology groups of the moduli space of stable bundles of rank 2 and odd degree with fixed determinant. This consists of a generalization of the result of Narasimhan and Ramanan that $H^0(N; T) = 0$ and $\dim H^1(N; T) = 3g - 3$ to consideration of the corresponding groups for the k-th symmetric power bundle $S^k T$. A consequence of this result is that the only globally defined holomorphic functions on the cotangent bundle of the moduli space, and polynomial in the fibres, are those generated by the functions defining the integrable system. Thus the *global* geometry of the moduli space determines the interesting symplectic geometry of its cotangent bundle.

The question which we have not treated here is to find explicitly the Hamiltonian differential equations which correspond to these systems. Since the moduli spaces are unirational (and conjectured to be rational) then ultimately these may be expressed as equations with rational coefficients. Finding some natural, concrete realization of the integrable systems which arise so naturally in this way may lead to an application in the other direction—from algebraic geometry to differential equations. This would be an agreeable outcome, and one consistent with Manin's view of the unity of mathematics.

§2. Stable bundles

2.1.   Let M be a Riemann surface and V a C^∞ complex vector bundle of rank m over M. By a *holomorphic structure A* on V we shall mean a differential operator

$$d_A'': \Omega^0(M; V) \to \Omega^{0,1}(M; V)$$

such that

$$d_A''(fs) = \bar{\partial} f \otimes s + f d_A'' s \tag{2.2}$$

for $s \in \Omega^0(M; V)$ and $f \in C^\infty(M)$.

The local sections s satisfying $d_A'' s = 0$ are defined to be holomorphic and with this definition one obtains holomorphic transition functions relating local holomorphic bases and consequently the usual definition of holomorphic bundle (see [2]).

2.3.   Let $\mathcal{G}$ denote the group of automorphisms of the vector bundle V. The group acts on the space of holomorphic structures by

$$d''_A \to g^{-1} d''_A g.$$

The space $\mathcal{A}$ of all holomorphic structures on V is an infinite dimensional affine space since from 2.1

$$d''_{A_1} - d''_{A_2} = B \in \Omega^{0,1}(M; \operatorname{End} V)$$

and $\mathcal{G}$ acts on $\mathcal{A}$ via affine transformations.

2.4.   If M is compact and of genus $g > 1$, then there is an open set $\mathcal{A}^s \subset \mathcal{A}$ of holomorphic structures, the set of *stable* structures, which is preserved by $\mathcal{G}$ and is such that the quotient $\mathcal{A}^s/\mathcal{G}$ is a smooth complex manifold whose tangent space at A is isomorphic to the sheaf cohomology group $H^1(M; \operatorname{End} V)$. The condition for stability is that for every proper subbundle U,

$$\frac{\deg(U)}{rkU} < \frac{\deg(V)}{rkV}. \tag{2.5}$$

If equality occurs, the bundle is *semistable*.

2.6.   The quotient space $\mathcal{N} = \mathcal{A}^s/\mathcal{G}$ is the *moduli space of stable vector bundles of rank m* over M. By the Riemann-Roch theorem (since stability implies $H^0(M; \operatorname{End} V) \cong \mathbb{C}$), the moduli space $\mathcal{N}$ has dimension

$$\dim \mathcal{N} = m^2(g - 1) + 1. \tag{2.7}$$

In the case where the rank m and Chern class $c_1(V)$ are mutually prime, then $\mathcal{N}$ is a compact projective variety.

2.8.   The notion of stability may be extended from vector bundles to principal bundles ([2], [9]). In this case the space of holomorphic structures on a principal bundle P over M with structure group a complex semisimple group G is an infinite dimensional affine space with group of translations $\Omega^{0,1}(M; \operatorname{ad} P)$, where ad P is the vector bundle associated to the adjoint representation. For a Lie group G, semistability of the principal bundle is equivalent to semistability of the holomorphic structure on the vector bundle ad P.

In this more general situation there is a group $\mathcal{G}$ of automorphisms of P, with Lie algebra given by $\Omega^0(M; \operatorname{ad} P)$ and a subspace $\mathcal{A}^s$ of stable holomorphic structures on P such that the quotient space $\mathcal{A}^s/\mathcal{G} = \mathcal{N}$ is a smooth complex quasi-projective variety with

$$\dim \mathcal{N} = \dim G(g - 1) \tag{2.9}$$

for a semisimple group G.

All these varieties have very special properties which have been investigated from many points of view ([2], [7], [8]). The point of view we adopt here is that of *symplectic geometry* in the holomorphic category. We shall consider the total space $T^*\mathcal{N}$ of the cotangent bundle of the moduli space of stable bundles as a *complex symplectic manifold*.

§3. Symplectic geometry.   We recall some basic facts from symplectic geometry.

3.1.   Let N be any manifold and T^*N its cotangent bundle with projection π: $T^*N \to N$. The tautological section of π^*T^*N defines a 1-form θ on T^*N such that $d\theta = \omega$ is a symplectic form on T^*N. If $(x_1, \ldots, x_n)$ are local coordinates on N and cotangent vectors are parametrized by the coordinates $(y_1, \ldots, y_n) \to \sum_{i=1}^{n} y_i \, dx_i$, then the form θ is defined by

$$\theta = \sum y_i \, dx_i$$

and

$$\omega = \sum dy_i \wedge dx_i.$$

3.2.   If f is a smooth function on a symplectic manifold M, then it defines a vector field X_f:

$$df = i(X_f)\omega.$$

The Poisson bracket of two functions f and g is defined by the function

$$\{f, g\} = X_f \cdot g = -X_g \cdot f = -\{g, f\}.$$

Two functions are said to Poisson-commute if $\{f, g\} = 0$. A particular case is if $\omega = \sum dy_i \wedge dx_i$; and f and g are functions of $(y_1, \ldots, y_n)$ alone, for then

$$df = \sum \frac{\partial f}{\partial y_i} \, dy_i = i(X_f)\left(\sum dy_i \wedge dx_i\right)$$

and so

$$X_f = -\sum \frac{\partial f}{\partial y_i} \frac{\partial}{\partial x_i}.$$

Thus $\{f, g\} = X_f \cdot g = -\Sigma(\partial f/\partial y_i)(\partial g/\partial x_i) = 0.$

3.3. If the vector fields $X_{f_1}, \ldots, X_{f_n}$ form the basis of a Lie sub-algebra $\mathfrak{g}$ of the Lie algebra of vector fields on a symplectic manifold M, then the functions define a map to the dual of $\mathfrak{g}$,

$$\mu \colon M \to \mathfrak{g}^*$$

given by

$$\mu(x) = \sum f_i(x)\xi_i$$

where $\{\xi_i\}$ is the dual basis in $\mathfrak{g}^*$.

When the vector fields integrate to give the action of a Lie group G on M, and if μ is equivariant, then μ is called a *moment map*.

Suppose G acts freely on the submanifold $\mu^{-1}(0)$. Then the symplectic form is degenerate along the orbits of G in $\mu^{-1}(0)$ and invariant by G. It defines a symplectic form on the quotient $\mu^{-1}(0)/G$ which is called the *Marsden-Weinstein* quotient.

3.4. A special case of the Marsden-Weinstein quotient is when a Lie group G acts freely on N, and hence on T^*N. A vector field X on N generated by this action defines a function from T^*N by contraction in the fibres: if $X = \sum a_i(\partial/\partial x_i)$, $f = \sum a_i y_i$. The vector field X_f on T^*N is the natural extension of X to the cotangent bundle.

In this case, the Marsden-Weinstein quotient is the cotangent bundle of the ordinary manifold quotient:

$$T^*(N/G) \cong \mu^{-1}(0)/G.$$

3.5. Let M be a symplectic manifold with the action of a Lie group G and moment map $\mu \colon M \to \mathfrak{g}^*$ defined by functions $f_1, \ldots, f_n$. If g is a G-invariant function on M then it defines by restriction a G-invariant function on $\mu^{-1}(0)$ and hence a function $\tilde{g}$ on the Marsden-Weinstein quotient $\mu^{-1}(0)/G$.

Let g, h be two such functions. Then $X_{f_i} g = 0$ since g is G-invariant. But then

$$X_g f_i = \{g, f_i\} = -X_{f_i} g = 0$$

so that X_g is tangential to $f_i^{-1}(0)$ for all i and thus to $\mu^{-1}(0)$.

If $\{g, h\} = 0$, then $X_g \cdot h = 0$ so h is constant along the orbits of X_g. But the projection of these orbits to $\mu^{-1}(0)/G$ are the orbits of $X_{\tilde{g}}$. Thus $\tilde{h}$ is constant on the orbits of $X_{\tilde{g}}$ and so $\{\tilde{g}, \tilde{h}\} = 0$.

Thus if g *and h Poisson-commute in M, so do $\tilde{g}$ and $\tilde{h}$ in $\mu^{-1}(0)/G$.*

3.6. A symplectic manifold M of dimension $2n$ is said to be *a completely integrable Hamiltonian system* if there exists functions $f_1, \ldots, f_n$ which Poisson-commute and for which $df_1 \wedge \cdots \wedge df_n$ is generically nonzero.

The map $f\colon M \to \mathbb{C}^n$ defined by these functions—the moment map for an abelian Lie algebra—has the property that a generic fibre is an n-dimensional submanifold with n linearly independent commuting vector fields $X_{f_1}, \ldots, X_{f_n}$. If this fibre is an open set in an abelian variety and the vector fields are linear, then we shall say that the system is *algebraically completely integrable*.

We shall see that the cotangent bundle of the moduli space $\mathcal{N}$ of stable bundles on a Riemann surface always has $n = \dim \mathcal{N}$ natural Poisson-commuting functions and, at least for the classical groups, these form an algebraically completely integrable Hamiltonian system.

The most trivial example is the case of the complex group $G = \mathbb{C}^*$. An equivalence class of holomorphic structures on a $\mathbb{C}^*$-bundle is defined by a point in the Picard group $H^1(M; \mathcal{O}^*)$, or fixing the degree by a point in the Jacobian $\mathrm{Jac}(M)$. The tangent bundle of the Jacobian is trivial and isomorphic to $H^1(M; \mathcal{O})$, and so by Serre duality the cotangent bundle is

$$T^*\mathcal{N} \cong \mathrm{Jac}(M) \times H^0(M; K)$$

where K is the canonical bundle.

The g functions obtained by projecting on the second factor Poisson-commute and every fibre of this map is a copy of the abelian variety which is the Jacobian. Every holomorphic vector field defined on a compact abelian variety is linear, so this is an algebraically completely integrable system, though not a very interesting one.

§4. Invariant polynomials. Let $\mathcal{N}$ be the moduli space of stable holomorphic structures on a principal G-bundle P. The tangent space at a point of $\mathcal{N}$ may be represented by the vector space $H^1(M; \mathrm{ad}\, P)$ and hence by Serre duality the cotangent space is isomorphic to $H^0(M; \mathrm{ad}\, P \otimes K)$.

The vector bundle $\mathrm{ad}\, P$ is a bundle of Lie algebras each isomorphic to $\mathfrak{g}$, and so if p is an invariant homogeneous polynomial on $\mathfrak{g}$ of degree d, it defines a map

$$p\colon H^0(M; \mathrm{ad}\, P \otimes K) \to H^0(M; K^d).$$

Let $p_1, \ldots, p_k$ be a *basis* for the ring of invariant polynomials on the Lie algebra $\mathfrak{g}$, then we obtain a map

$$p\colon H^0(M; \mathrm{ad}\, P \otimes K) \to \bigoplus_{i=1}^{k} H^0(M; K^{d_i}).$$

PROPOSITION 4.1. *Let G be a semisimple complex Lie group and P a stable holomorphic principal G-bundle over a compact Riemann surface M of genus $g > 1$.*

Then

$$\dim H^0(M; \operatorname{ad} P \otimes K) = \dim \bigoplus_{i=1}^{k} H^0(M; K^{d_i}).$$

Proof.   If P is stable, then $H^0(M; \operatorname{ad} P) = 0$ (see [2]) and so by Riemann-Roch, using the Killing form to identify $\operatorname{ad} P^* \cong \operatorname{ad} P$,

$$\dim H^0(M; \operatorname{ad} P \otimes K) = \dim G(g-1). \tag{4.2}$$

Since G is semisimple there are no linear invariant polynomials, so each $d_i > 1$ and therefore applying Riemann-Roch to the right hand side

$$\dim \bigoplus_{i=1}^{k} H^0(M; K^{d_i}) = \sum_{i=1}^{k} (2d_i - 1)(g-1). \tag{4.3}$$

But it is well known that

$$\dim G = \sum_{i=1}^{k} (2d_i - 1).$$

From a topological point of view, each invariant polynomial defines an invariant $(2d_i - 1)$-form on G and the exterior algebra generated by these is the cohomology $H^*(G; \mathbb{C})$ [3]. More algebraically, Kostant's principal three-dimensional subgroup breaks up the Lie algebra $\mathfrak{g}$ into representation spaces of dimension $2d_i - 1$ [6]. From either source, the proposition is proved.

From 4.1, we obtain a map

$$p: T^*\mathcal{N} \to \bigoplus_{i=1}^{k} H^0(M; K^{d_i}) \tag{4.4}$$

which provides n holomorphic functions f_i defined on the $2n$-dimensional symplectic manifold $T^*\mathcal{N}$.

PROPOSITION 4.5.   *The n functions f_i on $T^*\mathcal{N}$ Poisson-commute.*

Proof.   From 3.4 we shall represent $T^*\mathcal{N}$ as a Marsden-Weinstein quotient:

$$T^*\mathcal{N} = T^*(\mathcal{A}^s/\mathcal{G}) = \mu^{-1}(0)/\mathcal{G}$$

where

$$\mu: T^*\mathcal{A}^s \to \mathfrak{g}^*$$

is the moment map for the action of $\mathcal{G}$ on the affine space $\mathcal{A}$.

Now $T^*\mathscr{A}^s \cong \mathscr{A}^s \times \Omega^0(M; \text{ad } P \otimes K)$ with the canonical 1-form θ defined by

$$\theta(\dot{A}, \dot{\Phi}) = \int_M B(\dot{A} \wedge \dot{\Phi}) \tag{4.6}$$

where $\dot{A} \in \Omega^{0,1}(M; \text{ad } P)$ is a tangent vector to $\mathscr{A}^s$ at A and $\dot{\Phi} \in \Omega^0(M; \text{ad } P \otimes K)$ a tangent vector to $\Omega^0(M; \text{ad } P \otimes K)$ at Φ, and B is the Killing form of the group.

Since $\mathscr{G}$ acts on $\mathscr{A}^s$ by conjugation of the operator d_A'', the Lie algebra $\mathfrak{g} = \Omega^0(M; \text{ad } P)$ generates vector fields

$$\dot{A} = d_A''\psi \in \Omega^{0,1}(M; \text{ad } P)$$

for $\psi \in \Omega^0(M; \text{ad } P)$.

Consequently, from 3.4, $\mu(A, \Phi) = 0$ if and only if

$$0 = \int_M B(d_A''\psi \wedge \Phi) = \int_M B(\psi d_A''\Phi)$$

for all $\psi \in \Omega^0(M; \text{ad } P)$. Thus $\mu(A, \Phi) = 0$ if and only if $d_A''\Phi = 0$ i.e., $\Phi \in \Omega^0(M; \text{ad } P \otimes K)$ is holomorphic with respect to the holomorphic structure d_A''.

Taking a basis for the invariant polynomials on $\mathfrak{g}$ defines a map

$$q: \mathscr{A}^s \times \Omega^0(M; \text{ad } P \otimes K) \to \bigoplus_{i=1}^{k} \Omega^0(M; K^{d_i})$$

simply by evaluating on the second factor. On the submanifold $\mu^{-1}(0)$ (i.e., the points (A, Φ) satisfying $d_A''\Phi = 0$) it takes values in the *holomorphic* sections of K^{d_i} and, being invariant by $\mathscr{G}$, descends to $\mu^{-1}(0)/\mathscr{G} = T^*\mathscr{N}$ to define the map p in (4.4). Now any two of these functions certainly Poisson-commute in $T^*\mathscr{A}^s = \mathscr{A}^s \times \Omega^0(M; \text{ad } P \otimes K)$ since they are functions of $\Omega^0(M; \text{ad } P \otimes K)$ alone as in 3.2. Therefore, by 3.5, they commute also on the quotient $T^*\mathscr{N}$, proving the proposition.

We see then that the $2n$-dimensional symplectic manifold $T^*\mathscr{N}$ admits n Poisson-commuting functions, where $\mathscr{N}$ is the moduli space of stable holomorphic structures on a principal G-bundle, for G semisimple.

The same is true for $G = \text{GL}(m, \mathbb{C})$, i.e., the moduli space of stable bundles of rank m, the only difference being that the centre gives rise to a linear invariant, so that the dimension of the right hand side in 4.1 is $\Sigma_{i=2}^k(2d_i - 1)(g - 1) + g = (m^2 - 1)(g - 1) + g$, but by Riemann-Roch this is also the dimension of the left hand side. Restricting attention to vector bundles with fixed determinant bundle reduces this situation to that of the semisimple group $\text{SL}(m, \mathbb{C})$.

§5. Algebraic integrability. We shall show now that the n Poisson-commuting functions defined above make $T^*\mathcal{N}$ into an algebraically completely integrable Hamiltonian system for the classical groups $G = \mathrm{GL}(m, \mathbb{C})$, $\mathrm{Sp}(m, \mathbb{C})$, $\mathrm{SO}(2m, \mathbb{C})$ and $\mathrm{SO}(2m + 1, \mathbb{C})$. To do this we need to show that the generic fibre of the function p in (4.4) is n-dimensional (to obtain the functional independence of the functions f_i), and further that it is an open set in an abelian variety, with the vector fields X_{f_i} linear.

5.1. Let $G = \mathrm{GL}(m, \mathbb{C})$. A point of $T^*\mathcal{N}$ is given by a stable vector bundle V of rank m and a holomorphic section $\Phi \in H^0(M; \mathrm{End}\, V \otimes K)$.

A basis for the invariant polynomials on $\mathrm{GL}(m, \mathbb{C})$ is provided by the coefficients of the characteristic polynomial $\det(x - A)$ of a matrix $A \in \mathrm{gl}(m, \mathbb{C})$. There is one polynomial in each degree $\leqslant m$, so we obtain the map

$$p: T^*\mathcal{N} \to \bigoplus_{i=1}^{m} H^0(M; K^i).$$

To study the fibre of this map we take sections $a_i \in H^0(M; K^i)$ and consider the vector bundles V and sections $\Phi \in H^0(M; \mathrm{End}\, V \otimes K)$ for which

$$\det(x - \Phi) = x^m + a_1 x^{m-1} + \cdots + a_m. \tag{5.2}$$

Let X be the complex surface which is the total space of the cotangent bundle K_M of M. The pull-back of K_M to M has a tautological section λ. Consider the sections of the line bundle K_M^m on X of the form

$$s = \lambda^m + a_1 \lambda^{m-1} + \cdots + a_m,$$

for $a_i \in H^0(M; K^i)$.

As the a_i vary, the zero set of s forms a linear system of divisors on X, and its compactification $P(K_M \oplus 1)$. Any base point of this system must lie in the zero section of K_M since λ^m lies in the system. But then it is a base point for K_M^m on M since a_m vanishes there for all $a_m \in H^0(M; K^m)$. But K^m is without base points for any Riemann surface, so the above system has no base points either. By Bertini's theorem the generic divisor is a nonsingular curve S.

The genus of S is given by the adjunction formula

$$K_X \cdot S + S^2 = 2g(S) - 2.$$

Now $K_X = 0$ (X is a symplectic manifold!) and S is in the linear system mK_M. The zero section is in the linear system K_M and has self-intersection number $K_M^2 = c_1(K_M) = 2g - 2$, hence

$$2g(S) - 2 = 2m^2(g - 1)$$

and

$$g(S) = m^2(g - 1) + 1. \tag{5.3}$$

The curve S is an m-fold covering $\pi\colon S \to M$ of M. If $\det(x - \Phi) = x^m + a_1 x^{m-1} + \cdots a_m$, then pulling back the bundle V to S, we have $\Phi \in H^0(S; \operatorname{End} V \otimes K_M)$ satisfying $\det(\lambda - \Phi) = 0$. Thus on S, Φ has the eigenvalue $\lambda \in H^0(S; K_M)$. Generically the eigenvalues are distinct and so we obtain a one-dimensional eigenspace which defines a line bundle

$$L \subset \ker(\lambda - \Phi) \subset \pi^* V$$

on S and hence a point of the Jacobian $\operatorname{Jac}(S)$, an abelian variety of dimension $g(S) = m^2(g - 1) + 1 = \dim \mathcal{N}$ from (2.7).

Conversely, the line bundle L determines V and Φ on M. We consider the direct image sheaf $\pi_* L$ which is the sheaf of sections of a vector bundle of rank m on M. The fibre of the bundle $\pi_* L$ at $x \in M$ is by definition

$$(\pi_* L)_x \cong \mathcal{O}_S(L)/\mathscr{I}_{\pi^{-1}(x)}$$

where $\mathscr{I}_{\pi^{-1}(x)}$ is the ideal sheaf of $\pi^{-1}(x)$. If x is not a branch point then this is $\bigoplus_{y \in \pi^{-1}(x)} L_y$. If x is a branch point, then the fibre is

$$\bigoplus_{y \in \pi^{-1}(x)} J_{k(y)}(L)_y$$

a direct sum of jets of sections of L of degrees $k(y)$ given by the degree of ramification of y.

Each k-jet has a natural map to the 0-jet

$$J_k(L)_y \to L_y \to 0$$

and hence dually a map

$$0 \to L_y^* \to J_k(L)_y^*.$$

There is thus a natural sequence for all x

$$0 \to \bigoplus_{y \in \pi^{-1}(x)} L_y^* \to (\pi_* L)_x^* \to \bigoplus_{y \in \pi^{-1}(x)} J_{k(y)}^*(L)_y/L_y^* \to 0.$$

This defines a sheaf map on M

$$\mathcal{O}(\pi_* L)^* \overset{s}{\to} \mathscr{S} \to 0 \tag{5.4}$$

where $\mathscr{S}$ is a sheaf supported at the branch points of $\pi\colon S \to M$ in M. The kernel of s is a locally free sheaf $\mathcal{O}(W)$ of rank m.

If L was obtained from a vector bundle V and $\Phi \in H^0(M; \operatorname{End} V \otimes K)$ as above, then the vector bundle W is V^*, from the natural map $V_y^* \to \bigoplus_{y \in \pi^{-1}(x)} L_y^*$ induced by the inclusion $L \subset V$.

This way, we recover V from the line bundle L. The operation of multiplication by $\lambda \in H^0(S; K_M)$ on L yields a section of $\operatorname{End} \pi_* L \otimes K_M$ which takes W to $W \otimes K_M$ and so defines $\Phi \in H^0(M; \operatorname{End} V \otimes K_M)$. We see therefore that the generic fibre of p lies in an abelian variety of dimension $n = \dim \mathscr{N}$. Since p maps to an n-dimensional space this proves the functional independence of the Poisson-commuting functions $f_1, \ldots, f_n$ and shows that the fibre is an open set in the abelian variety. Not all vector bundles produced from line bundles on S this way will be stable and it is the stable ones which determine the open set.

What remains is to show that the vector fields X_{f_i} are linear on the Jacobian. This is equivalent to showing that they extend as holomorphic vector fields to the whole compact Jacobian.

Our construction of V and Φ from a line bundle L on S was, however, natural in the sense that if a holomorphic isomorphism takes V and Φ to V' and Φ', then it takes the corresponding line bundle L to L'. Thus the Jacobian is a space of equivalence classes of points in $\mathscr{A} \times \Omega^0(M; \operatorname{End} V \otimes K)$ with $d_A''\Phi = 0$, under the action of the group of automorphisms $\mathscr{G}$. This space is strictly larger than $T^*\mathscr{A}^s$ which we used in §4, but by the Marsden-Weinstein construction still gives rise to a symplectic quotient. The cotangent bundle $T^*\mathscr{N}$ therefore lies as an open set in a larger symplectic manifold which contains the whole Jacobian and hence allows the vector fields X_{f_i} to extend and therefore be linear. We shall consider this larger moduli space in more detail in the next section for the case $G = \mathrm{GL}(2, \mathbb{C})$.

The Hamiltonian system is thus in this case algebraically completely integrable.

Finally note that the sheaf map (5.4) gives rise to an exact sequence of sheaves:

$$0 \to \mathcal{O}(V^*) \to \mathcal{O}(\pi_* L)^* \to \mathscr{S} \to 0 \tag{5.5}$$

which determines the degree of the line bundle L.

From the Grothendieck-Riemann-Roch theorem,

$$\operatorname{ch}(\pi_* L)\operatorname{td}(M) = \pi_*(\operatorname{ch}(L)\operatorname{td}(S))$$

so

$$\operatorname{ch}(\pi_* L) = (1 + (g-1)[M])\pi_*(1 + (\deg L + (1 - g(S)))[S])$$

and therefore

$$\deg \pi_* L = m(g-1) + \deg L + (1 - g(S)). \tag{5.6}$$

On the other hand, the ramification divisor on S is defined by the points whose tangent space projects to zero on M and this is the divisor of a section of $K_S K_M^*$ which has degree

$$(2g(S) - 2) - m(2g - 2). \tag{5.7}$$

Thus from (5.5), since $\mathscr{S}$ is supported on the branch points,

$$\deg V^* = \deg(\pi_* L)^* - \deg \mathscr{S}$$

$$= m(g - 1) - (g(S) - 1) - \deg L$$

$$= -m(m - 1)(g - 1) - \deg L \quad \text{from (5.3).} \tag{5.8}$$

Hence

$$\deg L = -m(m - 1)(g - 1) - \deg V^*. \tag{5.9}$$

5.10.   Let $G = \mathrm{Sp}(m, \mathbf{C})$. A point of $T^*\mathcal{N}$ now consists of a stable vector bundle V of rank $2m$ with a symplectic form $\langle\ ,\ \rangle$, together with a holomorphic section $\Phi \in H^0(M; \mathrm{End}\, V \otimes K)$ which satisfies $\langle \Phi v, w \rangle = -\langle v, \Phi w \rangle$.

Suppose $A \in \mathrm{sp}(m, \mathbf{C})$ has distinct eigenvalues λ_i with eigenvectors v_i on $\mathbf{C}^{2m}$, then

$$\lambda_i \langle v_i, v_j \rangle = \langle Av_i, v_j \rangle = -\langle v_i, Av_j \rangle = -\lambda_j \langle v_i, v_j \rangle$$

so $\langle v_i, v_j \rangle = 0$ unless $\lambda_j = -\lambda_i$. Since $\langle v_i, v_i \rangle = 0$ it follows from the nondegeneracy of the symplectic inner product that if λ_i is an eigenvalue so is $-\lambda_i$.

Thus the characteristic polynomial is of the form

$$\det(x - A) = x^{2m} + a_2 x^{2m-2} + \cdots + a_{2m}.$$

The polynomials $a_2, \ldots, a_{2m}$ in A form a basis for the invariant polynomials on the Lie algebra $\mathrm{sp}(m, \mathbf{C})$.

As in the previous section, we consider the linear system of divisors of the sections $\lambda^{2m} + a_2 \lambda^{2m-2} + \cdots + a_{2m}$ of K_M^{2m} on the complex surface X and by Bertini's theorem see that the generic divisor is a nonsingular curve S of genus $g(S) = 4m^2(g - 1) + 1$ from (5.3). We obtain again a line bundle

$$L \subset \ker(\lambda - \Phi) \subset \pi^* V$$

and thus a point on the Jacobian which determines V and Φ.

In this case, however, the curve S with equation

$$\lambda^{2m} + a_2 \lambda^{2m-2} + \cdots + a_{2m} = 0$$

possesses the involution $\sigma(\lambda) = -\lambda$ with fixed points the zeros of $a_{2m} \in H^0(M; K^{2m})$ on the zero section $\lambda = 0$. The involution acts on the Jacobian of line bundles of degree zero. The subvariety of line bundles such that $\sigma^* L \cong L^*$ is an abelian variety—the *Prym variety*. Its dimension is $g(S) - g(S/\sigma)$.

Now σ has $4m(g-1)$ fixed points since they are the zeros of a section of K^{2m} on M, so

$$2 - 2g(S) = 2(2 - 2g(S/\sigma)) - 4m(g-1)$$

and thus

$$\dim \mathrm{Prym}(S) = g(S) - g(S/\sigma)$$

$$= g(S) - \tfrac{1}{2} - \tfrac{1}{2}g(S) + m(g-1)$$

$$= 2m^2(g-1) + m(g-1) \quad \text{from (5.3)}$$

$$= m(2m+1)(g-1). \tag{5.11}$$

In our situation the eigenspace bundle $L \subset \ker(\lambda - \Phi)$ actually defines a point in the Prym variety. This is because of the nondegenerate pairing above of eigenvectors corresponding to eigenvalues λ and $-\lambda$. If L is the eigenspace bundle corresponding to λ, then $\sigma^* L$ corresponds to $-\lambda$ and so the symplectic form defines a section of $L^* \otimes \sigma^* L^*$ which is nonvanishing if the eigenvalues are distinct i.e., away from the ramification locus of S.

Since V is symplectic, $\deg V = 0$ and so from (5.7) and (5.8),

$$\deg L = -\tfrac{1}{2}\deg(K_S K_M^*).$$

Thus

$$\deg(L^* \otimes \sigma^* L^*) = \deg(K_S K_M^*)$$

and since $L^* \otimes \sigma^* L^*$ has a section which vanishes on the ramification locus, a divisor of $K_S K_M^*$, we have

$$\sigma^* L \cong L^* \otimes (K_S K_M^*)^{-1}. \tag{5.12}$$

Choosing a holomorphic square root $(K_S K_M^*)^{1/2}$ of $K_S K_M^*$ we obtain

$$L \cong U(K_S K_M^*)^{-1/2} \tag{5.13}$$

where U is a line bundle such that $\sigma^* U \cong U^*$, i.e., U lies in the Prym variety.

Now from (5.11)

$$\dim \mathrm{Prym}(S) = m(2m+1)(g-1)$$

$$= \dim \mathrm{Sp}(m)(g-1)$$

$$= \dim \mathcal{N}.$$

Hence, the generic fibre of p is in this case an open set in an abelian variety—a *Prym variety*.

Note that in the converse construction of the vector bundle V from the line bundle L, the involution σ produces a bilinear form on $\pi_* L$ away from the branch locus by taking $v \in H^0(p^{-1}(U); L)$ and setting

$$\langle v, w \rangle s = \sigma^* v \cdot w \in H^0\big(p^{-1}(U), K_S K_M^*\big)$$

from (5.12), where $s \in H^0(S; K_S K_M^*)$ is the canonical section given by the derivative of $\pi\colon S \to M$.

The form is skew because σ has real codimension 2 fixed points and is therefore of *odd* type [1] i.e., it lifts to an action on $K_S^{1/2}$ of order 4. From (5.13) this means that $\sigma^2 = -1$ on L.

5.14. Let $G = \mathrm{SO}(2m, \mathbb{C})$. A point of $T^*\mathcal{N}$ is now a stable vector bundle V of rank $2m$ with a nondegenerate symmetric bilinear form $(\ ,\ )$, together with a holomorphic section $\Phi \in H^0(M; \mathrm{End}\, V \otimes K)$ which satisfies $(\Phi v, w) = -(v, \Phi w)$.

Now suppose $A \in \mathrm{so}(2m, \mathbb{C})$ has distinct eigenvalues λ_i with eigenvectors v_i on $\mathbb{C}^{2m}$. Then

$$\lambda_i(v_i, v_j) = \big(Av_i, v_j\big) = -\big(v_i, Av_j\big) = -\lambda_j(v_i, v_j)$$

so if $\lambda_i \neq 0$, $(v_i, v_i) = 0$ and, as in the previous case, if λ_i is an eigenvalue so is $-\lambda_i$. Thus the characteristic polynomial is of the form

$$\det(x - A) = x^{2m} + a_2 x^{2m-2} + \cdots + a_{2m}.$$

In this case, the polynomial $a_{2m} = \det A$ of degree $2m$ is the *square* of a polynomial p_m—*the Pfaffian*—of degree m. A basis for the invariant polynomials on $\mathrm{so}(2m, \mathbb{C})$ is then given by the coefficients $a_2, \ldots, a_{2m-2}$ and p_m.

Now let $p_m \in H^0(M; K^m)$ be a section with simple zeros and consider the linear system of divisors of the sections $\lambda^{2m} + a_2 \lambda^{2m-2} + \cdots + a_{2m-2}\lambda^2 + p_m^2$ for $a_{2i} \in H^0(M: K^{2i})$, $1 \leqslant i \leqslant m-1$. Each curve of this system has singularities where $\lambda = 0$ and $p_m = 0$. These are the base points of the system and so by Bertini's theorem a generic divisor has these as its only singularities. Since p_m is a section of K^m, there are $\deg K^m = 2m(g-1)$ singularities which are generically ordinary double points.

The curve S in X given by the generic divisor then has *virtual* genus given by the adjunction formula (5.3)

$$p = 4m^2(g - 1) + 1.$$

The genus of its nonsingular model $\hat{S}$ is thus

$$g(\hat{S}) = 4m^2(g - 1) + 1 - 2m(g - 1)$$

$$= 2m(2m - 1)(g - 1) + 1. \tag{5.15}$$

The involution $\sigma(\lambda) = -\lambda$ on S has as fixed points the singularities of S which are double points and so extends to an involution on $\hat{S}$ without fixed points.

We consider the Prym variety of $\hat{S}$. Its dimension is $g(\hat{S}) - g(\hat{S}/\sigma)$. Since σ has no fixed points,

$$2 - 2g(\hat{S}) = 2\big(2 - 2g(\hat{S}/\sigma)\big)$$

and thus

$$\dim \mathrm{Prym}(\hat{S}) = g(\hat{S}) - g(\hat{S}/\sigma)$$

$$= g(\hat{S}) - \tfrac{1}{2} - \tfrac{1}{2}g(\hat{S})$$

$$= m(2m - 1)(g - 1). \tag{5.16}$$

Because $\hat{S}$ is smooth we obtain an eigenspace bundle $L \subset \ker(\lambda - \Phi)$ inside the vector bundle V pulled back to $\hat{S}$, and as in the previous case, by applying the symmetric form, a section of $L^* \otimes \sigma^* L^*$ which is nonzero if the eigenvalues are distinct. It remains nonzero in fact when an eigenvalue tends to zero as may be seen by considering the eigenvectors of a 2×2 skew-symmetric matrix. Thus this section is nonvanishing away from the ramification points of the map $\pi \colon \hat{S} \to M$. Arguing analogously to the symplectic case, we find

$$\sigma^* L \cong L^* \otimes \big(K_{\hat{S}} K_M^*\big)^{-1}$$

and setting

$$L = U\big(K_{\hat{S}} K_M^*\big)^{-1/2}$$

we obtain a point of the Prym variety of $\hat{S}$.

Now from (5.16),

$$\dim \mathrm{Prym}(\hat{S}) = m(2m - 1)(g - 1)$$

$$= \dim \mathrm{SO}(2m)(g - 1)$$

$$= \dim \mathcal{N}.$$

We therefore obtain algebraic complete integrability in this case too. Since the involution σ on $\hat{S}$ now has no fixed points it is of *even* type and this is the reason that it provides a symmetric bilinear form on the corresponding vector bundle over M.

5.17. Let $G = SO(2m + 1, \mathbb{C})$. We now consider a stable vector bundle V over M of rank $(2m + 1)$ with a nondegenerate bilinear form $(,)$ and a holomorphic section $\Phi \in H^0(M; \text{End } V \otimes K)$ which satisfies $(\Phi v, w) = -(v, \Phi w)$.

By specializing from $SO(2m + 2)$, the characteristic polynomial of $A \in so(2m + 1)$ acting on $\mathbb{C}^{2m+1}$ is

$$\det(x - A) = x\left(x^{2m} + a_2 x^{2m-2} + \cdots + a_{2m}\right)$$

and the coefficients form a basis for the invariant polynomials on $SO(2m + 1)$.

As in the case of $Sp(m)$, we choose $a_{2i} \in H^0(M; K_M^{2i})$ for $1 \leqslant i \leqslant m$ such that the curve S defined by

$$\lambda^{2m} + a_2 \lambda^{2m-2} + \cdots + a_{2m} = 0$$

is nonsingular, and consider V and Φ such that

$$\det(\lambda - \Phi) = \lambda\left(\lambda^{2m} + a_2 \lambda^{2m-2} + \cdots + a_{2m}\right).$$

The vector bundle homomorphism $\Phi: V \to V \otimes K_M$ now always has an eigenvalue zero, so we automatically obtain a line bundle $V_0 \subset V$ defined by

$$V_0 \subset \ker \Phi.$$

The bundle V on M is therefore naturally an extension

$$0 \to V_0 \to V \to V_1 \to 0. \tag{5.18}$$

The zero eigenspace can be canonically identified for if we use the inner product on V to convert the skew map Φ to a holomorphic section $\phi \in H^0(M; \Lambda^2 V \otimes K_M)$, then

$$\phi^m \in H^0\left(M; \Lambda^{2m} V \otimes K_M^m\right) \cong H^0(M; V \otimes K_M^m)$$

defines a homomorphism from K_M^{-m} to V whose image is the line bundle V_0. Thus $V_0 \cong K_M^{-m}$ and hence from (5.18) since $V = V^*$, we have

$$\Lambda^{2m} V_1 \cong K_M^m. \tag{5.19}$$

Since V_0 lies in the kernel of Φ we obtain an induced homomorphism Φ:

$V_1 \to V_1 \otimes K_M$ and we proceed as in 5.1 to obtain a line bundle L on S defining the eigenspace of Φ.

The inner product $(\,,\,)$ on V defines a skew form on V_1 with values in K_M, by

$$\langle v, w \rangle = (\Phi v, w) \tag{5.20}$$

for representative vectors v and w in V. Since $(\Phi v, w) = -(v, \Phi w)$ the form is clearly well defined on V_1 but is singular when $\Phi \colon V_1 \to V_1 \otimes K_M$ is singular which is when $a_{2m} = 0$. It nevertheless plays a role analogous to the symplectic form in 5.10 and just as in that case, if we use the involution $\sigma(\lambda) = -\lambda$ on S, we obtain a section of

$$L^* \otimes \sigma^* L^* \otimes K_M$$

from the pairing of the eigenvectors with eigenvalues $\pm \lambda$. Now from (5.9),

$$\deg L = -2m(2m-1)(g-1) - \deg V_1^*$$

$$= -2m(2m-1)(g-1) + m \cdot 2m \cdot 2(g-1) \quad \text{from (5.19)}$$

$$= 2m(g-1)$$

$$= \tfrac{1}{2} \deg K_M.$$

Thus $L^* \otimes \sigma^* L^* \otimes K_M$ is of degree zero with a nontrivial section and is hence a trivial bundle. Therefore

$$\sigma^* L \cong L^* \otimes K_M.$$

Choosing a square root $K_M^{1/2}$ we obtain

$$L \cong U\!\left(K_M^{1/2}\right) \tag{5.21}$$

where U is a line bundle in the Prym variety of S. From (5.11)

$$\dim \operatorname{Prym} S = m(2m+1)(g-1)$$

$$= \dim \operatorname{SO}(2m+1)(g-1)$$

$$= \dim \mathcal{N}.$$

The methods of the previous cases allow us to recapture the rank $2m$ bundle V_1 from the line bundle L. It comes equipped with a skew form $\langle\,,\,\rangle$ with values in K_M and a homomorphism $\Phi \colon V_1 \to V_1 \otimes K_M$ satisfying $\langle \Phi v, w \rangle = -\langle v, \Phi w \rangle$. We need to reconstruct the vector bundle V of rank $(2m+1)$.

To do this we consider how V_1 arose and dualize the exact sequence (5.18) to obtain

$$0 \to V_1^* \to V \to V_0^* \to 0. \tag{5.22}$$

The composition $V_0 \to V \to V_0^*$ defines a section of $V_0^* \otimes V_0^* \cong K_M^{2m}$ which is easily seen to be the coefficient a_{2m}. On the zero set D of this we have an inclusion $V_0 \subset V_1^*$.

Now consider the exact sequence of sheaves

$$0 \to \operatorname{Hom}(V_0^*, V_1^*) \xrightarrow{\;a_{2m}\;} \operatorname{Hom}(V_0, V_1^*) \to \operatorname{Hom}(V_0, V_1^*)|_D \to 0$$

and its exact cohomology sequence. The extension (5.21) is defined by the coboundary map

$$\delta \colon H^0(D; \operatorname{Hom}(V_0, V_1^*)) \to H^1(M; \operatorname{Hom}(V_0^*, V_1^*))$$

applied to the inclusion $V_0 \subset V_1^*$. An analogous construction occurs in the twistor approach to magnetic monopoles [4].

The advantages of this construction of V is that it depends only on the bundle V_1^* and the symmetric form induced on it from V via the inclusion $V_1^* \subset V$. We consider a bundle V_1^* of rank $2m$ with a symmetric form. The form degenerates where the corresponding homomorphism $g \colon V_1^* \to V_1$ is singular i.e., on a divisor D of the line bundle $(\Lambda^{2m}V_1)$. Over D there is a 1-dimensional subbundle $V_0 \subset V_1^*$, the kernel of g. The bundle V_1^*/V_0 has a nondegenerate symmetric form and so over D,

$$V_0 \cong \Lambda^{2m}V_1^*.$$

In the exact sequence of sheaves

$$0 \to \operatorname{Hom}(\Lambda^{2m}V_1, V_1^*) \xrightarrow{\;\det g\;} \operatorname{Hom}(\Lambda^{2m}V_1^*, V_1^*) \to \operatorname{Hom}(\Lambda^{2m}V_1^*, V_1^*)|_D \to 0$$

the inclusion of $V_0 \cong \Lambda^{2m}V_1^*$ in V_1^* over D generates an extension

$$0 \to V_1^* \to V \to \Lambda^{2m}V_1 \to 0$$

by the coboundary map.

We use the above applied to the bundle V_1 arising from a point of the Prym variety of S. It has the skew form $\omega \in H^0(M; V_1^* \otimes V_1^* \otimes K_M)$ and the homomorphism $\Phi \in H^0(M; \operatorname{End} V_1 \otimes K_M)$. We obtain a symmetric form by considering

$$\omega^{2m-1} \in H^0\left(M; \Lambda^{2m-1}V_1^* \otimes \Lambda^{2m-1}V_1^* \otimes K_M^{2m-1}\right)$$

$$\cong H^0\left(M; V_1 \otimes V_1 \otimes \left(\Lambda^{2m}V_1^*\right)^2 \otimes K_M^{2m-1}\right)$$

$$= H^0\left(M; V_1 \otimes V_1 \otimes K_M^{-1}\right)$$

and applying Φ to obtain

$$\Phi\omega^{2m-1} \in H^0(M; V_1 \otimes V_1).$$

Hence, a generic fibre of the map p is an open set in the Prym variety of S and the system is integrable.

Remark. Note the similarity between the cases of $G = \mathrm{Sp}(m, \mathbb{C})$ and $\mathrm{SO}(2m + 1, \mathbb{C})$. In both cases the projection p maps to the same vector space of differentials on M and the generic fibre is the same Prym variety: locally the two systems are equivalent. Globally, the difference in identifying the fibres as Prym varieties (compare (5.13) and (5.21)) will affect the structure of the moduli spaces. This may be a convenient context in which to bring in the duality theory of Lie groups by means of which $\mathrm{Sp}(m, \mathbb{C})$ and $\mathrm{SO}(2m + 1, \mathbb{C})$ are viewed as dual groups.

§6. An application to moduli spaces.   We consider now an application of these symplectic methods—we use the integrable system to derive information about the sheaf cohomology groups of the moduli space $\mathcal{N}$ of stable vector bundles of rank 2 and odd degree over a Riemann surface of genus $g > 1$. This space is a compact smooth projective variety. If we fix the determinant line bundle $\Lambda^2 V$ of the stable vector bundle V, then it has complex dimension $3g - 3$.

In [7], Narasimhan and Ramanan proved that if $\mathcal{N}$ is the moduli space of stable bundles of rank m and degree k, with $(m, k) = 1$ and with fixed determinant bundle, then the sheaf cohomology groups $H^i(\mathcal{N}; T)$ of the holomorphic tangent bundle T vanish if $i \neq 1$ and $\dim H^1(\mathcal{N}; T) = 3g - 3$. In fact, the vanishing for $i > 1$ is a consequence of the vanishing theorem of Akizuki and Nakano, so the cases $i = 0$ and 1 are the only nontrivial ones. We shall generalize their result to deal with the *k-th symmetric power bundle $S^k T$*.

We restrict attention to the case of vector bundles of rank 2 because here, by analytical means, we have a precise description of the symplectic manifold alluded to in §5, in which $T^*\mathcal{N}$ lies as an open set [5]. We say that a pair $(A, \Phi) \in T^*\mathcal{A}$ consisting of a holomorphic structure A and a section $\Phi \in \Omega^0(M; \mathrm{ad}\, P \otimes K)$ satisfying $d_A''\Phi = 0$ is a *stable pair* if for each Φ-invariant holomorphic line bundle $L \subset V$, $\deg L < \frac{1}{2}\deg V$. Note from (2.5) that if the holomorphic structure is itself stable, then (A, Φ) is stable for all Φ.

The quotient of the set of stable pairs with fixed determinant by the group $\mathcal{G}$ of automorphisms of V of determinant one is, from [5], a complex symplectic manifold $\mathcal{M}$ of dimension $6g - 6$ with the following properties:

PROPOSITION 6.1.   (i) *$\mathcal{M}$ is a Kähler manifold.*

(ii) *The map* det: *$\mathcal{M} \to H^0(M; K^2)$ defined by applying the invariant polynomial* det *on* $\mathrm{sl}(2, \mathbb{C})$ *to Φ is proper and surjective.*

(iii) *The fibre of a point $q \in H^0(M; K^2)$ consisting of a section with simple zeros is isomorphic to the Prym variety of the double covering S of M defined by $\lambda^2 - q = 0$ in the total space $X = K_M$ of the cotangent bundle of M.*

(iv) *The cotangent bundle $T^*\mathcal{N}$ of the moduli space of stable vector bundles lies naturally in $\mathcal{M}$ as the complement of an analytic set of codimension at least g.*

Proof.   For parts (i) to (iii), see [5], §8. In fact (iii) is essentially the description for $\mathrm{SL}(2,\mathbb{C}) \cong \mathrm{Sp}(1,\mathbb{C})$ in 5.10. The fact that V is of odd degree only changes the isomorphism with the Prym variety.

For part (iv), if V is *unstable*, then it has a canonical subbundle $L \subset V$ with $\deg L > \frac{1}{2}\deg V$. Thus V is defined as an extension

$$0 \to L \to V \to L^* \otimes \Lambda^2 V \to 0$$

and hence by an element of $H^1(M; L^2 \otimes \Lambda^2 V^*)$.

Now the stability of the pair (V, Φ) implies from [5], (3.12), that

$$0 < \deg(L^2 \otimes \Lambda^2 V^*) \leqslant 2g - 2$$

(with strict inequality in this case since $\deg(V)$ is odd). This means that $\dim H^1(M; L^2 \otimes \Lambda^2 V^*) \leqslant g - 1$ with equality only if L is special.

We need now another property of the symplectic manifold $\mathcal{M}$. Recall that a submanifold of a symplectic manifold is *Lagrangian* if its tangent spaces are isotropic of maximal dimension. The moduli space $\mathcal{M}$ is foliated by Lagrangian submanifolds (of dimension $(3g - 3)$) each of which is obtained by fixing the equivalence class of the complex structure and allowing Φ to vary (see [5], §8). On the open set $T^*\mathcal{N} \subset \mathcal{M}$ these manifolds are just the fibres of the cotangent bundle.

The moduli space $\mathcal{M}$ is stratified by the degree of the destabilizing subbundle L, so each stratum (for degree > 0) is a bundle of $(3g - 3)$-dimensional spaces over equivalence classes of extensions of line bundles. Since L is special there is at most a $(g - 1)$-dimensional choice for L and since $\dim H^1(M; L^2 \otimes \Lambda^2 V^*)$ is less than g, at most a $(g - 2)$-dimensional family of extensions for a fixed L, giving the maximum dimension of stratum as $(3g - 3) + (g - 1) + (g - 2) = 5g - 6$, which is of codimension g in $\mathcal{M}$.

Using these facts we prove the following:

THEOREM 6.2.   *Let $\mathcal{N}$ be the moduli space of stable vector bundles with fixed determinant, rank 2 and odd degree over a compact Riemann surface of genus $g > 1$. If T is the tangent bundle of $\mathcal{N}$ and $S^k T$ the k-th symmetric power of T, then*

$$H^0(\mathcal{N}; S^k T) = 0 \ \text{if } k \text{ is odd}$$

$$H^0(\mathcal{N}; S^{2\ell} T) \cong S^\ell H^1(M; T_M).$$

Proof.   Let s be a holomorphic section of $S^k T$ on $\mathcal{N}$. By contraction it defines a holomorphic function $\hat{s}$ on the cotangent bundle $T^*\mathcal{N}$, which is a polynomial of degree k on each fibre.

Since $\mathcal{M}$, the moduli space of stable pairs, is obtained by adjoining an analytic set of codimension $g \geqslant 2$ to $T^*\mathcal{N}$, then $\hat{s}$ extends by Hartog's theorem to a holomorphic function on $\mathcal{M}$.

Consider the map det: $\mathcal{M} \to H^0(M; K^2)$. The generic fibre is compact and connected—a Prym variety—and hence, since det is proper and $H^0(M; K^2)$ smooth, every fibre is compact and connected. Thus the function $\hat{s}$ is constant on each fibre and so

$$\hat{s} = f \circ \det$$

for some holomorphic function $f\colon H^0(M; K^2) \to \mathbf{C}$.

Now the map $\Phi \to \lambda\Phi$ induces a $\mathbf{C}^*$ action on $\mathcal{M}$ and since s was homogeneous of degree k, we have

$$\hat{s}(\lambda \cdot m) = \lambda^k \hat{s}(m) \quad \text{for } m \in \mathcal{M}.$$

But $\det(\lambda \cdot m) = \lambda^2 \det(m)$, and so f satisfies

$$f(\lambda^2 x) = \lambda^k f(x). \tag{6.3}$$

Now a holomorphic function on a vector space W which is homogeneous of degree ℓ defines a holomorphic section of $\mathcal{O}(\ell)$ on $P(W)$ and hence a polynomial of degree ℓ. Since $\mathcal{O}(1)$ generates $H^1(P(W), \mathcal{O}^*)$ there are no homogeneous functions of noninteger degree, so if k is odd, $f \equiv 0$. If $k = 2\ell$, then f is a polynomial of degree ℓ on $W = H^0(M; K^2)$ and hence an element of the vector space $S^\ell W^*$. Thus, since $H^0(M; K^2)^* \cong H^1(M; K^{-1})$ by Serre duality, we have $\hat{s} = 0$ if M is odd and $\hat{s} = f \circ \det$ for $f \in S^\ell H^1(M; K^{-1})$ for $m = 2\ell$ and hence the theorem.

Remark 6.4. (1) Theorem 6.2 gives $H^0(\mathcal{N}; S^2 T) \cong H^1(M; K^{-1})$. This shows that the $3g - 3$ functions which make $T^*\mathcal{N}$ into an algebraically completely integrable Hamiltonian system are globally determined by $\mathcal{N}$ itself: they are the *only* holomorphic functions on $T^*\mathcal{N}$ which are homogeneous of degree 2 in the fibres. Thus if $\mathcal{N}$ is given in some way other than as the moduli space of stable vector bundles, it is still possible to describe $T^*\mathcal{N}$ as an integrable system simply by finding the global sections of $S^2 T$. An example would be the case of $g = 2$ where $\mathcal{N}$ is the intersection of two quadrics in P^5 [8].

(2) If we consider the ring of *all* holomorphic functions on $T^*\mathcal{N}$ which are polynomial in the fibres, then Theorem 6.2 shows that this ring is generated by the $3g - 3$ functions of degree 2. Since they Poisson-commute by 4.5, the whole Poisson algebra of polynomial functions is commutative.

In a similar manner we prove now the following:

THEOREM 6.5. *Let $\mathcal{N}$ be as in 6.2, and suppose that $g > 2$. Then*
(i) $H^1(\mathcal{N}; S^k T) = 0$ *if k is even,*
(ii) $H^1(\mathcal{N}; T) \cong H^1(M; T_M)$,
(iii) $H^1(\mathcal{N}; S^{2\ell+1} T) \cong H^1(\mathcal{N}; T) \otimes H^0(\mathcal{N}; S^{2\ell} T)$.

Proof.   We begin as in Theorem 6.2. Each cohomology class $\alpha \in H^1(\mathcal{N}; S^k T)$ defines by contraction a class $\hat{\alpha} \in H^1(T^*\mathcal{N}; \mathcal{O})$ which is homogeneous of degree k under the $\mathbb{C}^*$-action.

We may use the cohomological version of Hartog's theorem and extend this over an analytic set of codimension $\geqslant 3$, provided by Proposition 6.1 to a class $\alpha \in H^1(\mathcal{M}; \mathcal{O})$ [10]. This class is determined by its restriction to the complement of an analytic set of codimension $\geqslant 2$ and so is determined by its restriction to $T^*\mathcal{N}$ or even to the complement of the zero section in $T^*\mathcal{N}$. Thus *any* $\beta \in H^1(\mathcal{M}; \mathcal{O})$, homogeneous of degree k is determined by its restriction to the complement of the zero section in $T^*\mathcal{N}$ and hence by the class in $H^1(P(T^*\mathcal{N}); \mathcal{O}(k))$ which it defines, where $\mathcal{O}(1)$ is the tautological hyperplane bundle along the fibres.

Since $H^i(P^n; \mathcal{O}(k)) = 0$ for $i > 0$, the Leray spectral sequence identifies this group with $H^1(\mathcal{N}; S^k T)$. We may therefore simply consider classes $\beta \in H^1(\mathcal{M}; \mathcal{O})$ homogeneous of degree k.

We consider the map det: $\mathcal{M} \to H^0(M; K^2) = V$ which defines the integrable system and the first direct image sheaf $R^1\mathrm{det}_*\mathcal{O}$. Since det maps to a vector space V, all the higher cohomology groups of direct image sheaves vanish, and the Leray spectral sequence gives

$$H^1(\mathcal{M}; \mathcal{O}) \cong H^0(V; R^1\mathrm{det}_*\mathcal{O}). \tag{6.6}$$

We shall prove that, outside an analytic set of codimension 2 in V, the sheaf $R^1\mathrm{det}_*\mathcal{O}$ is isomorphic to the sheaf of holomorphic functions with values in $H^1(M; K^{-1})$. Since elements of $H^1(\mathcal{M}; \mathcal{O})$ are determined by their restriction to complements of codimension 2 sets this will, using Hartog's theorem, provide from (6.6) an isomorphism

$$H^1(\mathcal{M}; \mathcal{O}) \cong H^0(V; \mathcal{O}) \otimes H^1(M; K^{-1}). \tag{6.7}$$

To obtain this isomorphism we use the Kähler form to define a class w in $H^1(\mathcal{M}; T^*)$. Contraction with a holomorphic vector field gives a homomorphism from $H^0(\mathcal{M}; T)$ to $H^1(\mathcal{M}; \mathcal{O})$. We apply this to the vector space of Hamiltonian vector fields arising from linear functions on V: the basic fields of the integrable system. This defines a linear map from $V^* \cong H^1(M; K^{-1})$ to $H^1(\mathcal{M}; \mathcal{O})$.

Let $q \in H^0(M; K^2)$ have simple zeros, then the fibre $\mathcal{M}_q$ of det over q is isomorphic to a Prym variety, and the vector fields generated by V^* are tangent to $\mathcal{M}_q$ and span the space $H^0(\mathcal{M}_q; T)$ of all holomorphic vector fields on this abelian variety. Since the Kähler form restricts on $\mathcal{M}_q$ to a Kähler form it defines an *isomorphism*:

$$H^0(\mathcal{M}_q; T) \overset{w}{\to} H^1(\mathcal{M}_q; \mathcal{O}) \cong \mathbb{C}^{3g-3}. \tag{6.8}$$

This gives an isomorphism of $R^1\det_* \mathcal{O}$ with the functions with values in V^* on the open set of $q \in H^0(M; K^2)$ with simple zeros. This is the complement of a codimension 1 analytic set.

We need to consider more degenerate quadratic differentials. If $g \geqslant 3$, then the $q \in H^0(M; K^2)$ which have at most one double zero is the complement of a codimension 2 set, so we must consider the singular fibre $\mathcal{M}_q$ for q with a double zero at $x \in M$ and simple zeros elsewhere. We consider therefore a stable pair (V, Φ) on M such that $\det \Phi$ has a double zero at x.

The curve $S \subset X = K_M$ defined as in §5 now has an ordinary double point over x, but we obtain an eigenspace line bundle L over its desingularization $\hat{S}$, which has genus $(4g - 3) - 1 = (4g - 4)$. By arguments similar to those in (5.10), $L \cong U \otimes L_0$ for some fixed L_0 where U lies in the $(3g - 4)$-dimensional Prym variety of $\hat{S}$. If $\Phi(x)$ is nonzero, it is nilpotent and has a unique eigenspace in V_x. This provides an isomorphism between L_x and $L_{\sigma x}$ on $\hat{S}$ (and hence in fact a point on the Prym variety of the singular curve S).

Since $\sigma^* L \cong \sigma^* U \otimes L_0 \cong U^{-1} \otimes L_0 \cong L^* \otimes L_0^2$, an isomorphism between L_x and $L_{\sigma x}$ is a nonzero section of $(L_x^2 \otimes L_0^{-2})_x$. But the Poincaré bundle over $M \times \operatorname{Jac}(M)$ is trivial on $\{x\} \times \operatorname{Jac}(M)$, so if $\Phi(x)$ is nonzero, it corresponds to an open set of $\mathcal{M}_q$ isomorphic to $\mathbb{C}^* \times \operatorname{Prym}(\hat{S})$.

The remaining case is if $\Phi(x)$ vanishes. This must occur since the fibre of det is compact. In this case we can put $\Phi = s\Psi$ where s is a section of the line bundle $\mathcal{O}(x)$ vanishing at x and $\Psi \in H^0(M; \operatorname{ad} P \otimes K(-x))$. A repetition of 5.10 shows that such a stable pair is determined by a point in the Prym variety $\operatorname{Prym}(\hat{S})$, since the nonsingular curve $\hat{S}$ is the double cover of M branched over the simple zeros of $q = \det \Phi$. Thus each $\mathbb{C}^*$ is compactified by a single point and we find

$$\mathcal{M}_q \cong E \times \operatorname{Prym}(\hat{S})$$

where E is a rational curve with an ordinary double point: a degenerate elliptic curve.

Now $H^1(\mathcal{M}_q; \mathcal{O}) \cong H^1(E; \mathcal{O}) \oplus H^1(\operatorname{Prym}(\hat{S}); \mathcal{O}) \cong \mathbb{C}^{3g-3}$ so $R^1\det_* \mathcal{O}$ is locally free on the open set of q with at most one double zero. Also, the product with $w \in H^1(\mathcal{M}; T^*)$ maps the vector fields tangent to $\operatorname{Prym}(\hat{S})$ isomorphically to $H^1(\operatorname{Prym}(\hat{S}); \mathcal{O})$ as before, and by direct calculation certainly maps the unique holomorphic vector field on the singular curve E to a nonzero element in $H^1(E; \mathcal{O}) \cong \mathbb{C}$.

Thus the isomorphism of $R^1\det_* \mathcal{O}$ with the sheaf of functions extends to the complement of a codimension 2 set, and we obtain the isomorphism (6.7).

It remains to detect the classes in $H^1(\mathcal{M}; \mathcal{O})$ which are homogeneous of degree k. Now the canonical symplectic form ω on $T^*\mathcal{N}$ is of degree 1 (see (5.11)). Since a function f generates a Hamiltonian vector field X by $i(X)\omega = df$, then the functions $f \in V^* \cong H^1(M; K^{-1})$ which are homogeneous of degree 2 generate vector fields homogeneous of degree 1. It follows that the subspace of

$H^1(\mathcal{M}; \mathcal{O})$ of degree k is isomorphic in (6.7) with the tensor product of $H^1(M; K^{-1})$ with holomorphic functions on V of degree $\frac{1}{2}(k-1)$. The theorem then follows, using for part (iii) the result of Theorem 6.2.

Remarks. (1) The natural isomorphism $H^1(\mathcal{N}; T) \cong H^1(M; T_M)$ is almost certainly the Kodaira-Spencer map which associates to a deformation of the complex structure of the Riemann surface M, a deformation of the moduli space $\mathcal{N}$ (see [7]).

(2) The third part of the theorem says that $\oplus_k H^1(\mathcal{N}; S^kT)$ is a free module over $\oplus_k H^0(\mathcal{N}; S^kT)$ generated by $H^1(\mathcal{N}; T)$.

References

1. M. F. ATIYAH AND R. BOTT, *A Lefschetz fixed point formula for elliptic complexes: II.* Applications, Ann. of Math. **88** (1968), 451–491.
2. ______, *The Yang-Mills equations over Riemann surfaces*, Phil. Trans. R. Soc. Lond. **A308** (1982), 523–615.
3. C. CHEVALLEY, Proc. International Congress of Mathematicians, Cambridge, Mass. 1950, **II**, (1952) 21–24.
4. N. J. HITCHIN, *Monopoles and geodesics*, Commun. Math. Phys. **83** (1982), 579–602.
5. ______, *The self-duality equations on a Riemann surface*, to appear.
6. B. KOSTANT, *The principal three-dimensional subgroup and the Betti numbers of a complex simple Lie group*, Amer. J. Math. **81** (1959), 973–1032.
7. M. S. NARASIMHAN AND S. RAMANAN, *Deformations of the moduli space of vector bundles over an algebraic curve*, Ann. of Math. **101** (1975), 39–47.
8. P. E. NEWSTEAD, *Stable bundles of rank 2 and odd degree over a curve of genus 2*, Topology **I** (1968), 205–215.
9. S. RAMANAN, *The moduli spaces of vector bundles over an algebraic curve*, Math. Ann. **200** (1973), 69–84.
10. G. SCHEJA, *Riemannsche Hebbarkeitsatze für Cohomologieklassen*, Math. Ann. **144** (1961), 345–360.

ST. CATHERINE'S COLLEGE, OXFORD OX1 3UJ, ENGLAND.

Vol. 54, No. 1   DUKE MATHEMATICAL JOURNAL ©   1987

EXCEPTIONAL VECTOR BUNDLES ON PROJECTIVE SPACES

A. L. GORODENTSEV AND A. N. RUDAKOV

In this article we construct an infinite set of exceptional vector bundles on $\mathbb{P}^n$. We prove that all exceptional bundles on $\mathbb{P}^2$ are constructed in this way and their dimensions are Markov numbers. As a by-product we get some spectral sequences that are generalizations of the Beilinson spectral sequences.

In the paper we shall study algebraic coherent sheaves on $\mathbb{P}^n$. We shall call such a sheaf F an exceptional sheaf iff

$$\dim \operatorname{Ext}^0(F, F) = 1, \qquad \operatorname{Ext}^i(F, F) = 0 \quad \text{where } i > 0.$$

Our ground field is a complex number field but we believe that almost all our results are valid over an arbitrary field. An exceptional sheaf F on $\mathbb{P}^n$ must be locally free and the associated vector bundle we shall also call exceptional. For $n = 2$ such bundles were studied in [2] but there they had a different definition, an exceptional sheaf means a stable sheaf with discriminant less then $1/2$. We prove that for $\mathbb{P}^2$ this definition and ours are equivalent. It is an interesting question whether an exceptional sheaf on $\mathbb{P}^n$ for $n > 2$ is stable.

The first examples of exceptional sheaves are $\mathcal{O}(i)$. It turns out that other exceptional sheaves on $\mathbb{P}^n$ are naturally composed in infinite series (E_i), $i \in \mathbb{Z}$ and we call them helixes. The series $(\mathcal{O}(i))$ is an example of helix. We define mutations of helixes and via these mutations we obtain an infinite set of helixes on $\mathbb{P}^n$. We prove that for $\mathbb{P}^2$ this set consists of all helixes.

Each helix produces two spectral sequences in the manner of the Beilinson spectral sequences for the helix $(\mathcal{O}(i))$.

We find also that for each helix on $\mathbb{P}^2$ there is a corresponding integral solution of the Markov equation $x^2 + y^2 + z^2 = 3xyz$. It is known that integral

Received September 29, 1986

solutions of a Markov equation are vertexes of a Markov tree ([1]). One could form a graph with helixes as vertexes such that the correspondence became a morphism from "helix graph" to Markov tree. We hope to describe our investigations of these structures in subsequent papers.

We would like to thank F. Bogomolov, V. Danilov, Ju. Manin, I. Schafarevich, S. Zube for interesting and stimulating discussions and we wish especially to thank A. Tyurin; without his kind support and encouragement this paper would never have appeared.

1. Exceptional collections and helixes

1.1. *Definition.* A coherent sheaf E on $\mathbb{P}^n$ is exceptional iff $\mathrm{Ext}^0(E, E) \cong \mathbb{C}$ and $\mathrm{Ext}^i(E, E) = 0$ for $i \geqslant 1$.

It is easy to see that these cohomological conditions imply homogenity of E and thus E must be locally free. So we can talk about exceptional bundles instead of sheaves without any loss of generality.

1.2. *Definition.* An ordered collection of bundles $\varepsilon = (E_1, \ldots, E_k)$ is said to be an exceptional collection iff all its bundles are exceptional and whenever $1 \leqslant l < m \leqslant k$ it is

$$\mathrm{Ext}^i(E_l, E_m) = 0 \quad \text{for all } i \geqslant 1$$

$$\mathrm{Ext}^i(E_m, E_l) = 0 \quad \text{for all } i \geqslant 0.$$

It is trivial that if $\varepsilon = (E_1, \ldots, E_k)$ is an exceptional collection then the collections $\varepsilon^* = (E_k^*, \ldots, E_1^*)$ and $\varepsilon(i) = (E_1(i), \ldots, E_k(i))$ are the same.

1.3. *Mutations of pairs of bundles.* Let $\varepsilon = (A, B)$ be an ordered pair of bundles. Let there exist bundles A' and B' that there are exact sequences of bundles on $\mathbb{P}^n$:

$$0 \to B' \to A \otimes \mathrm{Hom}(A, B) \overset{\mathrm{can}}{\longrightarrow} B \to 0 \tag{1}$$

$$0 \to A \overset{\mathrm{can}}{\longrightarrow} \mathrm{Hom}(A, B)^* \otimes B \to A' \to 0 \tag{2}$$

where the morphism can $\in H^0(A^* \otimes \mathrm{Hom}(A, B)^* \otimes B) = \mathrm{Hom}(A, B)^* \otimes \mathrm{Hom}(A, B)$ is associated with the identity endomorphism of $\mathrm{Hom}(A, B)$.

In the above situation we call B' a left shift of B in ε and denote it by LB. Also we call A' a right shift of A in ε and denote it by RA. We will say that a pair $L\varepsilon = (B', A)$ is a left mutation of ε and $R\varepsilon = (B, A')$ is a right mutation of ε.

1.4. PROPOSITION. *Let $\varepsilon = (A, B)$ be an ordered pair of bundles on $\mathbb{P}^n$. Then $(L\varepsilon)^* = R\varepsilon^*$, $(R\varepsilon)^* = L\varepsilon^*$ and the left part of each equality is valid whenever the right one is.*

We leave the proof to the reader.

1.5. PROPOSITION.   *Let $\varepsilon = (A, B, E_1, \ldots, E_k)$ be an exceptional collection and there exists a left mutation (B', A) of pair (A, B). Then*
   (1) *a collection $\varepsilon' = (B', A, E_1, \ldots, E_k)$ is exceptional*
   (2) *there exists a right mutation $R(B', A)$ and it is equal to (A, B).*

Proof.   From (1) via a long exact sequence of functor $\mathrm{Hom}(*, E_j)$ we conclude that $\mathrm{Ext}^i(B', E_j) = 0$ for $i \geqslant 1$, remembering that ε is exceptional. With help of a long exact sequence for $\mathrm{Hom}(E_j, *)$ we also conclude that $\mathrm{Ext}^i(E_j, B') = 0$ for $i \geqslant 0$. In the same way we derive that $\mathrm{Ext}^i(A, B) = 0$ for $i \geqslant 0$ and $\mathrm{Ext}^i(B', A) = 0$ for $i \geqslant 1$ and that there is canonical isomorphism

$$\mathrm{Hom}(B', A) \cong \mathrm{Hom}(A, B)^*.$$

With the isomorphism in mind we can rewrite (1) in the form:

$$0 \to B' \to \mathrm{Hom}(B', A)^* \otimes A \to B \to 0.$$

This proves (2).

To prove (1) we need only show that B' is exceptional. Putting $\mathrm{Hom}(*, B')$ on (1) we get $\mathrm{Ext}^i(B', B') = \mathrm{Ext}^{i+1}(B, B')$ for $i \geqslant 0$. Then applying $\mathrm{Hom}(B', *)$ to (1) we see that $\mathrm{Ext}^1(B, B') = \mathbb{C}$ and $\mathrm{Ext}^i(B, B') = 0$ for $i \geqslant 2$, thereby proving the proposition.

1.6. COROLLARY.   *Let $\varepsilon = (E_1, \ldots, E_k, A, B)$ be an exceptional collection and there exists a right mutation (B, A') of a pair (A, B). Then:*
   (1) *a collection $\varepsilon' = (E_1, \ldots, E_k, B, A')$ is exceptional*
   (2) *there exists a left mutation $L(B, A')$ and it is equal to (A, B).*

Proof.   Apply 1.5 to collection ε^* and use 1.4.

1.7. *Mutations of collections.*   Let $\varepsilon = (\ldots A, B, \ldots)$ be an (ordered) collection of bundles and assume that mutations $L(A, B)$ and $R(A, B)$ exist. We define a mutation $L_B\varepsilon$ of ε to be the collection obtained by substituting $L(A, B)$ for (A, B) in ε. Appearing in $L_B\varepsilon$ is a new bundle we call a left shift of B in ε and denote $L_\varepsilon B$ or LB so it cannot lead to misunderstanding. In the same way we define $R_B\varepsilon$, $R_\varepsilon B$ or RB.

It is convenient to agree that $L_A^{(0)}\varepsilon = \varepsilon$ and $L_A^{(i)}\varepsilon$, $L_\varepsilon^{(i)}A$ are compositions $L_{LA}^{(i-1)}(L_A\varepsilon)$, $L_{L\varepsilon}^{(i-1)}(LA)$. Similar notations we shall use for compositions of right mutations.

1.8. *Infinite collections.*   Let $\tau = (E_i)$, $i \in \mathbb{Z}$ be an infinite ordered collection of bundles on $\mathbb{P}^n$. We say that it is periodic when $E_{i+n+1} = E_i(n + 1)$ for all $i \in \mathbb{Z}$.

A periodic collection is uniquely determined by each subcollection $(E_i, \ldots, E_{i+n})$ of $(n + 1)$ subsequent bundles. Such a subcollection we call a foundation of the collection.

A periodic collection we call exceptional iff each of its foundations is exceptional.

Let $\varepsilon = (A, B)$ be a pair of subsequent bundles from a periodic collection τ. If mutations $L\varepsilon$ and $R\varepsilon$ exist then they exist for $\varepsilon_k = \varepsilon(k(n+1))$. We define mutations $L_B\tau$ and $R_A\tau$ of a periodic collection τ to be the collection obtained by substituting $L\varepsilon_k$ and $R\varepsilon_k$ respectively for ε_k for every $k \in \mathbb{Z}$ in τ. It is clear that the mutations are also periodic. We keep notations of mutations as it was defined in 1.7.

1.9. *Definition.* An infinite periodic collection σ of bundles on $\mathbb{P}^n$ we shall call a helix iff every bundle E in it satisfies the condition

(hel) all collections $L_E^{(0)}\sigma, L_E^{(1)}\sigma, \ldots, L_E^{(n)}\sigma$ exist, they are exceptional and $L_E^{(n)}\sigma = \sigma$.

1.10. *Remark.* The condition (hel) is equivalent to the condition

(hel′) all collections $R_E^{(0)}\sigma, R_E^{(1)}\sigma, \ldots, R_E^{(n)}\sigma$ exist, they are exceptional and $R_E^{(n)}\sigma = \sigma$.

Mentioning 1.5.2) and 1.6.2) we see that either (hel) or (hel′) implies $R_E^{(i)}\sigma = L_E^{(n-i)}\sigma$ for $i = 0, 1, \ldots, n$ and from that it is easy to deduce its equivalence.

From 1.4 it is clear that σ is a helix iff σ^* is and iff $\sigma(m)$ is.

1.11. PROPOSITION. *The collection $\sigma_0 = (\mathcal{O}(i))$, $i \in \mathbb{Z}$, is a helix.*

Proof. It is sufficient to prove (hel′) for $\mathcal{O}$. Consider exterior powers of the Euler exact sequence:

$$0 \to \mathcal{O} \to V \otimes \mathcal{O}(1) \to \mathcal{T} \to 0$$

$$\ldots$$

$$0 \to \Lambda^{k-1}\mathcal{T} \to \Lambda^k V \otimes \mathcal{O}(k) \to \Lambda^k\mathcal{T} \to 0$$

$$\ldots$$

$$0 \to \Lambda^{n-1}\mathcal{T} \to \Lambda^n V \otimes \mathcal{O}(n) \to \mathcal{O}(n+1) \to 0.$$

It is easy to see that $\Lambda^k V = H^0(\Lambda^k\mathcal{T}(-k)) = \mathrm{Hom}(\mathcal{O}(k), \Lambda^k\mathcal{T})$ and that the morphisms in the above exact triples are canonical morphisms in the sense of 1.3. Thus $R_{\mathcal{O}}^{(k)}\sigma_0 = L_{\mathcal{O}}^{(n-k)}\sigma_0$ are valid and $R_{\mathcal{O}}^{(n)}\sigma_0 = \sigma_0$. To demonstrate an exceptionality of $R_{\mathcal{O}}^{(k)}\sigma_0$ it is sufficient to verify an exceptionality a foundations with $R^{(k)}\mathcal{O} = \Lambda^k\mathcal{T}$. Evidently any collection $(\mathcal{O}(i), \ldots, \mathcal{O}(i+n))$ is exceptional. Performing induction by k and $(n-k)$ and applying 1.5 and 1.6 we conclude that all collections $(\ldots, \mathcal{O}(k), \Lambda^k\mathcal{T})$ and $(\Lambda^k\mathcal{T}, \mathcal{O}(k+1), \ldots)$ are exceptional. After that it is easy to finish the proof.

2. Mutations of helixes

2.1. THEOREM. *Left and right mutations of a helix are helixes.*

Proof. From Proposition 1.4 and Remark 1.10 we deduce that it suffices to prove the theorem only for left mutation. We begin with two lemmas.

2.2. LEMMA. *Let $\varepsilon = (A, B, C)$ be an exceptional collection of bundles such that mutations $L_C^{(1)}\varepsilon = (A, C_1, B)$ and $L_C^{(2)}\varepsilon = (C_2, A, B)$ exist and $L_C^{(1)}\varepsilon$ is exceptional. If a collection $\varepsilon' = L_B\varepsilon$ exists then $L_C^{(1)}\varepsilon' = (B_1, C_1', A)$ and $L_C^{(2)}\varepsilon' = (C_2', B_1, A)$ exist also and they are exceptional and $C_2' = C_2$.*

Proof. First of all we observe that 1.5 implies that $\varepsilon' = (B_1, A, C)$ and $L_C^{(2)}\varepsilon = (C_2, A, B)$ are exceptional. By definition of C_1 there is an exact sequence

$$0 \to C_1 \to B \otimes \operatorname{Hom}(B, C) \to C \to 0. \tag{3}$$

As $\operatorname{Ext}^1(A, C_1) = 0$ then the sequence of vector spaces

$$0 \to \operatorname{Hom}(A, C_1) \to \operatorname{Hom}(A, B) \otimes \operatorname{Hom}(B, C) \to \operatorname{Hom}(A, C) \to 0$$

is exact. Defining sequences for B_1 and C_2 with (3) and (4) we can put in below the commutative diagram with exact rows and columns where unbroken arrows are canonical morphisms and dotted arrows are to prove to exist:

$$
\begin{array}{ccccccccc}
 & & 0 & & 0 & & 0 & & \\
 & & \downarrow & & \downarrow & & & & \\
0 & & C_2 & & B_1 \otimes \operatorname{Hom}(B, C) & & C_1' & & 0 \\
 & & \downarrow & & \downarrow & & & & \\
0 & \to & A \otimes \operatorname{Hom}(A, C_1) & \to & A \otimes \operatorname{Hom}(A, B) \otimes \operatorname{Hom}(B, C) & \to & A \otimes \operatorname{Hom}(A, C) & \to & 0 \\
 & & \downarrow & & \downarrow & & \downarrow_{\text{can}} & & \\
0 & \longrightarrow & C_1 & \longrightarrow & B \otimes \operatorname{Hom}(B, C) & \longrightarrow & C & \longrightarrow & 0 \\
 & & \downarrow & & \downarrow & & & & \\
 & & 0 & & 0 & & 0 & &
\end{array}
$$

$$\tag{5}$$

It is easy to see from the unbroken part of the diagram that a morphism can: $A \otimes \operatorname{Hom}(A, C) \to C$ is an epimorphism, thus C_1' exists. So we have all columns and then the first row is induced from the second one.

Applying $\operatorname{Hom}(*, C_2)$ to the defining exact sequence for B_1 we deduce that $\operatorname{Ext}^i(B_1, C_2) = 0$ for $i \geq 0$ on account of an exceptionality of (C_2, A, B). With it in mind we apply $\operatorname{Hom}(B_1, *)$ to the first row of (5) and derive that $\operatorname{Ext}^i(B_1, C_1') = 0$ for $i \geq 1$ and the isomorphism

$$\operatorname{Hom}(B_1, C_1') \cong \operatorname{Hom}(B, C).$$

So we can rewrite the first row in (5) as

$$0 \to C_2 \to \operatorname{Hom}(B_1, C_1') \otimes B_1 \to C_1' \to 0.$$

From that we see that the mutation $L_C^{(2)}\varepsilon' = (C_2', B_1, A)$ exists and $C_2' = C_2$, as was to be proven.

There remains to prove an exceptionality of collections (B_1, C_1', A) and (C_2, B_1, A). Because of 1.5 it suffices to consider only the first one. Applying $\operatorname{Hom}(*, B_1)$ to the right column of (5) we obtain that $\operatorname{Ext}^0(C_1', B_1)$ are trivial in consequence of an exceptionality of (B_1, A, C). Hence a pair (B_1, C_1') is exceptional. Pairs (B_1, A) and (C_1', A) are exceptional because of 1.5. This proves all our assertions.

2.3. COROLLARY.   *By the conditions of Lemma 2.2 there are canonical isomorphisms*:

$$\operatorname{Hom}(B, C) = \operatorname{Hom}(L_\varepsilon B, L_{\varepsilon'} C) = \operatorname{Hom}(L_\varepsilon^{(2)} C, L_\varepsilon^{(1)} B)^*.$$

2.4. LEMMA.   *Let τ be a periodic exceptional collection and let for some $B \in \tau$ the mutation $\tau' = L_B \tau$ exist and τ' is exceptional. If the condition* (hel) *is valid for some $C \in \tau$ not equal to B and not the first to the left to B in τ then the condition* (hel) *is valid for C in τ'.*

Proof.   We may assume that a foundation of τ is $(\ldots, A, B, \ldots, C)$ and the one of τ' is $(\ldots, B', A, \ldots, C)$. It is an easy consequence of 2.2 that bundles $C_i' = L_{\tau'}^{(i)} C$ exist and are equal to $C_i = L_\tau^{(i)} C$ in all cases except when C_i' is between B' and A. Hence $C_n' = C_n = C(-n-1)$.

To prove exceptionality in the cases when C_i' is not between B' and A we proceed inductively using 1.5 and 1.6 quite similarly to 1.11. In a remaining case the exceptionality is a consequence of 2.2.

2.5. COROLLARY.   *Let τ be a periodic exceptional collection and let for some $A \in \tau$ the mutation $\tau' = R_A \tau$ exists and τ' is exceptional. If the condition* (hel) *is valid for some $C \in \tau$ not equal to A and not the first to the right to A in τ then the condition* (hel) *is valid for C in τ'.*

Proof.   Clear on account of 1.4 and 1.10.

2.6. *Proof of Theorem 2.1.*   Let σ be a helix and $(A, B) \subset \sigma$ be a pair of two consecutive members of σ. By the definition of a helix $\sigma' = L_B \sigma$ exists and is exceptional. Let $L(A, B) = (B', A)$. We know that for B' in σ' the condition (hel) is valid and by Lemma 2.4 it is valid for all members except A. To prove it for A let us consider exceptional collections $R_B^{(i)} \sigma$, $i = 1, 2, \ldots, (n-1)$. They exist because of the definition of a helix. Proceeding inductively with 2.5 in mind we see that the condition (hel) is valid for A in all these collections. But $R_B^{(n-1)} \sigma = L_B^{(1)} \sigma = \sigma'$ so we have proved the theorem.

2.7. COROLLARY. *Let (A, B) be a pair of two consecutive members of a helix σ and $\omega = \omega(i) = L_A^{(i)}\sigma$, $i \geqslant 0$. Then there are canonical isomorphisms:*

$$\mathrm{Hom}(A, B) = \mathrm{Hom}\left(L_\sigma^{(i)}A, L_\omega^{(i)}B\right) = \mathrm{Hom}\left(L_\sigma^{(i+1)}B, L_\sigma^{(i)}A\right)^*.$$

Proof. For $i = 0$ see 1.5. Then proceed inductively with 2.3 in mind.

2.8. COROLLARY. *Let collections $(\ldots, E_1, \ldots, E_k, Q)$ and $(\ldots, F_1, \ldots, F_k, Q)$ be foundations of helixes σ and τ respectively on $\mathbb{P}^n$. Assume that τ can be made from σ by a sequence of mutations such that all members of these foundations except E_i and F_i, $i = 1, 2, \ldots, k$ are preserved. Then $L_\sigma^{(j)}Q = L_\tau^{(j)}Q$ for $k < j \leqslant n$.*

3. Spectral sequences associated with a helix

3.1. *Notations.* Consider $\mathbb{P}^n \times \mathbb{P}^n$ and let p_1 and p_2 be projection morphisms and $\Delta \subset \mathbb{P}^n \times \mathbb{P}^n$ be a diagonal. For bundles A and B on $\mathbb{P}^n$ we write $A \boxtimes B$ to denote the bundle $p_1^*A \otimes p_2^*B$ on $\mathbb{P}^n \times \mathbb{P}^n$.

3.2. *Definition.* A helix on $\mathbb{P}^n$ is said to be constructive iff it can be made from helix $\sigma_0 = (\mathcal{O}(i))$ be a sequence of right and left mutations.

3.3. PROPOSITION. *Let a helix $\sigma = (E_m)$ be constructive. Then for any $i \in \mathbb{Z}$ there are exact sequences of sheaves on $\mathbb{P}^n \times \mathbb{P}^n$:*

$$0 \to \mathcal{O}_{\mathbb{P}^n \times \mathbb{P}^n}|_\Delta \to E_i \boxtimes E_i^* \to E_{i+1} \boxtimes \left(L^{(1)}E_{i+1}\right)^* \to \cdots$$

$$\cdots \to E_{i+n-1} \boxtimes \left(L^{(n-1)}E_{i+n-1}\right)^* \to E_{i+n} \boxtimes \left(L^{(n)}E_{i+n}\right)^* \to 0 \quad (6)$$

$$0 \to \mathcal{O}_{\mathbb{P}^n \times \mathbb{P}^n}|_\Delta \to E_i \boxtimes E_i^* \to R^{(1)}E_{i-1} \boxtimes E_{i-1}^* \to \cdots$$

$$\cdots \to R^{(n-1)}E_{i-n+1} \boxtimes E_{i-n+1}^* \to R^{(n)}E_{i-n} \boxtimes E_{i-n}^* \to 0. \quad (7)$$

Proof. We shall investigate how the sequences are modified with the mutations of the helix.

3.4. LEMMA. *Let $\sigma = (E_m)$ be a helix on $\mathbb{P}^n$. There is a double complex $(Q_p^q; d: Q_p^q \to Q_p^{q+1}; \partial: Q_p^q \to Q_{p-1}^q)$ of bundles on $\mathbb{P}^n \times \mathbb{P}^n$ with exact rows and columns such that a cokernel for the morphism of complexes $d: Q_\cdot^{q-1} \to Q_\cdot^q$ is the complex*

$$0 \to E_q \boxtimes E_q^* \to R^{(1)}E_{q-1} \boxtimes E_{q-1}^* \to \cdots$$

$$\cdots \to R^{(n)}E_{q-n} \boxtimes E_{q-n}^* \to 0 \quad (8)$$

and the cokernel for the morphism of complexes $\partial: Q_{p+1}^\cdot \to Q_p^\cdot$ is

$$0 \to E_p \boxtimes E_p^* \to E_{p+1} \boxtimes \left(L^{(1)}E_{p+1}\right)^* \to \cdots$$

$$\cdots \to E_{p+n} \boxtimes \left(L^{(n)}E_{p+n}\right)^* \to 0. \quad (9)$$

Proof. Let $1 \leqslant (p - q) \leqslant n$. Then a pair $(E_q, R_\sigma^{(p-q)}E_p)$ is a pair of subsequent bundles in the helix $R_A^{(q-p)}\sigma$ for $A = E_p$ and by 2.7 we have a canonical isomorphism

$$\mathrm{Hom}\left(E_q, R^{(q-p)}E_p \right) \cong \mathrm{Hom}\left(L^{(q-p)}E_q, E_p \right).$$

In the following we shall denote this vector space by W_p^q and let $W_k^k = W_k^{k+n+1} = \mathbb{C}$, $k \in \mathbb{Z}$.

There are for right shifts of E_p in σ by definition an exact triple

$$0 \rightarrow R^{(k-1)}E_p \rightarrow W_p^{p+k} \otimes E_{p+k} \rightarrow R^{(k)}E_p \rightarrow 0.$$

From them one can make an exact sequence

$$0 \rightarrow E_p \rightarrow W_p^{p+1} \otimes E_{p+1} \rightarrow \cdots \rightarrow W_p^{p+n} \otimes E_{p+n} \rightarrow E_{p+n+1} \rightarrow 0.$$

Applying to the sequence a map p_1^* and multiplying by $p_2^* E_p^*$ we get an exact sequence of bundles on $\mathbb{P}^n \times \mathbb{P}^n$:

$$0 \rightarrow W_p^p \otimes E_p \boxtimes E_p^* \rightarrow W_p^{p+1} \otimes E_{p+1} \boxtimes E_p^* \rightarrow \cdots$$

$$\cdots \rightarrow W_p^{p+n+1} \otimes E_{p+n+1} \boxtimes E^* \rightarrow 0.$$

Let $Q_p^q = W_p^q \otimes E_q \boxtimes E_p^*$. We have made an exact complex $(Q_p^\cdot, d)$ and one immediately sees from their construction that

$$\mathrm{coker}\left(d \colon Q_p^{q-1} \rightarrow Q_p^q \right) = R^{(q-p)}E_p \boxtimes E_p^*.$$

Similarly for the left shifts of E_q we made an exact sequence

$$0 \rightarrow E_{q-n-1} \rightarrow \left(W_{q-n}^q \right)^* \otimes E_{q-n} \rightarrow \cdots$$

$$\cdots \rightarrow \left(W_{q-1}^q \right)^* \otimes E_{q-1} \rightarrow E_q \rightarrow 0.$$

Applying p_2^* to the dual sequence and multiplying by $p_1^* E_q$ we get an exact sequence of bundles on $\mathbb{P}^n \times \mathbb{P}^n$:

$$0 \rightarrow W_q^q \otimes E_q \boxtimes E_q^* \rightarrow W_{q-1}^q \otimes E_q \boxtimes E_{q-1}^* \rightarrow \cdots$$

$$\cdots \rightarrow W_{q-n-1}^q \otimes E_q \boxtimes E_{q-n-1}^* \rightarrow 0$$

or an exact complex $(Q_\cdot^q, \partial)$ such that

$$\mathrm{coker}\left(\partial \colon Q_{p+1}^q \rightarrow Q_p^q \right) = E_q \boxtimes \left(L^{(q-p)}E_q \right)^*.$$

Since morphisms d and ∂ are canonical then $d \circ \partial = \partial \circ d$ and so (Q_q^p, d, ∂) is a double complex.

3.5. COROLLARY. *Complexes* (8) *and* (9) *exist for any helix and their cohomologies are isomorphic and they are the same for all* i. *So if for some* i *one of the sequences* (6), (7) *is exact then they are exact for all* i.

3.6. *Remark.* Morphisms

$$E_{i+k} \boxtimes \left(L^{(k)}E_{i+k} \right)^* \to E_{i+k+1} \boxtimes \left(L^{(k+1)}E_{i+k+1} \right)^*$$

and

$$R^{(k)}E_{i-k} \boxtimes E_{i-k}^* \to R^{(k+1)}E_{i-k-1} \boxtimes E_{i-k-1}^*$$

are canonically associated with the identity endomorphisms of vector spaces by the isomorphisms

$$\mathrm{Hom}\left(E_{i+k} \boxtimes \left(L^{(k)}E_{i+k} \right)^*, E_{i+k+1} \boxtimes \left(L^{(k+1)}E_{i+k+1} \right)^* \right)$$

$$\cong H^0\left(\mathbb{P}^n \times \mathbb{P}^n, \left(E_{i+k}^* \otimes E_{i+k+1} \right) \boxtimes \left(\left(L^{(k+1)}E_{i+k+1} \right)^* \otimes L^{(k)}E_{i+k} \right) \right)$$

$$= \mathrm{Hom}\left(E_{i+k}, E_{i+k+1} \right) \otimes \mathrm{Hom}\left(L^{(k+1)}E_{i+k+1}, L^{(k)}E_{i+k} \right)$$

$$= W_{i+k}^{i+k+1} \otimes \left(W_{i+k}^{i+k+1} \right)^* = \mathrm{End}\, W_{i+k}^{i+k+1}$$

and similarly

$$\mathrm{Hom}\left(R^{(k)}E_{i-k} \boxtimes E_{i-k}^*, R^{(k+1)}E_{i-k-1} \boxtimes E_{i-k-1}^* \right) = \mathrm{End}\, W_{i-k-1}^{i-k}.$$

3.7. LEMMA. *Let* $\sigma = (E_m)$, $m \in \mathbb{Z}$ *be a helix on* $\mathbb{P}^n$, $A = E_k$ *such that* $0 < k < n$ *and let* $\tau = R_A\sigma = (E_m')$ *where* $E_0 = E_0'$. *Then the sequence* (6) *where* $i = 0$ *for* τ *is exact iff the same for* σ *is exact.*

Proof. From 2.8 for $i \neq k, (k+1)$ there are equalities $E_i' \boxtimes (L_\tau^{(i)}E_i')^* = E_i \boxtimes (L_\sigma^{(i)}E_i)^*$ therefore the sequences (6) for σ and τ distinguish in $(k-1)$-th and k-th terms. They are in τ:

$$E_k' \boxtimes \left(L_\tau^{(k)}E_k' \right)^* = E_{k+1} \boxtimes \left(L_\tau^{(k)}E_{K+1} \right)^*$$

and

$$E_{k+1}' \boxtimes \left(L_\tau^{(k+1)}E_{k+1}' \right)^* = R_\sigma^{(1)}E_k \boxtimes \left(L_\tau^{(k+1)}R_\sigma^{(1)}E_k \right)^* = R_\sigma^{(1)}E_k \boxtimes \left(L_\sigma^{(k)}E_k \right)^*.$$

One easily sees that $L_\tau^{(k)}E_{k+1} = L_\omega^{(k)}E_{k+1}$ where $\omega = L_A^{(k)}\sigma$. Applying 2.8 to σ and ω we get an equality $L_\omega^{(k+1)}E_{k+1} = L_\sigma^{(k+1)}E_{k+1}$ hence a canonical sequence that connects $L_\omega^{(k)}E_{k+1}$ and $L_\omega^{(k+1)}E_{k+1}$ in ω is

$$0 \to L_\sigma^{(k+1)}E_{k+1} \to \left(W_k^{k+1}\right)^* \otimes L_\sigma^{(k)}E_k \to L_\tau^{(k)}E_{k+1} \to 0.$$

Applying p_2^* to the dual sequence and multiplying by $p_1^* E_{k+1}$ we get on exact triple on $\mathbb{P}^n \times \mathbb{P}^n$:

$$0 \to E_{k+1} \boxtimes \left(L_\tau^{(k)}E_{k+1}\right)^* \to W_k^{k+1} \otimes E_{k+1} \boxtimes \left(L_\sigma^{(k)}E_k\right)^*$$

$$\to E_{k+1} \boxtimes \left(L_\sigma^{(k+1)}E_{k+1}\right)^* \to 0.$$

Applying also p_1^* to the canonical triple

$$0 \to E_k \to W_k^{k+1} \otimes E_{k+1} \to R_\sigma^{(1)}E_k \to 0$$

and multiplying by $p_2^*(L_\sigma^{(k)}E_k)^*$ we get another exact sequence

$$0 \to E_k \boxtimes \left(L_\sigma^{(k)}E_k\right)^* \to W_k^{k+1} \otimes E_{k+1} \boxtimes \left(L_\sigma^{(k)}E_k\right)^*$$

$$\to R_\sigma^{(1)}E_k\left(L_\sigma^{(k)}E_k\right)^* \to 0$$

on $\mathbb{P}^n \times \mathbb{P}^n$.

With these two sequences in mind we can put distinguish parts of the sequences (6) for σ and τ in a commutative diagram with exact diagonals:

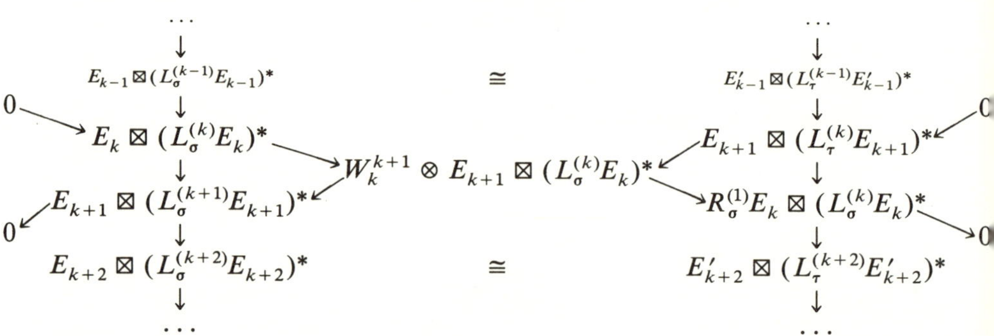

From the diagram it is easy to see that the first column is exact iff the last one is, as was to be shown.

3.8. *Proof of Proposition* 3.3. By Corollary 3.5 and Lemma 3.7 we see that if τ is made from σ by mutations then the assertion is valid for τ iff the one is valid for σ and it suffices to prove the exactness of (6) or (7) for some i.

We show that the sequence (6) where $i = 0$ is exact for σ_0. We can rewrite it as

$$0 \to \mathcal{O}|_\Delta \to \mathcal{O}_{\mathbb{P}^n \times \mathbb{P}^n} \to \mathcal{O}(1) \boxtimes \mathcal{T}(-1) \to \mathcal{O}(2) \boxtimes \Lambda^2\mathcal{T}(-2) \to \cdots$$

or:

$$0 \to \mathcal{O}|_\Delta \to \mathcal{O}_{\mathbb{P}^n \times \mathbb{P}^n} \xrightarrow{s} \mathcal{O}(1) \boxtimes \mathcal{T}(-1) \to \Lambda^2(\mathcal{O}(1) \boxtimes \mathcal{T}(-1)) \to \cdots$$

and the maps are induced by multiplication by a canonical section s in $\mathcal{O}(1) \boxtimes \mathcal{T}(-1)$ that arises from an identity endomorphism of a vector space V where $\mathbb{P}^n = \mathbb{P}(V)$ via isomorphisms $H^0(\mathbb{P}^n \times \mathbb{P}^n, \mathcal{O}(1) \boxtimes \mathcal{T}(-1)) = H^0(\mathbb{P}^n, \mathcal{O}(1)) \otimes H^0(\mathbb{P}^n, \mathcal{T}(-1)) = V^* \otimes V = \mathrm{End}\,V$ (see for example [4], Cap. 2, p. 3.1). The exactness of that complex is well known. So we have proved the proposition.

3.9. THEOREM. *Let* $\sigma = (E_m)$, $m \in \mathbb{Z}$ *be a constructive helix on* $\mathbb{P}^n$. *Then for any bundle Q on* $\mathbb{P}^n$ *there are spectral sequences* ${}^1E^{p,q}$ *and* ${}^2E^{p,q}$ *such that their terms are not trivial for* $0 \leqslant q \leqslant n$, $-n \leqslant p \leqslant 0$ *and their E_1-parts are*

$${}^1E_1^{p,q} = \mathrm{Ext}^q(E_{-p}, Q) \otimes L^{(-p)}E_{-p};$$

$${}^2E_1^{p,q} = \mathrm{Ext}^q(R^{(-p)}E_p, Q) \otimes E_p.$$

And they are convergent to

$${}^1E_\infty^i = {}^2E_\infty^i = \begin{cases} Q & \text{for } i = 0 \\ 0 & \text{for } i \neq 0 \end{cases}.$$

Proof. The assertion is easy to see as a generalisation of the well-known Beilinson theorem that arises if $\sigma = \sigma_0$, $E_i = \mathcal{O}(i)$. The proof of our theorem is quite similar to the Beilinson proof as in [4].

Multiplying the dual sequences to (6) and (7) by p_1^*Q we get locally free resolvents of a sheaf $Q|_\Delta = p_1^*Q|_\Delta$: $0 \to ({}^1C.) \to Q|_\Delta \to 0$ and $0 \to ({}^2C.) \to Q|_\Delta \to 0$, where ${}^1C_i = \mathcal{H}om(E_{-i}, Q) \boxtimes L^{(-i)}E_{-i}$; ${}^2C_i = \mathcal{H}om(R^{(-i)}E_i, Q) \boxtimes E_i$ and $i \leqslant 0$.

We are to count the higher direct image of $({}^1C.)$ and $({}^2C.)$ by p_2. There are standard spectral sequences with common limit for that and writing them explicitly we prove the theorem (see [4], Cap. 2, p. 3.1).

3.10. COROLLARY. *Let collection* $(E_0, E_1, \ldots, E_n)$ *be a foundation of a constructive helix on* $\mathbb{P}^n$. *If for the simple bundle Q* $\mathrm{Ext}^i(E_m, Q) = 0$ *where* $m = 1, 2, \ldots, n$ *and* $1 \leqslant m$ *then* $Q = E_0$.

Proof. From ${}^1E^{p,q}$ we immediately see that $Q = \mathrm{Hom}(E_0, Q) \otimes E_0$ so $Q = E_0$ since Q is simple.

4. Exceptional bundles on $\mathbb{P}^2$. In the following we consider only sheaves and bundles on $\mathbb{P}^2$. Let us denote a range of a bundle or a sheaf E by $r(E)$, its first Chern class by $c(E)$, its discriminant by $\Delta(E)$ and its slope by $\mu(E)$. When one can easily deduce from context what the sheaf in question is, we shall omit its notation in the above symbols. We mean here a stability by Mumford-Takemoto.

4.1. THEOREM. *If E is an exceptional sheaf on $\mathbb{P}^2$ then E is stable.*

Proof. A proof of the theorem depends on the next lemmas due to Mukai:

4.2. MUKAI LEMMAS. (1) *If $0 \to F \to E \to G \to 0$ is an exact sequence of sheaves and* $\mathrm{Hom}(F, G) = \mathrm{Ext}^2(G, F) = 0$ *then*

$$\dim \mathrm{Ext}^1(E, E) \geqslant \dim \mathrm{Ext}^1(F, F) + \dim \mathrm{Ext}^1(G, G).$$

(2) *If $0 \to F \to E \to G \to 0$ is an exact sequence of sheaves,* $\mathrm{Hom}(F, G) = 0$ *and E is exceptional then either $F = 0$ or $r(G) = 0$.*

Proof. Besides insignificant modifications the assertions and the proofs are the same as in [3].

4.3. *Proof of Theorem* 4.1. Considering restrictions on lines we see that slopes of subsheaves in E are bounded. Take subsheaves of E such that their factorsheaves are torsion free and their slopes are the maximal possible and choose from them a subsheaf F of maximal rank. To prove stability we have to show that $F = E$ ([4]). If $G = E/F$ then it is easily seen that $\mathrm{Hom}(F, G) = 0$ and because of 4.2 we have $r(G) = 0$. Since G is torsion free, $G = 0$.

4.4. PROPOSITION ([2]). *If E is an exceptional bundle (on $\mathbb{P}^2$) then $c(E)$ and $r(E)$ have no common divisors so they can be found from $\mu(E)$.*

Proof. It immediately derives from the Riemann-Roch theorem in the form suggested by Mukai in [3].

4.5. PROPOSITION ([2]). *If E is an exceptional bundle (on $\mathbb{P}^2$) then $\Delta(E) = (1 - 1/r(E)^2)/2$.*

Proof. By definition $\chi(E, E) = 1$. So by the Riemann-Roch Theorem $1 = r^2(1 - 2\Delta)$.

4.6. PROPOSITION ([2]). *If E_1 and E_2 are exceptional bundles (on $\mathbb{P}^2$) and their slopes are equal then $E_1 = E_2$.*

Proof. There is $\Delta(E_1) = \Delta(E_2)$ so $\chi(E_1, E_2) = 1$. Then either $\mathrm{Hom}(E_1, E_2) \neq 0$ or $\mathrm{Hom}(E_2, E_1(-3)) \neq 0$. In the second case there is therefore $\mathrm{Hom}(E_2, E_1) \neq 0$. And a homomorphism of the stable sheaf of the same slope must be an isomorphism.

5. Helixes on $\mathbb{P}^2$ and the Markov equation

5.1. PROPOSITION. *Let E_1 and E_2 be exceptional bundles (on $\mathbb{P}^2$), $\mu_1 = k_1/r_1$ and $\mu_2 = k_2/r_2$ their slopes. Assume that (E_1, E_2) is an exceptional pair. Then the integral numbers r_1, r_2 and $\widehat{\delta_{12}} = k_2 r_1 - k_1 r_2$ form a solution of the Markov* *equation*

$$x^2 + y^2 + z^2 = 3xyz.$$

Proof. By the definition of an exceptional pair $\chi(E_2, E_1) = 0$. But by the Riemann-Roch theorem we have

$$\chi(E_2, E_1) = r_1 r_2\left(1 + \tfrac{3}{2}(\mu_1 - \mu_2) + \tfrac{1}{2}(\mu_1 - \mu_2)^2 - \Delta_1 - \Delta_2\right).$$

Remembering 4.5 we obtain the assertion after simple calculations.

5.2. *Solutions of the Markov equation.* It is known (see [1]) that if $\bar{\mathbf{a}} = (a_1, a_2, a_3)$ is a solution of the Markov equation then

$$\pi_1 \bar{\mathbf{a}} = (a_1, 3a_1 a_2 - a_3, a_2)$$

and

$$\pi_r \bar{\mathbf{a}} = (a_2, 3a_2 a_3 - a_1, a_3)$$

are also. Passing from $\bar{\mathbf{e}} = (1,1,1)$ to $\pi_r \bar{\mathbf{e}} = (1,2,1)$ and to $\pi_1 \pi_r \mathbf{e} = (1,5,2)$ and so on, we can obtain exactly all the integral solutions of the Markov equation with the help of the transformations π_1 and π_r.

For any solution $\bar{\mathbf{a}} = (a_1, a_2, a_3)$ there exists a unique collection of integral numbers $k(a_1, \bar{\mathbf{a}})$, $k(a_2, \bar{\mathbf{a}})$, $k(a_3, \bar{\mathbf{a}})$ such that

$$k(a_1, \bar{\mathbf{a}}) \equiv a_2 a_3^{-1} \equiv -a_2^{-1} a_3 \pmod{a_1}, 0 \leqslant k(a_1, \bar{\mathbf{a}}) \leqslant a_1;$$

$$k(a_2, \bar{\mathbf{a}}) \equiv a_1 a_3^{-1} \equiv -a_1^{-1} a_3 \pmod{a_2}, 0 < k(a_2, \bar{\mathbf{a}}) \leqslant a_2;$$

$$k(a_3, \bar{\mathbf{a}}) \equiv a_1 a_2^{-1} \equiv -a_1^{-1} a_2 \pmod{a_3}, 0 < k(a_3, \bar{\mathbf{a}}) \leqslant a_3;$$

One knows that $k(a_1, \bar{\mathbf{a}}) = k(a_1, \pi_1 \bar{\mathbf{a}})$, $k(a_2, \bar{\mathbf{a}}) = k(a_2, \pi_1 \bar{\mathbf{a}})$ and the same is valid for π_r.

Definition. We shall call a solution $\bar{\mathbf{a}} = (a_1, a_2, a_3)$ conveniently written if $a_2 \geqslant a_1$, $a_2 \geqslant a_3$ and $0 \leqslant k(a_2, \mathbf{a})/a_2 \leqslant 1/2$.

One easily checks that the convenience is preserved under transformations and that for any solution there is a unique conveniently written solution made from the same elements.

PROPOSITION. *Let* $\bar{\mathbf{a}} = (a_1, a_2, a_3)$ *be a conveniently written solution and* $k_i = k(a_i, \bar{\mathbf{a}})$. *Then* $k_3 a_1 - k_1 a_3 = 3a_1 a_3 - a_2 > 0$ *and* $k_2 a_1 - k_1 a_2 = a_3$, $k_3 a_2 - k_2 a_3 = a_1$.

Proof. See in [1].

5.3. *Definition.* A foundation (E_1, E_2, E_3) of a helix on $\mathbb{P}^2$ we shall call convenient if $r_2 \geqslant r_1$, $r_2 \geqslant r_3$ and $0 \leqslant \mu(E_2) \leqslant 1/2$.

It is clear that from any helix multiplying by $\mathcal{O}(i)$, by dualising and renumbering of members if needed, one can make a unique helix with a foundation (E_1, E_2, E_3) which is convenient.

5.4. THEOREM. *Let $\varepsilon = (E_m)$ be a helix on $\mathbb{P}^2$. The numbers $r_1 = r(E_{3k+1})$, $r_2 = r(E_{3k+2})$, $r_3 = r(E_{3k+3})$ form a solution $\bar{r}(\varepsilon)$ of the Markov equation.*

Proof. Because of 5.3 one can assume that a foundation (E_1, E_2, E_3) is convenient. Let $\mu_i = \mu(E_i) = k_i/r_i$. By definition of a helix $\mathrm{Hom}(E_i, E_{i+1}) \neq 0$ hence because of 4.1 and 4.6, $\mu_i < \mu_{i+1}$ and we may use the next lemma with pairs (E_1, E_2) and (E_2, E_3):

5.5. LEMMA. *Under the conditions of 5.1 assume that $k_1/r_1 < k_2/r_2$. Then: if $r_1 < r_2$ and $0 \leqslant k_2/r_2 \leqslant 1/2$ then $\bar{a} = (r_1, r_2, \delta_{12})$ is a conveniently written solution and $k_1 = k(r_1, \bar{a})$, $k_2 = k(r_2, \bar{a})$; if $r_2 < r_1$ and $0 \leqslant k_1/r_1 \leqslant 1/2$ then $\bar{b} = (\delta_{12}, r_1, r_2)$ is a conveniently written solution and $k_1 = k(r_1, \bar{b})$, $k_2 = k(r_2, \bar{b})$.*

Proof. By the definition of δ_{12} there are

$$k_1 \equiv -\delta_{12} r_2^{-1} \ (\mathrm{mod} \ r_1) \quad \text{and} \quad k_2 \equiv \delta_{12} r_1^{-1} \ (\mathrm{mod} \ r_2).$$

Let $r_1 < r_2$. If $\delta_{12} < r_2$ then we prove the assertions by straightforward calculations. Otherwise $\delta_{12} > r_2$ and either (r_1, δ_{12}, r_2) is a conveniently written solution or (r_2, δ_{12}, r_1) is the same.

In the first case $k_1 \equiv -k(r_1, \bar{a}) \ (\mathrm{mod} \ r_1)$ hence because $k_1/r_1 < 1/2$ we get $k(r_1, \bar{a}) > r_1/2$ or $k(r_1, a)/r_1 > 1/2$. But because of proposition 5.2 we have $k(\delta_{12}, \bar{a})/\delta_{12} > k(r_1, \bar{a})/r_1$ which contradicts the definition of convenience.

In the second case we conclude that $k_1 = k(r_1, \bar{a})$, $k_2 = k(r_2, \bar{a})$ and by 5.2 we obtain $k_1/r_1 > k_2/r_2$ that contradicts the conditions of the lemma.

Considerations in the case when $r_2 < r_1$ are similar.

5.6. *Continuation of the proof of Theorem* 5.4. From Lemma 5.5 for exceptional pairs (E_1, E_2) and (E_2, E_3) we conclude that a solution $\bar{a} = (r_1, r_2, \delta_{12})$ is conveniently written, $k_2 = k(r_2, \bar{a})$, and $\bar{b} = (\delta_{23}, r_2, r_3)$ where $\delta_{23} = k_3 r_2 - k_2 r_3$ is the same and $k_2 = k(r_2, \bar{b})$.

Hence $k(r_2, \bar{a}) = k(r_2, \bar{b})$. But one easily checks from basic properties of solutions of the Markov equation that two conveniently written solutions with the same middle elements and the same middle associated numbers are the same. So $\bar{a} = \bar{b} = \bar{r}(\varepsilon)$.

5.7. COROLLARY. *If (E_1, E_2, E_3) is a convenient foundation of a helix $\varepsilon = (E_i)$ and $\mu_i = k_i/r_i$ then*

$$k_2 r_1 - k_1 r_2 = r_3$$

$$k_3 r_2 - k_2 r_3 = r_1$$

$$k_3 r_1 - k_1 r_3 = 3 r_1 r_3 - r_2.$$

Hence one can easily calculate the slopes of exceptional bundles on $\mathbb{P}^2$ with the help of induction by the Markov tree.

Proof. It has been demonstrated when proving that $k_i = k(r_i, \bar{\mathbf{r}}(\varepsilon))$.

5.8. PROPOSITION. *Let (E_1, E_2, E_3) be a convenient foundation of a helix ε. Then*

$$\bar{\mathbf{r}}(L_{E_1}\varepsilon) = \bar{\mathbf{r}}(R_{E_1}\varepsilon) = \pi_r(\bar{\mathbf{r}}(\varepsilon))$$

and

$$\bar{\mathbf{r}}(L_{E_3}\varepsilon) = \bar{\mathbf{r}}(R_{E_3}\varepsilon) = \pi_1(\bar{\mathbf{r}}(\varepsilon)).$$

Proof. The foundation of a helix $L_{E_1}\varepsilon$ is (E_2, E_1', E_3) where E_1' is defined by an exact sequence

$$0 \to E_1 \to \mathrm{Hom}(E_1, E_2)^* \otimes E_2 \to E_1' \to 0.$$

Since the pair (E_1, E_2) is exceptional there is

$$\dim \mathrm{Hom}(E_1, E_2) = \chi(E_1, E_2)$$

$$= r_1 r_2\left(1 + \tfrac{3}{2}(\mu_2 - \mu_1) + \tfrac{1}{2}(\mu_2 - \mu_1)^2 - \Delta_2 - \Delta_1\right) = 3r_3.$$

So $r(E_1') = 3r_2 r_3 - r_1$ and

$$\bar{\mathbf{r}}(L_{E_1}\varepsilon) = (r_2, 3r_2 r_3 - r_1, r_3) = \pi_r(\bar{\mathbf{r}}(\varepsilon)).$$

The equality $L_{E_1}\varepsilon = R_{E_1}\varepsilon$ is valid by the definition of mutations. The remaining cases are investigated quite similarly.

5.9. LEMMA ([2]). *If E and E' are exceptional and their slopes μ, μ' are such that $0 < (\mu - \mu') < 3$ then $\chi(E, E') \leqslant 0$.*

Proof. Remembering stability and the inequalities for slope we see that $\mathrm{Hom}(E, E') = 0$. Then by Serre duality and the same argument it is $\mathrm{Ext}^2(E, E') = \mathrm{Hom}(E', E(-3))^* = 0$. Therefore

$$\chi(E, E') = -\dim \mathrm{Ext}^1(E, E') \leqslant 0.$$

5.10. THEOREM. *Any exceptional bundle E' on $\mathbb{P}^2$ is an element in some helix. Any exceptional pair (E', E'') is a pair of subsequent elements in a unique helix.*

Proof. Begin from the first assertion and assume the opposite one is valid. Then it is clear that we may assume that $0 < \mu' < 1/2$ and by some mutations

one can make a helix $\varepsilon = (E_m)$ for which $0 \leqslant \mu_1 < \mu_2 < \mu_3 \leqslant 1/2$ and either
   (a) $\mu_1 < \mu' < \mu_2$, $r_2 > r'$, or
   (b) $\mu_2 < \mu' < \mu_3$, $r_2 > r'$.
Let $P(t) = 1 + \frac{3}{2}t + \frac{1}{2}t^2$: In the case (a) there are

$$P(\mu_1 - \mu_2) - \Delta_1 - \Delta_2 = 0$$

and

$$P(\mu_1 - \mu') - \Delta_1 - \Delta' \leqslant 0 \quad \text{by Lemma 5.9.}$$

And $\Delta' < \Delta_2$ because $r' < r_2$. Hence

$$P(\mu_1 - \mu') - P(\mu_1 - \mu_2) \leqslant \Delta' - \Delta_2 < 0$$

or $P(\mu_1 - \mu') < P(\mu_1 - \mu_2)$.

But $\mu_1 - \mu' > \mu_1 - \mu_2 > -1$ and it is easy to see that the function $P(t)$ is increasing for $t \geqslant -1$ so we come to contradiction.

We turn to the second assertion. We may assume for the pair (E', E) that a slope of its bundle which has maximal rank is in $[0, 1/2]$ and it suffices to prove the theorem for such pairs. Denoting such a pair by (E_1, E_2) and implying Lemma 5.5 we obtain that r_1 and r_2 are elements of a convenient solution $\bar{a}$ and $k_1 = k(r_1, \bar{a})$, $k_2 = k(r_2, \bar{a})$. But we know by 5.8 that any convenient solution is of a form $\bar{r}(\varepsilon)$ for some helix ε with a convenient foundation. And if $\bar{a} = \bar{r}(\varepsilon)$ then slopes of bundles from a convenient foundation for ε are $k(r_i, \bar{a})/r_i$ therefore we conclude that E_1 and E_2 are in this foundation.

5.11.   We assert also without proof two properties of exceptional bundles on a plane:

PROPOSITION.   (1) *Let E_1 and E_2 be exceptional bundles and $\mu_1 < \mu_2$. Then* $\mathrm{Hom}(E_1, E_2) \neq 0$ *and* $\mathrm{Ext}^1(E_1, E_2) = \mathrm{Ext}^2(E_1, E_2) = 0$.
   (2) *Let $\varepsilon = (E_m)$ be a helix. Then for any i and j* $\mathrm{Ext}^1(E_i, E_j) \cong \mathrm{Ext}^1(E_j, E_i) = 0$.

REFERENCES

1.   J. W. S. CASSELS, *An introduction to diophantine approximation*, "C. U. P.", Cambridge.
2.   J.-M. DREZET AND J. LE POTIER, *Fibrés stables et fibrés exceptionnels sur* $\mathbb{P}_2$, Ann. Ec. Norm. Sup., 1985, t. 18, p. 193–244.
2′.   J.-M. DRESET, *Fibrés exceptionnels et suite spectrale de Beilinson généralisée sur* $\mathbb{P}_2(\mathbb{C})$, Math. Ann., B.275, h.1, (1986)
3.   S. MUKAI, *On the moduli space of bundles on K3 surfaces I*, Preprint.
4.   CH. OKONEK, M. SCHNEIDER, AND H. SPINDLER, *Vector bundles on complex projective spaces*, Birkhäuser Boston, 1980.

DEPARTMENT OF MATHEMATICS, MOSCOW UNIVERSITY, MOSCOW 179-899

ENDOMORPHISMS AND TORSION OF
ABELIAN VARIETIES

YU. G. ZARHIN

To Yu. I. Manin, with admiration and gratitude

Introduction. Let K be a number field of finite degree over the field $\mathbb{Q}$ of rational numbers, $\overline{K}$ the algebraic closure of K and $G = \mathrm{gal}(\overline{K}/K)$ the Galois group. Let X be an Abelian variety over K and End X the ring of K-endomorphisms of X. The well-known Mordell-Weil theorem [4] asserts that $X(K)$ is a finitely generated commutative group. In particular, the torsion subgroup TORS $X(K)$ of $X(K)$ is finite.

Let p be a prime and K_p the subfield of $\overline{K}$ obtained by adjoining to K all p-power roots of unity in $\overline{K}$. B. Mazur conjectured that $X(K_p)$ is also finitely generated (see [6], [5]). In this connection J.-P. Serre and H. Imai [3] proved (independently) that the torsion subgroup of $X(K_p)$ is finite for each p. Moreover, let K^{cycl} be the compositum of all the K_p, i.e., the field obtained by adjoining to K all roots of unity in $\overline{K}$. K. A. Ribet [8] proved that the torsion subgroup of $X(K^{\mathrm{cycl}})$ is also finite.

Let $K^{\mathrm{ab}} \subset \overline{K}$ be the maximal abelian extension of K. The field K^{ab} contains K^{cycl}; if $K = \mathbb{Q}$, then $\mathbb{Q}^{\mathrm{cycl}} = \mathbb{Q}^{\mathrm{ab}}$ (the Kronecker-Weber theorem).

What can one say about the finiteness of the torsion subgroup TORS $X(K^{\mathrm{ab}})$ of $X(K^{\mathrm{ab}})$? The answer depends on the properties of the following endomorphism algebra of X:

$$\mathrm{End}^{\circ} X = \mathrm{End}\, X \otimes \mathbb{Q}.$$

Recall ([7], [10], [11]) that X is called of CM-type over K if the finite-dimensional semisimple $\mathbb{Q}$-algebra $\mathrm{End}^{\circ} X$ contains a semisimple commutative Q-subalgebra (i.e. the finite direct sum of number fields) of dimension $2 \dim X$. If X is a K-simple Abelian variety, then $\mathrm{End}^{\circ} X$ is a division algebra and the following conditions are equivalent; (1) X is of CM-type over K; (2) $\mathrm{End}^{\circ} X$ is a number field of degree $2 \dim X$; (3) $\mathrm{End}^{\circ} X$ contains a number field of degree $2 \dim X$.

If X is of CM-type over K, then all the torsion points of X are defined over K^{ab} [10, 11], i.e., the torsion subgroups of $X(K^{\mathrm{ab}})$ and $X(\overline{K})$ coincide. In particular, TORS $X(K^{\mathrm{ab}})$ is infinite if X is of CM-type over K.

If Y is an Abelian variety defined over K and K-isogenous to X, then TORS $X(K^{\mathrm{ab}})$ is finite if and only if TORS $Y(K^{\mathrm{ab}})$ is finite. If Y is K-isomor-

Received September 5, 1986.

phic to the product $A \times B$ of Abelian varieties A and B, then $\mathrm{TORS}\, Y(K^{\mathrm{ab}})$ is finite if and only if $\mathrm{TORS}\, A(K^{\mathrm{ab}})$ and $\mathrm{TORS}\, B(K^{\mathrm{ab}})$ are finite.

The main result of this paper is the following statement.

THEOREM 1. *Let X be a simple Abelian variety over a number field K. The torsion subgroup $\mathrm{TORS}\, X(K^{\mathrm{ab}})$ of $X(K^{\mathrm{ab}})$ is finite if and only if X is not of CM-type over K.*

COROLLARY. *Let X be an Abelian variety over a number field K. Let us choose simple abelian varieties $X_1, \dots, X_r$ over K such that X is K-isogenous to the product $X_1 \times \cdots \times X_r$. Then the torsion subgroup $\mathrm{TORS}\, X(K^{\mathrm{ab}})$ is finite if and only if all X_i are not of CM-type over K $(1 \leqslant i \leqslant r)$.*

In order to prove the theorem, we need some notations. For each positive integer n let X_n be the kernel of multiplication by n in $X(\overline{K})$. It is known [7] that X_n is a free Z/nZ-module of rank $2 \dim X$. Clearly, X_n is a finite Galois submodule of $X(\overline{K})$. For each prime p let

$$T_p(X) = \varprojlim X_{p^i}$$

be the Tate module [7] of X. (Here the transition map $X_{p^{i+1}} \to X_{p^i}$ is multiplication by p.) The Tate module $T_p(X)$ is a free $\mathbb{Z}_p$-module of rank $2 \dim X$. There is a natural structure of the Galois module on $T_p(X)$ ([10],[11]) such that the natural isomorphisms

$$T_p(X)/p^i T_p(X) \xrightarrow{\sim} X_{p^i}$$

are isomorphisms of Galois modules.

There are natural compatible embeddings

$$\mathrm{End}\, X \otimes \mathbb{Z}/n\mathbb{Z} \hookrightarrow \mathrm{End}_G X_n$$

and

$$\mathrm{End}\, X \otimes \mathbb{Z}_p \hookrightarrow \mathrm{End}_G T_p(X).$$

Let us put

$$V_p(X) = T_p(X) \bigotimes_{\mathbb{Z}_p} \mathbb{Q}_p.$$

Clearly, $V_p(X)$ is a vector space over $\mathbb{Q}_p$ and

$$\dim_{\mathbb{Q}_p} V_p(X) = 2 \dim X.$$

Extending the action of the Galois group by $\mathbb{Q}_p$-linearity on $V_p(X)$, one gets the

p-adic representation

$$\rho_p \colon G \to \operatorname{Aut} V_p(X),$$

and $V_p(X)$ becomes a Galois module. We also have the embedding

$$\operatorname{End} X \otimes \mathbb{Q}_p \hookrightarrow \operatorname{End}_G V_p(X),$$

extending the embedding of $\operatorname{End} X \otimes \mathbb{Z}_p$ into $\operatorname{End}_{\mathbb{Z}_p} T_p(X)$. The theorem of G. Faltings [2] asserts that the G-module $V_p(X)$ is semisimple and that

$$\operatorname{End}_G V_p(X) = \operatorname{End} X \otimes \mathbb{Q}_p.$$

It is also known [12] that for all but finitely many primes p the G-module X_p is semisimple and

$$\operatorname{End}_G X_p = \operatorname{End} X \otimes \mathbb{Z}/p\mathbb{Z}.$$

We refer to Serre [15, 16] for the excellent account of the Galois action on torsion points and Tate modules; see also [14].

Because of the structure of the torsion subgroup of $X(\overline{K})$ one sees easily that Theorem 1 is a corollary of the conjunction of two following statements.

THEOREM 2. *Let X be a simple Abelian variety over K which is not of CM-type. Let p be a prime, W a nonzero G-invariant $\mathbb{Q}_p$-vector subspace of $V_p(X)$ and G_W the image of G in $\operatorname{Aut}(W)$. Then the group G_W is not commutative.*

THEOREM 3. *Let X be a simple Abelian variety over K which is not of CM-type. Then for all but finitely many primes p the following conditions hold: Let W be a nonzero G-invariant $\mathbb{Z}/p\mathbb{Z}$-vector subspace of X_p and G_W the image of G in $\operatorname{Aut}(W)$. Then the group G_W is not commutative.*

Indeed, Theorem 3 asserts that the p-primary part of TORS $X(K^{\mathrm{ab}})$ vanishes for all but finitely many primes p and Theorem 2 asserts that this p-primary part is finite (the projective limit of nonempty finite sets is nonempty too).

The paper is organized as follows. Sections 1 and 2 contain auxiliary facts of linear algebra. Section 3 contains the proof of Theorems 2 and 3. In Section 4 we discuss torsion in noncyclotomic extensions.

This paper is dedicated to my teacher Yuri Ivanovich Manin on the occasion of his 50th birthday.

§1. Linear algebra. Let k be a perfect commutative field, H a finite-dimensional vector space over k, G a group and

$$\rho \colon G \to \operatorname{Aut}(H)$$

a semisimple representation of G, so that H is a semisimple G-module. Let us put

$$D = \operatorname{End}_G H,$$

and let E be a maximal commutative k-subalgebra of D, so that E coincides with its own centralizer in D. We have

$$E \subset D = \operatorname{End}_G H \subset \operatorname{End} H.$$

In particular, H has a natural structure of a faithful E-module. Let $\operatorname{End}_E H$ be the endomorphism algebra of the E-module H. Then, by the definition, $\operatorname{End}_E H$ coincides with the centralizer of E in the endomorphism algebra $\operatorname{End} H$ of the k-vector space H. Since E coincides with its own centralizer in D,

$$D \cap \operatorname{End}_E H = E.$$

Since E is commutative, $E \cap \operatorname{End}_E H$. Since $E \subset D = \operatorname{End}_G H$,

$$\rho(G) \subset \operatorname{End}_E H.$$

THEOREM 4. *Let us assume that E is a semisimple commutative k-algebra.*
(1) *Let $E(G) \subset \operatorname{End} H$ be the k-subalgebra generated by $\rho(G)$ and E. Then*

$$E(G) = \operatorname{End}_E H.$$

(2) *Let us assume that H is the free E-module of rank $r > 1$. Then the following conditions hold:*
Let W be a nonzero G-invariant k-vector subspace of H and G_W the image of G in the group $\operatorname{Aut}(W)$ of k-linear automorphisms of W. Then the group G_W is not commutative.

Proof. (1) Let $k(G) \subset \operatorname{End} H$ be the k-subalgebra generated by $\rho(G)$. Since H is the semisimple G-module, it becomes a faithful semisimple $k(G)$-module. In particular, $k(G)$ is a semisimple finite-dimensional k-algebra. We have

$$\operatorname{End}_{k(G)} H = \operatorname{End}_G H = D \subset \operatorname{End} H.$$

Clearly,

$$k(G) \subset E(G) \subset \operatorname{End}_E H$$

and the natural map

$$E \underset{k}{\bigotimes} k(G) \to E(G), \qquad e \otimes u \mapsto eu$$

is a surjective homomorphism of k-algebras. Since k is perfect, the finite-dimensional semisimple commutative k-algebra E splits into the direct sum

$$E = \bigoplus E_i$$

of commutative fields E_i, which are finite separable algebraic extensions of k. This implies that the tensor product $A = E \otimes_k k(G)$ is a semisimple k-algebra. Hence, $E(G)$ is also a semisimple k-algebra, because it is isomorphic to the quotient algebra of A.

Therefore, the faithful $E(G)$-module H is semisimple. By the definition of $E(G)$, the centralizer $\operatorname{End}_{E(G)}H$ of $E(G)$ in $\operatorname{End}(H)$ is equal to

$$\operatorname{End}_{k(G)}H \cap E = D \cap E = E.$$

By the Jacobson density theorem, $E(G)$ coincides with the centralizer of $\operatorname{End}_{E(G)}H$ in $\operatorname{End} H$, i.e., $E(G)$ coincides with the centralizer of E in $\operatorname{End} H$. But the centralizer of E in $\operatorname{End} H$ is equal to $\operatorname{End}_E H$. Thus

$$E(G) = \operatorname{End}_E H.$$

(2) Let us put

$$H_i = E_i H \subset H.$$

We have

$$H = \bigoplus H_i, \qquad E_i H_j = 0 \quad \text{while} \ \neq j.$$

Since H is the free E $(= \bigoplus E_i)$-module of rank r, all H_i are r-dimensional vector spaces over E_i.

We have

$$\operatorname{End} H \supset E(G) = \operatorname{End}_E H = \bigoplus \operatorname{End}_{E_i} H_i.$$

Here $\operatorname{End}_{E_i} H_i \subset E(G)$ kills H_j (while $i \neq j$) and acts on H_i in the obvious way. Clearly, all H_i are simple $E(G)$-submodules of H, and the image of $E(G)$ in the endomorphism algebra $\operatorname{End} H_i$ of the k-vector space E_i coincides with $\operatorname{End}_{E_i} H_i$. Reminding the definition of $E(G)$, we obtain the following statement.

Let G_i be the image of G in the automorphism group $\operatorname{Aut} H_i$ of the k-vector space H_i. Then

$$G_i \subset \operatorname{End}_{E_i} H_i.$$

Let

$$E_i(G_i) \subset \operatorname{End} H_i$$

be the k-subalgebra generated by E_i and G_i. Then

$$\operatorname{End}_{E_i} H_i = E_i(G_i).$$

Indeed, $E_i(G_i)$ coincides with the image of $E(G)$ in $\operatorname{End} H_i$. But this image is equal to $\operatorname{End}_{E_i} H_i$. Since

$$r = \dim_{E_i} H_i > 1,$$

the k-algebra

$$E_i(G_i) = \operatorname{End}_{E_i} H_i$$

is noncommutative. This implies that all $G_i = G_{H_i}$ are noncommutative groups. Notice that H_i constitute the list of all simple $E(G)$-submodules of H; in particular, each nonzero $E(G)$-submodule of H contains H_i (for some i).

LEMMA 1. *If W is a nonzero $E(G)$-submodule of H, then the group G_W is noncommutative.*

Proof of Lemma 1. If $H_i \subset W$ and G_W is commutative, then $G_{H_i} = G_i$ is also commutative, because G_{H_i} is isomorphic to the quotient group of G_W. Contradiction.

LEMMA 2. *Let W be a nonzero simple $k(G)$-submodule of H. Then G_W is a noncommutative group.*

Proof of Lemma 2. Let us assume that G_W is commutative. For each $e \in E \subset \operatorname{End} H$ let us put

$$eW = \{ ex \mid x \in W \} \subset H.$$

Clearly, eW is a $k(G)$-submodule of H and the map $x \mapsto ex$ is a surjective homomorphism $W \to eW$ of $k(G)$-modules. Since W is simple, either $eW = \{0\}$ or eW is the simple $k(G)$-module isomorphic to W. Let us put

$$V = \left\{ \sum_{e \in E} eW \right\} \subset H.$$

Clearly, V is a semisimple $k(G)$-submodule of H, isomorphic to the direct sum of some copies of W. In particular, G_V is isomorphic to G_W and, consequently, is a commutative group.

On the other hand, V is a nonzero E-invariant subspace of H, by the definition. This implies that V is a nonzero $E(G)$-submodule of H and, by Lemma 1, G_V is noncommutative. Contradiction.

End of the proof of Theorem 4 (Statement 2). Let W be a nonzero G-invariant vector k-subspace of H. Then W is the $k(G)$-module and there exists a simple

$k(G)$-submodule T of W. By Lemma 2, G_T is a noncommutative group. Since G_T is isomorphic to the quotient of G_W, the group G_W is also noncommutative.

§2. Maximal commutative subalgebras of semisimple algebras. Let T be a division algebra which is a finite-dimensional $\mathbb{Q}$-algebra and L an order in T. Let us choose a maximal commutative subfield F in T. Clearly, F coincides with its centralizer in T, i.e., if $x \in T$ commutes with F, then $x \in F$. Since F is the finite-dimensional $\mathbb{Q}$-algebra, it is a number field. Let us put

$$F_L = F \cap L \subset L.$$

The ring F_L is an order in F and the natural map $F_L \otimes \mathbb{Q} \to F$ is an isomorphism. Clearly, F_L coincides with its centralizer in L, i.e., if $x \in L$ commutes with F_L, then $x \in F_L$. In other words, F_L is a maximal commutative subring of L.

PROPOSITION 1. *For all primes p, $F \otimes_{\mathbb{Q}} \mathbb{Q}_p = F_L \otimes \mathbb{Q}_p$ is a maximal commutative $\mathbb{Q}_p$-subalgebra of $T \otimes_{\mathbb{Q}} \mathbb{Q}_p = L \otimes \mathbb{Q}_p$.*

PROPOSITION 2. *For all but finitely many primes p, F_L/pF_L is a maximal commutative $\mathbb{Z}/P\mathbb{Z}$-subalgebra of L/pL.*

PROPOSITION 3. *For all but finitely many primes p, F_L/pF_L is a semisimple $\mathbb{Z}/p\mathbb{Z}$-algebra.*

Proof of Proposition 3. Let O be the maximal order in the number field F, i.e. the ring of all the algebraic integers in F. Clearly, F_L is a subgroup of finite index, say m, in O. Then if prime p does not divide m and is unramified in F, then $F_L/pF_L = O/pO$ and O/pO is the direct sum of finite fields. Thus, for all but finitely many primes p, $F_L/pF_L = O/pO$ is semisimple.

The proof of Propositions 1 and 2 is based on the following obvious

LEMMA. *Let $u: A \to B$ be a homomorphism of free Abelian groups of finite rank. The homomorphism u induces the following obvious homomorphisms of vector spaces:*

$$u_{\mathbb{Q}}: A \otimes \mathbb{Q} \to B \otimes \mathbb{Q}, \qquad u_p: A \otimes \mathbb{Q}_p \to B \otimes \mathbb{Q}_p,$$

$$u \bmod p: A/pA \to B/pA.$$

The kernels of these homomorphisms enjoy the following properties:
(0) $\mathrm{Ker}(u_{\mathbb{Q}}) = \mathrm{Ker}(u) \otimes \mathbb{Q}$;
(1) for all primes p

$$\mathrm{Ker}(u_p) = \mathrm{Ker}(u) \otimes \mathbb{Q}_p$$

(2) for all but finitely many primes p

$$\mathrm{Ker}(u \bmod p) = \mathrm{Ker}\, u/p\, \mathrm{Ker}\, u.$$

Proof of Propositions 1 *and* 2.   Let us put

$$d = \dim_{\mathbf{Q}} F.$$

Clearly, d is equal to rank of the free abelian group F_L. Let us choose a basis $f_1, \ldots, f_d$ of F_L. Clearly, $f_1, \ldots, f_d$ is a basis of the $\mathbf{Q}$-vector space F and of the $\mathbf{Q}_p$-vector space

$$F_L \otimes \mathbf{Q}_p = F \otimes_{\mathbf{Q}} \mathbf{Q}_p.$$

The set

$$f_1 \bmod p, \ldots, f_d \bmod p \in F_L/pF_L$$

is a basis of the $\mathbf{Z}/p\mathbf{Z}$-vector space F_L/pF_L.

Let us apply Lemma to $A = L$, $B = L^d = L \oplus \cdots \oplus L$ (d times) and to u: $L \to L^d$, $x \mapsto ([x, f_1], \ldots, [x, f_d])$. Here $[\ ,\ ]$ is the commutator in L. Then, by the definition of u, $\mathrm{Ker}(u)$ is the centralizer of F_L in L, $\mathrm{Ker}(u_{\mathbf{Q}})$ is the centralizer of F in T, $\mathrm{Ker}(u_p)$ is the centralizer of

$$F_L \otimes \mathbf{Q}_p = F \underset{\mathbf{Q}}{\bigotimes} \mathbf{Q}_p$$

in

$$L \otimes \mathbf{Q}_p = T \underset{\mathbf{Q}}{\bigotimes} \mathbf{Q}_p$$

and $\mathrm{Ker}(u \bmod p)$ is the centralizer of F_L/pF_L in L/pL. Applying Lemma results in the centralizer of $F_L \otimes \mathbf{Q}_p$ in $L \otimes \mathbf{Q}_p$ coinciding with $F_L \otimes \mathbf{Q}_p$ for all primes p and the centralizer of F_L/pF_L in L/pL coinciding with F_L/pF_L for all but finitely many primes p.

§3. Proof of Theorems 1 and 2.   Let X be a simple Abelian variety over K which is not of CM-type and $L = \mathrm{End}\, X$. Then

$$T = \mathrm{End} \circ X = L \otimes \mathbf{Q}$$

is a division algebra. Clearly, L is an order in the finite-dimensional $\mathbf{Q}$-algebra T. Let us choose a maximal commutative field F in the division algebra T. The degree $[F: \mathbf{Q}]$ divides $2 \dim X$ [7]. Let us put

$$r = 2 \dim X/[F: \mathbf{Q}].$$

Since X is not of CM-type over K,

$$r > 1.$$

Let us put

$$F_L = F \cap L \subset L = \text{End } X.$$

Let G be the Galois group of K. Recall (see Introduction) that the following statements are valid.

Statement 1.   For all primes p the G-module $V_p(X)$ is semisimple and

$$\text{End}_G V_p(X) = \text{End } X \otimes \mathbb{Q}_p = L \otimes \mathbb{Q}_p.$$

Statement 2.   For all but finitely many primes p the G-module X_p is semisimple and

$$\text{End}_G X_p = \text{End } X \otimes \mathbb{Z}/p\mathbb{Z} = L/pL.$$

Notice that the results of K. A. Ribet ([9], Th. 2.1.1., Prop. 2.2.1) imply the following statements.

Statement 3.   For all primes p, $V_p(X)$ is the free $F_L \otimes \mathbb{Q}_p$-module of rank $r > 1$.

Statement 4.   For all but finitely many primes p, X_p is the free F_L/pF_L-module of rank $r > 1$.

Proof of Theorem 2.   One only has to apply Theorem 4 of Section 1 to $k = \mathbb{Q}_p$, $H = V_p(X)$, the Galois group G of K, $D = L \otimes \mathbb{Q}_p$ and $E = F_L \otimes \mathbb{Q}_p$. The validity of the assumption of Theorem 4 is guaranteed by Proposition 1 of Section 2 (with the same notations) and Statements 1 and 3.

Proof of Theorem 3.   For all but finitely many primes p the G-module X_p is semisimple and $\text{End}_G X_p = L/pL$ (Statement 2), F_L/pF_L is semisimple (Proposition 3 of Section 2) and coincides with its centralizer in L/pL (Proposition 2 of Section 2) and X_p is the free F_L/pF_L-module of rank $r > 1$ (Statement 4). One only has to apply Theorem 4 of Section 1 to $k = \mathbb{Z}/p\mathbb{Z}$, $H = X_p$, the Galois group G, $D = L/pL$ and $E = F_L/pF_L$.

Remark.   Using a technique similar to that described in [1], F. A. Bogomolov (Séminaire Délange-Pisot-Poitou, mai 1982, Paris) proved the following statement about the torsion in nonabelian extensions.

THEOREM OF BOGOMOLOV.   *Let $K' \subset \overline{K}$ be an infinite Galois extension of K such that $K' \cap K^{\text{ab}}$ is a finite algebraic extension of K. Then the torsion subgroup of $X(K')$ is finite for all Abelian varieties X over K.*

The rest of this Section contains auxiliary facts for the next one.

THEOREM 5. *Let k be a perfect commutative field, H a finite-dimensional k-vector space,*

$$\rho: G \to \mathrm{Aut}(H)$$

a semisimple representation of the Galois group G of K,

$$D = \mathrm{End}_G H \subset \mathrm{End}\, H$$

a finite-dimensional semisimple k-algebra and

$$R \subset D \subset \mathrm{End}\, H$$

the center of D. Let us split R into the direct sum

$$R = \bigoplus R_i$$

of commutative fields R_i which are finite separable algebraic extensions of k. Let us put

$$D_i = R_i D_i = R_i D_i R_i \subset D \subset \mathrm{End}\, H$$

and

$$H_i = R_i H = D_i H \subset H.$$

Then:

(a) *each D_i is a central simple algebra over R_i; $D = \bigoplus D_i$, $D_i = R_i D_i$ and $R_j D_i = \{0\}$ while $i \neq j$;*

(b) *each H_i is G-invariant and D_i-invariant; $H = \bigoplus H_i$, $H_i = R_i H_i = D_i H_i$ and $R_j H_i = D_j H_i = \{0\}$ while $i \neq j$. In particular, the natural map $D_i \to \mathrm{End}\, H_i$ is an embedding and identifies D_i with the centralizer of G in $\mathrm{End}\, H_i$, i.e.,*

$$D_i = \mathrm{End}_G H_i;$$

(c) *if $i \neq j$, then $\mathrm{Hom}_G(H_i, H_j) = 0$;*

(d) *each H_i is an isotypical G-module, i.e., it is isomorphic to the direct sum of some copies of a simple G-module;*

(e) *let W be a nonzero simple G-submodule of H. Then there exists an H_i such that $W \subset H_i$ and the G-module H_i is isomorphic to the direct sum of some copies of W*

(f) *let G' be an invariant subgroup of G. If the subspace*

$$H^{G'} = \{ x \in H \,|\, gx = x \text{ for } g \in G' \}$$

of G'-invariants does not vanish, then it contains some H_i

(g) *let us assume that* char $k \neq 2$. *Let M be a one-dimensional k-vector space and*

$$f \colon H \times H \to M$$

a nondegenerate skew-symmetric bilinear form such that

$$f(ax, y) = f(x, ay) \quad \text{for } a \in R \text{ and } x, y \in H.$$

Then H_i is orthogonal to H_j with respect to f while $i \neq j$. In particular, the restriction of f to H_i is nondegenerate for each H_i, and for each $m \in M$ there exist $x, y \in H_i$ with $f(x, y) = m$.

(h) *let us assume that M is a G-module and f is G-equivariant, i.e.,*

$$f(gx, gy) = gf(x, y) \quad \text{for } g \in G \text{ and } x, y \in H.$$

If the subspace $H^{G'}$ of G'-invariants does not vanish, then G' acts trivially on M, i.e.,

$$gm = m \quad \text{for } g \in G', m \in M.$$

The proofs of Theorem 5(a) and (b) are obvious.

Proof of (c). Let us extend a nonzero $u \in \operatorname{Hom}_G(H_i, H_j)$ to a homomorphism

$$u' \in \operatorname{Hom}_G(H, H_j) \subset \operatorname{Hom}_G(H, H) = D \subset \operatorname{End} H$$

by the formula

$$u' H_l = \{0\} \quad \text{while } l \neq i.$$

Since $i \neq j$, $u' \in D$ does not commute with $R_i \subset R$. But R is the center of D. Contradiction.

(d) By the assumption of the Theorem, H is a semisimple G-module. Since H_i is a G-submodule of H, it is also semisimple. By (a) and (b), the centralizer

$$D_i = \operatorname{End}_G H_i$$

is a simple central algebra. This implies that H_i is an isotypical G-module.

(e) Easy corollary of (c) and (d).

(f) Since G' is invariant, $H^{G'}$ is a G-submodule of H. Let W be a nonzero simple G-submodule of $H^{G'}$. Then, by (e), there exists an H_i such that the G-module H_i is isomorphic to the direct sum of some copies of W. Since G' acts trivially on W, it also acts trivially on H_i, i.e., $H_i \subset H^{G'}$.

(g) One only has to remark that by (a)

$$H_i = R_i H_i, \qquad R_i H_j = \{0\} \quad \text{while } i \neq j.$$

Indeed,

$$f(H_i, H_j) = f(R_i H_i, H_j) = f(H_i, R_i H_j) = f(H_i, \{0\}) = 0.$$

(h) By (f), $H^{G'}$ contains some H_i. By (g), for each $m \in M$ there exist $x, y \in H_i$ with

$$f(x, y) = m.$$

By G-equivariance of f, we have

$$gm = f(gx, gy) \quad \text{for all } g \in G.$$

Since

$$x, y \in H_i \subset H^{G'},$$

$gx = x$, $gy = y$ for each $g \in G' \subset G$. Hence, for all $g \in G'$

$$gm = f(gx, gy) = f(x, y) = m.$$

§4. Torsion in noncyclotomic extensions. Throughout this section G is the Galois group of K, $K' \subset \overline{K}$ is a finite or an infinite Galois extension of K, $G' = \mathrm{Gal}(\overline{K}/K')$ is the Galois group of K', μ_n is the group of n-th roots of unity in $\overline{K}$, X is an Abelian variety over K, $L = \mathrm{End}\, X$, C is the center of $\mathrm{End} \circ X$ and

$$C_L = C \cap L \subset L = \mathrm{End}\, X$$

is the center of $\mathrm{End}\, X = L$.

Results of this section are based on ideas of [13].

THEOREM 6.   *Let us assume that C is the direct sum of totally real number fields and K' contains only a finite number of roots of unity. Then the torsion subgroup* TORS $X(K')$ *is finite.*

Theorem 6 is a corollary of the conjunction of two following theorems.

THEOREM 7.   *Let p be a prime such that the p-primary part of* TORS $X(K')$ *is infinite. If C is the direct sum of totally real number fields, then K' contains K_p, i.e., K' contains all p-power roots of unity.*

Remark.   The p-primary part of TORS $X(K')$ is infinite if and only if the subspace of G'-invariants of $V_p(X)$ does not vanish.

THEOREM 8. *Let us choose a polarization*

$$\lambda\colon X \to \hat{X}$$

on X such that λ is defined over K. Here $\hat{X}$ is the dual of X. Let S be the set of primes defined as follows. A prime p does not lie in S if and only if it enjoys the following properties:
(a) *the G-module X_p is semisimple and*

$$\operatorname{End}_G X_p = L/pL;$$

(b) $C_L \otimes \mathbb{Z}/p\mathbb{Z}$ *is the center of L/pL;*
(c) p *is odd and prime to* $\deg \lambda$.
Then S is finite.

Let us assume that C is the direct sum of totally real number fields and the p-primary part of TORS $X(K')$ *does not vanish. If the prime p does not lie in S, then K' contains μ_p.*

Remark. Clearly, the following conditions are equivalent:
(a) the p-primary part of TORS $X(K')$ does not vanish;
(b) $X(K')$ contains a torsion point of order p;
(c) the subspace of G'-invariants of X_p does not vanish.

Remark. The finiteness of S is a corollary of Statement 2 of Section 3. Indeed, L is an order in the finite-dimensional (semisimple) $\mathbb{Q}$-algebra $\operatorname{End} \circ X$ and C_L is the center of L. In this case it is well known that $C_L \otimes \mathbb{Z}/p\mathbb{Z}$ is the center of L/pL for all but finitely many primes p. (One may easily check this assertion by applying the technique of Section 2.)

Notice that $C \otimes \mathbb{Q}_p$ is the center of

$$L \otimes \mathbb{Q}_p = \operatorname{End} \circ X \otimes_{\mathbb{Q}} \mathbb{Q}_p$$

for all primes p.

Remark. The set S does not depend on the choice of K'.

In order to prove Theorem 7, we shall apply Theorem 5 to $k = \mathbb{Q}_p$, $H = V_p(X)$, $D = \operatorname{End} X \otimes \mathbb{Q}_p$ and $R = C_L \otimes \mathbb{Q}_p$. In order to prove Theorem 8, we shall apply Theorem 5 to $k = \mathbb{Z}/p\mathbb{Z}$, $H = X_p$, $D = \operatorname{End} X \otimes \mathbb{Z}/p\mathbb{Z}$ and $R = C_L \otimes \mathbb{Z}/p\mathbb{Z}$. But in both cases we need the definition of M and f. We start with the definition of M. In the case of Theorem 8 let us put

$$M = \mu_p.$$

Clearly, μ_p is a one-dimensional $\mathbb{Z}/p\mathbb{Z}$-vector space provided with the structure of G-module; G' acts trivially on μ_p if and only if K' contains μ_p. In order to define M in the case of Theorem 7, let us consider the projective limit

$$\mathbb{Z}_p(1) = \varprojlim \mu_{p^i}$$

where the transition map $\mu_{p^{i+1}} \to \mu_{p^i}$ is raising to the p-th power. Clearly, $\mathbb{Z}_p(1)$ is a free $\mathbb{Z}_p$-module of rank 1 provided with the natural structure of the G-module and the natural isomorphisms

$$\mathbb{Z}_p(1) \otimes \mathbb{Z}/p^i\mathbb{Z} \xrightarrow{\sim} \mu_{p^i}$$

are isomorphisms of the G-modules. This implies that G' acts trivially on $\mathbb{Z}_p(1)$ if and only if K' contains μ_{p^i} for all i, i.e., K' contains K_p. Let us put

$$\mathbb{Q}_p(1) = \mathbb{Z}_p(1) \underset{\mathbb{Z}_p}{\otimes} \mathbb{Q}_p$$

and extend the action of G by $\mathbb{Q}_p$-linearity on $\mathbb{Q}_p(1)$. Clearly, $\mathbb{Q}_p(1)$ is a one-dimensional $\mathbb{Q}_p$-vector space, and G' acts trivially on $\mathbb{Q}_p(1)$ if and only if K' contains K_p. We put $M = \mathbb{Q}_p(1)$.

In order to define f, let us recall the properties of Weil pairing [7]

$$e_n \colon X_n \times \hat{X}_n \to \mu_n.$$

It is a nondegenerate G-equivariant bilinear pairing. The polarization λ defines the bilinear form

$$f_n \colon X_n \times X_n \to \mu_n, \qquad x, y \mapsto e_n(\lambda x, y).$$

The form f_n is G-equivariant and skew-symmetric; it is a nondegenerate if n is prime to deg λ, because in this case λ induces an isomorphism $X_n \xrightarrow{\sim} \hat{X}_n$. If C is the direct sum of totally real number fields, then the Rosati involution acts trivially on C [7] and

$$f_n(ax, y) = f_n(x, ay) \quad \text{for } a \in C_L; \ x, y \in X_n.$$

In the case of Theorem 8 we put $f = f_p$.

In order to define f in the case of Theorem 7, notice that the forms f_n are compatible for different n [7]. In particular, there exists a nondegenerate G-equivariant bilinear form (the so-called Riemann form [7])

$$f_{(p)} \colon T_p(X) \times T_p(X) \to \mathbb{Z}_p(1)$$

which is the "projective limit" of the forms f_{p^i}. If C is the direct sum of totally real number fields, then

$$f_{(p)}(ax, y) = f_{(p)}(x, ay) \quad \text{for } a \in C_L; \ x, y \in T_p(X).$$

Extending $f_{(p)}$ by $\mathbb{Q}_p$-linearity on $V_p(X)$, we obtain [7] the nondegenerate G-equivariant bilinear form

$$f_{(p)}: V_p(X) \times V_p(X) \to \mathbb{Q}_p(1).$$

Clearly,

$$f_{(p)}(ax, y) = f_{(p)}(x, ay) \quad \text{for } x, y \in V_p(X), \ a \in C_L \otimes \mathbb{Q}_p = C \underset{\mathbb{Q}}{\otimes} \mathbb{Q}_p$$

if C is the direct sum of totally real number fields. We put $f = f_{(p)}$.

Now, applying Theorem 5, we obtain $M = M^{G'}$ in both cases. As has been explained above, this means that K' contains K_p in the case of Theorem 7 and K' contains μ_p in the case of Theorem 8.

REFERENCES

1. F. A. Bogomolov, *Sur l'algébraicité des représentations l-adiques*, C. R. Acad. Sci. Paris **290** (1980), 701–704.
2. G. Faltings, *Endlichkeitssätze für Abelsche Varietäten über Zahlkörpern*, Inventiones Math. **73** (1983), 349–366.
3. H. Imai, *A remark on the rational points of abelian varieties with values in cyclotomic $\mathbb{Z}_p$ extensions*, Proc. Japan Acad. **51** (1975), 12–16.
4. Yu. I. Manin, *The Mordell-Weil theorem*. Appendix to [7].
5. ______, *Cyclotomic fields and modular curves*, Russian Math. Surveys **26** (1971), 7–78.
6. B. Mazur, *Rational points of abelian varieties with values in towers of number fields*, Inventiones Math. **18** (1972), 183–266.
7. D. Mumford, *Abelian Varieties*, Oxford University Press, 1974.
8. K. A. Ribet, *Torsion points of abelian varieties in cyclotomic extensions*, pp. 315–319. Appendix to N. M. Katz and S. Lang, *Finiteness theorems in geometric classified theory*, L'Enseignement Math. **27** (1981), 285–319.
9. ______, *Galois action on division points of abelian varieties with many real multiplications*, Amer. J. of Math. **98** (1976), 751–804.
10. J.-P. Serre and J. Tate, *Good reduction of Abelian varieties*, Ann. of Math. **88** (1968), 492–517.
11. G. Shimura, *Introduction to the arithmetic theory of automorphic functions*, Publ. Math. Soc. Japan **11** (1971), Tokyo–Princeton.
12. Yu. G. Zarhin, *A finiteness theorem for unpolarized Abelian varieties over number fields with prescribed places of bad reduction*, Inventiones Math. **79** (1985), 309–321.
13. ______, *Torsion of Abelian varieties in finite characteristic*, Math. Notes **22** (1978), 493–498.
14. Yu. G. Zarhin and A. N. Parshin, *Finiteness problems in Diophantine geometry*, pp. 369–438. Appendix to the Russian translation of S. Lang, *Fundamentals of Diophantine Geometry*, Mir, Moscow, 1986.
15. J.-P. Serre, *Résumé des cours de 1984–1985*, Annuaire du Collège de France, Paris, 1985.
16. ______, *Résumé des cours de 1985–1986*, Annuaire du Collège de France, Paris, 1986.

Research Computing Center of the USSR Academy of Sciences Pushchino Moscow Region, 142292 USSR

CONSTRUCTIVE HIGH-DIMENSIONAL
SPHERE PACKINGS

S. N. LITSYN AND M. A. TSFASMAN

To Yu. I. Manin

We use some ideas and methods of algebraic geometry, coding theory, and complexity theory to establish new constructive asymptotic existence bounds for the density of sphere packings (lattice and nonlattice) in Euclidean spaces. The main tools are: on one hand the beautiful construction of A. Bos, J. H. Conway and N. J. A. Sloane, on the other error-correcting codes obtained from algebraic curves. The results of this paper were announced in [1].

CONTENTS

§1. Preliminaries.   How should we place equal nonintersecting open spheres in $\mathbb{R}^N$ so as to obtain the densest possible packing? A sphere packing is a configuration of nonintersecting equal open spheres in $\mathbb{R}^N$. Let d be the diameter of the spheres; then the distance between any two sphere centers is at least d. Thus a *packing* is a set of points P in $\mathbb{R}^N$ such that the minimum distance between any two is at least d. If this set is an additive subgroup of $\mathbb{R}^N$ it is called a *lattice* or a *lattice packing*. For any packing P the *density* $\Delta(P)$ is defined as the fraction of space covered by spheres. More precisely, let $K_u = \{x = (x_0, \ldots, x_n) \mid |x_i| \leqslant u\}$ be a large cube centered at the origin, and $S = S(P)$ the set of points inside the spheres of the packing P; then

$$\Delta(P) = \limsup_{u \to \infty} \mu(S \cap K_u)/\mu(K_u),$$

where μ is the usual measure in $\mathbb{R}^N$ (cf. [2]). Of course $\Delta(P)$ does not depend on the fact that K_u is a cube.

Received November 10, 1986.

Let L be a lattice. Any choice of basis defines a map $L: \mathbb{Z}^N \to \mathbb{R}^N$ which, by abuse of notation, we also call L. The matrix of L is called a *generating matrix* of the lattice. For the diameter of spheres centered in L one can take $d(L) = \min_{v \in L\setminus\{0\}} |v|$. It is easy to check that

$$\Delta(L) = d(L)^N V_N / 2^N |\det L|, \tag{1}$$

where $V_N = \pi^{N/2}/\Gamma(N/2 + 1)$ is the volume of the unit sphere (in fact, the fundamental region F intersects with the spheres centered at its vertices in such a way that taken into the origin the pieces of spheres in F make up exactly one sphere; and $\mu(F) = |\det L|$).

For any packing $P \subset \mathbb{R}^N$ the ratio

$$\delta(P) = \Delta(P)/V_N \tag{2}$$

is called its *center density*.

Let $v(P) = \log \delta(P)$ (in this paper *all logarithms without subscript are binary*), and let

$$\lambda(P) = -(\log \Delta(P))/N \tag{3}$$

be the *density exponent*. Of course

$$\Delta(P) = 2^{-\lambda(P)N}. \tag{4}$$

There is a simple formula linking λ and v:

$$\lambda(P) = -v(P)/N - (\log \pi)/2 + (\log(\Gamma(N/2 + 1)))/N. \tag{5}$$

In this paper we are mostly interested in asymptotic problems (as $N \to \infty$). For these problems the density exponent is the most appropriate parameter of a packing. We give some additional definitions.

Let

$$\lambda(N) = \inf_{P \subset \mathbb{R}^N} \lambda(P) \tag{6}$$

and let

$$\tilde{\lambda} = \liminf_{N \to \infty} \lambda(N). \tag{7}$$

A *family of packings* $\{P_N\}$ is a set of packings $P_N \subset \mathbb{R}^N$, where N is some infinite subsequence of $\mathbb{N}$. Let

$$\lambda(\{P_N\}) = \liminf \lambda(P_N). \tag{8}$$

It is easy to check that

$$\tilde{\lambda} = \inf_{\{P_N\}} \lambda(P_N). \tag{9}$$

If we consider only lattice packings L we can give all the corresponding definitions:

$$\lambda_l(N) = \inf_{L \subset R^N} \lambda(L), \tag{6'}$$

$$\tilde{\lambda}_l = \liminf_{N \to \infty} \lambda_l(N). \tag{7'}$$

As above

$$\tilde{\lambda}_l = \inf_{\{L_N\}} \lambda(\{L_N\}). \tag{9'}$$

Analogous definitions can be given for $\nu(\{P_N\})$, $\nu(N)$, $\tilde{\nu}$, $\tilde{\nu}_l$, etc. For any family $\{P_N\}$, Stirling's formula applied to (5) gives

$$\lambda(\{P_N\}) = -\nu(\{P_N\})/N + \tfrac{1}{2}\log N + o(1). \tag{10}$$

We list some good lattices L in small dimensions and their parameters [3].

TABLE 1

N	L	$\nu(L)$	$\lambda(L)$
1	$\mathbb{Z}$	-1	0
2	A_2	-1.792	0.070
2	$\mathbb{Z}^2$	-2	0.174
4	D_4	-3	0.174
8	E_8	-4	0.247
16	Λ_{16}	-4	0.380
24	Λ_{24}	0	0.376
48	P_{48_q}	14.039	0.528

The lower bound for $\tilde{\lambda}$ was established by G. A. Kabatiansky and V. I. Levenstein [4], the upper (existence) bound by H. Minkowski [5]. In our terms

PROPOSITION 1. $0.599 \leqslant \tilde{\lambda} \leqslant \tilde{\lambda}_l \leqslant 1$.

Unfortunately, the proof of the Minkowski bound is purely an existence proof. The problem of how to construct packings explicitly is discussed in §3 below.

§2. Error-correcting codes.   There is a remarkable connection between packings and codes. Let us recall several important facts about codes; for additional information we refer to [6].

Fix a finite field $\mathbb{F}_q$, $q = p^b$. A *q-ary linear code* is a subspace $C \subseteq \mathbb{F}_q^n$; n is called the *length* of C, $k = \dim C$ its *dimension*, and $R = k/n$ its *rate*. The *minimal distance d* is the minimum *Hamming weight* (the number of nonzero coordinates) of a nonzero vector of C (or, which is the same, the number of coordinates in which any two elements of C differ); $\delta = d/n$ is called the *relative minimal distance*. We say that C is a $[n, k, d]_q$-code. A choice of basis in C defines a linear map $C \colon \mathbb{F}_q^k \to \mathbb{F}_q^n$; by abuse of notation we also denote it by C, and its matrix is called a *generating matrix* of the code. A set of codes $C_1 \subset C_2 \subset \cdots \subset C_m \subseteq \mathbb{F}_q^n$ is called a *nested family*.

Consider the set of pairs (δ, R) for all the linear q-ary codes. The limit points of this set form a domain U on the plane (δ, R). Yu. I. Manin [7] showed that there exists a continuous decreasing function $\alpha_q(\delta)$ such that $U = \{(\delta, R)|1 \geqslant \delta \geqslant 0,\ R \geqslant 0,\ R \leqslant \alpha_q(\delta)\}$. It is known that $\alpha_q(0) = 1$ and $\alpha_q(\delta) = 0$ for $(q - 1)/q \leqslant \delta \leqslant 1$. Any function $\beta_q(\delta)$ defined for $0 \leqslant \delta \leqslant (q - 1)/q$ is called an *upper bound* iff $\beta_q(\delta) \geqslant \alpha_q(\delta)$, and a *lower (existence) bound* iff $\beta_q(\delta) \leqslant \alpha_q(\delta)$.

Let $H_q(x) = x \log_q(q - 1) - x \log_q x - (1 - x)\log_q(1 - x)$ be the q-ary entropy function. Then the lower *Varshamov-Gilbert bound* is given by

$$\beta_q(\delta) = 1 - H_q(\delta). \tag{11}$$

This lower bound can be sharpened (for $q = p^{2m} \geqslant 49$) using algebraic geometry [8]. Here is a simple construction due to V. D. Goppa [9].

Let X be a smooth projective absolutely irreducible curve over $\mathbb{F}_q$ of genus g; let D be a divisor of degree $a \geqslant g - 1$, and $\mathscr{L}(D) = \{f \in \mathbb{F}_q(X)|(f) + D \geqslant 0\}$ the associated function space (cf. [10], Ch. 4). Suppose $P_1, \ldots, P_n$ are $\mathbb{F}_q$-points of X, $P_i \notin \operatorname{Supp} D$. Then there is a map $\varphi \colon \mathscr{L}(D) \to \mathbb{F}_q^n$, given by $\varphi(f) = (f(P_1), \ldots, f(P_n))$. Whenever $a < n$ this map is injective and its image is a linear q-ary codes of length n, dimension $k \geqslant a - g + 1$ (by the Riemann-Roch theorem), and distance $d \geqslant n - a$ (since the number of zeros of a function f cannot exceed the number of poles).

Now let $D = aP_0$, where P_0 is a $\mathbb{F}_q$-point of X, $P_0 \neq P_i$ for $i = 1, \ldots, n$. Then we get a nested family of codes C_a for $a = n - 1, n - 2, \ldots, g - 1$.

In the particular case $g = 0$ (i.e., X is a projective line) the algebraic-geometric codes were well known before. Such codes are called *Reed-Solomon codes*. Their parameters are $n = q$, $d = n - k - 1$, and they are nested.

Algebraic-geometric codes have good parameters when the ratio of the number of $\mathbb{F}_q$-points on the curve to its genus is high enough. The Vlăduţ-Drinfeld theorem [11] shows that asymptotically this ratio cannot exceed $\sqrt{q} - 1$. For $q = p^{2l}$ there exist families of modular curves (classical, Shimura, or Drinfeld) over $\mathbb{F}_q$ with such an asymptotic (cf. [12], [8], [13]); modular curves have many $\mathbb{F}_q$-points that correspond to supersingular elliptic curves or Drinfeld modules.

Modular curves provide the following lower bound [8] for $q = p^{2l}$:

$$\beta_q(\delta) = 1 - \left(\sqrt{q} - 1\right)^{-1} - \delta, \tag{12}$$

for $q \geqslant 49$ this bound (on the same segment $\delta_1 < \delta < \delta_2$) improves on the Varshamov-Gilbert bound.

We also need the following method of combining codes (concatenation). For a linear $[n, k, d]$-code C_0 over $\mathbb{F}_q$, and for a linear $[N, K, D]$-code C_1 over $\mathbb{F}_r$, where $r = q^k$, there exists a $[nN, kK, dD]$-code C over $\mathbb{F}_q$ constructed as follows:

$$\mathbb{F}_q^{kK} \simeq \mathbb{F}_r^{K} \xrightarrow{C_1} \mathbb{F}_r^{N} \simeq \left(\mathbb{F}_q^{k}\right)^{N} \xrightarrow{(C_0,\dots,C_0)} \mathbb{F}_q^{nN}. \tag{13}$$

C_0 is called the *inner* code, C_1 the *outer* code, C the *concatenated* code.

If there exists a $[n, k, d]_q$-code C (and k is even if q is an odd power of a prime), then we can concatenate this code with algebraic-geometric codes over $\mathbb{F}_{q^k}$, getting the bound (cf. [14]):

$$\beta_q(\delta) = \left(1 - \left(q^{k/2} - 1\right)^{-1}\right)k/n - \delta k/d. \tag{14}$$

When we make C vary we get an infinite polygon that "continues" the bound (12) to the points $(\delta, R) = (0, 1)$ and $((q - 1)/q, 0)$.

For example, if we take $C_0 = \mathbb{F}_q^n$, we get a $[nN, nK, D]$-code. Applying this to the lower bound (12) we get

$$\beta_q(\delta) = 1 - \left(q^{n/2} - 1\right)^{-1} - \delta/n \tag{15}$$

for any q and n, provided $q^{n/2}$ is an integer.

§3. Complexity.   When can we state that we can *construct* an infinite family of codes or sphere packings?

We know that there exist families of lattices $\{L_N\}$ on the Minkowski bound (i.e., with $N \to \infty$ and $\lambda(\{L_N\}) = 1$, cf. Proposition 1) and families of codes on the Varshamov-Gilbert bound (i.e., with parameters asymptotically given by (11)). More than that, it can be correctly stated and proved that almost all lattices lie on the Minkowski bound and almost all linear codes lie on the Varshamov-Gilbert bound. However, those codes and lattices are rather implicit and we should like to give explicit constructions of good families of codes and lattices even if their parameters are slightly worse.

There can be several different notions of what is meant by an explicit construction. As yet only one of them is mathematically precise (cf. [20]).

We first discuss the case of lattices and linear codes. Any lattice L (or linear code C) is determined by its basis. Therefore the minimal possible requirement to any explicit construction is the existence of an algorithm A giving some basis of L (resp. of C). Such an algorithm will be called a *generating algorithm*. Since we are interested in infinite families of packings and codes, for each family $\{L_N\}$ we must give a family of generating algorithms $A(N)$, and one can ask how these algorithms behave when $N \to \infty$.

Suppose given an infinite family of algorithms $B(N)$ for some sequence of integers $N \to \infty$. Then it is possible to define the *complexity* of this family. Suppose that $B(N)$ is described in terms of recursive functions, and let $f(N)$ be the length of recursion for $B(N)$. This function depends on the recursive description of $B(N)$, on the definition of the length of recursion, and on the fact that we have chosen recursive functions among several possible models of computations. Fortunately, any other sensible definition of complexity $g(N)$ differs from $f(N)$ polynomially, i.e., for some polynomial P we have $g(N) \leqslant f(P(N))$ and vice versa. Thus we get correct definitions of *polynomial* families of algorithms (for which $f(N)$ is bounded by some polynomial Q in N), *exponential* families (for which $f(N)$ is bounded by $\exp(Q(N))$ for some polynomial Q), doubly exponential (bounded by $\exp(\exp(Q(N))))$, and so on (see, for example, [14]).

We therefore define *constructive* families of lattices and linear codes as families for which the generating algorithm $A(N)$ is polynomial in N. This definition can be adopted to suit also nonlattice packings (and nonlinear codes); the difference is just that the generating algorithm produces not a basis but a polynomial algorithm enumerating one by one the coordinates of all sphere centers (resp., all vectors of the code).

There can be other definitions of constructive families. One can, for example, define constructivity as absence of direct search. There is another definition that, although imprecise, is rather dear to us: constructive packings and codes are those arising from some explicitly given and natural mathematical objects (such as algebraic-geometric codes or the number-theoretic lattices described in §7).

We now discuss lower bounds for parameters of codes. Let $C \subseteq \mathbb{F}_q^n$ be a linear $[n, k, d]_q$-code. The space $\mathbb{F}_q^n$ is a union of cosets of C. Any vector of minimal Hamming weight in the given coset is called a *coset leader*. If there exists a coset such that the weight of its leader is greater than d, this leader can be added to a basis of C, thus obtaining a $[n, k + 1, d]_q$-code C'. Repeating this procedure if possible, we get C'', etc. This method, called the *Varshamov procedure*, allows us to enlarge any code to a code C_0 with the same δ and R asymptotically (for a family $C = C(n)$, $n \to \infty$) on the Varshamov-Gilbert bound. Having obtained a code C_0 we can repeat this procedure for $d' < d$ starting from the code C_0, and get a set of nested codes $C_0 \subset C_1 \subset \cdots \subset C_a$, such that asymptotically all C_i lie on the Varshamov-Gilbert bound.

It is obvious that this algorithm is exponential in n. However there exist algebraic-geometric codes arising from Drinfeld modular curves that are polynomial; their parameters are asymptotically on the bound (12), cf. [13].

Therefore the bounds (14) and (15) are also polynomial.

Let us imitate the formulae (9) and (9′) and define

$$\tilde{\lambda}^{\mathrm{pol}} = \inf_{\{P_N\} \text{ polynomial}} \lambda(\{P_N\}), \tag{16}$$

$$\tilde{\lambda}_l^{\mathrm{pol}} = \inf_{\{L_N\} \text{ polynomial}} \lambda(\{L_N\}), \tag{16′}$$

$$\tilde{\lambda}^e = \inf_{\{P_N\} \text{ exponential}} \lambda(\{P_N\}), \tag{17}$$

$$\tilde{\lambda}_l^e = \inf_{\{L_N\} \text{ exponential}} \lambda(\{L_N\}). \tag{17′}$$

In §4 we shall recall the construction of packings and lattices from codes that goes back to E. S. Barnes, A. Bos, J. H. Conway, J. Leech and N. J. A. Sloane. In particular, These constructions lead to the estimate [16]:

$$\tilde{\lambda}^{\mathrm{pol}} \leqslant 6. \tag{18}$$

The results for lattices are less impressive; let us mention the existence of polynomial families [17] with

$$\lambda(\{L_N\}) \leqslant \log^* N + \mathcal{O}(I) \tag{19}$$

where $\log^* N = t$ means that t is the smallest integer such that $\underbrace{\log \ldots \log}_{t \text{ times}} N \leqslant 1$. Unfortunately (19) gives us no estimate on $\tilde{\lambda}_l^{\mathrm{pol}}$.

We present an algorithm generating packings on the Minkowski bound. This algorithm is exponential.

THEOREM 1. $\tilde{\lambda}^e \leqslant 1$.

Proof.   Let us construct the following family of packings. For each N consider the cube

$$K = \left\{ x = (x_1, \ldots, x_N) \in \mathbb{R}^N \mid |x_i| \leqslant N^{1/4}/2 \right\}$$

and the larger cube

$$K_1 = \left\{ x \in \mathbb{R}^N \mid |x_i| \leqslant N^{1/4}/2 + 1 \right\}.$$

Fix a grid $\mathscr{L} = (N^{-3/4}\mathbb{Z})^N$. We choose a set $U = \{u_1, \ldots, u_p\} \subset K \cap \mathscr{L}$ such that $|u_i - u_j| \geqslant 2$ for any two points of U and U is maximal with this property. To choose U we introduce a lexicographic order on $K \cap \mathscr{L}$, take the first point u_1, cross out of $K \cap \mathscr{L}$ all the points v such that $|v - u_1| \leqslant 2$, take the first point u_2 of the rest, and so on.

For each $v = (v_1, \ldots, v_N) \in K \cap \mathscr{L}$ such that $|u_1 - v| \leqslant 2$ consider the small cube $k_v = \{x = (x_1, \ldots, x_n)|0 \leqslant x_i - v_i < N^{-3/4}\}$, these cubes do not interest each other and their union lies inside a sphere of radius $2 + N^{-1/4}$ ($N^{-1/4}$ being the length of a diagonal of k_v). Comparing their total volume with the volume of this sphere we conclude that the number of such points v is less than $W = (2 + N^{-1/4})^N \cdot V_N \cdot N^{3N/4}$.

The procedure of choosing $u_1, u_2, \ldots$ will not stop before the number of its steps multiplied by W becomes greater than or equal to N^N (the total number of points in $K \cap \mathscr{L}$). Thus we have chosen at least N^N/W points $u_1, u_2, \ldots$. Place the center of a unit sphere into each of these points, these spheres do not intersect each other, and lie in the larger cube K_1. Fill up the whole space R^N with translates of K_1 containing its spheres.

The density of the packing P_N we obtain is at least

$$\Delta(P_N) \geqslant \left(V_N \cdot N^N/W\right)/\mu(K_1) = \left[\frac{N}{(2 + N^{-1/4})N^{3/4}(N^{1/4} + 2)}\right]^N$$

$$= \left[(2 + N^{-1/4})(1 + 2N^{-1/4})\right]^{-N}.$$

Thus

$$\lambda(P_N) \leqslant \log(2 + N^{-1/4}) + \log(1 + 2N^{-1/4}),$$

and $\lambda(\{P_N\}) \leqslant 1$.

To choose $u_1, u_2, \ldots$ we search through N^N points each time, and their number is less than N^N. Therefore the algorithm is exponential.   Q.E.D.

Remark 1. We do not know of any exponential algorithm generating lattices L_N with $\lambda(\{L_N\}) = 1$. Thus the best estimate for $\tilde{\lambda}_l^e$ we can give (Theorem 3) is greater than 1.

§4. Constructions. We recall here a construction of new packings (or lattices) from families of packings (lattices) and codes. This construction and particular cases of it were discovered and used by J. Leech and N. J. A. Sloane [3], A. Bos [18] (for packings), E. S. Barnes and N. J. A. Sloane [19], and by A. Bos, J. H. Conway, and N. J. A. Sloane [17] (for lattices). It is analogous to the construction of generalized concatenated codes (E. L. Bloch and V. V. Ziablov for linear codes, V. A. Zinoviev for nonlinear).

Let $P \subset \mathbb{R}^m$ be a packing, ν_0 its central density exponent, and $\rho_0 = \min_{x, y \in P, x \neq y} |x - y|^2 \geqslant 1$ the minimal norm. Suppose that P is the union of disjoint subsets: $P = \bigcup_{x \in \mathbb{F}_q^a} P_x$, where $q = 2^b$. For $v = (v_1, \ldots, v_i) \in \mathbb{F}_q^i$ ($i \leqslant a$) let $X_v = \{x = (x_1, \ldots, x_a) \in \mathbb{F}_q^a | (x_1, \ldots, x_i) = v\}$. Let $P_v^{(i)} = \bigcup_{x \in X_v} P_x$, and suppose that for any v the minimal norm of $P_v^{(i)}$ is greater than or equal to ρ_i. Let $C_1, \ldots, C_a \subset \mathbb{F}_q^n$ be a set of $[n, k_i, d_i]_q$-codes with rates $R_i = k_i/n$, such that $d_i \rho_i \geqslant 1$.

This data is enough to construct the following packing:

$$Q = \bigcup_{c_1 \in C_1} \bigcup_{c_2 \in C_2} \cdots \bigcup_{c_a \in C_a} P_{(c_1^{(1)}, c_2^{(1)}, \ldots, c_a^{(1)})} \times \cdots \times P_{(c_1^{(n)}, \ldots, c_a^{(n)})}, \qquad (20)$$

here $c_i = (c_i^{(1)}, \ldots, c_i^{(n)})$. This packing $Q \subset \mathbb{R}^{mn}$ has central density exponent

$$\nu_Q \geqslant n\nu_0 - b\left(\sum_{i=1}^{a} k_i\right). \qquad (21)$$

Now let $P = \Lambda \subset \mathbb{R}^m$ be a lattice with $\rho_0 \geqslant 1$, and let T be a similarity (i.e., an orthogonal transformation multiplied by a constant t) such that $T: \Lambda \to \frac{1}{2}\Lambda$, T^{-1} is an endomorphism of Λ, $2T = F(T^{-1}) \in \mathbb{Z}[T^{-1}]$, and $b = -m \log t$ is an integer. Then there is a sublattice $K \subseteq \Lambda$ of rank b such that $q = |K/(T^{-1}\Lambda \cap K)| = 2^b$. Define $\psi_0: \mathbb{F}_q \to \mathbb{R}^m$ as

$$\mathbb{F}_q \xrightarrow{\sim} \mathbb{F}_2^b \xrightarrow{\alpha} \mathbb{Z}^b \xrightarrow{\beta} K \subset \mathbb{R}^m$$

where

$$\alpha: \mathbb{F}_2 \xrightarrow{\sim} \{0, 1\} \subset \mathbb{Z},$$

β is a choice of basis in K. Let $\psi_i = T^i \psi_0$.

For $x = (x_1, \ldots, x_a) \in \mathbb{F}_q^a$ let

$$P_x = \Lambda + \sum_{i=1}^{a} \psi_i(x_i). \qquad (22)$$

Construction (20) applied to (22) and a family of nested $[n, k_i, d_i]_q$-codes $\mathbb{F}_q^n \supset C_1 \supset \cdots \supset C_a$ such that $d_i t^{-2i} \rho_0 \geqslant 1$ provides a lattice (sic!) with central density exponent given by (21).

We do not prove all these statements here, referring to [18], [19].

Let us sum up.

PROPOSITION 1.   *The above construction applied to a family of lattices $\Lambda_m \subset \mathbb{R}^m$ with density exponent $\lambda_0 = \lambda(\{\Lambda_m\})$ and to a family of sets of*

$$[4^l, k_i, 4^l - i + 1]_{2^m}\text{-codes } C_i \ (i = 1, \ldots, l)$$

gives a family of packings $Q_N \subset \mathbb{R}^N$, $N = m \cdot 4^l$, with

$$\lambda(\{Q_N\}) \leqslant \lambda_0 + \frac{1}{m}\log V_m - \frac{1}{2}\log\frac{2\pi e}{m} + \sum_i (1 - R_i) + o(1). \quad (23)$$

If the codes are nested, $Q_N = L_N$ are lattices.

Proof. Let T be a multiplication by $1/2$, then (21), (5), and Stirling's formula gives us (23).   Q.E.D.

We define a *T-lattice* to be a lattice L having a similarity T with all the above properties and $m = 2b$. We imitate (9'), (16'), and (17') to define $\tilde{\lambda}_T$, $\tilde{\lambda}_T^{\text{pol}}$, $\tilde{\lambda}_T^e$, etc.

PROPOSITION 2.   *The above construction applied to a family of T-lattices $\Lambda_m \subset \mathbb{R}^m$ with density exponent $\lambda_0 = \lambda(\{\Lambda_m\})$ and to a family of sets of $[2^l, k_i, 2^{l-i+1}]_{2^b}$-codes $C_i(i = 1, \dots, l)$ gives a family of packings $Q_N \subset \mathbb{R}^N$, $N = m \cdot 2^l$, with*

$$\lambda(\{Q_N\}) \leqslant \lambda_0 + \frac{1}{m}\log V_m - \frac{1}{2}\log\frac{2\pi e}{m} + \frac{1}{2}\sum_i (1 - R_i) + o(1). \quad (24)$$

If the codes are nested, $Q_N = L_N$ are T-lattices.

Proof. Choose any T with required properties and use (21), (5) and Stirling's formula. The resulted lattice L_N has $(T, \dots, T)$ for its similarity.   Q.E.D.

Remark 2.   All lattices of Table 1 except $\mathbb{Z}$ and A_2 are T-lattices (see [17], [19]).

Remark 3.   It follows from Stirling's formula that the second and the third term in (23) and (24) differ by $o(1)$ whenever $m \to \infty$.

§5. Decrease of complexity.   It is possible to get an improvement in complexity by slightly sacrificing the density.

THEOREM 2.

(a) $$\tilde{\lambda}_l^{\text{pol}} \leqslant \tilde{\lambda}_l^e + 4/3 \quad (25)$$

(b) $$\tilde{\lambda}_l^{\text{pol}} \leqslant \tilde{\lambda}_T^{\text{pol}} \leqslant \tilde{\lambda}_T^e + 1. \quad (26)$$

Proof. For every $\varepsilon > 0$ there exists a family of lattices $\{\Lambda_m \subset \mathbb{R}^m\}$ with $\lambda(\{\Lambda_m\}) \leqslant \tilde{\lambda}_l^e + \varepsilon$ and exponential complexity. Let us apply Proposition 1 to such a family and a family of sets of $[4^{m/2}, 4^{m/2} - 4^{m/2-i+1} + 1, 4^{m/2-i+1}]_{2^m}$ nested Reed-Solomon codes. By (23) we get a family of lattices L_N, $N = m \cdot 2^m$, with

$$\lambda(\{L_N\}) \leqslant \tilde{\lambda}_l^e + \varepsilon + \sum_i (1 - R_i) + o(1). \tag{27}$$

For the given codes $\lim_{l \to \infty} \sum_{i=1}^{l}(1 - R_i) = 4/3$.

The complexity of construction of L_m is exponential in m, hence it is polynomial in $N = m \cdot 2^m$; Reed-Solomon codes are also polynomially constructed [20], and the construction described in §4 is also polynomial (in the sense that the complexity of the resulting lattice is bounded by a polynomial in the complexities of the initial lattice and the codes).

Now letting $\varepsilon \to 0$ in (27) gives us (a).

If Λ_m are T-lattice (in this case m is always even) then using Proposition 2 and $[2^{m/2}, 2^{m/2} - 2^{m/2-i+1} - 1, 2^{m/2-i+1}]_{2^{m/2}}$ nested Reed-Solomon codes $(\lim_{l \to \infty} \sum_{i=1}^{l}(1 - R_i) = 1)$, we get (b).   Q.E.D.

§6. Asymptotically dense packings. Now we shall present several upper bounds. Let us arrange the lambdas we have introduced in the following diagram, where $\to$ means $\geqslant$.

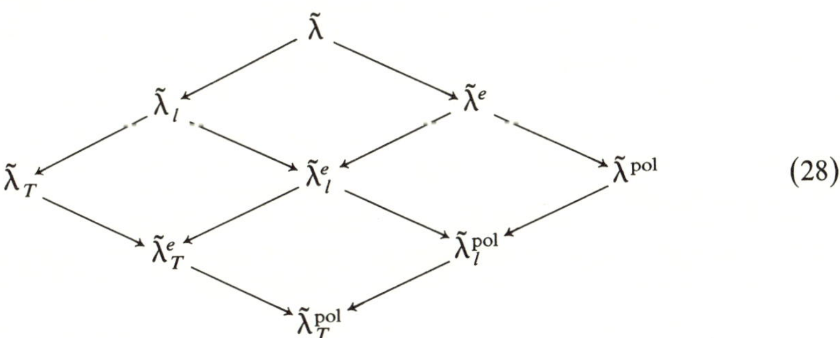

$$\tag{28}$$

THEOREM 3. $\tilde{\lambda}_l^e \leqslant 1.27$.

Proof.   We apply Proposition 1 to the lattice $\mathbb{Z} \subset \mathbb{R}^1$ and to a family of sets of exponentially constructed nested codes on the Varshamov-Gilbert bound. To evaluate λ we just take several first members and use the fact that

$$\sum_{i=a}^{\infty} H_2(x^i) \leqslant \frac{ax^a}{1 - x} - \frac{(a - 1)x^{a+1}}{(1 - x)^2} + \frac{1}{\ln 2} \frac{x^a}{(1 - x)}. \tag{29}$$

Q.E.D.

Remark 4.   It is easily seen from Theorems 2a and 4 that $\tilde{\lambda}_l^{\text{pol}} \leqslant 2.61$. This estimate will be sharpened in Corollary 1 below.

To estimate $\tilde{\lambda}_T^e$ we shall use algebraic-geometric codes arising from modular curves.

THEOREM 4. $\tilde{\lambda}^e_T \leqslant 1.30$.

Proof. We apply Proposition 2 to the Leech lattice Λ_{24} (it is a T-lattice) and to the following sets of 2^{12}-ary codes. For a given length n consider the largest d_0 such that for $1/2 \geqslant d \geqslant d_0$ algebraic-geometric codes are better than the Varshamov-Gilbert bound. Take a set of nested algebraic-geometric codes for $d \geqslant d_0$ and enlarge the largest of them by the Varshamov procedure for $d_0 > d \geqslant 1$. The constant 1.30 is obtained in the same way as in the proof of Theorem 3.
                                                                    Q.E.D.

COROLLARY 1. $\tilde{\lambda}^{\mathrm{pol}}_I \leqslant \tilde{\lambda}^{\mathrm{pol}}_T \leqslant 2.30$.

Proof. Follows directly from Theorems 2(b) and 4. Q.E.D.

For nonlattice packings we use concatenated codes from [14]:

THEOREM 5. $\tilde{\lambda}^{\mathrm{pol}} \leqslant 1.31$.

Proof. We apply Proposition 2 to the Leech lattice Λ_{24} and to the concatenations of algebraic-geometric codes. They are polynomial and the constant is estimated from (24) and (15). Q.E.D.

Remark 5. The Leech lattice is a remarkable phenomenon. We have tried several other lattices from Table 1, but the estimates we obtain appeared to be worse. Of course, if there existed codes on the McEliece-Rodemich-Ramsey-Welch bound which is highly doubtful, Proposition 1 applied to $\mathbb{Z}$ and such codes would have given a family of packings with $\lambda(\{P_N\}) \leqslant 0.90$, thus sharpening the Minkowski bound.

To present all the above estimates in one picture we give the following diagram (cf. (28)).

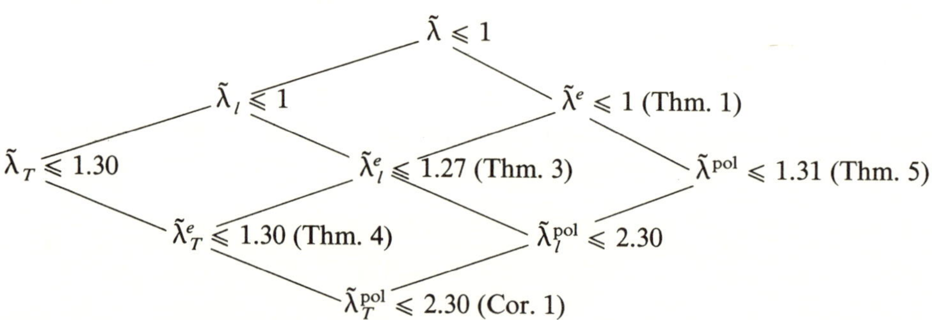

Recall that the density $\Delta = 2^{-\lambda N}$.

§7. Number-theoretic lattices. In this section we shall introduce another class of dense lattices–lattices of integers in algebraic number fields. The best asymptotic result presented here is a family of lattices $\{L_N\}$ with $\lambda(\{L_N\}) \leqslant 2.22$. The constructivity of these lattices from the point of view of §3 is rather low: it is not difficult to produce a triply exponential algorithm generating L_N, and as yet we know nothing better (through theorem 2 this gives rise to a family L'_N with $\lambda(\{L'_N\}) \leqslant 6.22$ and polynomial construction, which is much worse than 2.30 of corollary (1)). The reason we still think these lattices worth mentioning is the fact that they are extremely natural. They correspond to number fields in quite the same way as algebraic-geometric codes correspond to algebraic curves over finite fields.

Recall some facts and definitions (see, for example, [21] or [22]). Let k be a number field, $[k: Q] = N = s + 2t$ where s is the number of real embeddings $\sigma_i: k \hookrightarrow \mathbb{R}$ and t is the number of conjugate pairs of complex embeddings $\tau_j: k \to \mathbb{C}$. These embeddings give a natural map $\sigma: k \hookrightarrow \mathbb{R}^s \times \mathbb{C}^t \simeq \mathbb{R}^N$. For $x = (x_1, \ldots, x_s; y_1 + iz_1, \ldots, y_t + iz_t) \in \mathbb{R}^s \times \mathbb{C}^t$ let $N(x) = x_1 \cdot \cdots \cdot x_s \cdot (y_1^2 + z_1^2) \cdot \cdots \cdot (y_t^2 + z_t^2)$. Then $N(\sigma(a)) = N_{k/\mathbf{Q}}(a)$ is a norm of $a \in k$. Let $\mathcal{O} = \mathcal{O}_k = \{a \in k \mid N_{k/\mathbf{Q}}(a) \in \mathbb{Z}\}$ be the ring of integers, its discriminant $D_k = D_{\mathcal{O}} = \det(\mathrm{Tr}(\omega_i \cdot \omega_j)) \in \mathbb{Z}$. ($\omega_1, \ldots, \omega_N$ being a basis of $\mathcal{O}$). For $[K: k] = m$ let $D_{K/k} = \det(\mathrm{Tr}(\eta_i \cdot \eta_j)) \in \mathcal{O}_k$ ($\eta_1, \ldots, \eta_m$ being a basis of $\mathcal{O}_K$ over $\mathcal{O}_k$), then

$$D_K = D_k^m \cdot N_{k/\mathbf{Q}}(D_{K/k}). \tag{30}$$

Now, let k be totally complex (i.e. $N = 2t$). The image $L = \sigma(\mathcal{O})$ is a lattice in $\mathbb{R}^N \simeq \mathbb{C}^t$ and its determinant equals

$$|\det L| = 2^{-t} \cdot \sqrt{|D_k|} \tag{31}$$

(it is a direct calculation, see, for example, [22], Chapter 3).

For any nonzero $a \in \mathcal{O}$, let $\sigma(a) = (y_1 + iz_1, \ldots, y_t + iz_t)$; one has

$$t^{-1}|\sigma(a)|^2 = t^{-1} \sum_{i=1}^{t} (y_i^2 + z_i^2) \geqslant \left(\prod_{i=1}^{t} (y_i^2 + z_i^2) \right)^{1/t},$$

hence

$$|\sigma(a)| \geqslant \sqrt{t} \left(N_{k/\mathbf{Q}}(a) \right)^{1/2t} \geqslant \sqrt{t} = \sqrt{N/2}. \tag{32}$$

Combining (1), (31) and (32) we estimate the density

$$\Delta(L) \geqslant (N/2)^{N/2} \cdot V_N/2^N \cdot 2^{-N/2} \sqrt{|D_k|} = N^{N/2} \cdot V_N/2^N \sqrt{|D_k|}. \tag{33}$$

Thus

$$\lambda(L) \geqslant 1 - \frac{1}{2} \log N - \frac{1}{N} \log V_N + \frac{1}{2N} \log |D_k|. \tag{34}$$

For a nonramified extension $[K: k] = m$ one has $D_{K/k} = 1$, and (from (30)) $D_K = D_k^m$. The problem of existence of infinite towers of unramified extensions

$k \subset K_1 \subset K_2 \subset \cdots$ was positively solved by E. S. Golod and I. R. Shafarevich in [23]. We are interested in fields with infinite Golod-Shafarevich tower and $|D_k|^{1/N}$ as small as possible. The best such field known to us was discovered by J. Martinet [24] who discovered the field $k = \mathbb{Q}(\cos(2\pi/11), \sqrt{-46})$, $[k: Q] = 10$, $D_k = 2^{15}11^{8}23^{5}$, and showed that it has an infinite 2-tower.

THEOREM 6. *Let k be a totally complex field with infinite Golod-Shafarevich tower, $[k: Q] = N$. Then rings of integers of its unramified extensions give a family of lattice sphere packings $\{L_N\}$ with*

$$\lambda(\{L_N\}) \leqslant -\frac{1}{2}\log\frac{\pi e}{2} + \frac{1}{2N}\log|D_k| + o(1).$$

In particular, Martinet field $\mathbb{Q}(\cos(2\pi/11), \sqrt{-46})$ gives $\lambda(\{L_N\}) \leqslant 2.22$.

Proof. Let $L_N = \sigma(\mathcal{O}_K)$ for unramified extensions $[K: k] = m$ ($N = [K: \mathbb{Q}]$); the estimate for $\lambda(\{L_N\})$ follows from (34), Stirling's formula, and the fact that $D_K = D_k^m$. Q.E.D.

Remark 6. Here is a list of analogues (cf. [25]).

A number field k	A function field $\mathbb{F}_q(X)$ of a curve $X/\mathbb{F}_q$
The degree $[k: \mathbb{Q}]$	The number of $\mathbb{F}_q$-points on X
$\log|D_k|$	The genus of X
Lattice $\sigma(\mathcal{O}_k)$ and its sublattices	Algebraic-geometric codes

Remark 7. The lattice packings obtained in the above manner cannot be too good. Estimations of discriminant due to A. M. Odlyzko and J.-P. Serre (cf. [26]), which are analogues of Vlăduţ-Drinfeld inequality [11], show that if the tower is infinite, then $|D_k^{1/N}| \geqslant 22.3$ (and modulo the generalized Riemann conjecture, even $|D_k^{1/N}| \geqslant 44.7$). This gives $\lambda(\sigma(\mathcal{O}_k)) \geqslant 1.19$ (respectively, 1.69).

Remark 8. This construction can be also used for totally real fields, but the estimates one gets are worse.

Last but not the least our heartfelt thanks to Yu. I. Manin, G. L. Katsman, N. J. A. Sloane, and S. G. Vlăduţ, for their kind attention, and congratulations to Yu. I. Manin on his 50th birthday.

REFERENCES

1. S. N. LITSYN and M. A. TSFASMAN, *Algebro-geometric and number-theoretic packings of balls in $\mathbb{R}^N$*, Uspekhi Mat. Nauk, **40:2** (1985), 185–186; Russian Math. Surveys, **40:2** (1985), 219–220.
2. C. A. ROGERS, *Packing and Covering*, Cambridge University Press, Cambridge, 1964.
3. J. LEECH and N. J. A. SLOANE, *Sphere packings and error-correcting codes*, Can. J. Math., **23:4** (1971), 718–745.

4. G. A. KABATIANSKY and V. I. LEVENSHTEIN, *Bounds for packing on a sphere and in space*, Problemy Peredachi Informatsii **14:1** (1978), 3–25; Problems of Information Transmission, **14:1** (1978), 1–17.

5. H. MINKOWSKI, *Diskontinuitätsbereich für arithmetische Aequivallenz*, J. Reine Angew. Math. **129** (1905), 220–274.

6. F. J. MACWILLIAMS and N. J. A. SLOANE, *The theory of error-correcting codes*, North-Holland, Amsterdam, 3rd printing, 1981.

7. YU. I. MANIN, *What is the maximal number of points on a curve over* $\mathbb{F}_2$? J. Fac. Sci. Univ. Tokyo, Sec. IA, **28** (1982), 715–720.

8. M. A. TSFASMAN, S. G. VLĂDUȚ, and TH. ZINK, *Modular curves, Shimura curves, and Goppa codes, better than Varshamov-Gilbert bound*, Math. Nachrichten **109** (1982), 21–28.

9. V. D. GOPPA, *Codes on algebraic curves*, Dokl. Akad. Nauk SSSR **259:6** (1981), 1289–1290; Soviet Math. Doklady, **24** (1981), 170–172.

10. R. HARTSHORNE. *Algebraic Geometry*, Springer, New York-Heidelberg-Berlin, 1977.

11. S. G. VLĂDUȚ and V. G. DRINFELD, *Number of points on an algebraic curve*, Funk. Anal. i Ego Pril. **17:1** (1983), 68–69; English translation, Functional Analysis **17** (1983), 53–54.

12. Y. IHARA, *Some remarks on the number of rational points of algebraic curves over finite fields*, J. Fac. Sci. Univ. Tokyo, Sec. IA, **28** (1982), 721–724.

13. S. G. VLĂDUȚ and YU. I. MANIN, *Linear codes and modular curves*, Sovremennye Problemy Mat. **25** (1984), 209–257; J. Soviet Math **30** (1985) 2611–2643.

14. S. G. VLĂDUȚ, G. L. KATSMAN, and M. A. TSFASMAN, *Modular curves and codes with polynomial construction complexity*, Problemy Peredachi Informatsii, **20:1** (1984), 3–15; Problems of Information Transmission, **20:1** (1984).

15. YU. I. MANIN, *A Course in Mathematical Logic*, GTM 53, Springer-Verlag, 1977.

16. N. J. A. SLOANE, *Sphere packings constructed from BCH and Justesen codes*, Mathematika **19** (1972), 183–190.

17. A. BOS, J. H. CONWAY, and N. J. A. SLOANE, *Further lattice packings in high dimensions*, Mathematika **29** (1982), 171–180.

18. A. BOS, *Sphere packings in high-dimensional space*, preprint, 1980.

19. E. S. BARNES and N. J. A. SLOANE, *New lattice packings of spheres*, Can. J. Math., Vol. XXXV; **1** (1983), 117–130.

20. L. A. BASSALYGO, V. V. ZIABLOV, and M. S. PINSKER, *Complexity problems in the theory of error-correcting codes*, Problemy Peredachi Informatski, **13:3** (1977), 5–17; Problems of Information Transmission, **13** (1977), 166–175.

21. Z. I. BOREVICH and I. R. SHAFAREVICH, *Number Theory*, Nauka, Moscow, 1972; English translation by Academic Press, New York, 1966.

22. S. LANG, *Algebraic Numbers*. Addison-Wesley Publ. Comp., Reading, 1964.

23. E. S. GOLOD and I. R. SHAFAREVICH, *On the tower of class fields*, Izvestia Akad. Nauk SSSR, **28** (1964), 261–278.

24. J. MARTINET, *Tours de corps de classes et estimation de discriminants*, Invent. Math., 1978 **44:1** (1978), 65–73.

25. J.-P. SERRE, *Sur le nombre des points rationnels d'une courbe algébrique sur un corps fini*, C.R. Acad. Sci. Paris, sér. **1**, **296** (1983), 397–402.

26. G. POITOU, *Minorations de discriminants (d'après A. M. Odlyzko)*, Sém. Bourbaki 1975/1976, exp. 479, Lect. Notes Math., **567** (1977), 136–153.

LITSYN: PERM POLYTECHNICAL INSTITUTE, DEPT. AUTOMATICS AND TELEMECHANICS, 29 KOMSOMOLSKY PROSP., PERM, 614000, USSR

TSFASMAN: 10-75 UDALTSOVA ST., MOSCOW 117415, USSR

ANOTHER PROOF OF A CONJECTURE OF S. P. NOVIKOV ON PERIODS OF ABELIAN INTEGRALS ON RIEMANN SURFACES

ENRICO ARBARELLO AND CORRADO DE CONCINI

Introduction. In this paper we present a proof of a well known conjecture due to S. P. Novikov. The conjecture states that given a complex symmetric matrix τ, which is not in block form, and whose imaginary part is positive definite, then τ is the normalized period matrix for the abelian integrals of a compact connected Riemann surface if and only if the Riemann theta function $\theta(z, \tau)$ satisfies the so-called Kodomcev-Petviashvili equation, hereafter, K-P.

This conjecture has been recently proved by T. Shiota in [7]. The present note can be viewed as a shortcut to Shiota's proof. In our treatment we follow the spirit of Welters' paper [8] and of our previous paper [1]. We only use Riemann theta function, we never need the machinery of the abstract τ-function and wave operators. As a consequence we do not have to check properties of quasiperiodicity of formal solution to the K-P hierarchy. Moreover as we already found in our previous work, we only need to consider a particularly nice sub-set of equations in the K-P hierarchy.

More importantly, our point of view is to exploit a little bit more the geometry in the given abelian variety X. In fact a source of simplification in our argument is to restrict the differential equations we are interested in, to the subscheme of X defined by the vanishing of the theta function and of its D_1-derivative. Once this restriction is understood we can go back to Shiota's argument and see with clarity the path to follow. From Shiota's paper [7] we use the very skillful Lemma 7 which deals with the a priori possibility (which is, a posteriori, an impossibility) of the existence, inside the theta divisor, of integral curves for the D_1-flow.

It was in communicating with Gerald Welters that the ideas of §2 came to our minds. We thank him very much. We also thank Maurizio Cornalba for some useful hints concerning §3.

§1. The geometrical interpretation of the K-P hierarchy. In this section we shall introduce some notation and recall the basic results contained in [8] and [1].

Let $\mathscr{H}_g$ denote the Siegel upper-half phase of genus g. For $z \in \mathbb{C}^g$, $\tau \in \mathscr{H}_g$ and $n \in \frac{1}{2}\mathbb{Z}^g/\mathbb{Z}^g$ set

$$\theta\begin{bmatrix} n \\ 0 \end{bmatrix}(z, \tau) = \sum_{p \in \mathbb{Z}^g} \exp 2\pi i \left\{ \tfrac{1}{2}{}^t(p + n)\tau(p + n) + {}^t(p + n)z \right\}.$$

Received November 5, 1986.

The zero locus of Riemann's theta function

$$\theta(z) = \theta(z,\tau) = \theta\begin{bmatrix}0\\0\end{bmatrix}(z,\tau)$$

defines a principal polarization Θ on the complex torus $X = \mathbb{C}^g/(\mathbb{Z}^g + \tau\mathbb{Z}^g)$. The functions $\theta\begin{bmatrix}n\\0\end{bmatrix}(2z,2\tau)$ form a basis for the vector space of sections of $\mathcal{O}_X(2\Theta)$. We consider the map $\vec{\theta}\colon X \to \mathbb{P}^N$, $N = 2^g$, associated to the linear system $|2\Theta|$, so that

$$\vec{\theta}(\zeta) = \left(\ldots, \theta\begin{bmatrix}n\\0\end{bmatrix}(2\zeta,2\tau), \ldots\right).$$

Given

$$Y \hookrightarrow (X,0), \quad \text{with } Y \cong \operatorname{Spec}\mathbb{C}[\varepsilon]/\varepsilon^3,$$

Welters considers the "secant subscheme of X"

$$\tilde{V}_Y = \left\{\zeta \in X \colon \exists \text{ line } l \subset \mathbb{P}^N \text{ with } \vec{\theta}^{-1}(l) \supset \zeta + Y\right\} \subset X,$$

and, inspired by a criterion of Gunning [2], proves the following.

THEOREM 1 (Welters). *Let X be an irreducible principally polarized abelian variety. Then X is an irreducible Jacobian if and only if there exists a subscheme* $\operatorname{Spec}\mathbb{C}[\varepsilon]/\varepsilon^3 \cong Y \hookrightarrow (X,0)$, *such that* $\dim_0\tilde{V}_Y > 0$.

This criterion can be translated into a set of differential equations for θ in the following way. First of all let Y be given by

$$\varepsilon \mapsto 2W^{(1)}\varepsilon + 2W^{(2)}\varepsilon^2, \qquad W^{(i)} \in \mathbb{C}^g, \qquad i = 1,2,$$

and set $D_i = \sum_{j=1}^g W_j^{(i)}(\partial/\partial\zeta_j)$, $i = 1,2$ (the reason for the presence of the coefficient 2 is that, as it is easy to see, $\tilde{V}_Y$ coincides, up to second order with $\frac{1}{2}Y$, at the origin). Letting $\Delta_1 = D_1$, $\bar{\Delta}_2 = D_1^2 + D_2$ one sees that the natural subscheme structure $\tilde{V}_Y$ is given by

$$\tilde{V}_Y = \left\{\zeta \in X \colon \vec{\theta}(\zeta) \wedge \overrightarrow{\Delta_1\theta(\zeta)} \wedge \overrightarrow{\bar{\Delta}_2\theta(\zeta)} = 0\right\}.$$

The condition that $\tilde{V}_Y$ should be one-dimensional at the origin can then be translated by saying that there exists a formal curve

$$\varepsilon \mapsto \overrightarrow{\zeta(\varepsilon)} = \sum_{i \geqslant 1} 2W^{(i)}\varepsilon^i$$

such that $\overrightarrow{\zeta(\varepsilon)}$ satisfies the equations defining $\tilde{V}_Y$.

Let D_i be the constant vector field in $\mathbb{C}^g$ corresponding to the vector $W^{(i)}$ and recall the notation

$$e^{D(\varepsilon)} = e^{\sum D_i \varepsilon^i} = \sum_{s \geq 0} \Delta_s \varepsilon^s$$

where

$$\Delta_s = \Delta_s(D_1, \ldots, D_s) = \sum_{i_1 + 2i_2 + \cdots + si_s = s} \frac{1}{i_1! \cdots i_s!} D_1^{i_1} \ldots D_s^{i_s}.$$

With this notation we can write

$$f\left(\zeta + \overrightarrow{\zeta(\varepsilon)}\right) = e^{D(\varepsilon)} f(\zeta) = \sum_{s \geq 0} \Delta_s(f) \varepsilon^s.$$

Hence the condition that $\tilde{V}_Y$ be one-dimensional at the origin is translated into the existence of constant vector fields $D_3, D_4, \ldots$ such that

$$e^{D(\varepsilon)}\left(\vec{\theta} \wedge \Delta_1 \vec{\theta} \wedge \overline{\Delta}_2 \vec{\theta}\right)\Big|_{\zeta = 0} = 0.$$

In [1] p. 134 it is shown that this condition is equivalent to the existence of constant vector fields $D_3, D_4, \ldots$ and complex numbers $d_4, d_5, \ldots$ such that

$$\sum_{s \geq 0} \left[\left(\Delta_s \Delta_1 - \Delta_{s-1}\overline{\Delta}_2 + \sum_{i=3}^{s} d_{i+1}\Delta_{s-i}\right)\overrightarrow{\theta(\zeta)}\,\Big|_{\zeta=0}\right]\varepsilon^s = 0.$$

(This is in fact the only result from [1] that we need.) Using Riemann's quadratic relation we translate this equation into the following

$$(1) \qquad \sum_{s \geq 0}\left[\Delta_s \Delta_1 - \Delta_{s-1}\overline{\Delta}_2 + \sum_{i=3}^{s} d_{i+1}\Delta_{s-i}\right]\theta(z+\zeta)\theta(z-\zeta)\Big|_{\zeta=0}\;\varepsilon^s = 0.$$

For brevity we write (1) as

$$(2) \qquad \sum_{s \geq 0} P_s(z)\varepsilon^s = 0.$$

We then obtain the following translation of Welters' criterion.

THEOREM 2 ([1]). *Let X be an irreducible principally polarized abelian variety. Then X is an irreducible Jacobian if and only if there exist vector fields $D_1 \neq 0$, $D_2, D_3, \ldots$ and complex numbers $d_4, d_5, \ldots$ such that: $P_s(z) = 0$, $s \geq 1$.*

Notice that $P_1 = P_2 = 0$ and that the equation $P_3(z) = 0$ is the Kodomcev-Petviashvili equation

$$(3) \qquad D_1^4\theta \cdot \theta - 4D_1^3\theta \cdot D_1\theta + 3\left(D_1^2\theta\right)^2 - 3\left(D_2\theta\right)^2 + 3D_2^2\theta \cdot \theta$$

$$+ 3D_1\theta \cdot D_3\theta - 3D_1D_3\theta \cdot \theta + d_4\theta \cdot \theta = 0.$$

Our aim is to give a proof of the Novikov conjecture, namely, to prove the following.

THEOREM 3 (see also [7]). *Let X be an irreducible principally polarized abelian variety. Then X is an irreducible Jacobian if and only if there exist constant vector fields $D_1 \neq 0$, D_2, D_3 and a complex number d_4 such that the Riemann theta function of X satisfies the K-P equation (3).*

Notice that $P_s(z)$ only depends on $D_1, \ldots, D_s$ and $d_4, \ldots, d_{s+1}$. In symbols we write

$$P_s(z) = P_s(D_1, \ldots, D_s, d_4, \ldots, d_{s+1})(z).$$

In view of Theorem 2, the Novikov conjecture is then a trivial consequence of the following

THEOREM 4. *Let X be an irreducible principally polarized abelian variety. Assume there exist constant vector fields $D_1 \neq 0$, $D_2, \ldots, D_{s-1}$ and complex numbers $d_4, \ldots, d_s$ such that the Riemann theta function of X satisfies $P_i(z) = 0$, $3 \leqslant i \leqslant s - 1$. Then there exist a constant vector field D_s and a complex number d_{s+1} such that the equation $P_s(z) = 0$ is also satisfied.*

This is the theorem that we shall prove in the following sections, it will be an immediate consequence of Proposition 3 and Proposition 2, below.

We end this section by making a remark that we will need in the sequel. We would like to show that P_s/θ^2 is a polynomial in $D_1\log\theta, \ldots, D_s\log\theta$ with coefficients depending on $d_4, \ldots, d_{s+1}$. To see this set $f(z) = \log\theta(z)$ and $d(\varepsilon) = \sum_{i \geqslant 3} d_{i+1}\varepsilon^i$. Then

$$\sum P_s(z)\varepsilon^s = \sum\left(\Delta_s\Delta_1 - \Delta_{s-1}\overline{\Delta}_2 + \sum d_{i+1}\Delta_{s-i}\right)\theta(z + \zeta)\theta(z - \zeta)\Big|_{\zeta=0}\varepsilon^s$$

$$= e^{D(\varepsilon)}\left(\Delta_1 - \varepsilon\overline{\Delta}_2 + d(\varepsilon)\right)\theta(z + \zeta)\theta(z - \zeta)\Big|_{\zeta=0}$$

$$= e^{D(\varepsilon)}\left(\Delta_1 - \varepsilon\overline{\Delta}_2 + d(\varepsilon)\right)e^{f(z+\zeta)+f(z-\zeta)}\Big|_{\zeta=0}$$

$$= \left(\Delta_1 - \varepsilon\overline{\Delta} + d(\varepsilon)\right)e^{f(z+\zeta+\zeta(\varepsilon))+f(z-\zeta-\zeta(\varepsilon))}\Big|_{\zeta=0}.$$

Set now

$$(4) \qquad \cosh D(\varepsilon) = \frac{e^{D(\varepsilon)} + e^{-D(\varepsilon)}}{2}$$

$$\sinh D(\varepsilon) = \frac{e^{D(\varepsilon)} - e^{-D(\varepsilon)}}{2}.$$

An easy computation gives

$$\left(\Delta_1 - \varepsilon\overline{\Delta}_2\right) e^{f(z+\zeta+\zeta(\varepsilon)) + f(z-\zeta-\zeta(\varepsilon))}$$

$$= \Big(2\sinh D(\varepsilon)D_1 f - \varepsilon 2\sinh D(\varepsilon)D_2 f - \varepsilon 2\cosh D(\varepsilon)D_1^2 f$$

$$- \varepsilon\big(\sinh D(\varepsilon)D_1 f\big)^2\Big) e^{f(z+\zeta(\varepsilon)) + f(z-\zeta(\varepsilon))}.$$

On the other hand

$$e^{f(z+\zeta(\varepsilon)) + f(z-\zeta(\varepsilon))} = \theta(z)\exp\left(\sum_{s\geqslant 1}\left(\Delta_s + \tilde{\Delta}_s\right)f\varepsilon^s\right)$$

where $\tilde{\Delta}_s = \Delta_s(-D_1, \ldots, -D_s)$. In conclusion we have

$$(5) \qquad \sum_{s\geqslant 0}\frac{P_s(z)}{\theta^2}\varepsilon^s$$

$$= 2\Big(\sinh D(\varepsilon)D_1 f - \varepsilon\sinh D(\varepsilon)D_2 f - \varepsilon\cosh D(\varepsilon)D_1^2 f$$

$$- \varepsilon\big(\sinh D(\varepsilon)D_1 f\big)^2 + \tfrac{1}{2}d(\varepsilon)\Big)\exp\left(\sum_{s\geqslant 1}\left(\Delta_s + \tilde{\Delta}_s\right)f\varepsilon^s\right),$$

proving our contention.

§2. The first reduction. We make two observations. The first one is that, given any constant vector fields $D_1 \neq 0$, $D_2, \ldots, D_s$ and any complex numbers $d_4, \ldots, d_{s+1}$, the functions

$$P_s(z) = P_s(D_1, \ldots, D_s, d_4, \ldots, d_{s+1})(z)$$

defined in (2), are sections of $H^0(X, 2\Theta)$. This just depends on the fact that the quasi-periodicity factor of $\theta(z + \zeta)\theta(z - \zeta)$ is independent of ζ. The second is that

$$P_s(z) = P_s'(z) + 2D_1 D_s\theta(z) \cdot \theta(z) - 2D_1\theta(z) \cdot D_s\theta(z) + d_{s+1}\theta(z)^2$$

where

$$(6) \qquad P_s'(z) = P_s(D_1, \ldots, D_{s-1}, 0, d_4, \ldots, d_s, 0)(z).$$

PROPOSITION 1. *Let X be an irreducible principally polarized abelian variety. Let θ be its Riemann theta function. Suppose there exist constant vector fields $D_1 \neq 0$, $D_2, \ldots, D_{s-1}$, and complex numbers $d_4, \ldots, d_s$, $s \geqslant 4$ such that there exists a holomorphic solution $\varphi(z)$, in $\mathbf{C}^g$, to the equation*

$$(7) \qquad D_1\theta \cdot \varphi - D_1\varphi \cdot \theta = P_s'$$

then there exist a constant vector field D_s and a complex number d_{s+1} such that $P_s(z) = 0$.

Proof. Consider the exact sequences

$$(8) \qquad \mathcal{O}_X(\Theta) \xrightarrow{\cdot\theta} \mathcal{O}_X(2\Theta) \to \mathcal{O}_\Theta(2\Theta) \to 0$$

$$\mathcal{O}_\Theta(\Theta) \xrightarrow{\cdot D_1\theta} \mathcal{O}_\Theta(2\Theta) \to \mathcal{O}_{D_1\Theta}(2\Theta) \to 0$$

where $D_1\Theta$ is the subscheme of X defined by $\theta = D_1\theta = 0$. Also recall from [3] that a basis of $H^0(\mathcal{O}_\Theta(\Theta))$ is given by $\partial\theta/\partial\zeta_1, \ldots, \partial\theta/\partial\zeta_g$. Now equation (7) tells us that the restriction of the section $P_s'(z)$ to $D_1\Theta$ is zero. Therefore by the cohomology of the second sequence in (8), it follows that there exists a vector field D_s such that $P_s'(z) - 2D_1\theta \cdot D_s\theta$ is zero as a section of $H^0(\mathcal{O}_\Theta(2\Theta))$. This means that the section $P_s'(z) - 2D_1\theta \cdot D_s\theta + 2D_1D_s\theta \cdot \theta$ of $\mathcal{O}_X(2\Theta)$ vanishes when restricted to Θ. So that by the cohomology the first sequence in (8) there exists d_4 such that $P_s(z) = 0$ as a section of $\mathcal{O}_X(2\Theta)$.   Q.E.D.

By virtue of this Proposition the scheme of the proof of Theorem 4 is now the following. We must prove that given constant vector fields $D_1 \neq 0$, $D_2, \ldots, D_{s-1}$ and complex numbers $d_4, \ldots, d_s$, then there exists a holomorphic solution φ in $\mathbf{C}^g$ to the equation $D_1\theta \cdot \varphi - D_1\varphi \cdot \theta = P_s'$ (recall again that P_s' only depends on $D_1, \ldots, D_{s-1}, d_4, \ldots, d_s$).

§3. The second reduction. In the preceding section we reduced the proof of Novikov's conjecture to the problem of solving, globally in $\mathbf{C}^g$, equation (7). In this section we prove that it suffices to solve this equation only locally in a neighbourhood of each point of $\mathbf{C}^g$ which does not belong to the subscheme

$$V = \left\{ z \in \mathbf{C}^g;\ \theta(z) = D_1\theta(z) = 0 \right\}.$$

PROPOSITION 2. *Let X be an irreducible principally polarized abelian variety. Let θ be its Riemann theta function. Suppose there exist constant vector fields*

$D_1 \neq 0$, $D_2, \ldots, D_{s-1}$ *and complex numbers* $d_4, \ldots, d_s$ *such that* $P_i(z) = 0$, $3 \leqslant i$ $\leqslant s - 1$. *Suppose furthermore that near each point of* $\mathbb{C}^g \setminus V$ *there exists a meromorphic solution* h *to the equation*

$$(9) \qquad D_1 h = \frac{P_s'}{\theta^2}.$$

Then there exist a constant vector field D_s *and a complex number* d_{s+1} *such that* $P_s(z) = 0$.

Proof. In view of Proposition 1, it suffices to show that the local solutions of (9) lead to a global solution of (7). To prove this we need a Lemma which is a refinement of a Lemma due to J. Lee, T. Oda and A. Yukie. This lemma is also instrumental in Shiota's argument, ([7] Lemma 12(a)).

LEMMA. *Let* $\pi \colon X \to \mathbb{C}^g$ *be a smooth blow-up of* $\mathbb{C}^g$ *along a subscheme* W *of codimension at least equal to two. Let* D *be a nowhere vanishing vector field on* X. *Let* $Y \subset X$ *be a divisor which contains no integral curve of* D. *Let* f *be a global meromorphic function on* X *with poles contained in* Y. *Let* $Z \subset Y$ *be a subvariety of codimension at least one. Then a necessary and sufficient condition for the existence of a meromorphic solution to the equation* $Dh = f$ *in* $X \setminus Z$ *is that a local solution to the same equation exists at each point* $x \in X \setminus Z$. *Moreover the poles of* h *are contained in* Y.

Proof. Let $\mathcal{M}(Y)$ be the sheaf of meromorphic functions on $X \setminus Z$ with poles in Y and let $\mathcal{M}^0(Y)$ be the image of the sheaf map $D \colon \mathcal{M}(Y) \to \mathcal{M}(Y)$. The fact that Y contains no integral curve of D tells us that the kernel of D coincides with the kernel $\mathcal{O}'_{X \setminus Z}$ of the sheaf map $D \colon \mathcal{O}_{X \setminus Z} \to \mathcal{O}_{X \setminus Z}$.

Looking at the cohomology sequence for

$$0 \to \mathcal{O}'_{X \setminus Z} \to \mathcal{M}(Y) \xrightarrow{D} \mathcal{M}^0(Y) \to 0$$

it suffices to show that $H^1(\mathcal{O}'_{X \setminus Z}) = 0$. Clearly the map $D \colon H^0(\mathcal{O}_{X \setminus Z}) \to H^0(\mathcal{O}_{X \setminus Z})$ is surjective, so that we have an exact sequence

$$0 \to H^1(\mathcal{O}'_{X \setminus Z}) \to H^1(\mathcal{O}_{X \setminus Z}) \xrightarrow{D} H^1(\mathcal{O}_{X \setminus Z}).$$

We now proceed by induction on the dimension of Z. If $Z = \varnothing$ we follow the argument of Lemma 12 in [7]. We look at the spectral sequence of π. We get the exact sequence

$$\to H^1(\mathcal{O}_{\mathbb{C}^g}) \to H^1(\mathcal{O}_X) \to H^0\big(\mathbb{C}^g, R^1\pi_*(\mathcal{O}_X)\big).$$

Since X is smooth, $R^i\pi_*\mathcal{O}_X = 0$, $i > 0$, hence $H^1(\mathcal{O}_X) = H^1(\mathcal{O}_{\mathbb{C}^g}) = 0$. In general let S be the singular locus of Z and consider the spectral sequence of the inclusion $i: X \setminus Z \hookrightarrow X \setminus S$. The contribution to $H^1(\mathcal{O}'_{X \setminus Z})$ comes from $H^0(R^1 i_* \mathcal{O}'_{X \setminus Z})$ and from $H^1(i_* \mathcal{O}'_{X \setminus Z})$. By Hartogs' $i_* \mathcal{O}'_{X \setminus Z} = \mathcal{O}'_{X \setminus S}$, so that, by induction $H^1(i_* \mathcal{O}'_{X \setminus S}) = 0$. We are now reduced to show that if $\Delta = \{ z \in \mathbf{C}^g: |z| < 1 \}$ and $L = \{ z \in \Delta: z_1 = \cdot, \cdot = z_k = 0 \}$, $k \geqslant 2$, then the map $\partial / \partial z_1$: $H^1(\mathcal{O}_{\Delta \setminus L}) \to H^1(\mathcal{O}_{\Delta \setminus L})$ is injective, and this is easy (if $k \geqslant 3$, then $H^1(\mathcal{O}_{\Delta \setminus L}) = 0$, if $k = 2$, one explicitly computes the cohomology of $\mathcal{O}_{\Delta \setminus L}$).   Q.E.D.

Remark. Let X, Y, Z, W, f and h be as in Lemma 1. Assume that W is a subscheme of $\mathbf{C}^g$ invariant under the flow determined by a constant vector field D on $\mathbf{C}^g$, so that D lifts to a vector field on X which we still denote by D. Assume $f = r \circ \pi$ where r is a global meromorphic function on $\mathbf{C}^g$. Let φ be a meromorphic function on $\mathbf{C}^g$ whose polar divisor is contained in $\pi(Y)$ and therefore has no D-invariant component. Suppose that $D\varphi = r$. Then the polar divisor of $\varphi \circ \pi$ is contained in Y.

To see this we consider $\mathbf{C}^g \setminus W$ as an open subset of X and we remark that the open set A of $\mathbf{C}^g$ defined as $(\mathbf{C}^g \setminus W) \cap (X \setminus Z)$ is obtained from $\mathbf{C}^g$ by removing subvarieties of codimension at least equal to two. On A we have $D(\varphi - h) = 0$. On the other hand, by hypothesis, the polar divisor of $\varphi - h$ has no D-invariant component. Therefore the function $\varphi - h$ is holomorphic on A, and by Hartogs' has a unique holomorphic extension k to $\mathbf{C}^g$. It follows that $\varphi \circ \pi - h = k \circ \pi$ so that the polar divisor of $\varphi \circ \pi$ is contained in Y.

We are now going back to Proposition 2. In order to put ourselves in the hypotheses of the above Lemma we are going to use one of the important points in Shiota's argument, namely his Lemma 7. Consider the locus

$$(10) \qquad \Sigma = \{ \zeta \in \mathbf{C}^g: D_1^n \theta(z) = 0, \forall\, n \geqslant 0 \}.$$

In his Lemma Shiota proves that, if the Riemann theta function satisfies the K-P equation (3) then there exists a nonsingular blow-up

$$(11) \qquad X \xrightarrow{\pi} \mathbf{C}^g$$

along a subscheme with the same support as Σ, having the following properties. First of all it is equivariant under the flow determined by D_1, D_2 and D_3. Secondly the proper transform $\tilde{\Theta}$ of Θ under π, does not contain integral curves of D_1 (and coincides with the polar divisor of $D_i \theta(\pi(z)) / \theta(\pi(z))$, $i = 1, 2, 3$).

A consequence of Shiota's Lemma is the following

LEMMA 2.   *Let $X, \theta, D_1, \ldots, D_k$ $(k = s - 1)$ be as in Proposition 2. Assume $P_i(z) = 0$, $i \leqslant k$. Consider the blow-up (11). Then for any multiindex $(i_1, \ldots, i_k) \neq (0, \ldots, 0)$ the polar divisor of the function $(D^{i_1} \cdots D_k^{i_k} \log \theta) \circ \pi$ does not contain integral curves of D_1 and is in fact contained in $\tilde{\Theta}$.*

Proof. We proceed by induction on k. The case $k = 3$ is given by Shiota's Lemma. Assume Lemma 2 for $h \leqslant k - 1$ and let us prove it for $h = k$. We now proceed by induction on i_k. The case $i_k = 0$ is taken care of by the inductive hypothesis on k. We can then assume $i_k > 0$. Then

$$D_1\left(D_1^{i_1} \cdots D_k^{i_k}\log\theta\right)$$

$$= D_1^{i_1}\ldots D_{k-1}^{i_{k-1}}D_k^{i_k-1}\left(D_1 D_k\log\theta\right)$$

$$= D_1^{i_1}\ldots D_{k-1}^{i_{k-1}}D_k^{i_k-1}\left(\tfrac{1}{2}\frac{P_k'}{\theta^2}\right).$$

In fact the hypothesis $P_k(z) = 0$ means that $2D_1 D_k\log\theta = (P_k'/\theta^2) + d_k$. On the other hand recalling (5) and (6)

$$\frac{P_k'}{\theta^2} = \sum a_{j_1\ldots j_{k-1}}D_1^{j_1}\ldots D_{k-1}^{j_{k-1}}\log\theta,$$

with $(j_1,\ldots,j_{k-1}) \neq (0,\ldots,0)$. Hence, by the induction on i_k, $[D_1^{i_1}\ldots D_{k-1}^{i_{k-1}}D_k^{i_k-1}(\tfrac{1}{2}(P_k'/\theta^2))]\circ\pi$ has a polar divisor contained in $\tilde{\Theta}$. Therefore our claim follows from the above Remark.   Q.E.D.

We now start the proof of Proposition 2. Look at the function P_s'/θ^2. We know that this function is a linear combination of monomials of the type $D_1^{i_1}\ldots D_{s-1}^{i_{s-1}}\log\theta$. We then use Lemma 2, to conclude that the polar divisor of the function $f = (P_s'/\theta^2)\circ\pi$, on X, does not contain integral curves of D_1 and is in fact contained in $\tilde{\Theta}$, the proper transform of Θ under the blow-up (11). Let E be the exceptional divisor of this blow-up and denote by $\tilde{V}$ the proper transform of V. Set $Z = (\tilde{\Theta}\cap E)\cup\tilde{V}$. Clearly Z is a divisor on $\tilde{\Theta}$. We apply Lemma 1 with $\tilde{\Theta} = Y$. The only hypothesis of that Lemma that remains to be checked is that the equation $D_1 h = f$ can be locally solved in $X\setminus Z$. Write $X\setminus Z = [X\setminus E\cup \tilde{V}]\cup[(X\setminus Z)\cap E]$. Observe that $X\setminus E\cup\tilde{V}\cong \mathbb{C}^g\setminus V$ so that, by hypothesis local solutions exist in this open set. On the other hand in $(X\setminus Z)\cap E$ the function f is holomorphic and the local solution is given by integration. By Lemma 1 we can then conclude that h exists as a global meromorphic function on $X\setminus Z$. In this way we can define a meromorphic function in $\mathbb{C}^g\setminus V$, again denoted by h, which solves the equation $D_1 h = P_s'/\theta^2$. One then easily checks that locally the function $h\theta$ is holomorphic near any point of $\mathbb{C}^g\setminus V$. Therefore, by Hartogs' theorem, $h\theta$ extends to a holomorphic function φ in $\mathbb{C}^g$ which clearly solves the equation (7). We then apply Proposition 1 and conclude our proof.

§4. The local result.

By virtue of Proposition 2, the only thing that remains to be proved, in order to complete the proof of Novikov's conjecture is the following

PROPOSITION 3. *Let X be an irreducible principally polarized abelian variety. Let θ be its Riemann theta function. Suppose there exist constant vector fields*

$D_1 \neq 0$, $D_2, \ldots, D_{s-1}$ *and complex numbers* $d_4, \ldots, d_s$, *such that* $P_i(z) = 0$, $\zeta \leqslant i \leqslant s - 1$. *Let* $V = \{z \in \mathbf{C}^g : \theta(z) = D_1\theta(z) = 0\}$. *Then near each point of* $\mathbf{C}^g \setminus V$ *there exists a meromorphic solution* h_s *to the equation* $D_1 h_s = P_s'/\theta^2$.

Clearly it suffices to restrict our attention to points z in $\mathbf{C}^g \setminus V$ where $\theta(z) = 0$. Consider the differential operator

$$\Gamma = \theta^4\left(\tfrac{1}{3}D_1^4 + D_2^2 - D_1D_3 + 4D_1\left(D_1^2\log\theta\right)D_1\right)$$

the proof of Proposition 3 is an immediate consequence of the following Lemmas.

LEMMA 3. *The preceding Proposition is true under the additional hypothesis that* $\Gamma(P_s'/\theta^2) = 0$, *in* $\mathbf{C}^g \setminus V$.

LEMMA 4. *Under the hypothesis of Proposition* 3 *we have* $\Gamma(P_s'/\theta^2) = 0$ *in* $\mathbf{C}^g \setminus V$.

We start with Lemma 3. As we already remarked we can restrict our attention to points $z_0 \in \mathbf{C}^g$ for which $\theta(z_0) = 0$, $D_1\theta(z_0) \neq 0$. Therefore we can choose coordinates $z_1, \ldots, z_g$ in such a way that $z_0 = (0, \ldots, 0)$, $D_1 = \partial/\partial z_1$ and $\theta(z) = h(z_1 - b(z_2, \ldots, z_g))$ with $h(0, 0, \ldots, 0) \neq 0$ and $b(0, 0, \ldots, 0) = 0$. In a neighborhood of z_0 we can then write

$$(12) \qquad \frac{P_s'}{\theta^2} = \frac{H_{-1}}{(z_1 - b)^2} + \frac{H}{(z_1 - b)} + \sum_{i \geqslant 0} H_i(z_1 - b)^i = \frac{H}{z_1 - b} + K$$

where H_{-1}, H and the H_i's can be assumed to be independent of z_1. First of all we are going to prove that $\Gamma(K)$ is holomorphic in a neighborhood of z_0. To do so it clearly suffices to show that $\Gamma(H_{-1}/(z_1 - b)^2)$ is holomorphic near z_0. Set $G = -H_{-1}/(z_1 - b)$, then $\theta \cdot G$ is holomorphic in a neighborhood of z_0. In particular the function

$$f(z) = \left(\frac{1}{3}D_1^4 + D_2^2 - D_1D_3 + d_4\right)\theta(z + \zeta)\theta(z - \zeta)G(z - \theta)\Big|_{\zeta = 0}$$

is holomorphic near z_0. On the other hand

$$f(z) = G(z) \cdot \left(\frac{1}{3}D_1^4 + D_2^2 - D_1D_3 + d_4\right)\theta(z + \zeta)\theta(z - \zeta)\Big|_{\zeta = 0}$$

$$\cdot\,\theta(z)^2\left(\frac{1}{3}D_1^4 + D_2^2 - D_1D_3\right)G(z)$$

$$+ 4\left(\theta D_1^2\theta - D_1\theta \cdot D_1\theta\right)D_1^2 G(z).$$

The first summand in the above expression is zero since by hypothesis θ satisfies the K-P equation. We can then conclude that the function

$$\gamma(z) = \theta^2\left(\frac{1}{3}D_1^4 + D_2^2 - D_1 D_3\right)G + 4\theta^2\left(D_1^2 \log \theta\right)D_1^2 G$$

is holomorphic near z_0. Since $D_1 G = H_{-1}/(z_1 - b)^2$, the holomorphicity of $\Gamma(H_{-1}/(z_1 - b)^2)$ follows from the identity

$$\Gamma\left(\frac{H_{-1}}{(z_1 - b)^2}\right) = \theta^4 D_1\left(\frac{\gamma}{\theta^2}\right).$$

Going back to (12) we can conclude that $\Gamma(K)$ is holomorphic near z_0. Next we claim that $H = 0$. We have

$$0 = \Gamma\left(\frac{P_s'}{\theta^2}\right) = \Gamma\left(\frac{H}{z_1 - b}\right) + \Gamma(K).$$

By what we just proved we get the holomorphicity of $\Gamma(H/z_1 - b)$ near z_0. Recalling that $\theta = h(z_1 - b)$ near z_0 and expanding in terms of $(z_1 - b)$ one gets

$$\Gamma\left(\frac{H}{z_1 - b}\right) = -\frac{8H}{(z_1 - b)} + (\text{holo}).$$

Since H does not depend on z_1 we must have $H = 0$. The conclusion is that

$$\frac{P_s}{\theta^2} = \frac{H_{-1}}{(z_1 - b)^2} + \sum_{i \geqslant 0} H_i(z_1 - b)^i$$

where H_{-1} and the H_i do not depend on z_1. The meromorphic solution h_s of our equation is then given by

$$h_s = \frac{-H_{-1}}{z_1 - b} + \sum_{i \geqslant 0} \frac{H_i}{(i + 1)}(z_1 - b)^{i+1}.$$

Before proving Lemma 4 we need to introduce some notation. Let us go back to formula (5). Set

$$M = 2 \sinh D(\varepsilon), \qquad N = 2 \cosh D(\varepsilon).$$

Write $f = \log \theta$ and

$$(13) \qquad MD_1 f - \varepsilon MD_2 f - \varepsilon ND_1^2 f - \varepsilon(MD_1 f)^2 + d(\varepsilon) = \sum_{s \geqslant 0} T_s(z)\varepsilon^s.$$

Observe that

$$e^{\sum_{s \geqslant 1}(\Delta_s + \tilde{\Delta}_s)f\varepsilon^s} = 1 + a_1(z)\varepsilon + a_2(z)\varepsilon^2 + \cdots$$

is an invertible series in ε. Denote by $1 + \sum_{i \geqslant 1} b_i(z)\varepsilon^i$ its inverse. Formula (5) can then be written as

$$(14) \qquad \left(\sum_{s \geqslant 0} \frac{P_s}{\theta^2} \varepsilon^s \right) \cdot \left(1 + \sum_{i \geqslant 1} b_i \varepsilon^i \right) = \sum_{s \geqslant 0} T_s \varepsilon^s.$$

We now start the proof of Lemma 4. It will be an explicit computation. Recall from (6) the definition of P_s'. First of all we observe that given any constant vector field D_s and any complex number d_{s+1}, then

$$(15) \qquad \Gamma\left(\frac{P_s}{\theta^2} \right) = \Gamma\left(\frac{P_s'}{\theta^2} \right).$$

Clearly to prove (15) it suffices to prove that

$$\Gamma\left(\frac{2D_1 D_s \theta \cdot \theta - 2D_1\theta \cdot D_s\theta}{\theta^2} \right) = 0,$$

or, what is the same, that $\Gamma(2D_1 D_s f) = 0$. To show this first observe that the K-P equation for θ can be written in terms of f as

$$(16) \qquad \left(\frac{1}{3} D_1^4 + D_2^2 - D_1 D_3 \right) f = -2\left(D_1^2 f \right)^2 - d_4.$$

Using this equation we have

$$\Gamma(D_1 D_s f) = \left[\frac{1}{3} D_1^4 + D_2^2 - D_1 D_3 + 4D_1\left(D_1^2 f D_1 \right) \right](D_1 D_s f)$$

$$= -2D_1 D_s\left(D_1^2 f \right)^2 + 4D_1\left(D_1^2 f D_1^2 D_s f \right) = 0.$$

We then have to prove that

$$P_3 = \cdots = P_{s-1} = 0 \Rightarrow \Gamma\left(\frac{P_s}{\theta^2} \right) = 0.$$

In view of (14) this is equivalent to proving that

$$(17) \qquad \Gamma\left(\sum_{s \geqslant 0} T_s \varepsilon^s \right) \equiv 0 \quad \mathrm{mod} \sum_{s \geqslant 0} T_s \varepsilon^{s+1}.$$

To prove (17) we shall use the following remarks

$$(18) \qquad M \equiv N - 2 \equiv 0 \quad \mathrm{mod}\ \varepsilon,$$

$$(19) \qquad \begin{cases} M(\varphi\psi) = \dfrac{1}{2}(M\varphi N\psi + M\phi N\varphi) \\[2mm] N(\varphi\psi) = \dfrac{1}{2}(M\varphi M\psi + N\varphi N\psi). \end{cases}$$

Recalling (13) and the definition of Γ we write $\Gamma(\Sigma_{s \geqslant 0} T_s \varepsilon^s)$ as the sum of four terms

$$\Gamma\left(\sum_{s \geqslant 0} T_s \varepsilon^s\right) = \Gamma_1 + \Gamma_2 + \Gamma_3 + \Gamma_4,$$

$$\Gamma_1 = \left(\frac{1}{3}D_1^4 + D_2^2 - D_1 D_3\right)\left(MD_1 f - \varepsilon MD_2 f - \varepsilon ND_1^2 f\right)$$

$$\Gamma_2 = -\left(\frac{1}{3}D_1^4 + D_2^2 - D_1 D_3\right)\left(\varepsilon(MD_1 f)^2\right)$$

$$\Gamma_3 - 4D_1\left(D_1^2 f D_1\left(MD_1 f - \varepsilon MD_2 f - \varepsilon ND_1^2 f\right)\right)$$

$$\Gamma_4 = -4\varepsilon D_1\left(D_1^2 f D_1 (MD_1 f)^2\right), \qquad [\Gamma(d(\varepsilon)) = 0].$$

We develop the terms $\Gamma_1, \Gamma_2, \Gamma_3$ separately.

$$\Gamma_1 = \left(MD_1 - \varepsilon MD_2 - \varepsilon ND_1^2\right)\left(\left(\frac{1}{3}D_1 + D_2^2 - D_1 D_3\right)f\right)$$

$$(\text{using (16)}) = \left(-2MD_1 + 2\varepsilon MD_2 + 2\varepsilon ND_1^2\right)\left(D_1^2 f\right)^2$$

$$(\text{using (19)}) = -2MD_1^2 f ND_1^3 f - 2ND_1^2 f MD_1^3 f + 2\varepsilon MD_1^2 f ND_1^2 D_2 f$$

$$+ 2\varepsilon ND_1^2 f MD_1^2 D_2 f + 2\varepsilon ND_1^2 f ND_1^4 f$$

$$+ 2\varepsilon MD_1^2 f MD_1^4 f + 2\varepsilon ND_1^3 f ND_1^3 f + 2\varepsilon MD_1^3 f MD_1^3 f$$

$$\Gamma_2 = -2\varepsilon\left(\frac{1}{3}MD_1^5 f MD_1 f + \frac{4}{3}MD_1^4 f MD_1^2 f + MD_1^3 f MD_1^3 f + MD_1 D_2^2 f MD_1 f\right.$$

$$\left. + MD_1 D_2 f MD_1 D_2 f - MD_1^2 D_3 f MD_1 f - MD_1 D_3 f MD_1^2 f\right)$$

$$\Gamma_3 = 4D_1^3 f MD_1^2 f + 4D_1^2 f MD_1^3 f - 4\varepsilon D_1^3 f MD_1 D_2 f - 4\varepsilon D_1^2 f MD_1^2 D_2 f$$

$$- 4\varepsilon D_1^3 f ND_1^3 f - 4\varepsilon D_1^2 f ND_1^4 f.$$

Making the obvious cancellations and rearranging the terms we get

$$\Gamma_1 + \Gamma_2 + \Gamma_3 = -2\varepsilon MD_1\left(\frac{1}{3}D_1^4 + D_2^2 - D_1D_3\right)f \cdot MD_1f - 2\varepsilon(N-2)D_1^2f$$

$$\cdot D_1^2\left(MD_1 - \varepsilon MD_2 - \varepsilon ND_1^2\right)f - 2\varepsilon MD_1^2fM\left(\frac{1}{3}D_1^4 - D_1D_3\right)f$$

$$- 2\varepsilon(N-2)D_1^3fD_1\left(MD_1 - \varepsilon ND_1^2\right)f + 2\varepsilon MD_1^2fND_1^2D_2f$$

$$- 2\varepsilon MD_1D_2fMD_1D_2f - 4\varepsilon D_1^3fMD_1D_2f.$$

Recall that $N - 2 \equiv 0 \bmod \varepsilon$ so that we can use our relations to get

$$\Gamma_1 + \Gamma_2 + \Gamma_3 = 4\varepsilon MD_1\left(D_1^2f\right)^2 \cdot MD_1f - 2\varepsilon(N-2)D_1^2fD_1^2(MD_1f)^2$$

$$+ 4\varepsilon MD_1^2fM\left(D_1^2f\right)^2 + 2\varepsilon MD_1^2fMD_2^2f$$

$$- 2\varepsilon(N-2)D_1^3f \cdot D_1(MD_1f)^2 - 2\varepsilon(N-2)D_1^3fMD_1D_2f$$

$$+ 2\varepsilon MD_1^2fND_1^2D_2f - 2\varepsilon MD_1D_2fMD_1D_2f - 4\varepsilon D_1^3fMD_1D_2f.$$

We now have

$$2\varepsilon MD_1^2fMD_2^2f - 2\varepsilon(N-2)D_1^3fMD_1D_2f + 2\varepsilon MD_1^2fND_1^2D_2f$$

$$- 2\varepsilon MD_1D_2fMD_1D_2f - 4\varepsilon D_1^3fMD_1D_2f$$

$$= 2\left(MD_1^2fD_2\left(\varepsilon MD_2 - \varepsilon ND_1^2\right)f - D_1\left(\varepsilon ND_1^2f + \varepsilon MD_2f\right)MD_1D_2f\right).$$

Since $M \equiv 0 \bmod \varepsilon$, we can again use our relations and a straightforward computation shows that

$$MD_1^2fD_2\left(\varepsilon MD_2 - \varepsilon ND_1^2\right)f - D_1\left(\varepsilon ND_1^2f + \varepsilon MD_2f\right) \cdot MD_1D_2f = 0.$$

Thus:

$$\Gamma_1 + \Gamma_2 + \Gamma_3 = 4\varepsilon MD_1\left(D_1^2f\right)^2 MD_1f + 4\varepsilon MD_1^2f + M\left(D_1^2f\right)^2$$

$$- 2\varepsilon(N-2)D_1^2fD_1^2(MD_1f)^2 - 2\varepsilon(N-2)D_1^3fD_1(MD_1f)^2$$

$$= 2\varepsilon D_1\left(2MD_1fM\left(D_1^2f\right)^2 - (N-2)D_1^2fD_1(MD_1f)^2\right)$$

$$= 2\varepsilon D_1\left(2MD_1fMD_1^2fND_1^2f - ND_1^2fD_1(MD_1f)^2\right.$$

$$\left. + 2D_1^2fD_1(MD_1f)^2\right)$$

$$= 4\varepsilon D_1\left(D_1^2fD_1(MD_1f)^2\right) = -\Gamma_4. \hspace{2cm} \text{Q.E.D.}$$

§5. **Concluding remarks.** We would like to end this note by making some general remarks about the motivation of our proof and about a couple of related problems. Let us go back to the equations

$$P_s(z) = 0, \qquad s = 3, 4, \ldots$$

and to the subscheme $V = \{\theta = D_1\theta = 0\} \subset X$. There are two ideas in Proposition 1. First of all the equation $P_s|_V = 0$ makes sense even if one does not know D_s and d_{s+1}. Secondly

$$P_s|_V = 0 \Leftrightarrow \exists\, D_s, d_{s+1} \quad s.t. \ P_s = 0.$$

Now look, for instance, at the first equation, the K-P equation (3). Then one sees that

$$(20) \qquad P_3|_V = 3(D_1^2 + D_2)\theta \cdot (D_1^2 - D_2)\theta|_V = 0,$$

a particularly simple equation which says that the K-P equation is equivalent to the following geometrical statement

$$(21) \quad \{\theta = 0\} \cap \{D_1\theta = 0\} \subset \{(D_1^2 + D_2)\theta = 0\} \cup \{(D_1^2 - D_2)\theta = 0\}$$

this inclusion being in the sense of schemes. The geometrical statement (21) is the infinitesimal version of a criterion whose importance was brought to light by Mumford in [4], [5] and [6]. This criterion was also the object of Welters' paper [9].

Consider the totality of our equations

$$(22) \qquad \sum_{s \geq 0} P_s(z)\varepsilon^s = 0.$$

Now set

$$\sum P_s\left(z + \overrightarrow{\zeta(\varepsilon)}\right)\varepsilon^s = \sum R_s(z)\varepsilon^s.$$

Clearly the system (22) is equivalent to

$$\sum R_s(z)\varepsilon^s = 0$$

An easy computation shows that

$$(23) \qquad R_3 = \cdots = R_{s-1} = 0 \Rightarrow R_s|_V = (D_1^2 - D_2)\theta \cdot \overline{\Delta}_{s-1}\theta|_V$$

where $\overline{\Delta}_s = \Delta_s(2D_1, \ldots, 2D_s)$. This is a direct generalization of (20), and shows how simple these equations are when restricted to V. Therefore from (23) we conclude that to give a proof of Novikov's conjecture it suffices to show that

(24)

$$\left(D_1^2 - D_2\right)\theta \cdot \overline{\Delta}_{s-1}\theta \in (\theta, D_1\theta) = \{\text{the ideal generated by } \theta \text{ and } D_1\theta\} \quad \forall\, s.$$

Unfortunately, up to now the only thing that we can show (and this is very easy) is that

$$\left[\left(D_1^2 - D_2\right)\theta \cdot \overline{\Delta}_{s-1}\theta\right]^2 \in (\theta, D_1\theta).$$

The obstructions in proving (24) come from two sources. One is Shiota's subvariety Σ (see (10)). The other is the size of the singular locus of Θ. What we are able to show with simple algebraic computations is the following fact. Suppose that (a) the K-P equation is satisfied, (b) $\dim \Sigma \leqslant g - 3$, (c) $\dim \Theta_{\sin g} \leqslant g - 3$, then (24) holds, for every s. Thus another line of proof of the Novikov conjecture consists in proving (b) and (c) directly. The point (b) should perhaps be a refinement of Shiota's Lemma 7. Point (c) should actually be a consequence of the following general statement which seems to be quite reasonable; if X is an irreducible principally polarized abelian variety then $\dim \Theta_{\sin g} \leqslant g - 3$.

References

1. E. Arbarello and C. De Concini, *On a set of equations characterizing Riemann matrices,* Annals of Math. **120** (1984), 119–140.
2. R. C. Gunning, *Some curves in Abelian varieties,* Inv. Math. **66** (1982), 377–389.
3. J. Igusa, *Theta Functions,* Grundlehren der Math. Wiss, 194, Springer, 1972.
4. D. Mumford, *Curves and their Jacobians,* Ann Arbor, University of Michigan Press, 1975.
5. ______, *Tata Lectures on theta II,* Birkhaüser, Boston, 1983.
6. D Mumford and J. Fogarty, *Geometric Invariant Theory,* Ergebnisse der Math. 34, Springer 1982.
7. T. Shiota, *Characterization of Jacobian varieties in terms of soliton equations,* Inv. Math. **83** (1986), 333–382.
8. G. Welters, *A criterion for Jacobian varieties,* Annals of Math. **120** (1984), 497–504.
9. ______, *A characterization of non-hyperelliptic Jacobi varieties.*

Arbarello: Dipartimento di Matematica, "Guido Castelnuovo", Universita "La Sapienza" di Roma, Piazzale Aldo Moro, 5, 1-00185 Roma, Italia

De Concini: Dipartimento di Matematica, Università "Tor Vergata" Via Orazio Ruimondo 00173 Roma, Italia

SUR LES REPRÉSENTATIONS MODULAIRES DE DEGRÉ 2 DE Gal($\overline{\mathbf{Q}}/\mathbf{Q}$)

JEAN-PIERRE SERRE

à Yuri Ivanovich Manin, pour son 50e anniversaire

Le présent travail reprend, en la précisant, une *conjecture* faite en 1973, dont on trouvera un cas particulier dans [44], §3.

Il s'agit de représentations "modulaires" (au sens de Brauer), de degré 2, du groupe de Galois $G_\mathbf{Q} = \mathrm{Gal}(\overline{\mathbf{Q}}/\mathbf{Q})$.

Si $\rho\colon G_\mathbf{Q} \to \mathbf{GL}_2(\overline{\mathbf{F}}_p)$ est une telle représentation, supposée irréductible et de déterminant impair, la conjecture en question affirme que ρ est vraiment "modulaire", i.e. provient d'une forme modulaire parabolique mod p qui est fonction propre des opérateurs de Hecke.

Pour que cet énoncé soit à la fois utilisable et vérifiable numériquement, il est nécessaire de préciser le type de la forme modulaire correspondant à ρ: niveau N, poids k, caractère ε. En ce qui concerne N, les exemples connus suggèrent une réponse simple: N devrait être le *conducteur d'Artin* de ρ (n° 1.2); en particulier, il ne dépendrait que de la ramification de ρ en dehors de p. Une fois N connu, la classe de $k \bmod(p-1)$, et le caractère ε, s'obtiennent sans difficultés à partir du déterminant de ρ (n° 1.3). Il reste à déterminer la valeur exacte du *poids k* (ou plutôt sa valeur minimale). C'est là une question délicate, qui n'avait pas été abordée dans [44]. Il semble que k ne dépende que de la ramification de ρ en p (exposants des caractères de l'inertie modérée, inertie sauvage, etc); la recette précise que je propose est décrite aux n°s 2.2, 2.3 et 2.4.

Les définitions de N, k et ε esquissées ci-dessus font l'objet des §1 et §2. Le §3 contient l'énoncé principal, avec divers compléments. Le §4 explore les conséquences agréables qu'aurait cet énoncé, s'il était vrai: théorème de Fermat, conjecture de Taniyama-Weil, etc. Enfin, le §5 donne un certain nombre d'exemples numériques, pour $p = 2, 3$ et 7.

Ce texte doit beaucoup aux mathématiciens suivants, que j'ai plaisir à remercier:

—John Tate, pour ses nombreuses lettres (notamment en 1973 et 1974) relatives à la conjecture, ainsi qu'aux relations entre poids et inertie en p;

—Jean-Marc Fontaine, dont les résultats sur les représentations locales attachées à la cohomologie ont confirmé les idées de Tate, et ont permis de préciser la valeur du poids k attaché à une représentation;

Received December 5, 1986.

—Gerhard Frey, qui a eu l'idée fondamentale (cf. [17]) que la conjecture de Taniyama-Weil, convenablement complétée, entraîne le théorème de Fermat: i.e., "Weil + epsilon[1] $\Rightarrow$ Fermat";

enfin, et tout spécialement:

—Jean-François Mestre, qui a réussi à programmer et vérifier un nombre d'exemples suffisant pour me convaincre que la conjecture méritait d'être prise au sérieux.

Table des matières   pages

§1. Définitions de N, de ε, et de k mod ($p - 1$)

1.1. *Notations.* La lettre p désigne un nombre premier. On note $\overline{\mathbf{F}}_p$ une clôture algébrique du corps $\mathbf{F}_p$, et $\overline{\mathbf{Q}}$ une clôture algébrique du corps $\mathbf{Q}$. On pose $G_{\mathbf{Q}} = \mathrm{Gal}(\overline{\mathbf{Q}}/\mathbf{Q})$.

On se donne un homomorphisme continu

$$\rho \colon G_{\mathbf{Q}} \to \mathbf{GL}(V),$$

où V est un espace vectoriel de dimension 2 sur $\overline{\mathbf{F}}_p$. L'image de ρ est un groupe fini, noté G; par définition, ce groupe est isomorphe à un sous-groupe de $\mathbf{GL}_2(\mathbf{F}_q)$, où q est une puissance convenable de p. (Si $p \neq 2$, ou si ρ est irréductible, on peut prendre pour $\mathbf{F}_q$ le corps engendré par les *traces* des éléments de G.)

On se propose d'attacher à ρ des entiers positifs N et k, ainsi qu'un caractère de Dirichlet $\varepsilon \colon (\mathbf{Z}/N\mathbf{Z})^* \to \overline{\mathbf{F}}_p^*$.

1.2. *Définition de N.* L'entier N est simplement le *conducteur d'Artin* de ρ, défini comme en caractéristique 0 (cf. [1], [38]), à cela près que l'on se restreint aux places premières à p.

De façon plus précise, soit l un nombre premier $\neq p$. Choisissons une extension à $\overline{\mathbf{Q}}$ de la valuation l-adique de $\mathbf{Q}$, et soient

$$G_0 \supset G_1 \supset \cdots \supset G_i \supset \cdots$$

la suite des groupes de ramification de G correspondant à cette valuation ([38],

---

[1] Il semble que Ribet soit récemment parvenu à éliminer entièrement "epsilon"; d'où "Weil $\Rightarrow$ Fermat".

chap. IV). Notons V_i le sous-espace de V formé des éléments fixés par G_i, et posons

$$(1.2.1) \qquad n(l,\rho) = \sum_{i=0}^{\infty} \frac{1}{(G_0 : G_i)} \dim V/V_i.$$

On peut récrire (1.2.1) sous la forme:

$$(1.2.2) \qquad n(l,\rho) = \dim V/V_0 + b(V),$$

où $b(V)$ est "l'invariant sauvage" du G_0-module V, cf. [39], §19.3.

Ces formules entraînent:

(a) $n(l,\rho)$ est un entier $\geqslant 0$;

(b) $n(l,\rho) = 0$ si et seulement si $G_0 = \{1\}$, i.e., si et seulement si ρ est non ramifiée en l;

(c) $n(l,\rho) = \dim V/V_0$ si et seulement si $G_1 = \{1\}$, i.e., si et seulement si ρ est modérément ramifiée en l.

Il résulte de (a) et (b) que l'on peut définir un entier N par la formule

$$(1.2.3) \qquad N = \prod_{l \neq p} l^{n(l,\rho)}.$$

Nous appellerons N le *conducteur* de ρ; par construction, N est premier à p.

1.3. *Définition du caractère ε et de la classe de k* $\mathrm{mod}(p-1)$. Le *déterminant* de la représentation ρ est un homomorphisme

$$\det \rho \colon G_{\mathbf{Q}} \to \overline{\mathbf{F}}_p^*.$$

Son image est un sous-groupe cyclique fini de $\overline{\mathbf{F}}_p^*$, d'ordre premier à p. On peut donc regarder $\det \rho$ comme un caractère de $G_{\mathbf{Q}}$. Le conducteur de ce caractère divise pN: cela se voit, par exemple, en comparant les formules donnant les conducteurs de ρ et de $\det \rho$. On peut ainsi identifier $\det \rho$ à un homomorphisme de $(\mathbf{Z}/pN\mathbf{Z})^*$ dans $\overline{\mathbf{F}}_p^*$, ou, ce qui revient au même, à un couple d'homomorphismes

$$(1.3.1) \qquad \varphi \colon (\mathbf{Z}/p\mathbf{Z})^* \to \overline{\mathbf{F}}_p^*$$

et

$$(1.3.2) \qquad \varepsilon \colon (\mathbf{Z}/N\mathbf{Z})^* \to \overline{\mathbf{F}}_p^*.$$

Comme $(\mathbf{Z}/p\mathbf{Z})^*$ est cyclique d'ordre $p-1$, l'homomorphisme φ est de la forme

$$(1.3.3) \qquad x \mapsto x^h, \text{ avec } h \in \mathbf{Z}/(p-1)\mathbf{Z}.$$

182   JEAN-PIERRE SERRE

Ceci peut se récrire:

$$(1.3.4) \qquad\qquad \varphi = \chi^h,$$

où $\chi: G_{\mathbf{Q}} \to \mathbf{F}_p^*$ désigne le p-ième *caractère cyclotomique* de $G_{\mathbf{Q}}$ (celui qui donne l'action de $G_{\mathbf{Q}}$ sur les racines p-ièmes de l'unité).

On peut résumer ces formules en disant que, si l est un nombre premier ne divisant pas pN, et si $\mathrm{Frob}_{l,\rho}$ est l'élément de Frobenius correspondant dans G (défini à conjugaison près), on a

$$(1.3.5) \qquad\qquad \det(\mathrm{Frob}_{l,\rho}) = l^h \varepsilon(l) \quad \text{dans } \overline{\mathbf{F}}_p^*.$$

Au §2, nous définirons un certain entier k attaché à ρ et nous verrons (n° 2.5) que h est simplement la classe de $k - 1 \bmod(p - 1)$, de sorte que (1.3.5) peut se récrire:

$$(1.3.6) \qquad\qquad \det(\mathrm{Frob}_{l,\rho}) = l^{k-1} \varepsilon(l) \quad \text{dans } \overline{\mathbf{F}}_p^*.$$

Remarque. Soit c l'élément d'ordre 2 de $G_{\mathbf{Q}}$ donné par la conjugaison complexe (relativement à un plongement de $\overline{\mathbf{Q}}$ dans $\mathbf{C}$). L'image de c dans $(\mathbf{Z}/pN\mathbf{Z})^*$ est -1. On en conclut que:

$$(1.3.7) \qquad\qquad \det \rho(c) = (-1)^{k-1} \varepsilon(-1).$$

Dans la suite, on s'intéressera uniquement au cas où $\det \rho$ est *impair*, i.e.

$$(1.3.8) \qquad\qquad \det \rho(c) = -1,$$

ou encore:

$$(1.3.9) \qquad\qquad \varepsilon(-1) = (-1)^k \quad \text{dans } \overline{\mathbf{F}}_p^*.$$

Si $p = 2$, cette condition est automatiquement satisfaite, puisque $-1 = 1$. Si $p \neq 2$, elle signifie que $\rho(c)$ est conjuguée de la matrice $\begin{pmatrix} 1 & 0 \\ 0 & -1 \end{pmatrix}$.

§2. L'entier k. Le but de ce § est de définir l'entier k (le "poids") associé à une représentation ρ. Les n°s 2.1 à 2.4 contiennent la définition générale; les n°s 2.5 à 2.9 en donnent divers exemples.

2.1. *Préliminaires.* L'entier k ne dépend que de la restriction de la représentation ρ au groupe de décomposition en p (et même, en fait, au groupe d'inertie). Aussi, pour le définir, allons-nous partir d'une représentation "locale en p":

$$\rho_p: G_p \to \mathbf{GL}(V) \simeq \mathbf{GL}_2(\overline{\mathbf{F}}_p),$$

où $G_p = \mathrm{Gal}(\overline{\mathbf{Q}}_p/\mathbf{Q}_p)$.

Nous noterons I le groupe d'inertie de G_p, et I_p le plus grand pro-p-sous-groupe de I (groupe d'inertie *sauvage*). Le quotient $I_t = I/I_p$ est le *groupe d'inertie modérée* de G_p; il s'identifie à $\varprojlim \mathbf{F}_{p^n}^*$, cf. [41], prop. 2. Un caractère de I_t sera dit *de niveau n* s'il se factorise par $\mathbf{F}_{p^n}^*$, et ne se factorise par aucun $\mathbf{F}_{p^m}^*$, où m est un diviseur strict de n.

Si V^{ss} désigne le semi-simplifié de V vis-à-vis de l'action de G_p, le groupe I_p agit trivialement sur V^{ss} ([41], prop. 4), de sorte que I_t opère sur V^{ss}. Cette action de I_t est diagonalisable; elle est donnée par deux caractères

$$\varphi, \varphi' : I_t \to \overline{\mathbf{F}}_p^*.$$

PROPOSITION 1. *Les caractères φ et φ' donnant l'action de I_t sur V^{ss} sont de niveau 1 ou 2. S'ils sont de niveau 2, ils sont conjugués: on a $\varphi' = \varphi^p$ et $\varphi = \varphi'^p$.*

Soit s un élément de G_p dont l'image dans $G_p/I = \mathrm{Gal}(\overline{\mathbf{F}}_p/\mathbf{F}_p)$ soit l'automorphisme de Frobenius $x \mapsto x^p$. On vérifie facilement que, si $u \in I$, on a $sus^{-1} \equiv u^p$ (mod I_p): la conjugaison par s opère sur $I_t = I/I_p$ par $u \mapsto u^p$. Il en résulte que l'ensemble $\{\varphi, \varphi'\}$ est stable par l'opération de puissance p-ième. D'où deux cas:

(a) on a $\varphi^p = \varphi$, $\varphi'^p = \varphi'$ et les deux caractères φ et φ' sont de niveau 1;

(b) on a $\varphi^p = \varphi'$, $\varphi'^p = \varphi$, $\varphi \neq \varphi'$, et les deux caractères φ et φ' sont de niveau 2.

Cela démontre la prop. 1.

Nous allons maintenant nous occuper séparément de ces deux cas.

2.2 *Définition de k lorsque φ et φ' sont de niveau 2.*   Supposons que φ et φ' soient de niveau 2. La représentation V est alors *irréductible*, car si elle contenait un sous-espace stable de dimension 1, l'action de I_t sur ce sous-espace se ferait par un caractère prolongeable à G_p, donc de niveau 1. Appelons ψ et $\psi' = \psi^p$ les deux *caractères fondamentaux de niveau* 2 de I_t ([41], n° 1.7), autrement dit les deux caractères $I_t \to \mathbf{F}_{p^2}^* \to \overline{\mathbf{F}}_p^*$ correspondant aux deux plongements du corps $\mathbf{F}_{p^2}$ dans le corps $\overline{\mathbf{F}}_p$. On peut écrire φ de manière unique sous la forme

$$(2.2.1) \qquad \varphi = \psi^{a+pb} = \psi^a \psi'^b, \quad \text{avec} \quad 0 \leqslant a, b \leqslant p-1.$$

On a $b \neq a$, car sinon φ serait égal à $(\psi \psi')^a = \chi^a$, où χ est le caractère cyclotomique (ou plutôt sa restriction à I), et cela contredirait l'hypothèse que φ est de niveau 2. De plus, comme φ' est conjugué de φ, on a

$$(2.2.2) \qquad \varphi' = \psi^b \psi'^a.$$

Quitte à permuter φ et φ', on peut donc supposer que:

$$(2.2.3) \qquad 0 \leqslant a < b \leqslant p-1.$$

Ceci fait, l'entier k attaché à ρ_p est défini par:

$$(2.2.4) \qquad\qquad k = 1 + pa + b.$$

Remarques

(1) La plus petite valeur possible de k est $k = 2$, qui est obtenue lorsque $a = 0$, $b = 1$, c'est-à-dire lorsque φ et φ' sont égaux aux *caractères fondamentaux* ψ et ψ' de niveau 2.

(2) Dans le cas particulier $a = 0$, on a $(\varphi, \varphi') = (\psi^b, \psi'^b)$, avec $1 \leqslant b \leqslant p - 1$, et la définition de k se réduit à:

$$k = 1 + b \qquad \left(\text{d'où } 2 \leqslant k \leqslant p\right).$$

Le cas général se ramène à celui-là par "torsion". En effet, on peut écrire ρ_p sous la forme

$$\rho_p = \chi^a \otimes \rho_p',$$

où χ est le caractère cyclotomique (vu comme caractère de G_p, et pas seulement de I). Le couple (a, b) attaché à ρ_p' est alors $(0, b - a)$, et l'entier k correspondant est $k' = 1 + b - a$. On peut donc récrire (2.2.4) sous la forme

$$(2.2.5) \qquad\qquad k = k' + a(p + 1).$$

(Comparer avec la formule donnant la filtration de la forme modulaire "tordue" d'une forme donnée, cf. [24], [42].)

2.3. *Définition de k lorsque φ et φ' sont de niveau 1, et que I_p opère trivialement.* On suppose que l'action de I sur V est semi-simple, et qu'elle est donnée par deux caractères (φ, φ') qui sont des puissances χ^a et χ^b du caractère cyclotomique χ:

$$\rho_p | I = \begin{pmatrix} \chi^a & 0 \\ 0 & \chi^b \end{pmatrix}.$$

Les entiers a et b sont déterminés $\mathrm{mod}(p - 1)$. On les normalise de telle sorte que $0 \leqslant a, b \leqslant p - 2$. De plus, quitte à permuter φ et φ', on peut supposer que $a \leqslant b$. On a donc

$$(2.3.1) \qquad\qquad 0 \leqslant a \leqslant b \leqslant p - 2.$$

L'entier k est alors défini par:

$$(2.3.2) \qquad\qquad k = \begin{cases} 1 + pa + b & \text{si } (a, b) \neq (0, 0) \\ p & \text{si } (a, b) = (0, 0). \end{cases}$$

Remarques

(1) Ici encore, la plus petite valeur possible de k est $k = 2$, qui correspond à $\varphi = 1$, $\varphi' = \chi$.

(2) Le cas $(a, b) = (0, 0)$ est celui où I opère trivialement sur V, autrement dit où ρ_p est *non ramifiée*. La formule générale $k = 1 + pa + b$ donnerait alors $k = 1$. Comme les formes modulaires de poids 1 ont un comportement quelque peu exceptionnel, j'ai préféré les éviter, et "décaler" k par $p - 1$; d'où la valeur $k = p$ adoptée.

(3) Lorsqu'on tord ρ_p par les puissances successives $\chi, \chi^2,\ldots$ du caractère χ, les entiers k correspondants forment un *cycle de Tate*, cf. [21], [22].

2.4. *Définition de k lorsque I_p n'opère pas trivialement.* On suppose que I_p n'opère pas trivialement, i.e., que l'action de I n'est pas modérée. Les éléments de V fixés par I_p forment alors une droite D, qui est stable par G_p. L'action de G_p sur V/D (resp. sur D) se fait par un caractère θ_1 (resp. θ_2) de G_p:

$$(2.4.1) \qquad \rho_p = \begin{pmatrix} \theta_2 & * \\ 0 & \theta_1 \end{pmatrix}.$$

On peut écrire θ_1 et θ_2 de façon unique sous la forme

$$(2.4.2) \qquad \theta_1 = \chi^\alpha \varepsilon_1, \theta_2 = \chi^\beta \varepsilon_2, \qquad (\alpha, \beta \in \mathbf{Z}/(p-1)\mathbf{Z}),$$

où ε_1 et ε_2 sont des caractères non ramifiés de G_p à valeurs dans $\overline{\mathbf{F}}_p^*$. La restriction de ρ_p à I est donc:

$$\rho_p | I = \begin{pmatrix} \chi^\beta & * \\ 0 & \chi^\alpha \end{pmatrix}.$$

On normalise les exposants α et β par:

$$(2.4.3) \qquad 0 \leqslant \alpha \leqslant p - 2 \quad \text{et} \quad 1 \leqslant \beta \leqslant p - 1.$$

(Noter qu'ici χ^α et χ^β ne jouent pas des rôles symétriques.) On pose:

$$(2.4.4) \qquad a = \mathrm{Inf}(\alpha, \beta) \quad \text{et} \quad b = \mathrm{Sup}(\alpha, \beta).$$

Pour définir k, on distingue deux cas:

(i) *Le cas $\beta \neq \alpha + 1$* (*i.e.*, $\chi^\beta \neq \chi \cdot \chi^\alpha$). On pose alors, comme au n° 2.3:

$$(2.4.5) \qquad k = 1 + pa + b.$$

(Noter le cas où $\chi^\alpha = \chi^\beta = 1$, $p \geqslant 3$, où (2.4.3) impose $\alpha = 0$, $\beta = p - 1$, de sorte que (2.4.5) donne $k = p$, comme dans (2.3.2).)

(ii) *Le cas $\beta = \alpha + 1$* (*i.e.*, $\chi^\beta = \chi \cdot \chi^\alpha$).

La définition de k dépend alors du type de la ramification sauvage. Il y a deux types possibles, que j'appellerai respectivement *peu ramifié* et *très ramifié*. On les définit de la manière suivante:

Soit $K_0 = \mathbf{Q}_{p,\mathrm{nr}}$ l'extension non ramifiée maximale de $\mathbf{Q}_p$; on a $I = \mathrm{Gal}(\overline{\mathbf{Q}}_p/K_0)$. Le groupe $\rho_p(I)$ est le groupe de Galois d'une certaine extension totalement ramifiée K de K_0, et le groupe d'inertie sauvage $\rho_p(I_p)$ est le groupe de Galois de K/K_t, où K_t est la plus grande extension modérément ramifiée de K_0 contenue dans K.

$$
\begin{array}{c}
K \\
| \\
K_t \\
| \\
K_0
\end{array}
$$

Du fait que $\beta = \alpha + 1$, on déduit que $\mathrm{Gal}(K_t/K_0) = (\mathbf{Z}/p\mathbf{Z})^*$, donc que $K_t = K_0(z)$, où z est une racine primitive p-ième de l'unité. D'autre part, le groupe $\mathrm{Gal}(K/K_t) = \rho_p(I_p)$ est un groupe abélien élémentaire de type $(p, \ldots, p)$, représentable matriciellement par $\left(\begin{smallmatrix} 1 & * \\ 0 & 1 \end{smallmatrix} \right)$. De plus, l'hypothèse $\beta = \alpha + 1$ entraîne que l'action par conjugaison de $\mathrm{Gal}(K_t/K_0) = (\mathbf{Z}/p\mathbf{Z})^*$ sur $\mathrm{Gal}(K/K_t)$ est l'action évidente. Utilisant la théorie de Kummer, on en déduit que K peut s'écrire sous la forme

$$(2.4.6) \qquad K = K_t\big(x_1^{1/p}, \ldots, x_m^{1/p}\big), \quad \text{où } p^m = [K : K_t],$$

les x_i étant des éléments de K_0^*/K_0^{*p}. Si v_p désigne la valuation de K_0, normalisée par $v_p(p) = 1$, nous dirons que l'extension K (ou la représentation ρ_p) est *peu ramifiée* si

$$(2.4.7) \qquad v_p(x_i) \equiv 0 \,(\mathrm{mod}\; p) \quad \text{pour } i = 1, \ldots, m,$$

i.e., si les x_i peuvent être choisis parmi les *unités* de K_0. Dans le cas contraire, nous dirons que K et ρ_p sont *très ramifiées*.

Remarques

(1) Le cas très ramifié n'est possible que si les caractères ε_1 et ε_2 définis par (2.4.2) sont égaux, et l'on a alors $m = 1$ ou $m = 2$: cela se voit en utilisant l'action par conjugaison de G_p sur $\rho_p(I_p)$.

(2) Soit π une uniformisante de K_t, par exemple $\pi = 1 - z$ ou $\pi = p^{1/(p-1)}$. Si K/K_t est peu ramifiée, les $p^m - 1$ caractères d'ordre p associés à cette extension

sont tous de conducteur (π^2); dans le cas très ramifié, $p^m - p^{m-1}$ de ces caractères sont de conducteur $(\pi^{p+1}) = (p\pi^2)$ et les $p^{m-1} - 1$ autres sont de conducteur (π^2).

Nous pouvons maintenant définir l'entier k:
(ii$_1$) *Le cas* $\beta = \alpha + 1$, *peu ramifié*
La formule est la même que dans le cas $\beta \neq \alpha + 1$:

$$(2.4.8) \qquad k = 1 + pa + b = 2 + \alpha(p + 1).$$

(ii$_2$) *Le cas* $\beta = \alpha + 1$, *très ramifié*
On ajoute $p - 1$ (resp. 2 si $p = 2$) à ce que donnerait (2.4.8):

$$(2.4.9) \qquad k = \begin{cases} 1 + pa + b + p - 1 = (\alpha + 1)(p + 1) & \text{si } p \neq 2 \\ 4 & \text{si } p = 2. \end{cases}$$

Les formules (2.2.4), (2.3.2), (2.4.5), (2.4.8), et (2.4.9) fournissent une définition complète de l'entier k attaché à la représentation ρ_p considérée. Voici quelques propriétés qui résultent de cette définition:

2.5. *Classe de k* mod $(p - 1)$.

PROPOSITION 2.   *On a.*

$$(2.5.1) \qquad \det \rho_p | I = \chi^{k-1}.$$

(Comme χ est d'ordre $p - 1$, cette formule montre que la classe de k mod$(p - 1)$ est déterminée par $\det \rho_p$, et même seulement par la restriction de $\det \rho_p$ au groupe d'inertie I.)
Vérifions (2.5.1) dans le cas de hauteur 2 (cf. n° 2.2). On a alors

$$\det \rho_p | I = \varphi \cdot \varphi' = \left(\psi^a \psi'^b\right)\left(\psi^b \psi'^a\right) = \left(\psi\psi'\right)^{a+b} = \chi^{a+b}$$

$$= \chi^{k-1},$$

puisque $k - 1 = pa + b \equiv a + b \bmod (p - 1)$.
Les autres cas sont analogues.

On peut récrire (2.5.2) sous la forme

$$(2.5.2) \qquad \det \rho_p = \varepsilon_p \cdot \chi^{k-1},$$

où ε_p est un *caractère non ramifié* de G_p à valeurs dans $\overline{\mathbf{F}}_p^*$. Dans le cas où ρ_p provient d'une représentation globale ρ de $\mathrm{Gal}(\overline{\mathbf{Q}}/\mathbf{Q})$, le caractère ε_p n'est autre que la *p-composante* du caractère ε défini au n° 1.3; on a

$$(2.5.3) \qquad \varepsilon_p(\mathrm{Frob}_p) = \varepsilon(p),$$

où Frob_p est un élément de Frobenius de G_p.

2.6. *Valeurs de* k. Pour $p \neq 2$, les valeurs possibles de k sont les nombres de l'intervalle $[2, p^2 - 1]$ qui peuvent s'écrire sous la forme

$$k = 1 + a_0 + pa_1, \qquad 0 \leqslant a_0, a_1 \leqslant p - 1,$$

avec $a_1 \leqslant a_0 + 1$. Ainsi, pour $p = 3$, on a $k = 2, 3, 4, 5, 6$ ou 8.

Pour $p = 2$, on a $k = 2$ si l'action de I_p est triviale, ou peu ramifiée, et $k = 4$ si l'action de I_p est très ramifiée.

Exemple. Prenons $p = 2$. Soit $u: G_2 \to \mathbf{Z}/2\mathbf{Z}$ un homomorphisme surjectif, et soit $\rho_2: G_2 \to \mathbf{GL}_2(\mathbf{F}_2)$ la représentation définie par

$$s \mapsto \begin{pmatrix} 1 & u(s) \\ 0 & 1 \end{pmatrix}.$$

Soit $K/\mathbf{Q}_2$ l'extension quadratique correspondant au noyau de u. On a alors:

$$k = 2 \text{ si} \quad K/\mathbf{Q}_2 \text{ est non ramifiée, i.e., } K = \mathbf{Q}_2(\sqrt{5}\,);$$

$$k = 2 \text{ si} \quad \mathrm{discr}(K/\mathbf{Q}_2) = (4), \text{i.e., } K = \mathbf{Q}_2(\sqrt{-1}\,) \text{ ou } \mathbf{Q}_2(\sqrt{-5}\,);$$

$$k = 4 \text{ si} \quad \mathrm{discr}(K/\mathbf{Q}_2) = (8), \text{i.e., } K = \mathbf{Q}_2(\sqrt{2}\,), \mathbf{Q}_2(\sqrt{-2}\,), \mathbf{Q}_2(\sqrt{10}\,)$$

ou $\mathbf{Q}_2(\sqrt{-10}\,)$.

2.7. *Conditions pour que* $k \leqslant p + 1$, *lorsque* $p \neq 2$. Supposons $p \neq 2$. On a $k \leqslant p + 1$ si et seulement si l'une des deux conditions suivantes est satisfaite:

(2.7.1) Il existe un quotient V/D de V, de dimension 1, sur lequel I opère trivialement (i.e., *V a un quotient étale de dimension* 1); c'est le cas $a = 0$ des n°s 2.3 et 2.4.

(2.7.2) L'action de I sur V se fait par deux caractères modérés de la forme (ψ^b, ψ'^b), avec $1 \leqslant b \leqslant p - 1$, où ψ et ψ' sont les deux caractères fondamentaux de niveau 2 de I_t; c'est le cas $a = 0$ du n° 2.2.

Remarques

(1) On a $k = p + 1$ si et seulement si la restriction de ρ_p au groupe d'inertie I est de la forme $\begin{pmatrix} \chi & * \\ 0 & 1 \end{pmatrix}$ et est très ramifiée.

(2) Quelle que soit la représentation ρ_p, il existe une "tordue" $\chi^m \otimes \rho_p$ de ρ_p dont l'invariant k est $\leqslant p + 1$ (comparer à [44], th.3).

2.8. *Conditions pour que* $k = 2$.
L'énoncé suivant résulte immédiatement des définitions:

PROPOSITION 3. *Pour que l'invariant* k *de* ρ_p *soit égal à* 2, *il faut et il suffit que* $\rho_p | I$ *soit de l'un des deux types suivants:*

$$(2.8.1) \qquad\qquad \rho_p | I \simeq \begin{pmatrix} \psi' & 0 \\ 0 & \psi \end{pmatrix},$$

où $\psi, \psi'\colon I \to I_t \to \mathbf{F}_{p^2}^*$ *sont les deux caractères fondamentaux de I de niveau* 2;

$$(2.8.2) \qquad \rho_p|I \simeq \begin{pmatrix} \chi & 0 \\ 0 & 1 \end{pmatrix} \quad ou \quad \begin{pmatrix} \chi & * \\ 0 & 1 \end{pmatrix},$$

l'action du groupe d'inertie sauvage I_p étant, soit triviale, soit peu ramifiée.

On peut donner une autre caractérisation de ce cas, en termes de *schémas en groupes de type* (p, p). Pour l'énoncer, je me bornerai au cas où ρ_p prend ses valeurs dan $\mathbf{GL}_2(\mathbf{F}_p)$, donc définit un schéma en groupes (étale) de type (p, p) sur le corps $\mathbf{Q}_p$ (dans le cas général, il faudrait parler de "schémas en $\mathbf{F}_q$-vectoriels" au sens de Raynaud [35]). On peut se demander si ce schéma en groupes se prolonge en un schéma en groupes fini et plat sur $\mathbf{Z}_p$, cf. [35]; s'il en est ainsi, je dirai (cf. [48]) que la représentation ρ_p est *finie* en p.

PROPOSITION 4.   *On a $k = 2$ si et seulement si les deux conditions suivantes sont satisfaites*:

$$(2.8.3) \qquad \det \rho_p|I = \chi;$$

$$(2.8.4) \qquad \rho_p \text{ est finie en } p.$$

D'après le n° 2.5, la condition (2.8.3) équivaut à:

$$(2.8.5) \qquad k \equiv 2 \mod(p - 1).$$

Elle est donc nécessaire pour que k soit égal à 2. Montrons qu'elle est suffisante lorsque ρ_p est finie en p. D'après [35], cor. 3.4.4, chacun des caractères φ et φ' de I_t associés à ρ_p peut alors s'écrire sous la forme

$$\psi^n \psi'^{n'}, \quad \text{avec } 0 \leqslant n, n' \leqslant 1,$$

où ψ et ψ' sont comme ci-dessus les deux caractères fondamentaux de niveau 2. Cela fait quatre possibilités

$$1, \psi, \psi' \quad \text{et} \quad \psi\psi' = \chi$$

(qui se réduisent d'ailleurs à trois pour $p = 2$ puisque χ est alors égal à 1). Comme $\varphi\varphi' = \chi$ d'après (2.8.1), deux cas seulement sont possibles:

(i) $$\{\varphi, \varphi'\} = \{\psi, \psi'\}$$

et

(ii) $$\{\varphi, \varphi'\} = \{1, \chi\}.$$

Le cas (i) donne (2.8.1), d'où $k = 2$, comme annoncé. Occupons-nous du cas (ii), en nous bornant, pour simplifier, au cas $p \neq 2$ (le cas $p = 2$ est un peu différent, mais se traite de façon analogue). Soit J le schéma en groupes fini et plat sur $\mathbf{Z}_p$ prolongeant le schéma sur $\mathbf{Q}_p$ défini par ρ_p (d'après [35], prop. 3.3.2, ce schéma est unique). Il résulte de (ii) que ρ_p est réductible, et il en est de même de J. D'où l'existence d'une suite exacte de schémas en groupes finis et plats sur $\mathbf{Z}_p$:

$$(2.8.6) \qquad\qquad 0 \to A \to J \to B \to 0,$$

où A et B sont des schémas en groupes finis et plats d'ordre p. De plus, (ii) impose que l'un de ces schémas soit étale, et que l'autre soit de type multiplicatif. Il existe donc une extension finie étale R de $\mathbf{Z}_p$ sur laquelle A ou B devient isomorphe au schéma étale constant $\mathbf{Z}/p\mathbf{Z}$, et B ou A au schéma μ_p des racines p-èmes de l'unité. Sur R, la suite exacte (2.8.6) devient:

$$0 \to \mathbf{Z}/p\mathbf{Z} \to J \to \mu_p \to 0$$

où

$$0 \to \mu_p \to J \to \mathbf{Z}/p\mathbf{Z} \to 0.$$

Dans le premier cas, on voit facilement que l'extension J est *scindée* (utiliser la composante connexe de l'élément neutre), i.e., isomorphe sur R à $\mathbf{Z}/p\mathbf{Z} \oplus \mu_p$; d'où (2.8.2), avec action triviale de I_p, ce qui entraîne bien $k = 2$. Dans le second cas, on constate (suite exacte de Kummer) que la classe de l'extension J est donnée par un élément $u \in R^*/R^{*p}$, d'où

$$\rho_p | I \sim \begin{pmatrix} \chi & * \\ 0 & 1 \end{pmatrix},$$

et le corps K du n° 2.4 est égal à $K_t(u^{1/p})$; comme u est une unité, l'extension K/K_t est, soit non ramifiée, soit peu ramifiée, d'où encore $k = 2$ d'après (2.8.2). (Le fait que K/K_t ne soit pas très ramifiée peut aussi se déduire d'un résultat général de Fontaine, cf. [15], th. 1.)

Reste à prouver que $k = 2$ entraîne que ρ_p est finie en p. D'après la prop. 3, il y a deux cas à considérer:

(a) Celui où $\rho_p | I$ est donné par les deux caractères fondamentaux ψ et ψ'. Ce cas est traité dans Raynaud [35], th. 2.4.3.

(b) Celui où $\rho_p | I$ est de la forme $\begin{pmatrix} \chi & * \\ 0 & 1 \end{pmatrix}$, avec action de I_p triviale ou peu ramifiée. On fait alors une construction directe, basée sur la classification des extensions de $\mathbf{Z}/p\mathbf{Z}$ par μ_p, cf. ci-dessus (de façon un peu plus précise, on commence par remplacer $\mathbf{Z}_p$ par une extension finie étale R convenable, on construit l'extension en question sur R, et l'on descend ensuite à $\mathbf{Z}_p$).

2.9. *Exemple de calcul de k: points de p-division d'une courbe elliptique semi-stable.* Soit E une courbe elliptique sur $\mathbf{Q}_p$, d'invariant modulaire j_E, et soit E_p le groupe des points de p-division de E. L'action de G_p sur E_p définit une représentation

$$\rho_p \colon G_p \to \mathrm{Aut}(E_p) \simeq \mathbf{GL}_2(\mathbf{F}_p).$$

Comme $\det \rho_p = \chi$, l'invariant k associé à ρ_p satisfait à:

$$(2.9.1) \qquad\qquad k \equiv 2 \bmod (p-1).$$

Nous allons déterminer la valeur de k, en nous bornant au cas où E est *semi-stable*, i.e. a, soit bonne réduction, soit mauvaise réduction de type multiplicatif (cf. [41], n°s 1.11 et 1.12):

PROPOSITION 5.   (i) *Si E a bonne réduction, on a $k = 2$.*
(ii) *Si E a mauvaise réduction de type multiplicatif, on a*

$$k = \begin{cases} 2 & \textit{si } v_p(j_E) \textit{ est divisible par } p \\ p+1 & \textit{sinon.} \end{cases}$$

(Ici, et dans toute la suite, on note v_p la valuation p-adique, normalisée par la condition $v_p(p) = 1$.)

Lorsque E a bonne réduction, ρ_p est évidemment finie en p, et l'assertion (i) résulte de la prop. 4.

Lorsque E a mauvaise réduction de type multiplicatif, on utilise le modèle de Tate ([41], n° 1.12). Celui-ci montre que, après extension quadratique non ramifiée de $\mathbf{Q}_p$, on a une suite exacte de modules galoisiens

$$0 \to \mu_p \to E_p \to \mathbf{Z}/p\mathbf{Z} \to 0$$

d'où

$$\rho_p|I \simeq \begin{pmatrix} \chi & * \\ 0 & 1 \end{pmatrix}.$$

Soit de plus q_E l'élément de $\mathbf{Q}_p^*$ défini par l'identité

$$j_E = q_E^{-1} + 744 + 196884 q_E + \dots$$

On constate que l'extension K/K_t du n° 2.4 est $K = K_t(q_E^{1/p})$. Cette extension est donc très ramifiée si et seulement si $v_p(q_E)$ n'est pas divisible par p; comme $v_p(q_E) = -v_p(j_E)$, on en déduit bien (ii).

Remarques

(1) Supposons que l'on soit dans le cas (ii) avec $k = 2$, i.e. que E ait mauvaise réduction de type multiplicatif et que $v_p(j_E)$ soit divisible par p. Posons $m =$

$-v_p(j_E)/p$ et $u = p^{p^m} j_E$, de sorte que u est une unité p-adique et que q_E est égal au produit de u^{-1} par la puissance p-ème d'un élément de K_t. On a alors $K = K_t(u^{1/p})$ et l'on voit que:

(a) si $u^{p-1} \equiv 1 \pmod{p^2}$, on a $K = K_t$ et $\rho_p|I \simeq \begin{pmatrix} \chi & 0 \\ 0 & 1 \end{pmatrix}$;

(b) si $u^{p-1} \not\equiv 1 \pmod{p^2}$, on a $[K : K_t] = p$ et $\rho_p|I \simeq \begin{pmatrix} \chi & * \\ 0 & 1 \end{pmatrix}$.

Le cas (b) peut effectivement se présenter, contrairement à ce qui est affirmé dans [6], prop. 5.1.(3)(d).

(2) Des calculs analogues à ceux de la prop. 5 (mais plus compliqués) sont possibles lorsque E a mauvaise réduction de type additif. Je me borne à donner le résultat dans un cas particulier typique, celui où $p \equiv 1 \pmod 3$, et où l'équation minimale de E est de la forme

$$y^2 = x^3 + Ax + B,$$

avec $v_p(A) \geqslant 1$ et $v_p(B) = 1$ (type c_1 de Néron).
On trouve alors

$$\rho_p|I \simeq \begin{pmatrix} \chi^\beta & 0 \\ 0 & \chi^\alpha \end{pmatrix} \quad \text{ou} \quad \begin{pmatrix} \chi^\beta & * \\ 0 & \chi^\alpha \end{pmatrix},$$

avec $\alpha = (p - 1)/6$ et $\beta = (5p + 1)/6$.
Si $p > 7$, cela entraîne $k = 1 + p\alpha + \beta = 2 + (p - 1)(p + 5)/6$. Par contre, pour $p = 7$, on peut avoir, soit $k = 2 + (p - 1)(p + 5)/6 = 14$, soit $k = 2$, ce dernier cas survenant si $v_p(A) \geqslant 2$.

§3. Enoncé de la conjecture

3.1. *Rappels sur les formes paraboliques en caractéristique p.*   Soient:
N un entier $\geqslant 1$, premier à p;
k un entier $\geqslant 2$;
ε un caractère $(\mathbf{Z}/N\mathbf{Z})^* \to \overline{\mathbf{F}}_p^*$.
Supposons que:

$$(3.1.1) \qquad \begin{cases} (-1)^k = \varepsilon(-1) & \text{si } p \neq 2 \\ k \text{ est pair} & \text{si } p = 2. \end{cases}$$

Nous aurons à utiliser la notion de *forme parabolique de type* (N, k, ε) *à coefficients dans* $\overline{\mathbf{F}}_p$. Comme plusieurs définitions sont possibles (cf. [23] et [24] par exemple), il convient de préciser ce que nous entendons par là:
Identifions $\overline{\mathbf{Q}}$ à un sous-corps de $\mathbf{C}$, et choisissons une place de $\overline{\mathbf{Q}}$ au-dessus de p. Si $\overline{\mathbf{Z}}$ désigne l'anneau des entiers de $\overline{\mathbf{Q}}$, le choix de la place en question définit

un homomorphisme $\overline{\mathbf{Z}} \to \overline{\mathbf{F}}_p$ que nous noterons $z \mapsto \tilde{z}$. Notons enfin

$$\varepsilon_0 \colon (\mathbf{Z}/N\mathbf{Z})^* \to \overline{\mathbf{Z}}^*$$

le relèvement multiplicatif de ε, i.e., l'unique caractère à valeurs dans les racines de l'unité d'ordre premier à p tel que

$$\varepsilon_0(x)^{\sim} = \varepsilon(x) \quad \text{pour tout } x \in (\mathbf{Z}/N\mathbf{Z})^*.$$

D'après (3.1.1), on a $\varepsilon_0(-1) = (-1)^k$. On peut donc parler des *formes paraboliques de type* (k, ε_0) *sur* $\Gamma_0(N)$, au sens usuel. Rappelons (cf. par exemple [11]) qu'une telle forme est une série

$$(3.1.2) \qquad\qquad F = \sum_{n \geqslant 1} A_n q^n \qquad (q = e^{2\pi i z}),$$

convergeant dans le demi-plan $\mathrm{Im}(z) > 0$ et satisfaisant aux deux conditions suivantes:

(a) $F((az + b)/(cz + d)) = \varepsilon_0(d)(cz + d)^k F(z)$ pour tout $\begin{pmatrix} a & b \\ c & d \end{pmatrix} \in \Gamma_0(N)$ et tout $z \in \mathbf{C}$ tel que $\mathrm{Im}(z) > 0$;

(b) F s'annule aux pointes, i.e., pour tout $\begin{pmatrix} a & b \\ c & d \end{pmatrix} \in \mathbf{SL}_2(\mathbf{Z})$, la fonction

$$z \mapsto (cz + d)^{-k} F((az + b)/(cz + d))$$

a un développement en série du type (3.1.2), avec q remplacé par $q^{1/N}$.

Pour abréger, nous dirons qu'une telle forme F est *de type* (N, k, ε_0).

Nous pouvons maintenant définir la notion analogue en caractéristique p:

Définition.   *Une forme parabolique de type* (N, k, ε) *à coefficients dans* $\overline{\mathbf{F}}_p$ *est une série formelle*

$$f = \sum_{n \geqslant 1} a_n q^n, \qquad a_n \in \overline{\mathbf{F}}_p,$$

telle qu'il existe une forme parabolique

$$F = \sum_{n \geqslant 1} A_n q^n, \qquad A_n \in \overline{\mathbf{Z}},$$

de type (N, k, ε_0) *au sens rappelé ci-dessus, qui est telle que* $\tilde{F} = f$, *i.e., que* $\tilde{A}_n = a_n$ *pour tout* n.

(Au lieu de supposer que les A_n appartiennent à $\overline{\mathbf{Z}}$, on pourrait se borner à demander qu'ils appartiennent à l'*anneau local* de la place de $\overline{\mathbf{Q}}$ choisie au début. Cela ne changerait rien.)

Notons $S(N, k, \varepsilon)$ l'espace des f du type ci-dessus. Cet espace jouit des propriétés suivantes:

(3.1.3) $S(N, k, \varepsilon)$ ne dépend pas du choix de la place p-adique de $\overline{\mathbf{Q}}$ utilisée pour le définir. De plus, sa dimension sur $\overline{\mathbf{F}}_p$ est égale à la dimension de l'espace analogue $S(N, k, \varepsilon_0)$ sur $\mathbf{C}$.
Cela résulte de Shimura [51], th. 3.52 (voir aussi [11], prop. 2.7).

(3.1.4) $S(N, k, \varepsilon)$ est stable sous l'action des opérateurs de Hecke:

$$T_l: \sum a_n q^n \mapsto \sum a_{ln} q^n + \varepsilon(l) l^{k-1} \sum a_n q^{ln} \qquad (l \nmid pN),$$

$$U_l: \sum a_n q^n \mapsto \sum a_{ln} q^n \qquad\qquad (l \mid pN).$$

Pour les T_l et les U_l (l premier $\neq p$), cela résulte des propriétés analogues en caractéristique zéro. Pour U_p, il faut noter que c'est la réduction (mod p) de l'opérateur de Hecke

$$T_p: \sum a_n q^n \mapsto \sum a_{pn} q^n + \varepsilon_0(p) p^{k-1} \sum a_n q^{pn},$$

grâce à l'hypothèse $k \geqslant 2$.

(3.1.5) Les opérateurs de Hecke commutent entre eux. Si

$$f = \sum a_n q^n, \qquad f \neq 0,$$

est une fonction propre de ces opérateurs, on peut multiplier f par un scalaire non nul de telle sorte que $a_1 = 1$. Une fois f ainsi normalisée, on a $T_l(f) = a_l f$ pour $l \nmid pN$ et $U_l(f) = a_l f$ pour $l \mid pN$: les a_l sont les valeurs propres des T_l et U_l. De plus, la série formelle de Dirichlet

$$L_f(s) = \sum a_n n^{-s} \quad \left(\text{à coefficients dans } \overline{\mathbf{F}}_p\right)$$

est donnée par le produit eulérien usuel:

$$L_f(s) = \prod_{l \mid pN} \left(1 - a_l l^{-s}\right)^{-1} \prod_{l \nmid pN} \left(1 - a_l l^{-s} + \varepsilon(l) l^{k-1} l^{-2s}\right)^{-1}.$$

En particulier, f est déterminée par les a_l.

(3.1.6) Si $f = \sum a_n q^n$ est une fonction propre des opérateurs de Hecke normalisée comme ci-dessus, il existe une forme parabolique $F = \sum A_n q^n$ de type (N, k, ε_0) à coefficients dans $\overline{\mathbf{Z}}$, qui est fonction propre des $T_l(l \nmid N)$ et des $U_l(l \mid N)$ et vérifie:

$$A_1 = 1; \qquad \tilde{F} = f.$$

En effet, du fait que les opérateurs T_l et U_l commutent entre eux, tout système commun de valeurs propres de ces opérateurs dans $\overline{\mathbf{F}}_p$ se relève en caractéristique 0 (cf. par exemple [11], lemme 6.11). On déduit de là qu'il existe une forme parabolique $F = \Sigma\, A_n q^n$, de type (N, k, ε_0), fonction propre des T_l et des U_l, normalisée, et telle que $\tilde{A}_l = a_l$ pour tout nombre premier l. Il est alors immédiat que $\tilde{F} = f$.

(Bien entendu, il n'y a pas unicité de F: deux fonctions propres différentes en caractéristique 0 peuvent avoir la même réduction en caractéristique p.)

(3.1.7) Soit $f = \Sigma\, a_n q^n$ comme ci-dessus. D'après un théorème de Deligne ([11], th. 6.7), il existe une représentation continue semi-simple

$$\rho_f\colon G_\mathbf{Q} \to \mathbf{GL}_2\big(\overline{\mathbf{F}}_p\big)$$

caractérisée (à conjugaison près) par la propriété suivante:

(D). Pour tout nombre premier l ne divisant pas pN, la représentation ρ_f est non ramifiée en l, et, si l'on note $\rho_f(\mathrm{Frob}_l)$ l'élément de Frobenius correspondant (défini à conjugaison près), on a

$$(3.1.8) \qquad\qquad \mathrm{Tr}\,\rho_f(\mathrm{Frob}_l) = a_l$$

et

$$(3.1.9) \qquad\qquad \det\rho_f(\mathrm{Frob}_l) = \varepsilon(l) l^{k-1}.$$

La formule (3.1.9) peut être récrite avec les notations du n° 1.3 sous la forme

$$(3.1.10) \qquad\qquad \det\rho_f = \varepsilon \cdot \chi^{k-1}.$$

Compte tenu de (3.1.1) elle entraîne que $\det\rho_f(c) = -1$, autrement dit que $\det\rho_f$ est un caractère *impair*.

Remarque. J'ai supposé au début que le niveau est premier à p. En fait, ce n'est pas nécessaire: tous les résultats énoncés restent vrais dans le cas général. Toutefois, cette généralité accrue ne procure pas de "formes mod p" vraiment nouvelles; on sait en effet que toute forme parabolique à coefficients dans $\overline{\mathbf{F}}_p$ qui est de niveau $p^m N$ est aussi de niveau N, à condition de remplacer le poids par un poids plus grand. Un exemple typique est celui des formes de poids 2 et de niveau p, qui sont aussi de poids $p + 1$ et de niveau 1, cf. [43], th. 11.

3.2. *Les diverses variantes de la conjecture.* Revenons aux notations du §1, et soit

$$\rho\colon G_\mathbf{Q} \to \mathbf{GL}(V) \simeq \mathbf{GL}_2\big(\overline{\mathbf{F}}_p\big)$$

un homomorphisme continu, V étant un $\overline{\mathbf{F}}_p$-espace vectoriel de dimension 2. On suppose que:

(3.2.1) ρ *est irréductible*,

et

(3.2.2)   det ρ *est impair*, cf. (1.3.8).

La *conjecture* affirme que ρ est alors du type ρ_f de (3.1.7). Autrement dit:

(3.2.3$_?$) *Il existe une forme parabolique f* (de type convenable) *à coefficients dans $\overline{\mathbf{F}}_p$ qui est fonction propre des opérateurs de Hecke, et dont la représentation ρ_f associée est isomorphe à la représentation ρ donnée.*

Il convient de préciser (3.2.3$_?$) en donnant le type (N, k, ε) de f:

(3.2.4$_?$) *La forme parabolique f de* (3.2.3$_?$) *peut être choisie de type (N, k, ε), où N, k et ε sont les invariants de ρ définis au §1 et au §2.*

Si $f = \Sigma \, a_n q^n$ est normalisée ($a_1 = 1$), le fait que ρ_f soit isomorphe à ρ se traduit par les égalités

$$(3.2.5) \qquad \mathrm{Tr}(\mathrm{Frob}_{l,\rho}) = a_l \quad \text{et} \quad \det(\mathrm{Frob}_{l,\rho}) = \varepsilon(l) l^{k-1},$$

égalités qui doivent être valables pour tout nombre premier l ne divisant pas pN. (Il suffit d'ailleurs que la première égalité de (3.2.5) ait lieu pour un ensemble de l de densité 1.)

En ce qui concerne les a_l, pour l divisant pN, on peut conjecturer ceci:

(3.2.6$_?$) *Supposons que $f = \Sigma \, a_n q^n$ satisfasse à* (3.2.3$_?$) *et* (3.2.4$_?$) *et soit normalisée. Soit l un diviseur premier de pN. Alors:*

(a) *Si $a_l \neq 0$, il existe une droite D de V stable par le groupe de décomposition en l* (*relativement à une place l-adique donnée de $\overline{\mathbf{Q}}$*) *et telle que le groupe d'inertie en l opère trivialement sur V/D.* (*En d'autres termes, la restriction de ρ au groupe de décomposition en l a un quotient étale de dimension 1.*)

De plus, a_l est égal à la valeur propre de la substitution de Frobenius en l, opérant sur V/D.

(b) *Si $a_l = 0$, il n'existe aucune droite de V ayant les propriétés énoncées en* (a).

Remarques sur (3.2.6$_?$)

(1) Si l divise N, il existe au plus une droite D de V satisfaisant à (a). En effet, s'il en existait deux, ρ serait étale en l, et l ne diviserait pas le conducteur N.

On voit donc que, dans ce cas, a_l est déterminé sans ambiguïté par ρ.

[Il n'est pas difficile de prouver que D existe si et seulement si:

ou bien $v_l(N) = 1$, v_l désignant la valuation l-adique;

ou bien $v_l(N) = v_l(\text{cond. } \varepsilon) \geqslant 2$, où cond. ε désigne le conducteur du caractère ε.

De plus, dans le cas où $v_l(N) = 1$ et $v_l(\text{cond.} \varepsilon) = 0$, on peut montrer que la valeur propre λ de la substitution de Frobenius en l opérant sur V/D est telle que $\lambda^2 = \varepsilon_{\mathrm{prim}}(l)l^{k-2}$, où $\varepsilon_{\mathrm{prim}}$ est le caractère primitif défini par ε. D'après $(3.2.6_?)$ on aurait alors

$$a_l^2 = \varepsilon_{\mathrm{prim}}(l)l^{k-2},$$

en parfait accord avec [27], th. 3 (iii).]

(2) Si $l = p$ et si ρ est ramifiée en p, la situation est la même que si l divise N: la droite D, si elle existe, est unique; la valeur propre a_p est déterminée sans ambiguïté. D'où *l'unicité* de la forme f dans ce cas; ses coefficients appartiennent au corps de rationalité de ρ, et engendrent ce corps sur $\mathbf{F}_p$.

(3) Si $l = p$ et si ρ est non ramifiée en p (ce qui entraîne $k = p$ d'après nos conventions, cf. n° 2.3), la situation est différente. Il y a alors deux valeurs possibles pour a_p, à savoir les deux valeurs propres λ et μ de la substitution de Frobenius en p; on a d'ailleurs $\lambda\mu = \varepsilon(p)$. Bien entendu, on peut avoir $\lambda = \mu$, auquel cas a_p est déterminé sans ambiguïté. Lorsque $\lambda \neq \mu$, dans tous les cas que je connais, il y a *deux* formes paraboliques f distinctes telles que $\rho_f \sim \rho$, l'une avec $a_p = \lambda$ et l'autre avec $a_p = \mu$. On notera que λ et μ n'appartiennent pas nécessairement au corps de définition de ρ (qui est engendré par les a_l, pour $l \neq p$): ils peuvent être quadratiques sur ce corps; on en verra des exemples au n° 5.1.

(4) Il devrait être possible de préciser $(3.2.6_?)$ en déterminant l'action sur f des opérateurs de symétrie W_l $(l|N)$ d'Atkin-Lehner-Li [3]. Les pseudo-valeurs propres correspondantes (au sens de [3]) peuvent sans doute s'écrire en termes des *constantes locales* de ρ (Deligne [9], §6).

Remarques sur $(3.2.4_?)$

(5) Il est probable que N et k sont *minimaux* pour ρ, autrement dit que, si ρ est isomorphe à $\rho_{f'}$, avec f' de type (N', k', ε'), N' premier à p, $k' \geqslant 2$, alors N' est multiple de N et k' est $\geqslant k$. En particulier, si l'on écrit f sous la forme $\tilde{F}$ comme dans (3.1.6), F doit être une forme *primitive* ("newform" cf. [11], [27]) de type (N, k, ε_0).

(6) Au lieu de définir les formes paraboliques à coefficients dans $\overline{\mathbf{F}}_p$ par réduction à partir de la caractéristique 0, comme nous l'avons fait, nous aurions pu utiliser la définition de Katz [23], qui conduit à un espace *a priori* plus grand,[2] donc donne peut-être davantage de représentations ρ_f. Il serait intéressant de voir si les représentations supplémentaires ainsi obtenues peuvent être irréductibles; je n'en connais aucun exemple (pour $k \geqslant 2$), mais, si cela se produisait, il y aurait lieu de modifier $(3.2.4_?)$ et $(3.2.6_?)$. Il serait également intéressant d'étudier de ce

---

[2] La définition de Katz jouit de la propriété agréable suivante: toute forme de poids k est aussi de poids $k + p - 1$. Avec la définiton adoptée ici, cet énoncé est vrai pour $p \geqslant 5$, mais est faux pour $p = 2$ ou 3.

point de vue le cas $k = 1$, que nous avons exclu jusqu'ici; peut-être la définition de Katz donne-t-elle alors beaucoup plus de représentations ρ_f?

3.3 *Exemple $k = 2$*. Nous allons appliquer les conjectures du numéro précédent à une représentation

$$\rho\colon G_\mathbf{Q} \to \mathbf{GL}_2(\mathbf{F}_p)$$

telle que:

   (a) det $\rho = \chi$;
   (b) ρ est absolument irréductible (i.e., irréductible sur $\overline{\mathbf{F}}_p$);
   (c) ρ est finie en p, au sens du n° 2.8.

[Lorsque $p \neq 2$, on peut remplacer (b) par la condition suivante, en apparence plus faible:

   (b′) ρ est irréductible (sur $\mathbf{F}_p$).

En effet, (a) entraîne que det ρ est impair, donc que les valeurs propres de $\rho(c)$ sont $+1$ et -1; du fait que $p \neq 2$, ces valeurs propres sont distinctes. Si alors ρ se décomposait sur $\overline{\mathbf{F}}_p$ en somme directe de deux représentations de dimension 1, cette décomposition ne pourrait être que celle donnée par les vecteurs propres de $\rho(c)$, donc serait rationnelle sur $\mathbf{F}_p$, ce qui contredirait (b′).]

Soient N, k et ε les invariants de ρ. D'après le n° 1.3, on a $\varepsilon = 1$, et d'après la prop. 4 du n° 2.8, on a $k = 2$. La conjecture $(3.2.4_?)$ donne alors:

$(3.3.1_?)$ *Il existe une forme parabolique f de poids 2 et de niveau N, à coefficients dans $\overline{\mathbf{F}}_p$, qui est fonction propre des opérateurs de Hecke et dont la représentation ρ_f associée est isomorphe à ρ.*

D'après $(3.2.6_?)$, cette forme parabolique est à coefficients dans $\mathbf{F}_p$, sauf peut-être si ρ est non ramifiée en p (ce qui ne peut se produire que si $p = 2$).

On peut reformuler $(3.3.1_?)$ en termes de la jacobienne $J_0(N)$ de la courbe modulaire $X_0(N)$ associée au groupe $\Gamma_0(N)$:

$(3.3.2_?)$ *La représentation ρ intervient comme quotient de Jordan-Hölder dans la représentation de $G_\mathbf{Q}$ sur les points de p-division de $J_0(N)$.*

3.4. *Questions.*   En voici deux, l'une pour pessimistes, l'autre pour optimistes:

(1) Comment construire des contre-exemples aux conjectures du n° 3.2? J'ai fait de nombreux essais dans cette direction. Tous ces essais ont échoué, comme on le verra au §5.

(2) Peut-on reformuler ces conjectures dans le cadre d'une théorie des représentations (mod p) des groupes adéliques? Autrement dit, existe-t-il une "philosophie de Langlands modulo p", comme le demandent Ash et Stevens dans [2]? Si oui, cela permettrait peut-être:

   de donner une définition plus naturelle du poids k attaché à ρ;
   de remplacer $\mathbf{GL}_2$ par $\mathbf{GL}_N$, ou même par un groupe réductif;
   de remplacer $\mathbf{Q}$ par d'autres corps globaux.

§4. Applications.   Ces applications concernent:

l'équation de Fermat et ses variantes (n°s 4.1 à 4.3);

les discriminants des courbes elliptiques semi-stables (n° 4.4);

la structure des schémas en groupes de type (p, p) sur $\mathbf{Z}$ (n° 4.5);

la conjecture de Taniyama-Weil, et son extension aux variétés abéliennes à multiplications réelles (n°s 4.6 et 4.7);

la cohomologie des variétés projectives lisses sur $\mathbf{Q}$ ayant un nombre de Betti égal à 2 en dimension impaire (n° 4.8).

Excepté la dernière, ces applications n'utilisent la conjecture $(3.2.4_?)$ que dans le cas où $\varepsilon = 1$, $k = 2$, cf. n°. 3.3.

4.1. *Rappels sur certaines courbes elliptiques sur* $\mathbf{Q}$.   Soient A, B, C trois entiers non nuls, premiers entre eux deux à deux, et tels que

$$A + B + C = 0.$$

Choisissons des nombres entiers x_1, x_2, x_3 tels que

$$x_1 - x_2 = A, \, x_2 - x_3 = B, \, x_3 - x_1 = C.$$

La courbe elliptique d'équation

$$y^2 = (x - x_1)(x - x_2)(x - x_3)$$

est indépendante du choix des x_i, à isomorphisme près. Pour fixer les idées, nous prendrons $x_1 = A$, $x_2 = 0$, $x_3 = -B$, de sorte que l'équation ci-dessus s'écrit

$$(4.1.1) \qquad\qquad y^2 = x(x - A)(x + B).$$

Nous noterons $E_{A, B, C}$, ou simplement E, la courbe ainsi définie.

Remarque.   Une permutation de A, B, C de signature 1 (resp. -1) ne change pas E (resp. change E en sa "tordue" par l'extension quadratique $\mathbf{Q}(\sqrt{-1})/\mathbf{Q}$).

Donnons maintenant quelques propriétés de *mauvaise réduction* de E (cf. Frey [17]).

(4.1.2) *Mauvaise réduction en* $l \neq 2$.   Soit l un nombre premier $\neq 2$. La courbe E a mauvaise réduction en l si et seulement si l divise ABC, et cette mauvaise réduction est alors *de type multiplicatif*.

C'est immédiat sur (4.1.1). On notera également que cette équation fournit un *modèle minimal* de E en l, cf. Tate [4], p. 47.

(4.1.3) *Mauvaise réduction en* 2.   Nous nous bornerons au cas où:

$$(4.1.4) \qquad\qquad A \equiv -1 \, (\mathrm{mod} \, 4) \quad \text{et} \quad B \equiv 0 \, (\mathrm{mod} \, 32).$$

En faisant le changement de variables

$$x = 4X, \qquad y = 8Y + 4X,$$

on transforme (4.1.1) en l'équation

(4.1.5)

$$Y^2 + XY = X^3 + cX^2 + dX, \quad \text{avec } c = (B - 1 - A)/4, \ d = -AB/16,$$

dont la réduction (mod 2) est:

$$Y^2 + XY = \begin{cases} X^3 & \text{si } A \equiv 7 \ (\text{mod } 8) \\ X^3 + X^2 & \text{si } A \equiv 3 \ (\text{mod } 8). \end{cases}$$

On obtient ainsi une cubique sur $\mathbf{F}_2$ qui a un point double en $(0,0)$ à tangentes distinctes (ces tangentes étant rationnelles sur $\mathbf{F}_2$ si et seulement si $A \equiv 7 \ (\text{mod } 8)$). Il en résulte que E a *mauvaise réduction de type multiplicatif en* 2 (Tate, *loc. cit.*) et que (4.1.5) est une *équation minimale* en 2, donc aussi sur $\text{Spec}(\mathbf{Z})$ d'après ce qu'on vient de voir. Le discriminant Δ correspondant est:

(4.1.6) $$\Delta = 2^{-8}A^2B^2C^2.$$

Ainsi, E a partout, soit bonne réduction, soit mauvaise réduction de type multiplicatif: c'est une courbe *semi-stable*. Son *conducteur* est donné par:

(4.1.7) $$\text{cond}(E) = \text{rad}(ABC),$$

où $\text{rad}(X)$ désigne le produit des nombres premiers qui divisent X (i.e., le plus grand diviseur sans facteur carré de X).

L'invariant modulaire j_E de E est:

(4.1.8) $$j_E = 2^8(C^2 - AB)^3/A^2B^2C^2.$$

Si l divise ABC, on a:

(4.1.9) $$v_l(j_E) = -v_l(\Delta) = \begin{cases} -2v_l(ABC) & \text{si } l \neq 2 \\ 8 - 2v_2(ABC) & \text{si } l = 2. \end{cases}$$

Points de p-division de E. Soit p un nombre premier $\geqslant 5$. Nous nous intéresserons à la représentation

$$\rho_p^E \colon G_{\mathbf{Q}} \to \mathbf{GL}_2(\mathbf{F}_p)$$

fournie par les points de p-division de E.

On a tout d'abord:

PROPOSITION 6.   *La représentation ρ_p^E est irréductible.*

(Comme son déterminant est égal au caractère cyclotomique χ, cette représentation est même *absolument irréductible*, cf. n° 3.3.)

Supposons que ρ_p^E soit réductible, i.e. que E contienne un sous-groupe X d'ordre p qui soit rationnel sur $\mathbf{Q}$. Du fait que E est semi-stable, l'action de $G_{\mathbf{Q}}$ sur X se fait, soit par le caractère unité, soit par le caractère χ ([41], p. 307). Dans le premier cas, E a un point d'ordre p rationnel sur $\mathbf{Q}$; comme les points d'ordre 2 de E sont également rationnels sur $\mathbf{Q}$, l'ordre du groupe de torsion de $E(\mathbf{Q})$ est $\geqslant 4p \geqslant 20$, ce qui contredit un théorème de Mazur ([28], th. 8). Dans le second cas, la courbe $E' = E/X$ a un point d'ordre p rationnel sur $\mathbf{Q}$, et on lui applique le même argument que ci-dessus.

Remarque.   Au lieu d'utiliser le th. 8 de [28], on aurait pu se servir des résultats plus généraux de Mazur [29].

On va maintenant déterminer les *invariants* (N, k, ε) attachés à ρ_p^E:
(4.1.10) Comme $\det \rho_p^E = \chi$, on a $\varepsilon = 1$.
(4.1.11) *On a $k = 2$ si $v_p(\Delta)$ est divisible par p (i.e., si $v_p(ABC)$ est divisible par p), et $k = p + 1$ dans le cas contraire.* Cela résulte de la prop. 5 du n° 2.9, compte tenu du fait que E est semi-stable.
(4.1.12) *Le conducteur N de ρ_p^E est égal au produit des nombres premiers $l \neq p$ tels que $v_l(\Delta)$ ne soit pas divisible par p.*
C'est là une propriété générale des courbes semi-stables, qui se vérifie immédiatement sur les modèles de Tate "$\mathbf{G}_m/q^{\mathbf{Z}}$".

Remarque.   Vu (4.1.6), la condition "$v_l(\Delta)$ n'est pas divisible par p" est équivalente à:

$$(4.1.13) \qquad v_l(ABC) \not\equiv \begin{cases} 0 & (\mathrm{mod}\ p) & \text{si } l \neq 2 \\ 4 & (\mathrm{mod}\ p) & \text{si } l = 2. \end{cases}$$

4.2. *Le théorème de Fermat.*   Soit p un nombre premier $\geqslant 5$.

THÉORÈME 1.   *Admettons $(3.3.1_?)$. L'équation*

$$a^p + b^p + c^p = 0$$

n'a alors aucune solution avec $a, b, c \in \mathbf{Z}$, $abc \neq 0$.

Soit (a, b, c) une telle solution. Quitte à faire une homothétie, ainsi qu'une permutation, on peut supposer que a, b et c sont premiers entre eux, et que $b \equiv 0 \pmod 2$, $a \equiv -1 \pmod 4$. Si l'on pose

$$A = a^p, \qquad B = b^p, \qquad C = c^p,$$

les conditions (4.1.4) du n° 4.1 sont satisfaites. Soit $E = E_{A,B,C}$ la courbe elliptique correspondante, et soit ρ_p^E la représentation de $G_{\mathbf{Q}}$ fournie par ses points de p-division. Par construction, on a

$$v_l(ABC) \equiv 0 \pmod{p} \quad \text{pour tout } l \text{ premier.}$$

Il en résulte, d'après (4.1.11) et (4.1.13), que les invariants k et N attachés à ρ_p^E sont égaux à 2. De plus ρ_p^E est irréductible (prop. 6). La conjecture $(3.3.1_?)$ affirme alors que ρ_p^E est isomorphe à la représentation ρ_f associée à une forme parabolique normalisée f, de poids 2 et de niveau 2, à coefficients dans $\overline{\mathbf{F}}_p$. Mais une telle forme n'existe pas: la courbe modulaire $X_0(2)$ est de genre 0. D'où le théorème.

Remarque. La relation existant entre "solutions de l'équation de Fermat" et "points de p-division de certaines courbes elliptiques" figure déjà dans un travail de Hurwitz ([20]) datant de 1886.

Elle a été utilisée depuis par différents auteurs, notamment Hellegouarch [19], Vélu [54] et Frey [16], [17]. La méthode suivie ici est tirée de Frey [17].

4.3. *Variantes du théorème de Fermat.* Soit p un nombre premier $\geqslant 11$.

THÉORÈME 2. *Admettons* $(3.3.1_?)$. *Soit L un nombre premier $\neq p$ appartenant à l'ensemble*

$$S = \{3, 5, 7, 11, 13, 17, 19, 23, 29, 53, 59\},$$

et soit α un entier $\geqslant 0$. L'équation

$$(4.3.1) \qquad\qquad a^p + b^p + L^\alpha c^p = 0$$

n'a alors aucune solution avec $a, b, c \in \mathbf{Z}$ et $abc \neq 0$.

On procède comme pour le th. 1. Tout d'abord, on peut évidemment supposer que $0 < \alpha < p$. Soit alors (a, b, c) une solution de l'équation (4.3.1), avec a, b, c premiers entre eux. On prend pour A, B, C les trois entiers a^p, b^p, $L^\alpha c^p$ (qui sont premiers entre eux, comme on le vérifie tout de suite), à une permutation près choisie de telle sorte que B soit pair (donc divisible par 2^p et *a fortiori* par 32) et $A \equiv -1 \pmod{4}$. On considère la représentation ρ_p^E attachée à la courbe elliptique $E = E_{A,B,C}$. D'après (4.1.11) et (4.1.13) les invariants k et N de cette représentation sont $k = 2$ et $N = 2L$ (noter que L a été supposé distinct de p). D'après $(3.3.1_?)$, il existe alors une forme parabolique

$$f = q + a_2(f)q^2 + \cdots + a_n(f)q^n + \cdots$$

à coefficients dans $\overline{\mathbf{F}}_p$, de poids 2 et de niveau $2L$, fonction propre des opérateurs de Hecke, et telle que la représentation ρ_f associée soit isomorphe à ρ_p^E. Nous allons voir que ceci est impossible. C'est clair pour $L = 3, 5$ puisqu'aucune f n'existe dans ce cas: les courbes modulaires $X_0(6)$ et $X_0(10)$ sont de genre 0. On peut donc supposer $L \geqslant 7$.

LEMME 1. (a) *La forme f est la réduction en caractéristique p d'une forme primitive F de niveau $2L$ en caractéristique 0.*
(b) *On a $a_3(f) = 0$ ou ± 4.*
(c) *On a $a_5(f) = \pm 2$ ou ± 6.*

D'après (3.1.6), on a $f = \tilde{F}$, où F est une forme parabolique de poids 2 et de niveau $2L$, à coefficients dans $\overline{\mathbf{Z}}$, qui est fonction propre normalisée des opérateurs de Hecke. Si F n'était pas primitive, elle proviendrait du niveau L et la représentation ρ_f ne serait pas ramifiée en 2. Or ρ_p^E est ramifiée en 2, puisque son conducteur est $2L$. D'où (a).

Pour prouver (b) on distingue deux cas:
(b_1) *La courbe E a bonne réduction en 3, i.e., $ABC \not\equiv 0$ (mod 3).*
Soit $\tilde{E}$ la réduction de E en 3. C'est une courbe elliptique sur $\mathbf{F}_3$ dont les points d'ordre 2 sont rationnels. Le nombre de points rationnels de $\tilde{E}$ est donc divisible par 4. Comme ce nombre est compris entre $1 + 3 - 2\sqrt{3}$ et $1 + 3 + 2\sqrt{3}$, il est égal à 4. Cela signifie que la trace de l'endomorphisme de Frobenius de $\tilde{E}$ est égale à 0. D'où $a_3(f) = 0$ (dans $\mathbf{F}_p$), d'après (3.1.8).
(b_2) *La courbe E a mauvaise réduction en 3.*
On a vu que cette mauvaise réduction est de type multiplicatif. Si elle est déployée (i.e., si E est isomorphe sur $\mathbf{Q}_3$ à une courbe de Tate), le $G_{\mathbf{Q}_3}$-module galoisien E_p est extension de $\mathbf{Z}/p\mathbf{Z}$ par μ_p; les valeurs propres de l'endomorphisme de Frobenius en 3 sont donc 1 et 3; leur somme est 4. D'où $a_3(f) = 4$ dans ce cas. Lorsque la réduction est non déployée, il y a une "torsion" quadratique, et l'on obtient $a_3(f) = -4$.

La démonstration de (c) est analogue à celle de (b): on trouve que $a_5(f) = \pm 2$ lorsque E a bonne réduction en 5, et $a_5(f) = \pm 6$ sinon.

LEMME 2.   *Soit $L \in S$, avec $L \geqslant 7$, et soit*

$$F = q + A_2 q^2 + \cdots + A_n q^n + \cdots, \qquad A_n \in \overline{\mathbf{Z}},$$

une forme primitive normalisée de poids 2 et de niveau $2L$. On a alors:

$$A_3 = \pm 1, \pm 2 \ ou \ \pm 3 \ si \ L \neq 23$$

et

$$A_5 = 4 \ si \ L = 23.$$

Cela se vérifie cas par cas:

L	7	13	17	19	29	53	59
valeurs de A_3	-2	$1, -3$	-2	$1, -1$	$-1, -3$	$1, -1, 2, -2$	$-1, -1, 2, 2.$

($L = 11$ ne figure pas dans ce tableau, car il n'y a pas de forme primitive de poids 2 pour le niveau 22.)

On peut maintenant achever la démonstration du th. 2. Tout d'abord, pour $L = 23$, la comparaison des lemmes 1 et 2 montre que l'on a

$$\pm 2 \text{ ou } \pm 6 \equiv 4 \qquad (\text{mod } p),$$

ce qui est impossible pour $p \geqslant 7$. De même, si $L \neq 23$, $L \in S$ et $L \geqslant 7$, on a

$$0 \text{ ou } \pm 4 \equiv \pm 1, \pm 2 \text{ ou } \pm 3 \qquad (\text{mod } p),$$

ce qui est impossible pour $p \geqslant 11$.

Remarques

(1) L'hypothèse $p \neq L$ n'est pas essentielle; elle a seulement servi à assurer que le poids k est égal à 2, ce qui a permis d'appliquer $(3.3.1_?)$. Si $p = L$, on a $k = p + 1$, $N = 2$, et les arguments utilisés ci-dessus restent valables, à condition d'admettre $(3.2.4_?)$ pour $k = p + 1$ et non plus pour $k = 2$.

(2) Il est possible que le th. 2 reste vrai pour $p = 5$ et $p = 7$. La question devrait pouvoir se traiter, sans utiliser aucune conjecture, par les méthodes traditionnelles de factorisation et de descente (cf. par exemple Dénes [12]).

(3) La plus petite valeur de L ne figurant pas dans l'ensemble S du th. 2 est $L = 31$ (qui est un nombre de Mersenne–cf. ci-dessous). Pour cette valeur, la méthode suivie conduit à une représentation ρ_p^E qui pourrait, par exemple, être isomorphe à celle fournie par la forme primitive F de niveau 62 suivante:

$$F = q + q^2 + q^4 - 2q^5 + q^8 + \ldots$$

Je ne vois pas comment tirer de là une contradiction, d'autant plus que l'équation $a^5 + b^5 + 31c^5 = 0$ a effectivement une solution, à savoir $(1, -2, 1)$ [cette solution conduit à la courbe E d'équation $y^2 = x(x + 1)(x - 32)$, qui est une courbe de Weil de niveau 62 correspondant à F.]

Je ne vois pas non plus comment attaquer les équations

$$a^p + b^p + 15c^p = 0 \quad \text{et} \quad a^p + 3b^p + 5c^p = 0,$$

pour lesquelles le conducteur N est égal à 30.

(4) Si l'on fixe L, on peut se demander ce qui se passe pour p assez grand. Dans cette direction, Mazur m'a signalé le résultat suivant:

Admettons $(3.3.1_?)$. Soit L un nombre premier $\neq 2$ qui ne soit ni un nombre de Fermat ni un nombre de Mersenne (i.e. L ne peut pas s'écrire sous la forme

$2^n \pm 1$). *Il existe alors une constante C_L telle que, si $p \geqslant C_L$ et $\alpha \geqslant 0$, l'équation*

$$a^p + b^p + L^\alpha c^p = 0$$

n'a aucune solution avec $a, b, c \in \mathbf{Z}$ et $abc \neq 0$.

La démonstration est analogue à celle du th. 2; l'hypothèse faite sur L est utilisée pour montrer qu'il n'existe aucune courbe elliptique de conducteur $2L$ dont les trois points d'ordre 2 soient rationnels sur $\mathbf{Q}$.

4.4 *Discriminants des courbes elliptiques semi-stables.* La conjecture $(3.3.1_?)$ permettrait de répondre affirmativement à des questions de Brumer-Kramer ([6], §9):

PROPOSITION 7. *Admettons $(3.3.1_?)$. Soit E une courbe elliptique semi-stable sur $\mathbf{Q}$, et soit Δ le discriminant de son modèle minimal. Supposons que $|\Delta|$ soit une puissance p-ème. Alors E possède un sous-groupe d'ordre p rationnel sur $\mathbf{Q}$, et l'on a $p \leqslant 7$.*

Pour $p = 2$, on remarque que l'extension de $\mathbf{Q}$ engendrée par les points d'ordre 2 de E est non ramifiée en dehors de 2; son groupe de Galois ne peut donc être ni $\mathfrak{S}_3$ ni $\mathfrak{A}_3$, et cela montre que l'un de ces points est rationnel sur $\mathbf{Q}$. Pour $p = 3$, 5 ou 7, on applique un argument analogue (cf. [6], prop. 9.2). Il reste à prouver que le cas $p > 7$ est impossible. Or, si $p > 7$, la représentation ρ_p^E est irréductible (Mazur [29], th. 4). D'autre part, les hypothèses faites sur E entraînent que les invariants (N, k, ε) de ρ_p^E sont égaux à $(1, 2, 1)$. D'après $(3.3.1_?)$, ρ_p^E provient donc d'une forme parabolique normalisée de poids 2 et de niveau 1. Il y a contradiction: une telle forme n'existe pas.

PROPOSITION 8. *Admettons $(3.3.1_?)$. Soit E une courbe elliptique sur $\mathbf{Q}$ dont le conducteur est un nombre premier P. Soit $\Delta = \pm P^m$ le discriminant du modèle minimal de E. On a alors $m = 1$, sauf si E est une courbe de Setzer-Neumann, ou si $P = 11, 17, 19$ ou 37.*

Supposons $m > 1$. Il existe alors un nombre premier p qui divise m, et l'on peut appliquer la prop. 7. On en conclut d'abord que $p \leqslant 7$. Si $p = 2$, E a un point rationnel d'ordre 2, et c'est une courbe de Setzer-Neumann ([33], [50]), à moins que P ne soit égal à 17. Si $p = 3$, 5 ou 7, il existe une courbe isogène à E sur $\mathbf{Q}$ qui possède un point rationnel d'ordre p ([41], p. 307); d'après Miyawaki [32], cela est impossible pour $p = 7$ et cela entraîne $P = 11$ pour $p = 5$, et $P = 19$ ou 37 pour $p = 3$.

4.5. *Schémas en groupes de type (p, p) sur $\mathbf{Z}$.* Soit p un nombre premier $\geqslant 3$.

THÉORÈME 3. *Admettons $(3.3.1_?)$. Tout schéma en groupes fini et plat sur $\mathbf{Z}$, de type (p, p), est alors isomorphe à l'un des trois suivants:*

$$\mathbf{Z}/p\mathbf{Z} \oplus \mathbf{Z}/p\mathbf{Z}, \qquad \mathbf{Z}/p\mathbf{Z} \oplus \mu_p, \qquad \mu_p \oplus \mu_p.$$

Soit J un schéma en groupes fini et plat sur $\mathbf{Z}$, de type (p, p). On sait que J est étale sur $\mathrm{Spec}(\mathbf{Z}) - \{p\}$, donc définit une représentation

$$\rho\colon G_{\mathbf{Q}} \to \mathbf{GL}_2(\mathbf{F}_p)$$

qui est non ramifiée en dehors de p. Comme $p \neq 2$, la connaissance de ρ détermine celle de J (Raynaud [35], prop. 3.3.2).

LEMME 3.   *Si ρ est réductible, J est isomorphe à $\mathbf{Z}/p\mathbf{Z} \oplus \mathbf{Z}/p\mathbf{Z}$, $\mathbf{Z}/p\mathbf{Z} \oplus \mu_p$ ou $\mu_p \oplus \mu_p$.*

La réductibilité de ρ équivaut à l'existence d'une suite exacte

$$0 \to A \to J \to B \to 0,$$

où A et B sont des schémas en groupes finis et plats sur $\mathbf{Z}$, d'ordre p. D'après Oort-Tate [34], A et B sont isomorphes, soit à $\mathbf{Z}/p\mathbf{Z}$, soit à μ_p. Le lemme résulte alors de ce que toute extension de B par A est scindée (Fontaine [15], n° 3.4.3).

LEMME 4.   *Si ρ est irréductible, on a* $\det \rho = \chi$.

Le caractère $\det \rho\colon G_{\mathbf{Q}} \to \mathbf{F}_p^*$ est non ramifié en dehors de p, donc de la forme χ^i, avec $0 \leqslant i \leqslant p - 2$. Les résultats locaux de Raynaud [35] (cf. n° 2.8, démonstration de la prop. 4) montrent que les seules possibilités pour i sont $i = 0, 1$ et 2. De plus (*loc. cit.*) le cas $i = 0$ n'est possible que si J est étale en p, auquel cas ρ est non ramifiée partout, d'où $\rho = 1$ d'après Minkowski, ce qui contredit l'hypothèse que ρ est irréductible. De même, le cas $i = 2$ n'est possible que si le dual de J est étale en p, ce qui conduit à une contradiction par le même argument. On a donc $i = 1$, d'où le lemme.

Le th. 3 est maintenant immédiat. En effet, si ρ est réductible, on applique le lemme 3. Et, si ρ est irréductible, le lemme 4, joint à la prop. 4 du n° 2.8, montre que les invariants (N, k, ε) attachés à ρ sont $(1, 2, 1)$; d'où une contradiction avec $(3.3.1_?)$ par le même argument que celui utilisé dans la démonstration de la prop. 7.

Remarques

(1) Pour $p = 3, 5, 7, 11, 13$ ou 17, Fontaine [15] a démontré (sans utiliser aucune conjecture) un résultat plus général que le th. 3: tout schéma en groupes fini et plat sur $\mathbf{Z}$, de type $(p, \ldots, p)$, est somme directe de copies de $\mathbf{Z}/p\mathbf{Z}$ et de μ_p.

(2) Le th. 3 ne s'étend pas au cas $p = 2$: outre $\mathbf{Z}/2\mathbf{Z} \oplus \mathbf{Z}/2\mathbf{Z}$, $\mathbf{Z}/2\mathbf{Z} \oplus \mu_2$ et $\mu_2 \oplus \mu_2$, il y a une quatrième possibilité, à savoir une certaine extension non scindée de $\mathbf{Z}/2\mathbf{Z}$ par μ_2. La représentation ρ correspondante s'écrit

$$\rho = \begin{pmatrix} 1 & u \\ 0 & 1 \end{pmatrix}$$

où $u\colon G_{\mathbf{Q}} \to \mathbf{Z}/2\mathbf{Z}$ est l'homomorphisme de noyau $\mathrm{Gal}(\overline{\mathbf{Q}}/\mathbf{Q}(i))$. Ce schéma en groupes de type $(2,2)$ peut se réaliser comme celui des points de 2-division de la courbe elliptique

$$y^2 + xy + y = x^3 - x^2 - x - 14,$$

de conducteur 17 et de discriminant -17^4.

4.6. *La conjecture de Taniyama-Weil.*   Soit E une courbe elliptique sur $\mathbf{Q}$, soit j_E son invariant modulaire, et soit N son conducteur.

THÉORÈME 4.   *Admettons* $(3.3.1_?)$. *Alors E est une courbe de Weil de niveau N.*

(En particulier, E est isomorphe à un quotient de la jacobienne $J_0(N)$ de la courbe modulaire $X_0(N)$.)

Pour tout nombre premier p, notons $\rho_p^E\colon G_{\mathbf{Q}} \to \mathbf{GL}_2(\mathbf{F}_p)$ la représentation de $G_{\mathbf{Q}}$ fournie par les points de p-division de E. On a

(4.6.1) $\det \rho_p^E = \chi.$

De plus:

LEMME 5.   *Il existe une constante C_E telle que, pour tout $p \geqslant C_E$, on ait*:
(4.6.2) ρ_p^E *est irréductible*;
(4.6.3) *le conducteur de ρ_p^E est égal à N.*

C'est là un résultat connu. En effet, d'après Mazur [29], (4.6.2) est vrai dès que $p > 163$. D'autre part la définition du conducteur de E en termes de représentations l-adiques (cf. [18], [40], [49]) montre que le conducteur N_p de ρ_p^E *divise* N (ce qui d'ailleurs suffirait pour la suite). De plus, si $p \geqslant 5$, on vérifie que $N_p = N$ si et seulement si p satisfait aux deux conditions suivantes:

(a) p ne divise pas N;
(b) pour tout l tel que $v_l(N) = 1$, p ne divise pas $v_l(j_E)$.

(Noter, à propos de b), que l'hypothèse $v_l(N) = 1$ signifie que E a mauvaise réduction de type multiplicatif en l, et l'on a donc $v_l(j_E) < 0$.)

Bornons-nous maintenant aux nombres premiers $p \geqslant C_E$. D'après $(3.3.1_?)$, ρ_p^E est isomorphe à la représentation ρ_{f_p} associée à une forme parabolique de poids 2 et de niveau N

$$f_p = \sum a_{n,p} q^n,$$

à coefficients dans $\overline{\mathbf{F}}_p$, qui est fonction propre normalisée des opérateurs de Hecke.

D'après (3.1.6), f_p se relève en caractéristique 0: il existe une forme parabolique de poids 2 et de niveau N

$$F = \sum A_n q^n,$$

à coefficients dans $\overline{\mathbf{Z}}$, qui est fonction propre normalisée des opérateurs de Hecke, et telle que $\tilde{F} = f_p$. A priori, F dépend de p. Mais il n'y a qu'un nombre fini de F

possibles, puisque le poids et le niveau sont fixés. On en conclut qu'il existe un choix de F tel que l'on ait

$$\tilde{F} = f_p$$

pour tout $p \in P$, où P est un ensemble infini de nombres premiers. Soit alors l un nombre premier ne divisant pas N. La courbe E a bonne réduction en l. Soit a_l la trace de l'endomorphisme de Frobenius correspondant. On a

$$a_l \equiv a_{l,p} \pmod{p} \quad \text{pour tout } p \neq l.$$

Il en résulte que l'entier algébrique $A_l - a_l$ *a une image dans* $\overline{\mathbf{F}}_p$ *qui est égale à* 0 *pour tout* $p \in P$, $p \neq l$. Comme P est infini, cela entraîne

(4.6.4) $A_l = a_l \quad \text{pour tout } l \nmid N.$

En particulier, les A_l appartiennent à $\mathbf{Z}$. Ils définissent une *courbe de Weil* E_F de niveau un diviseur de N; d'après (4.6.4), les représentations l-adiques attachées à E et E_F sont isomorphes, et l'on sait (Faltings [13], [14]) que cela entraîne que E et E_F sont isogènes sur $\mathbf{Q}$. D'où le th. 4.

Remarques

(1) Le th. 4 m'a été suggéré par P. Colmez au Colloque de Luminy, en juin 1986. Jusque là, je ne m'étais pas rendu compte de l'étendue (à la fois intéressante et inquiétante) des conséquences des conjectures du §3.

(2) La forme F construite dans la démonstration ci-dessus est *primitive*; cela résulte d'un théorème de Carayol [8].

(3) La méthode suivie ici s'applique à d'autres questions du même genre. En voici un exemple, tiré de [52]:

Soit $K = \mathbf{Q}(\sqrt{D})$ un corps quadratique réel; notons σ l'involution de K. Soit E une courbe elliptique sur K, soit E^σ sa conjuguée, et soit $\lambda: E \to E^\sigma$ une isogénie telle que $\lambda^\sigma \circ \lambda = -c$, où c est un entier > 0. Shimura pose alors la question suivante ([52], p. 184): est-il vrai que E provienne (par la construction donnée dans [52]) d'une forme primitive de type $(N, 2, \varepsilon)$, où N est un entier convenable, et ε est le caractère quadratique associé à K? On peut montrer que la réponse est "oui" si l'on admet la conjecture $(3.2.4_?)$. La démonstration est analogue à celle du th. 4 (on travaille avec un système de représentations l-adiques qui est rationnel sur $\mathbf{Q}(\sqrt{-c})$, et dont le déterminant est le produit de ε et du caractère cyclotomique).

Pour d'autres exemples, voir les n°s 4.7 et 4.8 ci-après.

4.7. *Variétés abéliennes à multiplications réelles.* Soit X une variété abélienne sur $\mathbf{Q}$ de dimension $n \geqslant 1$. On dit que X est *à multiplication réelles* (cf. Ribet

[36]) si la **Q**-algèbre $K_X = \mathbf{Q} \otimes \mathrm{End}_{\mathbf{Q}}(X)$ est un corps de nombres algébriques totalement réel de degré n. On sait que de telles variétés apparaissent lorsqu'on décompose les jacobiennes $J_0(N)$ sous l'action des opérateurs de Hecke, cf. Shimura [51], §7.5. Réciproquement:

THÉORÈME 5. *Admettons* $(3.3.1_?)$. *Alors toute variété abélienne X sur* **Q**, *à multiplications réelles, de dimension n, est isomorphe à un quotient de $J_0(N)$, où N est la racine n-ème du conducteur de X.*

La démonstration est analogue à celle du th. 4 (que l'on retrouve pour $n = 1$). Je me bornerai à en indiquer les grandes lignes. Tout d'abord:

(4.7.1) *La variété abélienne X définit un "système de représentations λ-adiques" de $G_{\mathbf{Q}}$, de degré 2, rationnel sur K_X; le déterminant de ce système est le caractère cyclotomique.*

Cela est expliqué dans Ribet [36].

Si X a bonne réduction en l, on notera a_l la trace de l'endomorphisme correspondant (dans le système λ-adique ci-dessus); c'est un entier du corps K_X.

(4.7.2) *Le conducteur de X est de la forme N^n, avec N entier $\geqslant 1$.*

La définition du conducteur donnée dans [18], exposé IX, §4 (voir aussi [40], n° 2.1) fait intervenir certains caractères locaux de degré $2n$, à valeurs dans **Q**. Or on constate (comme pour (4.7.1) ci-dessus) que ces caractères peuvent s'écrire comme sommes des n conjugués de caractères de degré 2 à valeurs dans K_X. L'assertion (4.7.2) résulte facilement de là.

Fixons maintenant un plongement de K_X dans $\overline{\mathbf{Q}}$. Pour tout nombre premier p, on a choisi au n° 3.1 une place p-adique de $\overline{\mathbf{Q}}$, d'où une place λ_p de K_X. Si l'on suppose p *complètement décomposé* dans K_X, le corps résiduel de λ_p est $\mathbf{F}_p$; par réduction $(\mathrm{mod}\,\lambda_p)$, la représentation λ_p-adique correspondante définit une représentation

$$\rho_p^X \colon G_{\mathbf{Q}} \to \mathbf{GL}_2(\mathbf{F}_p).$$

Les ρ_p^X jouissent des propriétés suivantes:

(4.7.3) $$\det \rho_p^X = \chi.$$

Cela résulte de (4.7.1).

(4.7.4) *Si p est assez grand, ρ_p^X est irréductible.*

Cela résulte d'un théorème de Faltings ([14], p. 204), et cela peut aussi se voir par un argument élémentaire, analogue à celui que nous utiliserons au n° suivant pour démontrer le th. 6.

(4.7.5) *Si p est assez grand, le conducteur de ρ_p^X est N.*

Cela se vérifie au moyen des propriétés des modèles de Néron décrites dans [18], *loc. cit.* (Le fait que le conducteur de ρ_p^X soit un *diviseur* de N est nettement plus facile à démontrer, et serait suffisant pour la suite.)

(4.7.6) *Si p est assez grand, l'invariant k de ρ_p^X est égal à* 2.

Cela résulte de la prop. 4 du n° 2.8.

Une fois $(4.7.3), \ldots, (4.7.6)$ établis, on peut appliquer $(3.3.1_?)$. D'où, pour tout p assez grand et complètement décomposé dans K_X, une forme parabolique de poids 2 et de niveau N:

$$f_p = \sum a_{n,p} q^n,$$

à coefficients dans $\overline{\mathbf{F}}_p$, fonction propre normalisée des opérateurs de Hecke, telle que $\rho_p^X \simeq \rho_{f_p}$; on a en particulier

$$a_{l,p} = \tilde{a}_l \quad \text{pour tout } l \nmid N,\ l \neq p.$$

En relevant f_p en caractéristique zéro grâce à (3.1.6) on en déduit une forme parabolique de poids 2 et de niveau N:

$$F = \sum A_n q^n,$$

à coefficients dans $\overline{\mathbf{Z}}$, fonction propre normalisée des opérateurs de Hecke, et telle que $\tilde{F} = f_p$ pour tout $p \in P$, où P est un ensemble infini de nombres premiers complètement décomposés dans K_X. Si $l \nmid N$, on a donc

$$\tilde{A}_l = a_{l,p} = \tilde{a}_l \quad \text{pour tout } p \in P,\ p \neq l,$$

d'où $A_l = a_l$ puisque P est infini. Les systèmes de représentations λ-adiques définis par X et par F sont donc isomorphes. Le théorème en résulte d'après Faltings [13].

Remarque. Ici encore, F est *primitive*, cf. Carayol [8].

4.8. *Variétés projectives ayant un nombre de Betti égal à* 2 *en dimension impaire.* Soient:
X une variété projective lisse sur $\mathbf{Q}$;
$X_{\mathbf{C}} = X(\mathbf{C})$ la variété complexe définie par X;
m un entier impair ≥ 1;
$H^m(X_{\mathbf{C}}, \mathbf{C})$ le m-ème groupe de cohomologie de $X_{\mathbf{C}}$, à coefficients complexes.

Faisons les deux hypothèses suivantes:

(4.8.1) dim $H^m(X_{\mathbf{C}}, \mathbf{C}) = 2$ (i.e., le m-ème nombre de Betti de $X_{\mathbf{C}}$ est égal à 2);
(4.8.2) La décomposition de Hodge de $H^m(X_{\mathbf{C}}, \mathbf{C})$ est de type $(m, 0) + (0, m)$.

Choisissons un ensemble fini S de nombres premiers assez grand pour que X ait bonne réduction en dehors de S. Si $l \notin S$, on peut définir une réduction modulo l de X, qui est une variété $\tilde{X}_l$ lisse sur $\mathbf{F}_l$. Soient π_l et π_l' les valeurs propres de l'endomorphisme de Frobenius de $\tilde{X}_l$, opérant sur la cohomologie de dimension m. D'après Deligne, π_l et π_l' sont des entiers d'un corps quadratique imaginaire, et l'on a

$$(4.8.3) \qquad \pi_l' = \overline{\pi}_l \quad \text{et} \quad \pi_l \overline{\pi}_l = l^m.$$

On posera

$$(4.8.4) \qquad a_l(X) = \pi_l + \overline{\pi}_l.$$

On a $a_l(X) \in \mathbf{Z}$ et $|a_l(X)| \leqslant 2l^{m/2}$.

(Noter qu'il n'y a pas en général unicité de $\tilde{X}_l$, contrairement à ce qui se passe pour les variétés abéliennes. Toutefois, deux choix différents de $\tilde{X}_l$ conduisent au même $a_l(X)$, cf. [40], n° 1.2.)

THÉORÈME 6.   *Admettons* (3.2.4$_?$). *Il existe alors:*
(a) *un entier $N \geqslant 1$ dont tous les facteurs premiers appartiennent à S,*
(b) *une forme parabolique de type $(N, m + 1, 1)$:*

$$F = q + \cdots + A_n q^n + \dots,$$

fonction propre normalisée des opérateurs de Hecke,
tels que:

$$(4.8.5) \qquad A_l = a_l(X) \quad \text{pour tout } l \notin S.$$

(En d'autres termes, les $a_l(X)$ sont les valeurs propres associées à une forme de poids $m + 1$ dont le niveau ne fait intervenir que les nombres premiers de S.)

Il y a intérêt à reformuler le th. 6 en termes de représentations galoisiennes:
Soit $\overline{X}$ la $\overline{\mathbf{Q}}$-variété déduite de X par extension des scalaires de $\mathbf{Q}$ à $\overline{\mathbf{Q}}$, et soit $H^m_{\mathrm{et}}(\overline{X}, \mathbf{Q}_p)$ le m-ème groupe de cohomologie étale de $\overline{X}$ à coefficients dans $\mathbf{Q}_p$. Notons H_p le $\mathbf{Q}_p$-*dual* de $H^m_{\mathrm{et}}(\overline{X}, \mathbf{Q}_p)$. Le groupe $G_{\mathbf{Q}}$ opère sur H_p. On obtient ainsi une représentation p-adique de $G_{\mathbf{Q}}$ de dimension 2; son déterminant est la puissance m-ème du caractère cyclotomique $G_{\mathbf{Q}} \to \mathbf{Z}_p^*$. Lorsque p varie, ces représentations forment un système rationnel de représentations p-adiques compatibles entre elles, les traces des substitutions de Frobenius étant les a_l. (Noter qu'il s'agit ici de substitutions de Frobenius "arithmétiques", et non "géométriques"; c'est ce qui explique le passage au dual.) Le th. 6 équivaut à dire que *ce système de représentations est isomorphe à celui fourni par une forme parabolique de poids $k = m + 1$.*

Démonstration du th. 6.   On reprend la méthode utilisée pour le th. 4. Notons T l'ensemble des p tels que, soit $H^m_{\mathrm{et}}(\overline{X}, \mathbf{Z}_p)$, soit $H^{m+1}_{\mathrm{et}}(\overline{X}, \mathbf{Z}_p)$, ait une composante de torsion non nulle; c'est un ensemble fini. Si $p \notin T$, on a $\dim H^m_{\mathrm{et}}(\overline{X}, \mathbf{F}_p) = 2$; l'action de $G_{\mathbf{Q}}$ sur le dual de $H^m_{\mathrm{et}}(\overline{X}, \mathbf{F}_p)$ définit alors une représentation

$$\rho_p \colon G_{\mathbf{Q}} \to \mathbf{GL}_2(\mathbf{F}_p),$$

qui est non ramifiée en dehors de S et de p (c'est une réduction modulo p de la représentation de $G_{\mathbf{Q}}$ sur H_p considérée plus haut; en particulier, on a det $\rho_p = \chi^m$). Il est essentiel pour la suite de connaître le comportement de ρ_p en p, et, plus précisément, son invariant k au sens du §2. D'après un théorème de J.-M. Fontaine (démontré en utilisant certains de ses résultats récents obtenus en collaboration avec W. Messing), on a:

(4.8.6) (Fontaine–non publié). *Si p est assez grand, l'invariant k de la représentation ρ_p est égal à $m + 1$.*

(C'est ici que sert l'hypothèse faite sur la décomposition de Hodge de $H^m(X_{\mathbf{C}}, \mathbf{C})$.)

On va maintenant s'intéresser au conducteur N_p de ρ_p. Il est clair que N_p est de la forme

$$N_p = \prod_{l \in S} l^{n(l, p)}, \qquad \text{avec } n(l, p) \geqslant 0.$$

Il nous faut majorer les exposants $n(l, p)$, pour l fixé et p variable. Si l'on admettait la conjecture C_3 de [40], on saurait que les $n(l, p)$ sont *bornés* quand l varie (en fait, il est vraisemblable que $n(l, p)$, pour p assez grand, est *égal* à l'exposant du conducteur défini dans [40], formule (11)). Comme C_3 n'est pas démontrée, nous allons nous restreindre aux nombres premiers p satisfaisant aux congruences suivantes:

$$(4.8.7) \qquad \begin{cases} p \not\equiv \pm 1 \pmod{2^3} & \text{si } 2 \in S, \\ p \not\equiv \pm 1 \pmod{3^2} & \text{si } 3 \in S, \\ p \not\equiv \pm 1 \pmod{l} & \text{pour tout } l \in S, l \geqslant 5. \end{cases}$$

On peut alors majorer les $n(l, p)$:

(4.8.8) *Si p satisfait à (4.8.7), et si $l \in S$, $l \neq p$, on a:*

$$n(l, p) \leqslant 9 \quad \text{pour } l = 2,$$

$$n(l, p) \leqslant 5 \quad \text{pour } l = 3,$$

$$n(l, p) \leqslant 2 \quad \text{pour } l \geqslant 5.$$

En effet, soit $I_{l, p}$ le sous-groupe d'inertie en l de $\rho_p(G_{\mathbf{Q}})$. Comme det ρ_p n'est pas ramifiée en l, $I_{l, p}$ est contenu dans $\mathbf{SL}_2(\mathbf{F}_p)$, et son ordre est un diviseur de

$p(p^2 - 1)$. Si $l \geqslant 5$, l'hypothèse (4.8.7) entraîne que $I_{l,p}$ est d'ordre premier à l; la représentation ρ_p est donc modérée en l, et d'après le n° 1.2, on a $n(l, p) \leqslant 2$. Lorsque $l = 3$ (resp. $l = 2$), les l-sous-groupes de Sylow de $\mathbf{SL}_2(\mathbf{F}_p)$ sont cycliques d'ordre 3 (resp. quaternioniens d'ordre 8); en appliquant une majoration de conducteurs que l'on trouvera au n° 4.9 ci-après, on en déduit $n(3, p) \leqslant 5$ (resp. $n(2, p) \leqslant 9$).

Notons P l'ensemble des nombres premiers p satisfaisant aux conditions de (4.8.6) et (4.8.7). C'est un ensemble infini.

(4.8.9) *Si $p \in P$, et si p est assez grand, la représentation ρ_p est irréductible.*

Soit P' l'ensemble des $p \in P$ tels que ρ_p soit réductible. Si $p \in P'$, la semi-simplification de ρ_p est donnée par deux caractères

$$\alpha, \beta: G_{\mathbf{Q}} \to \mathbf{F}_p^*, \quad \text{avec } \alpha\beta = \chi^m.$$

Il résulte de (4.8.6) que l'un de ces caractères, disons α, est non ramifié en p. Le conducteur de α est alors un diviseur de N_p, et l'on a

(4.8.10) $$a_l(X) \equiv \alpha(l) + \alpha(l)^{-1} l^m \pmod{p}$$

pour tout $l \notin S$, $l \neq p$.

Soit $\alpha_0: (\mathbf{Z}/N_p\mathbf{Z})^* \to \overline{\mathbf{Z}}^*$ le relèvement multiplicatif de α, cf. n° 3.1. D'après (4.8.8), N_p ne prend qu'un nombre fini de valeurs. Il n'y a donc qu'un nombre fini de possibilités pour α_0. Si P' était infini, il y aurait alors un α_0 qui interviendrait pour un sous-ensemble infini P'' de P'. Si $l \notin S$, posons:

$$b_l = \alpha_0(l) + \alpha_0(l)^{-1} l^m.$$

D'après (4.8.10), $a_l(X)$ et b_l ont même image dans $\overline{\mathbf{F}}_p$ pour tout $p \in P''$, $p \neq l$. Comme P'' est infini, cela entraîne

$$a_l(X) = b_l \quad \text{pour tout } l \notin S,$$

donc aussi

$$\{\pi_l, \overline{\pi}_l\} = \left\{\alpha_0(l), \alpha_0(l)^{-1} l^m\right\},$$

ce qui est absurde. D'où (4.8.9).

En combinant (4.8.6), (4.8.8) et (4.8.9), on voit que l'on peut trouver un ensemble infini P_1 de nombres premiers, et un entier N de la forme $\prod_{l \in S} l^{n(l)}$, tels que, pour tout $p \in P_1$, la représentation ρ_p jouisse des propriétés suivantes:
(a) ρ_p est irréductible, de déterminant χ^m;
(b) le conducteur de ρ_p est égal à N;
(c) l'invariant k de ρ_p est égal à $m + 1$.

Comme m est impair, (a) entraîne que ρ_p est absolument irréductible si $p \in P_1$, $p \neq 2$. On peut alors appliquer $(3.2.4_?)$. D'où, pour tout $p \in P_1$, $p \neq 2$, l'existence d'une forme parabolique de poids $k = m + 1$ et de niveau N:

$$f_p = \sum a_{n,\,p} q^n,$$

à coefficients dans $\overline{\mathbf{F}}_p$, qui est fonction propre normalisée des opérateurs de Hecke, et telle que $\rho_p \simeq \rho_{f_p}$. On conclut alors comme dans la démonstration du th. 4, en relevant f_p en caractéristique 0, et en remarquant qu'il n'y a qu'un nombre fini de possibilités.

Remarques

(1) On trouvera dans Schoen [37] un exemple où les conditions (4.8.1) et (4.8.2) sont satisfaites, avec $m = \dim X = 3$, $k = 4$, $S = \{5\}$, $N = 5^2$. Il s'agit d'une variété X qui est une désingularisation de l'hypersurface de l'espace projectif $\mathbf{P}_4$ d'équation

$$X_0^5 + X_1^5 + X_2^5 + X_3^5 + X_4^5 - 5X_0 X_1 X_2 X_3 X_4 = 0.$$

On peut alors expliciter la forme parabolique F et démontrer la relation (4.8.5) sans utiliser aucune conjecture: il suffit d'appliquer la méthode de Faltings ([13], p. 362–363—voir aussi [47]) aux représentations 2-adiques définies par X et par F.

(2) Comme l'a remarqué S. Bloch [5], la conclusion du th. 6 peut aussi se déduire des conjectures "archimédiennes" (et non plus modulo p) sur les fonctions L attachées aux motifs (Deligne [10]), combinées avec la caractérisation des formes modulaires due à Weil [55]. De ce point de vue, l'hypothèse (4.8.2) sert à assurer que le facteur à l'infini de la fonction L est bien $(2\pi)^{-s}\Gamma(s)$.

(3) Si l'on supprime l'hypothèse (4.8.2) en question, la décomposition de Hodge de $H^m(X_{\mathbf{C}}, \mathbf{C})$ est de type $(m - r, r) + (r, m - r)$, avec $0 \leqslant r < m/2$. En admettant $(3.2.4_?)$, on peut alors démontrer l'existence d'une forme parabolique normalisée

$$F = \sum A_n q^n,$$

de poids $m - 2r$, telle que $a_l(X) = l^r A_l$ pour tout $l \notin S$: la représentation de $G_{\mathbf{Q}}$ sur H_p se déduit de celle associée à F par une "torsion de Tate" d'amplitude r. La démonstration est essentiellement la même.

4.9. *Une majoration de conducteurs.*   La question étant *locale*, nous utiliserons les notations standard suivantes:

K est un corps complet pour une valuation discrète;

$v_K\colon K^* \to \mathbf{Z}$ est la valuation normalisée de K;

$\overline{K}$ est une clôture algébrique de K;

$G_K = \mathrm{Gal}(\overline{K}/K)$ est le groupe de Galois de $\overline{K}$ sur K.

On suppose que K est de caractéristique 0, et que son corps résiduel est parfait de caractéristique $p > 0$. On note

$$e_K = v_K(p)$$

l'indice de ramification absolu de K.

(Attention au changement de notation: au n° précédent, la caractéristique résiduelle était notée l.)

Soit maintenant V un espace vectoriel de dimension finie sur un corps Ω de caractéristique $\neq p$, et soit $\rho : G_K \to \mathbf{GL}(V)$ un homomorphisme continu. *L'exposant du conducteur de* ρ est un entier $n(\rho) \geqslant 0$, que l'on définit comme au n° 1.2:

si $(G_i)_{i \geqslant 0}$ est la suite des groupes de ramification du groupe fini $G = \rho(G_K)$, on a

$$(4.9.1) \qquad n(\rho) = \sum_{i \geqslant 0} \frac{g_i}{g_0} \dim V/V_i,$$

où g_i est l'ordre de G_i, et V_i est le sous-espace de V fixé par G_i.

Il y a intérêt à récrire cette définition sous la forme

$$(4.9.2) \qquad n(\rho) = \dim V/V_0 + b(\rho),$$

où

$$b(\rho) = \sum_{i \geqslant 1} \frac{g_i}{g_0} \dim V/V_i$$

est *l'invariant sauvage* de ρ ([39], §19.3).

La majoration que nous avons en vue est la suivante:

PROPOSITION 9. *Soit p^c l'ordre du groupe d'inertie sauvage G_1, et soit N la dimension de V sur Ω. On a*

$$(4.9.3) \qquad b(\rho) \leqslant N e_K \left(c + \frac{1}{p-1} \right).$$

De plus, si G_1 n'est pas cyclique, cette inégalité est stricte.

Vu (4.9.2), ceci entraîne:

COROLLAIRE. *On a*

$$(4.9.4) \qquad n(\rho) \leqslant N(1 + e_K c + e_K/(p-1)),$$

avec inégalité stricte si G_1 n'est pas cyclique.

Démonstration de la prop. 9.   Soit I le plus grand indice $i \geqslant 1$ tel que $G_i \neq \{1\}$. On majore $\dim V/V_i$ par N si $i \leqslant I$, et par 0 si $i > I$. D'où:

$$(4.9.5) \qquad b(\rho) \leqslant \frac{N}{g_0}(g_1 + \cdots + g_I) \leqslant \frac{N}{g_0}\Big(I + \sum_{i \geqslant 1}(g_i - 1)\Big).$$

D'après un résultat élémentaire sur les groupes de ramification ([38], p. 79, exerc. 3), on a:

$$(4.9.6) \qquad\qquad I \leqslant g_0 e_K/(p - 1),$$

avec inégalité stricte si G_1 n'est pas cyclique.

D'autre part, l'entier

$$d = \sum_{i \geqslant 0}(g_i - 1)$$

est égal à la valuation de la différente de l'extension L/K de groupe de Galois G ([38], p. 72). D'après une majoration due à Hensel (reproduite dans [38], p. 67), on a

$$d \leqslant g_0 - 1 + g_0 e_K c,$$

d'où:

$$(4.9.7) \qquad\qquad \sum_{i \geqslant 1}(g_i - 1) \leqslant g_0 e_K c.$$

En combinant (4.9.5), (4.9.6) et (4.9.7), on obtient l'inégalité à démontrer (4.9.3), et l'on voit aussi que cette inégalité est stricte si G_1 n'est pas cyclique.

Remarque.   Lorsque G_1 est *abélien* d'exposant p^h, on peut montrer que

$$b(\rho) \leqslant Ne_K\left(h + \frac{1}{p - 1}\right).$$

Comme $h \leqslant c$, cela améliore (4.9.3).

Application à (4.8.8).   Dans la situation de (4.8.8), il y a deux cas à considérer:

(a) *Caractéristique résiduelle* 3. Avec les notations de la prop. 9 (qui diffèrent de celles du n° 4.8, comme on l'a signalé plus haut), on a $p = 3$, $N = 2$, $e_K = 1$ et $c \leqslant 1$, d'où $n(\rho) \leqslant 5$ d'après (4.9.4). Cette borne est optimale: il existe des courbes elliptiques de conducteur 3^5.

(b) *Caractéristique résiduelle* 2. On a alors $p = 2$, $N = 2$, $e_K = 1$ et $c \leqslant 3$, avec G_1 non cyclique si $c = 3$; d'où $n(\rho) \leqslant 9$ d'après (4.9.4). En fait, une analyse plus détaillée montre que l'on a même $n(\rho) \leqslant 8$, ce qui est une majoration optimale: il existe des courbes elliptiques de conducteur 2^8.

§5. Exemples. Ce § rassemble un certain nombre d'exemples sur lesquels on peut vérifier, au moins en partie, les conjectures du §3. La plupart des vérifications ont nécessité l'emploi d'un ordinateur; elles ont été programmées et réalisées par J-F. Mestre.

Les valeurs de p considérées sont:

$p = 2$ (n°s 5.1 et 5.2),

$p = 3$ (n°s 5.3 et 5.4),

$p = 7$ (n° 5.5).

5.1. *Exemples provenant de* $\mathbf{GL}_2(\mathbf{F}_2) \simeq \mathfrak{S}_3$. Soit K un corps cubique non abélien, et soit K^{gal} sa clôture galoisienne. Le groupe $\mathrm{Gal}(K^{\mathrm{gal}}/\mathbf{Q})$ est isomorphe au groupe symétrique $\mathfrak{S}_3$, lui-même isomorphe à $\mathbf{GL}_2(\mathbf{F}_2)$. On obtient ainsi une représentation

$$\rho^K\colon G_{\mathbf{Q}} \to \mathbf{GL}_2(\mathbf{F}_2),$$

qui est absolument irréductible, et à laquelle on peut appliquer les conjectures du §3.

Les invariants (N, k, ε) de ρ^K sont faciles à déterminer. Si l'on écrit le discriminant D du corps K sous la forme

$$D = \pm 2^m N, \text{ avec } N \text{ impair } > 0, \text{ et } m = 0, 2 \text{ où } 3,$$

on constate que:

le conducteur de ρ^K est égal à N;

le caractère ε est égal à 1;

le poids k de ρ^K est égal à 2 (resp. 4) si $m = 0, 2$ (resp. si $m = 3$).

La conjecture $(3.2.4_?)$ prédit alors l'existence d'une forme parabolique f à coefficients dans $\mathbf{F}_2$ (ou dans $\mathbf{F}_4$ si $m = 0$, i.e., si K est non ramifié en 2), de type $(N, k, 1)$, fonction propre normalisée des opérateurs de Hecke, et telle que ρ^K soit isomorphe à ρ_f. Le tableau suivant donne une liste de cas où ceci a été vérifié sur ordinateur:

$D < 0$	k = poids	N = niveau	$D > 0$	k = poids	N = niveau
-23	2	23	148	2	37
-31	2	31	229	2	229
-44	2	11	257	2	257
-59	2	59	316	2	79
-76	2	19			
-104	4	13			

(Dans les cas $D = -23$, $D = -31$ et $D = 257$, l'idéal (2) reste premier dans K, et la valeur propre de U_2 est une racine cubique primitive de l'unité, i.e., un élément de $\mathbf{F}_4 - \mathbf{F}_2$, conformément à $(3.2.6_?)$. Pour les autres valeurs de D, la valeur propre de U_2 est 0 ou 1, et tous les coefficients de f appartiennent à $\mathbf{F}_2$.)

Dans le cas général, je ne sais démontrer qu'un résultat un peu plus faible que $(3.2.4_?)$:

PROPOSITION 10. *Il existe une forme f de type $(N, k', 1)$, avec k' convenable, telle que ρ^K soit isomorphe à ρ_f.*

(En particulier, ρ^K satisfait à $(3.2.3_?)$.)

Démonstration. On utilise le plongement évident $\mathfrak{S}_3 \to \mathbf{GL}_2(\mathbf{Z})$, ce qui fournit une représentation

$$\rho_0^K \colon G_{\mathbf{Q}} \to \mathbf{GL}_2(\mathbf{C}),$$

qui "relève" ρ^K en caractéristique 0. Le déterminant de ρ_0^K est le caractère quadratique

$$\varepsilon_D \colon G_{\mathbf{Q}} \to \mathfrak{S}_3 \overset{\mathrm{sgn}}{\to} \{\pm 1\}$$

qui correspond au corps $\mathbf{Q}(\sqrt{D})$. Distinguons alors deux cas:

(i) $D < 0$, *i.e., K est un corps cubique imaginaire.*

Le caractère $\varepsilon_D = \det \rho_0^K$ est alors *impair*. Comme l'image de ρ_0^K est $\mathfrak{S}_3$, qui est un groupe diédral, on en conclut (cf. [11], [45]), que ρ_0^K est la représentation associée à une forme parabolique F_1 de poids 1, de caractère ε_D et de niveau $|D|$; on peut d'ailleurs écrire explicitement F en termes de fonctions thêta de formes quadratiques binaires de discriminant D. Soit E_D la série d'Eisenstein de poids 1 et de caractère ε_D (qui est aussi une fonction thêta). Le produit $F = F_1 \cdot E_D$ est une forme parabolique de poids 2, de caractère 1, et de niveau $|D|$. Si $f = \tilde{F}$ est la réduction (mod 2) de F, on a $f = \tilde{F}_1$, car $\tilde{E}_D = 1$. La forme f est la forme cherchée; en effet, par construction f est de type $(2^m N, 2, 1)$, donc aussi de type $(N, k', 1)$, pour k' convenable.

(Il devrait être possible de préciser cette démonstration, et d'obtenir la valeur exacte de k'. Je ne l'ai fait que pour $m = 0$, i.e., $D = -N$, où l'on obtient bien $k' = 2$, comme annoncé.)

(ii) $D > 0$, *i.e., K est un corps cubique totalement réel*

Le corps $\mathbf{Q}(\sqrt{D})$ est alors un corps quadratique réel, et la représentation ρ_0^K est induite par un caractère ψ d'ordre 3 de $\mathbf{Q}(\sqrt{D})$. Choisissons un caractère auxiliaire α de $\mathbf{Q}(\sqrt{D})$ ayant les propriétés suivantes:

(ii_1) l'ordre de α est une puissance de 2;

(ii_2) α a pour signatures $+$ et $-$ en les deux places à l'infini de $\mathbf{Q}(\sqrt{D})$;

(ii_3) α est non ramifié en toute place finie de $\mathbf{Q}(\sqrt{D})$ de caractéristique résiduelle $\neq 2$.

(L'existence d'un tel caractère est facile à démontrer.)

Soit $\rho_0' = \mathrm{Ind}(\psi\alpha)$ la représentation de $G_{\mathbf{Q}}$ *induite* par le caractère $\psi\alpha$ du corps $\mathbf{Q}(\sqrt{D})$. C'est une représentation irréductible de degré 2. D'après (ii_1), sa réduction en caractéristique 2 est isomorphe à $\mathrm{Ind}(\psi) \simeq \rho^K$. D'après (ii_2), son déterminant est impair, et d'après (ii_3) son conducteur est de la forme $2^M N$, avec M entier. On peut donc appliquer à ρ'_0 l'argument utilisé dans le cas (i) pour ρ_0^K: cette représentation est associée à une forme parabolique F' de poids 1 et de niveau $2^M N$; par réduction en caractéristique 2, F' donne la forme f cherchée. (Noter qu'ici F' est combinaison linéaire de fonctions thêta de formes binaires indéfinies.)

Remarque.   Le même genre d'argument s'applique à toute représentation

$$\rho_p \colon G_{\mathbf{Q}} \to \mathbf{GL}_2(\overline{\mathbf{F}}_p), \qquad p \neq 2,$$

à déterminant impair, et telle que l'image de $\rho_p(G_{\mathbf{Q}})$ dans $\mathbf{PGL}_2(\overline{\mathbf{F}}_p)$ soit un groupe *diédral*; en particulier, la conjecture faible $(3.2.3_?)$ est vraie pour une telle représentation.

5.2. *Exemples provenant de* $\mathbf{SL}_2(\mathbf{F}_4) \simeq \mathfrak{A}_5$.   Soit K un corps de degré 5 sur $\mathbf{Q}$ dont la clôture galoisienne K^{gal} ait pour groupe de Galois le groupe alterné $\mathfrak{A}_5$. Comme $\mathfrak{A}_5$ est isomorphe à $\mathbf{SL}_2(\mathbf{F}_4)$, on déduit de là un homomorphisme surjectif $G_{\mathbf{Q}} \to \mathbf{SL}_2(\mathbf{F}_4)$, d'où une représentation absolument irréductible

$$\rho^K \colon G_{\mathbf{Q}} \to \mathbf{GL}_2(\mathbf{F}_4),$$

avec $\det \rho^K = 1$.

Ici encore, on désire tester sur ρ^K les conjectures du §3. Comme le conducteur N de ρ^K est le plus souvent très grand, les calculs ne sont praticables que si N est un nombre premier, et si le poids k est égal à 2, car cela permet alors d'appliquer la "méthode des graphes" ([30], [31]). Le tableau suivant indique les différents cas étudiés par Mestre; on a noté D la racine carrée du discriminant de K, avec le signe $+$ si K est réel et le signe $-$ si K est imaginaire.

$D < 0$	N = niveau		$D > 0$	N = niveau
-2083	2083		$2^3 887$	887
-2707	2707		8311	8311
-3203	3203		$2^2 8447$	8447
-3547	3547		13613	13613
-4027	4027		$2^2 24077$	24077

Les exemples avec $D < 0$ sont extraits d'une table de J. Buhler ([7], p. 136–141); ceux avec $D > 0$ proviennent de [31], n° 4.2.

Remarques

(1) Dans chacun des cas considérés, Mestre obtient une forme parabolique f à coefficients dans $\mathbf{F}_4$ (ou, parfois, dans $\mathbf{F}_{16}$), du type $(N, 2, 1)$ voulu, fonction propre des opérateurs de Hecke $U_2, T_3, T_5, \ldots$, les valeurs propres des trois premiers opérateurs étant les bonnes. Il est donc vraisemblable que la représenta-

tion ρ_f associée à f est isomorphe à ρ^K; toutefois, une démonstration complète demanderait un travail considérable, qui n'a pas été entrepris.

(2) Le cas $D < 0$ n'est pas très surprenant. En effet, la représentation ρ^K peut se relever en caractéristique 0, son image étant alors une certaine extension centrale de $\mathfrak{A}_5$ par un groupe cyclique d'ordre une puissance de 2 (utiliser un plongement de $\mathfrak{A}_5$ dans $\mathbf{PGL}_2(\mathbf{C})$ et appliquer les résultats de Tate reproduits dans [45], §6). Si $D < 0$, cette représentation est de déterminant impair, et provient donc (si l'on admet la conjecture d'Artin sur les fonctions L) d'une forme parabolique F de poids 1. En réduisant F en caractéristique 2, on obtient une forme f telle que $\rho_f \simeq \rho^K$ (cf. démonstration de la prop. 10), ce qui montre que ρ^K satisfait à la conjecture faible (3.2.3$_?$).

Le cas $D > 0$ est plus étonnant: on ne voit *a priori* aucun moyen de rattacher ρ^K à une quelconque forme modulaire.

5.3. *Exemples provenant de* $\mathbf{GL}_2(\mathbf{F}_3) \simeq \tilde{\mathfrak{S}}_4$.   Le groupe $\mathbf{PGL}_2(\mathbf{F}_3)$ agit sur la droite projective $\mathbf{P}_1(\mathbf{F}_3)$, qui a 4 points, et cela définit un isomorphisme $\mathbf{PGL}_2(\mathbf{F}_3) \simeq \mathfrak{S}_4$. Comme le noyau de $\mathbf{GL}_2(\mathbf{F}_3) \to \mathbf{PGL}_2(\mathbf{F}_3)$ est $\{\pm 1\}$, on en conclut que $\mathbf{GL}_2(\mathbf{F}_3)$ est une extension centrale de degré 2 de $\mathfrak{S}_4$; en fait, c'est l'extension notée $\tilde{\mathfrak{S}}_4$ dans [46], n° 1.5.

Il est bien connu que $\tilde{\mathfrak{S}}_4$ peut se plonger dans $\mathbf{GL}_2(\mathbf{Z}[\sqrt{-2}\,])$, et que ce plongement donne par réduction (mod 3) l'isomorphisme $\tilde{\mathfrak{S}}_4 \simeq \mathbf{GL}_2(\mathbf{F}_3)$ ci-dessus. Ceci permet d'associer à toute représentation

$$\rho\colon G_{\mathbf{Q}} \to \mathbf{GL}_2(\mathbf{F}_3),$$

son *relèvement*

$$\rho_0\colon G_{\mathbf{Q}} \to \mathbf{GL}_2\big(\mathbf{Z}[\sqrt{-2}\,]\big) \subset \mathbf{GL}_2(\mathbf{C})$$

en caractéristique 0. Supposons que ρ satisfasse aux conditions du n° 3.2, i.e., soit irréductible et de déterminant impair. Il en est alors de même de ρ_0, et l'on peut appliquer les résultats de Langlands [26] et Tunnell [53]. On en déduit que ρ_0 provient d'une forme parabolique de poids 1 et de niveau le conducteur de ρ_0, conducteur que l'on peut écrire sous la forme $3^m N_0$, avec N_0 premier à 3. D'où, comme au n° 5.1:

PROPOSITION 11.   *Il existe une forme f de type* (N_0, k', ε), *avec k' convenable, telle que ρ soit isomorphe à ρ_f.*

(Ici, ε est le caractère $G_{\mathbf{Q}} \to \{\pm 1\}$ défini à partir de $\det \rho$ comme on l'a expliqué au n° 1.3.)

En particulier ρ *satisfait à la conjecture faible* (3.2.3$_?$).

Remarque.   Le conducteur $3^m N_0$ de ρ_0 est étroitement lié au conducteur N de ρ défini au §1. Si l'on pose

$$N = \prod_{l \neq 3} l^{n(l)} \quad \text{et} \quad N_0 = \prod_{l \neq 3} l^{n_0(l)},$$

on constate en effet que:

(5.3.1) Si le groupe d'inertie en l de $\rho(G_{\mathbf{Q}}) \simeq \rho_0(G_{\mathbf{Q}})$ est cyclique d'ordre 3, on a $n(l) = 1$ et $n_0(l) = 2$.

(5.3.2) Dans tout autre cas, on a $n(l) = n_0(l)$.

En particulier, N *divise* N_0, et les facteurs premiers de N et de N_0 sont les mêmes. La conjecture (3.2.4$_?$) affirme donc (entre autres choses) que le niveau N_0 intervenant dans la prop. 11 peut être abaissé à N. Voici deux exemples où cet abaissement a bien lieu:

Exemples tirés de courbes elliptiques.   Soit E une courbe elliptique sur $\mathbf{Q}$. Supposons qu'il y ait un nombre premier $l > 3$ en lequel E a mauvaise réduction de type c_3 ou c_6 au sens de Néron (types IV ou IV* de Kodaira). Avec les notations de [41], n° 5.6, cela équivaut à dire que E a potentiellement bonne réduction en l, et que le groupe Φ_l correspondant est cyclique d'ordre 3. Prenons pour ρ la représentation

$$\rho^E \colon G_{\mathbf{Q}} \to \mathbf{GL}_2(\mathbf{F}_3)$$

définie par les points de 3-division de E. D'après (5.3.1), l'exposant de l dans N (resp. N_0) est 1 (resp. 2). On doit donc constater un abaissement. Effectivement:

Exemple (5.3.3). *La courbe* 121_F (cf. [1], p. 97).   L'équation de E est

$$y^2 + xy = x^3 + x^2 - 2x - 7.$$

Il y a bonne réduction en dehors de $l = 11$, et mauvaise réduction de type c_3 en 11, d'où $N_0 = 11^2$ et $N = 11$. De plus, la représentation ρ^E est irréductible. La conjecture (3.2.4$_?$) prédit que ρ^E provient d'une forme de poids 2 et de niveau $N = 11$. Mais il n'y a qu'une telle forme (à homothétie près): celle qui correspond à la courbe E' de conducteur 11 et d'équation

$$y^2 + y = x^3 - x^2.$$

On en conclut que les représentations ρ^E et $\rho^{E'}$ doivent être isomorphes, ou encore que les traces a_l et a'_l de leurs endomorphismes de Frobenius doivent être telles que:

$$a_l \equiv a'_l \ (\mathrm{mod}\ 3) \quad \text{pour tout } l \neq 3, 11.$$

La table suivante (extraite de [4], p. 117–119) montre que c'est bien le cas, au moins pour $l < 50$:

l	2	5	7	13	17	19	23	29	31	37	41	43	47
a_l	1	1	-2	1	-5	6	2	9	-2	-3	-5	0	2
a'_l	-2	1	-2	4	-2	0	-1	0	7	3	-8	-6	8

Exemple (5.3.4). *La courbe* 147_I (cf. [4], p. 103).   L'équation de E est

$$y^2 + y = x^3 + x^2 - 114x + 473.$$

Son conducteur est $147 = 3.7^2$. Il y a mauvaise réduction de type multiplicatif en 3, et mauvaise réduction de type c_6 en 7, d'où $N_0 = 7^2$, $N = 7$. La représentation ρ^E a pour conducteur 7; comme elle est très ramifiée en 3, son poids k est égal à 4. La conjecture $(3.2.4_?)$ prédit que ρ^E provient d'une forme parabolique de poids 4 et de niveau 7. Or, ici encore, il n'y a qu'une seule telle forme (normalisée):

$$F = q + \sum_{n \geqslant 2} A_n q^n$$

$$= q - q^2 - 2q^3 - 7q^4 + 16q^5 + 2q^6 - 7q^7 + 15q^8 + \cdots .$$

(Pour le calcul des coefficients de F, voir ci-après.)

Si a_l désigne la trace de l'endomorphisme de Frobenius de E en l, on doit donc avoir

$$a_l \equiv A_l \pmod 3 \quad \text{pour tout } l \neq 3,7.$$

C'est bien ce qui se passe, au moins pour $l < 50$:

l	2	5	11	13	17	19	23	29	31	37	41	43	47
a_l	2	-2	-2	1	0	1	0	4	9	3	-10	5	-6
A_l	-1	16	-8	28	54	-110	48	-110	12	-246	182	128	324

Calcul de F.   Soit L l'anneau des entiers du corps $\mathbf{Q}(\sqrt{-7})$. Les séries

$$f_1 = \sum_{z \in L} q^{z\bar{z}} = 1 + 2q + 4q^2 + 6q^4 + 2q^7 + \cdots$$

$$f_2 = \frac{1}{2} \sum_{z \in L} z^2 q^{z\bar{z}} = q - 3q^2 + 5q^4 - 7q^7 - 3q^8 + \cdots$$

sont des formes modulaires de poids 1 et 3 respectivement, de niveau 7 et de caractère le caractère de Legendre mod 7. Leur produit $f_1 \cdot f_2$ est la forme F considérée plus haut; d'où le calcul des coefficients de F.

5.4. *Exemples provenant de* $\mathbf{SL}_2(\mathbf{F}_9) \simeq \tilde{\mathfrak{A}}_6$.   Soit G le sous-groupe de $\mathbf{GL}_2(\mathbf{F}_9)$ formé des éléments de déterminant ± 1. On a

$$G = \{ \pm 1, \pm i \} \cdot \mathbf{SL}_2(\mathbf{F}_9) = \mathbf{SL}_2(\mathbf{F}_9) \cup i \cdot \mathbf{SL}_2(\mathbf{F}_9),$$

où i désigne un élément d'ordre 4 de $\mathbf{F}_9^*$. L'image de ce groupe dans $\mathbf{PGL}_2(\mathbf{F}_9)$

est égale à $\mathbf{PSL}_2(\mathbf{F}_9)$, qui est isomorphe au groupe alterné $\mathfrak{A}_6$. D'où une projection $\varphi\colon G \to \mathfrak{A}_6$. Le couple $(\varphi, \det)$ définit un homomorphisme surjectif $G \to \mathfrak{A}_6 \times \{\pm 1\}$, de noyau $\{\pm 1\}$. On a donc une suite exacte:

$$(*) \qquad \{1\} \to \{\pm 1\} \to G \to \mathfrak{A}_6 \times \{\pm 1\} \to \{1\}.$$

Donnons-nous maintenant un corps K de degré 6 sur $\mathbf{Q}$, avec $\mathrm{Gal}(K^{\mathrm{gal}}/\mathbf{Q}) \cong \mathfrak{A}_6$, ainsi qu'un corps quadratique $\mathbf{Q}(\sqrt{D})$. On en déduit des homomorphismes

$$\alpha^K\colon G_\mathbf{Q} \to \mathfrak{A}_6 \quad \text{et} \quad \varepsilon_D\colon G_\mathbf{Q} \to \{\pm 1\},$$

d'où

$$\alpha\colon G_\mathbf{Q} \to \mathfrak{A}_6 \times \{\pm 1\}.$$

Cherchons à *relever* α en un homomorphisme

$$\rho\colon G_\mathbf{Q} \to G.$$

Vu $(*)$, il y a une *obstruction* à ce relèvement, qui est une classe de cohomologie

$$\mathrm{obs}(\alpha) \in H^2(G_\mathbf{Q}, \{\pm 1\}) \simeq \mathrm{Br}_2(\mathbf{Q}),$$

cf. [46], n° 1.1. Le lemme suivant donne un moyen de calculer cette classe:

LEMME 6. *Soit* $w \in \mathrm{Br}_2(\mathbf{Q})$ *l'invariant de Witt de la forme quadratique* $\mathrm{Tr}_{K/\mathbf{Q}}(x^2)$, *cf.* [46]. *On a*:

$$(5.4.1) \qquad \mathrm{obs}(\alpha) = w + (-1)(D).$$

(Rappelons, *loc. cit.*, que $(-1)(D)$ est l'élément de $\mathrm{Br}_2(\mathbf{Q})$ qui correspond à l'algèbre de quaternions $(-1, D)$.)

D'après le th. 1 de [46], w est l'obstruction à relever

$$\alpha^K\colon G_\mathbf{Q} \to \mathfrak{A}_6 \simeq \mathbf{PSL}_2(\mathbf{F}_9)$$

en un homomorphisme

$$G_\mathbf{Q} \to \tilde{\mathfrak{A}}_6 \simeq \mathbf{SL}_2(\mathbf{F}_9).$$

D'autre part, $(-1)(D)$ est l'obstruction à relever

$$\varepsilon_D\colon G_\mathbf{Q} \to \{\pm 1\}$$

en un homomorphisme:

$$G_\mathbf{Q} \to \{\pm 1, \pm i\}.$$

Le lemme résulte de ces deux faits, par un argument facile.

Fixons maintenant les choix de K et de D. Nous prendrons:

$D = -3$;

$K = $ corps sextique défini par une équation

$$X^6 + aX + b = 0, \qquad a, b \in \mathbf{Z},$$

le couple (a, b) étant choisi de telle sorte que l'équation soit irréductible et de groupe de Galois $\mathfrak{A}_6$.

[Voici quelques choix possibles de (a, b), obtenus par Mestre: $(a, b) = (24, -20)$; $(30, 25)$; $(240, 400)$; $(240, -400)$; $(48, -80)$; $(432, 720)$; $(480, -400)$.]

D'après [46], n° 3.3, le fait que K soit défini par une telle équation entraîne

$$w = (3)(-d) + (-1)(-1),$$

où d est le discriminant de K. Comme $\mathrm{Gal}(K^{\mathrm{gal}}/\mathbf{Q})$ est isomorphe à $\mathfrak{A}_6$, d est un carré, et l'on a

$$w = (3)(-1) + (-1)(-1) = (-1)(-3),$$

d'où

$$\mathrm{obs}(\alpha) = 0$$

en vertu du lemme 6. On peut donc relever α en un homomorphisme

$$\rho\colon G_{\mathbf{Q}} \to G \subset \mathbf{GL}_2(\mathbf{F}_9).$$

Bien entendu, la représentation ρ ainsi obtenue n'est pas unique: elle n'est définie qu'à torsion quadratique près. Comme dans la théorie de Tate (exposée dans [45], §6), on peut utiliser cette torsion pour rendre les invariants k et N de ρ aussi petits que possible; en particulier, on peut choisir ρ de telle sorte que $k = 2$ ou 4, et que N ne soit divisible que par les facteurs premiers du discriminant d qui sont $\neq 3$ (i.e., $l = 2$ et 5 dans les exemples donnés plus haut). Ceci fait, les conjectures du §3 affirment l'existence d'une forme parabolique $f = \Sigma a_n q^n$ de type $(N, k, 1)$, à coefficients dans $\mathbf{F}_9$, fonction propre normalisée des opérateurs de Hecke, et telle que $\rho \simeq \rho_f$. Cette dernière relation entraîne un lien étroit entre les a_l (pour $l \nmid 3N$), et la décomposition de l dans le corps K. De façon plus précise, notons $\mathrm{ord}(l)$ *l'ordre* de la substitution de Frobenius attachée à l dans $\mathrm{Gal}(K^{\mathrm{gal}}/\mathbf{Q}) \simeq \mathfrak{A}_6$. On doit avoir:

$$\mathrm{ord}(l) = 1 \text{ ou } 3 \quad \Leftrightarrow \quad a_l^2 = \left(\frac{l}{3}\right);$$

$$\mathrm{ord}(l) = 2 \qquad \Leftrightarrow \quad a_l = 0;$$

$$\mathrm{ord}(l) = 4 \qquad \Leftrightarrow \quad a_l^2 = -\left(\frac{l}{3}\right);$$

$$\mathrm{ord}(l) = 5 \qquad \Leftrightarrow \quad a_l^4 = -1.$$

(Rappelons que les a_l sont des éléments du corps $\mathbf{F}_9$.)

En particulier, si $l \neq 3$ ne divise pas le discriminant de $X^6 + aX + b$, le *nombre de solutions dans* $\mathbf{F}_l$ de la congruence

$$x^6 + ax + b \equiv 0 \;(\mathrm{mod}\; l)$$

doit être égal à 1 (resp. 2) si et seulement si a_l est un élément d'ordre 8 de $\mathbf{F}_9^*$ (resp. si $a_l = 0$).

La recherche d'une telle forme f a été effectuée par J.-F. Mestre dans chacun des cas $(a, b) = (24, -20), \ldots, (480, -400)$ cités plus haut, ainsi que dans quelques autres. Le conducteur N est alors égal à $2^m 5^n$, où m et n dépendent de (a, b). La détermination de n n'est pas difficile: lorsqu'il y a ramification sauvage en 5 (ce qui est le cas dans les exemples), n est égal à l'exposant de 5 dans $d^{1/2}$. Par contre, la détermination de m est un exercice dyadique que je n'ai pas fait; cela a obligé Mestre à essayer les différents niveaux possibles: 2.5^n, $2^2 5^n$, $2^3 5^n, \ldots$, jusqu'à ce qu'il en trouve un ayant une forme f du type voulu. Ses résultats sont résumés dans le tableau suivant:

a	b	$d^{1/2}$	$k = \text{poids}$	niveau
24	-20	$2^3 3^3 5^3$	2	$2^3 5^3 = 1000$
30	25	$2^3 3^3 5^4$	2	$\geqslant 20000\;?$
240	400	$2^2 3^3 5^4$	2	$2^2 5^4 = 2500$
240	-400	$2^3 3^3 5^4$	2	$2^3 5^4 = 5000$
48	-80	$2^3 3^3 5^3$	2	$2^3 5^3 = 1000$
432	720	$2^2 3^5 5^3$	4	$2^2 5^3 = 500$
480	-400	$2^3 3^2 5^4$	2	$2^3 5^4 = 5000$

Noter le cas $a = 30$, $b = 25$, où aucun niveau $\leqslant 10000$ ne convient: il semble que le conducteur N soit alors de la forme $2^m 5^4$, avec $m \geqslant 5$, d'où $N \geqslant 20000$, ce qui est un peu trop grand pour la méthode employée (basée sur la formule des traces d'Eichler-Selberg). Dans tous les autres cas, on trouve bien une forme parabolique ayant les propriétés cherchées, au moins pour l assez petit.

5.5. *Un exemple utilisant le groupe simple* $\mathbf{PSL}_2(\mathbf{F}_7)$ *d'ordre* 168. L'extension de $\mathbf{Q}$ de degré 7 définie par l'équation

$$(5.5.1) \qquad\qquad X^7 - 7X + 3 = 0$$

a pour groupe de Galois le groupe $\mathbf{PSL}_2(\mathbf{F}_7)$ (W. Trinks–cf. [25]). Nous allons l'utiliser pour construire une représentation de $G_{\mathbf{Q}}$ en caractéristique 7. La méthode est analogue à celle du n° précédent:

Soit G le sous-groupe de $\mathbf{GL}_2(\mathbf{F}_{49})$ défini par:

$$G = \{\pm 1, \pm i\} \cdot \mathbf{SL}_2(\mathbf{F}_7) = \mathbf{SL}_2(\mathbf{F}_7) \cup i \cdot \mathbf{SL}_2(\mathbf{F}_7),$$

où i est un élément d'ordre 4 de $\mathbf{F}_{49}^*$. On a det $G = \{\pm 1\}$, et l'image de G dans $\mathbf{PGL}_2(\mathbf{F}_{49})$ est égale à $\mathbf{PSL}_2(\mathbf{F}_7)$. D'où une suite exacte:

$$(*) \qquad \{1\} \to \{\pm 1\} \to G \to \mathbf{PSL}_2(\mathbf{F}_7) \times \{\pm 1\} \to \{1\}.$$

Soit K le corps de degré 7 défini par (5.5.1), et soit $\alpha^K \colon G_{\mathbf{Q}} \to \mathbf{PSL}_2(\mathbf{F}_7)$ l'homomorphisme correspondant. Soit d'autre part

$$\varepsilon \colon G_{\mathbf{Q}} \to \{\pm 1\}$$

le caractère quadratique associé au corps $\mathbf{Q}(\sqrt{-3})$. Le couple (α^K, ε) définit un homomorphisme

$$\alpha \colon G_{\mathbf{Q}} \to \mathbf{PSL}_2(\mathbf{F}_7) \times \{\pm 1\}.$$

Soit $\mathrm{obs}(\alpha) \in \mathrm{Br}_2(\mathbf{Q})$ l'obstruction à relever α en un homomorphisme

$$\rho \colon G_{\mathbf{Q}} \to G \subset \mathbf{GL}_2(\mathbf{F}_{49}).$$

Un calcul analogue à celui du lemme 6 montre que

$$\mathrm{obs}(\alpha) = w + (-1)(-3),$$

où w est l'invariant de Witt de la forme quadratique $\mathrm{Tr}_{K/\mathbf{Q}}(x^2)$. D'après [46], n° 3.3, on a $w = (-1)(-3)$, d'où $\mathrm{obs}(\alpha) = 0$. Cela démontre l'existence de la représentation

$$\rho \colon G_{\mathbf{Q}} \to \mathbf{GL}_2(\mathbf{F}_{49})$$

cherchée. Par construction, on a det $\rho = \varepsilon$.

Ici encore, on choisit ρ de telle sorte que son conducteur soit le plus petit possible. Le discriminant du polynôme $X^7 - 7X + 3$ est $3^8 7^8$ et celui du corps K est $3^6 7^8$. Il en résulte que le conducteur de ρ peut être choisi égal à 3^n, et un calcul de ramification en 3 montre que $n = 3$. D'autre part, l'étude de la ramification en 7 montre que l'action de l'inertie en 7 est:

$$\text{soit } \begin{pmatrix} \chi & * \\ 0 & \chi^{-1} \end{pmatrix}, \qquad \text{soit } \begin{pmatrix} \chi^4 & * \\ 0 & \chi^{-4} \end{pmatrix},$$

où χ est le caractère cyclotomique.

En faisant le produit tensoriel de ρ par χ, ou χ^4, on obtient une nouvelle représentation ρ' où l'action de l'inertie en 7 est donnée par:

$$\begin{pmatrix} \chi^2 & * \\ 0 & 1 \end{pmatrix},$$

ce qui conduit à un poids k égal à 3, cf. n°s 2.3 et 2.4. On a

$$\det \rho' = \varepsilon \cdot \chi^2.$$

[Noter que ρ' prend ses valeurs dans un groupe un peu plus grand que G: on a

$$\mathrm{Im}\, \rho' = \mathbf{GL}_2(\mathbf{F}_7) \cup i \cdot \mathbf{GL}_2(\mathbf{F}_7).]$$

Les conjectures du §3 affirment que ρ' est de la forme ρ_f, où $f = \Sigma a_n q^n$ est une forme parabolique de type $(3^3, 3, \varepsilon)$, à coefficients dans $\mathbf{F}_{49}$ et fonction propre normalisée des opérateurs de Hecke. Le lien entre les a_l $(l \neq 3, 7)$ et la décomposition de l dans K est le suivant:

si l'on note $\mathrm{ord}(l)$ *l'ordre* de la substitution de Frobenius attachée à l dans $\mathrm{Gal}(K^{\mathrm{gal}}/\mathbf{Q}) \simeq \mathbf{PSL}_2(\mathbf{F}_7)$, on doit avoir:

$$\mathrm{ord}(l) = 1 \text{ ou } 7 \quad \Leftrightarrow \quad a_l^2 = 4l^2\varepsilon(l) \quad \text{dans } \mathbf{F}_7$$

$$\mathrm{ord}(l) = 2 \quad\quad\quad \Leftrightarrow \quad a_l = 0 \quad\quad\quad \text{dans } \mathbf{F}_7$$

$$\mathrm{ord}(l) = 3 \quad\quad\quad \Leftrightarrow \quad a_l^2 = l^2\varepsilon(l) \quad\quad \text{dans } \mathbf{F}_7$$

$$\mathrm{ord}(l) = 4 \quad\quad\quad \Leftrightarrow \quad a_l^2 = 2l^2\varepsilon(l) \quad\quad \text{dans } \mathbf{F}_7,$$

avec $\varepsilon(l) = \left(\dfrac{l}{3}\right)$.

Effectivement, on trouve bien une forme f ayant ces propriétés, au moins pour l assez petit. C'est la réduction (mod 7) d'une forme parabolique primitive F en caractéristique 0:

$$F = q + \sum_{n \geqslant 2} A_n q^n$$

$$= 9 + 3iq^2 - 5q^4 - 3iq^5 + 5q^7 - 3iq^8 + \cdots$$

Cette forme est à coefficients dans $\mathbf{Z}[i]$. Elle se calcule sans difficulté, cf. ci-dessous. Le tableau suivant donne les valeurs des $\mathrm{ord}(l)$ et des A_l pour $l \leqslant 37$:

l	2	5	11	13	17	19	23	29	31	37
$\mathrm{ord}(l)$	7	7	7	4	3	3	3	7	7	4
A_l	$3i$	$-3i$	$-15i$	-10	$18i$	-16	$-12i$	$30i$	-1	20

(Par exemple, pour $l = 17$, on a $a_l^2 \equiv A_l^2 \equiv -2 \pmod 7$, $\varepsilon(l) = -1$, $l^2 \equiv 2 \pmod 7$, d'où $a_l^2 = l^2\varepsilon(l)$ dans $\mathbf{F}_7$, conformément au fait que $\mathrm{ord}(l) = 3$.)

Calcul de F. Soit θ_1 la fonction thêta associée au corps $\mathbf{Q}(\sqrt{-3})$:

$$\theta_1 = \sum_{x,\,y \in \mathbf{Z}} q^{x^2 + xy + y^2} = 1 + 6(q + q^3 + q^4 + 2q^7 + q^9 + \cdots).$$

C'est une série d'Eisenstein de poids 1, de niveau 3 et de caractère ε. Si l'on pose

$$\theta_2 = \theta_1(3z) = 1 + 6(q^3 + q^9 + q^{12} + \cdots)$$

$$\theta_3 = \theta_1(9z) = 1 + 6(q^9 + q^{27} + q^{36} + \cdots),$$

on obtient des formes de niveaux 3^2 et 3^3.

D'autre part, la série

$$g = q \prod_{n \geq 1} (1 - q^{3n})^2 (1 - q^{9n})^2 = q - 2q^4 - q^7 + 5q^{13} + \cdots$$

est l'unique forme parabolique normalisée de poids 2, de niveau 3^3 et de caractère unité (elle correspond à la courbe elliptique $y^2 + y = x^3 - 3$, de conducteur 3^3).

Les produits $g\theta_1$, $g\theta_2$ et $g\theta_3$ sont des formes de poids 3, de niveau 3^3 et de caractère ε. Ils forment une *base* de l'espace des formes paraboliques de type $(3^3, 3, \varepsilon)$. Les fonctions propres normalisées des opérateurs de Hecke s'obtiennent, par exemple, en diagonalisant l'opérateur T_2. On trouve:

$$F = \frac{1}{2} ig\theta_1 - \frac{1}{2}(1 + i)g\theta_2 + \frac{3}{2}g\theta_3 = q + 3iq^2 - 5q^4 + \cdots,$$

$$\overline{F} = -\frac{1}{2} ig\theta_1 - \frac{1}{2}(1 - i)g\theta_2 + \frac{3}{2}g\theta_3 = q - 3iq^2 - 5q^4 + \cdots,$$

$$G = g\theta_2 = q + 4q^4 - 13q^7 + \cdots$$

La série G est de type (CM): elle correspond à un caractère de Hecke du corps $\mathbf{Q}(\sqrt{-3})$.

La série F est la forme parabolique cherchée.

BIBLIOGRAPHIE

1. E. ARTIN, *Zur Theorie der L-Reihen mit allgemeinen Gruppencharakteren*, Hamb. Abh. **8** (1930), 292–306 (= Coll. P., 165–179).

2. A. ASH ET G. STEVENS, *Cohomology of arithmetic groups and congruences between systems of Hecke eigenvalues*, J. Crelle **365** (1986), 192–220.

3. A. O. L. ATKIN ET W. LI, *Twists of Newforms and Pseudo-Eigenvalues of W-Operators*, Invent. Math. **48** (1978), 221–243.

4. B. J. BIRCH ET W. KUYK (édit.), *Modular Forms of One Variable IV*, Lect. Notes in Math. **476**, Springer-Verlag, 1975.

5. S. BLOCH, *Algebraic cycles and values of L-functions*, II, Duke Math. J. **52** (1985), 379–397.

6. A. BRUMER ET K. KRAMER, *The rank of elliptic curves*, Duke Math. J. **44** (1977), 715–742.

7. J. P. BUHLER, *Icosahedral Galois Representations*, Lect. Notes in Math. **654**, Springer-Verlag, 1978.

8. H. CARAYOL, *Sur les représentations l-adiques associées aux formes modulaires de Hilbert*, Ann. Sci. E.N.S. **19** (1986), 409–468.

9. P. DELIGNE, *Les constantes des équations fonctionnelles des fonctions L*, Lect. Notes in Math. **349**, 501–597, Springer-Verlag, 1973.

10. ______, *Valeurs de fonctions L et périodes d'intégrales*, Proc. Symp. Pure Math. **33**, Amer. Math. Soc. (1979), vol. 2, 313–346.

11. P. DELIGNE ET J-P. SERRE, *Formes modulaires de poids 1*, Ann. Sci. E.N.S. **7** (1974), 507–530 (= J-P. Serre, *Oe.* 101).

12. P. DÉNES, *Über die Diophantische Gleichung $x^l + y^l = cz^l$*, Acta Math. **88** (1952), 241–251.

13. G. FALTINGS, *Endlichkeitssätze für abelsche Varietäten über Zahlkörpern*, Invent. Math. **73** (1983), 349–366; *Erratum*, ibid. **75** (1984), 381.

14. G. FALTINGS, G. WÜSTHOLZ ET AL., *Rational Points*, Vieweg, 1984.

15. J-M. FONTAINE, *Il n'y a pas de variété abélienne sur **Z***, Invent. Math **81** (1985), 515–538.

16. G. FREY, *Rationale Punkte auf Fermatkurven und getwisteten Modulkurven*, J. Crelle **331** (1982), 185–191.

17. ______, *Links between stable elliptic curves and certain Diophantine equations*, Ann. Univ. Saraviensis, Ser. Math. **1** (1986), 1–40.

18. A. GROTHENDIECK, *Groupes de Monodromie en Géométrie Algébrique (SGA 7 I)*, Lect. Notes in Math. **288**, Springer-Verlag, 1982.

19. Y. HELLEGOUARCH, *Courbes elliptiques et équation de Fermat*, Thèse, Besançon, 1972.

20. A. HURWITZ, *Über endliche Gruppen linearer Substitutionen, welche in der Theorie der elliptischen Transzendenten auftreten*, Math. Ann. **27** (1886), 183–233 (= Math. W. **XI**).

21. N. JOCHNOWITZ, *A study of the local components of the Hecke algebra mod l*, Trans. Amer. Math. Soc. **270** (1982), 253–267.

22. ______, *Congruences between systems of eigenvalues of modular forms*, Trans. Amer. Math. Soc. **270** (1982), 269–285.

23. N. KATZ, *p-adic properties of modular schemes and modular forms*, Lect. Notes in Math. **350**, 69–190, Springer-Verlag, 1973.

24. ______, *A result on modular forms in characteristic p*, Lect. Notes in Math. **601**, 53–61, Springer-Verlag, 1976.

25. S. LAMACCHIA, *Polynomials with Galois group PSL(2, 7)*, Comm. in Algebra **8** (1980), 983–992.

26. R. P. LANGLANDS, *Base Change for GL(2)*, Princeton Univ. Press, 1980.

27. W. LI, *Newforms and functional equations*, Math. Ann. **212** (1975), 285–315.

28. B. MAZUR, *Modular curves and the Eisenstein ideal*, Publ. Math. I.H.E.S. **47** (1977), 33–186.

29. ______, *Rational isogenies of prime degree*, Invent. Math. **44** (1978), 129–162.

30. J-F. MESTRE, *Courbes de Weil et courbes supersingulières*, Sém. de Théorie des Nombres, Bordeaux (1984–1985), exposé 23.

31. ______, *La méthode des graphes. Exemples et applications*, Taniguchi Symp., Kyoto, 1986, à paraître.

32. I. MIYAWAKI, *Elliptic curves of prime power conductor with **Q**-rational points of finite order*, Osaka Math. J. **10** (1973), 309–323.

33. O. NEUMANN, *Elliptische Kurven mit vorgeschriebenem Reduktionsverhalten* II, Math. Nach. **56** (1973), 269–280.

34. F. OORT ET J. TATE, *Group schemes of prime order*, Ann. Sci. E.N.S. **3** (1970), 1–21.

35. M. RAYNAUD, *Schémas en groupes de type $(p,\ldots,p)$*, Bull. S.M.F. **102** (1974), 241–280.

36. K. RIBET, *Galois action on division points of abelian varieties with real multiplications*, Amer. J. of Math. **98** (1976), 751–804.

37. C. SCHOEN, *On the geometry of a special determinantal hypersurface associated to the Mumford-Horrocks vector bundle*, J. Crelle **364** (1986), 85–111.

38. J-P. SERRE, *Corps Locaux*, 3ème édition, Hermann, Paris, 1980.

39. ______, *Représentations linéaires des groupes finis*, 3ème edition, Hermann, Paris, 1978.

40. ______, *Facteurs locaux des fonctions zêta des variétés algébriques* (*définitions et conjectures*), Sém. Delange-Pisot-Poitou 1969/1970, exposé 19 (= *Oe.* 87).

41. ______, *Propriétés galoisiennes des points d'ordre fini des courbes elliptiques*, Invent. Math. **15**, (1972), 259–331 (= *Oe.* 94).

42. ______, *Congruences et formes modulaires* (*d'après H.P.F. Swinnerton-Dyer*), Sém. Bourbaki 1971/1972, exposé 416 (= *Oe.* 95).

43. ______, *Formes modulaires et fonctions zêta p-adiques*, Lect. Notes in Math. **350**, 191–268, Springer-Verlag, 1973 (= *Oe.* 97).

44. ______, *Valeurs propres des opérateurs de Hecke modulo l*, Journées arith., Bordeaux, 1974, Astérisque **24–25** (1975), 109–117 (= *Oe.* 104).

45. ______, *Modular forms of weight one and Galois representations*, Algebraic Number Fields (A. Fröhlich édit.), Acad. Press, 1977, 193–268 (= *Oe.* 110).

46. ______, *L'invariant de Witt de la forme* $\mathrm{Tr}(x^2)$, Comm. Math. Helv. **59** (1984), 651–676 (= *Oe.* 131).

47. ______, *Résumé des cours de* 1984–1985, Annuaire du Collège de France (1985), 85–90.

48. ______, *Lettre à J-F. Mestre*, Arith. Alg. Geo. (K. Ribet édit.), Contemporary Math. Series, Amer. Math. Soc., 1986.

49. J-P. SERRE ET J. TATE, *Good reduction of abelian varieties*, Ann. of Math. **88** (1968), 492–517 (= J-P. Serre, *Oe.* 79).

50. C. B. SETZER, *Elliptic curves of prime conductor*, J. London Math. Soc. **10** (1975), 367–378.

51. G. SHIMURA, *Introduction to the arithmetic theory of automorphic functions*, Publ. Math. Soc. Japan, vol. **11**, Princeton Univ. Press, 1971.

52. ______, *Class fields over real quadratic fields and Hecke operators*, Ann. of Math. **95** (1972), 130–190.

53. J. TUNNELL, *Artin's conjecture for representations of octahedral type*, Bull. A.M.S. **5** (1981), 173–175.

54. J. VÉLU, *Courbes modulaires et courbes de Fermat*, Sém. de Théorie des Nombres, Bordeaux (1975–1976), exposé 16.

55. A. WEIL, *Über die Bestimmung Dirichletscher Reihen durch Funktionalgleichungen*, Math. Ann. **168** (1967), 149–156 (= *Oe. Sci.* [1967a]).

COLLÈGE DE FRANCE, F-75231 PARIS CEDEX 05, FRANCE.

INFINITE DETERMINANTS, STABLE BUNDLES AND CURVATURE.

S. K. DONALDSON

Introduction. Let (X, ω) be a compact Kahler manifold of complex dimension n and E a holomorphic r-plane bundle over X. For simplicity we suppose $c_1(E) = 0$, although this restriction could easily be removed from our discussion. We say E is $[\omega]$-stable if every subsheaf $S \subset \mathcal{O}(E)$ with torsion-free quotient $S/\mathcal{O}(E)$ satisfies:

$$c_1(S) \cdot [\omega]^{n-1} < 0.$$

This condition clearly implies that E is indecomposable. A Hermitian metric h on E determines a unique compatible connection, with curvature F_h. Let

$$\hat{F}_h = F_h \cdot \omega \in \Omega^0_X(\text{End } E).$$

We say that h is *Hermitian-Einstein* or Hermitian Yang-Mills if $\hat{F}_h = 0$ everywhere. In 1980 Hitchin and Kobayashi suggested that the existence of such a metric should be related to the $[\omega]$-stability of E. They conjectured:

PROPOSITION 1.   *If E is $[\omega]$-stable there exists a Hermitian Yang-Mills metric on E.*

(For the uniqueness and the converse see [3]). When X is a complex curve this is equivalent to a theorem of Narasimhan and Seshadri [8]. In [3] the author proved the proposition for bundles over complex algebraic surfaces. This paper extends the ideas in that reference in the light of various recent developments.

The most important development is the work of Uhlenbeck and Yau [10] who have proved Proposition 1 in full generality. Their proof is probably the most natural and displays clearly the role of stability. However it uses rather sophisticated analysis. In part III of this paper we extend the ideas of [3] to give an alternative proof of Proposition 1 for bundles over projective manifolds $X \subseteq \mathbb{CP}^N$ with a Hodge metric ω. By contrast this proof does not make explicit use of the definition of stability. It proceeds by induction on the dimension of X and uses the algebro-geometric work of Mehta and Ramanathan [7]. They prove:

PROPOSITION 2.   *Suppose $[\omega]$ is the hyperplane class of $X \subseteq \mathbb{CP}^N$. A bundle $E \to X$ is $[\omega]$-stable if and only if there is $d_0 > 0$ such that for generic smooth*

Received December 3, 1986.

intersections:

$$Y = X \cap S_d$$

with a hypersurface of degree $d \geq d_0$, the restriction of E to Y is $[\omega]$-stable.

This sharpens the earlier result of the same authors, used in [3], and enables us to give a proof of Proposition 1, in the algebraic setting, using only elementary analytical ideas. Even for algebraic surfaces this proof if simpler than that in [3]. In a sense, our method transfers the analytical difficulties to the algebrogeometric ones tackled by Mehta and Ramanathan. Perhaps one can think of the existence of Hermitian Yang-Mills metrics and the restriction theorem as roughly comparable results in that two important properties of stable bundles in high dimensions:

(i) the inequality $c_1^2 - 4c_2 \leq 0$,

(ii) the existence of a separated moduli space, can be easily deduced from either.

Our proof achieves less than that of Uhlenbeck and Yau. The main motivation for giving it is to exhibit the intricate geometry associated to these differential equations. We use the global functionals introduced in [3]. These fit into a circle of ideas introduced by Quillen, Ray and Singer. The functionals can be defined by norms in determinant line bundles. Thanks to the work of Bismut and Freed [2] we are able to identify these with norms defined by ζ-function determinants and so with spectral geometry. The fundamental formula we use for our inductive step in §III suggests a conjecture about the behavior of these determinants in exact sequences and it seems likely that this will fit in with the general theory being developed by Gillet and Soulé [5]. These ideas, which are discussed in the expository sections I, II, were one of the strands in Manin's Arbeitstagung manuscript [6]. They motivate Section III although are not actually needed in the proof there.

§1. Determinant lines. Let $\mathcal{A}$ be the space of unitary connections on a Hermitian bundle V over a Riemannian manifold M. There is a tautological connection $\mathbb{A}$ on the bundle $p_2^*(V)$ over $\mathcal{A} \times M$-flat in the $\mathcal{A}$ directions and equal to $p_2^*(A)$ on the slice $\{A\} \times M$. Suppose M has a fixed spin structure. There is then a family of Dirac operators D_A parametrised by $\mathcal{A}$ and acting on the positive V-valued spinors over M. Let L be the determinant bundle of this family; a complex line bundle over $\mathcal{A}$ whose fibres L_A are:

$$(3) \qquad\qquad L_A = \left(\Lambda^{\max} \mathrm{Ker}\, D_A \right)^{-1} \otimes \Lambda^{\max} \mathrm{Ker}\, D_A^*.$$

Bismut and Freed define a metric and unitary connection on L [2] (and on general families of operators). We recall their definitions. For each connection let $\{\lambda\}$ denote the spectrum of the Laplacian $D_A^* D_A$ or equally $D_A D_A^*$. $\mathcal{A}$ is covered by open sets U_l, for l in $\mathbb{R}^+$, on which the sums of eigenspaces $\mathcal{H}_l^+, \mathcal{H}_l^-$, associated to eigenvalues below l, form smooth vector bundles. D_A maps $\mathscr{H}_l^+$ to $\mathscr{H}_l^-$. Over U_l there is a canonical isomorphism:

$$(4) \qquad \phi_l: \left(\Lambda^{\max}\mathscr{H}_l^+\right)^{-1} \otimes \Lambda^{\max}\mathscr{H}_l^- \to L|_{U_l}.$$

Hilbert space geometry defines standard connections and metrics on $\mathscr{H}_l^{\pm}$, hence a connection $\nabla_{(l)}$ and metric on $L|_{U_l}$. Bismut and Freed modify these so that the modifications agree on the overlaps $U_l \cap U_m$. The L^2-metric on $L|_{U_l}$ is modified by multiplying by the "infinite determinant"

$$(5) \qquad \Pi\lambda = \exp(-\zeta'(0))$$

where $\zeta(s)$ is defined around 0 by analytic continuation from

$$(6) \qquad \zeta(s) = \text{Trace}(D_A^* D_A)^{-s}$$

for $\text{Re}(s) \gg 0$. The L^2 connection is modified to $\nabla = \nabla_{(l)} + \omega^{(l)}$ where the complex-valued l-form $\omega^{(l)}$ on U_l is:

$$(7) \qquad \omega_A^{(l)}(a) = \text{Reg.Tr}\big(\rho(a)_l D_A^{-1}\big).$$

Here a is a tangent vector in $\mathscr{A}$, $\rho(a)_l$ is the projection of the corresponding algebraic operator

$$\rho(a) = D_{A+a} - D_A$$

to $\mathscr{H}^- \ominus \mathscr{H}_l^-$ and D_A is inverted on this subspace. The trace is regularised by taking the principle value of:

$$\text{Tr}\big[(D_A D_A^*)^{-s}\rho(a)_l D_A^{-1}\big]$$

meromorphically continued to $s = 0$. The connection ∇ is unitary for the modified metric since the variation of ζ' is:

$$(8) \qquad \frac{\delta(-\zeta'(0))}{\delta A} a = \text{Reg.Tr}\big(\rho(a)_l D_A^{-1} + D_A^{*-1}\rho(a)_l^*\big)$$

$$= 2\,\text{Re}\,\omega_A(a).$$

Bismut and Freed's Theorem ([2]II, Thms. 1.14, 1.21) asserts that the curvature of this connection is the 2-form on $\mathscr{A}$:

$$(9) \qquad -2\pi i\left\{\int_M \text{ch}(\mathbf{A})\,\hat{A}(M)\right\}_{(2)}$$

(The candidate suggested by the index theorem).

Suppose now that $M = X$ is a Kahler manifold. A spin structure is equivalent to a choice of square root $K_X^{1/2}$ for the canonical bundle of X. Put $C^i = \Omega_X^{0,i}(V \otimes K_X^{1/2})$ and for each A in $\mathscr{A}$ form the sequence of operators:

$$(10) \qquad C^0 \xrightarrow{\bar{\partial}_A} C^1 \xrightarrow{\bar{\partial}_A} \cdots \xrightarrow{\bar{\partial}_A} C^n.$$

The Dirac operator D_A is identified with:

$$\bar{\partial}_A \oplus \bar{\partial}_A^* : \bigoplus C^{\text{even}} \to \bigoplus C^{\text{odd}}.$$

Let $\mathscr{A}^{(1,1)} \subset \mathscr{A}$ be the subset consisting of connections with $\bar{\partial}_A^2 = 0$, so (10) forms a complex. We define a complex structure on $\mathscr{A}$ by identifying a connection with its $\bar{\partial}$ operator [1], [3] so:

$$T\mathscr{A} \cong \Omega_X^1(\mathfrak{g}_V) \cong \Omega_X^{0,1}(\text{End } V)$$

where $\mathfrak{g}_V \subset \text{End } V$ is the bundle of skew adjoint endomorphisms. Then $\mathscr{A}^{(1,1)}$ is a complex analytic subvariety of $\mathscr{A}$ (we ignore the possible singularities in $\mathscr{A}^{(1,1)}$ since these do not affect our application).

Over $\mathscr{A}^{(1,1)}$ we form the family of cohomology groups H_A^i of the complexes $(C^*, \bar{\partial}_A)$. We define a line bundle $\mathscr{L} \to \mathscr{A}^{(1,1)}$ in a similar fashion to L; the fibres $\mathscr{L}_A$ are:

$$(11) \qquad \mathscr{L}_A = \bigotimes_i \left(\Lambda^{\max} H_A^i \right)^{(-1)^{i+1}}.$$

Locally in $\mathscr{A}^{(1,1)}$ we can find finite dimensional subcomplexes:

$$\left(U_A^i, \bar{\partial}_A \right) \subset \left(C^i, \bar{\partial}_A \right)$$

with U_A^i depending holomorphically on A and the inclusion defining an isomorphism on cohomology. The canonical isomorphism on the alternating tensor products of determinants exhibits the holomorphic structure of $\mathscr{L}$.

PROPOSITION 12. *There is an isomorphism* $j : \mathscr{L} \to L|_{\mathscr{A}}(1,1)$ *such that* $j^*(\nabla)$ *is compatible with the holomorphic structure of* $\mathscr{L}$.

(This is proved in [2] $I(k)$ for the case $n = 1$).

There are obvious isomorphisms i_A of the fibres L_A, $\mathscr{L}_A$ defined by (3), (11) and the Hodge isomorphism:

$$(12) \qquad \text{Ker } D_A \cong \bigoplus H_A^{\text{even}}, \qquad \text{Ker } D_A^* \cong \bigoplus H_A^{\text{odd}},$$

but this is not the map we need. Indeed the i_A do not define a continuous bundle map. This phenomenon can be seen already in the case of a single operator. If we form the determinant bundle L' over $\mathscr{A}$ using the operator D_A^* in place of D_A we clearly have canonical isomorphisms:

$$k_A : L_A^* \to L_A'.$$

But these do not vary continuously. Over an open set U_l the composite $\phi^* k_A \phi'$ mapping $(\Lambda^{\max}\mathscr{H}_l^+) \otimes (\Lambda^{\max}\mathscr{H}_l^-)^{-1}$ to itself is multiplication by:

$$\left(\prod_{\lambda < l}\lambda\right)^{-1},$$

which is discontinuous when an eigenvalue goes down to zero. To remedy this we multiply k_A by the regularised product $(\Pi\lambda)$, giving a smooth isomorphism. Similarly when we compare L and $\mathscr{L}$ we need to compensate for the different matching of the eigenvectors by $\bar{\partial}_A$ and $\bar{\partial}_A^*$. We split the positive spectrum $\{\lambda\}$ into two parts $\{\lambda'\}$, $\{\lambda''\}$ by restricting the Laplacian to $\operatorname{Im}\bar{\partial}_A$, $\operatorname{Im}\bar{\partial}_A^* \subset \bigoplus C^{\mathrm{even}}$ respectively. Then define $j_A(\alpha) = \pi_A \cdot i_A(\alpha)$, where π_A is the regularised product $(\Pi\lambda')^{-1}$. Then the j_A give a smooth isomorphism j from $\mathscr{L}$ to $L|_{\mathscr{A}(1,1)}$. We define this product with a combination of the ζ-functions:

$$\zeta_i(s) = \operatorname{Tr}\Delta_{A,i}^{-s}$$

where $\Delta_{A,i}$ is the Laplacian on C^i. Choose n_i so that:

$$n_i + n_{i-1} = 0 \qquad i \text{ odd}$$

$$= 1 \qquad i \text{ even}$$

and put:

$$\pi_A = \exp(\Sigma n_i \zeta_i'(0)).$$

The metric on L pulls back by j to the *Ray-Singer* metric on $\mathscr{L}$ [9]. For any holomorphic Hermitian bundle (E, h) this is the norm on the determinant line $\mathscr{L}_E$ of the cohomology defined by:

$$(13) \qquad \operatorname{Norm}_{h, X, \omega} = L^2 \text{ metric on harmonic forms } (\Pi\lambda')^{-1/2}(\Pi\lambda'')^{1/2}.$$

(The correction factor is the *analytic torsion* of the complex.)

To show that $j^*(\nabla)$ is compatible with the holomorphic structure we can work on an open set U where the cohomology has constant dimension. We choose l below the positive spectrum (locally) so that the bundles $\mathscr{H}_l^+$, $\mathscr{H}_l^-$ are the sums of harmonic form bundles $\mathscr{H}_i \subset C^i$. L^2 projection defines connections on the $\mathscr{H}_i$. The induced connection on L plainly agrees with $\nabla_{(l)}$. On the other hand the connection on $\mathscr{H}_i$ agrees with the holomorphic structure on the bundle

$H^i|_U$ under the Hodge isomorphism. So $i^*(\nabla_l)$ is compatible with $\mathscr{L}|_U$. Thus we must show that the correction terms $\omega^{(l)}$, π are compatible; i.e., that $-\bar{\partial} \log \pi$ is the $(0,1)$ part of ω. This follows from (8) when we put in the grading. We let:

$$\rho(a) = \left(\bar{\partial}_{A+a} - \bar{\partial}_A\right) + \left(\bar{\partial}_{A+a}^* - \bar{\partial}_A^*\right)$$
$$= \alpha + \alpha^*,$$

where α is the $(0,1)$ part of a. Then

$$\frac{\delta}{\delta A}\left(-\zeta_i'(0)\right)(a) = \mathrm{Reg.Tr}\left(\alpha\bar{\partial}_A^{-1} + \alpha_{i+1}^*\bar{\partial}_A^{*-1} + \bar{\partial}_A^{*-1}\alpha_i + \bar{\partial}_A^{-1}\alpha\right)$$

with $\alpha_i^*: C^i \to C^{i-1}$. So the $(0,1)$ form $-\bar{\partial} \log \pi$ maps a to:

$$n_i\left(\frac{\delta}{\delta A}\left(-\zeta_i'(0)\right)\right)_{0,1} a$$

$$= n_i \mathrm{Reg.Tr}\left(\alpha_{i+1}^*\bar{\partial}_A^{*-1} + \bar{\partial}_A^{*-1}\alpha_i^*\right)$$

$$= \left(n_i + n_{i-1}\right)\mathrm{Reg.Tr}\left(\alpha_i^*\bar{\partial}_A^{*-1}\right),$$

commuting inside the trace. But this is

$$\sum_{i \text{ even}} \mathrm{Reg.Tr}\left(\alpha_i^*\bar{\partial}_A^{*-1}\right)$$

which is the anti-holomorphic part of $\mathrm{Reg.Tr}(\rho(a)D^{-1})$, acting on $\bigoplus C^{\mathrm{odd}}$, as required.

Suppose now that X is an algebraic manifold and $H \to X$ an ample line bundle. Fix a metric on H with curvature form $-2\pi i\omega \in \Omega_X^{1,1}$, where ω is positive. We define a Kahler form on $\mathscr{A}$ by:

$$\Omega(a, b) = 2^{n-3}\pi^{-2} \int_X \mathrm{Tr}(a \wedge b) \wedge \omega^{n-1}$$

for $a, b \in \Omega_X^1(\mathfrak{y}_v)$. Bismut and Freed's theorem identifies a line bundle P over $\mathscr{A}$ with curvature form $-2\pi i\Omega$. P is the determinant of the Dirac family coupled to the virtual bundle:

$$(H - H^{-1})^{n-1} \otimes V$$

i.e., $P = \bigotimes_i L_i^{(-1)^i\binom{n-1}{i}}$ where L_i is the determinant line of $H^{n-1-i} \otimes V$.

PROPOSITION 14. *The curvature of the Bismut-Freed connection on P is $-2\pi i\Omega$.*

The curvature form $\mathbb{F}$ of $\mathbb{A}$ has as $(T\mathscr{A} \otimes TX)^*$ component the tautological pairing:

$$(a, v) \to a(v) \in \Omega_X^0(\mathfrak{y}v).$$

In the 2-dimensional part of

$$-\int_X ch \wedge ch(H - H^{-1})^{n-1}\hat{A}(X)$$

the only contribution comes from the term

$$-\frac{1}{2!}\int_X \mathrm{Tr}\left(\frac{i\mathbb{F}}{2\Pi}\right)^2 c_1(H - H^{-1})^{n-1}$$

$$= 2^{n-3}\Pi^{-2}\int_X \frac{1}{2}\mathrm{Tr}(\mathbb{F}^2)\omega^{n-1}$$

and

$$\mathrm{Tr}(\mathbb{F}^2)(a, b) = -2\,\mathrm{Tr}(a \wedge b).$$

Note that we can use any Kahler metric on X to define the Bismut-Freed connections, although it is natural to choose the metric with symplectic form ω.

§II. Secondary classes. Let $\mathscr{G}$ be the unitary gauge group of V and $\mathscr{G}^c$ its complexification, the group of complex linear automorphisms of V. The action of $\mathscr{G}$ on $\mathscr{A}$ clearly lifts to the determinant line bundles L_i and so to P, preserving the connection and metric. $\mathscr{G}^c$ also acts on $\mathscr{A}$ (and $\mathscr{A}^{(1,1)}$) via the $\bar{\partial}$-operators:

$$\bar{\partial}_{g(A)} = g \circ \bar{\partial}_A \circ g^{-1}, \qquad g \in \mathscr{G}^c.$$

The set of $\mathscr{G}^c$ orbits in $\mathscr{A}^{(1,1)}$ is identified with the set of equivalence classes of holomorphic bundles E, topologically equivalent to V. If o is an orbit the points of $o/\mathscr{G}$ correspond to metrics h on E ([1], [3]).

There is a holomorphic action of $\mathscr{G}^c$ on the bundle P over $\mathscr{A}^{(1,1)}$. We can see this in two ways. First, through the action of $\mathscr{G}^c$ on the cohomology of the twisted $\bar{\partial}$-complexes. Second, as the complexification of the $\mathscr{G}$ action, defined using the Bismut-Freed connection. By Proposition 2 these agree. The $\mathscr{G}^c$ action does not preserve the metric on P but we can identify explicitly the distortion it makes. Equivalently we see explicitly how the Ray-Singer norm $\mathrm{Norm}_{h, X, \omega}(\alpha)$ on elements α in the determinant line $\mathscr{P}_E$ of a virtual holomorphic bundle $E \otimes (H - H^{-1})^{n-1}$, depends on the metric h on E.

Let $\xi \in \mathrm{Lie}(\mathscr{G}) = \Omega^0_X(\mathfrak{h}_v)$ be an infinitesimal gauge transformation. For each point A in $\mathscr{A}$ let $im_A(\xi)$ be the vertical part of the action of ξ on P-measured by the connection. Then the map

$$\mu_1: \to \mathrm{Lie}(\mathscr{G})^* \cong \Omega^{2n}_X(\mathfrak{h}v)$$

$$(\mu_1(A), \xi) = \frac{1}{2\pi}m_A(\xi)$$

is a $\mathscr{G}$-equivariant *moment-map* for the action of $\mathscr{G}$ on $(\mathscr{A}, \Omega)$. This determines μ_1 uniquely up to the component in the centre of $\mathrm{Lie}(\mathscr{G})$, the constant scalars. We find this central component using the index theorem:

$$m_A(i\theta) = -2^{n-1}\big(c_1(V) \cdot \omega^{n-1}\big)\theta$$

since this is the weight of the action of the scalars on the fibres of P.

But we know a moment map for the action of $\mathscr{G}$, with the correct component in the centre. It is:

$$\mu_2(A) = 2^{n-3}\pi^{-2}F_A \wedge \omega^{n-1}$$

([1] p. 587). So $\mu_1 = \mu_2$. Now complexifying we have:

COROLLARY 15. *The vertical part of the action of $\eta \in \mathrm{Lie}(\mathscr{G}^c)$ on P, relative to the Bismut-Freed connection, is*:

$$i2^{n-2}\pi^{-1} \int_X \mathrm{Tr}\big(\eta \wedge F_A \wedge \omega^{n-1}\big)$$

This vertical part is zero for all η if and only if $F_A \wedge \omega^{n-1} = (1/n)\hat{F}_A \cdot \omega^n = 0$. So, fixing the holomorphic structure, we have:

COROLLARY 16. *Let $E \to X$ be a holomorphic bundle with $c_1(E) = 0$ and α be a basis element for the determinant line $\mathscr{P}_E$. A metric on E is Hermitian Yang-Mills if and only if it is an extremum for the functional*:

$$h \to \mathrm{Norm}_{h,X,\omega}(\alpha).$$

Now we recall from [3] §1.2 that for any two metrics h, k on E there is a Bott-Chern class:

$$R_2(h, k) \in \Omega_X^{1,1}/\mathrm{Im}\,\partial + \mathrm{Im}\,\bar{\partial}$$

such that:

(17)   (i) $R_2(k, k) = 0,\ R_2(h, k) = R_2(h, j) + R_2(j, k)$

(ii) The derivative with respect to h is

$$\delta R_2 = -2i\,\mathrm{Tr}\big(h^{-1}\delta h F_H\big)$$

(iii) $-i\bar{\partial}\partial R_2(h, k) = \mathrm{Tr}(F_h^2) - \mathrm{Tr}(F_k^2).$

We set:

(18) $$M_{X,\omega}(h, k) = \int_X R_2(h, k) \wedge \omega^{n-1}$$

(Of course we do not need X to be spin here).

Combining Property 7(ii) with Corollary 15 we have:

$$(19) \qquad \mathrm{Norm}_{h,X,\omega}(\alpha) = \mathrm{Norm}_{k,X,\omega}(\alpha)\exp\!\left(2^{n-4}\pi^{-1}M_{X,\omega}(h,k)\right).$$

So if the Kahler metric on X and form ω are fixed we can describe the Norm on the determinant line, up to an overall factor, using $M_{X,\omega}$.

Suppose $Y \subset X$ is a smooth hypersurface cut out by a section s of $H^{\otimes 2}$. Then the function $f = \frac{1}{4\Pi}\log|s|^2$ satisfies:

$$i\bar{\partial}\partial f = \omega - \tfrac{1}{2}T_Y$$

where T_Y is the δ-current of Y. Integrating by parts, using Property 7(iii), we get:

$$(20) \qquad M_{X,\omega}(h,k) = \tfrac{1}{2}M_{Y,\omega}(h,k) - \int_X f\,\mathrm{Tr}\!\left(F_h^2 - F_k^2\right)\omega^{n-2}.$$

This has a natural interpretation in terms of cohomology. The exact sequence

$$0 \to H^{-1} \overset{s}{\to} H \to H|_Y \to 0$$

combines with the adjunction isomorphism

$$K_Y \cong K_X \otimes H^2$$

to give:

$$0 \to H^{-1} \otimes K_X^{1/2} \to H \otimes K_X^{1/2} \to K_Y^{1/2} \to 0.$$

Tensoring with $E \otimes (H - H^{-1})^{n-2}$; the long exact cohomology sequence gives an isomorphism between the alternating products of the $\Lambda^{\max}(H^i(E \otimes K_X^{1/2} \otimes (H - H^{-1})^{n-1})$ and of the $\Lambda^{\max}(H^i(E|_Y \otimes K_Y^{1/2} \otimes (H - H^{-1})^{n-2})$. That is, an isomorphism

$$\hat{s} : \mathscr{P}_E \to \mathscr{P}_{E|Y}.$$

We have then

$$(21) \quad \mathrm{Norm}_{h,X,\omega}(\alpha) = c \cdot \mathrm{Norm}_{h,Y,\omega}(\hat{s}(\alpha))\exp\!\left(-2^{n-4}\pi^{-1}\int_X f\,\mathrm{Tr}\,F_h^2\right)$$

for a constant c, independent of h. It is reasonable to guess that c has the form:

$$c = \exp(\mathrm{rank}(E)C(X,Y,\omega,s)).$$

In §III we use the variational characterisation of Hermitian Yang-Mills metrics given by Corollary 6, but need only the relative version defined by the M's. The discussion above motivates our argument. We can pass between different dimensions because the line bundles P, whose geometry generates the differential equations, are essentially the same in every dimension.

§III. Proof of Proposition 1 for algebraic manifolds.   We argue by induction on dimension, assuming the result for complex curves. Observe first that, replacing H and ω by multiples and using Prop. 2, we can suppose that $H^{\otimes 2}$ is very ample and that the given stable bundle $E \to X$ remains stable on generic hypersurfaces Y in $|2H|$. We use the natural parabolic evolution equation for the problem:

$$(22) \qquad \frac{\partial h_t}{\partial t} = -2ih_t\hat{F}_t \qquad \left(\hat{F}_t = \hat{F}_{h_t}\right).$$

This has the following properties; the proofs in any dimension n are no different from those given in [3]. (For part (iv) use Lemma 19 of [3]).

PROPOSITION 23.   (i) *For any initial metric h_0 the equation* (22) *has a smooth solution, defined for $t \in [0, \infty)$.*

(ii) *If $\hat{e}_t = |\hat{F}_t|^2$ (where the norm is defined by h_t) then:*

$$\frac{d}{dt}\int_X \hat{e}_t = \frac{d}{dt}\int_X |F_t|^2 \leqslant 0,$$

and for any $\varepsilon > 0$

$$\sup_X \hat{e}_{t+\varepsilon} \leqslant c(\varepsilon)\int_X \hat{e}_t.$$

(iii) $\dfrac{d}{dt} M_{X,\omega}(h_t, h_0) = -8\displaystyle\int_X \hat{e} \geqslant 0.$

(iv) *If the C^0 norm of h_t (measured relative to h_0) is bounded for t in $[0, \infty)$ then some subsequence of the h_t converges (weakly in all L_2^p) to a smooth Hermitian Yang-Mills metric on E.*

We introduce a convenient "distance function" on the space of metrics. If h, k are two metrics on E with the same determinant, write:

$$h = k \cdot e^s,$$

so s is a trace-free section of End E, self-adjoint relative to h and k. Put $\overline{\lambda}_{h,k}(x) = $ maximum eigenvalue of $s(x)$. $\overline{\lambda}$ is a nonnegative Lipschitz function on X. We clearly have:

$$\text{const.}\|s\|_{C^0} \geqslant \sup_X \overline{\lambda}(h, k) \geqslant \text{const.}\|s\|_{C^0}$$

(norms calculated using k). By Proposition 13(iv) it suffices to show that the $\bar{\lambda}(h_t, k)$ are bounded, for some arbitrary fixed metric k.

To get such a bound on the h_t we argue in the following way. Restricting to a typical hypersurface Y, a Hermitian Yang Mills metric exists and, by calculus, the M functional dominates an integral norm of the metric. Using the maximum principle we can pass from integral norms to uniform norms. We know that $M_{X, \omega}(h_t, h_0)$ is decreasing and the comparison formula (20) gives the required bound on the M functional over X. Details are given in the next three lemmas.

LEMMA 24. *Suppose F is a stable bundle over a Kahler $(n - 1)$ manifold. (Y, ω) and that a Hermitian Yang-Mills metric k on F exists. Let $p = (2n - 2)/(2n - 3)$. There is a constant c_F such that*

$$M_{Y, \omega}(ke^s, k) \geq c_F\big(\|s\|_{L^{p}(Y)} - 1\big)$$

for all trace-free sections s of End F, *self-adjoint relative to k.*

Proof. Let $\nabla_k = \nabla'_k + \nabla''_k$ be the covariant derivative compatible with k. E is indecomposable so the only covariant constant sections of End E are constant scalars. For trace-free sections s we have the Sobolev inequality:

$$\|s\|_{L^{p}(Y)} \leq \text{const.}\|\nabla_k s\|_{L^1}$$

in real dimension $2(n - 1)$.

Consider, as in [3] Prop. 8, the family of metrics:

$$k_t = ke^{ts}, \qquad 0 \leq t \leq 1,$$

and put $M(t) = M_{Y, \omega}(k_t, k)$. So $M(0) = 0$, and $M'(0) = 0$ since k is Hermitian Yang-Mills. Now:

$$\frac{d^2M}{dt^2} = 2i\frac{d}{dt}\int_Y \text{Tr}\big(sF_{H_2}\big) \wedge \omega^{n-2}$$

$$= 2i\int_Y \text{Tr}\big(s\nabla''_{k_t}\nabla'_{k_t}s\big) \wedge \omega^{n-2}$$

$$= -2i\int_Y \text{Tr}\big(\nabla''_{k_t}s\nabla'_{k_t}s\big) \wedge \omega^{n-2}$$

$$= 2^{n-2}(n - 2)!\int_Y |\nabla''_k s|^2_{k_t}\, d\mu,$$

since $\nabla''_{k_t} = \nabla''_k$.

For each point y in Y let $U_y(t)$ be the solution of the differential equation:

$$U_y''(t) = \big(|\nabla''_k s|^2_{k_t}\big)_y$$

with $U_y(0) = U_y'(0) = 0$. So

$$M(1) = 2^{n-2}(n-2)! \int_Y U_y(1)\, d\mu.$$

We can compute U_y explicitly in a basis for the fibre F_y consisting of orthogonal eigenvectors of s. If $s = \mathrm{diag}(\lambda_i)$ and $(\nabla''s) = (\bar{\partial}s)_y = (\pi_{ij})$, a matrix of $(0,1)$ covectors, we have:

$$|\nabla''s|^2_{k_t} = \sum_{i,j} |\pi_{ij}|^2 e^{(\lambda_i - \lambda_j)t}.$$

So:

$$U_y(1) = \sum_{i,j} |\pi_{ij}|^2 \frac{\left(e^{(\lambda_i - \lambda_j)-(\lambda_i-\lambda_j)} - 1\right)}{(\lambda_i - \lambda_j)^2}.$$

(The summand is interpreted as $\tfrac{1}{2}|\pi_{ij}|^2$ if $\lambda_i = \lambda_j$.) Now for any μ in $\mathbb{R}$:

$$\frac{e^\mu - \mu - 1}{\mu^2} \geqslant \frac{1}{2\sqrt{\mu^2 + 1}}.$$

So

$$\frac{e^{(\lambda_i - \lambda_j)-(\lambda_i - \lambda_j)-1}}{(\lambda_i - \lambda_j)^2} \geqslant \frac{2c}{\sqrt{|s|^2_k + 1}},$$

where we can take $c = \sqrt{2}$. Hence

$$U_y(1) \geqslant \left(\frac{c}{\sqrt{|s|^2_k + 1}} |\nabla_k s|^2_k\right),$$

since $|\nabla_k s|^2 = 2|\nabla_k''s|^2$. Now write

$$\nabla_k s = \frac{\nabla_k s}{\left(|s|^2_k + 1\right)^{1/4}} \left(|s|^2_k + 1\right)^{1/4}$$

and apply the Cauchy inequality to get:

$$\left(\int_Y |\nabla_k s|\right)^2 \leqslant \int_Y \frac{|\nabla_k s|^2}{\left(s^2_k + 1\right)^{1/2}} \cdot \int_Y \left(|s|^2_k + 1\right)^{1/2}$$

$$\leqslant \text{const. } M(1)\left(1 + \|s\|^{L_1(X)}\right).$$

Write $I = \int_Y |\nabla_k s|$, and use the Sobolev embedding to get:

$$I^2 \leqslant \frac{\alpha}{2} M(1)(1 + I)$$

for some constant α. So if $I \geqslant 1$, I is bounded by $\alpha M(1)$. Hence:

$$I \leqslant \alpha M(1) + 1$$

and the result stated follows from the Sobolev inequality.

Now let h, k be two metrics on the bundle $E \to X$ and $\bar{\lambda} = \bar{\lambda}_{h,k}$ be the Lipschitz function defined above.

LEMMA 25.

$$\Delta \bar{\lambda} \leqslant 2\big(|\hat{F}_h|_h + |\hat{F}_k|_k\big)$$

in the weak sense:

$$\int_X \bar{\lambda} \, \Delta \phi \leqslant \int \phi \big(|\hat{F}_h| + |\hat{F}_k|\big)$$

for all smooth ϕ.

Proof. Let $h = ke^s$ and work initially in the open set Ω_s where s has distinct eigenvalues. We calculate in a local k-orthonormal frame of eigenvectors: so

$$s = \mathrm{diag}(\lambda_i)$$

and we suppose $\bar{\lambda} = \lambda_1$. The calculation is clearest if we temporarily adopt the point of view of Section II and use the "complex gauge transformation": $e^{s/2} \in \mathrm{End}\, E$. This takes the metric k to h. The pair (h, ∇_h) is equivalent to $(k, \tilde{\nabla})$ where

$$\tilde{\nabla}'' = e^{s/2} \nabla_k'' e^{-s/2}$$

$$\tilde{\nabla}' = e^{-s/2} \nabla_k' e^{s/2},$$

so $|F_h|_h = |F_{\tilde{\nabla}}|_k$. Now write $\nabla_k = \nabla + \alpha - \alpha^*$ in our orthonormal frame, with $\alpha = (\alpha_{ij})$ a matrix of $(0, 1)$ forms. So in this frame:

$$F_k = \partial \alpha - \bar{\partial} \alpha^* - \alpha \wedge \alpha^* - \alpha^* \wedge \alpha,$$

and

$$F_{\tilde{\nabla}} = \partial \beta - \bar{\partial} \beta^* - \beta \wedge \beta^* - \beta^* \wedge \beta$$

with

$$\beta = e^{s/2} \alpha e^{-s/2} - \tfrac{1}{2} \bar{\partial} s.$$

Consider the $(1, 1)$ entries in the curvature matrices:

$$(F_{\tilde{\nabla}})_{1,1} = (F_k)_{1,1} - \partial\bar{\partial}\lambda_1 - \sum_j (e^{\lambda_1 - \lambda_j} - 1)\alpha_{1j} \wedge \bar{\alpha}_{1j}$$

$$- \sum_j (1 - e^{(\lambda_j - \lambda_1)})\alpha_{j1} \wedge \bar{\alpha}_{j1}.$$

On a Kahler manifold

$$\Delta g = 2i(\bar{\partial}\partial g) \cdot \omega$$

on functions g. Also $i(\bar{\alpha} \wedge \alpha) \cdot \omega = |\alpha|^2$ for a $(0, 1)$ form α. Hence:

$$\Delta\lambda_1 = 2i(\hat{F}_{\tilde{\nabla}} - \hat{F}_k)_{1,1} - 2\sum_j (e^{\lambda_1 - \lambda_j} - 1)|\alpha_{ij}|^2$$

$$- 2\sum_j (1 - e^{\lambda_j - \lambda_i})|\alpha_{j1}|^2$$

$$\leqslant 2|\hat{F}_{\tilde{\nabla}}|_k + |\hat{F}_k|_k,$$

since $\lambda_1 \geqslant \lambda_j$.

This proves the inequality for the smooth function $\bar{\lambda} = \lambda_1$ on the open set Ω_s. To extend over the whole of X observe first that it suffices to prove the inequality for a C^2 dense set of metrics h. The set of self-adjoint matrices with coincident eigenvalues is a stratified space of codimension 3. So, by transversality we can suppose that the complement of Ω_s is a codimension 3 set in X. Hence the $(2n - 1)$ dimensional volume of the boundary ∂N_ε of an ε-neighbourhood N_ε of $X \setminus \Omega_s$ is $0(\varepsilon^2)$. Then:

$$\int_{X \setminus N_\varepsilon} \bar{\lambda}\, \Delta\phi = \int_{X \setminus N_\varepsilon} \Delta\bar{\lambda}\, \phi + \int_{\partial N_\varepsilon} \phi\frac{\partial\lambda}{\partial\nu} - \bar{\lambda}\frac{\partial\phi}{\partial\nu}$$

$$\leqslant \int_{X \setminus N_\varepsilon} 2(|\hat{F}_h| + |\hat{F}_k|) + \int_{\partial N_\varepsilon} \phi\frac{\partial\bar{\lambda}}{\partial\nu} - \lambda\frac{\partial\phi}{\partial\nu},$$

and since $\bar{\lambda}$ is Lipschitz the boundary terms go to zero with ε.

LEMMA 26. *Suppose $Y \subset X$ is a smooth hypersurface in $|2H|$. Let k be a metric on $E \to X$ and s a section of $H^{\otimes 2}$ cutting out Y, with $f = (1/4\pi)\log|s|^2$. There are constants c_1, c_2 depending on Y, f and c_3 depending on k and f such that for any* *other metric h on E*:

$$\tfrac{1}{2}M_{Y,\omega}(h,k) \leq M_{X,\omega}(h,k) + c_1\|F_h\|_{L^2(X)}$$

$$+ c_2\|\hat{F}_h\|_{L^\infty(X)} + c_3.$$

Proof. We have formula (20):

$$\tfrac{1}{2}M_{Y,\omega}(h,k) = M_{X,\omega}(h,k) + \int_X f\left(\operatorname{Tr} F_h^2 - \operatorname{Tr} F_k^2\right) \wedge \omega^{n-2}.$$

So we need to show that:

$$\int_X f \operatorname{Tr} F_h^2 \wedge \omega^{n-2} \leq c_1\|F_h\|_{L^2(X)} + c_2\|\hat{F}_h\|_{L^\infty(X)}.$$

If we write a real form θ in $\Omega_X^{1,1}$ as $\theta = (1/n)\hat{\theta}\omega + \theta^{\perp}$ where $\theta^{\perp}$ is orthogonal to ω, the "Hodge-Riemann" identity:

$$\theta \wedge \theta \wedge \omega^{n-2} = \left(\frac{1}{n^2}|\hat{\theta}|^2 - \frac{1}{8n(n-1)}|\theta^{\perp}|^2\right)\omega^n$$

holds. So if $F_h = (1/n)\hat{F}_h\omega + F_h^{\perp}$ we have:

$$\operatorname{Tr}\left(F_h^2\right) \wedge \omega^{n-2} = \left(-\frac{1}{n^2}|\hat{F}_h|^2 + \frac{1}{8n(n-1)}|F_h^{\perp}|^2\right)\omega^n.$$

Now $f = (1/4\pi)\log|s|^2$ is integrable on X and *bounded above*, so

$$\int_X f \operatorname{Tr} F_h^2 \wedge \omega^{n-2} = -\frac{1}{n^2}\int_X f|\hat{F}_h|^2 + \frac{1}{8n(n-1)}\int f|F_h^{\perp}|^2$$

$$\leq \frac{1}{n^2}\|\hat{F}_h\|_{L^\infty}\|f\|_{L^1} + \frac{1}{8n(n-1)}\operatorname{sup} f \cdot \|F^{\perp}\|_{L^2}$$

as required.

Completion of Proof. Since $c_1(E) = 0$ Hodge Theory shows that there exist metrics on E whose induced connection on $\Lambda^{\max}E$ is flat. This class is preserved by the flow (22) so we can suppose that all the metrics we consider have the same determinant.

We claim first that the bound on the h_t required to apply (23) (iv) will follow if we find a nonempty open set $U \subset X$ and a fixed metric k on E such that the

246   S. K. DONALDSON

integrals

$$\int_U \bar\lambda(h_t, k)^p$$

are bounded, where $p = 2n - 2/2n - 3$. For $\Delta\bar\lambda \leqslant (|\hat{F}_t| + |\hat{F}_k|) \leqslant c$ is bounded by Proposition (23) (ii) and Lemma 25. So if $V \Subset U$ we have:

$$\sup_V \bar\lambda \leqslant \text{const.}\left(\left(\int_U \bar\lambda^p\right)^{1/p} + 1\right)$$

by Moser iteration ([4], Chapter 8, especially Thm. 8.25). On the other hand if $W \Subset V$ is another nonempty open set there is a smooth function ψ on X such that $\Delta\psi = c$ on $X \setminus W$. Then the maximum principle, applied to $\bar\lambda - \psi$, gives:

$$\sup_{X \setminus W} \bar\lambda \leqslant \|\psi\|_{L^\infty} + \sup_{V \setminus W} \bar\lambda$$

as required.

The bundle $H^{\otimes 2}$ is assumed to be very ample and, for generic elements Y of $|2H|$, $E|_Y$ is stable. So there is a family Y_z of smooth hypersurfaces in $|2H|$, parametrised by $z \in \bar\Delta \subset \mathbb{C}$, over each of which E is stable; and an open set $U \cong \Delta \times B^{2n-2} \subset X$ foliated by the leaves $U_z = U \cap Y_z \cong B^{2n-2}$, for $z \in \Delta = \text{int } \bar\Delta$. Let k_z be the Hermitian Yang-Mills metric on $E|_{Y_z}$. Choose a metric k on E such that $k = k_z$ over U_z. This is possible since the solutions k_z vary smoothly with z. It suffices to get a bound on the integrals

$$\int_{U_z} \bar\lambda(h_t, k)^p$$

uniform in $z \in \Delta$. But for each z we know that $M_{Y_z, \omega}(h_t, k_z)$ is bounded, using Lemma (26), the bounds on $\hat{F}_t$ in (23, (ii)), and the fact that $M_{x, \omega}(h_t, h_0)$ is decreasing (23, (iii)). This implies that the integrals:

$$\int_{Y_z} \bar\lambda(h_t, k_z)^p$$

are bounded, using Lemma 24; and clearly

$$\int_{U_z} \bar\lambda(h_t, k)^p \leqslant \int_{Y_z} \bar\lambda(h_t, k_z)^p.$$

All the bounds are uniform in $z \in \Delta$, since the family k_z extends continuously to $\bar\Delta$.

References

1. M. F. Atiyah and R. Bott, *The Yang-Mills equations over Riemann surfaces*, Phil. Trans. Roy. Soc. London Series A, **308** (1982), 524–615.
2. J.-M. Bismut and D. S. Freed, *The analysis of elliptic families*, *I*, *II*, to appear in Commun. Math. Phys.
3. S. K. Donaldson, *Anti-self-dual Yang-Mills connections over complex algebraic surfaces and stable vector bundles*, Proc. London. Math. Soc. (3), **50** (1985), 1–26.
4. D. Gilbarg and N. S. Trudinger, *Elliptic partial differential equations of second order*, Springer-Verlag, Berlin, 1983.
5. H. Gillet and C. Soulé, *Classes caractéristiques en théorie d'Arakelov* C. R. Acad. Sc. Paris t. 301, Série I, No. 9, 1985.
6. Yu. I. Manin, "New dimensions in geometry," in *Proceedings of Arbeitstagung*, Bonn 1984. Lecture Notes in Math. IIII, Springer-Verlag Berlin, 1985.
7. V. B. Mehta and A. Ramanathan, *Restriction of stable sheaves and representations of the fundamental group*, Invent. Math. **77**, (1984), 163–172.
8. M. S. Narasimhan and C. S. Seshadri, *Stable and unitary vector bundles on Compact Riemann surfaces*, Ann. of Math. **82** (1965), 540–67.
9. D. B. Ray and I. M. Singer, *Analytic Torsion for complex manifolds*, Ann. of Math. **98** (1973), 154–177.
10. K. K. Uhlenbeck and S. T. Yau, *The existence of Hermitian Yang-Mills connections on stable bundles over Kahler manifolds*, preprint.

The Mathematical Institute, 24-29 St. Giles, Oxford, England.

p-ADIC K-THEORY OF ELLIPTIC CURVES

CHRISTOPHE SOULÉ

A Yu. I. Manin, avec admiration et sympathie.

Introduction. Let E be an elliptic curve over a number field F. A well-known conjecture of Birch and Swinnerton-Dyer asserts that the rank of the Mordell-Weil group $E(F)$ of rational points of E is equal to the order of vanishing of its L function $L(E; s)$ at $s = 1$. Bloch [2] and Beilinson [1] have proposed that the values of $L(E; s)$ at other integral points are related to other invariants of E, namely its higher K-groups $K_m(E)$, $m \in \mathbb{N}$ (notice that $K_0(E) = \mathbb{Z}^2 \oplus E(F)$). For instance, if we assume for simplicity that E has potentially good reduction, for any integer $i \geq 2$, the rank of $K_{2i-2}(E)$ should be equal to the degree $[F : \mathbb{Q}]$ of F over $\mathbb{Q}$. On the other hand, the order of vanishing of $L(E; s)$ at $s = 2 - i$ should be $[F : \mathbb{Q}]$ [14]. Furthermore the leading coefficient of $L(E; s)$ at $s = 2 - i$ is expected to be equal to a regulator defined using $K_{2i-2}(E)$, up to a rational number [1].

The descent theory (or Iwasawa theory) of elliptic curves with complex multiplication gives deep results about the conjecture of Birch and Swinnerton-Dyer [4], and provides it with p-adic analogs. In this paper we investigate how its methods and results can also be used, in some cases, to give p-adic analogs of Bloch and Beilinson's results.

Let us fix an odd prime p and consider the K-theory groups of E with coefficients in $\mathbb{Z}/p^n$, denoted $K_m(E; \mathbb{Z}/p^n)$. We shall be interested in the groups $K_m(E; \mathbb{Z}_p) = \lim_{\leftarrow n} K_m(E; \mathbb{Z}/p^n)$ and $K_m(E; \mathbb{Q}_p/\mathbb{Z}_p) = \lim_{\rightarrow n} K_m(E; \mathbb{Z}/p^n)$.

In Theorems 3.3.2 and 3.4, we use results of Yager [25] and Gross [7] to give examples of curves E and primes p such that the rank of $K_2(E, \mathbb{Q}_p/\mathbb{Z}_p)$ is equal to $[F : \mathbb{Q}]$. The idea of the proof is first to compare $K_2(E, \mathbb{Q}_p/\mathbb{Z}_p)$ with the étale cohomology group $H^2(E, \mathbb{Q}_p/\mathbb{Z}_p(2))$. This can be done using work of Merkurjev-Suslin [11] and Dwyer-Friedlander [5] (Proposition 3.2.). We then compute $H^2(E, \mathbb{Q}_p/\mathbb{Z}_p(2))$ using descent theory and some assumptions of regularity on the prime p.

In paragraph 4, given an elliptic curve with complex multiplication by an imaginary quadratic field K (and defined over K) we define, in some cases, a higher p-adic regulator map

$$r_i : K_{2i-2}\big(E, \mathbb{Z}_p\big) \to \mathbb{Z}_p^2$$

Received December 1, 1986.

(which factors through the p-adic K-theory of the curve E extended to the completion of K at a place dividing p). We introduce a subgroup V_i of $K_{2i-2}(E; \mathbb{Z}_p)$, using elliptic units in the tower of extension of K generated by the coordinates of p-torsion points in E. We then show (Theorem 4.6.) that the index of $r_i(V_i)$ inside $\mathbb{Z}_p^2$ is equal to p^{n_i}, where $n_i \leq +\infty$ is the p-adic valuation of the value at $s = 2 - i$ of a p-adic L-function analog to $L(E; s)$. The proof of this theorem relies upon a result of Yager [25].

The construction of V_i from units in a tower of fields is based on a general method to obtain elements in p-adic K-theory (1.4; see also [15] and [16]). In Theorem 2.1. we apply this method to the cyclotomic tower for any smooth projective variety X_S, over a ring of S-integers in a number field (with p invertible in this ring). Using a theorem of Thomason [24], we prove that, when m is large enough, the rank of $K_m(X_S; \mathbb{Z}_p)$ is greater or equal to the rank predicted by the Beilinson conjectures [1].

When doing this work the following people helped me a lot: S. Bloch, J. Coates, J.-L. Colliot-Thélène, B. Gross, K. Kato, B. Mazur, B. Perrin-Riou, W. Raskind, E. DeShalit, R. Thomason, J. Tilouine, and A. Wiles. Thanks are due to them, and to the Mathematics Scientific Research Institute of Berkeley where the final version was written.

1. Preliminaries

1.1. In this paper p is an *odd* prime integer.

When A is an abelian group and $q \geq 1$ an integer, we denote $_qA$ the kernel of multiplication by q in A, and A/q its cokernel. Let $Div(A)$ be the maximal divisible subgroup of A. Given a p-torsion group A, we write $dim(A)$ for the dimension over $\mathbb{F}_p$ of $_pDiv(A)$.

Given a (profinite) group G and a (topological) G-module M, we denote M^G the invariant subgroup of M, and M_G its quotient of coinvariants. Given a Galois Z_p-module M (or an étale sheaf) and $i \in \mathbb{Z}$, let $M(i)$ be the i-th Tate twist of M.

1.2. Let X be a noetherian separated scheme, $P(X)$ the exact category of locally free coherent $\mathcal{O}_X$-modules, and $BQP(X)$ the simplicial set attached to $P(X)$ by Quillen [13]. The *higher K-theory groups* of X are defined as the homotopy groups of $BQP(X)$:

$$K_m(X) = \pi_{m+1}BQP(X), \qquad m \geq 0.$$

For any integer $q \geq 1$ and $m \geq 1$, one defines the *K-theory of X with coefficients* $\mathbb{Z}/q$ using homotopy with coefficients [12]:

$$K_m(X; \mathbb{Z}/q) = \pi_{m+1}(BQP(X); \mathbb{Z}/q).$$

It fits in short exact sequences

$$(1.2.1) \qquad 0 \to K_m(X)/q \to K_m(X; \mathbb{Z}/q) \to_q K_{m-1}(X) \to 0.$$

When $m = 0$, define $K_0(X; \mathbb{Z}/q) = K_0(X)/q$.

The K-theory of X with coefficients in the ring $\mathbb{Z}_p$ of p-adic integers is, by definition,

$$K_m(X; \mathbb{Z}_p) = \lim_{\leftarrow_n} K_m(X; \mathbb{Z}/p^n),$$

where the transition maps $\varepsilon_n : K_m(X; \mathbb{Z}/p^n) \to K_m(X; \mathbb{Z}/p^{n-1})$ are the reduction of coefficients from $\mathbb{Z}/p^n$ to $\mathbb{Z}/p^{n-1}$.

We also consider

$$K_m(X; \mathbb{Q}_p) = K_m(X; \mathbb{Z}_p) \underset{\mathbb{Z}_p}{\bigotimes} \mathbb{Q}_p$$

and

$$K_m(X; \mathbb{Q}_p/\mathbb{Z}_p) = \lim_{\to_n} K_m(X; \mathbb{Z}/p^n).$$

We shall be mainly interested in evaluating $dim_{\mathbb{Q}_p} K_m(X; \mathbb{Q}_p)$ and $dim \, K_m(X; \mathbb{Q}_p/\mathbb{Z}_p)$. Notice that, when $K_m(X)$ and $K_{m-1}(X)$ are finitely generated abelian groups, these two numbers coincide with the rank of $K_m(X)$ (this follows from (1.2.1.)).

Most of the general properties of higher K-theory are expressed in terms of the space $BQP(X)$: products, localization exact-sequences, transfer maps (also called norm maps) etc. [13]. For this reason they are also valid for K-theory with finite coefficients, or with $\mathbb{Q}_p/\mathbb{Z}_p$ coefficients (since inductive limit is an exact functor).

1.3. Assume the prime p is invertible in $\mathcal{O}_X$. A good way to evaluate $K_m(X; \mathbb{Z}/p^n)$ is to compare it with the *étale K-theory* $K_m^{\acute{e}t}(X; \mathbb{Z}/p^n)$ introduced by Dwyer and Friedlander [5]. There are natural maps

$$\rho_m : K_m(X; \mathbb{Z}/p^n) \to K_m^{\acute{e}t}(X; \mathbb{Z}/p^n), \qquad m \geqslant 0,$$

which are compatible with products, norm maps and reduction of coefficients [5]. The groups $K_m^{\acute{e}t}(X; \mathbb{Z}/p^n)$ are defined for any integer $m \in \mathbb{Z}$ and can be computed as follows. For any integer $i \in \mathbb{Z}$, let $H^s(X, \mathbb{Z}/p^n(i))$ denote the s-th *étale cohomology* group of X with coefficients in the i-th Tate twist of $\mathbb{Z}/p^n$. Assume that the p-cohomological dimension $cd_p(X)$ of X is bounded. Then there is a spectral sequence $E_r^{st}(X)$ converging to $K_{-s-t}^{\acute{e}t}(X; \mathbb{Z}/p^n)$ whose E_2 term is given by

$$E_2^{st}(X) = \begin{cases} H^s(X, \mathbb{Z}/p^n(i)) & \text{if } t = -2i \\ 0 & \text{if } t \text{ is odd.} \end{cases}$$

We shall call it the *DF spectral sequence* [5].

The product structure in étale K-theory:

$$K_m^{\acute{e}t}(X; \mathbb{Z}/p^n) \otimes K_{m'}^{\acute{e}t}(X; \mathbb{Z}/p^n) \to K_{m+m'}^{\acute{e}t}(X; \mathbb{Z}/p^n)$$

is the abutment of a pairing of DF spectral sequences, given at the E_2 level by the cup-product

$$H^s(X, \mathbb{Z}/p^n(i)) \otimes H^s(X, \mathbb{Z}/p^n(i')) \to H^{s+s'}(X, \mathbb{Z}/p^n(i + i')).$$

To compute étale cohomology we shall often use the fact that, given a Galois covering $X' \to X$ with group G, the étale cohomology of X can be computed from the action of G on the étale cohomology of X'. More precisely, there is a Hochschild-Serre spectral sequence (or *HS spectral sequence*), such that

$$E_2^{st} = H^s(G, H^t(X'; \mathbb{Z}/p^n(i))),$$

which converges to $H^{s+t}(X; \mathbb{Z}/p^n(i))$ (for each value of i).

1.3. Assume that X is regular and has finite Krull dimension. Each group $K_m(X)$ can be endowed with a family of commuting endomorphisms ψ^k, $k \geq 1$, called the *Adams operations* [18]. These ψ^k also act upon K-theory with coefficients. Let $K_m(X, \cdot)^{(i)}$ be the subgroup of $K_m(X, \cdot)$ where each ψ^k acts by multiplication by k^i, $i \geq 0$. If $\alpha \in K_m(X, \cdot)^{(i)}$ we say that α has *weight i*.

1.4. In order to get elements in $K_m(X; \mathbb{Z}_p)$ we shall use the following construction. Let $X_n \to X_{n-1}$, $n \geq 1$, be a sequence of finite Galois coverings with $X_0 = X$. Assume all schemes X_n are regular and finite dimensional. Let $N_{n, n-1}$ be the norm map from X_n to X_{n-1}, j_n the corestriction from X_{n-1} to X_n, and ε_n the reduction of coefficients from $\mathbb{Z}/p^n$ to $\mathbb{Z}/p^{n-1}$.

Consider a sequence $\alpha = (\alpha_n)$ of elements $\alpha_n \in K_m(X_n; \mathbb{Z}/p^n)$, $n \geq 1$ (for a fixed value of m). We say that α has *property N* when

$$\varepsilon_n \circ N_{n, n-1}(\alpha_n) = \alpha_{n-1} \quad \text{in } K_m(X_{n-1}; \mathbb{Z}/p^{n-1}) \quad \text{for all } n \geq 2.$$

We say that α has *property R* when

$$j_n(\alpha_{n-1}) = \varepsilon_n(\alpha_n) \quad \text{in } K_m(X_n; \mathbb{Z}/p^{n-1}) \quad \text{for all } n \geq 2.$$

We denote by $\lim_{\leftarrow n} K_m(X_n; \mathbb{Z}/p^n)$ the set of α's having property N.

LEMMA 1.4. *If $\alpha = (\alpha_n)$ has property N and $\beta = (\beta_n)$ has property R, then $\alpha \cup \beta = (\alpha_n \cup \beta_n)$ has property N.*

Proof. Using the projection formula we get

$$\varepsilon_n N_{n, n-1}(\alpha_n \cup \beta_n) = N_{n, n-1}(\varepsilon_n(\alpha_n) \cup \varepsilon_n(\beta_n)) = N_{n, n-1}(\varepsilon_n(\alpha_n) \cup j_n(\beta_{n-1}))$$

$$= N_{n, n-1}(\varepsilon_n(\alpha_n)) \cup \beta_{n-1} = \alpha_{n-1} \cup \beta_{n-1} \qquad \text{q.e.d.}$$

Let N_n be the norm map from X_n to X. Given any sequence $\alpha = (\alpha_n)$ in $\lim_{\leftarrow n} K_m(X_n; \mathbb{Z}/p^n)$, the sequence $N(\alpha) = (N_n(\alpha_n))$ is an element in $K_m(X; \mathbb{Z}_p)$ (since $N_{n-1} \circ N_{n,n-1} = N_n$). Therefore, if $\beta_n \in K_{m'}(X_n; \mathbb{Z}/p^n)$ has property R, we can define an element $N(\alpha \cup \beta)$ in $K_{m+m'}(X; \mathbb{Z}_p)$. This is our basic construction. Notice that for any element g in the Galois group $\lim_{\leftarrow n} Gal(X_n/X)$ the following formula holds (because $N_n \circ g = N_n$):

$$N(g(\alpha) \cup g(\beta)) = N(\alpha \cup \beta).$$

A similar construction can be done in étale K-theory or in étale cohomology.

2. Varieties over number fields. In this paragraph as in paragraph 3 below we denote by X a smooth projective algebraic variety over a number field F. We also choose a finite set S of places in F, containing all archimedean ones, and a smooth projective model X_S of X over the ring of S-integers $\mathcal{O}_S$ in F.

2.1. Let $k \geqslant 1$ and $i \in \mathbb{Z}$ be some integers. Denote by $X(\mathbb{C})$ the set of complex points of the variety $X \otimes_{\mathbb{Q}} \mathbb{C}$ obtained by extension of scalars from $\mathbb{Q}$ to $\mathbb{C}$. The complex conjugation acts upon $X(\mathbb{C})$ and on its singular cohomology group with real coefficients $H^{k-1}(X(\mathbb{C}), \mathbb{R})$. Let $n_{i,k}(X)$ be the dimension of the real subvectorspace of $H^{k-1}(X(\mathbb{C}), \mathbb{R})$ where the complex conjugation acts by multiplication by $(-1)^{i-1}$. For example, when X is connected curve of genus g we have $n_{2,2}(X) = g[F:\mathbb{Q}]$, where $[F:\mathbb{Q}]$ is the absolute degree of F. Notice that $n_{i,k}(X)$ depends only on the parity of i.

Let $\overline{F}$ be an algebraic closure of F and $L_{k-1}(X; s)$ the L-function attached to the action of $Gal(\overline{F}/F)$ upon the étale cohomology of $X \otimes_F \overline{F}$ in degree $k - 1$. According to conjectures of Serre [14], when $i \geqslant k$ the integer $n_{i,k}(X)$ should be the order of vanishing of $L_{k-1}(X; s)$ at $s = k - i$. Let $d = dim(X)$ be the dimension of X over F. We assume that S contains all the places in F which divide p.

THEOREM 2.1. *Assume that $m \geqslant (8/3)(d + 2)(d + 3)(d + 4) + 2$ and that $m + k = 2i$. Then*

$$dim_{\mathbb{Q}_p} K_m(X_S; \mathbb{Q}_p)^{(i)} \geqslant n_{i,k}(X).$$

Proof. Let $F_n = F(\mu_{p^n}) \subset \overline{F}$ be the extension of F obtained by adding to F the roots of unity of order p^n, $\mathcal{O}_n$ the integral closure of $\mathcal{O}_S$ in F_n, and $X_n = X_S \otimes_{\mathcal{O}_S} \mathcal{O}_n$. According to [15] and [5] the projection $K_2(\mathcal{O}_n; \mathbb{Z}/p^n) \to {}_{p^n}K_1(\mathcal{O}_n) = \mu_{p^n}$ admits a natural splitting

$$\sigma : \mu_{p^n} \to K_2(\mathcal{O}_n; \mathbb{Z}/p^n).$$

Any element (ξ_n) in the Tate module $\mathbb{Z}_p(1) = \lim_{\leftarrow n} \mu_{p^n}$ defines a sequence $\beta = (\sigma(\xi_n))$ of elements in $K_2(\mathcal{O}_n, \mathbb{Z}/p^n)$, $n \geqslant 1$, which satisfies the property R

of 1.4. Therefore, by Lemma 1.4. we get a map

$$\varphi : \left(\varprojlim_n K_{m-2}(X_n; \mathbb{Z}/p^n) \right)(1) \to K_m(X_S; \mathbb{Z}_p)$$

sending $\alpha \otimes (\xi_n)$ to $N(\alpha \cup \beta)$.

The weight of β is one, therefore, using the compatibility of weights with products and norms [18] we see that φ gives a map

$$\varphi : \left(\varprojlim_n K_{m-2}(X_n, \mathbb{Z}/p^n)^{(i-1)} \right) \underset{\mathbb{Z}_p}{\otimes} \mathbb{Q}_p(1) \to K_m(X_S, \mathbb{Q}_p)^{(i)}.$$

When $m \geqslant \frac{8}{3}(d+2)(d+3)(d+4)+2$ Thomason has shown ([24], cor. 3.6.) that the morphism

$$\rho_{m-2} : K_{m-2}(X_n; \mathbb{Z}/p^n) \to K^{\acute{e}t}_{m-2}(X_n; \mathbb{Z}/p^n)$$

admits a natural splitting, compatible with norm maps and reduction of coefficients. Therefore the image of

$$\rho_m : K_m(X_S; \mathbb{Q}_p)^{(i)} \to K^{\acute{e}t}_m(X_S, \mathbb{Q}_p)^{(i)}$$

must contain the image of the map

$$\varphi^{\acute{e}t} : \left(\varprojlim_n K^{\acute{e}t}_{m-2}(X_n, \mathbb{Z}/p^n)^{(i-1)} \right) \underset{\mathbb{Z}_p}{\otimes} \mathbb{Q}_p(1) \to K^{\acute{e}t}_m(X_S; \mathbb{Q}_p)^{(i)},$$

where $\varphi^{\acute{e}t}$ is defined by the same method as φ.

According to [17], Theorem 1 and 2, the map $\varphi^{\acute{e}t}$ is the map

$$\varphi^{\acute{e}t} : \left(\varprojlim_n H^k(X_n, \mu_{p^n}^{\otimes i-1}) \right) \underset{\mathbb{Z}_p}{\otimes} \mathbb{Q}_p(1) \to H^k(X_S; \mathbb{Q}_p(i))$$

defined as φ using the cup-product and the fact that $H^0(X_n, \mu_{p^n})$ contains μ_{p^n}. We shall show that the image of $\varphi^{\acute{e}t}$ has $\mathbb{Q}_p$-rank greater or equal to $n_{i,k}(X)$.

Let

$$F_\infty = \bigcup_{n \geqslant 1} F_n, \ \mathcal{O}_\infty = \bigcup_{n \geqslant 1} \mathcal{O}_n, \ G = Gal(F_\infty/F),$$

$$G_n = Gal(F_\infty/F_n) \quad \text{and} \quad X_\infty = X_S \underset{\mathcal{O}_S}{\otimes} \mathcal{O}_\infty.$$

The *HS* spectral sequence for the Galois covering X_∞/X_n gives exact sequences

$$0 \to H^{k-1}(X_\infty, \mathbb{Z}/p^n(i))_{G_n} \to H^k(X_n, \mathbb{Z}/p^n(i)) \to H^k(X_\infty; \mathbb{Z}/p^n(i))^{G_n} \to 0.$$

The following diagram is commutative

$$H^{k-1}(X_\infty, \mathbb{Z}/p^n(i-1))_{G_n} \overset{\cup \xi_n}{\to} H^{k-1}(X_\infty, \mathbb{Z}/p^n(i))_{G_n} \to H^{k-1}(X_\infty, \mathbb{Z}/p^n(i))_G$$

$$\downarrow \qquad\qquad\qquad\qquad \downarrow \qquad\qquad\qquad\qquad \downarrow$$

$$H^{k-1}(X_n, \mathbb{Z}/p^n(i-1)) \overset{\cup \xi_n}{\to} H^{k-1}(X_n; \mathbb{Z}/p^n(i)) \overset{N_n}{\to} H^{k-1}(X_S; \mathbb{Z}/p^n(i))$$

where $\cup \xi_n$ is the cup-product by $\xi_n \in \mu_{p^n} \subset H^0(X_n, \mathbb{Z}/p^n(1))$. When ξ_n is primitive this cup-product is an isomorphism. Furthermore the diagrams above are compatible with each other when n varies. From this it follows that the image of $\varphi^{\acute{e}t}$ contains $H^{k-1}(X_\infty, \mathbb{Q}_p(i))_G$.

Let $\Delta = Gal(F_1/F)$ and $\Gamma = Gal(F_\infty/F_1)$. The cyclotomic character $\kappa: G \to \mathbb{Z}_p^*$ is injective. Therefore Δ has order prime to p, Γ is isomorphic to the additive group $\mathbb{Z}_p$, and G is the direct product of Δ and Γ. Let $\Lambda = \mathbb{Z}_p[[\Gamma]]$ be the topological group algebra of Γ, and M the Λ-submodule of $H^{k-1}(X_\infty, \mathbb{Z}_p)$ where Δ acts via the character κ^{-i}. The group $H^{k-1}(X_\infty, \mathbb{Z}_p(i))_G$ is isomorphic to $M(i)_\Gamma$.

The Λ-module M is finitely generated (its quotient by the maximal ideal in Λ is contained in the finite group $H^k(X_S; \mathbb{Z}/p(i))$). Let r be the rank of M over Λ. By the general structure theory of Λ-modules ([9], p. 132) we know that the $\mathbb{Z}_p$-rank of $M(i)_\Gamma$ is always greater or equal to r, and equal to r for almost all $i \in \mathbb{Z}$. Similarly we get that $H^k(X_\infty, \mathbb{Z}_p(i))^\Gamma$ is finite for almost all i. Therefore the $\mathbb{Z}_p$-rank of $H^k(X_S; \mathbb{Z}_p(i))$ is equal to r for almost all i. By [19] Lemma 1 and Theorem 1, i), this implies that $r \geq n_{i,k}(X)$. Therefore $H^{k-1}(X_\infty, \mathbb{Z}_p(i))_G$ has $\mathbb{Z}_p$-rank greater or equal to $n_{i,k}(X)$.   q.e.d.

3. Elliptic curves with complex multiplication

3.1. We keep the notations and §2 and assume that X has dimension one.

PROPOSITION 3.1. *The map*

$$\rho_2: K_2(X; \mathbb{Z}/p^n) \to K_2^{\acute{e}t}(X; \mathbb{Z}/p^n)$$

is an isomorphism.

Proof. We can assume that X is irreducible. Let $F(X)$ be its function field, X^1 its set of closed points, and $F(x)$ the residue field at $x \in X^1$. The localization exact sequence in K-theory with coefficients [13]

$$\cdots \bigoplus_{x \in X^1} K_3(F(x); \mathbb{Z}/p^n) \to K_3(X; \mathbb{Z}/p^n) \to K_3(F(X); \mathbb{Z}/p^n)$$

$$\to \bigoplus_{x \in X^1} K_2(F(x); \mathbb{Z}/p^n) \to K_2(X; \mathbb{Z}/p^n) \to K_2(F(X); \mathbb{Z}/p^n) \to \cdots$$

is mapped by ρ to an analogous exact sequence in étale K-theory [23]. Therefore we need only to check that, when $K = F(X)$ or $F(x)$, and $m = 2$ (resp. 3) the map

$$\rho_m : K_m(K; \mathbb{Z}/p^n) \to K_m^{\acute{e}t}(K; \mathbb{Z}/p^n)$$

is bijective (resp. surjective). When $m = 2$ this follows from a theorem of Merkurjev and Suslin, and when $m = 3$ it is due to Dwyer and Friedlander ([6], 2.5.).

Remark. The proof also shows that the map

$$\rho_3 : K_3(X; \mathbb{Z}/p^n) \to K_3^{\acute{e}t}(X; \mathbb{Z}/p^n)$$

is surjective.

3.2. Let J be the jacobian of X and $\overline{F}$ an algebraic closure of F. Denote by J_∞ the p-primary torsion subgroup of $J(\overline{F})$. Assume that S contains all the places dividing p, and let $j : Spec(F) \to Spec(\mathcal{O}_S)$ be the natural inclusion.

PROPOSITION 3.2.

(i) $dim\, K_2(X; \mathbb{Q}_p/\mathbb{Z}_p) = dim\, H^2(X; \mathbb{Q}_p/\mathbb{Z}_p(2)) = dim\, H^1(F, J_\infty(1))$

$$= dim\, H^1(Spec\, \mathcal{O}_S, j_* J_\infty(1)) \geqslant g[F : \mathbb{Q}].$$

(ii)   *Let F'/F be a finite Galois extension of F and $X' = X \otimes_F F'$. If*
$dim\, K_2(X'; \mathbb{Q}_p/\mathbb{Z}_p) = g[F' : \mathbb{Q}]$ *then* $dim\, K_2(X; \mathbb{Q}_p/\mathbb{Z}_p) = g[F : \mathbb{Q}]$.

Proof. (i). We can assume that X is irreducible. By Proposition 3.1. we know that

$$K_2(X; \mathbb{Q}_p/\mathbb{Z}_p) \simeq K_2^{\acute{e}t}(X; \mathbb{Q}_p/\mathbb{Z}_p).$$

The DF spectral sequence of X (cf. 1.3.) is such that (after taking the inductive limit on the coefficients) $E_2^{4,-6}(X) = H^4(X; \mathbb{Q}_p/\mathbb{Z}_p(3)) = H^2(F, \mathbb{Q}_p/\mathbb{Z}_p(2)) = 0$ and $E_2^{0,-2}(X) = H^0(X, \mathbb{Q}_p/\mathbb{Z}_p(1))$ is finite. Therefore,

$$dim\, K_2^{\acute{e}t}(X; \mathbb{Q}_p/\mathbb{Z}_p) = dim\, H^2(X; \mathbb{Q}_p/\mathbb{Z}_p(2)).$$

Let $\overline{X} = X \otimes_F \overline{F}$. The HS spectral sequence associated to the covering $\overline{X} \to X$ is such that

$$E_2^{rs} = H^r\!\left(F, H^s(\overline{X}; \mathbb{Q}_p/\mathbb{Z}_p(2))\right).$$

Therefore $E_2^{20} = H^2(F; \mathbb{Q}_p/\mathbb{Z}_p(2))$ and E_2^{02} are finite [21]. So

$$dim\, H^2(X, \mathbb{Q}_p/\mathbb{Z}_p(2)) = dim\, E_2^{11} = dim\, H^1\!\left(F, H^1(\overline{X}, \mathbb{Q}_p/\mathbb{Z}_p(2))\right)$$

$$= dim\, H^1(F, J_\infty(1)).$$

The equality $dim\, H^1(F, J_\infty(1)) = dim\, H^1(Spec\, \mathcal{O}_S, j_* J_\infty(1))$ is shown as in [19], Prop. 1. Let $b_k(\mathcal{O}_S) = dim\, H^k(Spec\, \mathcal{O}_S, j_* J_\infty(1))$. By [19], Prop. 2, we have

$$(3.2.1) \qquad\qquad g[F:Q] + b_0(\mathcal{O}_S) + b_2(\mathcal{O}_S) = b_1(\mathcal{O}_S).$$

This proves (i).

(ii). By enlarging S we can assume that F'/F is unramified outside S. From (3.2.1.) we see that the equality in (i) is equivalent to the vanishing of $b_0(\mathcal{O}_S)$ and $b_2(\mathcal{O}_S)$. Let $\mathcal{O}'_S$ be the integral closure of $\mathcal{O}_S$ in F'. A standard transfer argument shows that if $b_0(\mathcal{O}'_S)$ and $b_2(\mathcal{O}'_S)$ both vanish the same is true for $b_0(\mathcal{O}_S)$ and $b_2(\mathcal{O}_S)$. q.e.d.

Remark. If X_S is a smooth projective model of X over $\mathcal{O}_S$ (for any S) one can also prove ([20], Théorème 3) that

$$dim\, K_2(X; \mathbb{Q}_p/\mathbb{Z}_p) = dim_{\mathbb{Q}_p} K_2(X_S; \mathbb{Q}_p).$$

3.3. We shall now give examples of an elliptic curve E over a number field F such that $dim\, K_2(E; \mathbb{Q}_p/\mathbb{Z}_p) = [F:\mathbb{Q}]$, i.e., equality holds in Proposition 3.2(i) with $X = E$.

3.3.1. As a first example let $K = \mathbb{Q}(\sqrt{-d_K})$ be an imaginary quadratic field, E an elliptic curve over K with complex multiplication by the ring of integers $\mathcal{O}_K$ in K.

Let p be a prime number satisfying the following conditions: $p \neq 2, 3$, p splits in $\mathcal{O}_K$ as the product of two distinct primes: $(p) = \mathfrak{P}\mathfrak{P}^*$, and E has good reduction at $\mathfrak{P}$ and $\mathfrak{P}^*$. Choose a Weierstrass model of E: $y^2 = 4x^3 - g_2 x - g_3$ such that both g_2 and g_3 are in $\mathcal{O}_K$ and the discriminant is prime to p. Choose an imbedding of K into $\mathbb{C}$. The set of complex points $E(\mathbb{C})$ of E is then isomorphic to $\mathbb{C}/\Omega_\infty \mathcal{O}_K$, where $\Omega_\infty \in \mathbb{C}^*/\mathcal{O}_K^*$ denotes the period. Let ψ be the Hecke character attached to E (with values in K^*) and $\bar{\psi}$ its conjugate. Let $L(\bar{\psi}^k, s)$ be the primitive L function attached to powers of ψ($k \in \mathbb{Z}, s \in \mathbb{C}$). By Damerell's Theorem the complex numbers

$$L_\infty(\bar{\psi}^{k+j}, k) = \left(2\pi/\sqrt{d_K}\right)^j \Omega_\infty^{-(k+j)} L(\bar{\psi}^{k+j}, k)$$

lie in $\bar{K}$ when $k \geqslant 1$ and $j \geqslant 0$. Furthermore, if $0 \leqslant j \leqslant p - 1$ and $1 < k \leqslant p$ these are p-integral.

We shall say that p is *regular* (for E) when neither $\mathfrak{P}$ nor $\mathfrak{P}^*$ divides the numbers $L_\infty(\bar{\psi}^{k+j}, k)$ when $1 \leqslant j < p - 1$ and $1 < k \leqslant p$.

Let $\bar{K}$ be the algebraic closure of K in $\mathbb{C}$. For any α in $\mathcal{O}_K$ let E_α be the kernel of the endomorphism α of $E(\bar{K})$, and $K(E_\alpha)$ the field obtained by adding to K the coordinates of the points of E_α. The ideal $\mathfrak{P}$ (resp. $\mathfrak{P}^*$) is generated by

$\pi = \psi(\mathfrak{P})$ (resp. $\pi^* = \psi(\mathfrak{P}^*)$). According to a result of Yager ([25], Theorem 3), when p is regular for E the field $F = K(E_p)$ admits exactly one extension of degree p unramified outside $\mathfrak{P}$ (resp. $\mathfrak{P}^*$), namely $F(E_{\pi^2})$ (resp. $F(E_{\pi^*2})$).

3.3.2.   Theorem 3.3.2.   *When p is regular for E we have*

$$dim\, H^2\!\left(E \bigotimes_K F, \mathbb{Q}_p/\mathbb{Z}_p(2)\right) = [F:\mathbb{Q}].$$

Proof.   Let E_∞ be the p-primary torsion in $E(\overline{K})$ and $\mathcal{O}_F' = \mathcal{O}_F[1/p]$. By the criterion of Néron-Ogg-Shafarevic we know that $E \bigotimes_K F$ has everywhere good reduction. This implies by Proposition 3.2.(i) that

$$dim\, H^1\!\left(F, E_\infty(1)\right) = dim\, H^1\!\left(\mathcal{O}_F', E_\infty(1)\right)$$

(where we write $H^1(\mathcal{O}_F', E_\infty(1))$ instead of $H^1(Spec\,\mathcal{O}_F', j_*E_\infty(1)))$. Let $E_{\pi^\infty} = \bigcup_{n \geqslant 1} E_{\pi^n}$ and $E_{\pi^*\infty} = \bigcup_{n \geqslant 1} E_{\pi^*n}$. The group E_∞ decomposes as

$$E_\infty = E_{\pi^\infty} \oplus E_{\pi^*\infty}.$$

Let $F_n = F(E_{\pi^{n+1}})$, $F_\infty = \bigcup_{n \geqslant 1} F_n$, and let $\mathcal{O}_n'$ (resp. $\mathcal{O}_\infty'$) be the integral closure of $\mathcal{O}_F'$ in F_n (resp. F_∞). When p is regular F_{n+1} is the only extension of degree p of F_n which is unramified outside p [25] and this implies that the p-component of the class group $Pic(\mathcal{O}_n')$ of $\mathcal{O}_n'$ is trivial.

From this we deduce that $H^2(\mathcal{O}_\infty', \mathbb{Q}_p/\mathbb{Z}_p(1)) = 0$. In fact the Kummer exact sequence gives short exact sequences

$$0 \to Pic\left(\mathcal{O}_\infty'\right)/p^m \to H^2\left(\mathcal{O}_\infty', \mathbb{Z}/p^m(1)\right) \to {}_{p^m} H^2\left(\mathcal{O}_\infty', \mathbb{G}_m\right) \to 0 \quad \text{for all } m \geqslant 1.$$

But $Pic(\mathcal{O}_\infty')/p^m = \varinjlim_n Pic(\mathcal{O}_n')/p^m = 0$ and $H^2(\mathcal{O}_\infty', \mathbb{G}_m)$ is contained in $H^2(F_\infty, \mathbb{G}_m) = \varinjlim_n Br(F_n)$, whose p-torsion is trivial (when going from F_n to F_{n+1} the local invariants of the Brauer group are multiplied by p).

The Galois group $\Gamma = Gal(F_\infty/F)$ is isomorphic to $\mathbb{Z}_p$ and the HS spectral sequence for the extension $\mathcal{O}_\infty'/\mathcal{O}_F'$ proves that

$$H^2\left(\mathcal{O}_F', E_{\pi^\infty}(1)\right) = H^1\left(\Gamma, H^1(\mathcal{O}_\infty', E_{\pi^\infty}(1))\right) = \left(\mathcal{O}_\infty'^* \bigotimes_{\mathbb{Z}} E_{\pi^\infty}\right)_\Gamma.$$

But $\mathcal{O}_\infty'^*$ is a discrete Γ-module and Γ acts upon $E_{\pi^\infty} \simeq \mathbb{Q}_p/\mathbb{Z}_p$ by a character of infinite order. Using an argument of Tate ([21], Lemma) we conclude that $(\mathcal{O}_\infty'^* \bigotimes_{\mathbb{Z}} E_{\pi^\infty})_\Gamma$ vanishes. By the same argument we obtain that $H^2(\mathcal{O}_F', E_{\pi^*\infty}(1)) = 0$. Since $H^0(\mathcal{O}_F', E_\infty(1))$ is always a finite group, the proof of Proposition 3.2. shows that $dim\, H^1(\mathcal{O}_F', E_\infty(1)) = [F:\mathbb{Q}]$.

3.3.3. When p is regular the group $E(K)$ is finite [4]. For example, let $E_{\mathbb{Q}}$ be the elliptic curve defined over $\mathbb{Q}$ by the equation $y^2 = 4x^3 - 4x$. It has complex multiplication by the integers in $K = \mathbb{Q}(\sqrt{-1}\,)$. The prime $p = 5$ is regular for the corresponding curve over K (as was shown to me by Bernardi). Using Proposition 3.2. we get

$$dim\, K_2(E_{\mathbb{Q}}; \mathbb{Q}_5/\mathbb{Z}_5) = 1.$$

3.4. Let us now give another example, where the prime p, instead of being split, will be ramified. Choose $p > 3$ such that $K = \mathbb{Q}(\sqrt{-p}\,)$ has class number one (there are six possibilities: $p = 7, 11, 19, 43, 67$ or 163). It is known that there is a unique elliptic curve $A(p)$ defined over $\mathbb{Q}$, with complex multiplication by the ring of integers in K and minimal discriminant $-p^3$[7]. When $d \neq 0, 1$ is a square free integer, let $A(p)^d$ be the unique curve over $\mathbb{Q}$ which is isomorphic to $A(p)$ over $\mathbb{Q}(\sqrt{d}\,)$ but not over $\mathbb{Q}$(loc. cit.). Define $A(p)^1 = A(p)$.

Let $\pi = \sqrt{-p}$ be the unique prime factor of p in K, $E = A(p)^d$ and E_π the kernel of the multiplication by π in $E(\overline{\mathbb{Q}})$. Let $F = K(E_\pi)(\mu_p)$. The field F is abelian over $\mathbb{Q}$. Let ε be the character of $Gal(F/K)$ coming from its action on E_π. This character ε admits two extensions ε_+ and ε_- to $Gal(F/\mathbb{Q})$. We also denote by ε_+ and ε_- the corresponding primitive Dirichlet characters, and we assume that $\varepsilon_+(-1) = 1$ and $\varepsilon_-(-1) = -1$. The Weil pairing implies that $\varepsilon_+\varepsilon_- = \omega$ is the Teichmüller character.

For any nontrivial primitive Dirichlet character $\chi: \mathbb{Z}/m \to \mathbb{C}^*$ let

$$B_{1,\chi} = (1/m)\left(\sum_{a=0}^{m-1}\chi(a)a\right).$$

THEOREM 3.4.
(i) *Let* $E = A(p)^d$. *Assume that, when* $\chi = \varepsilon_-$ *or* $\chi = \varepsilon_-\omega^{-2}$, *either* p *does not divide* $B_{1,\chi}$ *or* $\chi = \omega$.
Then $dim\, H^2(E, \mathbb{Q}_p/\mathbb{Z}_p(2)) = 1$.
(ii) *Let* $p = 7, 11, 19, 43, 67$ *or* 163. *Then*

$$dim\, H^2\big(A(p), \mathbb{Q}_p/\mathbb{Z}_p(2)\big) = 1.$$

Proof.   (i) By Proposition 3.2.(i) we have

$$dim\, H^2\big(E, \mathbb{Q}_p/\mathbb{Z}_p(2)\big) = dim\, H^1(\mathbb{Q}, E_\infty(1)).$$

The Leray spectral sequence for the direct image functor of the inclusion $j: Spec(\mathbb{Q}) \to Spec(\mathbb{Z}[1/p])$ gives a short exact sequence

$$0 \to H^1(\mathbb{Z}[1/p], E_\infty(1)) \to H^1(\mathbb{Q}, E_\infty(1)) \to \bigoplus_{l \neq p} H^0\big(\mathbb{F}_l, H^1(T_l, E_\infty(1))\big),$$

where T_l is the maximal unramified extension of $\mathbb{Q}_l$. Since E has potentially good reduction everywhere, none of the groups $H^0(\mathbb{F}_l, H^1(T_l, E_\infty(1)))$, $l \neq p$, contains a divisible subgroup. Therefore

$$dim\, H^2\big(E, \mathbb{Q}_p/\mathbb{Z}_p(2)\big) = dim\, H^1(\mathbb{Z}[1/p], E_\infty(1)).$$

According to [7] the curve $E = A(p)^d$ is isogenous to $E^* = A(p)^{d^*}$, with $d^* = -pd$ (resp. $d = -pd^*$) when d is prime to p (resp. p divides d). The action of $-p$ on E is the composite of two isogenies

$$E \xrightarrow{\pi} E^* \xrightarrow{\pi} E,$$

so we get a short exact sequence of Galois modules

$$0 \to E_\pi(1) \to E_p(1) \to E_\pi^*(1) \to 0.$$

Since E and E^* become isomorphic over K we have $F = K(E_\pi)(\mu_p)$ and $\varepsilon_+, \varepsilon_-$ are given by the action of $\Delta = Gal(F/\mathbb{Q})$ upon E_π and E_π^*.

The group Δ maps injectively into $((\mathbb{Z}/p)^*)^2$ by $\varepsilon_+ \times \varepsilon_-$, therefore its order is prime to p. If $\mathcal{O}_F'$ denotes the integral closure of $\mathbb{Z}[1/p]$ inside F and M is the maximal abelian p extension of F unramified outside p we get

$$H^1\big(\mathbb{Z}[1/p], E_p(1)\big) = H^1\big(\mathcal{O}_F', E_p(1)\big)^\Delta = Hom\big(Gal(M/F), E_p(1)\big)^\Delta.$$

Therefore

$$dim_{\mathbb{F}_p} H^1\big(\mathbb{Z}[1/p], E_p(1)\big) = dim_{\mathbb{F}_p} Gal(M/F)^{\varepsilon_+\omega} + dim_{\mathbb{F}_p} Gal(M/F)^{\varepsilon_-\omega}.$$

Here, given a character χ of Δ and A a Δ-module, we denote A^χ the subgroup of A where Δ acts via the character χ. The Stickelberger theorem and Kummer theory prove the following ([9] or [7] p. 76):

When $\chi = 1$, $Gal(M/F)^\chi = \mathbb{Z}/p$

When $\chi(-1) = 1$, $\chi \neq 1$ and p does not divide $B_{1,\chi^{\omega-1}}$, then $Gal(M/F)^\chi = \mathbb{Z}/p$

When $\chi(-1) = -1$, $\chi \neq \omega$ and p does not divide $B_{1,\chi^{-1}}$, then $Gal(M/F)^\chi = 0$.

Using our hypotheses we see that when $\varepsilon_-\omega \neq 1$ we have $Gal(M/F)^{\varepsilon_-\omega} = \mathbb{Z}/p$ and $Gal(M/F)^{\varepsilon_+\omega} = 0$. When $\varepsilon_-\omega = 1$ then

$$Gal(M/F)^{\varepsilon_+\omega} = Gal(M/F)^{\varepsilon_-\omega} = E_p(1)^\Delta = \mathbb{Z}/p.$$

In all cases we get

$$dim_{\mathbb{F}_p} H^1\big(\mathbb{Z}[1/p], E_p(1)\big) = dim_{\mathbb{F}_p} H^0\big(\mathbb{Z}[1/p], E_p(1)\big) = 1,$$

from which $dim\, H^1(\mathbb{Z}[1/p], E_\infty(1)) = 1$ follows.

(ii) When $d = 1$ we have $\{\varepsilon_+, \varepsilon_-\} = \{\omega^{(p+1)/4}, \omega^{(3p-1)/4}\}$.
If k is even and $2 \leqslant k \leqslant p - 3$ then [9]

$$B_{1, \omega^{k-1}} \equiv B_k/k \qquad (\text{modulo } p),$$

with

$$t/(e^t - 1) = \sum_k B_k t^k/k!.$$

If $p = 7$ we find $\varepsilon_- = \omega^5$ and $\varepsilon_+ = \omega^2$. So

$$B_{1, \varepsilon_- \omega^{-2}} \equiv B_4/4 \not\equiv 0 \qquad (\text{modulo } 7).$$

If $p > 7$ then $\varepsilon_- = \omega^{(p+1)/4}$, $\varepsilon_+ = \omega^{(3p-1)/4}$, so

$$B_{1, \varepsilon_-} \equiv (4/(p + 5)) B_{(p+5)/4} \qquad (\text{modulo } p)$$

and

$$B_{1, \varepsilon_- \omega^{-2}} \equiv (4/(p - 3)) B_{(p-3)/4} \qquad (\text{modulo } p).$$

When $p = 11, 19, 43, 67$ or 163 none of these Bernoulli numbers is divisible
by p.   q.e.d.

4. Higher p-adic regulators

4.1. Let X be a smooth, projective, geometrically irreducible curve over a
field F. Assume that X has a rational point $P \in X(F)$ (chosen once for all). Let
J be the jacobian of X and p an even prime integer different from the
characteristic of F. We shall construct elements in $K_m(X; \mathbb{Z}_p)$ for all even integer
$m \geqslant 1$. The method we use is similar to 2.1. except that the tower of fields we
consider is the one generated by the p-torsion points of J (rather than the
cyclotomic tower). We first need the following result.

LEMMA 4.1.   *The composite morphism*

$$K_1(X; \mathbb{Z}/p^n)^{(1)} \to K_1(X; \mathbb{Z}/p^n) \to K_1^{\acute{e}t}(X; \mathbb{Z}/p^n) \to H^1(X; \mathbb{Z}/p^n(1))$$

is an isomorphism.

Proof.   Recall (1.3.) that $K_m(X; \mathbb{Z}/p^n)^{(i)}$ is the subgroup of $K_m(X; \mathbb{Z}/p^n)$ of weight i for the action of the Adams operations ψ^k, $k \geq 1$. The map $K_1^{\acute{e}t}(X; \mathbb{Z}/p^n) \to H^1(X; \mathbb{Z}/p^n(1))$ comes from the DF spectral sequence (1.2.) Consider the localization exact sequence

$$\bigoplus_{x \in X^1} K_1(F(x); \mathbb{Z}/p^n) \overset{u}{\to} K_1(X; \mathbb{Z}/p^n) \overset{v}{\to} K_1(F(X); \mathbb{Z}/p^n)$$

$$\overset{\delta}{\to} \bigoplus_{x \in X^1} K_0(F(x); \mathbb{Z}/p^n).$$

(see 3.1.) It gives a short exact sequence

$$0 \to Im(u) \to K_1(X; \mathbb{Z}/p^n) \to Im(v) \to 0.$$

According to [18] all elements of $Im(u)$ (resp. $Im(v)$) have weight 2 (resp. 1). When α has weight 1 and 2 we get $\psi^k(\alpha) = k\alpha = k^2\alpha$ for all $k \geq 1$, therefore $2\alpha = 0$ and $\alpha = 0$ since p is odd. Therefore the map $K_1(X, \mathbb{Z}/p^n)^{(1)} \to Im(v)$ is injective.

Let $\alpha \in Im(v)$ be equal to $v(\beta)$. Then $2\alpha = (4 - \psi^2)(\alpha) = v(\beta')$ with $\beta' = (4 - \psi^2)(\beta)$. But, for any $k \geq 1$, we have $(\psi^k - k)(\beta') = (4 - \psi^2)\beta''$ with $\beta'' = (\psi^k - k)(\beta)$. Hence $v(\beta'') = 0$, $\beta'' \in Im(u)$ and $(4 - \psi^2)(\beta'') = 0$. Therefore β' has weight one and 2α lies in the image of $K_1(X; \mathbb{Z}/(p^n)^{(1)}$. Since p is odd we see that the map $K_1(X; \mathbb{Z}(p^n)^{(1)} \to Im(v)$ is an isomorphism.

We shall now prove that $Im(v)$ is isomorphic to $H^1(X, \mathbb{Z}/p^n(1))$. Consider the localization exact sequence in étale cohomology [3]

$$0 \to H^1(X; \mathbb{Z}/p^n(1)) \overset{v^{\acute{e}t}}{\to} H^1(F(X); \mathbb{Z}/p^n(1)) \overset{\delta^{\acute{e}t}}{\to} \bigoplus_{x \in X^1} H^0(F(x), \mathbb{Z}/p^n).$$

By naturality of ρ_1 (1.2.) the diagram

$$\begin{array}{ccc}
K_1(X; \mathbb{Z}/p^n) & \overset{v}{\to} & K_1(F(X); \mathbb{Z}/p^n) \\
\downarrow & & \downarrow \\
H^1(X, \mathbb{Z}/p^n(1)) & \overset{v^{\acute{e}t}}{\to} & H^1(F(X); \mathbb{Z}/p^n(1))
\end{array}$$

commutes. Both groups on the right hand side are isomorphic to $F(X)^*/p^n$ (the map is the identity), and both δ and $\delta^{\acute{e}t}$ are given by the valuations of $F(X)$. Therefore the map $K_1(X; \mathbb{Z}/p^n) \to H^1(X; \mathbb{Z}/p^n(1))$ induces an isomorphism between $Im(v)$ and $Im(v^{\acute{e}t}) = H^1(X; \mathbb{Z}/p^n(1))$.   q.e.d.

4.2. Let $\overline{F}$ be an algebraic closure of F, $J_n = {}_{p^n}J(\overline{F})$ the group of points of order p^n in $J(\overline{F})$, and $F_n = F(J_n)$ the extension of F generated by the coordinates of these points ($n \geqslant 1$). Using the Weil pairing we know that F_n contains μ_{p^n}.

If $X_n = X \otimes_F F_n$ the Kummer exact sequence gives a short exact sequence

$$(4.2.1) \qquad 0 \to H^1(F_n, \mathbb{Z}/p^n(1)) \to H^1(X_n, \mathbb{Z}/p^n(1)) \to J_n \to 0.$$

The rational point P gives a splitting of (4.2.1). Combining this with Lemma 4.1. we get a morphism

$$\sigma' : J_n \to K_1(X_n; \mathbb{Z}/p^n).$$

Let (v_n) be an element in the Tate module $T_pJ = \lim_{\leftarrow n} J_n$. One checks that the sequence of elements $\alpha_n = \sigma'(v_n) \in K_1(X_n; \mathbb{Z}/p^n)$ has property R of 1.4..

Let us choose $(\xi_n) \in \lim_{\leftarrow n} \mu_{p^n} = \mathbb{Z}_p(1)$ and $(u_n) \in \lim_{\leftarrow n} F_n^*/p^n$ a sequence satisfying N. Let $\beta_n = \sigma(\xi_n) \in K_2(F_n; \mathbb{Z}/p^n)$ as in 2.1. The element

$$u_n \cup \beta_n^{i-2} \cup \alpha_n \in K_{2i-2}(X_n; \mathbb{Z}/p^n)$$

has property N (by Lemma 1.4). so we get an element $x = (N_n(u_n \cup \beta_n^{i-2} \cup \alpha_n))$ in $K_{2i-2}(X, \mathbb{Z}_p)$.

Let $F_\infty = \bigcup_{n \geqslant 1} F_n$ and $G_\infty = Gal(F_\infty/F)$. The map φ sending $(u_n) \otimes (\xi_n)^{\otimes i-2} \otimes v_n$ to x induces a morphism

$$\varphi : \left(\left(\lim_{\leftarrow n} F_n^*/p^n \right) \otimes T_pJ(i-2) \right)_{G_\infty} \to K_{2i-2}(X, \mathbb{Z}_p).$$

4.3. Assume that $cd_pF = 2$. The DF spectral sequence gives a map

$$H^4(X, \mathbb{Z}/p^n(i+1)) = H^2(F, \mathbb{Z}/p^n(i)) \to K_{2i-2}^{\acute{e}t}(X; \mathbb{Z}/p^n)$$

for any $i \in \mathbb{Z}$. This map is split by the rational point P. Therefore we get a projection

$$K_{2i-2}^{\acute{e}t}(X; \mathbb{Z}/p^n) \to H^2(X; \mathbb{Z}/p^n(i)).$$

The HS spectral sequence for the Galois covering $X \otimes_F \overline{F} \to X$ is also split by the choice of P, so we get a projection map

$$H^2(X; \mathbb{Z}/p^n(i)) \to H^1(F; J_n(i-1)).$$

Let us denote by $H^1(F, T_pJ(i-1))$ the projective limit $\lim_{\leftarrow n} H^1(F, J_n(i-1))$.

To study φ we shall compose it with the map

$$\psi : K_{2i-2}(X; \mathbb{Z}_p) \to H^1\big(F; T_p J(i-1)\big)$$

obtained by composing ρ_{2i-2} with the maps above. Let $J_\infty = \bigcup_{n \geqslant 1} J_n$ be the p-primary torsion subgroup of $J(\overline{F})$.

LEMMA 4.3. *Assume that F is a finite extension of $\mathbb{Q}_p$. Then the kernel (resp. cokernel) of $\psi \circ \varphi$ is contained in the Pontryagin dual of $H^2(G_\infty, J_\infty(1-i))$ (resp. $H^1(G_\infty, J_\infty(1-i))$.*

Proof. Since both the *DF* and the *HS* spectral sequences are compatible with products, the composite of the map

$$\big(F_n^*/p^n\big) \otimes \mu_{p^n}^{\otimes(i-2)} \otimes J_n \to K_{2i-2}\big(X_n, \mathbb{Z}/p^n\big)$$

defined in 4.2. with the map

$$K_{2i-2}\big(X_n; \mathbb{Z}/p^n\big) \to H^1\big(F_n, J_n(i-1)\big)$$

is the cup-product

$$H^1\big(F_n; \mathbb{Z}/p^n(1)\big) \otimes H^0\big(F_n, \mathbb{Z}/p^n(i-2)\big) \otimes H^0\big(F_n, J_n\big) \to H^1\big(F_n, J_n(i-1)\big).$$

This is an isomorphism because $Gal(\overline{F}/F_n)$ acts trivially upon μ_{p^n} and J_n. The local duality theorem [22] and the Weil pairing show that the norm map

$$H^1\big(F_n, J_n(i-1)\big) \to H^1\big(F, J_n(i-1)\big)$$

is the Pontryagin dual of the corestriction map

$$H^1\big(F; J_n(1-i)\big) \to H^1\big(F_n, J_n(1-i)\big).$$

The map $\psi \circ \varphi$ is therefore the Pontryagin dual of

$$H^1\big(F, J_\infty(1-i)\big) \to H^1\big(F_\infty, J_\infty(1-i)\big)^{G_\infty}.$$

The *HS* spectral sequence for the covering F_∞/F gives an exact sequence

$$0 \to H^1\big(G_\infty, J_\infty(1-i)\big) \to H^1\big(F, J_\infty(1-i)\big) \to H^1\big(F_\infty, J_\infty(1-i)\big)^{G_\infty}$$

$$\to H^2\big(G_\infty, J_\infty(1-i)\big),$$

from which the Lemma follows.

4.4. From now on let $E, K, p, \mathfrak{P}, \mathfrak{P}^{*}, \pi, \pi^{*}$ be as in 3.3. Define $F_n = K(E_{p^n})$, $F_\infty = \bigcup_{n \geq 1} F_n$, and $G_\infty = Gal(F_\infty/K)$. When v is a fixed place of F_∞ dividing $\mathfrak{P}$, let $G_{\infty, v}$ be its decomposition group in G_∞. Let $\tilde{E}$ be the reduction modulo $\mathfrak{P}$ of a Néron model of E.

LEMMA 4.4. *For all integers $r \geq 0$ and $i > 0$ the group $H^r(G_{\infty, v}; E_{p^\infty}(i))$ is finite. It is zero unless the following conditions are all satisfied*: $k \leq 1$, i *or* $i + 1$ *is divisible by* $p - 1$, *and* $\tilde{E}$ *has a point of order* p.

Proof. Let $\mathcal{O}_K$ be the ring of integers in K, and $\mathcal{O}_{\mathfrak{P}}$ (resp. $\mathcal{O}_{\mathfrak{P}^*}$) its completion at $\mathfrak{P}$ (resp. $\mathfrak{P}^*$). Call $\kappa_1 : G_\infty \to \mathcal{O}_{\mathfrak{P}}^* = \mathbb{Z}_p^*$ the character of G_∞ giving its action on

$$E_{\pi^\infty} = K_{\mathfrak{P}}/\mathcal{O}_{\mathfrak{P}} = \mathbb{Q}_p/\mathbb{Z}_p,$$

and $\kappa_2 : G_\infty \to \mathcal{O}_{\mathfrak{P}^*}^*$ the character attached to its action on $E_{\pi^*\infty}$. The direct product

$$\kappa_1 \times \kappa_2 : G_\infty \to \left(\mathbb{Z}_p^*\right)^2$$

is an isomorphism. The image of $G_{\infty, v}$ by $\kappa_1 \times \kappa_2$ is $\mathcal{O}_{\mathfrak{P}}^* \times \langle \pi \rangle$, where $\langle \pi \rangle$ is the closed subgroup of $\mathcal{O}_{\mathfrak{P}^*}^*$ generated by π. When $i > 0$, the action of G_∞ on $E_{\pi^\infty}(i)$ and $E_{\pi^*\infty}(i)$ is given by the nontrivial character $\kappa_1^{i+1}\kappa_2^i$ (resp. $\kappa_1^i\kappa_2^{i+1}$). Therefore

$$H^2\left(G_{\infty, v}, E_{p^\infty}(i)\right) = E_{p^\infty}(i)_{G_{\infty, v}} = 0$$

and the groups

$$H^1\left(G_{\infty, v}; E_{p^\infty}(i)\right) = H^1\left(\mathcal{O}_{\mathfrak{P}}^*, H^0\left(\langle \pi \rangle, E_{p^\infty}(i)\right)\right) \oplus H^0\left(\mathcal{O}_{\mathfrak{P}}^*, H^1\left(\langle \pi \rangle, E_{p^\infty}(i)\right)\right)$$

and $H^0(G_{\infty, v}; E_{p^\infty}(i))$ are finite.

Let $\Delta_{\mathfrak{P}}$ be the image of $G_{\infty, v}$ into $Gal(K(E_p)/K) = ((\mathbb{Z}/p)^*)^2$. Since $\Delta_{\mathfrak{P}}$ is a direct factor of $G_{\infty, v}$ with order prime to p, all groups $H^r(G_{\infty, v}; E_{p^\infty}(i))$ vanish unless $H^0(\Delta_{\mathfrak{P}}; E_{p^\infty}(i)) \neq 0$.

If $H^0(\Delta_{\mathfrak{P}}, E_\pi(i)) \neq 0$ and $\alpha \in (\mathbb{Z}/p)^*$ we have $\alpha^{i+1} = 1$ and $\pi^i \equiv 1$ (modulo π^*). In other words, $p - 1$ divides $i + 1$ and π^* divides $\pi - 1$. If $H^0(\Delta_{\mathfrak{P}}, E_{\pi^*}(i)) \neq 0$ we get similarly that $p - 1$ divides i and π^* divides $\pi - 1$.

The order of $\tilde{E}$ is equal to $(\pi - 1)(\pi^* - 1)$. It is divisible by p if and only if π^* divides $\pi - 1$.

4.5. We keep the notations of last paragraph and let

$$F_{\infty, \mathfrak{P}} = \prod_{v|\mathfrak{P}} F_{\infty, v} \left(\text{resp. } F_{n, \mathfrak{P}} = \prod_{v|\mathfrak{P}} F_{n, v}\right)$$

be the product of the completion of F_∞ (resp. F_n) at all places v dividing $\mathfrak{P}$

(there are finitely many of them [26], §2). Let U_n be the group of units in $F_{n,p}$ which are congruent to one modulo every maximal ideal, and let $U_\infty = \lim_{\leftarrow n} U_n$ be the projective limit of these groups for the norm maps. The group U_∞ is a compact $\mathbb{Z}_p[[G_\infty]]$-module. Let $K_{\mathfrak{P}}$ be the completion of K at $\mathfrak{P}$.

LEMMA 4.5. The composite morphism

$$\left(U_\infty \otimes T_p E(i-2)\right)_{G_\infty} \to K_{2i-2}\left(E \otimes_K K_{\mathfrak{P}}, \mathbb{Z}_p\right) \xrightarrow{\psi} H^1\left(K_{\mathfrak{P}}, T_p E(i-1)\right)$$

is injective and its cokernel is finite. It is an isomorphism of free $\mathbb{Z}_p$-modules of rank 2 except, when $\tilde{E}$ has a point of order p, for those i such that $p-1$ divides $i-1$ or $i-2$.

Proof. Fix a place v of F_∞ dividing $\mathfrak{P}$. When $n > 0$ let $U_{n,v}$ be the group of units in $F_{n,v}$ which are congruent to one modulo v, and $U_{\infty,v} = \lim_{\leftarrow n} U_{n,v}$. Since U_∞ is the G_∞-module induced from the $G_{\infty,v}$-module $U_{\infty,v}$, we get

$$\left(U_\infty \otimes T_p E(i-2)\right)_{G_\infty} = \left(U_{\infty,v} \otimes T_p E(i-2)\right)_{G_{\infty,v}}.$$

Consider the exact sequences

$$0 \to U_{n,v} \to F_{n,v}^* \to \mathbb{Z} \times k_n^* \to 0$$

where k_n is the residue field of $F_{n,v}$. The group k_n^* has order prime to p. When n is large, the extension $F_{n+1,v}/F_{n,v}$ has degree p^2 and ramification degree p. Therefore the norm map from the value group $\mathbb{Z}$ of $F_{n+1,v}$ to the value group $\mathbb{Z}$ of $F_{n,v}$ is the multiplication by p. With respect to these maps $\lim_{\leftarrow n} \mathbb{Z}/p^n = 0$. Consequently the map

$$U_{\infty,v} = \lim_{\leftarrow n} U_{n,v}/p^n \to \lim_{\leftarrow n} F_{n,v}^*/p^n$$

is an isomorphism, and the map

$$\left(U_\infty \otimes T_p E(i-2)\right)_{G_\infty} \to K_{2i-2}\left(E \otimes_K K_{\mathfrak{P}}, \mathbb{Z}_p\right)$$

is the map φ of 4.2. for the field $K_{\mathfrak{P}}$.

By Lemma 4.3. the kernel (resp. cokernel) of $\psi \circ \varphi$ is contained in the dual of $H^2(G_{\infty,v}; E_{p^\infty}(1-i))$ (resp. $H^1(G_{\infty,v}; E_{p^\infty}(1-i))$). By Lemma 4.4. this group is zero (resp. zero, or finite in the special case considered in the statement of Lemma 4.5.).   q.e.d.

4.6. Let I be the ring of integers in the maximal unramified extension of $\mathbb{Q}_p$. Since $K_{\mathfrak{P}} = \mathbb{Q}_p$ we get an inclusion $i_{\mathfrak{P}} : K \to I$. Let $i_\infty : K \to \mathbb{C}$ be a complex imbedding of K. According to Manin-Vishik [10] and Katz [8], given any pair of integers (i_1, i_2) modulo $(p-1)$, there exists a unique power series in two variables $G^{(i_1, i_2)}(T_1, T_2) \in I[[T_1, T_2]]$ such that

(4.6.1.) $\quad i_{\mathfrak{P}}^{-1}\Big(\Omega_{\mathfrak{P}}^{k_1 - k_2} G^{(i_1, i_2)}(u^{k_1} - 1, u^{k_2} - 1) \Big)$

$$= i_\infty^{-1}\Bigg((k_1 - 1)!\big(1 - \psi(\mathfrak{P})^{k_1 - k_2} p^{k_2 - 1}\big)\big(1 - \bar{\psi}(\mathfrak{P}*)^{k_1 - k_2} p^{-k_1}\big)$$

$$\times \left(\frac{2\pi}{\sqrt{d_K}} \right)^{-k_2} \Omega_\infty^{k_2 - k_1} L\big(\bar{\psi}^{k_1 - k_2}, k_1\big) \Bigg)$$

whenever $k_1 > -k_2 \geqslant 0$ and $(k_1, k_2) \equiv (i_1, i_2)$ modulo $(p - 1)$. Here $u = 1 + p$, $\Omega_{\mathfrak{P}}$ is a certain unit in I and $L(\bar{\psi}^k, s)$ is the Hecke L-series of $\bar{\psi}^k$ (not necessarily primitive) (cf. [26]).

When $i \geqslant 2$ the composite map

$$r_i: K_{2i-2}\big(E; \mathbb{Z}_p\big) \to K_{2i-2}\big(E \otimes_K K_{\mathfrak{P}}; \mathbb{Z}_p\big)$$

$$\overset{\varphi}{\to} H^1\big(K_{\mathfrak{P}}; T_p E(i - 1)\big)/\text{torsion} \simeq \mathbb{Z}_p^2$$

will be called a *higher p-adic regulator* map for the curve E.

THEOREM 4.6.   *For any $i \geqslant 2$ there exists a $\mathbb{Z}_p$-submodule V_i in $K_{2i-2}(E; \mathbb{Z}_p)$ with the following property. Let $p^{n_i}, 0 \leqslant n_i \leqslant +\infty$, be the index of $r_i(V_i)$ in $\mathbb{Z}_p^2$. Then n_i is the p-adic valuation of*

(4.6.2) $\quad c_i G^{(1-i, 2-i)}(u^{1-i} - 1, u^{2-i} - 1) G^{(2-i, 1-i)}(u^{2-i} - 1, u^{1-i} - 1),$

where $c_i \in I - \{0\}$. Furthermore c_i is a unit except, when $\tilde{E}$ has a point of order p, for those i such that $p - 1$ divides $i, i - 1$ or $i - 2$.

Proof.   Let C_n be the group of elliptic units in F_n which are congruent to one modulo the ideals dividing p (this group is denoted $C_{n-1, n-1}$ in [26], §9). By (4.2.) we have a map

$$\left(\lim_{\leftarrow n} (C_n/p^n) \otimes T_p E(i - 2) \right)_{G_\infty} \to K_{2i-2}\big(E; \mathbb{Z}_p\big).$$

Let V_i denote its image. We want to study the composite morphism

$$V_i \to K_{2i-2}\big(E \otimes_K K_{\mathfrak{P}}; \mathbb{Z}_p\big) \overset{\psi}{\to} H^1\big(K_{\mathfrak{P}}; T_p E(i - 2)\big).$$

Observe that C_n maps diagonally into U_n and that we have a commutative diagram

$$\left(\lim_{\leftarrow n}(C_n/p^n) \otimes T_p E(i - 2) \right)_{G_\infty} \overset{\alpha}{\to} \left(U_\infty \otimes T_p E(i - 2) \right)_{G_\infty}$$

$$\downarrow \qquad\qquad\qquad\qquad\qquad\qquad \downarrow$$

$$K_{2i-2}\big(E; \mathbb{Z}_p\big) \qquad\qquad \to K_{2i-2}\big(E \otimes_K K_{\mathfrak{P}}; \mathbb{Z}_p\big)$$

From Lemma 4.5. we know that we just need to compute the order of the cokernel of α. Remark that, if $\overline{C}_n$ is the closure of C_n inside U_n, the image of C_n/p^n inside U_n/p^n is equal to the image of $\overline{C}_n/p^n$. If $\overline{C}_\infty = \lim_{\leftarrow_n} \overline{C}_n$, the cokernel of α is then equal to $(Y_\infty \otimes T_p E(i-2))_{G_\infty}$, with $Y_\infty = U_\infty/\overline{C}_\infty$. The group G_∞ is isomorphic to $\Delta \times \Gamma$, where $\Delta = Gal(K(E_p)/K) = ((\mathbb{Z}/p)^*)^2$ and $\Gamma = Gal(F_\infty/K(E_p)) = ((1 + p\mathbb{Z}_p)^\times)^2$. Let χ_1 (resp. χ_2) be the character of Δ describing its action on E_π (resp. $E_{\pi*}$). Any $\mathbb{Z}_p[[G_\infty]]$-module M can be decomposed as a direct sum $M = \bigoplus_{(i_1, i_2)} M^{(i_1, i_2)}$, where i_1 and i_2 runs over integers modulo $p-1$, and Δ acts upon $M^{(i_1, i_2)}$ by $\chi_1^{i_2}\chi_2^{i_2}$. We have

$$\left(Y_\infty \otimes T_p E(i-2)\right)_{G_\infty} = \left(Y_\infty^{(1-i, 2-i)} \otimes T_\pi E(i-2)\right)_\Gamma$$

$$\otimes \left(Y_\infty^{(2-i, 1-i)} \otimes T_{\pi*} E(i-2)\right)_\Gamma.$$

Define an isomorphism $\lambda : \mathbb{Z}_p[[T_1, T_2]] \to \mathbb{Z}_p[\Gamma]$ by requiring that $\lambda(1 + T_1)$(resp. $\lambda(1 + T_2)$) lies in Γ and is mapped by $\kappa_1 \times \kappa_2$ to $(u, 1)$(resp. $(1, u)$). According to Yager ([26], Theorem 30), the Λ-module $Y_\infty^{(1-i, 2-i)}$ (resp. $Y_\infty^{(2-i, 1-i)}$) is in general isomorphic to $\Lambda/g\Lambda$, where $g \in \mathbb{Z}_p[T_1, T_2]$ is the product of $G^{(1-i, 2-i)}(T_1, T_2)$(resp. $G^{(2-i, 1-i)}(T_1, T_2)$) by a unit in $I[[T_1, T_2]]$. It follows that the order of $(Y_\infty^{(1-i, 2-i)} \otimes T_\pi E(i-2))_\Gamma$ is p^{mi}, where m_i is the p-adic valuation of $G^{(1-i, 2-i)}(u^{1-i} - 1, u^{2-i} - 1)$, and similarly for $(Y_\infty^{(2-i, 1-i)} \otimes T_{\pi*} E(i-2))_\Gamma$. In the special case of our statement, the same is true up to a constant ([26], Theorem 30 and Lemma 24).

4.7. According to (4.6.1.), the number (4.6.2.) can be viewed (up to non-vanishing constants and Euler factors) as an analog of

$$(4.7.1) \qquad\qquad L\left((\bar\psi)^{-1}, 1 - i\right) L(\bar\psi, 2 - i).$$

Since $\psi\bar\psi$ is the norm character, the functional equation for Hecke L-series relates (4.7.1.) to the value at $s = 2 - i$ of the L-function $L(E; s)$. In that sense, Theorem 4.6. is a p-adic analog of results of Bloch [2] and Beilinson [1].

One might expect (4.6.2.) to be nonzero for all $i \geqslant 2$. This is true when p is regular in the sense of 3.3.1.

REFERENCES

1. A. BEILINSON, *Higher regulators and values of L-functions, Sovremennye problemy mathematici,* VINITI **24** (1984), 181–238.
2. S. BLOCH, *Lectures on Algebraic Cycles,* Duke University Maths. Series **IV**, Durham N.C. (1980).
3. S. BLOCH AND A. OGUS, *Gersten's conjecture and the homology of schemes,* Ann. Sci. Ec. Norm. Sup. **7** (1974), 181–202.
4. J. COATES AND A. WILES, *On the conjecture of Birch and Swinnerton-Dyer,* Invent. Math. **39** (1977), 223–251.
5. W. DWYER AND E. FRIEDLANDER, *Algebraic and Etale K-theory,* to appear in Trans. Amer. Math. Soc.
6. W. DWYER, AND E. FRIEDLANDER, "Some remarks on the K-theory of Fields," in *Applications of Algebraic K-theory,* Contemporary Math., Vol. **55** Part I (1986), 149–158.

7. B. Gross, *Arithmetic of elliptic curves with complex multiplication*, Lecture Notes in Maths. 776 (1979), Springer-Verlag.

8. N. Katz, *p-adic interpolation of real analytic Einsenstein series*, Annals of Maths. **104** (1976), 459–571.

9. S. Lang, *Cyclotomic fields I*, Graduate Texts in Math. 59 (1978), Springer-Verlag.

10. Y. Manin and M. Vishik, *p-adic Hecke series for quadratic imaginary fields*, Math. Sb. **95** (1974), 357–383.

11. A. Merkurjev and A. Suslin, *K-cohomology of Severi-Brauer varieties and norm residue homomorphism*, Izv. Akad. Nauk SSSR Ser. Mat. **46** (1982), 1011–1046.

12. J. Neisendorfer, *Primary homotopy theory*, Memoirs AMS **232** (1980).

13. D. Quillen, *Higher Algebraic K-Theory I*, Lecture Notes in Math. 341 (1973), 85–147, Springer-Verlag.

14. J.-P. Serre, *Facteurs locaux des fonctions zêta des variétés algébriques (définitions et conjectures)*, Sém. Delange-Pisot-Poitou, Exposé **19** (1969–1970).

15. C. Soulé, *On higher p-adic regulators*, Lecture Notes in Maths. **854** (1981), 372–401, Springer-Verlag.

16. ______, *Eléments cyclotomiques en K-théorie*, to appear in Astérisque.

17. ______, *Operations on étale K-theory, Applications.* Lecture Notes in Maths. **966** (1982), 271–303, Springer-Verlag.

18. ______, *Opèrations en K-théorie algébrique*, Can. J. of Math. Vol 37, **3** (1985), 488–550.

19. ______, *On the rank of étale cohomology of varieties over p-adic or number fields*, Compositio Mathematica **53** (1984), 113–131.

20. ______, *K-théorie p-adique des courbes elliptiques*, preprint 1983.

21. J. Tate, *K_2 and Galois cohomology, letter to K. Iwasawa*, Lecture Notes in Maths. 342 (1973), 524–527, Springer-Verlag.

22. J. Tate, *Duality theorems in Galois cohomology over number fields*, Proc. Int. Congress, Stockholm (1962), 288–295.

23. R. Thomason, *Riemann-Roch for algebraic versus topological K-theory*, Journal of Pure and Applied Algebra **27** (1983), 87–109.

24. R. Thomason, "Bott stability in Algebraic K-theory," in *Applications of Algebraic K-Theory*, Contemporary Math., Vol. **55**, part I (1986), 389–406.

25. R. Yager, *A Kummer criterion for imaginary quadratic fields*, Compositio Math. **47** (1982), 31–42.

26. ______, *On two variable p-adic L-functions*, Annals of Maths. **115** (1982), 411–449.

U.E.R. de Mathématiques, Université Paris VII, Tour 45-55, 5ème étage, 2 Place Jussieu, 75251 Paris CEDEX 05, France

Vol. 54, No. 2   DUKE MATHEMATICAL JOURNAL ©   1987

ON THE RATIONALITY PROBLEM FOR CONIC BUNDLES

V. A. ISKOVSKIKH

To Yu. I. Manin on his 50th birthday

Introduction. Let V be a smooth irreducible projective threefold over $\mathbb{C}$. In this paper we are interested in the problem of characterizing rational threefolds V. For V to be rational, it is necessary that $q(V) := h^1(V, \mathcal{O}_V) = 0$ and $\kappa(V) = -\infty$ (κ stands for Kodaira dimension). For surfaces these two properties are also sufficient, but they are not in higher dimensions. However the Reid-Mori program for constructing minimal models (see, for example [11]) and results of Miyaoka [17] imply that every threefold V with $\kappa(V) = -\infty$ is birational to a variety W with at most terminal singularities of one of the following types:

(1) the anticanonical divisor $-K_W$ is ample and $\operatorname{Pic} W \simeq \mathbb{Z}$, i.e., W is a *minimal $\mathbb{Q}$-Fano variety*;

(2) There exists a morphism $\delta \colon W \to C$ onto a smooth curve C whose generic fiber is a del Pezzo surface and $\operatorname{Pic} W \simeq \delta^* \operatorname{Pic} C \oplus \mathbb{Z}$, i.e., W is a *minimal fiber space of del Pezzo surfaces*;

(3) There exists a morphism $\pi \colon W \to S$ onto a normal surface S, each fiber of π is isomorphic to a conic in $\mathbb{P}^2$, and $\operatorname{Pic} W \simeq \pi^* \operatorname{Pic} S \oplus \mathbb{Z}$, i.e., W is a *minimal conic bundle*. It is known that up to birational equivalence we can restrict ourselves to the case when W and S are nonsingular and projective (see [24], [16], [21]).

Thus the problem of characterizing rational threefolds is divided into three parts, each part requiring a description of rational varieties belonging to one of the three types. We note that if V is rational, then so is C or S in (2) or (3).

In the present paper we study the rationality problem for varieties of the third type, that is, conic bundles. The author's note [4] suggested a conjectural rationality criterion for such varieties, followed by some comments and a sketch of the proof. Here we discuss the problem in more detail.

In §1 we state Conjecture I which gives the rationality criterion for conic bundles, and we prove the sufficiency part (Theorem 1). Necessity is proved in Theorem 2 under the additional assumption that for a rational conic bundle V over S there exists a birational map $\chi \colon V \dashrightarrow \mathbb{P}^3$ which takes fibers of the morphism $\pi \colon V \to S$ to conics in $\mathbb{P}^3$. This assumption constitutes Conjecture II. By Theorems 1 and 2, Conjecture II is equivalent to the "only if" part of Conjecture I. We note that Kantor [10] gives a "proof" for Conjecture II, but this

Received November 5, 1986.

proof is beneath criticism; on the other hand, Millevoi [15] constructs a counter-example to this conjecture, but M. H. Gizatullin showed this also to be incorrect.

In §2 we propose another point of view on the problem of rationality of conic bundles via results of Sarkisov and the author [20], [21], [6], [7]. The main results of §2 are stated in Propositions 1 and 2 which, together with Conjecture III, give another approach to the proof of the rationality criterion for conic bundles involving the techniques of "untwisting" birational automorphisms. A first step in the proof of Conjecture III is given in Lemma 3.

The author would like to thank D. Markushevich for his kind help in the preparation of the English version of this paper. Thanks to J. Carlson, H. Clemens, and M. Reid for editing the manuscript.

In the following the ground field is the field of complex numbers $\mathbb{C}$.

§1. Conjecture on rationality of conic bundles.

Definition 1.   Let V be a smooth irreducible projective threefold and S a smooth rational surface. A subjective morphism $\pi\colon V \to S$ is called a *standard conic bundle* if the following properties are satisfied:

(i) for any point $s \in S$ the scheme-theoretic fiber $V_s = \pi^{-1}(s)$ is isomorphic (as a scheme) over the residue field $k(s)$ to a conic (possibly degenerate) in $\mathbb{P}^2_{k(s)}$;

(ii) (relative minimality) for any irreducible curve $D \subset S$ the surface $V_D = \pi^{-1}(D)$ is irreducible.

The following properties of a standard conic bundle $\pi\colon V \to S$ are well known.

LEMMA 1.   (i) *Let $C \subset S$ be the discriminant curve (possibly $C = \varnothing$). Then C is a reduced normal crossing divisor and*

$$V_s \simeq \begin{cases} P^1, & \textit{if } s \in S \setminus C; \\ P^1 \vee P^1, \textit{ two intersecting lines}, & \textit{if } s \in C \setminus \operatorname{Sing} C \\ 2P^1, \textit{ a double line}, & \textit{if } s \in \operatorname{Sing} C. \end{cases}$$

Here $\operatorname{Sing} C$ *is the set of double points of C (see* [2], [24], [21]).

(ii) *There is an isomorphism*

$$\operatorname{Pic} V \simeq \pi^*\operatorname{Pic} S + \begin{cases} \mathbb{Z} K_V & \textit{if } \pi\colon V \to S \textit{ has no rational sections} \\ \mathbb{Z} T & \textit{if } T \textit{ is the class of the image} \\ & \textit{of a rational section} \end{cases}$$

If $C \neq \varnothing$, then there is no rational section.

(iii) *Let $\tilde{C}$ be the curve parametrizing components of fibers in the ruled surface $V_C = \pi^{-1}(C)$, $\tilde{\pi}\colon \tilde{C} \to C$ the corresponding double covering, and $I\colon \tilde{C} \to \tilde{C}$ the natural involution. Then the restriction of $\tilde{\pi}$ to each irreducible component of $\tilde{C}$ is non-trivial and the following conditions are satisfied (the so-called Beauville conditions):*

tions):

$$(B) \qquad \tilde{\pi}(\operatorname{Sing} \tilde{C}) = \operatorname{Sing} C, \ \operatorname{Fix} I := \left\{ x \in \tilde{C} \,|\, I(x) = x \right\} = \operatorname{Sing} \tilde{C}.$$

(iv) *Conversely, let S be a rational surface, $C \subset S$ a reduced normal crossing curve, and $\tilde{\pi}: \tilde{C} \to C$ a double covering which is non-trivial on each irreducible component and which satisfies conditions (B). Then there exists a standard conic bundle $\pi: V \to S$ with the given $\tilde{\pi}: \tilde{C} \to C$, and all such standard conic bundles are birationally equivalent over S.*

Sketch proof of (iv). Consider the exact sequence for Brauer groups (see [1])

$$(A\text{-}M) \qquad 0 \to \operatorname{Br}(S) \to \operatorname{Br} k(S) \overset{\alpha}{\to}$$

$$\to \bigoplus_{\substack{\text{curves} \\ C_i \subset S}} H^1_{\text{ét}}(C_i, \mathbb{Q}|\mathbb{Z}) \overset{\beta}{\to} \bigoplus_{\substack{\text{points} \\ x \in C_i}} \mu_x^{-1} \overset{\gamma}{\to} \mu_x^{-1} \to 0,$$

where $\mu^{-1} = \bigcup_n \mu_n^{-1} = \bigcup_n \operatorname{Hom}(\mu_n, \mathbb{Q}|\mathbb{Z})$, and μ_n denotes the group of n-th roots of unity. Then the covering $\tilde{\pi}: \tilde{C} \to C$ specifies an element $a \in \bigoplus_{C_i \subset S} H^1_{\text{ét}}(C_i, \mathbb{Q}|\mathbb{Z})$ of order 2. From (B) it follows immediately that $\beta(a) = 0$, and hence there is an element $A \subset \operatorname{Br} k(S)$ such that $\alpha(A) = a$. As S is a rational surface we have $\operatorname{Br} S = 0$ so $A \in \operatorname{Br} k(S)$ has order 2. By virtue of the well-known Merkur'ev theorem [14] A is the product of classes representing quaternion algebras over the function field $k(S)$. Furthermore, since $k(S)$ is a C_2-field in Serre's classification, the product of classes of quaternion algebras is also represented by a quaternion algebra. Thus there exists a quaternion algebra $\mathscr{A}$ over $k(S)$ whose class in $\operatorname{Br} k(S)$ is equal to A. According to the classical theorem on central simple algebras the algebra $\mathscr{A}$ is uniquely determined up to isomorphism. As in [1] (see also [21]), one proves that the maximal orders of $\mathscr{A}$ over S are in 1–1 correspondence with standard conic bundles $\pi: V \to S$ with the given local invariants $\tilde{\pi}: \tilde{C} \to C$. All such standard conic bundles are birationally equivalent over S since their generic fibers are isomorphic conics over $k(S)$ associated to the quaternion algebra $\mathscr{A}$.

Definition 2. Conic bundles $\pi: V \to S$ and $\pi': V' \to S'$ are *birationally equivalent* if there exist birational maps $g: V \dashrightarrow V'$ and $h: S \dashrightarrow S'$ such that $\pi' \circ g = h \circ \pi$.

As we are concerned with the rationality questions we always assume S to be rational (indeed, if V is rational, then S is unirational, and hence rational). We also assume that C is connected. Otherwise V would be irrational by the theorem of Artin-Mumford [1]. Note also that there are no standard conic bundles with discriminant curve of arithmetic genus zero. Indeed, if $p_a(C) = 0$, then there exists an irreducible component with $p_a(C_1) = 0$ and $(C_1 \cdot C - C_1) = 1$. Then the resulting surface $V_{C_1} = \pi^{-1}(C_1)$ is reducible, since the curve $\tilde{C}_1$ parametrizing components of fibers of $V_{C_1} \to C_1$ is a double covering of C_1 with one branch point $= C_1 \cap \overline{C - C_1}$. This contradicts Definition 1, (ii).

We now state the main conjecture.

CONJECTURE I (compare [4]).   *Let* $\pi\colon V \to S$ *be a standard conic bundle with connected discriminant curve* $C \subset S$. *Then* V *is rational if and only if one of the following two properties is satisfied*:

(i) *there exists a standard conic bundle* $\pi'\colon V' \to S'$ *birationally equivalent to* $\pi\colon V \to S$ *and a base point free pencil* L' *of curves of genus zero on* S' *such that* $L'C' \leqslant 3$, *where* $C' \subset S'$ *is the discriminant curve*; *or*

(ii) *the standard conic bundle* $\pi\colon V \to S$ *is birationally equivalent to a standard conic bundle* $\pi_0\colon V_0 \to \mathbb{P}^2$ *with discriminant curve* $C_0 \subset \mathbb{P}^2$ *of degree 5, and the associated double covering* $\tilde{\pi}_0\colon \tilde{C}_0 \to C_0$ *is defined by an even theta-characteristic*.

Remark 1.   The sufficiency of these conditions is proved quite easily (Theorem 1 below). Necessity in general remains an open question. In the present paper we discuss some approaches to the solution of this problem. Note also that if $S = \mathbb{F}_N$ is a geometrically ruled surface or $S = \mathbb{P}^2$, then necessity is proved by V. V. Shokurov [22] by means of the Clemens-Griffiths intermediate Jacobian.

THEOREM 1 ([4], [19]).   *Either condition* (i) *or* (ii) *of Conjecture I is sufficient for the rationality of* V.

Proof.   (i) Let L be a base point free pencil of curves without fixed components on S and $L \cdot C \leqslant 3$. Consider the generic fiber L_ξ of L over the residue field $k(\xi)$ of the generic point $\xi \in \mathbb{P}^1$ over $k = \mathbb{C}$. The field $k(\xi)$ is isomorphic to the field of rational functions of one complex variable $\mathbb{C}(t)$, and hence by Tsen's theorem, L_ξ is isomorphic to $\mathbb{P}^1_{k(\xi)}$. Consider the generic surface $F_\xi = \pi^{-1}(L_\xi)$ over $k(\xi)$. Then the morphism $\pi_\xi\colon F_\xi \to L_\xi = \mathbb{P}_{k(\xi)}$ defines a standard conic bundle structure in the sense of [5] on the surface F_ξ over the non-closed field $k(\xi)$. The degeneracy divisor of the morphism $\pi_\xi\colon F_\xi \to L_\xi$ is $c_\xi = CL_\xi$ and $\deg c_\xi = CL \leqslant 3$. If K_{F_ξ} denotes the canonical divisor of F_ξ then $(K_{F_\xi} \cdot K_{F_\xi}) = 8 - \deg c_\xi \geqslant 5$. As $k(\xi) \simeq \mathbb{C}(t)$ is a field of type C_1 in the sense of Serre, Theorem 4.1 of [6] implies that F_ξ is rational over $k(\xi)$. Recall that F_ξ is the generic fiber of a pencil of surfaces on V with the base $\mathbb{P}^1$ so from the rationality of the fiber follows the rationality of V over $k = \mathbb{C}$.

(ii) This case was treated by I. Panin in [19]. Let $\pi_0\colon V_0 \to \mathbb{P}^2$ be a standard conic bundle, $C_0 \subset \mathbb{P}^2$ its discriminant curve. Suppose that C_0 is of degree 5 and that the double covering $\tilde{\pi}_0\colon \tilde{C}_0 \to C_0$ associated to π_0 corresponds to an even theta-characteristic. According to Lemma 1, (iv), such conic bundles exist and any two of them are birationally equivalent over $\mathbb{P}^2$. So it is enough to prove the rationality of any one of these. Following [19], we describe a construction giving such a conic bundle explicitly.

As $\tilde{\pi}_0\colon \tilde{C}_0 \to C_0$ satisfies the conditions (B) and corresponds to an even theta-characteristic (it is an odd pair in the sense of V. V. Shokurov [22]), the Prym variety $P(\tilde{C}_0, C_0)$ is (by [13] and [23]) isomorphic as a principally polarized Abelian variety to the Jacobian $J(Y)$ of a smooth curve of genus 5. The curve Y cannot be hyperelliptic, because otherwise C_0 would be hyperelliptic (see, for example, [22]). But C_0 is neither hyperelliptic, nor trigonal since its canonical model $\varphi_K(C_0) \subset \mathbb{P}^5$ lies in the Veronese surface. Similarly, Y is not trigonal; if C_0 is nonsingular this was proved in [13] and [23]. If C_0 is singular, one can obtain the proof either by generalizing the direct construction of [23] using quadrics in $\mathbb{P}^4$ defining the canonical curve $Y \subset \mathbb{P}^4$, or by a careful comparison of the singular locus of the theta-divisor of the Jacobian of a trigonal curve with the special types of singular loci described in [22], §5, under the assumption that the conditions (B) are satisfied.

Thus the canonical image $Y \subset \mathbb{P}^4$ is a complete intersection of three quadrics $Q_1 \cap Q_2 \cap Q_3$, so Y is the base locus of a net of quadrics $|2h - Y|$, where h is a hyperplane in $\mathbb{P}^4$. The invariant of this net in the sense of [23] is exactly the covering $\tilde{\pi}_0\colon \tilde{C}_0 \to C_0$.

Starting from this net of quadrics we build a standard conic bundle $\pi_0\colon V_0 \to \mathbb{P}^2$ with the given invariant $\tilde{\pi}_0\colon \tilde{C}_0 \to C_0$, and prove the rationality of V_0. Then by Lemma 1, (iv), any other standard conic bundle with the same $\tilde{\pi}_0$ will be rational.

Fix a point $0 \in Y \subset \mathbb{P}^4$. Consider the following family of conics in $\pi_0\colon V_0 \to \mathbb{P}^2$. For each point $s \in \mathbb{P}^2$ we define the fiber $V_s = \pi_0^{-1}(s)$ as the projectivized tangent cone to the quadric $Q_s \in |2h - Y|$ at the point 0. From [13], [23] it follows that none of the quadrics Q_s has singular points on Y, in particular at 0. Therefore, if the rank of Q_s equals 5, the maximal value, then the corresponding conic is irreducible; if $rk Q_s = 4$, then $rk V_s = 2$ and so $V_s \simeq P^1 \vee P^1$; and finally, if $rk Q_s = 3$, then $rk V_s = 1$ and V_s is a double line. Hence the degeneracies of $\pi_0\colon V_0 \to \mathbb{P}^2$ correspond precisely to the given covering $\tilde{\pi}_0\colon \tilde{C}_0 \to C_0$ from which the net of quadrics was built.

A more direct way to build the conic bundle $\pi_0\colon V_0 \to \mathbb{P}^2$ is as follows. Let

$$\mathscr{Q} \subset \mathbb{P}^2 \times \mathbb{P}^4$$
$$p_1 \downarrow$$
$$\mathbb{P}^2$$

be the family of quadrics of the net $|2h - Y|$, p_1 the projection, and let $P_0 = p_2^{-1}(0)$ be the rational section, where $p_2\colon \mathscr{Q} \to \mathbb{P}^4$ is the projection to the other factor. Put $\mathscr{V}_0 = \bigcup_{s \in \mathbb{P}^2}(T_{s,0} \cap Q_s)$, the union of the tangent cones to the quadrics Q_s, $s \in \mathbb{P}^2$ at the points $(s, 0) \in P_0$. Here $T_{s,0}$ denotes the tangent projective space of Q_s at the point $(s, 0) \in P_0$. Then V_0 is the union of the bases of these cones. The morphism $\pi_0\colon V_0 \to \mathbb{P}^2$ is induced by the projection $p_1\colon \mathscr{Q} \to p^2$.

The restriction of the projection p_2: $\mathscr{2} \to \mathbb{P}^4$ to the subvariety $\mathscr{V}_0 \subset \mathscr{2}$ is a birational morphism $p_{2,0}$: $\mathscr{V}_0 \to \mathbb{P}^4$. Indeed, it maps generators of cones from $\mathscr{V}_0$ onto lines in $\mathbb{P}^4$ passing through $0 \in Y$. It is clear that this map is surjective since each line in $\mathbb{P}^4$ passing through 0 lies on at least one quadric of the net $|2h - Y|$. Furthermore, the general such line is contained in exactly one quadric of the net $|2h - Y|$. Indeed, if a line $l \ni 0$ is contained in the intersection of two quadrics $Q_1, Q_2 \in |2h - Y|$, then any quadric $Q_3 \in |2h - Y|$ which is linearly independent on Q_1 and Q_2 intersects l not only in 0 but also in one more point $y \in Y$ (possibly $y = 0$) since $Q_1 \cap Q_2 \cap Q_3 = Y$. Hence l is either a chord, or a tangent line of Y, but this is not the case for a general line l. Thus we have shown that $p_2|_{\mathscr{V}_0}$: $\mathscr{V}_0 \to \mathbb{P}^4$ is a birational morphism, and moreover the projection p_0: $\mathbb{P}^4 \to \mathbb{P}^3$ from the point $0 \in Y$ induces the birational morphism χ: $V_0 \to \mathbb{P}^3$. The theorem is proved.

Remark 2.   Both rationality constructions for standard conic bundles satisfying conditions (i) or (ii) of Conjecture I suggested in the theorem lead to a birational map χ: $V \to \mathbb{P}^3$ which takes fibers V_s, $s \in S$ to conics in $\mathbb{P}^3$. In the case (i) the map χ is induced by the map from the generic surface F_ξ over the residue field $k(\xi)$. As was shown, the birational map χ_ξ: $F_\xi \to \mathbb{P}^2_{k(\xi)}$ is defined by the linear system $|-f_\xi - K_{F_\xi}|$, if $K^2_{F_\xi} = 8 - CL_\xi = 5$, and $|-f_\xi - K_{F_\xi} - x|$ if $K^2_{F_\xi} = 8 - CL_\xi = 6$, where f_ξ is a fiber of the morphism π_ξ: $F_\xi \to \mathbb{P}^1_{k(\xi)}$ and $x_\xi \in F_\xi$ is a $k(\xi)$-rational point in F_ξ. Since $-f_\xi K_{F_\xi} = V_s K_V = 2$, the image of a fiber V_s under such a map is a conic in $\mathbb{P}^3$.

In case (ii) the projection p_0: $\mathbb{P}^4 \to \mathbb{P}^3$ from the point $0 \in Y$ induces linear maps on fibers of π_0: $V_0 \to \mathbb{P}^2$ which are conics by construction. Thus the images of the fibers $V_{0,s}$ are conics in $\mathbb{P}^3$ intersecting the curve $Y' = p_0(Y)$ of genus 5 and degree 7 in $\mathbb{P}^3$ at 6 points. The two-dimensional family of conics in $\mathbb{P}^3$ (or the congruence of conics) obtained in this way is defined by a net $|3h - Y'|$ of cubics passing through Y' (h is the class of a plane in $\mathbb{P}^3$), the original conics appearing as residual curves of pairs of cubics which pass through Y'. Conversely if $Y' \subset \mathbb{P}^3$ is a nonsingular curve of degree 7 and genus 5, then the net of cubics $|3h - Y'|$ defines a rational map $\mathbb{P}^3 \dashrightarrow \mathbb{P}^2$ with a conic as the generic fiber, which becomes a standard conic bundle after one blow-up Y'.

In view of Remark 2 the following conjecture arises:

CONJECTURE II.   *For any rational standard conic bundle π: $V \to S$ there exists a birational map β: $V \dashrightarrow \mathbb{P}^3$ taking fibers V_s, $s \in S$ to conics in $\mathbb{P}^3$ (such a map is given by a subsystem of some complete linear system $|\pi^*D - K_V|$, $D \in \mathrm{Pic}\, S$). Equivalently, any 2-dimensional family (congruence) of rational curves in $\mathbb{P}^3$ of index 1 can be transformed into a 2-dimensional family of conics (or lines passing through a fixed point) by an appropriate birational transformation of $\mathbb{P}^3$.*

Remark 3.   M. H. Gizatullin pointed out to the author the paper [21] by Kantor, in which the positive answer to Conjecture II is stated. However, his proofs are hard to follow and seem not to be rigorous. M. H. Gizatullin pointed out also the paper by Millevoi [15] which claims a counter-example to Kantor's assertion (that is, an example of a congruence of index 1 of rational curves of degree 4 in $\mathbb{P}^3$ is given, which the author claims admits no birational transformation to a system of conics or lines). But this paper also proved to be mistaken.

We prove the following result.

THEOREM 2. *If Conjecture* II *is true, then for any standard conic bundle* $\pi\colon V \to S$ *with a rational* V *one of the properties* (i) *or* (ii) *of Conjecture* I *is satisfied.*

Proof. Let $\beta\colon V \dashrightarrow \mathbb{P}^3$ be a birational map taking fibers V_s, $s \in S$ to conics in $\mathbb{P}^3$, as in Conjecture II. (If the images of V_s are lines in $\mathbb{P}^3$, then $\pi\colon V \to S$ admits a section and $C = \varnothing$.) Let H_η be the generic plane in $\mathbb{P}^3$ over the field of rational functions of 3 variables $k(\eta) = \mathbb{C}(u_1, u_2, u_3)$, and $T_\eta = \beta^{-1}(H_\eta)$ the proper transform of H_η in V. Then $\pi|T_\eta\colon T_\eta \to S$ is a morphism of degree 2 at the generic point and there is a birational involution δ_η on T_η, which acts by interchanging the two points in the fibers of $\pi|T_\eta$. Let $G_\eta = \{id, \delta_\eta\} \simeq \mathbb{Z}/2$ be the group generated by δ_η; by classical results (see [5]; there the ground field is assumed to be algebraically closed, but all the arguments remain valid over a non-closed field) there exists a G_η-invariant pencil of rational curves or curves of genus 1 on T_η. In the case of curves of genus 1, T_η is birationally equivalent as a G-surface to a G_η-del Pezzo surface T_η^* with $\mathrm{Pic}_{k(\eta)}^{G_\eta} T_\eta^* \simeq \mathbb{Z}$ (see [5]).

We begin with the case in which there is a G_η-invariant pencil of rational curves. Choose any such pencil on T_η without fixed component, say M_η, and denote the image of M_η on S by L. Then L is a pencil of rational curves without fixed components on S. We can replace $\pi\colon V \to S$ if necessary by a birationally equivalent standard conic bundle using a sequence of elementary transformations as described in [20], §2–3, such that the following conditions (a), (b) are satisfied:

(a) T_η is not singular along any curve of V, except possibly for fibers of π.

Indeed, T_η moves in a linear system $|\pi^*D - K_V - \Sigma \nu_i B_i|$ of projective dimension 3, where the B_i are the base points or curves, the ν_i their multiplicities, and $D \in \mathrm{Pic}\, S$. By Bertini's theorem, the singular curves of a general element T are among the B_i. So, if $\dim \pi(B_i) \neq 0$, one can suppose according to [20], (3.14), that $\nu_i \leqslant 1$. This means that the surface T is non-singular at the generic point of B_i, as stated.

(b) M_η and L have no base points (on T_η and S resp.) and no fixed components. In particular, the generic element L_ξ of the pencil L is non-singular.

The argument here proceeds by a sequence of elementary transformations from [20], §2, with centers in the base locus of L.

Let $F_\xi = \pi^{-1}(L_\xi)$ be the surface over the generic fiber $L_\xi \simeq \mathbb{P}^1_{k(\xi)}$ of the pencil L. Since $\pi\colon V \to S$ is a standard conic bundle, we conclude that $\pi_\xi\colon F_\xi \to L_\xi \simeq \mathbb{P}^1_{k(\xi)}$ is also a standard surface with a pencil of rational curves over $k(\xi)$ in the sense of [5]; in particular, $\mathrm{Pic}\, F_\xi = \mathbb{Z} + \mathbb{Z} K_{F_\xi}$, where K_ξ is the canonical divisor. We want to show that $LC \leqslant 3$, to verify (i) of Conjecture I in the case under consideration. It is enough to show that $K_{F_\xi}^2 \geqslant 5$, since $K_{F_\xi}^2 = 8 - LC$ (see [5], [6]).

Let M be the restriction of M_η to F_ξ. Then M_ξ is a nonsingular rational curve varying in the linear system $|M_\xi| \sim af_\xi - K_{F_\xi} - \Sigma \nu_i x_i$, where $a \in \mathbb{Z}$, f is a fiber of $\pi\colon F_\xi \to \mathbb{P}^1_{k(\xi)}$, x_i are base points (possibly infinitely near), and $\nu_i \leqslant 1$ (according to (a)). Since $|M_\xi|$ is the restriction to F_ξ of the proper transform of the linear system of planes $|H|$ in $\mathbb{P}^3$, $\dim |H| = 3$. We now have two cases

Case 1. $\dim |M_\xi| \leqslant 2$. This means that the image of F_ξ in $\mathbb{P}^3$ lies in a plane (hence, coincides with it) so the surface F_ξ is rational over $k(\xi)$. From [5], [6] it follows that $K_{F_\xi}^2 \geqslant 5$, as was to be shown.

Case 2. $\dim |M_\xi| \geqslant 3$. Since $M_\xi \sim af_\xi - K_{F_\xi} - \Sigma \nu_i x_{i\xi}$ varies in a linear system of curves of genus zero, we have $a < 0$ by [6], (1.1). In this case $-K_{F_\xi}$ is ample, that is, F_ξ is a del Pezzo surface (see [6]) and, by the classification of such surfaces possessing a pencil of rational curves [5], [6], the inequality $\dim |M_\xi| \geqslant 3$ is possible only if $K_{F_\xi}^2 \geqslant 5$.

Consider now the case in which there exists a G_η-invariant system of curves of genus 1 on T_η, when T_η is birationally G_η-equivalent to a del Pezzo surface T_η^* with $\mathrm{Pic}^{G_\eta}_{k(\eta)} T_\eta^* = \mathbb{Z} K_{F_\eta}$. We have $1 \leqslant K_{T_\eta}^2 \leqslant 9$. We show first that if $K_{T_\eta^*}^2 \geqslant 3$, then there is a G_η-invariant pencil of rational curves on T_η^*, so such a pencil exists on T_η as well, and this brings us to the case considered earlier.

Since $\mathrm{Pic}^{G_\eta} T_\eta^* = \mathbb{Z} K_{T_\eta^*}$, the surface T_η^* is G_η^*-minimal in the sense of [5]. Moreover, T_η^* is rational over $k(\eta)$, since it is a birational image of $\mathbb{P}^2_{k(\eta)}$. Hence if $K_{T_\eta^*}^2 = 9$ or 8, then $T_\eta^* = \mathbb{P}^2_{k(\eta)}$, or a quadric $Q_\eta \subset \mathbb{P}^3_{k(\eta)}$. As G_η acts on these surfaces linearly, one can find a G_η-invariant pencil of rational curves in the linear system of hyperplane sections.

There are no G_η-minimal del Pezzo surfaces with $K_{T_\eta^*}^2 = 7$ (see [5]). Observe now that if $4 \leqslant K_{T_\eta^*}^2 \leqslant 6$ and if there exists a G_η-invariant point P on T_η^* rational over $k(\eta)$, then there exists a G_η-invariant pencil of rational curves as well. Indeed, $T_\eta^* \subset \mathbb{P}^d_{k(\eta)}$ is then a surface of degree d with $4 \leqslant d = K_{T_\eta^*}^2 \leqslant 6$, and one can choose a G_η-invariant pencil of rational curves in the linear system $|H_d - 2P|$ cut by hyperplanes H_d passing through the tangent plane to T_η^* at P. We now show that, in the case under consideration, a G_η-invariant point on T_η^* defined over $k(\eta)$ does exist. To see this, we note that, by construction, the quotient T_η^*/G_η is a rational surface defined over $\mathbb{C}$. Denote it by S^*. Since S^* is rational, G_η has fixed points on T_η^*. Let $D_\eta^* \subset T_\eta^*$ be the set of fixed points under the action of G_η (D_η^* may consist of isolated points and curves), and let $E^* \subset S^*$ be its image in S^*.

Since $E^* \neq 0$, the reduced geometrical inverse of any point $s^* \subset E^*$ over $\mathbb{C}$ is a $k(\eta)$-point of degree 1 on T_η^* which is fixed under G_η by construction.

We now turn to the case in which $K_{T_\eta^*}^2 = 3$. Here $T_\eta^* \subset \mathbb{P}^3_{k(\eta)}$ is a non-singular cubic surface, defined over $k(\eta)$. As it is rational over $k(\eta)$, it cannot be minimal over $k(\eta)$ (see [5], [6]). Let $Z_\eta \subset T_\eta^*$ be an exceptional curve of the first kind over $k(\eta)$; geometrically, it consists of one or more disjoint lines conjugate over some extension of $k(\eta)$. Let Z_η' be the image of Z_η under the non-trivial transformation of G_η. Since T_η^* is assumed to be G_η-minimal, $Z_\eta' \neq Z_\eta$, and thus Z_η' intersects Z_η at one point (use the fact that Z_η' and Z_η are lines on a cubic surface). We now have $(Z_\eta + Z_\eta')^2 = 0$ and $p_a(Z_\eta + Z_\eta') \leqslant 0$. Hence, Riemann-Roch implies that $Z_\eta + Z_\eta'$ varies in a pencil of curves of genus zero which, by construction, is G_η invariant. One can also show that the existence of the G_η-invariant curve $Z_\eta + Z_\eta'$ contradicts the assumption that $\operatorname{Pic}^{G_\eta} T_\eta^* \simeq \mathbb{Z} K_{T_\eta^*}$.

Thus we have proved that when $K_{T_\eta^*}^2 \geqslant 3$, there is a G_η-invariant pencil of rational curves on T_η^*. Consider now the two remaining cases: $K_{T_\eta^*}^2 = 2, 1$.

Case $K_{T_\eta^*}^2 = 2$. Here the quotient map $T_\eta^* \to T_\eta^*/G_\eta$ coincides with the anticanonical map $\varphi_{-K_{T_\eta^*}}: T_\eta^* \to \mathbb{P}_{\mathbb{C}}^2$ which takes the G_η-invariant 2-dimensional linear system of $|-K_{T_\eta^*}|$ of curves genus 1 to the linear system of lines in $\mathbb{P}_{\mathbb{C}}^2$. Just as in the case of a G_η-invariant pencil of rational curves, we can, by appropriate elementary transformations, change $\pi\colon V \to S$ to another birationally equivalent standard conic bundle $\pi'\colon V' \to S'$ with the following properties:

(a') Let T_η' be the proper transform of T_η on V'; then T_η' contains no singular curves other than components of fibers of π'.

(b') There is a 2-dimensional G_η-invariant base point free linear system $|M_\eta'|$ on T_η' of curves of genus 1, without fixed components, such that its image in S' is a 2-dimensional linear system $|L'|$ without base point or fixed components of curves of genus zero with $(L')^2 = 1$.

Indeed we can choose the proper transform of the system $|-K_{T_\eta^*}|$ as $|M_\eta'|$ (if necessary, resolve the singularities).

Let $F_\xi' = \pi'^{-1}(L_\xi')$ be the surface conic bundle over the generic element $L_\xi' \simeq \mathbb{P}_{k(\xi)}^1$ of the linear system $|L'|$ on S'. Then $\pi_\xi'\colon F_\xi' \to L_\xi' = \mathbb{P}_{k(\xi)}^1$ is a standard surface with a pencil of rational curves over $k(\xi)$. In particular, $\operatorname{Pic} F_\xi' = \mathbb{Z} + \mathbb{Z} K_{F_\xi'}$. The curve $M_\xi' \in |M_\eta'|$ varies on F_ξ' in the linear system $|M_\xi'| = |a'f' - K_{F_\xi'} - \Sigma v_i' x_{\xi i}'|$, where $a' \in \mathbb{Z}$, $x_{\xi i}'$ are base points (possibly infinitely near), and $v_i' \leqslant 1$ by (a'). Since $p_a(M_\xi')^* = 1$, $a' = 0$ by [6], $(1, 1)$ and similarly to the above we have the following two possibilities:

Case 1'. $\dim |M_\xi'| \leqslant 2$. The situation is the same as in Case 1 above. Thus we have $L'C' \leqslant 3$ for the discriminant curve $C' \subset S'$.

Case 2'. $\dim |M_\xi'| \geqslant 3$. Here $-K_{F_\xi'}$ is ample and, according to the classification of [5], [6] this is possible only if $K_{F_\xi'}^2 \geqslant 3$, i.e., if $L'C' = 8 - K_{F_\xi'}^2 \leqslant 5$.

If $L'C' \leqslant 4$, then using $\dim |L'| = 2$ we can find in $|L'|$ a subpencil with a base point on C', which after applying an elementary transformation satisfies condition (i) of Conjecture I.

Let $L'C' = 5$. Consider the birational morphism $\varphi_{|L'|}\colon S' \to \mathbb{P}^2$. Then $C_0 = \varphi_{|L'|}(C')$ is a curve of degree 5 in $\mathbb{P}^2$ birational to C'. If $p_a(C') < p_a(C_0) = 6$

then C_0 has a singular point. The pencil of lines passing through this point after the blow-up becomes a pencil satisfying condition (i) of Conjecture I.

If $p_a(C') = p_a(C_0) = 6$, then C_0 is a curve with normal crossings and the double covering $\tilde{\pi}'\colon \tilde{C}' \to C$ induces a covering $\tilde{\pi}_0\colon \tilde{C}_0 \to C_0$ which is "non-branched" (in the sense of [22], §3). By Lemma 1, (iv), to this covering we can attach a standard conic bundle $\pi_0\colon V_0 \to \mathbb{P}^2$ whose invariant is $\tilde{\pi}_0\colon \tilde{C}_0 \to C$ and which is birational to $\pi'\colon V' \to S'$, since their generic fibers correspond to quaternion algebras over $k(S') = \mathbb{C}(\mathbb{P}^2)$ with the same local invariants.

Now, if the covering $\tilde{\pi}_0\colon \tilde{C}_0 \to C_0$ corresponds to an odd theta-characteristic, then V_0, and hence V, is nonrational by Clemens-Griffiths, since the intermediate Jacobian $J(V_0)$, being the Prym variety of $\tilde{\pi}_0\colon \tilde{C}_0 \to C_0$, is not the Jacobian of a curve (see for example [2]). Thus, since by hypothesis of the theorem V is rational, $\tilde{\pi}_0\colon \tilde{C}_0 \to C_0$ corresponds to an even theta-characteristic, and condition (ii) of Conjecture I is satisfied.

Case $K^2_{T^*_\eta} = 1$. Here the quotient map $T^*_\eta \to T^*_\eta/G_\eta$ coincides with the morphism of degree two $\varphi_{|-2K_{T^*_\eta}|}\colon T^*_\eta \to Q^* \subset \mathbb{P}^3$ where Q^* is a quadratic cone. The G_η-invariant pencil of elliptic curves $|-K_{T^*}|$ lies over the pencil of generators of the cone Q^* and the G_η-invariant linear system $\varphi_{|-2K_{T^*_\eta}|}$ of curves of genus 2 over the linear system $|\mathcal{O}_{Q^*}(1)|$ of conics, hyperplane sections of Q^*.

Just as in the previous case, the standard conic bundle $\pi\colon V \to S$ admits a birational transformation to another standard conic bundle $\pi'\colon V' \to S'$ which satisfies the following properties.

(a'') Let T'_η be the proper transform of T_η on V'; then T'_η contains no singular curves other than components of fibers of π'.

(b'') There exists a G_η-invariant pencil of elliptic curves $|M'_\eta|$ without base points or fixed components on T'_η whose image on S' is a pencil $|L'|$ of curves of genus zero without base points and fixed components. Moreover there is a 3-dimensional G_η-invariant linear system $|R'_\eta|$ of curves of genus 2 without base points or fixed components on T'_η whose image on S' is a 3-dimensional linear system $|N'|$ of curves of genus zero without base points or fixed components, such that $(N')^2 = 2N'L' = 1$ and the linear system $|N'|$ defines the birational morphism $\varphi_{N'}\colon S' \to Q^* \subset \mathbb{P}^3$, taking L' to the pencil of generators of the cone Q^*.

Let $F'_\xi = \pi^{-1}(L'_\xi)$ be the surface conic bundle over the generic member $L'_\xi \simeq \mathbb{P}^1_{k(\xi)}$ of the pencil $|L'|$ and $G_\mu = \pi^{-1}(N'_\mu)$ the surface conic bundle over the generic member $N'_\mu \simeq \mathbb{P}^1_{k(\mu)}$ of the linear system $|N'|$. Then $\pi'_\xi\colon F'_\xi \to L'_\xi = \mathbb{P}^1_{k(\xi)}$ and $\pi'_\mu\colon G'_\mu \to N'_\mu \simeq \mathbb{P}^1_{k(\mu)}$ are standard surfaces with a pencil of rational curves in the sense of [5]. The curve $M'_\xi \in |M'_\eta|$ varies on F'_ξ in the linear system $|M'_\xi| = |-K_{F'_\xi} - \Sigma x'_{\xi i}|$. Similarly the curve $R'_\mu \in |R'_\eta|$ varies on G'_μ in $|R'_\mu| = |f'_\mu - K_{G'_\mu} - \Sigma y'_{\mu j}|$.

Now, if $L'C' \leqslant 3$ then condition (i) of Conjecture I is satisfied and all has been proved. We shall prove that $L'C' \geqslant 4$ is impossible. Assume that $L'C' \geqslant 4$. Then $K^2_{F'_\xi} = 8 - (L'C') \leqslant 4$. The linear system $|M'_\xi|$ defines a birational map from F'_ξ onto a surface $F''_\xi \subset \mathbb{P}^3$ of degree at least 3, since the image of $|M'_\xi|$ is cut out on

F_ξ'' by the linear system of planes $|H|$ and $p_a(M_\xi') = 1$. Let $D' = \varphi_{N'}^{-1}(q)$ be the curve on S' which is the geometric locus of the vertex q of the cone $Q^* \subset \mathbb{P}^3$, and let $E' = \pi^{-1}(D')$. Then $G' \sim 2F' + E'$. The linear system $|R_\mu'|$ defines the birational map of the surface G_μ^* to the surface G_μ^* in $\mathbb{P}^3$ whose degree is not less than $2 \deg F_\xi' \geqslant 6$. Hence,

$$6 \leqslant \left(R_\mu'\right)^2 = \left(f_\mu' - K_{G_\mu'} - \sum_j y_{\mu j}'\right)^2$$

$$= 4 + K_{G_\mu'}^2 - \sum_j \deg y_{\mu j}' \leqslant 4 + K_{G_\mu^*}^2 .$$

so $K_{G_\mu'}^2 \geqslant 2$ and $N'C' \leqslant 6$.

Let $\sigma: S'' \to Q^*$ be the blow-up of the vertex. Then the birational morphism $\varphi_{N'}: S' \to Q^*$ factorizes through the birational morphism $\varphi: S' \to S''$. Put $E'' = \sigma^{-1}(q)$ the exceptional curve on S'' with $(E'')^2 = -2$, $F'' = \varphi(F')$, the fiber of the ruled surface $S'' \to \mathbb{P}^1$, $N' = \varphi(N') \sim E'' + 2F''$ and $C'' = \varphi(C')$, the image of the discriminant curve C'. We have $C'' \sim cE'' + dF''$, where $c, d \in \mathbb{Z}$ and $c \geqslant 4$ (by hypothesis), and $d \leqslant 6$, since $6 \geqslant N''C'' = (E'' + 2F'' \cdot cE'' + dF'') = -2c + 2c + d = d$. Furthermore, $C''E'' = -2c + d \leqslant -2$. Hence E'' is a component of C'', and it appears in C'' with multiplicity 1, which follows from the fact that C', and consequently C'', is reduced. But this is impossible unless $c = 4$ and $d = 6$. It then follows that $C'' - E'' \cdot E'' = -2c + d + 2 = 0$, and therefore the curve C'' is disconnected. Hence the discriminant curve C'' of the standard conic bundle $\pi': V' \to S'$ is also disconnected, which contradicts the rationality of V', according to Artin-Mumford. The proof of Theorem 2 is now complete.

Conjecture II was used in the proof of Theorem 2 to ensure that the generic surface $T_\eta = \beta^{-1}(H_\eta)$ is a generically $2:1$ cover of the base S; this gives an action of $G_\eta \simeq \mathbb{Z}/2$ on T_η which fixes a pencil of rational or elliptic curves. More generally, for a birational map $\beta: V \dashrightarrow \mathbb{P}^3$ such that the covering $T_\eta = \beta^{-1}(H_\eta) \to S$ restricted to the generic point of T_η is a Galois covering with group G_η, there still exists a G_η-invariant pencil of rational or elliptic curves, and the analogue of Theorem 2 can be proved in just the same way as above. However, Galois coverings are quite rare among finite coverings of degree > 2. Possibly all the arguments go through even in the case when the covering $\pi_\eta: T_\eta \to S$ is not Galois, provided that there is a pencil of rational curves L on S whose inverse image is a pencil of rational or elliptic curves. This assumption also seems to be false in general.

Remark 5. Conjecture I is now equivalent to Conjecture II. Indeed by Theorems 1 and 2, I follows from II; conversely, by Theorem 1 and Remark 2, II follows from I.

§2. Another approach to the proof of Conjecture I. Let $\pi\colon V \to S$ be a standard conic bundle with rational V, $\beta\colon V \dashrightarrow \mathbb{P}^1 \times \mathbb{P}^2$ a birational map, $|H|$ the linear system of planes in $\mathbb{P}^2$, and $|M| = \beta^{-1}(|H|)$ the proper transform of the linear system $p_2^*|H|$ on V, where $p_2\colon \mathbb{P}^1 \times \mathbb{P}^2 \to \mathbb{P}^2$ denotes the projection to the second factor. If $C \neq \varnothing$ then $\operatorname{Pic} V = \pi^*\operatorname{Pic} S + \mathbb{Z} K_V$ and

$$|M| = \left| \pi^*D - bK_V - \sum_{i=0}^{N-1} \nu_i B_i \right| \tag{1}$$

where $D \in \operatorname{Pic} S$, K_V is the canonical divisor on V, b, $\nu_i \in \mathbb{Z}$, $b > 0$, $\nu_i \geq 0$, B_i are base curves or points (possibly infinitely near; for precise definitions of these notions see [8, 9]).

Let $\sigma\colon V_N \xrightarrow{\sigma_{N,N-1}} V_{N-1} \to \cdots \xrightarrow{\sigma_{1,0}} V_0 = V$ be a resolution of indeterminacies of β by monoidal transformation with nonsingular centers $B_i \subset V_i$ and $E_{i+1} = \sigma_{i+1,i}^{-1}(B_i)$ the exceptional divisors; let $\gamma\colon V_N \to \mathbb{P}^1 \times \mathbb{P}^2$ be a birational morphism resolving β and $M_N = \gamma^* \cdot p_2^*(H)$; then

$$M_N = \sigma^*\pi^*D - b\sigma^*K_V - \sum_{i=0}^{N-1} \nu_i \sigma_{N,i+1}^* E_{i+1}. \tag{2}$$

Put $X := \pi_*\sigma_*(M_N^2)$, (as a 1-cycle on S). Then for geometrical reasons of movability of $|H|$ it follows that X is movable on S and numerically effective (that is $XY \geq 0$ for any curve $Y \subset S$).

LEMMA 2 (V. G. Sarkisov [20], [21]). *Replacing the standard conic bundle* $\pi\colon V \to S$ *by an equivalent birational one if necessary, we may assume that formula* (2) *satisfies*:

(i) *Let* $A_i := \pi\sigma_{i,0}(B_i)$ *be the geometric image of* B_i *on* S. *Then* $C \cup \bigcup A_i$ *is a divisor with normal crossings, and the* A_i *are curves whose irreducible components are non-singular.*

(ii) $\nu_i \leq b$ *for all such that* $\dim \sigma_{i,0}(B_i) = 1$;

(iii) $h^0(S, \mathcal{O}_S(mD)) = 0$ *for* $m \geq 1$, *and the divisor* D *is not numerically effective* (*i.e., there is a curve* $Y \subset S$ *with* $DY < 0$);

(iv) $X = \pi_*\sigma_*(M_N^2)$ *is numerically effective and*

$$X = 4bD - b^2(4K_S + C) - \sum_{i=0}^{N-1} \nu_i^2 (\pi_i\sigma_{i,0})_* B_i \tag{3}$$

where C *is the discriminant curve and* K_S *the canonical divisor of* S.

Recall that under these assumptions we want to get conditions (i) or (ii) of Conjecture I. The following Conjecture III suggests an approach.

CONJECTURE III. *Under the hypotheses of Lemma 2, one of the following statements is true:*

(i) *there exists a pencil of curves of genus zero L on S without base points or fixed components such that $LD < 0$;*

(ii) *there exists a birational morphism $\varphi\colon S \to S_0$ to a del Pezzo surface S_0 with at most double singularities and with $\operatorname{Pic} S_0 = \mathbb{Z}$, such that the Weil divisor $\varphi(C) + 4K_{S_0}$ is negative.*

Remark 6.   There are only 27 types of surfaces S_0 in case (ii) and it is not difficult to describe them all explicitly. It follows from the condition $\varphi(C) + 4K_{S_0} < 0$ that the genus of C is bounded: we have a maximum for $S_0 = \mathbb{P}^2$ and $C \simeq \varphi(C)$, a curve of degree 11 with $p_a(C) = 45$. To get conditions (i) or (ii) of Conjecture I in this case one apparently has to consider quite a number of particular cases and to deal with the technique of the Clemens-Griffiths intermediate Jacobian.

In the case (i) of Conjecture III we prove the following:

PROPOSITION 1.   *Suppose that condition* (i) *of Conjecture* III *holds, and let $d = LC$. Then $d \leqslant 7$ and:*

(a) *if $5 \leqslant d \leqslant 7$, then there exists a birational automorphism $\tau\colon V \dashrightarrow V$ not preserving the conic bundle structure $\pi\colon V \to S$ and such that for the image $|M'| = |\tau(M)|$ of the linear system $|M|$ we have*

$$|M'| = |\pi^*D' - b'K_V - \Sigma v_j'B_j'| \tag{4}$$

with $b' < b$, where $D' \in \operatorname{Pic} S$;

(b) *if $d = 4$, then there exists another conic bundle structure on V such that its standard form $\pi'\colon V' \to S'$ and the image $|M'|$ on V' of the linear system $|M|$ satisfies* (4) *with $B' < B$.*

It is clear that if $d \leqslant 3$ then condition (i) *of Conjecture* I *is satisfied.*

Proof.   Let L_ξ be the generic fiber of the pencil L over the residue field $k(\xi)$ of the generic point $\xi \in \mathbb{P}^1$; then $F_\xi = \pi^{-1}(L_\xi)$, with $\pi_\xi\colon F_\xi \to L_\xi \simeq \mathbb{P}^1_{k(\xi)}$, is a standard surface with a pencil of rational curves in the sense of [5].

Suppose that $DL = a < 0$, and let $|M_\xi|$ be the restriction of the linear system $|M|$ to F_ξ. Since $L^2 = 0$, $K_{F_\xi} = K_V F_\xi$, we have

$$|M_\xi| = |af_\xi - bK_{F_\xi} - \Sigma v_i x_{\xi i}|,$$

where f_ξ is a fiber of the morphism $\pi_\xi\colon F_\xi \to \mathbb{P}^1_{k(\xi)}$, and $x_{\xi i} = (B_i F_\xi)$ are the restrictions of the B_i to F_ξ. The linear system $|M_\xi|$ is movable and has no fixed components, and by Lemma 2, (ii), $v_i \leqslant b$. Since $a < 0$, $-K_{F_\xi}$ is ample, that is, F_ξ is a del Pezzo surface over $k(\xi)$. Indeed, let $Z \subset F_\xi$ be any curve; if $(-K_{F_\xi} Z) \leqslant 0$, then $(af_\xi - bK_{F_\xi} \cdot Z) < 0$, and Z is a fixed component of the linear system $|M_\xi|$, which is impossible. We also have $0 \leqslant$

$(af_\xi - bK_{F_\xi})^2 = 4ab + b^2 K_{F_\xi}^2$. Hence $K_{F_\xi}^2 > 0$, since $a < 0$. Because $d = CL = 8 - K_{F_\xi}^2$, we get $d \leqslant 7$. Let us remark that by Lemma 2, (iv), the inequality $d \leqslant 7$ immediately follows from condition

$$0 \leqslant XL \leqslant 4bDL - b^2(4K_S + CL) < -b^2(4K_S + CL)$$

$$= b^2(8 - CL).$$

Case $d = 7$. In this case $K_{F_\xi}^2 = 1$ and F_ξ is a del Pezzo surface of degree 1 with a pencil of rational curves. Then the Bertini involution β_ξ acts biregularly on F_ξ and acts on Pic F_ξ by the formulas:

$$\beta_\xi(f_\xi) = -f_\xi - 4K_{F_\xi}, \qquad \beta_\xi(K_{F_\xi}) = K_{F_\xi}$$

(see [6]). Hence,

$$\beta_\xi(|M_\xi|) = |-af_\xi - (b + 4a)K_{F_\xi} - \Sigma \nu_i \beta(x_{\xi i})|.$$

The involution β_ξ extends to a birational involution $\tau \colon V \dashrightarrow V$ preserving the conic bundle structure and taking $|M|$ to $|M'|$, as in (4), with $B' = B + 4a < b$, since $a < 0$.

Case $d = 6$. Here F_ξ is a del Pezzo surface of degree $K_{F_\xi}^2 = 2$, on which the Geiser involution $\gamma_\xi \colon \gamma_\xi(f_\xi) = -f_\xi - 2K_{F_\xi}$, $\gamma_\xi(K_{F_\xi}) = K_{F_\xi}$ acts biregularly (see [6]). It extends to the birational involution $\tau \colon V \dashrightarrow V$ taking $|M|$ to $|M'|$ as in (4) with $B' = B + 2a < B$.

Case $d = 5$. As was shown in [6], §2, in this case $F_\xi \subset \mathbb{P}^3$ is a nonsingular cubic surface and there exists a birational automorphism $\delta_\xi \colon F_\xi \to F_\xi$, which acts on the linear system $|M_\xi|$ and diminishes the coefficient b:

$$\delta_\xi(|M_\xi|) = |a'f_\xi - b'K_{F_\xi} - \Sigma \nu_i' x_{\xi i}'|,$$

where $B' < B$. Again for its birational extension $\tau \colon V \dashrightarrow V$, the linear system $|M'| = \tau(|M|)$ has the smaller value of the coefficient b, namely $b' < b$. Thus (a) is proved.

Case $d = 4$. Here $F_\xi \subset \mathbb{P}^4$ is a del Pezzo surface of degree 4 with the pencil of rational curves $|f_\xi|$. Then (see [7]) there exists another pencil without base points or fixed components $|f_\xi'| = |-f_\xi - K_{F_\xi}|$. It defines the conic bundle $\pi_\xi'\colon F_\xi \to L_\xi' \simeq \mathbb{P}_\xi^1$. We have

$$|M_\xi| = |af_\xi - bK_{F_\xi} - \Sigma \nu_i x_{\xi i}| = |-af_\xi' - (b + a)K_{F_\xi} - \Sigma \nu_i x_{\xi i}|.$$

Here $b + a = b' < b$, that is, the linear system $|M_\xi|$ related to the conic bundle structure $\pi_\xi\colon V \dashrightarrow L'_\xi$ has coefficient $b' < b$. The morphism π_ξ extends to a rational map of V to a rational ruled surface, and, after transforming it to standard form, defines a standard conic bundle $\pi'\colon V' \to S'$. There exists a morphism $S' \to \mathbb{P}^1$, defined by the pencil of rational curves L' and, by construction, V' is birationally equivalent to V over $\mathbb{P}^1$ (both of them admit a surjective morphism to $\mathbb{P}^1$). Observe that if $C' \subset S'$ is the discriminant curve, then $L'C' = 4$, but the image $|M'|$ of the linear system $|M|$ on V' has $-K_V$ with coefficient $b' < b$. Thus (b) is proved, and with it the proposition.

The statement (b) is important because of the following fact.

PROPOSITION 2. *Under the hypotheses of Proposition* 1, (b), *conditions* (i) *or* (ii) *of Conjecture* I *hold for the standard conic bundle* $\pi\colon V \to S$ *if and only if they hold for the standard conic bundle* $\pi'\colon V' \to S'$.

Proof. Let $\psi_L\colon S \to \mathbb{P}^1$ be the morphism defined by the pencil L on S, and let $L_t = \sum_{i=1}^m \ell_i L_i$, $t \in \mathbb{P}^1$ be a singular fiber of the ruled surface $\varphi_L\colon S \to \mathbb{P}^1$, where L_i are irreducible components, with ℓ_i their multiplicities. Since the fiber L_t is singular, among the L_i there is at least one exceptional curve of the first kind, say L_1. Because $LC = 4$ we have $CL_1 \leqslant 4$ and $1 \leqslant \ell_1 \leqslant 4$. Let us show that there exists a birational morphism $\sigma\colon S \to \mathbb{F}_n$ over $\mathbb{P}^1$ to the geometrically ruled surface $\mathbb{F}_n \cdot\!\!- \mathbb{P}(\mathcal{O}_{p^1}(-n) \oplus \mathcal{O}_{p^1})$ for some integer $n \geqslant 0$, such that the curve $\sigma(C)$ has at most double points. The curve $C \subset S$ is a reduced divisor with normal crossings, that is, C has at most simple double points. Furthermore, if $CL_1 \leqslant 2$, then after contracting L_1, the image of the curve C will have at most double points. If $CL_1 = 3$ or 4, then the corresponding multiplicity ℓ_1 is equal to 1, and in this case there is at least one exceptional curve of the first kind, say L_2 among the components L_i, $i \neq 1$ of the degenerate fiber L_t.

Indeed, every component L_i in L_t has transversal intersection at one point with one or two other components, and its multiplicity ℓ_i is equal to the sum of the multiplicities of the adjoining components. Thus L_1 has nonempty intersection with only one component, for example, L_k and its multiplicity ℓ_k also is equal to 1. If $L_i^2 \leqslant -2$ for every $i \neq 1$, then after contracting L_1 there is a unique exceptional curve, the image of L_k. Since for every ruled surface there is a birational morphism to a relatively minimal model obtained by successively contracting exceptional curves of the first kind contained in the fibers, the image of L_k after the contraction of L_1 becomes an exceptional curve of the first kind. It is unique by the assumption $L_i^2 \leqslant -2$, $\forall i \neq 1$. Continuing the process of contraction, we come, after a finite number of steps, to the case of one component in the fiber with self-intersection index equal to 1, which is impossible for a fibration. This is a contradiction.

Consequently there is another exceptional curve L_2 of the first kind in L_t, but now $CL_2 \leqslant C(L_t - L_1) \leqslant 4 - 3 \leqslant 1$ and after contracting L_2, the image of C does not acquire new singularities. Introduction proves our assertion.

Now suppose that condition (i) of Conjecture I is satisfied for $\pi\colon V \to S$, and let

$$Z = \alpha f_n + \beta s_n - \Sigma \xi_i e_i$$

be the class of a pencil of rational curves with base points e_i and without fixed components, where the generators f_n (fiber), and s_n (the rigid section) correspond to the model $\mathbb{F}_n$ of S. If $C = \gamma f_n + 4 s_n - \Sigma \delta_i e_i$, then

$$(ZC) = 4\alpha + \beta\gamma - 4\beta n - \Sigma \delta_i \xi_i \leqslant 3,$$
$$\alpha, \beta, \gamma, \delta_i, \xi_i \in \mathbb{Z}, \qquad \beta > 0, \qquad \delta_i \geqslant 0, \qquad \xi_i \geqslant 0. \tag{5}$$

We use the classical technique of linear systems with assigned base conditions (see [12], Ch. V).

For the canonical class, we have

$$K = -(n + 2)f_n - 2s_n - \Sigma e_i$$

and

$$Z^2 = 2\alpha\beta - \beta^2 n - \Sigma \xi_i^2 = 0$$

$$-ZK = 2\alpha + 2\beta - \beta n - \Sigma \xi_i = 2, \tag{6}$$

since Z is the class of a pencil of curves of genus zero without fixed components. Because $\sigma(C)$ has at most double points on F_n, $\delta_i \leqslant 2$ for all i. Therefore, using $\Sigma \xi_i = 2\alpha + 2\beta - \beta n - 2$ and (5) we get

$$4\alpha + \beta\gamma - 4\beta n - 4\alpha - 4\beta + 2\beta n + 4 \leqslant 3,$$

or

$$\beta(\gamma - 2n - 4) \leqslant -1.$$

Since $\beta \geqslant 1$, we have $\gamma \leqslant 2n + 3$. On the other hand, $\sigma(C)$ is the image of a reduced curve, namely C, so the multiplicity of s_n in $\sigma(C)$ is $\leqslant 1$. Hence, either $\sigma(C)s_n \geqslant 0$ and $\gamma \geqslant 4n$, or $(\delta(C) - s_n)s_n \geqslant 2$ and $\gamma \geqslant 3n + 2$. In the second case, we have used the fact that each component of the discriminant curve intersects the union of the remaining components in an even number $\geqslant 2$ of points.

Thus, $2n + 3 \geqslant \gamma \geqslant 3n + 2$ or $4n$, and so $n = 0$ or 1.

Suppose $n = 0$. We have $\gamma = 2$ or 3. In either case, S has a pencil of curves of genus zero $Z = \sigma^{-1}(S_0)$ without base points or fixed components, with $ZC = 3$ or 2. Let Z_η be the generic fiber of the pencil Z and let $G_\eta = \pi^{-1}(Z_\eta)$. Then G_η is the generic fiber of a pencil of surfaces without base points or fixed components on V, and is itself a surface conic bundle with $K_G^2 = 8 - (Z \cdot C) = 5$ or 6. Since $Z \cdot L = 1$, $G_\eta \cap F_\xi$ is exactly one of the fibers of the morphism π, where $F_\xi = \pi^{-1}(L_\xi)$, and L_ξ is the generic fiber of the pencil L and S.

Then the rational map $T'_\eta: G_\eta \dashrightarrow S'$, induced by the other conic bundle structure $\pi': V \dashrightarrow S'$ has degree two. The linear system of curves of genus zero $|-f_\eta - K_{G_\eta}|$ on G_η, which is invariant under the birational involution of this double covering, is of dimension $K^2_{G_\eta} - 3 = 2$ or 3. Hence, its image $|Y'|$ on S' is a linear system of rational curves without fixed components and $\dim |Y'| = 2$ or 3. Let Y'_μ be the generic member of $|Y'|$, Y^*_μ the normalisation of Y'_μ, $H'_\mu = \pi'^{-1}(Y'_\mu)$, and H^*_μ a standard non-singular surface with a pencil of rational curves $\pi^*_\mu: H^* \to Y^*_\mu$ which is birational to $\pi'_\mu: H'_\mu \to Y'_\mu$ over the residue field $k(\mu)$. Then, for some integer a, G_η cuts out on H^*_μ a curve of genus zero, which varies in the pencil $|af^*_\mu - K_{H^*_\mu}|$. Hence (see [6]), $K^2_{H^*_\mu} \geqslant 4$. Therefore, the standard model $\pi^*: V^* \to S^*$ birational to $\pi': V' \to S'$, and for which the proper transform $|Y^*|$ on S^* of the linear system $|Y'|$ on S' has no base points, satisfies the inequality $Y^*C^* = 8 - K^2_{H^*_\mu} \leqslant 4$. Since $\dim |Y^*| = 2$ or 3, the pencil of curves of genus zero $|Y^* - s|$ obtained by fixing some point $s \in C^* \subset S^*$ satisfies $(Y^* - s)(C^* - s) \leqslant 3$. Hence in this case condition (i) of Conjecture I is satisfied by the standard conic bundle $\pi': V' \to S'$.

Now suppose $n = 1$; then $\gamma = 4$ or 5. If $\gamma = 4$, then the pencil of rational curves $Z = |f_1 + s_1 - e_1|$ obtained by fixing some point $e_1 \in C$ satisfies

$$(f_1 + s_1 - e_1 \cdot 4f_1 + 4s_1 + \Sigma\delta_i e_i) \leqslant 3.$$

Let Z_η be the generic fiber of this pencil and let $G_\eta = \pi^{-1}(Z_\eta)$. Then, repeating the preceding arguments just as in the case $n = 0$, we conclude that the standard conic bundle $\pi': V' \to S'$ satisfies condition (i) of Conjecture I.

If $\gamma = 5$, then two cases are possible:
(a) in the representation $C = 5f_1 + 4s_1 - \Sigma\delta_i e_i$ there is an i such that $\delta_i = 2$;
(b) $\delta_i \leqslant 1, \forall i$.

In case (a) there is a point, for example $e_1 \in C$ with $\delta_1 = 2$. Then for the pencil of rational curves $Z = f_1 + s_1 - e_1$ we have $(f_1 + s_1 - e_1 \cdot 5f_1 + 4s_2 - \Sigma\delta_i e_i) = 3$, so the above arguments work in this case as well.

In case (b) condition (i) of Conjecture I is not satisfied. Indeed, suppose there exists a pencil of rational curves $Z = \alpha f_n - \beta s_n - \Sigma\xi_i e_i$ as above. Then, taking in account that $\delta_i \leqslant 1$, we have the following inequality:

$$4\alpha + \beta\gamma - 4\beta n - 2\alpha - 2\beta + \beta n + 2 \leqslant 3.$$

Substituting $n = 1$ and $\gamma = 5$ we get the inequality $2\alpha \leqslant 1$. Hence $\alpha \leqslant 0$, which is impossible since

$$(\alpha f_1 + \beta s_1 - \Sigma\xi_i e_i)s_1 = \alpha - \beta < 0$$

because $\alpha \leqslant 0$ and $\beta > 0$.

Now assume that for $\pi: V \to S$ condition (ii) of Conjecture I is satisfied. Then the birational map $\tau: S \dashrightarrow P^2$ with $C_0 = \tau(C)$, $\deg C_0 = 5$ takes the pencil L on S with $LC = 4$ to the pencil of lines passing through some point $e_1 \in C_0 \subset P^2$.

Indeed, let $Z = d\ell - \Sigma \xi_i e_i$ be any pencil of rational curves without fixed components such that $(\alpha \ell - \Sigma \xi_i e_i \cdot C_0 - \Sigma \delta_i e_i) = 4$, where ℓ is the class of a line on P^2. Then since $\delta_i \leqslant 1 \; \forall i$, from the system of inequalities

$$(\alpha \ell - \Sigma \xi_i e_i \cdot C_0 - \Sigma e_i) = 5\alpha - \Sigma \xi_i$$

$$= 4 - KZ = 3\alpha - \Sigma \xi_i = 2$$

we get $\alpha = 1$. Hence $\xi_1 = 1$, and $\xi_i = 0$, for all $i \neq 1$, as was to be shown.

Let L_ξ be the generic fiber of the pencil $Z = l - e_1$. Then $F_\xi = \pi_0^{-1}(L_\xi)$ is embedded as a cubic surface in $P^3_{k(\xi)}$ and the fibers of the map $\pi': V' \dashrightarrow S'$ are embedded as conics cut by planes passing through one of the fibers $\pi_0^{-1}(e_1)$. From the explicit construction of the birational map $\chi: V_0 \to P^3$ (see Theorem 1 (ii) and Remark 2 (ii)), it is clear that the image of F_ξ in P^3 is a cubic surface, hence the fibers of π' map to conics in P^3. So Conjecture II is true for the conic bundle $\pi': V' \to S'$. Hence by Theorem 2 conditions (i) or (ii) for this conic bundle are also satisfied.

So we have proved that if conditions (i) or (ii) of Conjecture I are satisfied for $\pi: V \to S$, then they are also satisfied for $\pi: V' \to S'$. Substituting $\pi: V \to S$ for $\pi': V' \to S'$ we get the converse. The Proposition is proved.

Remark 7. In case (b) of the proof of Proposition 2, the birational morphism $\sigma: C \to F_1$ can be induced by a birational equivalence of the standard conic bundle $\pi: V \to S$ with a standard conic bundle $\pi_1: V_1 \to F_1$ having discriminant curve $C_1 = \sigma(C)$.

Proof. Let $L_t = \Sigma \ell_i L_i$ be a singular fiber of the morphism $\varphi_L: S \to P^1$ and let L_1 be an exceptional curve of the first kind. If condition (b) is satisfied, then the three following cases are the only ones possible:

(1) $L_1 C = 0$, $L_1 \not\subset C$;

(2) $L_1 C = 1$, $L_1 \not\subset C$;

(3) $L_1 C = 1$, $L_1 \subset C$ and $L_1(C - L_1) = 2$.

In case (1) $F_1 := \pi^{-1}(L_1)$ is a geometrically ruled surface over $L_1 \simeq P^1$. Let $Y_1 \subset F_1$ be an exceptional section. If $Y_1^2 = 0$ on F_1, then $F_1 \simeq P^1 \times P^1$ and since $L_1^2 = -1$ on S, $Y_1 F_1 = -1$ on V. Hence $F_1 \subset V$ satisfies the condition for the existence of a nonsingular contraction along the pencil $|Y_1|$. After contraction we get a standard conic bundle but with a smaller number of components in the degenerate fiber L_t. If $Y_1^2 = -n$, with $n > 0$ is an integer, then the normal sheaf $N_{Y_1/V}$ is isomorphic to $\mathcal{O}_{P^1}(-n) \oplus \mathcal{O}_{P^1}(-1)$. Hence, blowing up Y_1 and contracting the proper transform of F_1, we get a standard conic bundle $\pi': V' \to S$ with geometrically ruled surface $F_1' = \pi'^{-1}(L_1)$ and exceptional section $Y_1' \subset F_1'$, but with $(Y_1')^2 = -(n - 1)$. After n rearrangement steps of this kind we come to the case considered earlier.

In case (2), $F_1 = \pi^{-1}(L_1)$ is the blow-up of a nonsingular ruled surface at one point. Let $Y_1 \subset F_1$ be a section with minimal self-intersection number $Y_1^2 = -n$

on F_1. Then it is clear that $n \geqslant 1$ and that Y_1 intersects one of the components of the unique degenerate fiber of the ruled surface $F_1 \to L_1$. Let us denote this component by E_1. Suppose first that $n = 1$. Blowing up Y and then the proper transform of the curve E we obtain the same situation as we had when blowing up the fiber $V_s = \pi^{-1}(s)$, $s \in C \setminus \mathrm{Sing}\, C$, followed by resolution of the unique nondegenerate double singularity (see [20], Prop. 2.4). Hence the inverse procedure of contracting the inverse image of the curve E to a simple double point and then both components over L_1 gives a standard conic bundle $\pi': V' \to S'$, where S' is obtained from S by contracting L_1.

If $n \geqslant 2$, note that $\mathcal{N}_{Y_1/V} \simeq \mathcal{O}_{\mathbf{P}^1}(-n) \oplus \mathcal{O}_{\mathbf{P}^1}(-1)$. Blow up Y_1 on V, then the proper transform of E_1. The inverse image of E_1 is the ruled surface $\mathbf{P}^1 \times \mathbf{P}^1$, which can be contracted in the other direction. After the contraction the proper transform F_1 will also satisfy the contractibility condition. After contracting the proper transform F_1 we get a standard conic bundle $\pi': V' \to S'$, for which the blow-up of the ruled surface $F_1' = \pi^{-1}(L_1)$ at one point has exceptional section Y_1' with $(Y_1')^2 = -(n-1)$. Indeed, let F_1'' be the exceptional ruled surface of the blow-up of $Y_1 \subset V$. Then on F_1'' there is a section Y_1'' with $(Y_1'')^2 = -(n-1)$, since $\mathcal{N}_{Y_1/V} \simeq \mathcal{O}_{\mathbf{P}^1}(-n) \oplus \mathcal{O}_{\mathbf{P}^1}(-1)$. There is nothing to prove except that the center of the following blow-up, that is, the proper transform of the curve E_1 does not intersect Y_1''. But since the self-intersection number of the curve cut by the proper transform of F_1 on F_1'' is easy to calculate and turns out to be equal to $(n - 1)$, this curve does not intersect Z_1'', and F_1 lies on F_1. Thus, after $(n - 1)$ rearrangement steps of this kind, we recover the situation of $n = 1$ which we considered before.

In case (3) we obtain the result using a birational rearrangement inverse to that related to the blow-up of the point $s \in \mathrm{Sing}\, C \subset S$ (see [20], Prop. 2.4).

So by the birational rearrangements considered in (1), (2) and (3), we can reduce $\pi: V \to S$ to a standard conic bundle $\pi_1: V_1 \to \mathbf{F}_1$ with discriminant curve $C_1 \subset \mathbf{F}_1$, $C_1 \sim 5f_1 + 4s_1$. This proves Remark 7.

Observe that another birational rearrangement over S_1 can be effected to obtain a standard conic bundle $\pi_0: V_0 \to \mathbf{P}^2$ with discriminant curve $C_0 \subset \mathbf{P}^2$ of degree 5. Indeed, since $C_1 S_1 = 1$ we may use a rearrangement as in case (2) if $S_1 \not\subset C_1$, and as in (3) if $S_1 \subset C_1$.

Remark 8. If Conjecture III is true, then to prove Conjecture II (and hence Conjecture I) one must either give an explicit description of all rational varieties in case (ii) of Conjecture III, or find transformations similar to those of Propositions 1 and 2 to reduce the coefficient b in case (ii) also.

We can make the first step in the proof of Conjecture III using the following result:

LEMMA 3. *Let the hypotheses be as in Lemma 2. Let $Y \subset \bigcup A_i \cup C$ (see Lemma 2, (i)) be an irreducible nonsingular curve such that $Y^2 = -m < 0$ and $D \cdot Y < 0$. Then Y is a curve of genus zero and $m = 1$ or 2.*

Proof. First consider the case in which Y is not a component of the curve C. Then the surface $F = \pi^{-1}(Y)$ is nonsingular (since the intersection of Y with C is transverse by hypothesis) and has $r = C \cdot Y$ degenerate fibers, each isomorphic to $\mathbb{P}^1 \vee \mathbb{P}^1$. The linear system $|M| = |\pi^*D - bK_V - \sum_{i=0}^{N-1} v_i B_i|$ has no fixed components and $M \sim \pi^*D - bK_V$ in Pic V. Therefore the restriction $M_F = (\pi^*D - bK_V)_F$ is effective on F. Choose the curve-section Z on F with minimal self-intersection number $Z^2 = -e$. Let $\sigma\colon F \to F_0$ be the contraction of those components of the degenerate fibers which do not intersect the section Z. Then $F_0 \to Y$ is a geometrically ruled surface with invariant $e = -Z^2$ (see [3], Ch. V, §2) and $M_{F_0} := \sigma_*(M_F)$ is an effective divisor on F_0. We have:

$$M_{F_0} = \sigma_*(\pi^*D - bK_V)|_{F_0} \equiv (D \cdot Y)f - bK_{F_0} + b\sigma_* F^2$$

$$\equiv (D \cdot Y)f + b\big(2Z_0 + (2 - 2g(Y) + e - m)f\big),$$

where f is a fiber, $Z_0 = \sigma_* Z$ a section of the ruled surface $F_0 \to Y$, $g(Y)$ the genus of the curve Y, and $\equiv$ means numerical equivalence. Let us prove that $e \geq 0$. Assume that $e < 0$. Then the effectiveness of M_{F_0} gives the inequality $b(2 - 2g(Y) + e - m) + (D \cdot Y) \geq be$ (see [3], Ch. V, (2.21)). Since $(D \cdot Y) < 0$ and $m \geq 1$ it follows that $g(Y) = 0$. But then $e \geq 0$, which is a contradiction. Therefore $e \geq 0$, and since M_{F_0} is effective, we get the inequalities

$$2 - 2g(Y) + e - m > 0, \qquad e \geq 0. \tag{7}$$

Let $\tau\colon V' \to V$ be the blow up of the curve Z, $E' = \tau^{-1}(Z)$ the exceptional ruled surface over Z, e' its invariant, $Z' \subset E'$ the exceptional section $((Z')^2 = -e)$ and f' a fiber. The restriction $M_{E'}$ of the divisor $\tau^*\pi^*D - b\tau^*K_V - vE'$ to E' is effective, where v is the multiplicity of $|M|$ along Z. We have:

$$M_{F'} \equiv (\tau^*\pi^*D - b\tau^*K_V - vE')E'$$

$$\equiv (D \cdot Y)f' - bK_{E'} + (2b - v)E'^2$$

$$\equiv (D \cdot Y)f' + b\big(2Z' + (2 - 2g(Y) + e')f'\big)$$

$$+ (2b - v)\big(-Z' - \tfrac{1}{2}(e + e' + m)f'\big) \tag{8}$$

Indeed, $E'^3 = -\deg \mathscr{N}_{Z/V} = e + m$ and $E'^2 \equiv -Z' - \tfrac{1}{2}(e + e' + m)f'$.

Let $F' = \tau^*F - E'$ be the proper transform of the surface F and $Z'' = F' \cap E'$. We have: $\tau^*FZ'' = (E' + F')Z'' = (Z'')_{F'}^2 + (Z'')_{E'}^2 = -e + (Z'')_{E'}^2$. Consider separately the following two possible cases: $Z'' = Z'$ on E' and $Z'' \equiv Z' + rf'$, $r \in \mathbb{Z}$.

(a) $Z'' = Z'$. Here $(Z'')^2 = (Z')^2 = -e$. Taking into account the previous calculations we get $e + e' = m$. Hence

$$M_{E'} \equiv (D \cdot Y)f' + vZ' + b(2 - 2g(Y) + e' - 2m)f' + vmf'.$$

Since the divisor $M_{E'}$ is effective and $v \leqslant b$ (by Lemma 2 (ii)) the following inequality must hold: in the case $e' \geqslant 0$

$$2 - 2g(Y) + e' - m > 0 \qquad (9)$$

Using the fact that $e + e' = m$ and inequalities (7) and (9) we immediately get $g(Y) = 0$ and $m = 1$ or 2, as was to be proved. The case $e' < 0$ is impossible. Indeed, assume that $e' < 0$. Then the effectiveness of $M_{E'}$ gives the following inequality: $b(2 - 2g(Y) + e' - m) + (D \cdot Y) \geqslant \frac{1}{2}ve'$. Hence $b(2 - 2g(Y) - m) > (\frac{1}{2}v - b)e' > 0$. So again $g(Y) = 0$, which is impossible when $e' < 0$.

(b) $Z'' \equiv Z' + rf'$. In this case if $e' \geqslant 0$, then $r \geqslant e'$ and $e - m = (Z'')^2_{E'} \geqslant e'$ (see [3], Ch. V, (2.21)). As before, the following inequality must be true:

$$0 < 2 - 2g(Y) + \frac{e'}{2} - \frac{e}{2} - \frac{m}{2} \leqslant 2 - 2g(Y) - m.$$

Hence $g(Y) = 0$ and $m = 1$. The case $e' < 0$ is impossible. Indeed, assume that $e' < 0$. Then $r \geqslant 0$ (see [3], Ch. V, (2.21)), $e - m = (Z'')^2_{E'} \geqslant -e'$, so $e + e' \geqslant m$, and the following inequality must be true:

$$b\left(2 - 2g(Y) + \frac{e'}{2} - \frac{e}{2} - \frac{m}{2}\right) + (D \cdot Y) \geqslant \tfrac{1}{2}vc'.$$

Hence $2 - 2g(Y) - e/2 - m/2 > 0$ and again $g(Y) = 0$, which contradicts $e' < 0$.

Now consider the case when $Y = C_i$ is an irreducible component of the discriminant curve $C \subset S$. Assume $F = \pi^{-1}(Y)$; then F has double singular points along some exceptional curve $Z \simeq Y$, and has no other singularities. Let $\tau: V' \to V$ be the blow-up of Z, $E' = \tau^{-1}(Z)$ be a ruled surface over Z, e' the invariant of E', Z' the exceptional section of $E' \to Z$ with $(Z')^2_{E'} = -e'$, and f' a fiber. Then $\tau^* F = F' + 2E'$, where F' is a geometrically ruled surface over a curve which is isomorphic to $Z'' = F' \cap E'$ and covers Z with degree two. Let e be the invariant of this ruled surface. We shall show that the curve Z'' is its exceptional section. We have

$$-2m = \tau^* F Z'' = (F' + 2E')F'E' = (Z'')^2_{E'} + 2(Z'')^2_{F'}.$$

Since Z'' is an irreducible curve of the form $2Z' + rf$ on E', $(Z'')^2_{E'} \geqslant 0$. Hence, $2(Z'')^2_{F'} = -2m - (Z'')^2_{E'} < 0$, and so Z'' is the exceptional section on F' with the invariant $e = -(Z'')^2_{F'} > 0$. Let $n = (Z'')^2_{E'}$; then $n + 2m = 2e$ and $n = (2Z' + rf')^2_{E'} = -4e + 4r$.

The restriction $M_{E'}$ of the divisor $\tau^*\pi^* D - b\tau^* K_V - vE'$ to E' must be effective. (Here v is the multiplicity of the linear system $|M|$ along Z and $v \leqslant b$.)

We have

$$M_{E'} \equiv (D \cdot Y)f' + b(2Z' + (2 - 2g(Y) + e')f') + (2b - v)E'^2$$

Here $E'^2 = -Z' - \alpha f'$, where α can be found from the following equalities:

$$m = \tau^* F E'^2 = (F' + 2E')E'^2 = -e + 2E'^3$$

$$E'^3 = (-Z' - \alpha f')^2_{E'} = -e' + 2\alpha.$$

Thus

$$\alpha = (1/4)(m + e + e'),$$

and hence

$$M_{E'} \equiv (D \cdot Y)f' + vZ' + b\left(2 - 2g(Y) - \frac{e}{2} - \frac{m}{2}\right)f' + v\frac{m + e + 2e'}{4}f'.$$

If $e' \geqslant 0$, then the following inequality must hold:

$$2 - 2g(Y) + \frac{e'}{2} - \frac{e}{4} - \frac{m}{4} > 0 \tag{10}$$

In this case we have $Z'' \equiv 2Z' + rf'$ with $r' \geqslant 2e'$ and $-2m + 2e = n = -4e' + 4r \geqslant 4e'$. Hence $e' \leqslant (e - m)/2$. Using this fact in (10) we get the inequality $2 - 2g(Y) - m/2 > 0$. Consequently $g(Y) = 0$ and $m = 1$ or 2. The case $e' < 0$ is impossible. Indeed, assume that $e' < 0$, then the following inequality must hold:

$$b\left(2 - 2g(Y) - \frac{e}{2} - \frac{m}{2}\right) + v\frac{m + e + 2e'}{4} + (D \cdot Y) \geqslant \tfrac{1}{2}ve'$$

or

$$b\left(2 - 2g(Y) - \frac{e}{2} - \frac{m}{2}\right) + \frac{v(m + e)}{4} > 0.$$

Thus we obtain the inequality $2 - 2g(Y) - e/4 - m/4 > 0$, and hence $g(Y) = 0$, which contradicts the assumption $e' < 0$. The Lemma is now proved.

Remark 9. One can use the lemma to prove Conjecture III in the following way. By Lemma 2 (iv) there is a curve $Y \subset S$ with $D \cdot Y < 0$. We have

$$0 \leqslant X \cdot Y = 4bDY - b^2(4K_S + C)Y - \Sigma v_i^2(\pi \circ \delta_{i,0})_* B_i Y. \tag{11}$$

Hence, if $K_S Y \geqslant 0$, then Y is a component of the divisor $C + \Sigma A_i$ (see Lemma 2, (i)) and $Y^2 < 0$. Then, by Lemma 3, we have $p(Y) = 0$, $Y^2 = -1$ or -2, and so only one possible case remains, that is, $K_S Y = 0$ and $Y^2 = -2$. It is easy to show that there is either an extremal curve Z on S (in Mori's terminology [18]) with $DZ < 0$, or a curve Z with $p_a(Z) = 0$ and $Z^2 = -2$.

Using the classification of external curves on a surface (see [18]) we get:

(1) Z is an exceptional curve of the first kind;

(2) Z is a fiber of a geometrically ruled surface;

(3) Z is a line in $\mathbb{P}^2$.

In cases (2) and (3) S is a geometrically ruled surface or a plane, respectively, and so in these cases Conjecture III is true (as was remarked before, in these cases the criterion of rationality is also proved [22]).

In case (1) let $\sigma\colon S \to S'$ be the contraction of the curve Z, and let $X' = \sigma_* X$, $D' = \sigma_* D$, $C' = \sigma_* C$. Then the cycle X' as well as X is numerically effective for the same geometrical reasons, and the equality (3) can be rewritten as follows:

$$X' = 4bD' - b(4K_{S'} + C') - \Sigma v_i^2 \sigma_*(\pi \circ \sigma_{i,0})_* B_i.$$

We have $D = \sigma^* D' + aZ$, where $a = -DZ > 0$ and so $h^0(S', \mathcal{O}_{S'}(mD')) = 0$ for all $m \geqslant 1$. Using this fact and (12) with the Riemann-Roch theorem on the rational surface S', we see that D' as well as D is not numerically effective. To continue the process of contraction we need the analogue of Lemma 3 for the surface S'' (which we do not have). Nevertheless some remarks can be made.

Let Y' be a curve on S' with $D'Y' < 0$ and $Y'^2 = m' < 0$, and let Y be its proper transform on S. Let $r = YZ$; then $DY = (\sigma^* D' + aZ)Y < ar$ and $Y^2 = Y'^2 - r^2 = m' - r^2$. Since $(D'Y') < 0$ and $Y'^2 < 0$, either (1) $Y'K_{S'} \geqslant 0$, and then Y' is the exceptional curve of the first kind, hence the contraction may be continued, or (2) $Y'K_{S'} \geqslant 0$, and then Y is a component of $C + \Sigma A_i$ and hence the curve Y is non-singular. Repeating the arguments of Lemma 3 and using the inequality $DY < ar$ we get $g(Y) = 0$, $r = 1$ and $m' = -1$ or -2. Thus Y' is a smooth curve of genus zero and $Y'^2 = -1$ or -2. If $Y'^2 = -1$, then the contraction can be continued.

In the case of a curve $Z \subset S$ with $g(Z) = 0$, $Z^2 = -2$ and $DZ < 0$ there also exists a birational contraction $\sigma\colon S \to S'$ of Z, but now to a singular (double) point on S'. The previous arguments involving the attempt to find the analogue of Lemma 3 in this case fail.

REFERENCES

1. M. ARTIN, AND D. MUMFORD, *Some elementary examples of unirational varieties which are not rational*, Proc. London Math. Soc. **25** (1972), 79–95.

2. A. BEAUVILLE, *Variétés de Prym et jacobiennes intermédiaires*, Ann. Sci. École Norm. Sup. **10** (1977), 309–391.

3. R. HARTSHORNE, *Algebraic geometry*. Graduate Texts in Math. **52**, Springer-Verlag, New York, 1977.

4. V. A. Iskovskikh, *Congruences of conics in* $\mathbb{P}^3$, Vestnik Mosk. Univ., Matematika, **37**, No. **6** (1982), 57–62 = Moscow Univ. Math. Bulletin **37**, No. **6** (1982), 67–73.

5. ______, *Minimal models of rational surfaces over arbitrary fields*, Izv. Akad. Nauk SSSR **43** (1979), 19–43 = Math. USSR-Izv. **14** (1980), 17–39.

6. ______, *Rational surfaces with a pencil of rational curves and with positive square of the canonical class*, Mat. Sb. **83** (1970) = Math. USSR-Sb. **12** (1970), 91–117.

7. ______, *Birational properties of surfaces of degree* 4 *in* $\mathbb{P}^4_k$, Mat. Sb. **88** (1972), 31–37 = Math. USSR Sb. **17** (1972), 30–36.

8. ______, *Birational automorphisms of three-dimensional algebraic varieties*, Itogi Nauki: Sovremennye Problemy Mat. (Current problems of math) Vol. **12** VINITI, Moscow, 1979, 159–236 = J. Soviet Math. **13** (1980), 815–868.

9. V. A. Iskovskikh and Yu. I. Manin, *Three-dimensional quartics and counterexamples to the Lüroth problem*, Mat. Sb. **86** (1971), 140–166 = Math. USSR-Sb. **15** (1971), 141–166.

10. S. Kantor, *Die Typen der linearen Complexe rationaler Curven in* R_r, Amer. J. of Math. **23** (1901), 1–28.

11. Y. Kawamata, K. Matsuda, and K. Matsuki, *Introduction to the Minimal Model Problem.* preprint, 1986, 142 pp. in Proc. of Sendai Conference on algebraic geometry (1985), to appear in Advanced Studies in Pure Math.

12. Yu. I. Manin, *Cubic forms*, Nauka, Moscow, 1972 = *Cubic forms*; North Holland, 1974.

13. L. Masiewicki, *Prym varieties and the moduli space of curves of genus five*, Trans. Amer. Math. Soc. **222** (1976), 221–240.

14. A. S. Merkur'ev. *On the norm residue symbol of degree* 2, Dokl. Akad. Nauk. SSSR **261**, 3 (1981), 542–547 = Soviet Math-Doklady **24** (1981), 546–551.

15. T. Millevoi. *Sulla classificazione Cremoniana delle congruenze de curve razionali dello spazio.* Rend. Sem. Mat. Univ. Padova **30** (1960), **1**, 194–214.

16. M. Miyanishi, "Algebraic methods in the theory of algebraic threefolds," in *Algebraic Varieties and Analytic Varieties*, S. Iitaka ed., Advanced Studies in Pure Math. **1** (1983) Tokyo-Amsterdam, 69–99.

17. Y. Miyaoka, *On the Kodaira dimension of minimal threefolds*, preprint, 1986, to appear in Math. Ann.

18. S. Mori, *Threefolds whose canonical bundles are not numerically effective*, Ann. of Math. **116** (1982), 133–176.

19. I. A. Panin, "Rationality of conic bundles with the curve of degeneracy of degree 5 and even θ-characteristics," in *Modules and Linear Groups*, Sem. LOMI, Leningrad, **103** (1980), 100–105.

20. V. G. Sarkisov, *Birational automorphisms of conical fibrations*, Izv. Akad. Nauk SSSR **44** (1980), 918–945 = Math. USSR-Izv. **17** (1981).

21. ______, *On the structure of conic bundles*, Izv. Akad. Nauk SSSR **46** (1982), 371–408 = Math. USSR-Izv. **20** (1982) **2**, 355–390.

22. V. V. Shokurov, *Prym varieties: theory and applications*, Izv. Akad. Nauk SSSR **47** (1983), **4**, 785–855 = Math. USSR-Izv. **23** (1984), 83–147.

23. A. N. Tyurin, *On intersection of quadrics*, Uspekhi Mat. Nauk **30** (1975), 6, 51–99 = Russian Math. Surveys **30**:6 (1975), 51–105.

24. A. A. Zagorskii, *Three-dimensional conic bundles*, Mat. Zametki **21** (1977), 745–758 = Math. Notes Acad. Sci. USSR **21** (1977), 420–427.

Algebra Section, Department of Mathematics and Mechanics, Moscow State University, Moscow 117234, USSR

DE RHAM COHOMOLOGY AND CONDUCTORS
OF CURVES

SPENCER BLOCH

An *arithmetic surface* $\mathscr{X}$ is a regular scheme $\mathscr{X}$ which is flat and proper over Spec(Z) with fibre dimension 1. The zeta function of $\mathscr{X}$, $\zeta(\mathscr{X}, s)$, is defined by

$$\zeta(\mathscr{X}, s) = \prod_{\substack{x \in \mathscr{X} \\ \text{clsd. pt.}}} \left(1 - N(x)^{-s}\right)^{-1}$$

where $N(x)$ denotes the number of elements in the residue field at the closed point x [7]. One conjectures that $\zeta(\mathscr{X}, s)$, which is known to be analytic in a half plane, extends meromorphically to $\mathbf{C}$ and satisfies a functional equation. More precisely, one is *given* in addition to $\zeta(\mathscr{X}, s)$ a "Γ-factor" $\Gamma(\mathscr{X}, s)$ and a positive rational number A, *the conductor*, and one conjectures [6] that

$$\xi(\mathscr{X}, s) = A^{s/2}\zeta(\mathscr{X}, s)\Gamma(\mathscr{X}, s)$$

admits a meromorphic extension and satisfies

$$\xi(\mathscr{X}, s) = \pm\xi(\mathscr{X}, 2 - s).$$

The purpose of this note is to relate the conductor $A = A(\mathscr{X})$ to the complex of absolute de Rham differentials

$$\Omega_{\mathscr{X}}^{\cdot} = \Omega_{\mathscr{X}/\mathbf{Z}}^{\cdot} = \left\{ \mathcal{O}_{\mathscr{X}} \to \Omega_{\mathscr{X}/\mathbf{Z}}^{1} \to \Omega_{\mathscr{X}/\mathbf{Z}}^{2} \right\}.$$

Let $C^{\cdot}$ be a bounded complex of abelian sheaves on a scheme $\mathscr{X}$ proper over Spec(Z). Assume the C^i are Z-torsion and coherent sheaves. It follows that the hypercohomology $H^i(\mathscr{X}, C^{\cdot})$ are finite groups. We write

$$X(C^{\cdot}) = \prod_{i}\left(\#H^i(\mathscr{X}, C^{\cdot})\right)^{(-1)^i}$$

for the (multiplicative) Euler characteristic. Our first result was conjectured by K. Kato and proved by him in the equicharacteristic case for $\mathscr{X}$ of arbitrary dimension [4].

THEOREM 1. *Let $\mathscr{X}$ be an arithmetic surface, and let $\Omega_{\mathscr{X}, \text{tors}}^{\cdot} \subset \Omega_{\mathscr{X}}^{\cdot}$ be the subcomplex of Z-torsion differentials. Then the conductor $A(\mathscr{X}) = X(\Omega_{\mathscr{X}, \text{tors}}^{\cdot})$.*

Received October 31, 1986. Partially supported by the NSF.

As a consequence of this, it is also possible to interpret $A(\mathscr{X})$ in terms of a de Rham discriminant. For $\mathscr{X}$ an arithmetic surface, define

$$\Omega^{\cdot}_{\mathscr{X},\,\leqslant 1} = \mathcal{O}_{\mathscr{X}} \to \Omega^1_{\mathscr{X}/\mathbf{Z}}.$$

Poincaré duality gives a mapping in the derived category

$$\psi\colon R\Gamma(\mathscr{X},\Omega^{\cdot}_{\mathscr{X},\,\leqslant 1}) \to R\operatorname{Hom}(R\Gamma(\mathscr{X},\Omega^{\cdot}_{\mathscr{X},\,\leqslant 1}),\mathbf{Z})[-2]$$

which is an isomorphism when tensored with $\mathbf{Q}$. We define the *de Rham discriminant*

$$\operatorname{Disc}(H_{\mathrm{DR}}(\mathscr{X})) = \chi(\text{mapping cone}(\psi)).$$

THEOREM 2. $A(\mathscr{X}) = \operatorname{Disc}(H_{\mathrm{DR}}(\mathscr{X}))^{1/2}$.

Both the above results are corollaries of the basic theory developed in [1]. Hopefully this reference, which is currently in press, will have appeared before the present work. I should also like to acknowledge the influence of correspondence and conversation with K. Kato.

Finally, I am pleased and proud to be able to dedicate this paper to Y. Manin, whose work in arithmetic algebraic geometry has been a very great inspiration to me.

§1. The shape of the conductor. In this section I recall the conjectural functional equation for the Hasse-Weil L-function of a curve, as formulated by Serre in [6]. I then work out the minor modification in the conductor needed in order to pass from the Hasse-Weil L-function to the ζ-function associated to a particular global model of the curve (cf. [7]).

Let X be a smooth projective variety defined over a number field k. Let $m \geqslant 0$ be a fixed integer. The Hasse-Weil L-function $L(H^m(X),s)$ is the function (known to be analytic in a half-plane) defined as an indefinite product of local factors

$$L(H^m(X),s) = \prod_{\mathfrak{p}} L_{\mathfrak{p}}(H^m(X),s)$$

where $\mathfrak{p}$ runs through all non-archimedean primes of k.

Given $\mathfrak{p}$, let $\bar{\mathfrak{p}}$ be a prime of $\overline{\mathbf{Q}}$ lying over k. Let $G_{\bar{\mathfrak{p}}} \subset \operatorname{Gal}(\overline{\mathbf{Q}}/k)$ be the decomposition group at $\bar{\mathfrak{p}}$, and let $I_{\bar{\mathfrak{p}}} \subset G_{\bar{\mathfrak{p}}}$ be the inertia group. The quotient $G_{\bar{\mathfrak{p}}}/I_{\bar{\mathfrak{p}}} \cong \hat{\mathbf{Z}}$. We write $f_{\mathfrak{p}} \in \hat{\mathbf{Z}}$ for the geometric frobenius, which acts on $\overline{\mathsf{F}}_p$ by $x \to x^{1/N\mathfrak{p}}$, with $N\mathfrak{p} = $ number of elements in the residue field $\mathsf{F}(\mathfrak{p})$. The group $G_{\bar{\mathfrak{p}}}$ acts on the étale cohomology $H^m_{\text{ét}}(X_{\bar{k}},\mathbf{Q}_l)$, so there is an action of $\hat{\mathbf{Z}}$ on the inertia invariants $H^m_{\text{ét}}(X_{\bar{k}},\mathbf{Q}_l)^I$. By definition

$$L_{\mathfrak{p}}(H^m(X),s) = \det\left(1 - f_{\mathfrak{p}} \cdot N\mathfrak{p}^{-s} | H^m_{\text{ét}}(X_{\bar{k}},\mathbf{Q}_l)^I\right)^{-1}.$$

At each infinite (archimedean) place v of k there is defined a local Γ-factor $\Gamma_v(H^m(X), s)$. Write $\Gamma_\infty(H^m(X), s) = \Pi\Gamma_v(H^m(X), s)$. We have nothing constructive to say about these Γ-factors, and we simply refer to [6] for their definition.

Of more importance to us will be the *conductor* as defined by Serre in [op. cit.]. Let $K = \hat{k}_{\mathfrak{p}, n_r}$ be the completion of the maximal unramified extension of $k_\mathfrak{p}$, so $I = I_{\bar{\mathfrak{p}}} = \mathrm{gal}(\bar{K}/K)$. Let K' be a finite galois extension of K, $\Phi = I/I' = \mathrm{gal}(K'/K)$. Let $t \in K'$ be a uniformizing parameter, and let $v \colon K'^x \to \mathbf{Z}$ be the valuation, so $v(t) = 1$. The Swan character $\mathrm{sw}_{K'/K} \colon \Phi \to \mathbf{Z}$ is defined by

$$\mathrm{sw}_{K'/K}(g) = 1 - v(g(t) - t), \quad \text{if } g \neq e$$

$$\mathrm{sw}_{K'/K}(e) = - \sum_{g \neq e} \mathrm{sw}_{K'/K}(g).$$

The Swan character is supported on the wild inertia group $\Phi^{\mathrm{wild}} \subset \Phi$. Since the action of the pro-p group $I^{\mathrm{wild}} \subset I_{\bar{\mathfrak{p}}}$ on the l-adic vector space $H^m_{\text{ét}}(X_{\bar{k}}, \mathbf{Q}_l)$ factors through a finite quotient, one can define for K'/K sufficiently large,

$$\mathrm{sw}_{K'/K}\big(H^m_{\text{ét}}(X_{\bar{k}}, \mathbf{Q}_l)\big) = \dim \mathrm{Hom}_\Phi\big(\mathrm{sw}_{K'/K}, H^m_{\text{ét}}(X_{\bar{k}}, \mathbf{Q}_l)\big).$$

The Swan character is known to be the character of a projective $\mathbf{Z}_l[\Phi]$-module $SW(K'/K)$. If K''/K is galois with group Ψ such that $K' \subset K''$, one has

$$SW(K'/K) \cong SW(K''/K) \bigotimes_{\mathbf{Z}_l[\Psi]} \mathbf{Z}_l[\Phi].$$

It follows that $\mathrm{sw}_{K'/K}(H^m_{\text{ét}}(X_{\bar{k}}, \mathbf{Q}_l))$ is independent of the choice of K'. We write $\mathrm{sw}_\mathfrak{p}(H^m_{\text{ét}}(X_{\bar{k}}, \mathbf{Q}_l))$ for this number. The conductor at $\mathfrak{p}$ is by definition

$$f_\mathfrak{p}(H^m(X)) = \mathrm{sw}_\mathfrak{p}\big(H^m_{\text{ét}}(X_{\bar{k}}, \mathbf{Q}_l)\big) + \dim H^m_{\text{ét}}(X_{\bar{k}}, \mathbf{Q}_l) - \dim H^m_{\text{ét}}(X_{\bar{k}}, \mathbf{Q}_l)^I.$$

Note $f_\mathfrak{p} = 0$ for almost all $\mathfrak{p}$, so we many define the global conductor $f(H^m(X)) = \Sigma f_\mathfrak{p} \cdot \mathfrak{p}$. Finally, let $D = |d_{k/\mathbf{Q}}|$ be the absolute value of the field discriminant and let $B_m = \dim H^m_{\text{ét}}(X_{\bar{k}}, \mathbf{Q}_l)$. Define

$$A(H^m(X)) = D^{B_m} \cdot \prod_\mathfrak{p} N\mathfrak{p}^{f_\mathfrak{p}}.$$

Write

$$\xi(s) = A^{s/2} L(s) \Gamma_\infty(s).$$

Then the conjectured functional equation is

$$\xi(s) = \pm \xi(m + 1 - s).$$

Instead of selecting a particular m, one can work with the *Hasse-Weil ζ-function* of X, $\zeta_{\mathrm{HW}}(X, s)$ defined by

$$\zeta_{\mathrm{HW}}(X, s) = \prod_m L(H^m(X), s)^{(-1)^{m-1}}.$$

(I'm not certain anyone has used this terminology, although it seems natural.) The corresponding conductor is

$$A_{\mathrm{HW}}(X) = D^{\chi(X)} \cdot \prod_{\mathfrak{p}} N\mathfrak{p}^{f_{\mathfrak{p}}(X)}$$

where

$$f_{\mathfrak{p}}(X) = \sum (-1) m_{f_{\mathfrak{p}}}(H^m(X)) = \mathrm{sw}_{\mathfrak{p}}(\chi(X)) + \chi(X) - \chi(X)^{I_{\mathfrak{p}}}.$$

(We use the notation $\chi(X)$ interchangeably for the virtual galois representation $\Sigma(-1)^m H^m(X_{\bar{k}}, \mathbf{Q}_l)$ and for its dimension.) The Γ-factors $\Gamma_{\mathrm{HW}}(X, s)$ for $\zeta_{\mathrm{HW}}(X, s)$ are the alternating products of the Γ-factors for the $L(H^m(X), s)$, and the conjectured functional equation reads

$$\xi_{\mathrm{HW}}(X, s) = A(X)^{s/2} \cdot \zeta_{\mathrm{HW}}(X, s) \cdot \Gamma_{\mathrm{HW}}(X, s)$$

(*)

$$\xi_{\mathrm{HW}}(X, s) = \pm \xi_{\mathrm{HW}}(X, \dim X + 1 - s).$$

Suppose now we are given a global regular model $\mathscr{X}$ for X. By definition, $\mathscr{X}$ is a regular scheme flat and proper over $\mathrm{Spec}(\mathcal{O}_k)$ with generic fibre X. The ζ-function of $\mathscr{X}$, $\zeta(\mathscr{X}, s)$ is defined [7] by

$$\zeta(\mathscr{X}, s) = \prod_x (1 - Nx^{-s})^{-1}$$

where x runs through all closed points of $\mathscr{X}$, and Nx is the number of elements in the residue field at x. We have

$$\zeta(\mathscr{X}, s) = \prod_{\mathfrak{p}} \zeta(\mathscr{X}(\mathfrak{p}), s)^{-1}$$

where by means of the Weil conjectures, we can write

$$\zeta(\mathscr{X}(\mathfrak{p}), s) = \prod_m \det\left(1 - f_{\mathfrak{p}} \cdot N\mathfrak{p}^{-s} \mid H^m_{\text{ét}}(\mathscr{X}(\mathfrak{p}), \mathbf{Q}_l)\right)^{(-1)^{m+1}}.$$

Since at a good prime $\mathfrak{p}$, i.e. a prime over which $\mathscr{X}$ is smooth, we have $H^m_{\text{ét}}(\mathscr{X}(\mathfrak{p}), \mathbf{Q}_l) \cong H^m_{\text{ét}}(X_{\bar{k}}, \mathbf{Q}_l)^I = H^m_{\text{ét}}(X_{\bar{k}}, \mathbf{Q}_l)$, it follows that $\zeta(\mathscr{X}, s)$ and $\zeta_{\mathrm{HW}}(X, s)$ differ only by factors corresponding to bad primes. As the notation suggests, $\zeta(\mathscr{X}, s)$ depends on the *choice* of a global model $\mathscr{X}$, while $\zeta_{\mathrm{HW}}(X, s)$ depends only on X.

PROPOSITION (1.1). *Assume now that X is a curve which is geometrically connected over k, and that $\zeta_{\mathrm{HW}}(X, s)$ satisfies the function equation* (*). *Then $\zeta(\mathscr{X}, s)$ satisfies a functional equation of the same sort as* (*), *except that the conductor $f_{\mathfrak{p}}(X)$ is replaced by*

$$f_{\mathfrak{p}}(\mathscr{X}) = \mathrm{sw}_{\mathfrak{p}}(\chi(X)) + \chi(X) - \chi(\mathscr{X}(\mathfrak{p})).$$

Proof. We need to compare the virtual representations $\chi(\mathscr{X}(\mathfrak{p}))$ and $\chi(X)^I$ with $I = I_{\mathfrak{p}}$ the intertia group at $\mathfrak{p}$. To simplify notation, let $Y = \mathscr{X}(\mathfrak{p}) \times \bar{\mathsf{F}}(\mathfrak{p})$.

LEMMA (1.2). (i) $H^m_{\text{ét}}(X_{\bar{k}}, \mathsf{Q}_l)^I \cong H^m_{\text{ét}}(Y, \mathsf{Q}_l)$ *for $m = 0, 1$.*

(ii) *Let $M_{\mathfrak{p}}$ be the free abelian group spanned by the irreducible components of the geometric fibre Y. Since the individual components are not necessarily defined over $\mathsf{F}(\mathfrak{p})$, there is an action of $\hat{\mathsf{Z}}$ on $M_{\mathfrak{p}}$. Moreover, there is an exact sequence of $\hat{\mathsf{Z}}$-modules.*

$$0 \to \mathsf{Q}_l(-1) \to M_{\mathfrak{p}} \otimes \mathsf{Q}_l(-1) \to H^2_{\text{ét}}(Y, \mathsf{Q}_l) \to H^2_{\text{ét}}(X_{\bar{k}}, \mathsf{Q}_l) \to 0.$$

Proof of lemma. (Compare [8], p. 214). To simplify we write $H^m(\cdot)$ instead of $H^m_{\text{ét}}(\cdot, \mathsf{Q}_l)$. Also, let K be the completion of the maximal unramified extension of the local field $k_{\mathfrak{p}}$, and $\eta = \mathrm{Spec}(K)$. Let $\mathcal{O}_K$ be the ring of integers in K, and write $\mathscr{X}_{\mathfrak{p}} = \mathscr{X} \times \mathrm{Spec}(\mathcal{O}_K)$. There is a surjection $I = I_{\mathfrak{p}} \to \mathsf{Z}_l(1)$ with kernel profinite of (profinite) order prime to l, so $H^n(I, H^m_{\text{ét}}(\mathscr{X}_{\bar{\eta}})) = (0)$ for $n \geqslant 2$ and there are exact sequences

$$(1) \qquad 0 \to H^{m-1}(\mathscr{X}_{\bar{\eta}})_I \to H^m(\mathscr{X}_{\eta}) \to H^m(\mathscr{X}_{\bar{\eta}})^I \to 0$$

for all m. (The subscript I means coinvariants.) In addition there are localization sequences

$$\cdots \to H^m_Y(\mathscr{X}_{\mathfrak{p}}) \to H^m(\mathscr{X}_{\mathfrak{p}}) \to H^m(\mathscr{X}_{\eta}) \to H^{m+1}_Y(\mathscr{X}_{\mathfrak{p}}) \to \cdots.$$

One has isomorphisms

$$H^m(\mathscr{X}_{\mathfrak{p}}) \cong H^m(Y) \quad \text{(base change)}$$

$$H^m_Y(\mathscr{X}_{\mathfrak{p}}) \cong H^{4-m}(Y)^*(-2) \quad \text{(purity)}.$$

(In general, one does not know whether purity holds in this sort of mixed characteristic situation. However, in the case of codimension 1, one can apply [8]. Comparing the cohomology of Y and its normalization $\tilde{Y}$, one finds that the weights of the action of $\hat{Z}$ on $H^1(Y)$ are $\leqslant 1$, and

$$H^2(Y) \cong M_{\mathfrak{p}}^*(-1).$$

A piece of the localization sequence can be rewritten

$$(2) \qquad\qquad 0 \to H^1(Y) \to H^1(\mathscr{X}_\eta) \to M_{\mathfrak{p}}(-1) \underset{a}{\to} H^2(Y).$$

The map a is the twist by $\mathbf{Q}_l(-1)$ of the map associating to a divisor on $\mathscr{X}_{\mathfrak{p}}$ supported on Y its pullback class in $H^2(Y, \mathbf{Q}_l(1))$, and the kernel of a is one dimensional, generated by $[Y](-1)$ (Deligne, "Intersections sur les surfaces régulières", cor. 1.8). Applying (1) with $m = 1$, one finds

$$0 \to \mathbf{Q}_l(-1) \to H^1(\mathscr{X}_\eta) \to H^1(\mathscr{X}_{\bar{\eta}})^I \to 0.$$

Since weight$(\mathbf{Q}_l(-1)) = 2$, it follows from this sequence and (2) above that $H^1(Y) \cong H^1(\mathscr{X}_{\bar{\eta}})^I$. The analogous assertion for $m = 0$ is a consequence of Zariski connectedness. Finally, (ii) follows from (2) and the description of Ker(a), proving the lemma.

We now return to the proof of the proposition. It follows from the lemma that

$$\zeta(\mathscr{X}, s)/\zeta_{\mathrm{HW}}(X, s) = \prod_{\mathfrak{p}\ \mathrm{bad}} \det\left(1 - f_{\mathfrak{p}} N\mathfrak{p}^{-s} | M_{\mathfrak{p}}(-1)/\mathbf{Q}_l(-1)\right)^{-1}.$$

Write $G_{\mathfrak{p}}(s)$ for a factor of the right hand side, and let

$$F_{\mathfrak{p}}(s) = N\mathfrak{p}^{s \cdot (f_{\mathfrak{p}}(\mathscr{X}) - f_{\mathfrak{p}}(X))/2} \cdot G_{\mathfrak{p}}(s).$$

We have

$$\xi(\mathscr{X}, s)/\xi(X, s) = \prod F_{\mathfrak{p}}(s),$$

so it will suffice to show $F_{\mathfrak{p}}(s) = \pm F_{\mathfrak{p}}(2 - s)$. Note that $f_{\mathfrak{p}}(\mathscr{X}) - f_{\mathfrak{p}}(X) = 1 - \dim M_{\mathfrak{p}}$. If $M \cong \mathbf{Z}^{\oplus n}$ is a module on which $\hat{Z}$ acts by cyclic permutation of a basis, then $\det(1 - f_{\mathfrak{p}} N\mathfrak{p}^{-s} | M(-1)) = 1 - N\mathfrak{p}^{n(1-s)}$. Since

$$\left(N\mathfrak{p}^{-ns/2}\right) \cdot \left(1 - N\mathfrak{p}^{n(1-s)}\right)^{-1} = -\left(N\mathfrak{p}^{-n(2-s)/2}\right) \cdot \left(1 - N\mathfrak{p}^{n(s-1)}\right)^{-1},$$

the assertion follows easily after breaking up the $\hat{Z}$-action on the irreducible components of Y into orbits.   Q.E.D.

Remark (1.3).   The definitions of $A_{\mathrm{HW}}(X)$ and $A(\mathscr{X})$ are, in fact, independent of the choice of field k. Thus, for example, if $k' \subset k$ we have with obvious notation

$$D'^{\chi(X)} \cdot \prod_{\mathfrak{p}'} N\mathfrak{p}'^{f_{\mathfrak{p}'}(X)} = D^{\chi(X)} \cdot \prod_{\mathfrak{p}} N\mathfrak{p}^{f_{\mathfrak{p}}(X)}.$$

This follows from the fact that the $\mathrm{Gal}(\bar{k}/k')$-module structure on the étale cohomology of $X_{\bar{k}}$ is induced from the $\mathrm{Gal}(\bar{k}/k)$-structure, together with standard properties of the Swan conductor (cf. [5], VI §2, cor. to prop. 4). In particular, it will usually be convenient in the sequel to take $k = \mathbf{Q}$, so the factor $D^{\chi(X)}$ disappears from the formula for $A(\mathscr{X})$.

§2. The de Rham discriminant. Let $\mathscr{X}$ be an arithmetic surface, which we assume embeddable in a scheme P smooth over $\mathrm{Spec}(\mathbf{Z})$. Consider the exact sequence

$$(2.1) \qquad 0 \to N^*_{\mathscr{X}/P} \to \Omega^1_{P/\mathbf{Z}}|\mathscr{X} \to \Omega^1_{\mathscr{X}/\mathbf{Z}} \to 0.$$

Here $N^*_{\mathscr{X}/P}$ denotes the conormal bundle, and the map on the left is injective because it is so at the generic point of $\mathscr{X}$, and $N^*_{\mathscr{X}/P}$ is locally free. The canonical bundle $\omega_{\mathscr{X}/\mathbf{Z}}$ can be defined by

$$\omega_{\mathscr{X}/\mathbf{Z}} = \mathrm{Hom}\big(\det\big(N^*_{\mathscr{X}/P}\big), \det\big(\Omega^1_{P/\mathbf{Z}}|\mathscr{X}\big)\big).$$

It follows from (2.1) that there is a natural map

$$\rho\colon \Omega^1_{\mathscr{X}/\mathbf{Z}} \to \omega_{\mathscr{X}/\mathbf{Z}},$$

which is an isomorphism after tensoring with $\mathbf{Q}$.

We define the two term complex $\omega^\bullet_{\mathscr{X}/\mathbf{Z}}$ by

$$\omega^\bullet_{\mathscr{X}/\mathbf{Z}} \colon \mathcal{O}_{\mathscr{X}} \overset{\rho \cdot d}{\to} \omega_{\mathscr{X}/\mathbf{Z}}$$

so ρ defines a mapping

$$\Omega^\bullet_{\mathscr{X}, \,\leqslant 1} \to \omega^\bullet_{\mathscr{X}/\mathbf{Z}}.$$

LEMMA (2.2).   *There is a perfect pairing on $R\Gamma(\mathscr{X}, \omega^\bullet_{\mathscr{X}/\mathbf{Z}})$ fitting into a commutative diagram (which one can think of as defining ψ)*

$$R\Gamma(\mathscr{X}, \omega^\bullet_{\mathscr{X}/\mathbf{Z}}) \underset{\cong}{\to} R\,\mathrm{Hom}(R\Gamma(\mathscr{X}, \omega^\bullet_{\mathscr{X}/\mathbf{Z}}), \mathbf{Z})[-2]$$

$$R\Gamma(\rho)\uparrow \qquad\qquad\qquad \downarrow R\,\mathrm{Hom}(R\Gamma(\rho), 1_{\mathbf{Z}})[-2]$$

$$R\Gamma(\mathscr{X}, \Omega^\bullet_{\mathscr{X}, \,\leqslant 1}) \overset{\psi}{\underset{(\mathrm{P.D.})}{\longrightarrow}} R\,\mathrm{Hom}(R\Gamma(\mathscr{X}, \omega^\bullet_{\mathscr{X}, \,\leqslant 1}), \mathbf{Z})[-2]$$

Proof.   Indeed, standard calculations for the cohomology of coherent sheaves show the existence of a trace map

$$H^1\big(\mathscr{X}, \omega_{\mathscr{X}/\mathbf{Z}}\big) \to \mathbf{Z}$$

such that the cup products

$$H^1(\mathcal{X}, \mathcal{O}_\mathcal{X}) \times \Gamma(\mathcal{X}, \omega_{\mathcal{X}/\mathbf{Z}}) \to H^1(\mathcal{X}, \omega_{\mathcal{X}/\mathbf{Z}}) \to \mathbf{Z}$$

$$H^0(\mathcal{X}, \mathcal{O}_\mathcal{X}) \times H^1(\mathcal{X}, \omega_{\mathcal{X}/\mathbf{Z}}) \to \mathbf{Z}$$

are perfect pairings of free abelian groups. By de Rham theory, the differential
$\rho \cdot d$: $H^i(\mathcal{X}, \mathcal{O}_\mathcal{X}) \to H^i(\mathcal{X}, \omega_{\mathcal{X}/\mathbf{Z}})$ is zero, so

$$\mathsf{H}^0(\mathcal{X}, \omega^{\cdot}_{\mathcal{X}/\mathbf{Z}}) \cong H^0(\mathcal{X}, \mathcal{O}_\mathcal{X})$$

$$\mathsf{H}^2(\mathcal{X}, \omega^{\cdot}_{\mathcal{X}/\mathbf{Z}}) \cong H^1(\mathcal{X}, \omega_{\mathcal{X}/\mathbf{Z}})$$

and there is an exact sequence

$$0 \to H^0(\mathcal{X}, \omega_{\mathcal{X}/\mathbf{Z}}) \to \mathsf{H}^1(\mathcal{X}, \omega^{\cdot}_{\mathcal{X}/\mathbf{Z}}) \to H^1(\mathcal{X}, \mathcal{O}_\mathcal{X}) \to 0.$$

The assertions of the lemma follow easily.   (Q.E.D.)

As a consequence of the lemma, we have

$$\mathrm{Disc}(H_{\mathrm{DR}}(\mathcal{X})) = \chi(\text{mapping cone } \psi) = \chi(\text{mapping cone } \rho)^2.$$

In order to prove theorem 2, therefore, it will suffice to show:

THEOREM (2.3).   *Define the two term complex of sheaves $\mathscr{C}^{\cdot}$ on $\mathcal{X}$ in degrees
-1 and 0 by*

$$\Omega^1_{\mathcal{X}/\mathbf{Z}} \underset{\rho}{\to} \omega_{\mathcal{X}/\mathbf{Z}}.$$

Then $A(\mathcal{X}) = \chi(\mathscr{C}^{\cdot})$.

Remark.   The square root which appears in the statement of theorem 2 is
curious, but correct. Consider for example the case $\mathcal{X} = \mathbf{P}^1_\mathcal{O}$ with $\mathcal{O} = \mathcal{O}_K$ the ring
of integers in some finite extension K of $\mathbf{Q}$. Let $d_{K/\mathbf{Q}}$ be the absolute value of
the discriminant of K over $\mathbf{Q}$, and let $\mathscr{D} \subset \mathcal{O}$ be the different. One knows
$A(\mathcal{X}) = d^2_{K/\mathbf{Q}}$. The complex $\Omega^{\cdot}_{\mathcal{X}, \leqslant 1}$ is $\mathcal{O}_\mathcal{X}$ in degree 0 and $[\Omega^1_{\mathcal{O}/\mathbf{Z}} \otimes \mathcal{O}_\mathcal{X}] \oplus [\Omega^1_{\mathcal{X}/\mathcal{O}}]$
in degree 1. It follows that $H^0(\Omega^{\cdot}_{\mathcal{X}, \leqslant 1}) \cong \mathrm{Ker}(d\colon \mathcal{O} \to \Omega^1_{\mathcal{O}/\mathbf{Z}})$, $H^1 = \mathrm{Coker}(d\colon \mathcal{O} \to \Omega^1_{\mathcal{O}/\mathbf{Z}})$, and $H^2 \cong H^1(\Omega^1_{\mathcal{X}/\mathcal{O}}) \cong \mathcal{O}$. Since $\Omega^1_{\mathcal{O}/\mathbf{Z}} \cong \mathcal{O}/\mathscr{D}$, the de Rham discriminant in this case is the same as the discriminant on the rank two module
$\mathcal{O} \oplus \mathscr{D}$. The pairing $\mathscr{D} \otimes \mathcal{O} \to \mathbf{Z}$ has discriminant $d^2_{K/\mathbf{Q}}$, so the de Rham discriminant is $d^4_{K/\mathbf{Q}}$.

LEMMA (2.4).   *Fix an embedding $\mathcal{X} \to P$ with P smooth over $\mathrm{Spec}(\mathbf{Z})$, and let α:
$N^*_{\mathcal{X}/P} \to \Omega^1_{P/\mathbf{Z}}|\mathcal{X}$ be the natural map. Let $\mathscr{I} \subset \mathcal{O}_\mathcal{X}$ be the ideal generated by
determinants of maximal rank minors for α. Then*
   (i) $\mathscr{I}$ is independent of the choice of P.
   (ii) $\Omega^2_{\mathcal{X}/\mathbf{Z}}$ is an invertible $\mathcal{O}_\mathcal{X}/\mathscr{I}$-module.
   (iii) $\mathrm{Coker}(\rho)$ is an invertible $\mathcal{O}_\mathcal{X}/\mathscr{I}$-module.

Proof of lemma. The assertion in (i) is local, so it suffices to consider $\mathscr{X} \to P \to Q$ with P and Q smooth over Spec($\mathbf{Z}$) and show the embeddings of $\mathscr{X}$ in P and Q give rise to the same $\mathscr{I}$. This is clear because $\Omega^1_{P/\mathbf{Z}}|\mathscr{X}$ is a direct summand of $\Omega^1_{Q/\mathbf{Z}}|\mathscr{X}$.

For the proof of (ii) and (iii) we may localize and assume $P = \mathrm{Spec}(\mathbf{Z}[[t, u]])$ and $\mathscr{X}$: $f(t, u) = 0$. We have $\mathscr{I} = (\partial f/\partial t, \partial f/\partial u)$ and $\Omega^2_{\mathscr{X}/\mathbf{Z}} \cong \mathcal{O}_{\mathscr{X}}/(\partial f/\partial t, \partial f/\partial u)$ with generator $dt \wedge du$, so (ii) holds. Finally if we identify

$$N^*_{\mathscr{X}/P} \cong \mathcal{O}_{\mathscr{X}} \cdot df, \qquad \det\!\left(\Omega^1_{P/\mathbf{Z}}|\mathscr{X}\right) \cong \mathcal{O}_{\mathscr{X}} \cdot dt \wedge du,$$

the map

$$\rho \colon \Omega^1_{\mathscr{X}/\mathbf{Z}} \to \omega_{\mathscr{X}/\mathbf{Z}} \cong \mathrm{Hom}\!\left(N^*_{\mathscr{X}/P}, \det\!\left(\Omega^1_{P/\mathbf{Z}}|\mathscr{X}\right)\right)$$

becomes $\rho(\tau)(a\, df) = a\hat{\tau} \wedge df$. Here τ is a section of $\Omega^1_{\mathscr{X}/\mathbf{Z}}$ and $\hat{\tau}$ is a lifting to $\Omega^1_{P/\mathbf{Z}}|\mathscr{X}$. It is thus clear that $\mathrm{Coker}(\rho)$ is also isomorphic to $\mathcal{O}_{\mathscr{X}}/(\partial f/\partial t, \partial f/\partial u)$, proving the lemma.

Note that the map ρ is an isomorphism on the generic fibre, so $\ker(\rho) = \Omega^1_{\mathscr{X}/\mathbf{Z},\,\mathrm{tors}}$, the torson subsheaf of $\Omega^1_{\mathscr{X}/\mathbf{Z}}$.

LEMMA (2.5). *Let $x \in \mathscr{X}$ be a codimension* 1 *point. Then the stalks* $\ker(\rho)_x$ *and* $\mathrm{Coker}(\rho)_x$ *are* $\mathcal{O}_{\mathscr{X},\,x}$*-modules of finite length, and*

$$\mathrm{length}(\mathrm{Ker}(\rho)_x) = \mathrm{length}(\mathrm{Coker}(\rho)_x).$$

Proof. The question is local, so we may assume $\mathscr{X} \to P$ defined by $f = 0$ as above. Writing $T^1 = \mathrm{Hom}(\Omega^1, \mathcal{O})$, we get by dualizing the conormal bundle sequence

$$0 \to T^1_{\mathscr{X}/\mathbf{Z}} \to T^1_{P/\mathbf{Z}}|\mathscr{X} \xrightarrow{a} N_{\mathscr{X}/P} \to \mathrm{Ext}^1\!\left(\Omega^1_{\mathscr{X}/\mathbf{Z}}, \mathcal{O}_{\mathscr{X}}\right) \to 0.$$

Locally $N_{\mathscr{X}/P} \cong \mathcal{O}_{\mathscr{X}}$ and $\mathrm{Image}(a)$ is generated by $\partial f/\partial t$ and $\partial f/\partial u$ so by the previous lemma $\mathrm{Coker}(\rho) \cong \mathrm{Ext}^1(\Omega^1_{\mathscr{X}/\mathbf{Z}}, \mathcal{O}_{\mathscr{X}})$ (locally).

Write $\mathscr{E} = \mathrm{Ext}^1(\Omega^1_{\mathscr{X}/\mathbf{Z}}, \mathcal{O}_{\mathscr{X}})$ and $\mathscr{F} = \mathrm{Image}(a)$, so there is an exact sequence $0 \to \mathscr{F} \to N_{\mathscr{X}/P} \to \mathscr{E} \to 0$. Dualizing yields a diagram

$$
\begin{array}{ccc}
0 & & 0 \\
\downarrow & & \downarrow \\
N^*_{\mathscr{X}/P} & = & N^*_{\mathscr{X}/P} \\
\downarrow & & \downarrow \\
0 \to \mathrm{Hom}(\mathscr{F}, \mathcal{O}_{\mathscr{X}}) \to & \Omega^1_{P/\mathbf{Z}}|\mathscr{X} \to & (\Omega^1_{\mathscr{X}/\mathbf{Z}})^{**} \\
\downarrow & \downarrow & \| \\
0 \to \mathrm{Ext}^1(\mathscr{E}, \mathcal{O}_{\mathscr{X}}) \to & \Omega^1_{\mathscr{X}/\mathbf{Z}} \to & (\Omega^1_{\mathscr{X}/\mathbf{Z}})^{**} \\
\downarrow & & \downarrow \\
0 & & 0
\end{array}
$$

It follows that $\mathrm{Ext}^1(\mathscr{E}, \mathcal{O}_{\mathscr{X}}) \cong \Omega^1_{\mathscr{X}/\mathbf{Z}, \mathrm{tors}}$. Finally since $\mathcal{O}_{\mathscr{X}, x}$ is a discrete valuation ring and $\mathscr{E}$ is torsion, we have that

$$\mathrm{length}(\mathscr{E}_x) = \mathrm{length}\big(\mathrm{Ext}^1(\mathscr{E}, \mathcal{O}_{\mathscr{X}})_x\big).$$

This proves the lemma.   Q.E.D.

Let $s \in \mathrm{Spec}(\mathbf{Z})$ be a closed point and let $\mathscr{X}_s$ be the fibre of $\mathscr{X}$ over s. Since the complex $\mathscr{C}^{\cdot}$ is acyclic in $U - \mathscr{X}_s$ for some open neighborhood U of $\mathscr{X}_s$, the McPherson graph technique permits us (cf. [2], chap. 18) to define a chern character

$$[\mathscr{X}] \cap \mathrm{ch}_s(\mathscr{C}^{\cdot}) \in \mathrm{CH}_*(\mathscr{X}_s),$$

where $\mathrm{CH}_*(\mathscr{X}_s)$ denotes the Chow group of cycles mod rational equivalence on $\mathscr{X}_s$ as defined in (op. cit.).

LEMMA (2.6).   *The complex $\mathscr{C}^{\cdot}$ is acyclic on the generic fibre so we may define $\chi(\mathscr{C}^{\cdot})$. For $s \in \mathrm{Spec}(\mathbf{Z})$ write $N(s)$ for the number of elements in the residue field at s. Then*

$$\chi(\mathscr{C}^{\cdot}) = \prod_s N(s)^{\deg([\mathscr{X}] \cap \mathrm{ch}_{s,2}(\mathscr{C}^{\cdot}))}.$$

Proof.   It follows from the Riemann-Roch theorem (op. cit. thm. 18.2) that

$$\chi(\mathscr{C}^{\cdot}) = \prod_s N(s)^{\deg(Td(\mathscr{X}) \cap \mathrm{ch}_s(\mathscr{C}^{\cdot}))}.$$

Since $Td(\mathscr{X}) = [\mathscr{X}] + $ higher order terms (op. cit. 18.3.4) it suffices to show $\mathrm{ch}_{s,1}(\mathscr{C}^{\cdot}) = 0$, i.e. that $\mathrm{Coker}(\rho)$ and $\mathrm{Ker}(\rho)$ have the same lengths at generic points of $\mathscr{X}_s$. This is lemma 5.   Q.E.D.

LEMMA (2.7). $\mathrm{ch}_{s,2}(\mathscr{C}^{\cdot}) = c_{s,2}(\Omega^1_{\mathscr{X}/\mathbf{Z}})$, *where in the notation of* [1], §1

$$c_{s,2}\big(\Omega^1_{\mathscr{X}/\mathbf{Z}}\big) = c^{\mathscr{X}}_{2,\,Z}\big(\Omega^1_{\mathscr{X}/\mathbf{Z}}\big) \cap [\mathscr{X}]$$

with $Z = \mathscr{X}_s$.

Proof.   Write $\omega = \omega_{\mathscr{X}/\mathbf{Z}}$ and fix a resolution by locally free sheaves

$$0 \to E_1 \to E_0 \to \Omega^1_{\mathscr{X}/\mathbf{Z}} \to 0,$$

e.g. by embedding $\mathscr{X}$ in a P smooth over $\mathbf{Z}$ and taking the conormal sequence. Then $\mathscr{C}^{\cdot}$ is quasi-isomorphic to the complex

$$E_1 \to E_0 \to \omega.$$

Write G_1 (resp. G_0) for the Grassmann of subbundles of $E_1 \oplus E_0$ of rank = $\mathrm{rk}(E_1)$ (resp. subbundles of $E_0 \oplus \omega$ of rank = $\mathrm{rk}(E_0)$) and denote by pr_i the projection $G_i \to \mathcal{X}$. Let ζ_1 (resp. ζ_0) denote the tautological subbundle of $\mathrm{pr}_1^*(E_1 \oplus E_0)$ (resp. $\mathrm{pr}_0^*(E_0 \oplus \omega)$). Write $\xi = [\zeta_1] - [\zeta_0] + [\mathrm{pr}^*(\omega)] \in K_0(G)$ where $\mathrm{pr}: G = G_{1\mathcal{X}} \times G_0 \to \mathcal{X}$ is the structure map. Write $\bar{\xi} = [E_0] - [\zeta_1] \in K_0(G_1)$. Note that for $i = 1, 2$, $\mathrm{ch}_{s,i}(\mathscr{C}^{\,\cdot}) \cap (\cdot)$ is an operator from cycle classes on $\mathcal{X}$ to cycle classes on $\mathcal{X}_s$ and moreover, $\mathrm{ch}_{s,1}(\mathscr{C}^{\,\cdot}) \cap (\cdot) = 0$ by lemma 5. Let $\eta: G_{\mathcal{X}_s} \to \mathcal{X}_s$ be the projection. According to [2], chap. 18.1, we can associate to a cycle α on $\mathcal{X}$ a cycle $\gamma = \gamma(\alpha)$ on G such that

$$\mathrm{ch}_{s,2}(\mathscr{C}^{\,\cdot}) \cap \alpha = \eta_*(\mathrm{ch}_2(\xi) \cap \gamma).$$

On the other hand, by [1], §1,

$$c_{s,2}\big(\Omega^1_{\mathcal{X}/\mathbf{Z}}\big) \cap \alpha = \eta_{1*}\big(c_2(\bar{\xi}) \cap \gamma_1\big),$$

where γ_1 is the projection of γ on $G_{1, \mathcal{X}_s}$ and $\eta_1: G_{1, \mathcal{X}_s} \to \mathcal{X}_s$.
  Write

$$T = \mathrm{ch}_2(\xi) - c_2(E_0 - \zeta_1)$$

$$= \mathrm{ch}_2(\zeta_1) - \mathrm{ch}_2(E_0) - \mathrm{ch}_2(\zeta_0) + \mathrm{ch}_2(E_0) + \mathrm{ch}_2(\omega) - c_2(E_0 - \zeta_1).$$

Let $\mathscr{L}$ denote the tautological quotient line bundle on G_0. That is, $\mathscr{L} \cong \det(E_0) \otimes \omega \otimes \det(\zeta_0)^{-1}$. We have

$$T = \mathrm{ch}_2(\zeta_1) - \mathrm{ch}_2(E_0) + \mathrm{ch}_2(\mathscr{L}) - c_2(E_0 - \zeta_1)$$

$$= -c_2(\zeta_1) + c_1(\zeta_1)^2/2 + c_2(E_0) - c_1(E_0)^2/2 + c_1(\mathscr{L})^2/2 + c_2(\zeta_1)$$

$$- c_2(E_0) - c_1(\zeta_1)^2 + c_1(\zeta_1)c_1(E_0) = \big[c_1(\mathscr{L})^2 - (c_1(E_0) - c_1(\zeta_1))^2\big]/2.$$

Thus it will suffice to show that on the support of the cycle γ we have $\mathscr{L} \cong \det(E_0) \otimes \det(\zeta_1)^{-1}$. Substituting the definition of $\mathscr{L}$ and using the identify $\omega \cong \det(E_0) \otimes \det(E_1)^{-1}$, we are reduced to proving

$$(*) \qquad \det(E_1) \otimes \det(\zeta_1)^{-1} \cong \det(E_0) \otimes \det(\zeta_0)^{-1}$$

on $\mathrm{supp}(\gamma)$.
  We may assume α is effective, and then, replacing $\mathcal{X}$ by $\mathrm{supp}(\alpha)$, that $\alpha = [\mathcal{X}]$. (In fact, the case $\alpha = [\mathcal{X}]$ is the only one that we will need anyway.) With reference to the discussion on p. 341 of [2], let $d_1: E_1 \to E_0$ and $d_0: E_0 \to \omega$ be the maps in $\mathscr{C}^{\,\cdot}$. Let $\mathrm{rank}(E_1) = e$, so $\mathrm{rank}(E_0) = e + 1$. Define $\varphi: \mathcal{X} \times \mathbf{A}^1 \to G = G_1 \times G_0$ by

$$\varphi(x, t) = \{\text{graph of } t \cdot d_1(x) \subset E_1 \oplus E_0\} \times \{\text{graph of } t \cdot d_0(x) \subset E_0 \oplus \omega\}.$$

Let Z denote the Zariski closure of image(φ), and let $Z_\infty = Z \cap G \times \{\infty\}$. Then supp($\gamma$) $\subset Z_\infty$, so it will suffice to establish the isomorphism (*) in some neighborhood of Z_∞ on Z.

Since $\zeta_1 \subset E_1 \oplus E_0$ and $\zeta_0 \subset E_0 \oplus \omega$, there are projections $\zeta_i \to E_i$. Let θ_i: $\det(\zeta_i) \to \det(E_i)$ be the maps on determinant bundles. The θ_i are injective, and remain so when restricted to Z. Let t denote the standard parameter on $\mathbf{P}^1$. I claim that θ_1 vanishes to order $\geqslant e - 1$ at ∞, i.e. that $t^{e-1} \cdot \theta_1$: $\det(\zeta_1) \to \det(E_1)$, and that moreover the cycles [coker(θ_0)] and [coker($t^{e-1} \cdot \theta_1$)] have the same components and multiplicities in some neighborhood of Z_∞. Note that this suffices to verify (*) and complete the proof of lemma (2.7).

To check the claim, let Λ be a discrete valuation ring with uniformizing element π. Let $\eta, s \in \mathrm{Spec}(\Lambda)$ be the generic and closed points. Assume give f: $\mathrm{Spec}(\Lambda) \to Z$ with $f(\eta)$ lying over $(\mathscr{X} - \mathscr{X}_s) \times \mathbf{A}^1$ and $f(s)$ at ∞. We pull all bundles and maps of bundles back to $\mathrm{Spec}(\Lambda)$, but to simplify we omit f^* from the notation, writing E_0 in place of $f^*(E_0)$, etc. Choose bases $\varepsilon_1, \ldots, \varepsilon_e$ for E_1 and $\delta_1, \ldots, \delta_{e+1}$ for E_0 such that the map d_1 is given by

$$d_1(\varepsilon_i) = \delta_i, \quad 1 \leqslant i \leqslant e - 1; \qquad d_1(\varepsilon_e) = \pi^n \cdot \delta_e, \quad n \geqslant 0.$$

This is possible because we know that the fibre of coker(d_1) $\cong \Omega^1_{\mathscr{X}/\mathbf{Z}}$ has rank $\leqslant 2$ at every point.

It will suffice to show $t^{e-1} \cdot \theta_1$ integral and length coker($t^{e-1} \cdot \theta_1$) = length coker(θ_0). On Λ, $\zeta_1 \subset E_1 \oplus E_0$ is the subbundle generated by

$$\left\{ \left(t^{-1}\varepsilon_i, \delta_i \right) \right\}_{1 \leqslant i \leqslant e-1}$$

together with either $(t^{-1}\pi^{-n}\varepsilon_e, \delta_e)$ if $t^{-1}\pi^{-n} \in \Lambda$ or $(\varepsilon_e, t\pi^n\delta_e)$ if not. Thus the determinant θ_1 of the projection $\zeta_1 \to E_1$ has valuation $(1 - e) \cdot \mathrm{ord}(t) + \max(-n - \mathrm{ord}(t), 0)$, and $t^{e-1} \cdot \theta_1$ is integral with coker of length $= \max(-n - \mathrm{ord}(t), 0)$.

On the other hand, $\zeta_0 \subset E_0 \oplus \mathrm{Hom}(\det(E_1), \det(E_0))$ is generically the graph of the map $\varepsilon \to t(\Lambda^e d_1) \wedge \varepsilon$: $\det(E_1) \to \det(E_0)$. Over Λ, therefore, the image of ζ_0 in $E_0 \oplus \omega$ is generated by

$$\left\{ (\delta_i, 0) \right\}_{1 \leqslant i \leqslant e}$$

together with either

$$\left(\delta_{e+1}, t\pi^n \cdot \text{generator for } \omega \right) \text{ or } \left(t^{-1}\pi^{-n}\delta_{e+1}, \text{gen. for } \omega \right),$$

whichever is integral. Thus $\det(\theta_0) = \max(-n - \mathrm{ord}(t), 0) = t^{e-1} \cdot \det(\theta_1)$. This completes the proof of lemma (2.7).   Q.E.D.

Proof of theorem (2.3).   The fundamental result in [1] is that at a closed point $s \in \mathrm{Spec}(\mathbf{Z})$,

$$\deg\left\{ c_{s,2}\left(\Omega^1_{\mathscr{X}/\mathbf{Z}} \right) \cap [\mathscr{X}] \right\} = -\mathrm{sw}(\mathscr{X}/\mathbf{Z}) + \chi(\mathscr{X}_s) - \chi(\mathscr{X}_{\bar{\eta}}).$$

It follows from §1 that

$$A(\mathcal{X}) = \prod_s N(s)^{\deg\{c_{s,2}(\Omega^1_{\mathcal{X}/Z})\cap[\mathcal{X}]\}}.$$

From lemmas (2.6) and (2.7), we deduce

$$\chi(\mathcal{C}^{\cdot}) = A(\mathcal{X}),$$

proving theorem (2.3).

The proof of theorem 2 in the introduction had already been reduced to theorem (2.3), so the identity

$$A(\mathcal{X}) = \mathrm{disc}\big(H_{\mathrm{DR}}(\mathcal{X})\big)^{1/2}$$

is established.

§3. Torsion in the de Rham complex. Let $\mathcal{X}$ be an arithmetic surface, and let $\Omega^{\cdot}_{\mathcal{X},\mathrm{tors}} \subset \Omega^{\cdot}_{\mathcal{X}/Z}$ be the subcomplex of Z-torsion sections in the de Rham complex. Thus $\Omega^{\cdot}_{\mathcal{X},\mathrm{tors}}$ is 0 in degree 0, $\Omega^1_{\mathcal{X}/Z,\mathrm{tors}}$ in degree 1, and $\Omega^2_{\mathcal{X}/Z}$ in degree 2. Our objective is to prove theorem 1, viz.

$$A(\mathcal{X}) = \chi(\Omega^{\cdot}_{\mathcal{X},\mathrm{tors}}).$$

Note that $\Omega^1_{\mathcal{X}/Z,\mathrm{tors}}$ coincides with $\mathrm{Ker}(\rho)$ with ρ as in theorem (2.3), so it will suffice to compare $\mathrm{Coker}(\rho)$ and $\Omega^2_{\mathcal{X}/Z}$.

Let $\mathcal{I} \subset \mathcal{O}_{\mathcal{X}}$ be as in lemma (2.4), and let $Z \subset \mathcal{X}$ be the closed subscheme defined by $\mathcal{I}$. By that lemma, both $\mathrm{Coker}(\rho)$ and $\Omega^2_{\mathcal{X}/Z}$ are invertible $\mathcal{O}_Z$-modules, and there exists a line bundle $\mathcal{L}$ on Z such that

$$\mathrm{Coker}(\rho) \cong \Omega^2_{\mathcal{X}/Z} \otimes \mathcal{L}.$$

We want $\mathrm{ch}_{Z,2}(\mathrm{Coker}(\rho)) = \mathrm{ch}_{Z,2}(\Omega^2_{\mathcal{X}/Z})$. The sheaves $\mathrm{Coker}(\rho)$ and $\Omega^2_{\mathcal{X}/Z}$ can be filtered so that the supports of the graded pieces are reduced and irreducible, so it will suffice to show

LEMMA (3.1). *Let $Y \subset Z$ be an integral curve. Then $\mathcal{L}|Y \cong \mathcal{O}_Y$.*

Proof. Fix an embedding $\mathcal{X} \to P$ with P smooth over $\mathrm{Spec}(Z)$ of relative dimension m. Restrict the resulting conormal bundle sequence to Z, and define L to be the kernel:

$$(*) \qquad 0 \to L \to N^*_{\mathcal{X}/P}|Z \to \Omega^1_{P/Z}|Z \to \Omega^1_{\mathcal{X}/Z}|Z \to 0.$$

The same sort of local argument as in lemma 4 shows that $\Omega^1_{\mathcal{X}/Z}|Z$ is locally free of rank 2 over $\mathcal{O}_Z$, so L is invertible over $\mathcal{O}_Z$. Also

$$\Omega^2_{\mathcal{X}/Z} \cong \Omega^2_{\mathcal{X}/Z}|Z \cong \left(\Omega^m_{P/Z}|Z\right) \otimes \wedge^{m-2}\left(N^*_{\mathcal{X}/P}|Z/L\right)^*$$

$$\cong \Omega^m_{P/Z} \otimes \Gamma^{m-1}\left(N_{\mathcal{X}/P}\right) \otimes \wedge \cong \omega_{\mathcal{X}/Z} \otimes L.$$

Since $\mathrm{Coker}(\rho)$ is known to be invertible over $\mathcal{O}_Z$, however, it follows easily that $\mathrm{Coker}(\rho) \cong \omega \otimes \mathcal{O}_Z$. Thus $\mathscr{L} \cong L^*$.

Let $J \subset K \subset \mathcal{O}_p$ be the ideals of $\mathcal{X}$ and Y. I claim $JK = J \cap K^2$. Clearly, $JK \subset J \cap K^2$, and the question of equality is local. Since $\mathcal{X}$ is regular, it is defined locally in P by $f_1 = \cdots = f_n = 0$ with the f_i forming part of a system of parameters. Since $Y \subset \mathcal{X}$ is Cartier, we can choose g (locally) with $Y: g = f_1 = \cdots = f_n = 0$. An element $h \in J \cap K^2$ can be written $h = \Sigma h_i f_i$, $h = \Sigma h'_{ij} f_i f_j + \Sigma h'_i f_i g + h'' g^2$. Going mod J, we see $h'' \in J$, so $h \in JK$ as claimed.

Let $p \in \mathbf{Z}$ be the prime lying under Y. Since $\mathcal{X}$ is non-smooth over $\mathbf{Z}$ along Y, we have $p \in K^2 + J$, so there is a diagram

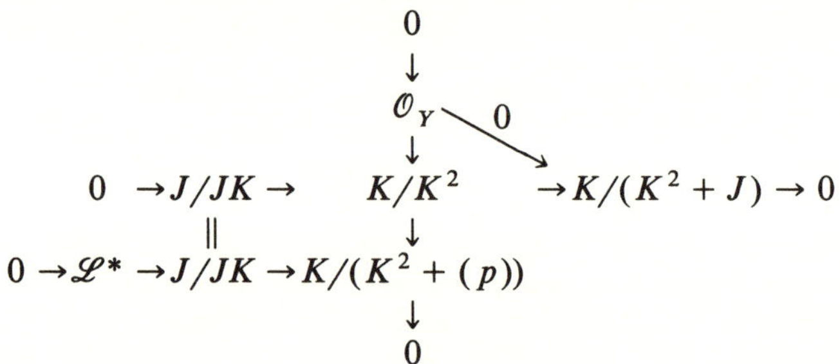

The bottom row is $(*)$ together with the remark that $K/(K^2 + (p))$ injects into $\Omega^1_{P/Z}|Z$. (Indeed, $K/(K^2 + (p))$ is the conormal bundle of Y in the closed fibre of P over $\mathrm{Spec}(\mathbf{Z}/p\mathbf{Z})$.) The lemma follows by a diagram chase.   Q.E.D.

Theorem 1 is now an easy consequence of theorem (2.3), since $\mathrm{Coker}(\rho)$ and $\Omega^2_{\mathcal{X}/Z}$ have the same ch_2 with supports in the singular fibres.   Q.E.D.

References

1.   S. BLOCH, *Cycles on arithmetic schemes and Euler characteristics of curves*, preprint, to appear in Bowdoin conference in Algebraic Geometry, Amer. Math. Soc. Proc. Symp. Pure Math.
2.   W. FULTON, *Intersection theory*, Springer-Verlag, Berlin, 1984.
3.   K. KATO, *Vanishing cycles, ramification of valuations, and class field theory*, preprint, University of Tokyo.
4.   ______, letter to the author.
5.   J.-P. SERRE, *Corps locaux*, Paris, Hermann, 1962.
6.   ______, *Facteurs locaux des fonctions zêta des variétés algébriques* (définitions et conjectures), Sém. Delange-Pisot-Poitou, 1969/70, no. 19.
7.   ______, article in *Conference on Arithmetical Algebraic Geometry*, Ed. by O. F. G. Schilling, N.Y., Harper and Row, 1965.
8.   P. DELIGNE, *La conjecture de Weil*. II, Publ. Inst. Hautes Études Sci. No. 43 (1974).
9.   ______, *Intersections sur les surfaces regulieres*, SGA 7 II, Springer Lecture Notes no. 340, Springer-Verlag, (1973).

DEPARTMENT OF MATHEMATICS, UNIVERSITY OF CHICAGO, CHICAGO, ILLINOIS 60637

Vol. 54, No. 2   DUKE MATHEMATICAL JOURNAL ©   1987

CORRESPONDANCE DE LANGLANDS GÉOMÉTRIQUE POUR LES CORPS DE FONCTIONS

GÉRARD LAUMON

A Yu. I. Manin
pour son 50ième Anniversaire

0. Introduction.   Soient $\mathbb{F}_q$ un corps fini et ℓ un nombre premier inversible dans $\mathbb{F}_q$. Soit X une courbe projective, lisse et géométriquement connexe sur $\mathbb{F}_q$. On note F le corps des fonctions de X, $\overline{F}$ une clôture algébrique de F et $\mathbf{A}$ l'anneau des adèles de F.

Langlands a conjecturé l'existence, pour tout entier $n \geqslant 1$, d'une correspondance entre les représentations ℓ-adiques de rang n du groupe de Galois de $\overline{F}$ sur F (ou mieux du groupe de Weil de $\overline{F}$ sur F) et les formes automorphes sur $GL_n(\mathbf{A})$ qui sont vecteurs propres des opérateurs de Hecke. En particulier, à une représentation ℓ-adique de rang n, géométriquement irréductible et partout non ramifiée, de $\mathrm{Gal}(\overline{F}/F)$ doit correspondre une fonction f sur $GL_n(\mathbf{A})$, invariante à gauche sous $GL_n(F)$ et à droite sous le compact maximal de $GL_n(\mathbf{A})$, géométriquement cuspidale et vecteur propre des opérateurs de Hecke, avec pour valeurs propres les traces des Frobenius dans la représentation galoisienne. Pour $n = 1$, cette correspondance n'est autre que la théorie du corps de classes abélien.

Cette correspondance de Langlands admet un analogue géométrique dégagé par Lang ($n = 1$) puis Drinfeld ($n = 2$). En particulier, à tout système local ℓ-adique géométriquement irréductible de rang $n \geqslant 1$ sur X doit correspondre un complexe de faisceaux constructibles ℓ-adiques sur l'espace de module $Fib_{X,\,n}$ des fibrés vectoriels de rang n sur X (ce complexe de faisceaux étant en un sens convenable géométriquement cuspidal et vecteur propre des opérateurs de Hecke). Inversement, compte-tenu du dictionnaire fonctions-faisceaux de Grothendieck et de l'interprétation adélique, due à Weil, des fibrés vectoriels sur X, cette correspondance géométrique doit impliquer la correspondance de Langlands.

L'objet principal de ce travail est de dégager une construction et une caractérisation (conjecturales) du complexe de faisceaux K_E sur $Fib_{X,\,n}$ associé à un système local de rang n, E, sur X. Pour motiver ces conjectures, nous donnons dans cet article une interprétation géométrique de la fonction de Whittaker associée à E et nous calculons la variété caractéristique du complexe de faisceaux K_E construit par Drinfeld pour $n = 2$.

Plus précisément, après avoir rappelé aux numéros 1 et 2 la conjecture de Langlands et son analogue géométrique, nous construisons au numéro 3 un faisceau pervers ℓ-adique W_E sur l'espace de module $Coh_{X,0}$ des $\mathscr{O}_X$-Modules

Received September 29, 1986. Unité de recherche associée au CNRS 752.

cohérents de torsion associé au système local E sur X. Cette construction est une globalisation de la construction de Springer des représentations des groupes de Weyl. Un calcul fibre par fibre de W_E permet de justifier la terminologie "faisceau de Wittaker associé à E" que nous adoptons pour W_E. Nous calculons aussi la variété caractéristique de W_E.

Les numéros 4 et 5 sont consacrés à notre construction conjecturale du complexe de faisceaux K_E. Cette construction s'inspire bien entendu de celle de Drinfeld pour $n = 2$ et de la théorie de Fourier globale pour les formes automorphes. Mais contrairement à ce qu'il se passe pour $n = 2$, la construction géométrique diffère très sensiblement de la construction arithmétique. Nous calculons à la fin du numéro 5 la variété caractéristique de K_E pour $n = 2$.

Au numéro 6, nous introduisons un analogue global du cône nilpotent dans l'algèbre de Lie duale d'un groupe réductif. Il s'agit d'un fermé lagrangien canonique Λ dans $T^*Fib_{X,n}$. Nous conjecturons que la variété caractéristique des K_E est toujours contenue dans Λ et que réciproquement les faisceaux pervers géométriquement irréductibles cuspidaux dont la variété caractéristique est contenue dans Λ sont tous de la forme K_E.

Dans l'appendice A, nous rappelons les résultats de Brylinski sur la transformation de Radon géométrique que nous utilisons aux numéros 4 et 5.

Nous utilisons librement dans cet article les notions de champs algébriques sur un corps k et de faisceaux pervers ℓ-adiques sur un tel champ. Les champs algébriques sur k considérés dans cet article sont tous localement de la forme X/G, où G est un k-groupe linéaire lisse et géométriquement connexe et où X est un k-schéma muni d'une action de G. Un faisceau pervers ℓ-adique sur le champ algébrique X/G n'est autre qu'un faisceau pervers ℓ-adique G-équivariant sur X.

Durant la préparation de cet article, j'ai bénéficié de nombreuses et très stimulantes discussions avec P. Deligne et D. Kazhdan. Madame Bonnardel a remarquablement effectué la frappe du manuscrit. Ce travail a été réalisé au cours de visites à l'IAS, à l'Université de Harvard, à l'IHES et au MSRI. J'ai reçu aussi un soutien partiel de la NSF.

1. Construction de la forme automorphe attachée à une représentation ℓ-adique partout non ramifiée du groupe de Galois d'un corps de fonctions: position du problème

(1.0). Soit X une courbe projective, lisse et géométriquement connexe de genre g sur un corps fini $k = \mathbb{F}_q$ de caractéristique p et soit F son corps des fonctions méromorphes. On identifie les places de F aux points fermés de X; on note $|X|$ l'ensemble de ces points fermés et, pour $x \in |X|$, on note F_x le complété de F en la place x, $\mathcal{O}_x$ l'anneau des entiers de F_x, v_x la valuation discrète de F_x, $\pi_x \in \mathcal{O}_x$ une uniformisante de F_x $(v_x(\pi_x) = 1)$, k_x le corps résiduel $\mathcal{O}_x/(\pi_x)$ et $\deg(x)$ le degré de k_x sur k (k_x a donc $q_x = q^{\deg(x)}$ éléments). On note $\mathbf{A}$ l'anneau des adèles de F et $\mathcal{O} = \prod_{x \in |X|} \mathcal{O}_x$ son sous-anneau compact maximal.

On fixe une clôture algébrique $\bar{F}$ de F, un nombre premier $\ell \neq p$, une clôture algébrique $\overline{\mathbf{Q}}_\ell$ de $\mathbf{Q}_\ell$ et un entier $n \geqslant 1$; on note $\bar{k}$ la clôture algébrique de k dans $\bar{F}$.

(1.1).   On considère d'une part l'ensemble $\mathscr{G}_n^0$ des classes d'isomorphie de représentations ℓ-adiques

$$\sigma\colon \mathrm{Gal}(\bar{F}/F) \to \mathrm{GL}_n(\overline{\mathbf{Q}}_\ell)$$

(i.e., des représentations définies sur une extension finie E_λ de $\mathbf{Q}_\ell$ dans $\overline{\mathbf{Q}}_\ell$ et continues pour la topologie de Krull sur $\mathrm{Gal}(\bar{F}/F)$ et la topologie λ-adique sur $\mathrm{GL}_n(E_\lambda)$) qui sont:

(i) *partout non ramifiées*

(ii) *géométriquement irréductibles* (i.e., telles que la restriction de σ à $\mathrm{Gal}(\bar{F}/\bar{k} \cdot F)$ soit irréductible).

On considère d'autre part l'ensemble $\mathscr{A}_n^0$ des fonctions non identiquement nulles

$$f\colon \mathrm{GL}_n(F) \backslash \mathrm{GL}_n(\mathbf{A})/\mathrm{GL}_n(\mathcal{O}) \to \overline{\mathbf{Q}}_\ell$$

qui sont:

(i) *cuspidales*, i.e., telles que, pour toute partition $\underline{n} = (n_1, \ldots, n_r)$ de n non triviale, déterminant le parabolique standard $P_{\underline{n}}$ de GL_n de radical unipotent $V_{\underline{n}}$, et pour toute mesure de Haar dv sur $V_{\underline{n}}(F) \backslash V_{\underline{n}}(\mathbf{A})$ à valeurs dans $\overline{\mathbf{Q}}_\ell$, on ait

$$\int_{V_{\underline{n}}(F)\backslash V_{\underline{n}}(\mathbf{A})} f(vg)\, dv = 0$$

pour tout $g \in \mathrm{GL}_n(\mathbf{A})$,

(ii) *vecteurs propres des opérateurs de Hecke T_x^i* ($x \in |X|$, $i = 1, \ldots, n$) définis par

$$\left(T_x^i f\right)(g) = \int_{M_n^i(\mathcal{O}_x)} f(g \cdot h_x)\, dh_x$$

où

$$M_n^i(\mathcal{O}_x) = \mathrm{GL}_n(\mathcal{O}_x) \cdot \mathrm{diag}\Big(\underbrace{\pi_x, \ldots, \pi_x}_{i}, 1, \ldots, 1\Big) \cdot \mathrm{GL}_n(\mathcal{O}_x)$$

$$\subset \mathrm{GL}_n(F_x) \subset \mathrm{GL}_n(\mathbf{A})$$

et où dh_x est la mesure de Haar sur $\mathrm{GL}_n(F_x)$ normalisée par

$$\int_{\mathrm{GL}_n(\mathcal{O}_x)} dh_x = 1.$$

La conjecture suivante fait partie de la "philosophie de Langlands" ([La]).

CONJECTURE (1.1.1) (Langlands).   *Pour toute $\sigma \in \mathcal{G}_n^0$, il existe une et une seule fonction (à un scalaire près dans $\overline{\mathbf{Q}}_\ell^\times$) $f_\sigma \in \mathcal{A}_n^0$ telle que, pour tout $x \in |X|$ et tout $i = 1, \ldots, n$, on ait*

$$T_x^i f_\sigma = q_x^{-i(i-1)/2} \mathrm{tr}\big(\Lambda^i \sigma(\mathrm{Frob}_x) \big) f_\sigma,$$

où $\mathrm{Frob}_x \in \mathrm{Gal}(\overline{F}_x/F_x) \subset \mathrm{Gal}(\overline{F}/F)$ *est n'importe quel relèvement du Frobenius géométrique de* $\mathrm{Gal}(\overline{k}_x/k_x)$.

Remarque.   Pour l'instant cette conjecture est démontrée pour $n = 1$, 2 et 3. Pour $n = 1$, c'est une conséquence de la théorie du corps de classes abélien. Pour $n = 2$ et 3, on utilise de plus l'interprétation cohomologique de la fonction L des représentations ℓ-adiques de $\mathrm{Gal}(\overline{F}/F)$ due à Grothendieck, la formule du produit pour la constante de l'équation fonctionnelle de ces fonctions L due à l'auteur et les théorèmes inverses de la théorie de Hecke dus à Jacquet, Langlands et Weil pour $n = 2$ et à Jacquet, Piatetski Shapiro et Shalika pour $n = 3$.

(1.2).   En fait, il est facile de construire la restriction $\tilde{f}_\sigma$ de f_σ à l'ensemble de doubles classes

$$Q_n(F) \backslash \mathrm{GL}_n(\mathbf{A}) / \mathrm{GL}_n(\mathcal{O})$$

où Q_n est le parabolique standard maximal de GL_n associé à la partition $(n - 1, 1)$, i.e., de Levi $\mathrm{GL}_{n-1} \times \mathrm{GL}_1$ inclus diagonalement dans GL_n.

Plus précisément, fixons une 1-forme différentielle méromorphe non nulle δ sur X ($\delta \in \Omega_{F/k}^1 - \{0\}$) et un caractère additif non trivial $\psi\colon k \to \overline{\mathbf{Q}}_\ell^\times$. On note B_n le Borel standard de GL_n (matrices triangulaires supérieures) et U_n son radical unipotent; δ et ψ définissent un caractère non trivial

$$\Psi\colon F \backslash \mathbf{A} \to \overline{\mathbf{Q}}_\ell^\times$$

par la formule

$$\Psi(\alpha) = \psi\bigg(\sum_{x \in |X|} \mathrm{tr}_{k_x/k}\big(\mathrm{Res}_x(\alpha_x \cdot \delta)\big) \bigg),$$

où $\mathrm{Res}_x\colon \Omega_{F_x/k_x}^1 - \{0\} \to k_x$ est l'application résidu, et un caractère

$$\Theta\colon U_n(F) \backslash U_n(\mathbf{A}) \to \overline{\mathbf{Q}}_\ell^\times$$

par la formule

$$\Theta(u) = \Psi\bigg(\sum_{i=1}^{n-1} u_{i,\,i+1} \bigg).$$

On note enfin Z_n le centre de GL_n.

Définition (1.2.1). Pour $\sigma \in \mathscr{G}_n^0$, on note W_σ et on appelle fonction de Whittaker associée à σ l'unique fonction

$$W_\sigma: Z_n(F)U_n(F) \backslash \mathrm{GL}_n(\mathbf{A})/\mathrm{GL}_n(\mathscr{O}) \to \overline{\mathbf{Q}}_\ell$$

qui vérifie les propriétés suivantes:

(i) $W_\sigma(ug) = \Theta(u) \cdot W_\sigma(g)$, $\forall u \in U_n(\mathbf{A})$ *et* $\forall g \in \mathrm{GL}_n(\mathbf{A})$,

(ii) $W_\sigma(g) = \prod_{x \in |X|} W_{\sigma,x}(g_x)$, *où* $W_{\sigma,x}$ *est la restriction de* W_σ *à* $\mathrm{GL}_n(F_x) \subset \mathrm{GL}_n(\mathbf{A})$,

(iii) *pour tout* $x \in |X|$ *et tout* $(a_1, \ldots, a_n) \in \mathbf{Z}^n$, *on a*

$$W_{\sigma,x}\big(\mathrm{diag}(\pi_x^{a_1}, \ldots, \pi_x^{a_n})\big)$$

$$= q_x^{\sum_{i=1}^{n-1}(i(i+1-2n)/2)d_i - (n(n-1)/2)a_n} \cdot \mathrm{tr}\big(R^{(d_1, \ldots, d_{n-1}; a_n)}(\sigma(\mathrm{Frob}_x))\big)$$

si tous les entiers

$$d_i = a_i - a_{i+1} + \mathrm{ord}_x(\delta) \qquad (i = 1, \ldots, n-1)$$

sont positifs ou nuls et

$$W_{\sigma,x}\big(\mathrm{diag}(\pi_x^{a_1}, \ldots, \pi_x^{a_n})\big) = 0$$

sinon, où on a noté

$$R^{(d_1, \ldots, d_{n-1}; a_n)}$$

la représentation algébrique de GL_n de plus haut poids $d_1\omega_1 + \cdots + d_{n-1}\omega_{n-1} + a_n\omega_n$ (ω_i étant le poids fondamental de GL_n correspondant à la puissance extérieure i-ième de la représentation tautologique de GL_n) et où $\mathrm{ord}_x(\delta)$ est l'ordre de δ en x ($\mathrm{ord}_x(fd\pi_x) = v_x(f)$ si $f \in F_x^\times$).

Remarques (1.2.2). (i) L'invariance de W_σ sous l'action de $Z_n(F)$ résulte du corps de classes abélien.

(ii) On a normalisé $W_{\sigma,x}$ par

$$W_{\sigma,x}\big(\mathrm{diag}(\pi_x^{-(n-1)\mathrm{ord}_x(\delta)}, \ldots, \pi_x^{-\mathrm{ord}_x(\delta)}, 1)\big) = 1.$$

On a alors, d'après [Sha] et [Shi], la formule suivante pour $\tilde{f}_\sigma$:

THÉORÈME (1.2.3) (Shalika, Shintani). *Pour $\sigma \in \mathscr{G}_n^0$, la fonction*

$$\tilde{f}_\sigma: Q_n(F) \backslash \mathrm{GL}_n(\mathbf{A})/\mathrm{GL}_n(\mathscr{O}) \to \overline{\mathbf{Q}}_\ell$$

définie par la formule

$$\tilde{f}_\sigma(g) = \sum_{\gamma \in \mathscr{U}_{n-1}(F)\backslash \mathrm{GL}_{n-1}(F)} W_\sigma\left(\begin{bmatrix} \gamma & 0 \\ 0 & 1 \end{bmatrix} g\right)$$

est l'unique fonction non identiquement nulle (à un scalaire près dans $\overline{\mathbf{Q}}_\ell^\times$) sur l'ensemble $Q_n(F) \setminus \mathrm{GL}_n(\mathbf{A})/\mathrm{GL}_n(\mathcal{O})$ qui est cuspidale et qui vérifie

$$T_x^i \tilde{f}_\sigma = q_x^{-i(i-1)/2} \mathrm{tr}(\Lambda^i)\sigma(\mathrm{Frob}_x))\tilde{f}_\sigma,$$

pour tous $x \in |X|$ et $i = 1, \ldots, n$.

Par suite, la conjecture (1.1.1) équivaut à la conjecture suivante:

CONJECTURE (1.2.4). *Pour tout $\sigma \in \mathscr{G}_n^0$, la fonction*

$$\tilde{f}_\sigma \colon Q_n(F) \setminus \mathrm{GL}_n(\mathbf{A})/\mathrm{GL}_n(\mathcal{O}) \to \overline{\mathbf{Q}}_\ell$$

définie en (1.2.3) *est en fait invariante à gauche par $\mathrm{GL}_n(F)$ tout entier.*

2. Traduction géométrique de la conjecture (1.1.1)

(2.0). Le point de départ de cette traduction géométrique est la remarque suivante de Weil: $\mathrm{GL}_n(F) \setminus \mathrm{GL}_n(\mathbf{A})/\mathrm{GL}_n(\mathcal{O})$ est en bijection avec l'ensemble des classes d'isomorphie de fibrés vectoriels de rang n sur X ($\mathrm{GL}_n(\mathbf{A})$ est en bijection avec l'ensemble des classes d'isomorphie de triplets

$$\left(\mathscr{L}, \mathscr{L}|\mathrm{Spec}(F) \simeq F^n, \left(\mathscr{L}|\mathrm{Spec}(\mathcal{O}_x) \simeq \mathcal{O}_x^n\right)_{x \in |X|}\right),$$

où $\mathscr{L}$ est un fibré vectoriel de rang n sur X).

En outre, cette traduction est basée sur le dictionnaire fonctions-faisceaux de Grothendieck: si Z est un schéma de type fini sur $\mathbf{F}_q$ et si $K \in \mathrm{ob}\, D_c^b(Z, \overline{\mathbf{Q}}_\ell)$, on peut associer à K la fonction $t_K \colon Z(\mathbf{F}_q) \to \overline{\mathbf{Q}}_\ell$ définie par

$$t_K(z) = \mathrm{tr}(\mathrm{Frob}_z, K).$$

(2.1). Soit X une courbe projective, lisse et géométriquement connexe de genre g sur un corps k de caractéristique $p \geqslant 0$ (on ne suppose plus k fini ni même de caractéristique strictement positive) et soit n un entier $\geqslant 1$. Soit ℓ un nombre premier distinct de p et soit $\overline{\mathbf{Q}}_\ell$ une clôture algébrique de $\mathbf{Q}_\ell$.

On considère d'une part la catégorie, que l'on notera encore $\mathscr{G}_n^0$, des $\overline{\mathbf{Q}}_\ell$-faisceaux *lisses* de rang n sur X qui sont *géométriquement irréductibles*, les flèches dans $\mathscr{G}_n^0$ étant les isomorphismes entre tels $\overline{\mathbf{Q}}_\ell$-faisceaux.

On considère d'autre part le champ algébrique lisse $Fib_{X,n}$ des fibrés vectoriels de rang n sur X; on a

$$Fib_{X,n} = \coprod_{\ell \in \mathbf{Z}} Fib_{X,n}^\ell$$

où $Fib_{X,n}^\ell$ est la composante connexe de $Fib_{X,n}$ des fibrés vectoriels de degré ℓ. On appellera simplement $\overline{\mathbf{Q}}_\ell$-faisceaux pervers sur $Fib_{X,n}$ les $K \in$

ob $D_c(Fib_{X,n}, \overline{\mathbf{Q}}_\ell)$ tels que, pour tout $\ell \in \mathbf{Z}$,

$$K^\ell \overset{\text{dfn}}{=} \left(K | Fib_{X,n}^\ell \right) \left[-(n-1)\ell + \frac{n(4n^2 - 3n + 5)}{6}(g-1) \right]$$

soit un $\overline{\mathbf{Q}}_\ell$-faisceau pervers au sens usuel sur $Fib_{X,n}^\ell$; on dira qu'un tel K est géométriquement irréductible si, pour tout $\ell \in \mathbf{Z}$, le $\overline{\mathbf{Q}}_\ell$-faisceau pervers K^ℓ sur $Fib_{X,n}^\ell$ l'est (en particulier un tel K est non identiquement nul sur chacune des composantes connexes $Fib_{X,n}^\ell$, $\ell \in \mathbf{Z}$). On note encore $\mathscr{A}_n^0$ la catégorie dont les objets sont les $\overline{\mathbf{Q}}_\ell$-faisceaux pervers géométriquement irréductibles sur $Fib_{X,n}$ (au sens ci-dessus) qui sont de plus:

(i) *cuspidaux*, i.e., tels que, pour toute partition $\underline{n} = (n_1, \ldots, n_r)$ de n non triviale, déterminant la correspondance

$$
\begin{array}{ccc}
 & Drap_{X,n} & \\
\varphi_{\underline{n}} \swarrow & & \searrow \pi_{\underline{n}}' \\
Fib_{X,n} & & Fib_{X,n_1} \times_k \cdots \times_k Fib_{X,n_r}
\end{array}
$$

où $Drap_{X,\underline{n}}$ est le champ algébrique lisse des drapeaux de fibrés vectoriels de type $\underline{n}$,

$$\mathscr{L}. = \left(\mathscr{L}_r \twoheadrightarrow \mathscr{L}_{r-1} \twoheadrightarrow \cdots \twoheadrightarrow \mathscr{L}_1 \twoheadrightarrow \mathscr{L}_0 \right),$$

avec $\mathscr{L}_j$ de rang $n_1 + \cdots + n_j$, et où

$$\varphi_{\underline{n}}(\mathscr{L}.) = \mathscr{L}_r$$

et

$$\pi_{\underline{n}}'(\mathscr{L}.) = \left(\text{Ker}(\mathscr{L}_r \twoheadrightarrow \mathscr{L}_{r-1}), \ldots, \text{Ker}(\mathscr{L}_1 \twoheadrightarrow \mathscr{L}_0) \right),$$

on ait

$$R\pi_{\underline{n}*}'\varphi_{\underline{n}}^* K = 0,$$

(ii) *vecteurs propres des opérateurs de Hecke* T^i ($i = 1, \ldots, n$) au sens où il existe $L_i \in \text{ob } D_c^b(X, \overline{\mathbf{Q}}_\ell)$ tel que

$$T^i(K) = L_i \boxtimes_k K,$$

l'opérateur T^i étant défini par la correspondance de Hecke de degré i

$$
\begin{array}{ccc}
 & Hecke_{X,n}^i & \\
\alpha_i \swarrow & & \searrow \beta_i \\
Fib_{X,n} & & X \times_k Fib_{X,n}
\end{array}
$$

et par

$$T^i(K) = R\beta_{i*}\alpha_i^* K,$$

où $Hecke_{X,n}^i$ est le champ algébrique des modifications supérieures de longueur i,

$$(\mathscr{L} \subset \mathscr{L}' \subset \mathscr{L}(x)),$$

long $(\mathscr{L}'/\mathscr{L}) = i$, et où

$$\alpha_i(\mathscr{L} \subset \mathscr{L}' \subset \mathscr{L}(x)) = \mathscr{L}'$$

et

$$\beta_i(\mathscr{L} \subset \mathscr{L}' \subset \mathscr{L}(x)) = (x, \mathscr{L}).$$

La traduction géométrique suivante de la conjecture (1.1.1) est due à Deligne et
Drinfeld.

CONJECTURE (2.1.1). *Pour tout* $E \in \mathrm{ob}\, \mathscr{G}_n^0$, *il existe* $K_E \in \mathrm{ob}\, \mathscr{A}_n^0$ *tel que,
pour* $i = 1, \ldots, n$, *on ait*

$$T^i(K_E) = (\Lambda^i E) \boxtimes_k K_E[i(i-1)]\left(\frac{i(i-1)}{2}\right).$$

De plus, K_E *est unique à tensorisation près par un* $\overline{\mathbf{Q}}_\ell$*-faisceau géométriquement
constant de rang* 1 *sur* $\mathrm{Fib}_{X,n}$ *et on peut normaliser* K_E *de telle sorte que* $E \mapsto K_E$
soit un foncteur de $\mathscr{G}_n^0$ *dans* $\mathscr{A}_n^0$ *qui commute aux extensions du corps de base* k.

Remarque. Pour $n = 1$, cette conjecture est une reformulation du corps de
classes abélien géométrique de Lang ([Se]). Pour $n = 2$, cette conjecture est
essentiellement démontrée par Drinfeld ([Dr]).

3. Le faisceau de Whittaker associé à un $\overline{\mathbf{Q}}_\ell$-faisceau lisse de rang n sur X

(3.0). L'objet de ce numéro est de traduire géométriquement les formules de
Shintani (1.2.1) (ii) et (iii) pour la fonction W_σ sur $Z_n(F) \backslash T_n(\mathbf{A})/T_n(\mathscr{O})$, où
$T_n \subset \mathrm{GL}_n$ est le tore maximal des matrices diagonales.

(3.1). On conserve les notations de (2.1). On considère de plus le champ
algébrique lisse $Coh_{X,0}$ des $\mathscr{O}_X$-Modules cohérents $\mathscr{M}$ de longueur finie (et donc
à support fini sur k); on a

$$Coh_{X,0} = \coprod_{n \in \mathbf{N}} Coh_{X,0}^m$$

où $Coh_{X,0}^m$ est la composante connexe de $Coh_{X,0}$ paramétrant les $\mathscr{O}_X$-Modules cohérents de longueur m. Rappelons que $GL_{m,k}$ agit naturellement sur le schéma de Grothendieck

$$\text{Quot}^m_{\mathcal{O}^m_X/X/k}$$

qui paramètre les $\mathcal{O}_X$-Modules quotients $q: \mathcal{O}^m_X \twoheadrightarrow \mathcal{M}$ tels que $\mathcal{M}$ soit de longueur m (cf. [Gr] 3) et que cette action préserve l'ouvert $\text{Quot}^{m,0}_{\mathcal{O}^m_X/X/k} \subset \text{Quot}^m_{\mathcal{O}^m_X/X/k}$ des $q: \mathcal{O}^m_X \twoheadrightarrow \mathcal{M}$ tels que

$$H^0(X, q): k^m = H^0(X, \mathcal{O}^m_X) \to H^0(X, \mathcal{M})$$

soit un isomorphisme; $Coh^m_{X,0}$ n'est autre que le champ algébrique quotient du k-schéma quasi-projectif et lisse $\text{Quot}^{m,0}_{\mathcal{O}^m_X/X/k}$ par cette action de $GL_{m,k}$.

On a une stratification de $Coh^m_{X,0}$ indexée par les partitions $\underline{m} = (m_1, \ldots, m_s)$, $m_1 \geqslant m_2 \geqslant \cdots \geqslant m_s$, de m. Pour chaque partition $\underline{m}$ de m, on a en fait un morphisme de champs algébriques

$$i_{\underline{m}}: X^{(m_1 - m_2)} \times_k \cdots \times_k X^{(m_{s-1} - m_s)} \times_k X^{(m_s)} \to Coh^m_{X,0}$$

d'image localement fermée qui envoie $(D_1, D_2, \ldots, D_s)$ sur le $\mathcal{O}_X$-Module cohérent de longueur m

$$\mathcal{O}_{X, D_1 + \cdots + D_s} \oplus \mathcal{O}_{X, D_2 + \cdots + D_s} \oplus \cdots \oplus \mathcal{O}_{X, D_s}$$

(on a noté $X^{(d)}$ la puissance symétrique d-ième sur k de X et identifié $X^{(d)}$ au schéma de modules des diviseurs effectifs D de degré d sur X, pour tout $d \in \mathbb{N}$; si $D \in X^{(d)}$, $\mathcal{O}_{X, D} = \mathcal{O}_X/\mathcal{O}_X(-D)$). Il est clair que $Coh^m_{X,0}$ est la réunion des images des $i_{\underline{m}}$ et que la strate ouverte est l'image de

$$i_{(m)}: X^{(m)} \to Coh^m_{X,0}.$$

La remarque suivante, due à Kazhdan, permet de mieux comprendre encore la structure de $Coh^m_{X,0}$ et de faire le lien entre notre construction des faisceaux de Whittaker et la construction de Springer des représentations des groupes de Weyl. Prenons pour X la droite affine $\mathbf{A}^1_k$ (X n'est plus projective), alors il revient au même de se donner un $\mathcal{O}_X$-Module cohérent $\mathcal{M}$ de longueur finie m ou de se donner le couple $(H^0(X, \mathcal{M}), \xi)$, où ξ est l'endomorphisme de $H^0(X, \mathcal{M})$ donné par l'action de la coordonnée standard de $\mathbf{A}^1_k$. Par suite,

$$\text{Quot}^{m,0}_{\mathcal{O}^m_X/X/k} = \text{Quot}^m_{\mathcal{O}^m_X/X/k}$$

s'identifie à $gl_{m,k}$ (l'algèbre de Lie de $GL_{m,k}$) et $Coh^m_{X,0}$ s'identifie au champ algébrique lisse quotient de $gl_{m,k}$ par l'action adjointe de $GL_{m,k}$. Chaque $\xi \in gl_m(k)$ admet des facteurs invariants $P_1(T), \ldots, P_s(T)$, qui sont des polynômes unitaires en T sur k non constants tels que $P_s(T)$ divise $P_{s-1}(T), \ldots$ et $P_2(T)$ divise $P_1(T)$; alors la classe de conjugaison de ξ est dans l'image de $i_{\underline{m}}$ pour la partition $\underline{m} = (m_1, \ldots, m_s)$ avec $m_i = \deg(P_i(T))$, $i = 1, \ldots, s$, de m.

(3.2). Considérons maintenant, pour chaque $m \in \mathbb{N}$, le champ algébrique lisse $\widetilde{Coh}^m_{X,0}$ paramètrant les drapeaux

$$\mathcal{M}. = \left(\mathcal{M}_m \twoheadrightarrow \mathcal{M}_{m-1} \twoheadrightarrow \cdots \twoheadrightarrow \mathcal{M}_1 \twoheadrightarrow \mathcal{M}_0 \right)$$

où chaque $\mathcal{M}_i$ est un $\mathcal{O}_X$-Module cohérent de longueur i. On a des projections naturelles

$$
\begin{array}{ccc}
\widetilde{Coh}^m_{X,0} & \xrightarrow{\ \rho\ } & X^m \\
{\scriptstyle \pi}\big\downarrow & & \\
Coh^m_{X,0} & &
\end{array}
$$

où

$$\pi(\mathcal{M}.) = \mathcal{M}_m$$

et

$$\rho(\mathcal{M}.) = (x_1, \ldots, x_m)$$

avec

$$x_i = \mathcal{N}_{X/k}\big(\mathrm{Ker}(\mathcal{M}_i \twoheadrightarrow \mathcal{M}_{i-1})\big)$$

($\mathcal{N}_{X/k}: Coh^1_{X,0} \to X$ est le morphisme norme, cf. [Gr] 6).

Le morphisme de champs algébriques $\pi: \widetilde{Coh}^m_{X,0} \to Coh^m_{X,0}$ est représentable et projectif; au-dessus de la strate ouverte $i_{(m)}(X^{(m)}) \subset Coh^m_{X,0}$, π est en fait un revêtement fini galoisien (ramifié si $m \geqslant 2$) de groupe de Galois le groupe symétrique $\mathfrak{S}_m$ et ρ induit un isomorphisme de champs algébriques

$$
\begin{array}{ccc}
X^{(m)} & & \widetilde{Coh}^m_{X,0} \xrightarrow{\ \sim\ } X^m \\
{\scriptstyle i_{(m)}}\!\!\searrow\!\!{\scriptstyle \times} & & \!\!\swarrow\!\!{\scriptstyle \pi} \\
& Coh^m_{X,0} &
\end{array}
$$

qui transforme la restriction de π à la strate ouverte en le revêtement fini canonique $r: X^m \to X^{(m)}, (x_1, \ldots, x_m) \mapsto x_1 + \cdots + x_m$.

Pour $X = \mathbf{A}^1_k$, le diagramme

$$
\begin{array}{ccc}
\widetilde{Coh}^m_{X,0} & \xrightarrow{\ \rho\ } & X^m \\
{\scriptstyle \pi}\big\downarrow & & \\
Coh^m_{X,0} & &
\end{array}
$$

s'identifie au diagramme

$$\begin{array}{ccc} \widetilde{\mathrm{gl}}_{m,k}/\mathrm{GL}_{m,k} & \xrightarrow{\rho} & \\ {\scriptstyle \pi}\downarrow & \searrow & \\ \mathrm{gl}_{m,k}/\mathrm{GL}_{m,k} & & t_{m,k} \end{array}$$

où

$$\widetilde{\mathrm{gl}}_{m,k} = \left\{ (\xi, gB_{m,k}) \in \mathrm{gl}_{m,k} \times_k (\mathrm{GL}_{m,k}/B_{m,k}) \,|\, g^{-1}\xi g \in b_{m,k} \right\},$$

où

$$\pi(\xi, gB_{m,k}) = \xi$$

$$\rho(\xi, gB_{m,k}) = \overline{g^{-1}\xi g}$$

et où $\mathrm{GL}_{m,k}$ agit sur $\widetilde{\mathrm{gl}}_{m,k}$ par

$$g' \cdot (\xi, gB_{m,k}) = \left(g'\xi g'^{-1}, g'gB_{m,k} \right)$$

($B_{m,k}$ est le sous-groupe de Borel de $\mathrm{GL}_{m,k}$ des matrices triangulaires supérieures, $b_{m,k}$ son algèbre de Lie, $t_{m,k} \subset b_{m,k}$ la sous-algèbre de Cartan des matrices diagonales et $\xi \mapsto \bar{\xi}$ la projection canonique de $b_{m,k}$ sur $t_{m,k}$).

(3.3). Soit maintenant E un $\overline{\mathbb{Q}}_\ell$-faisceau lisse de rang n sur la courbe X. Pour tout $m \in \mathbb{N}$, on peut former le complexe

$$R\pi_*\rho^*(E^{\boxtimes m}) \in \mathrm{ob}\, D_c^b\left(Coh^m_{X,0}, \overline{\mathbb{Q}}_\ell \right);$$

on a clairement

$$i^*_{(m)} R\pi_*\rho^*(E^{\boxtimes m}) = r_*(E^{\boxtimes m})$$

où $r\colon X^m \to X^{(m)}$ est le revêtement fini canonique.

THÉORÈME (3.3.1). *Le complexe $R\pi_*\rho^*(E^{\boxtimes m})$ est en fait un $\overline{\mathbb{Q}}_\ell$-faisceau pervers placé en degré 0. De plus, $R\pi_*\rho^*(E^{\boxtimes m})$ est le prolongement intermédiaire de sa restriction à $X^{(m)}$, i.e.,*

$$R\pi_*\rho^*(E^{\boxtimes m}) = i_{(m)!*}\left(r_*(E^{\boxtimes m})[m] \right).$$

Remarque (3.3.2). (i) Pour $X = \mathbf{A}^1_k$, ce résultat est dû à Lusztig (cf. [Lu 3] 3).

(ii) Comme $Coh^m_{X,0}$ est un champ algébrique lisse de dimension 0 sur k et que $i_{(m)}$ est un morphisme représentable de champs algébriques qui est lisse purement de dimension relative m, $i_{(m)!*}$ n'est à priori pas défini. Cependant, $i_{(m)}$ se comporte comme une immersion ouverte car $i_{(m)}$ admet une rétraction (!), à savoir la norme de Grothendieck

$$\mathcal{N}_{X/k}: \operatorname{Coh}^m_{X,0} \to X^{(m)}$$

qui, bien entendu, n'est pas représentable; en fait la différence de dimension entre $X^{(m)}$ et $\operatorname{Coh}^m_{X,0}$ vient uniquement des automorphismes.

Preuve du théorème. Pour toute partition $\underline{m} = (m_1, \ldots, m_s)$, $m_1 \geqslant m_2 \geqslant \cdots \geqslant m_s > 0$, de m et tout

$$(D_1, \ldots, D_s) \in X^{(m_1 - m_2)} \times_k \cdots \times_k X^{(m_s - m_{s+1})} \ (m_{s+1} = 0),$$

on vérifie facilement que

$$\dim\big(\pi^{-1}(i_{\underline{m}}(D_1, \ldots, D_s))\big) = \sum_{i=1}^{s} (m_i - m_{i+1}) \frac{i(i-1)}{2}$$

et que

$$\operatorname{codim}_{Coh^m_{X,0}}\big(i_{\underline{m}}\big(X^{(m_1 - m_s)} \times_k \cdots \times_k X^{(m_s - m_{s+1})}\big)\big) = \sum_{i=1}^{s} (m_i - m_{i+1})(i^2 - 1)$$

(la dimension du groupe des automorphismes de $\mathcal{O}_{X, D_1 + \cdots + D_s} \oplus \cdots \oplus \mathcal{O}_{X, D_s}$ est $\Sigma_{i=1}^{s}\deg(D_i)i^2$, pour tous diviseurs effectifs $D_1, \ldots, D_s$ sur X). Par suite, on a

$$\operatorname{codim}_{Coh^m_{X,0}}\big(i_{\underline{m}}\big(X^{(m_1 - m_2)} \times_k \cdots \times_k X^{(m_s - m_{s+1})}\big)\big)$$
$$> 2 \dim\big(\pi^{-1}(i_{\underline{m}}(D_1, \ldots, D_s))\big)$$

pour toute partition non triviale $\underline{m}$ de m et tout $(D_1, \ldots, D_s) \in X^{(m_1 - m_2)} \times_k \cdots \times_k X^{(m_s - m_{s+1})}$ et le théorème résulte de la caractérisation du prolongement intermédiaire (cf. [Go–Ma] 6.2).

Il résulte du théorème que le groupe symétrique $\mathfrak{S}_m$ agit sur $R\pi_*\rho^*(E^{\boxtimes m})$ et que

$$\big(R\pi_*\rho^*(E^{\boxtimes m})\big)^{\mathfrak{S}_m} = i_{(m)!*}\big(E^{(m)}[m]\big),$$

où $E^{(m)} = (r_*(E^{\boxtimes m}))^{\mathfrak{S}_m}$ est le $\overline{\mathbf{Q}}_\ell$-faisceau (constructible) sur $X^{(m)}$ puissance symétrique m-ième de E (cf. [De 1]).

Remarque (3.3.3). Pour $X = \mathbf{A}^1_k$ et $E = \overline{\mathbf{Q}}_{\ell, X}$, on retrouve la construction de Springer des représentations des groupes de Weyl (à une normalisation près).

DÉFINITION (3.3.4). *Soit E un $\overline{\mathbf{Q}}_\ell$-faisceau lisse sur X. On appellera faisceau de Whittaker associé à E le $\overline{\mathbf{Q}}_\ell$-faisceau pervers W_E sur $\operatorname{Coh}_{X,0}$ dont la restriction à $\operatorname{Coh}^m_{X,0}$ est égale à*

$$i_{(m)!*}\big(E^{(m)}[m]\big)$$

pour tout $m \in \mathbf{Z}$.

Pour justifier cette définition, nous allons calculer la restriction de W_E aux strates non ouvertes

$$i_{\underline{m}}\colon X^{(m_1-m_2)} \times_k \cdots \times_k X^{(m_s-m_{s+1})} \to Coh_{X,0}^m$$

de $Coh_{X,0}^m$ ($\underline{m} = (m_1, \ldots, m_s)$ partition non triviale de m; $m_{s+1} = 0$).

Pour cela, il nous faut quelques rappels sur les foncteurs polynomiaux (cf. [Mac]). Soit $\underline{m} = (m_1, \ldots, m_s)$, $m_1 \geqslant m_2 \geqslant \cdots \geqslant m_s > 0$, une partition de m et soient $d_i = m_i - m_{i+1}$ ($i = 1, \ldots, s$; $m_{s+1} = 0$); on note $\mathfrak{S}_{\underline{m}}$ le sous-groupe $\mathfrak{S}_{m_1} \times \cdots \times \mathfrak{S}_{m_s}$ du groupe symétrique $\mathfrak{S}_m$ et $\chi^{\underline{m}}$ le caractère irréductible (sur $\mathbf{Q}$) de $\mathfrak{S}_m$ associé à $\underline{m}$. On a

$$\mathrm{Ind}_{\mathfrak{S}_{\underline{m}}}^{\mathfrak{S}_m}(1) = \sum_{\underline{m}' \geqslant \underline{m}} K_{\underline{m}', \underline{m}} \chi^{\underline{m}'}$$

où 1 est le caractère trivial de $\mathfrak{S}_{\underline{m}}$, où $\underline{m}'$ parcourt les partitions de m, où

$$\underline{m}' \geqslant \underline{m} \Leftrightarrow \begin{cases} m_1' \geqslant m_1 \\ m_1' + m_2' \geqslant m_1 + m_2 \\ m_1' + m_2' + m_3' \geqslant m_1 + m_2 + m_3 \\ \cdots \end{cases}$$

et où $K_{\underline{m}', \underline{m}} \in \mathbb{N}$ est le nombre des $\underline{m}$-tableaux standard de type $\underline{m}'$ (nombre de Kostka) et en particulier

$$K_{\underline{m}, \underline{m}} = 1.$$

Par exemple, on a

$$\chi^{(m)} = 1$$

et

$$\chi^{(1^m)} = \mathrm{sgn}_{\mathfrak{S}_m}$$

(caractère signature).

On définit alors le foncteur polynomial $R^{(d_1, \ldots, d_s)}V$ en le $\overline{\mathbf{Q}}_\ell$-vectoriel V par

$$R^{(d_1, \ldots, d_s)}V = \left(V^{\otimes m} \otimes_{\mathbf{Q}} \chi^{\underline{m}}\right)^{\mathfrak{S}_m}$$

($\mathfrak{S}_m$ agit par permutation sur $V^{\otimes m}$, de manière naturelle sur $\chi^{\underline{m}}$ et diagonalement sur le produit tensoriel). On a $R^{(d_1, \ldots, d_s)}V = 0$ si $s > \dim_{\overline{\mathbf{Q}}_\ell}(V)$ et, sinon, $R^{(d_1, \ldots, d_s)}V$ est la représentation irréductible de $\mathrm{GL}(V)$ de plus haut poids

$$d_1\omega_1 + \cdots + d_s\omega_s$$

(par exemple,

$$R^{(0,\ldots,0,1)}V = \Lambda^s V \qquad (s = 1, 2, \ldots)$$

et

$$R^{(m)}V = S^m V).$$

En outre, comme

$$\bigotimes_{i=1}^{s} S^{m_i}V = \left(V^{\otimes m} \otimes_{\mathbf{Q}} \mathrm{Ind}_{\mathfrak{S}_{\underline{m}}}^{\mathfrak{S}_m}(1) \right)^{\mathfrak{S}_m},$$

on a

$$\bigotimes_{i=1}^{s} S^{m_i}V \simeq \bigotimes_{\underline{m}' \geqslant \underline{m}} \left(R^{(d'_1,\ldots,d'_{s'})}V \right)^{K_{\underline{m}',\underline{m}}}.$$

Les coefficients de Kostka ont des "q-analogues", à savoir les polynômes de Kostka-Foulkes (cf. [Mac] III 6)

$$K_{\underline{m}',\underline{m}}(q) = \sum_{a \geqslant 0} K_{\underline{m}',\underline{m},a} \cdot q^a,$$

qui se spécialisent en les coefficients de Kostka pour $q = 1$ ($K_{\underline{m}',\underline{m}}(1) = K_{\underline{m}',\underline{m}}$). Ce sont des polynômes unitaires de degré

$$\sum_{i=1}^{s} i m_i - \sum_{i'=1}^{s'} i' m'_{i'}$$

en q à coefficients entiers $\geqslant 0$ pour $\underline{m}' \geqslant \underline{m}$ partitions de m et en particulier

$$K_{\underline{m},\underline{m}}(q) \equiv 1.$$

Ces rappels étant faits, décrivons maintenant la fibre

$$\left(\widetilde{Coh}^m_{X,0} \right)_{(D_1,\ldots,D_s)}$$

de $\pi \colon \widetilde{Coh}^m_{X,0} \to Coh^m_{X,0}$ en $i_{\underline{m}}(D_1,\ldots,D_s)$ pour une partition $\underline{m} = (m_1,\ldots,m_s)$ de m et pour $(D_1,\ldots,D_s) \in X^{(d_1)} \times_k \cdots \times_k X^{(d_s)}$ ($d_i = m_i - m_{i+1}, m_{s+1} = 0$). On a

$$D_i = \sum_{x \in X} d_{i,x}[x] \; (i = 1,\ldots,s)$$

(on s'est placé ici sur une clôture algébrique $\overline{k}$ de k et par point d'un k-schéma on entend un $\overline{k}$-point) et on a le diagramme

$$(\widetilde{Coh}^m_{X,0})_{(D_1,\ldots,D_s)} \xrightarrow{\varphi} \prod_{x\in X} (\widetilde{Coh}^{m_x}_{X,0})_{(d_{1,x}[x],\ldots,d_{s,x}[x])}$$
$$\rho \downarrow$$
$$X^m$$

où on a posé $m_x = \sum_{i=1}^s id_{i,x}$, $\forall x \in X$, où

$$\varphi(\mathcal{M}_m \twoheadrightarrow \cdots \twoheadrightarrow \mathcal{M}_1 \twoheadrightarrow \mathcal{M}_0) = (\mathcal{M}_{x,m_x} \twoheadrightarrow \cdots \twoheadrightarrow \mathcal{M}_{x,1} \twoheadrightarrow \mathcal{M}_{x,0})_{x\in X}$$

correspond à la décomposition canonique

$$\mathcal{M} = \bigoplus_{x\in X} \mathcal{M}_x$$

d'un $\mathcal{O}_X$-Module cohérent $\mathcal{M}$ de longueur finie en ses composantes $\mathcal{M}_x$ à support dans $x \in X$ et où on a noté encore ρ la restriction de $\rho\colon \widetilde{Coh}^m_{X,0} \to X^m$ à $(\widetilde{Coh}^m_{X,0})_{(D_1,\ldots,D_s)}$. On vérifie alors facilement que l'image de ρ est l'ensemble des $(x_1,\ldots,x_m)$ avec

$$x_1 + \cdots + x_m = \sum_{i=1}^s iD_i = \sum_{x\in X} m_x[x],$$

que cette image est en bijection avec

$$\mathfrak{S}_m / \prod_{x\in X} \mathfrak{S}_{m_x}$$

et que, pour tout $(x_1,\ldots,x_m)$ dans cette image, φ envoie isomorphiquement $\rho^{-1}(x_1,\ldots,x_m)$ sur $\prod_{x\in X}(\widetilde{Coh}^{m_x}_{X,0})_{(d_{1,x}[x],\ldots,d_{s,x}[x])}$. Par suite, on a:

LEMME (3.3.5). (i) *Si* $\sigma^{\cdot}$ *(resp.* $\sigma_x^{\cdot}$) *est la représentation de* $\mathfrak{S}_m$ *(resp.* $\mathfrak{S}_{m_x}$) *sur*

$$H^{\cdot}\left((\widetilde{Coh}^m_{X,0})_{(D_1,\ldots,D_s)}, \overline{\mathbb{Q}}_\ell\right)$$

$\Big($*resp.*

$$H^{\cdot}\left((\widetilde{Coh}^{m_x}_{X,0})_{(d_{1,x}[x],\ldots,d_{s,x}[x])}, \overline{\mathbb{Q}}_\ell\right)\Big)$$

définie ci-dessus, on a

$$\sigma^{\cdot} = \operatorname{Ind}_{\prod_{x\in X} \mathfrak{S}_{m_x}}^{\mathfrak{S}_m}\left(\boxtimes_{x\in X} \sigma_x^{\cdot}\right).$$

(ii) *Pour tout* $\overline{\mathbf{Q}}_\ell$*-faisceau lisse E sur X, on a*

$$\mathcal{H}^{\cdot}(i_{\underline{m}}^* W_E)_{(D_1,\ldots,D_s)} = \bigotimes_{x\in X}\left(E_x^{\otimes m_x}\otimes\sigma_x^{\cdot}\right)^{\mathfrak{S}_{m_x}}.$$

Il reste à traiter le cas où $D_1,\ldots,D_s$ sont tous supportés par un même point $x \in X$, i.e., où $D_i = d_i[x]$ $(i = 1,\ldots,s)$. Dans ce cas, on peut supposer $X = \mathbf{A}^1_k$ et $x = 0$. Alors, par la remarque de Kazhdan, on est ramené au calcul de la représentation de Springer de $\mathfrak{S}_m$ sur $H^{\cdot}(\mathcal{B}_{\underline{m}},\overline{\mathbf{Q}}_\ell)$ où $\mathcal{B}_{\underline{m}}$ est la sous-variété de $\mathrm{GL}_{m,k}/B_{m,k}$ formée des Borel de $\mathrm{GL}_{m,k}$ fixés par la matrice unipotente standard de $\mathrm{GL}_{m,k}$ dont les blocs de Jordan sont de tailles $m_1,\ldots,m_s$. Ce calcul, dû à Macdonald (cf. [Ho-Sh] 8.5 et [Spr] 4.13), se traduit pour nous par le lemme suivant:

LEMME (3.3.6). *Pour tout $x \in X$ et tout $\overline{\mathbf{Q}}_\ell$-faisceau lisse E sur X,*

$$(i_{\underline{m}}^* W_E)_{(d_1[x],\ldots,d_s[x])}$$

n'a de cohomologie qu'en degrés pairs, compris entre 0 et $\sum_{i=1}^s d_i(i^2 - i)$ et, pour $a = 0,\ldots,\sum_{i=1}^s d_i i(i-1)/2$, on a un isomorphisme de $\overline{\mathbf{Q}}_\ell[\mathfrak{S}_m]$-modules, compatible à l'action de $\mathrm{Gal}(\overline{k}/k)$

$$\mathcal{H}^{2a}\left((i_{\underline{m}}^* W_E)_{(d_1[x],\ldots,d_s[x])}\right)(a) \simeq \bigoplus_{\underline{m}'\geqslant\underline{m}}\left(R^{(d_1',\ldots,d_s')}E_x\right)^{K_{\underline{m}',\underline{m},a}}$$

où $\underline{m}'$ parcourt les partitions de m avec $s' \leqslant \dim_{\overline{\mathbf{Q}}_\ell}(E_x)$ (et $\underline{m}' \geqslant \underline{m}$) et où $K_{\underline{m}',\underline{m},a}$ est le coefficient de q^a dans le polynôme de Kostka-Foulkes $K_{\underline{m}',\underline{m}}(q)$.

Remarque (3.3.7). On a donc un isomorphisme de $\overline{\mathbf{Q}}_\ell[\mathfrak{S}_m]$-modules, non compatible à l'action de $\mathrm{Gal}(\overline{k}/k)$,

$$\mathcal{H}^{\cdot}\left((i_{\underline{m}}^* W_E)_{(d_1[x],\ldots,d_s[x])}\right) \simeq \bigotimes_{i=1}^s S^{m_i}E_x.$$

En regroupant les résultats des lemmes (3.3.6) et (3.3.7), on obtient en particulier

THÉORÈME (3.3.8). *Pour tout $\overline{\mathbf{Q}}_\ell$-faisceau lisse E sur X, tout entier positif m et toute partition $\underline{m} = (m_1,\ldots,m_s)$ de m, le $\overline{\mathbf{Q}}_\ell$-faisceau de cohomologie en degré $\sum_{i=1}^s d_i(i^2 - i)$ (degré maximal) de $i_{\underline{m}}^* W_E$ sur $X^{(d_1)}\times_k\cdots\times_k X^{(d_s)}$ $(d_i = m_i - m_{i+1}; m_{s+1} = 0)$ a pour fibre en $(D_1,\ldots,D_s)$ le $\overline{\mathbf{Q}}_\ell$-vectoriel*

$$\bigotimes_{x\in X}\left(R^{(d_{1,x},\ldots,d_{s,x})}E_x\right)\left(-\sum_{i=1}^s d_{i,x}\frac{i(i-1)}{2}\right),$$

où on a posé $D_i = \sum_{x\in X}d_{i,x}[x]$ $(i = 1,\ldots,s)$.

Remarque (3.3.9).   Ce théorème justifie la terminologie "faisceau de Whittaker associé à E" et la notation W_E (comparer la formule ci-dessus à celle de Shintani, cf. (1.2.1) (ii) et (iii)).

Terminons ce numéro par deux propriétés importantes de W_E.

THÉORÈME (3.3.10).   *Si le $\overline{\mathbf{Q}}_\ell$-faisceau lisse E sur X est géométriquement irréductible, pour tout entier $m \geqslant 0$ le $\overline{\mathbf{Q}}_\ell$-faisceau pervers $W_E|Coh_{X,0}^m$ est géométriquement irréductible.*

Preuve.   Soit $X_{(1^m)}^{(m)}$ l'ouvert de $X^{(m)}$ formé des diviseurs $D = x_1 + \cdots + x_m$ avec $x_i \neq x_j$ si $i \neq j$ ($i, j = 1, \ldots, m$), alors $W_E|Coh_{X,0}^m$ s'obtient par prolongement intermédiaire à partir du $\overline{\mathbf{Q}}_\ell$-faisceau lisse $E^{(m)}|X_{(1^m)}^{(m)}$. Or, ce $\overline{\mathbf{Q}}_\ell$-faisceau lisse s'obtient par descente du $\overline{\mathbf{Q}}_\ell$-faisceau lisse $(E^{\boxtimes m})|r^{-1}(X_{(1^m)}^{(m)})$, où $r : X^m \to X^{(m)}$ est le revêtement fini galoisien canonique (cette remarque m'a été signalée par Gabber). Par suite, comme $E^{\boxtimes m}$ est géométriquement irréductible et que $r^{-1}(X_{(1^m)}^{(m)}) \subset X^m$ est un ouvert dense d'une k-variété lisse, $(E^{\boxtimes m})|r^{-1}(X_{(1^m)}^{(m)})$ est géométriquement irréductible donc $E^{(m)}|X_{(1^m)}^{(m)}$ aussi. D'où la conclusion.

La deuxième propriété importante de W_E est la nature remarquable de sa variété caractéristique. Celle-ci n'étant bien définie pour l'instant qu'en caractéristique nulle, nous supposerons que $\mathrm{car}(k) = p = 0$.

Commençons par décrire le fibré cotangent $T^*Coh_{X,0}$ au champ algébrique lisse $Coh_{X,0}$. C'est un champ algébrique muni d'une projection représentable

$$T^*Coh_{X,0} \to Coh_{X,0}.$$

dont la fibre au point $\mathcal{M}$ de $Coh_{X,0}$ s'identifie canoniquement au vectoriel

$$\mathrm{Hom}_{\mathcal{O}_X}\left(\mathcal{M}, \mathcal{M} \otimes_{\mathcal{O}_X} \Omega_X^1\right).$$

Plus précisément, on a

$$T^*Coh_{X,0} = \mathbb{V}(\mathcal{T}^1)$$

($\mathbb{V}(-)$ au sens de Grothendieck, cf. [Di-Gr]), où $\mathcal{T}^1$ est le $\mathcal{O}_{Coh_{X,0}}$-Module cohérent défini par

$$\mathcal{T}^1 = R^1 pr_{2*}\underline{R\mathrm{Hom}}_{\mathcal{O}_{X \times_k Coh_{X,0}}}(\tilde{\mathcal{M}}, \tilde{\mathcal{M}}),$$

avec $pr_2 \colon X \times_k Coh_{X,0} \to Coh_{X,0}$ la seconde projection canonique et $\tilde{\mathcal{M}}$ le $\mathcal{O}_{X \times_k Coh_{X,0}}$-Module cohérent universel.

Remarque (3.3.11).   Pour $X = \mathbf{A}_k^1$, on a $Coh_{X,0}^m = \mathfrak{gl}_{m,k}/\mathrm{GL}_{m,k}$ et

$$T^*Coh_{X,0}^m = \{(\xi, \xi') \in \mathfrak{gl}_{m,k} \times_k \mathfrak{gl}_{m,k}|[\xi, \xi'] = 0\}/\mathrm{GL}_{m,k},$$

326 GERARD LAUMON

où $\mathrm{GL}_{m,k}$ agit par $g.(\xi, \xi') = (g\xi g^{-1}, g\xi' g^{-1})$ et où la projection

$$T^*Coh^m_{X,0} \to Coh^m_{X,0}$$

est donnée par $(\xi, \xi') \mapsto \xi$. On voit sur ce cas particulier que, $m \geq 2$, $T^*Coh_{X,0}$ n'est pas lisse sur k.

Fixons maintenant un entier $n \geq 1$ et, pour toute suite $\underline{e} = (e_1, \ldots, e_n)$ d'entiers ≥ 0, notons

$$X^{(m)}_{(1^{e_1} \ldots n^{e_n})} \subset X^{(m)} \left(m = \sum_{i=1}^{n} i \cdot e_i \right)$$

la sous-variété localement fermée de $X^{(m)}$ formée des $D \in X^{(m)}$ de la forme

$$D = \sum_{i=1}^{n} i([x_{i1}] + \cdots + [x_{i, e_i}]),$$

où les $x_{i,j} \in X$ ($j = 1, \ldots, e_i$, $i = 1, \ldots, n$) sont deux à deux distincts. Alors,

$$Coh_{X,0,\underline{e}} = i_{(m)}\left(X^{(m)}_{(1^{e_1} \ldots n^{e_n})} \right)$$

est un sous-champ (lisse sur k) localement fermé de $Coh_{X,0}$.

DÉFINITION (3.3.12). *On note Λ_n le fermé lagrangien de $T^*Coh_{X,0}$ réunion des adhérences dans $T^*Coh_{X,0}$ des conormaux*

$$T^*_{Coh_{X,0,\underline{e}}} Coh_{X,0}$$

pour $\underline{e} = (e_1, \ldots, e_n) \in \mathbb{N}^n$.

On a $\Lambda_1 = T^*_{Coh_{X,0}} Coh_{X,0}$ (la section nulle du cotangent),

$$\Lambda_1 \subset \Lambda_2 \subset \cdots \subset \Lambda_n \subset \cdots \subset T^*Coh_{X,0}$$

et

$$\Lambda_m \cap T^*Coh^m_{X,0} = \Lambda_n \cap T^*Coh^m_{X,0}$$

pour tout $n \geq m$, de sorte que

$$\Lambda = \bigcup_{n \geq 1} \Lambda_n$$

est aussi un fermé lagrangien de $T^*Coh_{X,0}$.

THÉORÈME (3.3.13). (i) *Pour tout entier n et tout point $\mathcal{M}$ de $Coh_{X,0}$, la fibre de $\Lambda_n \to Coh_{X,0}$ en $\mathcal{M}$ s'identifie au cône dans $\mathrm{Hom}_{\mathcal{O}_X}(\mathcal{M}, \mathcal{M} \otimes_{\mathcal{O}_X} \Omega^1_X)$ formé des*

$$u: \mathcal{M} \to \mathcal{M} \otimes_{\mathcal{O}_X} \Omega^1_X$$

nilpotents de niveau $\leqslant n$, i.e., tels que l'homomorphisme composé

$$\mathcal{M} \overset{u}{\to} \mathcal{M} \otimes_{\mathcal{O}_X} \Omega^1_X \overset{u \otimes 1}{\to} \mathcal{M} \otimes_{\mathcal{O}_X} \left(\Omega^1_X\right)^{\otimes 2} \to ---\to \mathcal{M} \otimes_{\mathcal{O}_X} \left(\Omega^1_X\right)^{\otimes(n-1)}$$

$$\xrightarrow{u \otimes 1^{\otimes(n-1)}} \mathcal{M} \otimes_{\mathcal{O}_X} \left(\Omega^1_X\right)^{\otimes n}$$

soit identiquement nul (on a noté simplement 1 l'identité de Ω^1_X).

(ii) *Pour tout $\overline{\mathbf{Q}}_\ell$-faisceau lisse de rang n sur X, la variété caractéristique du $\overline{\mathbf{Q}}_\ell$-faisceau pervers W_E est égale à Λ_n et la multiplicité de l'adhérence de*

$$T^*_{Coh_{X,0,\underline{e}}} Coh_{X,0}$$

dans le cycle caractéristique de W_E est égale à

$$\prod_{k=1}^{n} \binom{n}{k}^{e_k}.$$

pour tout $\underline{e} = (e_1, \ldots, e_n) \in \mathbf{N}^n$.

Preuve. Le problème est local pour la topologie étale sur $Coh_{X,0}$ donc semi-local pour la topologie étale sur X. Par suite, il suffit de traiter le cas $X = \mathbf{A}^1_k$ et $E = (\overline{\mathbf{Q}}_{\ell,x})^n$.

Fixons un entier $m \geqslant 0$, alors, compte-tenu de la remarque (3.3.11), $\Lambda_n \cap T^*Coh^m_{X,0}$ admet la description suivante:

$$\Lambda_n \cap T^*Coh^m_{X,0} = \dot{\Lambda}^m_n / \mathrm{GL}_{m,k}$$

où

$$\dot{\Lambda}^m_n = \left\{ (\xi, \xi') \in \mathfrak{gl}_{m,k} \times_k \mathfrak{gl}_{m,k} \,\middle|\, [\xi, \xi'] = 0 \text{ et } \xi'^n = 0 \right\}.$$

Notons $\mathcal{O}_{\underline{e}} \subset \mathfrak{gl}_{m,k}$ l'orbite de la matrice de Jordan nilpotente ayant, pour chaque $i = 1, \ldots, n$, e_i blocs de Jordan de taille i ($\underline{e} = (e_1, \ldots, e_n) \in \mathbf{N}^n$ et $\Sigma_{i=1}^n i e_i = m$). Alors, on a

$$\dot{\Lambda}^m_n = \bigcup_{\underline{e}} \dot{\Lambda}^m_{n,\underline{e}} \text{ (réunion disjointe)},$$

où

$$\dot{\Lambda}^m_{n,\,\underline{e}} = \{(\xi,\xi')|[\xi,\xi'] = 0 \text{ et } \xi' = \mathcal{O}_{\underline{e}}\}$$

est une sous-variété localement fermée de $\mathrm{gl}_{m,\,k} \times_k \mathrm{gl}_{m,\,k}$ lisse et géométriquement irréductible sur k de dimension m^2 ($\dot{\Lambda}^m_{n,\,\underline{e}} \to \mathcal{O}_{\underline{e}}$, $(\xi,\xi') \mapsto \xi'$, est lisse, de fibre en ξ' l'algèbre de Lie du stabilisateur de ξ' dans $\mathrm{GL}_{m,\,k}$). De plus, si l'on note pour toute suite $\underline{e}$ comme ci-dessus

$$\mathrm{gl}_{m,\,k,\,\underline{e}} \subset \mathrm{gl}_{m,\,k}$$

la réunion des classes de $\mathrm{GL}_{m,\,k}$-conjugaison des matrices de Jordan de la forme

$$\mathrm{diag}\big(J(\alpha_{n1},n),\ldots,J(\alpha_{ne_n},n),\ldots,J(\alpha_{11},1),\ldots,J(\alpha_{1e_1},1)\big),$$

où $J(\alpha,i)$ est le bloc de Jordan de taille i

$$\begin{bmatrix} \alpha & 1 & & \\ & \alpha & \ddots & \\ & & \ddots & 1 \\ & & & \alpha \end{bmatrix}$$

et où $\alpha_{ij} \neq \alpha_{i'j'}$ si $(i,j) \neq (i',j')$, on vérifie facilement que

$$\{(\xi,\xi') \in \mathrm{gl}_{m,\,k,\,\underline{e}} \times_k \mathcal{O}_{\underline{e}}|[\xi,\xi'] = 0\}$$

est à la fois un ouvert dense de $\dot{\Lambda}^m_{n,\,\underline{e}}$ et du conormal de $\mathrm{gl}_{m,\,k,\,\underline{e}}$ dans $\mathrm{gl}_{m,\,k}$. Par suite, $\dot{\Lambda}^m_n$ est un fermé lagrangien de $T^*\mathrm{gl}_{m,\,k}$, égal à la réunion des adhérences des conormaux des $\mathrm{gl}_{m,\,k,\,\underline{e}}$ dans $\mathrm{gl}_{m,\,k}$. Ceci démontre l'assertion (i) du théorème.

Pour l'assertion (ii), commençons par remarquer que la variété caractéristique de W_E est contenue dans Λ. En effet, il suffit de montrer que la variété caractéristique du $\overline{\mathbf{Q}}_\ell$-faisceau pervers

$$R\dot{\pi}_*\overline{\mathbf{Q}}_{\ell,\,\widetilde{\mathrm{gl}}_{m,\,k}}[m^2],$$

où

$$\dot{\pi}\colon \widetilde{\mathrm{gl}}_{m,\,k} \to \mathrm{gl}_{m,\,k}, \qquad (\xi, gB_{m,\,k}) \mapsto \xi,$$

est contenue dans

$$\dot{\Lambda}^m = \{(\xi,\xi')|[\xi,\xi'] = 0 \text{ et } \xi' \text{ nilpotent}\}.$$

Or cela résulte immédiatement du fait suivant: pour tout $(\xi, gB_{m,\,k}) \in \widetilde{\mathrm{gl}}_{m,\,k}$, le noyau de

$$T_\xi^* \mathrm{gl}_{m,k} \to T^*_{(\xi,\, gB_{m,k})} \widetilde{\mathrm{gl}}_{m,k}$$

s'identifie canoniquement à

$$\left\{ \xi' \,|\, [\xi, \xi'] = 0 \text{ et } g^{-1}\xi' g \in u_{n,k} \right\},$$

où $u_{n,k} \subset b_{n,k}$ est le radical nilpotent (matrices triangulaires supérieures strictes).

Pour achever la preuve de l'assertion (ii), il reste à calculer la multiplicité de chaque composante irréductible de Λ dans le cycle caractéristique de W_E. Or, d'après ce qui précède, toute composante irréductible de $\Lambda \cap T^*Coh^m_{X,0}$ est l'adhérence de sa restriction à la strate ouverte $i_{(m)}(X^{(m)})$ de $Coh^m_{X,0}$ ($\mathrm{gl}_{m,k,\underline{e}} \subset \mathrm{gl}_{m,k,\mathrm{reg}}$, l'ouvert des éléments réguliers). Par suite, il suffit de calculer le cycle caractéristique de $E^{(m)}$ dans $T^*X^{(m)}$ ($m \geqslant 0$) pour achever la preuve du théorème. Nous supposerons toujours $E = (\overline{\mathbf{Q}}_{\ell,X})^n$ mais X sera de nouveau arbitraire (prendre $X = \mathbf{A}^1_k$ ne simplifie pas la fin de la démonstration du théorème). On sait déjà que la variété caractéristique de $E^{(m)}$ est contenue dans la variété lagrangienne

$$\bigcup_{\underline{e}} \text{adhérence}\left(T^*_{X^{(m)}_{(1_1^e \ldots\, m^e m)}} X^{(m)} \right) \subset T^*X^{(m)},$$

où $\underline{e} = (e_1, \ldots, e_m) \in \mathbf{N}^m$ et $\sum_{i=1}^m ie_i = m$, et il faut montrer que la multiplicité de la composante irréductible de cette variété lagrangienne correspondant à $\underline{e}$ dans le cycle caractéristique de $E^{(m)}$ est

$$\prod_{k=1}^m \binom{n}{k}^{e_k}$$

et en particulier nulle si $e_i > 0$ pour un $i > n$. Comme $E = (\overline{\mathbf{Q}}_{\ell,X})^n$, on a en fait

$$E^{(m)} = \bigoplus_{\substack{\mu_1 + \cdots + \mu_n = m \\ \mu_i \in \mathbf{N}}} \left(r^{(\mu_1,\ldots,\mu_n)} \right)_* \overline{\mathbf{Q}}_{\ell,\, X^{(\mu_1)} \times_k \cdots \times_k X^{(\mu_n)}},$$

où

$$r^{(\mu_1,\ldots,\mu_n)} \colon\; X^{(\mu_1)} \times_k \cdots \times_k X^{(\mu_n)} \to X^{(m)}$$

envoie $(\Delta_1, \ldots, \Delta_n)$ sur $\Delta = \Delta_1 + \cdots + \Delta_n$. De plus, le calcul du cycle caractéristique de

$$\left(r^{(\mu_1,\ldots,\mu_n)} \right)_* \overline{\mathbf{Q}}_{\ell,\, X^{(\mu_1)} \times_k \cdots \times_k X^{(p_n)}}$$

ne pose pas de difficultés: on trouve que la multiplicité de l'adhérence du conormal de $X^{(m)}_{(1^{e_1}\ldots m^{e_m})}$ dans ce cycle caractéristique est égale au nombre des $(\Delta_1,\ldots,\Delta_n) \in X^{(\mu_1)}_{(1^{\mu_1})} \times_k \cdots \times_k X^{(\mu_n)}_{(1^{\mu_n})}$ tels que

$$\Delta_1 + \cdots + \Delta_n = D$$

où D est un point fixé de $X^{(m)}_{(1^{e_1}\ldots m^{e_m})}$, i.e., au nombre des matrices

$$\left(\lambda_{ij}\right)_{\substack{i=1,\ldots,n\\ j=1,\ldots,e_1+e_2+\cdots+e_m}}$$

avec $\lambda_{ij} \in \{0,1\}$ pour tous i, j, $\sum_{j=1}^{e_1+\cdots+e_m}\lambda_{ij} = \mu_i$ pour $i = 1,\ldots,n$ et

$$\sum_{i=1}^{n}\lambda_{ij} = k \text{ si } e_1 + \cdots + e_{k-1} < j \leqslant e_1 + \cdots + e_k$$

pour $k = 1,\ldots, m$. Par suite, on trouve que la multiplicité de l'adhérence du conormal de $X^{(m)}_{(1^{e_1}\ldots m^{e_m})}$ dans le cycle caractéristique de $E^{(m)}$ est égale au nombre des matrices

$$\left(\lambda_{ij}\right)_{\substack{i=1,\ldots,n\\ j=1,\ldots,e_1+e_2+\cdots+e_m}}$$

avec $\lambda_{ij} \in \{0,1\}$ pour tous i, j et

$$\sum_{i=1}^{n}\lambda_{ij} = k \text{ si } e_1 + \cdots + e_{k-1} < j \leqslant e_1 + \cdots + e_k$$

pour $k = 1,\ldots, m$. D'où la conclusion.

4. Construction géométrique conjecturale de $\tilde{K}_E$

(4.0). ON GARDE LES NOTATIONS DE (2.1). Une fois que l'on a interprété géométriquement la fonction de Whittaker W_σ associée à une représentation ℓ-adique σ de rang n de $\mathrm{Gal}(\overline{F}/F)$ (partout non ramifiée et géométriquement irréductible), il est naturel d'essayer d'interpréter géométriquement la fonction

$$\tilde{f}_\sigma : Q_n(F) \backslash \mathrm{GL}_n(\mathbf{A})/\mathrm{GL}_n(\mathcal{O}) \to \overline{\mathbf{Q}}_\ell$$

définie par

$$\tilde{f}_\sigma(g) = \sum_{\lambda \in U_{n-1}(F)\backslash \mathrm{GL}_{n-1}(F)} W_\sigma\left(\begin{bmatrix} \gamma & 0 \\ 0 & 1 \end{bmatrix} g\right).$$

Pour $n = 1$, il n'y a rien à faire. Pour $n = 2$, cette interprétation géométrique est due à Drinfeld ([Dr]). Pour $n \geqslant 3$, nous proposons ci-dessous une interprétation géométrique conjecturale de $\tilde{f}_\sigma$.

(4.1). Commençons par réécrire en termes de fibrés vectoriels, diviseurs et extensions de fibrés vectoriels la définition de $\tilde{f}_\sigma$ pour mieux situer le problème.

On suppose $k = \mathbb{F}_q$ et on utilise librement les notations des numéros 1 et 2.

La première remarque est que pour toute partition $\underline{n} = (n_1, \ldots, n_r)$ de n, l'ensemble des doubles classes

$$P_{\underline{n}}(F) \backslash \mathrm{GL}_n(\mathbf{A})/\mathrm{GL}_n(\mathcal{O}),$$

qui est égal à l'ensemble des doubles classes

$$P_{\underline{n}}(F) \backslash P_{\underline{n}}(\mathbf{A})/P_{\underline{n}}(\mathcal{O}),$$

s'identifie à l'ensemble des classes d'isomorphisme d'objets de $Drap_{X, \underline{n}}(\mathbb{F}_q)$. Par exemple, pour $\underline{n} = (n_1, n_2)$, on associe à la matrice

$$\lambda_2 = \begin{pmatrix} \alpha_2 & \nu \\ 0 & \lambda_1 \end{pmatrix} \in P_{(n_1, n_2)}(\mathbf{A})$$

le drapeau

$$\mathscr{L}_2 \twoheadrightarrow \mathscr{L}_1$$

où $\mathscr{L}_i$ est le fibré vectoriel de rang $n_1 + \cdots + n_i$ associé à la double classe

$$\mathrm{GL}_{n_1 + \cdots + n_i}(F)\lambda_i \mathrm{GL}_{n_1 + \cdots + n_i}(\mathcal{O})$$

pour $i = 1, 2$; le noyau

$$\mathscr{A}_2 = \mathrm{Ker}(\mathscr{L}_2 \twoheadrightarrow \mathscr{L}_1)$$

est alors le fibré vectoriel de rang n_2 associé à la double classe

$$\mathrm{GL}_{n_2}(F)\alpha_2 \mathrm{GL}_{n_2}(\mathcal{O})$$

et l'extension (à un scalaire près) est déterminée par la classe de ν dans

$$\mathrm{Ext}^1_{\mathcal{O}_X}(\mathscr{L}_1, \mathscr{A}_2) = M_{n_2 \times n_1}(\mathbf{A})/\left(M_{n_2 \times n_1}(F)\lambda_1 + \alpha_2 M_{n_2 \times n_1}(\mathcal{O})\right).$$

Pour tout entier $i = 1, \ldots, n$, on désignera par $\mathscr{L}_i$ un fibré vectoriel de rang i sur X et par D_i un diviseur effectif sur X, on posera

$$\mathscr{A}_i = \left(\Omega^1_X\right)^{\otimes(i-1)}(D_i + D_{i+1} + \cdots + D_n)$$

et on notera ℓ_i le degré de $\mathscr{L}_i$, d_i le degré de D_i et

$$a_i = d_i + d_{i+1} + \cdots + d_n + (i-1)(2g-2)$$

le degré de $\mathscr{A}_i$. On posera aussi $\mathscr{L}_0 = 0$.

Pour $i = 1, \ldots, n$, considérons alors les ensembles suivants:

$$\mathscr{S}_{i-1} = \big\{ (\mathscr{L}_{i-1}; D_i, D_{i+1}, \ldots, D_n) \big\}$$

et

$$\mathscr{P}_i = \big\{ (0 \to \mathscr{A}_i \to \mathscr{L}_i \to \mathscr{L}_{i-1} \to 0; D_i, D_{i+1}, \ldots, D_n) \big\}$$

où

$$0 \to \mathscr{A}_i \to \mathscr{L}_i \to \mathscr{L}_{i-1} \to 0$$

est une extension de fibrés vectoriels (scindée ou non scindée), prise à un scalaire près.

On a le diagramme suivant

(4.1.1)

$$
\begin{array}{ccccccc}
\mathscr{P}_n & & \mathscr{P}_{n-1} & & \mathscr{P}_2 & \mathscr{P}_1 & \\
& \underset{\check{\pi}_{n-1}}{\searrow} \swarrow & \searrow \underset{\pi_{n-1}}{} & \!\!-\,-\,-\,\searrow \underset{\pi_2}{} & \searrow \underset{\check{\pi}_1}{} \swarrow \underset{\pi_1}{} & \searrow \underset{\check{\pi}_0}{} & \\
& \mathscr{S}_{n-1} & & & \mathscr{S}_2 & \mathscr{S}_1 & \mathscr{S}_0
\end{array}
$$

où $\check{\pi}_{i-1}$ et π_i envoient

$$(0 \to \mathscr{A}_i \to \mathscr{L}_i \to \mathscr{L}_{i-1} \to 0; D_i, \ldots, D_n)$$

sur

$$(\mathscr{L}_{i-1}; D_i, \ldots, D_n)$$

et sur

$$(\mathscr{L}_i; D_{i+1}, \ldots, D_n)$$

respectivement. En outre, pour $i = 2, \ldots, n$

$$
\begin{array}{ccc}
\mathscr{P}_i & & \mathscr{P}_{i-1} \\
\underset{\check{\pi}_{i-1}}{\searrow} & & \swarrow \underset{\pi_{i-1}}{} \\
& \mathscr{S}_{i-1} &
\end{array}
$$

sont en dualité au sens suivant: $\check{\pi}_{i-1}$ et π_{i-1} ont comme fibres en

$$(\mathscr{L}_{i-1}; D_i, \ldots, D_n)$$

l'ensemble quotient

$$\mathrm{Ext}^1_{\mathscr{O}_X}(\mathscr{L}_{i-1}, \mathscr{A}_i)/\mathbb{F}_q^{\times}$$

et l'espace projectif des droites dans

$$\operatorname{Hom}_{\mathcal{O}_X}\left(\mathscr{A}_i \otimes \left(\Omega_X^1\right)^{\otimes -1}, \mathscr{L}_{i-1}\right)$$

respectivement (en effet, si l'on a un homomorphisme non nul, donc injectif,

$$\mathscr{A}_i \otimes \left(\Omega_X^1\right)^{\otimes -1} \hookrightarrow \mathscr{L}_{i-1},$$

il existe un unique diviseur effectif D_i tel que cet homomorphisme se factorise en

$$\mathscr{A}_i \otimes \left(\Omega_X^1\right)^{\otimes -1} \hookrightarrow \mathscr{A}_i \otimes \left(\Omega_X^1\right)^{\otimes -1}(D_i) \hookrightarrow \mathscr{L}_{i-1}$$

et que $\mathscr{L}_{i-1}/\mathscr{A}_i \otimes (\Omega_X^1)^{\otimes -1}(D_i)$ soit un fibré vectoriel $\mathscr{L}_{i-2}$; on pose $\mathscr{A}_{i-1} = \mathscr{A}_i \otimes (\Omega_X^1)^{\otimes -1}(D_i)$). On a donc une correspondance d'incidence

(4.1.2)
$$\begin{array}{ccc} & \mathscr{Z}_{i-1} & \\ {}^{\check{p}_{i-1}}\swarrow & & \searrow^{p_{i-1}} \\ \mathscr{P}_i & & \mathscr{P}_{i-1} \end{array}$$

où $\mathscr{Z}_{i-1}$ est le sous-ensemble de $\mathscr{P}_i \times_{\mathscr{S}_{i-1}} \mathscr{P}_{i-1}$ formé des couples

$$((0 \to \mathscr{A}_i \to \mathscr{L}_i \to \mathscr{L}_{i-1} \to 0;\ D_i, \ldots, D_n),$$

$$(0 \to \mathscr{A}_{i-1} \to \mathscr{L}_{i-1} \to \mathscr{L}_{i-2} \to 0;\ D_{i+1}, \ldots, D_n))$$

tels que l'homorphisme composé

$$\mathscr{A}_i \otimes \left(\Omega_X^1\right)^{\otimes -1} \xrightarrow{D_{i-1}} \mathscr{A}_{i-1} \hookrightarrow \mathscr{L}_{i-1}$$

scinde l'extension

$$0 \to \mathscr{A}_i \to \mathscr{L}_i \to \mathscr{L}_{i-1} \to 0$$

i.e., se relève en

$$\begin{array}{ccc} & & \mathscr{L}_i \\ & \nearrow & \downarrow \\ \mathscr{A}_i \otimes (\Omega_X^1)^{\otimes -1} & \hookrightarrow & \mathscr{L}_{i-1}. \end{array}$$

Pour tout $\overline{\mathbf{Q}}_\ell$-faisceau lisse et géométriquement irréductible de rang n, E, sur X, on définit par récurrence sur i les fonctions

(4.1.3)
$$V_{E,i}: \mathscr{P}_i \to \overline{\mathbf{Q}}_\ell \qquad (i = 1, \ldots, n)$$

en posant

$$V_{E,1}(0 \to \mathscr{A}_1 \to \mathscr{L}_1 \to \mathscr{L}_0 \to 0; D_1, \ldots, D_n) = W_E(D_1, \ldots, D_n)$$

avec

$$W_E(D_1, \ldots, D_n) = q^{\sum_{i=1}^{n-1}(i(i+1-2n)/2)d_i - ((n-1)/2)a_n}$$

$$\cdot \prod_{x \in |X|} \mathrm{tr}\!\left(\mathrm{Frob}_x, R^{(d_{1,x}, \ldots, d_{n-1,x}; a_{n,x})}E_x\right),$$

où l'on a noté $d_{i,x}$ la multiplicité de D_i en x et $a_{n,x} = d_{n,x} + (n-1)\mathrm{ord}_x(\delta)(\forall x \in |X|)$ (cf. (1.2)), et en posant

$$V_{E,i} = q \cdot \left(\check{\rho}_{i-1*}\rho_{i-1}^{*}V_{E,i-1}\right) - \left(\check{\pi}_{i-1}^{*}\pi_{i-1*}V_{E,i-1}\right),$$

où, pour toute application $\varphi\colon A \to B$ à fibres finies et toutes fonctions $f\colon A \to \overline{\mathbf{Q}}_\ell$ et $g\colon B \to \overline{\mathbf{Q}}_\ell$, on note $\varphi_* f\colon B \to \overline{\mathbf{Q}}_\ell$ et $\varphi^* g\colon A \to \overline{\mathbf{Q}}_\ell$ les fonctions définies par

$$(\varphi_* f)(b) = \sum_{\substack{a \in A \\ \varphi(a) = b}} f(a)$$

et

$$(\varphi^* g)(a) = g(\varphi(a)).$$

Compte-tenu de l'identification ci-dessus entre $Drap_{X,(n-1,1)}(\mathbf{F}_q)$ et l'ensemble des doubles classes

$$Q_n(F) \backslash \mathrm{GL}_n(\mathbf{A})/\mathrm{GL}_n(\mathcal{O}) = Q_n(F) \backslash Q_n(\mathbf{A})/Q_n(\mathcal{O}),$$

on voit que $\mathscr{P}_n$ s'identifie à l'ensemble de doubles classes

$$Q_n^1(F) \backslash \left\{ g \in Q_n(\mathbf{A}) | v_x(g_{n,n}) \geqslant (n-1)\mathrm{ord}_x(\delta), \forall x \in |X| \right\}/Q_n(\mathcal{O})$$

où $Q_n^1 \subset Q_n$ est le sous-groupe formé des $g \in Q_n$ tels que $g_{n,n} = 1$ et où $\delta \in \Omega_F^1 - \{0\}$ a été fixée en (1.2).

PROPOSITION (4.1.4). *Soit σ la représentation ℓ-adique de $\mathrm{Gal}(\overline{F}/F)$ associée à E. Alors la fonction $V_{E,n}$ sur $\mathscr{P}_n$ est égale à la restriction de la fonction $\tilde{f}_\sigma$ à $\mathscr{P}_n$ (voir ci-dessus).*

Preuve. Il suffit de transcrire géométriquement [Sha] (5.9).

Dans loc. cit., Shalika définit par récurrence sur i les fonctions

$$V_{\sigma,i}\colon \mathrm{GL}_n(\mathbf{A}) \to \overline{\mathbf{Q}}_\ell \qquad (i = 1, \ldots, n)$$

vérifiant les propriétés suivantes:

(i) $V_{\sigma,i}$ est invariante à gauche sous $Q_i^1(F) \subset \mathrm{GL}_n(F)\left(\gamma \mapsto \begin{bmatrix} \gamma & 0 \\ 0 & 1_{n-i} \end{bmatrix}\right)$ et invariante à droite sous $\mathrm{GL}_n(\mathcal{O})$,

(ii) $V_{\sigma,i}(vg) = \Theta(v) \cdot V_{\sigma,i}(g)$, $\forall v \in V_{(i,1,\ldots,1)}(\mathbf{A}) \subset U_n(\mathbf{A})$, où $V_{(i,1,\ldots,1)}$ est le radical unipotent du parabolique standard $P_{(i,1,\ldots,1)}$ de Levi $\mathrm{GL}_i \times \mathrm{GL}_1 \times \cdots \times \mathrm{GL}_1$,

(iii) $V_{\sigma,1} = W_\sigma$ et $V_{\sigma,n} = \tilde{f}_\sigma$.

La relation de récurrence définissant $V_{\sigma,i}$ à partir de $V_{\sigma,i-1}$ $(i = 2,\ldots,n)$ est

$$V_{\sigma,i}(g) = \sum_{\gamma \in Q_{i-1}^1(F)\backslash \mathrm{GL}_{i-1}(F)} V_{\sigma,i-1}\left(\begin{bmatrix} \gamma & 0 \\ 0 & 1_{n-i+1} \end{bmatrix} g\right).$$

Il résulte des propriétés (i) et (ii) ci-dessus que $V_{\sigma,i}$ peut être considérée comme une fonction sur l'ensemble

$$\left(Q_i^1(F)\backslash Q_i(\mathbf{A})/Q_i(\mathcal{O})\right) \times (\mathbf{A}^\times/\mathcal{O}^\times)^{n-i}.$$

Comme le support de $V_{\sigma,1} = W_\sigma$ est contenu dans

$$\left\{(a_1,\ldots,a_n) \in (\mathbf{A}^\times/\mathcal{O}^\times)^n \,\middle|\, \mathrm{ord}_x\!\left(a_j a_{j+1}^{-1}\delta\right) \geqslant 0,\, \forall x \in |X| \text{ et } \forall j = 1,\ldots,n-1\right\}$$

(cf. (1.2)), on vérifie par récurrence sur i, que le support de $V_{\sigma,i}$ est contenu dans

$$\left\{\left(\left(Q_i^1(F)\begin{pmatrix} g_{i-1} & u \\ 0 & a_i \end{pmatrix}\right)Q_i(\mathcal{O}), a_{i+1},\ldots,a_n\right)\,\middle|\,\mathrm{ord}_x\!\left(a_j a_{j+1}^{-1}\delta\right) \geqslant 0,\right.$$

$$\left.\forall x \in |X| \text{ et } \forall j = i,\ldots,n-1\right\}.$$

Maintenant, ce dernier ensemble s'identifie à l'ensemble des

$$(0 \to \mathscr{A}_i \to \mathscr{L}_i \to \mathscr{L}_{i-1} \to 0;\; D_i,\ldots,D_n)$$

avec $\mathscr{A}_i, \mathscr{L}_i, \mathscr{L}_{i-1}, D_i,\ldots,D_n$ comme ci-dessus, à cela près qu'on n'impose plus à D_n d'être effectif ($\mathscr{A}_i \leftrightarrow a_i$, $\mathscr{L}_{i-1} \leftrightarrow g_{i-1}$, $\mathscr{L}_i \leftrightarrow g_i$, u définit l'extension de $\mathscr{L}_{i-1}$ par $\mathscr{A}_i$, $D_j = \mathrm{div}(a_j a_{j+1}^{-1}\delta)$ pour $j = i,\ldots,n-1$ et $D_n = \mathrm{div}(a_n\delta^{1-n})$). Par suite, on peut identifier $V_{\sigma,i}$ à une fonction de $(0 \to \mathscr{A}_i \to \mathscr{L}_i \to \mathscr{L}_{i-1} \to 0;\; D_i,\ldots,D_n)$ et donc restreindre $V_{\sigma,i}$ à S_i. On va voir en fait que $V_{E,i} = V_{\sigma,i}|\mathscr{P}_i$, pour $i = 1,\ldots,n$.

Un calcul algébrique facile utilisant le fait que, pour tout caractère additif non trivial $\psi\colon \mathbf{F}_q \to \overline{\mathbf{Q}}_\ell^\times$,

$$\sum_{\alpha \in \mathbf{F}_q^\times} \psi(\alpha\beta) = \begin{cases} -1 & \text{si } \beta \in \mathbf{F}_q^\times \\ q - 1 & \text{si } \beta = 0 \end{cases}$$

permet de réécrire la relation de récurrence définissant les $V_{\sigma,i}$ sous la forme

$$V_{\sigma,i}\left(\left[\begin{array}{cc|c} \begin{matrix} g_{i-1} & u \\ 0 & a_i \end{matrix} & & 0 \\ \hline 0 & & \begin{matrix} a_{i+1} & & \\ & \ddots & \\ & & a_n \end{matrix} \end{array}\right]\right)$$

$$= q \sum_{\substack{(v,v_{i-1})\in(F^{i-2}\times F^\times)/\mathbb{F}_q^\times \\ (v,v_{i-1})\cdot u=0}} V_{\sigma,i-1}\left(\left[\begin{array}{cc|c} \begin{pmatrix} 1_{i-2} & 0 \\ v & v_{i-1} \end{pmatrix}g_{i-1} & & 0 \\ \hline 0 & & \begin{matrix} a_i & & \\ & \ddots & \\ & & a_n \end{matrix} \end{array}\right]\right)$$

$$- \sum_{(v,v_{i-1})\in(F^{i-2}\times F^\times)/\mathbb{F}_q^\times} V_{\sigma,i-1}\left(\left[\begin{array}{cc|c} \begin{pmatrix} 1_{i-2} & 0 \\ v & v_{i-1} \end{pmatrix}g_{i-1} & & 0 \\ \hline 0 & & \begin{matrix} a_i & & \\ & \ddots & \\ & & a_n \end{matrix} \end{array}\right]\right).$$

Maintenant, achevons la preuve de la proposition. On remarque que

$$\left\{(v,v_{i-1})\in(F^{i-2}\times F^\times)/\mathbb{F}_q^\times \;\middle|\; \begin{pmatrix} 1_{i-2} & 0 \\ v & v_{i-1} \end{pmatrix}g_{i-1}\in\begin{pmatrix} g_{i-2} & w \\ 0 & a_{i-1} \end{pmatrix}\mathrm{GL}_{i-1}(\mathcal{O}) \text{ et}\right.$$

$$\left. \mathrm{ord}_x\left(a_{i-1}a_i^{-1}\delta\right)\geqslant 0, \forall x\in|X|\right\}$$

s'identifie à l'ensemble des droites dans

$$\mathrm{Hom}_{\mathcal{O}_X}\left(\mathscr{A}_i\otimes\left(\Omega_X^1\right)^{\otimes-1},\mathscr{L}_{i-1}\right),$$

de sorte que les relations de récurrence définissant $V_{\sigma,i}$ et $V_{E,i}$ coincident sur $\mathscr{P}_i$ et que $V_{E,i}=V_{\sigma,i}|\mathscr{P}_i$, $\forall i=1,\ldots,n$. D'où la proposition.

(4.2). Pour $n=2$, Drinfeld donne une interprétation géométrique de la valeur de $V_{E,2}$ en un point

$$(0\to\mathscr{A}_2\to\mathscr{L}_2\to\mathscr{L}_1\to 0;\, D_2)\in\mathscr{P}_2$$

avec $D_2 = 0$, $\mathscr{A}_2 = \Omega_X^1$, l'extension non scindée et $\ell_1 > 4g - 4$ (cf. [Dr]). Rappelons cette interprétation géométrique.

On suppose de nouveau k arbitraire et $g \geqslant 2$. On fixe un $\overline{\mathbf{Q}}_\ell$-faisceau lisse et géométriquement irréductible de rang 2, E, sur X et un entier $\ell_1 > 4g - 4 > 2g - 2$. Alors, d'après Riemann-Roch, l'application d'Abel-Jacobi

$$X^{(\ell_1)} \overset{\pi_1}{\to} \operatorname{Pic}_X^{\ell_1}, \ D_1 \mapsto \mathcal{O}_X(D_1) = \mathscr{L}_1,$$

où $\operatorname{Pic}_1^{\ell_1}$ est la composante de degré ℓ_1 de la variété de Picard de X, est un fibré projectif de rang $\ell_1 - g$: la fibre en $\mathscr{L}_1$ de π_1 est l'espace projectif des droites de

$$H^0(X, \mathscr{L}_1).$$

Le fibré projectif dual de π_1 est le fibré projectif

$$\operatorname{Drap}_X^{\ell_1} \overset{\check{\pi}_1}{\to} \operatorname{Pic}_X^{\ell_1},$$

où $\operatorname{Drap}_X^{\ell_1}$ est le schéma de module des extensions non scindées, prises à un scalaire près,

$$0 \to \Omega_X^1 \to \mathscr{L}_2 \to \mathscr{L}_1 \to 0,$$

et où

$$\check{\pi}_1\big(0 \to \Omega_X^1 \to \mathscr{L}_2 \to \mathscr{L}_1 \to 0\big) = \mathscr{L}_1 \in \operatorname{Pic}_X^{\ell_1}.$$

La fibre en $\mathscr{L}_1$ de $\check{\pi}_1$ est donc l'espace projectif des droites dans

$$\operatorname{Ext}_{\mathcal{O}_X}^1\big(\mathscr{L}_1, \Omega_X^1\big).$$

LEMME (4.2.1) (Deligne). *Pour tout $\overline{\mathbf{Q}}_\ell$-faisceau lisse et géométriquement irréductible de rang $n \geqslant 2$, E, sur X et pour tout entier $\ell_1 > n(2g - 2)$, on a*

$$R\pi_{1*}E^{(\ell_1)} = 0.$$

Remarque (4.2.2). Ce lemme traduit géométriquement le fait suivant: pour tout $\mathscr{L}_1 \in \operatorname{Pic}_X^{\ell_1}(\mathbf{F}_q)$ (ici on suppose de nouveau $k = \mathbf{F}_q$), on a

$$\sum_{\substack{D_1 \in X^{(\ell_1)}(\mathbf{F}_q) \\ \mathcal{O}_X(D_1) = \mathscr{L}_1}} W_E(D_1; 0) = 0$$

(la fonction L de E est un polynôme en $T = q^{-s}$ de degré $n(2g - 2)$).

Considérons alors le transformé de Radon géométrique (cf. (A.1))

$$(4.2.3) \qquad \mathscr{R}\big(E^{(\ell_1)}[\ell_1]\big) \in \mathrm{ob}\, D_c^b\big(\mathrm{Drap}_X^{\ell_1}, \overline{\mathbf{Q}}_\ell\big)$$

Comme $E^{(\ell_1)}[\ell_1]$ est un $\overline{\mathbf{Q}}_\ell$-faisceau pervers géométriquement irréductible d'après (3.3.10) et que $R\pi_{1*}(E^{(\ell_1)}[\ell_1])$ d'après (4.2.1), il résulte de (A.2) que:

PROPOSITION (4.2.4). *$\mathscr{R}(E^{(\ell_1)}[\ell_1])$ est un $\overline{\mathbf{Q}}_\ell$-faisceau pervers géométriquement irréductible sur* $\mathrm{Drap}_X^{\ell_1}$ *et on a*

$$R\breve{\pi}_{1*}\mathscr{R}\big(E^{(\ell_1)}[\ell_1]\big) = 0.$$

On normalise maintenant $\mathscr{R}(E^{(d_1)}[d_1])$ comme suit: soit $K_{\Lambda^2 E}$ le $\overline{\mathbf{Q}}_\ell$-faisceau lisse de rang 1 sur $Fib_{X,1}$ associé au $\overline{\mathbf{Q}}_\ell$-faisceau lisse de rang 1, $_{\Lambda^2 E}$, sur X par la théorie du corps de classes abélien géométrique ([Se]; voir aussi (5.1)) et soit $K_{\Lambda^2 E, \Omega_X^1}$ sa fibre en $\Omega_X^1 \in Fib_{X,1}^{2g-2}$; alors on pose

$$(4.2.5) \qquad \tilde{K}_E^{\ell_2} = K_{\Lambda^2 E, \Omega_X^1}(\ell_1 + 2g - 2) \otimes_{\overline{\mathbf{Q}}_\ell} \mathscr{R}\big(E^{(\ell_1)}[\ell_1]\big)(-1)$$

pour tout $\ell_1 > 4g - 4$, où $\ell_2 = \ell_1 + 2g - 2$.

Ce $\overline{\mathbf{Q}}_\ell$-faisceau pervers est l'analogue géométrique de $\tilde{f}_\sigma$ au sens où:

THÉORÈME (4.2.7) (Drinfeld). *Pour $k = \mathbb{F}_q$, la valeur de $\tilde{f}_\sigma$ en $(0 \to \Omega_X^1 \to \mathscr{L}_2 \to \mathscr{L}_1 \to 0; 0) \in \mathscr{P}_2$ est égale à*

$$\mathrm{tr}\big(\mathrm{Frob}_{(0 \to \Omega_X^1 \to \mathscr{L}_2 \to \mathscr{L}_1 \to 0)}, \tilde{K}_E^{\ell_2}\big)$$

dès que $\deg(\mathscr{L}_1) = \ell_1 > 4g - 4$ *(σ est la représentation galoisienne correspondant à E).*

Preuve. C'est une conséquence immédiate de la formule des traces de Grothendieck et des définitions.

(4.3). Soit toujours X une courbe projective, lisse et géométriquement connexe de genre $g \geqslant 2$ sur un corps k arbitraire et soit E un $\overline{\mathbf{Q}}_\ell$-faisceau lisse et géométriquement irréductible de rang $n \geqslant 2$ sur X. Nous allons donner maintenant une construction géométrique conjecturale d'un complexe de $\overline{\mathbf{Q}}_\ell$-faisceaux $\tilde{K}_E$ qui est l'analogue géométrique de $\tilde{f}_\sigma$ ou plus précisément de $V_{E,n}$.

Fixons un entier

$$\ell_1 \geqslant \frac{n(n+1)}{2}(2g - 2) + n$$

et posons, pour $i = 1, \ldots, n$

$$\ell_i = \ell_1 + \frac{i(i-1)}{2}(2g-2);$$

on a

$$\ell_i \geqslant i^2(2g-2) + i.$$

Pour $i = 1, \ldots, n$, notons $\mathscr{S}_i^0$ l'ouvert du champ algébrique lisse sur k, $Fib_{X,i}^{\ell_i}$, formé des fibrés vectoriels $\mathscr{L}_i$ sur X de rang i et de degré ℓ_i tels que

$$\mathrm{Hom}_{\mathcal{O}_X}\!\left(\mathscr{L}_i, \left(\Omega_X^1\right)^{\otimes i}(x)\right) = 0$$

pour tout $x \in X$. On remarque que $\mathscr{S}_i^0$ contient tous les fibrés vectoriels stables de rang i et de degré ℓ_i et remplit d'autant mieux $Fib_{X,i}^{\ell_i}$ que ℓ_1 est plus grand.

Toujours pour $i = 1, \ldots, n$ notons $\mathscr{P}_i^0$ le champ algébrique lisse sur k des homomorphismes injectifs de $\mathcal{O}_X$-Modules, à un scalaire près,

$$\left(\Omega_X^1\right)^{\otimes(i-1)} \hookrightarrow \mathscr{L}_i,$$

avec $\mathscr{L}_i \in \mathscr{S}_i^0$. On a une projection naturelle

$$\mathscr{P}_i^0 \xrightarrow{\ \pi_i\ } \mathscr{S}_i^0$$

qui est un fibré projectif de fibre en $\mathscr{L}_i$ l'espace projectif des droites dans

$$\mathrm{Hom}_{\mathcal{O}_X}\!\left(\left(\Omega_X^1\right)^{\otimes(i-1)}, \mathscr{L}_i\right)$$

(on a

$$\mathrm{Ext}^1_{\mathcal{O}_X}\!\left(\left(\Omega_X^1\right)^{\otimes(i-1)}, \mathscr{L}_i\right) = 0$$

pour $\mathscr{L}_i \in \mathscr{S}_i^0$).

Pour $i = 2, \ldots, n$, on a un ouvert dense

$$j_i \colon \mathscr{U}_i \hookrightarrow \mathscr{P}_i^0$$

formé des

$$\left(\Omega_X^1\right)^{\otimes(i-1)} \hookrightarrow \mathscr{L}_i$$

dont le conoyau $\mathscr{L}_{i-1}$ est un $\mathcal{O}_X$-Module cohérent sans torsion, i.e. un fibré vectoriel de rang $i-1$ et de degré $\ell_{i-1} = \ell_i - (i-1)(2g-2)$. De plus, $\mathscr{U}_i$

rencontre chaque fibre $\pi_i^{-1}(\mathscr{L}_i)$ suivant un ouvert dense $\big($on a

$$\dim \operatorname{Hom}_{\mathscr{O}_X}\!\Big(\big(\Omega_X^1\big)^{\otimes(i-1)}(x),\,\mathscr{L}_i\Big) = \dim \operatorname{Hom}_{\mathscr{O}_X}\!\Big(\big(\Omega_X^1\big)^{\otimes(i-1)},\,\mathscr{L}_i\Big) - i$$

pour tout $x \in X$, de sorte que

$$\operatorname{codim}_{\pi_i^{-1}(\mathscr{L}_i)}\!\Big(\pi_i^{-1}(\mathscr{L}_i) \cap \big(\mathscr{P}_i^0 - \mathscr{U}_i\big)\Big) \geqslant i - 1\big).$$

Toujours pour $i = 2, \ldots, n$, notons

$$\check{\mathscr{P}}_{i-1}^0 \xrightarrow{\check{\pi}_{i-1}} \mathscr{S}_{i-1}^0$$

le fibré projectif dual de π_{i-1}. La fibre de $\check{\pi}_{i-1}$ en $\mathscr{L}_{i-1} \in \mathscr{S}_{i-1}^0$ est l'espace projectif des droites dans

$$\operatorname{Ext}_{\mathscr{O}_X}^1\!\Big(\mathscr{L}_{i-1},\big(\Omega_X^1\big)^{\otimes(i-1)}\Big)$$

et $\check{\mathscr{P}}_{i-1}^0$ est donc le champ algébrique lisse sur k des extensions non scindées, à un scalaire près,

$$0 \to \big(\Omega_X^1\big)^{\otimes(i-1)} \to \mathscr{L}_i \to \mathscr{L}_{i-1} \to 0,$$

avec $\mathscr{L}_{i-1} \in \mathscr{S}_{i-1}^0$.
  Si

$$\Big(\big(\Omega_X^1\big)^{\otimes(i-1)} \hookrightarrow \mathscr{L}_i\Big) \in \mathscr{U}_i,$$

on vérifie aussitôt que le conoyau $\mathscr{L}_{i-1}$ de ce monomorphisme est dans $\mathscr{S}_{i-1}^0$ et que l'extension (à un scalaire près)

$$0 \to \big(\Omega_X^1\big)^{\otimes(i-1)} \to \mathscr{L}_i \to \mathscr{L}_{i-1} \to 0$$

est non scindée $\big($on a $H^0\big(X, \Omega_X^1\big) \neq 0$, de sorte que l'on peut injecter

$$\operatorname{Hom}_{\mathscr{O}_X}\!\Big(\mathscr{L}_{i-1},\big(\Omega_X^1\big)^{\otimes(i-1)}(x)\Big)$$

et

$$\operatorname{Hom}_{\mathscr{O}_X}\!\Big(\mathscr{L}_i,\big(\Omega_X^1\big)^{\otimes(i-1)}\Big)$$

dans

$$\operatorname{Hom}_{\mathscr{O}_X}\!\Big(\mathscr{L}_i,\big(\Omega_X^1\big)^{\otimes i}(x)\Big) = 0\big).$$

Par suite, on a un morphisme de champs algébriques sur k

$$j'_{i-1}\colon \mathscr{U}_i \to \check{\mathscr{P}}^0_{i-1}$$

qui envoie

$$\left(\Omega^1_X\right)^{\otimes(i-1)} \hookrightarrow \mathscr{L}_i$$

sur

$$0 \to \left(\Omega^1_X\right)^{\otimes(i-1)} \to \mathscr{L}_i \to \mathscr{L}_{i-1} \to 0;$$

ce morphisme est en fait une immersion ouverte d'image l'ouvert de $\check{\mathscr{P}}^0_{i-1}$ formé des extensions

$$0 \to \left(\Omega^1_X\right)^{\otimes(i-1)} \to \mathscr{L}_i \to \mathscr{L}_{i-1} \to 0$$

avec $\mathscr{L}_i \in \mathrm{ob}\ \mathscr{S}_i^0$. En particulier, on voit que

$$\dim \mathscr{P}_i^0 = \dim \check{\mathscr{P}}^0_{i-1}$$

pour tout $i = 2, \dots, n$, ce que l'on peut vérifier directement en montrant que

$$\dim \mathscr{P}_n^0 = \dim \check{\mathscr{P}}^0_{n-1} = \cdots = \dim \mathscr{P}_1^0 = \ell_1 - 1$$

(on rappelle que

$$\dim \mathscr{S}_i^0 = \dim Fib^{\ell_i}_{X,i} = i^2(g-1),$$

pour $i = 1, \dots, n$).

En résumé, on a le diagramme

$$(4.3.1)$$

$\mathscr{P}_3^0 \xleftarrow{j_3} \mathscr{U}_3 \xhookrightarrow{j'_2} \check{\mathscr{P}}^0_2 \qquad \mathscr{P}_2^0 \xleftarrow{j_2} \mathscr{U}_2 \xhookrightarrow{j'_1} \check{\mathscr{P}}^0_1 \qquad \mathscr{P}_1^0$

$\cdots\quad \Big\downarrow{\scriptstyle\pi_3} \qquad\quad \check{\pi}_2\searrow\ \swarrow{\scriptstyle\pi_2} \qquad\qquad\quad \check{\pi}_1\searrow\ \swarrow{\scriptstyle\pi_1}\ .$

$\mathscr{S}_3^0 \qquad\qquad\quad \mathscr{S}_2^0 \qquad\qquad\qquad\quad \mathscr{S}_1^0$

De plus, on a un morphisme de champs algébriques sur k

$$\mathscr{P}_1^0 \to X^{(\ell_1)}$$

qui envoie $(\mathscr{O}_X \hookrightarrow \mathscr{L}_1)$ sur le diviseur D_1 des zéros de cette section de $\mathscr{L}_1$.

Notons encore $E^{(\ell_1)}$ le $\overline{\mathbf{Q}}_\ell$-faisceau sur $\mathscr{P}_1^0$ image réciproque de $E^{(\ell_1)}$ sur $X^{(\ell_1)}$ par le morphisme $\mathscr{P}_1^0 \to X^{(\ell_1)}$ ci-dessus. Alors $E^{(\ell_1)}[\ell_1 - 1]$ est un $\overline{\mathbf{Q}}_\ell$-faisceau pervers géométriquement irréductible sur le champ $\mathscr{P}_1^0$ d'après (3.3.10) et comme $\ell_1 > n(2g - 2)$, on a

$$R\pi_{1*}E^{(\ell_1)} = 0$$

d'après (4.2.1).

On pose alors

(4.3.2)

$$V_{E,1}^{\ell_1} = K_{\Lambda^n E,\,(\Omega_X^1)^{\otimes(n-1)}}\big((n-1)\ell_1 + n(n-1)^2(g-1)\big) \otimes_{\overline{\mathbf{Q}}_\ell} E^{(\ell_1)}[\ell_1 - 1]$$

où $K_{\Lambda^n E}$ est le $\overline{\mathbf{Q}}_\ell$-faisceau lisse sur $Fib_{X,1}$ associé au $\overline{\mathbf{Q}}_\ell$-faisceau lisse de rang 1, $_{\Lambda^n E}$, sur X, par la théorie du corps de classes abélien géométrique ([Se]; voir aussi (5.1)) et où $K_{\Lambda^n E,\,(\Omega_X^1)^{\otimes(n-1)}}$ est le $\overline{\mathbf{Q}}_\ell$-espace vectoriel de dimension 1 fibre de $K_{\Lambda^n E}$ en $(\Omega_X^1)^{\otimes(n-1)} \in Fib_{X,1}^{(n-1)(2g-2)}$. $V_{E,1}^{\ell_1}$ est un $\overline{\mathbf{Q}}_\ell$-faisceau pervers géométriquement irréductible sur $\mathscr{P}_1^0$, on a

$$R\pi_{1*}V_{E,1}^{\ell_1} = 0$$

et, si $k = \mathbb{F}_q$, on a pour tout $D_1 \in X^{(\ell_1)}(\mathbb{F}_q)$

$$\mathrm{tr}\Big(\mathrm{Frob}_{(\mathscr{O}_X \hookrightarrow \mathscr{O}_X(D_1))},\, V_{E,1}^{\ell_1}\Big) = W_E(D_1, 0, \ldots, 0)$$

avec les notations de (4.1).

La conjecture suivante est motivée par la construction de Drinfeld pour $n = 2$ rappelée ci-dessus et par les résultats du numéro 3 concernant le faisceau de Whittaker.

CONJECTURE (4.3.3.).   *Pour tout entier*

$$\ell_1 \gg \frac{n(n+1)}{2}(2g - 2) + n$$

on peut construire par récurrence sur i des $\overline{\mathbf{Q}}_\ell$-faisceaux pervers géométriquement irréductibles $V_{E,1}^{\ell_1}, \ldots, V_{E,n}^{\ell_n}$ sur $\mathscr{P}_1^0, \ldots, \mathscr{P}_n^0$ respectivement de telle sorte que:
(i) $V_{E,1}^{\ell_1}$ soit le $\overline{\mathbf{Q}}_\ell$-faisceau pervers sur $\mathscr{P}_1^0$ défini en (4.3.2),
(ii) $R\pi_{i*}V_{E,i}^{\ell_i} = 0$ pour $i = 1, \ldots, n - 1$,
(iii) on ait la relation de récurrence

$$V_{E,i+1}^{\ell_{i+1}} = j_{(i+1)!*}\,j_i'^*\mathscr{R}_i\big(V_{E,i}^{\ell_i}\big)(-1),$$

pour $i = 1, \ldots, n - 1$, où

$$\mathscr{R}_i \colon D_c^b\big(\mathscr{P}_i^0, \overline{\mathbf{Q}}_\ell\big) \to D_c^b\big(\check{\mathscr{P}}_i^0, \overline{\mathbf{Q}}_\ell\big)$$

est la transformation de Radon géométrique pour le fibré projectif $\mathscr{P}_i^0 \xrightarrow{\pi_i} \mathscr{S}_i^0$.

En outre, si $k = \mathbb{F}_q$, pour tout $((\Omega_X^1)^{\otimes(i-1)} \to \mathscr{L}_i) \in \mathscr{P}_i^0(\mathbb{F}_q)$ qui se factorise par

$$\big(\Omega_X^1\big)^{\otimes(i-1)} \hookrightarrow \big(\Omega_X^1\big)^{\otimes(i-1)}(D_i) \hookrightarrow \mathscr{L}_i,$$

où $(\Omega_X^1)^{\otimes(i-1)}(D_i)$ est le sous-fibré vectoriel de $\mathscr{L}_i$ engendré par $(\Omega_X^1)^{\otimes(i-1)}$, on a

$$\mathrm{tr}\Big(\mathrm{Frob}_{((\Omega_X^1)^{\otimes(i-1)} \hookrightarrow \mathscr{L}_i)}, V_{E,i}^{\ell_i}\Big)$$

$$= V_{E,i}\Big(0 \to \big(\Omega_X^1\big)^{\otimes(i-1)}(D_i) \to \mathscr{L}_i \to \mathscr{L}_{i-1} \to 0; \, D_i, 0, \ldots, 0\Big)$$

avec les notations de (4.1).

En particulier, si l'on pose

$$(4.3.4) \qquad \tilde{K}_E^{\ell_n} = j_n^* V_{E,n}^{\ell_n} = j_{n-1}'^* \mathscr{R}_{n-1}\big(V_{E,n-1}^{\ell_{n-1}}\big)(-1),$$

ce $\overline{\mathbf{Q}}_\ell$-faisceau pervers sur $\mathscr{U}_n$ est l'analogue géométrique de $\tilde{f}_\sigma$, pour σ la représentation galoisienne attachée à E.

Remarque (4.3.5). Pour $n = 2$, la conjecture ci-dessus est une conséquence des résultats de Drinfeld ([Dr]).

5. Descente de $\tilde{K}_E$ en K_E et variété caractéristique de $\tilde{K}_E$

(5.0). On garde les notations de (2.1) et on fixe un $\overline{\mathbf{Q}}_\ell$-faisceau lisse et géométriquement irréductible de rang $n \geqslant 1$, E, sur X.

On va discuter dans ce numéro l'analogue géométrique de la conjecture (1.2.4) et étudier, pour k de caractéristique 0, la variété caractéristique de $\tilde{K}_E$ dans $T^* Drap_{X,(n-1,n)}$.

(5.1). Commençons par rappeler le cas $n = 1$ (voir [Se]). Fixons un entier $\ell_1 > 2g - 2$ et considérons l'application d'Abel-Jacobi

$$\pi_1 \colon X^{(\ell_1)} \to \mathrm{Pic}_X^{\ell_1}, \, D_1 \mapsto \mathscr{O}_X(D_1) = \mathscr{L}_1,$$

qui est un fibré projectif d'après Riemann-Roch.

Sur $X^{(\ell_1)}$, $E^{(\ell_1)}$ est un $\overline{\mathbf{Q}}_\ell$-faisceau lisse de rang 1 puisque E est de rang 1. Or, les espaces projectifs sont géométriquement simplement connexes. Par suite, il existe un unique $\overline{\mathbf{Q}}_\ell$-faisceau lisse de rang 1, $K_E^{\ell_1}$ sur $\mathrm{Pic}_X^{\ell_1}$ dont l'image réciproque

344   GERARD LAUMON

par π_1 soit isomorphe à $E^{(\ell_1)}$,

$$\pi_1^* K_E^{\ell_1} \simeq E^{(\ell_1)}.$$

On vérifie facilement que, pour tout $\ell_1 > 2g - 2$, on a

$$(5.1.1) \qquad\qquad \alpha_1^* K_E^{\ell_1+1} \simeq E \boxtimes_k K_E^{\ell_1},$$

où

$$\alpha_1 \colon X \times_k \operatorname{Pic}_X^{\ell_1} \to \operatorname{Pic}_X^{\ell_1+1}$$

est la correspondance de Hecke (voir (2.1)), et qu'il existe un unique $\overline{\mathbf{Q}}_\ell$-faisceau lisse de rang 1, $K_E^{\ell_1}$, sur $\operatorname{Pic}_X^{\ell_1}$ tel que (5.1.1) soit vérifiée quel que soit ℓ_1.

On a donc construit un $\overline{\mathbf{Q}}_\ell$-faisceau lisse de rang 1, K_E, sur Pic_X (ou si l'on préfère sur $Fib_{X,1}$) et la conjecture (2.1.1) est démontrée dans le cas $n = 1$.

Remarque. La présentation ci-dessus de la théorie du corps de classes abélien géométrique de Lang est due à ma connaissance à Deligne.

(5.2). Pour $n = 2$, Drinfeld s'inspire de la démonstration ci-dessus du cas $n = 1$. Le problème essentiel est de montrer que le $\overline{\mathbf{Q}}_\ell$-faisceau pervers géometriquement irréductible $\tilde{K}_E^{\ell_2}$ sur $\operatorname{Drap}_X^{\ell_1}$, construit en (4.2.5), se descend à $Fib_{X,2}^{\ell_2}$ par le morphisme

$$\pi_2 \colon \operatorname{Drap}_X^{\ell_1} \to Fib_{X,2}^{\ell_2}, \left(0 \to \Omega_X^1 \to \mathscr{L}_2 \to \mathscr{L}_1 \to 0\right) \mapsto \mathscr{L}_2,$$

où $\ell_2 = \ell_1 + 2g - 2$, dès que ℓ_1 est assez grand.

La difficulté majeure vient du fait que π_2 n'est pas propre bien que $\operatorname{Drap}_X^{\ell_1}$ soit une variété projective sur k ($Fib_{X,2}^{\ell_2}$ est non séparé). Cependant, si $\ell_1 > 6g - 4$ et si $\mathscr{L}_2 \in \mathscr{S}_2^0$ (cf. (4.3)), $\pi_2^{-1}(\mathscr{L}_2)$ est un ouvert d'un espace projectif. Par suite, pour montrer que les faisceaux de cohomologie de

$$\tilde{K}_E^{\ell_2} | \pi_2^{-1}(\mathscr{L}_2)$$

sont géométriquement constants, il suffit de montrer qu'ils sont lisses et qu'ils se prolongent en des $\overline{\mathbf{Q}}_\ell$-faisceaux lisses sur la complétion projective de $\pi_2^{-1}(\mathscr{L}_2)$. Par pureté, il suffit même de démontrer que ces faisceaux de cohomologie se prolongent en des faisceaux lisses sur le complémentaire d'un fermé de codimension $\geqslant 2$ de cette complétion projective de $\pi_2^{-1}(\mathscr{L}_2)$. C'est ce que fait Drinfeld dans [Dr].

(5.3). Pour n quelconque, on dispose conjecturalement, pour $\ell_1 \gg 0$, du $\overline{\mathbf{Q}}_\ell$-faisceau pervers et géométriquement irréductible $V_{E,n}^{\ell_n}$ sur $\mathscr{P}_n^0$ (cf. (4.3.3)) qui se restreint en $\tilde{K}_E^{\ell_n}$ sur $\mathscr{U}_n \xrightarrow{j_n} \mathscr{P}_n^0$. La conjecture suivante est motivée par les résultats de Lang et Drinfeld rappelés ci-dessus et par la conjecture (1.2.4).

CONJECTURE (5.3.1).   *Pour tout entier*

$$\ell_n \gg n^2(2g - 2) + n$$

il existe un unique $\overline{\mathbb{Q}}_\ell$*-faisceau pervers géométriquement irréductible* $K_E^{\ell_n}$ *sur* $Fib_{X,n}^{\ell_n}$ *tel que*

$$V_{E,n}^{\ell_n} = \pi_n^*\left(K_E^{\ell_n}|\mathscr{S}_n^0\right)\left[\ell_n - 1 - n(2n - 1)(g - 1)\right]$$

et donc tel que

$$\tilde{K}_E^{\ell_n} = j_n^*\pi_n^*\left(K_E^{\ell_n}|\mathscr{S}_n^0\right)\left[\ell_n - 1 - n(2n - 1)(g - 1)\right].$$

Remarques (5.3.2).   (i) L'entier

$$\ell_n - 1 - n(2n - 1)(g - 1)$$

est le rang du fibré projectif $\mathscr{P}_n^0 \xrightarrow{\pi_n} \mathscr{S}_n^0$ et l'on sait d'après [B–B–D] (4.2.5) que le foncteur $\pi_n^*(-)[\ell_n - 1 - n(2n - 1)(g - 1)]$ des $\overline{\mathbb{Q}}_\ell$-faisceaux pervers sur $\mathscr{S}_n^0$ dans les $\overline{\mathbb{Q}}_\ell$-faisceaux pervers sur $\mathscr{P}_n^0$ est pleinement fidèle.

(ii) Une fois acquis la conjecture (5.3.1) on peut utiliser l'opérateur de Hecke T^n pour définir un $\overline{\mathbb{Q}}_\ell$-faisceau pervers géométriquement irréductible K_n sur $Fib_{X,n}$ (au sens de (2.1)) qui vérifie

$$T^n(K_E) = (\Lambda^n E) \boxtimes_k K_E[n(n - 1)]\left(\frac{n(n - 1)}{2}\right)$$

et tel que

$$\left(K_E|Fib_{X,n}^{\ell_n}\right)\left[-(n - 1)\ell_n + \frac{n(4n^2 - 6n + 5)}{6}(g - 1)\right]$$

coincide avec le $\overline{\mathbb{Q}}_\ell$-faisceau pervers $K_E^{\ell_n}$ de (5.3.1) pour tout $\ell_n \gg n^2(2g - 2) + n$. Il n'est cependant pas clair que ce K_E vérifie (2.1.1) bien que cela soit certainement le cas. Cependant, si $k = \mathbb{F}_q$ et si l'on a démontré (4.3.3) ainsi que (5.3.1), on peut facilement en déduire la conjecture (1.2.4) avec

$$f_\sigma = t_{K_E}$$

(σ est la représentation galoisienne associée à E et l'on considère la fonction trace de Frobenius sur K_E,

$$t_{K_E}: Fib_{X,n}(\mathbb{F}_q) \to \overline{\mathbb{Q}}_\ell,$$

comme une fonction sur

$$\mathrm{GL}_n(F) \setminus \mathrm{GL}_n(\mathbf{A}) / \mathrm{GL}_n(\mathcal{O})).$$

(iii) Réciproquement, si l'on a démontré (4.3.3) et (1.2.4) pour un entier n (par exemple $n = 2$), on a aussi démontré (5.3.1): pour que $V_{E,n}^{\ell_n}$ (ou même $\tilde{K}_E^{\ell_n}$) se descende via $\mathcal{P}_n^0 \xrightarrow{\pi_n} \mathcal{S}_n^0$ (ou même $\mathcal{U}_n \xrightarrow{\pi_n \circ j_n} \mathcal{S}_n^0$), il suffit que, pour toute spécialisation de k en un corps fini, la fonction trace de Frobenius de $V_{E,n}^{\ell_n}$ (ou même de $\tilde{K}_E^{\ell_n}$) se descende à $\mathcal{S}_n^0$ (on utilise une variante du théorème de densité de Čebotarev qui assure qu'un $\overline{\mathbf{Q}}_\ell$-faisceau pervers géométriquement irréductible sur un $\mathbb{F}_q$-schéma lisse et géométriquement connexe est géométriquement constant dès que sa fonction trace de Frobenius est constante après tout changement de base de $\mathbb{F}_q$ à $\mathbb{F}_{q^m}$).

(5.4).　Le problème que pose la conjecture (5.3.1) est de nature topologique: pour vérifier qu'un $\overline{\mathbf{Q}}_\ell$-faisceau sur un espace projectif est géométriquement constant il suffit de vérifier qu'il est lisse. Ce problème se reformule très agréablement, si k est de caractéristique nulle, avec la notion de variété caractéristique. En effet, on a l'énoncé de descente suivant, dont la démonstration est laissée au lecteur.

PROPOSITION (5.4.1).　*Soient S une variété (ou un champ algébrique) lisse sur un corps de caractéristique 0 et $P \xrightarrow{\pi} S$ un fibré projectif de rang $r \geqslant 1$. Soit $\tilde{K}$ un $\overline{\mathbf{Q}}_\ell$-faisceau pervers géométriquement irréductible sur P. On suppose que la variété caractéristique de $\tilde{K}$ se descend à S, i.e, que*

$$|\mathrm{Car}(K)| \subset T^*S \times_S P \subset T^*P.$$

Alors, il existe un unique $\overline{\mathbf{Q}}_\ell$-faisceau pervers géométriquement irréductible K sur S tel que

$$\tilde{K} = \pi^*K[r].$$

De plus, les cycles caractéristiques de K et $\tilde{K}$ sont reliés par la relation

$$\mathrm{Car}(\tilde{K}) = \mathrm{Car}(K) \times_S P$$

dans $T^*S \times_S P \subset T^*P$.

Par suite, pour k de caractéristique nulle, on peut reformuler la conjecture (5.3.1) comme suit:

CONJECTURE (5.4.2).　*Pour tout entier*

$$\ell_n \gg n^2(2g - 2) + n$$

la variété caractéristique de $V_{E,n}^{\ell}$ est contenue dans

$$T^*\mathcal{S}_n^0 \times_{\mathcal{S}_n^0} \mathcal{P}_n^0 \subset T^*\mathcal{P}_n^0.$$

(5.5). A titre d'illustration de (5.4.2), calculons le cycle caractéristique de $\tilde{K}_E^{\ell_2}$ dans $T^*\mathrm{Drap}_X^{\ell_1}$ (on utilise librement les notations de (4.2); ce calcul correspond au "First vanishing cycle theorem" de [Dr]).

Nous supposerons bien entendu k de caractéristique 0 et $\ell_1 > 4g - 4$.

On a vu que le cycle caractéristique de $E^{(\ell_1)}[\ell_1]$ dans $T^*X^{(\ell_1)}$ est égal à

$$\mathrm{Car}\big(E^{(\ell_1)}[\ell_1]\big) = \sum_{\substack{e_1, e_2 \geqslant 0 \\ e_1 + 2e_2 = l_1}} 2^{e_1} \overline{\left[T^*_{X^{(\ell_1)}_{(1^{e_1}, 2^{e_2})}}\right]},$$

avec comme support l'ensemble des couples

$$\left(D, \mathscr{L}_1 \xrightarrow{\tilde{u}} \frac{\mathscr{L}_1}{\mathscr{O}_X} \otimes_{\mathscr{O}_X} \Omega_X^1\right),$$

où $\mathscr{L}_1 = \mathscr{O}_X(D)$, $\tilde{u}$ est $\mathscr{O}_X$-linéaire et l'homomorphisme induit par $\tilde{u}$,

$$\frac{\mathscr{L}_1}{\mathscr{O}_X} \xrightarrow{u} \frac{\mathscr{L}_1}{\mathscr{O}_X} \otimes_{\mathscr{O}_X} \Omega_X^1,$$

est nilpotent de niveau $\leqslant 2$, au sens où $(u \otimes 1) \circ u = 0$ $(\mathscr{L}_1/\mathscr{O}_X = \mathscr{O}_{X,D} \otimes_{\mathscr{O}_X} \mathscr{O}_X(D))$. Comme $\ell_1 > 4g - 4$, on a

$$\mathrm{Hom}_{\mathscr{O}_X}\big(\mathscr{L}_1, (\Omega_X^1)^{\otimes 2}\big) = 0$$

pour tout $\mathscr{L}_1 \in \mathrm{Pic}_X^{\ell_1}$. On en déduit facilement que

$$\big|\mathrm{Car}\big(E^{(\ell_1)}[\ell_1]\big)\big| \cap \big(T^*\mathrm{Pic}_X^{\ell_1} \times_{\mathrm{Pic}_X^{\ell_1}} X^{(\ell_1)}\big) \subset T^*_{X^{(\ell_1)}} X^{(\ell_1)},$$

i.e., que l'on est dans les conditions d'application de (A.3). On peut donc calculer le cycle caractéristique de $\tilde{K}_E^{\ell_2}$ en dehors de la section nulle en transférant simplement celui de $E^{(\ell_1)}[\ell_1]$ (en dehors de la section nulle) à $T^*\mathrm{Drap}_X^{\ell_1}$.

Plus précisément, commençons par remarquer que, pour tous $e_1 \geqslant 0$ et $e_2 > 0$ avec $e_1 + 2e_2 = \ell_1$, on a

$$\overline{T^*_{X^{(\ell_1)}_{(1^{e_1}, 2^{e_3})}}} = \left\{\left(D, \mathscr{L}_1 \xrightarrow{\tilde{u}} \frac{\mathscr{L}_1}{\mathscr{O}_X} \otimes_{\mathscr{O}_X} \Omega_X^1\right) \middle| \exists E_i \in X^{(e_i)} (i = 1, 2) \text{ avec} \right.$$

$$\left. D = E_1 + 2E_2 \text{ et } \mathrm{Im}(\tilde{u}) \subset \frac{\mathscr{L}_1(-E_1 - E_2)}{\mathscr{O}_X} \otimes_{\mathscr{O}_X} \Omega_X^1 \right\}.$$

Par suite, si l'on transfère par la dualité projective entre π_1 et $\check{\pi}_1$ ce fermé lagrangien homogène irréductible de $T^*X^{(\ell_1)}$ à $T^*\mathrm{Drap}_X^{\ell_1}$, on obtient, en dehors de la section nulle, le fermé lagrangien homogène irréductible formé des

$$\left(0 \to \Omega_X^1 \to \mathscr{L}_2 \to \mathscr{L}_1 \to 0, \ \mathscr{L}_2 \xrightarrow{\tilde{v}} \mathscr{L}_1 \otimes_{\mathcal{O}_X} \Omega_X^1\right) \in T^*\mathrm{Drap}_X^{\ell_1}$$

tels que l'homomorphisme composé

$$\Omega_X^1 \hookrightarrow \mathscr{L}_2 \xrightarrow{\tilde{v}} \mathscr{L}_1 \otimes_{\mathcal{O}_X} \Omega_X^1$$

soit non nul (et donc définisse un diviseur effectif D_1 de degré ℓ_1 avec $\mathscr{L}_1 = \mathcal{O}_X(D_1)$) et qu'il existe $E_1 \in X^{(e_1)}$ et $E_2 \in X^{(e_2)}$, avec

$$D_1 = E_1 + 2E_2$$

et

$$\mathrm{Im}(\tilde{v}) \subset \mathscr{L}_1(-E_1 - E_2) \otimes_{\mathcal{O}_X} \Omega_X^1.$$

Si l'on note

$$\mathrm{Drap}_{X, \, \varepsilon}^{\ell_1}$$

la projection de ce fermé lagrangien homogène irréductible sur $\mathrm{Drap}_X^{\ell_1}$, $\mathrm{Drap}_{X, \, \varepsilon}^{\ell_1}$ est le fermé irréductible de $\mathrm{Drap}_X^{\ell_1}$ formé des

$$\left(0 \to \Omega_X^1 \to \mathscr{L}_2 \to \mathscr{L}_1 \to 0\right)$$

tels qu'il existe un sous-fibré vectoriel $\mathscr{L}_1' \subset \mathscr{L}_2$ de rang 1 et de degré

$$\ell_1' \geqslant \ell_1 - e_2$$

et qu'il existe un diviseur effectif E_2'' de degré

$$e_2'' = \ell_1' - (\ell_1 - e_2),$$

avec

$$\mathrm{Hom}_{\mathcal{O}_X}\left(\mathscr{L}_1''(E_2''), \mathscr{L}_1'(-E_2'') \otimes_{\mathcal{O}_X} \Omega_X^1\right) \neq 0,$$

où on a posé $\mathscr{L}_1'' = \mathscr{L}_2/\mathscr{L}_1'$. En effet, pour $(0 \to \Omega_X^1 \to \mathscr{L}_2 \to \mathscr{L}_1 \to 0) \in \mathrm{Drap}_{X, \, \varepsilon}^{\ell_1}$ et $\mathscr{L}_1', E_2''$ comme ci-dessus, on a

$$\ell_1' \geqslant \ell_1 - e_2 \geqslant \frac{\ell_1}{2} > 2g - 2$$

et les flèches composées

$$\mathscr{L}_1' \hookrightarrow \mathscr{L}_2 \twoheadrightarrow \mathscr{L}_1$$

et

$$\Omega_X^1 \hookrightarrow \mathscr{L}_2 \to \mathscr{L}_1''$$

sont nécessairement injectives et définissent un même diviseur effectif E_2' de degré

$$e_2' = e_2 - e_2'';$$

on pose alors

$$E_2 = E_2' + E_2''$$

et on a

$$\mathrm{Hom}_{\mathcal{O}_X}\!\left(\Omega_X^1(E_2), \mathscr{L}_1(-E_2) \otimes_{\mathcal{O}_X} \Omega_X^1\right) \neq 0,$$

d'où l'existence d'un diviseur effectif E_1 de degré e_1 et d'un homomorphisme

$$\mathscr{L}_2 \xrightarrow{\ \tilde{v}\ } \mathscr{L}_1 \otimes_{\mathcal{O}_X} \Omega_X^1$$

tels que

$$\mathscr{L}_1 = \mathcal{O}_X(E_1 + 2E_2),$$

que la flèche composée

$$\Omega_X^1 \hookrightarrow \mathscr{L}_2 \xrightarrow{\ \tilde{v}\ } \mathscr{L}_1 \otimes_{\mathcal{O}_X} \Omega_X^1$$

soit non nulle (et définisse le diviseur $E_1 + 2E_2$) et que

$$\mathrm{Im}(\tilde{v}) \subset \mathscr{L}_1(-E_1 - E_2) \otimes_{\mathcal{O}_X} \Omega_X^1;$$

inversement, si l'on a E_1, E_2 et $\tilde{v}$, on retrouve $\mathscr{L}_1'$ et E_2'' en posant

$$\mathscr{L}_1' = \mathrm{Ker}(\tilde{v})$$

$$\mathscr{L}_1'' = \mathrm{Im}(\tilde{v})$$

et

$$\mathscr{L}_1(-E_1 - E_2) \otimes_{\mathcal{O}_X} \Omega_X^1 = \mathscr{L}_1''(E_2'').$$

En résumé, on a le diagramme commutatif

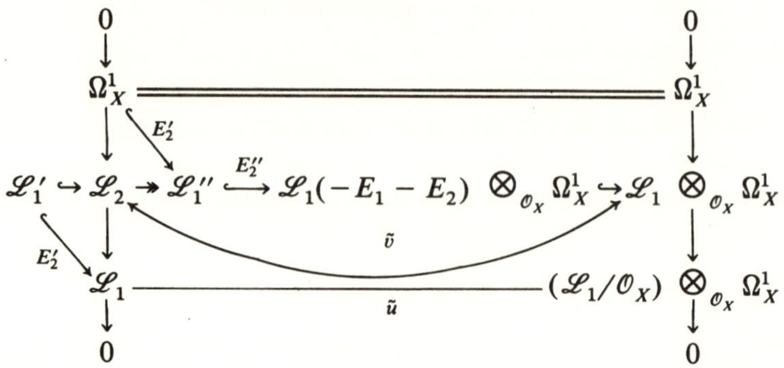

On a donc démontré:

PROPOSITION (5.5.1).   *Pour tous entiers $e_1 \geqslant 0$ et $e_2 > 0$ avec $\ell_1 = e_1 + 2e_2$,
soit* $\mathrm{Drap}^{\ell_1}_{X,\underline{e}}$ *le fermé irréductible de* $\mathrm{Drap}^{\ell_1}_X$ *formé des*

$$\left(0 \to \Omega^1_X \to \mathscr{L}_2 \to \mathscr{L}_1 \to 0\right) \in \mathrm{Drap}^{\ell_1}_X$$

tels que $\mathscr{L}_2$ ait la propriété suivante:
  $(*)$ *il existe un sous-fibré vectoriel $\mathscr{L}_1' \subset \mathscr{L}_2$ de rang 1 et de degré*

$$\ell_1' \geqslant \ell_2 - e_2 - 2g + 2$$

et il existe un diviseur effectif E_2'' sur X de degré

$$e_2'' = \ell_1' - (\ell_2 - e_2 - 2g + 2),$$

avec

$$\mathrm{Hom}_{\mathscr{O}_X}\!\left(\frac{\mathscr{L}_2}{\mathscr{L}_1'}(E_2''),\, \mathscr{L}_1'(-E_2'') \otimes_{\mathscr{O}_X} \Omega^1_X\right) \neq 0.$$

*Alors, pour tout $\overline{\mathbf{Q}}_\ell$-faisceau lisse et géométriquement irréductible de rang 2, E,
sur X, on a*

$$\mathrm{Car}\big(\tilde{K}^{\ell_2}_E\big) = m(E)\Big[T^*_{\mathrm{Drap}^{\ell_1}_X}\mathrm{Drap}^{\ell_1}_X\Big] + \sum_{\substack{e_1 \geqslant 0,\, e_2 > 0 \\ e_1 + 2e_2 = \ell_1}} 2^{e_1}\Big[T^*_{\mathrm{Drap}^{\ell_1}_{X,\underline{e}}}\mathrm{Drap}^{\ell_1}_X\Big].$$

pour un certain entier $m(E) \geqslant 0$.

Remarque (5.5.2).   On a

$$m(E) = (-1)^{\ell_1 - g - 1}\chi(H, E^{(\ell_1)})$$

où H est l'hyperplan générique de la fibre générique de $X^{(\ell_1)} \xrightarrow{\pi_1} \mathrm{Pic}_X^{\ell_1}$. Deligne a calculé cette caractéristique d'Euler-Poincaré et obtenu

$$\chi(H, E^{(\ell_1)}) = (-1)^{\ell_1 - g - 1} 2^{3g-3}$$

(cf. [De 2]). Par suite

$$m(E) = 2^{3g-3}$$

est indépendant de E (ce que l'on peut voir directement) et de ℓ_1 (ce qui est beaucoup plus surprenant).

Cette proposition est compatible avec la conjecture (5.4.2). En effet, pour $\mathrm{Drap}_X^{\ell_1} \xrightarrow{\pi_2} Fib_{X,2}^{\ell_2}$ la projection canonique, on a, pour tous entiers $e_1 \geqslant 0$ et $e_2 > 0$ avec $e_1 + 2e_2 = \ell_1 = \ell_2 - 2g + 2$,

$$\mathrm{Drap}_{X,\underline{e}}^{\ell_1} = \pi_2^{-1}\big(\pi_2\big(\mathrm{Drap}_{X,\underline{e}}^{\ell_1}\big)\big).$$

6. Cône nilpotent et variété caractéristique des K_E

(6.0).   Les considérations de ce numéro sont motivées d'une part par le calcul de la variété caractéristique de $\tilde{K}_E^{\ell_2}$ pour E de rang 2 effectué en (5.5) et par la théorie des "character sheaves" de Lusztig ([Lu 1] 13 et [Lu 2]).

(6.1).   Commençons par rappeler certains aspects de cette théorie de Lusztig. Soit G un groupe algébrique réductif connexe sur un corps algébriquement clos. Lusztig définit une classe de $\overline{\mathbb{Q}}_\ell$-faisceaux pervers irréductibles G-équivariants sur G (G agit par conjugaison sur lui-même) qu'il appelle "character sheaves".

Si k est la clôture algébrique d'un corps fini $\mathbb{F}_q$ et si G est défini sur $\mathbb{F}_q$, il n'y a qu'un nombre fini de ces "character sheaves" qui sont définis sur $\mathbb{F}_q$ et les fonctions centrales t_K sur $G(\mathbb{F}_q)$ pour ces derniers "character sheaves" K ($t_K(g) = \mathrm{tr}(\mathrm{Frob}_g, K)$) forment une base du $\overline{\mathbb{Q}}_\ell$-espace vectoriel des fonctions centrales sur $G(\mathbb{F}_q)$. Pour $G = \mathrm{GL}_{n,k}$, cette base (convenablement normalisée) coincide avec la base donnée par les caractères des $\overline{\mathbb{Q}}_\ell$-représentations irréductibles de $G(\mathbb{F}_q)$ (ce qui justifie la terminologie "character sheaf"). En général, ces deux bases diffèrent légèrement (L-indiscernabilité).

Pour $k = \mathbb{C}$, Lusztig propose dans [Lu 1] 13.8 une caractérisation topologie de ses "character sheaves" suggérée par le résultat suivant de Harish-Chandra ([H-C]). Soit G un groupe algébrique réductif connexe sur un corps local non archimédien d'inégale caractéristique et soit π une représentation admissible irréductible de G. Le caractère θ_π de π est une distribution G-invariante sur G et Harish-Chandra a montré que θ_π est lisse sur l'ouvert des éléments réguliers semi-simples, que θ_π est donnée sur G tout entier par intégration contre une fonction localement $\mathscr{L}^1$ et que la singularité de θ_π à l'origine de G peut être décrite de la façon suivante:

$$\theta_\pi(\exp X) = \sum_{\mathcal{O}^* \subset \mathcal{N}^*} c_{\mathcal{O}^*} \hat{\mu}_{\mathcal{O}^*}(X)$$

pour X voisin de l'origine dans l'algèbre de Lie $\mathfrak{g}$ de G, où $\mathcal{N}^*$ est le cône nilpotent dans l'algèbre de Lie duale de $\mathfrak{g}$, où $\mathcal{O}^*$ parcourt les orbites coadjointes nilpotentes, où $c_{\mathcal{O}^*} \in \mathbf{C}$ dépend de π et de $\mathcal{O}^*$ et où $\hat{\mu}_{\mathcal{O}^*}$ est le transformé de Fourier de "la" mesure G-équivariante sur $\mathfrak{g}^*$ de support $\mathcal{O}^*$.

On peut traduire plus algébriquement cette caractérisation "à la Harish–Chandra" des "characters sheaves" de la façon suivante. On suppose k de caractéristique 0 et on note

$$\Lambda \subset T^*G$$

le fermé réduit formé des

$$(g, \xi) \in G \times \mathfrak{g}^* = T^*G$$

tels que

$$\xi \in \mathcal{N}^*$$

et

$$g \in Z_G(\xi).$$

Alors Λ est lagrangien et les "characters sheaves" de Lusztig sont exactement les $\overline{\mathbf{Q}}_\ell$-faisceaux pervers irréductibles sur G tels que

$$|\mathrm{Car}(K)| \subset \Lambda$$

(les "character sheaves" de Lusztig sont définis par induction tordue à partir des $\overline{\mathbf{Q}}_\ell$-faisceaux lisses de rang 1, modérément ramifiés à l'infini et à monodromie finie; pour que la caractérisation ci-dessus soit raisonnable, il faut bien sûr supprimer cette condition "à monodromie finie").

(6.2). Soit de nouveau X une courbe projective, lisse et géométriquement connexe de genre g sur un corps k de caractéristique 0 et soit n un entier $\geqslant 1$.

Le fibré cotangent $T^*Fib_{X,n}$ au champ algébrique lisse sur k des fibrés vectoriels de rang n sur X s'identifie au champ algébrique sur k des couples

$$\left(\mathcal{L}, \mathcal{L} \xrightarrow{u} \mathcal{L} \otimes_{\mathcal{O}_X} \Omega^1_X \right),$$

où $\mathcal{L}$ est un fibré vectoriel de rang n et u est un homomorphisme de $\mathcal{O}_X$-Modules. On a

$$T^*Fib_{X,n} = \mathbb{V}(\mathcal{T}^1)$$

où

$$\mathscr{T}^1 = R^1 pr_{2*}\underline{\mathrm{End}}(\tilde{\mathscr{L}}),$$

avec $pr_2 \colon X \times_k Fib_{X,n} \to Fib_{X,n}$ la seconde projection canonique et $\tilde{\mathscr{L}}$ le fibré vectoriel universel sur $X \times_k Fib_{X,n}$; la fibre de la projection canonique

$$T^*Fib_{X,n} \to Fib_{X,n}$$

en $\mathscr{L}$ s'identifie au vectoriel

$$\mathrm{Hom}_{\mathscr{O}_X}\left(\mathscr{L}, \mathscr{L} \otimes_{\mathscr{O}_X} \Omega^1_X\right).$$

CONJECTURE (6.2.1). *Il existe un fermé réduit lagrangien canonique*

$$\Lambda \subset T^*Fib_{X,n}$$

tel que, pour tout corps $k' \supset k$, les objets de $\Lambda(k')$ soient exactement les couples

$$\left(\mathscr{L}', \mathscr{L}' \xrightarrow{u^*} \mathscr{L}' \otimes_{\mathscr{O}_X} \Omega^1_X\right),$$

où $\mathscr{L}'$ est un fibré vectoriel de rang n sur $X \otimes_k k'$ et où u' est un homomorphisme nilpotent de $\mathscr{O}_{X \otimes_k k'}$-Modules, i.e., tel que l'homomorphisme composé

$$\mathscr{L}' \xrightarrow{u'} \mathscr{L}' \otimes_{\mathscr{O}_X} \Omega^1_X \xrightarrow{u' \otimes 1} \mathscr{L}' \otimes_{\mathscr{O}_X} \left(\Omega^1_X\right)^{\otimes 2} \to ---\to \mathscr{L}' \otimes_{\mathscr{O}_X} \left(\Omega^1_X\right)^{\otimes(n-1)}$$

$$\xrightarrow{u' \otimes 1^{\otimes(n-1)}} \mathscr{L}' \otimes_{\mathscr{O}_X} \left(\Omega^1_X\right)^{\otimes n}$$

est identiquement nul (on a noté simplement 1 l'identité de Ω^1_X).

Remarque (6.2.2). (i) Pour $n = 1$, la conjecture est triviale: Λ est la section nulle $T^*Fib_{X,1}$.

(ii) Si G est n'importe quel groupe algébrique réductif connexe sur k, on peut formuler une conjecture analogue pour les fibrés principaux homogènes sous G de base X (la conjecture ci-dessus correspondant au cas particulier $G = \mathrm{GL}_{n,k}$).

(iii) Dans une lettre à Deligne, Drinfeld considère l'ouvert de $Fib_{X,n}$ formé des fibrés vectoriels qu'il appelle *très stables*, i.e., les fibrés vectoriels $\mathscr{L}$ pour lesquels le seul homomorphisme nilpotent $\mathscr{L} \to \mathscr{L} \otimes_{\mathscr{O}_X} \Omega^1_X$ est l'homomorphisme identiquement nul. Pour $n = 2$, il montre que le $\overline{\mathbf{Q}}_\ell$-faisceau pervers géométriquement irréductible K_E est lisse sur cet ouvert des fibrés très stables.

Pour $n = 2$, on peut vérifier à l'aide des calculs faits en (5.5) la conjecture. Plus précisément, on a:

PROPOSITION (6.2.2). *Pour tout couple d'entiers (λ_1, λ_2) avec*

$$\lambda_1 \leqslant \lambda_2 + 2g - 2,$$

notons

$$Fib_{X,2,(\lambda_1,\lambda_2)}$$

le fermé réduit de $Fib_{X,2}$ formé des fibrés vectoriels $\mathscr{L}$ de rang 2 et de degré $\lambda_1 + \lambda_2$ tels qu'il existe un fibré vectoriel quotient $\mathscr{L} \twoheadrightarrow \mathscr{L}_1$ de rang 1 et de degré $\ell_1 \leqslant \lambda_1$, un diviseur effectif E_1 sur X de degré $\lambda_1 - \ell_1$ et un homomorphisme de $\mathcal{O}_X$-Modules

$$\mathscr{L}_1(E_1) \to \mathrm{Ker}(\mathscr{L} \to \mathscr{L}_1)(-E_1) \otimes_{\mathcal{O}_X} \Omega^1_X$$

non identiquement nul.

Alors les fermés $Fib_{X,2,(\lambda_1,\lambda_2)}$ ainsi définis sont irréductibles et on a

$$\Lambda = \bigcup_{\substack{\lambda_1, \lambda_2 \in \mathbf{Z} \\ \lambda_1 \leqslant \lambda_2 + 2g - 2}} T^*_{Fib_{X,2,(\lambda_1,\lambda_2)}} Fib_{X,2}$$

(pour $\mathscr{Y}$ un fermé réduit irréductible d'un champ algébrique $\mathscr{X}$ lisse sur k, on note encore $T^*_{\mathscr{Y}}\mathscr{X}$ l'adhérence dans $T^*\mathscr{X}$ du conormal à l'ouvert de $\mathscr{Y}$ lisse sur k).

Remarque (6.2.3). (i) Si $\lambda_1 \leqslant \lambda_2 + g - 2$, on a simplement

$$Fib_{X,2,(\lambda_1,\lambda_2)} = \left\{ \mathscr{L} \in Fib^{\lambda_1+\lambda_2}_{X,2} \mid \ell_1(\mathscr{L}) \leqslant \lambda_1 \right\},$$

où $\ell_1(\mathscr{L})$ est la borne inférieure des degrés des fibrés vectoriels de rang 1 quotients de L, et par suite

$$Fib_{X,2,(\lambda_1,\lambda_2)} \supset Fib_{X,2(\lambda_1-1,\lambda_2+1)}.$$

On pose

$$Fib^0_{X,2,(\lambda_1,\lambda_2)} = Fib_{X,2,(\lambda_1,\lambda_2)} - Fib_{X,2,(\lambda_1-1,\lambda_2+1)}$$

$$= \left\{ \mathscr{L} \in Fib^{\lambda_1+\lambda_2}_{X,2} \mid \ell_1(\mathscr{L}) = \lambda_1 \right\}.$$

Pour $\lambda_1 < \lambda_2$, les strates localement fermées $Fib^0_{X,2,(\lambda_1,\lambda_2)}$ ont été introduites par Shatz; elles sont toutes lisses sur k et la codimension de $Fib^0_{X,2,(\lambda_1,\lambda_2)}$ dans $Fib_{X,2}$ est égale à $\lambda_2 - \lambda_1 + 2g - 2$ (cf. [Shat]).

Pour $\lambda_2 \leqslant \lambda_1 \leqslant \lambda_2 + g - 2$, les strates localement fermées $Fib^0_{X,2,(\lambda_1,\lambda_2)}$ ont été introduites par Lange et Narasimhan; elles ne sont plus lisses sur k en général mais la codimension de $Fib^0_{X,2,(\lambda_1,\lambda_2)}$ dans $Fib_{X,2}$ est toujours donnée par la formule $\lambda_2 - \lambda_1 + 2g - 2$ (cf. [La-Na]).

(ii) Pour $\lambda_2 + g - 1 \leqslant \lambda_1 \leqslant \lambda_2 + 2g - 2$, les strates fermées $Fib_{X,2,(\lambda_1,\lambda_2)}$ sont de codimension 1 dans $Fib_{X,2}$.

(iii) En rang n quelconque, on dispose encore des strates de Shatz paramétrées par les suites

$$(\lambda_1, \lambda_2, \ldots, \lambda_n) \in \mathbf{Q}^n,$$

où

$$\lambda_1 = \cdots = \lambda_{n_1} < \lambda_{n_1+1} = \cdots = \lambda_{n_1+n_2} < \cdots < \lambda_{n_1+\cdots+n_{r-1}+1} = \cdots = \lambda_{n_1+\cdots+n_r}$$

et où

$$n_j \lambda_{n_1+\cdots+n_{j-1}+1} = \cdots = n_j \lambda_{n_1+\cdots+n_j} \in \mathbf{Z}.$$

La strate localement fermée

$$Fib^0_{X,n,(\lambda_1,\ldots,\lambda_n)} \subset Fib_{X,n}$$

est formée des fibrés vectoriels de rang n et de degré $\lambda_1 + \cdots + \lambda_n$ qui admette un drapeau

$$\mathscr{L}. = \left(\mathscr{L} = \mathscr{L}_r \twoheadrightarrow \mathscr{L}_{r-1} \twoheadrightarrow \cdots \twoheadrightarrow \mathscr{L}_1 \twoheadrightarrow \mathscr{L}_0 = 0\right)$$

où, pour $j = 1, \ldots, r$, $\mathscr{L}_j$ est un fibré vectoriel de rang $n_1 + \cdots + n_j$ et où $\mathrm{Ker}(\mathscr{L}_j \to \mathscr{L}_{j-1})$ est semi-stable de pente

$$\lambda_{n_1+\cdots+n_{j-1}+1} = \cdots = \lambda_{n_1+\cdots+n_j}.$$

Un tel drapeau est nécessairement unique et est appelé le drapeau de Harder-Narasimhan.

Il n'est pas difficile de vérifier que pour toute suite $(\lambda_1, \ldots, \lambda_n)$ comme ci-dessus, on a

$$T^*_{Fib_{X,n,(\lambda_1,\ldots,\lambda_n)}} Fib_{X,n} \subset \Lambda,$$

où $Fib_{X,n,(\lambda_1,\ldots,\lambda_n)}$ et l'adhérence de $Fib^0_{X,n,(\lambda_1,\ldots,\lambda_n)}$ dans $Fib_{X,n}$. En effet, pour $\mathscr{L}$ dans la strate $Fib^0_{X,n,(\lambda_1,\ldots,\lambda_n)}$, de drapeau de Harder-Narasimhan $\mathscr{L}.$, le conormal de $Fib^0_{X,n,(\lambda_1,\ldots,\lambda_n)}$ dans $Fib_{X,n}$ a pour fibre en $\mathscr{L}$ le sous-espace vectoriel de $\mathrm{Hom}_{\mathscr{O}_X}(\mathscr{L}, \mathscr{L} \otimes \Omega^1_X)$ formé des u tels que

$$u\left(\mathrm{Ker}(\mathscr{L} \to \mathscr{L}_{j-1})\right) \subset \mathrm{Ker}(\mathscr{L} \to \mathscr{L}_j) \otimes_{\mathscr{O}_X} \Omega^1_X,$$

pour $j = 1, \ldots, r$. Mais ceci ne suffit pas à remplir Λ.

(6.3). On conserve les notations du numéro précédent. Si E est un $\overline{\mathbf{Q}}_\ell$-faisceau lisse et géométriquement irréductible de rang n sur X, on peut lui associer conjecturalement un $\overline{\mathbf{Q}}_\ell$-faisceau pervers géométriquement irréductible K_E sur $Fib_{X,n}$ (au sens de (2.1)).

CONJECTURE (6.3.1). *La variété caractéristique de K_E est contenue dans le fermé lagrangien canonique $\Lambda \subset T^*Fib_{X,n}$ de (6.2.1).*

Remarque (6.3.2). Pour $n = 1$, c'est trivial. Pour $n = 2$, cela résulte essentiellement de la construction de Drinfeld de K_E et du calcul de $|\mathrm{Car}(\tilde{K}_E^{\ell_2})|$ effectué en (5.5).

Enfin, on peut formuler la conjecture très optimiste suivante:

CONJECTURE (6.3.2). *Soit $K \in \mathrm{ob}\, D_c^b(Fib_{X,n}^0, \overline{\mathbf{Q}}_\ell)$ tel que*

$$K\left[\frac{n(4n^2 - 3n + 5)}{6}(g-1)\right]$$

soit un $\overline{\mathbf{Q}}$-faisceau pervers et géométriquement irréductible sur le champ algébrique (lisse et géométriquement connexe sur k) des fibrés vectoriels de rang n et de degré 0 sur X.

On suppose de plus que:
(i) $|\mathrm{Car}(K)| \subset \Lambda$,
(ii) K est cuspidal (au sens de (2.1)).
Alors, il existe un unique $\overline{\mathbf{Q}}_\ell$-faisceau lisse et géométriquement irréductible de rang n, E, sur X tel que

$$K \simeq K_E|Fib_{X,n}^0$$

(*avec les notations de (2.1.1)*).

Remarques (6.3.3). (i) Pour $n = 1$, la conjecture résulte du fait que le groupe fondamental géométrique de Pic_X^0 coincide avec l'abélianisé du groupe fondamental géométrique de X.

(ii) Les $\overline{\mathbf{Q}}_\ell$-faisceau pervers géométriquement irréductibles sur $Fib_{X,n}$ (au sens de (2.1)) dont la variété caractéristique est contenue dans Λ mais qui ne sont pas cuspidaux doivent correspondre aux $\overline{\mathbf{Q}}_\ell$-faisceaux lisses de rang n sur X qui ne sont pas géométriquement irréductibles. L'exemple le plus simple est le suivant: soit K le $\overline{\mathbf{Q}}_\ell$-faisceau pervers et géométriquement irréductible sur $Fib_{X,n}$ (au sens de (2.1)) dont la restriction à $Fib_{X,n}^\ell$ est

$$\overline{\mathbf{Q}}_{\ell,\,Fib_{X,n}^\ell}\left[(n-1)\ell - \frac{n(4n^2 - 9n + 5)}{6}(g-1)\right];$$

sa variété caractéristique est la section nulle de $T^*Fib_{X,n}$ et on a

$$T^1(K) = \left(\bigoplus_{j=0}^{n-1} \overline{\mathbf{Q}}_\ell[n-1-2j](-j)\right) \boxtimes_k K.$$

(iii) En tenant compte des phénomènes de L-indiscernabilité, il devrait être possible de formuler une conjecture semblable à (6.3.2) pour d'autres groupes réductifs connexes que GL_n, les K_E étant alors les analogues géométriques de combinaisons linéaires appropriées de formes automorphes dans un même L-paquet.

Appendice. Transformation de Radon géométrique d'après Brylinski (cf. [Br])

(A.0).   Soit S une variété lisse sur un corps k et soient $P \xrightarrow{\pi} S$ et $\check{P} \xrightarrow{\check{\pi}} S$ deux fibrés projectifs purement de dimension relative $r \geqslant 1$ en dualité parfaite. On note

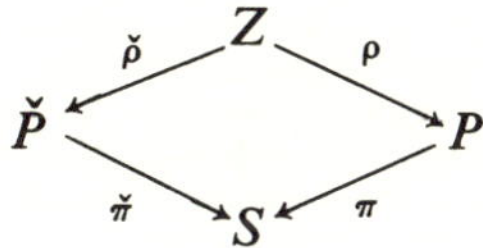

la variété d'incidence ($Z \subset \check{P} \times_S P$ et $Z_s \subset \check{P}_s \times_s P_s$ est le fermé formé des couples (H, x), où $x \in P_s$ et H est un hyperplan de P_s tels que $x \in H$, pour tout $s \in S$).

DÉFINITION (A.1).   *La transformation de Radon géométrique pour* $P \xrightarrow{\pi} S$ *est le foncteur*

$$\mathscr{R}(-) = R\check{\rho}_* \rho^*(-)[r-1]$$

de $D_c^b(P, \overline{\mathbf{Q}}_\ell)$ *dans* $D_c^b(\check{P}, \overline{\mathbf{Q}}_\ell)$.

THÉORÈME (A.2).   *Soit* K *un* $\overline{\mathbf{Q}}_\ell$*-faisceau pervers géométriquement irréductible tel que*

$$R\pi_* K = 0.$$

Alors, son transformé de Radon

$$K' = \mathscr{R}(K)$$

est aussi un $\overline{\mathbf{Q}}_\ell$*-faisceau pervers géométriquement irréductible tel que*

$$R\check{\pi}_* K' = 0.$$

De plus

$$K = \check{\mathscr{R}}(K')(r-1)$$

où

$$\check{\mathscr{R}}\colon D^b_c\big(\check{P},\overline{\mathbf{Q}}_\ell\big) \to D^b_c\big(\check{\check{P}},\overline{\mathbf{Q}}_\ell\big) = D^b_c\big(P,\overline{\mathbf{Q}}_\ell\big)$$

est la transformation de Radon géométrique pour $\check{P} \xrightarrow{\check{\pi}} \mathscr{S}$.

Dans ce qui suit, nous supposerons que le *corps k est de caractéristique* 0. On a un diagramme commutatif à carré cartésien (avec les notations évidentes)

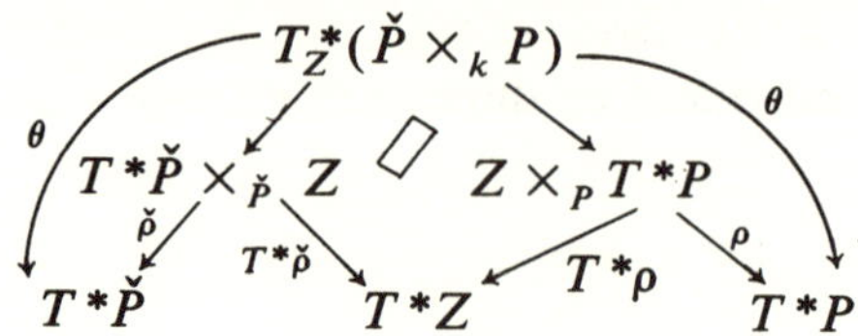

puisque $Z \subset \check{P} \times_k P$ est lisse sur k. De plus on a

$$T^*\check{P} - T^*S \times_S \check{P} \xleftarrow{\check{\theta}} T^*_Z\big(\check{P} \times_k P\big) - T^*_S(S \times_k S) \times_S Z$$

$$\xrightarrow{\theta} T^*P - T^*S \times_S P,$$

où on a identifié $T^*S \times_S P$ (resp. $T^*S \times_S \check{P}$, $T^*_S(S \times_k S) \times_S Z$) à une sous-variété de T^*P (resp. $T^*\check{P}$, $T^*_Z(\check{P} \times_k P)$) via $T^*\pi$ (resp. $T^*\check{\pi}$, $T^*(\check{\pi} \times_k \pi)$) qui est ici une immersion fermée; on note encore ψ l'isomorphisme

$$\psi = \check{\theta} \circ \theta^{-1}\colon T^*P - T^*S \times_S P \xrightarrow{\sim} T^*\check{P} - T^*S \times_S \check{P}.$$

THÉORÈME (A.3). *Soit K un faisceau pervers géométriquement irréductible sur P vérifiant*
(i) $R\pi_* K = 0$
(ii) $|\mathrm{Car}(K)| \cap (T^*S \times_S P) \subset T^*_P P$, *où* $\mathrm{Car}(K)$ *est le cycle caractéristique de K et* $|\mathrm{Car}(K)|$ *son support, i.e. la variété caractéristique de K.*
Soit $K' = \mathscr{R}(K)$ *le transformé de Radon de K. Alors, on a aussi*

$$|\mathrm{Car}(K')| \cap \big(T^*S \times_S \check{P}\big) \subset T^*_P \check{P}$$

et

$$\mathrm{Car}(K')|\big(T^*\check{P} - T^*S \times_S \check{P}\big) = \psi_*(\mathrm{Car}(K)|(T^*P - T^*S \times_S P)).$$

BIBLIOGRAPHIE

[B–B–D]   A. A. BEILINSON, J. BERNSTEIN ET P. DELIGNE, *Faisceaux pervers*, dans "Analyse et topologie sur les espaces singuliers", Astérisque 100 (1982).

[Br]   J. L. BRYLINSKI, *Transformations Canoniques, Dualité Projective, Théorie de Lefschetz, Transformations de Fourier et Sommes Trigonométriques*, Astérisque 140–141 (1986).

[De 1]   P. DELIGNE, "Cohomologie à supports propres (exposé XVII)," dans *Théorie des topos et cohomologie étale des schémas, dirigé par M. Artin, A. Grothendieck et J. L. Verdier*, Lecture Notes in Mathematics **305** (1973), 250–480.

[De 2]   ______, Notes manuscrites.

[Di–Gr]   J. A. DIEUDONNÉ ET A. GROTHENDIECK, *Eléments de Géométrie Algébrique* I, Springer-Verlag (1971).

[Dr]   V. G. DRINFELD, *Two dimensional ℓ-adic representations of the fundamental group of a curve over a finite field and automorphic forms on* GL(2), American Journal of Math. **105** (1983), 85–114.

[Go–Ma]   M. GORESKY AND R. MACPHERSON, *Intersection homology* II, Inventiones Math. **72** (1983), 77–130.

[Gr]   A. GROTHENDIECK, *Techniques de construction et théorèmes d'existence en géométrie algébrique* IV: *les schémas de Hilbert*, Séminaire Bourbaki 221 (1960/61), Benjamin (1966).

[H-C]   HARISH-CHANDRA, "The characters of reductive p-adic groups," in *Contribution to Algebra*, Academic Press (1977), 175–182.

[Ho–Sh]   R. HOTTA AND N. SHIMOMURA, *The fixed point subvarieties of unipotent transformations on generalized flag varieties and Green functions*, Math. Annalen **241** (1979), 193–208.

[La–Na]   H. LANGE AND M. S. NARASIMHAN, *Maximal subbundles of rank two vector bundles on curves*, Math. Annalen **266** (1983), 55–72.

[La]   R. P. LANGLANDS, "Problems in the theory of automorphic forms," in Lectures in Modern Analysis and Applications III, Lecture Notes in Math. **170** (1970), 18–86.

[Lu 1]   G. LUSZTIG, *Characters of reductive groups over a finite field*, Annals of Math. Studies **107** (1984).

[Lu 2]   ______, *Character sheaves* I, II, III, IV *and* V, Advances in Mathematics **56** (1985), 193–237; **57** (1985), 226–265, 266–315; **59** (1986) 1–63; and **61** (1986), 103–155.

[Lu 3]   ______, *Green polynomials and singularities of unipotent classes*, Advances in Mathematics **42** (1981), 169–178.

[Mac]   J. G. MACDONALD, *Symmetric Functions and Hall Polynomials*, Clarendon Press (1979).

[Se]   J. P. SERRE, *Groupes Algébriques et Corps de Classes*, Hermann (1959).

[Sha]   J. A. SHALIKA, The multiplicity one theorem for GL_n, Annals of Math. **100** (1974), 171–193.

[Shat]   S. SHATZ, *Algebraic families of vector bundles*, Compositio Mathematica **35** (1977), 163–187.

[Shi]   T. SHINTANI, *On an explicit formula for class* 1 "Whittaker functions" *on* GL_n *over p-adic fields*, Proc. Japan. Acad. **52** (1976), 180–182.

[Spr]   T. A. SPRINGER, *Quelques applications de la cohomologie d'intersection*, Séminaire Bourbaki **589** (1981/82), *Astérisque* **92–93** (1982), 249–273.

UNIVERSITÉ PARIS-SUD, 91405 ORSAY CEDEX, FRANCE

Ajoutée sur épreuve: J'ai démontré en mars 1987 la Conjecture (6.2.1).

CONGRUENCES FOR PERIODS OF MODULAR FORMS

NEAL KOBLITZ

To Yuri Ivanovich Manin on the occasion of his fiftieth birthday

§1. Introduction. Starting with Yu. I. Manin's paper [17], much attention has been devoted to p-adic congruences for the periods of modular forms of varying weights k. Our first purpose is to give an exposition of a part of this work. In addition, we give a congruence of the Ramanujan type for the odd periods of a cusp form for $SL_2,\mathbb{Z}$, in the six cases when the space of such cusp forms is 1-dimensional (Theorem 2). Finally, in §4 we discuss how p-adic congruences can be used as evidence of the existence of canonical square roots of the central critical Hecke L-series values $L(\psi^{k-1}, k/2)$, presumably connected with the arithmetic of the corresponding elliptic curve.

§2. Periods of cusp forms for $SL_2(\mathbb{Z})$. Let $\Phi(z) = \sum_{n=1}^{\infty} a_n e^{2\pi inz}$ be an element of the space S_k of cusp forms of weight k for the full modular group $SL_2(\mathbb{Z})$. The moments or *periods* of Φ are defined as follows:

$$r_m(\Phi) = \int_0^{i\infty} \Phi(z) z^m \, dz, \qquad m \geq 0. \tag{1}$$

Set $w = k - 2$. The periods of Φ for $m \leq w$ are closely related to the values at integer points in the critical strip of the Hecke L-series corresponding to Φ, which is defined in a right half-plane by $L_\Phi(s) = \sum a_n n^{-s}$. This will be described more precisely later. These L-series values are called "special values" or "critical values" of the L-function.

Eichler–Shimura's rationality result. If the coefficients a_n are real, then the periods $r_m(\Phi)$ with m odd are real, and the even periods are purely imaginary. One then separately considers the $(w/2)$-tuple $\mathbf{r}^-(\Phi)(r_1(\Phi), \ldots, r_{w-1}(\Phi)) \in \mathbf{R}^{w/2}$ of odd periods and the $(w/2 + 1)$-tuple $\mathbf{r}^+(\Phi) \in \mathbf{R}^{w/2+1}$ of even periods. Eichler and Shimura found a set of linear equations with rational integer coefficients which are satisfied by these period vectors for any $(\frac{1}{i}r_0(\Phi), \ldots, \frac{1}{i}r_w(\Phi))\Phi \in S_k$. Moreover, if $V^- \subset \mathbf{R}^{w/2}$ and $V^+ \subset \mathbf{R}^{w/2+1}$ denote the subspaces consisting of all vectors which satisfy these linear relations, then Eichler–Shimura proved that the map $S_k \to V^-$ given by $\Phi \mapsto \mathbf{r}^-(\Phi)$ is an isomorphism, and the map $S_k \to V^+$ given by $\Phi \mapsto \mathbf{r}^+(\Phi)$ is an isomorphism onto a subspace of codimension 1 in V^+. For a complete exposition, see [13].

Received October 1, 1986.

TABLE I

k	r_3/r_1	r_5/r_1	r_7/r_1	r_9/r_1	r_{11}/r_1
12	$-\dfrac{5^2}{2^4 \cdot 3}$	$\dfrac{5}{2^2 \cdot 3}$			
16	$-\dfrac{5 \cdot 7^2}{2^3 \cdot 3^2 \cdot 13}$	$\dfrac{7^2}{2^2 \cdot 3^2 \cdot 13}$	$-\dfrac{5 \cdot 7}{2^2 \cdot 3^2 \cdot 13}$		
18	$-\dfrac{11}{2^2 \cdot 3 \cdot 5}$	$\dfrac{1}{2^3 \cdot 3}$	$-\dfrac{1}{2^3 \cdot 3 \cdot 5}$		
20	$-\dfrac{5 \cdot 11 \cdot 13}{2^5 \cdot 3^2 \cdot 17}$	$\dfrac{11^2}{2^2 \cdot 3^2 \cdot 7 \cdot 17}$	$-\dfrac{5}{2^2 \cdot 3^2 \cdot 17}$	$\dfrac{1}{2^2 \cdot 3 \cdot 17}$	
22	$-\dfrac{7 \cdot 13^2}{2^2 \cdot 3^3 \cdot 5 \cdot 19}$	$\dfrac{7 \cdot 13}{2^5 \cdot 3^2 \cdot 19}$	$-\dfrac{41}{2^4 \cdot 3^2 \cdot 5 \cdot 19}$	$\dfrac{7}{2^4 \cdot 3^2 \cdot 5 \cdot 19}$	
26	$-\dfrac{17 \cdot 19}{2^2 \cdot 3^2 \cdot 5 \cdot 23}$	$\dfrac{5 \cdot 17}{2^3 \cdot 3^2 \cdot 7 \cdot 23}$	$-\dfrac{7}{2^3 \cdot 3^2 \cdot 5 \cdot 23}$	$\dfrac{1}{2^3 \cdot 3^2 \cdot 5 \cdot 23}$	$-\dfrac{1}{2^3 \cdot 3^2 \cdot 5 \cdot 7 \cdot 23}$

If, for example, S_k is 1-dimensional—i.e., if $k = 12$, 16, 18, 20, 22, or 26—then a consequence of the Eichler–Shimura result is that V^- is also 1-dimensional. Since V^- is defined by equations with rational coefficients, this means that it gives a rational projective point; in other words, in this case the ratio of any two odd periods is a rational number. Clearly this ratio is not affected by changing $\Phi \in S_k$ by a constant multiple. Table I shows these ratios for the indicated values of k. The missing values of $m \in \{1, 3, \ldots, w - 1\}$ are filled in using the symmetry property that $r_m(\Phi) = r_{w-m}(\Phi)$ for odd m.

Manin's rationality results. The Eichler–Shimura relations are proved by considering the action of the elements $\begin{pmatrix} 0 & -1 \\ 1 & 0 \end{pmatrix}$ and $\begin{pmatrix} 1 & -1 \\ 1 & 0 \end{pmatrix}$ on the integrand in (1). The resulting linear relations apply simultaneously for all $\Phi \in S_k$. In [17] Manin derived a whole series of additional linear relationships for the periods using the action of the Hecke operators on the integrand in (1). Manin's equations depend on the particular form $\Phi(z) = \Sigma a_n e^{2\pi i n z}$, and their coefficients are in the field generated by the a_n. For any fixed $\Phi \in S_k$ which is an eigenform for all of the Hecke operators, Manin proved that the corresponding linear equations satisfied by the periods uniquely determine a one-dimensional subspace of V^- or V^+. This implies

THEOREM 1 (MANIN [17]).   *Let $\Phi \in S_k$ be an eigenfunction for all of the Hecke operators. Let K denote the extension field of $\mathbf{Q}$ generated by the eigenvalues. Then there exist two constants ω^+ and ω^- such that the period vectors $(1/\omega^+)\mathbf{r}^+(\Phi)$ and $(1/\omega^-)\mathbf{r}^-(\Phi)$ have coordinates in K.*

The Eichler–Shimura equations suffice to give this conclusion only in the case of the odd period vector when S_k is 1-dimensional. Even for $k = 12$, 16, 18, 20,

22, 26, Manin's equations are needed to give rationality of the ratios r_m/r_0 for m even. The table analogous to our Table I for the even period ratios is given in [17].

Manin noticed that in all cases in his table the numerator of r_m/r_0 for $0 < m < w$ is divisible by the numerator of the k-th Bernoulli number. He used this fact to give a new proof of the Ramanujan congruence $\tau(n) \equiv \sigma_{11}(n)$ mod 691 for the coefficients of the weight 12 normalized cusp-form $\Delta(z)$ and the similar congruences that hold for the normalized form in S_k in the other 1-dimensional cases.

We shall now prove analogous congruences modulo the numerator of the k-th Bernoulli number for the odd periods in Table I.

Congruences for odd periods. If a and b are two rational numbers, we write $a \equiv b$ mod N if the numerator of $a - b$ is divisible by N. Let B_j denote the j-th Bernoulli number, i.e., the coefficient of $t^j/j!$ in $t/(e^t - 1)$, and let N_j denote the numerator of B_j. For example,

$$N_{12} = 691, \qquad N_{16} = \qquad 3617, \qquad N_{18} = 43867,$$

$$N_{20} = 283 \cdot 617, \qquad N_{22} = 131 \cdot 593, \qquad N_{26} = 657931.$$

We further denote $\tilde{B}_j = B_{j+1}/(j+1)$. Then

$$\tilde{B}_j = -\zeta(-j),$$

where ζ is the Riemann zeta-function.

THEOREM 2. *Let $k = 12, 16, 18, 20, 22,$ or 26, so that S_k is 1-dimensional. Let $\Phi \in S_k$. Let $w = k - 2$, and let $m < w$ be a positive odd integer. Then*

$$\frac{r_m(\Phi)}{r_1(\Phi)} \equiv \frac{\tilde{B}_m \tilde{B}_{w-m}}{\tilde{B}_1 \tilde{B}_{w-1}} \bmod N_k. \tag{2}$$

Proof. This congruence follows from an exact formula of Kohnen–Zagier [12] for the m-th period of a form $R_0 \in S_k$ which has the property that its Petersson scalar product with any $f \in S_k$ is equal to $r_0(f)$. Namely, since S_k is 1-dimensional, we can set $\Phi = cR_0$. Then $r_m(\Phi)/r_1(\Phi) = r_m(R_0)/r_1(R_0)$. If we multiply both sides of the formula in Theorem 1 of [12] by B_k and neglect terms on the right which are $\equiv 0$ mod N_k, we obtain (here we also take into account that r_m in [12] is defined with an extra factor of i^{m+1}):

$$B_k(-1)^k 2^{-w} w! r_m(R_0) \equiv kw! \tilde{B}_m \tilde{B}_{w-m} \bmod N_k,$$

from which the theorem follows.

Heuristics.   There is a heuristic argument which explains why one might expect a congruence such as (2) and provides a general approach to predicting many types of congruences for periods and other special values. If $\Phi(z) = \sum_{n=1}^{\infty} a_n e^{2\pi i n z}$ is a cusp form of weight $k = w + 2$, then term-by-term integration of the right side of (1) gives the relationship

$$r_m(\Phi) = \frac{m! i^{m+1}}{(2\pi)^{m+1}} L_\Phi(m+1), \qquad 0 \leqslant m \leqslant w. \tag{3}$$

Here L_Φ is the L-function associated to Φ, defined for $Re\ s > k - \frac{1}{2}$ by $L_\Phi(s) = \sum_{n=1}^{\infty} a_n n^{-s}$ and continued analytically onto the rest of the complex s-plane.

On the other hand, suppose we extend our definitions of r_m and the L-function to the Eisenstein series

$$G_k(z) = -\frac{B_k}{2k} + \sum_{n=1}^{\infty} \sigma_{k-1}(n) e^{2\pi i n z}, \quad \text{where } \sigma_{k-1}(n) = \sum_{d \mid n} d^{k-1},$$

by removing the constant term before inserting G_k in the integrand in (1) and also in the definition of L_{G_k}. Then we find that

$$L_{G_k}(s) = \sum_{n=1}^{\infty} \sigma_{k-1}(n) n^{-s} = \sum_{d,\, d'} d^{k-1} (dd')^{-s} = \zeta(s)\zeta(s - k + 1),$$

and that

$$r_m(G_k) = \frac{m! i^{m+1}}{(2\pi)^{m+1}} \zeta(m+1)\zeta(m - w).$$

Using the functional equation for ζ to relate $\zeta(m+1)$ to $\zeta(-m)$ and also the fact that $\zeta(-j) = -\tilde{B}_j$, we obtain:

$$r_m(G_k) = \tfrac{1}{2}\tilde{B}_m \tilde{B}_{w-m}. \tag{4}$$

Now in the case when S_k is 1-dimensional it is well known that the normalized cusp form $\Phi \in S_k$ and the Eisenstein series G_k, considered as power series in $\mathbf{Q}[[q]]$, where $q = e^{2\pi i z}$, satisfy the congruence

$$\Phi \equiv G_k \bmod N_k$$

(i.e., each coefficient satisfies this congruence). Thus, if we could justify using this congruence in the integrand of (1)—ignoring for a minute the impossibility of interchanging a congruence relation with the transcendental operation of integration—we could conclude that $r_m(\Phi) \equiv r_m(G_k) \bmod N_k$. However, this makes no sense, since the left side is (conjectured to be) transcendental. Nevertheless, if we remove the transcendental part ω^- of the period, or, equivalently, take ratios to a fixed choice of period, say $m = 1$, we can write the following congruence of two rational numbers:

$$\frac{r_m(\Phi)}{r_1(\Phi)} \equiv \frac{r_m(G_k)}{r_1(G_k)} \bmod N_k.$$

By (4), this is what we proved in Theorem 2.

One often encounters this type of situation. Suppose two power series or Dirichlet series satisfy a coefficient-by-coefficient congruence modulo N. After using some transcendental procedure to obtain some "special values," we remove a transcendental part to obtain two rational numbers. It frequently turns out that the two rational numbers are also congruent modulo N.

Manin's p-adic interpolation. An example of such a situation was studied by Manin in [17]–[18]. If $\Phi(z) = \sum_{n=1}^{\infty} a_n e^{2\pi i n z} \in S_k$ and if χ is a Dirichlet character, we let Φ_χ denote the series

$$\Phi_\chi(z) = \sum_{n=1}^{\infty} a_n \chi(n) e^{2\pi i n z}, \tag{5}$$

and we let $r_m(\Phi, \chi)$ denote the integral (1) with Φ replaced by Φ_χ. This "period twisted by χ" satisfies the same relation (3) as r_m, with $L_\Phi(m + 1)$ replaced by $L_\Phi(m + 1, \chi)$, where $L_\Phi(s, \chi)$ is defined by the same Dirichlet series as $L_\Phi(s)$ but with a_n replaced by $a_n \chi(n)$:

$$L_\Phi(s, \chi) = \sum_{n=1}^{\infty} a_n \chi(n) n^{-s} \quad \text{for } \operatorname{Re} s > k - \frac{1}{2}. \tag{6}$$

Suppose we choose two characters χ_1 and χ_2 and two integers m_1 and m_2 between 0 and w such that each term in the sum on the right in (6) for $L_\Phi(m_1 + 1, \chi_1)$ is congruent modulo N to the corresponding term in the sum for $L_\Phi(m_2 + 1, \chi_2)$. For example, suppose that $N = p$ is a prime dividing the conductor of χ_1 and we take $m_2 = m_1 + (p - 1)/2$ and $\chi_2(x) = (x/p)\chi_1(x)$, where $(\cdot/p)$ is the Legendre symbol. Then, heuristically speaking, we might expect that the algebraic parts of the L-series values, and hence of the periods, should be congruent modulo N. This is essentially what Manin proves in [17] and [18].

Manin actually shows that the (algebraic part of the) period can be written as an integral very similar in appearance to (1) but with the imaginary axis replaced by the set of p-adic integers and with $\Phi(z)\, dz$ replaced by a measure μ_Φ on the set of p-adic integers. The fact that any congruence modulo p^j between the terms in (6) carries over to a corresponding congruence between the algebraic parts of the periods then follows from an elementary p-adic argument. We now give a precise statement.

Suppose that $\Phi(z) = \sum_{n=1}^{\infty} a_n e^{2\pi i n z} \in S_k$ is a normalized (i.e., $a_1 = 1$) eigenfunction for all of the Hecke operators. For simplicity, we shall suppose that p is an odd prime such that the p-th coefficient a_p is prime to p, and that χ is a nontrivial Dirichlet character of conductor a power p^f of p. Then Manin constructs two measures μ_Φ^+ and μ_Φ^- on $\mathbf{Z}_p$ with values in the ring of integers of a certain finite extension of $\mathbf{Q}_p$ with the following property. Let m be an integer between 0 and $w = k - 2$. If χ has the same parity as m, i.e., if $\chi(-1) = (-1)^m$, let ω denote ω^+, the transcendental factor in the periods $r_m(\Phi)$ for m even, and let μ_Φ denote μ_Φ^+; if χ has the opposite parity from m, let $\omega = -\omega^-$ and $\mu_\Phi = \mu_\Phi^-$. Let $G(\chi) = \sum_{n=1}^{p^f} \chi(n) e^{2\pi i n/p^f}$ denote the Gauss sum, let ρ denote the p-adic unit root of the equation $\rho^2 - a_p \rho + p^{k-1} = 0$, and let $\mathbf{Z}_p^*$ denote the units of $\mathbf{Z}_p$.

THEOREM 3 (MANIN [18]). *With the above notation and assumptions*

$$\frac{1}{\omega} \frac{p^{f(m+1)}}{\rho^f G(\chi)} r_m(\Phi, \chi) = \int_{\mathbf{Z}_p^*} \chi^{-1}(z) z^m \, d\mu_\Phi(z). \tag{7}$$

Now if $\chi_1(n)/n^{m_1} \equiv \chi_2(n)/n^{m_2} \bmod p^j$ for every integer n prime to p, it follows that $\chi_1^{-1}(z) z^{m_1} \equiv \chi_2^{-1}(z) z^{m_2} \bmod p^j$ for every $z \in \mathbf{Z}_p^*$. In other words, the integrands on the right in (7) are congruent when χ and m are replaced by χ_1, m_1 and by χ_2, m_2. But whenever one integrates two functions which are congruent modulo p^j with respect to a p-adic integer valued measure, the results are also congruent modulo p^j. (This is the elementary p-adic fact that if two functions that are close to one another are integrated over a compact set, the results will be close to one another.)

Thus, Manin's theorem tells us in this case that the left hand sides of (7) for χ_1, m_1 and for χ_2, m_2—which turn out to be algebraic despite the very non-algebraic operation of complex integration on the left in (7)—are congruent modulo p^j.

§3. Periods of cusp forms associated to powers of Hecke characters. We now consider modular forms for congruence subgroups of $SL_2(\mathbf{Z})$ that are constructed in a very special way.

Hecke characters and their cusp forms. Let K be a quadratic imaginary field of discriminant $-d$, and let ψ be a Hecke character of the ring of integers O_K. For simplicity, we shall suppose that K has class number 1, and that ψ has infinity type $(1, 0)$ and takes values in K. That is, ψ is a character $x \mapsto \psi(x)$ from O_K to O_K which is trivial on the units of O_K and has the properties that $\psi(x)/x$ is a root of unity and $\psi(\bar{x}) = \bar{\psi}(x)$. Let u denote the number of roots of unity in

K, and let $\mathbf{N}(x)$ denote the norm from K to $\mathbf{Q}$ of an element $x \in K$. By the weight k cusp form associated to ψ we mean the series

$$\Phi(z) = \frac{1}{u} \sum_{x \in O_K} \psi^{k-1}(x) e^{2\pi i N(x)z}.$$

It is known (see [32, 23, 24, 22]) that Φ is a cusp form of weight k for the subgroup $\Gamma_0(N) = \left\{ \begin{pmatrix} a & b \\ c & d \end{pmatrix} \in SL_2(\mathbf{Z}) \mid c \equiv 0 \bmod N \right\}$, where the level N depends on ψ. N is a product of powers of primes dividing the discriminant $-d$ and the conductor of ψ. (The minimal level N may drop for certain k, because the conductor of ψ^{k-1} drops, i.e., the $(k-1)$-th power kills some of the ramification of ψ.) For χ a Dirichlet character we further denote

$$\Phi_\chi(z) = \frac{1}{u} \sum_{x \in O_K} \chi(\mathbf{N}(x)) \psi^{k-1}(x) e^{2\pi i N(x)z}.$$

In general, Φ_χ is a cusp form of level $(\operatorname{cond} \chi)^2 N$.

The forms Φ and Φ_χ are eigenforms for all of the Hecke operators. Their Fourier expansion coefficients are totally real algebraic integers in the field of values of χ and ψ. In particular, if χ is trivial or a quadratic character, then these coefficients are rational integers.

The L-series corresponding to the weight k cusp form associated to ψ will be denoted $L(\psi^{k-1}, s)$; it is the (analytic continuation of) the Dirichlet series

$$L(\psi^{k-1}, s) = \frac{1}{u} \sum_{x \in O_K} \psi^{k-1}(x) \mathbf{N}(x)^{-s}.$$

We let $L_{\psi, k}(\chi, s) = L(\chi \psi^{k-1}, s)$ denote the L-series which corresponds to Φ_χ.

These L-functions satisfy functional equations relating the value at s to the value at $k - s$. The region $0 < \operatorname{Re} s < k$ is called the "critical strip," and values of an L-function at integers of the critical strip are called "critical values."

The periods of these modular forms are related to the critical values of the corresponding L-series by (3), i.e.,

$$r_m(\Phi) = \int_0^{i\infty} \Phi(z) z^m \, dz = \frac{m! \, i^{m+1}}{(2\pi)^{m+1}} L(\psi^{k-1}, m+1), \qquad 0 \leqslant m \leqslant w = k - 2,$$

$$\tag{8}$$

and similarly with Φ replaced by Φ_χ and $L(\psi^{k-1}, m+1)$ replaced by $L_{\psi, k}(\chi, m+1)$.

Elliptic curves. Hecke characters ψ of the type under consideration arise from elliptic curves E with complex multiplication by O_K. Then the first Hecke L-series $L(\psi, s) = L_E(s)$ is the Hasse-Weil L-series of E, and the weight 2 cusp form associated to ψ gives the Weil parametrization of E.

Suppose now that p is a prime which splits in K: $(p) = (\rho)(\bar{\rho})$, where ρ and $\bar{\rho}$ are complex conjugate prime elements of O_K. Here we can choose the generator ρ so that $\rho = \psi(\rho)$. If we complete K at (ρ), obtaining a field $K_\rho \approx \mathbf{Q}_p$, then $\bar{\rho}$ is the p-adic unit reciprocal root of the congruence zeta-function of the corresponding elliptic curve E considered modulo p.

Let ω denote the transcendental period of E, i.e., ωO_K is the period lattice of E; explicitly, ω is a certain complete elliptic integral.

Example. Let $K = \mathbf{Q}(i)$, and let ψ be the Hecke character on the Gaussian integers defined by $x \mapsto i^j x$, where $j \in \{0, 1, 2, 3\}$ is chosen so that $i^j x \equiv 1 \bmod 2 + 2i$. Then E is the elliptic curve $y^2 = x^3 - x$ and $\omega = \int_1^\infty (dx/(\sqrt{x^3 - x})) = 2.622 \cdots$. The weight 2 cusp form corresponding to ψ is the unique normalized form in $S_2(\Gamma_0(32))$.

The Manin–Vishik interpolation. In [20], Manin and Vishik found p-adic congruences for the m-th period of the weight k cusp form $\Phi = \Phi_k$ associated to a Hecke character. More precisely, in the above situation they found a p-adic integral representation for the (algebraic) values

$$\left(\frac{\pi}{\omega}\right)^{k-1}\left(1 - \frac{p^m}{\bar{\rho}^{w+1}}\right)\left(1 - \frac{p^{w-m}}{\bar{\rho}^{w+1}}\right) r_m(\Phi_k) \tag{9}$$

for $k \geqslant 3, 0 \leqslant m \leqslant w = k - 2$.

Katz's interpolation. Then in [3] Katz used a completely different technique to obtain a more general result of this type. His point of view was to allow the elliptic curve E to vary and regard each period—equivalently, each L-series value in (8)—as a modular function of E. Katz expressed the resulting Eisenstein series as p-adic integrals with respect to modular-form-valued measures; if a specific elliptic curve meeting Manin–Vishik's conditions is plugged into Katz's formula, the result is the above period values. We shall give the p-adic integral formula for the periods in Katz's formulation.

We keep the above assumptions and notation; in particular, $p = \rho\bar{\rho}$ with $\bar{\rho}$ the p-adic unit reciprocal root of the congruence zeta-function of E. There exists a unit c in the p-adic completion of the unramified closure of $\mathbf{Q}_p$ which plays the role of a "p-adic period." It has the property that $c/\mathrm{Frobenius}(c) = \bar{\rho}$, which implies that $c^{1-p} \equiv \bar{\rho} \bmod p$.

Let N_0 be a positive integer, let χ be a character of $(O_K/N_0 O_K)^*$ (extended by zero to all of $O_K/N_0 O_K$), and let α be a level N_0 structure, i.e., an isomorphism from $\mathbb{Z}/N_0\mathbb{Z} \times \mathbb{Z}/N_0\mathbb{Z}$ to the group ${}_{N_0}E$ of points of order N_0 on E. Since $E \approx \mathbb{C}/\omega O_K$, multiplication by N_0/ω identifies ${}_{N_0}E$ with $O_K/N_0 O_K$. In this way α allows us to pull back χ to $\mathbb{Z}/N_0\mathbb{Z} \times \mathbb{Z}/N_0\mathbb{Z}$. Let χ' denote the character $x \mapsto \chi(\bar{x})$, and let χ^* denote the "partial Fourier transform" of χ, which is the function on $\mathbb{Z}/N_0\mathbb{Z} \times \mathbb{Z}/N_0\mathbb{Z}$ defined by $\chi^*(m, n) = \Sigma_{a, \bmod N_0} \chi(a, n) e^{2\pi i a m/N_0}$.

Let Φ_k denote the weight k cusp form associated to ψ.

In this situation, Katz p-adically interpolates the following algebraic-valued functions of $k = w + 2 \geqslant 4$ and $m = 0, \ldots, w$:

$$(iN_0)^{m+1}(-\sqrt{d})^{m-w}\left(1 - \frac{p^m}{\bar{\rho}^{w+1}\chi(\rho)}\right)\left(1 - \frac{p^{w-m}\chi(\bar{\rho})}{\bar{\rho}^{w+1}}\right)\left(\frac{2\pi}{\omega}\right)^{k-1} r_m(\Phi_k, \chi').$$

$$(10)$$

Katz constructs a modular-form-valued measure $\mu(x, y, u, v)$ on $\mathbb{Z}_p^* \times \mathbb{Z}_p^* \times \mathbb{Z}/N_0\mathbb{Z} \times \mathbb{Z}/N_0\mathbb{Z}$ whose moments essentially give (10). Namely, when evaluated at the specific elliptic curve E with level N_0 structure and complex multiplication as above, the value

$$\frac{1}{c^{w+1}} \int x^m y^{w-m} \chi'(u, v)\, d\mu_E \qquad (11)$$

is the p-integral algebraic number (10).

This immediately implies p-adic congruences for these numbers. For example, if we add $p - 1$ to the weight k then modulo p the number is multiplied by $\bar{\rho}$, because the integrand is unchanged modulo p while the factor in front is multiplied by $c^{1-p} \equiv \bar{\rho} \bmod p$.

Katz's results generalize to the situation when we also want to consider twistings of Φ_k by another character. For example, we can use an integral formula analogous to (11) to conclude that, if we insert the Legendre symbol $((N(\cdot))/p)$ and at the same time increase m by $(p - 1)/2$ (while keeping it within the range $0 \leqslant m \leqslant w$), then the expression (10) is unchanged modulo p. For more details, see [3] and [6].

§4. Square roots of periods. We now consider the periods r_m when m is in the center of its range of values, i.e., $m = w/2 = k/2 - 1$ (here the weight k is even); equivalently, we consider the "central critical values" of the L-series:

$$r_{w/2}(\Phi) = \frac{(w/2)! i^{k/2}}{(2\pi)^{k/2}} L_\Phi(k/2) \quad \text{and} \quad r_{w/2}(\Phi_\chi) = \frac{(w/2)! i^{k/2}}{(2\pi)^{k/2}} L_\Phi(k/2, \chi).$$

We shall also assume that the numbers which Manin, Manin–Vishik and Katz interpolate in (7), (9), and (10) are rational numbers, i.e., that Φ has rational Fourier coefficients and χ is either trivial or a quadratic character. In that case it turns out that, when suitably normalized (slightly differently than in (7), (9), (10)), the rational parts of the periods are squares of rational integers. See [11], [9] and [31] for proofs (but the details of the proper normalization have not yet been worked out in the case when the level is nonsquarefree).

The question then arises: Is there a natural way to choose the sign of the square root? Does that integer have some intrinsic meaning? In this section we discuss one type of evidence for the existence of an affirmative answer to this question.

We consider the situation in the last section, where for each Hecke character ψ satisfying the conditions described in §3 there is a two-parameter family of perfect squares: for each even weight k and for each quadratic character χ_D of conductor $|D|$ corresponding to $\mathbf{Q}(\sqrt{D}\,)$ (with k and D subject to the further condition that the root number be $+1$, so that the central critical value does not vanish) we may consider the rational part of the central period of

$$\Phi_{k,\chi_D}(z) = \frac{1}{u} \sum_{x \in O_K} \psi^{k-1}(x)\chi_D(\mathbf{N}(x))e^{2\pi i \mathbf{N}(x)z}.$$

Let $\pm s_{w/2}(\Phi_{k,\chi_D})$ denote the integer square roots of the properly normalized period $r_{w/2}(\Phi_{k,\chi_D})$. For example, if ψ is the Hecke character on the Gaussian integers in our example in §3, and if $w \equiv 0 \bmod 8$ and $D = 1$, as in Table II below, then we define

$$\left(s_{w/2}(\Phi_k)\right)^2 = -2i\left(\frac{\pi}{\omega}\right)^{w+1} r_{w/2}(\Phi_k).$$

For fixed k, once the sign of the square root is chosen for some D for which $r_{w/2}(\Phi_{k,\chi_D})$ does not vanish, this determines a natural choice of sign for all other D, because these integer values turn out to be the coefficients of a modular form of weight $k/2 + \frac{1}{2}$ (see [11], [9], [31]). But the question remains for the other

TABLE II

$w/2$	$s_{w/2}(\Phi_{w+2})$	$s_{w/2} \bmod 13$	$s_{w/2} \bmod 17$
0	1	3*	9*
4	2	2	2
8	-2^2	9	13
12	$2^3 \cdot 3 \cdot 17$	5	0
16	$2^4 \cdot 3 \cdot 283$	12	1
20	$-2^5 \cdot 3^2 \cdot 41 \cdot 71$	2	4
24	$2^8 \cdot 3^2 \cdot 149111$	4	9

* means after removal of an Euler factor

parameter k: For fixed D, having chosen the sign of $s_{w/2}(\Phi_{k,\chi_D})$ for some k for which this value is nonzero, is there a natural way to determine the sign for all other k? We now describe some evidence that there is a canonical relationship between the signs as k varies.

In §3 we saw that $(s_{w/2}(\Phi_k))^2$ satisfies p-adic congruences, because it can be written essentially as an integral of the form $c^{-w-1}\int(xy)^{w/2}\,d\mu$ where μ is a p-adic integer valued measure. Thus, for example, we have:

$$\left(s_{w/2+p-1}(\Phi_{k+2p-2})\right)^2 \equiv \bar{\rho}^2\left(s_{w/2}(\Phi_k)\right)^2 \bmod p. \tag{12}$$

(This is slightly inaccurate, because we must remove two Euler factors from the periods, as in (9); but both Euler factors are $\equiv 1 \bmod p$ as soon as $w > 0$.) It then trivially follows that for a suitable choice of sign we have

$$\pm s_{w/2+p-1}(\Phi_{k+2p-2}) \equiv \bar{\rho}s_{w/2}(\Phi_k)\bmod p. \tag{13}$$

Now suppose we arbitrarily choose the sign of, say, $s_0(\Phi_2)$, and then for some small prime p_1 that splits in O_K we let $w/2$ run through the multiples of $p_1 - 1$. If we want (13) with $p = p_1$ to hold with the plus sign on the left, then the signs of all of the $s_{w/2}(\Phi_k)$ not divisible by p_1 are determined. Let us choose the signs in that way.

Next suppose that we take a second prime p_2 which splits in O_k such that $p_2 - 1$ is a multiple of $p_1 - 1$. We can again examine (13) with $p = p_2$. If (13) holds with the plus sign on the left with the *same* choice of square roots $s_{w/2}$ that was determined by means of (13) with $p = p_1$, then we have evidence that the choice of square root is independent of the prime. That is, once $s_0(\Phi_2)$ is chosen, there is then an intrinsic choice of the other $s_{w/2}(\Phi_k)$.

Table II shows the results of one such series of tests in the case when ψ is the Hecke character of the Gaussian integers in the example in §3. Here we took $p_1 = 5$, in which case $\bar{\rho} = -1 + 2\sqrt{-1} \equiv 3 \bmod 5$. In the second column of Table II we arbitrarily took the sign of $s_0(\Phi_2)$ to be positive, and then determined the remaining signs by requiring that $s_{w/2+4}(\Phi_{w+10}) \equiv 3s_{w/2}(\Phi_{w+2})\bmod 5$. (We were fortunate in that none of the integers in the second column happens to be divisible by 5; also note that when $w = 0$ the Euler factor introduces a factor of -1.) We then took $p_2 = 13$ (for which $\bar{\rho} = 3 + 2\sqrt{-1} \equiv 6 \bmod 13$) and $p_3 = 17$ (for which $\bar{\rho} = 1 + 4\sqrt{-1} \equiv 2 \bmod 17$). The third and fourth columns in Table II show the least nonnegative residues modulo 13 and 17, respectively, of the integers $s_{w/2}(\Phi_{w+2})$ whose signs were chosen using congruences modulo 5. We see that in all cases in the table, we have

$$s_{w/2+p_j-1}\left(\Phi_{w+2p_j}\right) \equiv \bar{\rho}s_{w/2}(\Phi_{w+2})\bmod p_j, \qquad j = 2,3.$$

Thus, the choice of square roots in the second column is probably independent of

p. It is an open question to find an intrinsic interpretation for the choice of square roots.

Table II is based on a small part of the extensive computations of N. Stephens [29]. We made the same verification as in Table II for the other periods in Stephens' tables, and also for sequences of periods we computed in the case of a Hecke character for the field $\mathbf{Q}(\sqrt{-3})$. In all cases the evidence supports the existence of a canonical square root of the central period.

There is a second type of p-adic congruence that enables one to verify independence of p, in this case as we twist by quadratic characters at the same time as we change the weight. Namely, if in our integral of the form $c^{-w-1}\int(xy)^{w/2}\,d\mu$ we insert in the integrand the quadratic character $\chi_p = (\cdot/p)$ (suppose $p \equiv 1 \bmod 4$) and at the same time increase $w/2$ by $(p-1)/2$, then the integrand is unchanged modulo p, and the whole expression changes by the factor $c^{1-p} \equiv \bar{\rho} \bmod p$. This gives us another way to verify independence of p in our choice of square roots as both the weight and the quadratic character vary. For more details, see [6].

References

1. M. Eichler, *Eine Verallgemeinerung der Abelschen Integrale*, Math. Zeitschr. **67** (1957), 267–298.
2. B. Gross and D. Zagier, *On the critical values of Hecke L-series*, Soc. Math. de France, Mémoire No. 2 (1980), 49–54.
3. N. M. Katz, *p-adic interpolation of real analytic Eisenstein series*, Ann. Math. **104** (1976), 459–571.
4. N. Koblitz, *Non-integrality of the periods of cusp forms outside the critical strip*, Functional Anal. Appl. **9** (1975), No. 3, 224–226.
5. ______, *Introduction to Elliptic Curves and Modular Forms*, Springer, 1984.
6. ______, *p-adic congruences and modular forms of half integer weight*, Math. Annalen **274** (1986), 199–220.
7. W. Kohnen, *Kongruenzen für Hecke–Eisenstein'sche Riehen reell-quadratischer Körper*, Diplomarbeit, Bonn, 1977.
8. ______, *Beziehungen zwischen Modulformen halbganzen Gewichts und Modulformen ganzen Gewichts*, Bonner Mathematische Schriften **131**, Bonn, 1981.
9. ______, *Fourier coefficients of modular forms of half-integral weight*, Math. Annalen **271** (1985), 237–268.
10. ______, *p-adic distributions associated to Heegner points on modular curves*, preprint.
11. ______ and D. Zagier, *Values of L-series of modular forms at the center of the critical strip*, Invent. Math. **64** (1981), 175–198.
12. ______, *Modular Forms with Rational Periods*, in: Modular Forms, ed. by R. A. Rankin, Ellis Horwood Ltd., 1984, 197–249.
13. S. Lang, *Introduction to Modular Forms*, Springer, 1976.
14. Yu. I. Manin, *Cyclotomic fields and modular curves*, Russian Math. Surveys **26** (1971), No. 6, 7–78.
15. ______, *Parabolic points and zeta functions of modular curves*, Math. USSR Izvestija **6** (1972), 19–64.
16. ______, *Explicit formulas for the eigenvalues of Hecke operators*, Acta Arithmetica **XXIV** (1973), 239–249.
17. ______, *Periods of parabolic forms and p-adic Hecke series*, Math. USSR Sbornik **21** (1973), 371–393.

18. ______, *The values of p-adic Hecke series at integer points of the critical strip*, Math. USSR Sbornik **22** (1974), 631–637.

19. ______, *Non-archimedean integration and Jacquet-Langlands p-adic L-functions*, Russian Math. Surveys, **31** (1976), No. 1, 5–57.

20. ______, and M. M. Vishik, *p-adic Hecke series of imaginary quadratic fields*, Math. USSR Sbornik **24** (1974), No. 3, 345–371.

21. B. MAZUR, *On the arithmetic of special values of L-functions*, Invent. Math. **55** (1979), 207–240.

22. ______ and H. P. F. Swinnerton-Dyer, *Arithmetic of Weil curves*, Invent. Math. **25** (1974), 1–61.

23. A. OGG, *Elliptic curves and wild ramification*, Amer. J. Math. **89** (1967), 1–21.

24. ______, *Modular Forms and Dirichlet Series*, Benjamin, 1969.

25. M. RAZAR, *Values of Dirichlet series at integers in the critical strip*, Modular Functions of One Variable VI, Springer, 1977, 1–10.

26. K. RUBIN, *Congruences for special values of L-functions of elliptic curves with complex multiplication*, Invent. Math. **71** (1983), 339–364.

27. G. SHIMURA, *Sur les intégrales attachées aux formes automorphes*, J. Math. Soc. Japan **11** (1959), 291–311.

28. ______, *On modular forms of half-integral weight*, Ann. Math. **97** (1973), 440–481.

29. N. STEPHENS, *Computations of L (ψ_D^{2k-1}, k)*, Handwritten notes and computer printout, 1982.

30. G. STEVENS, *Arithmetic on Modular Curves*, Birkhäuser, 1982.

31. J.-L. WALDSPURGER, *Sur les coefficients de Fourier des formes modulaires de poids demi-entier*, J. Math. Pures Appl. **60** (1981), 375–484.

32. A. WEIL, *Über die Bestimmung Dirichletscher Reihen durch Funktoinalgleichungen*, Math. Ann. **168** (1967), 149–156. (= Collected Papers, III, [1967a])

MATHEMATICAL SCIENCES RESEARCH INSTITUTE, 1000 CENTENNIAL DR., BERKELEY, CALIFORNIA 94720, AND DEPARTMENT OF MATHEMATICS GN-50, UNIVERSITY OF WASHINGTON, SEATTLE, WASHINGTON 98195

Vol. 54, No. 2   DUKE MATHEMATICAL JOURNAL ©   1987

LA DESCENTE SUR LES VARIÉTÉS RATIONNELLES, II

JEAN-LOUIS COLLIOT-THÉLÈNE et JEAN-JACQUES SANSUC

à Yu. I. Manin

§0. Introduction. La méthode générale de la descente sur les variétés rationnelles a été introduite en 1976–77 dans une série de Notes ([17], [18], [19]) suscitées par des travaux de F. Châtelet [9], P. Swinnerton-Dyer, et Yu. I. Manin ([41], [42]) à propos de l'étude des points rationnels de certaines surfaces algébriques. Elle avait constitué notre première approche [17] de la description des points rationnels des tores [20]. Elle a été partiellement reprise et complétée en 1979 dans un exposé au congrès d'Angers [22] qui analyse davantage la problématique générale mais ne donne guère plus de démonstrations que les Notes précédentes.

Cette méthode a été appliquée depuis avec succès dans des situations nouvelles ([13], [27], [2], [15], [28], [16]) et l'intérêt arithmétique des résultats obtenus rend nécessaire la publication d'un texte de base, par nature quelque peu aride, qui rassemble les énoncés fondamentaux de la théorie, y compris certains obtenus récemment [25], et surtout en donne des démonstrations détaillées. Signalons que si les articles [13] et [2] évitaient le recours à la théorie générale, il n'en est pas de même de l'article [27], qui repose en grande partie sur le présent mémoire.

Des textes de présentation variés existent déjà: d'abord l'exposé [22] dont cet article est en quelque sorte la suite, mais aussi le §5 du rapport [43] de Manin et Tsfasman, le §2 de l'exposé [11] au congrès de Berkeley et divers textes de séminaire ou congrès ([29], [49], [51]).

Nous ne rappelons donc que brièvement la liste des problèmes algébriques, géométriques et arithmétiques auxquels on s'intéresse sur les variétés rationnelles, et le principe de la méthode de la descente dont l'objectif est d'aider à la solution de certains de ces problèmes.

Une *k-variété rationnelle* est une variété algébrique géométriquement intègre birationnelle à l'espace projectif sur une extension de son corps de définition k. On peut citer comme exemples les quadriques et les variétés de Severi-Brauer, les surfaces cubiques lisses, les surfaces fibrées en coniques sur la droite projective, les surfaces de Del Pezzo, les intersections complètes de deux quadriques, géométriquement intègres et sans point conique, dans l'espace projectif de dimension $\geqslant 4$, les espaces homogènes principaux sous un groupe algébrique linéaire connexe.

Received January 27, 1987.

Les *problèmes* auxquels on s'intéresse à propos de ces variétés sont de plusieurs types:

(a) problèmes k-birationnels: k-rationalité, k-rationalité stable, k-unirationalité (le corps des fonctions de la variété est-il transcendant pur sur k, le devient-il après adjonction de variables indépendantes, est-il un sous-corps d'une extension transcendante pure de k?);

(b) problèmes de points rationnels (existence d'un point rationnel, densité des points rationnels, répartition en classes pour la R-équivalence, paramétrage de l'ensemble des points rationnels, et, sur un corps de nombres, principe de Hasse et approximation faible, ...).

La méthode de *la descente*, inspirée de la descente sur les courbes de genre 1 (voir par exemple [7]), est conçue comme une méthode d'attaque de ces problèmes. Elle consiste à attacher à la variété rationnelle considérée un certain nombre de variétés auxiliaires, dites *variétés de descente*, de dimension plus grande mais de géométrie a priori plus simple, chacune équipée d'un k-morphisme dominant vers la variété initiale, tout point rationnel de celle-ci étant image d'un point rationnel de l'une de ces variétés auxiliaires (voir (2.7.2)), et le nombre de ces variétés auxiliaires étant fini lorsque k est un corps de nombres (théorème 2.7.3).

L'observation originale que nous avons faite, à la suite de questions de P. Swinnerton-Dyer en 1970, est que les exemples concrets de variétés de descente rencontrées dans la littérature (tant dans le travail original de Châtelet que dans les contre-exemples au principe de Hasse, tel celui de Swinnerton-Dyer [53] en 1962), variétés qui apparaissent après des "factorisations" adéquates, sont des torseurs sous des tores au-dessus de la variété initiale, non ramifiés sur les modèles projectifs et lisses de cette variété, et que ce fait explique les phénomènes rencontrés (telle la finitude ci-dessus, qui n'est autre qu'une variante du théorème faible de Mordell-Weil) et permet de définir le cadre le plus général où de telles méthodes s'appliquent (§2.7).

Si les variétés de descente sont bien choisies, ce qu'on peut imposer dans le cadre général des torseurs sous des tores quelconques ("torseurs universels" ou "torseurs admissibles"), les obstructions (dues à Manin) tant à la k-rationalité (§2.A) qu'au principe de Hasse (§3.1) s'évanouissent sur elles (§2.1).

Par ailleurs, si k est un corps de nombres, si la variété initiale a des points dans tous les complétés de k, et si l'obstruction au principe de Hasse définie par Manin au moyen du groupe de Brauer ne permet pas de conclure à l'absence de point rationnel, alors il existe une variété auxiliaire possédant des points dans tous les complétés de k et pour laquelle le groupe de Brauer est "trivial" (en fait mieux, voir §3.5).

La description de la méthode de descente et la justification des deux dernières phrases sont le premier objectif de cet article. Le résultat du §3.5 est fondamental dans les applications concrètes de la méthode.

Une fois établi que les obstructions usuelles s'évanouissent sur les variétés de descente d'un type convenable, il devient alors plus raisonnable de chercher à établir le principe de Hasse (lorsque k est un corps de nombres) et la k-rationalité de ces variétés auxiliaires (lorsqu'elles possèdent un point k-rationnel). Si ces deux propriétés sont satisfaites, on peut dire que l'arithmétique de la variété initiale est parfaitement connue (§2.8 et §3.8).

En général, nous ne disposons pas de procédé-miracle pour établir ces propriétés. De fait, à ce stade de la méthode, il faut de nouvelles idées pour analyser la géométrie et l'arithmétique des variétés auxiliaires. En l'absence de théorèmes généraux, on étudie de près ces variétés, et on essaye d'établir l'une ou l'autre des propriétés ci-dessus par des méthodes ad hoc ([13], [27], [28]). On ne sait le faire sans posséder des *équations concrètes* de ces variétés auxiliaires, variétés qui sont données a priori de façon très désincarnée. C'est le second objectif du présent article de donner de telles équations ("description locale des torseurs", §2.3) et de détailler en particulier le cas où la variété initiale est une surface fibrée en coniques sur la droite projective (§2.6).

Les résultats des §§ 2 et 3 sont fondés sur des propositions de cohomologie étale (ou plate) qui sont rassemblées dans le §1, paragraphe que nous invitons le lecteur à sauter en première lecture. Par ailleurs nous avons inclus une démonstration simple et valable en toute caractéristique d'un résultat de base sur le groupe de Picard des variétés rationnelles (§2.A), ainsi que la démonstration d'une remarque générale sur le calcul de ce groupe (§2.B). Enfin le résultat du §2.9 peut dans certains cas aider à établir la k-rationalité (stable) des torseurs universels.

Notons que si en dernière analyse (théorème 3.5.1), les obstructions "géométriques" au principe de Hasse définies par les torseurs sous des tores coïncident avec l'obstruction définie par Manin en termes de groupe de Brauer, des exemples montrent que la partition des points rationnels d'une k-variété rationnelle obtenue par accouplement avec des torseurs sous des tores peut être strictement plus fine que celle obtenue par accouplement avec le groupe de Brauer (cf. 2.7.11), et donc se rapprocher plus de la partition en classes pour la R-équivalence, telle que définie par Manin (cf. 2.7.2).

La méthode a complètement réussi dans le cas des surfaces de Châtelet généralisées [27] qui sont les surfaces fibrées en coniques non triviales les plus simples, et elle a partiellement réussi pour d'autres surfaces fibrées en coniques [28]. Autre exemple de réussite complète: le cas des espaces homogènes principaux sous des tores ([20], [22]). Complétée par la "méthode des sections hyperplanes", la méthode de la descente a permis de mieux comprendre l'arithmétique des intersections de deux quadriques [27] et aussi de certaines hypersurfaces cubiques [16]. Mais il convient de signaler que pour certaines variétés rationnelles très simples de dimension au moins 3, les torseurs universels ne sont pas (stablement) k-rationnels (cf. 2.8.18). Le cas des surfaces rationnelles est ouvert: le module galoisien défini par le groupe de Picard d'une surface rationnelle semble alors mieux contrôler la géométrie et l'arithmétique de la variété ([4], [10], [23]).

C'est avec grand plaisir que nous dédions ce travail à Yu. I. Manin, qui sut renouveler l'étude arithmétique des variétés rationnelles en y apportant le point de vue de la géométrie algébrique moderne.

Plan

§1. Torseurs sous des groupes de type multiplicatif

1.0. *Notations.* On utilise les abréviations usuelles pour les topologies suivantes sur un schéma X: *ét* pour étale, *Zar* pour Zariski, *pl* ou *fppf* pour fidèlement plat de présentation finie et enfin *fpqc* (voir [44] II §1 ou [31] Arcata I et II). Sauf mention explicite, la topologie considérée est la topologie plate *pl* et les faisceaux et leurs groupes de cohomologie sont relatifs à cette topologie. Si $\mathscr{F}$ et $\mathscr{G}$ sont deux faisceaux de groupes abéliens sur X, on désigne par $\mathscr{H}om_X(\mathscr{F}, \mathscr{G})$ le faisceau des homomorphismes de $\mathscr{F}$ dans $\mathscr{G}$. La notation G_m désigne le groupe multiplicatif $\operatorname{Spec}\mathsf{Z}[t, t^{-1}]$.

1.1. *Groupes de type multiplicatif.* Pour les notions générales sur les groupes de type multiplicatif, on renvoie à [32], chap. VIII–X. Voici néanmoins quelques rappels pour fixer la terminologie (on se place dans des situations moins générales que [32]).

Soit X un schéma. On note $\mathsf{G}_{m,X} := \mathsf{G}_m \times_{\mathsf{Z}} X$ le groupe multiplicatif sur X. Un X-schéma en groupes diagonalisable est un X-sous-schéma en groupes d'un $\mathsf{G}_{m,X}^n$. Un X-groupe de type multiplicatif (resp. un X-tore) est un X-schéma en groupes S localement diagonalisable (resp. localement isomorphe à un G_m^n) pour la topologie *fpqc* ([32] VIII 1.1, IX 1.1, 1.3). On se limite ici à des groupes de type fini, auquel cas on peut remplacer *fpqc* par *ét* ([32] X 4.5). Comme en outre X sera en général localement noethérien normal, un tel S sera même diagonalisable après revêtement étale (= morphisme étale fini surjectif) de X.

Comme exemples de groupes de type multiplicatif, on a le groupe $\mu_{n,X}$ des racines n-ièmes de l'unité, noyau de la multiplication par n dans $\mathsf{G}_{m,X}$, le tore descendu à la Weil $R_{X'/X}\mathsf{G}_m$ pour X'/X étale fini—X' non nécessairement connexe—le tore $R^1_{X'/X}\mathsf{G}_m$ noyau de la norme $N_{X'/X}\colon \mathsf{G}_{m,X'} \to \mathsf{G}_{m,X}$. Un X-tore est dit quasi-trivial s'il est isomorphe à un tore du type $R_{X'/X}\mathsf{G}_m$.

On appelle X-groupe constant tordu un X-schéma en groupes localement constant pour la topologie *fpqc* ([32] X 5.1). S'il est à engendrement fini, ce qui sera toujours le cas ici, il est même localement constant pour la topologie étale ([32] X 5.9). Si M est un groupe abélien de type fini, on note M_X le X-groupe constant $\coprod_X M$ qu'il définit.

On a une antiéquivalence de catégories, souvent appelée dualité, entre les X-groupes de type multiplicatif et les X-groupes constants tordus, via:

$$S \mapsto \hat{S} = \mathscr{H}om_{X\text{-gr}}(S, \mathsf{G}_{m,X})$$

$$M \mapsto D(M) = \mathscr{H}om_{X\text{-gr}}(M, \mathsf{G}_{m,X})$$

([32] X 5.1, 5.6 à 5.9). Lorsque X est connexe, S et M peuvent être trivialisés par un revêtement étale galoisien connexe $X' \to X$ de groupe $\mathfrak{g}$. On a alors une dualité entre les X-groupes de type multiplicatif trivialisés par X'/X et la catégorie des $\mathfrak{g}$-modules de type fini, via:

$$S \mapsto \hat{S}(X')$$

([32] X 1.1). A une suite exacte de $\mathfrak{g}$-modules correspond ainsi une "suite exacte" de X-groupes de type multiplicatif trivialisés par $X' \to X$ ([32] VIII 3.1) qui définit une suite exacte de faisceaux abéliens pour la topologie *pl*. Dans la dualité ci-dessus, S est lisse lorsque $\hat{S}(X')$ a une Z-torsion première aux caractéristiques résiduelles. C'est un tore lorsque $\hat{S}(X')$ est sans Z-torsion. Lorsque $X = \mathrm{Spec}\, k$ où k est un corps, on a même une dualité entre les k-groupes de type multiplicatif S et les $\mathfrak{g}$-modules continus discrets, de type fini, où $\mathfrak{g} = \mathrm{Gal}(\overline{k}/k)$ est le groupe de Galois d'une clôture séparable $\overline{k}$ de k, via:

$$S \mapsto \hat{S}(\overline{k})$$

$$M \mapsto D(M)$$

où $D(M) = \mathrm{Hom}_{\mathsf{Z}}(M, \overline{k}^*)$. On notera souvent $\hat{S}$ le $\mathfrak{g}$-module $\hat{S}(\overline{k})$.

Notons enfin que le dual de $\mathsf{G}_{m,\,X}$ est Z_X, celui de $\mu_{n,\,X}$ est $(\mathsf{Z}/n)_X$ et celui de $R^1_{X'/X}\mathsf{G}_m$ est le conoyau de la norme $\mathsf{Z} \to \mathsf{Z}[\mathfrak{g}/\mathfrak{h}]$, où $\mathfrak{g} = \mathrm{Gal}(X''/X)$ est le groupe de Galois d'un revêtement galoisien X''/X qui majore X'/X et $\mathfrak{h} = \mathrm{Gal}(X''/X')$.

1.2. *Torseurs.*   Soit S un X-groupe de type multiplicatif. On appelle *torseur* sur X sous S un espace principal homogène $\mathcal{T} \to X$ sur X sous S ([44] III §4). Autrement dit, $\mathcal{T} \to X$ est *fppf* et S agit sur $\mathcal{T}$ de façon compatible avec la projection, le morphisme canonique ainsi défini:

$$S \times_X \mathcal{T} \to \mathcal{T} \times_X \mathcal{T}$$

$$(s, t) \to (t, s \cdot t)$$

étant un isomorphisme.

Si $\mathrm{Tors}(X, S)$ désigne l'ensemble des classes d'isomorphisme de X-torseurs sous S,

$$\mathrm{Tors}(X, S) = H^1(X_{\mathrm{pl}}, S)$$

([44] III 4.7 et 2.10), et même, si S est lisse, par exemple si S est un tore,

$$\mathrm{Tors}(X, S) = H^1(X_{\text{ét}}, S)$$

([35] 11.7 ou [44] III 3.9). Autrement dit, un torseur $\mathcal{T}$ sur X sous S est dans ce dernier cas trivialisé par un recouvrement $\mathcal{U}$ pour la topologie étale: le 1-cocycle associé à $\mathcal{U} = \{U_i\}$ et à une famille $\{t_i\}$ de sections, $t_i \in H^0(U_i, \mathcal{T})$ est $\{t_{i,\,j}\} = \{t_j/t_i\} \in Z^1(\mathcal{U}, S)$ (cf. [31] Cycle). Si $S = \mathsf{G}_m$, on a même d'après le théorème 90 ([44] III 4.9):

$$\mathrm{Tors}(X, \mathsf{G}_m) = H^1(X_{\mathrm{Zar}}, \mathsf{G}_m) = \mathrm{Pic}\, X.$$

Voici quelques exemples de torseurs. Une suite exacte de groupes de type multiplicatif

$$1 \to M' \to M \to M'' \to 1$$

fait de M un torseur sur M'' sous M'. Une isogénie de courbes elliptiques $E' \to E$ fait de E' un torseur sur E sous le schéma en groupes fini noyau de l'isogénie. Mentionnons encore les λ-revêtements de courbes de genre 1 (cf. [7]). Si K/k est une extension séparable finie de corps, l'équation

$$N_{K/k}(\xi) = a$$

où $a \in k^*$ définit un torseur sur k sous le k-tore $R^1_{K/k}\mathbf{G}_m$ qui admet pour équation affine $N_{K/k}(\xi) = 1$. Un espace homogène G/H est la base d'un torseur, à savoir G, sous H.

1.3. *Lemmes de topologie.*

LEMME 1.3.1. *Soient* $q\colon X \to Y$ *un morphisme de schémas et* R *un* Y-*groupe constant tordu. Le morphisme naturel*

$$q^*R \to R_X$$

est un isomorphisme de faisceaux étales sur X.

Dans cet énoncé, q^*R désigne l'image réciproque de R en tant que faisceau étale, et le "morphisme naturel" est l'adjoint du morphisme "évident" $R \to q_*R_X$ de faisceaux étales sur Y. La vérification est locale pour la topologie considérée. Comme R est localement constant pour la topologie étale, il suffit de traiter le cas constant où $R = M_Y$ pour M un groupe abélien de type fini. On note qu'alors $M_Y \times_Y X = M_X$ et le lemme résulte de la suite d'égalités:

$$\mathrm{Hom}_X(M_X, \mathcal{F}) = \mathrm{Hom}_{\mathrm{gr}}(M, \mathcal{F}(X)) = \mathrm{Hom}_{\mathrm{gr}}(M, (q_*\mathcal{F})(Y))$$

$$= \mathrm{Hom}_Y(M_Y, q^*\mathcal{F}) = \mathrm{Hom}_X(q_*M_Y, \mathcal{F}).$$

LEMME 1.3.2. *Soit* S *un* X-*groupe de type multiplicatif. Le morphisme naturel*

$$S \to \mathscr{H}om_X(\hat{S}, \mathbf{G}_{m, X})$$

est un isomorphisme de faisceaux étales sur X.

En effet, $\mathscr{H}om_{X\text{-gr}}(\hat{S}, \mathbf{G}_{m, X}) \to \mathscr{H}om_X(\hat{S}, \mathbf{G}_{m, X})$ est un isomorphisme de faisceaux sur X.

LEMME 1.3.3.
(i) *Si* S *est un* X-*groupe de type multiplicatif,*

$$\mathscr{E}xt^i_{X_{\mathrm{pl}}}(\hat{S}, \mathbf{G}_{m, X}) = 0 \quad \textit{pour tout } i > 0.$$

(ii) *Si S est un X-groupe de type multiplicatif lisse,*

$$\mathcal{E}xt^{i}_{X_{\text{ét}}}\big(\hat{S}, \mathsf{G}_{m,X}\big) = 0 \quad \text{pour tout } i > 0.$$

La vérification est locale pour la topologie considérée. Comme $\hat{S}$ est localement constant pour la topologie étale, il suffit de traiter les cas constants $\hat{S} = \mathsf{Z}$ et $\hat{S} = \mathsf{Z}/n$. Let premier cas est immédiat pour toute topologie. On en déduit le second pour tout $i > 1$. Il reste la nullité pour $i = 1$ dans ce dernier cas: elle signifie que la multiplication par n

$$\mathsf{G}_{m,X} \xrightarrow{\ \times n\ } \mathsf{G}_{m,X}$$

est un épimorphisme de faisceaux, ce qui est toujours vrai pour la topologie *pl* et le reste pour la topologie étale si n est premier aux caractéristiques résiduelles de X, ce qui est le cas si S est lisse.

1.4. *Torseurs et extensions.*

PROPOSITION 1.4.1. *Soit S un X-groupe de type multiplicatif. On a des isomorphismes naturels*:

$$H^{i}\big(X_{\text{pl}}, S\big) \xrightarrow{\ \cong\ } \text{Ext}^{i}_{X_{\text{ét}}}\big(\hat{S}, \mathsf{G}_{m,X}\big) \quad \text{pour tout } i \geqslant 0.$$

Rappelons ([35] théorème 11.7) que, si S est lisse, $H^{i}(X_{\text{pl}}, S) = H^{i}(X_{\text{ét}}, S)$, ce qui permet dans ce cas de remplacer *pl* par *ét* dans l'énoncé ci-dessus.

La démonstration utilise la suite spectrale d'intégration ([1] V 6.1)

$$H^{p}\big(X_{\text{pl}}, \mathcal{E}xt^{q}\big(\hat{S}, \mathsf{G}_{m,X}\big)\big) \Rightarrow \text{Ext}^{p+q}_{X_{\text{pl}}}\big(\hat{S}, \mathsf{G}_{m,X}\big)$$

en topologie pl. Celle-ci dégénère complètement en raison de la nullité des faisceaux

$$\mathcal{E}xt^{i}_{X_{\text{pl}}}\big(\hat{S}, \mathsf{G}_{m,X}\big) \quad \text{pour tout } i > 0$$

(lemme 1.3.3). D'après le lemme 1.3.2, $\mathcal{H}om_{X}(\hat{S}, \mathsf{G}_{m,X}) = S$ comme faisceaux pour la topologie étale, ou pour une topologie plus fine. Les edges de la suite spectrale définnissent donc des isomorphismes

$$H^{i}\big(X_{\text{pl}}, S\big) \xrightarrow{\ \cong\ } \text{Ext}^{i}_{X_{\text{pl}}}\big(\hat{S}, \mathsf{G}_{m,X}\big).$$

Notons que, pour S lisse, on peut remplacer partout *pl* par *ét* dans les arguments ci-dessus, ce qui établit déjà la proposition dans ce cas-là. Sinon, il reste à prouver les égalités

$$\mathrm{Ext}^i_{X_{\text{ét}}}(\hat{S}, G_{m, X}) = \mathrm{Ext}^i_{X_{\text{pl}}}(\hat{S}, G_{m, X}).$$

Comme $\hat{S}$ est localement constant pour la topologie étale, il suffit, par descente (suite spectrale de Leray d'un recouvrement ([1] V 3.3)), de traiter les cas constants $\hat{S} = Z$ et $\hat{S} = Z/n$. Dans le premier cas, les égalités ci-dessus s'écrivent

$$H^i(X_{\text{ét}}, G_{m, X}) = H^i(X_{\text{pl}}, G_{m, X})$$

et sont vraies par lissité de $G_{m, X}$ ([35] théorème 11.7). Elles sont donc aussi vraies dans le second cas par le lemme des cinq.

PROPOSITION 1.4.2. *Soient X un schéma réduit, S un X-tore et $\mathcal{T} \xrightarrow{q} X$ un X-torseur sous S. On a une suite exacte naturelle de faisceaux étales sur X:*

$$(E_{\mathcal{T}}) \qquad\qquad 1 \to G_{m, X} \to q_* G_{m, \mathcal{T}} \to \hat{S} \to 0.$$

C'est une variante globale du lemme de Rosenlicht (cf. [46] VII 1.2): si Y est un schéma réduit et si S est un Y-tore, alors

$$G_m(S)/G_m(Y) = \hat{S}(Y).$$

Considérons d'abord le cas du torseur trivial $S \xrightarrow{q} X$. Un automorphisme φ de ce torseur est de la forme $s \mapsto ss_0$. Soit e la section unité. Le lemme de Rosenlicht affirme que toute fonction inversible f sur S s'écrit

$$f = \chi_f \cdot f(e)$$

où χ_f est un caractère de S et $f(e) = q^* e^* f$. On peut ainsi écrire une suite exacte de faisceaux sur X:

$$1 \to G_{m, X} \to q_* G_{m, S} \xrightarrow{\ \rho\ } \hat{S} \to 0$$

où $\rho(f) = \chi_f$. Un automorphisme φ induit un automorphisme φ^* de cette suite exacte qui est l'identité sur $G_{m, X}$, mais aussi sur $\hat{S}$, comme le montre le calcul suivant:

$$(\varphi^* f)(s) = f(ss_0) = \chi_f(ss_0) \cdot f(e) = \chi_f(s)\chi_f(s_0) \cdot f(e)$$

$$= \chi_f(s) \cdot f(s_0) = \chi_f(s) \cdot (\varphi^* f)(e).$$

Autrement dit: $\chi_{\varphi^* f} = \chi_f$, et ceci prouve que la projection ρ ne dépend pas de la trivialisation du torseur trivial. Ceci permet de définir, pour un torseur $\mathcal{T}$ quelconque, un morphisme de faisceaux étales sur X

$$\rho\colon q_*\mathsf{G}_{m,\,\mathcal{T}} \to \hat{S}$$

par recollement des morphismes ρ_U définis sans ambiguïté pour tout ouvert étale U qui trivialise $\mathcal{T}$. On obtient ainsi une suite naturelle de faisceaux étales sur X

$$1 \to \mathsf{G}_{m,\,X} \to q_*\mathsf{G}_{m,\,\mathcal{T}} \xrightarrow{\ \ \rho\ \ } \hat{S} \to 0.$$

Le fait que cette suite soit exacte se vérifie localement, et c'est alors le lemme de Rosenlicht déjà vu (il suffit d'ailleurs de le connaître pour $S = \mathsf{G}_{m,\,X}^n$).

Proposition 1.4.3.   *Soient X un schéma réduit et S un X-tore. L'isomorphisme naturel*

$$\varepsilon\colon H^1(X, S) \xrightarrow{\ \ \cong\ \ } \operatorname{Ext}^1_{X_{\text{ét}}}\!\left(\hat{S}, \mathsf{G}_{m,\,X}\right)$$

associe, au signe près, à la classe d'un torseur $\mathcal{T} \xrightarrow{q} X$ celle de l'extension de faisceaux étales $(E_{\mathcal{T}})$.

L'application naturelle ε est l'edge $E_2^{1,0} \to E^1$ de la suite spectrale

$$H^p\!\left(X_{\text{ét}}, \mathscr{E}xt^q\!\left(\hat{S}, \mathsf{G}_{m,\,X}\right)\right) \Rightarrow \operatorname{Ext}^{p+q}_{X_{\text{ét}}}\!\left(\hat{S}, \mathsf{G}_{m,\,X}\right)$$

d'intégration des $\mathscr{E}xt$ locaux (proposition 1.4.1).

Démonstration.   La vérification s'appuie sur la description générale de l'edge dans une suite spectrale du type ci-dessus donnée en appendice. Avec les notations de l'appendice, la suite spectrale ci-dessus n'est autre que la suite spectrale (1.A.1) pour les données suivantes:

$\mathscr{C} = \mathscr{C}'$ la catégorie des faisceaux de groupes abéliens sur $X_{\text{ét}}$ et $\mathscr{C}''$ la catégorie des groupes abéliens,

$$\Phi = \mathscr{H}om_X(\hat{S},\ ) \quad \text{et} \quad \Psi = H^0(X,\ ), \quad \text{de telle sorte que } \Psi\Phi = \operatorname{Hom}_X(\hat{S},\ ).$$

Etant donné un torseur $\mathcal{T} \xrightarrow{q} X$ sous S, on peut, d'après 1.3.3, appliquer le lemme 1.A.3 à l'extension de faisceaux $(E_{\mathcal{T}})$ sur $X_{\text{ét}}$. On a donc, au signe près, un carré commutatif

$$
\begin{array}{ccc}
\operatorname{Hom}_X(\hat{S}, \hat{S}) & \xrightarrow{\ \ \partial_0\ \ } & H^1(X, S) \\[4pt]
\Big\| & & \Big\downarrow{\scriptstyle \varepsilon} \\[4pt]
\operatorname{Hom}_X(\hat{S}, \hat{S}) & \xrightarrow{\ \ \partial_1\ \ } & \operatorname{Ext}^1_X(\hat{S}, \mathsf{G}_m)
\end{array}
$$

dans lequel ∂_0 est le bord déduit de la suite exacte de faisceaux $\Phi(E_{\mathscr{T}})$ par application de $\Psi = H^0(X, \ )$:

$$1 \to S \to \mathscr{H}om_X(\hat{S}, q_*\mathbf{G}_{m,\mathscr{T}}) \to \mathscr{H}om_X(\hat{S}, \hat{S}) \to 0$$

et ∂_1 est, au signe près, l'application pull-back de l'extension $(E_{\mathscr{T}})$. Ceci montre finalement qu'au signe près

$$[E_{\mathscr{T}}] = \partial_1(\mathrm{id}_{\hat{S}}) = \varepsilon\,\partial_0(\mathrm{id}_{\hat{S}}).$$

Il reste à voir qu'au signe près

$$[\mathscr{T}] = \partial_0(\mathrm{id}_{\hat{S}}).$$

Soit $\{U_i, t_i\}$ une trivialisation étale de $\mathscr{T}$, i.e. $\{U_i\}$ est un recouvrement étale de X et $t_i \in H^0(U_i, \mathscr{T})$. L'extension $(E_{\mathscr{T}})$ est trivialisée sur U_i par la section σ_i définie, avec des notations évidentes, par

$$(\sigma_i(\chi))(s \cdot t_i) = \chi(s).$$

Par suite, l'extension $\Phi(E_{\mathscr{T}})$ est trivialisée sur U_i par la section s_i définie par

$$((s_i\varphi)(\chi))(s \cdot t_i) = \varphi(\chi)(s).$$

On en déduit, pour $\varphi = \mathrm{id}_{\hat{S}}$,

$$\partial(\mathrm{id}_{\hat{S}})_{i,j}(\chi)(t) = \chi(s_j(t))/\chi(s_i(t)) = \chi(s_j(t)/s_i(t)) = \chi(t_i/t_j)$$

$$\text{où } t = s_i(t) \cdot t_i = s_j(t) \cdot t_j \text{ et } t_i/t_j := s_j(t_i).$$

et, par suite en "oubliant" χ:

$$\partial(\mathrm{id}_{\hat{S}})_{i,j} = t_i/t_j.$$

Or, $\{t_{i,j} := t_i/t_j\}$ est, au signe près, le 1-cocycle du recouvrement $\{U_i\}$ à valeurs dans S associé à la trivialisation du torseur $\mathscr{T}$ définie par les sections $\{t_i\}$. Ceci prouve $[E_{\mathscr{T}}] = \partial_0(\mathrm{id}_{\hat{S}})$ et achève donc la démonstration.

1.5. *Une suite exacte fondamentale.*

1.5.0. Fixons d'abord quelques notations qui seront constamment utilisées dans la suite. Si k est un corps, on désigne par $\bar{k}$ une clôture séparable et par $\mathfrak{g}$ le groupe de Galois de $\bar{k}/k$. Si X est une k-variété, on note $\bar{X} = X \times_k \bar{k}$. Si K/k est une extension quelconque, on note $K[X]$ l'anneau des fonctions régulières sur X_K et, si X_K est intègre, on note $K(X)$ le corps des fonctions rationnelles sur X_K.

Soit enfin F un fermé de Zariski de X. On note $\mathrm{Div}\, X$ (resp. $\mathrm{Div}_F X$) le groupe des diviseurs de Cartier (resp. à support dans F) de X, et $\mathrm{Pic}\, X$ le groupe de Picard de X. On note

$$\mathrm{Br}_1 X = \ker\big(H^2(X, \mathsf{G}_m) \to H^2(\overline{X}, \mathsf{G}_m)\big) = \ker(\mathrm{Br}\, X \to \mathrm{Br}\, \overline{X})$$

le sous-groupe du groupe de Brauer-Grothendieck de X formé des éléments tués par passage à $\overline{k}$ (voir le début de 3.1). Pour X géométriquement intègre,

$$\mathrm{Br}_1 X = \ker\big(H^2(\mathfrak{g}, \overline{k}(X)^*) \to H^2(\mathfrak{g}, \mathrm{Div}\, \overline{X})\big)$$

(cf. [20] lemme 14, p. 213). On a la suite exacte naturelle (cf. [41], [20] p. 214):

$$(1.5.0) \quad 0 \to H^1(\mathfrak{g}, \overline{k}[X]^*) \to \mathrm{Pic}\, X \to (\mathrm{Pic}\, \overline{X})^{\mathfrak{g}} \to H^2(\mathfrak{g}, \overline{k}[X]^*)$$

$$\to \mathrm{Br}_1 X \xrightarrow{\chi} H^1(\mathfrak{g}, \mathrm{Pic}\, \overline{X}) \to H^3(\mathfrak{g}, \overline{k}[X]^*)$$

qui est la suite des termes de bas degré de la suite spectrale de Leray pour $X \to \mathrm{Spec}\, k$ et le faisceau étale G_m:

$$H^p\big(\mathfrak{g}, H^q(\overline{X}, \mathsf{G}_m)\big) \Rightarrow H^{p+q}(X, \mathsf{G}_m).$$

La suite exacte qui fait l'objet de ce paragraphe généralise simplement le début de (1.5.0) pour un k-groupe de type multiplicatif quelconque à la place de G_m.

Après les résultats préliminaires indiqués plus haut, venons-en à cette suite exacte, essentielle pour les paragraphes ultérieurs:

THÉORÈME 1.5.1. *Soient k un corps, X une k-variété algébrique et S un k-groupe de type multiplicatif. Soit $\mathfrak{g} = \mathrm{Gal}(\overline{k}/k)$ où $\overline{k}$ est une clôture séparable de k. On a une suite exacte naturelle:*

$$(1.5.1) \quad 0 \to \mathrm{Ext}^1_{\mathfrak{g}}(\hat{S}, \overline{k}[X]^*) \xrightarrow{i_1} H^1(X, S) \xrightarrow{\chi} \mathrm{Hom}_{\mathfrak{g}}(\hat{S}, \mathrm{Pic}\, \overline{X})$$

$$\xrightarrow{\partial} \mathrm{Ext}^2_{\mathfrak{g}}(\hat{S}, \overline{k}[X]^*) \xrightarrow{i_2} H^2(X, S).$$

Dans cet énoncé, $\hat{S}$ désigne le $\mathfrak{g}$-module $\hat{S}(\overline{k})$ et $H^i(X, S) := H^i(X_{\mathrm{pl}}, S)$.

La suite exacte du théorème est la suite des termes de bas degré de la suite spectrale

$$\mathrm{Ext}^p_{k_{\text{ét}}}\big(\hat{S}, R^q p_* \mathsf{G}_{m, X}\big) \Rightarrow \mathrm{Ext}^{p+q}_{X_{\text{ét}}}\big(p^* \hat{S}, \mathsf{G}_{m, X}\big)$$

où $p\colon X \to \operatorname{Spec} k$ est le morphisme structural. Il s'agit de la suite spectrale

$$(L)_{p,\mathscr{F},\mathscr{G}} \qquad \operatorname{Ext}^p_{Y_{\text{ét}}}(\mathscr{F}, R^q p_* \mathscr{G}) \Rightarrow \operatorname{Ext}^{p+q}_{X_{\text{ét}}}(p^*\mathscr{F}, \mathscr{G})$$

associée à un morphisme $p\colon X \to Y$ et à deux faisceaux, $\mathscr{F}$ sur $Y_{\text{ét}}$ et $\mathscr{G}$ sur $X_{\text{ét}}$ ([1] V 5.5), considérée pour $Y = \operatorname{Spec} k$, $\mathscr{F} = \hat{S}$ et $\mathscr{G} = \mathsf{G}_{m,X}$. Autrement dit, avec les notations de l'appendice, c'est la suite spectrale (1.A.4) pour les données suivantes:

$\mathscr{C}$ (resp. $\mathscr{C}'$) = la catégorie des faisceaux de groupes abéliens sur $X_{\text{ét}}$ (resp. $Y_{\text{ét}}$) et $\mathscr{C}''$ = la catégorie des groupes abéliens,

$\Phi = p_*$ et $\Psi = \operatorname{Hom}_Y(\mathscr{F}, \ )$, de telle sorte que, par adjonction, $\Psi\Phi(G) = \operatorname{Hom}_X(p^*\mathscr{F}, \mathscr{G})$, et que la condition (1.4.0) est satisfaite, p_* admettant pour adjoint à gauche p^* qui est exact. La suite des termes de bas degré s'écrit en général

$$0 \to \operatorname{Ext}^1_Y(\mathscr{F}, p_*\mathscr{G}) \to \operatorname{Ext}^1_X(p^*\mathscr{F}, \mathscr{G}) \to \operatorname{Hom}_Y(\mathscr{F}, R^1 p_*\mathscr{G})$$

$$\to \operatorname{Ext}^2_Y(\mathscr{F}, p_*\mathscr{G}) \to \operatorname{Ext}^2_X(p^*\mathscr{F}, \mathscr{G}).$$

Si S est un Y-groupe de type multiplicatif, si $\mathscr{F} = \hat{S}$ et $\mathscr{G} = \mathsf{G}_{m,X}$, on a $p^*\hat{S} = \hat{S}_X$ (lemme 1.3.1) et cette suite s'écrit

$$0 \to \operatorname{Ext}^1_Y(\hat{S}, p_*\mathsf{G}_{m,X}) \to \operatorname{Ext}^1_X(\hat{S}, \mathsf{G}_{m,X}) \to \operatorname{Hom}_Y(\hat{S}, \mathscr{P}\!ic_{X/Y})$$

$$\to \operatorname{Ext}^2_Y(\hat{S}, p_*\mathsf{G}_{m,X}) \to \operatorname{Ext}^2_X(\hat{S}, \mathsf{G}_{m,X})$$

où $\mathscr{P}\!ic_{X/Y} = R^1 p_* \mathsf{G}_{m,X}$. Si en outre $Y = \operatorname{Spec} k$, on identifie faisceaux étales sur $\operatorname{Spec} k$ et $\mathfrak{g}$-modules continus discrets ([1] VIII 2.2), via

$$\mathscr{F} \mapsto \varinjlim_{k'/k} \mathscr{F}(k') = \mathscr{F}(\bar{k}),$$

ce qui identifie $p_*\mathsf{G}_{m,X}$ au $\mathfrak{g}$-module $\bar{k}[X]^*$, et $R^1 p_* \mathsf{G}_{m,X}$ au $\mathfrak{g}$-module $\operatorname{Pic} \overline{X}$. La proposition 1.4.1 permet de remplacer $\operatorname{Ext}^i_X(\hat{S}, \mathsf{G}_{m,X})$ par $H^i(X_{\text{pl}}, S)$, ou même, si S est lisse, par exemple si c'est un tore, par $H^i(X_{\text{ét}}, S)$.

Voici quelques propriétés de cette suite exacte, qui nous seront utiles dans la suite:

PROPOSITION 1.5.2. *La suite exacte* (1.5.1) *du théorème précédent a les propriétés suivantes:*
  (i) *Elle est fonctorielle en* (X, S), *covariante en* S *et contravariante en* X.
  (ii) *Le morphisme* χ *s'obtient par fonctorialité de la façon suivante:*

$$H^1(X, S) \times \hat{S}(\bar{k}) \to H^1(\overline{X}, \overline{S}) \times \operatorname{Hom}_{\text{gr}}(\overline{S}, \mathsf{G}_{m,\bar{k}}) \to H^1(\overline{X}, \mathsf{G}_m) = \operatorname{Pic} \overline{X}.$$

(iii) *Si S est un tore, χ se décrit également comme suit: si $\mathcal{T}$ est un X-torseur sous S, l'extension $(E_{\mathcal{T}})$ associée définit, par application de $R^{\cdot}p_{*}$, un bord $\partial_{\mathcal{T}}$: $\hat{S}(\overline{k}) \to \operatorname{Pic} \overline{X}$, et, au signe près:*

$$\chi([\mathcal{T}]) = \partial_{\mathcal{T}}.$$

(iv) *Le morphisme ∂ est, au signe près, le cup-produit par la classe de l'extension*

$$1 \to \overline{k}[X]^{*} \to \overline{k}(X)^{*} \to \operatorname{Div} \overline{X} \to \operatorname{Pic} \overline{X} \to 0.$$

Autrement dit, il associe à un $\mathfrak{g}$-morphisme λ: $\hat{S} \to \operatorname{Pic} \overline{X}$ la classe, au signe près, de la 2-extension de $\hat{S}$ par $\overline{k}[X]^{*}$ obtenue par pull-back à partir de l'extension ci-dessus.

(v) *i_1 et i_2 sont les morphismes naturels*

$$\operatorname{Ext}^{i}_{k}(\hat{S}, p_{*}\mathsf{G}_{m, X}) \to \operatorname{Ext}^{i}_{X}(\hat{S}_{X}, \mathsf{G}_{m, X})$$

pour $i = 1, 2$.

Démonstration.

(i) La suite spectrale $(L)_{p, \mathscr{F}, \mathscr{G}}$ est covariante en $\mathscr{G}$ et contravariante en $\mathscr{F}$. De plus, étant donné un morphisme j: $X' \to X$ de Y-schémas, on a un morphisme naturel de suites spectrales

$$j^{*}\colon (L)_{p, \mathscr{F}, \mathscr{G}} \to (L)_{p \circ j, \mathscr{F}, j^{*}\mathscr{G}}.$$

On en déduit aussitôt la covariance en S de la suite

$$0 \to \operatorname{Ext}^{1}_{Y}(\hat{S}, p_{*}\mathsf{G}_{m, X}) \to \operatorname{Ext}^{1}_{X}(\hat{S}, \mathsf{G}_{m, X}) \to \operatorname{Hom}_{Y}(\hat{S}, \mathscr{P}ic_{X/Y})$$

$$\to \operatorname{Ext}^{2}_{Y}(\hat{S}, p_{*}\mathsf{G}_{m, X}) \to \operatorname{Ext}^{2}_{X}(\hat{S}, \mathsf{G}_{m, X})$$

et en particulier celle de (1.5.1).

Pour la variance en X, soient j: $X' \to X$ un Y-morphisme et $p' = p \circ j$. Le morphisme j définit des morphismes de suites spectrales:

$$(L)_{p, \hat{S}, \mathsf{G}_{m, X}} \to (L)_{p', \hat{S}, j^{*}\mathsf{G}_{m, X}} \to (L)_{p', \hat{S}, \mathsf{G}_{m, X'}}.$$

Le premier traduit la fonctorialité de (L) en X, et le second est défini par le morphisme $j^{*}\mathsf{G}_{m, X} \to \mathsf{G}_{m, X'}$ adjoint du morphisme naturel $\mathsf{G}_{m, X} \to j_{*}\mathsf{G}_{m, X'}$.

(ii) Par fonctorialité de la suite spectrale, on se ramène au cas particulier

$$k = \overline{k} \quad \text{et} \quad S = \mathsf{G}_{m, k},$$

auquel cas on vérifie que $\chi = \operatorname{id}_{\operatorname{Pic} \overline{X}}$.

(iii) On va montrer plus généralement que, pour la suite spectrale $(L)_{p,\mathscr{F},\mathscr{G}}$, le morphisme

$$\chi: \operatorname{Ext}^1_X(p^*\mathscr{F}, \mathscr{G}) \to \operatorname{Hom}_Y(\mathscr{F}, R^1p_*\mathscr{G})$$

se décrit comme suit: étant donné une extension de faisceaux sur $X_{\text{ét}}$

$$0 \to \mathscr{G} \to \mathscr{E} \to p^*\mathscr{F} \to 0,$$

elle définit un bord $\partial: p_*p^*\mathscr{F} \to R^1p_*\mathscr{G}$, et $\chi(\mathscr{E})$ n'est autre, au signe près, que le morphisme composé

$$\mathscr{F} \to p_*p^*\mathscr{F} \xrightarrow{\;\partial\;} R^1p_*\mathscr{G}$$

avec le morphisme naturel $\mathscr{F} \to p_*p^*\mathscr{F}$. La démonstration s'appuie sur le lemme 1.A.2 de l'appendice. Si l'on plonge $\mathscr{G}$ dans un injectif $I^0_{\mathscr{G}}$, on obtient un diagramme commutatif à lignes exactes

$$
\begin{array}{ccccccccc}
0 & \to & \mathscr{G} & \to & \mathscr{E} & \to & p^*\mathscr{F} & \to & 0 \\
  &   & \| & & \downarrow & & \downarrow{\scriptstyle\beta} & & \\
0 & \to & \mathscr{G} & \to & I^0_{\mathscr{G}} & \to & Z^1_{\mathscr{G}} & \to & 0,
\end{array}
$$

d'où l'on tire le suivant:

$$
\begin{array}{ccccccc}
\operatorname{Ext}^1_X(p^*\mathscr{F}, \mathscr{G}) \xleftarrow{\;\partial\;} \operatorname{Hom}_X(p^*\mathscr{F}, p^*\mathscr{F}) = \operatorname{Hom}_Y(\mathscr{F}, p_*p^*\mathscr{F}) \xrightarrow{\;\partial\;} \operatorname{Hom}_Y(\mathscr{F}, R^1p_*\mathscr{G}) \\
\| \qquad\qquad\qquad \downarrow{\scriptstyle\beta} \qquad\qquad\qquad \downarrow{\scriptstyle\beta} \qquad\qquad\qquad \| \\
\operatorname{Ext}^1_X(p^*\mathscr{F}, \mathscr{G}) \xleftarrow{\;\partial\;} \operatorname{Hom}_X(p^*\mathscr{F}, Z^1_{\mathscr{G}}) = \operatorname{Hom}_Y(\mathscr{F}, p_*Z^1_{\mathscr{G}}) \xrightarrow{\;\partial\;} \operatorname{Hom}_Y(\mathscr{F}, R^1p_*\mathscr{G}).
\end{array}
$$

D'après le lemme 1.A.2, le morphisme χ est donné par la deuxième ligne. Or, sur la première ligne, le morphisme canonique $\iota: \mathscr{F} \to p_*p^*\mathscr{F}$ donne à gauche la classe de l'extension $\mathscr{E}$, et à droite le composé avec ∂. D'où: $\chi(\mathscr{E}) = \partial\iota$, ce qui achève de prouver (iii).

(iv) Dans cet énoncé, on suppose X géométriquement intègre. Le résultat est un cas particulier du suivant: si, dans la suite spectrale $(L)_{p,\mathscr{F},\mathscr{G}}$, on a une extension

$$0 \to \mathscr{G} \to \mathscr{E} \to \mathscr{D} \to 0$$

de faisceaux étales sur X, avec $R^1p_*\mathscr{E} = 0$, alors le morphisme

$$\partial: \operatorname{Hom}_Y(\mathscr{F}, R^1p_*\mathscr{G}) \to \operatorname{Ext}^2_Y(\mathscr{F}, p_*\mathscr{G})$$

est donné, au signe près, par le cup-produit avec la 2-extension de faisceaux

$$0 \to p_* \mathcal{G} \to p_* \mathcal{E} \to p_* \mathcal{D} \to R^1 p_* \mathcal{G} \to 0.$$

C'est un corollaire immédiat du lemme 1.A.4. Il s'applique à $\mathcal{G} = \mathbf{G}_{m,\,X}$ grâce à la suite exacte

$$1 \to \mathbf{G}_{m,\,X} \to i_* \mathbf{G}_{m,\,\eta} \to \mathscr{D}iv_X \to 0$$

de faisceaux sur $X_{\text{ét}}$, où $i: \eta \to X$ désigne l'inclusion du point générique η de X. On a effectivement

$$\left(R^1 p_*\right)\left(i_* \mathbf{G}_{m,\,\eta}\right) = 0,$$

car, si $r = p \circ i$, c'est un sous-faisceau de $R^1 r_* \mathbf{G}_{m,\,\eta}$ qui vaut 0 par le théorème 90. Dans l'énoncé, $Y = \text{Spec } k$.

(v) Là aussi, on se place dans le cadre plus général de la suite spectrale $(L)_{p,\,\mathcal{F},\,\mathcal{G}}$. On va voir que l'edge

$$\text{Ext}_Y^n(\mathcal{F}, p_* \mathcal{G}) \to \text{Ext}_X^n(p^* \mathcal{F}, \mathcal{G})$$

se décrit comme suit en termes de n-extensions: étant donné une n-extension de $\mathcal{F}$ par $p_* \mathcal{G}$, on lui applique p^*, ce qui donne une n-extension de $p^* \mathcal{F}$ par $p^* p_* \mathcal{G}$ qu'on pousse par le morphisme naturel $p^* p_* \mathcal{G} \to \mathcal{G}$. Cette assertion revient au lemme suivant qu'on pourrait d'ailleurs énoncer dans un cadre plus général:

LEMME 1.5.3.   *Soit* $p: X \to Y$. *Soient* $\mathcal{F}, \mathcal{F}'$ *deux faisceaux sur* Y *et* $\mathcal{G}$ *un faisceau sur* X. *Soient enfin* $\lambda: \mathcal{F}' \to p_* \mathcal{G}$ *et* $\mu: p^* \mathcal{F}' \to \mathcal{G}$ *deux morphismes adjoints. On a alors un diagramme commutatif:*

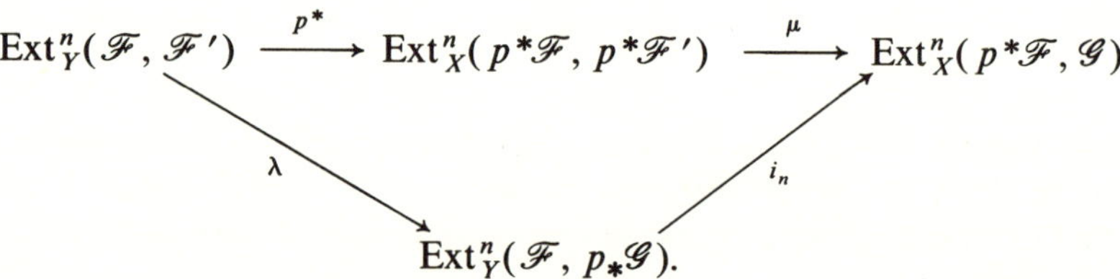

On applique ce lemme pour $\mathcal{F}' = p_* \mathcal{G}$, et pour λ l'identité de $p_* \mathcal{G}$ et μ son adjointe. La démonstration du lemme s'appuie sur la description générale de l'edge

$$i_n: E_2^{n,0} \to E^n$$

donnée au lemme 1.A.1. On considère des résolutions injectives

$$0 \to \mathcal{F}' \to D^{\cdot}, \qquad 0 \to \mathcal{G} \to I^{\cdot}, \qquad 0 \to p_* \mathcal{G} \to J^{\cdot}$$

respectivement de $\mathscr{F}'$, $\mathscr{G}$ et $p_*\mathscr{G}$. Il existe alors des morphismes de complexes γ et θ rendant commutatif le diagramme

$$
\begin{array}{ccc}
0 \to \mathscr{F}' \to & D^{\cdot} \\
\lambda \downarrow \qquad & \downarrow \gamma \\
0 \to p_*\mathscr{G} \to & J^{\cdot} \\
\| \qquad & \downarrow \theta \\
0 \to p_*\mathscr{G} \to & p_*I^{\cdot}
\end{array}
$$

Soit $\xi = \mathrm{adj}(\theta\gamma)$ l'adjointe de $\theta\gamma$. On obtient, par adjonction, le diagramme commutatif

$$
\begin{array}{ccc}
0 \to p^*\mathscr{F}' \to & p_*D^{\cdot} \\
\mu \downarrow \qquad & \downarrow \xi \\
0 \to \quad \mathscr{G} \quad \to & I^{\cdot}
\end{array}
$$

Le lemme 1.A.1 montre que le diagramme de l'énoncé est simplement l'homologie de degré n du diagramme de complexes:

$$
\mathrm{Hom}_Y(\mathscr{F}, D^{\cdot}) \xrightarrow{\ p^*\ } \mathrm{Hom}_X(p^*\mathscr{F}, p^*D^{\cdot}) \xrightarrow{\ \xi\ } \mathrm{Hom}_X(p^*\mathscr{F}, I^{\cdot})
$$

avec γ et θ:

$$
\mathrm{Hom}_Y(\mathscr{F}, J^{\cdot}) \xrightarrow{\ \theta\ } \mathrm{Hom}_Y(\mathscr{F}, p_*I^{\cdot}).
$$

Or ce diagramme est commutatif, car, si $\omega \in \mathrm{Hom}_Y(\mathscr{F}, D^{\cdot})$,

$$
(\mathrm{adj} \circ \theta\gamma)(\omega) = \mathrm{adj}(\theta\gamma) \circ p^*(\omega) = \xi \circ p^*(\omega).
$$

1.6. *Comportement par restriction à un ouvert.* Le comportement de la suite exacte fondamentale (1.5.1) par restriction à un ouvert de Zariski est donné par l'énoncé suivant:

THÉORÈME 1.6.1. *Soient X une variété algébrique lisse sur le corps k et S un k-groupe de type multiplicatif. Soit U un ouvert de Zariski non vide de X. On note F le fermé complémentaire de U et on considère le $\mathfrak{g}$-module $P_{\overline{X},\overline{U}}$ défini par la suite exacte*

$$(1.6.1) \qquad 0 \to P_{\overline{X},\overline{U}} \to \mathrm{Pic}\,\overline{X} \to \mathrm{Pic}\,\overline{U} \to 0.$$

Sous ces hypothèses, on a le diagramme suivant, dont les lignes et les colonnes sont exactes, et qui commute à des signes près:

$$
\begin{array}{ccccccc}
\mathrm{Hom}_{\mathfrak{g}}(\hat{S}, \overline{k}[U]^*) & \longrightarrow & H^0(U, S) & & 0 & & \mathrm{Ext}^1_{\mathfrak{g}}(\hat{S}, \overline{k}[U]^*) \\
\downarrow & & \downarrow & & \downarrow & & \downarrow \\
0 \to \mathrm{Hom}_{\mathfrak{g}}(\hat{S}, \overline{k}[U]^*/\overline{k}[X]^*) & \longrightarrow & \mathrm{Hom}_{\mathfrak{g}}(\hat{S}, \mathrm{Div}_{\overline{F}}\overline{X}) & \to & \mathrm{Hom}_{\mathfrak{g}}(\hat{S}, P_{\overline{X},\overline{U}}) & \to & \mathrm{Ext}^1_{\mathfrak{g}}(\hat{S}, \overline{k}[U]^*/\overline{k}[X]^*) \\
\downarrow & & \downarrow & & \downarrow & & \downarrow \\
0 \longrightarrow \mathrm{Ext}^1_{\mathfrak{g}}(\hat{S}, \overline{k}[X]^*) & \longrightarrow & H^1(X, S) & \longrightarrow & \mathrm{Hom}_{\mathfrak{g}}(\hat{S}, \mathrm{Pic}\,\overline{X}) & \longrightarrow & \mathrm{Ext}^2_{\mathfrak{g}}(\hat{S}, \overline{k}[X]^*) \\
\downarrow & & \downarrow & & \downarrow & & \downarrow \\
0 \longrightarrow \mathrm{Ext}^1_{\mathfrak{g}}(\hat{S}, \overline{k}[U]^*) & \longrightarrow & H^1(U, S) & \longrightarrow & \mathrm{Hom}_{\mathfrak{g}}(\hat{S}, \mathrm{Pic}\,\overline{U}) & \longrightarrow & \mathrm{Ext}^2_{\mathfrak{g}}(\hat{S}, \overline{k}[U]^*) \\
\downarrow{\scriptstyle \pi} & & \downarrow & & & & \\
\end{array}
$$

$$
\mathrm{Hom}_{\mathfrak{g}}(\hat{S}, P_{\overline{X},\overline{U}}) \to \mathrm{Ext}^1_{\mathfrak{g}}(\hat{S}, \overline{k}[U]^*/\overline{k}[X]^*) \to \mathrm{Ext}^1_{\mathfrak{g}}(\hat{S}, \mathrm{Div}_{\overline{F}}\overline{X}).
$$

Dans ce diagramme, les suites exactes horizontales sont des morceaux de $(1.5.1)_X$, *de* $(1.5.1)_U$ *et de la suite exacte longue de cohomologie tirée de la suite exacte courte de* $\mathfrak{g}$*-modules*

$$
(1.6.3) \qquad 1 \to \overline{k}[U]^*/\overline{k}[X]^* \to \mathrm{Div}_{\overline{F}}\overline{X} \to P_{\overline{X},\overline{U}} \to 0.
$$

Les suites exactes verticales sont des morceaux des suites longues de cohomologie tirées des deux suites exactes de $\mathfrak{g}$*-modules*

$$
(1.6.1) \qquad 0 \to P_{\overline{X},\overline{U}} \to \mathrm{Pic}\,\overline{X} \to \mathrm{Pic}\,\overline{U} \to 0
$$

$$
(1.6.4) \qquad 1 \to \overline{k}[X]^* \to \overline{k}[U]^* \to \overline{k}[U]^*/\overline{k}[X]^* \to 1
$$

ainsi que de la suite exacte de faisceaux étales définissant le faisceau de droite

$$
(1.6.5) \qquad 1 \to \mathsf{G}_{m,X} \to j_*\mathsf{G}_{m,U} \to \mathscr{D}iv_F X \to 0,
$$

où j *est l'immersion ouverte de* U *dans* X, *par application du foncteur* $\mathrm{Hom}_X(\hat{S},\ )$.

Pour préciser ce dernier point, il est bon d'isoler le lemme suivant:

LEMME 1.6.2. *Sous les hypothèses du théorème, on a la suite exacte naturelle*

$$
(1.6.6) \qquad 0 \to H^0(X, S) \to H^0(U, S) \to \mathrm{Hom}_{\mathfrak{g}}(\hat{S}, \mathrm{Div}_{\overline{F}}\overline{X})
$$

$$
\to H^1(X, S) \to H^1(U, S) \to \mathrm{Ext}^1_{\mathfrak{g}}(\hat{S}, \mathrm{Div}_{\overline{F}}\overline{X}).
$$

Démonstration du lemme. La suite exacte de faisceaux étales sur X

$$
(1.6.5) \qquad 1 \to \mathsf{G}_{m,X} \to j_*\mathsf{G}_{m,U} \to \mathscr{D}iv_F X \to 0
$$

définit le faisceau étale $\mathscr{D}iv_F X$ des diviseurs de X à support dans F. On en déduit, par application du foncteur $\mathrm{Hom}_{X_{\mathrm{\acute{e}t}}}(\hat{S}, \cdot)$, une longue suite exacte de cohomologie étale dont le début s'écrit

$$(1.6.7) \quad 0 \to \mathrm{Hom}_X(\hat{S}, \mathsf{G}_{m, X}) \to \mathrm{Hom}_X(\hat{S}, j_*\mathsf{G}_{m, U}) \to \mathrm{Hom}_X(\hat{S}, \mathscr{D}iv_F X)$$

$$\to \mathrm{Ext}^1_X(\hat{S}, \mathsf{G}_{m, X}) \to \mathrm{Ext}^1_X(\hat{S}, j_*\mathsf{G}_{m, U}) \to \mathrm{Ext}^1_X(\hat{S}, \mathscr{D}iv_F X).$$

Dans cette suite, $\mathrm{Ext}^i_X := \mathrm{Ext}^i_{X_{\mathrm{\acute{e}t}}}$. D'après 1.3.2 et 1.4.1,

$$\mathrm{Hom}_X(\hat{S}, \mathsf{G}_{m, X}) = H^0(X, S) \quad \text{et} \quad \mathrm{Ext}^1_X(\hat{S}, \mathsf{G}_{m, X}) = H^1(X, S).$$

Par adjonction et par 1.3.2,

$$\mathrm{Hom}_X(\hat{S}, j_*\mathsf{G}_{m, U}) = H^0(U, S).$$

D'autre part, $R^1 j_* \mathsf{G}_{m, U} = 0$ pour $j = j_{\mathrm{\acute{e}t}} \colon U_{\mathrm{\acute{e}t}} \to X_{\mathrm{\acute{e}t}}$; ceci se vérifie par réduction au cas strictement local X_ξ, auquel cas $\mathrm{Pic}\, U_\xi = 0$ par lissité de X. La suite spectrale de Leray pour $j = j_{\mathrm{\acute{e}t}}$:

$$(1.6.8) \qquad \mathrm{Ext}^p_{X_{\mathrm{\acute{e}t}}}(\hat{S}, R^q j_* \mathsf{G}_{m, U}) \Rightarrow \mathrm{Ext}^{p+q}_{U_{\mathrm{\acute{e}t}}}(\hat{S}, \mathsf{G}_{m, U})$$

donne alors, compte tenu de 1.4.1, l'identification

$$\mathrm{Ext}^1_{X_{\mathrm{\acute{e}t}}}(\hat{S}, j_*\mathsf{G}_{m, U}) = \mathrm{Ext}^1_{U_{\mathrm{\acute{e}t}}}(\hat{S}, \mathsf{G}_{m, U}) = H^1(U, S).$$

Pour interpréter enfin les termes en $\mathscr{D}iv_F X$, on considère la suite spectrale de Leray pour le morphisme structural $p = p_{\mathrm{\acute{e}t}} \colon X_{\mathrm{\acute{e}t}} \to \mathrm{Spec}\, k_{\mathrm{\acute{e}t}}$

$$(1.6.9) \qquad \mathrm{Ext}^p_{k_{\mathrm{\acute{e}t}}}(\hat{S}, R^q p_* \mathscr{D}iv_F X) \Rightarrow \mathrm{Ext}^{p+q}_{X_{\mathrm{\acute{e}t}}}(\hat{S}, \mathscr{D}iv_F X).$$

Nous allons établir les deux assertions suivantes:
  (i) le faisceau étale $p_*\mathscr{D}iv_F X$ s'identifie au $\mathfrak{g}$-module $\mathrm{Div}_{\overline{F}}\overline{X}$;
  (ii) $R^1 p_* \mathscr{D}iv_F X = 0$.
On commence par noter que la suite exacte (1.6.5) est aussi une suite exacte de faisceaux sur X_{Zar}, auquel cas on retrouve la définition même du faisceau des diviseurs de Cartier à support dans F. Pour l'établir, soit $r \colon X_{\mathrm{\acute{e}t}} \to X_{\mathrm{Zar}}$ le morphisme canonique. Le théorème 90 assure que $\mathrm{Pic}\, X_\xi = 0$ pour tout localisé de X, d'où $R^1 r_* \mathsf{G}_{m, X} = 0$, et le résultat annoncé. Cette remarque vaut tout autant sur $\overline{X}$. Dans la correspondance entre faisceaux étales sur $\mathrm{Spec}\, k$ et modules galoisiens, $p_*\mathscr{D}iv_F X$ correspond au $\mathfrak{g}$-module $H^0(\overline{X}, \mathscr{D}iv_{\overline{F}}\overline{X})$ et $R^1 p_*\mathscr{D}iv_F X$ au $\mathfrak{g}$-module $H^1(\overline{X}, \mathscr{D}iv_{\overline{F}}\overline{X})$, où $\mathscr{D}iv_{\overline{F}}\overline{X}$ est le faisceau des diviseurs de Cartier de $\overline{X}$ à support dans $\overline{F}$. On a donc $H^0(\overline{X}, \mathscr{D}iv_{\overline{F}}\overline{X}) = \mathrm{Div}_{\overline{F}}\overline{X}$. D'où (i). D'autre part,

$$H^1(\overline{X}, \mathscr{D}iv_{\overline{F}}\overline{X}) = 0,$$

car X est lisse et $H^1(Y, \mathbf{Z}) = 0$ pour tout schéma Y normal intègre ([1] IX 3.6(iii)). D'où (ii). La suite exacte des termes de bas degré de la suite spectrale (1.6.9)

$$0 \to \mathrm{Ext}^1_{k_{\text{ét}}}\big(\hat{S},\, p_* \mathscr{D}iv_F X\big) \to \mathrm{Ext}^1_{X_{\text{ét}}}\big(\hat{S},\, \mathscr{D}iv_F X\big) \to \mathrm{Hom}_k\big(\hat{S},\, R^1 p_* \mathscr{D}iv_F X\big)$$

définit donc, d'après (i) et (ii), un isomorphisme canonique

(iii) $\mathrm{Ext}^1_{X_{\text{ét}}}(\hat{S},\, \mathscr{D}iv_F X) = \mathrm{Ext}^1_{\mathfrak{g}}(\hat{S},\, \mathrm{Div}_{\overline{F}}\overline{X})$.

Remarque 1.6.3. La suite exacte du lemme prouve que, si S est un k-tore flasque (cf. [20] §1 ou [26] §0), la restriction $H^1(X, S) \to H^1(U, S)$ est surjective (voir aussi [26] théorème 2.2).

Démonstration du théorème. Il suffit d'appliquer le lemme 1.A.5 avec les données suivantes:

$\mathscr{C}$ (resp. $\mathscr{C}'$) = la catégorie des faisceaux de groupes abéliens sur $X_{\text{ét}}$ (resp. $(\mathrm{Spec}\,k)_{\text{ét}}$) et $\mathscr{C}'' =$ la catégorie des groupes abéliens,

$\Phi = p_*$ et $\Psi = \mathrm{Hom}_k(\hat{S},\ )$, de telle sorte que, par adjonction, d'après 1.3.1, $\Psi\Phi(\mathscr{G}) = \mathrm{Hom}_X(\hat{S}, \mathscr{G})$, et que la condition (1.4.0) est satisfaite, la suite exacte

(1.6.5) $1 \to \mathsf{G}_{m,X} \to j_* \mathsf{G}_{m,U} \to \mathscr{D}iv_F X \to 0$

jouant le rôle de (1.A.15), ce qui est possible car l'hypothèse $R^1 p_* \mathscr{D}iv_F X = 0$ est vérifiée comme on l'a vu plus haut, et alors, avec les notations du lemme 1.A.5,

$$P = \ker(\mathrm{Pic}\,\overline{X} \to \mathrm{Pic}\,\overline{U}) = P_{\overline{X},\overline{U}},$$

par identification de $R^1 p_* \mathsf{G}_{m,X}$ avec le module galoisien $\mathrm{Pic}\,\overline{X}$, et de $(R^1 p_*) j_* \mathsf{G}_{m,U}$ avec le module galoisien $\mathrm{Pic}\,\overline{U}$: cette dernière identification passe par l'égalité $(R^1 p_*) j_* \mathsf{G}_{m,U} = R^1(pj)_* \mathsf{G}_{m,U}$, due au fait que $R^1 j_* \mathsf{G}_{m,U} = 0$ (voir démonstration du lemme 1.6.2).

LEMME 1.6.4. *Le cheminement ci-dessous, extrait du diagramme du théorème précédent, est cohérent, au signe près:*

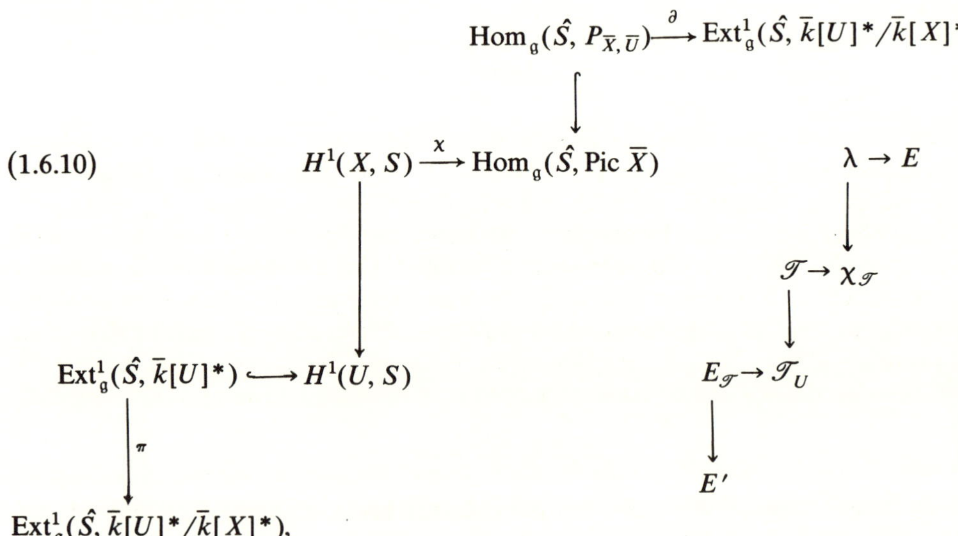

autrement dit, si la classe du torseur $\mathcal{T}$ dans $H^1(X, S)$ est reliée à

$$E \text{ et } E' \in \mathrm{Ext}^1_{\mathfrak{g}}(\hat{S}, \overline{k}[U]^*/\overline{k}[X]^*)$$

comme indiqué sur le diagramme ci-dessus, alors, au signe près, $E = E'$.

Démonstration. Il suffit d'appliquer le lemme 1.A.6 de l'appendice avec les mêmes données que dans l'application du lemme 1.A.5 au théorème précédent. On notera que $[\mathcal{T}] \in H^1(X, S)$ est reliée à E et E' comme indiqué dans l'énoncé, si et seulement si $\chi(\mathcal{T}_U) = 0$.

LEMME 1.6.5. *Soit $p\colon X \to \mathrm{Spec}\, k$ un k-schéma. Soient*

$$(1.6.11) \qquad 1 \to S \to M \to T \to 1$$

une suite exacte de k-groupes de type multiplicatif et

$$(1.6.12) \qquad 0 \to \hat{T} \to \hat{M} \to \hat{S} \to 0$$

la suite exacte duale de groupes constants tordus. On a alors le diagramme commutatif suivant, où les cobords ∂ sont définis par (1.6.11) et (1.6.12), où ε est l'edge (1.4.1), et où η est l'edge de la suite spectrale (1.5.2):

(1.6.13)

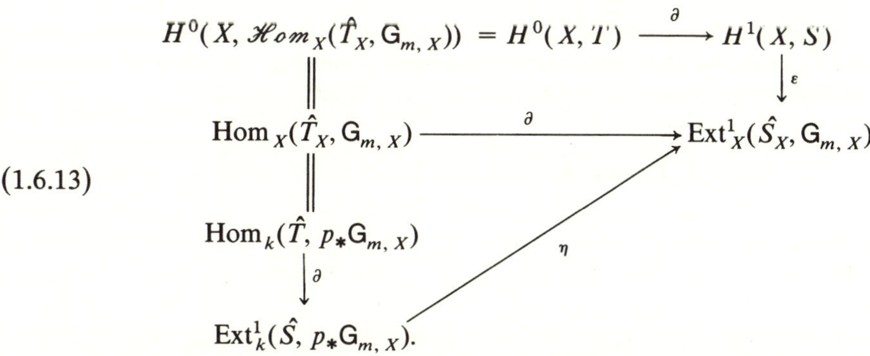

Démonstration. On est en topologie *pl*. Le diagramme ci-dessus s'écrit

1. Montrons d'abord la commutativité du carré supérieur. Si

$$0 \to \mathbf{G}_{m, X} \to I^0 \to I^1$$

est le début d'une résolution injective de $\mathbf{G}_{m, X}$, elle définit le diagramme commutatif suivant, à lignes et colonnes exactes:

$$
\begin{array}{ccccccccc}
0 & \to & \mathscr{H}\!om_X(\hat{S}, \mathbf{G}_{m, X}) & \to & \mathscr{H}\!om_X(\hat{M}, \mathbf{G}_{m, X}) & \to & \mathscr{H}\!om_X(\hat{T}, \mathbf{G}_{m, X}) & \to & 0 \\
 & & \downarrow & & \downarrow & & \downarrow & & \\
0 & \to & \mathscr{H}\!om_X(\hat{S}, I^0) & \to & \mathscr{H}\!om_X(\hat{M}, I^0) & \to & \mathscr{H}\!om_X(\hat{T}, I^0) & \to & 0 \\
 & & \downarrow & \ast & \downarrow & \ast & \downarrow & & \\
0 & \to & \mathscr{H}\!om_X(\hat{S}, I^1) & \to & \mathscr{H}\!om_X(\hat{M}, I^1) & \to & \mathscr{H}\!om_X(\hat{T}, I^1) & \to & 0.
\end{array}
$$

Le bord

$$\partial : \mathrm{Hom}_X\!\left(\hat{T}, \mathbf{G}_{m, X}\right) \to H^1\!\left(X, \mathscr{H}\!om_X\!\left(\hat{S}, \mathbf{G}_{m, X}\right)\right)$$

est le morphisme de liaison défini par le diagramme $H^0(X, \ast)$ où $\ast$ désigne le diagramme formé par les deux dernières lignes ci-dessus, qui forment le début d'une résolution acyclique de la première ligne. Quant au bord

$$\partial : \mathrm{Hom}_X\!\left(\hat{T}, \mathbf{G}_{m, X}\right) \to \mathrm{Ext}^1_X\!\left(\hat{S}, \mathbf{G}_{m, X}\right)$$

c'est aussi directement le morphisme de liaison défini par le même diagramme $H^0(X, \ast)$. On en déduit la commutativité de **1**, compte tenu de l'interprétation de l'edge ε.

2. Si $\lambda \in \mathrm{Hom}_k(\hat{T}, p_*\mathbf{G}_{m, X})$, il lui correspond, par l'isomorphisme d'adjonction, le composé μ des morphismes

$$\hat{T}_X \xrightarrow{\;p^*\lambda\;} p^*p_*\mathbf{G}_{m, X} \xrightarrow{\;\omega\;} \mathbf{G}_{m, X}.$$

On a les diagrammes commutatifs successifs:

$$
\begin{array}{ccccccccc}
0 & \longrightarrow & \hat{T} & \longrightarrow & \hat{M} & \to & \hat{S} & \to & 0 \\
 & & \lambda \downarrow & & \downarrow & & \| & & \\
0 & \to & p_*\mathbf{G}_{m, X} & \to & E & \to & \hat{S} & \to & 0
\end{array}
$$

dont la dernière ligne n'est autre que l'extension $\partial(\lambda)$, et

$$
\begin{array}{ccccccccc}
0 & \longrightarrow & p^*\hat{T} & \longrightarrow & p^*\hat{M} & \rightarrow & p^*\hat{S} & \rightarrow & 0 \\
 & & {\scriptstyle p^*\lambda}\downarrow & & \downarrow & & \| & & \\
0 & \rightarrow & p^*p_*\mathbf{G}_{m,\,X} & \rightarrow & p^*E & \rightarrow & p^*\hat{S} & \rightarrow & 0 \\
 & & {\scriptstyle \omega}\downarrow & & \downarrow & & \| & & \\
0 & \longrightarrow & \mathbf{G}_{m,\,X} & \longrightarrow & F & \longrightarrow & p^*\hat{S} & \rightarrow & 0.
\end{array}
$$

La description de η donnée dans la proposition 1.5.2(v) montre d'après les deux dernières lignes que l'extension F a pour classe $\eta(\partial(\lambda))$. D'autre part, d'après les deux lignes extrêmes, la classe de F est aussi $\partial(\omega \circ p^*\lambda) = \partial(\mu)$. Ceci prouve la commutativité de **2** et achève la démonstration.

1.A. *Appendice d'algèbre homologique.* On renvoie à [6] ou [36] pour les généralités d'algèbre homologique; voir aussi [44] (appendice B). Pour les diverses définitions ou interprétations des Ext^i, voir par exemple [36], IV 7–9.

1.A.0. On introduit, pour tout cet appendice, les hypothèses et notations suivantes:

Soient $\mathscr{C}, \mathscr{C}', \mathscr{C}''$ trois catégories abéliennes, et $\Phi\colon \mathscr{C} \to \mathscr{C}'$, $\Psi\colon \mathscr{C}' \to \mathscr{C}''$ deux foncteurs additifs exacts à gauche. On suppose que $\mathscr{C}$ et $\mathscr{C}'$ ont assez d'injectifs et que Φ transforme injectif en Ψ-acyclique. On fait même souvent l'hypothèse plus forte suivante:

$$(1.\mathrm{A}.0) \qquad \Phi \text{ possède un adjoint à gauche exact,}$$

auquel cas Φ transforme injectif en injectif. Sous ces hypothèses, on a la suite spectrale de dérivation du foncteur composé $\Psi \circ \Phi$:

$$(1.\mathrm{A}.1) \qquad E_2^{p,\,q} = R^p\Psi\big(R^q\Phi(A)\big) \Rightarrow E^{p+q} = R^{p+q}(\Psi\Phi)(A),$$

dont la suite des termes de bas degré commence par

$$(1.\mathrm{A}.2) \qquad 0 \to (R^1\Psi)(\Phi A) \to (R^1\Psi\Phi)(A) \to \Psi\big((R^1\Phi)(A)\big)$$

$$\to (R^2\Psi)(\Phi A) \to (R^2\Psi\Phi)(A),$$

et dont les edges s'écrivent

$$(1.\mathrm{A}.3) \qquad E^n = (R^n\Psi\Phi)(A) \to E_2^{0,\,n} = \Psi(R^n\Phi)(A)$$

$$(1.\mathrm{A}.4) \qquad E_2^{n,\,0} = (R^n\Psi)(\Phi A) \to E^n = (R^n\Psi\Phi)(A).$$

Rappelons comment on peut décrire la suite spectrale (1.A.1). Soient $A \to I_A^{\cdot}$ une résolution injective de A, puis $\Phi A \to I^{\cdot}_{\Phi A}$ une résolution injective de ΦA, enfin $\Phi I^{\cdot}_A \to L^{\cdot\cdot}$ une résolution de Cartan-Eilenberg de $\Phi I^{\cdot}_A$. On obtient ainsi le carré commutatif:

$$(1.A.5) \qquad \begin{array}{ccc} \Phi I^{\cdot}_A & \xrightarrow{\ \alpha\ } & L^{\cdot\cdot} \\[2mm] \uparrow & \nwarrow{\scriptstyle \theta} & \uparrow{\scriptstyle \beta} \\[2mm] \Phi A & \longrightarrow & I^{\cdot}_{\Phi A} \end{array}$$

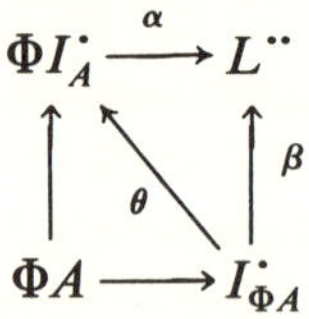

qu'on complète, sous (1.A.0), par la diagonale θ, telle que le triangle inférieur soit commutatif, tandis que β et $\alpha\theta$ sont homotopes. L'existence de θ, sous (1.A.0), tient à ce que $I^{\cdot}_{\Phi A}$ est acyclique et $\Phi I^{\cdot}_A$ injectif. Par application de Ψ, on obtient ainsi le diagramme

$$(1.A.6) \qquad \begin{array}{ccc} \Psi\Phi I^{\cdot}_A & \xrightarrow{\ \Psi(\alpha)\ } & \Psi L^{\cdot\cdot} \\[2mm] \uparrow & \nwarrow{\scriptstyle \Psi(\theta)} & \uparrow{\scriptstyle \Psi(\beta)} \\[2mm] \Psi\Phi A & \longrightarrow & \Psi I^{\cdot}_{\Phi A} \end{array}$$

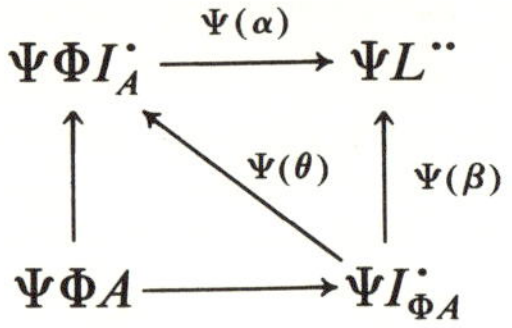

dans lequel chaque horizontale $\Psi I^q_A \to \Psi L^{\cdot q}$ est une résolution de ΨI^q_A. La suite spectrale (1.A.1) est la suite spectrale du complexe bigradué $\Psi(L^{\cdot\cdot})$ filtré par le premier exposant p. Par dégénérescence de la suite spectrale du même complexe filtré par le second exposant q, l'aboutissement de (1.A.1) s'identifie via $H^n(\Psi(\alpha))$ à $H^n(\Psi\Phi(I^{\cdot}_A)) = (R^n\Psi\Phi)A$, et les termes initiaux "de côté" sont d'une part $H^n(\Psi I^{\cdot}_{\Phi A}) = R^n\Psi(\Phi A)$ via $H^n(\Psi(\alpha))$ et d'autre part $\Psi(H^n(\Phi(I^{\cdot}_A))) = \Psi(R^n\Phi A)$. D'où le lemme:

LEMME 1.A.1. *Avec les notations introduites ci-dessus, la suite spectrale (1.A.1) admet pour edge $E^n = (R^n\Psi\Phi)(A) \to E^{0,n}_2 = \Psi(R^n\Phi)A$ le morphisme naturel $H^n(\Psi\Phi I^{\cdot}_A) \to \Psi(H^n(\Phi I^{\cdot}_A))$, et pour edge $E^{n,0}_2 = (R^n\Psi)\Phi A \to E^n = (R^n\Psi\Phi)A$ le morphisme $H^n(\Psi(\theta))$: $H^n(\Psi I^{\cdot}_{\Phi A}) \to H^n(\Psi\Phi I^{\cdot}_A)$.*

Démonstration. Vérification facile à partir de ce qu'on a dit précédemment.

Pour le début de la suite des termes de bas degré on a le lemme suivant:

LEMME 1.A.2. *Soit*

$$(1.A.7) \qquad 0 \to A \to I^0_A \to Z^1_A \to 0$$

une suite exacte dans $\mathscr{C}$ avec I^0_A injectif. Soit

$$(1.A.8) \qquad 0 \to \Phi A \to \Phi I^0_A \to \Phi Z^1_A \to R^1\Phi A \to 0$$

la 2-extension qu'on en déduit par application de Φ et qu'on peut couper en deux extensions

$$(1.\text{A}.9) \qquad 0 \to \Phi A \to \Phi I_A^0 \to Z_{\Phi A}^1 \to 0,$$

$$(1.\text{A}.10) \qquad 0 \to Z_{\Phi A}^1 \to \Phi Z_A^1 \to R^1\Phi A \to 0.$$

On a alors le diagramme de suites exactes suivant, commutatif à des signes près:

$$(1.\text{A}.11)$$

$$
\begin{array}{ccc}
\Psi\Phi I_A^0 & == & \Psi\Phi I_A^0 \\
\downarrow & & \downarrow \\
\end{array}
$$

$$0 \longrightarrow \Psi Z_{\Phi A}^1 \longrightarrow \Psi\Phi Z_A^1 \longrightarrow \Psi(R^1\Phi A) \to (R^1\Psi)Z_{\Phi A}^1 \to (R^1\Psi\Phi)Z_A^1$$

$$0 \to (R^1\Psi)\Phi A \longrightarrow (R^1\Psi\Phi)(A) \to \Psi(R^1\Phi A) \to (R^2\Psi)\Phi A \to (R^2\Psi\Phi)A$$

dont l'horizontale médiane est le début de la longue suite exacte dérivée $R^{\cdot}\Psi$ (1.A.10) suivi de l'edge

$$(R^1\Psi)\Phi Z_A^1 \to (R^1\Psi\Phi)Z_A^1$$

et dont les bords ∂ dérivent de la suite exacte courte (1.A.9).

Le diagramme (1.A.11) donne une description "concrète" du début de la suite des termes de bas degré (1.A.2). La vérification se fait à partir des diagrammes (1.A.5) et (1.A.6). Sous l'hypothèse (1.A.0), le début de la résolution injective $\Phi A \to I_{\Phi A}^{\cdot}$ peut être pris égal à $\Phi A \to \Phi I_A^0 \to \Phi I_A^1$, auquel cas on peut prendre pour θ^0 et θ^1 les morphismes "identité." Sinon, la notation $Z_{\Phi A}^1$ est quelque peu abusive, mais l'hypothèse de Ψ-acyclicité des ΦI_A^q, laquelle implique par exemple $(R^1\Psi)\Phi I_A^0 = 0$, permet d'obtenir encore les mêmes présentations. On notera que le morphisme $\partial\colon \Phi Z_A^1 \to R^1\Phi A$ de (1.A.8) est l'opposé du morphisme naturel de définition de $R^1\Phi A$, résultat qu'on utilise souvent implicitement ici; plus généralement, avec des notations évidentes, le bord itéré $\partial\colon \Phi Z_A^n \to R^n\Phi A$, défini en tronquant la résolution $I_A^{\cdot}$ au niveau n, est égal au morphisme naturel de définition de $R^n\Phi A$ au signe $(-1)^{n(n+1)/2}$ près ([6] V 7.1).

On a également les énoncés particuliers suivants:

LEMME 1.A.3. *Si $R^1\Phi A = 0$, toute extension*

$$(1.\text{A}.12) \qquad 0 \to A \to E \to B \to 0$$

définit un carré commutatif, au signe près:

$$(1.A.13) \qquad \begin{array}{ccc} \Psi\Phi(B) & \xrightarrow{\;\partial\;} & (R^1\Psi)\Phi A \\[2pt] \Big\| & & \Big\downarrow{\scriptstyle i_1} \\[2pt] \Psi\Phi(B) & \xrightarrow{\;\partial\;} & (R^1\Psi\Phi)A \end{array}$$

où i_1 *est l'edge* (1.A.4) *et où les bords* ∂ *sont déduits de* Φ(1.A.12) *par application de* Ψ *et de* (1.A.12) *par application de* $\Psi\Phi$.

Démonstration. Conséquence de l'existence, par l'injectivité de I_A^0, d'un diagramme commutatif

$$\begin{array}{ccccccccc} 0 & \to & A & \to & E & \to & B & \to & 0 \\[2pt] & & \Big\| & & \Big\downarrow & & \Big\downarrow & & \\[2pt] 0 & \to & A & \to & I_A^0 & \to & Z_A^1 & \to & 0. \end{array}$$

LEMME 1.A.4. *Le morphisme*

$$\Psi(R^1\Phi A) \to (R^2\Psi)\Phi A$$

est, au signe près, le bord itéré défini par application de Ψ *à la 2-extension* (1.A.8). *Si*

$$(1.A.12) \qquad\qquad 0 \to A \to E \to B \to 0$$

est une extension telle que $R^1\Phi E = 0$, *c'est aussi, au signe près, le bord itéré défini par application de* Ψ *à la 2-extension*

$$(1.A.14) \qquad\qquad 0 \to \Phi A \to \Phi E \to \Phi B \to R^1\Phi A \to 0.$$

Démonstration. La première partie résulte du lemme 1.A.2. Pour la seconde assertion, on procède comme au lemme 1.A.3, ce qui permet d'obtenir un isomorphisme (1.A.14) → (1.A.8) de 2-extensions.

LEMME 1.A.5. *On garde les hypothèses et notations initiales. On considère en outre, dans la catégorie $\mathscr{C}$, une suite exacte*

$$(1.A.15) \qquad\qquad 0 \to A \to B \to C \to 0$$

telle que $R^1\Phi C = 0$. *Soit P le noyau du morphisme* $R^1\Phi A \to R^1\Phi B$. *On a ainsi les* *trois suites exactes courtes*

$$(1.A.16) \qquad 0 \to \Phi A \to \Phi B \to \Phi B/\Phi A \to 0$$

$$(1.A.17) \qquad 0 \to \Phi B/\Phi A \to \Phi C \to P \to 0$$

$$(1.A.18) \qquad 0 \to P \to R^1\Phi A \to R^1\Phi B \to 0.$$

On a alors le diagramme suivant, formé de lignes et de colonnes exactes, et commutatif à des signes près:

$$(1.A.19)$$

$$
\begin{array}{ccccccc}
\Psi\Phi B & \!=\!=\!= & \Psi\Phi B & & 0 & & (R^1\Psi)\Phi B \\
\downarrow & {}^* & \downarrow & & \downarrow & & \downarrow \\
0 \to \Psi(\Phi B/\Phi A) & \longrightarrow & \Psi\Phi C & \longrightarrow & \Psi P & \longrightarrow & (R^1\Psi)(\Phi B/\Phi A) \\
\downarrow & {}^{\mathbf{1}} & \downarrow & {}^{\mathbf{2}} & \downarrow & {}^{\mathbf{3}} & \downarrow \\
0 \to (R^1\Psi)\Phi A & \longrightarrow & (R^1\Psi\Phi)A & \to & \Psi(R^1\Phi)A & \longrightarrow & (R^2\Psi)\Phi A \\
\downarrow & {}^* & \downarrow & {}^* & \downarrow & {}^* & \downarrow \\
0 \to (R^1\Psi)\Phi B & \longrightarrow & (R^1\Psi\Phi)B & \to & \Psi(R^1\Phi)B & \longrightarrow & (R^2\Psi)\Phi B \\
& & & & \downarrow & & \\
& & & & (R^1\Psi\Phi)C & & \\
& & {}^{\mathbf{4}} & & \uparrow {}^{\cong} & & \\
\end{array}
$$

$$\Psi(P) \to (R^1\Psi)(\Phi B/\Phi A) \to (R^1\Psi)\Phi C.$$

Dans ce diagramme, les horizontales sont les suites $(1.A.2)_A$, $(1.A.2)_B$, *et un morceau de la longue suite de cohomologie* $(R^{\cdot}\Psi)$ $(1.A.17)$, *les verticales sont des morceaux des longues suites de cohomologie* $(R^{\cdot}\Psi)$ $(1.A.16)$, $(R^{\cdot}\Psi\Phi)$ $(1.A.15)$ *et* $(R^{\cdot}\Psi)$ $(1.A.18)$. *La "soudure"*

$$(R^1\Psi)\Phi C \xrightarrow{\ \cong\ } (R^1\Psi\Phi)C$$

est donnée par l'edge de la suite spectrale $(1.A.4)_C$.

Démonstration. On vérifie successivement les commutativités des carrés $*$, puis **1** à **4**.

$*$. La commutativité du carré supérieur $*$ est évidente; celle des trois autres carrés marqués $*$ résulte de la fonctorialité en A de la suite spectrale $(1.A.1)$.

1. Avec les notations du lemme 1.A.2, on a, par injectivité de I_A^0, un diagramme

$$(1.A.20) \qquad \begin{array}{ccccccccc} 0 & \to & A & \to & B & \to & C & \to & 0 \\ & & \| & & \downarrow & & \downarrow & & \\ 0 & \to & A & \to & I_A^0 & \to & Z_A^1 & \to & 0, \end{array}$$

d'où l'on déduit par application de Φ les deux diagrammes suivants de suites exactes:

$$(1.A.21) \qquad \begin{array}{ccccccccc} 0 & \to & \Phi A & \to & \Phi B & \to & \Phi B/\Phi A & \to & 0 \\ & & \| & & \downarrow & & \downarrow & & \\ 0 & \to & \Phi A & \to & \Phi I_A^0 & \to & Z_{\Phi A}^1 & \to & 0, \end{array}$$

$$(1.A.22) \qquad \begin{array}{ccccccccc} 0 & \to & \Phi B/\Phi A & \to & \Phi C & \to & R^1\Phi A & & \\ & & \downarrow & & \downarrow & & \| & & \\ 0 & \to & Z_{\Phi A}^1 & \to & \Phi Z_A^1 & \to & R^1\Phi A & \to & 0. \end{array}$$

On déduit de ce dernier diagramme le carré commutatif

$$\begin{array}{ccc} \Psi(\Phi B/\Phi A) & \to & \Psi\Phi C \\ \downarrow & & \downarrow \\ \Psi Z_{\Phi A}^1 & \longrightarrow & \Psi\Phi Z_A^1, \end{array}$$

qui, d'après le lemme 1.A.2 et le diagramme (1.A.21) ci-dessus, induit le carré **1** après passage au quotient par l'image de $\Psi\Phi I_A^0$ sur la ligne du bas.

2. Il s'agit de voir que le triangle

$$(1.A.23) \qquad \begin{array}{ccc} & \Psi\Phi C & \\ \partial \downarrow & & \searrow \partial \\ (R^1\Psi\Phi)A & \to & \Psi(R^1\Phi A) \end{array}$$

commute. Or, dans le diagramme ci-dessous:

$$\begin{array}{ccc} \Psi\Phi C & \xrightarrow{\ \partial\ } & (R^1\Psi\Phi)A \\ \downarrow & \nearrow & \downarrow \\ \Psi\Phi Z_A^1 & \xrightarrow[\ \partial\ ]{} & \Psi(R^1\Phi A) \end{array}$$

les deux triangles commutent, celui du bas d'après le lemme 1.A.2. Le carré commute donc, et si l'on trace l'autre diagonale:

$$\begin{array}{ccc}
\Psi\Phi C & \xrightarrow{\ \partial\ } & (R^1\Psi\Phi)A \\
\downarrow & \searrow & \downarrow \\
\Psi\Phi Z_A^1 & \xrightarrow{\ \partial\ } & \Psi(R^1\Phi A)
\end{array}$$

le triangle inférieur commute comme image par Ψ d'un triangle commutatif, ce qui montre finalement que le triangle supérieur, qui n'est autre que (1.A.23), commute aussi.

3. Le diagramme (1.A.20) donne le diagramme commutatif de 2-extensions:

$$(1.A.24)\qquad\begin{array}{ccccccccc}
0 & \to & \Phi A & \to & \Phi B & \to & \Phi C & \to & P & \to & 0 \\
& & \| & & \downarrow & & \downarrow & & \downarrow & & \\
0 & \to & \Phi A & \to & \Phi I_A^0 & \to & \Phi Z_A^1 & \to & R^1\Phi A & \to & 0,
\end{array}$$

d'où le carré commutatif

$$(1.A.25)\qquad\begin{array}{ccc}
\Psi P & \xrightarrow{\ \partial_2\ } & (R^2\Psi)\Phi A \\
\downarrow & & \| \\
\Psi(R^1\Phi A) & \xrightarrow{\ \partial\ } & (R^2\Psi)\Phi A
\end{array}$$

où, d'après le lemme 1.A.4, le morphisme ∂ coïncide avec la flèche correspondante dans la partie 3 de (1.A.19). Quant à ∂_2, c'est la composée des applications

$$\Psi P \to (R^1\Psi)(\Phi B/\Phi A) \to (R^2\Psi)(\Phi A)$$

de la même partie **3** du diagramme (1.A.19). D'où la commutativité de **3**.

4. On peut considérer un diagramme commutatif de suites exactes

$$(1.A.26)\qquad\begin{array}{ccccccccc}
0 & \to & B & \to & I_B^0 & \to & Z_B^1 & \to & 0 \\
& & \downarrow & & \downarrow & & \downarrow & & \\
0 & \to & C & \to & I_C^0 & \to & Z_C^1 & \to & 0,
\end{array}$$

où I_B^0 et I_C^0 sont injectifs. On en tire successivement les diagrammes commutatifs suivants:

$$(1.A.27) \qquad \begin{array}{ccccccc}
0 & \longrightarrow & \Phi B & \longrightarrow & \Phi I_B^0 & \longrightarrow & Z_{\Phi B}^1 & \longrightarrow & 0 \\
& & \downarrow & & \downarrow & & \downarrow & & \\
0 & \to & \Phi B/\Phi A & \to & \Phi I_C^0 & \to & Z_{\Phi B/\Phi A}^1 & \to & 0 \\
& & \downarrow & & \| & & \downarrow & & \\
0 & \longrightarrow & \Phi C & \longrightarrow & \Phi I_C^0 & \longrightarrow & Z_{\Phi C}^1 & \longrightarrow & 0,
\end{array}$$

dont les lignes sont exactes, et dont la partie droite se prolonge par

$$\begin{array}{ccc}
Z_{\Phi B}^1 & \hookrightarrow & \Phi Z_B^1 \\
\downarrow & & \downarrow \\
Z_{\Phi C}^1 & \xrightarrow{\;\cong\;} & \Phi Z_C^1,
\end{array}$$

compte tenu de $R^1 \Phi C = 0$, puis

$$(1.A.28) \qquad \begin{array}{ccc}
\Psi Z_{\Phi B}^1 & \longrightarrow & \Psi \Phi Z_B^1 \\
& & \downarrow \\
& & \Psi \Phi Z_C^1 \\
\downarrow & & \uparrow \cong \\
\Psi Z_{\Phi B/\Phi A}^1 & \to & \Psi Z_{\Phi C}^1,
\end{array}$$

et, d'après le lemme 1.A.2, ce dernier diagramme induit le carré **4** de l'énoncé par passage au quotient, par $\Psi \Phi I_B^0$ sur la ligne du haut et par $\Psi \Phi I_C^0$ sur le reste. Ceci achève la démonstration.

LEMME 1.A.6. *Avec les hypothèses et notations du lemme précédent, le cheminement ci-dessous, extrait du diagramme* (1.A.19), *est cohérent, au signe près:*

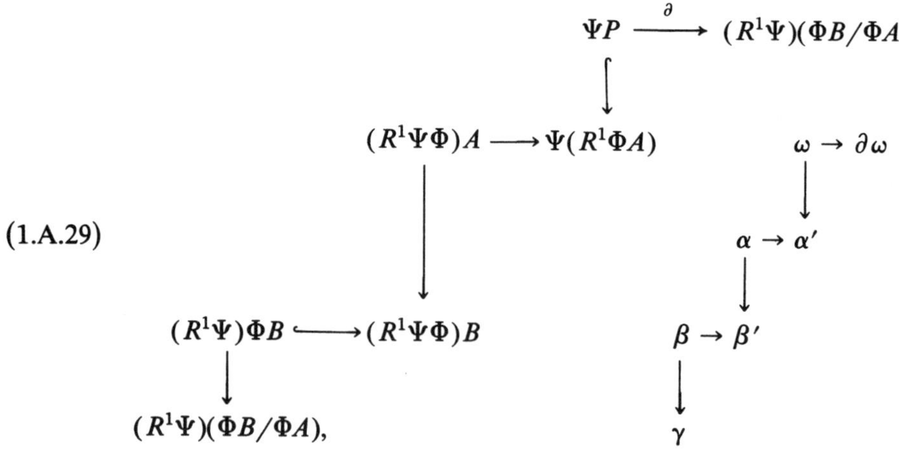

(1.A.29)

autrement dit, si α est relié à $\partial\omega$ et γ comme indiqué ci-dessus, alors, au signe près,

$$\gamma = \partial\omega \in (R^1\Psi)(\Phi B/\Phi A).$$

Démonstration. On notera d'abord qu'un élément α de $(R^1\Psi\Phi)A$ est relié à des éléments $\partial\omega$ et γ comme indiqué dans l'énoncé, si et seulement si c'est un élément du noyau de $(R^1\Psi\Phi)A \to \Psi(R^1\Phi B)$. On part d'un diagramme commutatif

$$
\begin{array}{ccccccccc}
0 & \to & A & \to & B & \to & C & \to & 0 \\
& & \downarrow & & \downarrow & & \downarrow & & \\
0 & \to & I_A^0 & \to & I_B^0 & \to & I_C^0 & \to & 0,
\end{array}
$$

dont la ligne du bas est exacte et formée d'injectifs I_A^0, I_B^0, I_C^0, et on note Z_A^1, Z_B^1 et Z_C^1 les conoyaux respectifs des trois morphismes verticaux.

(i) Montrons d'abord que, pour établir la cohérence du cheminement (1.A.29), il suffit d'établir celle du cheminement défini de façon analogue à partir du diagramme naturel suivant:

(1.A.30)

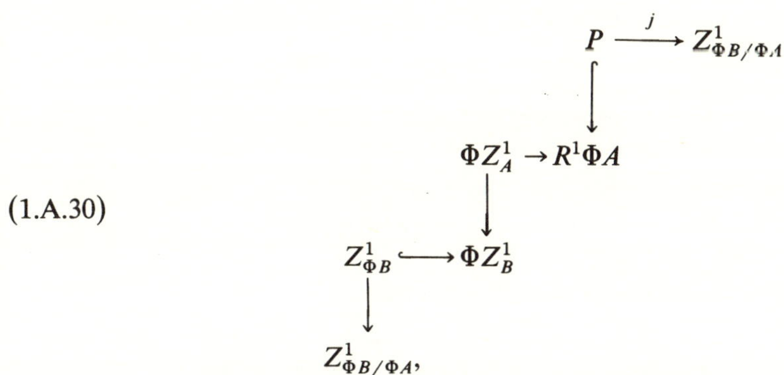

dans lequel, par définition, j rend commutatif le diagramme de suites exactes

$$
\begin{array}{ccccccccc}
0 & \to & \Phi B/\Phi A & \to & \Phi C & \overset{\partial}{\longrightarrow} & P & \longrightarrow & 0 \\
& & \| & & \downarrow & & \downarrow j & & \\
0 & \to & \Phi B/\Phi A & \to & \Phi I_C^0 & \to & Z_{\Phi B/\Phi A}^1 & \to & 0,
\end{array}
$$

tandis que les autres flèches sont les flèches naturelles évidentes. Introduisons

$$N := \ker\big(\Phi Z_A^1 \to R^1\Phi B\big).$$

Le diagramme (1.A.30) se complète alors comme suit:

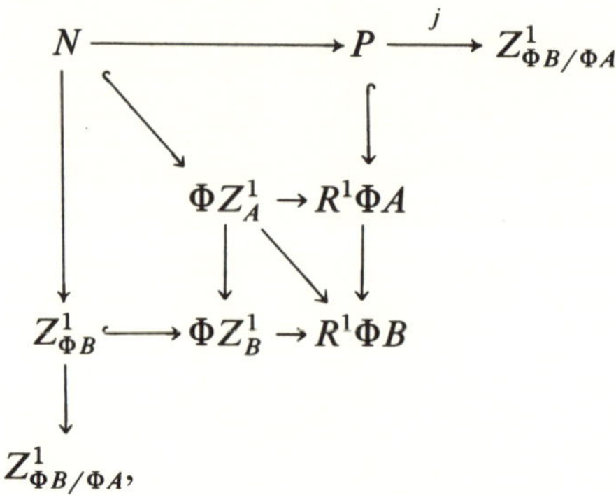

et dire que le cheminement défini par ce diagramme est cohérent, c'est dire que les deux morphismes

$$N \to Z^1_{\Phi B/\Phi A}$$

ainsi définis, l'un via P, l'autre via $Z^1_{\Phi B}$, coïncident.

La cohérence du cheminement (1.A.29) s'interprète de façon analogue, en introduisant

$$N_\Psi := \ker\big((R^1\Psi\Phi)A \to \Psi(R^1\Phi B)\big)$$

et en complétant (1.A.29) comme on l'a fait pour (1.A.30). On obtient ainsi sur ce diagramme complété deux applications $N_\Psi \to (R^1\Psi)(\Phi B/\Phi A)$ dont on veut voir qu'elles coïncident, au signe près.

Or, on a les identifications canoniques suivantes:

$$(R^1\Psi\Phi)A = \Psi\Phi Z^1_A/\Psi\Phi I^0_A, \qquad (R^1\Psi\Phi)B = \Psi\Phi Z^1_B/\Psi\Phi I^0_B,$$

$$(R^1\Psi)\Phi B = \Psi Z^1_{\Phi B}/\Psi\Phi I^0_B, \qquad (R^1\Psi)(\Phi B/\Phi A) = \Psi Z^1_{\Phi B/\Phi A}/\Psi\Phi I^0_C,$$

et enfin $N_\Psi = \Psi N/\Psi\Phi I^0_A$.

Ces identifications montrent que le complété du diagramme (1.A.29) est défini par passage au quotient de l'image par Ψ du complété du diagramme (1.A.30). Pour obtenir le lemme, il suffit donc d'établir la cohérence du cheminement défini par (1.A.30).

(ii) Comme on le voit sur (1.A.27), on a une surjection naturelle $Z^1_{\Phi B/\Phi A} \overset{\eta}{\to} Z^1_{\Phi C}$, et on montre aisément qu'on a le diagramme commutatif suivant, à lignes et colonnes exactes:

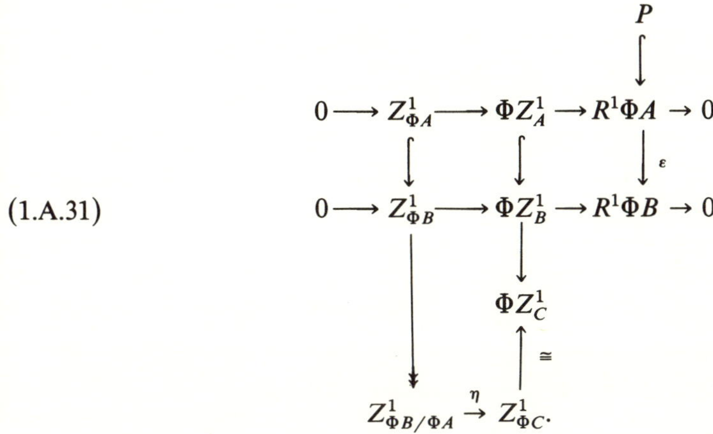

$$(1.A.31)$$

Ce diagramme définit un morphisme de liaison $j'\colon P \to Z^1_{\Phi B/\Phi A}$ qui, mis à la place de j dans le cheminement (1.A.30), le rend évidemment cohérent.

Considérons alors le diagramme commutatif suivant:

$$(1.A.32)$$

Le diagramme formé des deux dernières lignes définit un morphisme de liaison induisant un isomorphisme

$$j''\colon \ker \eta \xrightarrow{\ \cong\ } P = \ker\!\left(R^1\Phi A \xrightarrow{\ \varepsilon\ } R^1\Phi B \right).$$

Il est immédiat que j'' est l'inverse de l'isomorphisme $j'\colon P = \ker \varepsilon \xrightarrow{\cong} \ker \gamma$, défini par (1.A.31). Le diagramme formé par les lignes extrêmes définit un morphisme de liaison

$$\ker \zeta = \Phi C \to R^1 \Phi A = \operatorname{coker} \iota,$$

qui n'est autre que le bord ∂ déduit de la suite (1.A.17) par application de Φ. Il résulte ainsi de (1.A.32) que

$$\partial = j'' \circ (\pi|_{\Phi C}).$$

Ainsi, j coïncide avec $j''^{-1}\colon P \to Z^1_{\Phi B/\Phi A}$, et finalement $j = j'$, ce qui achève la démonstration.

§2. Géométrie des torseurs.

2.0. *On considère la situation suivante*:

(2.0.1) k est un corps et X une k-variété algébrique lisse géométriquement intègre telle que

$$\bar{k}^* = \bar{k}[X]^*,$$

ce qui est le cas si X est propre sur k.

Sous ces hypothèses, la suite exacte (1.5.1) se réécrit

(2.0.2)

$$0 \to H^1(k, S) \xrightarrow{i_1} H^1(X, S) \xrightarrow{\chi} \mathrm{Hom}_{\mathfrak{g}}(\hat{S}, \mathrm{Pic}\,\overline{X}) \xrightarrow{\delta} H^2(k, S) \xrightarrow{i_2} H^2(X, S).$$

On dit qu'un torseur $\mathcal{T}$ sur X sous S est de type

$$\lambda \in \mathrm{Hom}_{\mathfrak{g}}(\hat{S}, \mathrm{Pic}\,\overline{X})$$

si l'on a $\chi([\mathcal{T}]) = \lambda$. On note $\mathrm{Tors}(X, S, \lambda)$ l'ensemble des classes d'isomorphisme de torseurs de type λ. S'il est non vide, c'est un espace principal homogène sous le groupe $H^1(k, S)$.

Supposons de plus que X satisfasse la condition

(2.0.3)   le $\mathfrak{g}$-module $\mathrm{Pic}\,\overline{X}$ est de type fini.

On note alors

$$S_0 := D(\mathrm{Pic}\,\overline{X})$$

le k-groupe de type multiplicatif dual. Vu l'appendice 2.A, la condition (2.0.3) est vérifiée pour les variétés rationnelles projectives et lisses, et S_0 est alors un tore, appelé tore de Néron-Severi de X. On dit qu'un torseur $\mathcal{T}$ sur X sous S est *universel* si l'on a

(2.0.4) $S = S_0$ et le type de $\mathcal{T}$ est l'identité de $\mathrm{Hom}_{\mathfrak{g}}(\mathrm{Pic}\,\overline{X}, \mathrm{Pic}\,\overline{X})$.

En pratique, on considère aussi des torseurs qui, sans être universels, en sont suffisamment proches pour en avoir certaines des propriétés. On dit qu'un torseur

$\mathscr{T}$ sur X sous un k-groupe de type multiplicatif S est *admissible* si son type

$$\lambda: \hat{S} \to \mathrm{Pic}\,\overline{X}$$

est une injection dont le conoyau est un $\mathfrak{g}$-module de permutation.

De façon plus générale, on peut aussi considérer une sous-extension galoisienne K/k de $\overline{k}/k$ telle que, si l'on note $\mathfrak{h} = \mathrm{Gal}(\overline{k}/K)$ et $G = \mathrm{Gal}(K/k) = \mathfrak{g}/\mathfrak{h}$, on ait

(2.0.5)   le $\mathfrak{h}$-module $\mathrm{Pic}\,\overline{X}$ stablement de permutation;

cette condition est en particulier vérifiée si X est K-rationnelle (appendice 2.A). Un torseur $\mathscr{T}$ sur X est dit *quasi-universel* s'il a pour type l'injection naturelle

$$i: \mathrm{Pic}\,X_K \to \mathrm{Pic}\,\overline{X}$$

pour K/k vérifiant (2.0.5); plus généralement, un torseur $\mathscr{T}$ sur X sous le k-groupe de type multiplicatif S, trivialisé par K, est dit *quasi-admissible* s'il a pour type l'application composée

$$\hat{S} \xrightarrow{\quad j \quad} \mathrm{Pic}\,X_K \xrightarrow{\quad i \quad} \mathrm{Pic}\,\overline{X}$$

d'une injection j à conoyau G-inversible (= facteur direct d'un G-module de permutation) et de l'injection naturelle i, où K/k vérifie (2.0.5).

2.1. *Propriétés globales des torseurs sous des tores.*

PROPOSITION 2.1.1.   *Soient X une k-variété algébrique lisse et géométriquement intègre, S un k-tore et $\mathscr{T}$ un torseur sur X sous S de type λ. Soient K/k une sous-extension de $\overline{k}/k$ et $\mathfrak{h} = \mathrm{Gal}(\overline{k}/K)$. On a alors la suite exacte de groupes abéliens*:

(2.1.1) $1 \to K[X]^* \to K[\mathscr{T}]^* \to \hat{S}^{\mathfrak{h}} \to \mathrm{Pic}\,X_K \to \mathrm{Pic}\,\mathscr{T}_K \to H^1(K, \hat{S}),$

fonctorielle en K. Lorsque $K = \overline{k}$, le morphisme

$$\hat{S} \to \mathrm{Pic}\,\overline{X}$$

ainsi obtenu est, au signe près, le type de $\mathscr{T}$.

Démonstration.   Soit $q: \mathscr{T} \to X$ le morphisme structural. D'après la proposition 1.4.2, on a l'extension de faisceaux étales sur X:

$(E_{\mathscr{T}})$ $1 \to \mathbf{G}_{m,X} \to q_* \mathbf{G}_{m,\mathscr{T}} \to \hat{S} \to 0.$

Il est clair que la longue suite exacte de cohomologie obtenue à partir de $(E_{\mathscr{T}})$

par application de $H^0(X_K, \cdot)$ est fonctorielle en K. Il suffit donc de traiter le cas $K = k$. La longue suite de cohomologie commence alors par

$$1 \to k[X]^* \to k[\mathscr{T}]^* \to H^0\big(X, \hat{S}_X\big) \to \mathrm{Pic}\, X \to H^1\big(X, q_*\mathsf{G}_{m,\mathscr{T}}\big) \to H^1\big(X, \hat{S}_X\big)$$

Comme X est géométriquement connexe, on a l'isomorphisme de faisceaux étales sur $\mathrm{Spec}\, k$:

$$\hat{S} \xrightarrow{\;\cong\;} p_*\hat{S}_X$$

comme on le voit en passant à $\bar{k}$, par réduction au cas $\hat{S} = \mathsf{Z}$. On en déduit l'isomorphisme

$$\hat{S}^{\mathfrak{g}} = H^0(k, \hat{S}) \cong H^0\big(k, p_*\hat{S}_X\big) = H^0\big(X, \hat{S}_X\big).$$

Par ailleurs,

$$R^1 p_* \hat{S}_X = 0,$$

comme on le voit en calculant la fibre géométrique de ce faisceau:

$$H^1\big(\overline{X}, \hat{S}_{\overline{X}}\big) = 0$$

car $H^1(Y, \mathsf{Z}) = 0$ pour tout schéma normal intègre Y ([1] IX 3.6(ii)). Compte tenu de l'isomorphisme $\hat{S} \cong p_*\hat{S}_X$, la suite spectrale

$$H^p\big(k, R^q p_* \hat{S}_X\big) \Rightarrow H^{p+q}\big(X, \hat{S}_X\big)$$

donne alors les isomorphismes

$$H^1(k, \hat{S}) \cong H^1\big(k, p_*\hat{S}_X\big) \cong H^1\big(X, \hat{S}_X\big).$$

Il reste à calculer le terme $H^1(X, q_*\mathsf{G}_{m,\mathscr{T}})$. La suite spectrale

$$H^p\big(X, R^q q_*\mathsf{G}_{m,\mathscr{T}}\big) \Rightarrow H^{p+q}\big(\mathscr{T}, \mathsf{G}_{m,\mathscr{T}}\big)$$

donne la suite exacte

$$1 \to H^1\big(X, q_*\mathsf{G}_{m,\mathscr{T}}\big) \to H^1\big(\mathscr{T}, \mathsf{G}_{m,\mathscr{T}}\big) \to H^0\big(X, R^1 q_*\mathsf{G}_{m,\mathscr{T}}\big).$$

Montrons que le faisceau $R^1 q_*\mathsf{G}_{m,\mathscr{T}}$ est nul. Il suffit (cf. [1] VIII 5.2) de montrer que sa fibre en un localisé strict $Y = \mathrm{Spec}\, \mathscr{O}_{X,\xi}$ en un point géométrique ξ de X est nulle. Compte tenu de [1], VII 5.9 et 5.7, cette fibre est égale à $H^1(\mathscr{T}\times_X Y, \mathsf{G}_m)$. Au-dessus de Y, le tore $S_X \times_X Y$ est isomorphe à un $\mathsf{G}^n_{m,Y}$ et, comme Y est le spectre d'un anneau local, $\mathrm{Pic}\, Y = 0$, le torseur $\mathscr{T}\times_X Y$ est donc trivial, et ainsi isomorphe à $\mathsf{G}^n_{m,Y}$. Comme X est normal, Y l'est aussi et la flèche naturelle

$$\mathrm{Pic}\, Y \to \mathrm{Pic}\big(\mathsf{G}^n_{m,Y}\big)$$

est un isomorphisme ([33] Errata 4ème partie, 21.4.13), d'où finalement $\mathrm{Pic}(\mathscr{T}\times_X Y) = 0$, ce qui établit la nullité de $R^1 q_* \mathsf{G}_{m,\mathscr{T}}$. En résumé:

$$H^1(X, q_* \mathsf{G}_{m,\mathscr{T}}) \xrightarrow{\ \simeq\ } H^1(\mathscr{T}, \mathsf{G}_{m,\mathscr{T}}).$$

On a ainsi obtenu la suite exacte (2.1.1) pour $K = k$. On vérifie aisément que la flèche $\mathrm{Pic}\, X \to \mathrm{Pic}\, \mathscr{T}$ est la flèche naturelle. Lorsque $K = \bar{k}$, il reste à identifier la flèche

$$\hat{S} \to \mathrm{Pic}\, \overline{X}.$$

Or on a vu (proposition 1.5.2(iii)) que la flèche composée

$$\hat{S} \to p_* \hat{S}_X \to R^1 p_* \mathsf{G}_{m,X},$$

une fois identifiée au morphisme de $\mathfrak{g}$-modules

$$\hat{S} \to \mathrm{Pic}\, \overline{X},$$

coïncide, au signe près, avec le type du torseur $\mathscr{T}$. Or, la flèche bord

$$\partial: p_* \hat{S}_X \to R^1 p_* \mathsf{G}_{m,X},$$

traduite en termes de $\mathfrak{g}$-modules, n'est autre que la flèche bord

$$\partial: H^0(\overline{X}, \hat{S}_{\overline{X}}) \to H^1(\overline{X}, \mathsf{G}_{m,\overline{X}})$$

déduite de la suite exacte initiale.

THÉORÈME 2.1.2. *Soit X une k-variété algébrique vérifiant les conditions (2.0.1) et (2.0.3). Soient $\mathscr{T}$ un torseur sur X sous un k-tore S et $\mathscr{T}^c$ une k-compactification lisse de $\mathscr{T}$.*

(a) Si $\mathscr{T}$ est un torseur universel, ou même un torseur admissible, le $\mathfrak{g}$-module $\mathrm{Pic}\, \overline{\mathscr{T}}^c$ est de permutation.

(b) Si $\mathscr{T}$ est un torseur quasi-admissible, relativement à une extension algébrique K/k telle qu'on ait (2.0.5) et $X(K) \neq \varnothing$, le G-module $\mathrm{Pic}\, \mathscr{T}_K^c$ est inversible, et

$$H^1(\mathfrak{g}, \mathrm{Pic}\, \overline{\mathscr{T}}^c) = 0.$$

Démonstration. On démontre (a) et (b) en parallèle, le cas (a) correspondant aux hypothèses particulières: $K = \bar{k}$, $G = \mathfrak{g}$ et le conoyau de λ est de permutation. On commence par montrer les deux résultats suivants:

(i) $K^* = K[\mathscr{T}]^*$

(ii) le G-module $\mathrm{Pic}\, \mathscr{T}_K$ est inversible.

Comme S_K est un K-tore trivial, on a

$$\hat{S}^{\flat} = \hat{S} \quad \text{et} \quad H^1(K, \hat{S}) = 0,$$

et la suite exacte (2.1.1) donne donc l'égalité (i) et la suite exacte

$$0 \to \hat{S} \to \operatorname{Pic} X_K \to \operatorname{Pic} \mathscr{T}_K \to 0,$$

ce qui établit (ii) puisque $\mathscr{T}$ est supposé quasi-admissible relativement à K/k; de plus, ceci donne que, dans le cas (a), le $\mathfrak{g}$-module $\operatorname{Pic} \bar{\mathscr{T}}$ est de permutation. Soit alors Y le fermé complémentaire de $\mathscr{T}$ dans $\mathscr{T}^c$. La suite exacte de G-modules

$$0 \to K[\mathscr{T}]^*/K[\mathscr{T}^c]^* \to \operatorname{Div}_{Y_K} \mathscr{T}_K^c \to \operatorname{Pic} \mathscr{T}_K^c \to \operatorname{Pic} \mathscr{T}_K \to 0$$

se réduit d'après (i) et (ii) à l'extension suivante de G-modules

$$0 \to \operatorname{Div}_{Y_K} \mathscr{T}_K^c \to \operatorname{Pic} \mathscr{T}_K^c \to \operatorname{Pic} \mathscr{T}_K \to 0$$

qui est scindée, car c'est l'extension d'un module inversible par un module de permutation. Ceci prouve que le G-module $\operatorname{Pic} \mathscr{T}_K^c$ est inversible, et achève la preuve de (a). On va ensuite établir

(iii) $\operatorname{Pic} \mathscr{T}_K^c = (\operatorname{Pic} \bar{\mathscr{T}}^c)^{\flat}$.

Ceci vaut plus généralement pour une K-variété Z telle que $Z(K) \neq \varnothing$ et que $K[Z]^* = K^*$. En effet la suite exacte des termes de bas degré de la suite spectrale de Leray pour $p: Z \to \operatorname{Spec} K$

$$H^p(K, R^q p_* \mathbf{G}_{m, Z}) \Rightarrow H^{p+q}(Z, \mathbf{G}_{m, Z})$$

s'écrit

$$0 \to H^1(\mathfrak{h}, \bar{k}[Z]^*) \to \operatorname{Pic} Z \to (\operatorname{Pic} \bar{Z})^{\flat} \to H^2(\mathfrak{h}, \bar{k}[Z]^*) \to H^2(Z, \mathbf{G}_m).$$

Or l'hypothèse et le théorème 90 de Hilbert impliquent

$$H^1(\mathfrak{h}, \bar{k}[Z]^*) = H^1(\mathfrak{h}, \bar{k}^*) = 0$$

et l'injectivité de l'application

$$H^2(K, \mathbf{G}_m) = H^2(\mathfrak{h}, \bar{k}^*) = H^2(\mathfrak{h}, \bar{k}[Z]^*) \to H^2(Z, \mathbf{G}_m)$$

puisque tout point rationnel de Z en définit une rétraction. Il reste à voir que $\mathscr{T}_K^c$ vérifie les hypothèses indiquées pour Z: comme S_K est un K-tore trivial, la fibre du torseur $\mathscr{T}_K$ en un point de $X(K) \neq \varnothing$ est triviale (théorème 90 de Hilbert), et donc $\mathcal{T}_K(K)$ est non vide. A fortiori $\mathcal{T}_K^c(K) \neq \varnothing$, et évidemment $K[\mathcal{T}_K^c]^* = K^*$, ce qui achève de prouver (iii).

On établit enfin

(iv) le $\mathfrak{h}$-module $\operatorname{Pic} \bar{\mathcal{T}}^c$ est stablement de permutation.

Comme le torseur $\mathcal{T}_K$ est localement trivial sur X_K pour la topologie de Zariski, puisque S_K est un K-tore trivial (théorème 90 de Hilbert), la K-variété $\mathcal{T}_K$ est K-birationnelle à $X_K \times_K \mathsf{P}_K^n$ pour un certain entier n: on en déduit (cf. appendice 2.A et [20] lemme 11) l'existence d'un isomorphisme de $\mathfrak{h}$-modules

$$P_1 \oplus \operatorname{Pic} \bar{\mathcal{T}}^c \cong P_2 \oplus \operatorname{Pic}\left(\bar{X} \times_{\bar{k}} \mathsf{P}_{\bar{k}}^n \right) \cong P_2 \oplus \mathsf{Z} \oplus \operatorname{Pic} \bar{X}$$

où P_1 et P_2 sont des modules de permutation, ce qui d'après l'hypothèse (2.0.5) prouve (iv). La suite exacte de restriction-inflation

$$0 \to H^1\left(G, (\operatorname{Pic} \bar{\mathcal{T}}^c)^{\mathfrak{h}} \right) \to H^1\left(\mathfrak{g}, \operatorname{Pic} \bar{\mathcal{T}}^c \right) \to H^1\left(\mathfrak{h}, \operatorname{Pic} \bar{\mathcal{T}}^c \right) = 0$$

donne alors la conclusion

$$H^1\left(\mathfrak{g}, \operatorname{Pic} \bar{\mathcal{T}}^c \right) = 0.$$

Remarque 2.1.3.   *Le résultat* (b) *s'applique aux k-variétés* X, *propres, lisses et géométriquement intègres telles que* X_K *soit K-rationnelle.* Il est intéressant de noter que, même dans ce cas, le $\mathfrak{g}$-module $\operatorname{Pic} \bar{\mathcal{T}}^c$ n'est pas forcément de permutation, ni même inversible, contrairement à la situation décrite en (a). De fait, il peut se produire qu'on ait $H^1(L, \operatorname{Pic} \bar{\mathcal{T}}^c) \neq 0$ pour une extension convenable L/k, comme on le voit en utilisant la proposition 6.1 de [13] et la description locale des torseurs donnée ci-après, alors que néanmoins $H^1(k, \operatorname{Pic} \bar{\mathcal{T}}^c) = 0$!

Remarque 2.1.4.   Pour X propre sur k, tout torseur $\mathcal{T}$ sur X sous un k-tore S admet une k-compactification lisse, même en caractéristique > 0: J.-L. Brylinski [5] a en effet établi l'existence d'une k-compactification lisse équivariante $S \to S^c$ pour tout k-tore S, et il suffit de prendre le produit contracté

$$\mathcal{T}^c = \mathcal{T} \times^S S^c,$$

ce qui compactifie dans les fibres de $\mathcal{T} \to X$.

2.2. *L'obstruction élémentaire: condition nécessaire d'existence de torseurs d'un type donné.*

Définition 2.2.1.   Soit X une k-variété algébrique lisse et géométriquement intègre. On appelle *obstruction élémentaire* la classe dans $\operatorname{Ext}^1_{\mathfrak{g}}(\bar{k}(X)^*/\bar{k}^*, \bar{k}^*)$ de l'extension définie par la suite exacte naturelle

$$(2.2.1) \qquad\qquad 1 \to \bar{k}^* \to \bar{k}(X)^* \to \bar{k}(X)^*/\bar{k}^* \to 1.$$

La proposition qui suit montre que la non-nullité de cette classe est une obstruction à l'existence d'un point rationnel pour X. Si l'obstruction élémentaire est nulle, on dira aussi qu'elle est " vide".

PROPOSITION 2.2.2. *Soit X une k-variété algébrique lisse et géométriquement intègre.*

(a) *Si X possède un point k-rationnel, la suite exacte de $\mathfrak{g}$-modules (2.2.1) est scindée; il en est de même, si U est un ouvert de Zariski non vide de X, de la suite exacte de $\mathfrak{g}$-modules*

$$(2.2.2) \qquad 1 \to \overline{k}^* \to \overline{k}[U]^* \to \overline{k}[U]^*/\overline{k}^* \to 1.$$

(b) *Si k est parfait et si X est quasi-projective, les extensions (2.2.1) et (2.2.2) sont triviales dès que X possède un 0-cycle de degré 1.*

Démonstration. Soit $P \in X(k)$. Soit $\mathcal{O}_{\overline{X}, P}$ l'anneau local de $\overline{X}$ en P. Soit $\mathrm{Div}_P \overline{X}$ le groupe abélien libre sur les points de codimension 1 de $\overline{X}$ dont l'adhérence contient P: c'est un $\mathfrak{g}$-module somme directe de $\mathfrak{g}$-modules du type $\mathbf{Z}[\mathfrak{g}/\mathfrak{h}]$, pour $\mathfrak{h}$ sous-groupe ouvert de $\mathfrak{g}$, i.e. c'est un $\mathfrak{g}$-module de permutation, mais non de type fini sur $\mathbf{Z}$. Comme X est lisse en P, on a la suite exacte de $\mathfrak{g}$-modules

$$1 \to \mathcal{O}^*_{\overline{X}, P} \to \overline{k}(X)^* \to \mathrm{Div}_P \overline{X} \to 0.$$

Pour tout sous-groupe ouvert $\mathfrak{h}$ de $\mathfrak{g}$, une variante du théorème 90 de Hilbert donne

$$H^1\big(\mathfrak{h}, \mathcal{O}^*_{\overline{X}, P}\big) = 0,$$

d'où par le lemme de Shapiro

$$\mathrm{Ext}^1_{\mathfrak{g}}\big(\mathbf{Z}[\mathfrak{g}/\mathfrak{h}], \mathcal{O}^*_{\overline{X}, P}\big) = 0,$$

ce qui prouve que la suite exacte de $\mathfrak{g}$-modules ci-dessus est scindée (pour une variante de cet argument, voir [21]). Autrement dit, le $\mathfrak{g}$-morphisme naturel

$$\mathcal{O}^*_{\overline{X}, P} \to \overline{k}(X)^*$$

admet une $\mathfrak{g}$-rétraction. Comme le point P définit évidemment une $\mathfrak{g}$-rétraction de la flèche naturelle

$$\overline{k}^* \to \mathcal{O}^*_{\overline{X}, P},$$

on en conclut que le morphisme naturel

$$\overline{k}^* \to \overline{k}(X)^*$$

admet une $\mathfrak{g}$-rétraction; il en est a fortiori de même des morphismes

$$\bar{k}^* \to \bar{k}[U]^*$$

pour tout ouvert non vide U de X, ce qui achève la preuve de (a). Le point (b) s'établit de façon analogue en considérant l'anneau semi-local de $\bar{X}$ aux points du support d'un 0-cycle de degré 1.

Remarque 2.2.3.   L'assertion (a) reste vraie pourvu que $X_{\text{lisse}}(k) \neq \varnothing$. L'obstruction élémentaire est donc une obstruction à l'existence d'un point rationnel lisse, qui, par sa définition même, présente un caractère k-birationnel, et qui, d'après (b), ne peut faire la différence entre les k-variétés qui possèdent un 0-cycle de degré 1 et celles qui possèdent un point rationnel.

PROPOSITION 2.2.4.   *Soit X une k-variété algébrique vérifiant* (2.0.1). *L'obstruction élémentaire est nulle si et seulement si la classe de la 2-extension naturelle*

$$(2.2.4) \qquad 1 \to \bar{k}^* \to \bar{k}(X)^* \to \operatorname{Div}\bar{X} \to \operatorname{Pic}\bar{X} \to 0$$

est nulle dans $\operatorname{Ext}^2_{\mathfrak{g}}(\operatorname{Pic}\bar{X}, \bar{k}^*)$.

Démonstration.   La suite exacte (2.2.4) donne aussitôt la suite exacte d'Ext

$$\operatorname{Ext}^1_{\mathfrak{g}}(\operatorname{Div}\bar{X}, \bar{k}^*) \to \operatorname{Ext}^1_{\mathfrak{g}}(\bar{k}(X)^*/\bar{k}^*, \bar{k}^*) \xrightarrow{\partial} \operatorname{Ext}^2_{\mathfrak{g}}(\operatorname{Pic}\bar{X}, \bar{k}^*).$$

Comme ∂ est le cup-produit par la classe de l'extension

$$1 \to \bar{k}(X)^*/\bar{k}^* \to \operatorname{Div}\bar{X} \to \operatorname{Pic}\bar{X} \to 0,$$

l'obstruction élémentaire a pour image par ∂ la classe de l'extension (2.2.4). Comme X est lisse, $\operatorname{Div}\bar{X}$ est un $\mathfrak{g}$-module de permutation, d'où par application du lemme de Shapiro et du théorème 90 de Hilbert:

$$\operatorname{Ext}^1_{\mathfrak{g}}(\operatorname{Div}\bar{X}, \bar{k}^*) = 0,$$

ce qui permet de conclure.

PROPOSITION 2.2.5.   *Soit X une k-variété algébrique lisse et géométriquement intègre telle que l'obstruction élémentaire soit nulle. Si S est un k-groupe de type multiplicatif, les applications naturelles*

$$H^i(k, S) \to H^i(X, S) \quad et \quad H^i(k, S) \to H^i(k(X), S)$$

sont injectives pour $i = 0, 1$ et 2. En particulier, on a des injections naturelles

$$\operatorname{Br}k \to \operatorname{Br}X \quad et \quad \operatorname{Br}k \to \operatorname{Br}k(X).$$

Démonstration. Noter que, si $X(k) \neq \varnothing$, les résultats relatifs à X sont immédiats. En général, il suffit d'établir le résultats relatifs à $k(X)$. Le cas $i = 0$ est trivial. Pour tout ouvert non vide U de X, on a la suite exacte (1.5.1):

$$0 \to \mathrm{Ext}^1_{\mathfrak{g}}\big(\hat{S}, \bar{k}[U]^*\big) \to H^1(U, S) \to \mathrm{Hom}_{\mathfrak{g}}\big(\hat{S}, \mathrm{Pic}\,\overline{U}\big)$$

$$\to \mathrm{Ext}^2_{\mathfrak{g}}\big(\hat{S}, \bar{k}[U]^*\big) \to H^2(U, S),$$

qui, par passage à la limite inductive sur U, donne la suite exacte

$$0 \to \mathrm{Ext}^1_{\mathfrak{g}}\big(\hat{S}, \bar{k}(X)^*\big) \to H^1(k(X), S) \to 0 \to \mathrm{Ext}^2_{\mathfrak{g}}\big(\hat{S}, \bar{k}(X)^*\big) \to H^2(k(X), S).$$

On en déduit le résultat par fonctorialité (proposition 1.5.2(i)) de (1.5.1) appliquée à $X \to \mathrm{Spec}\,k$. L'hypothèse de nullité de l'obstruction élémentaire assure que toute $\mathfrak{g}$-rétraction σ de $\bar{k}^* \to \bar{k}(X)^*$ induit une $\mathfrak{g}$-rétraction de chaque morphisme naturel $\mathrm{Ext}^i_{\mathfrak{g}}\big(\hat{S}, \bar{k}(X)^*\big) \to H^i(k(X), S)$. On a donc pour $i = 1$ le résultat supplémentaire suivant: toute $\mathfrak{g}$-rétraction σ définit une $\mathfrak{g}$-rétraction du morphisme naturel

$$H^1(k, S) \to H^1(k(X), S), \quad \text{donc aussi de } H^1(k, S) \to H^1(X, S).$$

L'énoncé pour le groupe de Brauer correspond au cas $S = \mathsf{G}_{m,\,k}$.

Exemple 2.2.6. Soit X une k-variété lisse et géométriquement intègre et soit K/k une extension finie séparable. S'il existe $\alpha \in k^*$ tel que

$$\alpha \notin N_{K/k}K^*, \quad \text{mais } \alpha \in N_{K(X)/k(X)}K(X)^*,$$

alors $X(k) = \varnothing$. Il suffit d'appliquer la proposition précédente en considérant le tore $S = R^1_{K/k}\mathsf{G}_m$. On notera que si $\alpha = N_{K(X)/k(X)}(g)$ avec $g \in K(X)^*$, il est immédiat que X ne possède pas de point rationnel en dehors du diviseur de g, mais la proposition assure qu'elle n'en possède pas non plus sur le support de $\mathrm{div}(g)$.

Remarque 2.2.7. L'obstruction élémentaire est nulle pour une k-variété vérifiant (2.0.1) et contenant un ouvert de Zariski U non vide tel que

$$\bar{k}^* = \bar{k}[U]^* \quad \text{et} \quad \mathrm{Pic}\,\overline{U} = 0.$$

En effet, on dispose alors de la suite exacte de $\mathfrak{g}$-modules

$$1 \to \bar{k}^* \to \bar{k}(U)^* \to \mathrm{Div}\,\overline{U} \to 0,$$

qui est $\mathfrak{g}$-scindée (lemme de Shapiro et théorème 90).

La proposition suivante montre la relation entre l'obstruction élémentaire et l'existence de torseurs d'un type donné:

PROPOSITION 2.2.8. *Soit X une k-variété algébrique vérifiant (2.0.1). Soient S un k-groupe de type multiplicatif,*

$$\lambda \in \mathrm{Hom}_{\mathfrak{g}}(\hat{S}, \mathrm{Pic}\,\overline{X}) \quad et \quad \partial \colon \mathrm{Hom}_{\mathfrak{g}}(\hat{S}, \mathrm{Pic}\,\overline{X}) \to H^2(k, S)$$

le bord dans la suite exacte (1.5.1).

(i) *La non-nullité de $\partial(\lambda) \in H^2(k, S)$ est une obstruction $(\mathscr{O}'_\lambda)$ à l'existence d'un point k-rationnel sur X, et même, si k est parfait, à l'existence d'un 0-cycle de degré 1.*

(ii) *Cette obstruction est nulle si, et seulement si, X possède un torseur de type λ. C'est le cas si l'obstruction élémentaire est nulle.*

(iii) *Si $\mathrm{Pic}\,\overline{X}$ est de type fini sur $\mathbf{Z}$ et si*

$$\lambda_0 = \mathrm{id}_{\mathrm{Pic}\,\overline{X}},$$

l'obstruction $(\mathscr{O}'_{\lambda_0})$ est la plus fine de ces obstructions $(\mathscr{O}'_\lambda)$, et elle coïncide avec l'obstruction élémentaire.

(iv) *Si l'obstruction élémentaire est nulle, toute $\mathfrak{g}$-rétraction σ du morphisme évident*

$$\overline{k}^* \to \overline{k}(X)^*$$

définit un scindage de la suite exacte

$$(2.2.5) \qquad 0 \to H^1(k, S) \to H^1(X, S) \to \mathrm{Hom}_{\mathfrak{g}}(\hat{S}, \mathrm{Pic}\,\overline{X}) \to 0.$$

(v) *Soit U un ouvert de Zariski non vide de X. Soient*

$$P_{\overline{X}, \overline{U}} \coloneqq \ker(\mathrm{Pic}\,\overline{X} \to \mathrm{Pic}\,\overline{U})$$

et

$$\lambda \colon P_{\overline{X}, \overline{U}} \to \mathrm{Pic}\,\overline{X}$$

l'injection naturelle. L'obstruction $(\mathscr{O}'_\lambda)$ est nulle si et seulement si la suite évidente de $\mathfrak{g}$-modules

$$1 \to \overline{k}^* \to \overline{k}[U]^* \to \overline{k}[U]^*/\overline{k}^* \to 1$$

est scindée.

Démonstration.

(i) D'après 1.5.2(iv), $\partial(\lambda)$ est la classe dans $H^2(k, S) = \mathrm{Ext}^2_{\mathfrak{g}}(\hat{S}, \overline{k}^*)$ de la 2-extension déduite de

$$(2.2.4) \qquad 1 \to \overline{k}^* \to \overline{k}(X)^* \to \mathrm{Div}\,\overline{X} \to \mathrm{Pic}\,\overline{X} \to 0$$

via λ. Si $\partial(\lambda) \neq 0$, la 2-extension ci-dessus est a fortiori non triviale, d'où la conclusion d'après 2.2.4.

(ii) Compte tenu de la suite exacte (2.0.2), c'est une conséquence immédiate des définitions et de (i).

(iii) L'hypothèse permet de considérer le k-groupe de type multiplicatif

$$S_0 := D(\mathrm{Pic}\,\overline{X}).$$

Par fonctorialité de la suite exacte (2.0.2), $\partial(\lambda)$ est l'image de $\partial(\lambda_0)$ par le morphisme $D(\lambda)\colon S_0 \to S$. La nullité de $\partial(\lambda_0)$ implique donc celle de tous les $\partial(\lambda)$. D'autre part (cf. (i)), d'après 1.5.2(iv), $\partial(\lambda_0)$ n'est autre que la classe de la 2-extension (2.2.4). La conclusion suit alors de la proposition 2.2.4.

(iv) Par définition même, la nullité de l'obstruction élémentaire équivaut à l'existence d'une $\mathfrak{g}$-rétraction

$$\sigma\colon \overline{k}(X)^* \to \overline{k}^*,$$

qui induit donc (cf. 2.2.5) des rétractions de i_1 et i_2 dans la suite exacte (2.0.2); celle-ci donne alors la suite exacte scindée

$$0 \to H^1(k, S) \to H^1(X, S) \to \mathrm{Hom}_{\mathfrak{g}}(\hat{S}, \mathrm{Pic}\,\overline{X}) \to 0.$$

(v) Considérons le k-groupe de type multiplicatif

$$S := D(P_{\overline{X}, \overline{U}})$$

et l'injection naturelle $\lambda\colon P_{\overline{X}, \overline{U}} \to \mathrm{Pic}\,\overline{X}$. On a donc $\lambda \in \mathrm{Hom}_{\mathfrak{g}}(\hat{S}, \mathrm{Pic}\,\overline{X})$. La description de $\partial(\lambda)$ montre alors que c'est la classe dans $H^2(k, S) = \mathrm{Ext}^2_{\mathfrak{g}}(P_{\overline{X}, \overline{U}}, \overline{k}^*)$ de la 2-extension

$$1 \to \overline{k}^* \to \overline{k}[U]^* \to \mathrm{Div}_{\overline{Y}}\overline{X} \to P_{\overline{X}, \overline{U}} \to 0,$$

où Y désigne le fermé complémentaire de U dans X (c'est le pull-back de (2.2.4) par λ). L'extension

$$1 \to \overline{k}[U]^*/\overline{k}^* \to \mathrm{Div}_{\overline{Y}}\overline{X} \to P_{\overline{X}, \overline{U}} \to 0$$

définit la suite exacte

$$0 = \mathrm{Ext}^1_{\mathfrak{g}}\big(\mathrm{Div}_{\overline{Y}}\overline{X}, \overline{k}^*\big) \to \mathrm{Ext}^1_{\mathfrak{g}}\big(\overline{k}[U]^*/\overline{k}^*, \overline{k}^*\big) \xrightarrow{\delta} \mathrm{Ext}^2_{\mathfrak{g}}\big(P_{\overline{X}, \overline{U}}, \overline{k}^*\big)$$

dont le premier termes est nul par lissité de X (lemme de Shapiro et théorème 90 de Hilbert). D'après ce qui précède, $\partial(\lambda)$ est, au signe près, l'image par δ de la classe de l'extension évidente

$$1 \to \overline{k}^* \to \overline{k}[U]^* \to \overline{k}[U]^*/\overline{k}^* \to 1,$$

dont la non-nullité équivaut donc à celle de $\partial(\lambda)$, i.e. à celle de $(\mathcal{O}'_\lambda)$.

Remarque 2.2.9. Si l'obstruction élémentaire est nulle, par exemple si $X(k) \neq \varnothing$, il existe des torseurs de tout type λ. Inversement, si Pic $\overline{X}$ est de type fini, l'existence de torseurs universels entraîne la nullité de l'obstruction élémentaire.

Si k est un corps $\mathfrak{p}$-adique, les théorèmes de dualité du corps de classes local entraînent que l'obstruction élémentaire équivaut, dans ce cas, au fait que la flèche naturelle

$$\mathrm{Br}\, k \to \mathrm{Br}\, X$$

est injective.

Exemple 2.2.10. Soient k un corps de caractéristique $\neq 2$ et U la k-variété lisse et géométriquement intègre définie par l'équation

$$0 \neq y^2 - az^2 = P(x)$$

où x, y, z sont les variables, où $a \in k^*$ est donné et où $P \in k[x]$ est un polynôme séparable non constant donné. Soit $X \supset U$ une k-compactification lisse de U. Soient c le coefficient dominant de P et A la k-algèbre finie séparable produit de $k[t]/(t^2 - a)$ et de $k[x]/(P)$. On vérifie, via la proposition 2.2.8(v), que l'obstruction élémentaire pour X est vide si, et seulement si, c est une norme d'un élément de A^*. Elle est en particulier toujours vide si P est de degré impair.

Exemples 2.2.11. Ce sont des exemples où l'obstruction élémentaire est la seule obstruction à l'existence d'un point rationnel pour la k-variété algébrique X (on suppose k parfait):

(a) les variétés de Severi-Brauer (formes sur k de l'espace projectif);

(b) les formes sur k d'une surface rationnelle propre et lisse, relativement minimale sur $\overline{k}$ (formes sur k de $\mathsf{P}^2, \mathsf{P}^1 \times \mathsf{P}^1$ ou F_n, pour $n \geqslant 2$);

(c) les k-compactifications lisses d'espaces homogènes principaux de tores;

(d) les surfaces de Del Pezzo de degré 6 (formes sur k de l'éclaté de P^2 en trois points non alignés).

On a en fait plus, car, dans les cas (a), (b), (d), si $X(k) \neq \varnothing$, alors X est k-rationnelle. Par ailleurs, si E^c est une k-compactification lisse d'un espace principal homogène E sous un k-tore, on peut montrer en utilisant 2.2.8(v) que, si l'obstruction élémentaire est nulle, alors E est trivial. Ainsi, $E^c(k) \neq \varnothing$ implique $E(k) \neq \varnothing$.

Exemples 2.2.12. Voici au contraire des exemples où l'obstruction élémentaire est nulle et où pourtant $X(k) = \varnothing$:

(a) une quadrique lisse dans $\mathbf{P}_{\mathbf{R}}^n$ sans point réel, pour $n \geqslant 4$;

(b) une k-compactification lisse d'un espace homogène principal non trivial sous $\mathrm{SL}(D)$ où D est une k-algèbre centrale simple pour laquelle la norme réduite Nrd: $D^* \to k^*$ n'est pas surjective;

(c) la surface cubique X d'équation

$$x^3 + 2y^3 + 4z^3 + 9t^3 = 0$$

considérée sur $k = \mathbf{Q}_3$: tout d'abord $X(k) = \varnothing$ par congruences mod 9; ensuite Pic $X = (\mathrm{Pic}\,\overline{X})^{\mathfrak{g}}$ car, X étant k-minimale (cf. [42] III 4.12 (= III 21.10)), $(\mathrm{Pic}\,\overline{X})^{\mathfrak{g}} = \mathbf{Z}\omega_X = \mathrm{Pic}\,X$ où ω_X est la classe canonique ([42] IV 6.1 (= IV 23.6)); il résulte alors de (2.0.2) pour $S = \mathbf{G}_{m,\,k}$ que Br k s'injecte dans Br X, ce qui équivaut, pour un corps $\mathfrak{p}$-adique, à la nullité de l'obstruction élémentaire (cf. remarque 2.2.9);

(d) certaines surfaces rationnelles sur $\mathbf{Q}$ ou $\mathbf{Q}_p$ [12] possédant un 0-cycle de degré 1 sans posséder de point rationnel.

2.3. *Description locale des torseurs.* Soit X une k-variété algébrique qui vérifie (2.0.1), c'est-à-dire qui soit lisse, géométriquement intègre et telle que $\overline{k}[X]^* = \overline{k}^*$. Soient S un k-groupe algébrique de type multiplicatif et

$$\lambda \colon \hat{S} \to \mathrm{Pic}\,\overline{X}$$

un $\mathfrak{g}$-homomorphisme. Soit alors U un ouvert de Zariski non vide de X tel que λ se factorise par

$$P_{\overline{X},\,\overline{U}} := \ker(\mathrm{Pic}\,\overline{X} \to \mathrm{Pic}\,\overline{U}).$$

Soit en outre

$$(2.3.1) \qquad 1 \to S \to M \to R \to 1 \qquad\qquad (\aleph)$$

une suite exacte de k-groupes algébriques de type multiplicatif où M soit un k-tore quasi-trivial et telle qu'il existe un diagramme commutatif de $\mathfrak{g}$-modules

$$(2.3.2) \quad \begin{array}{ccccccccc} 1 & \to & \overline{k}[U]^*/\overline{k}^* & \to & \mathrm{Div}_{\overline{Y}}\overline{X} & \to & P_{\overline{X},\,\overline{U}} & \to & 0 \\ & & {\scriptstyle\rho}\uparrow & & \uparrow & & \uparrow{\scriptstyle\lambda} & & \\ 0 & \longrightarrow & \hat{R} & \longrightarrow & \hat{M} & \longrightarrow & \hat{S} & \longrightarrow & 0 \end{array}$$

où Y désigne le fermé $X - U$. La suite exacte (2.3.1) fait de M un torseur $(\aleph)$ sur R sous S. Tout k-morphisme $\phi \colon U \to R$ définit un $\mathfrak{g}$-morphisme

$$\hat{\phi} \colon \hat{R} \to \overline{k}[U]^*,$$

et inversement, autrement dit:

$$H^0(U, R) = \mathrm{Hom}_{\mathfrak{g}}(\hat{R}, \overline{k}[U]^*).$$

Tout k-morphisme $\phi\colon U \to R$ définit par pull-back de l'extension (2.3.1) un torseur $\phi^*(\aleph)$ sur U sous S modelé sur le torseur-type $(\aleph)$.

Ayant ainsi fixé λ, U, $(\aleph)$ et ρ, considérons l'ensemble

$$(2.3.3) \qquad\qquad \mathrm{Tors}(U, S; \aleph, \rho)$$

des classes d'isomorphisme de tels torseurs $\phi^*(\aleph)$ lorsque le comorphisme

$$\hat{\phi}\colon \hat{R} \to \overline{k}[U]^* \text{ est un relèvement de } \rho.$$

On note

$$(2.3.4) \qquad\qquad \mathrm{Tors}(X, S; \lambda)$$

l'ensemble des classes d'isomorphisme de torseurs sur X sous S de type λ.

THÉORÈME 2.3.1. *On garde les hypothèses et notations ci-dessus. La restriction de X à U définit une surjection*

$$\gamma\colon \mathrm{Tors}(X, S; \lambda) \to \mathrm{Tors}(U, S; \aleph, \rho).$$

C'est une bijection si, et seulement si, la flèche naturelle

$$\mathrm{Ext}^1_{\mathfrak{g}}(\hat{S}, \overline{k}^*) \to \mathrm{Ext}^1_{\mathfrak{g}}(\hat{S}, \overline{k}[U]^*)$$

est une injection, ce qui est le cas si le morphisme $\overline{k}^* \to \overline{k}[U]^*$ admet une $\mathfrak{g}$-rétraction.

Cet énoncé a au moins trois aspects:

(a) il *décrit* localement la restriction à U d'un torseur $\mathscr{T}$ de type λ comme un produit fibré d'un torseur-type $(\aleph)$ par un morphisme $\phi\colon U \to R$, ce qui fournit, si R est aussi quasi-trivial, une description d'un *modèle* de $\mathscr{T}$ par des équations explicites;

(b) il assure que certains torseurs définis par produit fibré d'un torseur-type sur certaines k-variétés "ouvertes" U *se prolongent*, parfois même de façon unique, à toute k-compactification lisse X;

(c) il caractérise l'existence d'un torseur de type λ par l'existence d'un $\mathfrak{g}$-relèvement $\hat{R} \to \overline{k}[U]^*$ de la flèche $\rho\colon \hat{R} \to \overline{k}[U]^*/\overline{k}^*$.

Démonstration. Elle s'appuie sur le diagramme du théorème 1.6.1 et sur la propriété de cohérence du lemme 1.6.4.

On va commencer par montrer l'égalité

(2.3.5) $$\mathrm{Tors}(U, S; \aleph, \rho) = \pi^{-1}(\partial(\lambda)) \subset H^1(U, S)$$

sur la portion suivante du diagramme du théorème 1.6.1:

$$
\begin{array}{ccc}
\mathrm{Ext}^1_{\mathfrak{g}}(\hat{S}, \overline{k}[U]^*) & \longleftarrow\!\!\!\longrightarrow & H^1(U, S) \\
\downarrow{\scriptstyle \pi} & & \\
\mathrm{Hom}_{\mathfrak{g}}(\hat{S}, P_{\overline{X}, \overline{U}}) \xrightarrow{\ \partial\ } & \mathrm{Ext}^1_{\mathfrak{g}}(\hat{S}, \overline{k}[U]^*/\overline{k}^*).
\end{array}
$$

(α) Pour établir l'égalité ci-dessus, on commence par factoriser le diagramme commutatif (2.3.2) en introduisant l'extension intermédiaire E qui est à la fois le pull-back par λ de la suite du haut et le push-out par ρ de celle du bas:

$$
\begin{array}{ccccccccc}
1 & \to & \overline{k}[U]^*/\overline{k}^* & \to & \mathrm{Div}_{\overline{Y}}\overline{X} & \to & P_{\overline{X}, \overline{U}} & \to & 0 \\
& & \| & & \uparrow & & \uparrow{\scriptstyle \lambda} & & \\
1 & \to & \overline{k}[U]^*/\overline{k}^* & \longrightarrow & E & \longrightarrow & \hat{S} & \longrightarrow & 0 \\
& & \uparrow{\scriptstyle \rho} & & \uparrow & & \| & & \\
0 & \longrightarrow & \hat{R} & \longrightarrow & \hat{M} & \longrightarrow & \hat{S} & \longrightarrow & 0.
\end{array}
$$

On a donc, au signe près,

$$\partial(\lambda) = [E] = \partial(\rho),$$

où les bords

$$\partial : \mathrm{Hom}_{\mathfrak{g}}\left(\hat{S}, P_{\overline{X}, \overline{U}}\right) \to \mathrm{Ext}^1_{\mathfrak{g}}\left(\hat{S}, \overline{k}[U]^*/\overline{k}^*\right)$$

et

$$\partial : \mathrm{Hom}_{\mathfrak{g}}\left(\hat{R}, \overline{k}[U]^*/\overline{k}^*\right) \to \mathrm{Ext}^1_{\mathfrak{g}}\left(\hat{S}, \overline{k}[U]^*/\overline{k}^*\right)$$

sont respectivement définis par les suites exactes du haut et du bas. L'égalité (2.3.5) équivaut donc à

$$\mathrm{Tors}(U, S; \aleph, \rho) = \pi^{-1}(\partial(\rho)) \subset H^1(U, S)$$

où π et ∂ sont les morphismes naturels du diagramme commutatif

(2.3.6)
$$
\begin{array}{ccccc}
\mathrm{Hom}_{\mathfrak{g}}(\hat{R}, \overline{k}[U]^*) & \xrightarrow{\ \partial\ } & \mathrm{Ext}^1_{\mathfrak{g}}(\hat{S}, \overline{k}[U]^*) & \longleftarrow\!\!\!\longrightarrow & H^1(U, S) \\
\downarrow{\scriptstyle \pi} & & & & \downarrow{\scriptstyle \pi} \\
\rho \in \mathrm{Hom}_{\mathfrak{g}}(\hat{R}, \overline{k}[U]^*/\overline{k}^*) & \xrightarrow{\ \partial\ } & \mathrm{Ext}^1_{\mathfrak{g}}(\hat{S}, \overline{k}[U]^*/\overline{k}^*).
\end{array}
$$

(β) La suite exacte (2.3.1) définit, d'après le lemme 1.6.5, le diagramme commutatif

$$\begin{array}{ccc}
\operatorname{Hom}_{\mathfrak{g}}(\hat{R}, \overline{k}[U]^*) & = & H^0(U, R) \\
\partial \downarrow & & \partial' \downarrow \\
\operatorname{Ext}^1_{\mathfrak{g}}(\hat{S}, \overline{k}[U]^*) & \to & H^1(U, S)
\end{array}$$

dont les flèches ∂' et ∂ sont les bords déduits de la suite exacte (2.3.1) et de la suite exacte duale. Dans ce diagramme, l'image de $\phi \in H^0(U, R)$ par ∂' est la classe du torseur déduit de (2.3.1) par pull-back par le k-morphisme $\phi: U \to R$, autrement dit le produit fibré $M \times_R U$. D'après la définition (2.3.3), un élément de $H^1(U, S)$ appartient donc à $\operatorname{Tors}(U, S; \aleph, \rho)$ si, et seulement si, c'est l'image par ∂' d'un $\phi \in H^0(U, R)$ dont le comorphisme $\hat{\phi} \in \operatorname{Hom}_{\mathfrak{g}}(\hat{R}, \overline{k}[U]^*)$ a pour image $\rho \in \operatorname{Hom}_{\mathfrak{g}}(\hat{R}, \overline{k}[U]^*/\overline{k}^*)$ par le morphisme naturel π. Ainsi, $\operatorname{Tors}(U, S; \aleph, \rho)$ s'identifie, sur le diagramme (2.3.6), à l'ensemble des classes d'extensions

$$[E'] \in \operatorname{Ext}^1_{\mathfrak{g}}(\hat{S}, \overline{k}[U]^*)$$

de la forme $[E'] = \partial(\hat{\phi})$ avec $\pi(\hat{\phi}) = \rho$. Autrement dit:

$$\operatorname{Tors}(U, S; \aleph, \rho) = \partial\big(\pi^{-1}(\rho)\big).$$

(γ) Avec les notations de (2.3.6), l'égalité (2.3.5) équivaut donc à la suivante:

$$\partial\big(\pi^{-1}(\rho)\big) = \pi^{-1}(\partial(\rho)).$$

Comme (2.3.6) commute, l'inclusion $\partial(\pi^{-1}(\rho)) \subset \pi^{-1}(\partial(\rho))$ est évidente. Considérons alors (2.3.6) dans le diagramme commutatif suivant, formé de suites exactes naturelles évidentes:

$$
\begin{array}{ccccccc}
 & & & & & & \operatorname{Ext}^1_{\mathfrak{g}}(\hat{M}, \overline{k}^*) = 0 \\
 & & & & & & \downarrow \\
\operatorname{Hom}_{\mathfrak{g}}(\hat{M}, \overline{k}[U]^*) & \to & \operatorname{Hom}_{\mathfrak{g}}(\hat{R}, \overline{k}[U]^*) & \overset{\partial}{\to} & \operatorname{Ext}^1_{\mathfrak{g}}(\hat{S}, \overline{k}[U]^*) & \to & \operatorname{Ext}^1_{\mathfrak{g}}(\hat{M}, \overline{k}[U]^*) \\
\downarrow & & \downarrow \pi & & \downarrow \pi & & \downarrow \\
\operatorname{Hom}_{\mathfrak{g}}(\hat{M}, \overline{k}[U]^*/\overline{k}^*) & \to & \operatorname{Hom}_{\mathfrak{g}}(\hat{R}, \overline{k}[U]^*/\overline{k}^*) & \overset{\partial}{\to} & \operatorname{Ext}^1_{\mathfrak{g}}(\hat{S}, \overline{k}[U]^*/\overline{k}^*) & \to & \operatorname{Ext}^1_{\mathfrak{g}}(\hat{M}, \overline{k}[U]^*/\overline{k}^*) \\
\downarrow & & & & & & \\
\operatorname{Ext}^1_{\mathfrak{g}}(\hat{M}, \overline{k}^*) = 0. & & & & & &
\end{array}
$$

Dans ce diagramme, M étant un tore quasi-trivial,

$$\mathrm{Ext}^1_{\mathfrak{g}}(\hat{M}, \overline{k}^*) = H^1(k, M) = 0$$

d'après le théorème 90 de Hilbert et le lemme de Shapiro. L'inclusion

$$\pi^{-1}(\partial(\rho)) \subset \partial(\pi^{-1}(\rho))$$

résulte aisément du diagramme ci-dessus: si $\pi([E']) = \partial(\rho)$, la partie droite du diagramme montre que

$$[E'] = \partial(\hat{\phi}) \quad \text{avec } \partial(\pi(\hat{\phi}) - \rho) = 0;$$

la partie gauche du diagramme permet alors de modifier $\hat{\phi}$ de sorte que $\pi(\hat{\phi}) = \rho$.

(δ) L'égalité (2.3.5) étant ainsi établie, montrons que la restriction de X à U définit une application

$$\gamma: \mathrm{Tors}(X, S; \lambda) \to \mathrm{Tors}(U, S; \aleph, \rho).$$

Cela résulte du diagramme du théorème 1.6.1 et du lemme de cohérence 1.6.4. Il s'agit de vérifier sur le diagramme que la restriction $H^1(X, S) \to H^1(U, S)$ induit une application $\mathrm{Tors}(X, S; \lambda) \to \pi^{-1}(\partial(\lambda))$. Soit $\mathcal{T}$ un torseur sur X sous S de type λ. Comme

$$\lambda \in \mathrm{Hom}_{\mathfrak{g}}(\hat{S}, P_{\overline{X}, \overline{U}}),$$

le diagramme du théorème 1.6.1, plus précisément la portion

$$\begin{array}{ccc}
H^1(X, S) & \to & \mathrm{Hom}_{\mathfrak{g}}(\hat{S}, \mathrm{Pic}\,\overline{X}) \\
\downarrow & & \downarrow \\
0 \to \mathrm{Ext}^1_{\mathfrak{g}}(\hat{S}, \overline{k}[U]^*) \to H^1(U, S) & \to & \mathrm{Hom}_{\mathfrak{g}}(\hat{S}, \mathrm{Pic}\,\overline{U}),
\end{array}$$

montre aussitôt que l'image de $[\mathcal{T}]$ dans $H^1(U, S)$ vient d'une classe d'extensions

$$\alpha \in \mathrm{Ext}^1_{\mathfrak{g}}(\hat{S}, \overline{k}[U]^*).$$

La cohérence du cheminement (1.6.10) affirme alors que

$$\pi(\alpha) = \partial(\lambda) \quad \text{dans } \mathrm{Ext}^1_{\mathfrak{g}}(\hat{S}, \overline{k}[U]^*/\overline{k}^*).$$

(ε) La surjectivité de γ résulte aussi du diagramme du théorème 1.6.1 et du lemme 1.6.4. Soit en effet

$$\alpha \in \mathrm{Ext}^1_{\mathfrak{g}}(\hat{S}, \overline{k}[U]^*)$$

tel que $\pi(\alpha) = \partial(\lambda)$. Le diagramme du théorème 1.6.1 montre aussitôt que son image dans $H^1(U, S)$ est la restriction d'un $\mathscr{T} \in H^1(X, S)$; on trouve en outre que son type $\chi(\mathscr{T})$ appartient à $\mathrm{Hom}_{\mathfrak{g}}(\hat{S}, P_{\overline{X}, \overline{U}})$ et même, d'après le lemme 1.6.4, qu'il a même image que λ dans $\mathrm{Ext}^1_{\mathfrak{g}}(\hat{S}, \overline{k}[U]^*/\overline{k}^*)$. On peut donc modifier $\mathscr{T}$ par l'image d'un élément venant de $\mathrm{Hom}_{\mathfrak{g}}(\hat{S}, \mathrm{Div}_{\overline{Y}}\overline{X})$ de telle sorte qu'il garde la même restriction à U et qu'il ait pour type λ.

(ζ) D'après le diagramme du théorème 1.6.1, deux classes de torseurs $[\mathscr{T}]$ et $[\mathscr{T}'] \in H^1(X, S)$ ont même type λ et même restriction à U si, et seulement si, elles diffèrent par l'image d'un élément de

$$\mathrm{Hom}_{\mathfrak{g}}(\hat{S}, \overline{k}[U]^*/\overline{k}^*).$$

Or, d'après le théorème 1.6.1, ce groupe a pour image 0 dans $H^1(X, S)$ si, et seulement si, la flèche

$$\mathrm{Ext}^1_{\mathfrak{g}}(\hat{S}, \overline{k}^*) \to \mathrm{Ext}^1_{\mathfrak{g}}(\hat{S}, \overline{k}[U]^*)$$

est injective, ou, de façon équivalente, si le morphisme

$$\mathrm{Hom}_{\mathfrak{g}}(\hat{S}, \overline{k}[U]^*) \to \mathrm{Hom}_{\mathfrak{g}}(\hat{S}, \overline{k}[U]^*/\overline{k}^*)$$

est surjectif. Ceci achève la démonstration du théorème.

Exemple 2.3.2. Un cas particulier important est celui où

$$\lambda = \lambda_U := \mathrm{id}_{P_{\overline{X}, \overline{U}}}.$$

On peut alors prendre pour suite exacte ($\aleph$) la suite exacte de k-tores

$$(2.3.7) \qquad 1 \to S_U \to M \to R_U \to 1$$

duale de celle de $\mathfrak{g}$-modules

$$(2.3.8) \qquad 1 \to \overline{k}[U]^*/\overline{k}^* \to \mathrm{Div}_{\overline{Y}}\overline{X} \to P_{\overline{X}, \overline{U}} \to 0,$$

et pour flèches verticales du diagramme (2.3.2) les flèches "identité", et il existe un torseur de ce type si, et seulement si, la projection

$$\bar{k}[U]^* \to \bar{k}[U]^*/\bar{k}^*$$

admet une g-section σ (on retrouve ainsi la proposition 2.2.8(v)). D'après le théorème 2.3.1, la flèche de restriction $\mathrm{Tors}(X, S; \lambda) \to \mathrm{Tors}(U, S; \aleph, \rho)$ est alors bijective.

Une telle section σ définit un torseur $\mathscr{T}^\sigma$ sur X sous S de type λ: il est caractérisé par le fait que sa restriction à U est le pull-back de (2.3.8) par le morphisme $\phi_\sigma\colon U \to R_U$ dont le comorphisme

$$\hat{\phi}_\sigma\colon \bar{k}[R_U] = \bar{k}[\hat{R}_U] \to \bar{k}[U]$$

est précisément défini par

$$\sigma\colon \hat{R}_U = \bar{k}[U]^*/\bar{k}^* \to \bar{k}[U]^*.$$

On obtient ainsi toutes les classes de torseurs sur X de type λ.

Si l'obstruction élémentaire est nulle, toute g-section σ de la projection $\bar{k}(X)^* \to \bar{k}(X)^*/\bar{k}^*$ en induit une pour $\bar{k}[U]^* \to \bar{k}[U]^*/\bar{k}^*$ et définit donc, comme indiqué ci-dessus, un torseur $\mathscr{T}^\sigma$ dont la classe appartient à

$$\mathrm{Tors}(X, S, \lambda) \xrightarrow{\ \cong\ } \pi^{-1}([E]) = \mathrm{Tors}(U, S; \aleph, \mathrm{id}) \subset \mathrm{Ext}^1_{\mathfrak{g}}\big(\hat{S}, \bar{k}[U]^*\big)$$

et donne 0 par la rétraction $\mathrm{Ext}^1_{\mathfrak{g}}(\hat{S}, \bar{k}[U]^*) \to \mathrm{Ext}^1_{\mathfrak{g}}(\hat{S}, \bar{k}^*)$ définie par σ. Comme on l'a vu en 2.2.5 et 2.2.8(iv), la section σ définit des rétractions pour les morphismes $H^1(k, S) \to H^1(X, S)$ et $H^1(k, S) \to H^1(U, S)$, toutes les deux induites par la rétraction définie de façon évidente par σ:

$$H^1(k(X), S) = \mathrm{Ext}^1_{\mathfrak{g}}\big(\hat{S}, \bar{k}(X)^*\big) \to \mathrm{Ext}^1_{\mathfrak{g}}\big(\hat{S}, \bar{k}^*\big) = H^1(k, S).$$

Ces diverses rétractions r, r' et r'' sont donc compatibles: elles rendent commutatif le diagramme

$$
\begin{array}{ccc}
\mathrm{Ext}^1_{\mathfrak{g}}(\hat{S}, \bar{k}^*) = H^1(k, S) & \xleftarrow{\ r'\ } & H^1(X, S) \\[2mm]
\big\uparrow{\scriptstyle r} & {\scriptstyle r''}\nwarrow & \big\downarrow \\[2mm]
\mathrm{Ext}^1_{\mathfrak{g}}(\hat{S}, \bar{k}[U]^*) & \longrightarrow & H^1(U, S),
\end{array}
$$

ce qui prouve que, si l'on considère le scindage de la suite exacte (2.2.5) défini par σ, la classe de $\mathscr{T}^\sigma$ est aussi l'élément de $H^1(X, S)$ de type λ et "fibre triviale en σ".

Exemple 2.3.3. *les torseurs universels.*

Supposons

(2.0.3) le $\mathfrak{g}$-module Pic $\overline{X}$ de type fini.

On peut alors considérer l'exemple précédent dans le cas particulier où Pic $\overline{U} = 0$, auquel cas

$$\lambda = \lambda_0 := \mathrm{id}_{\mathrm{Pic}\,\overline{X}}.$$

On retrouve alors la description locale des torseurs universels donnée dans ([22], assertion 1, p. 227):

COROLLAIRE 2.3.4. *Soit X une k-variété algébrique vérifiant les conditions (2.0.1) et (2.0.3), et soit U un ouvert de Zariski non vide de X, tel que* Pic $\overline{U} = 0$. *Une telle variété possède un torseur universel si et seulement si la projection $\overline{k}[U]^* \to \overline{k}[U]^*/\overline{k}^*$ possède une $\mathfrak{g}$-section σ.*

Une telle section σ définit un torseur universel $\mathscr{T}_0^\sigma$ dont la restriction à U est le pull-back par ϕ_σ de la suite exacte de k-tores

$$(2.3.9) \qquad\qquad 1 \to S_0 \to M \to R_U \to 1$$

duale de celle de $\mathfrak{g}$-modules

$$(2.3.10) \qquad\qquad 1 \to \overline{k}[U]^*/\overline{k}^* \to \mathrm{Div}_{\overline{Y}}\overline{X} \to \mathrm{Pic}\,\overline{X} \to 0,$$

le morphisme $\phi_\sigma\colon U \to R_U$ étant défini par dualité à partir de $\sigma\colon \hat{R}_U \to \overline{k}[U]^*$.

On peut également caractériser $\mathscr{T}_0^\sigma$ par le fait qu'il a pour image 0 par la rétraction (cf. 2.2.8(iv))

$$H^1(X, S) \to H^1(k, S)$$

définie par σ; autrement dit, par le fait que $\mathscr{T}_0^\sigma$ vaut 0 en σ.

Dans le cas où $U(k) \neq \varnothing$, si $P \in U(k)$, on peut prendre pour section

$$\sigma_P\colon f \bmod \overline{k}^* \to f/f(P).$$

On note alors $\mathscr{T}_0^P$ le torseur universel défini par la section σ_P: c'est le torseur universel de fibre triviale en P.

Les résultats énoncés résultent de la discussion de l'exemple précédent. On notera qu'ici toute $\mathfrak{g}$-section de $\overline{k}[U]^* \to \overline{k}[U]^*/\overline{k}^*$ provient d'une $\mathfrak{g}$-section de $\overline{k}(X)^* \to \overline{k}(X)^*/\overline{k}^*$.

2.4. *Exemples d'application au prolongement de torseurs.* Le théorème 2.3.1 contient la plupart des énoncés de prolongement de torseurs que nous avons rencontrés. Joint aux propositions 2.1.1 et 2.1.2, il permet aussi d'établir la trivialité d'invariants liés au groupe de Picard sur tout compactifié lisse de certains torseurs, sans avoir à calculer explicitement une telle compactification lisse.

2.4.1. Soit X une k-variété algébrique vérifiant la condition (2.0.1). Soit n un entier > 0. Soit $f \in k(X)^*$ une fonction dont le diviseur soit un multiple de n. Il est clair que f définit un torseur sous μ_n au-dessus de l'ouvert U où f est inversible, par pull-back par f du torseur-type

$$1 \to \mu_{n,k} \to \mathsf{G}_{m,k} \xrightarrow{\times n} \mathsf{G}_{m,k} \to 1.$$

L'hypothèse sur le diviseur de f assure que ce torseur se prolonge, de façon unique, en un torseur sur X tout entier sous μ_n: il suffit d'appliquer le théorème 2.3.1 pour le diagramme

$$
\begin{array}{ccccccccc}
1 & \to & \overline{k}[U]^*/\overline{k}^* & \to & \mathrm{Div}_{\overline{Y}}\overline{X} & \to & P_{\overline{X},\overline{U}} & \to & 0 \\
& & \uparrow & & \uparrow & & \uparrow & & \\
0 & \longrightarrow & \mathsf{Z} & \longrightarrow & \mathsf{Z} & \longrightarrow & \mathsf{Z}/n & \to & 0
\end{array}
$$

où la verticale de gauche expédie 1 sur la classe de f et la verticale médiane expédie 1 sur D tel que $\mathrm{div}(f) = nD$ (on notera que k^*/k^{*n} s'injecte dans $k[U]^*/k[U]^{*n} \subset \mathrm{Ext}^1_{\mathfrak{g}}(\mathsf{Z}/n, \overline{k}[U]^*)$).

2.4.2. Soit X qui vérifie (2.0.1). Soit K/k une extension finie et séparable de corps. Soit $f \in k(X)^*$ une fonction dont le diviseur est une norme pour K/k: $\mathrm{div}(f) = N_{K/k}(D)$ pour $D \in \mathrm{Div}\, X_K$. Soit U l'ouvert complémentaire du plus petit fermé Y de X tel que Y_K contienne le support de D. Il est clair que f définit sur U un torseur sous le k-tore $R^1_{K/k}\mathsf{G}_m$ par pull-back par f du torseur-type

$$1 \to R^1_{K/k}\mathsf{G}_m \to R_{K/k}\mathsf{G}_m \xrightarrow{N_{K/k}} \mathsf{G}_{m,k} \to 1.$$

L'hypothèse sur $\mathrm{div}(f)$ assure que ce torseur se prolonge, de façon unique si k^*/NK^* s'injecte dans $k[U]^*/NK[U]^*$, en un torseur sur X sous le k-tore $S := R^1_{K/k}\mathsf{G}_m$: il suffit d'appliquer le théorème 2.3.1 au diagramme

$$
\begin{array}{ccccccccc}
1 & \to & \overline{k}[U]^*/\overline{k}^* & \to & \mathrm{Div}_{\overline{Y}}\overline{X} & \to & P_{\overline{X},\overline{U}} & \to & 0 \\
& & \uparrow & & \uparrow & & \uparrow & & \\
0 & \longrightarrow & \mathsf{Z} & \xrightarrow{N} & \mathsf{Z}[\mathfrak{g}/\mathfrak{h}] & \longrightarrow & \hat{S} & \to & 0
\end{array}
$$

où

$$\mathfrak{g} = \mathrm{Gal}(\overline{k}/k), \qquad \mathfrak{h} = \mathrm{Gal}(\overline{k}/K), \qquad N = N_{\mathfrak{g}/\mathfrak{h}},$$

et où la verticale de gauche envoie 1 sur la classe de f et la verticale médiane envoie 1 sur D.

Pour un exemple de la même veine, mais plus complexe, voir [27] §12 Step IV(f).

2.4.3. *Le cas des torseurs universels.* On peut interpréter le corollaire 2.3.4 comme un énoncé de prolongement (unique) à tout X du torseur sur U sous S_0 défini explicitement comme le pull-back par $\phi_\sigma\colon U \to R_U$ du torseur sur R_U sous S_0 défini par la suite exacte (2.3.9).

2.4.4. *Le cas des tores.* Le case particulier de 2.4.3 où X est une k-compactification lisse d'un k-tore T est particulièrement intéressant. On peut prendre $U = T$. On trouve alors $R_U = T$. Si l'on prend $\sigma = \sigma_O$ où O est l'élément neutre de $T(k)$, le torseur sur T sous S_0 ainsi obtenu n'est autre que celui donné par la suite exacte de k-tores

$$1 \to S_0 \to M \to T \to 1$$

déduite par dualité de la suite exacte de $\mathfrak{g}$-modules

$$0 \to \hat{T} \to \mathrm{Div}_{\overline{Y}}\overline{X} \to \mathrm{Pic}\,\overline{X} \to 0,$$

et la proposition assure qu'il se prolonge, de façon unique, en un torseur sur X sous S_0. Comme M est un k-tore quasi-trivial, en particulier une variété k-rationnelle, on obtient ainsi les principaux résultats de [20]. De façon plus générale, la première hypothèse sur les torseurs universels (cf. §2.8 ci-après) est ici satisfaite, ce qui fait le succès de la méthode de la descente dans ce cas. Pour plus de détails, voir [22], §3.

2.5. *Exemples de calcul d'équations locales de torseurs universels.*

2.5.1. Soit k un corps de caractéristique $\neq 3$, contenant une racine cubique de l'unité $\theta \neq 1$. Soit $a \in k^* \setminus k^{*3}$. La k-*surface cubique* X d'équation homogène

$$T_0^3 + T_1^3 + T_2^3 + aT_3^3 = 0,$$

dont on sait qu'elle n'est pas k-rationnelle (Shafarevitch), a été étudiée par Manin [42] du point de vue de l'équivalence de Brauer. On va appliquer la proposition pour trouver des équations locales pour les torseurs universels sur cette surface cubique sur l'ouvert complémentaire de la réunion des 27 droites (en fait même seulement de 9 d'entre elles). On note

$$\alpha := \sqrt[3]{a}, \qquad K = k(\alpha), \qquad G = \mathrm{Gal}(K/k) \cong \mathbf{Z}/3.$$

Avec les notations de Manin ([42] chap. VI, 5.4 (= 45.4)), la surface $\overline{X}$ contient les droites suivantes, définies sur K:

$$A_i = L_2(1, i): \qquad T_0 + \theta T_1 = 0 \qquad T_2 + \theta^i \alpha T_3 = 0$$

$$B_i = L_1(0, i): \qquad T_0 + T_2 = 0 \qquad T_1 + \theta^i \alpha T_3 = 0$$

$$C_i = L_2(0, i): \qquad T_0 + T_1 = 0 \qquad T_2 + \theta^i \alpha T_3 = 0,$$

où $i = 0, 1, 2$.

LEMME 2.5.2. *Chacun des deux systèmes suivants définit une base du groupe abélien libre* Pic $\overline{X}$:

$$(A_0, A_1, B_0, B_1, C_0, C_1, C_2),$$

$$(A_0 - C_0, A_1 - C_1, B_0 - C_0, B_1 - C_1, C_0, C_1, C_2).$$

Démonstration. Le déterminant de Gram du premier système vaut 1.

Soit U l'ouvert de X complémentaire de la réunion des 9 droites A_i, B_i et C_i. On a Pic $\overline{U} = 0$. On peut donc appliquer le théorème 2.3.1 au diagramme commutatif de suites exactes

$$
\begin{array}{ccccccccc}
1 & \to & \overline{k}[U]^*/\overline{k}^* & \longrightarrow & \mathrm{Div}_{\overline{Y}}\overline{X} & \longrightarrow & \mathrm{Pic}\,\overline{X} & \to & 0 \\
& & \cong \uparrow & & \delta \uparrow & & \lambda \uparrow & & \\
0 & \longrightarrow & \mathbf{Z}^2 & \xrightarrow{\,N\,} & \mathbf{Z}[G]^2 & \longrightarrow & \hat{S} & \longrightarrow & 0
\end{array}
$$

défini comme suit: $N = (N_G, N_G)$; l'isomorphisme de gauche est donné par les classes des fonctions

$$f = (T_0 + \theta T_1)/(T_0 + T_1) \quad \text{et} \quad g = (T_0 + T_2)/(T_0 + T_1);$$

l'injection δ est donnée par

$$\delta(1,0) = A_0 - C_0 \quad \text{et} \quad \delta(0,1) = B_0 - C_0,$$

et on a effectivement

$$\mathrm{div}(f) = N(\delta(1,0)) \quad \text{et} \quad \mathrm{div}(g) = N(\delta(0,1)),$$

d'où, d'après le lemme:

$$\mathrm{Div}_{\overline{Y}}\overline{X} = \mathrm{im}(\delta) \oplus \mathbf{Z}[G] \cdot C_0 \quad \text{et} \quad \mathrm{Pic}\,\overline{X} \cong \hat{S} \oplus \mathbf{Z}[G] \cong (\mathbf{Z}[G]/N_G \cdot \mathbf{Z})^2 \oplus \mathbf{Z}[G].$$

Noter qu'on en déduit aussitôt

$$H^1(k, \operatorname{Pic} \overline{X}) \cong (\mathbf{Z}/3)^2.$$

Ainsi, λ est un type admissible et l'application de la proposition au diagramme ci-dessus donne les équations au-dessus de l'ouvert U de X d'un torseur admissible quelconque $\mathcal{T}^{\gamma}$ de type λ:

$$0 \neq f(T_0, T_1, T_2, T_3) = \gamma_1 \cdot N_{K/k}\big(x_0 + \alpha x_1 + \alpha^2 x_2\big)$$

$$0 \neq g(T_0, T_1, T_2, T_3) = \gamma_2 \cdot N_{K/k}\big(y_0 + \alpha y_1 + \alpha^2 y_2\big)$$

$$T_0^3 + T_1^3 + T_2^3 + aT_3^3 = 0,$$

en les variables $(T_0, T_1, T_2, T_3; x_0, x_1, x_2; y_0, y_1, y_2)$, avec $\gamma = (\gamma_1, \gamma_2) \in k^{*2}$ quelconque. Le torseur-type est alors donné par la suite exacte de k-tores

$$1 \to R^1_{K/k}\mathbf{G}_m^2 \to R_{K/k}\mathbf{G}_m^2 \xrightarrow{N_{K/k}} \mathbf{G}_{m,k}^2 \to 1$$

et l'application $\phi\colon U \to \mathbf{G}_{m,k} \times \mathbf{G}_{m,k}$ est donnée par (f, g). Tout torseur universel est, sur l'ouvert U, le produit direct d'un torseur admissible $\mathcal{T}^{\gamma}$ et du facteur "trivial" $R_{K/k}\mathbf{G}_m$.

2.5.3.   *La surface cubique diagonale X d'équation*

$$aT_0^3 + bT_1^3 + cT_2^3 + dT_3^3 = 0,$$

en les variables T_0, T_1, T_2, T_3, avec $a, b, c, d \in k^*$. On fait sur k les mêmes hypothèses qu'en 2.5.1, mais on suppose qu'aucun des nombres ab/cd, ni des a/b n'est un cube dans k^*, ce qui exclut le cas précédent. On considère alors le corps

$$K = k(\alpha, \gamma) \quad \text{où } \alpha = \sqrt[3]{b/a} \text{ et } \gamma = \sqrt[3]{ad/bc}\,.$$

La surface X est K-rationnelle et on construit dans [15], §10, les torseurs quasi-universels relativement à K/k, i.e. de type l'injection naturelle

$$i\colon \operatorname{Pic} X_K \to \operatorname{Pic} \overline{X}.$$

Contrairement au cas précédent, ces torseurs ne sont pas admissibles, le conoyau de i n'étant pas un $\mathfrak{g}$-module de permutation. Il n'empêche qu'en vertu du théorème 2.1.2(b), ces torseurs peuvent être aussi intéressants que les torseurs admissibles.

Nous renvoyons à [15] pour le détail des calculs: voir en particulier le lemme 17 et le diagramme (143) de [15], qui permet d'appliquer le théorème 2.3.1 et d'obtenir ainsi des équations locales pour les torseurs de type i. On note

$$K_1 = k(\alpha), \ K_2 = k(\beta), \ L_1 = k(\gamma), \ L_2 = k(\delta) \quad \text{où } \beta = \alpha\gamma \text{ et } \delta = \alpha/\gamma.$$

Le torseur-type est donné par la suite exacte de k-tores

$$1 \to S \to \mathsf{G}_{m,k} \times R_{K/k}\mathsf{G}_m \to S_1 \to 1$$

où S et S_1 sont respectivement les conoyau de ρ et noyau de π dans la suite exacte

$$1 \to \mathsf{G}_{m,k} \xrightarrow{\iota} R_{L_1/k}\mathsf{G}_m \times R_{L_2/k}\mathsf{G}_m \xrightarrow{\rho} \mathsf{G}_{m,k} \times R_{K/k}\mathsf{G}_m \xrightarrow{\nu}$$

$$R_{K_1/k}\mathsf{G}_m \times R_{K_2/k}\mathsf{G}_m \xrightarrow{\pi} \mathsf{G}_{m,k} \to 1$$

où $\iota(x) = (x, x^{-1})$, $\rho(x, y) = (N_{L_1/k}(x) \cdot N_{L_2/k}(y), 1/xy)$,

$\nu(x, y) = \big(x \cdot N_{K/K_1}(y), x \cdot N_{K/K_2}(y)\big)$ et enfin $\pi(x, y) = N_{K_1/k}(x) \cdot N_{K_2/k}(y)$.

Quant au k-morphisme $\phi \colon U \to S_1$, il est défini par

$$\phi(T_0, T_1, T_2, T_3) = \big((T_0 + \alpha T_1)/\varepsilon_\alpha T_3, (T_2 + \beta T_3)/\varepsilon_\beta T_3\big),$$

avec $\varepsilon_\alpha \in K_1^*$ et $\varepsilon_\beta \in K_2^*$ tels que $N_{K/L_1}(\varepsilon_\alpha/\varepsilon_\beta) = -c/a$. On obtient ainsi, au-dessus d'un ouvert U non vide de X convenable, comme équations des torseurs quasi-universels, relativement à K/k, les équations suivantes, où $c_i \in K_i$:

$$aT_0^3 + bT_1^3 + cT_2^3 + dT_3^3 = 0,$$

$$(T_0 + \alpha T_1)/T_3 = c_1 u \cdot N_{K/K_1}(\xi) \neq 0$$

$$(T_2 + \beta T_3)/T_3 = c_2 u \cdot N_{K/K_2}(\xi) \neq 0,$$

en les variables $(T_0, T_1, T_2, T_3; u; \xi)$ dans $\mathsf{P}_k^3 \times \mathsf{G}_{m,k} \times R_{K/k}\mathsf{G}_m \hookrightarrow \mathsf{P}_k^3 \times \mathsf{A}_k^{10}$, avec

$$(c_1, c_2) \in K_1^* \times K_2^* \quad \text{tel que } a \cdot N_{K_1/k}(c_1) = -c \cdot N_{K_2/k}(c_2).$$

On peut montrer (cf. [15] §10 proposition 11) qu'un tel torseur admet pour modèle un cône intersection dans A_k^9 de deux hypersurfaces cubiques d'un type géométrique très particulier.

2.6. *Le cas des surfaces fibrées en coniques sur la droite projective.* De nombreux exemples de calcul d'équations locales de torseurs universels ou admissibles sur de telles surfaces rationnelles ont déjà été exposés dans d'autres articles, e.g. [22], [13], [50], [2], [27], dans des situations particulières variées. Nous traitons ici le cas général, selon la méthode annoncée dans [25], avant de mentionner rapidement un certain nombre de cas particuliers donnant parfois lieu à des simplifications.

On suppose k parfait de caractéristique $\neq 2$. On note (u, v) les coordonnées homogènes sur la droite projective P^1_k et $x = u/v$ la coordonnée sur la droite affine $\mathsf{A}^1_k = \mathsf{P}^1_k - \{\infty\}$. Soit

$$X \xrightarrow{\ \pi\ } \mathsf{P}^1_k$$

une k-surface projective et lisse, fibrée en coniques sur P^1_k (par quoi l'on entend que chaque fibre est une conique réduite). Une telle surface vérifie (2.0.1) et (2.0.3), et S_0 est même un tore. On suppose la fibre à l'infini lisse. Soient $\{e_i\}_{i=1,\ldots,s}$ les points fermés de P^1_k en lesquels les fibres de π sont dégénérées, $K_i = k(e_i)$ le corps résiduel en e_i et K'_i/K_i l'algèbre quadratique séparable définie par les composantes irréductibles géométriques de la fibre de e_i. Soient

$$P(x) = \prod_i N_{K_i/k}(x - e_i) \quad \text{et} \quad A = k[x]/(P) = \prod_i K_i.$$

On note $\theta = \{e_i\}$ la classe de x dans A. L'algèbre $A' = \prod_i K'_i$ est quadratique étale sur A et peut s'écrire

$$A' = k\left[x, \sqrt{a(x)}\,\right]/(P(x)) = A\left[\sqrt{a(\theta)}\,\right]$$

avec $a(x)$ séparable (si $b \in B$, on désigne par la notation $B[\sqrt{b}\,]$ l'algèbre $B[T]/((T^2 - b)))$. Ainsi:

$$K'_i = K_i\left[\sqrt{a_i}\,\right] \quad \text{avec } a_i = a(e_i).$$

On note $r = \deg P$ le nombre de fibres géométriques dégénérées de π. La fibration π définit une $k(x)$-algèbre de quaternions $\mathscr{A}$ dont la classe dans $\operatorname{Br} k(x) = \operatorname{Br} k(\mathsf{P}^1_k)$ a pour résidu en e_i la classe de a_i dans K_i^*/K_i^{*2}.

Soient U'_0 l'ouvert de P^1_k complémentaire de la réunion des e_i et $U_0 = U'_0 - \{\infty\}$. On note U l'ouvert $\pi^{-1}(U_0)$ de X. De la suite exacte de $\mathfrak{g}$-modules

$$1 \to \overline{k}[U]^*/\overline{k}^* \to \operatorname{Div}_{\overline{Y}}\overline{X} \to \operatorname{Pic}\overline{X} \to \operatorname{Pic}\overline{U} \to 0$$

on tire (voir [23] (2.1)) des suites exactes naturelles de k-tores:

(2.6.1) $$1 \to \mathsf{G}_{m,k} \to S_0 \to S \to 1$$

(2.6.2) $$1 \to S \to \mathsf{G}_{m,k} \times \prod_i R_{K'_i/k}\mathsf{G}_m \xrightarrow{\ \nu\ } \prod_i R_{K_i/k}\mathsf{G}_m \to 1,$$

où $\nu(t, t_i') = \{ t^{-1} \cdot N_i(t_i') \}$ avec $N_i := N_{K_i'/K_i}$; on réécrit celle-ci

$$(2.6.3) \qquad 1 \to S \to \mathsf{G}_{m,\,k} \times R_{A'/k}\mathsf{G}_m \xrightarrow{\ \nu\ } R_{A/k}\mathsf{G}_m \to 1,$$

ou encore

$$(2.6.4) \qquad 1 \to S \to M \to R \to 1.$$

La suite exacte (2.6.1) vient, par dualité, de celle de $\mathfrak{g}$-modules:

$$(2.6.5) \qquad 0 \to \hat{S} \xrightarrow{\ \lambda\ } \operatorname{Pic} \overline{X} \xrightarrow{\ \omega\ } \mathsf{Z} \to 0,$$

où ω est la restriction à la fibre générique, ou, ce qui revient au même, à U. Ainsi,

$$P_{\overline{X},\,\overline{U}} = \hat{S}.$$

De plus, le $\mathfrak{g}$-module $\overline{k}[U]^*/\overline{k}^*$ est de permutation, son dual est le k-tore $R = R_{A'/A}\mathsf{G}_m = \prod_i R_{K_i/k}\mathsf{G}_m$ et la famille de fonctions $\{(x - e_i) \in K_i(U)\}$, ou, ce qui revient au même, la fonction $x - \theta \in A[U]$, définit une $\mathfrak{g}$-section de la projection $\overline{k}[U]^* \to \overline{k}[U]^*/\overline{k}^*$.

Il existe donc toujours (proposition 2.2.8(v)) des torseurs de type λ sur X et la description locale des torseurs donnée par le théorème 2.3.1 s'applique sur l'ouvert U: pour chaque relèvement $\alpha \in A^* = \prod_i K_i^*$ de

$$\operatorname{cl}(\alpha) = \operatorname{cl}((\alpha_i)) \in H^1(k, S) = A^*/\operatorname{im}(\nu(k)) = \left(\prod_i K_i^*\right)/\operatorname{im}(\nu(k)),$$

la fonction $\varphi_\alpha \colon U \to R = R_{A/k}\mathsf{G}_m = \prod_i R_{K_i/k}\mathsf{G}_m$ définie par $x \mapsto \alpha(x - \theta) = \{\alpha_i(x - e_i)\}$ a pour comorphisme un relèvement de l'identité de $\hat{R} = \overline{k}[U]^*/\overline{k}^*$. Le torseur $\mathscr{T}^\alpha$ de type λ est donc sur U le pull-back par φ_α du torseur-type (2.6.2): c'est la situation de l'exemple 2.3.2. Comme φ_α se factorise clairement par U_0, on a le diagramme suivant dont les deux carrés sont cartésiens:

$$(2.6.6) \qquad \begin{array}{ccccc} \mathscr{T}_U^\alpha & \longrightarrow & W_0 & \longrightarrow & M \\ \downarrow & & \downarrow & & \downarrow \\ U & \xrightarrow{\ \pi\ } & U_0 & \longrightarrow & R. \end{array}$$

Les équations de W_0 sont donc données dans $\mathsf{A}_k^2 \times R_{A'/k}\mathsf{A}^1 = \mathsf{A}_k^2 \times \prod_i R_{K_i'/k}\mathsf{A}^1$ par:

$$(2.6.7) \qquad \alpha(x - \theta) = t^{-1} \cdot N_{A'/A}(t') \neq 0,$$

ou encore

$$(2.6.8) \qquad \alpha_i(x - e_i) = t^{-1} \cdot N_i(t_i') \neq 0 \qquad i = 1, \ldots, s,$$

où $(x, t; t') = (x, t; t_i')$ sont les coordonnées, t' étant à valeurs dans A' et les t_i' dans K_i'. Considérons alors la k-variété W' définie dans le même espace, de coordonnées $(u, v; t') = (u, v; t_i')$ par l'équation

$$(2.6.9) \qquad \alpha(u - \theta v) = N_{A'/A}(t') \neq 0,$$

ou encore par les équations

$$(2.6.10) \qquad \alpha_i(u - e_i v) = N_i(t_i') \neq 0 \qquad i = 1, \ldots, s.$$

Le changement de variables $x = u/v$ et $t = v$ définit une k-immersion ouverte $W_0 \to W'$ compatible avec les projections p_0 sur $U_0 \subset \mathsf{A}_k^1$ et p' sur $U_0' \subset \mathsf{P}_k^1$ données respectivement par x et par (u, v). On peut ensuite plonger W' dans la k-variété $W = W_\alpha$ définie dans $\mathsf{A}_k^2 \times R_{A'/k}\mathsf{A}^1$ par l'équation

$$(2.6.11) \qquad u - \theta v = \alpha^{-1} \cdot N_{A'/A}(t') \qquad (u, v) \neq (0,0),$$

où $\alpha = \alpha(\theta) = \{\alpha_i\} \in A^*$, ou encore, dans $\mathsf{A}_k^2 \times \prod_i R_{K_i/k}\mathsf{A}^2$ par les équations $(i = 1, \ldots, s)$:

$$(2.6.12) \qquad u - e_i v = \alpha_i^{-1} \cdot N_i(t_i') \qquad (u, v) \neq (0,0),$$

de façon compatible avec les projections sur U_0' et P_k^1 respectivement. Soit $\Omega := \mathsf{A}_k^2 - \{0,0\}$ et soit $\kappa\colon \Omega \to \mathsf{P}_k^1$ le morphisme canonique. Soit Ω' l'image réciproque de U_0' dans Ω. Avec ces notations, on a le diagramme d'inclusions et k-morphismes:

$$(2.6.13)$$

$$
\begin{array}{ccccc}
W & \xrightarrow{\ q\ } & \Omega & \xrightarrow{\ \kappa\ } & \mathsf{P}_k^1 \\
\cup & & \cup & & \cup \\
W' & \longrightarrow & \Omega' & \longrightarrow & U_0' \\
\cup & & & & \cup \\
W_0 & & \longrightarrow & & U_0.
\end{array}
$$

Dans ce diagramme, on note $p\colon W \to \mathsf{P}_k^1$ la restriction $\kappa \circ q$ de κ à W, et les projections $p'\colon W' \to U_0'$ et $p_0\colon W_0 \to U_0$ sont les restrictions de p à W' et W_0 respectivement.

Dans ce qui suit on utilise la notion de groupe de Brauer-Grothendieck d'une variété. Pour les notations et propriétés utilisées ici, voir 1.5.0 et le début de 3.1. En particulier, si Y est une k-variété algébrique, $\mathrm{Br}_1 Y$ désigne le noyau de la restriction $\mathrm{Br}\, Y \to \mathrm{Br}\, \overline{Y}$, et, si $\overline{k}[Y]^* = \overline{k}^*$, on a la suite exacte

$$(1.5.0) \quad 0 \to \mathrm{Pic}\, Y \to (\mathrm{Pic}\, \overline{Y})^{\mathfrak{g}} \to \mathrm{Br}\, k \to \mathrm{Br}_1 Y \to H^1(\mathfrak{g}, \mathrm{Pic}\, \overline{Y}) \to H^3(\mathfrak{g}, \overline{k}^*).$$

LEMME 2.6.1.   *Avec les notations et hypothèses ci-dessus*:

$$\overline{k}[W]^* = \overline{k}^*, \qquad \mathrm{Pic}\,\overline{W} = 0, \qquad \mathrm{Br}_1 W = \mathrm{Br}\,k.$$

Démonstration.   L'égalité $\mathrm{Br}_1 W = \mathrm{Br}\,k$ résulte des deux premières et celles-ci se vérifient aisément, comme on le voit ci-après, en observant que sur $\overline{k}$ les équations (2.6.12) s'écrivent plus simplement:

$$u - e_j v = u_j' v_j' \qquad (u, v) \neq (0, 0),$$

où les u_j', v_j' sont des variables "ordinaires" ($j = 1, \ldots, r$). Il est immédiat que $\overline{W}'$ est le produit de $\overline{\Omega}'$ et d'un tore $\cong \mathsf{G}_{m,\overline{k}}^r$. On a donc:

$$\mathrm{Pic}\,\overline{W}' = \mathrm{Pic}\,\overline{\Omega}' = 0 \quad \text{et} \quad \overline{k}[W']^*/\overline{k}^* = \overline{k}[\overline{\Omega}']^*/\overline{k}^* \times \mathsf{Z}^r,$$

où Z^r admet pour base les classes des u_j'. Autrement dit,

$$\overline{k}[W']^* \text{ est engendré par } \overline{k}[\Omega]^* \text{ et par les } u_j'.$$

Soient δ_j (resp. δ_j') les sous-variétés irréductibles de codimension 1 définies dans $\overline{W}$ par $u_j' = 0$ (resp. $v_j' = 0$). D'après ce qui précède, une fonction inversible sur $\overline{W}$ s'écrit

$$f = g \cdot \prod_j u_j'^{n_j}$$

avec $g \in \overline{k}[\Omega']^*$. Ce groupe $\overline{k}[\Omega']^*$ est engendré mod $\overline{k}^*$ par les classes des fonctions $g_j = u - e_j v$. Comme $\mathrm{div}(u - e_j v) = \delta_j + \delta_j'$, on voit que, si, à une constante près,

$$g = \prod_j g_j^{m_j},$$

alors

$$\mathrm{div}(f) = \sum_j \big((n_j + m_j)\delta_j + m_j \delta_j'\big).$$

Comme f est inversible, $\mathrm{div}(f) = 0$ et $m_j = 0$ pour chaque j, donc aussi $n_j = 0$, et f est donc constante, ce qui prouve

$$\overline{k}[W]^* = \overline{k}^*.$$

Le groupe des diviseurs de $\overline{W}$ à support hors de $\overline{W}'$ admet pour base $\{\delta_j, \delta_j'\}$. Comme chaque δ_j ou δ_j' est un diviseur principal et que $\mathrm{Pic}\,\overline{\Omega}' = 0$, on en déduit

$$\mathrm{Pic}\,\overline{W} = 0.$$

A fortiori, $H^1(\mathfrak{g}, \mathrm{Pic}\,\overline{W}) = 0$ et, comme $\overline{k}[X]^* = \overline{k}^*$, on obtient ainsi $\mathrm{Br}_1 W = \mathrm{Br}\,k$.

Remarque 2.6.2.   Le lemme 2.6.1 assure que, pour toute k-compactification lisse Z de W, le $\mathfrak{g}$-module $\operatorname{Pic} \overline{Z}$ est de permutation et les obstructions standard à la k-rationalité (voir appendice 2.A ci-après) et au principe de Hasse ([41], [42] chap. VI, et §3 ci-après) disparaissent donc sur Z.

PROPOSITION 2.6.3.   *Soit* $\alpha = (\alpha_i) \in \prod_i K_i^*$ *pour* $i = 1, \ldots, s$. *Soit* $W = W_\alpha$ *la k-variété lisse définie dans* $\mathsf{A}_k^2 \times \prod_i R_{K_i/k}\mathsf{A}^2$ *par les équations* $(i = 1, \ldots, s)$

$$(2.6.14) \qquad \alpha_i(u - e_i v) = u_i^2 - a_i v_i^2 \qquad (u, v) \neq (0, 0),$$

en les variables $(u, v; u_i, v_i)$, *à valeurs dans* $k \times k \times \prod_i (K_i \times K_i)$.

(i) *La k-variété W_α s'identifie à un ouvert d'un cône épointé ayant pour base une intersection complète* $V = V_\alpha$ *de* $r - 2$ *quadriques dans* P_k^{2r-1}, *qui est géométriquement intègre et n'est pas un cône.*

(ii) *Tout modèle k-birationnel, propre et lisse, Z de W, ou de V, vérifie*

$$H^1(\mathfrak{g}, \operatorname{Pic} \overline{Z}) = 0.$$

Démonstration.   L'assertion (ii) est un corollaire du lemme 2.6.1 et de (i) comme indiqué dans la remarque 2.6.2. Les équations (2.6.12) donnent (2.6.14) en posant

$$t_i' = u_i + v_i\sqrt{a_i}\,.$$

Si l'on pose formellement

$$t' = \left(u_1 + u_2\theta + \cdots + u_r\theta^{r-1}\right) + \left(v_1 + v_2\theta + \cdots + v_r\theta^{r-1}\right)\sqrt{a(\theta)}\,,$$

où les variables u_j, v_j sont à valeurs dans k, il vient:

$$\alpha^{-1} \cdot N_{A'/A}(t') = Q_1(\mathbf{u}, \mathbf{v}) + Q_2(\mathbf{u}, \mathbf{v})\theta + \cdots + Q_r(\mathbf{u}, \mathbf{v})\theta^{r-1},$$

où $\mathbf{u} = (u_1, \ldots, u_r)$ et $\mathbf{v} = (v_1, \ldots, v_r)$, et l'équation (2.6.12) équivaut ainsi au système de $r - 2$ équations quadratiques homogènes:

$$Q_3(\mathbf{u}, \mathbf{v}) = \cdots = Q_r(\mathbf{u}, \mathbf{v}) = 0,$$

en $2r$ variables "ordinaires," à coefficients dans k, avec $(Q_1(\mathbf{u}, \mathbf{v}), Q_2(\mathbf{u}, \mathbf{v})) \neq (0, 0)$, ce qui prouve (i).

THÉORÈME 2.6.4.   *Soit X une surface fibrée en coniques sur la droite projective sur un corps parfait k de caractéristique* $\neq 2$. *Soit r le nombre de fibres géométriques dégénérées et soit*

$$\lambda \colon \hat{S} \to \operatorname{Pic} \overline{X}$$

le type admissible considéré en (2.6.5), *noyau de la restriction à la fibre générique.*

(i) *X possède toujours des torseurs $\mathcal{T}^\alpha$ de type λ.*

(ii) *Sous chacune des hypothèses suivantes, tout tel torseur $\mathcal{T}^\alpha$ admet pour modèle k-birationnel le produit par une conique et par P_k^1 d'une intersection complète V_α de $r-2$ quadriques dans P_k^{2r-1}:*

(a) *k est un corps local, ou un corps de nombres, ou un corps C_1;*

(b) *la fibration X/P_k^1 possède une section sur une extension quadratique de k;*

(c) *$W_\alpha(k) \neq \varnothing$;*

(d) *$\mathcal{T}^\alpha$ possède un point k-rationnel non situé au-dessus d'un point d'une fibre singulière de la fibration;*

(e) *$\mathrm{car}(k) = 0$ et $\mathcal{T}^\alpha(k) \neq \varnothing$.*

(iii) *Soit $\mathcal{T}$ un torseur universel sur X; sous (a), sous (b), ou si $\mathcal{T}$ satisfait l'une des conditions (d) ou (e), alors $\mathcal{T}$ admet pour modèle k-birationnel le produit d'une variété V_α par une conique et par P_k^2.*

Voici d'abord deux remarques avant la démonstration.

Remarque 2.6.5.   Si aucune des hypothèses (a), (b), (c), (d) n'est vérifiée, on peut au moins affirmer que $\mathcal{T}^\alpha$ est stablement (i.e. quitte à faire son produit par l'espace projectif d'une dimension convenable) k-birationnellement le produit de V_α par une k-variété de Severi-Brauer.

Remarque 2.6.6.   L'assertion (iii) n'affirme pas l'existence automatique d'un torseur universel, mais n'a d'intérêt que s'il en existe!

Démonstration.   (i) a déjà été établi au début de la section 2.6. D'après la description locale qu'on a déjà donnée, le torseur $\mathcal{T}^\alpha$ a pour restriction sur l'ouvert U de X le produit fibré de U et de l'ouvert W_0 de $W = W_\alpha$ au-dessus de l'ouvert U_0 de A_k^1. Il admet donc pour modèle le pull-back par $W \to \mathsf{P}_k^1$ de la fibration en coniques $X \to \mathsf{P}_k^1$, qui est lisse au-dessus de U_0', et ramifiée en $\{e_i\}$ tout comme le morphisme $W \to \mathsf{P}_k^1$. Le point-clé de la démonstration est que la ramification "mange" la ramification ("lemme d'Abhyankar"): le passage de P_k^1 à W a pour effet de tuer la ramification de la fibration en coniques, comme on va le voir.

D'après le théorème de Tsen, la classe de la $k(x)$-algèbre de quaternions $\mathcal{A}$ associée à la conique générique de la fibration π est trivialisée par passage à $\bar{k}$. Elle se prolonge en une algèbre d'Azumaya, encore notée $\mathcal{A}$, au-dessus de U_0', et sa classe $[\mathcal{A}]$ appartient donc à $\mathrm{Br}_1 U_0'$. Son image réciproque $p'^*([\mathcal{A}])$ sur W' par $p': W' \to U_0'$ appartient donc à $\mathrm{Br}_1 W'$. Or, on a la suite exacte (cf. [34], §1) et le diagramme commutatif:

$$
\begin{array}{ccccc}
0 \to \mathrm{Br}_1 W \to \mathrm{Br}_1 W' & \xrightarrow{\;\{\partial_{2,w}\}\;} & \displaystyle\bigoplus_w H^1(k(w), \mathsf{Q}/\mathsf{Z}) \\[2ex]
\Big\uparrow{\scriptstyle p'^*} & & \Big\uparrow{\scriptstyle p^*} \\[2ex]
\mathrm{Br}_1 U_0' & \xrightarrow{\;\{\partial_{2,i}\}\;} & \displaystyle\bigoplus_i H^1(k(x_i), \mathsf{Q}/\mathsf{Z}),
\end{array}
$$

où w parcourt l'ensemble des points de codimension 1 de $W - W'$, où $i = 1, \ldots, s$, où $\partial_{2,w}$ désigne le "second résidu" en w et $\partial_{2,i}$ le second résidu en e_i, et où $k(w)$ désigne le corps résiduel de w et $k(e_i)$ celui de e_i. Dans le diagramme ci-dessus, l'application p^* est exactement induite par les inclusions naturelles des $k(e_i)$ dans $k(w)$, pour tous les couples (e_i, w) avec w au-dessus de e_i. L'absence de coefficients de ramification tient au fait que la fibre de p en chacun des e_i est un diviseur réduit. Fixons (e_i, w) avec w au-dessus de e_i. La fibration en coniques $X \to \mathsf{P}^1_k$ admet, au voisinage de e_i, une écriture locale du type

$$y^2 - a(x)z^2 = b(x) \cdot P_i(x)t^2 \quad \text{où } P_i(x) = N_{K_i/k}(x - e_i),$$

et où a et b sont des unités dans l'anneau local de e_i. Au voisinage de e_i, l'algèbre $\mathscr{A}$ équivaut donc à l'algèbre de quaternions $(a(x), b(x) \cdot P_i(x))$ et, par suite,

$$\partial_{2,i}(\mathscr{A}) \text{ est la classe de } a_i := a(e_i) \text{ dans } k(e_i)^*/k(e_i)^{*2} \cong H^1(k(e_i), \mathsf{Z}/2).$$

On en déduit que

$$\partial_{2,w}(p'^*(\mathscr{A})) \text{ est la classe de } a_i \text{ dans } k(w)^*/k(w)^{*2} \cong H^1(k(w), \mathsf{Z}/2).$$

Or, compte tenu du fait que u_i et v_i sont inversibles en w, l'équation

$$u - e_i v = \alpha_i\left(u_i^2 - a_i v_i^2\right)$$

montre que

$$a_i = u_i^2/v_i^2 \quad \text{dans } k(w)^*,$$

ce qui prouve finalement que

$$\partial_{2,w}(p'^*(\mathscr{A})) = 1.$$

Ainsi, la classe de l'algèbre d'Azumaya $p'^*(\mathscr{A})$ appartient à $\mathrm{Br}_1 W$, autrement dit l'algèbre $p'^*(\mathscr{A})$ est "non ramifiée" sur la variété W tout entière. D'après le lemme 2.6.1, $\mathrm{Br}_1 W = \mathrm{Br}\, k$. Il existe donc une algèbre constante $\mathscr{A}_0$ semblable à $p'^*(\mathscr{A})$ dans $\mathrm{Br}\, k(W)$. Comme $\mathscr{A}$ est une algèbre de quaternions, sa classe appartient au sous-groupe ${}_2\mathrm{Br}\, k(x)$ des éléments annulés par 2 dans $\mathrm{Br}\, k(x)$. Ainsi $p'^*(\mathscr{A})$ appartient à ${}_2\mathrm{Br}_1 W \subset {}_2\, \mathrm{Br}\, k(W)$, et la classe de $\mathscr{A}_0$ appartient à ${}_2\mathrm{Br}\, k$.

Pour établir l'assertion (ii), il suffit de prouver que, dans chacun des cas (a) à (d), on a:

(ii') la classe de $\mathscr{A}_0$ est celle d'une k-algèbre de quaternions, c'est-à-dire celle d'une k-conique C. En effet les deux fibrations en coniques $\mathscr{T}_U^\alpha \to W_0$ et $C \times_k W_0 \to W_0$ ont alors des fibres génériques isomorphes.

Dans le cas (a), tout élément de ${}_2\mathrm{Br}\, k$ est la classe d'une algèbre de quaternions, qui est même nulle dans le cas C_1.

Sous l'hypothèse (b), la classe de $\mathscr{A}$ dans $\mathrm{Br}\, k(x)$ est trivialisée par passage à une extension quadratique $k(\sqrt{a}\,)/k$. Il en est donc de même de la classe de $\mathscr{A}_0$ dans $\mathrm{Br}\, k$. D'où (ii').

Le cas (c) résulte immédiatement du lemme suivant appliqué à l'anneau local d'un point k-rationnel de la k-variété lisse W_α:

LEMME 2.6.7. *Soit k un corps de caractéristique $\neq 2$. Soit A une k-algèbre locale régulière de corps des fractions K et de corps résiduel κ. Si un élément $\beta \in \mathrm{Br}\, k$ a pour image dans $\mathrm{Br}\, K$ la classe d'une conique, il en est de même de son image dans $\mathrm{Br}\, \kappa$.*

Démonstration. Supposons d'abord A de valuation discrète complet. Comme toute classe d'algèbre de quaternions dans $\mathrm{Br}\, K$, la restriction de β à K peut s'écrire comme un symbole $(a, b)_2$ avec $a, b \in A^*$, ou bien $a \in A^*$ et b uniformisante. Dans le premier cas, comme $\mathrm{Br}\, A$ s'injecte dans $\mathrm{Br}\, K$, l'image de β dans $\mathrm{Br}\, A$ coïncide avec $(a, b)_2$. Par spécialisation à κ, on obtient le lemme. Dans le second cas, la classe de a dans κ^*/κ^{*2} est triviale, car c'est le résidu de l'algèbre non ramifiée $(a, b)_2 = \beta_K$. Ainsi, A étant complet, a est un carré dans A^* et la classe de $(a, b)_2$ dans $\mathrm{Br}\, K$ est nulle, donc aussi celle de β dans $\mathrm{Br}\, A$, puis, par spécialisation, dans $\mathrm{Br}\, k$.

Le cas où A est de valuation discrète résulte du précédent par complétion. Le cas général s'obtient par récurrence. En effet, si t est un paramètre régulier de A, le corps résiduel du localisé $A_{(t)}$, qui est un anneau de valuation discrète, est le corps des fractions de la k-algèbre locale régulière A/t, qui est de dimension $\dim(A) - 1$.

Considérons le cas (d). Vu (a), on peut se limiter à k infini. Quitte à faire un automorphisme de P^1_k, on peut alors supposer que $\mathscr{T}^\alpha$ possède un point k-rationnel dont la projection sur P^1_k appartient, avec les notations précédentes, à U_0, donc dont la projection M sur X appartient à $W_0(k)$. La fibration en coniques $\mathscr{T}^\alpha_U \to W_0$ est lisse, de fibre triviale en M. Par spécialisation en M, on voit donc que la classe de $\mathscr{A}_0$ dans $\mathrm{Br}\, k$ est nulle, si bien que $\mathscr{T}^\alpha$ est k-birationnel à $\mathsf{P}^1_k \times_k W_\alpha$.

Dans (e), la caractéristique de k étant nulle, on peut considérer une k-compactification lisse W_1 de W_0. On dispose alors d'une k-application rationnelle de la k-variété lisse $\mathscr{T}^\alpha$ dans la k-variété propre W_1. D'après le lemme de Nishimura ([45], [13] lemme 3.1) l'hypothèse $\mathscr{T}^\alpha(k) \neq \varnothing$ entraîne $W_1(k) \neq \varnothing$. On peut alors appliquer le lemme 2.6.7 aux données suivantes: A l'anneau local d'un k-point de W_1, β la classe de $\mathscr{A}_0$ dans $\mathrm{Br}\, k$ et la classe dans $\mathrm{Br}\, k(W_0) = \mathrm{Br}\, k(W_1)$ de la fibre générique de la fibration $\mathscr{T}^\alpha_U \to W_0$.

Il reste (iii). L'application $H^1(X, S_0) \to H^1(X, S)$ transforme, par produit contracté via (2.6.1), un torseur universel $\mathscr{T}$ en un torseur $\mathscr{T}^\alpha$ de type λ et on a ainsi, d'après (2.6.1) et le théorème 90 de Hilbert, une fibration $\mathscr{T} \to \mathscr{T}^\alpha$ localement triviale de groupe $\mathsf{G}_{m,\, k}$, d'où (iii).

Remarque 2.6.8.   Notons qu'on peut prouver que la classe de $p'{}^*(\mathscr{A})$ vient de k sans passer par le lemme 2.6.1. L'argument est essentiellement dû à P. Salberger. On prouve, par un calcul de résidus sur Ω analogue à celui de la démonstration précédente, que, dans $\mathrm{Br}\,\Omega$,

$$\mathscr{B}_0 := \kappa^*(\mathscr{A}) - \sum_i N_{K_i/k}\big(a_i, \alpha_i(u - e_iv)\big) \in {}_2\,\mathrm{Br}_1\Omega = {}_2\mathrm{Br}\,k,$$

la dernière égalité résultant, d'après (1.5.0), de la nullité de $\mathrm{Pic}\,\overline{\Omega}$, puis on montre (voir (2.6.14)) que dans $\mathrm{Br}\,W'$:

$$q^*\bigg(\sum_i N_{K_i/k}\big(a_i, \alpha_i(u - e_iv)\big)\bigg) = 0.$$

Ainsi, $\mathscr{A}_0 := p^*(\mathscr{A}) = q^*(\mathscr{B}_0) \in {}_2\mathrm{Br}\,k$. On trouve de plus, en spécialisant en $(u, v) = (1, 0)$:

$$\mathscr{A}_0 = \mathscr{A}(\infty) - \sum_i N_{K_i/k}\big((a_i, \alpha_i)\big) \in {}_2\mathrm{Br}\,k.$$

Remarque 2.6.9.   Ce théorème généralise aux surfaces fibrées en coniques quelconques sur $\mathbf{P}_k^1$ la description des torseurs universels obtenue dans le cas des surfaces de Châtelet "généralisées"

$$y^2 - az^2 = P(x),$$

avec $a \in k^*$ et $P \in k[x]$ séparable de degré 4 ([27],[2]). Plus précisément, le théorème 2.6.4(ii)(b) explique la "séparation de variables" qui se produit lors des calculs explicites dans ce cas particulier ([27], (7.14) et (7.15)).

Exemple 2.6.10.   C'est le cas d'une surface fibrée en coniques d'équation

$$y^2 - az^2 = P(x) \quad \text{avec } a \in k^* \text{ non carré et } P \text{ séparable de degré impair.}$$

Ce cas est traité en détail dans ([2], §2, remarques 5 et 6). L'hypothèse sur le degré de P implique que la fibre à l'infini de la fibration $X \to \mathbf{P}_k^1$ définie par x est singulière et possède en particulier un point k-rationnel. Ce cas est plus simple que le cas général; on peut en effet alors décrire directement les torseurs universels, car il existe dans ce cas une suite exacte de $\mathfrak{g}$-modules

$$0 \to P_1 \to P_2 \to \mathrm{Pic}\,\overline{X} \to 0,$$

où les P_i sont de permutation, coïncidant avec la suite (2.3.10). Une telle suite n'existe pas dans le cas général, comme on le voit sur l'exemple de Tsfasman (cf. [50],[40]) sur $k = \mathbf{Q}$:

$$y^2 - 221z^2 = (x^2 - 13)(x^2 - 17).$$

Pour cette surface, $H^1(\mathbf{Q}, \operatorname{Pic} \overline{X}) = \mathbf{Z}/2$, alors que, X étant $\mathbf{Q}_v$-rationnelle pour chaque place v de $\mathbf{Q}$, on a $H^1(\mathbf{Q}_v, \operatorname{Pic} \overline{X}) = 0$ pour toute place v, ce qui, vu le théorème de Tchebotarev, est incompatible avec l'existence d'une suite de $\mathfrak{g}$-modules du type ci-dessus.

Exemple 2.6.11.   Si, dans le cas précédent, P admet une racine dans k, on peut donner une description encore plus simple des torseurs universels. Le cas où P est scindé sur k est traité dans ([22], §4); ce cas inclut l'exemple des surfaces de Châtelet originelles

$$y^2 - az^2 = (x - e_1)(x - e_2)(x - e_3).$$

Exemple 2.6.12.   Le cas d'une surface fibrée en coniques d'équation

$$y^2 - az^2 = P_1(x) \cdots P_s(x)$$

où les P_i sont tous de degré pair et irréductibles sur $k' := k(\sqrt{a})$.

Dans ce cas, comme X est k'-rationnelle, on a considéré dans [13] et [24] des torseurs quasi-universels relativement à k'/k. On construit alors des variétés de descente intermédiaires:

$$P_i(x) = \alpha_i\left(u_i^2 - av_i^2\right) \qquad \alpha_i \in k^* \text{ et } \alpha_1 \ldots \alpha_s = 1$$

par simple factorisation sur le corps de base. Cette méthode a l'avantage de fournir des variétés dont on arrive à trouver plus facilement des équations. L'exemple

$$y^2 - az^2 = P(x) \quad \text{avec } P \text{ irréductible}$$

montre les limites de cette méthode qui, en l'occurrence, ne donne rien, alors que la considération des véritables torseurs universels ou admissibles peut donner des informations supplémentaires (cf. [27]).

2.7. *Accouplements avec les points rationnels.*   On s'intéresse ici à une k-variété algébrique X telle que $X(k) \neq \varnothing$, qui sera souvent rationnelle ou proche d'une variété rationnelle, et à la "différence" qui peut exister entre X et l'espace projectif de même dimension.

En particulier, faute de pouvoir paramétrer de façon essentiellement biunivoque les points rationnels de X, on aimerait trouver une partition de $X(k)$, indexée par $i \in I$, en classes paramétrées par des "cartes k-rationnelles," i.e. par des k-morphismes dominants

$$\Psi_i \colon Z_i \to X$$

où Z_i soit une variété k-rationnelle. On aimerait même en trouver une optimale, i.e. avec $\#I$ minimal.

A défaut, on aimerait au moins calculer la R-équivalence sur $X(k)$. Cette relation, introduite par Manin ([42] chap. II, §4 (= II.14)), est ainsi définie: deux points P_0 et $P_1 \in X(k)$ sont directement R-équivalents s'il existe une k-application rationnelle $\gamma\colon \mathbf{P}_k^1 \cdots \to X$ définie en 0 et ∞, telle que $\gamma(0) = P_0$ et $\gamma(\infty) = P_1$, et ils sont R-équivalents s'il existe une chaîne finie de points directement R-équivalents, qui relie P_0 à P_1.

2.7.0. Une approche de ces questions est fournie par la considération des partitions de $X(k)$ définies par les accouplements avec des torseurs sur X sous un k-groupe algébrique de type multiplicatif S. Ces accouplements

$$(2.7.1) \qquad X(k) \times H^1(X, S) \to H^1(k, S)$$

$$(P, [\mathcal{T}]) \mapsto \mathcal{T}(P) := [\mathcal{T}_P]$$

où $\mathcal{T}_P = \mathcal{T} \times_X \operatorname{Spec} k(P)$, et où $\mathcal{T}(P)$ désigne la classe de ce torseur dans $H^1(k(P), S)$, dérivent de la fonctorialité contravariante en X du groupe $H^1(X, S)$ et sont additifs à droite et fonctoriels en S. On note $q_{\mathcal{T}}\colon \mathcal{T} \to X$ le morphisme structural du torseur $\mathcal{T}$. On a le lemme suivant, évident, mais essentiel:

LEMME 2.7.1. *Avec les notations ci-dessus, les conditions suivantes, qui ne dépendent que de la classe $[\mathcal{T}]$ de $\mathcal{T}$ dans $H^1(X, S)$ sont équivalentes:*
(i) $\mathcal{T}(P) = 0$ *dans* $H^1(k, S)$;
(ii) $\mathcal{T}_P(k) \neq \varnothing$;
(iii) P *appartient à* $q_{\mathcal{T}}(\mathcal{T}(k))$.

Compte tenu de ce lemme, il nous arrivera, dans l'étude des points rationnels de X, de ne pas faire la distinction entre un torseur $\mathcal{T}$ et sa classe $[\mathcal{T}] \in H^1(X, S)$.

La *partition* définie par un torseur $\mathcal{T}$ est constituée des fibres de l'application

$$\theta_{\mathcal{T}}\colon X(k) \to H^1(k, S)$$

qu'il définit par $\theta_{\mathcal{T}}(P) = \mathcal{T}(P)$. Pour $\alpha \in H^1(k, S)$, notons $q_\alpha\colon \mathcal{T}^\alpha \to X$ un torseur de classe $[\mathcal{T}] - \alpha$ dans $H^1(X, S)$. La *partition* définie par $\mathcal{T}$ s'écrit

$$(2.7.2) \qquad X(k) = \bigcup_{\alpha \in \operatorname{im} \theta_{\mathcal{T}}} q_\alpha(\mathcal{T}^\alpha(k)).$$

L'ensemble $\operatorname{Tors}(X, S)$ des classes de torseurs sur X sous un S donné définit une partition, a priori plus fine, formée des fibres de l'application

$$X(k) \to \operatorname{Hom}(H^1(X, S), H^1(k, S))$$

$$P \to ([\mathcal{T}] \mapsto \mathcal{T}(P)).$$

Comme on le voit ci-après, si X est propre sur k, ces partitions sont a priori plus grossières que celle de la R-équivalence, mais, dans certains cas, il y a coïncidence, ce qui donne alors un excellent moyen d'étude de la R-équivalence. D'autre part, si X est propre et lisse sur k de caractéristique 0, et si l'une de ces partitions est non triviale, X ne peut être k-rationnelle ([20] proposition 10), ce qui donne parfois un test très commode et calculatoire de non-k-rationalité.

PROPOSITION 2.7.2. *Si X est une k-variété algébrique propre et S un k-groupe de type multiplicatif, l'accouplement (2.7.1) passe au quotient par la R-équivalence:*

$$(X(k)/R) \times H^1(X, S) \to H^1(k, S).$$

Autrement dit, l'équivalence définie par les torseurs sous S est plus grossière que la R-équivalence. Pour le démontrer, il suffit de considérer deux points P_0 et $P_1 \in X(k)$ directement R-équivalents, ce qui, par fonctorialité en X, X étant propre, revient à traiter le cas $X = \mathsf{P}^1_k$. Il suffit même de savoir traiter le cas, non propre, $X = \mathsf{A}^1_k$. Or, le théorème 1.5.1 et la proposition 1.4.3 montrent que, si Y est une k-variété qui vérifie

$$(2.7.3) \qquad \overline{k}[Y]^* = \overline{k}^* \quad \text{et} \quad \operatorname{Pic} \overline{Y} = 0,$$

ce qui est le cas pour A^1_k, alors

$$H^1(k, S) = H^1(Y, S),$$

ce qui achève la preuve.

THÉORÈME 2.7.3. *Soient X une k-variété algébrique propre et $\mathscr{T}$ un torseur sur X sous un k-groupe de type multiplicatif S. Si k est de type fini sur le corps premier, l'image de l'application*

$$X(k) \xrightarrow{\ \mathscr{T}\ } H^1(k, S)$$

$$P \mapsto \mathscr{T}(P),$$

est finie.

Autrement dit, la partition de $X(k)$ définie par le torseur $\mathscr{T}$ est *finie*. La démonstration donnée dans ([22], proposition 2) dans le cas où S est un tore vaut aussi bien pour S de type multiplicatif.

Si X est une k-variété algébrique propre géométriquement intègre telle que $X(k) \neq \varnothing$, le début de la suite exacte (2.0.2) forme une suite exacte dont tout point $P \in X(k)$ définit un scindage:

$$(2.2.5) \qquad 0 \to H^1(k, S) \to H^1(X, S) \to \operatorname{Hom}_{\mathfrak{g}}(\hat{S}, \operatorname{Pic} \overline{X}) \to 0.$$

On en déduit que deux torseurs de même type λ diffèrent par un élément de $H^1(k, S)$ et définissent donc la même partition de $X(k)$. Parmi toutes les partitions de $X(k)$ ainsi définies—toutes plus grossières que la R-équivalence—celle définie par un torseur universel, lorsqu'il y en a, est la plus fine:

PROPOSITION 2.7.4. *Soit X une k-variété algébrique propre et géométriquement intègre. On suppose $X(k) \neq \varnothing$ et le $\mathfrak{g}$-module $\operatorname{Pic} \overline{X}$ de type fini. La relation d'équivalence définie sur $X(k)$ par un torseur universel est plus fine que celle définie par tout autre torseur sous un k-groupe de type multiplicatif.*

Démonstration. Les hypothèses entraînent l'existence de torseurs universels et leur groupe structural S_0 est un k-groupe de type multiplicatif. Deux torseurs d'un même type

$$\lambda: \hat{S} \to \operatorname{Pic} \overline{X},$$

par exemple deux torseurs universels, définissent la même partition de $X(k)$. Soit $P \in X(k)$. Soient $\mathcal{T}_0^P$ le torseur universel de fibre triviale en P, et $\mathcal{T}^P$ le torseur de type λ, de fibre triviale en P. Soit $D(\lambda)$ le k-morphisme $S_0 \to S$ dual de λ. Il induit une application $H^1(X, S_0) \to H^1(X, S)$ qui, par fonctorialité des scindages des suites (2.2.5) définis par P, envoie $\mathcal{T}_0^P$ sur $\mathcal{T}^P$. L'application composée

$$X(k) \xrightarrow{\ \mathcal{T}_0^P\ } H^1(k, S_0) \xrightarrow{\ D(\lambda)\ } H^1(k, S)$$

est donc l'application définie par $\mathcal{T}^P$, ce qui prouve la proposition.

PROPOSITION 2.7.5. *Soit X une k-variété algébrique intègre de corps des fonctions $k(X)$, et soit S un k-groupe de type multiplicatif. Soit $[\mathcal{T}] \in H^1(X, S)$ la classe d'un torseur sur X sous S. Si $P \in X(k)$ est lisse, la fibre $\mathcal{T}(P) = [\mathcal{T}_P] \in H^1(k, S)$ de $\mathcal{T}$ en P ne dépend que de l'image de $\mathcal{T}$ dans $H^1(k(X), S)$.*

Démonstration. Soit $\mathcal{O}_{X, P}$ l'anneau local de X en P. L'application $H^1(X, S) \to H^1(k(X), S)$ se factorise en

$$H^1(X, S) \to H^1(\mathcal{O}_{X, P}, S) \to H^1(k(X), S).$$

La fibre $\mathcal{T}(P)$ de $\mathcal{T}$ en P ne dépend évidemment que de l'image de $\mathcal{T}$ dans $H^1(\mathcal{O}_{X, P}, S)$. Pour établir la proposition, il suffit de montrer que, $\mathcal{O}_{X, P}$ étant régulier, donc géométriquement localement factoriel, l'application $H^1(\mathcal{O}_{X, P}, S) \to H^1(k(X), S)$ est injective (voir aussi [21] et [26] §4). Par passage à la limite dans le diagramme (1.6.2), ceci se ramène à établir l'injectivité de l'application.

$$\operatorname{Ext}^1_{\mathfrak{g}}\left(\hat{S}, \mathcal{O}^*_{\overline{X}, P}\right) \to \operatorname{Ext}^1_{\mathfrak{g}}\left(\hat{S}, \overline{k}(X)^*\right).$$

Or l'extension de $\mathfrak{g}$-modules $1 \to \mathcal{O}^*_{\overline{X}, P} \to \overline{k}(X)^* \to \mathrm{Div}_{\{P\}}\overline{X} \to 0$ est scindée, car $\mathrm{Div}_{\{P\}}\overline{X}$ étant un $\mathfrak{g}$-module de permutation, le théorème 90 de Hilbert implique

$$\mathrm{Ext}^1_{\mathfrak{g}}\left(\mathrm{Div}_{\{P\}}\overline{X}, \mathcal{O}^*_{\overline{X}, P}\right) = 0.$$

On déduit de cette proposition une version "fonctionnelle" et "calculatoire" de l'accouplement de $X(k)$ avec $H^1(X, S)$ en introduisant, sous les hypothèses de 2.7.5, le groupe

$$(2.7.4) \qquad D^S(X) := \mathrm{im}\left(H^1(X, S) \to H^1(k(X), S)\right).$$

La proposition assure l'existence, pour X géométriquement intègre et lisse, d'un accouplement

$$(2.7.5) \qquad X(k) \times D^S(X) \to H^1(k, S),$$

induit par (2.7.1), linéaire à droite. Le lemme qui suit "calcule" $D^S(X)$ de façon "fonctionnelle":

LEMME 2.7.6. *Si X est une k-variété algébrique géométriquement intègre et S un k-groupe de type multiplicatif,*

$$D^S(X) = \ker\left(\mathrm{Ext}^1_{\mathfrak{g}}(\hat{S}, \overline{k}(X)^*) \to \mathrm{Ext}^1_{\mathfrak{g}}(\hat{S}, \mathrm{Div}\,\overline{X})\right).$$

Démonstration. D'après la proposition 1.4.3, le morphisme naturel $H^1(X, S) \to H^1(k(X), S)$ s'écrit encore

$$\mathrm{Ext}^1_{X_{\text{ét}}}(\hat{S}, \mathsf{G}_{m, X}) \to \mathrm{Ext}^1_{\eta_{\text{ét}}}(\hat{S}, \mathsf{G}_{m, \eta})$$

où η désigne le point générique de X. L'edge de la suite spectrale de Leray pour le morphisme $i : \eta \to X$ donne une injection

$$\mathrm{Ext}^1_{X_{\text{ét}}}(\hat{S}, i_*\mathsf{G}_{m, \eta}) \to \mathrm{Ext}^1_{\eta_{\text{ét}}}(\hat{S}, \mathsf{G}_{m, \eta}),$$

si bien que $D^S(X)$ est encore l'image de l'application naturelle

$$\mathrm{Ext}^1_{X_{\text{ét}}}(\hat{S}, \mathsf{G}_{m, X}) \to \mathrm{Ext}^1_{X_{\text{ét}}}(\hat{S}, i_*\mathsf{G}_{m, \eta}).$$

La suite exacte de faisceaux étales sur X

$$1 \to \mathsf{G}_{m, X} \to i_*\mathsf{G}_{m, \eta} \to \mathscr{D}iv_X \to 0,$$

qui définit le faisceau étale $\mathscr{D}iv_X$, identifie donc $D^S(X)$ au noyau de

$$\mathrm{Ext}^1_{X_{\text{ét}}}(\hat{S}, i_*\mathsf{G}_{m, \eta}) \to \mathrm{Ext}^1_{X_{\text{ét}}}(\hat{S}, \mathscr{D}iv_X).$$

Soit alors $p\colon X \to \operatorname{Spec} k$ le morphisme structural. L'edge de la suite spectrale

$$\operatorname{Ext}^p_{k_{\text{ét}}}(\hat{S}, R^q p_* \mathscr{G}) \Rightarrow \operatorname{Ext}^{p+q}_{X_{\text{ét}}}(p^*\hat{S}, \mathscr{G}),$$

où $\mathscr{G}$ est un faisceau étale sur X, donne, pour $\mathscr{G} = i_* \mathsf{G}_{m,\eta}$ et $\mathscr{G} = \mathscr{D}iv_X$, le carré commutatif

$$\begin{array}{ccc}
\operatorname{Ext}^1_{k_{\text{ét}}}(\hat{S}, p_* i_* \mathsf{G}_{m,\eta}) & \overset{\delta}{\longrightarrow} & \operatorname{Ext}^1_{k_{\text{ét}}}(\hat{S}, p_* \mathscr{D}iv_X) \\
{\scriptstyle\cong}\downarrow & & \uparrow \\
\operatorname{Ext}^1_{X_{\text{ét}}}(\hat{S}, i_* \mathsf{G}_{m,\eta}) & \longrightarrow & \operatorname{Ext}^1_{X_{\text{ét}}}(\hat{S}, \mathscr{D}iv_X),
\end{array}$$

dont la verticale de droite est injective—c'est un edge i_1—et dont la verticale de gauche est un isomorphisme parce que $R^1 p_*(i_* \mathsf{G}_{m,\eta}) = 0$, par application du théorème 90 de Hilbert. Finalement, $D^S(X) = \ker(\delta)$. La suite exacte définissant $\mathscr{D}iv_X$ restant exacte par restriction à X_{Zar} puisque $H^1(A, \mathsf{G}_m) = \operatorname{Pic} A = 0$ pour un anneau local A, le faisceau $p_* \mathscr{D}iv_X$ s'identifie au module galoisien $\operatorname{Div} \bar{X}$, ce qui achève la démonstration.

Remarque 2.7.7.   Si S est déployé par une extension galoisienne K/k, on peut remplacer dans le lemme $\bar{k}/k$ par K/k, puis $\mathfrak{g} = \operatorname{Gal}(\bar{k}/k)$ par $G = \operatorname{Gal}(K/k)$, et $\bar{k}(X)$ par $K(X)$, enfin $\operatorname{Div} \bar{X}$ par $\operatorname{Div} X_K$.

Exemples 2.7.8.

(a) $S = \mu_{n,k}$ pour n entier > 0. On trouve

$$D^S(X) = \ker\big(k(X)^*/k(X)^{*n} \to \operatorname{Div}(X)/n\big),$$

i.e. le groupe des fonctions rationnelles, dont le diviseur est un multiple de n, modulo celles qui sont des puissances n-ièmes. Ce groupe apparaît naturellement dans l'étude des isogénies de courbes elliptiques.

(b) $S = R^1_{K/k}\mathsf{G}_m$ pour K/k une extension de corps, séparable et finie. On trouve

$$D^S(X) = \ker\big(k(X)^*/N_{K/k}K(X)^* \to \operatorname{Div}(X)/N_{K/k}(\operatorname{Div} X_K)\big),$$

i.e. le groupe des fonctions rationnelles, dont le diviseur est une norme d'un diviseur de X_K, modulo celles qui sont des normes de $K(X)^*$.

(c) $S = R_{K/k}\mathsf{G}_m/\mathsf{G}_{m,k}$ pour K/k une extension finie de corps, galoisienne de groupe G. On trouve

$$D^S(X) = \ker\big(H^2(G, K(X)^*) \to H^2(G, \operatorname{Div}(X_K))\big),$$

qui n'est autre que le groupe $\operatorname{Br}(X, K)$ considéré par Manin dans ([42], VI 1.3

($= $ VI 41.3)). Pour X lisse sur k,

$$\mathrm{Br}(X, K) = \ker(\mathrm{Br}\, X \to \mathrm{Br}\, X_K)$$

(cf. [20], lemme 14, p. 213).

Exemples 2.7.9. Nous allons expliciter, pour X/k lisse, l'accouplement (2.7.5) dans chacun des exemples 2.7.8. La méthode de calcul utilise la proposition 2.7.5 et le lemme 2.7.6.

(a) L'accouplement

$$X(k) \times D^S(X) \to k^*/k^{*n}$$

s'obtient ainsi: soient $P \in X(k)$ et $\alpha \in D^S(X) = \ker(k(X)^*/n \to \mathrm{Div}(X)/n)$, représenté par $f \in k(X)^*$, il existe $g \in \mathcal{O}^*_{X, P}$ telle que $g/f \in k(X)^{*n}$, et alors $\langle P, \alpha \rangle = \mathrm{cl}(g(P)) \in k^*/k^{*n}$.

(b) L'accouplement

$$X(k) \times D^S(X) \to k^*/N(K^*)$$

s'obtient ainsi: soient $P \in X(k)$ et $\alpha \in D^S(X) = \ker(k(X)^*/N(K(X)^*) \to \mathrm{Div}(X)/N(\mathrm{Div}\, X_K))$, représenté par $f \in k(X)^*$, il existe $g \in \mathcal{O}^*_{X, P}$ telle que $g/f \in N(K(X)^*)$, et alors $\langle P, \alpha \rangle = \mathrm{cl}(g(P)) \in k^*/N(K^*)$.

(c) L'accouplement

$$X(k) \times D^S(X) \to \mathrm{Br}(k, K) = \ker(\mathrm{Br}\, k \to \mathrm{Br}\, K)$$

s'obtient ainsi: soient $P \in X(k)$ et $\alpha \in D^S(X) = \ker(H^2(G, K(X)^*) \to H^2(G, \mathrm{Div}\, X_K))$; on peut représenter α par un 2-cocycle a de G à valeurs dans $\mathcal{O}^*_{X_K, P}$, et $\langle P, \alpha \rangle$ est alors la classe dans $\mathrm{Br}(k, K)$ de $a(P)$. On retrouve ainsi l'accouplement avec des algèbres d'Azumaya considéré par Manin ([42] VI 1.7 et 1.8 ($=$ VI 41.7 et 41.8)). Ceux-ci apparaissent donc comme des cas particuliers d'accouplements définis par des torseurs sous des tores.

(d) Pour d'autres calculs explicites, faisant intervenir d'autres groupes S, voir par exemple [50], §3, et, pour le cas des surfaces de Châtelet d'équation $y^2 - az^2 = (x - e_1)(x - e_2)(x - e_3)$, [22] (§IV), et [30] (proposition 4.7).

La proposition suivante sera utilisée de façon cruciale au §3 dans la démonstration du théorème 3.5.1.

PROPOSITION 2.7.10. *Soit X une k-variété algébrique, lisse et géométriquement intègre, telle que*

$$\bar{k}[X]^* = \bar{k}^*.$$

Soit σ une $\mathfrak{g}$-rétraction de la flèche naturelle $\bar{k}^* \to \bar{k}(X)^*$. Soit S un k-groupe de type multiplicatif. Soit $\mathcal{T}^\sigma$ le torseur sur X sous S, de type λ, trivial "en σ". Soit *enfin*

$$t_\sigma\colon H^1(\mathfrak{g}, \operatorname{Pic} \overline{X}) \to \operatorname{Br}_1 X$$

la section de la flèche naturelle, définie par σ. Le diagramme suivant est commutatif, au signe près:

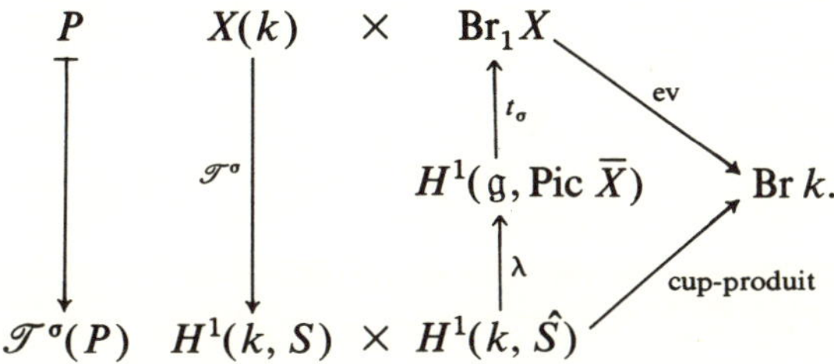

Démonstration. Par définition, l'existence d'une $\mathfrak{g}$-rétraction σ signifie que l'obstruction élémentaire est nulle. On sait en outre, d'après la proposition 2.2.8(iv)—voir aussi 2.3.4—que σ définit un torseur $\mathcal{T}^\sigma$ de type λ, trivial "en σ," et un seul dans $H^1(X, S)$. Dans le diagramme commutatif suivant, les verticales sont des suites exactes naturelles: l'exactitude de celle de droite résulte de $H^1(\mathfrak{g}, \operatorname{Div} \overline{X}) = 0$ (lissité de X et lemme de Shapiro), pour celle du milieu voir 1.5.0:

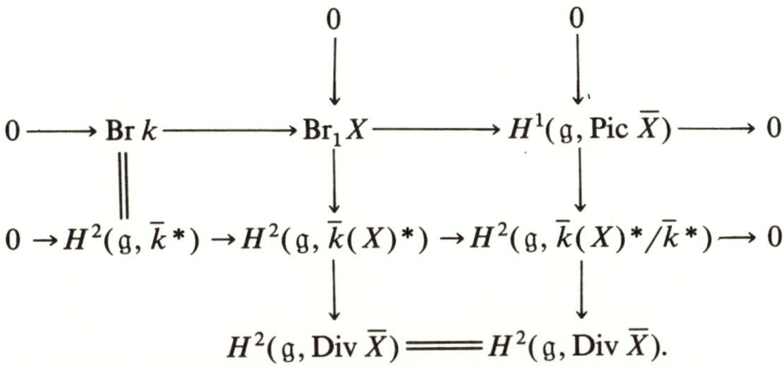

La nullité de l'obstruction élémentaire assure que, dans ce diagramme, dont la première ligne est extraite de (1.5.0), la deuxième suite exacte horizontale est scindée par σ, ce qui induit la rétraction t_σ et établit l'exactitude de la première.

Soit $P \in X(k)$. La flèche $\operatorname{Br}_1 X \to H^2(\mathfrak{g}, \overline{k}(X)^*)$ se factorise par $H^2(\mathfrak{g}, \mathcal{O}^*_{\overline{X}, P})$, et l'évaluation $\operatorname{ev}_P\colon \operatorname{Br}_1 X \to \operatorname{Br} k$ en P est induite par l'évaluation $\operatorname{ev}_P\colon H^2(\mathfrak{g}, \mathcal{O}^*_{\overline{X}, P}) \to H^2(\mathfrak{g}, \overline{k}^*)$.

Considérons l'extension naturelle de $\mathfrak{g}$-modules

$$1 \to \mathcal{O}^*_{\overline{X}, P}/\overline{k}^* \to \operatorname{Div}_{\overline{X}-\{P\}} \overline{X} \to \operatorname{Pic} \overline{X} \to 0,$$

et celle qui s'en déduit en poussant par la section $\mathcal{O}_{\overline{X}, P}^{*}/\overline{k}^{*} \to \mathcal{O}_{\overline{X}, P}^{*}$ définie par σ et en prenant le pull-back par $\lambda\colon \hat{S} \to \mathrm{Pic}\ \overline{X}$, soit:

$$1 \to \mathcal{O}_{\overline{X}, P}^{*} \to E \to \hat{S} \to 0.$$

En passant à la limite sur les ouverts U contenant P dans le diagramme (1.6.2) et dans le lemme 1.6.4, on vérifie sans peine que la restriction de $\mathcal{T}^{\sigma}$ à $\mathrm{Spec}\ \mathcal{O}_{X, P}$, qui appartient à $H^{1}(\mathcal{O}_{X, P}, S) = \mathrm{Ext}_{\mathfrak{g}}^{1}(\hat{S}, \mathcal{O}_{\overline{X}, P}^{*})$ s'interprète, au signe près, comme l'extension E ci-dessus. La valeur $\mathcal{T}^{\sigma}(P)$ de $\mathcal{T}^{\sigma}$ en P appartient à $H^{1}(k, S) = \mathrm{Ext}_{\mathfrak{g}}^{1}(\hat{S}, \overline{k}^{*})$ et s'obtient donc, en tant qu'extension, en poussant l'extension E par l'évaluation ev_{P} en $P\colon \mathcal{O}_{\overline{X}, P}^{*} \to \overline{k}^{*}$. On a ainsi le diagramme commutatif

$$
\begin{array}{ccccccccc}
1 & \to & \mathcal{O}_{\overline{X}, P}^{*} & \to & E_{0} & \to & \mathrm{Pic}\ \overline{X} & \to & 0 \\
 & & \parallel & & \uparrow & & \uparrow \lambda & & \\
1 & \to & \mathcal{O}_{\overline{X}, P}^{*} & \to & E & \longrightarrow & \hat{S} & \longrightarrow & 0 \\
 & & \downarrow & & \downarrow & & \parallel & & \\
1 & \to & \overline{k}^{*} & \longrightarrow & E_{P} & \longrightarrow & \hat{S} & \longrightarrow & 0,
\end{array}
$$

et, étant donné $\beta \in H^{1}(k, \hat{S})$, le cup-produit $\mathcal{T}^{\sigma}(P) \cup \beta$ coïncide au signe près, d'après la définition du cup-produit à la Yoneda, avec $\partial_{E_{P}}(\beta)$, donc aussi avec l'image de β par l'application composée:

$$H^{1}(k, \hat{S}) \to H^{1}(\mathfrak{g}, \mathrm{Pic}\ \overline{X}) \xrightarrow{\ \partial_{E_{0}}\ } H^{2}\big(\mathfrak{g}, \mathcal{O}_{\overline{X}, P}^{*}\big) \xrightarrow{\ \mathrm{ev}_{P}\ } H^{2}\big(\mathfrak{g}, \overline{k}^{*}\big).$$

Or, par définition même de t_{σ}, on a le diagramme commutatif:

$$
\begin{array}{ccc}
H^{1}(\mathfrak{g}, \mathrm{Pic}\ \overline{X}) & \xrightarrow{\ t_{\sigma}\ } & \mathrm{Br}_{1} X \\
\partial_{0} \downarrow & & \downarrow \\
H^{2}(\mathfrak{g}, \mathcal{O}_{\overline{X}, P}^{*}/\overline{k}^{*}) & \xrightarrow{\ \sigma\ } & H^{2}(\mathfrak{g}, \mathcal{O}_{\overline{X}, P}^{*}).
\end{array}
$$

D'où, finalement, en utilisant $\sigma\partial_{0} = \partial_{E_{0}}$, les égalités:

$$\big(t_{\sigma}\lambda(\beta)\big)(P) = \big(\sigma\big(\partial_{0}(\lambda(\beta))\big)\big)(P) = \mathrm{ev}_{P}\big(\partial_{E_{0}}(\lambda(\beta))\big) = \mathcal{T}^{\sigma}(P) \cup \beta,$$

ce qui achève la démonstration.

Remarque 2.7.11. Il est facile de déduire de ce qui précède les résultats suivants pour une k-variété algébrique X, lisse et géométriquement intègre, telle que $X(k) \neq \varnothing$:

(i) l'équivalence de Brauer sur $X(k)$ définie par $\mathrm{Br}_1 X$ est moins fine que celle définie par les accouplements avec les torseurs sur X sous les k-tores (utiliser 2.7.5, 2.7.6 et 2.7.8(c));

(ii) si $\overline{k}[X]^* = \overline{k}^*$ et si $\mathrm{Pic}\,\overline{X}$ est un groupe abélien de type fini, l'équivalence de Brauer associée à $\mathrm{Br}_1 X$ est moins fine que celle définie par un torseur universel (résulte de (i) et de 2.7.4);

(iii) sous les hypothèses de (ii), si, en outre, k est un corps $\mathfrak{p}$-adique, l'équivalence de Brauer et celle définie par un torseur universel coïncident.

On doit cependant noter que ce dernier résultat (iii), qui résulte du fait que, pour k $\mathfrak{p}$-adique, le cup-produit

$$H^1(k, S_0) \times H^1\big(k, \hat{S}_0\big) \to \mathrm{Br}(k) = \mathsf{Q}/\mathsf{Z}$$

est une dualité parfaite de groupes finis ([52] II.40) et de la compatibilité établie dans la proposition précédente, n'est plus vrai en général. On trouvera dans [20], proposition 21, p. 224, et dans [50] V.2.c, des exemples de k-variétés rationnelles X propres et lisses pour lesquelles l'équivalence de Brauer sur $X(k)$ est triviale, alors que celle définie par un torseur universel ne l'est pas.

Il se trouve que dans le cas d'un modèle projectif et lisse X de la k-surface d'équation affine

$$y^2 - az^2 = P(x),$$

où $P \in k[x]$ est un polynôme *scindé* sur k, ou dans le cas de la surface cubique X d'équation homogène

$$x^3 + y^3 + z^3 + at^3 = 0, \qquad \sqrt[3]{1} \in k,$$

l'équivalence de Brauer et celle définie par un torseur universel coïncident, cela étant dû au fait que dans ces cas S_0 est, à des facteurs quasi-triviaux près, un produit de k-tores $R^1_{K/k}\mathsf{G}_m$ pour K/k cyclique, auquel cas $R^1_{K/k}\mathsf{G}_m \cong R_{K/k}\mathsf{G}_m/\mathsf{G}_{m,k}$, ce qui ramène à l'équivalence de Brauer (cf. 2.7.8(c)). C'est le cas en particulier pour les surfaces de Châtelet d'équation affine

$$y^2 - az^2 = (x - e_1)(x - e_2)(x - e_3),$$

ce qui explique le rôle particulier que joue dans ce cas-là l'équivalence de Brauer, comme introduite et utilisée par Manin ([42] chap. VI). Mais dans beaucoup d'autres situations, l'équivalence de Brauer est trop grossière et celle définie par un torseur universel peut alors la remplacer de façon utile (cf. [20], [40], [50]).

2.8. *La première hypothèse sur les torseurs universels.* Soit X une k-variété rationnelle propre et lisse définie sur un corps k de caractéristique zéro, de clôture algébrique $\overline{k}$. On note $\mathfrak{g} = \mathrm{Gal}(\overline{k}/k)$ et S_0 le k-tore dual du $\mathfrak{g}$-module

Z-libre de type fini Pic $\overline{X}$ (voir appendice 2.A). Dans ce paragraphe, nous discutons l'hypothèse suivante et ses conséquences:

(H1) *Les torseurs universels $\mathcal{T}$ sur X qui possèdent un point k-rationnel sont des variétés k-rationnelles.*

Il y a aussi intérêt à considérer l'hypothèse plus faible:

(H1′) *Les torseurs universels $\mathcal{T}$ sur X qui possèdent un point k-rationnel sont des variétés stablement k-rationnelles* (i.e. qui deviennent k-rationnelles après multiplication avec un k-espace projectif convenable).

De fait, il est également utile de considérer les hypothèses analogues pour les torseurs d'autres types $\lambda \in \mathrm{Hom}_{\mathfrak{g}}(\hat{S}, \mathrm{Pic}\,\overline{X})$ pour S un k-tore quelconque.

(H1)$_\lambda$ *Les torseurs $\mathcal{T}$ sur X de type λ qui possèdent un point k-rationnel sont des variétés k-rationnelles.*

(H1′)$_\lambda$ *Les torseurs $\mathcal{T}$ sur X de type λ qui possèdent un point k-rationnel sont des variétés stablement k-rationnelles.*

Notons que $\mathcal{T}$ est une k-variété rationnelle: ceci résulte du fait que c'est un torseur sous un k-tore sur une k-variété rationnelle: sur $\overline{k}$, $\overline{\mathcal{T}}$ est un torseur sur $\overline{X}$ sous un tore $\mathbf{G}_m^n$, donc localement trivial et $\overline{\mathcal{T}}$ est $\overline{k}$-birationnellement équivalent au produit de $\overline{X}$ et de $\mathbf{G}_m^n$.

L'hypothèse (H1) est motivée par le théorème 2.1.2(a), qui assure que, pour toute k-compactification lisse $\mathcal{T}^c$ de $\mathcal{T}$, le $\mathfrak{g}$-module Pic $\overline{\mathcal{T}}^c$ est de permutation. Ainsi l'obstruction à la k-rationalité décrite dans l'appendice 2.A disparaît-elle sur la k-variété rationnelle $\mathcal{T}^c$. Il en est de même pour les torseurs admissibles.

Dans les énoncés conditionnels ci-après, on fait les hypothèses (H1)$_\lambda$ ou (H1′)$_\lambda$. Mais c'est seulement pour les torseurs universels, ou pour les torseurs admissibles, que le théorème 2.1.2(a) donne quelque espoir de validité pour (H1)$_\lambda$ ou (H1′)$_\lambda$. Même pour ces torseurs, ces hypothèses ne valent pas en général (voir 2.8.18 ci-après) mais on ne connaît pas de contre-exemple lorsque X est une k-surface rationnelle projective et lisse.

L'hypothèse (H1) a été vérifiée lorsque X est une surface de Châtelet généralisée [27] ou une surface fibrée en coniques avec 4 fibres géométriques dégénérées [28], c'est-à-dire que (H1) vaut pour les surfaces rationnelles non-triviales les plus simples (du point de vue de la classification d'Enriques-Manin-Iskovskih, cf. [43] §3). Cette hypothèse vaut aussi pour X une k-compactification lisse d'un k-tore ([20]; [22], §III). Il n'est donc pas déraisonnable de décrire des conséquences concrètes de (H1) et (H1′).

PROPOSITION 2.8.1. *Si X vérifie* (H1′)$_\lambda$, *et si $X(k)$ est non vide, X est k-unirationnelle et $X(k)$ est Zariski-dense dans X.*

Démonstration. Il existe alors un torseur $\mathscr{T}$ de type λ sur X, de fibre triviale en un k-point M de X, donc tel que $\mathscr{T}(k)$ soit non vide (lemme 2.7.1).

Remarque 2.8.2. La première de ces conséquences impliquerait que toute variété algébrique complexe de dimension trois fibrée en coniques sur le plan projectif complexe est unirationnelle, ce qui est un problème ouvert bien connu. La seconde de ces conséquences, plus faible, est inconnue déjà pour des surfaces rationnelles très simples sur $\mathbb{Q}$ (voir [24], où dans certains cas l'on déduit cette conséquence d'une hypothèse bien connue en théorie des nombres). La k-unirationalité a récemment été établie par Yančevskiĭ [56] dans le cas où X est une surface fibrée en coniques sur la droite projective sur un corps k $\mathfrak{p}$-adique (le cas $k = \mathbb{R}$ avait été obtenu par Iskovskikh).

PROPOSITION 2.8.3. *Si X vérifie* (H1$'$) *et $X(k) \neq \varnothing$, et si le $\mathfrak{g}$-module* $\operatorname{Pic} \overline{X}$ *est stablement de permutation, resp. est facteur direct d'un module de permutation, la k-variété X est stablement k-rationnelle, resp. est facteur direct k-birationnel d'une k-variété k-rationnelle.*

Démonstration. Il existe un k-tore quasi-trivial S_1, resp. un k-tore S_1, tel que le k-tore $S_2 = S_0 \times_k S_1$ soit quasi-trivial. Le théorème 90 de Hilbert sous la forme de Grothendieck implique qu'alors tout torseur sous S_0 est localement trivial pour la topologie de Zariski. Si donc $\mathscr{T}$ est un torseur universel sur X, la k-variété $\mathscr{T}$ est k-birationnelle au produit $X \times_k S_0$. Sous (H1$'$), ceci suffit à conclure que X est facteur direct k-birationnel d'une variété k-rationnelle. Par ailleurs $\mathscr{T} \times_k S_1$ est k-birationnel à $X \times_k S_2$. Si S_1 et S_2 sont quasi-triviaux, ce qui implique que ce sont des variétés k-rationnelles, (H1$'$) implique que la k-variété X est stablement k-rationnelle.

Remarque 2.8.4. La réciproque de cette proposition résulte de l'appendice 2.A et de [20] lemme 11, p. 188.

PROPOSITION 2.8.5. *Si X vérifie l'hypothèse* (H1)$_\lambda$, *chaque classe pour la R-équivalence sur $X(k)$ est paramétrée par les k-points k-rationnels d'une k-variété lisse k-rationnelle, qu'on peut supposer propre.*

Démonstration. Soit M dans $X(k)$, et soit $\mathscr{T}$ le torseur de type λ sur X de fibre triviale en M. D'après la proposition 2.7.2, la fibre de $\mathscr{T}$ en tout point N de $X(k)$ qui est R-équivalent à M est triviale. Ainsi l'image de la projection naturelle $\mathscr{T}(k) \to X(k)$ contient la classe pour la R-équivalence de M. Montrons que cette image coïncide avec la classe de M pour la R-équivalence. Tout d'abord il existe une k-compactification lisse $\mathscr{T}^c$ de $\mathscr{T}$ telle que la projection q: $\mathscr{T} \to X$ se prolonge en un k-morphisme propre q^c: $\mathscr{T}^c \to X$; comme la caractéristique de k est nulle, ceci résulte du théorème d'Hironaka, mais cela résulte déjà, en toute caractéristique, de l'existence, due à Brylinski [5], de k-compactifications toriques lisses S^c d'un k-tore S, ce qui permet, par produit contracté, de compactifier q dans ses fibres. Si (H1)$_\lambda$ vaut, la k-variété $\mathscr{T}^c$ est k-rationnelle. Il résulte alors de l'hypothèse $\operatorname{car}(k) = 0$ et du théorème d'Hironaka (voir [20], prop. 10 p. 195) que deux points quelconques de $\mathcal{T}^c(k)$ sont R-équivalents, et donc que leurs projections par q^c sur X sont deux points R-équivalents de $X(k)$. Ceci établit que la R-classe de M sur $X(k)$ coïncide avec $q(\mathcal{T}(k))$ d'une part, avec $q^c(\mathcal{T}^c(k))$ d'autre part.

Remarque 2.8.6. Dans la proposition 2.8.5 ci-dessus, l'hypothèse $\mathrm{car}(k) = 0$ sert uniquement pour assurer que les points d'une R-classe sont *exactement* paramétrés par les k-points d'une k-variété k-rationnelle (que sous $(\mathrm{H1})_\lambda$ les k-points d'une R-classe proviennent tous des k-points d'une k-variété k-rationnelle est clair; ce qui ne l'est pas, c'est que si Y est une k-variété lisse k-rationnelle ouverte, et si $f\colon Y \to X$ est un k-morphisme vers une k-variété propre, alors $f(Y(k))$ est constitué de points R-équivalents).

Proposition 2.8.7. *Le même énoncé vaut sous l'hypothèse* $(\mathrm{H1'})_\lambda$.

Démonstration. Soit n un entier tel que la k-variété propre et lisse $Y = \mathcal{T}^c \times_k \mathsf{P}^n_k$ soit k-rationnelle. On considère alors la projection composée $r\colon Y \to \mathcal{T}^c \to X$. Il est clair que la R-classe de M, qui est contenue dans l'image de $\mathcal{T}^c(k)$ (voir ci-dessus), est contenue dans l'image de $Y(k)$. Par ailleurs tous les points de $Y(k)$ sont R-équivalents sur Y (même argument que dans la proposition 2.8.5), et donc l'image de $Y(k)$ par r est formée de points R-équivalents sur X.

Remarque 2.8.8. Cette démonstration montre que sous $(\mathrm{H1'})_\lambda$ il est encore vrai que chaque classe pour la R-équivalence coïncide avec l'image $q(\mathcal{T}(k))$ des k-points d'un k-torseur convenable $\mathcal{T}$ de type λ par la projection structurale $q\colon \mathcal{T} \to X$.

Remarque 2.8.9. Cette démonstration permet de renforcer la proposition 2.8.1: sous $(\mathrm{H1'})_\lambda$, toute R-classe sur $X(k)$ est Zariski-dense dans X.

Proposition 2.8.10. *Si X vérifie* $(\mathrm{H1'})_\lambda$ *et $\mathcal{T}$ est un torseur de type λ sur X, l'application de $X(k)$ dans $H^1(k, S)$ qui à un k-point M associe la fibre de $\mathcal{T}$ en M induit une injection*

$$i\colon X(k)/R \to H^1(k, S).$$

Démonstration. Nous avons déjà vu (proposition 2.7.2) que l'application fibre induit une application $i\colon X(k)/R \to H^1(k, S)$. Si deux k-points de X ont même image par cette application, il existe un torseur de type λ sur X de fibre triviale en chacun de ces k-points. On conclut alors comme dans la proposition précédente.

Proposition 2.8.11. *Sous chacune des conditions suivantes, la R-équivalence sur $X(k)$ est triviale:*
   (i) *X vérifie $(\mathrm{H1'})$ et $\mathrm{Pic}\ \overline{X}$ est un facteur direct d'un $\mathfrak{g}$-module de permutation;*
   (ii) *X vérifie $(\mathrm{H1'})_\lambda$ et la dimension cohomologique $\mathrm{cd}\ k$ de k est au plus 1.*

Démonstration. Dans le premier cas, il existe un k-tore S_1 tel que le produit $S_0 \times_k S_1$ soit un tore quasi-trivial. Il résulte donc du théorème 90 de Hilbert et du lemme de Shapiro que le groupe $H^1(k, S_0)$ est nul, ce qui implique (i) d'après la proposition 2.8.10. Dans le cas (ii), pour tout k-tore S, le groupe $H^1(k, S)$ est nul, ce qui permet encore de conclure.

PROPOSITION 2.8.12. *Si X vérifie* $(H1')_\lambda$, *l'application naturelle*

$$X(k)/R \to \mathrm{CH}_0(X)$$

des classes de points rationnels sur X modulo la R-équivalence dans le groupe de Chow de dimension zéro est injective: deux k-points qui sont rationnellement équivalents sont R-équivalents.

Démonstration. Supposons les k-points M et N rationnellement équivalents sur X. Si $\mathscr{T}$ est un torseur sur X sous un k-tore, les fibres de $\mathscr{T}$ en M et N sont égales ([20], prop. 12, p. 198). Si $\mathscr{T}$ est le torseur sur X de type λ trivial en M, sa fibre est aussi triviale en N, et M et N appartiennent tous deux à l'image de $\mathscr{T}(k)$ par la projection naturelle. En raisonnant comme dans les propositions 2.8.5 et 2.8.7, on voit qu'alors M et N sont R-équivalents sur X.

PROPOSITION 2.8.13. *Si k est un corps de type fini sur $\mathbf{Q}$, et si X vérifie l'hypothèse $(H1')_\lambda$, alors la R-équivalence sur $X(k)$ est finie ($X(k)/R$ est fini), et il existe un nombre fini de k-morphismes*

$$q_i\colon X_i \to X,$$

où chaque X_i est une k-variété lisse k-rationnelle, tels que $X(k)$ soit la réunion des $q_i(X_i(k))$.

Démonstration. Soit $\mathscr{T}$ un torseur de type λ sur X. L'image de l'application $X(k) \to H^1(k, S_0)$ qui à un k-point M associe la fibre de $\mathscr{T}$ en M est finie (proposition 2.7.3). L'énoncé résulte alors des propositions 2.8.7 et 2.8.10.

Remarque 2.8.14. En termes vagues mais plus imagés, cette proposition dit que les points k-rationnels de X peuvent être décrits au moyen d'un nombre fini de paramétrisations, chacune d'entre elles utilisant en général plus de paramètres que la dimension de X. On notera que la partie de la proposition qui ne porte pas sur la R-équivalence vaut plus généralement pour k de type fini sur le corps premier (cf. remarque 2.8.6).

PROPOSITION 2.8.15. *Si k est un corps local ($\mathfrak{p}$-adique ou réel) et si X vérifie $(H1')_\lambda$, la R-équivalence sur $X(k)$ est finie, chaque R-classe est ouverte et fermée dans $X(k)$ pour la topologie induite par celle de k. Si $k = \mathbf{R}$, les classes pour la R-équivalence coïncident avec les composantes connexes de l'espace topologique $X(\mathbf{R})$.*

Démonstration. Que la R-équivalence soit finie résulte de la proposition 2.8.10 et de la finitude du groupe $H^1(k, S)$ pour k local et S un k-tore quelconque. D'après la remarque 2.8.8, chaque R-classe peut s'écrire $q(\mathscr{T}(k))$, et le théorème des fonctions implicites implique que cet ensemble est ouvert dans $X(k)$. Chaque R-classe est donc ouverte dans $X(k)$, et par suite fermée. Pour $k = \mathbf{R}$, ceci implique que chaque R-classe est union de composantes connexes réelles de $X(\mathsf{R})$. Mais deux R-points qui sont R-équivalents sur la R-variété propre X appartiennent à la même composante connexe de $X(\mathsf{R})$ (puisque $\mathsf{P}^1(\mathsf{R})$ est connexe), ce qui suffit à conclure.

Remarque 2.8.16.   Les R-surfaces fibrées en coniques sur P^1_R admettent des modèles explicites très simples (modèle affine $y^2 + z^2 = \prod_i(x - e_i)$) sur lesquels on vérifie aisément la coïncidence des classes pour la R-équivalence avec les composantes connexes.

Remarque 2.8.17.   Toujours sous $(\mathrm{H1}')_\lambda$, on peut voir que, si k est un corps local, la R-équivalence coïncide avec l'équivalence de Brauer sur $X(k)$: il suffit d'utiliser les propositions 2.7.10, 2.8.10 et le fait que, pour k local, le cup-produit

$$H^1(k, S_0) \times H^1(k, \operatorname{Pic} \overline{X}) \ \xrightarrow{\ \cup\ } \ \operatorname{Br} k \to \mathsf{Q}/\mathsf{Z}$$

est une dualité parfaite de groupes abéliens finis ([52] II.40).

Remarque 2.8.18.   Pour $k = \mathsf{R}$, il résulte alors de la proposition 2.8.10 que le nombre de composantes connexes réelles de $X(\mathsf{R})$ est borné par l'ordre de $H^1(\mathsf{R}, S_0)$, et en particulier au plus égal à 1 si $\operatorname{Pic} \overline{X}$ est un facteur direct d'un module de permutation, ce qui permet de construire des contre-exemples à $(\mathrm{H1}')$ lorsque la dimension de X est au moins 3: on peut facilement écrire les équations d'une intersection lisse X de deux quadriques dans P^5_R (alors $\operatorname{Pic} \overline{X} = \mathsf{Z}$ avec action triviale de $\mathfrak{g} = \operatorname{Gal}(\mathsf{C}/\mathsf{R})$) telle que $X(\mathsf{R})$ possède deux composantes connexes réelles ([22], p. 228). De tels exemples n'existent pas en dimension deux. Soit en effet X une R-*surface* rationnelle propre et lisse, soit $O \in X(\mathsf{R})$ et soit $\mathscr{T}$ le torseur universel sur X de fibre triviale en O. De façon générale (cf. [23]), la flèche

$$i\colon X(k)/R \to H^1(k, S_0)$$

définie par $\mathscr{T}$ peut s'écrire comme la composée de l'application naturelle de $X(k)/R$ dans $A_0(X)$, induite par l'application qui à un point M associe la classe de $(M - O)$ dans le groupe $A_0(X)$ des 0-cycles de degré zéro sur X, et d'une application

$$\Phi\colon A_0(X) \to H^1(k, S_0),$$

qui est injective si $X(k)$ est non vide [10]. Si donc deux R-points de X ont même image par i, ils sont rationnellement équivalents sur X. Mais ceci implique [14] qu'ils appartiennent à la même composante connexe de la R-variété X. En utilisant les mêmes arguments, on peut montrer que pour toute compactification lisse $\mathscr{T}^c$ d'un torseur universel $\mathscr{T}$ sur X, $\mathscr{T}^c(\mathsf{R})$ est soit vide, soit connexe, et l'on peut aussi voir que la décomposition en composantes connexes est donnée par l'équivalence de Brauer sur $X(\mathsf{R})$.

PROPOSITION 2.8.19.   *Si k est un corps de nombres et si X vérifie $(\mathrm{H1}')_\lambda$ et possède un point k-rationnel, il existe un ensemble fini Σ de places de k tel que pour toute place v non dans Σ, tous les points de $X(k_v)$ soient R-équivalents sur la k_v-variété $X \times_k k_v$.*

Démonstration. Soit M dans $X(k)$, et soit $\mathcal{T}$ le torseur sur X de type λ et de fibre triviale en M. Il existe un ouvert affine $\operatorname{Spec} A$ du spectre de l'anneau des entiers de k tel que $(X, S, \mathcal{T})$ possède un modèle $(\tilde{X}, \tilde{S}, \tilde{\mathcal{T}})$ sur $\operatorname{Spec} A$, avec $\tilde{X}/A$ propre et lisse, $\tilde{S}$ un A-tore et $\tilde{\mathcal{T}}$ un torseur sur $\tilde{X}$ sous $\tilde{S}$. Pour toute place finie v de l'anneau des entiers de k correspondant à un idéal premier de l'anneau A, notons A_v le complété de A en ce premier, k_v le corps des fractions de A_v et κ_v le corps résiduel de A ou A_v en v. On a le diagramme commutatif:

$$
\begin{array}{ccc}
X(k_v) & \to & H^1(k_v, S) \\
\uparrow & & \uparrow \\
\tilde{X}(A_v) & \to & H^1(A_v, \tilde{S}),
\end{array}
$$

où les flèches horizontales sont obtenues par accouplement avec $\mathcal{T}$ et $\tilde{\mathcal{T}}$. La flèche verticale de gauche est une bijection car $\tilde{X}$ est propre sur l'anneau A. Par ailleurs, le lemme de Hensel dit que le groupe $H^1(A_v, \tilde{S})$ s'identifie au groupe $H^1(\kappa_v, \tilde{S} \times_A \kappa_v)$, et ce groupe est nul car $\tilde{S} \times_A \kappa_v$ est un tore sur le corps fini κ_v (théorème de Lang). Ainsi l'image de $X(k_v)$ par le flèche horizontale supérieure est réduite à l'élément 0, c'est-à-dire que l'application $\mathcal{T}(k_v) \to X(k_v)$ induite par la projection structurale est surjective. Sous $(\mathrm{H1'})_\lambda$, on conclut $X(k_v)/R = \{M\}$, en utilisant les mêmes arguments que dans les propositions 2.8.5 et 2.8.7.

PROPOSITION 2.8.20. *Soit k un corps local. Soit X une k-surface rationnelle qui vérifie* $(\mathrm{H1'})_\lambda$. *Si k est un corps $\mathfrak{p}$-adique et X a bonne réduction, ou si $k = \mathsf{R}$ et $X(\mathsf{R})$ est connexe, tous les points de $X(k)$ sont R-équivalents.*

Démonstration. Soit O un k-point de X, et soit $\mathcal{T}$ le torseur sur X de type λ et de fibre triviale en O. D'après la proposition 2.8.12 et la remarque 2.8.18, l'application $X(k)/R \to H^1(k, S)$ définie par $\mathcal{T}$ se factorise par le groupe $A_0(X)$ des zéro-cycles de degré zéro modulo l'équivalence rationnelle, l'application $i\colon X(k)/R \to A_0(X)$ associant à un k-point P la classe de $(P - O)$, et, sous $(\mathrm{H1'})_\lambda$, cette application i est injective. Sous chacune des hypothèses de l'énoncé, le groupe $A_0(X)$ est nul (pour le cas $\mathfrak{p}$-adique, voir [4] et [10], pour $k = \mathsf{R}$, voir [14]).

2.9. *Comportement des torseurs universels par transformations k-birationnelles.* La proposition suivante permet d'établir la k-rationalité stable des torseurs universels sur une k-variété rationnelle X qui possède un k-point rationnel, lorsqu'on la connaît pour un torseur universel particulier et que la k-variété X possède suffisamment d'automorphismes k-birationnels pour passer d'un k-point à un autre. Cette dernière propriété est par exemple satisfaite par les surfaces de Del Pezzo de degré 4 (intersections lisses de deux quadriques dans P^4) et les surfaces cubiques lisses dans P^3 (voir [42] II 6.1 (= II 16.1)).

PROPOSITION 2.9.1. *Soit k un corps de caractéristique zéro. Soient X et Y deux k-variétés rationnelles, propres et lisses. Si $f\colon X \to Y$ est une application k-birationnelle qui induit un k-isomorphisme d'un voisinage ouvert de $P \in X(k)$ avec un* *voisinage ouvert de $Q \in Y(k)$, le torseur universel sur X de fibre triviale en P est stablement k-birationnellement équivalent au torseur universel sur Y de fibre triviale en Q.*

Démonstration. Considérons d'abord le cas où f est un k-morphisme birationnel. Puisque X est rationnelle, et donc Pic $\overline{X}$ de type fini sur $\mathbf{Z}$ (appendice 2.A), il existe des voisinages ouverts U de P dans X et V de Q dans Y tels que f induise un isomorphisme de U sur V et que de plus Pic $\overline{U}$ = Pic $\overline{V}$ = 0. Puisque X est propre et Y est lisse, il y a un ouvert Y_1 de Y qui contient tous les points de codimension 1 et tel que l'inverse f^{-1} de f définisse un k-morphisme de Y_1 vers X. On a alors le diagramme commutatif de g-modules:

$$1 \to \overline{k}[U]^*/\overline{k}^* \to \mathrm{Div}_{\overline{X}-\overline{U}}\overline{X} \to \mathrm{Pic}\ \overline{X} \to 0$$
$$\cong \Big\downarrow \qquad\qquad \Big\downarrow \qquad\qquad \Big\downarrow$$
$$1 \to \overline{k}[V]^*/\overline{k}^* \to \mathrm{Div}_{\overline{Y}-\overline{V}}\overline{Y} \to \mathrm{Pic}\ \overline{Y} \to 0,$$

où les flèches verticales sont définies au moyen de $(f^{-1})^*$ (puisque Y est lisse, les groupes des diviseurs de Cartier sur Y et Y_1 coïncident ainsi que les groupes de Picard). Dans ce diagramme tous les g-modules sont $\mathbf{Z}$-libres de type fini. Nous pouvons donc le réécrire comme un diagramme de groupes de caractères de k-tores:

$$0 \to \hat{T}_1 \to \hat{M}_1 \to \hat{S}_1 \to 0$$

(2.9.1) $\qquad\qquad \cong \Big\downarrow \qquad \Big\downarrow \qquad \Big\downarrow$

$$0 \to \hat{T}_2 \to \hat{M}_2 \to \hat{S}_2 \to 0.$$

Remarquons que le torseur M_1 sur $T_1 \cong T_2$ se déduit du torseur M_2 sur T_2 par le changement de groupe structural $\rho\colon S_2 \to S_1$, lui-même dual de l'application $(f^{-1})^*\colon \mathrm{Pic}\ \overline{X} \to \mathrm{Pic}\ \overline{Y}$. On a aussi le diagramme commutatif de g-modules:

$$
\begin{array}{ccc}
0 & & 0 \\
\Big\downarrow & & \Big\downarrow \\
1 \to \overline{k}[V]^*/\overline{k}^* \to \mathrm{Div}_{\overline{Y}-\overline{V}}\overline{Y} \to & \mathrm{Pic}\ \overline{Y} & \to 0 \\
f^* \Big\downarrow \qquad\qquad f^* \Big\downarrow \qquad\qquad & f^* \Big\downarrow & \\
1 \to \overline{k}[U]^*/\overline{k}^* \to \mathrm{Div}_{\overline{X}-\overline{U}}\overline{X} \to & \mathrm{Pic}\ \overline{X} & \to 0 \\
\Big\downarrow \qquad\qquad & \Big\downarrow \alpha & \\
\hat{P} & =\!=\!=\!= & \hat{P} \\
\Big\downarrow & & \Big\downarrow \\
0 & & 0,
\end{array}
$$

que l'on peut réécrire:

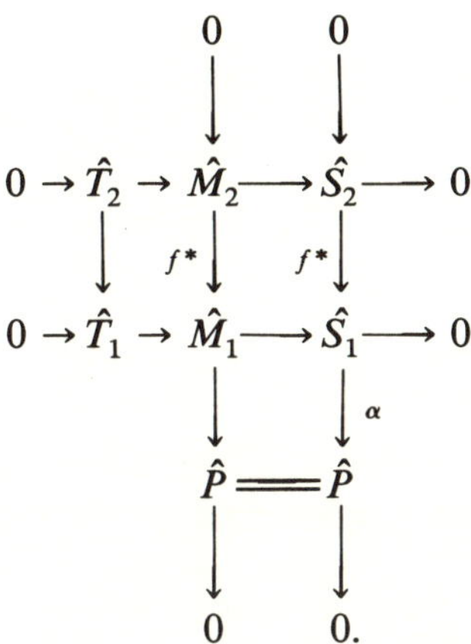

L'application $(f^{-1})^*$ définit une rétraction de chaque application f^* dans les suites exactes verticales (il s'agit là d'un fait simple mais essentiel). En particulier, le $\mathfrak{g}$-module $\hat{P}$ est $\mathbf{Z}$-libre de type fini, c'est le module des caractères d'un k-tore P qui vérifie:

$$(2.9.2) \qquad\qquad M_1 \cong M_2 \times_k P$$

$$(2.9.3) \qquad\qquad S_1 \cong S_2 \times_k P,$$

ce dernier isomorphisme étant donné sur les modules des caractères par l'application:

$$\hat{S}_1 \to \hat{S}_2 \oplus \hat{P}$$

$$\chi \mapsto \left((f^{-1})^*(\chi), \alpha(\chi)\right).$$

On vérifie alors que le composé de l'application $\rho\colon S_2 \to S_1$ avec l'isomorphisme (2.9.3) est l'application évidente $x \to (x, 1)$.

Soit alors $\mathscr{T}$ le torseur universel sur X de fibre triviale en P, et soit $\mathscr{T}'$ le torseur universel sur Y de fibre triviale en Q. Il résulte de la description locale des torseurs universels (proposition 2.3.1) que la restriction de $\mathscr{T}$ à U et celle de $\mathscr{T}'$ à V sont données par les produits fibrés respectifs:

$$\begin{array}{ccc}
\mathscr{T}_U \to M_1 & & \mathscr{T}'_V \to M_2 \\
\downarrow \quad \downarrow & \text{et} & \downarrow \quad \downarrow \\
U \to T_1 & & V \to T_2,
\end{array}$$

où les applications $U \to T_1$ et $V \to T_2$ correspondent aux g-sections des projections

$$\overline{k}[U]^* \to \overline{k}[U]^*/\overline{k}^* \quad \text{et} \quad \overline{k}[V]^* \to \overline{k}[V]^*/\overline{k}^*$$

définies respectivement par les points k-rationnels P et Q. Maintenant les remarques suivant le diagramme (2.9.1) impliquent qu'il y a un isomorphisme de torseurs sur U sous S_1:

$$\mathcal{T}_U \cong f^*(\mathcal{T}'_V) \times^{S_2} S_1,$$

où le changement de groupe structural est donné par ρ. Cet isomorphisme et l'isomorphisme (2.9.3) donnent naissance à un isomorphisme de k-variétés

$$\mathcal{T}_U \cong f^*(\mathcal{T}'_V) \times_k P,$$

et donc, puisque f est un isomorphisme de U avec V, à un k-isomorphisme

$$\mathcal{T}_U \cong \mathcal{T}'_V \times_k P.$$

On peut maintenant utiliser (2.9.2) et obtenir le k-isomorphisme

$$\mathcal{T}_U \times_k M_2 \cong \mathcal{T}'_V \times_k M_1.$$

Puisque $\hat{M}_1$ et $\hat{M}_2$ sont des modules de permutation, les k-tores M_1 et M_2 sont quasi-triviaux, donc sont des k-variétés k-rationnelles, et ceci achève la démonstration de la proposition 2.9.1 quand f est un k-morphisme.

Pour établir le résultat dans le cas général, on utilise l'hypothèse $\mathrm{car}(k) = 0$ et le théorème de désingularisation d'Hironaka pour insérer l'application k-birationnelle f dans un diagramme commutatif de k-applications rationnelles

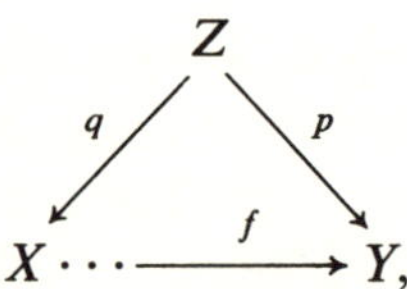

où Z est propre et lisse, et où il existe un point $M \in Z(k)$ et des voisinages ouverts U, V, W de P, Q, M, tels que p, resp. q, induise un k-isomorphisme de W avec U, resp. V. L'argument donné ci-dessus montre que le torseur universel sur X de fibre triviale en P est stablement k-birationnellement équivalent au torseur universel sur Z de fibre triviale en M, lui-même stablement k-birationnellement équivalent au torseur universel sur Y de fibre triviale en Q, ce qui achève d'établir la proposition 2.9.1.

De fait, on peut établir, essentiellement par les mêmes arguments, l'énoncé suivant:

PROPOSITION 2.9.2. *Soit k un corps de caractéristique zéro, et soient X et Y deux k-variétés rationnelles propres et lisses k-birationnellement équivalentes, et soit $\mathscr{T}$ un torseur universel sur X. Il existe alors un torseur universel $\mathscr{T}'$ sur Y qui est k-stablement birationnellement équivalent à $\mathscr{T}$.*

Démonstration. Voici simplement l'idée de la démonstration. Soient U, resp. V, des ouverts de X, resp. Y, qui sont k-isomorphes et tels que Pic $\overline{U}$ = Pic $\overline{V}$ = 0. L'existence de $\mathscr{T}$ assure (corollaire 2.3.4) que le $\mathfrak{g}$-homomorphisme naturel $\overline{k}^* \to \overline{k}[U]^*$ admet une $\mathfrak{g}$-section, qui définit ainsi une $\mathfrak{g}$-section de l'homomorphisme $\overline{k}^* \to \overline{k}[V]^*$. Comme Pic $\overline{V}$ = 0, l'obstruction élémentaire à l'existence d'un k-point sur Y est vide et plus précisément on peut considérer le torseur universel $\mathscr{T}'$ sur Y associé à la $\mathfrak{g}$-section ci-dessus par le corollaire 2.3.4. On raisonne alors exactement comme dans la proposition 2.9.1 précédente.

2.A. *Appendice: Invariance k-birationnelle stable du $\mathfrak{g}$-module Pic $\overline{X}$.* La proposition suivante a été obtenue pour les surfaces par Manin [42] et en dimension quelconque, mais en caractéristique 0, par Voskresenskiĭ ([55] théorème 4.35).

PROPOSITION 2.A.1. *Soient X et Y deux k-variétés propres, lisses et géométriquement intègres, k-birationnellement équivalentes. Soit K/k une extension galoisienne, finie ou non, de groupe G. Il existe deux G-modules de permutation P_1 et P_2 et un isomorphisme de G-modules:*

$$\text{Pic } X_K \oplus P_1 \cong \text{Pic } Y_K \oplus P_2.$$

Avant d'indiquer la démonstration, dégageons deux conséquences, dont la première est immédiate.

COROLLAIRE 2.A.2. *Si X est une k-variété propre, lisse, géométriquement intègre et $\overline{k}$-rationnelle, alors le $\mathfrak{g}$-module Pic $\overline{X}$ est un groupe abélien libre, de type fini, et le groupe $H^1(\mathfrak{g}, \text{Pic } \overline{X})$ est fini.*

COROLLAIRE 2.A.3. *Si X est une k-variété propre, lisse, géométriquement intègre et K/k est une extension galoisienne finie, ou non, de groupe G, si X_K est K-rationnelle et si $H^1(G, \text{Pic } X_K) = 0$, alors*

$$H^1(\mathfrak{g}, \text{Pic } \overline{X}) = 0.$$

Démonstration. Il suffit d'écrire la suite exacte d'inflation-restriction, d'utiliser la proposition 2.A.1 et d'observer que Pic $X_K = (\text{Pic } \overline{X})^{\mathfrak{h}}$, où $\mathfrak{h} = \text{Gal}(\overline{k}/K)$, ce qui résulte du fait que X_K est propre, géométriquement intègre et possède un K-point.

Démonstration de la proposition (suggérée par L. Moret-Bailly). Il existe une k-variété propre, géométriquement intègre et normale Z et des k-morphismes propres k-birationnels $p: Z \to X$ et $q: Z \to Y$: il suffit pour cela de définir Z

comme la normalisation du graphe d'une k-application k-birationnelle de X vers Y. Considérons le k-morphisme p. Comme Z est normal et que p est k-birationnel et propre, il existe un ouvert U de X qui contienne l'ensemble $X^{(1)}$ des points de codimension 1 de X et qui soit tel que p induise un k-isomorphisme entre $V = p^{-1}(U)$ et U. Considérons alors le diagramme:

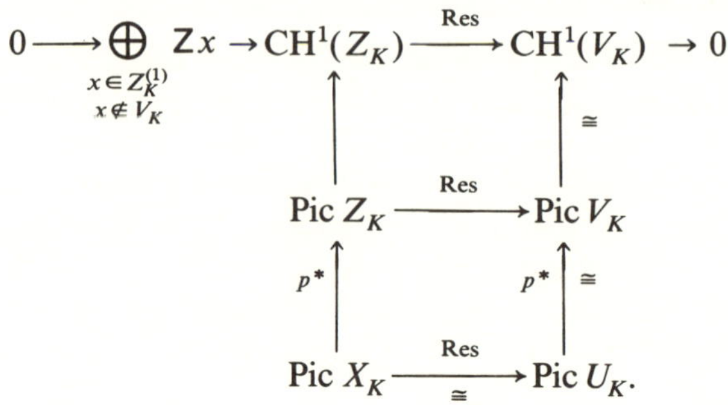

Dans ce diagramme, les groupes CH^1 sont les groupes de Chow de codimension 1, i.e. les quotients par l'équivalence rationnelle des groupes abéliens libres ayant pour base l'ensemble des points de codimension 1. La ligne horizontale supérieure est une suite exacte. L'injectivité à gauche est le seul point à vérifier. Soit $\Sigma_x n_x x$ un cycle de codimension 1 dans ce noyau. Il existe alors une fonction f dans le corps des fonctions rationnelles $K(Z)$ de Z_K dont le diviseur $\mathrm{div}(f)$ est égal à $\Sigma_x n_x x$. Comme V_K est isomorphe à U_K et que U_K contient tous les points de codimension 1 de la K-variété propre et géométriquement intègre X_K, ceci implique que f, vue comme fonction sur X_K, est une constante dans K^*. Il en est donc aussi ainsi sur Z_K, et $\Sigma_x n_x x = 0$. Les flèches verticales $\mathrm{Pic} \to CH^1$ sont les flèches naturelles. Sur V_K, qui est lisse, cette flèche est un isomorphisme. Que la restriction en bas soit un isomorphisme résulte du fait que X_K est lisse et que l'ouvert U_K contient tous les points de codimension 1. Il résulte alors immédiatement du diagramme que la suite exacte de G-modules apparaissant à la première ligne est G-scindée, et qu'on a un isomorphisme de G-modules:

$$\mathrm{Pic}\, X_K \oplus P_1 \cong CH^1(Z_K),$$

où P_1 est le G-module de permutation défini par le terme de gauche de la suite exacte horizontale. Comme on obtient un isomorphisme similaire

$$\mathrm{Pic}\, Y_K \oplus P_2 \cong CH^1(Z_K)$$

en analysant le morphisme $q: Z \to Y$, ceci établit la proposition.

2.B. *Appendice: Calcul du $\mathfrak{g}$-module* $\mathrm{Pic}\, \overline{X}$ *en l'absence de point rationnel.* On observe souvent que lorsqu'une classe de k-variétés rationnelles propres et lisses satisfait la propriété:

(P) *si une variété de cette famille possède un point k-rationnel, alors elle est k-rationnelle,*
la même classe satisfait le principe de Hasse.

Compte tenu de l'appendice 2.A, l'énoncé suivant, qui a été annoncé dans ([27], proposition 3.5) et utilisé dans ce même article, indique au moins que lorsque la propriété (P) est satisfaite, l'obstruction de Brauer-Manin au principe de Hasse (voir §3) s'annule.

THÉORÈME 2.B.1. *Soit k un corps de caractéristique zéro. Soit $\mathscr{C}$ une classe, stable par extension arbitraire du corps de base, de variétés X/K pour K extension de k, propres, lisses, géométriquement intègres, et satisfaisant $H^1(X, \mathcal{O}_X) = 0$. Si, pour X/K dans $\mathscr{C}$ avec $X(K) \neq \varnothing$, le $\mathrm{Gal}(\overline{K}/K)$-module $\mathrm{Pic}\, X_{\overline{K}}$ est de permutation, resp. stablement de permutation, resp. facteur direct d'un module de permutation, resp. satisfait $H^1(\mathrm{Gal}(\overline{K}/K), \mathrm{Pic}\, X_{\overline{K}}) = 0$, alors il en est de même pour tout X/K dans $\mathscr{C}$.*

Démonstration. Donnons d'abord la démonstration sous la première hypothèse. Soit X/K dans $\mathscr{C}$. Quitte à changer les notations, on peut supposer $K = k$. Soient $E = k(X)$ le corps des fonctions de X, puis $F = \overline{k}(X)$ le corps des fonctions de $\overline{X} = X \times_k \overline{k}$, et enfin $\overline{F}$ une clôture algébrique de F. Soient

$$G = \mathrm{Gal}(\overline{k}/k) = \mathrm{Gal}(F/E), \quad \text{puis } G_1 = \mathrm{Gal}(\overline{F}/E) \text{ et } H_1 = \mathrm{Gal}(\overline{F}/F).$$

Le groupe H_1 est distingué dans G_1, et $G = G_1/H_1$. La E-variété X_E possède un E-point rationnel, à savoir celui donné par le point générique de X. Par hypothèse, le G_1-module $\mathrm{Pic}\, X_{\overline{F}}$ est un G_1-module de permutation. Comme H_1 est distingué dans G_1, on en déduit que le G-module $(\mathrm{Pic}\, X_{\overline{F}})^{H_1}$ est un G-module de permutation. De $X(F) \neq \varnothing$ (de fait $X(\overline{k}) \neq \varnothing$), on déduit $\mathrm{Pic}\, X_F = (\mathrm{Pic}\, X_{\overline{F}})^{H_1}$. Ainsi le G-module $\mathrm{Pic}\, X_F$ est de permutation. L'hypothèse $H^1(X, \mathcal{O}_X) = 0$ assure que l'injection naturelle

$$\mathrm{Pic}\, \overline{X} \to \mathrm{Pic}\, X_F$$

est un isomorphisme de G-modules. Ainsi, $\mathrm{Pic}\, \overline{X}$ est-il un G-module de permutation. La démonstration sous les deuxième et troisième hypothèses est essentiellement la même. Sous la quatrième hypothèse, il convient d'utiliser la suite exacte d'inflation-restriction, qui, en utilisant les arguments ci-dessus, donne une injection

$$H^1(G, \mathrm{Pic}\, \overline{X}) \hookrightarrow H^1(G_1, \mathrm{Pic}\, X_{\overline{F}}),$$

et donc la nullité annoncée.

§3. Arithmétique des torseurs. Pour une variété algébrique X définie sur un corps de nombres k, le groupe de Brauer-Grothendieck de X permet de définir des obstructions au principe de Hasse (Manin [41], [42]) et à l'approximation faible. L'objet de ce paragraphe est de montrer que la théorie de la descente fournit, pour les variétés rationnelles, une interprétation géométrique de ces obstructions. On peut ainsi montrer, pour certaines classes de variétés, que les obstructions définies au moyen du groupe de Brauer sont les seules. Ceci donne alors des critères effectifs pour décider si une variété X de la classe considérée a un point rationnel ou non, ou pour décrire l'adhérence de $X(k)$ dans le produit des $X(k_v)$.

3.0. *Notations et rappels.* Dans toute la suite, sauf mention contraire, k désigne un corps de nombres, Ω_k l'ensemble des places de k et k_v le complété de k en la place v. On note $\mathcal{O}_k$ l'anneau des entiers de k, et $\mathcal{O}_{(v)}$ son localisé en v et $\mathcal{O}_v$ l'anneau des entiers de k_v, complété du précédent, enfin $\kappa(v)$ le corps résiduel en v. On note $\bar{k}$ une clôture algébrique de k, puis $\mathfrak{g} = \mathrm{Gal}(\bar{k}/k)$ le groupe de Galois de $\bar{k}/k$ et $\mathfrak{g}^v$ le groupe de décomposition d'une place de k prolongeant v, groupe qui s'identifie à $\mathrm{Gal}(\bar{k}_v/k_v)$. On note enfin

$$\mathrm{inv}_v \colon \mathrm{Br}\, k_v \to \mathbf{Q}/\mathbf{Z}$$

l'invariant local en v.

On dit d'une k-variété algébrique X qu'elle *ne vérifie pas le principe de Hasse* si elle a un point dans chaque complété k_v de k sans en avoir dans k:

$$(3.0.1) \qquad X(k) = \varnothing \quad \text{et pourtant} \prod_{v \in \Omega_k} X(k_v) \neq \varnothing.$$

On connaît de multiples exemples de telles variétés: il y en a parmi les courbes elliptiques (Reichardt-Lind, cf. [7]), les tores algébriques (Hasse), les surfaces rationnelles (Swinnerton-Dyer [53], Iskovskih [39], Cassels et Guy [8], Birch et Swinnerton-Dyer [3]). Au congrès international de Nice en 1970, Manin [41] a introduit une obstruction au principe de Hasse définie au moyen du groupe de Brauer-Grothendieck de X et censée englober les diverses obstructions particulières utilisées dans les contre-exemples considérés auparavant.

On dit d'une k-variété algébrique X qui a un point rationnel qu'elle *vérifie l'approximation faible* si

$$(3.0.2)$$

$X(k)$ est dense, par l'application diagonale, dans le produit $\prod_{v \in \Omega_k} X(k_v)$.

Là aussi on connaît de multiples exemples de variétés qui ne vérifient pas l'approximation faible.

On dit d'une classe $\mathscr{C}$ de k-variétés algébriques qu'elle vérifie *le principe de Hasse*, s'il n'y a pas dans $\mathscr{C}$ de variété qui vérifie (3.0.1), autrement dit, si pour X dans $\mathscr{C}$,

$$(3.0.3) \qquad \prod_{v \in \Omega_k} X(k_v) \neq \varnothing \Rightarrow X(k) \neq \varnothing.$$

On dit que $\mathscr{C}$ vérifie *l'approximation faible* si, pour X dans $\mathscr{C}$,

$$(3.0.4) \qquad X(k) \neq \varnothing \Rightarrow X(k) \text{ dense dans le produit } \prod_{v \in \Omega_k} X(k_v).$$

On peut évidemment associer les deux propriétés: $\mathscr{C}$ vérifie *le principe de Hasse et l'approximation faible*, si pour tout X dans $\mathscr{C}$,

$$(3.0.5) \qquad X(k) \text{ est dense dans le produit } \prod_{v \in \Omega_k} X(k_v).$$

Il s'agit là d'une condition très simple et naturelle. Néanmoins, on préfère souvent en pratique dissocier les deux conditions (3.0.3) et (3.0.4).

Voici enfin quelques rappels sur la dualité de Tate-Nakayama (cf. [54], [55]), et quelques notations supplémentaires. Tout d'abord, si A est un groupe abélien fini, on note

$$\tilde{A} = \mathrm{Hom}_{\mathbf{Z}}(A, \mathbf{Q}/\mathbf{Z})$$

son dual. Si M est un $\mathfrak{g}$-module continu et Σ une partie de Ω_k, on note

$$\text{III}_{\Sigma}^{i}(k, M) := \ker\!\left(H^i(k, M) \to \prod_{v \notin \Sigma} H^i(k_v, M) \right)$$

le noyau de l'application naturelle de restriction: on écrit $H^i(k, M)$ pour $H^i(\mathfrak{g}, M)$, et $H^i(k_v, M)$ pour $H^i(\mathfrak{g}^v, M)$. On pose

$$\text{III}^{i}(k, M) := \text{III}_{\varnothing}^{i}(k, M) = \ker\!\left(H^i(k, M) \to \prod_{v \in \Omega_k} H^i(k_v, M) \right).$$

Etant donné un k-tore S, on note

$$\text{Ч}^{1}(k, S) := \mathrm{coker}\!\left(H^1(k, S) \to \bigoplus_{v \in \Omega_k} H^1(k_v, S) \right).$$

Pour un tel S, on désigne de plus par τ_v l'application duale de la restriction de groupes abéliens finis

$$(3.0.6) \qquad H^1(k_v, S) \to H^1(k, \hat{S})^{\sim},$$

obtenue en composant la dualité de Tate-Nakayama locale

$$H^1(k_v, S) \cong H^1(k_v, \hat{S})^{\sim}$$

avec l'application duale de la restriction

$$H^1(k, \hat{S}) \to H^1(k_v, \hat{S}).$$

Les dualités de Tate-Nakayama, locales et globale, définissent alors la suite exacte de groupes abéliens finis (cf. [20] §8):

$$(3.0.7) \quad H^1(k, S) \to \bigoplus_v H^1(k_v, S) \xrightarrow{\Sigma \tau_v} H^1(k, \hat{S})^{\sim} \longrightarrow H^2(k, S)$$
$$\to \bigoplus_v H^2(k_v, S),$$

qui donne en particulier la suite exacte de groupes abéliens finis

$$(3.0.8) \qquad 0 \longrightarrow Ш^1(k, S) \xrightarrow{\Sigma \tau_v} H^1(k, \hat{S})^{\sim} \longrightarrow Ш^2(k, S) \longrightarrow 0.$$

3.1. *L'obstruction de Manin au principe de Hasse.* Voici d'abord quelques rappels sur le groupe de Brauer-Grothendieck (cf. [34], [35], [41], [42], chap. VI, voir aussi [20], §7, et [48]). Soient k un corps *quelconque*, $\overline{k}$ une clôture séparable et $\mathfrak{g}$ le groupe de Galois de $\overline{k}/k$. Soient X une k-variété algébrique géométriquement intègre et $\overline{X} = X \times_k \overline{k}$. Le groupe de Brauer Br X formé des classes d'algèbres d'Azumaya sur X est un sous-groupe du groupe de Brauer cohomologique Br$'X := H^2_{\text{ét}}(X, \mathsf{G}_m)$. On note Br$_1 X$ le noyau de la restriction Br $X \to$ Br $\overline{X}$, et de même Br$_1'X$ pour le groupe de Brauer-Grothendieck Br$'$. On démontre (cf. [20] lemme 14 et [42], VI 41.7 ($=$ VI 1.7)) que

$$(3.1.1) \qquad \mathrm{Br}_1 X = \mathrm{Br}_1'X = \ker\big(H^2(\mathfrak{g}, \overline{k}(X)^*) \to H^2(\mathfrak{g}, \mathrm{Div}\,\overline{X})\big).$$

On en déduit que, pour k de caractéristique 0 et X *rationnelle*, complète et lisse, auquel cas Br$'\overline{X} = 0$,

$$(3.1.2) \qquad \mathrm{Br}\,X = \mathrm{Br}'X = \ker\big(H^2(\mathfrak{g}, \overline{k}(X)^*) \to H^2(\mathfrak{g}, \mathrm{Div}\,\overline{X})\big).$$

Si k est un corps de nombres ou un corps local complété d'un tel corps de nombres,

$$H^3(k, \mathsf{G}_m) = H^3(\mathfrak{g}, \overline{k}^*) = 0.$$

Si X est une variété algébrique *complète* et géométriquement intègre sur un tel corps k, la suite exacte (1.5.0) s'écrit:

$$(3.1.3) \quad 0 \to \mathrm{Pic}\,X \to (\mathrm{Pic}\,\overline{X})^{\mathfrak{g}} \to \mathrm{Br}\,k \to \mathrm{Br}_1 X \xrightarrow{\chi} H^1(\mathfrak{g}, \mathrm{Pic}\,\overline{X}) \to 0.$$

Rappelons qu'au signe près l'application χ est définie comme suit: soit $\mathscr{A} \in \mathrm{Br}_1 X$; on peut la représenter par un 2-cocycle $f = (f_{s,t}) \in Z^2(\mathfrak{g}, \bar{k}(X)^*)$, de telle sorte que $\mathrm{div}(f) = \partial D$ où $D = (D_s) \in C^1(\mathfrak{g}, \mathrm{Div}\,\bar{X})$; au signe près, $\chi(\mathscr{A})$ est la classe de D dans $Z^1(\mathfrak{g}, \mathrm{Pic}\,\bar{X})$.

Soit X une variété algébrique définie sur le corps de nombres k, et telle que

$$\prod_{v \in \Omega_k} X(k_v) \neq \varnothing \, .$$

On appelle point adélique $x = \{x_v\}_{v \in \Omega_k}$ de X un élément du produit $\prod_{v \in \Omega_k} X(k_v)$ pour lequel il existe un ouvert non vide U de $\mathrm{Spec}\,\mathcal{O}_k$ au-dessus duquel X se prolonge en un U-schéma $\tilde{X}$ et, pour toute place v définie par un point fermé de U, le point x_v se prolonge en un point $\tilde{x}_v \in \tilde{X}(\mathcal{O}_v)$. On note $X(\mathbf{A}_k)$ l'ensemble des points adéliques de X. Si X est *complète*,

$$X(\mathbf{A}_k) = \prod_{v \in \Omega_k} X(k_v).$$

On a un accouplement naturel, introduit par Manin:

$$X(\mathbf{A}_k) \times \mathrm{Br}'X \longrightarrow \mathbf{Q}/\mathbf{Z}$$

défini par

$$(3.1.4) \qquad \langle \{x_v\}, \mathscr{A} \rangle = \sum_v \mathrm{inv}_v(\mathscr{A}(x_v)),$$

et noté également

$$i_{\mathscr{A}}(\{x_v\}) := \langle \{x_v\}, \mathscr{A} \rangle.$$

Cette définition a un sens car, pour presque tout v, toute la situation se prolonge au-dessus de $\mathcal{O}_v$, et $\mathscr{A}(x_v)$ vient donc de $\mathrm{Br}'\mathcal{O}_v$, qui vaut 0.

Définition 3.1.1. Soit B un sous-groupe de $\mathrm{Br}'X$. On appelle *obstruction de Manin associée à B* la condition suivante:

$(3.1.5)$

pour tout $\{x_v\} \in X(\mathbf{A}_k)$, il existe $\mathscr{A} \in B$ telle que $\sum_v \mathrm{inv}_v(\mathscr{A}(x_v)) \neq 0$,

et on appelle *obstruction de Manin au principe de Hasse* celle associée à $\mathrm{Br}'X$ tout entier. Chacune de ces conditions est une obstruction à l'existence d'un point k-rationnel pour X. Le fondement même de cette obstruction est la loi de réciprocité:

$$\mathscr{A} \in \mathrm{Br}\,k \Rightarrow \sum_v \mathrm{inv}_v(\mathscr{A}(x_v)) = 0.$$

Toute classe d'algèbres d'Azumaya $\mathcal{A} \in \mathrm{Br}\, X$ définit un diagramme commutatif

$$(3.1.6) \qquad \begin{array}{ccc} X(k) & \longrightarrow & X(\mathbf{A}_k) \\ {\scriptstyle\mathcal{A}}\downarrow & & {\scriptstyle\mathcal{A}}\downarrow \quad\searrow^{i_{\mathcal{A}}} \\ \mathrm{Br}\, k & \longrightarrow & \displaystyle\bigoplus_{v \in \Omega_k} \mathrm{Br}\, k_v \xrightarrow{\;\Sigma_v \mathrm{inv}_v\;} \mathbf{Q}/\mathbf{Z}, \end{array}$$

dont la suite du bas est un complexe—en fait exact—de sorte que $i_{\mathcal{A}}$ vaut 0 sur $X(k)$. Ainsi:

$$(3.1.7) \qquad X(k) \subset \bigcap_{\mathcal{A}} \ker i_{\mathcal{A}} \quad \text{où } \mathcal{A} \text{ parcourt } \mathrm{Br}'X.$$

Avec ces notations, comme $i_{\mathcal{A}}$ ne dépend que de la classe de $\mathcal{A}$ modulo $\mathrm{Br}\, k$, i.e. modulo l'image $\mathrm{Br}_0 X$ de $\mathrm{Br}\, k$ dans $\mathrm{Br}'X$, *l'obstruction de Manin*—ou de Brauer-Manin—*au principe de Hasse* est la condition suivante:

$$(3.1.8) \qquad \bigcap_{\mathcal{A}} \ker i_{\mathcal{A}} = \varnothing \quad \text{où } \mathcal{A} \text{ parcourt } \mathrm{Br}'X/\mathrm{Br}_0 X.$$

On peut aussi bien l'écrire

$$(3.1.9) \qquad\qquad\qquad \ker i_{\mathrm{Br}} = \varnothing,$$

où

$$i_{\mathrm{Br}}\colon X(\mathbf{A}_k) \to \mathrm{Hom}(\mathrm{Br}'X/\mathrm{Br}_0 X, \mathbf{Q}/\mathbf{Z})$$

est définie par la collection des $i_{\mathcal{A}}$. Pour une variété rationnelle X propre et lisse sur le corps de nombres k,

$$\mathrm{Br}'X/\mathrm{Br}_0 X = H^1(\mathfrak{g}, \mathrm{Pic}\, \overline{X}),$$

et c'est ainsi un groupe abélien fini. Si $\{\mathcal{A}_1, \ldots, \mathcal{A}_r\}$ engendre $\mathrm{Br}'X/\mathrm{Br}_0 X$, on peut se limiter dans (3.1.8) à $\mathcal{A} = \mathcal{A}_1, \ldots, \mathcal{A}_r$. On voit qu'alors cette obstruction de Manin est théoriquement calculable de façon effective (cf. [42], chap. VI, voir aussi corollaire 3.4.5 ci-après). Elle est, de fait, très simple à calculer dans certains cas, comme celui des surfaces de Châtelet généralisées [13], mais son calcul est déjà délicat pour les surfaces cubiques diagonales (cf. [15]).

On dira que l'obstruction de Manin est " vide ", si la condition (3.1.8) n'est pas vérifiée, i.e. si

$$\bigcap_{\mathcal{A}} \ker i_{\mathcal{A}} \neq \varnothing,$$

autrement dit s'il existe $\{x_v\} \in X(\mathbf{A}_k)$ tel que $i_{\mathscr{A}}(\{x_v\}) = 0$ pour tout $\mathscr{A} \in \mathrm{Br}'X$, i.e. si l'application

$$i_{\mathrm{Br}} = (i_{\mathscr{A}})_{\mathscr{A} \in \mathrm{Br}'X} \text{ atteint la valeur } 0.$$

3.2. *Obstructions au principe de Hasse définies par des torseurs sous des tores.* On commence par quelques préliminaires.

3.2.0. Si $[\mathscr{T}]$ est la classe dans $H^1(X, S)$ d'un torseur $\mathscr{T}$ sur X sous un k-tore S, et si K/k est une extension quelconque, l'expression $[\mathscr{T}](K) \neq \varnothing$ a un sens: elle signifie $\mathscr{T}(K) \neq \varnothing$, cette condition étant indépendante du représentant $\mathscr{T}$ choisi. Si $\alpha \in H^1(k, S)$, on note $\mathscr{T}^\alpha$ un torseur sur X sous S dont la classe vérifie (pour i_1 comme en (2.0.2)):

$$(3.2.1) \qquad\qquad [\mathscr{T}^\alpha] = [\mathscr{T}] - i_1(\alpha) \text{ dans } H^1(X, S).$$

Supposons que X vérifie (2.0.1). La définition même, en 2.0, d'un torseur $\mathscr{T}$ sur X de type λ et la suite exacte (2.0.2) montrent le résultat suivant, sur lequel repose essentiellement ce paragraphe:

PROPOSITION 3.2.1. *Soit X une k-variété algébrique vérifiant (2.0.1). Soit λ un type donné de torseur sur X sous S.*

(i) $\mathrm{Tors}(X, S, \lambda)$ *est soit vide, soit affine ($=$ principal homogène) sous le groupe* $H^1(k, S)$.

(ii) $\{q_{\mathscr{T}}(\mathscr{T}(k))\}$ *pour* $[\mathscr{T}] \in \mathrm{Tors}(X, S, \lambda)$ *est une partition de* $X(k)$.

(iii) *Si $\mathscr{T}$ est un torseur de type λ, $\{q_\alpha(\mathscr{T}^\alpha(k))\}$ pour $\alpha \in H^1(k, S)$ est une partition de $X(k)$.*

Démonstration. Précisons d'abord que $q_{\mathscr{T}}: \mathscr{T} \to X$ est le morphisme structural, qu'on note simplement q_α pour $\mathscr{T} = \mathscr{T}^\alpha$. L'assertion (i) résulte aussitôt de (2.0.2). Si $P \in X(k)$, il existe une classe et une seule $[\mathscr{T}^P] \in \mathrm{Tors}(X, S, \lambda)$ telle que $\mathscr{T}^P(P) = 0 \in H^1(k, S)$, ce qui prouve (ii), et (iii) résulte de (i) et (ii).

3.2.2. Etant donné une variété algébrique X définie sur le corps de nombres k, et supposée complète, un k-tore S et un torseur $\mathscr{T}$ sur X sous S, on appelle *place de bonne réduction* de $\mathscr{T}$ toute place finie v de k telle que $(X, S, \mathscr{T})$ admette, au-dessus de $\mathrm{Spec}\, \mathscr{O}_{(v)}$, un prolongement $(\tilde{X}, \tilde{S}, \tilde{\mathscr{T}})$ où $\tilde{X}$ est un $\mathscr{O}_{(v)}$-schéma propre, $\tilde{S}$ un $\mathscr{O}_{(v)}$-tore et $\tilde{\mathscr{T}}$ un torseur sur $\tilde{X}$ sous $\tilde{S}$. D'après ([33], §8) l'ensemble des places finies de k où $\mathscr{T}$ n'a pas bonne réduction est *fini*. On note $\Sigma(\mathscr{T})$ l'ensemble des places finies de mauvaise réduction de $\mathscr{T}$ et des places à l'infini.

LEMME 3.2.3. *Sous les hypothèses ci-dessus, si $\mathscr{T}$ a bonne réduction en v, et si $P_v \in X(k_v)$, alors*

$$\mathscr{T}(P_v) = 0 \quad \text{dans } H^1(k_v, S).$$

Démonstration. Soit en effet $(\tilde{X}, \tilde{S}, \tilde{\mathscr{T}})$ un prolongement de $(X, S, \mathscr{T})$ au-dessus de $\mathcal{O}_{(v)}$. Soit

$$(X', S', \mathscr{T}') = (\tilde{X}, \tilde{S}, \tilde{\mathscr{T}}) \times_{\mathrm{Spec}\, \mathcal{O}_{(v)}} \mathrm{Spec}\, \mathcal{O}_v.$$

Comme X' est propre sur $\mathcal{O}_v$, qui est un anneau de valuation discrète, le point $P_v \in X(k_v)$ se prolonge en $P'_v \in X'(\mathcal{O}_v)$. Ainsi $\mathscr{T}(P_v)$ vient de $\mathscr{T}'(P'_v) \in H^1(\mathcal{O}_v, S')$ qui est égal à $H^1(\kappa(v), \tilde{S}_{\kappa(v)})$ par bonne réduction et par le lemme de Hensel, et vaut donc 0 par le théorème de Lang.

COROLLAIRE 3.2.4. *Pour $(X, S, \mathscr{T})$ comme ci-dessus, il existe un ensemble fini Σ de places tel que, pour toute place $v \notin \Sigma$, et tout $P_v \in X(k_v)$, on ait:*

$$\mathscr{T}(P_v) = 0 \quad dans\ H^1(k_v, S).$$

3.2.5. Pour simplifier, on suppose *désormais*

(3.2.2) X est une variété algébrique lisse sur le corps de nombres k, complète et rationnelle.

On note alors S_0 le k-tore dual du $\mathfrak{g}$-module Pic $\overline{X}$ et λ_0 l'identité du $\mathfrak{g}$-module $\hat{S}_0 = \mathrm{Pic}\,\overline{X}$. Soient S un k-tore, $\lambda \in \mathrm{Hom}_\mathfrak{g}(\hat{S}, \hat{S}_0)$ et $D(\lambda)$: $S_0 \to S$ le k-morphisme dual.

PROPOSITION-DÉFINITION 3.2.6. *Soit X une k-variété vérifiant les hypothèses (3.2.2) et soit λ comme ci-dessus. On suppose $X(\mathbf{A}_k) \neq \varnothing$. Chacune des conditions suivantes est une obstruction à l'existence d'un point k-rationnel de X:*

$\mathcal{O}_\lambda$: *pour tout torseur $\mathscr{T}$ de type λ, il existe une place v telle que $\mathscr{T}(k_v) = \varnothing$;*

$\mathcal{O}'_\lambda$: *il n'existe pas de torseur de type λ;*

$\mathcal{O}''_\lambda$: *il existe un torseur $\mathscr{T}$ de type λ, et, pour tout $\alpha \in H^1(k, S)$, il existe une place v telle que*

$$\mathscr{T}^\alpha(k_v) = \varnothing.$$

De plus:

$$\mathcal{O}_\lambda \Leftrightarrow \mathcal{O}'_\lambda \ ou\ \mathcal{O}''_\lambda.$$

L'obstruction $\mathcal{O}'_{\lambda_0}$ est la plus fine des obstructions $\mathcal{O}'_\lambda$; elle est à valeurs dans $\mathrm{III}^2(k, S_0)$, et on l'appelle la première obstruction au principe de Hasse. *L'obstruction $\mathcal{O}_{\lambda_0}$ est la plus fine des obstructions $\mathcal{O}_\lambda$; on l'appelle* l'obstruction de Picard au principe de Hasse.

Démonstration. Si $P \in X(k)$, le torseur $\mathscr{T}^P$ de type λ et fibre nulle en P a, par définition même, un point dans k, au-dessus de P: $\mathscr{T}^P(k) \neq \varnothing$ (proposition

3.2.1). A fortiori $\mathcal{T}^P(k_v) \neq \varnothing$ quelle que soit v. La condition $\mathcal{O}_\lambda$ est donc une obstruction à $X(k) \neq \varnothing$.

L'équivalence $\mathcal{O}_\lambda \Leftrightarrow \mathcal{O}'_\lambda$ ou $\mathcal{O}''_\lambda$ est immédiate, car l'ensemble des torseurs de type λ est soit vide, soit un espace affine sous $H^1(k, S)$: si $\mathcal{T}$ est un torseur de type λ, tout autre torseur de type λ est de la forme $\mathcal{T}^\alpha$ pour $\alpha \in H^1(k, S)$ et inversement.

D'après la proposition 2.2.8, l'obstruction $\mathcal{O}'_\lambda$ se traduit par $\partial(\lambda) \neq 0$. De l'hypothèse $X(\mathbf{A}_k) \neq \varnothing$ résulte que cette obstruction est à valeurs dans $\text{III}^2(k, S)$. Toujours d'après cette proposition, l'obstruction $\mathcal{O}'_{\lambda_0}$ est la plus fine des obstructions $\mathcal{O}'_\lambda$ et elle coïncide avec l'obstruction élémentaire étudiée au §2.2.

Que l'obstruction $\mathcal{O}_{\lambda_0}$ soit la plus fine des obstructions $\mathcal{O}_\lambda$ est une conséquence de la remarque suivante:

SCHOLIE 3.2.7. *Si l'obstruction de Picard est vide, il existe un torseur universel $\mathcal{T}_0$ qui a des points dans tous les complétés de k:*

$$\prod_{v \in \Omega_k} \mathcal{T}_0(k_v) \neq \varnothing,$$

et, de même, pour tout type λ, il existe un torseur de ce type qui a des points dans tous les complétés de k.

Démonstration. C'est immédiat sur la définition de $\mathcal{O}_\lambda$: la négation de $\mathcal{O}_\lambda$ signifie qu'il existe un torseur $\mathcal{T}$ de type λ qui a des points dans chaque complété de k. Si $\mathcal{T}_0$ est un torseur universel ayant cette propriété, et si λ est un type quelconque, on peut alors prendre pour $\mathcal{T}$ le produit contracté de $\mathcal{T}_0$ et S sous l'action de S_0 agissant sur $\mathcal{T}_0$ comme groupe du torseur et sur S via $D(\lambda)$: $S_0 \to S$.

3.3. *La première obstruction et l'obstruction de Manin associée à* Б. On suppose toujours que X est une variété rationnelle propre et lisse sur un corps de nombres k et que l'on a $X(\mathbf{A}_k) \neq \varnothing$.

PROPOSITION 3.3.1. *La première obstruction est nulle dans chacun des cas suivants*:

    (i) $\text{III}^2(k, S_0) = 0$;
    (ii) $\text{III}^1(k, \hat{S}_0) = 0$;
    (iii) $H^1(k, \text{Pic}\,\overline{X}) = 0$.

Démonstration. Le cas (i) résulte immédiatement de la proposition 3.2.6. La condition (i) équivaut à (ii) d'après la dualité de Poitou-Tate [52], qui affirme que les deux groupes abéliens finis $\text{III}^2(k, S_0)$ et $\text{III}^1(k, \hat{S}_0)$ sont en dualité. Quant à (iii), elle implique évidemment (ii).

PROPOSITION 3.3.2. *La première obstruction au principe de Hasse coïncide avec l'obstruction de Manin associée au sous-groupe*

$$Б(X) := \text{III}(\text{Br}_a X),$$

formé des classes d'algèbres localement constantes.

Autrement dit, la trivialité de l'obstruction de Manin associée à $Б(X)$ équivaut à l'existence d'un torseur universel sur X.

Dans l'énoncé ci-dessus, $\mathrm{Br}_a X$ désigne $\mathrm{Br}_1 X/\mathrm{Br}_0 X$ et $Б(X)$ est l'image dans $\mathrm{Br}_a X$ du noyau B de la restriction

$$\mathrm{Br}_1 X \longrightarrow \prod_{v \in \Omega_k} \left(\mathrm{Br}_1 X_{k_v}/\mathrm{Br}\, k_v \right),$$

noyau formé des classes d'algèbres localement constantes.

Démonstration. Comme indiqué au lemme 6.2 de [48], l'accouplement de Manin (3.1.4) définit un homomorphisme

$$(3.3.1) \qquad\qquad \gamma \colon Б(X) \to \mathbf{Q}/\mathbf{Z}$$

par

$$\gamma(\mathscr{A}) = \sum_v \mathrm{inv}_v(\mathscr{A}(P_v))$$

où $\{P_v\}$ désigne un point quelconque de $X(\mathbf{A}_k)$, l'expression $\mathrm{inv}_v(\mathscr{A}(P_v))$ étant indépendante de P_v puisque, par hypothèse, l'image $\mathscr{A}_v$ de $\mathscr{A}$ dans $\mathrm{Br}\, X_v$ vient de $\mathrm{Br}\, k_v$. Autrement dit, si $\mathscr{A} \in Б(X)$, l'application $i_{\mathscr{A}}$ est constante. L'obstruction de Manin associée à $Б(X)$ est la condition suivante:

$$(3.3.2) \qquad\qquad \text{il existe } \mathscr{A} \in Б(X) \text{ telle que } \gamma(\mathscr{A}) \neq 0.$$

La dualité de Poitou-Tate est donnée par l'accouplement

$$\mathrm{III}^2(k, S_0) \times \mathrm{III}^1(k, \hat{S}_0) \xrightarrow{\ \tau\ } \mathbf{Q}/\mathbf{Z}$$

défini par

$$(3.3.3) \qquad\qquad \tau(\beta, \alpha) = \sum_v \mathrm{inv}_v(\varepsilon_v),$$

où $\varepsilon_v \in \mathrm{Br}\, k_v$ est défini de la façon suivante (cf. [48] p. 58): on représente α et β par des cocycles a et b, puis, compte tenu de $H^3(k, \mathbf{G}_{m,k}) = 0$, on écrit

$$\alpha \cup \beta = dh \quad \text{où } h \in C^2(\mathfrak{g}, \overline{k}^*);$$

pour chaque place v, on peut écrire

$$b = d\xi_v \quad \text{dans } Z^2(k_v, S_0) \text{ où } \xi_v \in C^1(k_v, S_0);$$

on vérifie aussitôt que, pour chaque place v,

$$\xi_v \cup a - h \in Z^2(\mathfrak{g}^v, \overline{k}_v^*),$$

ce qui permet de poser

$$\varepsilon_v = \mathrm{cl}\big((\xi_v \cup a) - h\big) \in \mathrm{Br}\, k_v.$$

Comme $H^3(k, \mathsf{G}_{m,k}) = 0$, l'application

$$\chi \colon \mathrm{Br}_1 X \to H^1(k, \mathrm{Pic}\,\overline{X})$$

induit des isomorphismes

$$\overline{\chi}\colon \mathrm{Br}_a X \xrightarrow{\ \cong\ } H^1(k, \hat{S}_0) \quad \text{et} \quad \overline{\chi}\colon \mathrm{B}(X) \xrightarrow{\ \cong\ } \mathrm{III}^1(k, \hat{S}_0).$$

Lemme 3.3.3. *Si $\mathscr{A}$ est une algèbre dont la classe appartienne à $\mathrm{B}(X)$, on a l'égalité*

$$\gamma(\mathscr{A}) = \tau\big(\partial(\lambda_0), \chi(\mathscr{A})\big).$$

Autrement dit, le triangle ci-dessous

$$
\begin{array}{ccc}
\mathrm{B}(X) & \xrightarrow[\ \overline{\chi}\ ]{\ \cong\ } & \mathrm{III}^1(k, \hat{S}_0) \\
& \gamma \searrow \quad \swarrow \tau' & \\
& \mathbb{Q}/\mathbb{Z} &
\end{array}
$$

où τ' est définie par $\tau'(\alpha) = \tau(\partial(\lambda_0), \alpha)$, est un triangle commutatif.

Ce lemme étant admis, la proposition résulte du fait que τ est une dualité parfaite de groupes finis; on a en effet les équivalences:

$$\partial(\lambda_0) \neq 0 \Leftrightarrow \tau' \neq 0 \Leftrightarrow \gamma \neq 0;$$

autrement dit, la première obstruction est non vide si, et seulement si, l'obstruction de Manin (3.3.2) est non vide. Pour achever la démonstration de la proposition, il reste à démontrer le lemme.

Démonstration du lemme. Elle se fait en plusieurs étapes.

1. On représente $\mathscr{A}$ par un 2-cocycle $f = (f_{s,t}) \in Z^2(\mathfrak{g}, \overline{k}(X)^*)$ tel que

$$\mathrm{div}(f) = dD \quad \text{où } D = (D_s) \in C^1(\mathfrak{g}, \mathrm{Div}\,\overline{X}).$$

On sait qu'alors

$$\chi(\mathscr{A}) = \text{classe dans } H^1(k, \hat{S}_0) \text{ de } \mathrm{cl}(D) \in Z^1(\mathfrak{g}, \mathrm{Pic}\,\overline{X}),$$

autrement dit, avec les notations ci-dessus, on peut représenter $\alpha := \chi(\mathscr{A})$ par le 1-cocycle $a = \mathrm{cl}(D)$.

2. Soit $P \in X(\bar{k})$ tel que les fonctions $f_{s,t}$ soient toutes définies en P, ce qui est possible, X étant lisse. On sait que $\partial(\lambda_0)$ est la classe de la 2-extension de $\mathfrak{g}$-modules

$$1 \to \bar{k}^* \to \bar{k}(X)^* \to \operatorname{Div} \bar{X} \to \operatorname{Pic} \bar{X} \to 0.$$

Comme P est lisse, il existe un scindage de l'extension de groupes abéliens

$$1 \to \bar{k}^* \to \bar{k}(X)^* \to \operatorname{Div}_0 \bar{X} \to 0$$

provenant de la rétraction

$$\mathcal{O}^*_{\bar{X}, P} \to \bar{k}^*$$

donnée par $g \mapsto g(P)$. On a ainsi une section de groupes abéliens $\sigma_P \colon \operatorname{Div}_0 \bar{X} \to \bar{k}(X)^*$ dont le bord $d\sigma_P$ est à valeurs dans $\bar{k}^*$ et s'étend à $\operatorname{Div} \bar{X}$ en un morphisme de groupes abéliens ψ:

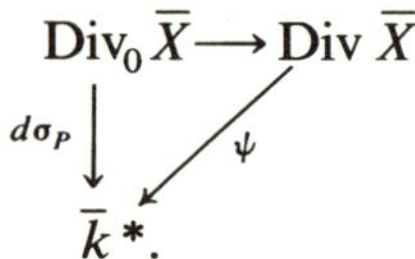

L'obstruction élémentaire $\partial(\lambda_0)$ peut alors être représentée par

$$b := d\psi \in Z^2(\mathfrak{g}, \operatorname{Pic} \bar{X}).$$

3. On continue à appliquer la méthode indiquée en (3.3.3) pour calculer $\tau(\partial(\lambda_0), \chi(\mathscr{A}))$. Montrons qu'on peut prendre pour $h \in C^2(\mathfrak{g}, \bar{k}^*)$ tel que

$$b \cup a = d\psi \cup D = dh$$

la 2-cochaîne $h = \psi \cup D + e_P \cup f$ où $e_P \in \operatorname{Hom}(\bar{k}(X)^*, \bar{k}^*)$ est définie par $e_P(g) = g/\sigma_P(g)$.

En effet: $de_P = d\sigma_P$ et $dh = (d\psi \cup D) - (\psi \cup dD) + (d\sigma_P \cup f)$

$$= (d\psi \cup D) - (\psi \cup f) + (d\sigma_P \cup f) = b \cup a.$$

4. Pour chaque place v, on peut fixer $P_v \in X(k_v)$ de telle sorte que le 2-cocycle f soit défini en P_v: c'est possible d'après la lissité de X, par application du théorème des fonctions implicites local. Soit

$$\theta_v = \sigma_P/\sigma_{P_v} \colon \operatorname{Div}_0 \bar{X} \to \bar{k}^*;$$

c'est un morphisme de groupes abéliens tel que, si $g \in \bar{k}(X)^*$ est une fonction définie en P et en P_v:

$$\theta_v(\operatorname{div} g) = g(P_v)/g(P).$$

On suppose, ce qui est possible, car $\operatorname{Div}_{\{P_v\}} \bar{X}$ est un $\mathfrak{g}^v$-module de permutation, que σ_{P_v} est un $\mathfrak{g}^v$-morphisme. Dès lors

$$d\theta_v = d\sigma_P \in B^1\big(\mathfrak{g}^v, \operatorname{Hom}(\operatorname{Div}_0 \bar{X}, \bar{k}^*)\big).$$

On peut prolonger θ_v en un morphisme de groupes abéliens μ_v:

$$\operatorname{Div}_0 \bar{X} \longrightarrow \operatorname{Div} \bar{X}$$
$$\theta_v \Big\downarrow \quad \swarrow \mu_v$$
$$\bar{k}^*,$$

et on peut alors prendre

$$\xi_v = \psi - d\mu_v \in C^1(\mathfrak{g}^v, S_0),$$

car, d'une part, $(\psi - d\mu_v)|_{\operatorname{Div}_0 \bar{X}} = d\sigma_P - d\theta_v = 0$, d'autre part $d\xi_v = d\psi = b$ dans $Z^2(\mathfrak{g}^v, S_0)$.

5. Ces divers choix étant faits, ε_v est la classe dans $\operatorname{Br} k_v$ du 2-cocycle

$$\xi_v \cup a - h = \psi \cup D - d\mu_v \cup D - \psi \cup D - e_P \cup f$$
$$= -d(\mu_v \cup D) + \mu_v \cup dD - e_P \cup f.$$

Comme $\mu_v \cup D$ appartient à $C^1(\mathfrak{g}^v, \bar{k}^*)$, ε_v est aussi la classe du 2-cocycle

$$\mu_v \cup f - e_P \cup f = (\theta_v/e_P) \cup f = f(P_v).$$

C'est donc $\mathscr{A}(P_v)$, ce qui montre que les termes locaux intervenant dans le calcul de $\gamma(\mathscr{A})$ et dans celui de $\tau(\partial(\lambda_0), \chi(\mathscr{A}))$ sont les mêmes, ce qui achève la démonstration du lemme, et donc celle de la proposition.

A propos de cette première obstruction, voir aussi [41], ([42], chap. VI, 1.23 et 1.24) et [48].

3.4. *La deuxième obstruction: la traduction fonctionnelle.* Soit X comme en 3.3, soit S un k-tore, et soit $\lambda \in \operatorname{Hom}_{\mathfrak{g}}(\hat{S}, \operatorname{Pic} \bar{X})$.

Définition 3.4.1. S'il existe un torseur de type λ, on a également une application

$$i_\lambda \colon X(\mathbf{A}_k) \to H^1(k, \hat{S})^{\tilde{}}$$

définie par

$$i_\lambda(\{P_v\}) = \sum_{v \in \Omega_k} \tau_v(\mathcal{T}(P_v)),$$

et cette application ne dépend pas du choix du torseur $\mathcal{T}$ de type λ, car deux tels torseurs diffèrent par un élément $\alpha \in H^1(k, S)$, et pour un tel élément on a encore, d'après la suite exacte (3.0.7), une loi de réciprocité $\Sigma_v\tau_v(\alpha) = 0$.

PROPOSITION 3.4.2. *On suppose* Tors$(X, S, \lambda) \neq \varnothing$. *Chacune des conditions suivantes équivaut à* $\mathcal{O}_\lambda''$:

(i) *pour tout torseur $\mathcal{T}$ de type λ, il existe une place v telle que $\mathcal{T}(k_v) = \varnothing$*;

(ii) *pour un torseur $\mathcal{T}$ de type λ, pour tout $\alpha \in \text{III}^1_{\Sigma(\mathcal{T})}(k, S)$, il existe $v \in \Sigma(\mathcal{T})$ telle que $\mathcal{T}^\alpha(k_v) = \varnothing$*;

(iii) $\ker i_\lambda = \varnothing$, *i.e. i_λ ne prend pas la valeur 0*;

(iv) *pour un torseur $\mathcal{T}$ de type λ, et pour tout $\{P_v\} \in X(\mathbf{A}_k)$*,

$$\sum_v \tau_v(\mathcal{T}(P_v)) \neq 0.$$

Remarque 3.4.3. Dans (ii) et (iv), l'expression "pour un torseur $\mathcal{T}$ de type λ" signifie aussi bien "il existe un torseur $\mathcal{T}$ de type λ" que "pour tout torseur $\mathcal{T}$ de type λ" ou encore "étant donné un torseur $\mathcal{T}$ de type λ", la validité des conditions (ii) et (iv) ne dépendant pas du choix de $\mathcal{T}$ dans Tors(X, S, λ).

Le fait que dans la somme figurant dans la condition (iv) presque tous les termes soient nuls résulte de l'énoncé de bonne réduction 3.2.3 comme indiqué ci-après au pas n° 3 de la démonstration du lemme 3.4.4.

La condition (i) constitue la version géométrique de $\mathcal{O}_\lambda''$, la condition (ii) en est la version géométrique "effective", les expressions (iii) et (iv) en sont les versions algébriques qui permettent de faire le lien, dans les exemples, avec les versions "calculatoires" fonctionnelles.

Démonstration. L'énoncé (i) est la définition même de $\mathcal{O}_\lambda''$. Comme à la proposition 3.2.1, il équivaut à

(i$'$) *pour un torseur $\mathcal{T}$ de type λ, et pour tout $\alpha \in H^1(k, S)$, il existe v telle que $\mathcal{T}^\alpha(k_v) = \varnothing$*.

La définition même de i_λ montre que l'énoncé (iii) équivaut à (iv). L'équivalence de (i$'$), (ii), et (iv) résulte alors du lemme suivant:

LEMME 3.4.4. *Soit $\mathcal{T}$ un torseur de type λ. Les conditions suivantes sont équivalentes*:

(a) *Pour tout $\alpha \in H^1(k, S)$, il existe une place v telle que $\mathcal{T}^\alpha(k_v) = \varnothing$.*

(b) *Pour tout $\alpha \in \text{III}^1_{\Sigma(\mathcal{T})}(k, S)$, il existe une place $v \in \Sigma(\mathcal{T})$ telle que $\mathcal{T}^\alpha(k_v) = \varnothing$.*

(c) *Pour tout $\{P_v\} \in X(\mathbf{A}_k)$, on a*: $\Sigma_v\tau_v(\mathcal{T}(P_v)) \neq 0$.

Démonstration.

1. Notons d'abord que (a) équivaut de façon immédiate à

(a′) *Pour tout* $\alpha \in H^1(k, S)$, *il existe une place v telle que, pour tout $P_v \in X(k_v)$, on ait*

$$\mathcal{T}^\alpha(P_v) \neq 0.$$

2. Ensuite, compte tenu de la suite exacte

$$H^1(k, S) \longrightarrow \bigoplus_{v \in \Omega_k} H^1(k_v, S) \xrightarrow{\Sigma_v \tau_v} H^1(k, \hat{S})^{\sim}$$

rappelée en (3.0.7), on voit que la condition (c) équivaut à celle-ci: pour tout $\{P_v\} \in X(\mathbf{A}_k)$, la famille $\{\mathcal{T}(P_v)\}$ n'appartient pas à l'image de $H^1(k, S)$, autrement dit, pour tout $\alpha \in H^1(k, S)$, il existe une place v telle que $\mathcal{T}(P_v) \neq \mathrm{res}_v(\alpha)$, où res_v désigne la restriction de $\mathfrak{g}$ à $\mathfrak{g}^v$; ainsi, la condition (c) équivaut à

(c′) *Pour tout* $\{P_v\} \in X(\mathbf{A}_k)$ *et pour tout* $\alpha \in H^1(k, S)$, *il existe une place v telle que* $\mathcal{T}^\alpha(P_v) \neq 0$.

3. On va montrer le résultat auxiliaire suivant:

(br) *Pour tout* $\alpha \in \text{III}^1_{\Sigma(\mathcal{T})}(k, S)$ *et tout* $w \notin \Sigma(\mathcal{T})$, *on a* $\mathcal{T}^\alpha(k_w) \neq \varnothing$.

Soit $w \notin \Sigma(\mathcal{T})$. Par hypothèse, il existe $P_w \in X(k_w)$. Comme w est une place de bonne réduction pour $\mathcal{T}$, on a $\mathcal{T}(P_w) = 0$ d'après 3.2.3. De plus, les hypothèses sur α et sur w impliquent $\mathrm{res}_w(\alpha) = 0$ dans $H^1(k_w, S)$. Finalement

$$\mathcal{T}^\alpha(P_w) = \mathcal{T}(P_w) - \mathrm{res}_w(\alpha) = 0,$$

ce qui établit le résultat annoncé.

4. (a) $\Rightarrow$ (b). Conséquence immédiate du résultat de bonne réduction (br) ci-dessus: la place v de (a) appartient nécessairement à $\Sigma(\mathcal{T})$.

5. (b) $\Rightarrow$ (a). Pour vérifier (a), il suffit de considérer $\alpha \notin \text{III}^1_{\Sigma(\mathcal{T})}(k, S)$. Il existe alors une place de bonne réduction w telle que $\mathrm{res}_w(\alpha) \neq 0$ dans $H^1(k_w, S)$. Quel que soit $P_w \in X(k_w)$, on a donc, w étant de bonne réduction, $\mathcal{T}(P_w) = 0$ d'après 3.2.3, et par suite:

$$\mathcal{T}^\alpha(P_w) = \mathcal{T}(P_w) - \mathrm{res}_w(\alpha) = -\mathrm{res}_w(\alpha) \neq 0.$$

Ceci implique $\mathcal{T}^\alpha(k_w) = \varnothing$.

6. (a′) $\Rightarrow$ (c′). Soient $\{P_v\} \in X(\mathbf{A}_k)$ et $\alpha \in H^1(k, S)$. D'après (a′), il existe une place v telle que, en tout point de $X(k_v)$, en particulier en P_v, la valeur de $\mathcal{T}^\alpha$ soit $\neq 0$.

On note (a″) la négation de (a′) et (c″) celle de (c′).

7. (a″) $\Rightarrow$ (c″). Par hypothèse, il existe $\alpha \in H^1(k, S)$ ayant la propriété suivante: pour toute place v de k, il existe $P_v \in X(k_v)$ tel que $\mathcal{T}^\alpha(P_v) = 0$. On a ainsi trouvé $\{P_v\} \in X(\mathbf{A}_k)$ et $\alpha \in H^1(k, S)$ tels que, pour toute place v,

$$\mathscr{T}^{\alpha}(P_v) = 0.$$

Autrement dit, la conditioon (c″) est vérifiée.

On a ainsi fini de prouver l'équivalence (a) ⇔ (c), ce qui achève la démonstration du lemme, et donc celle de la proposition.

COROLLAIRE 3.4.5. *Il suffit d'un nombre* fini *d'opérations élémentaires pour déterminer si l'obstruction* $\mathcal{O}_{\lambda}''$ *est vide ou non.*

On a conservé les mêmes hypothèses sur X et suppose $\mathrm{Tors}(X, S, \lambda) \neq \varnothing$, i.e. $\partial(\lambda) = 0$. Pour déterminer si $\mathcal{O}_{\lambda}''$ est vide ou non, on part d'un torseur $\mathscr{T}$ de type λ *quelconque*. On détermine un ensemble *fini* Σ de places contenant $\Sigma(\mathscr{T})$, autrement dit tel que $\mathscr{T}$ ait bonne réduction en dehors de Σ. Comme le groupe $\mathrm{III}^1_{\Sigma}(k, S)$ est *fini*, on doit considérer un nombre *fini* de torseurs $\mathscr{T}^{\alpha}$, à savoir pour $\alpha \in \mathrm{III}^1_{\Sigma}(k, S)$, et vérifier pour chacune de ces variétés, si elle a un point dans k_v pour $v \in \Sigma$, ou non. Par application du lemme de Hensel, ceci conduit à un nombre fini d'opérations "élémentaires", qui peuvent être évidemment difficiles à mettre en oeuvre dans les exemples.

COROLLAIRE 3.4.6. *Si* $Ч^1(k, S) = 0$, *l'obstruction* $\mathcal{O}_{\lambda}''$ *est vide.*

Démonstration. C'est une conséquence de la version 3.4.2(iii) de l'obstruction, car d'après la suite exacte (3.0.7), l'application $i_{\lambda}(\{P_v\}) = \Sigma_v \tau_v(\mathscr{T}(P_v))$ se factorise par $Ч^1(k, S)$ et vaut donc nécessairement 0.

COROLLAIRE 3.4.7. *Soit X une k-variété rationnelle, complète et lisse, telle que* $\Pi_v X(k_v) \neq \varnothing$. *Si $H^1(\mathfrak{g}, \mathrm{Pic}\, \overline{X}) = 0$, l'obstruction de Picard au principe de Hasse est vide. Autrement dit, il existe un torseur universel $\mathscr{T}$ tel que $\Pi_v \mathscr{T}(k_v) \neq \varnothing$, et de même il existe un torseur de tout type λ ayant la même propriété.*

Démonstration. Par hypothèse, le terme médian de la suite exacte (3.0.8)

$$0 \to Ч^1(k, S_0) \to H^1\big(k, \hat{S}_0\big)^{\widetilde{}} \to \mathrm{III}^2(k, S_0) \to 0$$

est nul. On a donc $\mathrm{III}^2(k, S_0) = 0$, ce qui assure l'existence d'un torseur universel $\mathscr{T}$, et $Ч^1(k, S_0) = 0$, ce qui assure que la deuxième obstruction est également vide.

3.5. *Obstruction de Picard et obstruction de Manin au principe de Hasse.* Le résultat principal du §3 est le suivant:

THÉORÈME 3.5.1. *Soient k un corps de nombres et X une k-variété rationnelle, complète et lisse, telle que* $\Pi_v X(k_v) \neq \varnothing$. *L'obstruction de Picard au principe de Hasse coïncide avec celle de Manin. En particulier, si l'obstruction de Manin est* *vide, il existe un torseur universel $\mathscr{T}$ sur X tel que*

$$\mathscr{T}(k_v) \neq \varnothing \quad \textit{pour toute place } v \textit{ de } k,$$

et de même pour tout autre type λ de torseur.

Démonstration. Elle utilise essentiellement la proposition 3.3.2 sur la première obstruction et la proposition 2.7.10.

S'il n'existe pas de torseur universel, l'obstruction de Picard est "non vide" par définition même, et celle de Manin est "non vide" d'après la proposition 3.3.2.

On peut donc supposer qu'il existe un torseur universel. D'après la proposition 2.2.8, l'obstruction élémentaire est nulle et le g-morphisme naturel $\overline{k}^* \to k(X)^*$ admet ainsi une rétraction σ. Il est alors utile de dériver de la proposition 2.7.10 l'énoncé suivant:

LEMME 3.5.2. *Soient k un corps quelconque et X une k-variété rationnelle, complète et lisse. Etant donnés un torseur universel $\mathscr{T}$ sur X et $\mathscr{A} \in \mathrm{Br}\, X$, il existe $a_0 \in \mathrm{Br}\, k$ tel que, pour toute extension K/k, on ait, pour tout $P \in X(K)$, l'égalité*

$$(3.5.1) \qquad \mathscr{T}(P) \cup \chi_K(\mathscr{A}) = \mathscr{A}(P) + \mathrm{Res}_k a_0 \quad \textit{dans } \mathrm{Br}\, K.$$

Démonstration du lemme. Dans cet énoncé, $\mathscr{T}(P) \in H^1(K, S_K)$, et la notation Res_K désigne la restriction $\mathrm{Br}\, k \to \mathrm{Br}\, K$, tandis que

$$\chi_K \colon \mathrm{Br}\, X_K \to H^1(K, \mathrm{Pic}\, X_{\overline{K}})$$

désigne l'application naturelle donnée par (1.5.0).

Comme deux torseurs universels $\mathscr{T}$ et $\mathscr{T}'$ diffèrent par un élément $\alpha \in H^1(k, S_0)$, les expressions $\mathscr{T}(P) \cup \chi_K(\mathscr{A})$ et $\mathscr{T}'(P) \cup \chi_K(\mathscr{A})$ diffèrent par $\mathrm{Res}_K(a)$ où $a = \alpha \cup \chi(\mathscr{A})$ est un élément fixe de $\mathrm{Br}\, k$. Il suffit donc de montrer le lemme pour un torseur universel particulier. Soit $\mathscr{T}_0 := \mathscr{T}_0^\sigma$ le torseur universel trivial par σ. On peut appliquer la proposition 2.7.10 en prenant $S = S_0$ et $\lambda = \lambda_0$. Le diagramme de 2.7.10 s'écrit alors au niveau de K:

$$\begin{array}{ccc}
X(K) & \times & \mathrm{Br}\, X_K \\
\downarrow{\scriptstyle \mathscr{T}_0} & & \chi_K \downarrow\!\!\uparrow{\scriptstyle t_\sigma} \quad\searrow \\
H^1(K, S_0) & \times & H^1(K, \mathrm{Pic}\, \overline{X}) \quad \xrightarrow{\ \cup\ } \mathrm{Br}\, K.
\end{array}$$

D'après la proposition 2.7.10,

$$\mathscr{T}_0(P) \cup \beta = \big(t_\sigma(\beta)\big)(P) \quad \textit{pour tout } \beta \in H^1(K, \mathrm{Pic}\, \overline{X}).$$

Comme

$$t_\sigma \chi(\mathscr{A}) - \mathscr{A} =: a_0 \in \mathrm{Br}\, k,$$

on obtient, pour $\beta = \chi(\mathscr{A})$:

$$\mathscr{T}_0(P) \cup \chi_K(\mathscr{A}) = \mathscr{A}(P) + \mathrm{Res}_K a_0.$$

COROLLAIRE 3.5.3. *Sous les mêmes hypothèses,*

$$(3.5.2) \qquad i_{\mathscr{A}} = \left\langle i_{\lambda_0}, \chi(\mathscr{A}) \right\rangle \quad \text{pour tout } \mathscr{A} \in \mathrm{Br}\, X,$$

autrement dit

$$(3.5.3) \qquad i_{\mathrm{Br}} = \left\langle i_{\lambda_0}, \chi \right\rangle.$$

De plus,

$$(3.5.4) \qquad \ker i_{\mathrm{Br}} = \ker i_{\lambda_0}.$$

La notation $\langle\ ,\ \rangle$ désigne l'accouplement

$$H^1\!\left(k, \hat{S}_0\right)^{\sim} \times H^1(k, \mathrm{Pic}\, \overline{X}) \to \mathbf{Q}/\mathbf{Z},$$

et $i_{\mathrm{Br}} = \{i_{\mathscr{A}}\}\colon X(\mathbf{A}_k) \to \mathrm{Hom}(\mathrm{Br}\, X, \mathbf{Q}/\mathbf{Z})$. Autrement dit, le corollaire affirme qu'on a un triangle commutatif

$$(3.5.5) \qquad \begin{array}{ccc} X(\mathbf{A}_k) & \xrightarrow{\ i_{\mathrm{Br}}\ } & \mathrm{Hom}(\mathrm{Br}\, X, \mathbf{Q}/\mathbf{Z}) \\ & \searrow{\scriptstyle i_{\lambda_0}} \quad \nearrow{\scriptstyle \overline{\chi}} & \\ & H^1(k, \mathrm{Pic}\, \overline{X})^{\sim}. & \end{array}$$

Démonstration du corollaire. Il s'agit de voir que, pour tout $\{P_v\} \in X(\mathbf{A}_k)$,

$$\left\langle i_{\lambda_0}, \chi(\mathscr{A}) \right\rangle(P_v) = i_{\mathscr{A}}(P_v),$$

autrement dit, si $\mathscr{T}_0$ est un torseur universel:

$$(3.5.6) \qquad \sum_v \left(\tau_v(\mathscr{T}_0(P_v))\right)(\chi(\mathscr{A})) = \sum_v \mathrm{inv}_v(\mathscr{A}(P_v)).$$

Par définition de τ_v,

$$(3.5.7) \qquad \left(\tau_v(\mathscr{T}_0(P_v))\right)(\chi(\mathscr{A})) = \mathrm{inv}_v(\mathscr{T}_0(P_v) \cup \chi(\mathscr{A})).$$

Or, d'après le lemme 3.5.2, étant donnés $\mathscr{T}_0$ et $\mathscr{A}$, il existe $a_0 \in \mathrm{Br}\, k$ telle que, pour tout $P_v \in X(k_v)$:

$$(3.5.8) \qquad \mathscr{T}_0(P_v) \cup \chi(\mathscr{A}) = \mathscr{A}(P_v) + \mathrm{Res}_v a_0 \quad \text{dans } \mathrm{Br}\, k_v.$$

Comme, par la loi de réciprocité, $\sum_v \mathrm{inv}_v(a_0) = 0$, les formules (3.5.8) et (3.5.7) établissent (3.5.6), donc les assertions (3.5.2) et (3.5.3) du corollaire. Comme enfin χ est surjective, $\tilde{\chi}$ est injective et l'égalité (3.5.4) résulte alors aussitôt du triangle commutatif (3.5.5).

Fin de la démonstration du théorème. Par définition, l'obstruction de Manin équivaut à $\ker i_{\mathrm{Br}} = \varnothing$. Comme on a supposé la première obstruction "vide", l'obstruction de Picard équivaut à la seconde obstruction, autrement dit à $\ker i_{\lambda_0} = \varnothing$. L'équivalence des obstructions de Manin et de Picard est alors une conséquence immédiate de l'égalité (3.5.4) de ces deux noyaux dans $X(\mathbf{A}_k)$. Notons enfin que, par définition même, l'obstruction de Picard signifie qu'il n'y a pas de torseur universel ayant des points dans chaque complété de k. Si au contraire il en existe un $\mathcal{T}_0$ et si λ est un type quelconque, le torseur $\mathcal{T}$ sur X sous S obtenu par produit contracté de $\mathcal{T}_0$ et S sous S_0 agissant sur $\mathcal{T}_0$ comme groupe du torseur et sur S via $D(\lambda)$: $S_0 \to S$ a évidemment aussi des points dans chaque complété de k.

Remarque 3.5.4. On vérifie facilement que si $\lambda \in \mathrm{Hom}_{\mathfrak{g}}(\hat{S}, \hat{S}_0)$ est un type quasi-admissible, la flèche naturelle $H^1(k, \hat{S}_0)^{\sim} \to H^1(k, \hat{S})^{\sim}$ est injective, d'où il suit que, si la première obstruction au principe de Hasse est vide, l'obstruction $\mathcal{O}_\lambda$ équivaut à l'obstruction $\mathcal{O}_{\lambda_0}$. On peut aussi montrer en toute généralité que si λ est un type admissible, l'obstruction $\mathcal{O}_\lambda$ équivaut à l'obstruction $\mathcal{O}_{\lambda_0}$, et donc à l'obstruction de Manin.

3.6. *Exemples.*

PROPOSITION 3.6.1. *Les variétés suivantes vérifient le principe de Hasse*:
(a) *les variétés de Severi-Brauer*;
(b) *les k-formes de $\mathsf{P}^1 \times \mathsf{P}^1$, en particulier les quadriques lisses dans P^3_k*;
(c) *les k-formes des surfaces F_n $(n \geqslant 2)$*;
(d) *les surfaces de del Pezzo de degré 6.*

Démonstration. Pour chacune de ces variétés, la propriété $X(k) \neq \varnothing$ entraîne que X est k-rationnelle. D'après l'appendice 2.B, le $\mathfrak{g}$-module $\mathrm{Pic}\,\overline{X}$ est stablement de permutation, d'où $\mathrm{III}^2(k, S_0) = 0$. Si $\prod_v X(k_v) \neq \varnothing$, l'obstruction élémentaire est donc nulle pour X. Or, pour ces variétés-là, l'obstruction élémentaire est, sur un corps quelconque, la seule obstruction à l'existence d'un point rationnel (cf. 2.2.11).

Le principe de Hasse pour les espaces homogènes principaux des groupes semi-simples simplement connexes sans facteur de type E_8 (Kneser-Harder) permet de déduire aisément de [48] le résultat suivant:

THÉORÈME 3.6.2. *Pour un k-compactifié lisse d'un espace homogène principal sous un groupe algébrique linéaire connexe sans facteur de type E_8, la première obstruction au principe de Hasse est la seule.*

Exemples 3.6.3.   Il s'agit d'exemples d'obstructions $\mathcal{O}_\lambda''$ associées à certains tores. Ce sont souvent ces obstructions qui apparaissent dans les contre-exemples au principe de Hasse qu'on trouve dans la littérature, sans relation a priori avec les obstructions attachées aux torseurs universels.

Soit K/k une extension galoisienne finie, de groupe de Galois G et soit $S = R_{K/k}\mathbf{G}_m/\mathbf{G}_{m,k}$. De la suite exacte

$$1 \to \mathbf{G}_{m,k} \to R_{K/k}\mathbf{G}_m \to S \to 1$$

on déduit une surjection

$$\rho\colon H^1(X, S) \to \mathrm{Br}(X, K).$$

On laisse au lecteur le soin de démontrer la proposition suivante (voir aussi 2.7.8(c)), qui fournit une autre démonstration du fait que l'obstruction de Manin équivaut à l'obstruction de Picard (théorème 3.5.1).

PROPOSITION 3.6.4.   *Soient k un corps de nombres et X une k-variété rationnelle propre et lisse satisfaisant $X(\mathbf{A}_k) \neq \varnothing$, et soient K et S comme ci-dessus.*

(a)   *Soit $\mathcal{T}$ un torseur sur X sous S de type $\lambda \in \mathrm{Hom}_{\mathfrak{g}}(\hat{S}, \hat{S}_0)$. Soit $\mathcal{A} = \rho(\mathcal{T}) \in \mathrm{Br}(X, K)$. L'obstruction $\mathcal{O}_\lambda''$ est équivalente à l'obstruction de Manin associée à $\mathcal{A}$ (i.e. au sous-groupe de $\mathrm{Br}_a X$ engendré par $\mathcal{A}$).*

(b)   *L'obstruction de Manin pour $\mathrm{Br}(X, K)$ équivaut à la condition: il existe un entier m et $\lambda \in \mathrm{Hom}_{\mathfrak{g}}(\hat{S}^m, \hat{S}_0)$ tels qu'on ait $\mathcal{O}_\lambda''$.*

(c)   *Soit $(\lambda_1, \ldots, \lambda_m)$ un système générateur du groupe $\mathrm{Hom}_{\mathfrak{g}}(\hat{S}, \hat{S}_0)$. Soit*

$$\mu = (\lambda_1, \ldots, \lambda_m) \in \mathrm{Hom}_{\mathfrak{g}}\bigl(\hat{S}^m, \hat{S}_0\bigr).$$

Si l'obstruction élémentaire $\partial(\lambda_0)$ est nulle, l'obstruction de Manin associée au groupe $\mathrm{Br}(X, K)$ est équivalente à $\mathcal{O}_\mu''$.

(d)   *Supposons l'obstruction élémentaire nulle, ou $H^3(G, K^*) = 0$, ce qui est loisible quitte à agrandir K, et supposons que X_K est K-rationnelle. Alors, pour μ comme en (c), l'obstruction $\mathcal{O}_\mu''$ équivaut à l'obstruction de Manin associée à $\mathrm{Br}\, X$.*

Soit toujours K/k une extension finie galoisienne de corps de nombres, de groupe de Galois G, et considérons maintenant le k-tore $S = R^1_{K/k}\mathbf{G}_m$. On a ici la suite exacte de k-tores

$$1 \to S \to R_{K/k}\mathbf{G}_m \xrightarrow{\ N\ } \mathbf{G}_{m,k} \to 1,$$

où $N = N_{K/k}$ désigne la norme de K à k. On en tire d'abord $Ш^2(k, S) = 0$. La suite exacte du corps de classes (3.0.7) donne alors l'isomorphisme

$$Ш^1(k, S) = \mathrm{coker}\biggl(k^*/NK^* \to \bigoplus_v k_v^*/NK_w^*\biggr) \xrightarrow[\cong]{\ \Sigma_v i_v\ } G/G',$$

où G' désigne le groupe dérivé de G, et où chaque $i_v\colon k_v^*/NK_w^* \to G/G'$ vient de l'isomorphisme du corps de classes local $k_v^*/NK_w^* \cong G_v/G_v'$, isomorphisme donné, lorsque v est non ramifiée dans K, par l'application $\alpha \to F_v^{v(\alpha)}$, où F_v désigne la classe de conjugaison des Frobenius en v.

PROPOSITION 3.6.5. *Soit S comme ci-dessus, et soit X une k-variété rationnelle propre et lisse telle que $X(\mathbf{A}_k) \neq \varnothing$. Soit $\lambda \in \mathrm{Hom}_{\mathfrak{g}}(\hat{S}, \hat{S}_0)$. L'obstruction $\mathcal{O}_\lambda'$ est vide: il existe toujours des torseurs de type λ. L'existence d'une obstruction $\mathcal{O}_\lambda''$ équivaut à celle d'une fonction $f \in k(X)^*$ dont le diviseur soit une norme d'un diviseur de X_K, satisfaisant les conditions équivalentes suivantes, où U désigne l'ouvert où f est inversible:*

(i) *pour tout $\alpha \in k^*$, il existe v telle que, pour tout $P_v \in U(k_v)$, on ait $\alpha \cdot f(P_v) \notin NK_w^*$;*

(ii) *pour tout $\{P_v\} \in \Pi_v U(k_v)$, on a: $\Pi_v i_v(f(P_v)) \neq 1$.*

Démonstration. Que $\mathcal{O}_\lambda'$ soit vide résulte de $\mathrm{III}^2(k, S) = 0$. La seconde partie de l'énoncé est une simple traduction de la proposition 3.4.2, compte tenu de l'exemple 2.7.8 (b): un argument de continuité permet de se limiter aux points de l'ouvert U (cf. [13], §4).

Remarque 3.6.6. La proposition ci-dessus permet de rendre compte de beaucoup des contre-exemples au principe de Hasse pour des variétés rationnelles que l'on trouve dans la littérature. Ainsi les exemples de [39], [53], [3] §4, [13] §7 et [37] relèvent-ils tous de cette proposition, dans le cas particulièrement simple où K/k est une extension cyclique. On notera que dans ce dernier cas, on dispose d'isomorphismes

$$R_{K/k}^1\mathsf{G}_m \cong R_{K/k}\mathsf{G}_m/\mathsf{G}_{m,\,k},$$

ce qui permet de passer de l'obstruction étudiée en 3.6.5 à celle étudiée en 3.6.4, et donc à l'obstruction de Manin. En pratique cependant, c'est plutôt la version calculatoire 3.6.5, en termes de fonctions (dont les diviseurs sont des normes), qui apparaît naturellement.

Il y a cependant un certain nombre d'exemples qui ne rentrent pas dans le moule simple de la proposition 3.6.5. Nous laissons au lecteur le plaisir de découvrir les tores S et les types λ sous-jacents au §3 de [3] et à l'exemple de Cassels et Guy [8] (pour ce dernier exemple, voir [15]).

Remarque 3.6.7. En principe, étant donné une k-variété X, lisse, complète et rationnelle, on peut en un nombre *fini* d'opérations "élémentaires" déterminer si l'obstruction de Picard au principe de Hasse, ou celle de Manin, est "vide" ou non. Néanmoins, en pratique, ces calculs peuvent devenir rapidement inextricables, et même dans des cas apparemment élémentaires posent des problèmes délicats.

Le cas a priori le plus simple est celui où X est rationnelle sur une extension quadratique—ou plus généralement cyclique—K/k. Le cas des surfaces de

Châtelet généralisées (cf. [42], chap. VI, [13]) est particulièrement simple à traiter. Cependant, même dans ce cadre, il convient parfois de passer à des extensions non cycliques pour calculer *effectivement* l'obstruction de Manin. Pour des exemples où on utilise une extension biquadratique, voir [38] et [50]. Dans le cas des surfaces cubiques diagonales, on utilise une extension bicubique, voir [15].

3.7. *Les obstructions à l'approximation faible.* Tout comme pour le principe de Hasse, le groupe de Brauer et les torseurs permettent de définir des obstructions à l'approximation faible.

LEMME 3.7.1. *Soit X une variété algébrique propre et lisse sur le corps de nombres k. Soient S un k-tore, $\mathscr{A} \in \operatorname{Br} X$ et*

$$\lambda \in \operatorname{Hom}_{\mathfrak{g}}(\hat{S}, \operatorname{Pic} \overline{X}).$$

Les applications

$$i_{\mathscr{A}}\colon X(\mathbf{A}_k) \to \mathbf{Q}/\mathbf{Z} \quad et \quad i_{\lambda}\colon X(\mathbf{A}_k) \to \mathbf{Q}/\mathbf{Z}$$

sont continues et triviales sur l'image de $X(k)$ par l'application diagonale. De même

$$i_{\mathrm{Br}}\colon X(\mathbf{A}_k) \to \operatorname{Hom}(\operatorname{Br} X, \mathbf{Q}/\mathbf{Z}) \quad et \quad i_{\lambda_0}\colon X(\mathbf{A}_k) \to H^1(\mathfrak{g}, \operatorname{Pic} \overline{X})^{\sim},$$

en supposant pour cette dernière que $\operatorname{Pic} \overline{X}$ est $\mathbf{Z}$-libre de type fini.

Démonstration. Voir [41], [42], [48] p. 36.

COROLLAIRE 3.7.2. *Chacune des conditions*

$$i_{\mathscr{A}} \neq 0, \ i_{\lambda} \neq 0, \ i_{\mathrm{Br}} \neq 0 \ et \ i_{\lambda_0} \neq 0$$

est une obstruction à l'approximation faible pour X. Plus précisément:

$$(3.7.1) \qquad \overline{X(k)} \subset \ker i_{\mathrm{Br}} = \ker i_{\lambda_0} \subset X(\mathbf{A}_k).$$

Démonstration. En effet, ces diverses applications sont continues et triviales sur l'image diagonale de $X(k)$, donc sur son adhérence.

Définition 3.7.3 [19]. On appelle *obstruction de Manin—ou de Brauer—à l'approximation faible* pour X la condition $i_{\mathrm{Br}} \neq 0$, autrement dit la condition:

$$(3.7.2) \quad \text{il existe } \mathscr{A} \in \operatorname{Br} X \text{ et } \{P_v\} \in X(\mathbf{A}_k) \text{ tels que } \sum_v \operatorname{inv}_v(\mathscr{A}(P_v)) \neq 0.$$

Définition 3.7.4. Etant donné un type λ, on appelle *obstruction $\mathscr{O}_{\lambda}^a$ à l'approximation faible* pour X la condition $i_{\lambda} \neq 0$, définie par les torseurs de type λ, autrement dit la condition:

(3.7.3)

étant donné un torseur $\mathcal{T}$ de type λ, il existe $\{P_v\} \in X(\mathbf{A}_k)$

$$\text{tel que } \sum_v \tau_v(\mathcal{T}(P_v)) \neq 0.$$

Définition 3.7.5. Supposons que Pic $\overline{X}$ est $\mathbf{Z}$-libre de type fini. On appelle *obstruction de Picard à l'approximation faible pour X* la condition $\mathcal{O}_{\lambda_0}^a$, autrement dit la condition:

(3.7.4)

étant donné un torseur universel $\mathcal{T}_0$, il existe $\{P_v\} \in X(\mathbf{A}_k)$

$$\text{tel que } \sum_v \tau_v(\mathcal{T}_0(P_v)) \neq 0.$$

On notera que cette obstruction est à valeurs dans $Ш^1(k, S_0) \subset H^1(\mathfrak{g}, \operatorname{Pic} \overline{X})^{\sim}$.

THÉORÈME 3.7.6. *L'obstruction de Picard à l'approximation faible a les propriétés suivantes*:
  (i) *c'est la plus fine des obstructions $\mathcal{O}_\lambda^a$*;
  (ii) *elle coïncide avec celle de Brauer-Manin*;
  (iii) *elle est nulle si $Ш^1(k, S_0) = 0$, en particulier si $H^1(\mathfrak{g}, \operatorname{Pic} \overline{X}) = 0$.*

Démonstration. Les assertions (i) et (iii) sont immédiates; quant à (ii), c'est une conséquence de (3.7.1), en fait du diagramme (3.5.5).

De façon générale, si $\mathcal{T}$ est un torseur de type λ, et si Σ_0 est un ensemble fini de places en dehors duquel $\mathcal{T}$ a bonne réduction, on a la factorisation suivante de $i_{\mathcal{T}} = i_\lambda$:

$$(3.7.5)$$

$$
\begin{array}{ccccccc}
\mathcal{T}(k) & \xrightarrow{q} & X(k) & \xrightarrow{\partial} & Ш_{\Sigma_0}^1(k, S) & \hookrightarrow & H^1(k, S) \\
\downarrow{\scriptstyle \delta_{\mathcal{T}}} & & \downarrow{\scriptstyle \delta_X} & & \downarrow{\scriptstyle \iota} & & \downarrow \\
\mathcal{T}(\mathbf{A}_k) & \xrightarrow{\{q_v\}} & X(\mathbf{A}_k) & \xrightarrow{\{\partial_v\}} & \bigoplus_{v \in \Sigma_0} H^1(k_v, S) & \hookrightarrow & \bigoplus_{v \in \Omega_k} H^1(k_v, S) \\
& & & {\scriptstyle i_{\mathcal{T}}} \searrow & \downarrow & & \downarrow \\
& & & & Ш_{\Sigma_0}^1(k, S) & \longrightarrow & Ш^1(k, S) \\
& & & & & & \downarrow \\
& & & & & \searrow & H^1(k, \hat{S})^{\sim}.
\end{array}
$$

En particulier, l'obstruction $\mathcal{O}_\lambda^a$ disparaît dès que $Ш^1(k, S) = 0$.

Remarque 3.7.7. On vérifie aisément que si λ est un type quasi-admissible, l'obstruction $\mathcal{O}_\lambda^a$ coïncide avec l'obstruction de Picard $\mathcal{O}_{\lambda_0}^a$.

Exemples 3.7.8. Comme pour les contre-exemples au principe de Hasse, la plupart des contre-exemples à l'approximation faible pour des variétés rationnelles que l'on trouve dans la littérature peuvent être analysés du point de vue ci-dessus en utilisant des tores S du type $R_{K/k}^1 \mathbf{G}_m$. Il en est ainsi des contre-exemples de [53], [19] et [13] §7.

3.8. *Deux hypothèses arithmétiques sur les torseurs universels.* Dans ce paragraphe, k est toujours un corps de nombres, et X une k-variété rationnelle, propre et lisse.

Nous indiquons les conséquences des deux hypothèses suivantes sur les torseurs universels sur X. On notera que (H3) est un affaiblissement de l'hypothèse (H1') introduite au §2.8:

(H2) *Les torseurs universels sur X satisfont le principe de Hasse*: si un tel torseur $\mathcal{T}$ a des points dans chaque complété k_v de k, il a aussi un point dans k.

(H3) *Les torseurs universels sur X satisfont l'approximation faible dès qu'ils possèdent un point dans k*: si Σ est un ensemble (fini ou non) de places de k, et $\mathcal{T}$ un torseur universel sur X avec $\mathcal{T}(k) \neq \varnothing$, alors $\mathcal{T}(k)$ est dense dans le produit topologique des $\mathcal{T}(k_v)$ pour v dans Σ.

De fait, il est également utile de considérer les hypothèses analogues pour les torseurs d'autres types $\lambda \in \mathrm{Hom}_{\mathfrak{g}}(\hat{S}, \mathrm{Pic}\,\overline{X})$:

(H2)$_\lambda$ *Les torseurs sur X de type λ satisfont le principe de Hasse.*

(H3)$_\lambda$ *Les torseurs sur X de type λ satisfont l'approximation faible dès qu'ils possèdent un point dans k.*

Les meilleurs candidats pour vérifier le principe de Hasse ou l'approximation faible sont a priori les torseurs universels, et plus généralement les torseurs quasi-admissibles relativement à une extension galoisienne K/k sur laquelle X devient K-rationnelle. En effet, sur les compactifiés lisses de tous ces torseurs, les obstructions de Brauer-Manin au principe de Hasse et à l'approximation faible disparaissent d'après le théorème 2.1.2.

Les hypothèses (H2) et (H2)$_\lambda$ pour certains types λ on été établies pour les surfaces de Châtelet généralisées ([27], voir aussi [13] où on étudie des torseurs quasi-admissibles), et par voie de conséquence pour les modèles de certaines surfaces de Del Pezzo singulières [30]. Elles ont été aussi établies pour de nombreuses autres surfaces fibrées en coniques d'invariant 4 [47]. Elles sont également établies pour certaines hypersurfaces cubiques singulières dans [16]. Quant à (H3), elle a été établie dans certains cas comme conséquence de (H1) (voir §2.8). Pour $k = \mathbf{Q}$, l'hypothèse H de Schinzel implique (H2)$_\lambda$ et (H3)$_\lambda$ pour certains types λ quasi-admissibles pour une assez large classe de surfaces fibrées en coniques (voir [24]).

On notera que, pour λ quasi-admissible et $X(\mathbf{A}_k) \neq \varnothing$, l'application

$$H^1\big(k, \hat{S}_0\big)^{\sim} \xrightarrow{\ \tilde{\lambda}\ } H^1(k, \hat{S})^{\sim}$$

induite par λ est injective, ce qui entraîne

$$\ker i_\lambda = \ker i_{\lambda_0}.$$

THÉORÈME 3.8.1. *Supposons que X vérifie* (H2)$_\lambda$. *Alors l'obstruction de Manin au principe de Hasse pour X est la seule. Sous la même hypothèse, si $H^1(\mathfrak{g}, \mathrm{Pic}\,\overline{X}) = 0$, le principe de Hasse vaut pour X.*

Démonstration. Si l'obstruction de Manin disparaît sur X, par exemple si $H^1(\mathfrak{g}, \mathrm{Pic}\,\overline{X}) = 0$, il existe d'après le théorème 3.5.1 un torseur $\mathscr{T}$ de type λ qui a des points dans chaque complété de k, donc dans k d'après l'hypothèse, ce qui, par projection sur X, implique $X(k) \neq \varnothing$.

PROPOSITION 3.8.2. *Si X vérifie* (H2)$_\lambda$ *et possède un k-point*,

$$\#X(k)/R \geqslant \#\mathrm{III}^1(k, S) = \#\mathrm{III}^2(k, \hat{S}).$$

Démonstration. Soient $O \in X(k)$ et $\mathscr{T}$ le torseur de type λ sur X et de fibre triviale en O. Considérons, pour tout $\alpha \in \mathrm{III}^1(k, S)$, un torseur $\mathscr{T}^\alpha$ sur X sous S obtenu par torsion de $\mathscr{T}$ par un 1-cocycle de $Z^1(k, S)$ de classe $-\alpha$. La fibre de $\mathscr{T}^\alpha$ en O est un torseur sur Spec k qui est trivial sur chaque Spec k_v. Ainsi $\mathscr{T}^\alpha$ possède un k_v-point pour chaque v et donc, d'après (H2)$_\lambda$, $\mathscr{T}^\alpha$ possède un k-point. D'où l'énoncé de la proposition, puisque les projections de chacun des $\mathscr{T}^\alpha(k)$ sur $X(k)$ correspondent à des classes différentes pour la R-équivalence.

Remarque 3.8.3. Dans [23] nous avons donné une conjecture (Conjecture A, op cit. p. 443) sur l'ordre du groupe fini $A_0(X)$ pour X une surface rationnelle définie sur un corps de nombres. Il vaut la peine de noter que les arguments développés aux pages 444 et 445 du même article montrent plus généralement le résultat suivant:

PROPOSITION 3.8.4. *Si X est une surface fibrée en coniques sur la droite projective $\mathbf{P}_k^1$, avec au plus 5 fibres géométriques dégénérées, si $X(k)$ est non vide et si* (H2) *vaut pour X, alors la conjecture A vaut pour X.*

PROPOSITION 3.8.5. *Si X vérifie les hypothèses* (H1$'$)$_\lambda$ *et* (H2)$_\lambda$, *l'ordre de $X(k)/R$ est fini et divisible par l'ordre de $\mathrm{III}^1(k, S)$.*

Démonstration. Soit i l'application $X(k)/R \to H^1(k, S)$, d'image finie (proposition 2.7.3) définie par un torseur $\mathscr{T}$ de type λ sur X. La démonstration de la proposition 3.8.2 montre que si γ est dans l'image de i, et si X vérifie (H2)$_\lambda$, pour tout élément α dans $\mathrm{III}^1(k, S)$, l'élément $\gamma - \alpha$ appartient à l'image de i. Comme X vérifie (H1$'$)$_\lambda$, l'application i est injective (proposition 2.8.10), d'où l'énoncé.

488 COLLIOT-THÉLÈNE ET SANSUC

Remarque 3.8.6. Sous les hypothèses de la proposition, l'ordre de $X(k)/R$ est en principe déterminable de façon effective. On part de l'application i définie par un torseur $\mathcal{T}$ de type λ et de fibre triviale en un point $O \in X(k)$. On commence par déterminer un ensemble fini Σ de places de k tel que l'image de i soit contenue dans le groupe fini $\text{III}^1_\Sigma(k, S)$ (noyau de la restriction de $H^1(k, S)$ à la somme directe, pour $v \notin \Sigma$, des $H^1(k_v, S)$). Pour chaque $\alpha \in \text{III}^1_\Sigma(k, S)$, on considère un torseur $\mathcal{T}^\alpha$ comme plus haut. Chacun de ces torseurs a des k_v-points pour $v \notin \Sigma$. On élimine les α pour lesquels il existe une place $v \in \Sigma$ avec $\mathcal{T}^\alpha(k_v) = \varnothing$. L'ordre de $X(k)/R$ est égal au nombre des α qui subsistent.

PROPOSITION 3.8.7. *Supposons* $X(k) \neq \varnothing$. *Si* X *vérifie* (H2)$_\lambda$ *et* (H3)$_\lambda$, *l'adhérence de* $X(k)$ *dans* $X(\mathbf{A}_k)$ *est ouverte et elle coïncide avec le noyau de l'application* i_λ.

Démonstration. C'est une conséquence du diagramme (3.7.5). D'après (3.7.1), on sait déjà que l'adhérence de $X(k)$ est contenue dans le noyau de i_λ. Inversement, soit $\{P_v\} \in X(\mathbf{A}_k)$ tel que $i_\lambda(P_v) = 0$. Il existe $\alpha \in \text{III}^1_{\Sigma_0}(k, S)$ tel que $\iota(\alpha) = \{\partial_v(P_v)\}$. Considérons alors le torseur $\mathcal{T}' := \mathcal{T}^\alpha$. Il a encore bonne réduction en dehors de Σ_0, et on peut considérer le diagramme (3.7.5) pour $\mathcal{T}'$ au lieu de $\mathcal{T}$. On trouve alors, pour toute place v:

$$\partial'_v(P_v) = \partial_v(P_v) - \iota_v(\alpha) = 0.$$

Ainsi, $\mathcal{T}'(P_v) = 0$, autrement dit il existe $Q_v \in \mathcal{T}'(k_v)$ tel que $q_v(Q_v) = P_v$. En particulier, $\Pi_v \mathcal{T}'(k_v) \neq \varnothing$ et, puisque, par hypothèse, les torseurs de type λ vérifient le principe de Hasse et l'approximation faible, $\mathcal{T}'(k)$ est dense dans $\mathcal{T}'(\mathbf{A}_k)$. Etant donné un voisinage ouvert de $\{P_v\}$, il existe donc $Q \in \mathcal{T}'(k)$ tel que $\{q_v \delta_{\mathcal{T}}(Q)\}$ appartienne à ce voisinage ouvert, i.e. $\delta_X q(Q)$ appartient à ce voisinage ouvert, et $\{P_v\}$ appartient à l'adhérence de $\delta_X(X(k))$.

COROLLAIRE 3.8.8. *Supposons* $X(k) \neq \varnothing$ *et que les torseurs d'un certain type* λ *vérifient le principe de Hasse et l'approximation faible. Alors,* X *vérifie l'approximation faible sous l'une quelconque des hypothèses suivantes:*

(i) *L'obstruction à l'approximation faible définie par les torseurs de type* λ *est vide.*

(ii) *L'obstruction de Brauer-Manin à l'approximation faible est vide.*

(iii) $H^1(k, \hat{S}) = 0$.

Démonstration. L'assertion dans le cas (i) dérive de la proposition précédente 3.8.7, et elle implique l'assertion dans le cas (iii); d'après le théorème 3.7.6, l'hypothèse (ii) implique l'hypothèse (i).

PROPOSITION 3.8.9. *Si* k *est un corps de nombres et si* X *vérifie* (H3)$_\lambda$ *et possède un* k-*point, il existe un ensemble fini* Σ_0 *de places de* k *tel que pour tout ensemble fini* Σ *de places de* k *ne rencontrant pas* Σ_0, *l'image de* $X(k)$ *par l'application diagonale est dense dans le produit topologique, pour* v *dans* Σ, *des* $X(k_v)$.

Démonstration. Le même argument de bonne réduction qu'à la proposition 2.8.19 montre qu'il existe un torseur $\mathcal{T}$ de type λ sur X avec $\mathcal{T}(k) \neq \varnothing$ et un ensemble fini Σ_0 de places de k (contenant les places réelles), tel que, pour tout ensemble Σ de places de k ne rencontrant pas Σ_0, l'application naturelle

$$\prod_{v \in \Sigma} \mathcal{T}(k_v) \rightarrow \prod_{v \in \Sigma} X(k_v)$$

est surjective. Mais, si X vérifie $(H3)_\lambda$, l'approximation faible vaut pour la k-variété lisse $\mathcal{T}$, c'est-à-dire que $\mathcal{T}(k)$ est dense dans tout produit topologique fini de $\mathcal{T}(k_v)$, ce qui permet de conclure.

PROPOSITION 3.8.10. *Si k est un corps de nombres, si X vérifie $(H3)_\lambda$, et si Σ est un ensemble fini de places de k, l'adhérence de $X(k)$ plongé par l'application diagonale dans le produit topologique, pour v dans Σ, des $X(k_v)$ (où k_v désigne la complétion de k en la place v) est ouverte et fermée dans ce produit. En particulier, si Σ est composé de places réelles, cette adhérence est une union de composantes connexes du produit.*

Démonstration. On peut supposer $X(k) \neq \varnothing$. Soit alors $\mathcal{T}$ un torseur sur X de type λ, et considérons le diagramme commutatif

$$
\begin{array}{ccc}
X(k) & \xrightarrow{\ \ i\ \ } & H^1(k, S) \\
\downarrow{\scriptstyle \rho} & & \downarrow{\scriptstyle \sigma} \\
\displaystyle\prod_{v \in \Sigma} X(k_v) & \xrightarrow{\ \ j\ \ } & \displaystyle\prod_{v \in \Sigma} H^1(k_v, S)
\end{array}
$$

où i et j associent à un point la fibre de $\mathcal{T}$ en ce point, et où les flèches verticales sont les flèches diagonales. Le théorème des fonctions implicites implique que j est une application continue de l'espace topologique produit de gauche dans le groupe abélien fini discret de droite. Ainsi chaque fibre de j est ouverte et fermée. Soit $\{\alpha_1, \ldots, \alpha_n\}$ l'image finie (proposition 2.7.3) de i. Nous allons montrer que l'adhérence de $\rho(X(k))$, qui est clairement contenue dans le fermé réunion des $j^{-1}(\sigma(\alpha_i))$, coïncide avec cette réunion, ce qui suffira à établir la proposition. Soit $\mathcal{T}_i$ un torseur universel sur X dont la classe dans $H^1(X, S_0)$ se déduise de celle de $\mathcal{T}$ par soustraction de α_i. Comme α_i est dans l'image de $X(k)$ par i, $\mathcal{T}_i(k)$ est non vide. L'image par la projection structurale q_i du produit des $\mathcal{T}_i(k_v)$ dans le produit des $X(k_v)$ n'est autre que $j^{-1}(\sigma(\alpha_i))$. Sous l'hypothèse $(H3)_\lambda$, $\mathcal{T}_i(k)$ est dense dans le produit des $\mathcal{T}_i(k_v)$, donc $q_i(\mathcal{T}_i(k))$ est dense dans $j^{-1}(\sigma(\alpha_i))$.

Remarque 3.8.11. On peut établir directement le résultat de cette proposition pour une surface cubique lisse ou une intersection lisse de deux quadriques dans $\mathbf{P}^4$ en utilisant la k-unirationalité de telles k-variétés X lorsque $X(k) \neq \varnothing$, et l'existence dans ce cas de beaucoup de k-automorphismes birationnels, en utilisant la même idée que dans ([42], II. 16.1).

PROPOSITION 3.8.12. *Soit k un corps de nombres. Soit X une k-surface rationnelle qui possède un k-point et vérifie* (H3)$_\lambda$. *Si Σ est un ensemble fini de places de k tel que, pour v dans Σ, X a bonne réduction en v si v est non archimédienne et $X(k_v)$ est connexe si v est archimédienne, alors $X(k)$ est dense dans le produit des $X(k_v)$ pour v dans Σ.*

Démonstration. Soit $O \in X(k)$ et soit $\mathcal{T}$ le torseur de type λ de fibre triviale en O. Les mêmes arguments que dans la proposition 2.8.20 montrent que la projection structurale $q\colon \mathcal{T} \to X$ induit une surjection

$$\prod_{v \in \Sigma} \mathcal{T}(k_v) \to \prod_{v \in \Sigma} X(k_v).$$

Sous (H3)$_\lambda$, $\mathcal{T}(k)$ est dense dans le produit de gauche, et donc $X(k)$ est dense dans le produit de droite.

RÉFÉRENCES

1. M. ARTIN, A. GROTHENDIECK, ET J.-L. VERDIER, *Théorie des topos et cohomologie étale des schémas*, Séminaire de géométrie algébrique du Bois-Marie SGA 4, Lecture Notes in Math. **269, 270, 305**, Springer, Berlin-Heidelberg-New York 1972–1973.

2. A. BEAUVILLE, J.-L. COLLIOT-THÉLÈNE, J.-J. SANSUC, ET SIR PETER SWINNERTON-DYER, *Variétés stablement rationnelles non rationnelles*, Annals of Math. **121** (1985), 283–318.

3. B. J. BIRCH ET H. P. F. SWINNERTON-DYER, *The Hasse problem for rational surfaces*, J. reine angew. Math. **274** (1975), 164–174.

4. S. BLOCH, *On the Chow groups of certain rational surfaces*, Ann. scient. Éc. norm. sup. (4) **14** (1981), 41–59.

5. J.-L. BRYLINSKI, *Décomposition simpliciale d'un réseau, invariante par un groupe fini d'automorphismes*, C.R. Acad. Sci. Paris **288**, série A (1979), 137–139.

6. H. CARTAN ET S. EILENBERG, *Homological Algebra*, Princeton Univ. Press, Princeton, 1956.

7. J. W. S. CASSELS, *Diophantine equations with special reference to elliptic curves*, J. London Math. Soc. **41** (1966), 193–291; Corrigenda: ibid **42** (1967), 183.

8. J. W. S. CASSELS ET M. J. T. GUY, *On the Hasse principle for cubic surfaces*, Mathematika **13** (1966), 111–120.

9 F. CHÂTELET, *Points rationnels sur certaines courbes et surfaces cubiques*, Enseign. Math. (2) **5** (1959), 153–170.

10. J.-L. COLLIOT-THÉLÈNE, *Hilbert's theorem* 90 *for* K_2, *with application to the Chow groups of rational surfaces*, Invent. Math. **71** (1983), 1–20.

11. ______, "Arithmétique des variétés rationnelles et problèmes birationnels," in *Actes du congrès intern. math. Berkeley 1986* (à paraître).

12. J.-L. COLLIOT-THÉLÈNE ET D. CORAY, *L'équivalence rationnelle sur les points fermés des surfaces rationnelles fibrées en coniques*, Compositio Math. **39** (1979), 301–332.

13. J.-L. COLLIOT-THÉLÈNE, D. CORAY, ET J.-J. SANSUC, *Descente et principe de Hasse pour certaines variétés rationnelles*, J. reine angew. Math. **320** (1980), 150–191.

14. J.-L. COLLIOT-THÉLÈNE ET F. ISCHEBECK, *L'équivalence rationnelle sur les cycles de dimension zéro des variétés algébriques réelles*, C.R. Acad. Sci. Paris **292**, série I (1981), 723–725.

15. J.-L. COLLIOT-THÉLÈNE, D. KANEVSKY, ET J.-J. SANSUC, *Arithmétique des surfaces cubiques diagonales*, Springer Lecture Notes in Math., éd. G. Wüstholz (à paraître) (preprint Orsay 85 T 46, 1985).

16. J.-L. COLLIOT-THÉLÈNE ET P. SALBERGER, *Arithmetic on singular cubic hypersurfaces* (en préparation).

17. J.-L. Colliot-Thélène et J.-J. Sansuc, *Torseurs sous des groupes de type multiplicatif; applications à l'étude des points rationnels de certaines variétés algébriques*, C.R. Acad. Sci. Paris **282**, série A (1976), 1113–1116.

18. ______, *Variétés de première descente attachées aux variétés rationnelles*, C.R. Acad. Sci. Paris **284**, série A (1977), 967–970.

19. ______, *La descente sur une variété rationnelle définie sur un corps de nombres*, C.R. Acad. Sci. Paris **284**, série A (1977), 1215–1218.

20. ______, *La R-équivalence sur les tores*, Ann. scient. Éc. norm. sup. (4) **10** (1977), 175–229.

21. ______, *Cohomologie des groupes de type multiplicatif sur les schémas réguliers*, C.R. Acad. Sci. Paris **287**, série A (1978), 449–452.

22. ______, "La descente sur les variétés rationnelles," in *Journées de géométrie algébrique d'Angers 1979*, éd. A. Beauville, Sijthoff & Noordhoff, Alphen aan den Rijn 1980, 223–237.

23. ______, *On the Chow groups of certain rational surfaces: a sequel to a paper of S. Bloch*, Duke Math. J. **48** (1981), 421–447.

24. ______, *Sur le principe de Hasse et l'approximation faible, et sur une hypothèse de Schinzel*, Acta Arith. **41** (1982), 33–53.

25. ______, *La descente sur les surfaces rationnelles fibrées en coniques*, C.R. Acad. Sci. Paris **303**, série I (1986), 303–306.

26. ______, *Principal homogeneous spaces under flasque tori, applications*, J. of Algebra **106** (1987), 148–205.

27. J.-L. Colliot-Thélène, J.-J. Sansuc, et Sir Peter Swinnerton-Dyer, *Intersections of two quadrics and Châtelet surfaces*, J. reine angew. Math. **373** (1987), 37–107 & **374** (1987), 72–168 (voir aussi: *Intersections de deux quadriques et surfaces de Châtelet*, C.R. Acad. Sci. Paris **298**, série I (1984), 377–380).

28. J.-L. Colliot-Thélène et A. N. Skorogobatov, *R-equivalence on conic bundles of degree 4*, this volume.

29. D. F. Coray, "The Hasse principle for pairs of quadratic forms," in *Number theory days*, Exeter 1980, London Math. Soc. Lecture note series **56**, Cambridge Univ. Press, Cambridge-New York 1982, 237–246.

30. D. F. Coray et M. A. Tsfasman, *Arithmetic on singular Del Pezzo surfaces*, Proc. London Math. Soc. (à paraître).

31. P. Deligne, *Cohomologie étale*, Séminaire de géométrie algébrique du Bois-Marie SGA 4 1/2, Lecture Notes in Math. **569**, Springer, Berlin-Heidelberg-New York 1977.

32. M. Demazure et A. Grothendieck, *Schémas en groupes*, Séminaire de géométrie algébrique du Bois-Marie SGA 3, Lecture Notes in Math. **151**, **152**, **153**, Springer, Berlin-Heidelberg-New York 1970.

33. A. Grothendieck et J. Dieudonné, *Eléments de géométrie algébrique*, *IV*, Publ. math. I.H.E.S. **20**, **24**, **28**, **32**, Bures sur Yvette 1964–65–66–67.

34. A. Grothendieck, "Le groupe de Brauer, I: Algèbres d'Azumaya et interprétations diverses; II: Théorie cohomologique." Exposés Bourbaki **290** & **297** (1965), in *Dix exposés sur la cohomologie des schémas*, Masson-North-Holland, Amsterdam 1968.

35. A. Grothendieck, "Le groupe de Brauer, III: Exemples et compléments," in *Dix exposés sur la cohomologie des schémas*, Masson-North-Holland, Amsterdam 1968.

36. P. Hilton et U. Stammbach, *A Course in Homological Algebra*, Springer, Berlin-Heidelberg-New York 1971.

37. W. Hürlimann, "Brauer group and diophantine geometry: a cohomological approach," in *Brauer Groups in Ring Theory and Algebraic Geometry*, Antwerp 1981, ed. F. van Oystaeyen & A. Verschoren, Lecture Notes in Math. **917**, Springer, Berlin-Heidelberg-New York 1982.

38. ______, H^3 *and rational points on biquadratic bicyclic norm forms*, Arch. Math. **47** (1986), 113–116.

39. V. A. Iskovskih, *Un contre-exemple au principe de Hasse pour un système de deux formes quadratiques en cinq variables* (en russe), Mat. Zametki **10** (1971), 253–257 (= Math. Notes **10** (1971), 575–577).

40. B. È. KUNYAVSKIĬ ET M. A. TSFASMAN, *Zero-cycles on rational surfaces and Néron-Severi tori*, Izv. Akad. Nauk SSSR **48** (1984), 631–654 (= Math. USSR Izvestiya **24** (1985), 583–603).

41. YU. I. MANIN, "Le groupe de Brauer-Grothendieck en géométrie diophantienne," *Actes du congrès intern. math. Nice* **1** (1970), 401–411, Gauthier-Villars, Paris 1971.

42. ______, *Formes cubiques: algèbre, géométrie, arithmétique* (en russe), Nauka, Moscou 1972 (trad. angl.: *Cubic forms: algebra, geometry, arithmetic*, North-Holland, Amsterdam 1974 et 1986, 2ème éd.).

43. YU. I. MANIN ET M. A. TSFASMAN, *Rational varieties: Algebra, geometry and arithmetic*, Uspekhi Mat. Nauk **41** (1986), 43–94 (= Russian Math. Surveys **41** (1986), 51–116).

44. J. S. MILNE, *Etale Cohomology*, Princeton Univ. Press, Princeton, 1980.

45. H. NISHIMURA, *Some remark on rational points*, Mem. Coll. Sci. Kyoto **29**, Ser. A (1955), 189–192.

46. M. RAYNAUD, *Faisceaux amples sur les schémas en groupes et les espaces homogènes*, Lecture Notes in Math. **119**, Springer, Berlin-Heidelberg-New York 1970.

47. P. SALBERGER, "On the arithmetic of conic bundle surfaces," in *Séminaire de théorie des nombres de Paris 1985–1986* (à paraître) (voir aussi: *Sur l'arithmétique de certaines surfaces de del Pezzo*, C.R. Acad. Sci. Paris **303**, série I (1986), 273–276).

48. J.-J. SANSUC, *Groupe de Brauer et arithmétique des groupes algébriques linéaires sur un corps de nombres*, J. reine angew. Math. **327** (1981), 12–80.

49. ______, Descente et principe de Hasse pour certaines variétés rationnelles," in *Séminaire de théorie des nombres (DPP)*, Paris 1980–1981, 253–271, Birkhäuser, 1982.

50. ______, "A propos d'une conjecture arithmétique sur le groupe de Chow d'une surface rationnelle," in *Séminaire de théorie des nombres de Bordeaux 1981–1982*, exp. n° 33, 1–38.

51. ______, "Principe de Hasse, surfaces cubiques et intersections de deux quadriques, in *Journées arithmétiques de Besançon 1985*, Astérisque **147–148** (1987), 183–207.

52. J.-P. SERRE, *Cohomologie Galoisienne*, Lecture Notes in Math. **5**, Springer, Berlin-Heidelberg-New York 1965.

53. H. P. F. SWINNERTON-DYER, *Two special cubic surfaces*, Mathematika **9** (1962), 54–56.

54. J. T. TATE, *The cohomology groups of tori in finite Galois extensions of number fields*, Nagoya Math. J. **27** (1966), 709–719.

55. V. E. VOSKRESENSKIĬ, *Tores algébriques* (en russe), Nauka, Moscou 1977.

56. V. I. YANČEVSKIĬ, *k-unirationalité des fibrés en coniques et corps de décomposition des algèbres simples centrales* (en russe), Doklady Akad. Nauk BSSR **29** (1985), 1061–1064.

JEAN-LOUIS COLLIOT-THÉLÈNE: MATHÉMATIQUES, BÂT. 425, UNIVERSITÉ DE PARIS-SUD, F-91405 ORSAY FRANCE

JEAN-JACQUES SANSUC: ÉCOLE NORMALE SUPERIEURE, 45 RUE D'ULM, F-75230 PARIS CEDEX 05 FRANCE, UNIVERSITÉ PARIS VII, MATHÉMATIQUES, 2 PLACE JUSSIEU, F-75251 PARIS CEDEX 05 FRANCE

Vol. 54, No. 2   DUKE MATHEMATICAL JOURNAL ©   1987

SERRE DUALITY ON COMPLEX SUPERMANIFOLDS

CARL HASKE AND R. O. WELLS, JR.

In 1955 J. P. Serre published his well-known duality theorem (Serre, 1955), which has had many consequences in complex geometry. In particular, as pointed out in his original paper, the classical Riemann-Roch theorem for Riemann surfaces is a very quick consequence. This duality theorem has had many applications and has had a number of generalizations over the past 30 years, and more recently there have been generalizations to the context of supermanifolds (Penkov, 1983), and applications of this in the context of supermanifolds have been discussed in the paper of Ogijevetsky and Penkov (1984). In particular the problems of geometric realizations of representations of super Lie algebras (generalizations of the Bott-Borel-Weil Theorem) are contexts in which this duality theory is useful (see Penkov, 1985).

The recent book of Manin (1984) gives a beautiful introduction to the ideas of supergeometry. This is the study of manifolds with both commuting and anti-commuting coordinates, which was inspired by pioneering work in theoretical physics over the past two decades. In this paper we want to outline a new proof of Serre duality on complex supermanifolds which is a generalization of the original proof of Serre (1955). The proof of Penkov (1983) uses the theory of $\mathcal{D}$-modules, and is a generalization of the proof of Hartshorne (1966) in which he has an extension of Serre duality using the Ext functor. Our proof uses the natural duality between differentiable functions and distributions, and uses the underlying differentiable structure of the complex supermanifold, just as in the original Serre paper. The details will be published elsewhere.

The basic Serre duality theorem includes as a special case the following result. If V is a holomorphic vector bundle on a complex manifold M of complex dimension n, then there is a natural duality between $H^q(M, \Omega^p(V))$ and $H^{n-q}(M, \Omega^{n-p}(V^*))$. This result has been generalized by Hartshorne (1966) to a duality between $H^q(M, \mathcal{F})$ and $\mathrm{Ext}^{n-q}(M, \mathcal{F}, \Omega^n)$, where F is a coherent $\mathcal{O}$-module. This is a generalization, since for a locally free $\mathcal{O}$-module $\mathcal{F}$ there is an isomorphism

$$\mathrm{Ext}^k(M, \mathcal{F}, \Omega^n) \cong H^k(M, \Omega^n(\mathcal{F}^*)).$$

Using the theory of $\mathcal{D}$-modules, Serre duality was proved for supermanifolds (Penkov, 1983) in the following form: for a compact complex supermanifold $(M, \mathcal{O})$ of dimension $m|n$, there is a duality between $\mathrm{Ext}^{m-q}(M, \mathcal{F}, \mathcal{B}\mathit{er}_\mathcal{O})$ and

Received January 5, 1987.

$H^q(M, \mathcal{F})$, where $\mathcal{F}$ is a coherent $\mathcal{O}$-module and $\mathcal{B}er_{\mathcal{O}}$ is the Berezinian sheaf on M, which is the analogue in the supermanifold context of the classical canonical sheaf on an ordinary complex manifold (we shall use Leites (1980) and Manin (1984) as references for supermanifold concepts and terminology which we use in this paper). Letting $\mathcal{F} = \mathcal{O}$, we obtain the simplest form of Serre duality between $H^q(M, \mathcal{O})$ and $H^{m-q}(M, \mathcal{B}er_{\mathcal{O}})$. More generally, if $\mathcal{F}$ is a locally free $\mathcal{O}$-module, the duality is between $H^q(M, \mathcal{F})$ and $H^{m-q}(M, \mathcal{B}er_{\mathcal{O}} \otimes \mathcal{F}^*)$. This is the precise analogue of Serre's original duality between $H^q(M, \mathcal{O}(V))$ and $H^{n-q}(M, \Omega^n(V^*))$ for a complex manifold M.

Our goal is to give a new proof of this version of Serre duality on supermanifolds, and as we mentioned above, it is in the spirit of Serre's original proof. Let us analyze the main tools used by Serre. First and simplest is the fact that a complex manifold has an underlying differentiable structure which allows for differentiable analysis on a complex manifold. Second are the Dolbeault resolutions of holomorphic forms, one by sheaves of smooth (p, q)-forms, the other by sheaves of (p, q)-currents, allowing for two distinct realizations of $H^q(M, \Omega^p(V))$ as the Dolbeault group of smooth (p, q)-forms, and (p, q)-currents. Lastly, we have the functional analysis apparatus which can be used in conjunction with the above data to generate the required duality theorem.

We shall develop the analogous analytic tools for a complex supermanifold $(M, \mathcal{O})$, and, having done this, we shall see that the functional analysis used by Serre may be applied exactly as in the original proof.

We summarize the results in the following four constructions which will be described in more detail below:

1. To every complex supermanifold $(M, \mathcal{O})$ of dimension $m|n$ one may associate a smooth supermanifold $(M, \mathcal{E})$ of dimension $2m|2n$ and a natural map of ringed spaces $(i, i^{\#})\colon (M, \mathcal{E}) \to (M, \mathcal{O})$ such that $i\colon M \to M$ is the identity and $i^{\#}\colon \mathcal{O} \to \mathcal{E}$ is an inclusion.

2. If $(M, \mathcal{E})$ is the smooth supermanifold associated to a complex supermanifold $(M, \mathcal{O})$, then

$$0 \to \Omega^p \to \mathcal{E}^{p,0} \to \mathcal{E}^{p,1} \to \cdots$$

is a resolution of the "superholomorphic" forms Ω^p by the super (p, q)-forms $\mathcal{E}^{p, \bullet}$ and Dolbeault's theorem holds in this context.

3. There is a differential sheaf $\{\mathcal{E}_{p, \bullet}\}$ whose cohomology sheaves vanish except in degree 0, where $\mathcal{H}^0(\mathcal{E}_{p, \bullet}) = \mathcal{B}er_{\mathcal{O}} \otimes (\Omega^{m-p})^*$. This will yield a realization of $H^q(M, \mathcal{B}er_{\mathcal{O}})$ in terms of the global section cohomology of $\mathcal{E}_{m, \bullet}$.

4. There is a sequence of sheaves of (p, q)-currents $\mathcal{D}^{p, \bullet}$ which resolves Ω^p.

5. The global sections can be given a suitable topology so that

$$\Gamma_c(M, \mathscr{D}^{p,q}),$$

the compactly supported (p, q)-currents on M, is the topological dual of

$$\Gamma(M, \mathscr{E}_{m-p, m-q}).$$

We begin with step 1. We use the concept of the "conjugate" space of a real vector space with a complex structure. If V is a real vector space with complex structure J, then its complexification $V_c = V \otimes_{\mathbf{R}} \mathbf{C}$ splits into $V^{1,0} \oplus V^{0,1}$, the $+i$ and $-i$ eigenspaces of J, respectively. Here V may be complex-linearly identified with $V^{1,0}$ and conjugate-linearly identified with $V^{0,1}$, hence we write the above splitting as $V_c = V \oplus \bar{V}$, where $\bar{V}$ is the *conjugate* space of V. This gives the notion of the conjugate $\bar{v} \in \bar{V}$ of a vector $v \in V$.

Recall that for a complex supermanifold $(M, \mathcal{O})$, locally $\mathcal{O} \cong \mathcal{O}_{\mathrm{rd}} \otimes \Lambda(\mathbf{R}^{a^n})$. If $\{U_\alpha\}$ is an open cover of M trivializing $\mathcal{O}$, then for each α, we have a notion of the conjugate space $\overline{\mathcal{O}_{\mathrm{rd}}|_{U_\alpha} \otimes \Lambda(\mathbf{C}^n)}$ of $\mathcal{O}_{\mathrm{rd}}|_{U_\alpha} \otimes \Lambda(\mathbf{C}^n)$ which is just the anti-holomorphic functions $\overline{\mathcal{O}}_{\mathrm{rd}}|_{U_\alpha}$ tensored with the conjugate space $\overline{\Lambda(\mathbf{C}^n)}$ of $\Lambda(\mathbf{C}^n)$. The cocycles $\tau_{\alpha\beta}$ induced from the isomorphisms $\phi_\alpha \colon \mathcal{O}|_{U_\alpha} \xrightarrow{\sim} \mathcal{O}_{\mathrm{rd}}|_{U_\alpha} \otimes \Lambda(\mathbf{C}^n)$ define cocycles $\bar{\tau}_{\alpha\beta}$ on $\overline{\mathcal{O}}_{\mathrm{rd}}|_{U_{\alpha\beta}} \otimes \overline{\Lambda(\mathbf{C}^n)}$ and by gluing we obtain the *conjugate sheaf* $\overline{\mathcal{O}}$ of $\mathcal{O}$. If $\theta^1, \ldots, \theta^n$ are local odd coordinates for $\mathcal{O}$, then $\bar{\theta}^1, \ldots, \bar{\theta}^n$ are local odd coordinates for $\overline{\mathcal{O}}$.

If (z, θ) are local coordinates for $\mathcal{O}|_U$, we define

$$x = \tfrac{1}{2}(z + \bar{z}), \quad y = 1/2i(z - \bar{z}), \quad \alpha = \tfrac{1}{2}(\theta + \bar{\theta}), \quad \beta = 1/2i(\theta - \bar{\theta})$$

as coordinates for $C^\infty|_U \otimes \Lambda(\mathbf{R}^{2n})$, where $(x, y) = (x^1, y^1, \ldots, x^m, y^m)$ are even coordinates, and $(\alpha, \beta) = (\alpha^1, \beta^1, \ldots, \alpha^n, \beta^n)$ are odd coordinates. The cocycles $\tau_{\alpha\beta}$ and $\bar{\tau}_{\alpha\beta}$ induce cocycles $\tilde{\tau}_{\alpha\beta}$ on $C^\infty|_{U_\alpha \cap U_\beta} \otimes \Lambda(\mathbf{R}^{2n})$ by

$$\tilde{\tau}(x) = \frac{1}{2}\big(\tau(z) + \bar{\tau}(\bar{z})\big),$$

$$\tilde{\tau}(y) = \frac{1}{2i}\big(\tau(z) - \bar{\tau}(\bar{z})\big),$$

$$\tilde{\tau}(\alpha) = \frac{1}{2}\big(\tau(\theta) + \bar{\tau}(\bar{\theta})\big),$$

$$\tilde{\tau}(\beta) = \frac{1}{2i}\big(\tau(\theta) - \bar{\tau}(\bar{\theta})\big).$$

By gluing with these cocycles we obtain a $\mathbf{Z}_2$-graded sheaf $\mathscr{E}$. One can easily see that $\mathscr{E}$ is independent of the covering used. The ringed space $(M, \mathscr{E})$ is a smooth supermanifold of dimension $2m|2n$ and is called the *associated smooth supermanifold* of $(M, \mathcal{O})$. It is clear by construction that we have an inclusion of sheaves $\mathcal{O} \hookrightarrow \mathcal{E}$ inducing a morphism of ringed spaces $i: (M, \mathcal{E}) \to (M, \mathcal{O})$.

The sheaf of forms $\mathcal{E}^r = \wedge^r \mathcal{T}_\mathcal{E}^*$, where $\mathcal{T}_\mathcal{E}$ is the tangent sheaf $\mathcal{D}er\mathcal{E}$ of $(M, \mathcal{E})$, decomposes into $\mathcal{E}^r = \oplus_{p+q=r} \mathcal{E}^{p,q}$ where $\mathcal{E}^{p,q}$ is the subsheaf of $\mathcal{E}^r$ which is locally generated by forms $dz^I d\bar{z}^J d\theta^\mu d\bar\theta^\nu$, $|I| + |\mu| = p$, $|J| + |\nu| = q$, over $\mathcal{E}$. The property of being (p, q) is readily seen to be independent of local coordinates since $\tilde\tau_{\alpha\beta}$ is induced from the "superholomorphic" coordinate change τ.

Define $\bar\partial$: $\mathcal{E}^{p,q} \to \mathcal{E}^{p,q+1}$ and ∂: $\mathcal{E}^{p,q} \to \mathcal{E}^{p+1,q}$ by $\bar\partial = \pi_{p, q+1} \circ d$ and $\partial = \pi_{p+1, q} \circ d$, where $\pi_{r, s}$ denotes the projection from $r + s$ to (r, s)-forms. It is clear that $d = \bar\partial + \partial$ and $\bar\partial^2 = \partial^2 = \partial\bar\partial + \bar\partial\partial = 0$. We have the following analogue of the Cauchy-Riemann equations.

LEMMA 1. *Let $f \in \Gamma(U, \mathcal{E})$, then $\partial/\partial\bar{z}^i$ and $\partial/\partial\bar\theta^j$ annihilate f for all i, j if and only if $f \in \Gamma(U, \mathcal{O})$, where U is an open set in M.*

The sheaf of superholomorphic p-forms Ω^p is the subsheaf of $\mathcal{E}^{p,0}$ whose sections have coefficients in $\mathcal{O}$. Clearly, $\Omega^p = \ker \bar\partial$: $\mathcal{E}^{p,0} \to \mathcal{E}^{p,1}$ by Lemma 1. The following is the analogue of the Dolbeault resolution in this context.

THEOREM 2. *The sequence $0 \to \Omega^p \to \mathcal{E}^{p,\bullet}$ is exact.*

Proof. The question is of a local nature. If we let (z, θ) be local complex coordinates then we have $\mathcal{E}^{p,\bullet} = \oplus_{r+s=p} \mathcal{E}_{\mathrm{rd}}^{r,\bullet} \otimes \mathcal{E}_{\mathrm{odd}}^{s,\bullet}$, where $\mathcal{E}_{\mathrm{odd}}^{s,\bullet}$ is the complex of odd forms over the subalgebra of $\mathcal{E}$ generated by 1 and $(\theta, \bar\theta)$. It is evident that $\mathcal{E}_{\mathrm{rd}}^{r,\bullet}$ and $\mathcal{E}_{\mathrm{odd}}^{s,\bullet}$ are stable under $\bar\partial$, hence by the Künneth formula,

$$H^*(\mathcal{E}^{p,\bullet}) = \bigoplus_{r+s=p} H^*(\mathcal{E}_{\mathrm{rd}}^{r,\bullet}) \otimes H^*(\mathcal{E}_{\mathrm{odd}}^{s,\bullet}).$$

Now $H^0(\mathcal{E}_{\mathrm{rd}}^{r,\bullet}) = \Omega_{\mathrm{rd}}^r$ and $H^k(\mathcal{E}_{\mathrm{rd}}^{r,\bullet}) = 0$, $k > 0$ by the classical Dolbeault theorem. To compute $H^*(\mathcal{E}_{\mathrm{odd}}^{s,\bullet})$ we have the following lemma.

LEMMA 3. *Let $S = \oplus S^p$ be the symmetric algebra over a field k with n generators $\{e_1, \ldots, e_n\}$ and $\wedge = \oplus \wedge^p$ be the Grassmann algebra over k with n generators $\{\theta^1, \ldots, \theta^n\}$. Consider the complex $S^\bullet \otimes \wedge$ with boundary $d: S^p \otimes \wedge \to S^{p+1} \otimes \wedge$ given by*

$$d(\omega\theta^\mu) = \sum (-1)^{p+j-1} \omega e_{\mu_j} \theta^{\mu_1} \ldots \hat\theta^{\mu_j} \ldots \theta^{\mu_l}.$$

Then $H^0(S \otimes \wedge) = k$ and

$$H^i(S^\bullet \otimes \wedge) = 0$$

for $i > 0$.

Using this lemma one can see that $H^0(\mathcal{E}^{s,\bullet}_{\mathrm{odd}}) = \Omega^s_{\mathrm{odd}}$ and $H^i(\mathcal{E}^{s,\bullet}_{\mathrm{odd}}) = 0$, for $i > 0$. Hence $H^i(\mathcal{E}^{p,\bullet}) = 0$, for $i > 0$ and

$$H^0(\mathcal{E}^{p,\bullet}) = \bigoplus_{r+s=p} \Omega^r_{\mathrm{rd}} \otimes \Omega^s_{\mathrm{odd}} = \Omega^p.$$

Now define the *adjoint sheaf* $\mathcal{E}_{p,q}$ of $\mathcal{E}^{m-p,m-q}$ by

$$\mathcal{E}_{p,q} := \mathcal{B}er_{\mathcal{E}} \otimes \bigwedge^{m-p,m-q} \mathcal{T}_{\mathcal{E}}$$

where $\bigwedge^{m-p,m-q} \mathcal{T}_{\mathcal{E}} = (\mathcal{E}^{m-p,m-q})^*$. The map $\bar{\partial}' \colon \mathcal{E}_{p,q} \to \mathcal{E}_{p,q+1}$ is defined so that it is the adjoint of $\bar{\partial}$ with respect to the natural pairing

$$\Gamma(\mathcal{E}_{p,q}) \times \Gamma(\mathcal{E}^{m-p,m-q}) \to \mathbf{C}$$

given by

$$(\omega, \eta) \to \int_M \langle \omega, \eta \rangle,$$

when this is defined. Note that the values of the bilinear pairing before integration are sections of the Berezinian sheaf, and hence, this integration pairing is defined, in particular, if either factor has compact support (Leites (1980) and Manin (1984)). Integration for noncompact support has been defined by Rothstein (1986). We have the following analogue of the Grothendieck lemma which occurs in the classical Dolbeault resolution.

THEOREM 4. *The cohomology sheaves of $\mathcal{E}_{m-p,\bullet}$ are*

$$\mathcal{H}^k(\mathcal{E}_{m-p,\bullet}) = \begin{cases} \mathcal{B}er_{\mathcal{O}} \otimes_{\mathcal{O}} \bigwedge^p \mathcal{T}_{\mathcal{O}} & k = 0 \\ 0 & k \neq 0. \end{cases}$$

Proof. It is possible to construct a homotopy to show that any $\bar{\partial}'$-closed $(m-p, q)$ adjoint form is exact for $q \neq 0$. Let

$$\frac{\partial}{\partial \bar{z}} = \frac{\partial}{\partial \bar{z}^1} \frac{\partial}{\partial \bar{z}^2} \cdots \frac{\partial}{\partial \bar{z}^m}.$$

For $p = 0$, the only closed $(m, 0)$-forms which are not exact are of the form, at a point $x = (z, \theta)$, $\rho f \bar{\theta}^1 \ldots \bar{\theta}^n (\partial/\partial \bar{z})$, where $f \in \mathcal{O}_x$ and $\rho \in \mathcal{B}er_{\mathcal{E},x}$ is nonzero (i.e., is a local generator for this rank one sheaf). It remains to show that under a complex change of coordinates these forms transform as

$$\rho f \bar{\theta}^1 \ldots \bar{\theta}^n \frac{\partial}{\partial \bar{z}} = \mathrm{Ber}\left(\frac{\partial(z, \theta)}{\partial(\omega, \eta)} \right) \rho f \bar{\eta}^1 \ldots \bar{\eta}^n \frac{\partial}{\partial \bar{\omega}} + \text{closed terms.}$$

This is a direct calculation using the fact that

$$\text{Ber } J(\bar{\tau}) = \text{Ber } J(\tau)\overline{\text{Ber } J(\tau)}$$

for a complex change of coordinates τ and its associated smooth change of coordinates $\tilde{\tau}$, and where $J(\bullet)$ denotes the Jacobian matrix of partial derivatives of the change of variables. Hence $\mathscr{H}^0(\mathscr{E}_{m,\bullet}) = \mathscr{B}er_{\mathscr{O}}$. The proof for $p \neq 0$ is similar.

The first consequence of the above Dolbeault lemma is that

$$\mathbb{H}^*(M, \mathscr{B}er_{\mathscr{O}}) \cong \mathbb{H}^*(M, \mathscr{E}_{m,\bullet}).$$

But

$$H^*(M, \mathscr{B}er_{\mathscr{O}}) = \mathbb{H}^*(M, \mathscr{B}er_{\mathscr{O}})$$

and the fact that $\mathscr{E}_{p,q}$ are $\mathscr{E}$-modules and $\mathscr{E}$ is fine imply that

$$\mathbb{H}^*(M, \mathscr{E}_{m,\bullet}) = H^*_{\bar{\partial}'}\big(H^0(M, \mathscr{E}_{m,\bullet})\big).$$

We summarize these results in the following analogues of Dolbeault's theorem.

THEOREM 5. *Let M be a complex supermanifold, then there are canonical isomorphisms*

$$H^*(M, \Omega^p) \cong H^*_{\bar{\partial}}\big(H^0(M, \mathscr{E}^{p,\bullet})\big),$$

$$H^*(M, \mathscr{B}er_{\mathscr{O}}) \cong H^*_{\bar{\partial}'}\big(H^0(M, \mathscr{E}_{m,\bullet})\big).$$

Note that there is a natural pairing

$$H^q(M, \mathscr{O}) \leftrightarrow H^{m-q}(M, \mathscr{B}er_{\mathscr{O}})$$

induced by the natural pairing

$$H^0(M, \mathscr{E}^{0,q}) \leftrightarrow H^0(M, \mathscr{E}_{m,m-q}).$$

We shall see that this pairing is nondegenerate.

Consider the vector space of smooth superfunctions on M, denoted by $\Gamma(M, \mathscr{E})$ and the vector space of compactly supported smooth superfunctions on M, denoted by $\Gamma_c(M, \mathscr{E})$. These sections are locally vector valued-valued smooth functions, and we can use the usual topology on smooth functions and smooth functions with compact support (see, e.g., Rudin, 1974) to define natural topologies on these spaces, in which they become complete topological vector spaces. The space $\Gamma(M, \mathscr{E})$ will be a Frechet space, and $\Gamma_c(M, \mathscr{E})$ will be an inductive limit of Frechet spaces. These topologies extend to the sections of the sheaves $\mathcal{E}^{p,q}$ and $\mathcal{E}_{r,s}$. We define the sheaf of (p, q)-currents $K^{p,q}$ to be the sheaf whose sections are continuous linear maps on the space of compactly supported sections of $\mathcal{E}_{m-p,m-q}$. The sheaf of 0-currents $K^{0,0}$, which we denote by $\mathcal{D}'$, is called the sheaf of distributions on $(M, \mathcal{E})$, and the global sections of $\mathcal{D}'$ is the dual space of the space of compactly supported sections of $\mathcal{B}e\imath_{\mathcal{E}}$. Locally $\mathcal{D}'$ is isomorphic to $\mathcal{D}'_{\mathrm{rd}} \otimes \Lambda(\mathbf{R}^{2n})$. A straightforward calculation in local coordinates shows that $K^{p,q} \cong \mathcal{E}^{p,q} \otimes \mathcal{D}'$; the space of (p, q)-forms with distribution coefficients. Distributions and currents are $\mathbf{Z}_2$-graded by considering $\mathbf{C}$ as $\mathbf{Z}_2$-graded with zero odd part.

As usual, we may define ∂ on currents by $\langle \partial T, \phi \rangle = \langle T, -\partial\phi \rangle$. The analogue for regularity of the $\bar{\partial}$-operator acting on distributions holds in this context. Namely, let $\mathcal{O}(U) \otimes \Lambda(\mathbf{C}^n)$ be a complex superdomain with associated smooth superdomain $C^\infty(U) \otimes \Lambda(\mathbf{R}^{2n})$, where U is a subdomain of $\mathbf{C}^n$, and $\mathcal{O}$ is the usual sheaf of holomorphic functions on U.

LEMMA 6. *Let T be a distribution acting on $C_c^\infty(U) \otimes \Lambda(\mathbf{R}^{2n})$. If $\bar{\partial}T = 0$, then $T \in \mathcal{O}(U) \otimes \Lambda(\mathbf{C}^n)$.*

This lemma is reduced to the classical one in a straightforward manner.
Now we see that the classical Grothendieck lemma generalizes to this case.

THEOREM 7. *The sequence*

$$0 \to \Omega^p \to K^{p,0} \to K^{p,1} \to \cdots$$

is exact.

We now have all the analogues of the data which Serre used. Applying the functional analysis used in Serre (1955) we obtain the following duality theorem, where we let H_c denote cohomology with compact supports.

THEOREM 8. *Let $(M, \mathcal{O})$ be a complex supermanifold, and let $\mathcal{F}$ be a locally free $\mathcal{O}$-module, then*

$$H^p(M, \mathcal{F})$$

is canonically dual to

$$H_c^{m-p}(M, \mathcal{B}e\imath_{\mathcal{O}} \otimes \mathcal{F}^*).$$

From this we have the immediate corollary.

COROLLARY 9. *Let $(M, \mathcal{O})$ be a compact complex supermanifold, and let $\mathcal{F}$ be a locally free $\mathcal{O}$-module, then*

$$H^p(M, \mathcal{F})$$

is canonically dual to

$$H^{m-p}(M, \mathcal{B}e\imath_{\mathcal{O}} \otimes \mathcal{F}^*).$$

REFERENCES

1. P. A. GRIFFITHS AND J. HARRIS, *Principles of Algebraic Geometry*, Wiley, New York, 1978.
2. ROBIN HARTSHORNE, *Residues and Duality*, Lecture Notes in Mathematics No. 20, Springer-Verlag, Heidelberg, 1966.
3. D. A. LEITES, *Introduction to the theory of supermanifolds*, Russian Math Surveys **35:1** (1980), 1–64.
4. YU. I. MANIN, *Gauge Fields and Complex Geometry*, Nauka, Moscow, 1984.
5. O. V. OGIJEVETSKY AND I. B. PENKOV, *Serre duality for projective supermanifolds*, Funct. Anal. and Appl. **18** (1984), 68–70.
6. I. B. PENKOV, *D-modules on supermanifolds*, Invent. Math. **71** (1983), 501–512.
7. ______ ,"An introduction to geometric representation theory of complex simple Lie superalgebras," *Differential Geometric Methods in Theoretical Physics*, Shumen, 1984," World Scientific Publ., Singapore, 1984, pp. 89–106.
8. MITCHELL ROTHSTEIN, *Integration in supermanifolds*, Trans. Amer. Math. Soc. **299** (1987), 387–396.
9. WALTER RUDIN, *Functional Analysis*, McGraw-Hill, New York, 1973.
10. J.-P. SERRE, *Un theorème de dualité*, Comment. Math. Helv. **29** (1955), 1–26.
11. R. O. WELLS, JR., *Differential Analysis on Complex Manifolds*, 2nd ed., Springer-Verlag, Berlin-New York, 1980.

DEPARTMENT OF MATHEMATICS, RICE UNIVERSITY, HOUSTON, TEXAS 77251

TORSION POINTS ON FERMAT JACOBIANS, ROOTS OF CIRCULAR UNITS AND RELATIVE SINGULAR HOMOLOGY

GREG W. ANDERSON

Dedicated to Yu. I. Manin on the occasion of his fiftieth birthday

Introduction.   Our object is to analyze certain relationships among galois modules arising as relative singular homology groups. The theory developed in this paper permits one to prove, for example, the following result: Let N be a positive integer. Let U denote the affine Fermat curve

$$x^N + y^N = 1$$

over $\mathbb{Q}$. Let $\overline{U}$ denote the projective closure of U. Let ∞ denote the squarefree effective divisor on $\overline{U}$ of support the complement of U in $\overline{U}$. Let S denote the generalized jacobian of $\overline{U}$ of conductor ∞. Then S is an extension of an abelian variety by a torus and is defined over $\mathbb{Q}$. Let b denote the $\mathbb{Q}$-rational point of S corresponding to the difference of the points $(0, 1)$ and $(1, 0)$ of U.

THEOREM 0.   *The numberfield generated by the coordinates of the Nth roots of the point b in the group $S(\overline{\mathbb{Q}})$ contains the numberfield generated by the roots of the equation*

$$1 - (1 - x^N)^N = 0.$$

If N is prime, the asserted containment becomes an equality.

A similar result was obtained by Greenberg [G].

The paper consists of two parts. Let X be a smooth quasi-projective $\mathbb{Q}$-scheme, D an effective divisor on X with singularities no worse than normal crossings and N a positive integer. The purpose of Part I is to explain how to equip relative singular homology groups of the form $H_*(X(\mathbb{C}), D(\mathbb{C}); \mathbb{Z}/N\mathbb{Z})$ with a natural action of

$$G(\mathbb{Q}) \stackrel{\text{def}}{=} \text{the galois group over } \mathbb{Q} \text{ of the algebraic}$$
$$\text{closure of } \mathbb{Q} \text{ in } \mathbb{C}.$$

We employ the methods of étale homotopy theory [AM] to do this. There is no novelty to be claimed for Part I other than for the approach, which is to proceed

Received September 29, 1986.

directly to the construction of a $G(\mathbf{Q})$-action in relative singular homology *without* first considering a free-standing relative étale homology theory. This simplifies matters and suffices for our purpose, which is to set up a calculus of $G(\mathbf{Q})$-modules.

In Part II we take up the study of some special relative singular homology groups and distinguished elements thereof. For each positive integer k, we consider the group $H_k(U_k(\mathbf{C}), Y_k(\mathbf{C}); \mathbf{Z}/N\mathbf{Z})$ where U_k is the affine Fermat hypersurface

$$x_0^N + \cdots + x_k^N = 1,$$

Y_k is the divisor on U_k defined by the equation

$$x_0 \ldots x_k = 0,$$

and we consider the homology class to which the singular k-simplex

$$(t_0, \ldots, t_k) \mapsto \left(\sqrt[N]{t_0} \geq 0, \ldots, \sqrt[N]{t_k} \geq 0\right) : \left\{(t_0, \ldots, t_k) \in \mathbf{R}^{k+1} \,|\, t_j \geq 0, \Sigma t_j = 1\right\}$$
$$\to U_k(\mathbf{C})$$

belongs. We study also the group $H_1(V(\mathbf{C}), Z(\mathbf{C}); \mathbf{Z}/N\mathbf{Z})$ where V is the largest affine open subscheme of $\mathrm{Spec}(\mathbf{Q}[t])$ on which

$$\sum_{j=0}^{N-1} t^j$$

is invertible, Z is the closed subscheme of V defined by the equation

$$t(1-t) = 0,$$

and we consider the homology class to which the 1-simplex

$$t \mapsto t \colon [0,1] \to V(\mathbf{C}) = \mathbf{C} \setminus \left(\mu_N(\mathbf{C}) \setminus \{1\}\right)$$

belongs. A key observation is that $H_1(V(\mathbf{C}), Z(\mathbf{C}); \mathbf{Z}/N\mathbf{Z})$ is a $G(\mathbf{Q})$-subquotient of $H_1(U_1(\mathbf{C}), Y_1(\mathbf{C}); \mathbf{Z}/N\mathbf{Z})$ in a fashion compatible with the specifications made above of distinguished homology classes; it is this fact which, after some computation, implies the containment asserted in Theorem 0.

This then raises the question of the extent to which $H_1(U_1(\mathbf{C}), Y_1(\mathbf{C}); \mathbf{Z}/N\mathbf{Z})$ may be reconstructed on the basis of knowledge of $H_1(V(\mathbf{C}), Z(\mathbf{C}); \mathbf{Z}/N\mathbf{Z})$. The answer to this question is provided by the main result of the paper, Theorem 10 of §9. In particular, for N prime, it turns out that complete reconstruction is possible, whence the equality of numberfields asserted in Theorem 0 after some computation.

The Fermat *surface* has an important role to play in the proof of our main result: By exhibiting $H_2(U_2(\mathbf{C}), Y_2(\mathbf{C}); \mathbf{Z}/N\mathbf{Z})$ as a $G(\mathbf{Q})$-quotient of the tensor square of $H_1(U_1(\mathbf{C}), Y_1(\mathbf{C}); \mathbf{Z}/N\mathbf{Z})$ and taking into account the S_3-symmetry of the pair (U_2, Y_2), we obtain a severe constraint on the size of the image of $G(\mathbf{Q})$ in $\operatorname{Aut}(H_1(U_1(\mathbb{C}), Y_1(\mathbb{C}); \mathbb{Z}/N\mathbb{Z}))$. This constraint is naturally expressed as a sort of 2-cocycle which, when split, yields a mod N analogue of the classical Γ-function, an object of some interest in its own right.

Prior to the author's having obtained the proof of Theorem 10 given in this paper, a p-adic analytic proof of a somewhat weakened version of Theorem 10 was obtained jointly with R. Coleman. Coleman's ideas in this line remain valuable and, indeed, become even more valuable when run in reverse and used, in *conjunction* with Theorem 10, to gain insight into the local behavior of the ℓ-adic representations arising from Fermat curves and, more generally, the motives considered in [A1]. A joint paper on this topic is planned.

The main inspiration for this paper is the recent work of Ihara [I], together with a conjecture of Ihara [IKY], proven by Coleman (unpublished) and independently by Ihara, Kaneko and Yukinari [IKY]. Also figuring as a motivation are the congruences of Miki [M] for Gauss sums. The author sought to link and to generalize these results by extending the factorization studied in [A1] into new territory. Superficially, the theory which resulted from this effort and is detailed here little resembles any of the works cited above but, nonetheless, is closely linked to all. These connections, not apparent at level N, become so in the limit $N \to \infty$. In a sequel devoted to the study of the mod N Γ-function in the limit $N \to \infty$ and to the proof of the results announced in [A2], we hope to clarify these connections.

Acknowledgements. This paper would never have been written had I not enjoyed a very lively exchange of ideas with R. Coleman and Y. Ihara. I am deeply grateful to both of them. I thank H. Miki for a valuable discussion. I thank N. Katz and B. Mazur for perceptive remarks and criticism. Support during much of the period of preparation of this paper was provided by the Miller Institute for Basic Research in Science at the University of California at Berkeley.

TABLE OF CONTENTS

PART I. THE GALOIS ACTION IN RELATIVE SINGULAR HOMOLOGY

§1. Homological apparatus

1.1. A positive integer N is fixed throughout the paper. We set

$$A \overset{\text{def}}{=} \mathbb{Z}/N\mathbb{Z}.$$

1.2. Given an abelian group G and a set X, we denote by $G \otimes X$ the free abelian group on the collection of symbols

$$\{ g \otimes x \mid g \in G, x \in X \}$$

modulo the subgroup generated by the collection

$$\{ g \otimes x + g' \otimes x - (g + g') \otimes x \mid g, g' \in G, x \in X \}.$$

Set

$$d^0(G \otimes X) \overset{\text{def}}{=} \text{kernel}(g \otimes x \mapsto g \colon G \otimes X \to G).$$

1.3. Let Δ denote the category the objects of which are the sets

$$[n] \overset{\text{def}}{=} \{ j \in \mathbb{Z} \mid 0 \leqslant j \leqslant n \},$$

n ranging over the set of nonnegative integers, and the morphism sets of which are given by the rule

$$\text{Hom}_\Delta([m], [n]) \overset{\text{def}}{=} \{ f \colon [m] \to [n] \mid f \text{ is monotone.} \}.$$

For each nonnegative integer n, let $\Delta_{\leqslant n}$ denote the full subcategory of Δ consisting of the objects $[j]$ for $0 \leqslant j \leqslant n$. Given nonnegative integers m and n, $i \in [m + 1]$ and $j \in [n]$, let $\delta_m^i \colon [m] \to [m + 1]$ denote the unique one-to-one monotone map omitting the value i and let $\sigma_n^j \colon [n + 1] \to [n]$ denote the unique onto monotone map taking the value j twice.

1.4. Let $\mathscr{C}$ be a category. A *simplicial object* of $\mathscr{C}$ is by definition a contravariant functor $\Delta \to \mathscr{C}$, a *bisimplicial object* of $\mathscr{C}$ a contravariant functor $\Delta \times \Delta \to \mathscr{C}$, and an *$n$-truncated simplicial object* of $\mathscr{C}$ a contravariant functor $\Delta_{\leqslant n} \to \mathscr{C}$. In each case, a morphism of such objects is defined to be a natural transformation of contravariant functors.

1.5. Let $\mathscr{C}$ be a category, X a simplicial object of $\mathscr{C}$ and Y a bisimplicial object of $\mathscr{C}$. In order to expedite calculations, the following abbreviations will be employed:

$$X([n]) \mapsto X_n, \qquad Y([p],[q]) \mapsto Y_{pq},$$

$$X(\delta_m^i) \mapsto \partial_i, \qquad X(\sigma_n^j) \mapsto s_j,$$

$$Y(\delta_m^i, id_{[q]}) \mapsto \partial_i', \qquad Y(id_{[p]}, \delta_m^i) \mapsto \partial_i'',$$

$$Y(\sigma_n^j, id_{[q]}) \mapsto s_j', \qquad Y(id_{[p]}, \sigma_n^j) \mapsto s_n''.$$

1.6. Let $\mathscr{C}$ and $\mathscr{D}$ be categories, $F: \mathscr{C} \to \mathscr{D}$ a functor, G a functor with source $\mathscr{C}$ and target the category of simplicial objects of $\mathscr{D}$, X and X' simplicial objects of $\mathscr{C}$, and Y and Y' bisimplicial objects of $\mathscr{C}$. We set

$$X_{\leqslant n} \overset{\text{def}}{=} \text{the restriction of } X \text{ to } \Delta_{\leqslant n},$$
$$\text{an } n\text{-truncated simplicial object of } \mathscr{C},$$

$$Y^{op} \overset{\text{def}}{=} ([p],[q]) \mapsto Y_{qp}, \text{ a bisimplicial object of } \mathscr{C},$$

$$/Y \overset{\text{def}}{=} [n] \mapsto Y_{nn}, \text{ a simplicial object of } \mathscr{C},$$

$$F(X) \overset{\text{def}}{=} [n] \mapsto F(X_n), \text{ a simplicial object of } \mathscr{D},$$

$$G(X) \overset{\text{def}}{=} ([p],[q]) \mapsto G(X_q)_p, \text{ a bisimplicial object of } \mathscr{D}.$$

Provided that $\mathscr{C}$ is closed under the formation of cartesian products, we set

$$X \times X' \overset{\text{def}}{=} [n] \mapsto X_n \times X_n', \text{ a simplicial object of } \mathscr{C},$$

$$Y \times Y' \overset{\text{def}}{=} ([p],[q]) \mapsto Y_{pq} \times Y_{pq}', \text{ a bisimplicial object of } \mathscr{C},$$

$$X \boxtimes X' \overset{\text{def}}{=} ([p],[q]) \mapsto X_p \times X_q', \text{ a bisimplicial object of } \mathscr{C}.$$

1.7. Let n be a nonnegative integer and $\mathscr{C}$ a category. The left (respectively, right) adjoint of the functor $X \mapsto X_{\leqslant n}$ from the category of simplicial objects of $\mathscr{C}$ to the category of n-truncated simplicial objects of $\mathscr{C}$ is denoted by $sk_n^{\mathscr{C}}$ (respectively, $\cos k_n^{\mathscr{C}}$) provided that this adjoint functor exists.

1.8. A *chain complex* is understood to be a graded abelian group vanishing in negative degree equipped with a differential (always denoted by ∂) of degree -1.

A *filtered chain complex* is understood to be a chain complex X equipped in each degree n with a filtration

$$\cdots \subseteq \mathrm{Fil}_p X_n \subseteq \mathrm{Fil}_{p+1} X_n \subseteq \cdots$$

such that

$$\partial\left(\mathrm{Fil}_p X_n\right) \subseteq \mathrm{Fil}_p X_{n-1},$$

$$\mathrm{Fil}_{-1} X_n = 0,$$

$$\mathrm{Fil}_n X_n = X_n.$$

1.9. Given a filtered chain complex X, let $E^r_{pq}(X)$ denote the general term of the necessarily convergent spectral sequence of X. Let $E^1_{*q}(X)$ denote the chain complex which in degree p is the term $E^1_{pq}(X)$ of the spectral sequence of X, the differential $\partial\colon (E^1_{*q}(X))_p \to (E^1_{*q}(X))_{p-1}$ of which coincides with the differential $E^1_{pq}(X) \to E^1_{p-1,q}(X)$ of the spectral sequence of X.

1.10. Given a simplicial abelian group X, let $\mathit{tot}(X)$ denote the chain complex coinciding in degree $n \geqslant 0$ with X_n, the differential $\partial\colon \mathit{tot}(X)_n \to \mathit{tot}(X)_{n-1}$ of which is given by the usual rule

$$\partial \overset{\mathrm{def}}{=} \sum_{j=0}^{n} (-1)^j \partial_j.$$

LEMMA. *Let X and $\mathit{tot}(X)$ be as above. Let D denote the graded group*

$$D_n \overset{\mathrm{def}}{=} \left. \begin{array}{ll} \sum_{j=0}^{n-1} s_j X_{n-1} \subseteq X_n & \text{if } n > 0 \\ 0 & \text{if } n \leqslant 0 \end{array} \right\}.$$

Then D is a subcomplex of $\mathit{tot}(X)$ and

$$H_*(D) = 0.$$

Proof. Well known. $\qquad\qquad\qquad\qquad\qquad\qquad\qquad\qquad\qquad\qquad\square$

1.11. Let X be a bisimplicial abelian group. Given integers p and q, set $X_{pq} \overset{\mathrm{def}}{=} 0$ if $p < 0$ or $q < 0$. Given $x \in X_{pq}$, set

$$\partial' x \overset{\mathrm{def}}{=} \sum_{i=0}^{p} (-1)^i \partial'_i x, \qquad \partial'' x \overset{\mathrm{def}}{=} \sum_{j=0}^{q} (-1)^{p+j} \partial''_j x.$$

Then

$$(\partial')^2 = 0, \qquad (\partial'')^2 = 0, \qquad \partial' \partial'' + \partial'' \partial' = 0.$$

Let $\mathit{bitot}(X)$ denote the chain complex given in degree n by

$$\mathit{bitot}(X)_n \overset{\text{def}}{=} \bigoplus_{p+q=n} X_{pq},$$

equipped with the differential

$$\partial \overset{\text{def}}{=} \partial' + \partial'' \colon \mathit{bitot}(X)_n \to \mathit{bitot}(X)_{n-1}.$$

Let $\mathit{bitotfil}(X)$ denote the filtered chain complex the underlying chain complex of which is $\mathit{bitot}(X)$, filtered by the rule

$$\mathrm{Fil}_p\,\mathit{bitotfil}(X)_n \overset{\text{def}}{=} \bigoplus_{\substack{i+j=n \\ j \leqslant p}} X_{ij}.$$

Let $H_q^{\mathrm{horiz}}(X)$ denote the chain complex given in degree p by

$$H_q^{\mathrm{horiz}}(X)_p \overset{\text{def}}{=} \ker\!\big(\partial' \colon X_{qp} \to X_{q-1,p}\big)/\partial' X_{q+1,p},$$

the differential $\partial \colon H_q^{\mathrm{horiz}}(X)_p \to H_q^{\mathrm{horiz}}(X)_{p-1}$ of which is induced by $\partial'' \colon X_{qp} \to X_{q,p-1}$. Note that

$$E^1_{*q}\big(\mathit{bitotfil}(X)\big) = H_q^{\mathrm{horiz}}(X).$$

1.12. Let

$$\mathit{ez} \colon \mathit{bitot}(?) \to \mathit{tot}(/?)$$

denote a natural transformation of functors of bisimplicial abelian groups taking values in the category of chain complexes inducing the identity transformation in degree zero. Such a natural transformation exists by the acyclic models theorem and has, moreover, the property that for all bisimplicial abelian groups X the induced map

$$\mathit{ez}(X)_* \colon H_*\big(\mathit{bitot}(X)\big) \to H_*\big(\mathit{tot}(/X)\big)$$

is an isomorphism. Let

$$\mathit{edch} \colon \mathit{bitot}(?) \to H_0^{\mathrm{horiz}}(?)$$

denote the natural transformation of functors of bisimplicial abelian groups taking values in the category of chain complexes such that for all bisimplicial abelian groups X and $x \in X_{pq}$,

$$\mathscr{edch}(X)(x) \stackrel{\text{def}}{=} \left. \begin{array}{ll} x \bmod \partial' X_{1q} & \text{if } p = 0 \\ 0 & \text{otherwise} \end{array} \right\}.$$

1.13. Let X be a bisimplicial abelian group. Identifying the abutment $H_*(\mathscr{bitot}(X))$ of the spectral sequence deduced from the filtered chain complex $\mathscr{bitotfil}(X)$ with $H_*(\mathscr{tot}(/X))$ under the isomorphism induced by the chain map $\mathscr{ez}(X)$, we obtain the *Eilenberg-Zilber spectral sequence*

$$E^2_{pq} = H_p\big(H_q^{\text{horiz}}(X)\big) \Rightarrow H_{p+q}(\mathscr{tot}(/X))$$

of the bisimplicial abelian group X.

1.14. Given a simplicial set X, let $\pi_0(X)$ denote the set of equivalence classes in X_0 under the equivalence relation $R(X) \subseteq X_0 \times X_0$ obtained by intersecting all the equivalence relations $R' \subseteq X_0 \times X_0$ such that

$$R' \supseteq \{(\partial_0 x, \partial_1 x) \in X_0 \times X_0 | x \in X_1\}.$$

Note that

$$A \otimes \pi_0(X) = H_0(\mathscr{tot}(A \otimes X)).$$

1.15. Given a bisimplicial set X, let $\pi_0^{\text{horiz}}(X)$ denote the simplicial set which in degree q is the set of equivalence classes in X_{0q} under the equivalence relation $R_q(X) \subseteq X_{0q} \times X_{0q}$ obtained by intersecting all equivalence relations $R' \subseteq X_{0q} \times X_{0q}$ such that

$$R' \supseteq \{(\partial_0' x, \partial_1' x) \in X_{0q} \times X_{0q} | x \in X_{1q}\},$$

the face maps ∂_i and degeneracies s_j of which are induced by the face maps ∂_i'' and degeneracies s_j'' of X. We denote by

$$\mathscr{ed}(X) \colon /X \to \pi_0^{\text{horiz}}(X)$$

the natural map given by the rule that for all $x \in X_{qq}$,

$$\mathscr{ed}(X)(x) \stackrel{\text{def}}{=} (\partial_0')^q x \bmod R_q(X).$$

Note that

$$\mathscr{tot}\big(A \otimes \pi_0^{\text{horiz}}(X)\big) = H_0^{\text{horiz}}(A \otimes X).$$

Note also that the diagram

$$\begin{array}{ccc}
\textit{bitot}(A \otimes X) & \xrightarrow{\textit{ez}(A \otimes X)} & \textit{tot}(A \otimes /X) \\
\searrow{\scriptstyle \textit{edch}(A \otimes X)} & & \swarrow{\scriptstyle \textit{ed}(X)_*} \\
& \textit{tot}(A \otimes \pi_0^{\text{horiz}}(X)) &
\end{array}$$

commutes up to chain homotopy natural in X by the acyclic models theorem.

1.16. Let $Y \xrightarrow{f} X$ be a morphism of chain complexes. Let $Y(-1)$ denote the chain complex given in degree n by

$$Y(-1)_n \overset{\text{def}}{=} Y_{n-1},$$

the differential $\partial: Y(-1)_n \to Y(-1)_{n-1}$ of which is by definition $-\partial: Y_{n-1} \to Y_{n-2}$. Let $\textit{cone}(Y \xrightarrow{f} X)$ denote the chain complex given in degree n by

$$\textit{cone}\left(Y \xrightarrow{f} X\right)_n \overset{\text{def}}{=} X_n \oplus Y_{n-1},$$

the differential $\partial: \textit{cone}(Y \xrightarrow{f} X)_n \to \textit{cone}(Y \xrightarrow{f} X)_{n-1}$ of which is given by

$$\partial(x \oplus y) \overset{\text{def}}{=} x - f(y) \oplus - \partial y.$$

We have at our disposal the *cone exact sequence*

$$0 \to X \xrightarrow{x \mapsto x \oplus 0} \textit{cone}\left(Y \xrightarrow{f} X\right) \xrightarrow{x \oplus y \mapsto y} Y(-1) \to 0$$

and the *cone resolution homomorphism*

$$x \oplus y \mapsto x \bmod f(Y): \textit{cone}\left(Y \xrightarrow{f} X\right) \to X/f(Y).$$

If $Y \xrightarrow{f} X$ is injective, the cone resolution homomorphism induces an isomorphism

$$H_*\left(\textit{cone}\left(Y \xrightarrow{f} X\right)\right) \xrightarrow{\sim} H_*(X/f(Y))$$

and, moreover, the square diagram

$$\begin{array}{ccc}
H_n(\textit{cone}(Y \xrightarrow{f} X)) & \xrightarrow{\;\;①\;\;} & H_n(X/f(Y)) \\
\Big\downarrow{\scriptstyle ②} & & \Big\downarrow{\scriptstyle ③} \\
H_n(Y(-1)) & \xrightarrow[④]{} & H_{n-1}(f(Y))
\end{array}$$

commutes where ① is induced by the cone resolution, ② is induced by the corresponding homomorphism of the cone exact sequence, ③ is the boundary map extracted from the long exact sequence associated to the short exact sequence

$$0 \to f(Y) \to X \to X/f(Y) \to 0,$$

and ④ is the isomorphism induced by f.

1.17. Let $Y \xrightarrow{f} X$ be a morphism of filtered chain complexes. Let $\mathit{conefil}\,(Y \xrightarrow{f} X)$ denote the filtered chain complex the pth filtrand of which in degree n is given by

$$\mathrm{Fil}_p \mathit{conefil}\left(Y \xrightarrow{f} X\right)_n \overset{\mathrm{def}}{=} \mathrm{Fil}_p X_n \oplus \mathrm{Fil}_{p-1} Y_{n-1},$$

the differential $\partial: \mathit{conefil}\,(Y \xrightarrow{f} X)_n \to \mathit{conefil}\,(Y \xrightarrow{f} X)_{n-1}$ of which is given by the rule

$$\partial(x \oplus y) \overset{\mathrm{def}}{=} \partial x - f(y) \oplus - \partial y.$$

We then have

$$E^1_{*q}\left(\mathit{conefil}\left(Y \xrightarrow{f} X\right)\right) = \mathit{cone}\left(E^1_{*q}(Y) \xrightarrow{f_*} E^1_{*q}(X)\right),$$

as one verifies by some diagram chasing.

1.18. Let X and Y be chain complexes. Let $X \otimes Y$ denote the chain complex given in degree n by

$$(X \otimes Y)_n \overset{\mathrm{def}}{=} \bigoplus_{p+q=n} X_p \otimes Y_q,$$

the differential $\partial: (X \otimes Y)_n \to (X \otimes Y)_{n-1}$ of which is determined by the rule that for all $x \in X_p$ and $y \in Y_{n-p}$,

$$\partial(x \otimes y) \overset{\mathrm{def}}{=} \partial x \otimes y + (-1)^p x \otimes \partial y.$$

1.19. Let $Y \xrightarrow{f} X$ and $Y' \xrightarrow{f'} X'$ be morphisms of simplicial sets. We denote by $(Y \xrightarrow{f} X) \wedge (Y' \xrightarrow{f'} X')$ or simply $f \wedge f'$ the morphism of simplicial sets with source the push-out of the diagram

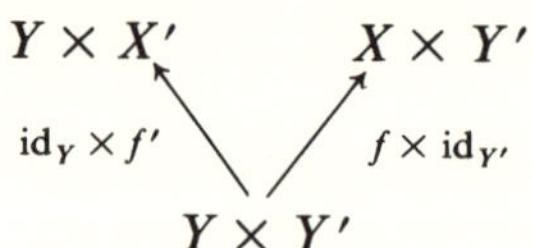

and target $X \times X'$ which is induced by the commutative square

$$
\begin{array}{ccc}
Y \times X' & \xrightarrow{f \times \mathrm{id}_{X'}} & X \times X' \\
\uparrow{\scriptstyle \mathrm{id}_Y \times f'} & & \uparrow{\scriptstyle \mathrm{id}_X \times f'} \\
Y \times Y' & \xrightarrow[f \times \mathrm{id}_{Y'}]{} & X \times Y'
\end{array}
$$

of morphisms of simplicial sets. Note that in the case $Y \subseteq X,\, Y' \subseteq X'$ and the morphisms f and f' are the respective inclusions, $(Y \xrightarrow{f} X) \wedge (Y' \xrightarrow{f} X')$ becomes the inclusion of $Y \times X' \cup X \times Y'$ into $X \times X'$.

1.20. Let

$$
(Q) \qquad
\begin{array}{ccc}
W & \xrightarrow{a} & X \\
\uparrow{\scriptstyle b} & & \uparrow{\scriptstyle c} \\
Y & \xrightarrow[d]{} & Z
\end{array}
$$

be a commutative square of morphisms of chain complexes. Set

$$
\mathscr{I}(Q) \stackrel{\mathrm{def}}{=} \mathrm{image}(\, y \mapsto -b(y) \oplus d(y) \colon Y \to W \oplus Z).
$$

Let $\mathit{bicone}(Q)$ denote the chain complex given in degree n by

$$
\mathit{bicone}(Q)_n \stackrel{\mathrm{def}}{=} X_n \oplus ((W \oplus Z)/\mathscr{I}(Q))_{n-1},
$$

the differential $\partial \colon \mathit{bicone}(Q)_n \to \mathit{bicone}(Q)_{n-1}$ of which is given by

$$
\partial\big(x \oplus (w \oplus z \bmod \mathscr{I}(Q)) \big)
$$

$$
\stackrel{\mathrm{def}}{=} \partial x - a(w) - c(z) \oplus (-\partial w \oplus -\partial z \bmod \mathscr{I}(Q)).
$$

1.21. Let $Y \xrightarrow{f} X$ and $Y' \xrightarrow{f'} X'$ be morphisms of chain complexes. Let the commutative square diagram

$$
\begin{array}{ccc}
Y \otimes X' & \xrightarrow{f \otimes \mathrm{id}_{X'}} & X \otimes X' \\
\uparrow{\scriptstyle \mathrm{id}_Y \otimes f'} & & \uparrow{\scriptstyle \mathrm{id}_X \otimes f'} \\
Y \otimes Y' & \xrightarrow[f \otimes \mathrm{id}_{Y'}]{} & X \otimes Y'
\end{array}
$$

512   GREG W. ANDERSON

be denoted by $Q_{cc}(f, f')$. We define a natural chain map

$$\textit{join}(f, f'): \textit{cone}(f) \otimes \textit{cone}(f') \to \textit{bicone}(Q_{cc}(f, f'))$$

by the rule that for all $x \oplus y \in \textit{cone}(f)_p = X_p \oplus Y_{p-1}$ and $x' \oplus y' \in \textit{cone}(f')_q = X'_q \oplus Y'_{q-1}$,

$$\textit{join}(f, f')((x \oplus y) \otimes (x' \oplus y'))$$

$$\overset{\text{def}}{=} x \otimes x' \oplus \left(y \otimes x' \oplus (-1)^p x \otimes y' \bmod \mathscr{I}(Q_{cc}(f, f')) \right).$$

1.22. Let $Y \xrightarrow{f} X$ and $Y' \xrightarrow{f'} X'$ be morphisms of simplicial sets. Let f_*, f'_* and $(f \wedge f')_*$ denote the morphisms of chain complexes obtained by applying the functor $\textit{tot}(A \otimes \,?)$ to the morphisms f, f' and $f \wedge f'$, respectively. We shall presently define a chain map

$$\textit{ezrel}(f, f'): \textit{cone}(f_*) \otimes \textit{cone}(f'_*) \to \textit{cone}((f \wedge f')_*)$$

natural in pairs $(Y \xrightarrow{f} X, Y' \xrightarrow{f'} X')$ of morphisms of simplicial sets. Let $Q_{biss}(f, f')$ denote the commutative square

$$
\begin{array}{ccc}
Y \boxtimes X' & \xrightarrow{f \boxtimes \text{id}_{X'}} & X \boxtimes X' \\
\uparrow{\scriptstyle \text{id}_Y \boxtimes f} & & \uparrow{\scriptstyle \text{id}_X \boxtimes f'} \\
Y \boxtimes Y' & \xrightarrow{f \boxtimes \text{id}_{Y'}} & X \boxtimes Y'
\end{array}
$$

of morphisms of bisimplicial sets. Let $Q_{\text{bitot}}(f, f')$ and $Q_{\text{tot}}(f, f')$ denote, respectively, the commutative squares of chain complexes obtained by applying the functors $\textit{bitot}(A \otimes \,?)$ and $\textit{tot}(A \otimes /?)$, respectively, to the square $Q_{\text{biss}}(f, f')$. Let $\textit{ez}(A \otimes Q_{\text{biss}}(f, f'))$ denote the morphism $Q_{\text{bitot}}(f, f') \to Q_{\text{tot}}(f, f')$ in the category of commutative squares of morphisms of chain complexes obtained by applying the natural transformation $\textit{ez}: \textit{bitot}(?) \to \textit{tot}(/?)$ to the square $Q_{\text{biss}}(f, f')$. We define $\textit{ezrel}(f, f')$ to be the composite chain map diagrammed below:

$$
\begin{array}{l}
\textit{cone}(f_*) \otimes \textit{cone}(f'_*) \\
\quad\downarrow{\scriptstyle \textit{join}(f, f')} \\
\textit{bicone}(Q_{cc}(f_*, f'_*)) = \textit{bicone}(Q_{\text{bitot}}(f, f')) \\
\qquad\qquad\qquad\qquad\downarrow{\scriptstyle \textit{ez}(A \otimes Q_{\text{biss}}(f, f'))_*} \\
\textit{cone}((f \wedge f')_*) \quad = \textit{bicone}(Q_{\text{tot}}(f, f'))
\end{array}
$$

Note that when f, f' and therefore also $f \wedge f'$ are one-to-one, the diagram

$$
\begin{array}{ccc}
\mathit{cone}(f_*) \otimes \mathit{cone}(f'_*) & \xrightarrow{\;①\;} & \mathit{cone}((f \wedge f')_*) \\
\Big\downarrow{②} & & \Big\downarrow{③} \\
\mathrm{coker}(f_*) \otimes \mathrm{coker}(f'_*) & \xrightarrow[④]{} & \mathrm{coker}((f \wedge f')_*)
\end{array}
$$

commutes, where ① is the map $\mathit{exrel}(f, f')$ just defined, ② is the tensor product of the cone resolution homomorphisms associated to f_* and f'_*, ③ is the cone resolution homomorphism associated to $(f \wedge f')_*$, and ④ is induced by the chain map

$$
\mathit{ez}\big(A \otimes (X \boxtimes X')\big) \colon \mathit{bitot}\big(A \otimes (X \boxtimes X')\big) \to \mathit{tot}\big(A \otimes (X \times X')\big).
$$

1.23. Given a morphism $Y \xrightarrow{f} X$ of simplicial sets, set

$$
H_n\!\left(Y \xrightarrow{f} X\right) \overset{\mathrm{def}}{=} H_n\!\left(\mathit{cone}\!\left(\mathit{tot}(A \otimes Y) \xrightarrow{f_*} \mathit{tot}(A \otimes X)\right)\right).
$$

In the case $Y - \varnothing$ we may also write simply $II_n(X)$.

We have at our disposal an exact sequence

$$
\cdots \to H_n(Y) \xrightarrow{f_*} H_n(X) \to H_n\!\left(Y \xrightarrow{f} X\right) \xrightarrow{\partial} H_{n-1}(Y) \to \cdots
$$

functorial in $Y \xrightarrow{f} X$ deduced from the cone exact sequence. We refer to ∂ as the *boundary homomorphism*. Given a pair $Y \xrightarrow{f} X$ and $Y' \xrightarrow{f'} X'$ of morphisms of simplicial sets we have at our disposal a natural *external product*

$$
\kappa \colon H_p\!\left(Y \xrightarrow{f} X\right) \otimes H_q\!\left(Y' \xrightarrow{f'} X'\right) \to H_{p+q}\!\left(\left(Y \xrightarrow{f} X\right) \wedge \left(Y' \xrightarrow{f'} X'\right)\right),
$$

namely that induced by the chain map $\mathit{exrel}(f, f')$ defined in ¶1.22.

1.24. Given a pair $X \supseteq Y$ of simplicial sets, we define

$$
H_n(X/Y) \overset{\mathrm{def}}{=} H_n(\mathit{tot}(A \otimes X/A \otimes Y)).
$$

When $Y = \varnothing$, we may also write simply $H_n(X)$. We have at our disposal a long exact sequence

$$
\cdots \to H_n(Y) \to H_n(X) \to H_n(X/Y) \xrightarrow{\partial} H_{n-1}(Y) \to \cdots
$$

natural in pairs $X \supseteq Y$. We refer to ∂ as the *boundary homomorphism* of the pair $X \supseteq Y$. Given a pair $X \supseteq Y$ and $X' \supseteq Y'$ of pairs of simplicial sets, we have at our disposal an *external product*

$$\kappa : H_p(X/Y) \otimes H_q(X'/Y') \to H_{p+q}(X \times X'/X \times Y' \cup Y \times X'),$$

namely that induced by the chain map $ez(A \otimes (X \boxtimes X'))$ defined in ¶1.12.

1.25. Given a pair $X \supseteq Y$ of simplicial sets, denote by $Y \to X$ the inclusion morphism and let

$$c_n(X/Y): H_n(Y \to X) \overset{\sim}{\to} H_n(X/Y)$$

denote the isomorphism induced by the cone resolution homomorphism. When $Y = \varnothing$, $c_n(X/Y)$ reduces to the identity map $H_n(X) \to H_n(X)$. The commutativity of the square diagram of ¶1.16 insures that the diagram

$$
\begin{array}{ccc}
H_n(X, Y) & & \\
\ \ \downarrow c_n(X/Y) & \searrow^{\partial} & \\
 & & H_{n-1}(Y) \\
 & \nearrow_{\partial} & \\
H_n(X/Y) & &
\end{array}
$$

commutes. The commutativity of the square diagram at the conclusion of ¶1.22 insures that for all pairs $X \supseteq Y$ and $X' \supseteq Y'$ of pairs of simplicial sets, the diagram

$$
\begin{array}{ccc}
H_p(Y \to X) \otimes H_q(Y' \to X') & \overset{\kappa}{\longrightarrow} & H_{p+q}((Y \to X) \wedge (Y' \to X')) \\
\downarrow{\scriptstyle c_p(X/Y) \otimes c_q(X'/Y')} & & \downarrow{\scriptstyle c_{p+q}(X \times X'/X \times Y' \cup Y \times X')} \\
H_p(X/Y) \otimes H_q(X'/Y') & \overset{\kappa}{\longrightarrow} & H_{p+q}(X \times X'/X \times Y' \cup Y \times X')
\end{array}
$$

commutes.

1.26. Let $Y \overset{f}{\to} X$ be a morphism of bisimplicial sets. We define the *relative Eilenberg-Zilber spectral sequence* associated to $Y \overset{f}{\to} X$ to be the spectral sequence of the filtered chain complex

$$conefil\left(bitotfil(A \otimes Y) \overset{f_*}{\to} bitotfil(A \otimes X) \right)\left(\overset{def}{=} K, \text{ for brevity}\right),$$

the abutment

$$H_*\left(cone\left(bitot(A \otimes Y) \overset{f_*}{\to} bitot(A \otimes X) \right)\right)$$

of which we *identify* with

$$H_*\left(/Y \overset{f_*}{\to} /X\right)$$

under the isomorphism induced by the commuting square

$$
\begin{array}{ccc}
\textit{bitot}(A \otimes Y) & \xrightarrow{\ ez(A \otimes Y)\ } & \textit{tot}(A \otimes /Y) \\
\big\downarrow{\scriptstyle f_*} & & \big\downarrow{\scriptstyle f_*} \\
\textit{bitot}(A \otimes X) & \xrightarrow[\ ez(A \otimes X)\ ]{} & \textit{tot}(A \otimes /X)
\end{array}
$$

of chain maps. We set

$$E_{pq}^r\left(Y \overset{f}{\to} X\right) \overset{\text{def}}{=} E_{pq}^r(K), \qquad E_{*q}^1\left(Y \overset{f}{\to} X\right) \overset{\text{def}}{=} E_{*q}^1(K).$$

The early terms of the relative Eilenberg-Zilber spectral sequence admit the following interpretation:

$$E_{*q}^1\left(Y \overset{f}{\to} X\right) \overset{\text{def}}{=} \textit{cone}\left(H_q^{\text{horiz}}(A \otimes Y) \overset{f_*}{\to} H_q^{\text{horiz}}(A \otimes X)\right),$$

$$E_{p0}^2\left(Y \overset{f}{\to} X\right) \overset{\text{def}}{=} H_p\left(\pi_0^{\text{horiz}}(Y) \overset{f_*}{\to} \pi_0^{\text{horiz}}(X)\right).$$

We define

$$\varepsilon_p\left(Y \overset{f}{\to} X\right): H_p\left(/Y \overset{f_*}{\to} /X\right) \to H_p\left(\pi_0^{\text{horiz}}(Y) \overset{f_*}{\to} \pi_0^{\text{horiz}}(X)\right)$$

to be the homomorphism induced by the commuting square

$$
\begin{array}{ccc}
/Y & \xrightarrow{\ ed(Y)\ } & \pi_0^{\text{horiz}}(Y) \\
\big\downarrow{\scriptstyle f_*} & & \big\downarrow{\scriptstyle f_*} \\
/X & \xrightarrow[\ ed(X)\ ]{} & \pi_0^{\text{horiz}}(X)
\end{array}
$$

of morphisms of simplicial sets. In view of the natural homotopy commutativity of the triangular diagram of ¶1.15, it follows that that "edge map"

$$H_p\left(/Y \overset{f_*}{\to} /X\right) \to E_{p0}^2\left(Y \overset{f}{\to} X\right)$$

of the relative Eilenberg-Zilber spectral sequence coincides with $\varepsilon_p(Y \xrightarrow{f} X)$. This circumstance gives us a useful bound on the size of the kernel and cokernel of $\varepsilon_p(Y \xrightarrow{f} X)$ in terms of the relative Eilenberg-Zilber spectral sequence.

1.27. Let X be a bisimplicial set which we may, rather pedantically, view as a morphism $\varnothing \to X$ of bisimplicial sets, and to which, therefore, we may apply the constructions of ¶1.26. The resulting spectral sequence will be referred to as the *absolute Eilenberg-Zilber spectral sequence* associated to X and we write $E^r_{pq}(X)$, $E^1_{*q}(X)$, $\varepsilon_p(X)$ and so on.

§2. Relative singular homology of q.p.q.s. Q-schemes

2.1. A *good pair* (X, Y) of **Q**-schemes consists by definition of a smooth quasi-projective **Q**-scheme X and a reduced closed subscheme Y of X enjoying the following two properties: (i) Each irreducible component of $Y \otimes \mathbb{C}$ is smooth and of codimension one in $X \otimes \mathbb{C}$. (ii) For each closed point x of $X \otimes \mathbb{C}$ there exists in the local ring at x a system of parameters $t_1, \ldots, t_{\dim X}$ and there exists an integer $0 \leqslant k \leqslant \dim X$ such that in any sufficiently small neighborhood of x, the subscheme $Y \otimes \mathbb{C}$ of $X \otimes \mathbb{C}$ is defined by the single equation

$$\prod_{i=1}^{k} t_i = 0.$$

The following examples come under close scrutiny in Part II of this paper.

Example 2.1.1.   Let $n > 0$ be given. Set

$$X \stackrel{\text{def}}{=} \operatorname{Spec}\left(\mathbb{Q}[x_1, \ldots, x_n] \middle/ \left(1 - \sum_{j=1}^{n} x_j^N \right) \right),$$

$$Y \stackrel{\text{def}}{=} \operatorname{Spec}\left(\mathbb{Q}[x_1, \ldots, x_n] \middle/ \left(1 - \sum_{j=1}^{n} x_j^N, \prod_{j=1}^{n} x_j \right) \right).$$

Then (X, Y) is a good pair of **Q**-schemes.

Example 2.1.2.   Set

$$X \stackrel{\text{def}}{=} \operatorname{Spec}\left(\mathbb{Q}\left[x, \left(\sum_{j=0}^{N-1} x^j \right)^{-1} \right] \right),$$

$$Y \stackrel{\text{def}}{=} \operatorname{Spec}\left(\mathbb{Q}\left[x, \left(\sum_{j=0}^{N-1} x^j \right)^{-1} \right] \middle/ (x(1-x)) \right).$$

Then (X, Y) is a good pair of **Q**-schemes.

2.2. We say that a $\mathbf{Q}$-scheme Z is *quasi-projective quasi-smooth* (hereafter abbreviated to *q.p.q.s.*) if there exists a collection

$$\{(X_j, Y_j)\}_{j=1}^{n}$$

of good pairs of $\mathbf{Q}$-schemes such that Z is isomorphic to

$$\overset{n}{\underset{j=1}{\text{\Large X}}} Y_j.$$

A *q.p.q.s. pair* (Z, Z') of $\mathbf{Q}$-schemes consists by definition of a $\mathbf{Q}$-scheme Z and a closed subscheme Z' of Z such that both Z and Z' are q.p.q.s. $\mathbf{Q}$-schemes.

2.3. Given a topological space X, we denote by $S(X)$ the *total singular complex* of X, a simplicial set which, in degree n, is the set of continuous maps of the *standard n-simplex* (the convex hull in $\mathbb{R} \otimes [n]$ of the set $1 \otimes [n]$) into X.

2.4. Let X be a q.p.q.s. $\mathbf{Q}$-scheme. We denote the *normalization* of X by X^{norm}. We denote the classically topologized set of complex points of X by X^{top}. Set

$$\bar{S}(X) \overset{\mathrm{def}}{=} \mathrm{image}\big(S\big((X^{\mathrm{norm}})^{\mathrm{top}} \to S(X^{\mathrm{top}})\big)\big),$$

$$\bar{\pi}_0(X) \overset{\mathrm{def}}{=} \pi_0(\bar{S}(X)).$$

2.5. The slightly complicated definition of the functor $\bar{S}$ is convenient for the following reason: Let (X', Y') and (X'', Y'') be good pairs of $\mathbf{Q}$-schemes and set

$$(X, Y) \overset{\mathrm{def}}{=} (X' \times X'', X' \times Y'' \cup Y' \times X''),$$

obtaining once again a good pair of $\mathbf{Q}$-schemes. Then the diagram

$$
\begin{array}{ccc}
\bar{S}(X' \times Y'') & \to & \bar{S}(Y) \\
\uparrow & & \uparrow \\
\bar{S}(Y' \times Y'') & \to & \bar{S}(Y' \times X'')
\end{array}
$$

is cocartesian. Equivalently, in the notation of ¶1.19, we have

$$\bar{S}(Y) \to \bar{S}(X) = (\bar{S}(Y') \to \bar{S}(X')) \wedge (\bar{S}(Y'') \to \bar{S}(X'')).$$

It follows that the diagram of sets

$$\begin{array}{ccc}
\bar{\bar{\pi}}_0(X' \times Y'') & \to & \bar{\bar{\pi}}_0(Y) \\
\uparrow & & \uparrow \\
\bar{\bar{\pi}}_0(Y' \times Y'') & \to & \bar{\bar{\pi}}_0(Y' \times X'')
\end{array}$$

is cocartesian.

2.6. Let (X, Y) be a q.p.q.s. pair of $\mathbf{Q}$-schemes and set

$$H_n(X, Y) \stackrel{\text{def}}{=} H_n(\bar{S}(X)/\bar{S}(Y)),$$

a finite abelian group killed by N and which vanishes for $n \gg 0$. We have at our disposal a long exact sequence

$$\cdots \to H_n(X, Y) \stackrel{\partial}{\to} H_{n-1}(Y, \emptyset) \to H_{n-1}(X, \emptyset) \to H_{n-1}(X, Y) \to \cdots$$

functorial in (X, Y). We call ∂ the *boundary homomorphism* of the pair (X, Y). (Cf. ¶1.24 above.)

2.7. Let (X', Y') and (X'', Y'') be good pairs of $\mathbf{Q}$-schemes, p and q integers and set

$$(X, Y) \stackrel{\text{def}}{=} (X' \times X'', X' \times Y'' \cup Y' \times X''),$$

obtaining another good pair of $\mathbf{Q}$-schemes. We have at our disposal an *external product*

$$\kappa: H_p(X', Y') \otimes H_q(X'', Y'') \to H_{p+q}(X, Y),$$

namely the external product defined in ¶1.24.

2.8. Let X be a q.p.q.s. $\mathbf{Q}$-scheme. We define the *simplicial normalization* X^{snrm} of X to be the simplicial $\mathbf{Q}$-scheme which in degree n is the $(n+1)$-fold fiber product of X^{norm} over X. Note that X_n^{snrm} is smooth for all n.

LEMMA. *With X as above, the evident map $/\bar{S}(X^{snrm}) \to \bar{S}(X)$ induces an isomorphism $H_*(/\bar{S}(X^{snrm})) \stackrel{\sim}{\to} H_*(\bar{S}(X))$.*

Proof. Since

$$\bar{S}_q(X_p^{snrm}) = \bar{S}_q\left(\left(\coprod_X\right)_{j \in [p]} X^{\text{norm}}\right)$$

$$= \left(\coprod_{\bar{S}_q(X)}\right)_{j \in [p]} \bar{S}_q(X^{\text{norm}}),$$

it follows that

$$\pi_0^{\text{horiz}}\left(\bar{S}(X^{snrm})^{\text{op}}\right) = \bar{S}(X),$$

$$H_q^{\text{horiz}}\left(\bar{S}(X^{snrm})^{\text{op}}\right) = 0 \quad \text{if } q > 0.$$

Therefore the absolute Eilenberg-Zilber spectral sequence associated to the bisimplicial set $\bar{S}(X^{snrm})^{\text{op}}$ (cf. ¶1.27 above) degenerates at E^1, whence the desired result. $\qquad\square$

Therefore we have at our disposal the *normalization spectral sequence* of the q.p.q.s. $\mathbb{Q}$-scheme X

$$E_{pq}^1 = H_q\left(X_p^{snrm},\emptyset\right),\ E_{p0}^2 = H_p\left(\bar{\pi}_0(X^{snrm})\right) \Rightarrow H_{p+q}(X,\emptyset)$$

obtained by identifying the abutment $H_*(/\bar{S}(X^{snrm}))$ of the absolute Eilenberg-Zilber spectral sequence of the bisimplicial set $\bar{S}(X^{snrm})$ with $H_*(X,\emptyset)$ under the isomorphism provided by the lemma above.

2.9. Given a topological space X and closed subspaces $Y_1,\ldots,Y_n$ of X, we set

$$H_*(X; Y_1,\ldots,Y_n) \overset{\text{def}}{=} H_*\left(tot\left(A \otimes S(X)/A \otimes \left(\bigcup_{j=1}^{n} S(Y_j)\right)\right)\right).$$

This notion will prove useful in calculating relative homology groups of good pairs in part II.

§3. The main spectral sequence

3.1. Let X be a noetherian scheme. We denote by $X^{\text{ét}}$ the category the objects of which are étale X-schemes of finite type and the morphisms of which are étale morphisms of X-schemes, equipped with a Grothendieck topology by declaring each surjective morphism of $X^{\text{ét}}$ to be a covering. An *étale hypercovering U of X* is by definition a simplicial object of $X^{\text{ét}}$ such that (i) the structure morphism

$$U_0 \to X$$

is a covering and (ii) for each nonnegative integer n the adjunction morphism

$$U \to \cos k_n^{X^{\text{ét}}} U_{\leqslant n}$$

induces a covering morphism

$$U_{n+1} \to \left(\cos k_n^{X^{\text{ét}}} U_{\leqslant n}\right)_{n+1}$$

of $X^{\text{ét}}$. A *morphism* $U \to V$ of étale hypercoverings is by definition a morphism $U \to V$ of simplicial objects of $X^{\text{ét}}$. For all nonnegative integers j and n set

$$I(j, n) \stackrel{\text{def}}{=} \{ f \in \text{Hom}_\Delta([n], [j]) | f \text{ is onto.} \}.$$

It can be shown that for each étale hypercovering U of X and for each $j \geq 0$ there exists a unique open and closed subscheme U_j^{new} of U_j such that for all $n \geq 0$ the evident map

$$\coprod_{j=0}^{n} \coprod_{f \in I(j, n)} U_j^{\text{new}} \to U_n$$

is an isomorphism. Given a positive integer n and a morphism $U \to V$ of étale hypercoverings of X, we say that $U \to V$ is an *n-refinement* of étale hypercoverings of X if i) the induced map $U_{\leq n-1} \to V_{\leq n-1}$ of truncations is an isomorphism and ii) the diagram

$$\begin{array}{c} U_n^{\text{new}} \to U_n \\ \downarrow \\ V_n^{\text{new}} \to V_n \end{array}$$

admits a (necessarily unique) completion to a cartesian square

$$\begin{array}{ccc} U_n^{\text{new}} & \to & U_n \\ \downarrow & & \downarrow \\ V_n^{\text{new}} & \to & V_n \end{array}$$

of $X^{\text{ét}}$. In some contexts it is convenient to refer to a morphism of étale hypercoverings as a *0-refinement* of étale hypercoverings.

REFINEMENT LEMMA. *Let X be a noetherian scheme, V an étale hypercovering of X, n a nonnegative integer and $W \to V_n^{\text{new}}$ an étale surjective morphism of finite type. Then there exists an n-refinement $U \to V$ of étale hypercoverings of X such that the V^{new}-scheme $U_n^{\text{new}} \to V_n^{new}$ is isomorphic to $W \to V_n^{\text{new}}$.*

Proof. See §8 of [AM] for generalities concerning hypercoverings in sites. □

3.2. Let X be a q.p.q.s. **Q**-scheme, U an étale hypercovering of X.

RESOLUTION LEMMA. *With X and U as above, the evident map*

$$/\bar{S}(U) \to \bar{S}(X)$$

of simplicial sets induces an isomorphism

$$H_*(/\bar{S}(U)) \xrightarrow{\sim} H_*(\bar{S}(X)) = H_*(X, \emptyset).$$

Proof. A straightforward adaptation of the argument proving Theorem 12.1 of [AM] suffices for the purpose of proving that the evident map

$$/S(U^{\text{top}}) \to S(X^{\text{top}})$$

induces a weak homotopy equivalence

$$|/S(U^{\text{top}})| \to |S(X^{\text{top}})|$$

of geometric realizations and, in particular, an isomorphism

$$H_*(/S(U^{\text{top}})) \xrightarrow{\sim} H_*(S(X^{\text{top}})).$$

Under the additional assumption that X is smooth, $S(X^{\text{top}}) = \bar{S}(X), /S(U^{\text{top}}) = /\bar{S}(U)$ and we are done. It remains only to show that by assuming X to be smooth, we lose no generality. For general X, consider the bisimplicial sets

$$K \overset{\text{def}}{=} ([p],[q]) \mapsto \bar{S}_p(U_q \underset{X}{\times} X_p^{snrm}),$$

$$L \overset{\text{def}}{=} ([p],[q]) \mapsto \bar{S}_p(U_p \underset{X}{\times} X_q^{snrm}),$$

noting that

$$/K = /L.$$

Let $K \to \bar{S}(U)$ and $L \to \bar{S}(X^{snrm})$ denote the evident maps of bisimplicial sets. Consider the commutative square

$$
\begin{array}{ccc}
/K = /L & \overset{\textcircled{1}}{\longrightarrow} & /\bar{S}(U) \\
{\scriptstyle\textcircled{3}}\downarrow & & \downarrow{\scriptstyle\textcircled{4}} \\
/\bar{S}(X^{snrm}) & \underset{\textcircled{2}}{\longrightarrow} & \bar{S}(X)
\end{array}
$$

of morphisms of simplicial sets. The arrow $\textcircled{2}$ induces an isomorphism in homology by Lemma 2.5.1 and the same lemma guarantees that (in the notation of ¶1.21)

$$E^2_{pq}(K \to \bar{S}(U)) = 0,$$

whence the conclusion that $\textcircled{1}$ likewise induces an isomorphism in homology.

Now for all $n \geqslant 0$ the induced map

$$H_*\!\left(/\overline{S}\!\left(U \underset{X}{\times} X_n^{snrm}\right)\right) \rightarrow H_*\!\left(\overline{S}(X^{snrm})\right)$$

is an isomorphism by the "smooth" case of the Resolution Lemma already proven. Therefore

$$E_{pq}^2\!\left(L \rightarrow \overline{S}(X^{snrm})\right) = 0$$

and in particular ③ induces an isomorphism in homology. Since ①, ② and ③ induce isomorphisms in homology, so also does ④. $\square$

3.3. Let (X, Y) be a q.p.q.s. pair of **Q**-schemes, U an étale hypercovering of X, n an integer. We define

$$r_n(X, Y; U)\colon H_n\!\left(/\overline{S}\!\left(U \underset{X}{\times} Y\right) \rightarrow /\overline{S}(U)\right) \rightarrow H_n\!\left(\overline{S}(Y) \rightarrow \overline{S}(X)\right)$$

to be the homomorphism induced by the commutative square

$$\begin{array}{ccc}
/\overline{S}(U \underset{X}{\times} Y) & \rightarrow & /\overline{S}(U) \\
\downarrow & & \downarrow \\
\overline{S}(Y) & \rightarrow & \overline{S}(X)
\end{array}$$

of morphisms of simplicial sets which, by virtue of the Resolution Lemma and the 5-Lemma, is an isomorphism. The *main spectral sequence* associated to the data of the good pair (X, Y) and the étale hypercovering U of X is defined to be the relative Eilenberg-Zilber spectral sequence (cf. ¶1.26 above) of the morphism $\overline{S}(U \underset{X}{\times} Y) \rightarrow \overline{S}(U)$ of bisimplicial sets. By definition, the abutment of the main spectral sequence is $H_*(/\overline{S}(U \underset{X}{\times} Y) \rightarrow /\overline{S}(U))$. We denote by $E_{pq}^r(X, Y; U)$ the general term of the main spectral sequence and by $E_{*q}^1(X, Y; U)$ the chain complex of E^1-terms with fixed complementary degree q. The homomorphism

$$\varepsilon_n(X, Y; U)\colon H_n(X, Y) \rightarrow H_n\!\left(\overline{\pi}_0\!\left(U \underset{X}{\times} Y\right) \rightarrow \overline{\pi}_0(U)\right)$$

is defined to be that which renders the diagram

commutative. Note that the morphisms on the left edge of the diagram are isomorphisms. The early terms of the main spectral admit the following interpretation:

$$E^1_{*q}(X, Y; U) = \operatorname{cone}\left(\operatorname{tot}\left(H_q\left(U \underset{X}{\times} Y, \emptyset\right)\right) \to \operatorname{tot}\left(H_q(U, \emptyset)\right)\right),$$

$$E^2_{p0}(X, Y; U) = H_p\left(\bar{\pi}_0\left(U \underset{X}{\times} Y\right) \to \bar{\pi}_0(U)\right).$$

Remark. The point of view to be exploited in what follows is that the homomorphism $\varepsilon_p(X, Y; U)$ gives an "approximation" to $H_p(X, Y)$ in terms of $H_p(\bar{\pi}_0(U \underset{X}{\times} Y) \to \bar{\pi}_0(U))$ the "degree of accuracy" of which is usefully gauged by the main spectral sequence.

§4. Formulation of the principal results of Part I

4.1. Let $\overline{\mathbf{Q}}$ denote the algebraic closure of $\mathbf{Q}$ in $\mathbf{C}$. Given any subfield k of $\overline{\mathbf{Q}}$, let $G(k)$ denote the group of automorphisms of $\overline{\mathbf{Q}}$ over k. We say that $G(\mathbf{Q})$ acts *admissibly* on a set X if for all $x \in X$ there exists a subfield k of $\overline{\mathbf{Q}}$ of finite degree over $\mathbf{Q}$ such that $\{\sigma \in G(\mathbf{Q}) | \sigma x = x\} = G(k)$.

4.2. Let π_0^{zar} denote the functor assigning to each scheme the set of its Zariski connected components. Given now a q.p.q.s. $\mathbf{Q}$-scheme X, let

$$\nu(X): \bar{\pi}_0(X) \to \pi_0^{\mathrm{zar}}(X \otimes \overline{\mathbf{Q}})$$

denote the map assigning each path component $Z \subseteq X^{\mathrm{top}}$ to the unique Zariski connected component $Z' \subseteq X \otimes \overline{\mathbf{Q}}$ such that

$$\operatorname{Hom}_{\operatorname{Spec}(\overline{\mathbf{Q}})}(\operatorname{Spec}(\mathbf{C}), Z') \supseteq Z.$$

LEMMA. *The map $\nu(X)$ is a bijection.*

Proof. We may assume X to be projective without loss of generality. Let X^{an} denote the analytic space underlying $X \otimes \mathbf{C}$. It will be enough to show that the maps

$$H^0(X \otimes \mathbf{C}, \mathcal{O}_{X \otimes \mathbf{C}}) \overset{①}{\to} H^0(X^{\mathrm{an}}, \mathcal{O}_{X^{\mathrm{an}}}),$$

$$H^0(X \otimes \overline{\mathbf{Q}}, \mathcal{O}_{X \otimes \overline{\mathbf{Q}}}) \underset{\overline{\mathbf{Q}}}{\bigotimes} \mathbf{C} \overset{②}{\to} H^0(X \otimes \mathbf{C}, \mathcal{O}_{X \otimes \mathbf{C}})$$

induced, respectively, by the evident maps

$$X^{\mathrm{an}} \to X \otimes \mathbf{C}, \quad X \otimes \mathbf{C} \to X \otimes \overline{\mathbf{Q}}$$

of ringed spaces are *isomorphisms*. But ① is an isomorphisms by [GAGA] and ②
an isomorphism since coherent sheaf cohomology commutes with flat base-change
(Prop. 9.3, p. 255 of [Ha]).　　　　　　　　　　　　　　　　　　　　　□

Therefore, the set $\bar{\bar{\pi}}_0(X)$ is endowed, via the natural transformation ν, with the
structure of finite admissible $G(\mathbf{Q})$-set. In turn, if X is part of a good pair (X, Y)
of $\mathbf{Q}$-schemes and one is given an étale hypercovering U of X, the group

$$E^2_{p0}(X, Y; U) = H_p\big(\bar{\bar{\pi}}_0(U \underset{X}{\times} Y) \to \bar{\bar{\pi}}_0(U)\big)$$

is endowed, via ν, with the structure of finite admissible $G(\mathbf{Q})$-module functorially in étale hypercoverings U of X.

4.3. The principal results of Part I are the following.

THEOREM 1. *Given a good pair (X, Y) of $\mathbf{Q}$-schemes and a nonnegative integer
n, there exists a unique way to equip the finite abelian group $H_n(X, Y)$ with the
structure of admissible $G(\mathbf{Q})$-module relative to which, for all étale hypercoverings
U of X, the homomorphism*

$$\varepsilon_n(X, Y; U)\colon H_n(X, Y) \to H_n\big(\bar{\bar{\pi}}_0(U \underset{X}{\times} Y) \to \bar{\bar{\pi}}_0(U)\big)$$

*becomes $G(\mathbf{Q})$-equivariant, the target of $\varepsilon_n(X, Y; U)$ being equipped with admissi-
ble $G(\mathbf{Q})$-module structure via the natural transformation ν of the preceding
paragraph.*

Let us define the ε-*action* of $G(\mathbf{Q})$ upon $H_n(X, Y)$, where (X, Y) and n are as
above, to be that specified by Theorem 1.

COROLLARY 2. *Let X be an étale $\mathbf{Q}$-scheme of finite type. Then the evident
isomorphism*

$$H_0(X, \emptyset) \xrightarrow{\sim} A \otimes X(\overline{\mathbf{Q}})$$

is compatible with the ε-action of $G(\mathbf{Q})$ upon $H_0(X, \emptyset)$.

COROLLARY 3. *Let (X, Y) and (X', Y') be good pairs of $\mathbf{Q}$-schemes, n a
nonnegative integer and $f\colon X' \to X$ a morphism of $\mathbf{Q}$-schemes such that $f(Y') \subseteq Y$.
Then the induced map*

$$f_*\colon H_n(X', Y') \to H_n(X, Y)$$

commutes with the ε-action of $G(\mathbf{Q})$.

COROLLARY 4. *Let (X, Y) be a good pair of $\mathbf{Q}$-schemes such that Y is smooth.
Let n be a positive integer. Then the boundary map*

$$\partial\colon H_n(X, Y) \to H_{n-1}(Y, \emptyset)$$

commutes with the ε-action of $G(\mathbf{Q})$.

COROLLARY 5. *Let (X', Y') and (X'', Y'') be good pairs of $\mathbf{Q}$-schemes, p and q nonnegative integers. Set*

$$(X, Y) \overset{\text{def}}{=} (X' \times X'', X' \times Y'' \cup Y' \times X''),$$

obtaining once again a good pair of $\mathbf{Q}$-schemes. Then the external product

$$\kappa \colon H_p(X', Y') \otimes H_q(X'', Y'') \to H_{p+q}(X, Y)$$

commutes with the ε-action of $G(\mathbf{Q})$.

The proofs of these results will be completed in §6 after an investigation of the main spectral sequence is carried out in §5. In Part II we shall drop the terminology of ε-actions, tacitly regarding relative singular homology groups as $G(\mathbf{Q})$-modules.

§5. Degeneration in the limit of the main spectral sequence

5.1. Let F and G be functors from a category $\mathscr{C}$ to the category $\mathscr{ABGP}$ of abelian groups. We say that F is *null in the limit* if for every object X of $\mathscr{C}$ there exists a morphism $X' \to X$ in $\mathscr{C}$ such that the induced map $F(X') \to F(X)$ is the zero map. We say that F *dominates* G if there exists a positive integer n such that for every sequence

$$X_n \to X_{n-1} \to \cdots \to X_1 \to X_0$$

of n morphisms of $\mathscr{C}$, the vanishing for $j = 1, \ldots, n$ of the induced maps

$$F(X_j) \to F(X_{j-1})$$

is a sufficient condition for the vanishing of the map

$$G(X_n) \to G(X_0)$$

induced by the composite $X_n \to X_0$ of the n given morphisms. The following trivial observation encapsulates a style of argument which we employ repeatedly in what follows.

LEMMA. *Let $G \colon \mathscr{C} \to \mathscr{ABGP}$ be a functor, $\{F_j \colon \mathscr{C} \to \mathscr{ABGP}\}_{j=1}^n$ a finite collection of functors. If $\oplus F_j$ dominates G, then a sufficient condition for G to be null in the limit is that F_j be null in the limit for each $j = 1, \ldots, n$.*

5.2. Let k be a subfield of $\mathbb{C}$ and X a smooth quasi-projective k-scheme. We write $(X/k)^{\text{man}}$ to denote the set of $\operatorname{Spec}(k)$-morphisms $\operatorname{Spec}(\mathbb{C}) \to X$, equipped with the structure of complex manifold with which it is naturally endowed.

5.3. Let $f: X \to Y$ be a morphism of complex manifolds. We say that f is *étale* if for each point x of X, the induced map $f_*: T_x(X) \to T_{f(x)}(Y)$ of complex tangent spaces is an isomorphism.

5.4. Let X be a complex manifold. It is convenient to define a *proper étale X-manifold Z* to be a complex manifold equipped with the additional structure of a proper étale map $Z \to X$ of complex manifolds. A *morphism* of proper étale X-manifolds is a map of complex manifolds rendering the evident triangle commutative.

5.5. Let X be a complex manifold. We denote by $H_*(X)$ the singular homology of X with coefficients in A. We say that X is *good* if for all proper étale morphisms $U \to X$ of complex manifolds there exists a proper étale surjective morphism $U' \to U$ of complex manifolds such that for all n > 0 the induced map $H_n(U') \to H_n(U)$ is the zero map.

5.6. Let X be a smooth quasi-projective $\overline{\mathbb{Q}}$-scheme. For each $\sigma \in G(\mathbb{Q})$ let X^σ denote the conjugate of X relative to σ. We say that X is *absolutely good* if for all $\sigma \in G(\mathbb{Q})$ the complex manifold $(X^\sigma/\overline{\mathbb{Q}})^{\mathrm{man}}$ is good.

LEMMA. *With X as above, there exists a finite open cover $\{U_i\}$ of X such that each open U_i of the cover is absolutely good.*

Proof. This is a consequence of Artin's good neighborhood theorem, XI.3.3 of [SGA 4]. □

5.7. Let X again be a smooth quasi-projective $\overline{\mathbb{Q}}$-scheme. Let $\mathscr{FET}(X)$ denote the category of finite étale X-schemes, $\mathscr{FET}_{/\mathbb{C}}(X)$ the category of finite étale $X \otimes_{\overline{\mathbb{Q}}} \mathbb{C}$-schemes, and $\mathscr{COV}(X)$ the category of proper étale $(X/\overline{\mathbb{Q}})^{\mathrm{man}}$-manifolds.

LEMMA. *With X as above, the functor $\mathscr{FET}(X) \to \mathscr{COV}(X)$ induced by $(?/\overline{\mathbb{Q}})^{\mathrm{man}}$ is an equivalence of categories.*

Proof. The functor $(?/\mathbb{C})^{\mathrm{man}}$ induces an equivalence of categories $\mathscr{FET}_{/\mathbb{C}}(X) \xrightarrow{\sim} \mathscr{COV}(X)$ be the generalized Riemann existence theorem [GR]. The functor $\mathscr{FET}(X) \to \mathscr{FET}_{/\mathbb{C}}(X)$ induced by the base-change functor $? \otimes_{\overline{\mathbb{Q}}} \mathbb{C}$ is fully faithful by IX.3.4 of [SGA 1]. It will therefore be enough to prove that the base change functor $\mathscr{FET}(X) \to \mathscr{FET}_{/\mathbb{C}}(X)$ is essentially surjective. Let W be an arbitrary object of $\mathscr{FET}_{/\mathbb{C}}(X)$. Without loss of generality we may assume that there exists an affine scheme T of finite type and smooth over $\mathrm{Spec}(\overline{\mathbb{Q}})$, a finite étale morphism $Z \to X \underset{\overline{\mathbb{Q}}}{\times} T$, and a point t_0 of T in $\mathbb{C}$ such that the object of $\mathscr{FET}_{/\mathbb{C}}(X)$ deduced from $Z \to X \underset{\overline{\mathbb{Q}}}{\times} T$ by base-change along $\mathrm{id}_X \underset{\overline{\mathbb{Q}}}{\times} t_0$ is isomorphic to W. Given an arbitrary point t of T in $\mathbb{C}$, let Z_t denote the object of $\mathscr{FET}_{/\mathbb{C}}(X)$ deduced via base-change along $\mathrm{id}_X \underset{\overline{\mathbb{Q}}}{\times} t$ from the morphism $Z \to X \underset{\overline{\mathbb{Q}}}{\times} T$. Then the isomorphism class of the object $(Z_t/\mathbb{C})^{\mathrm{man}}$ of $\mathscr{COV}(X)$

depends only on the connected component of $(T/\overline{\mathbb{Q}})^{\mathrm{man}}$ to which t belongs. In particular, for t close enough to t_0 in the classical topology, the objects $(Z_t/\mathbb{C})^{\mathrm{man}}$ and $(W/\mathbb{C})^{\mathrm{man}}$ of $\mathscr{COV}(X)$ are isomorphic, and hence, by the generalized Riemann existence theorem cited above, the objects Z_t and W of $\mathscr{FET}_{/\mathbb{C}}(X)$ are isomorphic. Selecting now a point t_1 of T in $\overline{\mathbb{Q}}$ close enough to t_0 in the sense above, we obtain from the morphism $Z \to X \underset{\overline{\mathbb{Q}}}{\times} T$ via base-change along $\mathrm{id}_X \underset{\overline{\mathbb{Q}}}{\times} t_1$ an object W' of $\mathscr{FET}(X)$ such that $W' \otimes_{\mathbb{Q}} \mathbb{C}$ is isomorphic in $\mathscr{FET}_{/\mathbb{C}}(X)$ to W, as required. $\qquad\square$

5.8. Let X be a smooth quasi-projective $\mathbb{Q}$-scheme.

LEMMA. *With X as above, there exists an étale covering $U \to X$ of finite type such that for all $n > 0$ the induced map*

$$H_n(U,\emptyset) \to H_n(X,\emptyset)$$

is the zero map.

Proof. Let $\{W_i\}$ be a finite open cover of $X \otimes \overline{\mathbb{Q}}$ such that each W_i is absolutely good; such a cover exists by the lemma of ¶5.6. For a suitable subfield k of $\overline{\mathbb{Q}}$ of finite degree over $\mathbb{Q}$, the covering $\{W_i\}$ of $X \otimes \overline{\mathbb{Q}}$ may be recovered via base-change from an open covering $\{W_i'\}$ of $X \otimes k$. Set

$$W' \overset{\mathrm{def}}{=} \coprod_i W_i'.$$

Then W' is an étale covering of X of finite type and the complex manifold

$$(W'/\mathbb{Q})^{\mathrm{man}} = \coprod_i \coprod_{\sigma \in G(\mathbb{Q})/G(k)} \left(W_i^{\sigma}/\overline{\mathbb{Q}}\right)^{\mathrm{man}}$$

is good. Replacing X by W', we may as well assume that $(X/\mathbb{Q})^{\mathrm{man}}$ is good. Then there exists a proper étale surjective morphism $Z \to (X/\mathbb{Q})^{\mathrm{man}}$ of complex manifolds such that for all $n > 0$ the induced map

$$H_n(Z) \to H_n\left((X/\mathbb{Q})^{\mathrm{man}}\right) = H_n(X,\emptyset)$$

is the zero map. By the lemma of ¶5.7, there exists a finite étale $X \otimes \overline{\mathbb{Q}}$-scheme W such that the proper étale $(X/\mathbb{Q})^{\mathrm{man}}$-manifold $(W/\overline{\mathbb{Q}})^{\mathrm{man}}$ is isomorphic to Z. For a suitable subfield k of $\overline{\mathbb{Q}}$ of finite degree over $\mathbb{Q}$ and a suitable finite étale $X \otimes k$-scheme W', W is isomorphic to $W' \otimes_k \overline{\mathbb{Q}}$. Let U be the object of $X^{\mathrm{ét}}$ representing the functor

$$\mathrm{Hom}_{(X \otimes k)^{\mathrm{ét}}}(? \otimes k, W').$$

Then $U \to X$ is an étale covering of X of finite type and one can check that the proper étale $(X/\mathbf{Q})^{\mathrm{man}}$-manifold $(U/\mathbf{Q})^{\mathrm{man}}$ is isomorphic to the fiber product over $(X/\mathbf{Q})^{\mathrm{man}}$ of the proper étale $(X/\mathbf{Q})^{\mathrm{man}}$-manifolds $(W^{\sigma}/\overline{\mathbf{Q}})^{\mathrm{man}}$, where the index σ runs over a set of representatives for the cosets of $G(k)$ in $G(\mathbf{Q})$. In particular, the map $(U/\mathbf{Q})^{\mathrm{man}} \to (X/\mathbf{Q})^{\mathrm{man}}$ factors through $Z \to (X/\mathbf{Q})^{\mathrm{man}}$. Consequently, for all $n > 0$, the induced map

$$H_n(U,\emptyset) \to H_n(X,\emptyset)$$

is the zero map, as required. $\square$

5.9. Let X be a smooth quasi-projective $\mathbf{Q}$-scheme, Y a smooth closed subscheme of X.

LEMMA. *With X and Y as above, there exists an étale covering $U \to X$ of finite type such that for all $n > 0$, the induced map*

$$H_n\!\left(U \underset{X}{\times} Y,\emptyset\right) \to H_n(Y,\emptyset)$$

is the zero map.

Proof. It will be enough, given an étale covering $V \to Y$ of finite type to construct an étale covering $U \to X$ of finite type such that there exists an étale morphism $U \underset{X}{\times} Y \to V$ over Y. We omit the details of this construction, as it is straightforward once one takes into account the local description of étale morphisms provided by I.7.6 of [SGA 1]. $\square$

5.10. Let X' and X'' be smooth quasi-projective $\mathbf{Q}$-schemes, Y' and Y'' smooth closed subschemes of X' and X'', respectively. Set

$$X \overset{\mathrm{def}}{=} X' \times X'', \; Y \overset{\mathrm{def}}{=} Y' \times Y''.$$

LEMMA. *Notation as above, there exist étale covers $U' \to X'$ and $U'' \to X''$ of finite type such that for all positive integers n the induced map*

$$H_n\!\left((U' \times U'') \underset{X}{\times} Y,\emptyset\right) \to H_n(Y,\emptyset)$$

is the zero map.

Proof. Let $\mathscr{C}'$ denote the subcategory of $X^{\mathrm{\acute{e}t}}$ consisting of objects with surjective structure morphisms and the surjective morphisms from one such object to another; let $\mathscr{C}''$ denote the analogous subcategory of $X''^{\mathrm{\acute{e}t}}$. Set

$$\mathscr{C} \overset{\mathrm{def}}{=} \mathscr{C}' \times \mathscr{C}''.$$

Consider the functors $\mathscr{C} \to \mathscr{A}\mathscr{B}\mathscr{G}\mathscr{P}$ defined as follows:

$$F_n \overset{\text{def}}{=} (U', U'') \mapsto H_n\big((U' \times U'') \underset{X}{\times} Y\big)$$

$$F_p' \overset{\text{def}}{=} (U', U'') \mapsto H_p\big(U' \underset{X'}{\times} Y'\big)$$

$$F_q'' \overset{\text{def}}{=} (U', U'') \mapsto H_q\big(U'' \underset{X''}{\times} Y''\big)$$

$$\Phi_{pq} \overset{\text{def}}{=} (U', U'') \mapsto H_p\Big(\mathit{tot}\Big(H_q\big(U'' \underset{X''}{\times} Y''\big) \otimes \bar{S}\big(U' \underset{X'}{\times} Y'\big)\Big)\Big).$$

Our task is to show that $\oplus_{n>0} F_n$ is null in the limit. Since for all $n \gg 0$ the functor F_n vanishes identically, it will suffice to fix a positive integer n and to prove that F_n is null in the limit. Consideration of the absolute Eilenberg-Zilber spectral sequence of the bisimplicial set

$$\bar{S}\big(U' \underset{X'}{\times} Y'\big) \boxtimes \bar{S}\big(U'' \underset{X''}{\times} Y''\big)$$

leads to the conclusion that

$$\underset{\substack{p+q=n \\ p,q \geqslant 0}}{\bigoplus} \Phi_{pq} \quad \text{dominates } F_n.$$

The universal coefficients theorem provides a spectral sequence the existence of which guarantees that

$$\underset{\substack{p+q+r=n \\ p,q,r \geqslant 0}}{\bigoplus} \operatorname{Tor}_r^A(F_p', F_q'') \quad \text{dominates} \quad \underset{\substack{p+q=n \\ p,q \geqslant 0}}{\bigoplus} \Phi_{pq}.$$

Taking into account the facts that i) both F_p' and F_q'' dominate $\operatorname{Tor}_r^A(F_p', F_q'')$ and that ii) $\operatorname{Tor}_n^A(F_0', F_0'')$ vanishes identically (the functors F_0' and F_0'' take values that are *free A*-modules) we find that

$$\Big(\underset{p>0}{\bigoplus} F_p'\Big) \oplus \Big(\underset{q>0}{\bigoplus} F_q''\Big) \quad \text{dominates } F_n.$$

Thus F_n is dominated by a functor which, by the lemma of ¶5.9, is null in the limit. $\square$

5.11. Let (X, Y) be a good pair of $\mathbf{Q}$-schemes. Then $G(\mathbf{Q})$ operates admissibly on the set of Zariski connected components of $X \otimes \overline{\mathbf{Q}}$ and the orbit space for this $G(\mathbf{Q})$-action is in natural one-to-one correspondence with the set of Zariski connected components of X. Also $G(\mathbf{Q})$ operates admissibly on the set of irreducible components of $Y \otimes \overline{\mathbf{Q}}$ and the orbit space for this $G(\mathbf{Q})$-action is in natural one-to-one correspondence with the set of irreducible components of Y. We say that (X, Y) is *split* if for all connected components X_i of $X \otimes \overline{\mathbf{Q}}$, irreducible components Y_j of $Y \otimes \overline{\mathbf{Q}}$ lying in X_i and $\sigma \in G(\mathbf{Q})$, X_i is fixed by σ only if Y_j is fixed by σ. Clearly there exists for any good pair (X, Y) a subfield k of $\overline{\mathbf{Q}}$ of finite degree over $\mathbf{Q}$ such that $(X \otimes k, Y \otimes k)$ is split. Supposing that we can write

$$(X, Y) = (X' \times X'', X' \times Y'' \cup Y' \times X'')$$

where (X', Y') and (X'', Y'') are good pairs of $\mathbf{Q}$-schemes, (X, Y) is split only if (X', Y') and (X'', Y'') are both split. If (X, Y) is split, then each irreducible component of Y is smooth and Y^{norm} is the disjoint union of the irreducible components of Y. More generally, given a nonnegative integer n and writing

$$Y_n^{snrm} = \coprod_{i \in I} Z_i, \quad \text{each } Z_i \text{ connected,}$$

if (X, Y) is split, then for all $i \in I$ the evident map $Z_i \to X$ is a closed immersion.

5.12. Let (X, Y) be a good pair of $\mathbf{Q}$-schemes. We say that (X, Y) is *untangled* under either of the following equivalent conditions:

(5.12.1) For all nonnegative integers n, the natural map

$$\overline{\pi}_0\left(\left(\underset{X}{\overset{n}{\underset{j=0}{\mathsf{X}}}}\right) Y^{\mathrm{norm}}\right) \to \left(\underset{\overline{\pi}_0(X)}{\overset{n}{\underset{j=0}{\mathsf{X}}}}\right) \overline{\pi}_0(Y^{\mathrm{norm}})$$

is a bijection.

(5.12.2) Given any connected component X_i of $X \otimes \overline{\mathbf{Q}}$ and collection $\{Y_j\}_{j \in J \neq \varnothing}$ of irreducible components of $Y \otimes \overline{\mathbf{Q}}$ lying in X_i, the intersection $\bigcap_{j \in J} Y_j$ is nonempty and connected.

LEMMA 5.12.3. *With* (X, Y) *as above, if* (X, Y) *is untangled, then for all positive integers* n,

$$H_n(\overline{\pi}_0(Y^{snrm})) = 0.$$

Proof.   Follows directly from (5.12.1).   □

LEMMA 5.12.4. *With* (X, Y) *as above, if there exist good pairs* (X', Y') *and* (X'', Y'') *of* $\mathbf{Q}$-schemes *such that*

$$(X, Y) = (X' \times X'', X' \times Y'' \cup Y' \times X''),$$

then a sufficient condition for (X, Y) to be untangled is that (X', Y') and (X'', Y'') be untangled.

Proof. Follows directly from (5.12.2). $\qquad\square$

LEMMA 5.12.5. *With (X, Y) as above, there exists an étale cover $U \to X$ of finite type such that $(U, U \underset{X}{\times} Y)$ is untangled.*

Proof. Let $\{Y_i\}_{i \in I}$ be the irredundantly indexed collection of irreducible components of $Y \otimes \overline{\mathbf{Q}}$, x a closed point of $X \otimes \overline{\mathbf{Q}}$ and J a subset of I. Set

$$Y_J \overset{\text{def}}{=} \left.\begin{cases} \bigcap_{j \in J} Y_j & \text{if } J \neq \varnothing \\[2mm] X & \text{otherwise} \end{cases}\right\},$$

$$V_{x, J} \overset{\text{def}}{=} \begin{array}{l} \text{the complement in } X \otimes \overline{\mathbf{Q}} \text{ of the union of} \\ \text{the connected components of } Y_J \text{ not} \\ \text{containing } x, \end{array}$$

$$V_x \overset{\text{def}}{=} \bigcap_{J \subseteq I} V_{x, J}.$$

Then

$$\#\pi_0^{\text{zar}}(V_x \cap Y_J) = \left.\begin{cases} 1 & \text{if } Y_J \ni x \\ 0 & \text{otherwise} \end{cases}\right\}.$$

Therefore we can find a finite open covering $\{V_i\}_{i=1}^m$ of $X \otimes \overline{\mathbf{Q}}$ such that, setting

$$V \overset{\text{def}}{=} \coprod_{i=1}^m V_i,$$

we obtain an étale covering $V \to X \otimes \overline{\mathbf{Q}}$ of finite type such that for all nonnegative integers n the evident map

$$\pi_0^{\text{zar}}\left(V \underset{X}{\times} \left(\underset{X}{\boldsymbol{\mathsf{X}}}\right)_{j=0}^n Y^{\text{norm}}\right) \to \left(\underset{\pi_0^{\text{zar}}(V)}{\boldsymbol{\mathsf{X}}}\right)_{j=0}^n \pi_0^{\text{zar}}\left(V \underset{X}{\times} Y^{\text{norm}}\right)$$

is a bijection. For a suitable subfield k of $\overline{\mathbf{Q}}$ of finite degree over $\mathbf{Q}$, $V \to X \otimes \overline{\mathbf{Q}}$ is the base-change of an étale covering $U \to X \otimes k$ of finite type. The composite $U \to X$ of $U \to X \otimes k$ and the evident map $X \otimes k \to X$ is then an étale covering of X of finite type with the following property: As an X-scheme (albeit not, in general, as an $X \otimes \overline{\mathbf{Q}}$-scheme) $U \otimes \overline{\mathbf{Q}}$ is isomorphic to the disjoint union of $[k: \mathbf{Q}]$ copies of V. Consequently $(U, U \underset{X}{\times} Y)$ verifies the analogue of (5.12.1) in which the functor $\overline{\pi}_0$ is replaced by the functor $\pi_0^{\text{zar}}(? \otimes \overline{\mathbf{Q}})$ and hence, by the lemma of ¶4.2, $(U, U \underset{X}{\times} Y)$ verifies (5.12.1) and is untangled. $\qquad\square$

5.13. Let (X', Y') and (X'', Y'') be good pairs of $\mathbf{Q}$-schemes and set

$$(X, Y) \stackrel{\text{def}}{=} (X' \times X'', \, X' \times Y'' \cup Y' \times X''),$$

obtaining another good pair of $\mathbf{Q}$-schemes. Let n be a positive integer.

LEMMA. *Notation as above, there exist étale covers $U' \to X'$ and $U'' \to X''$ of finite type such that the induced map*

$$H_n\big((U' \times U'') \underset{X}{\times} Y, \emptyset\big) \to H_n(Y, \emptyset)$$

is the zero map.

Proof. Let the category $\mathscr{C}$ be defined just as in the proof of the lemma of ¶5.10. Consider the functors $\mathscr{C} \to \mathscr{ABGP}$ defined as follows:

$$F \stackrel{\text{def}}{=} (U', U'') \mapsto H_n\big((U' \times U'') \underset{X}{\times} Y, \emptyset\big)$$

$$\Phi_j \stackrel{\text{def}}{=} (U', U'') \mapsto H_n\big((U' \times U'') \underset{X}{\times} Y_{n-j}^{snrm}\big) \qquad (0 < j \leqslant n)$$

$$\Phi_0 \stackrel{\text{def}}{=} (U', U'') \mapsto H_n\big(\bar{\pi}_0\big((U' \times U'') \underset{X}{\times} Y^{snrm}\big)\big)$$

It will suffice to show that the functor F is null in the limit. By virtue of the existence of the normalization spectral sequence discussed in ¶2.8, F is dominated by $\bigoplus_{j=0}^{n} \Phi_j$. The functor Φ_0 is null in the limit by Lemmas 5.12.3, 4, 5. Assuming without loss of generality that (X', Y') and (X'', Y'') are split, for $0 < j \leqslant$ n, the functor Φ_j is a finite direct sum of functors of the form

$$(U', U'') \mapsto H_j\big((U' \times U'') \underset{X}{\times} (Z' \times Z'')\big) \colon \mathscr{C} \to \mathscr{ABGP},$$

where Z' and Z'' are smooth closed subschemes of X' and X'', respectively, and therefore, by the lemma of ¶5.9, Φ_j is null in the limit. $\qquad\square$

5.14. Let (X', Y') and (X'', Y'') be good pairs of $\mathbf{Q}$-schemes, p a nonnegative integer, q a positive integer, V' an étale hypercovering of X' and V'' an étale hypercovering of X''. Set

$$(X, Y) \stackrel{\text{def}}{=} (X' \times X'', \, X' \times Y'' \cup Y' \times X''),$$

obtaining again a good pair of $\mathbf{Q}$-schemes. Let $V' \times V''$ denote the simplicial object

$$[n] \mapsto V_n' \times V_n''$$

of $X^{\text{êt}}$ which, as one can easily check, is an étale hypercovering of X.

LEMMA 5.14.1. *Notation as above, there exist morphisms $U' \to V'$ and $U'' \to V''$ of étale hypercoverings of X' and of X'', respectively, such that the induced map*

$$E_{pq}^2(X,\emptyset; U' \times U'') \to E_{pq}^2(X,\emptyset; V' \times V'')$$

is the zero map.

Proof. Let $\mathscr{C}'$ denote the category of étale hypercoverings of X', $\mathscr{C}''$ the category of étale hypercoverings of X'' and set

$$\mathscr{C} \overset{\text{def}}{=} \mathscr{C}' \times \mathscr{C}''.$$

For each nonnegative integer n, let $\mathscr{C}_n'$ denote the subcategory of $\mathscr{C}'$ consisting of the same objects but in which a morphism must be an n-refinement; let $\mathscr{C}_n''$ denote the analogously defined subcategory of $\mathscr{C}''$. For all nonnegative integers i and j, set

$$\mathscr{C}_{ij} \overset{\text{def}}{=} \mathscr{C}_i' \times \mathscr{C}_j''.$$

Consider the functors

$$F \overset{\text{def}}{=} (U',U'') \mapsto E_{pq}^2(X,\emptyset; U' \times U''): \mathscr{C} \to \mathscr{ABGP},$$

$$\Phi_{ij} \overset{\text{def}}{=} (U',U'') \mapsto H_q\big(U_i'^{\,\text{new}} \times U_j''^{\,\text{new}}\big): \mathscr{C}_{ij} \to \mathscr{ABGP},$$

the functor

$$G \overset{\text{def}}{=} (U',U'') \mapsto H_q(U' \boxtimes U'')$$

with source $\mathscr{C}$ and target the category of bisimplicial abelian groups and the functors Φ_{ij} associating to each object (U',U'') of $\mathscr{C}$ the E_{ij}^2-term of the Eilenberg-Zilber spectral sequence of $G(U',U'')$. Our task is to prove that F is null in the limit. The functor Φ_{ij} is null in the limit by the Refinement Lemma combined with the lemma of ¶5.10 and, by the lemma of ¶1.10, dominates the restriction of Γ_{ij} to the subcategory $\mathscr{C}_{ij}$ of $\mathscr{C}$. Hence, *a fortiori*, the functor Γ_{ij} is null in the limit. Since $F = H_p(/G)$,

$$\bigoplus_{i=0}^{p} \Gamma_{i,\,p-i} \text{ dominates } F.$$

Hence F is null in the limit. $\square$

LEMMA 5.14.2. *Notation as at the head of the paragraph, there exist morphisms $U' \to V'$ and $U'' \to V''$ of étale hypercoverings of X' and of X'', respectively, such that the induced map*

$$E^2_{pq}\big(Y, \emptyset; (U' \times U'') \underset{X}{\times} Y\big) \to E^2_{pq}\big(Y, \emptyset; (V' \times V'') \underset{X}{\times} Y\big)$$

is the zero map.

Proof. Let the categories $\mathscr{C}$ and $\mathscr{C}_{ij}$ be defined just as in the proof of the preceding lemma. Consider the functors

$$F \overset{\text{def}}{=} (U', U'') \mapsto E^2_{pq}\big(Y, \emptyset; (U' \times U'') \underset{X}{\times} Y\big) : \mathscr{C} \to \mathscr{ABGP},$$

$$\Phi_{ij} \overset{\text{def}}{=} (U', U'') \mapsto H_q\big((U_i'^{\text{new}} \times U_j''^{\text{new}}) \underset{X}{\times} Y\big) : \mathscr{C}_{ij} \to \mathscr{ABGP},$$

the functor

$$G \overset{\text{def}}{=} (U', U'') \mapsto H_q\big((U' \times U'') \underset{X}{\times} Y\big)$$

with source the category $\mathscr{C}$ and target the category of bisimplicial abelian groups and the functors Γ_{ij} associating to each object (U', U'') of $\mathscr{C}$ the E^2_{ij}-term of the Eilenberg-Zilber spectral sequence of $G(U', U'')$. By the Refinement Lemma combined with the lemma of ¶5.13, the functors Φ_{ij} are null in the limit. The remainder of the proof is formally identical to the proof of the preceding lemma. $\square$

LEMMA 5.14.3. *Notation as at the head of the paragraph, there exist morphisms $U' \to V'$ and $U'' \to V''$ of étale hypercoverings of X' and of X'', respectively, such that the induced map*

$$E^2_{pq}(X, Y; U' \times U'') \to E^2_{pq}(X, Y; V' \times V'')$$

is the zero map.

Proof. Let $\mathscr{C}$ be defined just as in the proof of Lemma 5.14.1. Since we have at our disposal an exact sequence

$$E^2_{pq}(X, \emptyset; U' \times U'') \to E^2_{pq}(X, Y; U' \times U'')$$

$$\to E^2_{p-1,\,q}\big(Y, \emptyset; (U' \times U'') \underset{X}{\times} Y\big)$$

functorial in objects (U', U'') of $\mathscr{C}$, the functor

$$(U', U'') \mapsto E_{pq}(X, \emptyset; U' \times U'')$$

$$\oplus\, E_{p-1, q}\big(Y, \emptyset; (U' \times U'') \underset{X}{\times} Y\big)\colon \mathscr{C} \to \mathscr{ABGP}$$

dominates the functor

$$(U', U'') \mapsto E_{pq}(X, Y; U' \times U'')\colon \mathscr{C} \to \mathscr{ABGP}. \qquad \square$$

5.15. Let (X, Y) be a good pair of $\mathbf{Q}$-schemes, p a nonnegative integer, q a positive integer, and V an étale hypercovering of X. We record here some special cases of the lemmas of the preceding paragraph for the sake of convenience.

LEMMA 5.15.1. *Notation as above, there exists a morphism $U \to V$ of étale hypercoverings of X such that the induced map*

$$E_{pq}^2(X, Y; U) \to E_{pq}^2(X, Y; V)$$

is the zero map.

LEMMA 5.15.2. *Notation as above, there exists a morphism $U \to V$ of étale hypercoverings of X such that the induced map*

$$E_{pq}^2\big(Y, \emptyset; U \underset{X}{\times} Y\big) \to E_{pq}^2\big(Y, \emptyset; V \underset{X}{\times} Y\big)$$

is the zero map.

§6. Completion of the proofs

6.1. *Proof of Theorem 1.* Let $\mathscr{C}$ denote the category of étale hypercoverings of X. Set

$$F \overset{\mathrm{def}}{=} U \mapsto \mathrm{kernel}\big(\varepsilon_n(X, Y; U)\big)\colon \mathscr{C} \to \mathscr{ABGP},$$

$$G \overset{\mathrm{def}}{=} U \mapsto \mathrm{cokernel}\big(\varepsilon_n(X, Y; U)\big)\colon \mathscr{C} \to \mathscr{ABGP}.$$

The functor

$$U \mapsto \underset{\substack{p+q=n \\ p \geqslant 0,\, q > 0}}{\bigoplus} E_{pq}^2(X, Y; U)\colon \mathscr{C} \to \mathscr{ABGP}$$

dominates F and the functor

$$U \mapsto \underset{\substack{p+q=n-1 \\ p \geqslant 0,\, q > 0}}{\bigoplus} E_{pq}^2(X, Y; U)\colon \mathscr{C} \to \mathscr{ABGP}$$

dominates G. By Lemma 5.15.1, the functors F and G are null in the limit. In particular, we can find a sequence $W^2 \to W^1 \to W^0$ of morphisms of étale hypercoverings of X such that the induced maps $F(W^1) \to F(W^0)$ and $G(W^2) \to G(W^1)$ are zero maps and, consequently, such that the map

$$(6.1.1) \qquad H_n(X, Y) \to \text{image}\big(E^2_{n0}(X, Y; W^2) \to E^2_{n0}(X, Y; W^1)\big)$$

induced by $\varepsilon_n(X, Y; W^1)$ is an isomorphism. Let V be an arbitrary étale hypercovering of X. It will suffice now to show that, once $H_n(X, Y)$ is equipped with the unique admissible $G(\mathbb{Q})$-action rendering (6.1.1) $G(\mathbb{Q})$-equivariant, the map $\varepsilon_n(X, Y; V)$ becomes $G(\mathbb{Q})$-equivariant. There certainly exist morphisms $U^0 \to V$ and $U^0 \to W^2$ of étale hypercoverings of X with a common source U^0, e.g. take U^0 to be the simplicial X-scheme $[n] \mapsto V_n \underset{X}{\times} W_n^2$.

Since F and G are null in the limit, there exists a sequence of morphisms $U^2 \to U^1 \to U^0$ of étale hypercoverings of X such that the induced maps $F(U^1) \to F(U^0)$ and $G(U^2) \to G(U^1)$ are zero maps. Consider the commutative diagram

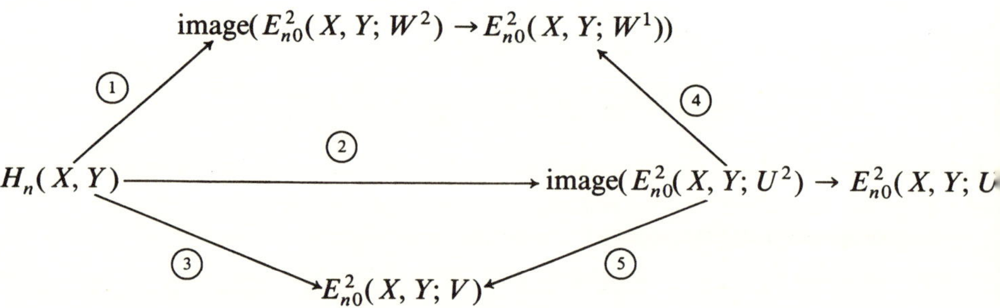

wherein ① is induced by $\varepsilon_n(X, Y; W^1)$, ② induced by $\varepsilon_n(X, Y; U^1)$, ③ equals $\varepsilon_n(X, Y; V)$, ④ is induced by the sequence of morphisms $U^2 \to U^1 \to U^0 \to W^2 \to W^1$ and ⑤ is induced by the sequence of morphisms $U^2 \to U^1 \to U^0 \to V$. Then ① and ② are isomorphisms, ④ and ⑤ $G(\mathbb{Q})$-equivariant and, therefore, ③ becomes $G(\mathbb{Q})$-equivariant if $H_n(X, Y)$ is equipped with the unique admissible $G(\mathbb{Q})$-action rendering ① $G(\mathbb{Q})$-equivariant. $\qquad \square$

6.2. *Proof of Corollary* 2. The "evident isomorphism" in question is none other than the homomorphism $\varepsilon_0(X, \emptyset; U)$, where U is taken to be the constant étale hypercovering $[n] \mapsto X$ of X. $\qquad \square$

6.3. *Proof of Corollary* 3. Let U be an arbitrary étale hypercovering of X. View X' and Y' as X-schemes via the morphism f. It suffices to exhibit a $G(\mathbb{Q})$-equivariant map $E^2_{n0}(X', Y'; U \underset{X}{\times} X') \to E^2_{n0}(X, Y; U)$ functorial in U

rendering the diagram

$$
\begin{array}{ccc}
H_n(X', Y') & \xrightarrow{\;\;f_*\;\;} & H_n(X, Y) \\
{\scriptstyle \varepsilon_n(X', Y';\, U \underset{X}{\times} X')} \Big\downarrow & & \Big\downarrow {\scriptstyle \varepsilon_n(X, Y;\, U)} \\
E_{n0}^2(X', Y';\, U \underset{X}{\times} X') & \longrightarrow & E_{n0}^2(X, Y;\, U)
\end{array}
$$

commutative. We leave the details to the reader. $\qquad\qquad\square$

6.4. *Proof of Corollary* 4. Let U be an arbitrary étale hypercovering of X. Consider the square diagram

$$
\begin{array}{ccc}
H_n(X, Y) & \xrightarrow{\;\;\partial\;\;} & H_{n-1}(Y, \emptyset) \\
{\scriptstyle \varepsilon_n(X, Y;\, U)} \Big\downarrow & & \Big\downarrow {\scriptstyle \varepsilon_{n-1}(Y, \emptyset;\, U \underset{X}{\times} Y)} \\
E_{n0}^2(X, Y;\, U) & \longrightarrow & E_{n-1,0}^2(Y, \emptyset;\, U \underset{X}{\times} Y)
\end{array}
$$

in which the lower horizontal arrow is the boundary map

$$
\partial: H_n\!\left(\bar{\pi}_0\!\left(U \underset{X}{\times} Y\right) \to \bar{\pi}_0(U)\right) \to H_{n-1}(\bar{\pi}_0(U))
$$

defined in ¶1.23. We claim that this square diagram commutes. We leave the task of verifying this claim to the reader; the observations of ¶1.23, ¶1.24 and ¶1.25 are intended to expedite this task. Now the lower horizontal arrow of the square diagram is clearly $G(\mathbf{Q})$-equivariant, and the vertical arrows so by definition. It remains only to exhibit an étale hypercovering U of X such that $\varepsilon_{n-1}(Y, \emptyset;\, U \underset{X}{\times} Y)$ is injective. Let $\mathscr{C}$ denote the category of étale hypercoverings of X. Set

$$
F \overset{\text{def}}{=} U \mapsto \operatorname{kernel}\!\left(\varepsilon_{n-1}\!\left(Y, \emptyset;\, U \underset{X}{\times} Y\right)\right): \mathscr{C} \to \mathscr{ABGP}.
$$

Since the functor

$$
U \mapsto \bigoplus_{\substack{p+q=n-1 \\ p\geqslant 0,\, q>0}} E_{pq}^2\!\left(Y, \emptyset;\, U \underset{X}{\times} Y\right): \mathscr{C} \to \mathscr{ABGP}
$$

dominates F, by Lemma 5.12.2 we conclude that F is null in the limit. Selecting an arbitrary étale hypercovering V of X and a morphism $U \to V$ of étale hypercoverings of X such that the induced map $F(U) \to F(V)$ is the zero map, we obtain an étale hypercovering U of X such that, as required, $\varepsilon_{n-1}(Y, \emptyset;\, U \underset{X}{\times} Y)$ is injective. $\qquad\qquad\square$

$$H_p(X',Y') \otimes H_q(X'',Y'') \xrightarrow{\quad\kappa\quad} H_{p+q}(X,Y)$$

$$H_p(\overline{S}(Y') \to \overline{S}(X')) \otimes H_q(\overline{S}(Y'') \to \overline{S}(X'')) \xrightarrow{\quad\kappa\quad} H_{p+q}(\overline{S}(Y) \to \overline{S}(X))$$

$$H_p(/\overline{S}(U' \underset{X'}{\times} Y') \to /\overline{S}(U')) \otimes H_q(/\overline{S}(U'' \underset{X''}{\times} Y'') \to /\overline{S}(U'')) \xrightarrow{\;\;\textcircled{1}\;\;} H_{p+q}(/\overline{S}((U' \times U'') \underset{X}{\times} Y) \to /\overline{S}(U' \times U''))$$

$$H_p(\overline{\pi}_0(U' \underset{X'}{\times} Y') \to \overline{\pi}_0(U')) \otimes H_q(\overline{\pi}_0(U'' \underset{X''}{\times} Y'') \to \overline{\pi}_0(U'')) \xrightarrow{\;\;\textcircled{2}\;\;} H_{p+q}(\overline{\pi}_0((U' \times U'') \underset{X}{\times} Y) \to \overline{\pi}_0(U' \times U''))$$

FIGURE 1

6.5. *Proof of Corollary* 5. Let U' be an arbitrary étale hypercovering of X', U'' an arbitrary étale hypercovering of X''. Consider Figure 1, in which the unlabelled arrows are the evident ones (cf. ¶3.3 above) and the arrows labelled ① and ② will be defined presently. By the observations of ¶2.5,

$$\left(\overline{\pi}_0\!\left(U' \underset{X'}{\times} Y'\right) \to \overline{\pi}_0(U') \right) \wedge \left(\overline{\pi}_0\!\left(U'' \underset{X''}{\times} Y''\right) \to \overline{\pi}_0(U'') \right)$$

$$= \overline{\pi}_0\!\left((U' \times U'') \underset{X}{\times} Y\right) \to \overline{\pi}_0(U' \times U''),$$

$$\left(/\overline{S}\!\left(U' \underset{X'}{\times} Y'\right) \to /\overline{S}(U') \right) \wedge \left(/\overline{S}\!\left(U'' \underset{X''}{\times} Y''\right) \to /\overline{S}(U'') \right)$$

$$= /\overline{S}\!\left((U' \times U'') \underset{X}{\times} Y\right) \to /\overline{S}(U' \times U'').$$

We define the arrows ① and ② to be the external product maps obtained by the procedure of ¶1.23. Having so defined ① and ②, Figure 1 is made commutative. Consequently, there exists a commutative square (obtained by collapsing Figure 1)

$$
\begin{array}{ccc}
H_p(X',Y') \otimes H_q(X'',Y'') & \xrightarrow{\;\kappa\;} & H_{p+q}(X,Y) \\[1ex]
{\scriptstyle \varepsilon_p(X',Y';U') \otimes \varepsilon_q(X'',Y'';U'')} \downarrow & & \downarrow {\scriptstyle \varepsilon_{p+q}(X,Y;U' \times U'')} \\[1ex]
E^2_{p0}(X',Y';U') \otimes E^2_{q0}(X'',Y'';U'') & \to & E^2_{p+q,0}(X,Y;U' \times U'')
\end{array}
$$

the lower horizontal arrow of which is $G(\mathbf{Q})$-equivariant and functorial in the pair (U', U''). It remains only to exhibit étale hypercoverings U' of X' and U'' of X'' such that the map $\varepsilon_{p+q}(X,Y;U' \times U'')$ is injective. Let $\mathscr{C}$ denote the product of the categories of étale hypercoverings of X' and étale hypercoverings of X''. Consider the functor

$$F \stackrel{\text{def}}{=} (U',U'') \mapsto \mathrm{kernel}\!\left(\varepsilon_{p+q}(X,Y;U' \times U'')\right) \colon \mathscr{C} \to \mathscr{ABGP}.$$

Then F is dominated by the functor

$$(U', U'') \mapsto \bigoplus_{\substack{i+j=p+q \\ i \geq 0,\, j > 0}} E^2_{ij}(X, Y; U' \times U''): \mathscr{C} \to \mathscr{ABGP}.$$

By Lemma 5.14.3, the functor F is null in the limit. Selecting arbitrary étale hypercoverings V' of X and V'' of X'', and morphisms $U' \to V'$ and $U'' \to V''$ of étale hypercoverings of X' and of X'', respectively, such that the induced map $F(U', U'') \to F(V', V'')$ is the zero map, we obtain a pair (U', U'') such that the map $\varepsilon_{p+q}(X, Y; U' \times U'')$ is injective, as required. $\qquad\square$

PART II. THE FERMAT k-MOTIVE AND THE CIRCULAR 1-MOTIVE

§7. The homology of the Fermat k-motive

7.1. For each nonnegative integer k, let $\mathbf{A}^{k+1}$ denote affine space of dimension $k + 1$ over $\mathbf{Q}$ and let $x_0, \ldots, x_k$ denote the standard coordinates on $\mathbf{A}^{k+1}$.

7.2. For each positive integer k, let U_k be the hypersurface in $\mathbf{A}^{k+1}$ defined by the equation

$$\sum_{j-0}^{k} x_j^N = 1,$$

and $Y_k \subseteq U_k$ the divisor defined by the equation

$$\prod_{j=0}^{k} x_j = 0.$$

Then (U_k, Y_k) is a good pair of $\mathbf{Q}$-schemes which we shall call the *Fermat k-motive*. Given a nonnegative integer $j \leq k$, we denote by Y_{kj} the divisor on U_k defined by the equation

$$x_j = 0.$$

7.3. For each nonnegative integer k, let G_k denote the free abelian group on the symbols

$$[\zeta]_i \quad (0 \leq i \leq k, \zeta \in \mu_N(\mathbf{C}))$$

modulo the relations

$$[\zeta]_i [\zeta']_i = [\zeta\zeta']_i \quad (\zeta, \zeta' \in \mu_N(\mathbf{C}), 0 \leq i \leq k).$$

Set

$$\Lambda_k \stackrel{\text{def}}{=} \text{the group ring of } G_k \text{ over } A.$$

We consider G_k to act upon $\mathbf{A}^{k+1} \otimes \mathbf{C}$ according to the rule

$$(x_j \otimes 1) \circ [\zeta]_i \overset{\text{def}}{=} x_j \otimes \zeta^{\delta_{ij}} \quad (0 \leqslant i, j \leqslant k, \zeta \in \mu_N(\mathbf{C})),$$

where δ_{ij} is the Kronecker delta. This action of G_k stabilizes $U_k \otimes \mathbf{C} \subseteq \mathbf{A}^{k+1} \otimes \mathbf{C}$ and each $Y_{kj} \otimes \mathbf{C} \subseteq U_k \otimes \mathbf{C}$, hence induces the structure of Λ_k-module in $H_*(U_k, Y_k)$.

7.4. For each nonnegative integer k, let S_k be the symmetric group on the set $\{0, \ldots, k\}$. We consider S_k to act upon $\mathbf{A}^{k+1}$ by exchange of coordinates thus:

$$x_i \circ s \overset{\text{def}}{=} x_{s^{-1}(i)} \quad (s \in S_k, 0 \leqslant i \leqslant k).$$

This action of S_k stabilizes $U_k \subseteq \mathbf{A}^{k+1}$ and exchanges the divisors Y_{kj} hence induces an action of S_k upon $H_*(U_k, Y_k)$.

7.5. Hence three groups, namely G_k, S_k and (as was explained in Part I) $G(\mathbf{Q})$ operate on the homology group $H_*(U_k, Y_k)$. Note that for all $s \in S_k$, $\zeta \in \mu_N(\mathbf{C})$, $0 \leqslant i \leqslant k$, $\sigma \in G(\mathbf{Q})$ and $c \in H_*(U_k, Y_k)$,

$$s[\zeta]_{s(i)}c = [\zeta]_i sc,$$

$$\sigma[\zeta]_i c = [\sigma\zeta]_i \sigma c,$$

$$\sigma s c = s \sigma c.$$

7.6. The family of Fermat k-motives has an *inductive structure* (cf. Katsura-Shioda [KS]). Let $sq_k \colon \mathbf{A}^2 \times \mathbf{A}^{k+1} \to \mathbf{A}^{k+2}$ be the map defined by the relation

$$x_j \circ sq_k \overset{\text{def}}{=} \left.\begin{array}{ll} x_0 \otimes 1 & \text{if } j = 0 \\ x_1 \otimes x_{j-1} & \text{if } 0 < j \leqslant k+1 \end{array}\right\},$$

where k is any positive integer. Then sq_k induces a morphism

$$(U_1 \times U_k, Y_1 \times U_k \cup U_1 \times Y_k) \to (U_{k+1}, Y_{k+1})$$

which we again denote by sq_k. Then for all ζ, $\zeta' \in \mu_N(\mathbf{C})$, $i \in \{0,1\}$, $0 \leqslant j \leqslant k$, $c \in H_*(U_1, Y_1)$ and $c' \in H_*(U_k, Y_k)$,

$$sq_{k*}\kappa([\zeta]_i c \otimes [\zeta']_j c')$$

$$= \left.\begin{array}{ll} [\zeta]_0 sq_{k*}\kappa(c \otimes c') & \text{if } i = 0 \\ [\zeta']_{j+1}\left(\displaystyle\prod_{i=1}^{k+1}[\zeta]_i\right) sq_{k*}\kappa(c \otimes c') & \text{if } i = 1 \end{array}\right\}.$$

7.7. For each positive integer k, let $\beta_k \in H_k(U_k, Y_k)$ denote the homology class to which the singular k-simplex

$$(t_0, \ldots, t_k) \mapsto \left(\sqrt[N]{t_0} \geqslant 0, \ldots, \sqrt[N]{t_k} \geqslant 0\right):$$

$$\left\{(t_0, \ldots, t_k) \in \mathbb{R}^{k+1}_{\geqslant 0} \,\middle|\, \sum t_j = 1\right\} \to U_k^{\text{top}} \subseteq \mathbb{C}^{k+1},$$

the jth face of which lies in Y_{kj}^{top}, belongs. We have

(7.7.1) $$\sigma\beta_k = (-1)^\sigma \beta_k \quad (\sigma \in S_k),$$

(7.7.2) $$sq_{k*}\kappa(\beta_1 \otimes \beta_k) = \beta_{k+1}.$$

The remainder of §7 is devoted to proving

THEOREM 6. *For each positive integer k, the homology group $H_*(U_k, Y_k)$ is concentrated in dimension k and is a rank one free module over Λ_k for which β_k is a basis.*

7.8. For each $k > 0$ and $j \geqslant 0$, set

$$P_k \stackrel{\text{def}}{=} \left\{(t_0, \ldots, t_k) \in \mathbb{R}^{k+1} \,\middle|\, \sum t_j = 1\right\},$$

$$Q_{kj} \stackrel{\text{def}}{=} \left\{(t_0, \ldots, t_k) \in P_k \,\middle|\, t_j = 0\right\},$$

$$Q_k \stackrel{\text{def}}{=} \left\{(t_0, \ldots, t_k) \in P_k \,\middle|\, \prod t_j = 0\right\}.$$

Given also $e = (e_0, \ldots, e_k) \in \{0, 1\}^{k+1}$, set

$$P_k^e \stackrel{\text{def}}{=} \left\{(t_0, \ldots, t_k) \in P_k \,\middle|\, (-1)^{e_i} t_i \geqslant 0, \text{ for } 0 \leqslant i \leqslant k\right\},$$

$$Q_{kj}^e \stackrel{\text{def}}{=} P_k^e \cap Q_{kj},$$

$$Q_k^e \stackrel{\text{def}}{=} P_k^e \cap Q_k.$$

Set

$$E \stackrel{\text{def}}{=} (z_0, \ldots, z_k) \mapsto \left(z_0^N, \ldots, z_k^N\right): \mathbb{C}^{k+1} \to \mathbb{C}^{k+1}.$$

We consider the following homology groups (defined as in ¶2.9 above).

$$W_k^e \stackrel{\text{def}}{=} H_*(P_k^e; Q_{k0}^e, \ldots, Q_{kk}^e),$$

$$M_k^e \stackrel{\text{def}}{=} H_*(E^{-1}(P_k^e); E^{-1}(Q_{k0}^e), \ldots, E^{-1}(Q_{kk}^e)),$$

$$M_k \stackrel{\text{def}}{=} H_*(E^{-1}(P_k); E^{-1}(Q_{k0}), \ldots, E^{-1}(Q_{kk})).$$

LEMMA. *The evident map $M_k^0 \to M_k$ is an isomorphism.*

Proof. Clearly

$$M_k^0 = \bigoplus_s s_* W_k^0,$$

the direct sum extended over all continuous sections s of E over P_k^0. Moreover,

$$M_k = \bigoplus_{(e,s)} s_* W_k^e,$$

the direct sum extended over all pairs (e, s) consisting of $e \in \{0,1\}^{k+1}$ and continuous sections s of E over P_k^e. But $W_k^e = 0$ when $e \neq 0$. $\qquad\square$

7.9. We turn to the construction of homotopies.

LEMMA. *There exists a continuous function $\Phi\colon \mathbf{C} \times [0,1] \to \mathbf{C}$ such that for all $z \in \mathbf{C}$ and $t \in [0,1]$,*

$$\Phi(z,0) = z, \ \Phi(z,t)^N = (1-t)z^N + t\,Re(z^N).$$

Proof. Let

$$\varphi\colon \{z \in \mathbf{C} \,|\, \mathrm{Im}(z) \geqslant 0\} \to \mathbf{C}$$

be the unique continuous function such that

$$\varphi(z)^N = z, \qquad t > 0 \Rightarrow \varphi(t) > 0.$$

Then φ maps the closed upper half-plane onto the closed sector bounded by the positive real axis and the ray issuing from the origin at the angle π/N. For each integer n, set

$$R_n \stackrel{\text{def}}{=} \{z \in \mathbf{C} \,|\, z = |z|\exp(\pi i n/N)\},$$

$$S_n \stackrel{\text{def}}{=} \{z \in \mathbf{C} \,|\, z = |z|\exp(\pi i(n+t)/N) \text{ for some } 0 \leqslant t \leqslant 1\},$$

and consider the map $\Phi_n\colon S_n \times [0,1] \to \mathbb{C}$ given by the rule

$$\Phi_n(z,t) \overset{\text{def}}{=} \left.\begin{array}{ll} \exp(\pi i n/N)\varphi\big((1-t)z^N + t\,\mathrm{Re}(z^N)\big) & \text{if } n \text{ even} \\ \exp(\pi i(n+1)/N)\bar{\varphi}\big((1-t)\bar{z}^N + t\,\mathrm{Re}(z^N)\big) & \text{if } n \text{ odd} \end{array}\right\}.$$

Then

$$\Phi_{n+2N} = \Phi_n, \quad \Phi_n\big|_{R_n \times [0,1]} = \Phi_{n+1}\big|_{R_n \times [0,1]}.$$

Thus the maps Φ_n patch together to given a map $\Phi\colon \mathbb{C} \times [0,1] \to \mathbb{C}$ which clearly has the desired properties. $\qquad\square$

7.10. *Proof of Theorem 6.* Consider the map

$$F \overset{\text{def}}{=} (z_0,\ldots,z_k,t) \mapsto (\Phi(z_0,t),\ldots,\Phi(z_k,t))\colon \mathbb{C}^{k+1} \times [0,1] \to \mathbb{C}^{k+1},$$

where Φ is the map provided by the lemma of ¶7.9. Then F is a retraction of U_k^{top} onto $E^{-1}(P_k)$ and, for $j = 0,\ldots,k$, a retraction of Y_{kj}^{top} onto $E^{-1}(Q_{kj})$. It follows that the evident map

$$M_k \to H_*(U_k, Y_k)$$

is an isomorphism. In view of the lemma of ¶7.8, we are done. $\qquad\square$

§8. Splitting of cocycles

8.1. Consider the diagram

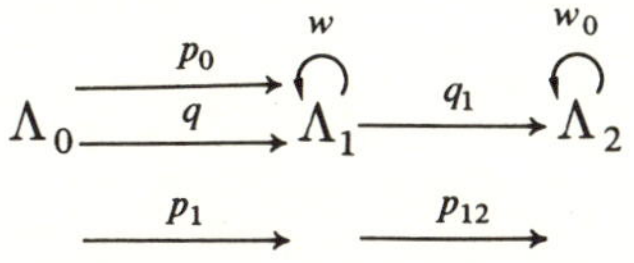

$$
\begin{array}{ccccc}
& & \overset{w}{\curvearrowright} & & \overset{w_0}{\curvearrowright} \\
& \overset{p_0}{\longrightarrow} & & \overset{q_1}{\longrightarrow} & \\
\Lambda_0 & \overset{q}{\longrightarrow} & \Lambda_1 & \longrightarrow & \Lambda_2 \\
& \underset{p_1}{\longrightarrow} & & \underset{p_{12}}{\longrightarrow} &
\end{array}
$$

of ring homomorphisms where the homomorphisms are uniquely determined by the following relations in which ζ is an arbitrary complex N^{th} root of unity:

$$p_0[\zeta]_0 = [\zeta]_0,\ q[\zeta]_0 = [\zeta]_0[\zeta]_1,\ p_1[\zeta]_0 = [\zeta]_1,$$

$$w[\zeta]_0 = [\zeta]_1,\ w[\zeta]_1 = [\zeta]_0,$$

$$q_1[\zeta]_0 = [\zeta]_0,\ q_1[\zeta]_1 = [\zeta]_1[\zeta]_2,$$

$$p_{12}[\zeta]_0 = [\zeta]_1,\ p_{12}[\zeta]_1 = [\zeta]_2,$$

$$w_0[\zeta]_0 = [\zeta]_1,\ w_0[\zeta]_1 = [\zeta]_0,\ w_0[\zeta]_2 = [\zeta]_2.$$

Given any abelian group-valued functor F of commutative rings with unit, set

$$K_F^1 \overset{\text{def}}{=} F(\Lambda_0),$$

$$K_F^2 \overset{\text{def}}{=} \text{kernel}(F(w) - 1: F(\Lambda_1) \to F(\Lambda_1)),$$

$$K_F^3 \overset{\text{def}}{=} F(\Lambda_2),$$

$$d_F' \overset{\text{def}}{=} F(p_0) + F(p_1) - F(q),$$

$$d_F'' \overset{\text{def}}{=} (F(w_0) - 1)(F(q_1) + F(p_{12})),$$

thereby defining a complex

$$K_F^{\boldsymbol{\cdot}}: \quad \cdots \to 0 \to K_F^1 \overset{d_F'}{\to} K_F^2 \overset{d_F''}{\to} K_F^3 \to 0 \to \cdots$$

concentrated in degrees 1, 2 and 3.

8.2. For each $\sigma \in G(\mathbf{Q})$, let $B_{\sigma, N} \in \Lambda_1^{\times}$ be defined by the relation

$$\sigma\beta_1 \overset{\text{def}}{=} B_{\sigma, N}\beta_1 \in H_1(U_1, Y_1).$$

The existence and uniqueness of $B_{\sigma, N}$ is guaranteed by Theorem 6.

THEOREM 7. $B_{\sigma, N} \in \text{kernel}(d_{\mathbf{G}_m}'') \subseteq K_{\mathbf{G}_m}^2$.

Proof.   Let $s \in S_1$ be the transposition $0 \mapsto 1, 1 \mapsto 0$. Then

$$B_{\sigma, N}\beta_1 = -\sigma s\beta_1$$

$$= -s\sigma\beta_1$$

$$= -(wB_{\sigma, N})s\beta_1$$

$$= (wB_{\sigma, N})\beta_1,$$

whence the conclusion

$$B_{\sigma, N} \in K_{\mathbf{G}_m}^2.$$

Let $s_0 \in S_2$ be the transposition $0 \mapsto 1, 1 \mapsto 0, 2 \mapsto 2$. Then

$$
\begin{aligned}
H_2(U_2, Y_2) \ni \sigma\beta_2 &= \sigma\big(sq_{1*}(\kappa(\beta_1 \otimes \beta_1))\big) \\
&= sq_{1*}(\kappa(\sigma\beta_1 \otimes \sigma\beta_1)) \\
&= sq_{1*}(\kappa(B_{\sigma,N}\beta_1 \otimes B_{\sigma,N}\beta_1)) \\
&= (q_1 B_{\sigma,N})(p_{12} B_{\sigma,N})\beta_2 \\
&= -\sigma s_0 \beta_2 \\
&= -s_0 \sigma \beta_2 \\
&= -s_0(q_1 B_{\sigma,N})(p_{12} B_{\sigma,N})\beta_2 \\
&= (w_0 q_1 B_{\sigma,N})(w_0 p_{12} B_{\sigma,N})\beta_2,
\end{aligned}
$$

whence the desired conclusion. $\qquad\square$

For each commutative ring R, let $\Omega(R)$ denote the module of Kähler differentials of F, $d\colon R \to \Omega(R)$ the universal derivation and $d\log\colon R^* \to \Omega(R)$ the logarithmic derivative map.

COROLLARY 8. $\quad d\log B_{\sigma,N} \in \mathrm{kernel}(d''_\Omega) \subseteq K^2_\Omega$.

8.3. For each $p \in \mathrm{Spec}(A)$, i.e., for each prime number p dividing N, set

$$
A_p \overset{\mathrm{def}}{=} \text{the local ring of } \mathrm{Spec}(A) \text{ at } p
$$

$$
= \mathbb{Z}/p^{\mathrm{ord}_p(N)}\mathbb{Z},
$$

$$
A_p^{sh} \overset{\mathrm{def}}{=} \text{the strict henselization of } A_p,
$$

$$
A^{sh} \overset{\mathrm{def}}{=} \prod_{p \in \mathrm{Spec}(A)} A_p^{sh}.
$$

Let W denote the ring-valued functor of rings $R \mapsto A^{sh} \otimes R$. For each $p \in \mathrm{Spec}(A)$, let $e_p \in A \subseteq A^{sh}$ denote the corresponding idempotent.

PROPOSITION 8.3.1. *The homomorphism*

$$
\delta \overset{\mathrm{def}}{=} \left((g_p)_{p \in \mathrm{Spec}(A)} \mapsto \sum_{p \in \mathrm{Spec}(A)} e_p g_p\right)\colon G_0^{\mathrm{Spec}(A)} \to H^1\big(K^{\cdot}_{\mathbf{G}_m \circ W}\big)
$$

is an isomorphism.

Proof. The homomorphism in question is clearly injective; we have only to establish surjectivity. Let $a \in H^1(K^{\cdot}_{\mathbf{G}_m \circ W})$ be given and write

$$a \overset{\text{def}}{=} \sum_{\zeta \in \mu_N(\mathbb{C})} a_\zeta [\zeta]_0, \quad a_\zeta \in A^{sh}.$$

Necessarily the coefficients a_ζ satisfy the relations

$$\sum_\zeta a_\zeta \in (A^{sh})^\times,$$

$$\left(\sum_\zeta a_\zeta [\zeta]_0 \right)\left(\sum_\zeta a_\zeta [\zeta]_1 \right) = \sum_\zeta a_\zeta [\zeta]_0 [\zeta]_1.$$

It follows that for all $\zeta, \zeta' \in \mu_N(\mathbb{C})$,

$$a_\zeta a_{\zeta'} = \left.\begin{matrix} a_\zeta & \text{if } \zeta = \zeta' \\ 0 & \text{if } \zeta \neq \zeta' \end{matrix}\right\}.$$

$$\sum_\zeta a_\zeta = 1 \in A^{sh}.$$

It follows in turn that for each $p \in \operatorname{Spec}(A)$ there exists a unique $\zeta \in \mu_N(\mathbb{C})$ such that $a_\zeta e_p = e_p$. This proves the proposition. $\qquad\square$

PROPOSITION 8.3.2. *Let*

$$(*) \qquad\qquad 1 \to \mathbf{G}_m \overset{i}{\to} E \overset{\pi}{\to} \mu_N \to 1$$

be an exact sequence of commutative affine group schemes over $\operatorname{Spec}(A^{sh})$. *Suppose that there exists a section* $\sigma\colon \mu_N \to E$ *of the morphism of* $\operatorname{Spec}(A^{sh})$-*schemes underlying* π. *Then* $(*)$ *splits.*

Proof. Let $E[N]$ denote the subgroup of E killed by N. Consider the sequence

$$(**) \qquad\qquad 1 \to \mu_N \overset{i'}{\to} E[N] \overset{\pi'}{\to} \mu_N \to 1$$

deduced from $(*)$. We claim that $(**)$ is an exact sequence of finite flat group schemes over $\operatorname{Spec}(A^{sh})$; in order to verify the claim it will be enough to show that π' is finite and faithfully flat. Let t be an indeterminate, set

$$R_0 \overset{\text{def}}{=} A^{sh}[t]/(t^N - 1),$$

and regard t as an R_0-valued point of μ_N. Then there exists unique $y \in R_0^\times$ such that

$$i(y) \stackrel{\text{def}}{=} \sigma(t)^N.$$

Let x be an indeterminate, set

$$R_1 \stackrel{\text{def}}{=} R_0[x]/(x^N - y),$$

and, regarding x as an R_1-valued point of $\mathbf{G}_m$, set

$$e \stackrel{\text{def}}{=} i(x^{-1})\sigma(t) \in E[N](R_1).$$

Then for all A^{sh}-algebras R, the map

$$f \mapsto E(f)(e) \colon \operatorname{Hom}_{A^{sh}-\text{alg.}}(R_1, R) \to E[N](R)$$

is a bijection, hence R_1 an isomorphic copy of the coordinate ring of E. As R_1 is a finite and faithfully flat R_0-algebra, π' is finite and faithfully flat. This establishes the claim. Consider the exact sequence

$$(***) \qquad\qquad 1 \to \mathbb{Z}/N\mathbb{Z} \stackrel{\hat{\pi}'}{\to} E[N]^{\widehat{}} \stackrel{\hat{i}'}{\to} \mathbb{Z}/N\mathbb{Z} \to 1$$

Cartier dual to (**). Since the groups on the left and on the right are *constant*, $E[N]^{\widehat{}}$ must be *étale* over $\operatorname{Spec}(A^{sh})$. Moreover, for each $p \in \operatorname{Spec}(A)$, $E[N]^{\widehat{}}$ must be *constant* over $\operatorname{Spec}(A_p^{sh}) \subseteq \operatorname{Spec}(A^{sh})$ and since $E[N]^{\widehat{}}$ is killed by N, (***) must *split* over $\operatorname{Spec}(A_p^{sh})$. Hence (***) splits over $\operatorname{Spec}(A^{sh})$, hence (**) splits, hence (*) splits. $\qquad\square$

THEOREM 9. $H^2(K_{\mathbf{G}_m \circ W}^{\cdot}) = \{1\}.$

Proof. Let ζ be a primitive complex N^{th} root of unity and let $b \in \operatorname{kernel}(d_{\mathbf{G}_m \circ W}'') \subseteq K_{\mathbf{G}_m \circ W}^2$ be given. Write

$$b \stackrel{\text{def}}{=} \sum_{i,j \in A} b_{ij}[\zeta^i]_0[\zeta^j]_1, \qquad b_{ij} \in A^{sh},$$

$$b^{-1} \stackrel{\text{def}}{=} \sum_{i,j \in A} b_{ij}'[\zeta^i]_0[\zeta^j]_1, \qquad b_{ij}' \in A^{sh},$$

$$(q_1 b)(p_{12} b) \stackrel{\text{def}}{=} \sum_{i,j,k \in A} b_{ijk}''[\zeta^i]_0[\zeta^j]_1[\zeta^k]_2, \qquad b_{ijk}'' \in A^{sh}.$$

Let R be an arbitrary A^{sh}-algebra. Given $\mu_1, \mu_2 \in \mu_N(R)$, note that

$$\left(\sum_{i,\,j \in A} b'_{ij}\mu_1^i\mu_2^j \right)\left(\sum_{i,\,j \in A} b_{ij}\mu_1^i\mu_2^j \right) = 1 \in R,$$

hence

$$\sum_{i,\,j \in A} b_{ij}\mu_1^i\mu_2^j \in R^x.$$

Set

$$E(R) \stackrel{\mathrm{def}}{=} R^\times \times \mu_N(R).$$

Given $(t_1, \mu_1), (t_2, \mu_2) \in E(R)$, set

$$(t_1, \mu_1) \stackrel{*}{_R} (t_2, \mu_2) \stackrel{\mathrm{def}}{=} \left(t_1 t_2 \left(\sum_{i,\,j \in A} b_{ij}\mu_1^i\mu_2^j \right), \mu_1\mu_2 \right),$$

thereby defining a composition law $\stackrel{*}{_R}\colon E(R) \times E(R) \to E(R)$ functorial in A^{sh}-algebras R. Now by hypothesis,

$$b_{ij} = b_{ji},$$

and hence $\stackrel{*}{_R}$ is *commutative*. Note that

$$(t_1, \mu_1) \stackrel{*}{_R} \left((t_2, \mu_2) \stackrel{*}{_R} (t_3, \mu_3) \right)$$

$$= \left(t_1 t_2 t_3 \left(\sum_{i,\,j,\,k \in A} b''_{ijk}\mu_1^i\mu_2^j\mu_3^k \right), \mu_1^i\mu_2^j\mu_3^k \right).$$

By hypothesis,

$$b''_{ijk} = b''_{jik}.$$

It follows that $\stackrel{*}{_R}$ is *associative*. Since for all $(t, \mu) \in E(R)$,

$$((t, \mu) \stackrel{*}{_R} (1, 1)) \stackrel{*}{_R} (1, 1) = \left(t \left(\sum_{i,\,j \in A} b_{ij}\mu^i \right)^2, \mu \right),$$

$$(t, \mu) \stackrel{*}{_R} ((1, 1) \stackrel{*}{_R} (1, 1)) = \left(t \left(\sum_{i,\,j \in A} b_{ij}\mu^i \right)\left(\sum_{i,\,j \in A} b_{ij} \right), \mu \right),$$

associativity implies that

$$\sum_{i,\,j\in A} b_{ij}\mu^i = \sum_{i,\,j\in A} b_{ij},$$

and that, in particular,

$$1_{E(R)} \stackrel{\text{def}}{=} \left(\left(\sum_{i,\,j\in A} b_{ij} \right)^{-1}, 1 \right) \in E(R)$$

is a neutral element for the composition law $\underset{R}{*}$. For all $(t,\mu) \in E(R)$,

$$(t,\mu)\left(t^{-1}\left(\sum_{i,\,j\in A} b_{ij} \right)^{-1}\left(\sum_{i,\,j\in A} b_{ij}\mu^i\mu^{-j} \right)^{-1}, \mu^{-1} \right) = 1_{E(R)}.$$

Hence every element of $E(R)$ has an inverse with respect to $\underset{R}{*}$. In short, $E(R)$ is a *commutative group* under the composition $\underset{R}{*}$ and the functor E of A^{sh}-algebras, equipped with the natural composition $*$, is an *affine group scheme* over A^{sh}. Set

$$i \stackrel{\text{def}}{=} t \mapsto (t,1)\colon \mathbf{G}_m \to E,$$

$$\pi \stackrel{\text{def}}{=} (t,\mu) \mapsto \mu\colon E \to \mu_N,$$

$$\sigma \stackrel{\text{def}}{=} \mu \mapsto (1,\mu)\colon \mu_N \to E.$$

Then

(8.3.3) $$1 \to \mathbf{G}_m \xrightarrow{i} E \xrightarrow{\pi} \mu_N \to 1$$

is an exact sequence of commutative affine group schemes over A^{sh} and $\sigma\colon \mu_N \to E$ is a scheme theoretic section of π. By Prop. 8.3.2, there exists a homomorphism $\gamma\colon \mu_N \to E$ splitting (8.3.3). Let x and y be indeterminates, set

$$R_0 \stackrel{\text{def}}{=} A^{sh}[x,y]/(x^N - 1, y^N - 1),$$

and regard x and y as R_0-valued points of μ_N. Then there exists for each $j \in A$ a unique coefficient $g_j \in A^{sh}$ such that

$$\gamma(x) = \left(\sum_{j\in A} g_j x^j, x \right) \in E(R_0).$$

Set

$$g \stackrel{\mathrm{def}}{=} \sum_{j \in A} g_j [\zeta^j]_0 \in A^{sh} \otimes \Lambda_0.$$

Then since

$$\sum_{j \in A} g_j x^j \in R_0^{\times}$$

it follows that

$$g \in \left(A^{sh} \otimes \Lambda_0 \right)^{\times} = K^1_{\mathbf{G}_m \circ W}.$$

From the identity

$$\gamma(x) \stackrel{*}{R_0} \gamma(y) = \gamma(xy)$$

follows the relation

$$\left(\sum_{j \in A} g_j x^j \right) \left(\sum_{j \in A} g_j y^j \right) = \left(\sum_{j \in A} g_j x^j y^j \right) \left(\sum_{j, k \in A} b_{jk} x^j y^k \right)$$

which in turn clearly implies

$$d'_{\mathbf{G}_m \circ W}(g) = b. \qquad \square$$

8.4. By Theorems 7 and 9, there exists for all $\sigma \in G(\mathbf{Q})$ unique

$$\Gamma_{\sigma, N} \in \left(A^{sh} \otimes \Lambda_0 \right)^{\times} / \delta \left(G_0^{\mathrm{Spec}(A)} \right),$$

where δ is as defined in Prop. 8.3.1, such that

$$B_{\sigma, N} \stackrel{\mathrm{def}}{=} \left(W(p_1) \Gamma_{\sigma, N} \right) \left(W(p_2) \Gamma_{\sigma, N} \right) / \left(W(q) \Gamma_{\sigma, N} \right),$$

where W is the functor defined in ¶8.3. The object $\Gamma_{\sigma, N}$ is a mod N analogue of the classical gamma function.

8.5. Set

$$\varepsilon \stackrel{\mathrm{def}}{=} \exp(2\pi i / N),$$

a primitive complex N^{th} root of unity. Note that for each nonnegative integer k, $\Omega(\Lambda_k)$ is a free Λ_k-module of rank $k + 1$, $\Omega(A^{sh} \otimes \Lambda_k)$ a free $A^{sh} \otimes \Lambda_k$-module of rank $k + 1$ and that the collection

$$\left\{ d \log[\varepsilon]_j \right\}_{j=0}^{k}$$

forms a basis of the former over Λ_k, and a basis of the latter over $A^{sh} \otimes \Lambda_k$. We therefore may and do identify $\Omega(\Lambda_k)$ with an A-submodule of $\Omega(A^{sh} \otimes \Lambda_k)$ under the map induced by the evident inclusion $\Lambda_k \hookrightarrow A^{sh} \otimes \Lambda_k$.

PROPOSITION 8.5.1. $H^1(K_{\dot\Omega}) = Ad \log[\varepsilon]_0$, $H^1(K_{\dot\Omega \circ W}) = A^{sh}d \log[\varepsilon]_0$.

Proof. It suffices to prove the latter equality. Let $\omega \in H^1(K_{\dot\Omega \circ W})$ be given and write

$$\omega \stackrel{\text{def}}{=} \sum_{j \in A} \omega_j [\varepsilon^j]_0 d \log[\varepsilon]_0, \ \omega_j \in A^{sh}.$$

Then, by definition,

$$\left(\sum_{j \in A} \omega_j [\varepsilon^j]_0 d \log[\varepsilon]_0 \right) + \left(\sum_{j \in A} \omega_j [\varepsilon^j]_1 d \log[\varepsilon]_1 \right)$$

$$= \sum_{j \in A} \omega_j [\varepsilon^j]_0 [\varepsilon^j]_1 (d \log[\varepsilon]_0 + d \log[\varepsilon]_1),$$

whence the conclusion that $\omega \in H^1(K_{\dot\Omega \circ W})$ if and only if

$$\sum_{j \in A} \omega_j [\varepsilon^j]_0 (1 - [\varepsilon^j]_1) = 0. \qquad \square$$

Let $h: \Omega(A^{sh} \otimes \Lambda_1) \to (A^{sh} \otimes \Lambda_0)$ be the unique A^{sh}-linear homomorphism such that for all $j, k \in A$,

$$h\left([\varepsilon^j]_0[\varepsilon^k]_1 d \log[\varepsilon]_0\right) \stackrel{\text{def}}{=} 0,$$

$$h\left([\varepsilon^j]_0[\varepsilon^k]_1 d \log[\varepsilon]_1\right) \stackrel{\text{def}}{=} - [\varepsilon^j]_0 d \log[\varepsilon]_0.$$

Note that

$$(8.5.2) \qquad h\left(\Omega(\Lambda_1)\right) \subseteq \Omega(\Lambda_0) \subseteq \Omega\left(A^{sh} \otimes \Lambda_0\right).$$

By an easy calculation, one finds that

$$(8.5.3) \qquad d'_{\dot\Omega \circ W} \circ h \circ d'_{\dot\Omega \circ W} = d'_{\dot\Omega \circ W}.$$

For all $\sigma \in G(\mathbf{Q})$, by definition of $\Gamma_{\sigma, N}$,

$$(8.5.4) \qquad d'_{\dot\Omega \circ W}(d \log \Gamma_{\sigma, N}) = d \log B_{\sigma, N}.$$

Combining (8.5.3) and (8.5.4), we obtain

PROPOSITION 8.5.5. $d'_{\Omega}(h(d \log B_{\sigma, N})) = d \log B_{\sigma, N}.$ □

Remark 8.5.6.   Since clearly

$$h(d \log B_{\sigma, N}) \in \Omega(\Lambda_0) \subseteq \Omega(A^{sh} \otimes \Lambda_0),$$

it follows that

$$d \log \Gamma_{\sigma, N} \in \left(\Omega(\Lambda_0) + A^{sh}d \log[\varepsilon]_0\right)/Ad \log[\varepsilon]_0.$$

§9. Approximating the Fermat *l*-motive by the circular 1-motive

9.1. Let $\tilde{V}$ denote the largest open subscheme of $\mathbf{A}^1$ on which $x_0^N - 1$ is invertible. Let $\lambda_0 \in H_1(V, \emptyset)$ denote the homology class to which the 1-simplex with vanishing boundary

$$t \mapsto 1 + \theta \exp(2\pi i t) \colon [0,1] \to \tilde{V}^{\mathrm{top}}\left(= \mathbf{C} \setminus \mu_N(\mathbf{C})\right)$$

belongs for all sufficiently small positive values of the parameter θ. The action of G_0 upon $\mathbf{A}^1 \times \mathrm{Spec}(\mathbf{C})$ defined in ¶2.1 induces an action of Λ_0 upon $H_1(\tilde{V}, \emptyset)$ relative which $H_1(\tilde{V}, \emptyset)$ is free of rank one and λ_0 a basis. Let $\lambda \colon \Omega(\Lambda_0) \xrightarrow{\sim} H_1(\tilde{V}, \emptyset)$ denote the unique isomorphism of Λ_0-modules such that

$$\lambda(d \log[\varepsilon]_0) \stackrel{\mathrm{def}}{=} \lambda_0.$$

9.2. Let V denote the largest open subscheme of $\mathbf{A}^1$ on which

$$\sum_{j=0}^{N-1} x_0^j$$

is invertible. Note that $\tilde{V} \subseteq V$; let $\tilde{v} \colon \tilde{V} \to V$ denote the inclusion. Note that the sequence

$$(9.2.1) \qquad\qquad 0 \to Ad \log[\varepsilon]_0 \to \Omega(\Lambda_0) \xrightarrow{\tilde{v}_* \circ \lambda} H_1(V, \emptyset) \to 0$$

is *exact*. Let Z denote the closed subscheme of V on which $x_0(1 - x_0)$ vanishes. We call (V, Z) the *circular 1-motive*. Let $\psi \in H_1(V, Z)$ denote the homology class to which the 1-simplex

$$t \mapsto t \colon [0,1] \to V^{\mathrm{top}}\left(= \mathbf{C} \setminus \left\{\zeta \in \mathbf{C} \mid \sum_{j=0}^{N-1} \zeta^j = 0\right\}\right)$$

with boundary in $Z^{top} = \{0,1\}$ belongs. Note that for all $\sigma \in G(\mathbf{Q})$

$$(\sigma - 1)\psi \in H_1(V,\emptyset) \subseteq H_1(V, Z),$$

since σ preserves the boundary map $\partial\colon H_1(V, Z) \to H_0(Z,\emptyset)$ and

$$(\sigma - 1)\,\partial\psi = 0;$$

let $\Psi_{\sigma, N} \in \Omega(\Lambda_0)/Ad\,\log[\varepsilon]_0$ be defined by the relation

$$\tilde{v}_*(\lambda(\Psi_{\sigma, N})) \overset{\text{def}}{=} (\sigma - 1)\psi.$$

Our goal in §9 is to prove

THEOREM 10.   *Assumptions and notation as above,*

$$d\,\log \Gamma_{\sigma, N} \equiv \Psi_{\sigma, N}\,\mathrm{mod}\,A^{sh}d\,\log[\varepsilon]_0.$$

9.3. Let $\tilde{U}$ denote the largest open subscheme of U_1 on which $x_0^N - 1$ is invertible, $\tilde{u}\colon \tilde{U} \to U_1$ the inclusion, and $\tilde{p}\colon \tilde{U} \to \tilde{V}$ the map induced by the coordinate $x_0\colon \mathbf{A}^2 \to \mathbf{A}$. Let $\rho\colon\{z \in \mathbf{C}\,||z| \leqslant 1\} \to \mathbf{C}$ denote the unique continuous function such that

$$\rho(0) = 1, \qquad \rho(z)^N = 1 - z.$$

For each $\zeta \in \mu_N(\mathbf{C})$, let $c(\zeta)$ denote the homology class in $H_1(\tilde{U},\emptyset)$ to which the 1-cycle

$$t \mapsto \big(\zeta\rho\big(\theta^N\exp(2\pi iNt)\big), \theta\exp(2\pi it)\big)\colon [0,1] \to \tilde{U}^{top} \subseteq \mathbf{C}^2$$

belongs for all sufficiently small positive values of the parameter θ. More intuitively, if less precisely, we may describe $c(\zeta)$ as the homology class in $H_1(\tilde{U},\emptyset)$ of any loop in $\tilde{U}^{top}$ winding once counterclockwise about and sufficiently close to the point $(\zeta, 0) \in U_1^{top} \subseteq \mathbf{C}^2$. The classes $c(\zeta)$ generate the kernel of $\tilde{u}_*\colon H_1(\tilde{U},\emptyset) \to H_*(U_1,\emptyset)$ and belong to the kernel of $\tilde{p}_*\colon H_1(\tilde{U},\emptyset) \to H_1(\tilde{V},\emptyset)$. Accordingly, there exists a unique homomorphism $\tilde{h}\colon H_1(U_1,\emptyset) \to H_1(\tilde{V},\emptyset)$ rendering the triangle

$$
\begin{array}{ccc}
 & H_1(\tilde{U},\emptyset) & \\
{}^{\tilde{u}_*}\swarrow & & \searrow^{p_*} \\
H_1(U_1,\emptyset) & \xrightarrow[\tilde{h}]{} & H_1(\tilde{V},\emptyset)
\end{array}
$$

commutative; clearly $\tilde{h}$ is $G(\mathbf{Q})$-equivariant.

9.4. The action of G_1 upon $\mathbf{A}^2 \times \text{Spec}(\mathbf{C})$ defined in ¶7.3 induces an action of G_1 upon $H_1(\tilde{U},\emptyset)$ relative to which $\tilde{u}_*$ is G_1-equivariant and such that for all $\zeta \in \mu_N(\mathbf{C})$,

$$\tilde{p}_* \circ [\zeta]_0 = [\zeta]_0 \circ \tilde{p}_*, \, \tilde{p}_* \circ [\zeta]_1 = \tilde{p}_*.$$

Consequently the map $\tilde{h}: H_1(U_1,\emptyset) \to H_1(\tilde{V},\emptyset)$ satisfies

$$(9.4.1) \qquad\qquad \tilde{h} \circ [\zeta]_0 = [\zeta]_0 \circ \tilde{h}, \, \tilde{h} \circ [\zeta]_1 = \tilde{h},$$

a relation analogous to the relation

$$(9.4.2) \qquad\qquad h \circ [\zeta]_0 = [\zeta]_0 \circ h, \, h \circ [\zeta]_1 = h$$

satisfied by $h: \Omega(\Lambda_1) \to \Omega(\Lambda_0)$, as an immediate consequence of the definition of h.

9.5. The homology class $(1 - [\varepsilon]_0)(1 - [\varepsilon]_1)\beta_1 \in H_1(U_1, Y_1)$ has vanishing boundary, hence belongs to $H_1(U_1,\emptyset)$. We shall calculate the image under $\tilde{h}$ of the class $(1 - [\varepsilon]_0)(1 - [\varepsilon]_1)\beta_1$. Let $\tilde{Y}$ denote the closed subscheme of $\tilde{U}$ on which x_0 vanishes. Let $\tilde{\beta} \in H_1(\tilde{U}, \tilde{Y})$ denote the homology class to which the 1-simplex

$$t \mapsto \left(\rho(\exp(2\pi i t)), \exp(2\pi i t/N)\right): [0,1] \to \tilde{U}^{\text{top}}$$

with boundary in $\tilde{Y}^{\text{top}} = \{0\} \times \mu_N(\mathbf{C}) \subseteq \mathbf{C}^2$ belongs. Now $\tilde{u}(\tilde{Y}) \subseteq Y_1$, hence u induces a homomorphism $\tilde{u}_*: H_*(\tilde{U}, \tilde{Y}) \to H_*(U_1, Y_1)$, and

$$\tilde{u}_*(\tilde{\beta}) = (1 - [\varepsilon]_1)\beta_1,$$

hence

$$\tilde{u}_*\left((1 - [\varepsilon]_0)\tilde{\beta}\right) = (1 - [\varepsilon]_0)(1 - [\varepsilon]_1)\beta_1.$$

Since $\partial(1 - [\varepsilon]_0)\tilde{\beta} = 0, (1 - [\varepsilon]_0)\tilde{\beta}$ belongs to $H_1(\tilde{U},\emptyset)$ and

$$\tilde{p}_*\left((1 - [\varepsilon]_0)\tilde{\beta}\right) = (1 - [\varepsilon]_0)\lambda_0 \in H_1(\tilde{V},\emptyset).$$

We conclude that

$$(9.5.1) \qquad\qquad \tilde{h}\left((1 - [\varepsilon]_0)(1 - [\varepsilon]_1)\beta_1\right) = (1 - [\varepsilon]_0)\lambda_0.$$

9.6. A homology class

$$c \stackrel{\text{def}}{=} \sum_{j, k \in A} c_{jk}[\varepsilon^j]_0[\varepsilon^k]_1\beta_1 \in H_1(U_1, Y_1), \qquad c_{jk} \in A,$$

satisfies

$$\partial c = 0$$

if and only if for all $k \in A$,

$$\sum_{j \in A} c_{jk} = 0 = \sum_{j \in A} c_{kj}.$$

Consequently,

$$H_1(U_1, \emptyset) = I_1 \beta_1 \subseteq H_1(U_1, Y_1),$$

where

$$I_1 \overset{\mathrm{def}}{=} (1 - [\varepsilon]_0)(1 - [\varepsilon]_1) \Lambda_1,$$

a principal ideal of Λ_1. In particular, it follows that relations (9.4.1) and (9.5.1) uniquely determine $\tilde{h}$.

9.7. We claim that for all $r \in I_1$,

$$(9.7.1) \qquad \lambda(h(dr)) = \tilde{h}(r\beta_1)$$

In order to verify the claim we calculate thus: For all $j, k \in A$,

$$\tilde{h}\big([\varepsilon^j]_0[\varepsilon^k]_1(1 - [\varepsilon]_0)(1 - [\varepsilon]_1)\beta_1\big) = [\varepsilon^j]_0(1 - [\varepsilon]_0)\lambda_0,$$

$$\lambda\big(h\big(d([\varepsilon^j]_0[\varepsilon^k]_1(1 - [\varepsilon]_0)(1 - [\varepsilon]_1))\big)\big)$$

$$= \lambda\big((-k[\varepsilon^j]_0 + k[\varepsilon^{j+1}]_0 + (k+1)[\varepsilon^j]_0 - (k+1)[\varepsilon^{j+1}]_0)d\log[\varepsilon]_0\big)$$

$$= [\varepsilon^j]_0(1 - [\varepsilon]_0)\lambda_0.$$

This establishes the claim. Suppose given $r, r' \in I_1$ satisfying

$$(1 + r)(1 + r') = 1.$$

Then

$$(9.7.2) \qquad h(d\log(1 + r)) = h(dr + r'dr) = h(dr),$$

since, by (9.4.2), h vanishes on $I_1 \Omega(\Lambda_1)$.

9.8. Let U' be the largest open subscheme of U_1 on which

$$\sum_{j=0}^{N-1} x_0^j$$

is invertible, let u': $U' \to U_1$ denote the inclusion, and let p': $U' \to V$ denote the map induced by the coordinate x_0: $\mathbf{A}^2 \to \mathbf{A}^1$. Let Y' denote the unique reduced closed subscheme of U_1 such that

$$Y'(\mathbf{C}) \stackrel{\text{def}}{=} \{(1,0),(0,1)\} \subseteq U_1(\mathbf{C}) \subseteq \mathbf{C}^2,$$

a closed subscheme of U' as well. Clearly $p'(Y') \subseteq Z$. By an evident modification of the argument presented in ¶4.3, there exists a unique homomorphism a: $H_1(U_1, Y') \to H_1(V, Z)$ rendering the diagram

$$
\begin{array}{ccc}
H_1(U', Y') & \xrightarrow{\;u'_*\;} & H_1(U_1, Y') \\
{\scriptstyle p'_*}\big\downarrow & \swarrow{\scriptstyle a} & \\
H_1(V, Z) & &
\end{array}
$$

commutative. It follows that the diagram

(9.8.1)
$$
\begin{array}{ccc}
 & & H_1(U_1, Y') \supseteq H_1(U_1,\varnothing) \\
 & {\scriptstyle a}\swarrow & \big\downarrow{\scriptstyle \tilde{h}} \\
H_1(V, Z) \supseteq H_1(V,\varnothing) & \xleftarrow{\;\tilde{v}_*\;} & H_1(\tilde{V},\varnothing)
\end{array}
$$

commutes.

9.9. Fix $\sigma \in G(\mathbf{Q})$ arbitrarily. Since the boundary of β_1 lies in Y',

$$\beta_1 \in H_1(U_1, Y') \subseteq H_1(U_1, Y_1).$$

Clearly,

$$a\beta_1 = \psi.$$

Since a is $G(\mathbf{Q})$-equivariant,

$$(\sigma - 1)\psi = a((\sigma - 1)\beta_1) = a((B_{\sigma, N} - 1)\beta_1).$$

Since

$$(\sigma - 1)\,\partial\beta_1 = 0,$$

it follows that

$$(\sigma - 1)\beta_1 \in H_1(U_1,\varnothing) \subseteq H_1(U_1, Y'),\quad B_{\sigma, N} - 1 \subseteq I_1 \subseteq \Lambda_1.$$

Since (9.8.1) commutes,

$$(\sigma - 1)\psi = \tilde{v}_*\big(\tilde{h}((B_{\sigma, N} - 1)\beta_1)\big).$$

By (9.7.1),

$$(\sigma - 1)\psi = \tilde{v}_*\big(\lambda\big(h\big(d\big(B_{\sigma, N} - 1\big)\big)\big)\big).$$

By (9.7.2),

$$(\sigma - 1)\psi = \tilde{v}_*\big(\lambda\big(h\big(d \log B_{\sigma, N}\big)\big)\big).$$

In view of Prop. 8.5.5 and the definitions, the proof of Theorem 10 is now complete. $\qquad\square$

§10. The proof of Theorem 0

10.1. First we consider the following general situation: Let X be a smooth quasi-projective curve over $\mathbb{Q}$. Let x_0 and x_1 be distinct $\mathbb{Q}$-rational points of X. Set

$$P \stackrel{\text{def}}{=} \{x_0, x_1\}.$$

Let E be an extension of an abelian variety by a torus, defined over $\mathbb{Q}$. Let $\Phi\colon X \to E$ be a morphism of $\mathbb{Q}$-schemes such that the induced map $\Phi_*\colon H_1(X, \emptyset) \to H_1(E, \emptyset)$ is an isomorphism. Let $E[N]$ denote the kernel of multiplication by N in E.

LEMMA. *Assumptions and notation as above, the $G(\mathbb{Q})$-modules*

$$E[N](\overline{\mathbb{Q}})$$

and

$$H_1(X, \emptyset)$$

are isomorphic. Moreover, the $G(\mathbb{Q})$-sets

$$\{c \in H_1(X, P)\,|\,\partial c = 1 \otimes x_1 - 1 \otimes x_0 \in A \otimes P = H_0(P, \emptyset)\}$$

and

$$\{e \in E(\overline{\mathbb{Q}})\,|\,Ne = \Phi(x_1) - \Phi(x_0)\}$$

are isomorphic.

Proof. Given any E-scheme Q, let Q' denote the E-scheme deduced by base-change relative to the map $e \mapsto Ne\colon E \to E$. Regard X as an E-scheme via

Φ. Let $\varphi\colon \mathbb{Z} \otimes P \to E(\mathbb{Q})$ denote the $\mathbb{Z}$-linear extension of the structure map $P \to E$, $\varphi'\colon \mathbb{Z} \otimes P'(\overline{\mathbb{Q}}) \to E(\overline{\mathbb{Q}})$ the $\mathbb{Z}$-linear extension of the structure map $P' \to E$. Set

$$I \overset{\text{def}}{=} \{(N\delta, \varphi(\delta)) \in d^{\circ}(\mathbb{Z} \otimes P) \times E(\mathbb{Q}) \mid \delta \in d^{\circ}(\mathbb{Z} \otimes P)\}.$$

$$M \overset{\text{def}}{=} \{(\delta, e)) \in d^{\circ}(\mathbb{Z} \otimes P) \times E(\overline{\mathbb{Q}}) \mid Ne = \varphi(\delta)\}/I.$$

Then M is a finite $G(\mathbb{Q})$-module killed by N. Let

$$\alpha\colon d^{\circ}\left(A \otimes P'(\overline{Q})\right) \to M$$

be the unique homomorphism such that for all $\delta' \in d^{\circ}(\mathbb{Z} \otimes P'(\overline{\mathbb{Q}}))$ and $\delta \in d^{\circ}(\mathbb{Z} \otimes P)$ such that $\delta' \mapsto \delta$ under the map induced by the projection $P' \to P$,

$$\alpha(\delta' \bmod N) = ((\delta, \varphi'(\delta')) \bmod I).$$

Note that α is surjective and $G(\mathbb{Q})$-equivariant. Now the induced map $\Phi_{*}\colon H_1(X,\emptyset) \to H_1(E,\emptyset)$ is an isomorphism by hypothesis, hence the induced map

$$H_1(X',\emptyset) \to H_1(X,\emptyset)$$

the zero map. Accordingly, there exists a unique homomorphism

$$\gamma\colon d^{\circ}\left(A \otimes P'(\overline{\mathbb{Q}})\right) \to H_1(X, P)$$

rendering the diagram

$$\text{kernel}(H_0(P',\emptyset) \to H_0(X',\emptyset)) = d^{\circ}(A \otimes P'(\overline{\mathbb{Q}}))$$

$$\partial \uparrow \qquad\qquad\qquad\qquad\qquad \downarrow \gamma$$

$$H_1(X', P') \longrightarrow H_1(X, P)$$

commutative. Note that γ is surjective and $G(\mathbb{Q})$-equivariant. It is not difficult to show that

$$\text{kernel}(\alpha) = \text{kernel}(\gamma).$$

Therefore we have at our disposal an isomorphism $H_1(X, P) \overset{\sim}{\to} M$ of $G(\mathbb{Q})$-modules from which the desired isomorphisms of $G(\mathbb{Q})$-modules and $G(\mathbb{Q})$-sets can readily be deduced. $\square$

10.2. We now place ourselves in the situation of Theorem 0.

LEMMA. *Let $\sigma \in G(\mathbb{Q})$ be given. Then σ fixes every N^{th} root of b in $S(\overline{\mathbb{Q}})$ if and only if σ fixes the Nth roots of unity and $B_{\sigma, N} = 1$.*

Proof. ($\Rightarrow$) Let J denote the abelian variety of which S is an extension by a torus. Then σ fixes every N-torsion element of $J(\overline{\mathbf{Q}})$. By virtue of the existence and nondegeneracy of the Weil pairing, σ must fix each Nth root of unity in $\overline{\mathbf{Q}}$. (This argument breaks down when $\dim J = 0$, but that only happens when $N = 1$ or $N = 2$ in which case we already have N^{th} roots of unity in $\mathbf{Q}$.) By the lemma of ¶10.1,

$$\sigma\beta_1 = \beta_1 \in H_1(U_1, Y_1),$$

hence

$$B_{\sigma, N} = 1.$$

($\Leftarrow$) We have, by hypothesis,

$$\sigma\beta_1 = \beta_1 \in H_1(U_1, Y_1).$$

Since, by hypothesis, σ fixes the Nth roots of unity, σ operates Λ_1-linearly in $H_1(U_1, Y_1)$ and hence fixes *every* element of $H_1(U_1, Y_1)$. Invoking the lemma of ¶10.1 a second time, we conclude that σ fixes every N^{th} root of b in $S(\overline{\mathbf{Q}})$. □

10.3. Let T denote the torus over $\mathbf{Q}$ which, as a group-valued functor of $\mathbf{Q}$-algebras R is given by the rule

$$R \mapsto \left\{ t: \mu_N(\overline{\mathbf{Q}}) \setminus \{1\} \to (R \otimes \overline{\mathbf{Q}})^{\times} | t \text{ is } G(\mathbf{Q})\text{-equivariant.} \right\}.$$

Let $f: V \to T$ denote the map which on R-valued points is given by the rule

$$\begin{array}{c} R \\ \cup\mathsf{I} \\ v \mapsto (\zeta \mapsto 1 - \zeta^{-1}v): \ V(R) \to T(R). \end{array}$$

Set

$$u \overset{\text{def}}{=} f(1)/f(0) = (\zeta \mapsto 1 - \zeta^{-1}) \in T(\mathbf{Q}).$$

Then one can easily check that the field generated by the Nth roots of u in $T(\overline{\mathbf{Q}})$ is precisely that generated by the roots in $\overline{\mathbf{Q}}$ of the equation

$$1 - (1 - x^N)^N = 0$$

and that, moreover, the induced map $f_*: H_1(V, \emptyset) \to H_1(T, \emptyset)$ is an isomorphism.

LEMMA.   *Let $\sigma \in G(\mathbf{Q})$ be given. Then σ fixes every N^{th} root of u in $T(\overline{\mathbf{Q}})$ if and only if σ fixes every N^{th} root of unity and $\Psi_{\sigma, N} = 0$.*

Proof.   ($\Rightarrow$) By the lemma of ¶10.1,

$$\sigma\psi = \psi \in H_1(V, Z),$$

hence

$$\Psi_{\sigma, N} = 0.$$

Clearly σ fixes every N^{th} root of unity. ($\Leftarrow$) By hypothesis,

$$\sigma\psi = \psi \in H_1(V, Z).$$

Invoking the first part of the lemma of ¶10.1, σ fixes every element of $H_1(V, \emptyset)$ since the N-torsion of $T(\overline{\mathbf{Q}})$ is rational over the field of Nth roots of unity. Since

$$H_1(V, Z) = A\psi + H_1(V, \emptyset),$$

it follows that σ fixes every element of $H_1(V, Z)$. Invoking the second part of the lemma of ¶10.1, it follows that σ fixes every N^{th} root of u in $T(\overline{\mathbf{Q}})$. $\qquad\square$

10.4. By Theorem 10, given any $\sigma \in G(\mathbf{Q})$,

$$B_{\sigma, N} = 1 \Rightarrow d \log B_{\sigma, N} = 0 \Rightarrow \Psi_{\sigma, N} = 0.$$

By the lemmas of ¶10.2 and ¶10.3, we have already established the containment of fields asserted in Theorem 0.

10.5. We now suppose that N is prime and turn to the proof of the equality of fields asserted in Theorem 0. Let $\sigma \in G(\mathbf{Q})$ fix all the roots of the equation $1 - (1 - x^N)^N = 0$, or, equivalently, fix the Nth roots of unity and satisfy $\Psi_{\sigma, N} = 0$. Then, by Theorem 10, $d \log B_{\sigma, N} = 0$, or, equivalently,

$$(10.5.1) \qquad\qquad\qquad\qquad dB_{\sigma, N} = 0.$$

Now it has already been observed in ¶9.6 that

$$(10.5.2) \qquad\qquad\qquad B_{\sigma, N} - 1 \in (1 - [\varepsilon]_0)(1 - [\varepsilon]_1)\Lambda_1.$$

Write

$$B_{\sigma, N} - 1 \overset{\text{def}}{=} \sum_{i, j \in A} b_{ij}[\varepsilon^i]_0[\varepsilon^j]_1, \qquad b_{ij} \in A.$$

Then for all $i, j \in A$, by (10.5.1),

$$(10.5.3) \qquad\qquad\qquad\qquad ib_{ij} = jb_{ij} = 0.$$

For all $i \in A$, by (10.5.2),

$$(10.5.4) \qquad\qquad\qquad \sum_{j \in A} b_{ij} = \sum_{j \in A} b_{ji} = 0.$$

Together $(10.5.3, 4)$ imply that for all $i, j \in A$,

$$b_{ij} = 0.$$

This concludes the proof of Theorem 0 and the paper. $\square$

REFERENCES

[A1]   G. ANDERSON, *Cyclotomy and an extension of the Taniyama group*, Comp. Math. **57** (1986), 153–217.

[A2]   ______ , The hyperadelic gamma function: a précis, to appear in Advanced Studies in Pure Math., Vol. 12.

[AM]   M. ARTIN AND B. MAZUR, *Etale Homotopy*, Lecture Notes in Math. **100** (1969), Springer, New York.

[G]   R. GREENBERG, *On the jacobian variety of some algebraic curves*, Comp. Math. **42** (1981), 345–359.

[GAGA]   J.-P. SERRE, *Géométrie algébrique et géometrie analytique*, Ann. Inst. Fourier **6** (1956), 1–42.

[GR]   H. GRAUERT AND R. REMMERT, *Komplex Räume*, Math. Ann. **136** (1958), 245–318.

[Ha]   R. HARTSHORNE, *Algebraic Geometry*, Graduate Texts in Math. **50**, Springer, New York 1977.

[I]   Y. IHARA, *Profinite braid groups, Galois representations, and complex multiplications*, Ann. of Math. **123** (1986), 43–106.

[IKY]   Y. IHARA, M. KANEKO AND A. YUKINARI, *On some properties of the universal power series for Jacobi sums*, to appear in Advanced Studies in Pure Math., Vol. **12**.

[KS]   T. KATSURA AND T. SHIODA, *On Fermat varieties*, Tohoku Math. J. **31** (1979), 97–115.

[M]   H. MIKI, *On the ℓ-adic expansion of certain Gauss sums and its applications*, to appear in Advanced Studies in Pure Math., Vol. **12**.

[SGA 1]   A. GROTHENDIECK ET AL., *Séminaire de Géométrie Algébrique du Bois Marie 1960/61*, Lecture Notes in Math. **224**, Springer, New York 1971.

[SGA 4]   ______ , *Théorie des Topos et Cohomologie Etale des Schémas*, Lecture Notes in Math. **269, 270, 305**, Springer, New York 1972–73.

SCHOOL OF MATHEMATICS, UNIVERSITY OF MINNESOTA, MINNEAPOLIS, MINNESOTA 55455

Vol. 54, No. 2   DUKE MATHEMATICAL JOURNAL ©   1987

A NEW INTERPRETATION OF GELFAND–TZETLIN BASES

I. V. CHEREDNIK

To Yuri Ivanovich Manin on his 50th birthday

Introduction. In this paper we determine q-analogues of Gelfand-Tzetlin bases for a q-deformation $\mathcal{D}_N^q$ of the universal enveloping algebra $\mathcal{U}(\widetilde{\mathfrak{gl}}_N)$ for the Lie algebra $\widetilde{\mathfrak{gl}}_N = \mathfrak{gl}_N \otimes_{\mathbb{C}} \mathbb{C}[[x]]$. The definition of $\mathcal{D}_N^q$ (and its generalization) was given by V. G. Drinfeld and M. Jimbo (see [1, 2]). We introduce $\mathcal{D}_N^q$ (§1) in a somewhat different way by means of R-matrix technique, developed by L. D. Faddeev and his collaborators (see e.g. [3]). This technique was applied to construct first some "elliptic" deformations of $\mathcal{U}(\mathfrak{gl}_2)$ [4], $\mathcal{U}(\mathfrak{gl}_N)$ [5] and then to introduce the "elliptic" R-algebras of level n [6]. The latter are deformations of $\widetilde{\mathfrak{gl}}_N \mathrm{mod}(x^n)$ and also natural generalizations of $\mathcal{D}_N^q$ as $n \to \infty$.

We use the results and methods of [6, 7] in this paper but can prove our "branching" theorems only in the special case of $\mathcal{D}_N^q$ since the branching properties for the elliptic R-algebras are of a more complicated nature than for $\mathcal{D}_N^q$. These properties are analoques of the well-known Young–Gelfand–Tzetlin rules for decomposing irreducible finite-dimensional representations of $\mathfrak{gl}_N$ or S_n under the action of $\mathfrak{gl}_K \subset \mathfrak{gl}_N$ or $S_k \subset S_n$ and are most important in our paper. It was shown recently by G. I. Olshansky that the Yangians, which were introduced by V. G. Drinfeld and are the limits of $\mathcal{D}_N^q$ as $q \to 1$, play an essential role in calculating the centralizers of $\mathcal{U}(\mathfrak{gl}_K)$ in $\mathcal{U}(\mathfrak{gl}_N)$. For S_n the degenerated Hecke algebras from [8] arise when we find the centralizers of $\mathbb{C}[S_k]$ in $\mathbb{C}[S_n]$ (cf. [6, 9]). We develop these results, generalizing the classic branching properties and extending them to $\mathcal{D}_N^q$ and affine Hecke algebras in §2, 3.

We apply the structural theorems for $\mathcal{D}_N^q$ obtained in this way to their irreducible representations, connected with the so-called skew Young diagrams. This class includes q-analogues of finite-dimensional irreducible representations of $\mathfrak{gl}_N$, constructed independently in [10, 11] for $\mathcal{D}_N^q$ of level $= 1$ and in [6, 7] for its elliptic generalization. As we prove (for sufficiently large N) at the end of the paper, only representations of the above type have the classic branching properties. We do not detail here the structure of the considered representations (e.g., do not describe the action of generators of $\mathcal{D}_N^q$ on Gelfand–Tzetlin bases). For further information we refer to [11] (where the latter problem was solved for level $= 1$) and the author's paper on the second Weyl character formula to be published soon (see also [6, 9] for Hecke algebras).

Received October 17, 1986.

Our technique of studying the representations of $\mathscr{D}_N^q$ (or more general R-algebras) has its origin in the "fusion process" by P. P. Kilish, N. Yu. Reshetikhin, E. K. Sklyanin and V. E. Korepin and in the notion of the quantum determinant (see [12]). We refer to accounts [12, 15], and recent papers by Drinfeld and Jimbo papers (see also [6] and, last but not least, new works of Leningrad's School of Faddeev) for the various results on R-matrices. The fusion process is similar to some constructive methods in the theory of representations of Hecke algebras by A. V. Zelevinsky and J. D. Rogawski (see [13]) since there exists a generalization of Weil duality for Yangians [8] and for $\mathscr{D}_N^q$ (see §3 below) with Hecke algebras instead of $\mathbb{C}[S_n]$. Another remarkable property of $\mathscr{D}_N^q$ is as follows.

We can obtain from $\mathscr{D}_N^q$ an algebra d_N of commutative polynomials with some Poisson bracket by some passage to limit. There is a maximal commutative subalgebra in $\mathscr{D}_N^q$, diagonalized by Gelfand-Tzetlin bases in any "good" representations. Its limit is a maximal involutive subalgebra $a_N \subset d_N$ relative to the Poisson bracket. Sufficiently general fibers of the morphism Specm $d_N \to$ Specm a_N of the corresponding affine (ind-) varieties are some open subsets of Jacobians of appropriate algebraic curves. These fibers are also orbits of the collection of flows with elements of a_N as Hamiltonians. These two statements (and their variants—see [14, 15]) generalize some classic results on integration of the equation of motion of the "rigid body" and many results on "finite-zoned" integration of soliton equations. Thus we have new and surprising connections between algebraic geometry and representation theory.

In studying the algebraic aspects of the Hamiltonian formalism and the "finite-zoned" integration, [16] and further works by Yu. I. Manin on these problems play an important role.

The author thanks G. I. Olshansky for useful discussions and I. M. Gelfand for his kind interest in this work.

§1. Trigonometric R-algebras.

Let I^{lm} be the matrices with matrix elements $(I^{lm})_{ij} = \delta_{il}\delta_{jm}$ which form a basis in the algebra M_N of all $N \times N$-matrices over $\mathbb{C}$ (δ_{ij} is the Kronecker symbol). One defines inclusions of M_N, $M_N^{\otimes 2} = M_N \otimes M_N$ into $M_N^{\otimes n}$ by formulas ${}^iA = I^{\otimes(i-1)} \otimes A \otimes I^{\otimes(n-i)}$,

$$ {}^{ij}(A \otimes B) = {}^iA\,{}^jB \quad \text{for } A, B \in M_N, 1 \leqslant i \neq j \leqslant n \quad \text{and} \quad I = \sum_{l=1}^{N} I^{ll} $$

being the matrix unit. Let us denote by $\mathscr{L}_N(u)$ the vector space of all functions

$$ L(x) = \sum_{\substack{0 \leqslant s \leqslant n \\ 1 \leqslant i,\,j \leqslant N}} x^s \tilde{E}_{ij}^s I^{ji} \prod_{i=1}^{n} (x - q^{u_i})^{-1}, $$

where $u = (u_i) \in \mathbb{C}^n$, $\tilde{E}_{ij}^s \in \mathbb{C}$ and $\tilde{E}_{ij}^n = 0 = \tilde{E}_{ji}^0$ for $i < j$. Here and below the parameter $q^{1/2} \in \mathbb{C} \setminus \{1\}$ is "sufficiently close" to $q^{1/2} = 1$ (in fact further results are valid for almost all q).

Let us introduce

$$R(x) = R_N(x; q)$$

$$= \sum_{i=1}^{N} I^{ii} \otimes \left(I + (q^{1/2} - 1)I^{ii}\right) + \frac{q^{1/2} - q^{-1/2}}{x - 1} \sum_{i \geqslant j} I^{ij} \otimes I^{ji}$$

$$+ \frac{q^{1/2} - q^{-1/2}}{1 - x^{-1}} \sum_{i < j} I^{ij} \otimes I^{ji}$$

$$\in \mathscr{L}(0) \otimes_{\mathbf{C}} M_N$$

(see [10, 15]). We define the quotient-algebra $\tilde{\mathscr{D}}_N^q(u)$ of the tensor algebra $T(\mathscr{L}_N^*(u))$ for $\mathscr{L}_N^*(u) = \mathrm{Hom}_{\mathbf{C}}(\mathscr{L}_N(u), \mathbf{C}) = \oplus \mathbf{C}\,\tilde{E}_{ij}^s$ by Yang-Baxter-Faddeev relations

$$^{12}R\left(x_1 x_2^{-1}\right)^{1}L(x_1)^{2}L(x_2) - {}^{2}L(x_2)^{1}L(x_1)^{12}R\left(x_1 x_2^{-1}\right) = 0. \tag{1}$$

To be more exact $\tilde{\mathscr{D}}_N^q(u) = T(\mathscr{L}_N^*(u))/\tilde{K}_N^q(u)$, where $\tilde{K}_N^q(u)$ is the two-sided ideal generated by all matrix elements of the left-hand side of (1) for arbitrary values of $x_1, x_2 \in \mathbf{C}$. The subalgebra $\mathscr{D}_N^q(u)$ of a proper localization of $\tilde{\mathscr{D}}_N^q(u)$ generated by $\tilde{E}_{ij}^s(\tilde{E}_{ii}^n)^{-1} \overset{\text{def}}{=} E_{ij}^s$ is called the R-**algebra** (of level n), corresponding to $R(x; q)$. Later on $\mathscr{D}_N^q(\varnothing) = \mathbf{C} = \mathscr{D}_0^q(u)$.

We have to pass from $\tilde{\mathscr{D}}_N^q(u)$ to $\mathscr{D}_N^q(u)$ to eliminate the symmetry $^{1}\tilde{h}^{2}\tilde{h}R = R^{1}\tilde{h}^{2}\tilde{h}$ of $R(x; q)$, where $\tilde{h} = \sum_{i=1}^{N} q^{h_i} I^{ii}$, $h_i \in \mathbf{C}$. This symmetry involves an action $\tau_h\colon L \to L\tilde{h} \Leftrightarrow \tau_h(\tilde{E}_{ij}^s) = q^{h_i}\tilde{E}_{ij}^s$ of the group $\mathbf{C}^N \ni (h_i) = h$ on the space $\mathscr{L}_N(u)$, the algebra $\tilde{\mathscr{D}}_N^q(u)$ and its localization $\tilde{\mathscr{D}}_N^q(u)'$ by $\{\tilde{E}_{ii}^n\}$.

PROPOSITION 1.1. *The algebra of invariants of $\tilde{\mathscr{D}}_N^q(u)'$ under the action τ of $\mathbf{C}^N$ is $\mathscr{D}_N^q(u)$.*

This statement follows from the identities

$$\tilde{E}_{ii}^n \tilde{E}_{jk}^s = q^{(\delta_{ij} - \delta_{ik})/2} \tilde{E}_{jk}^s \tilde{E}_{ii}^n$$

proved by a consideration of (1) at $x_1 = \infty$. $\qquad\square$

PROPOSITION 1.2. *The algebra $\mathscr{D}_N^q(u)$ is generated by E_{ij}^s with the relations*

$$E_{ij}^r E_{kl}^s q^{(\delta_{ij} - \delta_{il})/2} - E_{kl}^s E_{ij}^r q^{(\delta_{ij} - \delta_{jk})}$$

$$= \left(q^{1/2} - 1\right)\left(E_{kl}^s E_{ij}^r \delta_{ik} - E_{ij}^r E_{kl}^s \delta_{jl}\right)$$

$$+ \left(q^{1/2} - q^{-1/2}\right)\left(\left(\theta_{sr} + \theta_{jl} - 1\right)E_{il}^r E_{kj}^s + \left(1 - \theta_{sr} - \theta_{ki}\right)E_{il}^s E_{kj}^r\right)$$

$$+ \left(q^{1/2} - q^{-1/2}\right) \sum_{\substack{a < r < b}}^{a + b = r + s} \left(E_{il}^a E_{kj}^b - E_{il}^b E_{kj}^a\right), \tag{2}$$

where $1 \leqslant i, j \leqslant N$, $E_{ij}^n = 0 = E_{ji}^0$ *for* $i < j$, $E_{ij}^s = 0$ *for* $s < 0, s > n$, $E_{ii}^n = 1$, $\theta_{ij} = 1, 0 \overset{\text{def}}{\Leftrightarrow} i \geqslant j, i < j$. $\qquad\qquad\qquad\qquad\qquad\qquad\qquad\qquad\qquad\qquad\qquad\qquad$ $\square$

Let us enumerate some simple properties of $\mathscr{D}_N^q(u)$.

PROPOSITION 1.3. (i) *The mapping*

$$L(x) \to L(xq^d) \prod_{i=1}^{n} \frac{x - q^{u_i - d}}{x - q^{v_i}}$$

induces an isomorphism $\pi_u^v(d): \mathscr{D}_N^q(u) \to \mathscr{D}_N^q(v)$ *for* $d \in \mathbb{C}$, $u, v \in \mathbb{C}^n$. (ii) *The natural inclusion* $\mathscr{L}_N(u') \to \mathscr{L}_N(u)$ *for* $u' = (u_i, i \leqslant n' < n) \in \mathbb{C}^{n'}$ *induces a surjective homomorphism of* $\mathscr{D}_N^q(u)$ *onto* $\mathscr{D}_N^q(u')$. (iii) *The mapping* $E_{ij}^1 \to E_{ij}^n$, $E_{ij}^0 \to E_{ij}^0$ *can be prolonged to an isomorphism of* $\mathscr{D}_N^q(0) \overset{\text{def}}{=} \mathscr{D}_N^q$ $(u = 0 \in \mathbb{C})$ *onto some subalgebra* $\mathscr{D}_N^q(u)_0 \subset \mathscr{D}_N^q(u)$. $\qquad\qquad\qquad$ $\square$

We will give below a posteriori reasons for such a definition for arbitrary u. As for $\mathscr{D}_N^q(0)$, if in (2) we put $q^{1/2} = 1 + \delta/2$, $E_{ij}^{0,1} = \delta e_{ij} - \delta_{ij}$ except $E_{ii}^1 = 1$, where $\delta^2 = 0$, then e_{ij} must obey the relations $[e_{ij}, e_{kl}] = e_{il}\delta_{jk} - e_{kj}e_{il}$. Hence, $\mathscr{D}_N^q(0)$ is a deformation of the universal enveloping algebra $U(\mathfrak{gl}_N)$ of $\mathfrak{gl}_N$. Moreover this deformation is flat (this statement requires a special proof—see Theorem 2.2).

Let us construct a $N \times N$-matrix

$$L_u^q(x) = {}^1 L_u^q(x) = {}^{11'}R(xq^{-u_1}) \cdots {}^{1n'}R(xq^{-u_n})$$

with elements in ${}^{1'}M_N \cdots {}^{n'}M_N \simeq M_N^{\otimes n}$ (we numerate by $1', \ldots, n'$ some extra components of tensor products, which are different from the components without primes). The main property of $\mathscr{D}_N^q(u)$ is as follows.

PROPOSITION 1.4. *The matrix* $L_u^q(x)$ *satisfies* (1) *and therefore determines a* $\mathscr{D}_N^q(u)$-*module* $V_u^q \simeq (\mathbb{C}^N)^{\otimes n}$ *after the passage from* $\tilde{\mathscr{D}}_N^q(u)$ *to* $\mathscr{D}_N^q(u)$.

The proposition is easily deduced from its particular case $n = 1$ (see e.g., [3, 12]). For $n = 1$ one must verify the Yang–Baxter identity

$$ {}^{12}R(x_1 x_2^{-1}) {}^{13}R(x_1 x_3^{-1}) {}^{23}R(x_2 x_3^{-1}) = {}^{23}R(x_2 x_3^{-1}) {}^{13}R(x_1 x_3^{-1}) {}^{12}R(x_1 x_2^{-1}). $$

To compare $R(x; q)$ with the analogous R-matrix from [15], §3, let us note that the last equality is valid for a more general $\tilde{R}$ with $I^{ii} \otimes I^{jj}$ multiplied by $\rho_{i-j} = \rho_{j-i}^{-1} \in \mathbb{C}$ and $I^{ij} \otimes I^{ji}$ multiplied by $x^{a(i-j)}$, $a \in \mathbb{C}$ for $i \neq j$ in addition to the coefficients of R. $\qquad\qquad\qquad\qquad\qquad\qquad$ $\square$

Finally, we will introduce some "good" irreducible representations of $\mathscr{D}_N^q(u)$. See [6] for the general case.

The **skew Young diagram** $\mu = \{\alpha, \beta\}$ of degree n is given by two sequences $\alpha = (\alpha_i)$, $\beta = (\beta_i) \subset \mathbf{C}$ coinciding for almost all $i \in \mathbf{Z}$ with the properties (a) $\beta_i - \alpha_i \in \mathbf{Z}_+ \overset{\text{def}}{=} \{m \in \mathbf{Z} \,|\, m \geqslant 0\}$, (b) $i \leqslant k < j$, $\alpha_i - \alpha_j \in \mathbf{Z} \Rightarrow \alpha_k - \alpha_j$, $\beta_k - \beta_j \in \mathbf{Z}_+$. Let us construct by μ some ordered finite set $\tilde{\mu} = \{(i_k, y_k)\} \subset \mathbf{Z} \times \mathbf{C}$, where (a) $1 \leqslant k \leqslant n \overset{\text{def}}{=} = \Sigma_i(\beta_i - \alpha_i)$, (b) $\beta_{i_k} - \alpha_{i_k} > \beta_{i_k} - y_k \in \mathbf{Z}_+$, (c) if $l > k$ then $i_l > i_k$ or $y_l - y_k > 0 = i_l - i_k$. If $i \geqslant 1$, $\{\alpha_i, \beta_i\} \subset \mathbf{Z}$ then we obtain the standard definition. We put $u_k = i_k - y_k$, $\tilde{u}_k = u_k + (i_k - 1)v$, $v \in \mathbf{C}$, $R_{\tilde{u}} = \prod_{i < j} {}^{ij}R(q^{\tilde{u}_i - \tilde{u}_j})$, where $1 \leqslant i$, $j \leqslant n$ and (i, j) is before (i', j') in the product iff $i > i'$ or $j - j' > 0 = i - i'$. For this u we will sometimes write $u \sim \mu$.

PROPOSITION 1.5. (see [6, 7]). (i) *There exists a limit* $\lim_{v \to 0} R_{\tilde{u}} \overset{\text{def}}{=} R_\mu$. (ii) *The subspace* $V_\mu^q = R_\mu V_u^q$ *of* V_u^q *is an irreducible* $\mathscr{D}_N^q(u)$*-submodule, corresponding to* $L_\mu^q(x) = L_u(x)P_\mu$ *for a projector* P_μ *onto* V_μ^q. (iii) *The algebra* $\mathscr{D}_N^q(u)$ *acts on* V_μ^q *through its quotient-algebra* $\mathscr{D}_N^q(a)$, *where* $a = (\alpha_i + 1 - i \,|\, \alpha_{i-1} \neq \alpha_i \neq \beta_i)$. $\qquad\square$

If we restrict our representations to $\mathscr{D}_N^q(u)_0 \simeq \mathscr{D}_N^q(0)$ and degenerate $\mathscr{D}_N^q(0)$ to $u(\mathfrak{gl}_N)$ by making $q^{1/2}$ tend to 1 (see above) then V_u^1 as $\mathfrak{sl}_N$-module will be the usual n-th tensor power of the fundamental N-dimensional representation and $V_\mu^1 = R_\mu(q = 1)V_u$ will be isomorphic to the $\mathfrak{sl}_N$-module arising from μ in the classic representation theory. If $\alpha_1 \neq \beta_1$, $\alpha_i = \beta_i$ for $i \leqslant 0$ (one can get this by a common shift of i, α_i, β_i) and $\alpha_i = 0$ for each i then V_μ^1 is the irreducible $\mathfrak{gl}_N$-module with the highest weight $(\beta_1, \beta_2, \ldots, \beta_i, \ldots)$. We further put:

$$L_{\{\alpha, \alpha\}}^q(x) = \mathbf{1} = L_\phi^q(x), \qquad V_{\{\alpha, \alpha\}}^q = \mathbf{C} = V_\phi^q.$$

§2. Structural theorems. We begin with the notion of "**quantum determinant**" of L in the spirit of [12] (see also [15, 10, 7]). For the universal solution L of (I) with $\tilde{E}_{ij}^s$ from $\tilde{\mathscr{D}}_N^q(u)$ (or for a concrete L according to the context) let us define

$$H^N(x) = Sp_{1,\ldots,N}\left({}^N L(xq^{-1})\ldots {}^1 L(xq^{-N})\mathscr{Q}_N^N\right) \Big/ \prod_{i=1}^{N} \tilde{E}_{ii}^n. \tag{3}$$

Here $Sp_{1,\ldots,K}$ denotes the matrix spur in $M_N^{\otimes K} \simeq M_{NK}$ over any algebra of coefficients, $\mathscr{Q}_N^K \overset{\text{def}}{=} 1/k! R_w$ for $w = (1, \ldots, K) \in \mathbf{C}^K$ (see §1), $Q_1^1 = Q_0^0 = 1$. A priori $H^N \in \mathscr{L}_1(u \uparrow N) \otimes_{\mathbf{C}} \mathscr{D}_N^q(u)$, where $u \uparrow K \overset{\text{def}}{=} (u_i + j, 1 \leqslant i \leqslant n, 1 \leqslant j \leqslant K) \in \mathbf{C}^{nK}$. We will use later some properties of Q.

PROPOSITION 2.1. (i) *If* $q \to 1$ *then* Q_N^K *tends to the "antisymmetrizer"—the canonical projector onto the exterior power* $\Lambda^K \mathbf{C}^N \subset (\mathbf{C}^N)^{\otimes K}$; $2Q_N^2 = \Sigma_{i \neq j} I^{ii} \otimes I^{jj} - q^{1/2}\Sigma_{i > j} I^{ij} \otimes I^{ji} - q^{-1/2}\Sigma_{i < j} I^{ij} \otimes I^{ji}$, $rk(Q_N^K) = \dim_{\mathbf{C}} \Lambda^K \mathbf{C}^N$, $(Q_N^N)^2 =$

Q_N^N. (ii) *If $C_i = I^{i'i'}$ for some*

$$1 \leqslant 1', \ldots, K' \leqslant K \quad then \quad Sp_{1,\ldots,K}\left(^{1\cdots N}Q_N^{N1}C_1 \ldots {}^KC_K\right)$$

is equal to

$$\frac{(N-K)!}{N!}{}^{K+1\cdots N}Q_{N-K}^{N-K} \quad if \; i \neq j \Rightarrow i' \neq j'$$

and otherwise it is equal to zero; the analogous is true for $Sp_{K+1,\ldots,N}$ and $K < i' \leqslant N$. (iii) The image of Q_N^N is the space $\mathbf{C}\Sigma_\sigma(-q^{1/2})^{l(\sigma)}e_{\sigma(1)} \otimes \cdots \otimes e_{\sigma(N)}$, where e_i is the first column of I^{i1}, $\sigma \in S_N$, $l(\sigma)$—the length of σ (see next §3);

$$\prod_{i=1}^{N} \tilde{E}_{ii}^n \cdot H^N(x)$$

$$= \sum_{\sigma,r} \tilde{E}_{\sigma(N)N}^{r_1} \tilde{E}_{\sigma(N-1)N-1}^{r_2} \cdots \tilde{E}_{\sigma(1)1}^{r_N}(-q^{1/2})^{l(\sigma)} \prod_{j=1}^{N}\left(xq^{-j}\right)^{r_j} \prod_{i=1}^{n}\left(xq^{-j} - q^{u_i}\right)^{-1},$$

where $0 \leqslant r_i \leqslant n$ (cf. [11], Prop. 6.3). □

The symmetry of R (see §I) involves the following branching properties of $\mathscr{D}_N^q(u)$. The functions $[L]^K \overset{\text{def}}{=} \Sigma_{i,j=1}^K I^{ii}LI^{jj}$, $[L]_K = \Sigma_{i,j=N-K+1}^N I^{ii}LI^{jj}$ satisfy (1) and hence determine inclusions ρ^K: $\mathscr{D}_K^q(u) \to \mathscr{D}_N^q(u) \leftarrow \mathscr{D}_K^q(u)$: ρ_K. Let us denote by H^K, $H_K \in \mathscr{L}_1(u \uparrow K) \otimes_{\mathbf{C}} \rho(\mathscr{D}_K^q(u))$ the H-functions (3) constructed for $N := K$ and $[L]^K$, $[L]_K$ in place of L. Lataer on $H^0 = H_0 = 1$.

THEOREM 2.2 (cf. [6, 7]). (i) *The subspace $[\mathscr{D}_N^q(u)]_p$ of $\mathscr{D}_N^q(u)$, linearly generated by monomials of degree $\leqslant p$ in E_{ij}^r, is isomorphic to the space of all polynomials in nN^2 variables of the same degree.*

(ii) *The coefficients of the principal parts of $H^1(\bar{x}), \ldots, H^N(x)$ (in their expansions at points $u_i + j$ from $u \uparrow N$) are all algebraically independent and generate a maximal commutative subalgebra $A_N^q(u)$ in $\mathscr{D}_N^q(u)$. The analogue is true for $H_1, \ldots, H_N$.*

(iii) *The coefficients of $H^N = H_N$ generate the centre $Z_N^q(u)$ of $\mathscr{D}_N^q(u)$.*

Proof scheme. For every p one can find a skew diagram μ such that the homomorphism $\mathscr{D}_N^q(u) \to \text{End}_{\mathbf{C}}V_\mu^q$ is injective on $[\mathscr{D}_N^q(u)]_p$. To do this we tend $q^{1/2}$ to 1 and construct a flat degeneration of $\mathscr{D}_N^q(u)$ (from a punctured neighbourhood of $q^{1/2} = 1$) to some quotient-algebra $\tilde{U}'$ of $\tilde{U} \overset{\text{def}}{=} U(\mathfrak{gl}_N[t]/(t^n))$ (see above and [6, 7]). We have: $\dim_{\mathbf{C}}[\tilde{U}']_p = \dim_{\mathbf{C}}[\mathscr{D}_N^q(u)]_p$. Next we use the representation theory of $U(\mathfrak{gl}_N)$ to find μ with the above property for $\tilde{U}$ and V_μ^1. This μ necessarily suit $\mathscr{D}_N^q(u)$ too. We have proved (i) and also the coincidence of $\tilde{U}$ and $\tilde{U}'$ since p is arbitrary. We remind the reader that $q^{1/2}$ is close to 1 here and everywhere.

It is not difficult to deduce the identities $[H^k(x), H^m(y)] = 0 = [H^N(x), E_{ij}^r]$, $x, y \in \mathbf{C}$ from Proposition 2.1, i) and (1) (see [12], [15] §3, [5, 6, 7]). To complete the proof of (ii), (iii) we pass to a quasi-classical limit. First we put $q^{1/2} = 1$ in (2), §1 and will come to the commutative algebra $d_N(u) \overset{\text{def}}{=} \mathbb{C}[E_{ij}^r]$. Next, by the formal substitution $q^{1/2} \to 1 + \delta/2$, $[E_{ij}^r, E_{kl}^s] \to \delta\{E_{ij}^r, E_{kl}^s\}$ in (2) and making $\delta = 0$ (after reducing by δ) one can define the brackets $\{E_{ij}^r, E_{kl}^s\}$ and then extend them to a Poisson bracket $\{\,,\,\}$ on $d_N(u)$. Let us denote $H^K(x; q = 1)$, $H_K(x; q = 1)$ by $h^K(x)$, $h_K(x)$. We have:

$$h^K(x) = \det\big([L(x)]^K\big)/\prod_{i=1}^{K} \tilde{E}_{ii}^n, \quad h_K(x) = \det\big([L(x)]_K\big)/\prod_{i=N-K+1}^{N} \tilde{E}_{ii}^N.$$

Let $q_{K'+1}, q_{K'+2}, \dots, q_{K'+nK}$ be the coefficients of the principal part of h^K, which are put in any order ($1 \leqslant K \leqslant N$, $K' = nK(K-1)/2$). The subalgebra $a_N(u) \overset{\text{def}}{=} \mathbb{C}[q_1, \dots, q_{(N+1)'}]$ is commutative (involutive) relative to $\{\,,\,\}$.

LEMMA 2.3. (i) *The mapping* $L \to (h^1, \dots, h^N)$ *determines a regular surjective morphism* α: $\mathbb{C}^{N^2 n} \to \mathbb{C}^{N(N+1)n/2}$ *where* $\mathbb{C}^{N^2 n} = \operatorname{Specm} d_N(u)$, $\mathbb{C}^{N(N+1)n/2} = \operatorname{Specm} a_N(u)$. (ii) *Sufficiently general fibers of* α *are isomorphic to open submanifolds of some Jacobians of algebraic curves, which are the orbits of the entire collection of flows (differential equations)*

$$\partial E_{ij}^s/\partial t = \{h, E_{ij}^s\},$$

commuting with one another, h being an element of $a_N(u)$. All this holds good for $h_1, \dots, h_N$.

Statement (i) is verified directly. As for (ii) we refer the reader to [14, 15]. $\square$

The remaining part of the Theorem is deduced from the Lemma in the following way. If $h \in d_N(u)$ and $\{h, h^i(x)\} = 0$ for all i, x then h has to be constant along general fibers of α and hence is a polynomial in (q_i) (since α is surjective). By computation of the dimensions we obtain that sufficiently general fibers of the morphism ξ: $\mathbb{C}^{N^2 n} \to \mathbb{C}^{Nn} = \operatorname{Specm} z_N(u)$ induced by the mapping $L \to h^N$ have non-degenerate Poisson structures for $z_N(u) = \lim Z_N^q(u)$. Therefore the kernel of $\{\,,\,\}$ coincides with $z_N(u) = \mathbb{C}[q_{N'+1}, \dots, q_{(N+1)'}]$. The subsequent lifting to $\mathscr{D}_N^q(u)$ uses statement (i) of the Theorem. $\square$

Our next task is to calculate the centralizer $Z(\operatorname{Im} \rho^K)$ of $\mathscr{D}_K^q(u)$ in $\mathscr{D}_N^q(u)$. In solving this problem L-functions and R-algebras of arbitrary level n display their usefulness and add some new features to the classical results.

If L is our universal solution of (1) then $L(x)^{-1}$, $L^*(x) = H^N(xq)L(x)^{-1}$ and $\hat{L}(x) = L^*(x)/H^N(x)$ satisfy (1) for $\hat{R}(x) = R(x; q^{-1})$ (an easy computation). It follows (cf. [12]) from the definition and properties of H and Proposition 2.1, (ii) that

$$^N L^*(x) \prod_{i=1}^{N} \tilde{E}_{ii}^n = S_{p_1, \dots, N-1}\big(^{N-1}L(xq^{-1}) \dots {}^1 L(xq^{-N+1}) Q_N^N\big).$$

Hence $L^* \in \mathcal{L}_N(u \uparrow N - 1) \otimes_{\mathbf{c}} \mathcal{D}_N^q(u)$ determines homomorphisms

$$\rho_K^*: \mathcal{D}_K^{q-1}(u \uparrow N - 1) \to \mathcal{D}_N^q(u)$$

corresponding to the mappings $L \to [L^*]_K$ (after the passage from $\tilde{\mathcal{D}}$ to $\mathcal{D}$). Let us define $\hat{H}_K$ by means of (3) for $\hat{L}(x)$ with $\hat{Q}_N^N$ constructed for $\hat{R}$ instead of R. One can prove with the aid of Proposition 2.1, (ii) the identities

$$H^K(x) = \hat{H}_{N-K}(xq^{-1}) H^N(x), \tag{4}$$

which become the well-known relations for principal minors for $q = 1$.

THEOREM 2.4.   *The centralizer* $Z(\operatorname{Im} \rho^K)$ *coincides with*

$$Z_N^q(u) \cdot \rho_{N-K}^* \left(\mathcal{D}_{N-K}^{q^{-1}}(u \uparrow N - 1) \right).$$

Conversely, $Z(\operatorname{Im} \rho_{N-K}^*) = Z_N^q(u) \cdot \rho^K(\mathcal{D}_K^q(u))$.

Proof.   We obtain from (1) the identity

$$^{12}R^t\left(x_1 x_2^{-1}\right)^1 L^t(x_1)^2 L^*(x_2) = {}^2 L^*(x_2)^1 L^t(x_1)^{12} R^t\left(x_1 x_2^{-1}\right),$$

where A^t is the transpose of $A \in M_N$, $(A \otimes B)^t = A^t \otimes B$. A moment's consideration shows that matrix elements of $[L]^K$ commute pairwise with matrix elements of $[L^*]_{N-K}$ (use the above identity and the special structure of R^t). To complete the proof we put $q = 1$ everywhere in the same way as above. We keep all the notation of Theorem 2.2.

For some sufficiently general stratum $F_c = \{ q_{N'+i} = c_i \}$, $1 \leqslant i \leqslant Nn$ and a point $l \in F_c$ one can supplement $(q_1, \ldots, q_{N'})$ with such local functions $(p_1, \ldots, p_{N'})$ that $\tilde{q}_i = q_i - q_i(l)$, $\tilde{p}_j = p_j - p_j(l)$ form coordinates in a neighbourhood of l in F_c and $\{ q_i, p_j \} = \delta_{ij}$, $\{ p_i, p_j \} = 0$. We remind the reader that $\{ q_i, q_j \} = 0$ and $(q_{N'+i})$ generate the kernel $z_N(u)$ of $\{ , \}$—see Theorem 2.2. The functions $p_1, \ldots, p_{(K+1)'}$ belong necessarily (modulo $(q_{N'+i})$) to some formal completion of a localization of the algebra $\operatorname{Im} \rho^K \simeq d_K(u)$. The analogous case is true for $p_{(K+1)'+1}, \ldots, p_{N'}$ and $\operatorname{Im} \rho_{N-K}^*$ since $h^{K+i}(x) = \hat{h}_{N-K-i}(x) h^N(x)$ (see (4)). Here we use the fact that ρ^K and ρ_{N-K}^* preserve the appropriate Poisson brackets (i.e. are canonical maps).

Any local function f on F_c at l is a series of powers of $\tilde{q}_i$, $\tilde{p}_i$. Hence, $\{ f, p_i \} = 0 = \{ f, q_i \}$ for $1 \leqslant i \leqslant K'$ iff f does not depend on $p_1, \ldots, p_{K'}$ and belongs to a completion of a localization of the algebra $\operatorname{Im} \rho_{N-K}^* \cdot Z_N(u)$. Now let us consider the regular morphism $\mathbf{C}^{N^2 n} \to \mathbf{C}^{(N-K)^2(N-1)n} \times \mathbf{C}^{Nn} = \operatorname{Specm} d_{N-K}(u \uparrow N - 1) \times \operatorname{Specm} z_N(u)$ induced by $\rho_{N-K}^* \times \xi$. Its image is a normal affine variety. By means of this we eliminate the words "completion and localization" if the above f is from $d_N(u)$ and prove the quasi-classical version of the Theorem (the converse statement is quite analogous). The lifting of this limited statement to a general one with $q^{1/2} \neq 1$ uses Theorem 2.2, (i). $\square$

We begin the description of the branching rules for representations of $\mathcal{D}_N^q(u)$ with the following proposition that is a formal consequence of the definition of

L_μ^q, $\hat{L}$. We remind the reader that L_μ^q are determined non-uniquely and depend on the choice of P_μ.

PROPOSITION 2.5. (i) *If we put* $\hat{\chi} = (\hat{\chi}_i)$, $\hat{\chi}_i = -\chi_{-i} - 1$ *for a set* $\chi = (\chi_i)$ *then*

$$\hat{L}^q_{\{\alpha,\beta\}}(x) = L^{q^{-1}}_{\{\hat{\beta},\hat{\alpha}\}}(x). \qquad \square$$

Let us define $\tilde{H}^K$ *by* (3) *with* Q_K^K *and* $\tilde{L}^K(x) \overset{\text{def}}{=} ([L^{-1}]_K)^{-1} H^{N-K}(x)/H^{N-K}(xq)$ *($\tilde{L}^K$ satisfies* (1) *for* $R_K(x; q)$). *One can easily deduce from* (4) *that*

$$H^N(x) = H^{N-K}(x)\tilde{H}^K(x).$$

The mapping $L_\mu \to ([L_\mu]^K, \tilde{L}_\mu^{N-K})$ *gives us an action of* $\mathscr{D}_K^q(u) \times \mathscr{D}_{N-K}^q(u)$ *on* V_μ.

THEOREM 2.6. (i) *There exists such a basis* $\{\xi_\lambda\} \subset V_\mu$ *that*

$$H^K(x)\xi_\lambda = \xi_\lambda \prod_i \left(x - q^{i-\lambda_i^{(K)}}\right)/\left(x - q^{i-\alpha_i}\right),$$

where λ *ranges over the sequence of all collections* $\lambda = \{\lambda^{(0)}, \ldots, \lambda^{(N)}\}$, $\lambda^{(K)} = (\lambda_i^{(K)} \in \mathbb{C})$, $\lambda^{(0)} = \alpha$, $\lambda^{(N)} = \beta$ *with the following properties*: a) $\lambda_i^{(K)} - \lambda_i^{(K-1)} \in \mathbb{Z}_+$, b) $\mathbb{Z}_+ \ni \lambda_i^{(K)} - \lambda_{i-1}^{(K)} \Rightarrow \lambda_i^{(K-1)} - \lambda_{i-1}^{(K)} \in \mathbb{Z}_+$. *The vectors* ξ_λ *are unique up to scalars.*

(ii) *If we put* $W_\gamma = \bigoplus \mathbb{C}\xi_\lambda$, *where* $\gamma = (\gamma_i)$, $\lambda_i^{(K)} = \gamma_i$, *then* $V_\mu^q = \bigoplus_\gamma W_\gamma$ *and* $W_\gamma \simeq V_{\{\alpha,\gamma\}}^q \otimes V_{\{\gamma,\beta\}}^q$ *as* $\mathscr{D}_K^q(u) \times \mathscr{D}_{N-K}^q(u)$-*modules.*

To prove the Theorem we can restrict ourselves to the case $K = N - 1$ and then use the induction and Proposition 2.5. We note that statement (i) easily follows from (ii) and the formula for the action of H^N (see [6, 7, 11]), and the Theorem is verified by a direct computation for $n = 1$.

One can generalize to V_u^q considered as $\mathscr{D}_{N-1}^q(u)$-module the classic decomposition

$$(\mathbb{C}^N)^{\otimes n} = \bigoplus_{k=0}^n \mathbb{C}[S_n]V_k, \quad \text{where } V_k = e_N^{\otimes k} \otimes \left(\bigoplus_{i=1}^{N-1} \mathbb{C}e_i\right)^{\otimes(n-k)},$$

S_n acts naturally on $(\mathbb{C}^N)^{\otimes n}$, e_i are from Proposition 2.1. The Hecke algebra $H_n^q \subset M_N^{\otimes n}$ plays the role of $\mathbb{C}[S_n]$ in its q-analogue (see next §3). The branching properties of Hecke algebras (ibid.) allow us to decompose V_μ^q when N is "indefinite" (we note that the projector P_μ of §1 can be chosen from H_n^q). The restrictions for $\lambda_i^{(K)}$ (see (i)) arise from Proposition 2.5. and the following vanishing theorem:

$$V_\mu^q \neq \{0\} \Leftrightarrow i_k - i_l < N \quad \text{if } y_k = y_l.$$

We would like to outline another way of proving the Theorem. One can degenerate $\mathcal{D}_N^q(u)$ to some Yangian, while tending $q^{1/2}$ to 1, and put $n = 1$ to apply the classic results or the recent results by G. I. Olshansky about $U(\mathfrak{gl}_N)$ and Yangians. $\square$

§3. Hecke algebras. The Hecke algebra H_n^q is generated over $\mathbb{C}$ by $1, T_1, \ldots, T_{n-1}$ with the relations $[T_i, T_j] = 0$ whenever $i \neq j \pm 1$, $T_i T_{i+1} T_i = T_{i+1} T_i T_{i+1}$ and $(T_i - q)(T_i + 1) = 0$. As above $q \in \mathbb{C} \setminus \{1\}$ is sufficiently close to 1. Let S_n be the symmetric group acting on any set with n elements; $s_i = (i, i + 1)$, s_0—the unit of S_n. Given any function f of arguments $x_1, \ldots, x_n$ permutations $w \in S_n$ act on f by the formula $({}^w f)(x_1, \ldots, x_n) = f(w^{-1}(x_1, \ldots, x_n))$. We denote by $l(w)$ the length of $w \in S_n$ (the length of a decomposition of w in a product of s_i of minimal length).

By adjoining to H_n^q pairwise commuting independent $X_1, \ldots, X_n$ with the relations $[X_i, T_j] = 0$ for $i \neq j + 1, j$ and

$$X_{i+1} T_i - T_i X_i = (q - 1) X_{i+1} = T_i X_{i+1} - X_i T_i \tag{5}$$

we shall get the **affine Hecke algebra** $\mathfrak{H}_n^q$ (see e.g. [13]). The localization of $\mathfrak{H}_n^q$ by $X_1, \ldots, X_n$ (generated by $\mathfrak{H}_n^q$ and $X_1^{-1}, \ldots, X_n^{-1}$) will be denoted by $\overline{\mathfrak{H}}_n^q$.

PROPOSITION 3.1 [13]. (i) *The collection of H_n^q-valued functions $\varphi_i = \varphi_{s_i} = T_i + (q - 1)/(x_i x_{i+1}^{-1} - 1)$ in $x_1, \ldots, x_n$ extends uniquely to a homogeneous 1-cocycle $\varphi_w(x_1, \ldots, x_n)$ with the following property: $\varphi_{ab} = \varphi_b{}^{b-1}\varphi_a$ if $ab \geqslant b \overset{\mathrm{def}}{\Leftrightarrow} l(ab) = l(a) + l(b)$ for $a, b \in S_n$. (ii) The set of elements $\Upsilon_i = \Upsilon_{s_i} = T_i + (q - 1) \times (X_i X_{i+1}^{-1} - 1)^{-1} \in \overline{\mathfrak{H}}_n^q$ extends uniquely to $\{\Upsilon_w, w \in S_n\}$ with the following properties: $X_i \Upsilon_w = \Upsilon_w X_{w(i)}, \Upsilon_{ab} = \Upsilon_b \Upsilon_a$ if $ab \geqslant b \in S_n$.* $\square$

In the notations of §1 given a skew μ let us define the permutation $\omega = \omega_\mu$ by means of the following rule: if $0 < y_{\omega(k)} - y_{\omega(l)} \in \mathbb{Z}_+$ or $i_{\omega(k)} - i_{\omega(l)} < 0$ when $y_{\omega(k)} - y_{\omega(l)} \in \{0, \mathbb{C} \setminus \mathbb{Z}\}$ then $l > k$. This ω reorders in a natural way the two-dimensional set $\tilde{\mu}$ when it has been reflected in a line $i + y = \mathrm{const}$. We put $\tilde{\varphi}_w = \varphi_w(q^{\tilde{u}_1}, \ldots, q^{\tilde{u}_n})$, $\tilde{u}_k = u_k + (i_k - 1)v$, $u_k = i_k - y_k$. There exist limits $\lim_{v \to 0} \tilde{\varphi}_w \overset{\mathrm{def}}{=} f_w \in H_n^q$ for $w \geqslant \omega$ (see [6,9]). Let $f_\omega = 1$ when $\mu = \{\alpha, \alpha\}$.

For an arbitrary $u = (u_i) \in \mathbb{C}^n$ we will denote below by $J_u^q \simeq H_n^q$ the only right $\mathfrak{H}_n^q$-module defined by relations $\langle z \rangle h = zh, \langle 1 \rangle X_i = q^{u_i}$, where $z, h, 1 \in H_n^q$, h takes z into $\langle z \rangle h$. Let $\mathfrak{H}_n^q(a) = \mathfrak{H}_n^q/A$, where $a = (a_i) \in \mathbb{C}^m$, A is the two-sided ideal $\mathfrak{H}_n^q(\prod_{i=1}^m (X_n - q^{a_i}))\mathfrak{H}_n^q$. One can prove that $\mathfrak{H}_n^q(a) = \oplus X_1^{i_1} \ldots X_n^{i_n} H_n^q$, where $0 \leqslant i_k < m$.

PROPOSITION 3.2 [6,9]. (i) *The subspace $J_\mu^q \overset{\mathrm{def}}{=} f_w H_n^q \subset J_u^q$ for $u \sim \mu$ does not depend on $w \geqslant \omega$ and is an irreducible $\mathfrak{H}_n^q$-module and is, in fact, a $\mathfrak{H}_n^q(a)$-module, where $a = (\alpha_i + 1 - i | \alpha_{i-1} \neq \alpha_i \neq \beta_i)$. (ii) The vectors $\{f_w, w \geqslant \omega\}$ form a basis of J_μ^q with the following property: $\langle f_w \rangle X_i = q^{w-1}(u_i)f_w$. The last formulas determine each f_w uniquely up to a factor of proportionality.* $\square$

Let us denote below by $\hat{\mathfrak{H}}_k^q$ the subalgebra in $\mathfrak{H}_n^q$ generated by $1, T_{n-k+1}, \ldots, T_{n-1}, X_{n-k+1}, \ldots, X_n, 0 \leqslant k \leqslant n$ and by $\tilde{\mathfrak{H}}_k^q$ the algebraic span of $1, T_1, \ldots, T_{k-1}, X_1, \ldots, X_k$ ($\hat{\mathfrak{H}}_k^q, \tilde{\mathfrak{H}}_k^q \simeq \mathfrak{H}_k^q$). We can connect with an arbitrary $w \in S_n$ two subsets $[w]^k = (w(n-k+1), \ldots, w(n))$, $[w]_k = (w(1), \ldots, w(k))$ of $(1, \ldots, n)$ and two permutations $\hat{w}^k, \hat{w}_k \in S_{n-k}$ by throwing away respectively $[w]^k, [w]_k$ (e.g. $\hat{w}_k$: $(1, \ldots, \widehat{w(i_1)}, \ldots, \widehat{w(i_k)}, \ldots, n) \rightarrow (\widehat{w(1)}, \ldots, \widehat{w(k)}, w(k+1), \ldots)$, where omitted terms are marked off by "$^\wedge$", $1 \leqslant i_l \leqslant k$). The following Theorem is deduced from [6, 9].

THEOREM 3.3. (i) *Let* $\gamma = (\gamma_i)$, $\alpha_i \leqslant \gamma_i \leqslant \beta_i$ *and both* $\{\alpha, \gamma\}$ *and* $\{\gamma, \beta\}$ *be skew diagrams of degree k and $n - k$ respectively. Define two disjoint sets* $\gamma^* = \{1 \leqslant l \leqslant n \,|\, (i_l, y_l) \in \{\alpha, \gamma\}\}$, $\gamma_* = \{1 \leqslant l \leqslant n \,|\, (i_l, y_l) \in \{\gamma, \beta\}\}$ *and a subset* $[\gamma] = \{g \in S_n \,|\, [g]^k = \gamma^*\} = \{g \,|\, [g]_{n-k} = \gamma_*\}$. *Then* $J_\mu^q = \bigoplus_\gamma I_\gamma$, *where* $I_\gamma = \bigoplus_{g \in [\gamma]} \mathbb{C} f_g$.

(ii) *We have:* $g^k \geqslant \omega_{\{\alpha, \gamma\}}$, $g_{n-k} \geqslant \omega_{\{\gamma, \beta\}}$ *for* $g \in [\gamma]$. *The linear map taking f_g,* $g \in [\gamma]$ *into* $f_{g^k} \otimes f_{g_{n-k}}$ *is an isomorphism of the* $\tilde{\mathfrak{H}}_k^q \times \hat{\mathfrak{H}}_{n-k}^q$*-modules I_γ (which is invariant under the action of* $\hat{\mathfrak{H}}_k^q \cdot \tilde{\mathfrak{H}}_{n-k}^q$*) and* $\mathscr{J}_{\{\alpha, \gamma\}}^q \otimes \mathscr{J}_{\{\gamma, \beta\}}^q$. $\qquad\square$

THEOREM 3.4. (i) *The subalgebra* $\mathbb{C}[X_1, \ldots, X_n]$ *is a maximal commutative one in* $\mathfrak{H}_n^q$. *The centre* Z_n^q *of* $\mathfrak{H}_n^q$ *is generated by symmetric polynomials in* $X_1, \ldots, X_n$. *The centralizer* $Z(\hat{\mathfrak{H}}_k^q)$ *of* $\hat{\mathfrak{H}}_k^q \subset \mathfrak{H}_n^q$ *coincides with* $Z_n^q \tilde{\mathfrak{H}}_{n-k}^q$; *conversely,* $Z(\tilde{\mathfrak{H}}_k^q) = Z_n^q \hat{\mathfrak{H}}_{n-k}^q$.

(ii) *Statements* (i) *are true for* $\mathfrak{H}_n^q(0) = \mathfrak{H}_n^q / \mathfrak{H}_n^q (X_n - 1) \mathfrak{H}_n^q \simeq H_n^q$ *and for the images* $\tilde{\mathfrak{H}}_k^q(0)$ *of* $\tilde{\mathfrak{H}}_k^q$ *in* $\mathfrak{H}_n^q(0)$, *where* $X_1, \ldots, X_n$ *are replaced by their images*

$$X_n = 1, \qquad X_{n-1} = qT_{n-1}^{-2}, \ldots, \qquad X_i = q^{n-i}\left(T_i^{-1} \ldots T_{n-1}^{-1}\right)\left(T_{n-1}^{-1} \ldots T_i^{-1}\right), \ldots \,.$$

Proof. It follows from Proposition 3.1, (ii), that $\bar{\mathfrak{H}}_n^q$ is a simple algebra of dimension $(n!)^2$ over its centre $\mathbb{C}(Z_n^q)$. This allows us to prove (i) for $\mathfrak{H}_n^q$. The passage to $\mathfrak{H}_n^q$ is not difficult. Part (ii) follows from Theorem 3.3, since H_n^q is semi-simple (for q being sufficiently close to 1). The latter is a well-known fact. The formulas for X_i are the consequence of (5). $\qquad\square$

Conjecture 3.5. Statements (i) of Theorem 3.4 are valid for arbitrary $\mathfrak{H}_n^q(a)$. $\qquad\square$

Next we explain the similarity of the results of §2 and §3 and prove an analogue of Hermann Weyl duality and Drinfeld's theorem on Yangians from [8]. We discuss below the simplest version of this duality only. In a more functorial form (as in [8]) it can be proved by the same techniques.

According to M. Jimbo $q^{1/2}R(x)P = (T + q - 1)/(x - 1)I \otimes I$, where

$$T = qI \otimes I - \sum_{i>j}\left(qI^{ii} \otimes I^{jj} + I^{jj} \otimes I^{ii}\right) + q^{1/2}\sum_{i \neq j} I^{ij} \otimes I^{ji}$$

satisfies the equality

$$(T - q)(T + 1) = 0, \qquad P = \sum_{i,j=1}^{N} I^{ij} \otimes I^{ji}$$

is the permutation matrix. Let us further identify $^{ii+1}T, {}^{ii+1}P$ with $T_i \in H_{n+1}^q$, $s_i \in S_{n+1}$ for $1 \leqslant i \leqslant n$. Then we have $^{12}R(xq^{-u_1}) \ldots {}^{1n+1}R(xq^{-u_n}) = \varphi_{s_n \ldots s_1}(x_0, x_1, \ldots, x_n)(s_n \ldots s_1)$ for $x_0 = x$, $x_i = q^{u_i}$ (by the way, $R_\mu = f_{w_0} w_0$ for $u \sim \mu$ and $w_0 \colon (1, \ldots, n) \to (n, \ldots, 1)$). Here and below we change the indices $1, 1', \ldots, n'$ in the definition of L_u, V_u (see §1) for $1, 2, \ldots, n + 1$.

THEOREM 3.6. *Any $\mathcal{D}_N^q(u)$-invariant submodule of the $\mathcal{D}_N^q(u)$-module V_u^q for some arbitrary $u \in \mathbf{C}^n$ can be described as follows. If B is a right ideal in H_n^q which becomes $\mathfrak{H}_n^q$-invariant after H_n^q has been identified with the $\mathfrak{H}_n^q$-module $\mathscr{J}_u^q$ then BV_u^q is a $\mathcal{D}_N^q(u)$-submodule of V_u^q for the inclusion $T_i \to {}^{ii+1}T$ of H_n^q into $M^{\otimes_N^n}$.*

Proof. We put

$$\varphi_{s_n \ldots s_1}(x, x_1, \ldots, x_n) \prod_{i=0}^{n-1}(x_i - x_{i+1})$$

$$= \left(xT_1 - qx_1 T_1^{-1}\right) \ldots \left(xT_{n-1} - qx_n T_n^{-1}\right) \stackrel{\text{def}}{=} F$$

$$= \sum_{i=0}^{n} x^i F_{n-i}(-q)^{n-i}.$$

One has $F_0 = T_1 \ldots T_n$, $F_n = \prod_{i=1}^{n} x_i T_1^{-1} \ldots T_n^{-1}$, $F_1 = x_1 T_1^{-1} T_2 \ldots T_n + x_2 T_1 T_2^{-1} T_3 \ldots T_n + \cdots + x_n T_1 \ldots T_{n-1} T_n^{-1}$ and so on. Let us define a projection $h \to \operatorname{res} h$ of $H_{n+1}^q \ni h$ onto $\hat{H}_n^q = H_{n+1}^q \cap \hat{\mathfrak{H}}_n^q$ ($\hat{\mathfrak{H}}_n^q$ is from Theorem 3.4) by the relations: $\operatorname{res}(\hat{H}_n^q T_n \hat{H}_n^q) = 0 = \operatorname{res} h - h$ for $h \in \hat{H}_n^q$. We need two lemmas on Hecke algebras.

LEMMA 3.7. *If $\operatorname{res}(hT_n^{-1} \ldots T_{k+1}^{-1} T_{k-1} \ldots T_1) = 0$ for $h \in H_{n+1}^q$ and $k = 1, \ldots, n + 1$ then $h = 0$.*

Proof. The equivalent condition for h is as follows: for such a sufficiently general $u' \in \mathbf{C}^{n+1}$ that $\mathscr{J}_{u'}^q = \oplus_{k=1}^{n+1} E_k$, where $E_k \stackrel{\text{def}}{=} \{ z \in \mathscr{J}_{u'}^q | \langle z \rangle X_k = u_1' z \}$, the projections of h onto E_k vanish for $1 \leqslant k \leqslant n + 1$. $\qquad \square$

LEMMA 3.8. *For each $h \in \hat{H}_n^q$ we have the identities $(q - 1)\langle h \rangle(X_{k+1} - x_{k+1}) = q^{1-k}\operatorname{res}((hF_1 - F_1 h')T_n^{-1} \ldots T_{k+1}^{-1} T_{k-1} \ldots T_1)$, $h' = F_0^{-1} h F_0 = F_n^{-1} h F_n$, where the isomorphism $h \to h'$ from $\hat{H}_n^q$ onto $\tilde{H}_n^q = \tilde{\mathfrak{H}}_h^q \cap H_{n+1}^q$ transfers T_{k+1} to T_k for $1 \leqslant k \leqslant n$.* $\qquad \square$

Any invariant submodule of V_u^q has the form SV_u^q for some $S = S^2 \in {}^2M_N {}^3M_N \ldots {}^{n+1}M_N \simeq M_N^{\otimes_n}$ with the following property: $SFS' = FS'$, where

$S' = P_0^{-1} S P_0$ for $P_0 = s_1 \ldots s_n$. In particular we have: $SF_0 S' = F_0 S'$, $SF_n S' = F_n S'$. These two relations are equivalent to the invariance of SV_u^q under the action of $\mathcal{D}_N^q(u)_0$ (see Proposition 1.3, (iii)). But $\mathcal{D}_N^q(u)_0 \simeq \mathcal{D}_N^q(0)$ is a deformation of $U(\mathfrak{gl}_N)$ (§I) and the $\mathcal{D}_N^q(u)_0$-decomposition of V_u is like that for $U(\mathfrak{gl}_N)$ and $(\mathbb{C}^N)^{*n}$ (see the end of §1). Hence, by means of H. Weyl's duality and Lemma 3.8 (cf. [10], Theorem 1) we can assume S to belong to $\hat{H}_n^q$ and reduce the problem of describing S to Hecke algebras.

Next, by using the identities of Lemma 3.8 we obtain that if $SF_1 S' = F_1 S'$ for $S^2 = S \subset \hat{H}_n^q$ then $\langle S \rangle \hat{\mathfrak{S}}_n^q \subset S\hat{H}_n^q$. The converse statement is valid too, which is proved with the aid of the same Lemma, Lemma 3.7 and the formula

$$\operatorname{res}\big((SF_1 - F_1 S')T_n \ldots T_1\big) = (S - 1)\operatorname{res}\big(F_1 F_n^{-1}\big)S = 0.$$

The analogous implications are true for F_{n-1} in place of F_1.

Thus we obtain that a subspace of V_u^q has the form BV_u^q for a $\hat{\mathfrak{S}}_n^q$-invariant B iff it is invariant under the action of all matrix elements of F_0, F_1, F_{n-1}, F_n (we would remind the reader that F_i are $N \times N$-matrices with their elements in $M_N^{\otimes n} \simeq {}^2 M_N \ldots {}^{n+1} M_N$). But $\{E_{ij}^s,\ s = 0, 1, n - 1, n\}$ generate $\mathcal{D}_N^q(u)$ (see Proposition 1.2). $\qquad\qquad\square$

If for some skew μ, μ' the sequence (u_k') can be obtained from (u_k) by a permutation of such segments $(u_{a_i}, u_{a_i+1}, \ldots, u_{b_i})$ that $u_{a_i-1} - u_{a_i}$, $u_{b_i} - u_{b_i+1} \notin \mathbb{Z}$, $1 \leqslant a_i \leqslant b_i \leqslant n$, then $V_\mu^q \simeq V_{\mu'}^q$ and we will call μ, μ' to be equivalent. It follows from Theorem 3.6 that isomorphism classes of the $\mathcal{D}_N^q(u)$-modules $V_\mu^q \neq \{0\}$ up to $\pi_u^{u'}(0)$ from Proposition 1.3 are in one-to-one correspondence with the classes of μ modulo equivalence. This can be deduced from the results of §2 too. The automorphism $\pi_u^u(d)$ (ibid.) corresponds to the translation $u_k \to u_k + d$, $1 \leqslant k \leqslant n$.

Let us also use homomorphisms of the type (ii), (ibid.) to compare V_μ^q with μ of different degrees and identify representations modulo multiplications of their L-functions by rational scalar functions in x. Then if $V\mu$ is not one-dimensional it is equivalent to some V_μ^q with the following μ': $y_k' = y_l' \Rightarrow i_k' - i_l' < N - 1$ (cf. Theorem 2.6). The mapping $\{\mu'\} \to \{V\mu'\}$ of equivalence classes is a one-to-one correspondence.

The following Theorem exposes to some extent the place of modules V_μ^q among all irreducible representations of $\mathcal{D}_N^q(u)$.

THEOREM 3.9. *In the notations of §2 we suppose the subalgebra $A_N^q(u)$, generated by the coefficients of $H^k(x)$, to act semi-simply and to have only simple eigenvalues on some irreducible $\mathcal{D}_N^q(u)$-composition factor V of V_u^q, $u \in \mathbb{C}^n$ (i.e. V is a quotient-module of a submodule of V_u^q). Then $V \simeq V\mu$ for some skew μ and the initial u is a permutation of $u = (i_k - y_k) \sim \mu$ if N/n is sufficiently large (e.g. $N > n$).*

Sketch of Proof.   The next lemmas are valid for an arbitrary u, N.

LEMMA 3.10.   (i) *In the notations of Theorem 2.6 $V_u = \oplus_{k=0}^{n} H_n^q V_k$, where V_k and $H_n^q V_k = \mathbf{C}[S_n]V_k$ are $\mathscr{D}_{N-1}^q(u)$-modules under the action of $\mathscr{D}_{N-1}^q(u) = \rho^{N-1}(\mathscr{D}_{N-1}^q(u))$. (ii) There exists a filtration of $H_n^q V_k$ by some $\mathscr{D}_{N-1}^q(u)$-modules with the modules $V_{u'}$ as its composition factors, where u' ranges over all the subsets $u' \in \mathbf{C}^k$ of u.*   □

LEMMA 3.11 (cf. Theorem 3.6).   *Every $\mathscr{D}_{N-1}^q(u)$-submodule of $H_n^q V_k$ has the form $\hat{B}V_k$ for $\hat{B} \subset H_n^q$, which is a submodule under the action of $\hat{\mathfrak{H}}_k^q \subset \mathfrak{H}_n^q$ on $H_n^q \simeq \mathscr{J}_u^q$ (any such $\hat{B}V_k$ is a $\mathscr{D}_{N-1}^q(u)$-submodule).*   □

The proof of the Theorem is by induction on n/N with the case $1/(N - n + 1)$ being clear. Firstly, we can suppose V to be not only a composition factor but a submodule of V_u^q (we use Theorem 3.6 and the theory of Hecke algebras). All the irreducible composition factors of $V' = V \cap H_n^q V_{N-1}$ are pairwise non-isomorphic $\mathscr{D}_{N-1}^q(u)$-submodules. It follows from the assumption of the Theorem. Therefore, if $V = BV_u^q$ for some $\mathfrak{H}_n^q$-submodule B of $H_n^q \simeq \mathscr{J}_u^q$ then the irreducible composition factors B_i of B as an $\hat{\mathfrak{H}}_{n-1}^q$-module are pairwise non-isomorphic too. This is deduced from some corollary of Lemma 3.11 and implies a semi-simplicity of the action of X_1 on $B([X_1, \hat{\mathfrak{H}}_{n-1}^q] = 0)$. Next we apply the Lemmas and the statement of the Theorem for $(n - 1)/(N - 1)$ to show that each B_i corresponds to a skew diagram of $\deg = n - 1$. Hence, other $X_2, \ldots, X_n$ act semi-simply on B and we can use Theorem 5 from [6].   □

We conjecture this Theorem to be true for any V, N modulo the above equivalence of representations (by homomorphisms of Proposition 1.3 and the proportionality of L-functions). To prove (or disprove) it in general one needs, to begin with, a description of irreducible representations of $\mathscr{D}_N^q(u)$ in terms of their "highest weights" like that of V. G. Drinfeld, V. O. Tarasov for Yangians [17] to show that every irreducible representation belongs to some V_u^q. Then one can use Theorem 3.6 and the classification of irreducible $\mathfrak{H}_n^q$-modules (A. V. Zelevinsky, J. D. Rogawski—see [13]) and can try to generalize the above branching properties of $\mathfrak{H}_n^q$.

Appendix: Yangians.   We explain briefly how to specialize the above results to Yangians (see [8] for their general definition). We have to put $x = q^\lambda$, $\tilde{E}_{ii}^n = 1$ and tend $q^{1/2}$ to 1 in all the notations and formulas. After some substitutions we can degenerate

$$L(x) \text{ to } L(\lambda) = I + \sum_{s=1}^{n} \lambda^{n-s} e_{ij}^{s-1} I^{ji} \Big/ \prod_{i=1}^{n} (\lambda - u_i), \qquad 1 \leqslant i, j \leqslant n.$$

The relations of Proposition 1.2 will become as follows:

$$[e_{ij}^r, e_{kl}^s] = \delta_{jk} e_{il}^{r+s} - \delta_{li} e_{kj}^{r+s} + \sum_{\substack{a+b=r+s-1 \\ a<r\leqslant b}}^{a<r\leqslant b} \left(e_{ki}^a e_{il}^b - e_{kj}^b e_{il}^a \right).$$

The corresponding algebra $Y_N(u)$ is the Yangian of level n for $\mathfrak{gl}_N$ (note that $Y_N(0) \simeq U(\mathfrak{gl}_N)$). Drinfield's Yangian $Y(\mathfrak{sl}_N)$ is the quotient-algebra of $\lim_{\leftarrow u} Y_N(u)$ by the relation $H^N(\lambda) = 1$, where we replace xq^a by $\lambda + a$ in formula (3) for H^N and in any expressions with L, H-functions. For $u = 0$ Theorems 2.4, 2.6 are deduced from the results by G. I. Olshanski, Theorem 2.2 is a version of the classic theorems.

One obtains the degenerated Hecke algebra $\mathfrak{H}_n^1$ [8] by putting $X_i = q^{\chi_i}$, $T_i = s_i$ for $q \to 1$. This algebra is generated by $\mathbb{C}[S_n]$ and χ_i with relations (5) in the following form: $\chi_{i+1} s_i - s_i \chi_i = 1 = s_i \chi_{i+1} - \chi_i s_i$. Theorems 3.3, 3.4, (ii) (for χ_i instead of X_i) are deduced from [9]. The degenerated homomorphism $\mathfrak{H}_n^1 \to \mathfrak{H}_n^1(0) \simeq \mathbb{C}[S_n]$ was defined by Drinfeld. He conjectured in [8] the existence of q-versions of this homomorphism and some Weil-type duality (see Theorem 3.6) for $q \neq 1$.

References

1. V. G. Drinfeld, *Hopf algebras and the quantum Yang-Baxter equation*, Doklady Akad. Nauk SSSR, **283** (1985), N 5, 1060–1064.

2. M. Jimbo, *A q-difference analoque of $U(\mathfrak{g})$ and the Yang-Baxter equation*, Lett. Math. Phys. **10** (1985), 63–69.

3. L. D. Faddeev, *Quantum completely integrable models in field theory*, Sov. Sci. Reviews. Sec. C., **1** (1980), 107–155. Harwood Acad. Publishers, Chur (Switzerland)-New York.

4. E. K. Sklyanin, *On some algebraic structures connected with the Yang-Baxter equation*, Funct. Anal. and Appl. **16** (1982), N 4, 27–34.

5. I. V. Cherednik, *On some finite-dimensional representations of generalized Sklyanin algebras*, Funct. Anal. and Appl., **19** (1985), N 1, 89–90.

6. ______, *On R-matriz quantization of formal loop groups*, In: *Group Theoretic Methods in Physics*. Nauka, Moskow, 1986 (VNU Publ. B.V., 1986).

7. ______, *On "quantum" irreducible finite-dimensional representations of $\mathfrak{gl}_N$*, Doklady Akad. Nauk SSSR, **287** (1986), N 5, 1076–1079.

8. V. G. Drinfeld, *Degenerated affine Hecke algebras and Yangians*, Funct. Annal. and Appl., **20** (1986), N 1, 69–70.

9. I. V. Cherednik, *On special bases of irreducible representations of the degenerated affine Hecke algebra*. Funct. Annal. and Appl., **20** (1986), N 1, 87–89.

10. M. Jimbo, *A q-analogue of $U(\mathcal{H}(N+1))$, Hecke algebra and the Yang-Baxter equation*, Preprint RIMS-517 (1985), Kyoto University.

11. M. Jimbo, *Quantum R-matrix related to the generalized Toda system: an algebraic approach*, Preprint RIMS-521 (1985), Kyoto University.

12. P. P. Kulish, E. K. Sklyanin, Quantum spectral transform method. Recent developments, In: *Integrable quantum field theories* (Lect. Notes in Phys. N 151), p. 61–119. Springer-Verlag, Berlin-Heidelberg-New York, 1982.

13. J. D. Rogawski, *On modules over the Hecke algebra of a p-adic group*. Invent. Math, **79** (1985), 443–465.

14. I. V. Cherednik, *Hamilton theory of steady-state differential equations with an elliptic pencil*, Doklady Akad. Nauk SSSR, **271** (1984) N 1, 51–55.

15. I. V. Cherednik, *Elliptic curves and matrix soliton differential equations*, Itogi nauki i techniki VINITI. Algebra, Topology, Geometry, **22** (1984), 205–265.

16. Yu. I. Manin, *Algebraic aspects of nonlinear differential equations*, Itogi nauki i techniki VINITI. Modern problems of Math. **11** (1978), 5–152.

17. V. G. Drinfeld, *New realization of Yangians and quantized affine algebras*, Preprint FTINT AN USSR, N 17-86, Kharkov, 1986.

Belozersky Laboratory, Moscow State University, Moscow 119899

Vol. 54, No. 2   DUKE MATHEMATICAL JOURNAL ©   1987

QUANTUM FIELD THEORIES IN ONE AND TWO DIMENSIONS

WERNER NAHM

1. Introduction.   All known physical forces with the exception of gravity have been described by quantum field theories. This is probably the most important achievement of theoretical physics in the past fifty years. However, for several decades this success led to an estrangement between physics and mathematics, as we have not yet learnt to analyse those complex theories with methods which are both efficient and rigorous. In recent years, however, physicists found uses for simpler systems of that kind, mainly as tools for the description of surface phenomena in statistical physics and in attempts to understand the relation of gravity to the other forces of nature, and mathematicians discovered that these systems could be adapted to their purposes. A striking discovery along these lines was the realization that one of the simplest possible quantum field theories has the biggest sporadic simple group F_1 (or FG) as automorphism group.

Good introductions to quantum field theory are available (Streater and Wightman 1978, with further references), but they are mostly geared towards the treatment of more complex systems. Here I will give an introduction to the simplest quantum field theories and some new results. I hope that this will contribute to the new upsurge of interactions between the mathematics and physics communities, to which Yu. I. Manin has contributed so much.

Section 2 introduces a set of rather general quantum field theories, which should include the experimentally relevant ones in four dimensions. However, features not used in sections 3, 4 are not discussed. The main concern is the dimension of local fields, for which I found no sufficient discussion elsewhere. In section 3 several important examples in one dimension are discussed, and section 4 gives some new results for particularly simple chiral theories in two dimensions. The examples in section 3 include the one with symmetry group F_1. The results of section 4 apparently are related to the correspondence between the ADE series of simple Lie algebras and the finite subgroups of SU(2).

2. Axioms for quantum field theories.   Quantum field theories model the causal structure of space-time, the states of a physical system, the results of measurements performed on the states, and the interrelationships between these structures.

Space-time is modelled by a differentiable manifold M of dimension d equipped with a causal structure.

Received January 24, 1987.

Definition 1.1.   A pseudo-Riemannian metric on M is a non-degenerate quadratic form of signature $(1, d - 1)$ in the tangent bundle of M. A tangent vector v is called timelike, if $v^2 > 0$. A timelike curve is a differentiable curve with only timelike tangent vectors.

Definition 1.2.   Two pseudo-Riemannian metrics on M are conformally equivalent, if they yield the same sets of timelike tangent vectors. A conformal structure on M is a class of conformally equivalent pseudo-Riemannian metrics on M.

Definition 1.3.   A causal structure of M is defined as follows: For $d = 1$, two open sets $\sigma, \sigma' \subset M$ are causally unrelated, if their intersection is empty. For $d \geqslant 2$ let M have a conformal structure and no closed timelike curves. Two points $m, m' \in M$ are causally related, if they can be connected by a timelike curve. Two open sets $\sigma, \sigma' \subset M$ are causally unrelated, if no two points $m \in \sigma$, $m' \in \sigma'$ are causally related.

The set of states of a physical system is modelled by a topological space $\hat{V}$, which may have components $\hat{V}_i$ called superselection sectors. In quantum field theories the $\hat{V}_i$ are dense subspaces $\hat{V}_i = V_i/\mathbb{C}^*$ of projective Hilbert spaces $P\mathcal{H}_i = \mathcal{H}_i/\mathbb{C}^*$. States usually are represented by vectors in $\mathcal{H} = \oplus \mathcal{H}_i$. The hermitean form on $\mathcal{H}$ will be written $\langle \, , \, \rangle$.

Measurements both change the states and yield real numbers as results. Measurable quantities are represented by self-adjoint operators A. The spectral decomposition

$$A = \int a \, d\pi(a)$$

yields a projector valued measure $d\pi(a)$ on $\mathbb{R}$. For any state v the $\mathbb{R}$-valued measure $\langle v, d\pi(a)v \rangle / \langle v, v \rangle$ models the probability distribution of the results of the measurement of A in v. Self-adjoint operators generate unitary transformations of $\mathcal{H}$, and these can be used to describe the change of v due to the measurement.

Measurements take place in finite domains of space-time. Those performed in causally unrelated domains should not influence each other. This can be modelled by a co-sheaf $\mathcal{M}$ of unitary transformation groups of $\mathcal{H}$. The generators of the group $\mathcal{M}(\sigma)$ model the experiments which can be done in σ. The groups $\mathcal{M}(\sigma)$, $\mathcal{M}(\sigma')$ commute whenever σ, σ' are causally unrelated. Moreover, all $\mathcal{M}(\sigma)$ should respect the decomposition of v into superselection sectors.

Local quantum field theory uses the concept that measurable quantities can be written additively in terms of contributions from arbitrarily small spacetime domains. In other words, there should be distributions ϕ over M such that $\phi[f]$ is a generator of $\mathcal{M}(\sigma)$ whenever the support of f lies in σ. Such ϕ are called local fields. Examples are the electromagnetic fields and currents and the densities of energy and momentum.

Other interactions also have been described successfully in terms of operator valued distributions and their expectation values $\langle v, \phi[f]v \rangle$. These $\phi[f]$ can be assumed to be symmetric operators, but not much is known about their selfadjointness. Moreover, exponentiation of $\phi[f]$ has no direct physical interpretation, only a time ordered version of it has.

Thus at least temporarily one is led to consider co-sheaves of operator algebras and not the potentially more elegant groups. However, in the one- and two-dimensional cases considered later, time ordering is inessential, and certain groups do appear, namely subgroups of $\mathrm{Diff}(M)$ and $\mathrm{Map}(M, G)$, where G is a compact Lie group. The interpretation of their tangent spaces in terms of local quantum fields is natural. Much bigger unitarily representable groups with nice tangent spaces might be generated in this way, but here this question will not be pursued further.

Definition 2.1.   Let V be a dense subspace of the Hilbert space $\mathcal{H}$ and $O(V)$ the algebra of those linear operators $A: V \to V$ which have an adjoint $A^+: V \to V$, equipped with the weak topology. Let $\mathcal{D}(M)$ be the space of C^∞ functions from M to $\mathbf{C}$ with compact support, with the standard topology and the involution $f \to \bar{f}$ given by complex conjugation. Let $M^n = M \times \cdots \times M(n$ times$)$ and let

$$\mathcal{D}(M)^x = \bigoplus_{n=0}^{\infty} \mathcal{D}(M^n)$$

be the graded tensor algebra of test functions in $\coprod_n M^n$, with involution

$$\bar{f}(x_1,\ldots, x_n) = \overline{f(x_n,\ldots, x_1)} \quad \text{for } f \in \mathcal{D}(M^n).$$

An algebra of multilocal fields belonging to $(M, \mathcal{H}, V)$ is a graded algebra F^x of continuous homomorphisms $\phi: \mathcal{D}(M)^x \to O(V)$ with the natural grading

$$F^x = \bigoplus_{n=0}^{\infty} F^{(n)}$$

and with an involution $+$ given by

$$\phi^+[f] = (\phi[\bar{f}])^+.$$

The elements of $F^{(1)}$ are called local fields. Real fields are defined by $\phi^+ = \phi$. By abuse of language we will call F^x an algebra of local fields, if it is generated by $F^{(1)}$. We write $F = F^{(1)}$.

Remarks.   Often it is useful to take spaces of paths in M as testfunction spaces for operator valued distributions. This concept will not be used here.

Note that $(a + A)^+(a + A)$ has self-adjoint closure for all $A \in O(V)$, $a \in \mathbf{C}$. All elements of $O(V)$ can be written as linear combinations of elements of this form.

The action of ϕ on f will be written as $\phi[f]$. After fixing a measure $d\mu$ on M we also use the physicist's notation

$$\phi[f] = \int_{M^n} \phi(x) f(x)\, d\mu^n, \qquad \phi \in F^{(n)}, f \in \mathcal{D}(M^n).$$

For diffeomorphisms $g\colon M^n \to M^n$ we may then write

$$\int_{M^n} \phi(gx) f(x)\, d\mu^n = \int_{M^n} \phi(x) f(g^{-1}x)|g'|^{-1}\, d\mu^n,$$

where g' is the Jacobian of g.

Definition 2.2. A system $(M, \mathcal{H}, V, F^x)$ is called decomposable into superselection sectors, if any subspace V' of V which is closed in V and mapped by $F^x[\mathcal{D}(M)^x]$ into itself has an orthogonal complement in V. Such subspaces V' are called superselection sectors of the system.

Because of the adjoint operation in F, the orthogonal complement of V' also is a superselection sector.

To relate the superselection sectors we use the following concept of coherence.

Definition 2.3. For any subalgebra $\mathfrak{U}$ of $O(V)$ let $\mathfrak{U}'$ be the centralizer of $\mathfrak{U}$ in $O(V)$. The system $(M, \mathcal{H}, V, F^x)$ is coherent, if there is an open covering

$$M = \bigcup_i \sigma_i$$

of M such that for any σ_i only the trivial superselection sectors $\{0\}, V$ are mapped by $F^x[\mathcal{D}(\sigma_i)^x]'$ into themselves.

Definition 2.4. Let M be a differentiable manifold with causal structure, $\mathcal{H}$ a separable Hilbert space, V a dense subspace. Let F^x be an algebra of local fields belonging to $(M, \mathcal{H}, V)$, such that $(M, \mathcal{H}, V, F^x)$ is decomposable into superselection sectors and coherent. The system $Q = (M, \mathcal{H}, V, F)$ is a quantum field theory, if for causally unrelated open sets σ, σ' of M one has

$$\phi[f]\phi'[f'] = \phi'[f']\phi[f] \text{ for all } \phi, \phi' \in F,$$

$$f \in \mathcal{D}(\sigma), \qquad f' \in \mathcal{D}(\sigma'). \tag{1}$$

In short, $F[\mathcal{D}(\sigma)]$ and $F[\mathcal{D}(\sigma')]$ commute.

Definition 3. A field theory $(M, \mathcal{H}, V, F)$ is trivial, if $F = 0$.

Note that triviality implies that $V = \mathcal{H}$, as all closed subspaces of V are superselection sectors and have orthogonal complements, in particular the space orthogonal to any $v \in \mathcal{H}$.

Definition 4.   The product of two field theories $Q_i = (M, \mathcal{H}_i, V_i, F_i)$, $i = 1, 2$, is the field theory $Q_1 \times Q_2 = (M, \mathcal{H}_1 \otimes \mathcal{H}_2, V_1 \otimes V_2, F_1 \oplus F_2)$.

LEMMA 1.   *Let $\phi \in F$ and $\phi V' = 0$ for some nontrivial superselection sector V'. Then ϕ vanishes identically.*

Proof.   This follows from coherence (def. 2.3). Let $f_i \in \mathcal{D}(\sigma_i)$ and consider the maximal superselection sector on which $\phi[f_i]$ vanishes. By coherence this must be all of V. By a partition of unity argument, $\phi[f] = 0$ for all $f \in \mathcal{D}(M)$.   □

Thus for arbitrary Q_1, Q_2 there is no natural way to define a Q with $\mathcal{H} = \mathcal{H}_1 \oplus \mathcal{H}_2$.

Definition 5.   Let $Q = (M, \mathcal{H}, V, F)$ and $Q' = (M, \mathcal{H}, V, F')$ be two quantum field theories. Q' is an extension of Q, if $F \subset F'$. Q is maximal, if it has no proper extensions. A superselection sector V' of Q is natural, if it remains a superselection sector for any extension of Q.

It happened several times that seemingly unrelated field theories were found to have isomorphic extensions, to the surprise of the physics community. A famous conjecture in high energy physics states that in the correct quantum field theory of the world all superselection sectors are natural ("anything which can go, will go, e.g., proton decay").

LEMMA 2.   *Let h be a C^∞ vector field on M. Then F can be extended such that $hF \subset F$.*

Proof. $h\phi[f] = -\phi[hf]$ defines a continuous map $h\phi : \mathcal{D}(M) \to O(V)$ and therefore a natural extension of F^x as an algebra of local fields belonging to $(M, \mathcal{H}, V)$. The superselection sectors do not change. Eq. (1) is satisfied as the support of hf is contained in the support of f.   □

One also can enlarge the theory by including $\mathcal{D}(M)$ in a bigger space of test functions. When M is an abelian group manifold, it has technical advantages to work with the Schwartz space $\mathcal{S}(M)$ of C^∞ functions of rapid decay.

Symmetries of quantum field theories are transformation groups with a projective unitary action on $\mathcal{H}$ which transform superselection sectors into superselection sectors and have a prescribed induced action on the local fields. Using $P\mathcal{H} = \mathcal{H}/\mathbb{C}^*$ they are usually represented by a unitary representation of a central extension of the transformation group on $\mathcal{H}$, where the extension is by a subgroup of $\mathbb{C}^*$.

Definition 6.   Let M be a connected abelian group manifold with invariant causal structure. Let R_x be the group of invariant vector fields on M, and R_p the dual group of R_x. Let tR_x, tR_p, be the sets of nonzero vector fields in R_x, R_p for $d = 1$ and of timelike vector fields in R_x, R_p for $d \geqslant 2$. Let

$$tR_x = R_x^+ \cup R_x^-, \qquad tR_p = R_p^+ \cup R_p^-$$

their decomposition into connected parts such that $ph > 0$ for $p \in R_p^+$, $h \in R_x^+$.

For $m \in M$ let $m_*: \mathcal{D}(M) \to \mathcal{D}(M)$ be defined by $(m_* f)(x) = f(mx)$ and the operator valued distributions $m^* \phi$ by $m^* \phi[f] = \phi[m_* f]$. A quantum field theory $Q = (M, \mathcal{H}, V, F)$ is translationally invariant, if there is an extension M' of M by a cyclic group and a representation U of M' in such that

— $U(m')V' = V'$ for all $m' \in M'$ and all superselection sectors V' of Q
— $hF \subset F$ for all $h \in R_x$.
— $U(m')\phi U(m')^{-1} = (\pi^M m')^* \phi$ for all $\phi \in F$, $m' \in M'$
— the spectrum of $U^*(R_x)$ considered as a subset of R_p is contained in $p_0 + R_p^+$ for some $p_0 \in R_p$
— for any $h \in R_x^+$ and any real number E one has $\pi^h(E)\mathcal{H} \subset V$.

Here $\pi^M: M' \to M$ is the natural projection, U^* is the map from the generators of M' into the selfadjoint operators in $\mathcal{H}$ induced by U and

$$U^*(h) = \int_{E \geq p_0 h} E \, d\pi^h(E)$$

is the spectral decomposition of $U^*(h)$.

The following two lemmata are immediate consequences of this definition.

LEMMA 3. *For any compact domain $X \subset R_p$ let $\pi^*(X)$ be the corresponding spectral projector of $U^*(R_x)$. The operator valued distribution $\pi^*(X_1)\phi\pi^*(X_2)$ has a Fourier transform with support in the subset $\{p_1 - p_2 | p_1 \in X_1, p_2 \in X_2\}$ of R_p.*

Proof. This is just a restatement of the transformation law of local fields under M'. $\square$

LEMMA 4. *Let IM be an index set labelling the irreducible representations of the cyclic group $\ker \pi^M$ and $\mathcal{H} = \bigoplus_{\nu \in IM} \mathcal{H}_\nu$ the corresponding decomposition of $\mathcal{H}$. Let $V_\nu = \mathcal{H}_\nu \cap V$. Then $V = \bigoplus_{\nu \in IM} V_\nu$. The V_ν are superselection sectors of any translationally invariant extension of Q.*

Proof. The V_ν are superselection sectors, as $\ker \pi^M$ commutes with all elements of F^x. As $\pi^h(E)\mathcal{H} \subset V$ for any E their direct sum is dense in $\mathcal{H}$. As V decomposes into superselection sectors, the sum is equal to V. $\square$

Definition 7. Let $\mathcal{S}(M)$ be the Schwartz space of test functions of rapid decay, $\tilde{f}$ the Fourier transform of $f \in \mathcal{S}(M)$. The dimension $d(\phi)$ of a local field ϕ is the lower bound of the real numbers r for which

$$\|\pi^*(p_1 + B)\phi[f]\pi^*(p_2 + B)\| = O\left((p_1 h + p_2 h)^{r - dc/2} \max_{p \in p_1 - p_2 + 2B} |\tilde{f}(p)|\right)$$

for some $h \in R_x^+$, all $f \in \mathcal{S}(M)$ and all balls B around the origin in R_p. Here $d_c = d - 1$ for $d \geq 2$ and $d_c = 1$ for $d = 1$. If no such r exists, we put $d(\phi) = \infty$.

Remarks. Our assumption about the spectrum of $U^*(R_x)$ implies that $d(\phi)$ is independent of the choice of h. The value of d_c arises as the dimension of the space of null vectors in R_p. For constant local fields, which have support in $\{0\} \subset R_p$, one should put $d_c = 0$, but for technical reasons this adjustment of $d(\ )$ will be done later.

LEMMA 5. $d(\phi^+) = d(\phi)$, $d(\phi_1 + \phi_2) \leqslant \max(d(\phi_1), d(\phi_2))$, $d(h\phi) \leqslant d(\phi) + 1$ *for* $h \in R_x$.

Proof. Obvious. $\qquad\qquad\qquad\qquad\qquad\qquad\qquad\qquad\qquad\qquad\qquad\qquad\Box$

LEMMA 6. *Let* ϕ *be a field of finite dimension,* $\overline{\phi[f]}$ *the closure of* $\phi[f]$ *in* $\mathscr{H}$, *and for some* $h \in R_x^+$, $\Delta > 0$

$$V_0 = \left\{ v \in \mathscr{H} \mid \|(\pi^h(E + \Delta) - \pi^h(E))v\| = O(E^{-n}) \text{ for all } n \in Z \right\} \quad (2)$$

Then $\overline{\phi[f]}$ *is defined on* V_0 *and* $\overline{\phi[f]}\, V_0 \subset V_0$.

Proof. Write $V = \lim_{E \to \infty} \pi^h(E)V$. Use that $\phi[f]$ is defined on $\pi^h(E)V$ and the definition of $d(\phi)$, together with the rapid decrease of f. $\qquad\qquad\Box$

PROPOSITION 1. *Let* $Q = (M, \mathscr{H}, V_0, F)$ *be a translationally invariant quantum field theory, with* V_0 *given by eq. (2). Let* ϕ *admit an extension to a continuous map* $\hat{\phi}\colon \mathscr{S}(M) \to O(V_0)$. *Then* ϕ *has finite dimension.*

Proof. By lemma 5 we may assume ϕ to be real. Let for some B

$$\Phi(p', p) = \pi^*(p' + B)\phi\pi^*(p + B).$$

We have $\Phi(p', p)^+ = \Phi(p, p')$. Assume that $d(\phi) = \infty$. First we shall construct a real $f \in \mathscr{S}(M)$ and a sequence (p_i', p_i) with $p_i'h \geqslant p_i h$ such that $p_i h \to \infty$ and $\rho_i = \|\Phi(p_i', p_i)[f]\|$ increases faster than any power of $p_i'h$. Our assumption means that there is such a sequence and a $g \in \mathscr{S}(M)$ such that

$$\tilde{\rho}_i = \|\Phi(p_i', p_i)[g]\| \Big/ \max_{p \in p_i' - p_i + 2B} |\tilde{g}(p)|$$

increases faster than any power of hp_i'. If infinitely many of the displaced balls $p_i' - p_i + 2B$ contain a common ball $p_\infty + B$, we just take the corresponding subsequence and $f = g$, $\tilde{\rho} = \rho$. Otherwise the sequence of balls contains a subsequence of separated balls, in which case one may construct f by interpolation of the following values in those balls,

$$\tilde{f}(p) = \tilde{g}(p)\tilde{\rho}_i^{-1/2} \Big/ \max_{p \in p_i' - p_i + 2B} |\tilde{g}(p)|$$

and use

$$\rho = \tilde{\rho}^{-1/2}.$$

Now we choose v_i of unit norm such that

$$\|\Phi(p_i', p_i)[f]v_i\| \geq \rho_i/2 \text{ for all } i \in \mathbb{Z}_+$$

Try to define iteratively a subsequence by $\iota\colon \mathbb{Z}_+ \to \mathbb{Z}_+$ such that for all $j \in \mathbb{Z}_+$

$$\sum_{n=\iota(j)}^{\infty} |\Phi(p_{\iota(j-1)}', p_n)[f]|\rho_n^{-1} \leq \frac{1}{6}$$

$$\sum_{k=0}^{j-1} |\Phi(p_{\iota(j)}', p_{\iota(k)})[f]|\rho_{\iota(k)}^{-1} \leq \frac{1}{6}$$

If such a subsequence exists, an easy estimate yields

$$v = \sum_{j=0}^{\infty} \rho_{\iota(j)}^{-1} v_{\iota(j)} \in V_0$$

$$\hat{\phi}[f]v \notin V_0.$$

If no such subsequence exists, there is a fixed E such that $\pi^h(E')\phi[f]\pi^h(E)$ cannot be bounded by any power of E'. As $\pi^h(E)\mathscr{H}$ is closed, it contains a v such that $\hat{\phi}[f]v \notin V_0$. Thus in any case $\hat{\phi}[f]$ does not belong to $O(V_0)$ and $d(\phi) = \infty$ leads to a contradiction. $\qquad\square$

Perhaps the finiteness of the local field dimensions can be deduced from much weaker assumptions. No candidates for local fields of infinite dimension are known to me. Experimentally accessible fields have low dimensions.

Even if one considers $Q = (\mathbb{R}^d, \mathscr{H}, V, F)$ it is often useful to deal first with related theories $Q_c = (M_c \times \mathbb{R}, \mathscr{H}_c, V_c, F_c)$ for $d \geq 2$ or $(S^1, \mathscr{H}_c, V_c, F_c)$ for $d = 1$, such that M_c has dimension $d - 1$ and is compact and Riemannian, whereas $F[\mathscr{D}(\sigma_i)]$ and $F_c[\mathscr{D}(\sigma_i)]$ yield isomorphic operator algebras for small isomorphic subsets σ_i^c and σ_i of $M_c \times \mathbb{R}$ or S^1 and $\mathbb{R}^d$. The compact theories are easier to study and the limit towards infinite extension may be taken later. Moreover, the compact systems also have important direct applications. From now on we shall only consider such cases.

Definition 8. A translationally invariant quantum field theory $Q = (M_c \times \mathbb{R}, \mathscr{H}, V_0, F)$ with M_c compact or $Q = (S^1, \mathscr{H}, V_0, F)$ is of standard type, if V_0 is the space defined in lemma 6 and $\exp(-U^*(h))$ is of trace class for all $h \in R_x^+$. The map $Z\colon (R_x + iR_x^+) \to \mathbb{C}$, $Z(h) = \operatorname{tr}\exp(iU^*(h))$ is the partition function of Q.

All experimentally relevant translationally invariant quantum field theories are assumed to be of standard type, when they are restricted to a finite space volume. The partition function is the basic quantity in thermodynamics.

In theories of standard type the spectrum of $U^*(R_x)$ consists of isolated points with finite multiplicities, which often allows a treatment which concentrates on algebra and not on functional analysis.

3. One dimensional quantum field theories. Let $Q = (S^1, \mathcal{H}, V, F)$ be a translationally invariant quantum field theory. On S^1 we use the coordinate function $x \in \mathbb{R} \bmod 2\pi \mathbb{Z}$ and the measure dx. The generator

$$L_0 = U^*(\partial_x), \qquad \partial_x = \partial/\partial_x \in R_x^+$$

of $U(M')$ has the spectral decomposition

$$L_0 = \sum_{\nu \in IM} \sum_{n=0}^{\infty} (n + n_\nu)\pi(n + n_\nu)$$

such that $\pi(n_\nu)\mathcal{H}$ lies in the superselection sector V_ν. In terms of the Fourier coefficients $\phi_n = \phi[e^{inx}]$ of $\phi \in F$, translational invariance implies

$$[L_0, \phi_n] = n\phi_n, \quad \text{where } [AB] = AB - BA.$$

According to def. 7 the dimension of ϕ now is the lower bound of the real numbers r for which

$$|\phi_n \pi(m)| = O\big((n + m)^{r-1/2}\big).$$

The criterion of proposition 1 for the finiteness of $d(\phi)$ also simplifies, as $\mathscr{S}(S^1) = \mathscr{D}(S^1) = C^\infty(S^1)$. Thus all $d(\phi)$ are finite, when we use for V the space V_0 defined in eq. (2.2).

THEOREM 1. *Any translationally invariant quantum field theory* $(S^1, \mathcal{H}, V, \tilde{F})$ *has an extension* $(S^1, \mathcal{H}, V, F)$ *in which*

$$[\phi(x)\chi(y)] = \sum_{k=0}^{[d(\phi)+d(\chi)-1]} W(\phi, \chi, k)(x)(-i\partial_x)^k \delta(x - y),$$

with $W(\phi, \chi, k) \in F$ *for arbitrary* $\phi, \chi \in F$.

Proof. Due to eq. (1) for any $v, v' \in V$ the distribution $\langle v', [\phi(x)\chi(y)]v \rangle$ on $C^\infty(S^1 \times S^1)$ must have support in the diagonal of $S^1 \times S^1$. Thus we may write

$$[\phi(x)\chi(y)] = \sum_{k=0}^{\infty} W(\phi, \chi, k)(x)(-i\partial_x)^k \delta(x - y),$$

where for $\phi, \chi \in \tilde{F}$ the $W(\phi, \chi, k)$, $k \in \mathbb{Z}$ are a family of quadratic forms on $\mathcal{H}$

with values in the distributions on $C^\infty(S^1)$, such that for any v only a finite number of the distributions $\langle v, W(\phi, \chi, k)v\rangle$ is different from zero. Let $g_k \in C^\infty(S^1)$ be equal to $(ix)^k/k!$ in some neighbourhood of $x = 0$. Then

$$W(\phi, \chi, k) = \int g_k(z)[\phi(x)\chi(x + z)]\, dz.$$

Thus for all $f \in C^\infty(S^1)$ the operator $W(\phi, \chi, k)[f]$ lies in $O(V)$ and maps all superselection sectors into themselves. As the support of g_k may be chosen to be arbitrarily small, eq. (1) is satisfied, too. Thus there are no obstacles for the extension. The rest of the theorem is a part of the following lemma.

LEMMA 7.

$$d(W(\phi, \chi, k)) \leq d(\phi) + d(\chi) - k - 1/2$$

$$W(\phi, \chi, k) = 0 \quad \text{for } k > d(\phi) + d(\chi) - 1.$$

Proof. We have

$$[\phi_{n-m}\chi_m] = \sum_{k=0}^{\infty} W(\phi, \chi, k)_n m^k$$

and

$$W(\phi, \chi, k)_n = \sum_{m \in \mathbb{Z}} c_k(n, m)[\phi_{n-m}\chi_m]$$

where c_k is any function for which

$$\sum_{m \in \mathbb{Z}} c_k(n, m)m^r = \delta_{kr}, \qquad r \in \mathbb{Z}_+.$$

Let $d(\phi)$ and $d(\chi)$ be finite, as otherwise there is nothing to prove. Let the function $c_k': \mathbb{R} \to \mathbb{R}$ have support in $\{0, 1, \ldots, [d(\phi) + d(\chi)] - 1\}$ and satisfy

$$\sum_{m \in \mathbb{Z}} c_k'(m)m^r = \delta_{kr} \quad \text{for } r \in \{0, 1, \ldots, [d(\phi) + d(\chi)] - 1\}.$$

For any $\varepsilon > 0$ and any sufficiently small $\eta > 0$ one may choose c_k to have the form

$$c_k(n, m) = c_k'(m/n)n^{-k} + c_k^\varepsilon(n, m)$$

with

$$c_k^\varepsilon(n, m) = 0 \text{ for } m \leq n, \quad c_k^\varepsilon(n, m) < \varepsilon|m|^{-d(\phi)-d(\chi)-\eta} \quad \text{for all } n, m.$$

The statements of the lemma now follow from the definition of the dimension and from the fact that c_k' vanishes identically for $k > d(\phi) + d(\chi) - 1$. $\qquad \square$

LEMMA 8.

$$W(\chi, \phi, k) = (-)^{k+1} \sum_{n=0}^{\infty} \binom{n+k}{k} (i\partial_x)^n W(\phi, \chi, n+k)$$

$$W(\phi, \partial_x\chi, k) = -iW(\phi, \chi, k-1)$$

Proof. Obvious.   □

LEMMA 9. *If $\phi v = 0$ for some eigenstate v of L_0, then $\phi = 0$.*

Proof. We show that $\phi \chi_n v = 0$ for any local field χ and any n and use induction. Let $v, v' \in V$ be eigenstates of L_0 with eigenvalues n_0, n_1. Then

$$\langle v', [\phi(x)\chi(y)]v\rangle$$

$$= \langle v', \phi(x)\chi(y)v\rangle$$

$$= \sum_{n \in n_\nu + \mathbf{Z}_+} \exp(i(n-n_0)(x-y) - i(n_1-n_0)x)\langle v', \phi_{n_1-n}\chi_{n-n_0}v\rangle.$$

Thus $\exp i(n_1 - n_0)x\langle v', \phi(x)\chi(y)v\rangle$ is the boundary value of a function of $\exp i(x-y)$ which is meromorphic in the interior of the unit circle of the complex plane and vanishes in an interval (in fact everywhere except at 1). Thus it vanishes identically. By induction one finds that $\phi\psi[f]v = 0$ where $\psi \in F^x$ is arbitrary and f is any exponential function in $\mathscr{D}(M)^x$. The linear combinations of such $\psi[f]v$ are dense in some superselection sector V', such that $\phi V' = 0$. By lemma 1, $\phi = 0$. This is a standard argument in quantum field theories For more details see (Streater and Wightman 1978, chapter 4, sect. 2).   □

The Fourier transforms of the local fields form the infinite dimensional graded Lie algebra

$$[\phi_{n-m}\chi_m] = \sum_{k=0}^{[d(\phi)+d(\chi)-1]} W(\phi, \chi, k)_n m^k \tag{3}$$

with the involution $\phi_n^+ = (\phi_{-n})^+$. From now on we shall assume that $Q = (M, \mathscr{H}, V_0, F)$ is of standard type, such that in particular all fields have finite dimension and the spectrum of L_0 has finite multiplicities.

PROPOSITION 2. *Let F_k be the vector space of local fields of dimension $\leqslant k$. Then*

$$\dim F_k \leqslant 1 + 2 \sum_{n=1}^{[k]} \operatorname{tr} \pi(n + n_\nu) \quad \text{for any } \nu \in IM.$$

Proof. Otherwise for a state $v \in \pi(n_\nu)$ one can construct at least two linearly independent real fields $\phi \in F_k$ such that $\phi_n v = 0$ for $1 \leqslant n \leqslant [k]$. According to eq. (3), $\langle v, [\phi_{-n}\phi_n]v\rangle$ is an odd polynomial of degree $\leqslant 2[k] - 1$. Thus it has to vanish identically, which yields $\phi_n v = 0$ for $n > 0$. Thus the proposition follows from

LEMMA 10. *Let $\phi_m v$ vanish for almost all $m > 0$. Then ϕ is a multiple of the constant field 1_F defined by $1_F[f] = \int f\,dx \cdot 1_0$, where 1_0 is the identity in $O(V_0)$.*

Proof. For a sufficiently high power n one has $(e_x^i \partial_x)^n \phi v = 0$. According to lemma 9 the local field $e^{-inx}(e^{ix}\partial_x)^n \phi$ vanishes. By eq. (2), ϕ commutes with all local fields, such that for any $f \in C^\infty(S^1)$ the eigenspaces of $\phi[f]$ define superselection sectors. By lemma 1, the eigenvalues in all superselection sectors are the same. Fourier decomposition shows that $\phi[f]$ is proportional to $\int f\,dx$. $\square$

LEMMA 11. *For any eigenvector v of L_0 one has*

$$\langle v, W(\phi^+, \phi, k)v\rangle_0 \geqslant 0 \quad \text{for } k = [2d(\phi) - 1] \quad \text{and any } \phi \in F$$

$$\langle v, W(\phi, \phi, k)v\rangle_0 \geqslant 0 \quad \text{for } k = 2[d(\phi)] - 1 \quad \text{and any real } \phi \in F.$$

Proof. Use eq. (3) with $n = 0$, $m \to \infty$, and lemma 10. $\square$

Proposition 2 has an immediate corollary: Any Q of standard type has maximal extensions, when the extensions are required to be of standard type, too. One just has to extend Q sequentially such that $F_1, F_2, \ldots$ cannot be extended further and to take the new F to be the union of the F_k generated by this process.

Note that proposition 2 implies that all fields of dimension < 1 are trivial. If the representation on some superselection sector $V_{0\nu}$ has a real structure, such that $\phi_n v = 0$ for $v \in \pi(n_\nu)$ implies $\phi_n^+ v = 0$, the estimate can be sharpened to

$$\dim F_k \leqslant 1 + \sum_{n=1}^{[k]} \operatorname{tr} \pi(n + n_\nu) \tag{4}$$

Quantum field theories for which the inequality (4) is saturated in some superselection sector are from many points of view the best behaved ones. After one redefines the dimension of 1_F to be zero all dimensions $d(\phi)$ are integers. Extending the terminology for affine Kac-Moody algebras, the representation of the Lie algebra (3) on the superselection sector for which the lowest eigenvalue of L_0 has multiplicity 1 may be called the basic representation.

The standard main tool for extending the space of local fields is the normal ordered product $:\phi\chi:$ of two fields. One has

PROPOSITION 3. *For any two local fields ϕ, χ the limit*

$$:\phi\chi:(x) = \lim_{\varepsilon \to 0} \int \varepsilon^{-1} g\left(\frac{y}{\varepsilon}\right)\left(\phi(x)\chi(x + y) - \pi_y^+[\phi(x)\chi(x + y)]\right) dy$$

exists. Q can be extended such that $:\phi\chi: \in F$ for any $\phi, \chi \in F$. Here g is some C^∞

function with support in $S^1 - \{\pi\}$ for which $\int g\,dy = 1$, and π_y^+ projects a distribution to its positive frequency part

$$\pi_y^+\psi = \sum_{n=1}^{\infty} e^{-iny}\psi_n$$

The limit is independent of the choice of g.

Proof. By direct calculation

$$:\!\phi\chi\!:_n = \sum_{m\leqslant 0} \phi_{n-m}\chi_m + \sum_{m>0} \chi_m\phi_{n-m}$$

such that $:\!\phi\chi\!:$ is independent of the choice of g and well defined on any eigenspace of L_0. Its dimension is bounded by $d(\phi) + d(\chi) + 1/2$. By the argument of lemma 6 one has $:\!\phi\chi\!:[f]V_0 \subset V_0$ for all $f \in C^\infty(S^1)$. It is easy to check that there are no obstacles to the extension of Q. In particular, eq. (1) follows from the definition of $:\!\phi\chi\!:$ and the Jacobi identity. $\qquad\square$

By using derivatives, commutators, and normal ordered products, one can construct a large class of local fields from any finite set. For arbitrary Lie algebras of the type given by eq. (3) one expects few linear relations in this class, such that $\dim(F_k)$ would grow exponentially with k. By proposition 2 the same would be true of the dimensions of the eigenspaces of L_0. Thus $\exp(-cL_0)$ would not be of trace class for some $c > 0$ and Q not standard. It seems to be difficult to find useful criteria for the growth of the dimensions, though a related problem has been treated successfully (Kac 1980).

As long as this problem is unsolved, one must use somewhat tentative conditions to obtain objects which have a nice structure. This certainly should be the case when all dimensions are integral and when the bilinear forms considered in lemma 11 are positive definite. I have not proceeded very far into this direction, but the following results show that it is promising:

LEMMA 12. *Let all fields have integral dimension. Consider the family of projections P: $F_k \to F_k/F_{k-1}$. We write $\hat\phi = P\phi$. Define*

$$\hat{W}(,,k)\colon PF_m \times PF_n \to PF_{n+m-k-1}$$

by

$$\hat{W}(\hat\phi, \hat\chi, k) = PW(\phi, \chi, k)$$

Then

$$[\hat\phi(x)\hat\chi(y)] = \sum_{k=0}^{d(\hat\phi)+d(\hat\chi)-1} \hat{W}(\hat\phi, \hat\chi, k)(x)(-i\partial_x)^k\delta(x-y)$$

is a consistent algebra of distributions with values in an abstract operator algebra with involution. These distributions have normal ordered products $:\hat{\phi}\hat{\chi}:$ *such that* $d(:\hat{\phi}\hat{\chi}:) = d(\hat{\phi}) + d(\hat{\chi})$.

Proof. By lemma 7 one has $d(W(\phi, \chi, k)) \leqslant d(\phi) + d(\chi) - k - 1$, such that $\hat{W}$ is well defined. As now $d(:\phi\chi:) \leqslant d(\phi) + d(\chi)$, the same is true for $:\hat{\phi}\hat{\chi}:$. All linear relations in the Jacobi identities for the algebra remain valid, as

$$\phi(x)\delta(x - y)\delta(x - z) = \phi(y)\delta(x - y)\delta(x - z) = \phi(z)\delta(x - y)\delta(x - z)$$

and its derivatives are compatible with the projection. In fact lemma 5 yields $\partial_x P = P \partial_x$. According to the same lemma, the involution $^+$ acts on PF_k, $k \in \mathbb{Z}$. $\qquad\square$

PROPOSITION 4. *Under the conditions of lemma 12, let*

$$\left[\hat{\phi}_{n-m}\hat{\chi}_m\right] = \sum_{k=0}^{d(\hat{\phi})+d(\hat{\chi})-1} \hat{W}(\hat{\phi}, \hat{\chi}, k)_n m^k$$

be the Lie algebra of the Fourier coefficients of the projected local fields. Let the bilinear forms $N_k : PF_k \times PF_k \to \mathbb{C}$ *defined by*

$$N_k(\hat{\phi}, \hat{\chi}) = \langle v, W(\phi, \chi, 2k - 1)_0 v\rangle \quad for\ \phi, \chi \in F_k, v \in \pi(n_v)\mathscr{H}\ fixed,$$

be nondegenerate. Then the Lie algebra is invariant under an SU(1, 1) *group with generators* L_{-1}, L_0, L_1, *where* L_0 *gives the grading of the algebra.*

Proof. Using the bilinear forms, which by lemma 11 are positive, we may write

$$PF_k = PF_k' \oplus \partial_x PF_{k-1}$$

such that PF_k' and $\partial_x PF_{k-1}$ are orthogonal. Call the elements of PF_k', $k \in \mathbb{Z}$, non-derivative. We define the action of the L on the Fourier coefficients of the non-derivative fields by

$$\left[L_n\hat{\phi}_m\right] = \left(m - n\left(d(\hat{\phi}) - 1\right)\right)\hat{\phi}_{n+m}, \qquad n \in \{-1, 0, 1\},$$

such that L_0 yields the grading. Write the algebra in terms of such Fourier coefficients and consider the Jacobi identities for $\hat{\phi}_n$, $\hat{\chi}_n$, $\hat{\psi}_n$, $n \in \mathbb{Z}$. For definiteness, let $d(\hat{\chi}), d(\hat{\phi}) \leqslant d(\hat{\psi})$. If $d(\hat{\phi}) + d(\hat{\chi}) > d(\hat{\psi})$, the Jacobi identities yield just enough linearly independent relations to show that the L_n are derivations. Otherwise one checks first the action of the L_n on normal ordered products of $\hat{\phi}$, $\hat{\chi}$, and their derivatives. Then one shows by use of the Jacobi identities that $[\hat{\phi}(x)\hat{\psi}(y)]$ and $[\hat{\chi}(x)\hat{\psi}(y)]$ vanish, if $N_{d(\hat{\psi})}(\hat{\psi}, :\partial_x^n\hat{\phi}\,\partial_x^m\hat{\chi}:) = 0$ for $d(\hat{\phi}) + d(\hat{\chi}) + n + m = d(\hat{\psi})$. $\qquad\square$

As the proof is straightforward, but purely calculational and somewhat tedious, I omit the details. It is very likely that the result can be made much stronger, probably under mild additional conditions: The $SU(1, 1)$ symmetry should act on $\mathscr{H}$ and should be embedded into a central extension of $\mathrm{Diff}(S^1)$ which acts similarly. Moreover there should be a field $T \in F_2$ with $T_n = L_n$.

In fact, all quantum field theories considered below have this property. These theories are called conformally invariant and have been studied extensively. (Belavin et al. 1984, Friedan et al. 1984, Knizhnik et al. 1984, Cardy 1986, Witten 1984, Gepner and Witten 1986.) They all have basic representations for which the inequality (4) is saturated. On the other hand, preliminary calculations indicate that the saturation of this inequality is a sufficient condition for the conformal invariance of the theory.

The quantum field theories of physics usually are defined in terms of functional integrals. I have not mentioned this very important tool so far, as it has been made rigorous only in a restricted set of examples, and not even all quantum field theories of standard type can be described in this way. If they can, many beautiful structures appear. In particular the partition function of conformally invariant theories is invariant under the action of the modular group $SL(2, Z)/Z_2$ on the complex upper half plane $R_x + iR_x^+$. This arises, as the partition function is given in terms of an integral over conformally invariant functions on a Riemann surface of genus one and only depends on the corresponding point in moduli space.

Now let me give some examples. All are conformally invariant and of standard type.

Example 1. The Lie algebra of $\mathrm{Diff}(S^1)$ has a central extension, which is called Virasoro algebra by the physicists. Its generators have a natural representation as values of a local field T of dimension 2 for which

$$i[T(x)T(y)] = (T(x) + T(y))\delta'(x - y) + \frac{c}{12}\delta'''(x - y)$$

for some constant c. The inclusion of the group S^1 into $\mathrm{Diff}(S^1)$ yields a natural generator of translations. Thus translationally invariant field theories with $F = \langle T \rangle$ just are graded unitary highest weight representations of the Virasoro algebra. The representations need not be irreducible, but the constant c by coherence has to be the same in all irreducible parts. Moreover, $L_0 = T_0$ implies that given irreducibles representations have finite multiplicities, as otherwise Q would not be standard.

For $c < 1$ the partition function of such a quantum field theory (Feigin and Fuchs 1983, Rocha-Caridi 1984) is invariant under a subgroup of the modular group of finite index, but not for $c \geqslant 1$. For $c > 1$ it is the ordinary partition function of number theory. Fixing the lowest eigenvalue of L_0 to be $-1/24$, it becomes a meromorphic modular form, but even this is incompatible with a description of the theory by a functional integral.

Using the real structure of the basic representation for fixed c and ineq. (4) one can show easily that a maximal space of local fields is spanned by the normal ordered products of T and its derivatives.

Example 2. Let F consist of fields of dimension 1 only. If one excludes the constant field, $W(\,,\,,1)$ yields a nondegenerate form and may be diagonalized. Let $\mathfrak{h}$ be the Lie algebra of the Fourier coefficients J_0^α of the local fields J^α, which must be reductive. Let $f_{\alpha\beta}^\gamma$ be the structure constants of $\mathfrak{h}$. For some constant k one obtains the algebra

$$i\left[J^\alpha(x)J^\beta(y)\right] = \sum_\gamma f_\gamma^{\alpha\beta}J^\gamma(x)\delta(x-y) + k\delta^{\alpha\beta}\delta'(x-y).$$

The fields J^α are called currents by the physics community and this algebra is called a current algebra. When one evaluates the distributions on $C^\infty(S^1)$ it yields a central extension of the loop algebra $\mathrm{Map}(S^1, \mathfrak{h})$.

The quantum field theories of this type are just graded unitary highest weight representations of affine Kac-Moody algebras and for the abelian case Heisenberg algebras. The constant k, called the level of the representation must be the same in all superselection sectors. In the basic representation, ineq. (4) is saturated. Derivatives and normal ordered products again yield a maximal space of local fields. The fields of dimension 2 are spanned by $\partial_x J^\alpha$ and $:J^\alpha J^\beta:$. Among them there is a field T which generates $\mathrm{Diff}(S^1)$ and in particular the translations. For semisimple $\mathfrak{h}$ the partition functions of finitely reducible representations are invariant under subgroups of the modular group of finite index. Some are invariant under the whole modular group. In particular, the basic representation of E_8 yields a partition function which is invariant up to a multiplicative third root of unity.

If $\mathfrak{h}$ has an abelian component, the partition function again becomes meromorphic modular form. The reason for the distinction is that in this case the values of the currents only exponentiate to the connected component of a central extension of $\mathrm{Map}(S^1, G)$, where G is the compact Lie group generated by the J_0^α. When one considers all of this loop group, one again obtains a modular invariant partition function (Segal 1981).

Later we will need a few facts about quantum field theories given by projective irreducible unitary highest weight representations of $\mathrm{Map}(S^1, G)$, G compact, which can be found in (Kac 1983, Segal 1981). These representations can be labelled by the irreducible representation of G in $\pi(n_\nu)\mathcal{H}$ and the level k. The latter is normalized to take arbitrary values in $\mathbb{Z}_+$.

Let us first state how we label the representations of G. After choosing a Cartan subgroup, one may fix a set of positive roots $\Delta_+(G)$ of G. This defines a partial ordering of the weight lattice of G, and one may label an irreducible representation by its highest weight. For us a slightly different and less arbitrary labelling is more convenient, namely the one by the free orbits $W(G)\lambda$ of the

Weyl group $W(G)$ in the weight lattice. Recall that the weight lattice minus those elements which are invariant under some non-trivial $w \in W(G)$ decomposes into cones and that

$$\rho^{(G)} = \frac{1}{2} \sum_{\alpha \in \Delta_+(G)} \alpha$$

sits at the tip of one of those cones. To obtain the highest weight corresponding to a free orbit $W(G)\lambda$ choose its representative $w\lambda$ in this particular cone and subtract $\rho(G)$. For example, $W(G)\rho(G)$ labels the trivial representation.

In the abelian case, we define that $W(G) = 1$, $\rho(G) = 0$ and the weight lattice is the dual group. Then all statements trivially remain true.

Let us use the notation $\mathscr{L}G$ for the central extension of $\mathrm{Map}(S^1, G)$. We define the character of a highest weight representation of $\mathscr{L}G$ in $\mathscr{H}_\nu$ by

$$\chi[e^h] = \mathrm{tr}_{\mathscr{H}}(e^h q^{L_0 - n_\nu})$$

where h is a generator of a fixed Cartan subgroup of G, and q is the standard exponential map from the upper complex half plane (identified with $iR_x^+ + R_x$) to the interior of the unit circle in C, such that

$$Z = q^{n_\nu}\chi[1]$$

is the partition function. The character can be written as a formal sum over elements e^α, where the α are weights of G and

$$e^\alpha[e^h] = e^{\alpha(h)}.$$

The Weyl group of $\mathscr{L}G$ is the semidirect product $W(G) \ltimes T(G)$, where $T(G)$ is the lattice spanned by the long roots of G. We define the theta functions

$$\theta_n(T(G), \lambda) = \sum_{\alpha \in nT(G)} e^{\lambda + \alpha} q^{(\lambda + \alpha)^2 / 2n} \quad \text{for } n \in \mathbb{Z}_+$$

using the standard metric on the weight lattice. Let g be the dual Coxeter number of T. Then the Weyl-Kac character formula takes the form

$$\chi(\lambda, k) = D(\mathscr{L}G)^{-1} \sum_{w \in W(G)} \varepsilon(w) \theta_{g+k}(T(G), w\lambda),$$

where $\varepsilon: W(G) \to \{1, -1\}$ is the signature function, the denominator $D(\mathscr{L}G)$ is given by

$$D(\mathscr{L}G) = \sum_{\alpha \in \Delta_+(G)} \left(e^{\alpha/2} - e^{-\alpha/2} \right)$$

$$\times \prod_{l=1}^{\infty} \left[(1 - q^l)^{\mathrm{rank}\, G} \prod_{\alpha \in \Delta_+(G)} (1 - e^{\alpha}q^l)(1 - e^{-\alpha}q^l) \right]$$

and $\Delta_+(G)$ is chosen such that $\lambda\alpha > 0$ for all $\alpha \in \Delta_+(G)$ to avoid an overall minus sign. We indiscriminantly use the labelling λ, k or $W(G)\lambda, k$ for the representation. For the trivial representation, which yields a trivial field theory, we have $k = 0$, $\lambda = \rho(G)$, $\chi[1] = 1$. This yields the Weyl-Kac denominator formula

$$D(\mathscr{L}G) = \sum_{w \in W(G)} \varepsilon(w)\theta_g(T(G), \rho(G)).$$

For $U(1)$ one needs $g = 0$, $T(U(1)) = \mathbb{Z}$ and

$$D(\mathscr{L}U(1)) = \prod_{l=1}^{\infty} (1 - q^l).$$

Moreover, we allow a double covering of $U(1)$ and use the weight lattice $\mathbb{Z}/2$.

At level k the number of distinct irreducible unitary highest weight representations of $\mathscr{L}G$ is equal to the number of free orbits of $W(G)$ in the weight lattice modulo $(g + k)T(G)$. In particular, it is finite. For $\mathscr{L}U(1)$ this yields $2k$ representations. In this case one also can make sense out of levels $k \in \mathbb{Z}_+ + 1/2$, but we shall not need that.

Example 3. Let W be a finite dimensional euclidean vector space with bilinear form $(\,,)$. Let S, S' be the spaces of those functions in $C^{\infty}(R, W)$ for which

$$f(x + 2n) = f(x), \qquad f(x + 2\pi) = -f(x),$$

resp. Let

$$S = \bigoplus_{n \in \mathbb{Z}} S_n, \qquad S' = \bigoplus_{r \in \mathbb{Z}+1/2} S_r'$$

be the gradings given by the Fourier decompositions and put

$$S_{\pm} = \bigoplus_{n \gtrless 0} S_n, \qquad S'_{\pm} = \bigoplus_{r \gtrless 0} S_r'$$

On both S and S' one can define a Clifford algebra by

$$\{fg\} = \int_0^{2\pi} (f(x), g(x))\, dx.$$

Let $C(S_0)$ be the standard $2^{\lfloor \dim W/2 \rfloor}$ dimensional irreducible Clifford module of S_0. The Clifford algebras over S, S' have natural representations ρ, ρ' on

$$\mathcal{H}_p = \overline{\Lambda S_+} \otimes C(S_0), \qquad \mathcal{H}_a = \overline{\Lambda S_+'} \quad \text{resp.,}$$

where the bar denotes completion and the metric is the natural one. These spaces inherit a grading from S, S' and can be used as the Hilbert spaces of translationally invariant field theories on S^1.

For $v \in W$ we define the operator valued distributions $\psi_v : C^\infty(S^1) \to \rho(S)$ by

$$\psi_v[f] = \pi(S^1)^* f v$$

where $\pi(S^1)$ is the projection from R to S^1. Thus

$$\psi_v(x)\psi_w(y) + \psi_w(y)\psi_v(x) = (v, w)\delta(x - y).$$

For functions with support S^1-{point}, the same procedure works for S', but involves an arbitrary sign choice. The ψ_v are not local fields in the sense of this paper, but share many of their properties. They have dimension $1/2$. Their operator products

$$J^{v \wedge w}(x) = \lim_{\varepsilon \to 0} \int_0^{2\pi} \varepsilon^{-1} g(y/\varepsilon) \big[\psi_v(x)\psi_w(x + y) - \pi_y^+\big(\psi_v(x)\psi_w(x + y)$$

$$+ \psi_w(x + y)\psi_v(x)\big)\big]\, dy$$

have dimension 1 and generate $\mathcal{L}\,\mathrm{SO}(W)$, or more precisely $\mathcal{L}(\mathrm{Spin}(W)/z)$, where the subgroup z of the center of $\mathrm{Spin}(W)$ is different on the various superselection sectors. Note that bilinear expressions like $J^{v \wedge w}$ do not depend on the sign choice or the suppressed point of S^1 used to define ψ on $\mathcal{H}_a$.

We now have a special case of example 2, where the group G is fixed and the level is 1. However, the ψ_v now yield natural operators $\psi_v[f]$ which commute with the $J^{v' \wedge v''}[g]$ when the supports of f, g do not intersect. Thus they can be used to show the coherence of superselection sectors in the sense of def. 2.3.

$\mathcal{H}_a$ splits into two superselection sectors $\mathcal{H}_0$ and $\mathcal{H}_1$ which can be characterized by $1 \in \mathcal{H}_0$, $\rho'(S')\mathcal{H}_0 \subset \mathcal{H}_1$, $\rho'(S')\mathcal{H}_1 \subset \mathcal{H}_0$, or by the fact that the currents yield $\mathcal{L}(\mathrm{SO}(W)/Z_2)$ on $\mathcal{H}_0$ and $\mathcal{L}\,\mathrm{SO}(W)$ on $\mathcal{H}_1$. If the dimension of W is odd, $\mathcal{H}_p$ is irreducible and the currents yield $\mathcal{L}\,\mathrm{Spin}(W)$. If $\dim W$ is even, $C(S_0)$ splits into two irreducible representations of $\mathrm{Spin}(W)$

$$C(S_0) = C^+(S_0) \oplus C^-(S_0)$$

and this induces the decomposition $\mathcal{H}_p = \mathcal{H}^+ \oplus \mathcal{H}^-$, with loop groups $\mathcal{L}\,\mathrm{Spin}^+(W)$ and $\mathcal{L}\,\mathrm{Spin}^-(W)$. More details can be found in (Kac and Peterson

1981), and also in most physics books on quantum field theories under the heading Fock space.

Operators which intertwine between $\mathcal{H}_a$ and $\mathcal{H}_p$ also can be constructed. They are closely related to the Ising field of statistical mechanics and particularly elegant and obvious for $\dim W = 8$, where $\mathcal{H}_1, \mathcal{H}^+, \mathcal{H}^-$ are related by triality and the simplest intertwining operators have dimension $1/2$ (Goddard, Olive and Schwimmer 1985). In general, their operator dimension is $\dim W/16$. For $\dim W = 16$, they are currents and enlarge SO(16) to E_8.

In a fancier language, the preceding construction may be described as follows. S, S' should be considered as spaces of maps from S^1 to a superspace of $\dim W$ odd dimensions and no even one. They have a complex structure given by the splitting into positive and negative frequency parts and a Kähler form given by the Clifford structure. $\mathcal{H}_0$ and $\mathcal{H}_1$ are the spaces of holomorphic maps from S' to a superspace of one even or one odd complex dimension resp. The tricky point in this description is the discrete analogue of a holomorphic function needed to understand the occurrence of $C(S_0)$. This feature will reappear in the examples 4 and 5 below.

The constructions of this example made it possible to solve the following problem. Let G be a semisimple Lie group, H a compact subgroup, $\mathcal{H}$ the representation space of an irreducible highest weight representation of $\mathcal{L}G$. For the simple factors G_j of G let $k(G_j)$ be the level of the representation of $\mathcal{L}G_j$ in $\mathcal{H}$. For any simple or one-dimensional factor H_i of H let $k(H_i)$ be the level of the induced representation of $\mathcal{L}H_i$ in $\mathcal{H}$. One has

$$k(H_i) = \sum_j n_{ij} k(G_j),$$

where n_{ij} only depends on the homomorphism from H_i to G_j given by the embedding. For a trivial homomorphism one has $n_{ij} = 0$, for a regular embedding $n_{ij} = 1$. For the natural embedding of H into SO(Lie H) one finds $n = g(H)$, where Lie H is the Lie algebra of H and $g(H)$ its dual Coxeter number.

Of particular interest are the cases where $\mathcal{H}$ splits into a finite number of irreducible representations of $\mathcal{L}H$. This can only happen, if $H = \otimes H_j'$, so that H_j' is embedded in G_j and if $k(G_j) = 1$ for all j (Bais and Bouwknegt 1986, Schellekens and Warner 1986). Thus we may assume G to be simple and $\mathcal{L}G$ to have level 1. For classical G one has

THEOREM (Goddard, Nahm, Olive 1985). *Let G be the orthogonal, special unitary, or symplectic structure group of some real, hermitean, or quarternionic symmetric space resp. In other words, let G be of the form* SO($T\tilde{G}/H$), SU($T_c\tilde{G}/(H \times U(1))$) *or* Sp($T_H\tilde{G}/(H \times SU(2))$), *where* T, T_c, T_H *denote the real, complex or quaternionic tangent spaces. Then the natural embedding of H into G induces splittings of the level 1 irreducible unitary highest weight representations of $\mathcal{L}G$ into a finite number of irreducible representations of $\mathcal{L}H$ and (up to isomorphisms) these are the only embeddings into nonexceptional G for which this happens.*

We gave the proof for orthogonal G, but due to the standard embeddings $U(N) \to SO(2N)$ and $Sp(N) \times Sp(1) \to SO(4N)$ given by the interpretation of $U(N + 1)/U(N)$ and $Sp(N + 1)/(Sp(N) \times Sp(1))$ as real symmetric spaces the generalization to unitary and symplectic G is immediate. For exceptional G, the analogous embeddings are known, but their interpretation is not clear (Bais and Bouwknewgt 1986, Schellekens and Warner 1986).

The proof of the theorem uses the following construction. The currents of H can be written as linear combinations of normal ordered products : $\psi_v \psi_w$:, as discussed above. Now construct the field $T \in F_2$ by $T = \Sigma_\alpha : J^\alpha J_\alpha$: (up to a constant factor). Finite reducibility is obtained exactly when this expression for T reduces to the form $\Sigma_v : \psi_v \, \partial_x \psi_v$:, where the sum is over an orthonormal basis, as this is the field yielding the representation of $\text{Diff}(S^1)$ for all of $\mathcal{H}$. Explicit calculation shows that $\Sigma_\alpha : J^\alpha J_\alpha$: reduces to $\Sigma_v : \psi_v \partial_x \psi_v$: in the cases described above, but the deeper reasons for the occurrence of the symmetric spaces are still mysterious.

Example 4 (Frenkel and Kac 1980, Segal 1981). We now consider the analogue of example 3 for an ordinary space, not a superspace. Let S be the space of C^∞ mappings from S^1 into a torus $\mathbb{R}^r/L$, for some lattice L. As before, we write

$$S = S_+ \oplus S_0 \oplus S_-$$

but now S_0 is more complicated. It consists of the functions

$$f(x) = f_0 + \lambda x/2\pi \bmod L, \quad \text{where } \lambda \in L.$$

On $S_+ \oplus S_-$ we have the Kähler form

$$\langle f, g \rangle = i \int_0^{2\pi} (\partial_x f, g) \, dx,$$

and one factor of the Hilbert space just is given by the holomorphic functions on S_+, with action of S_- and hermitean form given by the Kähler form. The functions on S_0 yield the space $(L^v \times L) \times \mathbb{C}$, where L^v is the lattice dual to L, which describes the functions of f_0. The discrete analogue of the holomorphic function is a function with support on $(\lambda, \lambda) \in (L^v \times L)$. For self-dual lattices this construction yields a representation of $\mathscr{L}\mathbb{R}^r/L$ otherwise a representation of $\mathscr{L}\mathbb{R}^r/(L \cap L^v)$. The connected part of this group is represented by the currents coming from S itself, with the algebra

$$[J^v(x) J^w(y)] = -i(v, w)\delta'(x - y)$$

coming from the Kähler form. As in example 2 the representation has the character

$$\chi = \prod_{l=0}^{\infty} (1 - q^l)^{-r} \theta_1(L \cap L^v, 0)$$

where the level of the representation is 1. Again one can construct enough local fields to saturate ineq. (4) for this representation. The r-th power of the classical partition function corresponds to normal ordered products of the currents and their derivatives. For each $\lambda \in L \cap L^v$ there are additional local fields V_λ called vertex operators, which according to ineq. (4) should have dimensions $\lambda^2/2$, and indeed they have. The local fields are spanned by expressions of the form

$$: (J)^{n_1} (\partial_x J)^{n_2} (\partial_x^2 J)^{n_3} \cdots V_\lambda :$$

which have dimension $(n_1 + 2n_2 + 3n_3 + \cdots) + \lambda^2/2$, with obvious correspondence to the terms in the character.

When the lattice $L^v \cap L$ has elements of length squared 2, the corresponding vertex operators are currents. Those elements span the root lattice of a Lie algebra, and the currents have the current algebra given by this Lie algebra.

Again the main point to be clarified at a conceptual level is the concept of something like holomorphic functions on S_0. The actual construction of the vertex operators uses the triviality of $\mathrm{Ext}(L, \mathbb{Z}_2)$ to resolve this difficulty.

Example 5.   This example is an interpretation of (Frenkel et al. 1984). Let L be the Leech lattice in $\mathbb{R}^{24}$ and consider the space S of C^∞ maps from S^1 to $(\mathbb{R}^{24}/L)/\mathbb{Z}_2$, where $\mathbb{Z}_2$ acts by reflection at the origin. S now has two components with an intersection of codimension ∞, given by $S_p/\mathbb{Z}_2$ and $S_a/\mathbb{Z}_2$, where

$$S_p = \left\{ f | f(x) = \sum_{n \in \mathbb{Z}} f_n e^{inx} + \lambda x/2\pi \bmod L, \lambda \in L \right\}$$

$$S_a = \left\{ f | f(x) = \sum_{r \in \mathbb{Z}+1/2} f_r e^{irx} + \lambda/2 \bmod L, \lambda \in L \right\}.$$

The action of $\mathbb{Z}_2$ takes f_n, λ, f_r to their negatives. As in examples 3 and 4 one can construct Hilbert spaces $\mathscr{H}_p, \mathscr{H}_a$ of holomorphic maps on S_p, S_a. This yields a quantum field theory $(S^1, \mathscr{H}_p \oplus \mathscr{H}_a, V_p \oplus V_a, F_p)$ with superselection sectors V_p, V_a which are dense in $\mathscr{H}_p, \mathscr{H}_a$. The restriction $(S^1, \mathscr{H}_p, V_p, F_p)$ of this theory is just one of the maximally extended theories considered in example 4 and has the partition function

$$Z_p = q^{-1} \prod_{l=1}^{\infty} (1 - q^l)^{-24} \theta_1(L, 0)[1]$$

$$= j(q) - 720 = q^{-1} + 24 + 196884q + \cdots$$

Inequality (4) is saturated. In particular the constant term in Z_p corresponds to the currents of the translation group $\mathbb{R}^{24}/L$.

The partition function of the restriction $(S^1, \mathcal{H}_a, V_a, F_p)$ is

$$Z_a = 2^{12}q^{1/2} \prod_{r \in \mathbb{Z}_+ + 1/2} (1 - q^r)^{-24},$$

as the discrete analogue of the holomorphic functions on $L/2L$ is a 2^{12} dimensional space. Its construction involves a Clifford algebra, as in example 3.

Now, the holomorphic functions on $S_p/\mathbb{Z}_2$ and $S_a/\mathbb{Z}_2$ are assumed to lift to even functions on S_p and to odd functions on S_a.

This yields subspaces $\mathcal{H}_p^+$, $\mathcal{H}_a^-$, F_p^+ of $\mathcal{H}_p$, $\mathcal{H}_a$, F_p and a quantum field theory $Q = (S^1, \mathcal{H}, V, F_p^+)$ with

$$\mathcal{H} = \mathcal{H}_p^+ \oplus \mathcal{H}_a^-, \quad V = V_p^+ \oplus V_a^-$$

$$V_p^+ = V_p \cap \mathcal{H}_p^+, \quad V_a^- = V_a \cap \mathcal{H}_a^-.$$

The restriction $(S^1, \mathcal{H}_p^+, V_p^+, F_p^+)$ of Q has the partition function

$$Z_p^+ = \frac{1}{2}q^{-1} \prod_{l=1}^{\infty} (1 - q^l)^{-24}(\theta_1(L,0)[1] - 1)$$

$$+ \frac{1}{2}q^{-1}\left(\prod_{l=1}^{\infty} (1 - q^l)^{-24} + \prod_{l=1}^{\infty} (1 + q^l)^{-24} \right)$$

$$= q^{-1} + 98580q + \cdots$$

Inequality (4) is still saturated. Note that the translation currents of $\mathbb{R}^{24}/L$ are odd under $\mathbb{Z}_2$ and do not belong to F_p^+. The partition function of the restriction $(S^1, \mathcal{H}_a^-, V_a, F_p^+)$ of Q is

$$Z_a^- = 2^{11}q^{1/2}\left(\prod_{r \in \mathbb{Z}_+ + 1/2} (1 - q^r)^{-24} - \prod_{r \in \mathbb{Z}_+ + 1/2} (1 + q^r)^{-24} \right)$$

$$= 24 \cdot 2^{12}q + \cdots$$

The partition function of Q itself is

$$Z = Z_p^+ + Z_a^- = j(q) - 744$$

$$= q^{-1} + 196884q + \cdots$$

The problem now is to extend Q to a maximal translationally invariant quantum field theory $(S^1, \mathcal{H}, V, F)$. In fact there is a unique natural extension which saturates ineq. (4).

To do this one does not need any currents. In fact ineq. (4) shows that none can exist. But we need 196884 local fields of dimension 2 and only 98580 are present. Thus one has to extend the theory by $24 \cdot 2^{12}$ local fields which intertwine $\mathcal{H}_a^-$ and $\mathcal{H}_p^+$, such that only one superselection sector exists in $\mathcal{H}$. This problem is solved by Frenkel, Lepowsky and Meurman by techniques which seem to be related to the construction of L from a lattice in R^{26} with a metric of signature $(25, 1)$. Instead one might have used the fields of Ising type mentioned in example 3.

The 196884 resulting fields of dimension 2 form a space F_2 with a splitting $\langle T \rangle \oplus F_2'$, where T yields $\mathrm{Diff}(S_1)$, as before. We now use the terminology of eq. (2). For $\phi, \chi \in F_2$ the term $W(\phi, \chi, 3)$ is proportional to the constant field and yields a nondegenerate bilinear form on F_2. The term $W(\phi, \chi, 1)$ yields a commutative map from $F_2 \times F_2$ to F_2, which is the Griess algebra. The Fischer-Griess sporadic simple group FG is the symmetry group of the algebra of all local fields, or equivalently of the Lie algebra of their Fourier coefficients. Because of the term $W(\phi, \chi, 0)$, which yields local fields of dimension 3, the Fourier coefficients of fields in F_2 alone do not form a Lie algebra.

F_3 decomposes into $\partial_x F_2$ and F_3'. The latter yields an irreducible representation of FG. It is convenient to decompose $\mathcal{H}$ into irreducible representations of $\mathrm{Diff}(S^1) \times FG$ to control the representations of FG at higher grades or equivalently on local fields of higher dimensions. At least up to dimension 10 and grade 9 no one-dimensional representations of FG appear except the obvious ones given by the action of the T_n on the state of the lowest grade.

One of the difficult steps in the construction of FG is the proof that it is finite. I conjecture that all grade conserving continuous symmetries of a quantum field theory Q of standard type are given by the zeroth Fourier coefficients of currents. In our case this would imply the finiteness of the automorphism group of the Griess algebra, as one easily can show that F is generated from F_2 by normal ordered products and derivatives, such that this group is a symmetry group of Q.

4. Chiral quantum field theories in two dimensions. Let us consider translationally invariant quantum field theories of the form $Q = (\mathbb{R} \times S^1, \mathcal{H}, V, F)$. On $\mathbb{R} \times S^1$ we use the coordinates $t \in \mathbb{R}$, $x \in S^1$ and the causal structure given by the pseudo-Riemannian metric $dt^2 - dx^2$. Now R_x contains the zero norm vector fields

$$P_R = \frac{\partial}{\partial t} + \frac{\partial}{\partial x}, \qquad P_L = \frac{\partial}{\partial t} - \frac{\partial}{\partial x}$$

We write the partition function in the form $Z \colon \mathfrak{H} \times \overline{\mathfrak{H}} \to \mathbb{C}$

$$Z(w, w') = \mathrm{tr}\, \exp\!\left(iwU^*(P_R) - iw'U^*(P_L)\right)$$

where $\mathfrak{H}$, $\overline{\mathfrak{H}}$ are the upper and lower complex half planes.

Definition 9.   Local fields in the spaces

$$F_R = \{\, \phi \in F \,|\, P_R\phi = 0 \,\}$$

$$F_L = \{\, \phi \in F \,|\, P_L\phi = 0 \,\}$$

are called right-handed resp. left-handed. All such fields are called chiral.
Note that a field which is both right- and left-handed, is constant.

LEMMA 13.   *Let $\phi \in F_R$, $\chi \in F_L$. Then $[\phi\chi] = 0$.*

Proof.   Let M_R, M_L be the subgroups of $M = \mathbb{R} \times S^1$ which are generated by
P_R, P_L. For $m_R \in M_R$, $m_L \in M_L$ one has

$$[\phi[m_{R*}f]\chi[m_{L*}g]] = [\phi[f]\chi[g]].$$

If the supports of f, g are contained in sufficiently small open sets, it is easy to
find m_R, m_L such that $m_R\mathrm{supp}(f)$ and $m_L\mathrm{supp}(g)$ are casually unrelated. The
result then follows from a partition of unity argument.   □

LEMMA 14.   *Let $\phi, \chi \in F_R$, $\pi_R\colon M \to M/M_R$ the natural projection. If the sets
$\pi_R\mathrm{supp}(f)$ and $\pi_R\mathrm{supp}(g)$ do not intersect, then $\phi[f]\chi[g] = \chi[g]\phi[f]$, and
analogously for F_L.*

Proof.   As in lemma 13.   □

Note that the fields in F_R, F_L satisfy algebraic relation of the form of eq. (2).

LEMMA 15.   *Let $Q = (S^1, \mathscr{H}, V, F)$ be a translationally invariant field theory.
Let us identify S^1 with M/M_R or M/M_L resp. Then $\iota_+Q = (\mathbb{R} \times S^1, \mathscr{H}, V, F_+)$
and $\iota_-Q = (R \times S^1, \mathscr{H}, V, F_-)$ are translationally invariant field theories, when
one defines F_+, F_- by the maps: $\iota_\pm\colon F \to F_\pm$*

$$(\iota_\pm\phi)[f] = \phi[\iota_{\pm*}f], \qquad (\iota_{\pm*}f)(x) = \int_0^{2\pi} f(x \pm y, y)\, dy$$

*and the action of the translation group by the projection from M to S^1. The maps $\iota_\pm$
conserve the dimensions of the fields.*

The proof is straightforward.

Definition 10. $Q = (\mathbb{R} \times S^1, \mathscr{H}, V, F)$ is called chiral, if its restriction
$(\mathbb{R} \times S_1, \mathscr{H}, V, F_R + F_L)$ contains only finitely many superselection sectors.

LEMMA 16.   *The minimal superselection sectors of a chiral Q can be written in
the form $\iota_+Q' \times \iota_-Q''$, where Q', Q'' are one-dimensional quantum field theories.*

Proof.   By the second remark after definition 2.1, the image of $F_R^x \otimes F_L^x$ in
$O(V)$ contains sufficiently many selfadjoint operators to decompose the Hilbert

space of any superselection sector in the form $\mathcal{H}_R \otimes \mathcal{H}_L$, where F_R only acts on $\mathcal{H}_R$ and F_L on $\mathcal{H}_L$. $\square$

Of particular interest are those Q, for which F_R, F_L contain local fields T_R, T_L of dimension 2 which generate $\mathrm{Diff}(M/M_R) \times \mathrm{Diff}(M/M_L)$. This is essentially $\mathrm{Diff}_c(\mathbb{R} \times S^1)$, i.e., the subgroup of $\mathrm{Diff}(\mathbb{R} \times S^1)$ which leaves the causal structure invariant. More precisely one has $\mathrm{Diff}_c(\mathbb{R} \times S^1) = (\mathrm{Diff}(M/M_R) \times \mathrm{Diff}(M/M_L))/\mathbb{Z}$, such that the eigenvalues of $T_{0L} - T_{0R}$ are restricted to be integral. These Q are called conformally invariant.

If Q is conformally invariant and realized by a functional integral, the partition function is invariant under simultaneous modular transformations of w, w'

$$Z\left(\frac{aw + b}{cw + d}, \frac{aw' + b}{cw' + d}\right) = Z(w, w'), \qquad \begin{pmatrix} a & b \\ c & d \end{pmatrix} \in SL_2(\mathbb{Z})$$

as for $w = \bar{w}'$ one can relate $Z(w, w')$ to a conformally invariant integral over functions on a compact Riemann surface of genus 1 and modulus w.

The modular invariance of the partition function in two dimensions is more basic than the corresponding phenomenon in one dimension, as the Riemann surface is just another section of a two dimensional complex space in which $\mathbb{R} \times S^1$ is embedded. The modular invariance of one dimensional partition functions arises from the realization of $\iota_+ Q \times \iota_- Q$ in terms of functional integrals, where Q is the one dimensional theory.

The partition function of a chiral quantum field theory can be written in the form

$$Z(w, w') = \sum_i Z_i(w) Z_i'(w')$$

where the sum is over superselection sectors. If this partition function is modular invariant, the space spanned by the Z_i yields a representation of the modular group.

CONJECTURE. *Let* $Q = (\mathbb{R} \times S^1, \mathcal{H}, V, F)$ *be a conformally invariant quantum field theory such that* $F_R \times F_L$ *contains the currents of* $H_R \times H_L$, *with* H_R *and* H_L *isomorphic. Let*

$$\mathcal{H} = \bigoplus_{\alpha, \beta} \left(\mathcal{H}_{\alpha L} \otimes \mathcal{H}^{\alpha \beta} \otimes \mathcal{H}_{\beta R} \right)$$

be the splitting of $\mathcal{H}$ *given by the irreducible representations of* $\mathcal{L} H_R \times \mathcal{L} H_L$. *Let the partition function of* Q *be invariant under modular transformations. Then* $Q' = (\mathbb{R} \times S^1, \mathcal{H}', V', F')$ *with*

$$\mathcal{H}' = \bigoplus_{\alpha} \mathcal{H}^{\alpha \alpha},$$

the corresponding V', and

$$F' = \left\{ \phi \in F \,|\, \left[\phi J_R^\alpha\right] = \left[\phi J_L^\beta\right] = 0 \text{ for all currents } J_R^\alpha, J_L^\beta \text{ of } \mathscr{L}H_R \times \mathscr{L}H_L \right\}$$

is a conformally invariant quantum field theory with modular invariant partition function.

I have a proof of this conjecture which is satisfactory from a phycisist's point of view, but not rigorous: When one couples a Yang-Mills field to the currents, L splits into the conformally invariant part and another part which changes continuously with the coupling constant and can be shifted to arbitrarily high eigenvalues of $T_{0L} + T_{0R}$.

THEOREM 2. *For the Q of this conjecture, let F be given by the currents of $\mathscr{L}G_R \times \mathscr{L}G_L'$ and let $\mathscr{H}$ be finitely reducible as a representation of this group. Then the statement of the conjecture is true.*

Proof. Let Z, Z_α be the partition functions of $\mathscr{H}$, $\mathscr{H}_\alpha$ and let χ, χ_α be the corresponding characters of $\mathscr{L}G_R \times \mathscr{L}G_L'$. Let $\gamma, \bar{\gamma}$ be transformations of the complex plane and the space of generators of the complexified Cartan subgroup of $G \times G'$ given by

$$\gamma(w) = \frac{aw + b}{cw + d}, \quad \bar{\gamma}(w, h) = \left(\frac{aw + b}{cw + d}, \frac{h}{cw + d} \right)$$

and $\gamma_*, \bar{\gamma}_*$ the induced maps on functions of w and w, e^h resp. Then (Kac 1983, chapter 13)

$$\gamma_* Z_\alpha = \sum_\beta U_{\alpha\beta}(\gamma) Z_\beta, \qquad \bar{\gamma}_* \chi_\alpha = f(\gamma) \sum_\beta U_{\alpha\beta}(\gamma) \chi_\beta,$$

where $U(\gamma)$ is a unitary matrix and $f(\gamma)$ is a function which does not depend on α. Using the analogous result for $\mathscr{L}G_R \times \mathscr{L}G_L'$, one sees that Z is invariant under γ^* if χ is invariant under $\bar{\gamma}^*$ up to such a factor $f(\gamma)$. In fact the splitting of a finitely reducible representation of some $\mathscr{L}G$ can be read off from its partition function, as the irreducible representations yield partition functions which are independent over $\mathbb{Z}$. Thus up to a given factor, χ is invariant under $\bar{\gamma}^*$. A fortiori the character of $\mathscr{L}H_R \times \mathscr{L}H_L$ on $\mathscr{H}$ has the same invariance, which yields

$$\sum_{\alpha, \beta} \bar{\gamma}_* \left(\bar{\chi}_{\alpha L} Z^{\alpha\beta} \chi_{\beta R} \right) = |f(\gamma)|^2 \sum_{\alpha, \beta} \bar{\chi}_{\alpha L} Z^{\alpha\beta} \chi_{\beta R}$$

$$= |f(\gamma)|^2 \sum_{\alpha, \beta, \delta, \varepsilon} \bar{\chi}_{\alpha L} U(\gamma)_{\alpha\beta}^{-1} \left(\gamma_* Z^{\beta\delta} \right) U(\gamma)_{\delta\varepsilon} \chi_{\varepsilon R},$$

where $Z^{\alpha\beta}$ is the partition function of $\mathscr{H}^{\alpha\beta}$. As the χ_α are independent over the power series in q, one obtains

$$\gamma_* Z^{\alpha\beta} = \sum_{\delta,\,\varepsilon} U(\gamma)_{\alpha\delta} Z^{\delta\varepsilon} U(\gamma)_{\varepsilon\beta}^{-1},$$

which yields the modular invariance of the partition function of Q'. The local fields yielding the action of $\mathrm{Diff}_c(\mathbb{R} \times S^1)$ on $\mathscr{H}'$ are $T_R - T_R(G_R)$, $T_L - T_L(G_L)$, where $T_R(G_R)$, $T_L(G_L)$ are the Virasoro fields of section 3, example 2, which are given by the normal ordered products $:J_R J_R:$, $:J_L J_L:$. $\square$

Due to the physics background referred to in the conjecture, we call this procedure the gauging of H_{L+R}. It was inspired by (Goddard, Kent and Olive 1985). It can be used to construct many finitely reducible representations of $\mathscr{L}G \times \mathscr{L}G$ with modular invariant partition functions. We write the Hilbert space of such a representation in the form

$$\mathscr{H}(\delta) = \bigoplus_{\alpha,\,\beta \in I(\mathscr{L}G,\,k)} \left(\mathscr{H}_{\alpha L} \otimes \mathbb{C}^{\delta(\alpha,\,\beta)} \otimes \mathscr{H}_{\beta R} \right)$$

where $I(\mathscr{L}, G, k)$ is the index set for the irreducible representations of $\mathscr{L}G$ at level k and the dimensions $\delta(\alpha, \beta)$ are finite. The modular invariance of the partition function imposes linear constraints on δ. The unitarity of the transformation matrix $U(\gamma)$ used before implies that for any G, k one solution is given by

$$\delta_0^{G,\,k}(\alpha, \beta) = \delta_{\alpha\beta}, \qquad \alpha, \beta \in I(\mathscr{L}G, k).$$

In fact one can construct a quantum field theory $(\mathbb{R} \times S^1, \mathscr{H}(\delta_0), V, F)$ with F given by $\mathscr{L}G_R \times \mathscr{L}G_L$ (Witten 1984, Knizhnik and Zamolodchikov 1984). Now we consider embeddings $\iota\colon \mathscr{L}H \times \mathscr{L}H' \to \mathscr{L}G$ for which the $\mathscr{H}_\alpha$ decompose like

$$\mathscr{H}_\alpha = \bigoplus_{\substack{\gamma \in I(\mathscr{L}H,\,k(H)) \\ \gamma' \in I(\mathscr{L}H',\,k(H'))}} \mathscr{H}_\gamma^H \otimes \mathscr{H}_{\gamma'}^{H'} \otimes \mathbb{C}^{n(\alpha,\,\gamma,\,\gamma')}$$

with finite $n(\alpha, \gamma, \gamma')$. This requires $k = 1$, as discussed in section 3, example 3. The gauging of H'_{L+R} then yields the Hilbert space $\mathscr{H}(\iota\delta)$ with

$$\iota\delta(\gamma, \varepsilon) = \sum_{\substack{\alpha,\,\beta \in I(\mathscr{L}G,\,k) \\ \gamma' \in \mathscr{I}(\mathscr{L}H',\,k(H'))}} n(\alpha, \gamma, \gamma') \delta(\alpha, \beta) n(\beta, \varepsilon, \gamma')$$

which according to theorem 2 has a modular invariant partition function.

For $H = \mathrm{SU}(2)$ we use the labelling

$$\alpha = \mathrm{tr}\,\pi(n_\alpha) \in \{1, \ldots, k+1\} = I(\mathscr{L}\mathrm{SU}(2), k)$$

of the representation spaces $\mathscr{H}_\alpha$. Arguments supported by extensive computer studies (Cappelli et al. 1986, Gepner 1986) led to the conjecture that at level k the cone of δ which yield modular invariant partition functions has generators $\delta[\mathfrak{h}]$, where $\mathfrak{h}$ labels the simple Lie algebras of ADE type (the simply laced ones) with Coxeter number $k + 2$. In particular, $\delta_0 = \delta[A_{k+1}]$. The value $\delta[\mathfrak{h}](i, i)$ is the number of exponents of which are equal to i. Indeed $I(\mathscr{L}\mathrm{SU}(2), k)$ coincides with the set of exponents of A_{k+1}. In other cases the sets of exponents are

$$
\begin{aligned}
&1, 3, 5, \ldots, 2m - 3, m - 1 &&\text{for } D_m \\
&1, 4, 5, 7, 8, 11 &&\text{for } E_6 \\
&1, 5, 7, 9, 11, 13, 17 &&\text{for } E_7 \\
&1, 7, 11, 13, 17, 19, 23, 29 &&\text{for } E_8.
\end{aligned}
$$

These facts have remained mysterious so far. Here I will give at least a construction of $\delta[\mathfrak{h}]$ in terms of $\mathfrak{h}$.

For any Lie group G there is up to multiple coverings a unique quaternionic space $G/(\mathrm{SU}(2) \times H)$. Let $W = TG/(\mathrm{SU}(2) \times H)$ be its real tangent space, which yields an embedding $\iota_G : \mathscr{L}\,\mathrm{SU}(2) \times \mathscr{L}H \to \mathscr{L}\,\mathrm{SO}(W)$. For level 1 of $\mathscr{L}\,\mathrm{SO}(W)$ one finds in the case of a $U(1)$ factor of H

$$
k(U(1)) = \dim W/2
$$

and for simple factors H_1 of H

$$
k(H_1) = g(G) - g(H_1),
$$

i.e. the difference of the dual Coxeter numbers. Similarly

$$
k(\mathrm{SU}(2)) = g(G) - 2.
$$

This follows from the fact the embedding ι_G is regular and the remarks made in section 3, example 3.

THEOREM 3.

$$
\iota_G \delta_0^{\mathrm{SO}(W), 1} = 2\big(\mathrm{rank}(G)\delta_0^{\mathrm{SU}(2), k} + \delta[\mathrm{Lie}\,G]\big)
$$

for G of type D, E

$$
\iota_G \delta_0^{\mathrm{SO}(W), 1} = 2(\mathrm{rank}(G) - 1)\delta_0^{\mathrm{SU}(2), k}
$$

for G of type A, where always $k = g(G) - 2$. The case distinction is due to peculiarities of the hermitean symmetric spaces.

Proof. We calculate the characters explicitly, using a procedure invented by (Parthasarathy 1972) for ordinary Lie groups and generalized by (Kac and

Peterson 1981) to affine Kac-Moody groups. However, the result given in the latter reference is wrong as it stands and given without derivation, so we supply here the details.

Let us first consider a finite dimensional symmetric space G/H with rank $(G) = \text{rank}(H)$. The tangent space TG/H has a natural metric, so we can give it the structure of a Clifford algebra with standard module $C(TG/H)$, which splits as in example 3, section 3. Consider the character difference

$$d\chi(G, H) = \chi(C^+(TG/H)) - \chi(C^-(TG/H)).$$

One finds

$$d\chi(G, H) = D(G)/D(H)$$

where

$$D(G) = \prod_{\alpha \in \Delta_+(G)} \left(e^{\alpha/2} - e^{-\alpha/2} \right)$$

and analogously for $D(H)$. The Weyl denominator formula

$$D(G) = \sum_{w \in W(G)} \varepsilon(w)\exp(w\rho(G))$$

and the Weyl character formula

$$\chi(\lambda) = D(H)^{-1} \sum_{w \in W(H)} \varepsilon(w)\exp(w\lambda)$$

yield

$$d\chi(G, H) = \sum_{s \in W(G)/W(H)} \varepsilon(s)\chi(s\rho(G)).$$

The labelling of the representations used here is again that by the free orbits of the Weyl group $W(G)$, as discussed in example 2, section 3.

In the special case $G = H \times H$ one obtains

$$d\chi(H \times H, H) = D(H).$$

The same procedure works for $\mathscr{L}G$, one just has to use the Weyl-Kac character and denominator formulas given in that example. Using example 3, one finds the difference of level 1 characters of $\mathscr{L}\,\mathrm{Spin}(W)$

$$d\chi(\mathscr{L}G, \mathscr{L}H) = \chi(\mathscr{H}^+) - \chi(\mathscr{H}^-) = D(\mathscr{L}G)/D(\mathscr{L}H).$$

If H is semisimple, on has $T(H) \subset T(G)$ and obtains

$$d\chi(\mathscr{L}G, \mathscr{L}H)$$

$$= \sum_{t \in g(G)(T(G)/T(H))} \sum_{s \in W(G)/W(H)} \varepsilon(s)\chi(t + s\rho(G), g(G) - g(H)).$$

In the reference cited above, the sum over $T(G)/T(H)$ was omitted. In the only explicit example given there, namely $G/H = \mathrm{Sp}(N + M)/\mathrm{Sp}(N) \times \mathrm{Sp}(M))$ this does not matter, as $T(G) = \mathbf{Z}^{N+M} = T(H)$. In general, the contribution of the trivial element in $T(G)/T(H)$ just yields those representations of $\mathscr{L}H$ for which the highest weight lies in the subspace of lowest grade in $\mathscr{H}$, as can be seen by comparison with the ordinary Lie group case. For the quaternionic symmetric spaces of our construction one has for semisimple H

$$g(G)(T(G)/T(H)) = \{0, t_0\}.$$

The representations given by $(0, \gamma)$ and (t_0, γ) are related by an exterior automorphism of H, which does not respect the grading. From the character formula one sees that such exterior automorphisms occur, when t lies in a non-trivial coset of the dual of the root lattice modulo $T(H)$. For H of type ADE the root lattice and $T(H)$ coincide. The dual lattice is the weight lattice, such that the exterior automorphisms correspond naturally to the representations of the center of the covering group of H.

In the case of hermitean symmetric spaces one must write

$$d\chi(\mathscr{L}G, \mathscr{L}H) = \sum_{t \in IT} \sum_{s \in IS} \varepsilon(s)\chi(t + s\rho(G), g(G) - g(H))$$

$$IT = g(G)(T(G)/(T(G) \cap T(H)))$$

$$IS = (W(G) \ltimes (T(G) \cap T(H)))/(W(H) \ltimes T(H)),$$

where the sum over t again is a sum over representations which are transformed into each other by outer automorphisms.

Using the quaternionic structure of our symmetric space one can factor the embedding $\mathscr{L}(\mathrm{SU}(2) \times H) \to \mathscr{L}\,\mathrm{Spin}(W)$ in the form

$$\mathscr{L}\,\mathrm{SU}(2) \times \mathscr{L}H \to \mathscr{L}\,\mathrm{SU}(2) \times \mathscr{L}\,\mathrm{Sp}(T_H G/(H \times \mathrm{SU}(2))) \to \mathscr{L}\,\mathrm{Spin}(W).$$

Level 1 representations of $\mathscr{L}\,\mathrm{Spin}(W)$ induce level 1 representations of $\mathscr{L}\,\mathrm{Sp}(T_H G/(H \times \mathrm{SU}(2))) \simeq \mathscr{L}\,\mathrm{Sp}(n)$ and level n representations of $\mathscr{L}\,\mathrm{SU}(2)$, where

$$n = \dim(T_H G/(H \times \mathrm{SU}(2))) = g(G) - 2.$$

Let

$$\Lambda^j \mathbf{C}^{2n} \oplus \Lambda^{j-1}\mathbf{C}^{2n} \simeq \bigoplus_{i=0}^{j} \Lambda^{j,n}, \qquad j = 0, 1, \ldots, n$$

be the splitting of the exterior products of the defining representation of $\mathrm{Sp}(n)$ into irreducible representations. The representation spaces of $\mathrm{Sp}(n)$ which occur at the lowest grade of irreducible level 1 representations of $\mathscr{L}\,\mathrm{Sp}(n)$ are just isomorphic to the $\Lambda^{i,n}$, $i = 0, 1, \ldots, n$. Accordingly, we use the index set

$$I(\mathcal{L}\,\mathrm{Sp}(n), 1) = \{0, 1, \ldots, n\}.$$

With this notation

$$\chi(\mathcal{H}_0) - \chi(\mathcal{H}_1) = \sum_{i=0}^{n} (-)^i \chi_{i+1}[\mathcal{L}\,\mathrm{SU}(2)]\chi_{i+1}[\mathcal{L}\,\mathrm{Sp}(n)].$$

We still have to calculate the decomposition of the other two irreducible representations of $\mathcal{L}\,\mathrm{Spin}(W)$ of level 1, for which we use the notation $\mathcal{H}_0$ and $\mathcal{H}_1$ as in example 3, section 3. For them, Kac and Peterson gave no formula. Using an outer automorphism of $\mathcal{L}\,\mathrm{Spin}(W)$ one finds

$$\chi(\mathcal{H}_0) - \chi(\mathcal{H}_1) = \sum_{i=0}^{n} (-)^i \chi_{i+1}[\mathcal{L}\,\mathrm{SU}(2)]\chi_i[\mathcal{L}\,\mathrm{Sp}(n)].$$

Thus the terms coming from $\mathcal{H}_0$, $\mathcal{H}_1$ yield exactly the same contribution to $\iota_G\delta$ as the terms coming from $\mathcal{H}^+$, $\mathcal{H}^-$, which explains the factor 2 in the formula for $\iota_G\delta$.

For the decomposition of $\Lambda^{i,n}$ into irreducible representations of the Lie group H, with G an exceptional Lie group, I used a computer. It is easy to calculate the diagonal terms of $\delta[\mathfrak{h}]$, as one just has to count the number of irreducible representations in $\Lambda^{i,n}$. For each exponent i of $\mathfrak{h}$, an extra term appears in $\Lambda^{i-1,n}$, and Poincaré duality corresponds to the outer automorphisms used above. To calculate the off-diagonal terms of $\delta[\mathfrak{h}]$, one has to find representations of H which appear in different level 1 representations of $\mathcal{L}\,\mathrm{Sp}(n)$. Their highest weights are invariant under the outer automorphisms which are induced from those of $\mathcal{L}G$ and correspond to the center of G, as discussed above.

In the case of $G = \mathrm{SO}(2k + 4)$, $H = \mathrm{SU}(2) \times \mathrm{SO}(2k)$, such a term appears in the $\mathcal{L}\,\mathrm{Sp}(2k)$ representation spaces $\mathcal{H}_i$, $\mathcal{H}_j$, $i \neq j$, iff $i + j = 2k$ and $k - i$ is even. The corresponding highest weight of $\mathrm{SU}(2)$ is the one of the k-fold symmetric tensor product of the fundamental representation, and the highest weight of $\mathrm{SO}(2k)$ is the highest weight in $\Lambda^{(k-i)/2}R^{2k} \otimes \Lambda^{(k+i)/2}R^{2k}$, where $i < k$.

For $G = E_6, E_7, E_8$, $H = \mathrm{SU}(6)/Z_3$, $\mathrm{Spin}^+(12)$, E_7, the corresponding terms are listed in Table 1.

Table 2 lists the generating functions for the resulting $\delta[\mathfrak{h}]$, which earlier had been found on the computer. $\square$

It is not yet clear what are the maximal superelection sectors of the quantum field theories obtained in this way. Obviously a trivial factor with $\mathcal{H} = \mathbf{C}^2$ should be split off. Quantum field theories with a Hilbert space with $\delta = \delta[\mathfrak{h}]$ have been constructed in all cases, except for $\mathfrak{h} = E_7$ (Gepner and Witten 1986, Bouwknegt and Nahm 1986). We also obtained a theory with $\delta = 2\delta[E_7] + \delta[D_{10}]$. Our construction only can yield δ for which the generating function can be written as a sum of squares. It is interesting that for all the root lattices which yield even

TABLE 1

Numbering of the vertices of the extended Dynkin diagrams for A_5, D_6, E_7:

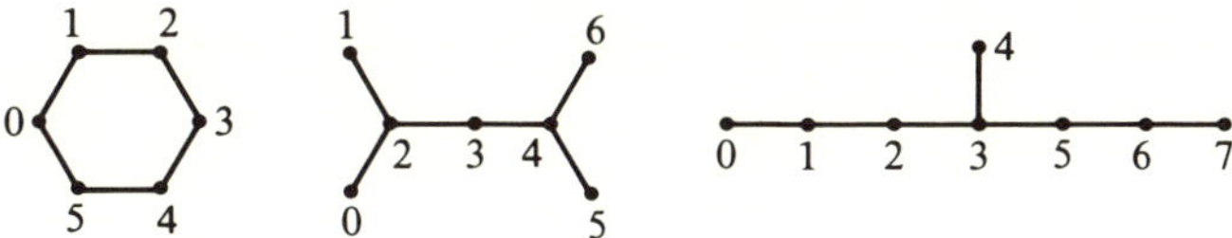

Lie $G = E_6$, Lie $H = A_5$:

highest weight	appearing in $\chi_i[\mathscr{L} \, \mathrm{Sp}(10)]$ with
020202	$i = 0, 6$
202020	$i = 4, 10$
111111	$i = 3, 7$

Lie $G = E_7$, Lie $H = D_6$

highest weight	appearing in $\chi_i[\mathscr{L} \, \mathrm{Sp}(16)]$ with
0004000	$i = 0, 16$
2010102	$i = 2, 8$
0210120	$i = 8, 14$
0020200	$i = 4, 12$
1110111	$i = 6, 10$

Lie $G = E_8$, Lie $H = E_7$

highest weight	appearing in $\chi_i[\mathscr{L} \, \mathrm{Sp}(28)]$ with
00006000	$i = 0, 28$
00400000	$i = 0, 18$
40020000	$i = 0, 10$
00000400	$i = 10, 28$
00020004	$i = 18, 28$
02010020	$i = 10, 18$
00030000	$i = 6, 22$
03000200	$i = 6, 16$
22010010	$i = 6, 12$
00200030	$i = 12, 22$
01010022	$i = 16, 22$
11100111	$i = 12, 16$

self-dual lattices in 24 dimension the corresponding sum $\sum_i \delta[\mathfrak{y}_i]$ can be written as a sum of squares, like for the case $D_{10} E_7^2$ just mentioned. So far, I cannot explain this fact.

There is an indication that the result of theorem 3 is related to the correspondence between finite subgroups of SU(2) and simple Lie algebras of ADE type. The decomposition of the symmetric algebra over $\mathbf{C}^2$ under these finite subgroups is described by a generating function with parameter t which is proportional to $(1 - t^a)^{-1}(1 - t^b)^{-1}$ for certain positive integers a, b, where $ab/2$ is the order of the finite subgroup. These integers can be interpreted in terms of the

WERNER NAHM

TABLE 2

The functions $\sum\limits_{ij} \delta[\mathfrak{y}](i, j)\xi_i\bar{\xi}_j$

for $\mathfrak{y} = A_k$:	$\displaystyle\sum_{i=1}^{k} \lvert\xi_i\rvert^2$
for $\mathfrak{y} = D_{2k}$:	$\displaystyle\frac{1}{2}\sum_{i=0}^{2k-2} \lvert\xi_{2i+1} + \xi_{4k-3-2i}\rvert^2$
for $\mathfrak{y} = D_{2k+1}$:	$\displaystyle\sum_{i=0}^{2k-1} \lvert\xi_{2i+1}\rvert^2 + \sum_{i=1}^{2k-1} \xi_{2i}\bar{\xi}_{4k-2i}$
for $\mathfrak{y} = E_6$:	$\lvert\xi_1 + \xi_7\rvert^2 + \lvert\xi_4 + \xi_8\rvert^2 + \lvert\xi_5 + \xi_{11}\rvert^2$
for $\mathfrak{y} = E_7$:	$\lvert\xi_1 + \xi_{17}\rvert^2 + \lvert\xi_5 + \xi_{13}\rvert^2 + \lvert\xi_7 + \xi_{11}\rvert^2$ $+ \lvert\xi_9\rvert^2 + \xi_9(\bar{\bar{\xi}}_3 + \bar{\xi}_{25}) + \bar{\xi}_9(\xi_3 + \xi_{15})$
for $\mathfrak{y} = E_8$:	$\lvert\xi_1 + \xi_{11} + \xi_{19} + \xi_{29}\rvert^2 + \lvert\xi_7 + \xi_{13} + \xi_{17} + \xi_{23}\rvert^2$

root systems (Kostant 1984). Here we just note the following coincidence: For the finite group which corresponds to G, we have

$$a = g(G) - g(H_1)$$

$$b = g(H_1) + 2,$$

where $G/(\mathrm{SU}(2) \times H)$ is the quaternionic symmetric space used above, $H = H_1$ for simple H, and $H = H_1 \times H_2$ otherwise. In the latter case, switching H_1 and H_2 switches a and b.

Quantum field theories obviously have deep connections to many branches of mathematics, and already the very simplest ones show a surprising richness of structure.

Acknowledgement.   Most of this work was done at the Physics Department of the University of California at Davis, which I thank for its hospitality. Much of it was inspired by discussions with mathematicians and physicists at the Max-Planck-Institut for mathematics at Bonn. I wish to thank L. Breen, E. Corrigan, G. Mackey, D. Zagier and in particular P. Goddard and D. Olive for explanations and discussions.

REFERENCES

F. A. BAIS AND P. G. BOUWKNEGT, Nucl. Phys. **B279** (1987), 561.

P. G. BOUWKNEGT AND W. NAHM, *Realizations of the exceptional modular invariant $A_1^{(1)}$ partition functions*, Phys. Lett. **184B** (1987), 349.

A. Cappelli, C. Itzykson, and J.-B. Zuber, *Modular invariant partition functions in two dimensions*, Nucl. Phys. **B280** (1987), 445.

J. L. Cardy, Nucl. Phys. **B270** (1986), 186.

B. L. Feigin and D. B. Fuchs, Funct. Anal. Appl. **17** (1983), 241.

I. B. Frenkel and V. G. Kac, Invent. Math. 62 (1980) 23.

I. B. Frenkel, J. Lepowsky and A. Meurman, Proc. Nat. Acad. Sci. USA (1984), 3256.

D. Friedan, Z. Qiu, and S. Shenker, Phys. Rev. Lett. **52** (1984), 1575.

D. Gepner, *On the spectrum of 2d conformal field theories*, Princeton preprint, June 1986.

D. Gepner and E. Witten, Nucl. Phys. **B278** (1986), 493.

P. Goddard, A. Kent, and D. Olive, Phys. Lett. **152B** (1985), 88.

P. Goddard, W. Nahm, and D. Olive, Phys. Lett. **160B** (1985), 111.

P. Goddard, D. Olive, and A. Schwimmer, Phys. Lett. **157B** (1985), 393.

R. L. Griess, Inv. Math. **69** (1982), 1.

V. G. Kac, Bull. Amer. Math. Soc. **2** (1980), 311.

V. G. Kac and D. H. Peterson, Proc. Nat. Acad. Sci. USA **78** (1981), 3308.

V. G. Kac, *Infinite dimensional Lie algebras*, Birkhäuser, 1983.

V. Knizhnik and A. B. Zamolodchikov, Nucl. Phys. **B247** (1984), 83.

B. Kostant, Proc. Nat. Acad. Sci. USA **81** (1984), 5275.

R. Parthasarathy, Ann. Math. **96** (1972), 1.

A. Rocha-Caridi, in *Vertex Operators in Mathematics and Physics*, Springer 1984, J. Lepowski, S. Mandelstam, I. M. Singer eds. p. 451.

A. N. Schellekens and N. P. Warner, Phys. Rev. **D34** (1986), 3092.

G. Segal, Comm. Math. Phys. **80** (1981), 301.

R. F. Streater and A. S. Wightman, *PCT, Spin and Statistics, and all that*, 2nd print, Benjamin 1978.

E. Witten, Comm. Math. Phys. **92** (1984), 455.

Department of Physics, State University of New York, Stony Brook, New York 11794

Vol. 54, No. 2   DUKE MATHEMATICAL JOURNAL ©   1987

RAMIFIED TORSION POINTS ON CURVES

ROBERT F. COLEMAN

Dedicated to Yu I. Manin on the occasion of his fiftieth birthday

Introduction. Let K be a field and C a smooth complete connected curve over K. Let $\overline{K}$ denote an algebraic closure of K. If $P, Q \in C(\overline{K})$ we write $P \sim Q$ if a positive integral multiple of the divisor $P - Q$ is principal. Then " $\sim$ " is an equivalence relation. With respect to this relation, we call an equivalence class a torsion packet.

Using Abel's addition theorem, the Manin-Mumford conjecture proven by Raynaud [R-2] is equivalent to:

THEOREM A.   *If* $\operatorname{char}(K) = 0$ *and the genus of C is at least two then every torsion packet in $C(\overline{K})$ is finite.*

We propose the following conjecture:

CONJECTURE B.   *Suppose K is a number field and T is a torsion packet in $C(\overline{K})$ stable under* $\operatorname{Gal}(\overline{K}/K)$. *Suppose* $\mathfrak{p}$ *is a prime of K satisfying*

(i) char $\mathfrak{p} > 3$.
(ii) $K/\mathbb{Q}$ *has good reduction at* $\mathfrak{p}$.
(iii) C *has good reduction at* $\mathfrak{p}$.
If the genus of C is at least two, the extension $K(T)/K$ is unramified above $\mathfrak{p}$.

We have proven this conjecture under any of the following additional hypotheses:

(a) char $\mathfrak{p} > 2g$.
(b) C has ordinary reduction at $\mathfrak{p}$ (i.e., the Hasse-Witt matrix at $\mathfrak{p}$ is invertible).
(c) C has superspecial reduction at $\mathfrak{p}$ (i.e., the Hasse-Witt matrix at $\mathfrak{p}$ is zero).
(d) C is an abelian branched covering of $\mathbb{P}^1_K$ over K unbranched outside $\{0, 1, \infty\}$, and T is the torsion packet on C containing the inverse image of $\{0, 1, \infty\}$.

In this paper we will prove the conjecture under any of the hypotheses (a)–(c). This generalizes the first part of Theorem A of [C-1]. In [C-2] we will prove the conjecture under hypothesis (d). (See also [C-3] for additional evidence.) We will also give a new proof, of the Manin-Mumford conjecture, which we will now sketch.

First, by standard arguments it suffices to prove Theorem A when K is a number field. Second, Bogomolov [B-1] has proven.

Received March 14, 1987.

615

THEOREM C (Bogomolov). *Let $f\colon C \to A$ be a morphism over K into an Abelian variety such that the image of C is not contained in a one-dimensional subgroup of A. Then $\#(f(C) \cap A[N^{\infty}])(\overline{K}) < \infty$ for any positive integer N.*

(Note: One-dimensional subgroups are not necessarily connected.)

Let T be a torsion packet in $C(\overline{K})$. After replacing K by a finite extension field we may suppose there is a $P \in T \cap C(K)$. Let $f\colon C \to A$ be the Albanese morphism of C into its Jacobian such that $P \mapsto 0$. Then

$$T = f^{-1}\big(C(K) \cap A_{\mathrm{Tor}}\big)(\overline{K})$$

by Abel's addition theorem. Also T is stable under $\mathrm{Gal}(\overline{K}/K)$. Hence the Manin-Mumford conjecture will follow from Theorem C and Conjecture B under hypothesis (a) together with

THEOREM D. *Let N be a positive integer and K_N the maximal extension of K in $\overline{K}$ unramified outside N. Then the subgroup of $A(K_N)$ consisting of torsion points of order prime to N is finite.*

The main idea of this paper is to use the observation of [C-1] that a torsion packet containing a point P is the set of common zeros of the set of p-adic Abelian integrals of the first kind which vanish at P. We study these zeros residue class by residue class, for on a residue class the integrals may be expanded in formal power series; moreover, the zeros of a power series may be studied by means of its Newton polygon. Hence we undertake an analysis of the Newton polygons of the formal expansions of p-adic Abelian integrals on a residue class. This is accomplished by using the action of Frobenius on the de Rham cohomology of a p-adic curve (Sections 1 and 2). In Section 3, we obtain information about the common zeros of the integrals in a residue class by obtaining information on the common slopes of their Newton polygons. We draw our main conclusions in Section 4 and give our proof of the Manin-Mumford conjecture in Section 5.

Throughout this article we will employ the following notations: for a rational prime p, $\mathbb{Q}_p$ will denote the field of p-adic numbers and $\mathbb{C}_p$ the completion of a fixed algebraic closure of $\mathbb{Q}_p$. The completion in $\mathbb{C}_p$ of the maximal unramified algebraic extension of $\mathbb{Q}_p$ will be denoted K_p^u while R_p^u and $\mathbb{F}_p^u$ will denote its ring of integers and residue field respectively. The absolute Frobenius automorphisms of K_p^u and $\mathbb{F}_p^u$ will be denoted by the same letter, σ. We use v_p to denote the unique continuous valuation map $v_p\colon \mathbb{C}_p \to \mathbb{Q} \cup \{\infty\}$ such that $v_p(p) = 1$.

Suppose K is a complete subfield of K_p^u, $R = R_p^u \cap K$, and $\mathbb{F}$ the residue field of K. By a curve over R we mean a proper smooth connected one-dimensional scheme over R. We use the notation $\tilde{C}$ to denote the special fiber of C. By a *residue class* of C we mean both a point in $C(\mathbb{F}_p^u)$ and its inverse image in $C(\mathbb{C}_p)$ considered as an analytic space.

We are grateful to Ken Ribet for acquainting us with his exposition of Bogomolov's unpublished paper [B-2].

§1. Remarks on F-crystals. We now fix a rational prime p and allow ourselves to drop the subscripts "p" from our notation when convenient.

Let K be a complete subfield of K^u and $R = K \cap R^u$ its ring of integers. Also let $\mathbb{F} \subseteq \mathbb{F}^u$ denote the residue field of K. A quadruple $H = (H, W, F, V)$ will be called an F-crystal over R if it is a free R-module of finite rank, W is a direct summand of H, F is an σ-linear endomorphism of H, and V is a σ^{-1}-linear endomorphism of H such that

$$(1) \qquad\qquad FV = VF = p$$

and

$$(2) \qquad\qquad VH = W + pH.$$

It follows that

$$(3) \qquad\qquad FH \supseteq pH \supseteq FW.$$

We set

$$\tilde{H} = H/pH, \qquad \tilde{W} = \frac{W + pH}{pH} \subseteq \tilde{H}$$

as vector spaces over $\mathbb{F}$. For $h \in H$, $\tilde{h}$ denotes its image in $\tilde{H}$. We also set

$$m(h) = \max\{n : h \in p^n H\} \in \mathbb{N} \cup \{\infty\}$$

and

$$h^* = \begin{cases} \left(p^{-m(h)}h\right)^{\tilde{}}, & h \neq 0 \\ \tilde{0}, & h = 0. \end{cases}$$

LEMMA 1. *Suppose $h \in H$, $\tilde{h} \neq 0$ is such that $V(h)^* \in \tilde{W}$, then $V(h)^{\tilde{}} = V(h)^* \neq 0$.*

Proof. By (2) there exists $k \in H$ such that $V(k)^{\tilde{}} = V(h)^*$. Suppose $V(h)^{\tilde{}} = 0$, then $V(h)^* = (p^{-1}V(h))^{\tilde{}}$ and $V(pk) \equiv V(h) \bmod p^2 H$. So $V(h - pk) \equiv 0 \bmod p^2 H$. Applying F to this we get $p(h - pk) \equiv 0 \bmod p^2 H$, so $h \equiv 0 \bmod pH$, which contradicts our hypotheses. Thus $V(h)^{\tilde{}} \neq 0$ so $V(h)^{\tilde{}} = V(h)^*$, which proves our lemma. $\qquad\square$

We call H *ordinary* if

$$(4) \qquad\qquad VW \equiv W \bmod pH$$

and *superspecial* if

$$(5) \qquad\qquad\qquad FH = W + pH.$$

We call *H extraordinary* otherwise.

LEMMA 2. *Suppose H is superspecial; then*
   (i) $VW \equiv 0 \bmod pH$;
   (ii) $\dfrac{V(W)}{p} \oplus W = H$;
   (iii) $\dfrac{V^2(W)}{p} \equiv W \bmod pH$.

Proof. Applying V to (5) we obtain,

$$pH = VW + pVH$$

which implies (i). We also obtain

$$H = \frac{V}{p} W + VH = \frac{V}{p} W + W + pH$$

using (2). Now (ii) will follow once we show,

$$(6) \qquad\qquad\qquad V(W) \cap W = (0).$$

Suppose $w \in W$ such that $w \neq 0$ and $V(w) \in W$. We may suppose $\tilde{w} \neq 0$. Then

$$F\!\left(\frac{V(w)}{p} \right) = w \notin pH$$

but $p^{-1}V(W) \in W$ which contradicts (3), which establishes (6) and (ii). Now applying V to (ii) we get

$$W + pH = VH = \frac{V^2(W)}{p} \oplus V(W).$$

This, together with (i), implies (iii). $\qquad\qquad\qquad\qquad\qquad\qquad\square$

As an immediate corollary we have:

COROLLARY 2.1. *H is superspecial iff* $\mathrm{rk}_R H = 2\,\mathrm{rk}_R W$ *and* $VW \equiv 0 \bmod pH$.

The following lemma will be crucial in the proofs of the main theorems.

LEMMA 3. *Suppose* $n \geqslant 0$ *such that* $V^n(w)^* \in \tilde{W}$ *for all* $w \in W$. *Then* $\{V^n(w)^* : w \in W\} = \tilde{W}$.

Proof. Let $m = \max\{m(V^n(w)): w \in W, \tilde{w} \neq 0\}$. For $k \in \mathbb{N}$ let $L_k = \{w \in W: m(V^n w)) \geqslant k\}$. Clearly L_k is an R submodule of W with maximal rank.

Claim. $L_{m+1} = pL_m$.

Indeed, if $w \in L_{m+1}$ then $w \in pW$ by the definition of m. Let $w' = p^{-1}w$. Then $m(V^n(w')) = m(V^n(w)) - 1 \geqslant m$. So $w' \in L_m$ which yields the claim.

It follows that the map $w \mapsto V^n(w)$ induces a σ^{-n}-linear injection

$$L_m/pL_m \to \frac{p^m H}{p^{m+1} H} \xrightarrow{\sim} \tilde{H}.$$

By hypothesis this map factors through $\tilde{W}$. Since

$$\dim_{\mathbf{F}}(L_m/pL_m) = \mathrm{rk}_R L_m = \mathrm{rk}_R W = \dim_{\mathbf{F}} \tilde{W}$$

we deduce that our map

$$L_m/pL_m \to \tilde{W}$$

is an isomorphism, which proves the lemma. $\qquad\square$

We call H, PCM, for pseudo-complex-multiplication, if W is given as the direct sum of rank one submodules, $\{B_i\}_{i \in I}$, for some index set I, such that if $b \in B = \bigcup_{i \in I} B_i$, $n \in \mathbb{N}$ such that $V^n(b)^* \in \tilde{W}$ then $V^n(b) \in B$.

LEMMA 4. *Suppose H is PCM. If $V^n(W) \subseteq W$ then*

$$\bigoplus_{i \in I} \left(V^n(B_i) \otimes \mathbb{Q}_p\right) = W \otimes \mathbb{Q}_p.$$

Proof. If $i \neq j$ then $V^n(B_i) \cap V^n(B_j) = (0)$ since V is injective. The lemma follows immediately. $\qquad\square$

LEMMA 5. *Suppose H is PCM, extraordinary, and $\mathrm{rk}\, H = 2\,\mathrm{rk}\, W$. Then $VB \subsetneq W$ and $VB \cap W \neq (0)$.*

Proof. $VB \subsetneq W$ since H is not ordinary.

Now since H is not superspecial there exists $w \in W$ such that $V(w)\tilde{} \neq 0$ and lies in $\tilde{W}$. By Corollary 2.1 we may write $w = \sum_{i \in I} b_i$, where $b_i \in B_i$. Since $V(w)\tilde{} \neq 0$ we must have $V(b_i)\tilde{} \neq 0$ for some i. By Lemma 1 it follows that $V(b_i)\tilde{} \in \tilde{W}$ and hence by the definition of PCM, $V(b_i) \in W$. This completes the proof. $\qquad\square$

LEMMA 6. *Suppose $w \in W$, $w \neq 0$, then $\lim_{n \to \infty}(n - m(V^n(w))) = \infty$.*

Proof. Let $U = \bigcap_{n \geqslant 0} F^n H$. Then $FU = U$ and it follows that

$$(7) \qquad\qquad m(F(u)) = m(u)$$

for $u \in U$. Now suppose $n - m(V^n(w)) = k$ for all $n \gg 0$. Then

$$V^n(w) = p^{n-k}h_n$$

for some $h_n \in H$. Hence applying F^n we get $p^k w = F^n h_n$ for all $n \geq w$. It follows that $p^k w \in U$. But

$$m(F(p^k w)) = k + m(F(w)) = k + m(w) + 1 = m(p^k w) + 1.$$

This contradicts (7) and proves the lemma. $\qquad\qquad\square$

We call U the unit root subspace of H. From the proof we have

COROLLARY 6a.　*The conclusion of Lemma 6 holds for $w \notin U$.*

§2. Newton polygons of Abelian integrals.　First we establish some general notations for Newton polygons. Let $\mathbb{C}_p[[x]]$ denote the ring of power series over $\mathbb{C}_p$. If

$$L(x) = \sum_{n=0}^{\infty} a_n x^n \in \mathbb{C}_p[[x]]$$

we define the Newton polygon of L, $N(L)$ to be the lower convex hull in $\mathbb{R}^2$ of the points $\{(n, v(a_n)): a_n \neq 0\}$.

By a *slope* of $N(L)$ we mean the slope of one of its sides, or $-\infty$ if $a_0 = 0$ and $L \neq 0$. By a *minus* slope we mean the additive inverse of a slope, by the *run* of a slope, the length of the projection of the corresponding side onto the x-axis, and by a *vertex* of $N(L)$, an endpoint of a side. We will have more to say about Newton polygons in the next section.

Let C be a curve of genus g over R. Let

$$H^1_{dR} = H^1_{dR}(C/R) = \mathbb{H}^1(C, \Omega^*_{C/R})$$

be the first de Rham cohomology group of C over R. It is a free R-module of rank $2g$. Set

$$\Omega = \Omega(C) = H^0(C, \Omega^1_{C/R}),$$

to be the module of holomorphic differentials on C over R. Then Ω is a free R-module of rank g which injects naturally into H^1_{dR} as a direct summand. We will identify Ω with its image in H^1_{dR}. The context should make it clear about which manifestation of Ω we are speaking. There exist natural σ-linear and σ^{-1}-linear endomorphisms of H^1_{dR}, F and V, such that

$$H^1_{dR} = \left(H^1_{dR}, \Omega, F, V\right)$$

is an F-crystal over R, which we shall describe presently. We say C is *ordinary*, *superspecial* or *extraordinary* if the corresponding statement is true of H^1_{dR}.

Regard C now as a formal scheme so that $\mathcal{O}_C$ and $\Omega^1_{C/R}$ denote the formal structure sheaf and the sheaf of formal differentials. Let $\mathcal{C}$ be an open covering of C by formal affines. A collection of pairs $\omega = \{(U, \omega_U)\}_{U \in \mathcal{C}}$ such that $\omega_U \in \Omega^1_{C/R}(U)$ is called a $\mathcal{C}$-Čech one-cocycle if for all $U, V \in \mathcal{C}$,

$$\omega_U|_{U \cap V} - \omega_V|_{U \cap V} \in d\mathcal{O}_C(U \cap V).$$

The cocycle ω is said to be exact if $\omega_U \in d\mathcal{O}_C(U)$ for all $U \in \mathcal{C}$. Let $H(\mathcal{C})$ denote the classes of $\mathcal{C}$-Čech one-cocycles modulo exact $\mathcal{C}$-Čech one-cocycles. Then the natural map

$$H^1_{dR} \to \varinjlim_{\mathcal{C}} H(\mathcal{C}),$$

where the direct limit runs over formal affine open covers $\mathcal{C}$ of C, is an isomorphism [B-0]. In fact we need only to consider formal affine coverings whose members are formal completions along Zariski affine open subschemes of $\tilde{C}$. We may think of such formal opens as affinoides [R-1] and we call them *Zariski affinoides*. In the following we will employ the notation of [C-1, §1].

We may now describe F. Let $\mathcal{C}$ be a covering of C by Zariski affinoides. For each $U \in \mathcal{C}$ we let

$$\phi_U \colon U \to U^\sigma$$

be a lifting of the absolute Frobenius morphism from $\tilde{U}$ to $\tilde{U}^\sigma$ as in [K] or [C-1]. Let $\omega = \{(U, \omega_U)\}_{U \in \mathcal{C}}$ be a $\mathcal{C}$-Čech one-cocycle. Then

$$\omega' = \{(U, \phi_U^* \omega_U^\sigma)\}$$

is also a $\mathcal{C}$-Čech one-cocycle and its class in H^1_{dR} is independent of the choices of ϕ_U. If $[\omega]$ denotes the class of ω in H^1_{dR} then

$$F[\omega] = [\omega'].$$

PROPOSITION 6. *Let X and Y be two affinoides over K with good reduction. Let $f, g \colon X \to Y$ be two morphisms such that $\tilde{f} = \tilde{g}$. Let ω be a closed continuous one-form on Y in the image of $\Omega^1_{A^0(Y)/R}$. Then*

$$f^*\omega - g^*\omega \in p\, dA^0(X).$$

Proof. This is an immediate corollary of the proof of Proposition 1.2 of [C-1]. $\square$

Definition. Suppose X is a Zariski subaffinoide of C and $\eta \in \Omega^1_{X/R}$. Then we say η is of the *second kind* on C if there exists a formal affine covering $\mathcal{C}$ of C by Zariski-affinoides and a $\mathcal{C}$-Čech one-cocycle $\omega = \{(U, \omega_U)\}_{U \in \mathcal{C}}$ on C over R such that $(X, \eta) \in \omega$.

Remark. This notion depends on the embedding of X in C_K.

We let $\mathcal{D}_C^{sk}(X)$ denote the R-module of differentials of the second kind on X. Note that

$$(1) \qquad\qquad \lim_{\substack{\longrightarrow \\ \mathcal{C}}} H(\mathcal{C}) \cong H(\mathcal{C}_0)$$

if $\mathcal{C}_0$ is Zariski-affinoide covering.

PROPOSITION 7. $H_{dR}^1 = \mathcal{D}_C^{sk}(X)/dA^0(X)$ *under the natural map.*

Proof. In view of (1), the only thing to check is injectivity, but this follows from the following lemma.

LEMMA 8. *Suppose X is a connected one dimensional affinoide with good reduction and $\omega \in \Omega^1_{A^0(X)/R}$ which lies in $dA^0(Y)$ for some nonempty Zariski subdomain Y of X. Then ω is exact on X.*

Proof. First note that $X - Y$ is a finite union of residue disks. By an inductive argument we may suppose $X - Y = U$ is a single residue disk. Let $T \in A^0(X)$ be a uniformizing parameter on U, i.e., $T\colon U \overset{\sim}{\to} B(0,1)$, the open unit ball. Let X' denote the Zariski subaffinoide of X whose reduction is $U \cup (\tilde{X}\text{-locus}$ of $\tilde{T}\,d\tilde{T})$. Let $Y' = Y \cap X'$. Then $\omega = f\,dT$ on X' for some $f \in A^0(X')$. Suppose $g \in A^0(Y')$. Then we may write uniquely

$$(2) \qquad\qquad g = h + \sum_{n=1}^{\infty} a_n T^{-m}$$

where $h \in A^0(X')$, $a_n \in R$, and $a_n \to 0$ as $n \to \infty$. Suppose now $g \in A^0(Y)$ such that $dg = \omega$ on Y. Express $g|_{Y'}$ as in (2). Then

$$f\,dT = dh - \sum_{n=1}^{\infty} na_n T^{-(n+1)}\,dT$$

on Y'. Hence as $f \in A^0(X)$ we must have $a_n = 0$ for all $n \geqslant 1$. Hence $g|_{Y'} = h|_{X'}$, so $g|_{Y'}$ analytically continues to X'. Since it already continues to an element of $A^0(Y)$ it must continue uniquely to an element of $A^0(Y \cup X') = A^0(X)$. Finally since $dg - \omega = 0$ on Y, by the uniqueness of analytic continuation (X is connected), $dg - \omega = 0$ on X. This completes the proof. $\square$

If $\omega \in \mathcal{D}_C^{sk}(X)$ we let $[\omega]$ denote its class in H_{dR}^1.

PROPOSITION 9. *Suppose $\omega \in \mathcal{D}_C^{sk}(X)$, such that $[\omega] \in \Omega + pH_{dR}^1$. Suppose $\phi\colon X \to X^\sigma$ is a lifting of absolute Frobenius. Then $\phi^*\omega^\sigma/p \in \mathcal{D}_C^{sk}(X)$.*

Proof. Let $\mathcal{C}$ be a cover of C by Zariski subdomains. Suppose $\nu \in \Omega$ and η is a $\mathcal{C}$-Čech one-cocycle such that $X \in \mathcal{C}$ and $\omega = \nu + p\eta_X$. For each $U \in \mathcal{C}$ let

$\phi_U \colon U \to U^\sigma$ be a lifting of absolute Frobenius such that $\phi_X = \phi$. For $U \in \mathscr{C}$ let $\alpha_U = p^{-1}\phi_U^*(\nu^\sigma + p\eta_U^\sigma)$.

Claim. $\alpha = \{(U, \alpha_U) \colon U \in \mathscr{C}\}$ is a $\mathscr{C}$-Čech one-cocycle.

First, $\alpha_U \in \Omega_C^1(U)$, obviously. Second,

$$\alpha_U - \alpha_V = p^{-1}(\phi_U^*\nu^\sigma - \phi_V^*\nu^\sigma) + \phi_U^*\eta_U^\sigma - \phi_V^*\eta_V^\sigma$$

$$= p^{-1}(\phi_U^*\nu^\sigma - \phi_V^*\nu^\sigma) + (\phi_U^*\eta_U^\sigma - \phi_V^*\eta_U^\sigma) + \phi_V^*(\eta_U^\sigma - \eta_V^\sigma)$$

which lies in $d\mathscr{O}_C(U \cap V)$ using Proposition 5 and the fact that η is a $\mathscr{C}$-Čech one-cocycle. This completes the proof. $\square$

As an immediate consequence we have

COROLLARY 9.1. *Let* $\omega, \nu \in \mathscr{D}_C^{sk}(X)$ *such that* $[\nu] = V[\omega]$. *Then for some* $\omega = \phi^*\nu^\sigma/p + df$, $f \in A^0(X)$.

Now let $\nu \in \mathscr{D}_C^{sk}(X)$. Let $\omega_i \in \mathscr{D}_C^{sk}(X)$ such that

$$\omega_0 = \nu$$

$$[\omega_i] = V^i[\omega]$$

$$\omega_i \in p^{m_i}\mathscr{D}_C^{sk}(X)$$

where $m_i = m([\omega_i])$ and if $[\omega_i]^* \in \tilde{\Omega}$, $\omega_i = \nu + p\eta$ for some $\nu \in \Omega$, $\eta \in \mathscr{D}_C^{sk}(X)$. Let U be a residue disk of X and $Q \in U(R^u)$. Let $T \in A^0(X)$ be a uniformizing parameter on U such that $T(Q) = 0$. Let Y denote the Zariski subaffinoide of X whose reduction is $\tilde{X}$-(locus of $d\tilde{T}$). It follows from Theorem 1.1 of [C-1] that there exists a lifting of absolute Frobenius

$$\phi \colon Y \to Y^\sigma$$

such that

$$\phi(Q) = Q^\sigma \quad \text{and} \quad \phi^*T^\sigma = T^p.$$

Hence by Corollary 9.1 there exist $f_i \in p^{m_i}A^0(X)$ such that

$$\omega_i = \frac{\phi^*\omega_{i+1}}{p} + df_i.$$

Let $0 < n_0 < n_1 < \cdots < n_i < \cdots$ be the sequence of positive integers such that $\tilde{\omega}_i \in \tilde{\Omega}$. If $[\nu]$ is not in the unit root subspace, (e.g., if $\nu \in \Omega$), this sequence is infinite by Lemma 6. Then by Lemma 1, $m_k = k - i$ for $n_i \leq k < n_{i+1}$. Let

$$\nu_i = \frac{\omega_{n_i}}{p^{n_i - i}}, \qquad a_i = n_{i+1} - n_i.$$

It follows that

$$v_i = \frac{\left((\phi^{a_i})^* v_{i+1}^{\sigma^{a_i}}\right)}{p^{a_i}} + dg_i$$

for some $g_i \in A^0(Y)$. Let $h_i = g_i - g_i(Q) \in A^0(Y \hat{X} R^u)$. Let $h_i(T)$ denote the expansion of h_i at Q in a series in T so that $h_i(T) \in R^u[[T]]$. It follows that if $L(T) \in K^u[[T]]$ such that $dL(T) = \omega$ on U then

$$(3) \qquad L(T) = c + \sum_{i=0}^{\infty} \frac{h_i^{\sigma^{n_i}}(T^{p^{n_i}})}{p^i}$$

for some $c \in K^u$.

We now come to the main result of this section. Let $k_i = \operatorname{ord}_{\tilde{U}} \tilde{v}_i + 1$ and $Q_i = (k_i p^{n_i}, -i) \in \mathbb{R}^2$, $i \geq 0$.

THEOREM 10. *Suppose $\tilde{v} \in \tilde{\Omega}$, $\tilde{v} \neq 0$. Suppose $k_i \leq p - 1$ for all i. Let $M = \max\{0, 1 - v(c)\}$. Then*

(i) The vertices of $N(L)$ with x-coordinate at least $k_M p^{n_M}$ lie among $\{Q_i : i \geq M\}$

(ii) If $M > 0$ then the vertices of $N(L)$ lie among the points $\{(0, 1 - M)\} \cup \{Q_i : i \geq M\}$.

(iii) If $i \geq M$ and Q_i is not a vertex of $N(L)$, then Q_{i+1} is a vertex of $N(L)$, $n_{i+1} = n_i + 1$, $k_{i+1} = 1$, and $k_i \geq p/2$.

Proof. Let $L(T) = \sum_{i=0}^{\infty} c(i) T^i$. It follows from our hypotheses and (3) that $v(c(j)) > v(c(k_i p^{n_i})) = -i$ for $1 \leq j < k_i p^{n_i}$. From this, (i) and (ii) follow immediately.

Now suppose $i \geq M$ and Q_i does not lie on $N(L)$. Let P and Q_j, $j > i$, be the vertices of $N(L)$ to the immediate left and right of Q_i. Then we must have that

$$(4) \qquad \operatorname{slope}\left(\overrightarrow{PQ_i}\right) \geq \operatorname{slope}\left(\overrightarrow{PQ_j}\right).$$

By (i), either $x(P) < k_M p^{n_M}$ or $P = Q_h$ for some $0 \leq h < i$.

Case (i). $x(P) < k_M p^{n_M}$. First suppose $M = 0$. Then $n_M = 0$. Hence $P = (h, s)$ where $0 \leq h < k_0$ and $s > 0$. Then (4) becomes

$$\frac{-i - s}{k_0 - h} \geq \frac{-j - s}{k_j p^{n_j} - h}$$

or

$$(5) \qquad (i + s)\left(k_j p^{n_j} - h\right) \leq (j + s)(k_0 - h).$$

As $k_0 - h \leqslant p - 1$, $h \leqslant p - 2$ and $k_j \geqslant 1$ this implies

$$\left(p^{t+r} - (p-2) \right) \leqslant \left(\frac{t}{l} + 1 \right)(p-1)$$

where $l = i + s$, $t = j - i$ and $r = n_j - t$. From this and the facts that $l \geqslant 1$, $t > 0$, and $r \geqslant 0$, we see that $r = 0$, $t = 1$. Hence $n_j = 1$ which implies $j = 1$, $i = 0$, and Q_{i+1} is a vertex. Plugging this back into (5) we get

$$s(k_1 p - h) \geqslant (1 + s)(k_0 - h)$$

or

$$0 \leqslant h \leqslant (1 + s)k_0 - spk_1$$

so as $k_0 \leqslant p - 1$, $s > 0$ and $k_1 \geqslant 1$, this implies

$$k_1 = 1 \quad \text{and} \quad k_0 \geqslant \frac{sp}{1 + s} \geqslant \frac{p}{2}.$$

This completes the proof when $M = 0$.

Suppose $M > 0$, then $P = (0, 1 - M)$ by (ii) and a similar argument to the above completes the proof in this case.

Case (ii). $P = Q_h$. Then we have

$$\frac{-i + h}{k_i p^{n_i} - k_h p^{n_h}} \geqslant \frac{-j + h}{k_j p^{n_j} - k_h p^{n_h}}$$

or equivalently,

$$(6) \qquad (i - h)\left(k_j p^{n+r} - k_h \right) \leqslant (j - h)\left(k_i p^n - k_h \right)$$

where $n = n_i - n_h$ and $r = n_j - n_i$. Set $i - h = l > 0$. Since $n \geqslant i - h$, $r \geqslant j - i$ and $1 \leqslant k_s \leqslant p - 1$ for all s, (6) implies

$$\left(p^{n+r} - (p-1) \right) \leqslant (r/l + 1)((p-1)p^n - 1).$$

From this we deduce that $r = 1$. Hence $n_j = n_i + 1$ so $j = i + 1$. Moreover, (6) now becomes

$$l\left(k_{i+1} p^{n+1} - k_h \right) \leqslant (l + 1)\left(k_i p^n - k_h \right)$$

so

$$p^h\left((l + 1)k_i - lpk_{i+1} \right) \geqslant k_h > 0.$$

This occurs only if $k_{i+1} = 1$ and $k_i > pl/(l + 1) \geqslant p/2$, since $k_i \leqslant p - 1$ and $k_{i+1} \geqslant 1$. This completes the proof. $\qquad\square$

LEMMA 11. *Suppose $p|k_i$. Then $n_{i+1} = n_i + 1$ and $pk_{i+1} = k_i$.*

Proof. This follows immediately from the fact that $V \bmod p$ is the Cartier operator. □

PROPOSITION 12. *Suppose $\tilde{\nu} \in \tilde{\Omega}$, $\tilde{\nu} \neq 0$, $\mathrm{ord}_Q \nu_i = \mathrm{ord}_Q \tilde{\nu}_i$ for all $i \geq 0$ and $c = 0$. Then the vertices of $N(L)$ lie in the set $\{Q_i: (k_i, p) = 1\}$.*

Proof. This follows easily from (3) and Lemma 11. □

§3. On the zeros of a de Rham F-crystal. We say a pair (H, W) of submodules of $H^1_{dR}(C)$ is a de Rham F-crystal on C if H is stable under F, W is a direct summand of $\Omega(C)$ and the quadruple $H = (H, W, F, V)$ is an F-crystal. (Note: H need not be a direct summand of $H^1_{dR}(C)$.) We fix such an H throughout this section. We fix a residue class U on C. As in §2 we choose a point $Q \in U(R^u)$ and a uniformizing parameter T on U at Q. We also suppose we have an R-linear homomorphism

$$W \rightarrow K^u[[T]]$$

$$\omega \mapsto L_\omega(T)$$

such that

$$dL_\omega(T) = \omega$$

on U. (Note: If $f \in K[[T]]$ is a series convergent on the open unit ball then we regard $f(T)$ as a rigid analytic function on U.)

For $\omega \in W$ let $0 = n_0(\omega) < n_1(\omega) < \cdots < n_i(\omega) < \cdots$ be the sequence of integers such that

$$V^{n_i(\omega)}[\omega]^* \in \tilde{\Omega}.$$

Let $\omega_i \in W + p\mathscr{D}^{sk}(X)$ such that

$$p^{n_i(\omega)-i}[\omega_i] = V^{n_i(\omega)}[\omega].$$

Then $\tilde{\omega}_i \in \tilde{\Omega}$, $\tilde{\omega}_i \neq 0$. Let $k_i(\omega) = \mathrm{ord}_U \omega_i + 1$ and $Q_i(\omega) = (k_i(\omega)p^{n_i(\omega)}, -i) \in \mathbb{R}^2$. Let $W^* = W - pW$.

PROPOSITION 13. *Suppose U is not a base point for $\tilde{W}$. Suppose $\mathrm{ord}_U \omega \leq p - 2$ for all $\omega \in \tilde{W}^*$ and there exists a $\nu \in \tilde{W}^*$ and $0 < \mathrm{ord}_U \nu < p - 2$. Suppose b is a common zero of $\{L_\omega(T): \omega \in W\}$. Then $b \in U(K^u)$.*

Proof. First note that the hypotheses imply $p > 2$. Suppose $\omega \in W^*$. Set $k_i = k_i(\omega)$, $n_i = n_i(\omega)$, $Q_i = Q_i(\omega)$, and $M = M(\omega)$. Let i_0 be the minimal integer such that Q_{i_0} is a vertex of $N(L_\omega)$. It follows from Theorem 10 that the minus slopes of $N(L_\omega)$ are among the following types:

Type (*a*). $\dfrac{1}{k_{i+1}p^{n_{i+1}} - k_i p^{n_i}}$ if $i \geqslant i_0$ and both Q_i and Q_{i+1} are vertices of $N(L_\omega)$.

Type (*b*). $\dfrac{2}{p^{n_{i+2}} - k_i p^{n_i}}$ if $i \geqslant i_0$, Q_i is a vertex, Q_{i+1} is not a vertex of $N(L_\omega)$ and $k_{i+2} = 1$.

Type (*c*). $\dfrac{r}{t}$, where $r \geqslant 1$, $1 \leqslant t \leqslant p$ lie in $\mathbb{Z}$ and if $t > k_0$ then $r \geqslant 2$ and $k_0 = 1$. These slopes occur only if $M = 0$, in which case they are slopes of sides of $N(L_\omega)$ to the left of Q_{i_0}.

Type (*d*). $\dfrac{1}{k_M p^{n_M}}$ if $M > 0$ and $i_0 = M$.

Type (*e*). $\dfrac{2}{p^{n_M+1}}$ if $M > 0$ and $i_0 = M + 1$.

Now let $s = v(T(b))$. By the theory of the Newton polygon, s is a minus slope of $N(L_\omega)$ for all $v \in W$, $v \neq 0$. Let D denote the subset of $\{a, b, c, d, e\}$ consisting of elements x for which there exists a $v \in W^*$ such that s is of type (x) for v.

LEMMA 13.1.   *D is contained in one of the following four sets*:

$$\{a, d\}, \{a, c\}, \{b\}, \{e\}.$$

Proof.   Case (a): $a \in D$. Then since $p > 2$, $b, e \notin D$.

Case (b): $b \in D$. Then, by case (a), we already know $a \notin D$. We see that the denominator of s in lowest terms, since $n_{i+2}(v) - n_i(v) > 2$ and $0 < k_i(v) < p$ for all $v \in W^*$, is a power of p times some integer prime to p and greater than p. This implies c, d and e are not in D.

Case (c): $c \in D$. By case (b), we already know $b \notin D$. Let $s = r/t$ in lowest terms. If d lies in D then t must be divisible by p. But then r must be greater than one by the conditions of type (c); but this contradicts the conditions of type (d). Finally, e is not in D since the denominator of a slope of type (e) is at least p^2.

These three arguments only allow D to be contained in one of the four sets listed in the statement of the lemma. $\square$

To finish the proof of the proposition, we must eliminate each of the above possibilities for D. Let $\omega, v \in W^*$. We set $n_i = n_i(\omega)$ and $n'_i = n_i(v)$, etc.

Case (i): $D \subseteq \{a, d\}$. Suppose first that s is of type (a) for ω, with respect to i as above. If s is of type (a) for v then we see easily that

(1) $$n_i = n'_j$$

and

$$(2) \qquad\qquad k_i = k'_j$$

for some $j \geq 0$. If s is of type (d) for ν, then

$$(3) \qquad\qquad n_i = n'_{M'}$$

and

$$k_{i+1} p^{n_{i+1} - n_i} - k_i = k'_{M'}$$

but since both k_i and $k'_{M'}$ are less than p, $n_{i+1} - n_i > 1$ and $k_{i+1} \geq 1$, we deduce that

$$(4) \qquad\qquad p - k_i = k'_{M'}.$$

It follows from (1) and (3) that

$$v^{n_i}[\eta]^* \in \tilde{W}$$

for all $\eta \in \tilde{W}^*$. Hence, by Lemma 3 combined with (2) and (4), this implies

$$1 + \text{ord}_U \eta = k_i \quad \text{or} \quad p - k_i$$

for all $\eta \in \tilde{W}^*$. Since U is not a base point for $\tilde{W}$, this implies $\text{ord}_U \eta$ equals 0 or $p - 2$ for all such η. But this contradicts our hypotheses.

Next suppose $D = \{d\}$. Then we see that

$$n_M = n'_{M'} \quad \text{and} \quad k_M = k'_{M'}.$$

Using Lemma 3 again, we see that this contradicts our hypotheses as before.

Case (ii): $D \subset \{a, c\}$. Since U is not a base point for $\tilde{W}$, we may suppose $k_0 = 1$. First suppose s is of type (a) for ω. If s is of type (c) for ν then $n_i = 0$, $n_{i+1} = 1$ and $k_{i+1} = 1$, so that $i = 0$. As $k_0 = 1$ we must have $s = 1/(p - 1)$ and $k'_0 = p - 1$. By case (i), we know $c \in D$. Hence, from (2) and the above we see that $k_0(\eta)$ equals 1 or $p - 1$ for all $\eta \in \tilde{W}^*$, a contradiction.

Now suppose s is of type (c) for ω. Then $k_0 = 1$ and as $p > 2$, we conclude from Theorem 10 that $i_0 = 0$. Hence s is a positive integer and the run of the side of $N(L_\nu)$ with slope $-s$ is 1. It follows that L_ν has only one zero in $U(\mathbb{C}_p)$ with valuation s and so b is fixed by $\text{Gal}(\mathbb{C}_p/K^u)$ and so lies in $U(R^u)$.

Case (iii): $D = \{b\}$. Suppose s is of type (b) for ω with respect to i. We see easily that $n_i = n'_j$ and $k_i = k'_j$ for some j. It follows from Lemma 3 again that $\text{ord}_U \eta = k_i - 1$ for all $\eta \in \tilde{W}^*$, in contradiction to our hypotheses.

Case (iv): $D = \{e\}$. By our usual arguments, we see that $\text{ord}_U \eta = 0$ for all $\eta \in \tilde{W}^*$, which is a contradiction. $\qquad\square$

The proof of Proposition 13 also yields information when U is a base point. In fact, an examination of the proof yields

COROLLARY 13.2. *Suppose $k_0(\omega) \leqslant p - 1$ for all $\omega \in W^*$. Let $z = \mathrm{Min}\{k_0\omega\}: \omega \in W^*\}$ and suppose $z \leqslant (p - 1)/2$. Finally suppose that there exists a $v \in W^*$ such that $k_0(v) \neq z$ or $p - z$. Then if b is a common zero of $\{L_\omega(T): \omega \in W\}$, b has degree at most z over K^u and in particular is at worst tamely ramified over K^u.*

Suppose for the rest of this section that U is not a base point for $\tilde{W}$.

COROLLARY 13.3. *Suppose instead of the hypotheses of the proposition, that H is extraordinary, rk $H = 2$ rk W, $\mathrm{ord}_U v < p/2 - 1$ for all $v \in \tilde{W}^*$ and there exists a $\omega \in W^*$ such that $M(\omega) = 0$. Then if $b \in U(\mathbb{C}_p)$ is a common zero of $\{L_\omega(T): \omega \in W\}$, $b \in U(K^u)$.*

Proof. Let $s = v(T(b))$. By Theorem 10, under the hypotheses of the corollary, $\{Q_i(\omega): i \geqslant M\}$ are all vertices of $N(L_\omega)$ for $\omega \in W^*$. Hence s cannot be of type (b) or (e) for any $\omega \in W$. Furthermore, if D is as in the proof of the proposition, the hypothesis, $\mathrm{ord}_U v < p/2 - 1$, combined with the arguments in the proof of the proposition, shows that D must equal $\{a\}$, $\{c\}$ or $\{d\}$. Finally, the hypothesis that there exists an ω such that $M(\omega) = 0$ implies that D cannot equal $\{d\}$.

If $D = \{c\}$, then an argument as in the proof of the proposition, using the hypothesis that U is not a base point, shows that $b \in U(K^u)$.

Now suppose $D = \{a\}$. Let ω, v, n_i, n_i', etc., be as in the proof of the proposition. If

$$s = \frac{1}{k_{i+1}p^{n_{i+1}} - k_i p^{n_i}}$$

there is a j so that

$$s = \frac{1}{k_{j+1}' p^{n_{j+1}'} - k_j' p^{n_j'}}.$$

It follows that

(5) $$n_i = n_j', \qquad n_{i+1} = n_{j+1}'$$

so

$$V^{n_i}[v]^* \in \tilde{W}$$

for all $v \in W$, $v \neq 0$. So using Lemma 3 again we conclude

$$VW \equiv W \bmod pH$$

if $n_{i+1} - n_i = 1$ and $VW \equiv 0 \bmod pH$ if $n_{i+1} - n_i > 1$. But this contradicts the hypotheses that H is extraordinary and rk $H = 2$ rk W, using Corollary 2.1. $\qquad \square$

COROLLARY 13.4. *Under the hypotheses of either Proposition 13 or Corollary 13.1 if $M(\omega) > 0$ for some $\omega \in W$ then the elements of $\{L_\omega\colon \omega \in W\}$ have no common zeros.*

Proof. If $M(\omega) > 0$ then no slopes of $N(L_\omega)$ are natural numbers. Hence L_ω has no zeros in $U(K^u)$. $\qquad\square$

We need to make some more observations about Newton polygons. If S is a subset of $\mathbb{C}_p[[T]]$ then we define the Newton polygon of S, $N(S)$, to be the lower convex hull of $\{N(s)\colon s \in S\}$. We define a slope of $N(S)$ and the run of a slope just as we did in Section 2 for the Newton polygon of a single power series.

LEMMA 14. *Suppose a is a common zero of S. Then $-v(a)$ is a slope of $N(S)$ and S has at most the run of $-v(a)$ common zeros of valuation equal to $v(a)$.*

Proof. This follows easily from the standard facts about Newton polygons of single power series. $\qquad\square$

Now suppose H is not extraordinary. We set $t = 1$, $q = p$, $\phi = \sigma$ if H is ordinary and $t = 2$, $q = p^2$ and $\phi = \sigma^2$ if H is superspecial. We also define a ϕ^{-1}-linear map

$$\mathscr{C}\colon \tilde{W} \to \tilde{W}$$

as follows:

$$\mathscr{C} = \tilde{V}|_{\tilde{W}}$$

if H is ordinary and

$$\mathscr{C} = \left(\frac{V^2}{p}\right)^{\tilde{}}\Big|_{\tilde{W}}$$

if H is superspecial. It follows from the definition of ordinary and Lemma 2 that $\mathscr{C}$ is bijective. If H is ordinary, $\mathscr{C}$ is just the Cartier operator.

We now use the notation of [C-1], §IV. Let $j = j(U)$ denote the element of $\mathbb{P}(\tilde{W})$ represented by the linear map

$$w \in \tilde{W} \mapsto \left(\frac{\omega}{d\tilde{T}}\right)(\tilde{Q}).$$

This is just the image of U under the canonical map $\tilde{C} \to \mathbb{P}(\tilde{W})$ given by the linear system $\tilde{W}$ (see [C-1], §4).

Let $A = \min_{\omega \in W} v(L_\omega(0))$. If $A \neq \infty$, let $e = e(a)$ denote the point

$$\omega \in \tilde{W} \mapsto \left(p^{-A}L(0)\right)^{\tilde{}}$$

of $\mathbb{P}(\tilde{W})$. As in [C-1], $\mathscr{C}$ induces an endomorphism $\mathbb{P}(\mathscr{C})\colon \mathbb{P}(\tilde{W}) \to \mathbb{P}(\tilde{W})$. Let $M = \max_{\omega \in W} M(\omega) = \max\{0, 1 - A\}$.

PROPOSITION 15. *Suppose as above that H is not extraordinary. Suppose $b \in U(\mathscr{C}_p)$ is a common zero of $S = \{ L_\omega(T): \omega \in W \}$. Then one of the following occurs: Let $s = v(T(b))$.*

(i) $b \in U(K^u)$, $s = A > 0$.

(ii) $s = q^{-M}$, $M > 0$. If $q \geqslant 3$ then $\mathbb{P}(\mathscr{C})^M(j) = e$ and every element of $\tilde{W}$ which vanishes at $\tilde{Q}$ vanishes $q - 1$ times. Moreover, S has at most q^M common zeros with valuation s.

(iii) $s = 1/q^n(q - 1)$ with $n \geqslant M$. Then $\mathbb{P}(\mathscr{C})(j) = j$ and every differential at $\tilde{Q}$ vanishes $q - 2$ times if $n = M$ and $q - 1$ times if $n > M$. Moreover S has at most $q^n(q - 1)$ common zeros with valuation s.

Proof. For all $\omega \in W$, $\omega \neq 0$, $n_i(\omega) = it$ and if $\tilde{\omega} \neq 0$, $\tilde{\omega}_i = \mathscr{C}^i\tilde{\omega}$, since $\mathscr{C}$ is bijective, $\tilde{W} = \{ \tilde{\omega}_i: \omega \in W \}$. Since U is not a base point of $\tilde{W}$ it follows from Proposition 10 that $N(S)$ is the lower convex hull of $\{(0, A)\} \cup \{(q^i, -i):$ $i \geqslant M \}$. An easy estimate shows that all these points are vertices of $N(S)$ except $(0, A)$ when $A = \infty$ and $(1, 0)$ when $p = 2$, $A = 1$.

It follows from Lemma 14 that s is of one of the following types:

Type (a): $s = A > 0$, which can occur only if $M = 0$.

Type (b): $s = q^{-M}$ which can occur only if $M > 0$.

Type (c): $s = 1/q^n(q - 1)$, $n \geqslant M$.

Suppose s is of type (a). Then the run of the slope $-s$ is 1 when $q > 2$ or $s > 1$, and the run is 2 when $q = 2$ and $s = 1$. From this it follows immediately in the first case and by the arguments in the proof of Theorem 4.1 of [C-1] in the second that $b \in U(R^u)$.

Now as in §2, for $\omega \in W$, we may write

$$L_\omega(T) = L_\omega(0) + \sum_{i=0}^{\infty} \frac{g_{\omega,i}^{\phi^i}(T^{q^i})}{p^i}$$

for some $g_{\omega,i}(T) \in R^u[[T]]$ such that

$$dg_{\omega,i} \equiv \omega_i \bmod T^{q-1} dT$$

$$\equiv \mathscr{C}^i\omega \bmod(p, T^{q-1} dT).$$

Let

$$C_i(\omega) = (g_{\omega,i}(0))^{\phi^i}$$

which is congruent to

$$((\mathscr{C}^i(\omega)/dT)|_U)^{\phi^i}$$

modulo p. Then setting $a = T(b)$, we have

$$L_\omega(T)(b) = L_\omega(a) \equiv L(0) + \frac{C_M(\omega) a^{q^M}}{p^M} \bmod p^A a$$

if s is of type (b) and $q \geqslant 3$. Also,

$$L_\omega(T)(b) = L_\omega(a) \equiv \frac{g_{\omega,n}^{\phi^n}(a^{q^n})}{p^n} + \frac{C_{n+1}(\omega)a^{q^{n+1}}}{p^{n+1}} \bmod p^{1-n}a_n$$

if $s = 1/(q^n(q-1))$ is of type (c), where $a_n = a^{q^n}$ if $n > M$ and $a_M = 1$. Now the rest of the proof follows the same lines as that of Theorem 4.2.1 of [C-1].

Remark. One can also adapt this argument to the case when U is a base point.

Now suppose H is PCM. Let $W = \bigoplus_{i \in I} B_i$ be the corresponding direct sum decomposition. Let $B = \bigcup B_i$.

PROPOSITION 16. *Suppose H is extraordinary and no nonzero element of $\tilde{B}$ vanishes at U. Suppose also that $M(\omega) = 0$ for all $\omega \in B$. If b is a common zero of $\{L_\omega(T): \omega \in W\}$, then $b \in U(R^u)$.*

Proof. If $\omega \in B$, $\tilde{\omega} \neq 0$, then we may choose K, $\omega_i \in B$. If $A = v(L_\omega(0))$, then the Newton polygon of L is the lower convex hull of $\{(0, A)\} \cup \{(p^{n_i}, -i): i \geqslant 0\}$ where $n_i = n_i(\omega)$, etc. Let $s = V(T(b))$. As usual s is of one of the following types:

Type (a): $s = A > 0$.

Type (b): $s = \dfrac{1}{p^{n_{i+1}} - p^{n_i}}$, $i \geqslant 0$.

As usual, if s is of type (a) then $b \in U(K^u)$. Otherwise if s is of type (b) and if $\omega' \in B$, $\tilde{\omega}' \neq 0$, then

$$s = \frac{1}{p^{n'_{i+1}} - p^{n'_j}} \quad \text{for some } j.$$

It follows that

$$(6) \qquad\qquad n_i = n'_j \quad \text{and} \quad n_{i+1} = n'_{j+1}.$$

Hence $V^{n_i}[v]^* \in B$ for all $v \in B$. From this, Lemma 4, and (6) it follows that $VB \cap W = (0)$ if $n_{i+1} - n_i > 1$ and $VB \subseteq W$ if $n_{i+1} - n_i = 1$. This contradicts Lemma 5.

§4. General conclusions. In this section we fix a de Rham F-crystal $H = (H, W)$ on C such that $\check{W}$ has no base points. We also fix an R-linear homomorphism

$$\omega \mapsto \lambda_\omega$$

of W into the locally analytic functions on $C(\mathbb{C}_p)$ such that λ_ω is a primitive of in the sense of [C-1], §2, and such that

1)
$$\lambda_\omega(Q^\tau) = \lambda_\omega(Q)^\tau$$

for all $\tau \in I \overset{\mathrm{def}}{=} \mathrm{Aut}_{\mathrm{cont}}^n(\mathbb{C}_p/K^u)$, $Q \in C(\mathbb{C}_p)$. For example, if $P \in C(K^u)$ then we may set

$$\lambda_\omega(Q) = \int_P^Q \omega$$

in the sense of [C-1], §2.

LEMMA 17. *The restriction of λ_ω to each residue class is analytic over K^u.*

Proof. This follows immediately from Corollary 2.1a of [C-1] and (1). $\qquad\square$

Let $T = \{Q \in C(\mathbb{C}): \lambda_\omega(Q) = 0 \text{ for all } \omega \in W\}$.

Let U be a residue class of C and let $\omega \in M$. Let

$$M_U(\omega) = \max\{0, 1 - \nu(\lambda_\omega(P)): P \in U(K^u)\}.$$

THEOREM 18. *Suppose H is extraordinary. Suppose $Q \in T - C(K^u)$. Then*
(i) *$M_{\tilde{Q}}(\omega) > 0$ for all $\omega \in W^*$ or*
(ii) *There exists a $\nu \in W^*$ such that $\mathrm{ord}_{\tilde{Q}}\nu \geqslant p/2 - 1$.*

Proof. This follows from Lemma 17 and Corollary 13.1. $\qquad\square$

Remark. When $[K: \mathbf{Q}_p] < \infty$, $\tilde{W}$ is finite and so, when $p > 2$, condition (ii) allows only finitely many possibilities for $\tilde{Q}$. Moreover, although we do not give the details here, condition (i) can often be eliminated.

We suppose $K = K^u$ for the rest of this section.

THEOREM 19. *Suppose $\mathrm{rk}_R W \geqslant 2$ and $g < p/2$. Then $T \subseteq C(K)$.*

Proof. Let $U \in C(\mathbb{F})$. Then $\mathrm{ord}_U \omega \leqslant 2g - 2 < p - 2$ for all $\omega \in \tilde{W}$, $\omega \neq 0$. Since $\dim_\mathbb{F} \tilde{W} = \mathrm{rk}_R W \geqslant 2$ and $\mathbb{F} = \mathbb{F}^u$ is algebraically closed, there exists a $\nu \in \tilde{W}$ such that $\mathrm{ord}_U \nu > 0$. Hence H satisfies all the hypotheses of Proposition 10 for each residue class of C and so the theorem follows from this and Lemma 17. $\qquad\square$

Definition. If A is an Abelian scheme over R and $X \subseteq A$ is an irreducible curve over R then we say X generates A if A_K is the smallest Zariski-closed subgroup scheme of A_K containing X_K. I.e., X is not contained in the translate of a proper Abelian subscheme by a point of finite order.

COROLLARY 19.1. *Suppose C is as in the theorem and $f: C \to A$ is a smooth morphism over R of C into an Abelian scheme over R of dimension at least two whose image generates A. Then $f^{-1}(f(C) \cap A_{Tor})(\mathbb{C}_p) \subseteq C(K)$.*

Proof.   Set

$$H = f^* H^1_{dR}(A/R)$$

$$W = f^* H^0\left(A, \Omega^1_{A/R}\right)$$

Then $H = (H, W)$ is a de Rham F-crystal on C by the functoriality of Frobenius together with the smoothness of f. Smoothness also implies that W has no base points. If $\mathrm{rk}\, W < 2$ then $f(C)$ is contained in the translate of a proper Abelian subscheme B of A. Since $f(C)$ generates A, $C \subseteq B + t$ for some point t of infinite order. It follows that $(C \cap A_{\mathrm{Tor}})(\mathbb{C}_p) = \varnothing$. Hence we may suppose that $\mathrm{rk}\, W \geqslant 2$.

Now let D be a divisor on $C(K)$ such that

$$f(D) = 0$$

(Note: $f(D)$ denotes the sum on A of the image of D under f.) Set

$$\lambda_\omega(Q) = \int_{Q-D} \omega$$

in the sense of [C-1], §2. Then for $\tau \in I$, $Q \in C(\mathbb{C}_p)$,

$$\lambda(Q^\tau) = \int_{Q^\tau - D} \omega$$

$$= \int_{Q^\tau - D^\tau} \omega^\tau = \lambda_\omega(Q)^\tau$$

by Corollary 2.1e of [C-1]. As

$$\{Q \colon \lambda_\omega(Q) = 0,\, \omega \in W\} = f^{-1}(f(C) \cap A_{\mathrm{Tor}})(\mathbb{C}_p)$$

by Proposition 3.1 of [C-1], the corollary is now an immediate consequence of the theorem.   □

Remark.   If we had systematically dealt with base points in §3 we could obtain a slightly weaker result than this when $\tilde{f}$ is only supposed to be separable.

THEOREM 20.   *Suppose $(H, W) = (H^1_{dR}(C), \Omega(C))$ and H is not extraordinary. Let $q = p$ if H is ordinary and $q = p^2$ if H is superspecial. Suppose $g \geqslant 2$. If either*
   (i) $q \geqslant 5$ or
   (ii) $q = 2$ or 4 and $\tilde{C}$ is not hyperelliptic, or
   (iii) $q = 3$, $g \geqslant 3$ and $\tilde{C}$ is hyperelliptic,
then $T \subseteq C(K)$.

Proof. We may apply the conclusions of Proposition 19. Suppose that $Q \in T - C(K)$. Then it follows from Proposition 15 that every differential in $\tilde{W}$ which vanishes at $\tilde{Q}$ vanishes at least $q - 2$ times. This and Riemann-Roch imply $g \leqslant 1$ if $q \geqslant 5$ and $\tilde{C}$ is hyperelliptic if $q = 4$.

If $q = 2$ or 3 then

$$\mathbb{P}(\mathscr{C})\big(j(\tilde{Q})\big) = j(\tilde{Q})$$

by Proposition 15, using the notation of [C-1], §4, which is impossible under hypotheses (ii) or (iii) by the part of the proof of Theorem 5.5(ii) on pages 146–148. $\square$

COROLLARY 20.1. *Assume the initial hypotheses of the theorem. Suppose $q = 3$, then $T - C(K)$ contains at most two points in each residue class U of C, such that $v(\lambda_\omega(Q)) > 0$ for all $\omega \in W$, $Q \in U(K)$.*

Proof. This follows from Proposition 15 together with the proof of Theorem 5.5(iii) of [C-1]. $\square$

COROLLARY 20.2. *Suppose $p/2 > g$, or C satisfies the hypotheses of Theorem 20. Let T be a torsion packet in $C(\mathbb{C}_p)$ rational over K. Then $T \subseteq C(K)$.*

Proof. We may suppose $T \neq 0$. Let $P \in T$. For $\omega \in \Omega$ set $Q \in C(\mathbb{C}_p)$, $\lambda_\omega(Q) = \int_P^Q \omega$. Then for $\tau \in I$, $\lambda_\omega(Q^\tau) = \int_{P^\tau}^{Q^\tau} \omega + \int_P^{P^\tau} \omega = \lambda_\omega(Q)^\tau$. by Proposition 2.4 and Theorem 2.9 of [C-1], since $P^\tau \in T$ and so some multiple of $P^\tau - P$ is principal.

By Proposition 3.1 of [C-1] again, $T = \{Q \in C(\mathbb{C}_p): \lambda_\omega(Q) = 0, \omega \in \Omega\}$. Hence the corollary follows from Theorems 19 and 20. This implies Conjecture B of the Introduction under any of the hypotheses (a)–(c). $\square$

Now suppose H is PCM and $W = \bigoplus_{i \in I} B_i$ is the corresponding direct sum decomposition. Let $B = \bigcup_{i \in I} B_i$. Let

$$S = \big\{ U \in C(\mathbb{F}): \tilde{\omega}(U) = 0 \quad \text{for some } \omega \in B, \tilde{\omega} \neq 0 \big\}.$$

THEOREM 21. *Suppose $H \neq 0$ is extraordinary and $v(\lambda_\omega(P)) > 0$ for all $P \in C(R)$. Then if $Q \in T - C(R)$, $\tilde{Q} \in S$.*

Proof. This follows immediately from Proposition 16. $\square$

COROLLARY 21.1. *Suppose $f: C \to A$ is a smooth morphism over R into an Abelian scheme with complex multiplication by a ring $\mathcal{O}$ such that the image of C generates A. Suppose A has extraordinary reduction. Then if Q belongs to $f^{-1}((f(C) \cap A_{Tor})(\mathbb{C}_p))$, either $Q \in C(K)$ or $(f^*\omega)\tilde{\,}(\tilde{Q}) = 0$ for some $\omega \in H^0(A, \Omega^1_{A/R})$, $\tilde{\omega} \neq 0$, which is an eigenvector for the action of $\mathcal{O}$.*

Proof. Let

$$H = f^* H^1_{dR}(C/R)$$

$$W = f^* H^0\big(A, \Omega^1_{A/R}\big).$$

As in the proof of Corollary 20.1, we may suppose $f^*\colon H^1_{dR}(A/R) \to H^1_{dR}(C/R)$ is an injection. Then $H = (H, W)$ is a de Rham F-crystal on C and $\tilde{W}$ has no base points on $\tilde{C}$ as explained in the proof of Corollary 15.1. Also, through its action on A, $\mathcal{O}$ acts on H and stabilizes W. Since A has CM by $\mathcal{O}$,

$$(3) \qquad\qquad W = \bigoplus_{i \in I} B_i$$

where the B_i are the eigenspaces for the action of $\mathcal{O}$ on W. They are all one-dimensional, again using the CM of A. By the functoriality of Frobenius, $F \circ \alpha = \alpha \circ F$ for $\alpha \in \mathcal{O}$ considered as an endomorphism of H. It follows from this that H is PCM with respect to the direct sum decomposition (3).

Now let D be a divisor of degree one on $C(R)$ such that

$$f(D) = 0$$

and set for $\omega \in W$,

$$\lambda = \int_{P-D} \omega$$

for $\omega \in W$, $P \in C(\mathbf{C}_p)$, as in the proof of Corollary 19.1. By Lemma 17, λ_ω is analytic over K on each residue class of C. As $f^{-1}(f(C) \cap A_{\mathrm{Tor}})(d_p) = \{Q\colon \lambda_\omega(Q) = 0,\ \omega \in W\}$, the corollary will follow at once from the theorem once we know $V(\lambda_\omega(P)) > 0$ for all $P \in C(R)$. But this follows from Lemma 3.2 of [C-1]. ($\Lambda(U(R)) \subseteq pR$, even when $p = 2$, if U is the kernel of reduction) and the following lemma.

LEMMA 22.   *The map $A(R)_{\mathrm{Tor}} \to A(\mathbb{F})$ is a surjection if A has CM by $\mathcal{O}$ over R.*

Proof.   It suffices to show $V_p(A(R)) \overset{\sim}{\to} V_p(A(\mathbb{F}))$ where for an Abelian group B, $V_p(B)$ is the p-Tate module of B over $\mathbf{Q}_p$. Since A has CM by $\mathcal{O}$, $\mathcal{O} \times \mathbf{Q}_p$ is a semi-simple $\mathbf{Q}_p$-algebra and $V_p = V_p(A(\mathbf{C}_p))$ is a principal $\mathcal{O} \otimes \mathbf{Q}_p$-module. Let A^0 denote the formal group of A. Then

$$V^0 = V_p\big(A^0(\mathbf{C}_p)\big)$$

is an $\mathcal{O} \otimes \mathbf{Q}_p$ submodule of V_p. Hence there exists a unique $\mathcal{O} \otimes \mathbf{Q}_p$ submodule M of V_p such that

$$(4) \qquad\qquad M \oplus V^0 = V_p.$$

The sequence

$$0 \to V^0 \to V_p \to V_p(A(\mathbb{F})) \to 0$$

is an exact sequence of I-modules. By (4) it has a canonical splitting. As $(V_p^0)^I = (0)$, $V_p(A(R)) = M \cong V_p(A/\mathbb{F})$. This completes the proof.

§5. The Manin-Mumford Conjecture. In this section we will use Corollary 19.1 and Bogomolov's theorem (Theorem C of the Introduction) to give a new proof of the Manin-Mumford conjecture. As explained in the Introduction, it suffices to prove Theorem D.

Let N and e be integers. Let $F_{N,e}$ be the maximal Galois extension of F, in a fixed algebraic closure $\overline{F}$ of F, with Galois group G such that the inertia group in G^{ab} at any prime of F outside N has exponent dividing e. The following theorem implies Theorem D and is useful in the study of arbitrary morphisms of curves to Abelian varieties (see Corollary 13.2).

THEOREM 23. *The subgroup of $A(F_{N,e})$ consisting of torion of order prime to N is finite.*

As an immediate corollary, we have

THEOREM 24 (Bogomolov, [B-2]). *Let L be a Galois extension of F whose maximal Abelian subextension is of finite degree over K. Then $A(L)_{\mathrm{Tor}}$ is finite.*

Proof of Theorem 23. After passing to a finite extension of F we may suppose A has semi-stable reduction. For each prime l of $\mathbf{Q}$ let T_l denote the l-Tate module of $A(\overline{F})$ and M_l the group of points of order l in $A(\overline{F})$. Let $G = \mathrm{Gal}(\overline{F}/F)$ and $H = \mathrm{Gal}(\overline{F}/F_{N,e})$. It suffices to prove

$$(1) \qquad\qquad T_l^H = 0 \quad \text{for all } l, (l, N) = 1$$

and

$$(2) \qquad\qquad M_l^H = 0 \quad \text{for all but finitely many } l.$$

Suppose now $(l, N) = 1$ and $\mathrm{rk}_{\mathbf{Z}_l} T_l^H = d > 0$. Let $D = \Lambda^d T_l^H$, the maximal exterior power of T_l^H. Hence D is a rank one $\mathbf{Z}_l$-module and a G-module. We get a representation $r\colon G \to \mathbf{Z}_l^*$, such that the image of the inertia group of each prime outside N has exponent dividing e. Since $(N, l) = 1$ this representation factors through a finite subgroup of $\mathbf{Z}_l^*$. But D is a submodule of $\Lambda^d T_l$ which by the Riemann hypothesis is pure of weight $d/2$. Thus we obtain a contradiction. This establishes (1). Next we need:

LEMMA 25. *Suppose D is a one-dimensional $\mathbf{F}_l$-subspace of $\Lambda^d M_l$ stable under G. Then the representation*

$$r\colon G \to \mathrm{Aut}(D) = \mathbf{F}_l^*$$

is unramified outside l.

Proof. Let $\mathfrak{p}$ be a prime of $\overline{F}$, $(\mathfrak{p}, l) = 1$. Let $I_{\mathfrak{p}}$ denote the inertia group of $\mathfrak{p}$. Since A has semistable reduction, the action of $I_{\mathfrak{p}}$ on $\Lambda^d M_l$ is unipotent. The lemma follows.

End of proof of Theorem 23.   If (2) is false then there exist a positive integer d and an infinite set S of rational primes S not dividing N such that for $l \in S$, $\Lambda^d M_l$ contains a one-dimensional representation r_l of G unramified outside N. It follows from Lemma 25 that r_l factors through the Galois group of the maximal Abelian extension of K unramified outside l and whose inertia subgroups have exponent dividing e. The exponent of this group is bounded by $m = eh$ where h is the class number of K. Let $\mathfrak{p}$ be a prime of K at which A has good reduction. Let $f(T)$ denote the characteristic polynomial of Frobenius of $\mathfrak{p}$ on $\Lambda^d T_l$, $(l, \mathfrak{p}) = 1$. Then $f(T) \in \mathbb{Z}[T]$ is independent of l. The above argument shows that it has a root congruent to an m-th root of unity for infinitely many l. This implies $f(T)$ has a root which is an m-th root of unity in contradiction to the Riemann hypothesis. This completes the proof of the theorem. $\square$

As an immediate consequence of the proof of the theorem, we have

COROLLARY 25.1.   *Let A be an Abelian variety over a number field F and e a positive integer. Suppose l is a rational prime and $P \in A(\overline{F})$ is a point of order l. Then for l sufficiently large, the extension $F(P)/F$ is ramified above l. Moreover, the exponent of the inertia group at some prime above l in the abelianization of the Galois group of the Galois hull of $F(P)/F$ is greater than e.*

Remark.   Theorem 23 is similar to the following theorem of Ribet [K-L].

THEOREM 26 (Ribet).   *Let L be the cyclotomic extension of F. Then $A(L)_{\mathrm{Tor}}$ is finite.*

Final Remarks

(1) The results of Sections 1–3 lead one to attach an infinite Newton polygon to an F-crystal (not to be confused with the finite Newton polygon of Frobenius). More precisely, let notations be as in §1. Let (H, W, F, V) be an F-crystal. We define the Newton polygon of H, $N(H)$ to be the lower convex hull of the points:

$$(p^n, i), \qquad n \geq 0$$

where n and i are such that there exists a $w \in W$, $w \neq 0$, such that the set of integers $0 < j \leq n$ such that

$$V^j(w)^* \in \tilde{W}$$

has exactly i elements in it including n. Using the results of §2 and §3 one can show:

THEOREM 27.   *Let U be a residue class of C and $P \in U(R^u)$. Let T be a parameter on U centered at P. Let $S = \{ L \in K[[T]]: dL \in \Omega|_U, \ L(0) = 0\}$. Then*

    (i) *$N(S)$ depends only on U.*
    (ii) *$N(S)$ lies above $N(H^1_{dR}(C))$.*
    (iii) *Moreover, the equality*

$$(1) \qquad\qquad N(S) = N\big(H^1_{dR}(C)\big)$$

always holds if C is ordinary or superspecial and holds in general for almost all residue classes of C.

We call the residue classes, where equality (1) fails, the rise points of C. We will show in [C-2] that the rise points of the extraordinary Fermat quotients are precisely the cuspidal residue classes. Also, note that ordinary or superspecial curves have no rise points; however, Proposition 15 shows that there are still special residue classes, namely, the solutions of

$$\mathbb{P}(\mathscr{C})(j(U)) = j(U).$$

Perhaps these should be called higher rise points.

(2) One should be able to adapt the methods in this paper to the cases of a ramified base field and of bad reduction (stable reduction would be enough). However, the most obvious generalization of Conjecture B is false. For example, let $m \in \mathbb{Z}$, such that $(m, p) = 1$ and let $\pi \in \mathbb{C}_p$ such that $\pi^{p-1} = -p$. Let Y be the smooth complete curve defined over $\mathbb{Q}_p(\pi)$ with affine equation:

$$y^p = 1 + \pi p x^m.$$

Then Y has genus at least 2 for m large and has good reduction over $\mathbb{Q}_p(\pi)$, as one can see by making the change of variables $x = x'$ and $y = \pi y' + 1$. The branch points of Y as a cover of the x-line lie in a torsion packet rational over $\mathbb{Q}_p(\pi)$, but as may be easily seen, they generate a ramified extension of $\mathbb{Q}_p(\pi)$.

(3) Suppose ω is a differential on C of the first kind over R^u. Then it follows from [C-3] that the values of λ_ω on $C(R^u)$ modulo pR^u are closely related to the Hodge-Tate periods of C.

(4) The proof of the Manin-Mumford conjecture, given in this paper, differs from Raynaud's [R-2] in that we study ramified torsion points and Raynaud studies unramified torsion points, specifically points of order prime to p. Together the two proofs should provide a method of obtaining estimates as in [C-1] for the number of torsion points on a curve with good reduction over K^u.

(5) Ekedahl has studied superspecial curves [E] and found that one must have $g \leqslant p(p-1)/2$. Moreover, the Fermat curve of degree $p+1$ is superspecial at p and $g = p(p-1)/2$.

REFERENCES

[B-0]   P. BERTHELOT AND A. OGUS *F-isocrystals and De Rham cohomology*, I. Invent. Math. **72** (1983), 159–199.

[B-1]   F. BOGOMOLOV, *Sur l'algébricité des représentations χ-adiques*, C.R. Acad. Sci. **290** (1980), 701–704.

[B-2]   ______, *Torsion points on an abelian variety*, (unpublished).

[C-1]   R. COLEMAN, *Torsion points on curves and p-adic Abelian integrals*, Annals of Math. **121** (1985), 111–168.

[C-2]   ______, *Torsion points on Abelian coverings of* $\mathbb{P}^1 - \{0, 1, \infty\}$, to appear.

[C-3]   ______, *Hodge-Tate periods and p-adic Abelian integrals*, Invent. Math. **78** (1984), 351–379.

[C-4]   ______, *Torsion points on curves*, to appear.

[E]   T. EKEDAHL, *On supersingular curves and abelian varieties*, Orsay preprint.

[K]   N. KATZ, *Travaux de Dwork*, sém. Bourbaki **409** (1972), 1–34.

[K-L]   N. KATZ, AND S. LANG, *Finiteness theorems in geometric class field theory*, L'Enseign. Math. **27**, fasc. 3–4 (1981), 285–319.

[R-1]   M. RAYNAUD, *Géométrie analytique rigide d'après Tate, Kiehl*,... Bull. Soc. Math. France Memoire **39–40** (1974), 319–327.

[R-2]   ———, *Courbes sur une variété abelienne et points de torsion*, Invent. Math. **71** (1983), 207–233.

DEPARTMENT OF MATHEMATICS, UNIVERSITY OF CALIFORNIA, BERKELEY, CALIFORNIA 94720

Vol. 54, No. 2   DUKE MATHEMATICAL JOURNAL ©   1987

CYCLIC HOMOLOGY OF DIFFERENTIAL OPERATORS

MARIUSZ WODZICKI

To Dear Yuri Ivanovich on His Fiftieth Birthday

1. Let $\mathscr{D}(X)$ denote the k-algebra of differential operators on a smooth manifold X in one of the following categories: algebraic, holomorphic or C^∞. In the first case X has to be an affine variety over the ground field k of characteristic zero, in the second case a Stein manifold ($k = \mathbb{C}$), assumed, for simplicity, to possess finitely many connected components, and in the last case a compact C^∞-manifold (possibly with boundary or nonorientable; $k = \mathbb{R}$ or $\mathbb{C}$). The purpose of this article is to determine Hochschild and cyclic homology of $\mathscr{D}(X)$ denoted, respectively, $H_*(\mathscr{D}(X), \mathscr{D}(X))$ and $HC_*(\mathscr{D}(X))$. In the holomorphic and C^∞ settings, $\mathscr{D}(X)$ is naturally a locally convex algebra with respect to $\hat{\otimes}_\pi$-tensor product, and the groups above mean the corresponding *topological* homology groups. For basic definitions and properties of cyclic homology see [5] and for basics on locally convex homological algebra consult [4] and [7].

2. Theorem.

$$H_q(\mathscr{D}(X), \mathscr{D}(X)) \simeq H_{\mathrm{DR}}^{2n-q}(X) \qquad (q \in \mathbb{N}; \, n = \dim X). \tag{1}$$

3. Theorem.

$$HC_q(\mathscr{D}(X)) \simeq H_{\mathrm{DR}}^{2n-q}(X) \oplus H_{\mathrm{DR}}^{2n-q+2}(X) \oplus H_{\mathrm{DR}}^{2n-q+4}(X) \oplus \cdots$$

$$(q \in \mathbb{N}). \tag{2}$$

4. Remark. In proof of the holomorphic case of Theorem 3 we shall assume, for simplicity, that $H_{\mathrm{DR}}^*(X)$ is finite-dimensional; the similar condition automatically holds in the two remaining cases.

The isomorphisms in (1) are canonical and functorial with respect to embeddings of codimension zero. The proof of Theorem 3 which is presented below will provide similarly functorial isomorphisms in (2), for $q \geqslant 2n - 1$. The existence of *canonical* isomorphisms in the "unstable" range $q < 2n - 1$ can be proved as well, at least in C^∞ case, but requires stronger means (cf. Remarks 8 and 13.1 below).

Received December 19, 1986. Supported by the National Science Foundation Grant No. DMS-8610730(1).

5. Recall that for a general algebra $\mathscr{A}$ the groups $H_*(\mathscr{A}, \mathscr{A})$ and $HC_*(\mathscr{A})$ are related to each other by Connes long exact sequence (cf., e.g., [5, Thm. 1.6])

$$\cdots \to H_q(\mathscr{A}, \mathscr{A}) \xrightarrow{I} HC_q(\mathscr{A}) \xrightarrow{S} HC_{q-2}(\mathscr{A}) \xrightarrow{B} H_{q-1}(\mathscr{A}, \mathscr{A}) \to \cdots .$$

$$(3)$$

Assume that $\dim_k H_*(\mathscr{A}, \mathscr{A}) < \infty$. Then the following simple lemma holds.

6. LEMMA. *One has*

$$\dim HC_q(\mathscr{A}) \leqslant \sum_{i=0}^{[q/2]} \dim H_{q-2i}(\mathscr{A}, \mathscr{A}) \qquad (4)$$

for every q. If there is equality in (4) *for $q \gg 0$ (suffices for just one q) sequence* (3) *splits into the short exact sequences*

$$0 \to H_k(\mathscr{A}, \mathscr{A}) \xrightarrow{I} HC_k(\mathscr{A}) \xrightarrow{S} HC_{k-2}(\mathscr{A}) \to 0 \qquad (k \in \mathbb{N}),$$

and equality holds in (4) *for all q.*

7. Recall that $HC_*(\mathscr{A})$ can be obtained as homology of $\operatorname{Tot} \mathscr{B}_{**}(\mathscr{A})$ where $(\mathscr{B}_{**}(\mathscr{A}), b, B)$ is Connes double complex ([5, (1.8)]). One has $\mathscr{B}_{kl}(\mathscr{A}) = \mathscr{C}_{l-k}(\mathscr{A}, \mathscr{A})$ $(k, \ l \geqslant 0)$ where $\mathscr{C}_j(\mathscr{A}, \mathscr{A}) = \mathscr{A} \hat{\otimes} \cdots \hat{\otimes} \mathscr{A}$ $(j+1$ times). Here $\hat{\otimes}$ denotes $\hat{\otimes}_\pi$ if $\mathscr{A}$ is a locally convex $\hat{\otimes}_\pi$-algebra, and the ordinary (inductive) tensor product if $\mathscr{A}$ has no topology.

Assume that $\mathscr{A}$ is filtered: $\{0\} \subset \mathscr{A}^0 \subset \cdots \subset \mathscr{A}^p \subset \cdots$ $(p \in \mathbb{N})$, and $\mathscr{A}^i \mathscr{A}^j \subset \mathscr{A}^{i+j}$. This induces the filtrations $\{0\} \subset \mathscr{C}_j^0 \subset \cdots \subset \mathscr{C}_j^p \subset \cdots$ on spaces $\mathscr{C}_j(\mathscr{A}, \mathscr{A})$ which are defined by images in $\mathscr{C}_j(\mathscr{A}, \mathscr{A})$ of the spaces $\oplus(\mathscr{A}^{i_0} \hat{\otimes} \cdots \hat{\otimes} \mathscr{A}^{i_j})$ (summation over $i_0 + \cdots + i_j = p$). It is clear from the corresponding definitions that both Hochschild and Connes boundary maps (denoted above b and B respectively) preserve these filtrations.

If $\mathscr{A}$ is topologized we shall assume in addition that the canonical maps $\varinjlim_p \mathscr{C}_j^p \to \mathscr{C}_j(\mathscr{A}, \mathscr{A})$ are isomorphisms for all j. Then the spectral sequence E_{pq}^r associated with the considered filtration on $\operatorname{Tot} \mathscr{B}_{**}(\mathscr{A})$ *converges to* $HC_{p+q}(\mathscr{A})$. Similarly, the spectral sequence $'E_{pq}^r$ associated with the filtration on $(\mathscr{C}_*(\mathscr{A}, \mathscr{A}), b)$ *converges to* $H_{p+q}(\mathscr{A}, \mathscr{A})$.

Reduction of Theorem 3 to Theorem 2. The above remarks apply to $\mathscr{A} = \mathscr{D}(X)$ filtered by order of operator in any of the three cases quoted at the beginning. The spectral sequence $E_{pq}^r \Rightarrow HC_{p+q}(\mathscr{D}(X))$ is a priori located in the region $(p \geqslant 0$ and $p + q \geqslant 0)$. In fact, we shall see that E_{pq}^r $(r \geqslant 1)$ vanish outside the

region shown below

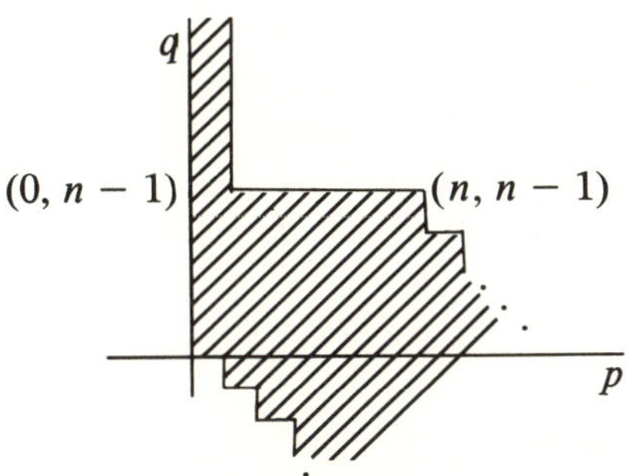

i.e., $E_{pq}^r = 0$ if either $p \geq 1$ and $q \geq n$ or $p \geq 1$ and $p + q \geq 2n$.

Indeed, $\mathrm{gr}\, \mathcal{D}(X)$ is the graded algebra $\mathcal{O} = \oplus_{p=0}^{\infty} \mathcal{O}(p)$ of functions on $\mathcal{T}^*X$ polynomial along fibres of $\mathcal{T}^*X \to X$ and algebraic, holomorphic, or C^∞ in the X-direction (depending on the case). Thus $E_{pq}^1 = HC_{p+q}(\mathcal{O})(p) = H_{p+q}(\mathrm{Tot}\,\mathcal{B}_{**}(\mathcal{O})(p))$. One can show that, for every $p \in \mathbb{N}$, the first spectral sequence (corresponding to filtering by columns) of the double complex $\mathcal{B}_{**}(\mathcal{O})(p)$ degenerates at E^2-term (for an algebraic X this is Theorem 2.9 of [5] plus the considerations which precede it). This yields, in particular, that

$$HC_{p+q}(\mathcal{O})(p) \simeq \Omega_{\mathcal{O}}^{p+q}(p)/d\Omega_{\mathcal{O}}^{p+q-1}(p) \qquad (p \geq 1) \qquad (5)$$

(for brevity, $\Omega_{\mathcal{O}}^*$ will denote $\Omega_{\mathcal{O}/k}^*$). Since the right-hand side of (5) vanishes for $q \geq n$, as well as for $p + q \geq 2n$, we obtain the required location region for nonvanishing terms of E_{pq}^1.

The same spectral sequence for $\mathcal{B}_{**}(\mathcal{O})(p)$ yields, if $p = 0$, that

$$E_{0q}^1 = HC_q(\mathcal{O})(0) = HC_q(\mathcal{O}(0)) = H_{\mathrm{DR}}^{(\tilde{q})}(X) \qquad (q \geq n)$$

where $\tilde{q} = $ parity of q and $H^{(\varepsilon)}(X)$ denotes the standard $\mathbb{Z}/2$-grading of $H_{\mathrm{DR}}^*(X)$. In view of the shape of E_{**}^1, the terms E_{0q}^1 $(q \geq 2n - 1)$ survive to infinity and we have $HC_q(\mathcal{D}(X)) \simeq H_{\mathrm{DR}}^{(\tilde{q})}(X)$ $(q \geq 2n - 1)$. By comparing this with the statement of Theorem 2, we obtain equality in (4) for $q \geq 2n - 1$. An application of Lemma 6 then finishes the proof of Theorem 3.

8. Remark. As a corollary of the proof, we obtain that the natural embedding of the algebra $\mathcal{O}(X)$ of functions on X viewed as differential operators of order zero induces an isomorphism in cyclic homology $HC_q(\mathcal{O}(X)) \xrightarrow{\sim} HC_q(\mathcal{D}(X))$ for $q \geq 2n - 1$. This is an "additive analog" of the theorem of D. Quillen saying that $\mathcal{O}(X) \hookrightarrow \mathcal{D}(X)$ induces an isomorphism in algebraic K-theory (for X smooth affine). In the unstable range $q < 2n - 1$ the maps $HC_q(\mathcal{O}(X)) \to HC_q(\mathcal{D}(X))$

are always surjective, and usually with nontrivial kernel, as shows an easy inductive argument (q going from the stable range to zero) which involves comparing Connes exact sequences for $\mathcal{O}(X)$ and $\mathcal{D}(X)$.

Note that our proof does not necessitate information about the behavior of the spectral sequence *inside* the marked region. This additional information, however, can be useful, e.g., in proving that the surjections $HC_q(\mathcal{O}(X)) \to HC_q(\mathcal{D}(X))$ split or, equivalently, that the short exact sequences

$$0 \to H_q(\mathcal{D}(X), \mathcal{D}(X)) \to HC_q(\mathcal{D}(X)) \to HC_{q-2}(\mathcal{D}(X)) \to 0$$

admit a canonical splitting (cf. Remark 13.i below), for all q, therefore giving *canonical* identifications in (2).

Theorem 2 is a corollary of the proposition describing the term E^1 of the spectral sequence $'E_{pq}^r \Rightarrow H_{p+q}(\mathcal{D}(X), \mathcal{D}(X))$ of subsection 7 above.

9. **PROPOSITION.** *There is a natural identification* $'E_{pq}^1 \simeq \Omega_{\mathcal{O}}^{2n-p-q}(n-q)$ *($p \geqslant 0$ and $p + q \geqslant 0$). Under this identification* d_{pq}^1 *corresponds to de Rham differential* d_{DR}: $\Omega_{\mathcal{O}}^{2n-p-q}(n-q) \to \Omega_{\mathcal{O}}^{2n-p-q+1}(n-q)$.

10. **COROLLARY.** $'E_{pq}^2 \simeq H_{\mathrm{DR}}^{2n-p-q}(X)$, *if* $q = n$, *otherwise* $'E_{pq}^2 = 0$. *In particular,* $'E_{pq}^r$ *degenerates at* E^2-term, *and* $H_k(\mathcal{D}(X), \mathcal{D}(X)) \simeq 'E_{k-n, n}^2 \simeq H_{\mathrm{DR}}^{2n-k}(X)$.

(Notice that $\Omega_{\mathcal{O}}^{2n-p-q}(n-q) = 0$ for $p < 0$.)

We shall need one general construction. For a given algebra $\mathcal{A}$ and an $\mathcal{A}$-bimodule $\mathcal{M}$ let $\mathfrak{a}$ denote $\mathcal{A}$ regarded as a Lie algebra, and $\mathfrak{m}$ denote $\mathcal{M}$ as a *right* $\mathfrak{a}$-module (i.e. $m \cdot a \equiv ma - am$). *Let* $(C_*(\mathfrak{a}; \mathfrak{m}), \partial)$ *be the standard Koszul chain complex:* $C_q(\mathfrak{a}; \mathfrak{m}) = \mathfrak{m} \hat{\otimes} \wedge^q \mathfrak{a}$, *and*

$$\partial\left(m \otimes a_1 \wedge \cdots \wedge a_q \right)$$

$$= \sum_{1 \leqslant i < j \leqslant q} (-1)^{i+j} m \otimes [a_i, a_j] \wedge a_1 \wedge \cdots \wedge \widehat{a_i} \wedge \cdots \wedge \widehat{a_j} \wedge \cdots \wedge a_q$$

$$+ \sum_{1 \leqslant i \leqslant q} (-1)^{i-1} m \cdot a_i \otimes a_1 \wedge \cdots \wedge \widehat{a_i} \wedge \cdots \wedge a_q. \tag{6}$$

Define a set-theoretic embedding η: $C_*(\mathfrak{a}; \mathfrak{m}) \to \mathscr{C}_*(\mathcal{A}, \mathcal{M})$ by

$$m \otimes a_1 \wedge \cdots \wedge a_q \mapsto \sum_{\sigma \in \mathscr{S}_q} (\operatorname{sign} \sigma) m \otimes a_{\sigma(1)} \otimes \cdots \otimes a_{\sigma(q)}, \tag{7}$$

and let $\mathscr{L}_*(\mathcal{A}, \mathcal{M}) = \eta C_*(\mathfrak{a}; \mathfrak{m})$.

11. LEMMA. *One has $b\eta = \eta\partial$, i.e., Hochschild boundary induces Lie algebra boundary* (6) *on $\mathscr{L}_*(\mathscr{A}, \mathscr{M})$.*

12. Remark. The induced map $\eta\colon H_*(\mathfrak{a}; \mathrm{ad}) \to H_*(\mathscr{A}, \mathscr{A})$ coincides with the edge homomorphism of a certain spectral sequence $''E^2_{pq} = H_p(\mathscr{A}, H_q(\mathfrak{a}; \mathscr{A} \hat{\otimes} \mathscr{A})) \Rightarrow H_{p+q}(\mathfrak{a}; \mathrm{ad})$ where $\mathscr{A} \hat{\otimes} \mathscr{A}$ is an $\mathscr{A}$-bimodule via the action: $a'(a_1 \otimes a_2)a'' = a'a_1 \otimes a_2a''$, and a right $\mathfrak{a}$-module via: $(a_1 \otimes a_2) \cdot a = a_1a \otimes a_2 - a_1 \otimes aa_2$. There is a similar interpretation of $\eta\colon H_*(\mathfrak{a}; \mathfrak{m}) \to H_*(\mathscr{A}, \mathscr{M})$ for a general $\mathscr{A}$-bimodule $\mathscr{M}$.

Proof of Proposition 9. We have $'E^1_{pq} = H_{p+q}(\mathscr{O}, \mathscr{O})(p)$—the component of weight p of the Hochschild homology of the graded algebra $\mathscr{O}$. Let $\omega \in \Omega^2_{\mathscr{O}}(1)$ be the canonical symplectic form. For $f \in \mathscr{O}$ we denote by i_f interior product with the Hamiltonian vector field corresponding to f. Then the correspondence

$$f_0 \otimes f_1 \otimes \cdots \otimes f_j \mapsto \frac{(-1)^j}{j!} f_0 i_{f_1} \cdots i_{f_j} \omega^n \qquad (j \in \mathbb{N})$$

defines surjections $\tau^p_j\colon \mathscr{C}_j(\mathscr{O}, \mathscr{O})(p) \to \Omega^{2n-j}_{\mathscr{O}}(p - j + n)$. One verifies directly that $\tau b = 0$. One can also show that $\operatorname{Ker} \tau^p_j$ consists entirely of boundaries (in the algebraic case this is essentially the dual formulation of the Hochschild-Kostant-Rosenberg theorem [3, Thm. 3.1]; the other two cases are more subtle but not difficult). Thus τ^p_{p+q} induce identifications $'E^1_{pq} \simeq \Omega^{2n-p-q}_{\mathscr{O}}(p)$. In these terms d^1_{pq} are determined as follows. Notice that already the restriction of τ^p_{p+q} to $\mathscr{L}_{p+q}(\mathscr{O}, \mathscr{O})(p)$ is a surjection onto $\Omega^{2n-p-q}_{\mathscr{O}}(n - q)$. Denote this by $\bar{\tau}^p_{p+q}$. One has $\mathscr{L}_*(\mathscr{O}, \mathscr{O}) = \operatorname{gr} \mathscr{L}_*(\mathscr{D}(X), \mathscr{D}(X))$, and by Lemma 11 we know that Hochschild boundary induces on $\mathscr{L}_*(\mathscr{D}(X), \mathscr{D}(X))$ the Lie algebra boundary (6). Recall that commutator of two operators of orders, say, l and m modulo operators of order $l + m - 2$ induces on $\mathscr{O}$ the structure of the graded Poisson algebra (denote it by $\mathscr{P}$). Now, it should be clear that d^1_{pq} on $\Omega^{2n-p-q}_{\mathscr{O}}(n - q)$ is the corresponding homogeneous component of the projection of the boundary homomorphism in the Koszul complex $C_*(\mathscr{P}; \mathrm{ad})$, under the composite map $C_*(\mathscr{P}; \mathrm{ad}) \xrightarrow{\eta} \mathscr{L}_*(\mathscr{O}, \mathscr{O}) \xrightarrow{\bar{\tau}} \Omega_{\mathscr{O}, *}$ where $\Omega_{\mathscr{O}, *}$ denotes de Rham complex with the dual grading $\Omega_{\mathscr{O}, k} \equiv \Omega^{2n-k}_{\mathscr{O}}$ and η is the identification (7). This composite map is, in fact, a morphism of complexes (we proved this, e.g., in a slightly different context in [8, 1.25 and 1.19]; mind there the opposite sign convention for the boundary in Koszul complex). In particular, d^1_{pq} identifies with de Rham differential $d_{\mathrm{DR}}\colon \Omega^{2n-p-q}_{\mathscr{O}}(n - q) \to \Omega^{2n-p-q+1}_{\mathscr{O}}(n - q)$. ∎

13. Final remarks. (i) It follows from the proof of Theorem 2 that the homomorphism $H_q(\mathscr{O}(X), \mathscr{O}(X)) \to H_q(\mathscr{D}(X), \mathscr{D}(X))$ induced by the natural embedding $\mathscr{O}(X) \hookrightarrow \mathscr{D}(X)$ is zero except $q = n$ where it corresponds to the projection $\Omega^n_{\mathscr{O}} \to H^n_{\mathrm{DR}}(X)$.

It is worth noting that Proposition 9 plus a rather substantial amount of diagram chasing plus an argument involving "counting dimensions" similar to that of Lemma 6 yield together an almost complete description of the spectral sequence $E^r_{pq} \Rightarrow HC_{p+q}(\mathscr{D}(X), \mathscr{D}(X))$ used previously to prove Theorem 3.

This can be summarized as follows. The term E^2 is given by

$$E^2_{pq} \simeq \begin{cases} H^q_{\mathrm{DR}} \oplus H^{q-2}_{\mathrm{DR}} \oplus \cdots, & p = 0, q \geqslant n, \\ H^{q-2}_{\mathrm{DR}} \oplus H^{q-4}_{\mathrm{DR}} \oplus \cdots, & p = 0, q < n, \\ H^{q-p+1}_{\mathrm{DR}}, & 2 \leqslant p \leqslant n, q < n, \\ 0, & \text{otherwise} \end{cases}$$

($H^*_{\mathrm{DR}} \equiv H^*_{\mathrm{DR}}(X)$). The only nontrivial differentials d^r_{pq} ($r \geqslant 2$) are $d^p_{pq} \colon E^p_{pq} \to E^p_{0, p+q-1}$ which inject $E^p_{pq} = E^2_{pq} = H^{q-p+1}_{\mathrm{DR}}(X)$ into $E^p_{0, p+q-1}$. In particular, E^r_{**} never stops earlier than at E^{n+1} (the "last" differential being $d^n_{n, n-1} \colon H^0_{\mathrm{DR}}(X) = E^2_{n, n-1} = E^n_{n, n-1} \twoheadrightarrow E^n_{0, 2n-2}$), and the term $E^\infty_{**} = E^{n+1}_{**}$ vanishes except $E^\infty_{0, *}$. A slightly more precise information about differentials[1] is necessary to conclude that the composition of the inclusion

$$H^{2n-q}_{\mathrm{DR}} \oplus H^{2n-q+2}_{\mathrm{DR}} \oplus \cdots \hookrightarrow H^{(\tilde{q})}_{\mathrm{DR}} \simeq E^2_{0q} \qquad (q \geqslant n)$$

with the canonical projection $E^2_{0q} \to E^\infty_{0q}$ is, in fact, an isomorphism. Since $E^2_{0q} \equiv HC_q(\mathscr{O}(X))$, this gives, in equivalent terms, canonical splitting of the surjections $HC_q(\mathscr{O}(X)) \to HC_q(\mathscr{D}(X))$ and hence, canonical isomorphisms in (2) (cf. the discussion in Remarks 8 and 4 above).

For a fuller treatment of the presented results, their extension to noncompact C^∞ manifolds and yet another proof of Theorem 2 in the C^∞ case (actually chronologically prior; based on a certain kind of "noncommutative" Poincaré lemma) the reader will be referred to [9]. One may add that the point-of-view adopted here has its source in the author's work on noncommutative residue.

(ii) Homology of $\mathscr{D}(X)$ for an affine space $X = \mathbf{A}^n$ reduces to the single nonvanishing group $H_{2n}(\mathscr{D}(\mathbf{A}^n), \mathscr{D}(\mathbf{A}^n)) \simeq \ell$ (this was found by B. B. Feigin and B. B. Tsygan, and also by T. Masuda (cf. [2], [6])).

Our approach gives immediately what the corresponding generator is. Denote by ∂_j, the partial derivative $\partial/\partial x_j$, and let $e_j = 1 \otimes 1 \otimes 1 - 1 \otimes X_j \otimes \partial_j + 1 \otimes \partial_j \otimes X_j$. Then $e = e_1 \times \cdots \times e_n$ is a nontrivial $2n$-cycle (the symbol $\times$ denotes the shuffle product that identifies $H_{2n}(\mathscr{D}(\mathbf{A}^n), \mathscr{D}(\mathbf{A}^n))$ with the n-th tensor power of $H_2(\mathscr{D}(\mathbf{A}^1), \mathscr{D}(\mathbf{A}^1)))$. For a general smooth affine variety (in characteristic 0), Hochschild homology of $\mathscr{D}(X)$ was determined also by Ch. Kassel and C. Mitschi (as is reported by J.-L. Brylinski in [1]; we do not know, however, what

---

[1] Which is known to the author at present in C^∞ case and seems very likely in the two remaining cases.

their method is[2]). The evidence towards the existence of isomorphisms like (1) and (2) was given also in the preprint of Brylinski mentioned above (the author thanks Ezra Getzler for drawing his attention to it).

REFERENCES

1. J.-L. BRYLINSKI, *A differential complex for Poisson manifolds*, preprint IHES, 1986.
2. B. B. FEIGIN AND B. B. TSYGAN, *Cohomology of Lie algebras of generalized Jacobi matrices* (in Russian), Funct. Anal. and Appl. **17**: 2 (1983), 86–87.
3. G. HOCHSCHILD, B. KOSTANT AND A. ROSENBERG, *Differential forms on regular affine algebras*, Trans. Amer. Math. Soc. **102** (1962), 383–408.
4. A. YA. KHELEMSKIĬ, *Homological methods in the holomorphic calculus of several operators in a Banach space, according to Taylor* (in Russian), Sov. Math. Surveys **36**: 1 (1981), 127–172.
5. J.-L. LODAY AND D. QUILLEN, *Cyclic homology and the Lie algebra homology of matrices*, Comment. Math. Helv. **59** (1984), 565–591.
6. T. MASUDA, *Duality for differential crossed product and its periodic cyclic homology*, preprint, IHES, 1985.
7. J. L. TAYLOR, *Homology and cohomology for topological algebras*, Adv. Math. **9** (1972), 137–182.
8. M. WODZICKI, "Noncommutative residue. Chapter I. Fundamentals," in *Arithmetical Geometry*, ed. by Yu. I. Manin, Lecture Notes in Math., USSR Subseries, Springer-Verlag, Berlin and New York, 1987.
9. ______,"Noncommutative residue. Chapter IV. Homology of algebras of differential operators and symbols" (in preparation).

THE INSTITUTE FOR ADVANCED STUDY, PRINCETON, NJ 08540

---

[2]Cf., however, the recent preprint by J.-L. Brylinski, *Some examples of Hochschild and cyclic homology*, Fall 1986 (note added in proof).

Vol. 54, No. 2   DUKE MATHEMATICAL JOURNAL ©   1987

SELECTED PROBLEMS OF SUPERMANIFOLD THEORY

D. LEITES

For Yu. I. Manin's fiftieth birthday

> ... being at times somewhat of a humorist myself I know that it is hard to have an amusing story to tell and find no listener. S. Maugham.

Supermanifold theory is very popular nowadays. Its fantastic applications in physics (unified theory of all the known forces) and some not so spectacular results in mathematics, discussed and reviewed in thousands of papers and dozens of books, screen somewhat its shortcomings, the most tantalizing mysteries.

The best introductions to supermanifold theory that give you its specific character are at present [1–3]; see also [4–6] and references therein. The most extensive exposition of supermanifold theory (calculus, differential geometry, representation theory, mechanics, quantization, etc.) is contained in the workouts of my "Seminar on Supermanifolds" (1977–86), which while in preparation to be published by D. Reidel are preprinted by Stockholm University. The vague hints at some results that the reader will find in what follows refer to "Seminar."

I will start with the most exciting problems. They also seem to be rather difficult. Simpler ones, to keep you busy for a couple of evenings, are contained in "Seminar"'s Problems.

(1) *Generalize the notion of supermanifold.* (a) Supergroups are better than groups because they provide us with a bigger stock of automorphisms of the same entities, which is the whole point. In addition, it turns out that by introducing nilpotents and noncommutative algebras instead of algebras of functions in this particular way (via supermanifolds and their rings of functions), we are able to retain all the notions of differential geometry! And even more so, on supermanifolds practically all of them have at least two versions.

But after introducing such a nice generalization of groups, we suddenly confine ourselves only to automorphisms of *homogeneous* objects, and those automorphisms themselves are only *even*. This does not seem fair, and from the mathematical point of view looks arbitrary. Physicists may refer to Pauli's principle on the relation of spin and statistics, but is it not an artifact of the constraints

Received November 26, 1986.

imposed on the apparatus we work with, the same as with divergences of the field theories before supergroups came into play?

A trifle more precisely, I mean here the following: Let C be a commutative superalgebra (or better, an algebra satisfying $[x,[y, z]] = 0$, where $[\cdot, \cdot]$ is the *ordinary commutator*, not supercommutator; see 16). The functor that assigns to C the group $\mathrm{Aut}_C C[x, \xi]$ of automorphisms of the algebra of "functions" over C in even variables $x = (x_1, \ldots, x_n)$ and odd variables $\xi = (\xi_2, \ldots, \xi_m)$ is presented by a supergroup (pseudo, if $n \neq 0$) whose Lie superalgebra is $\underline{w}(n, m)$, if we consider even automorphisms. If we allow all automorphisms, then a transformation is of the form

$$\xi_i \mapsto \underline{\xi_i'(\xi, x)}(1 + \zeta(\xi, x)), \text{ where } \zeta \text{ is odd}$$

$$x_j \mapsto \underline{x_j'(\xi, x)} + f_j(x)\xi_1, \ldots, \xi_n(1 - \varepsilon), \text{ where } \varepsilon = 0 \text{ if } n \text{ is odd, 1 otherwise,}$$

and the underlined terms correspond to an even automorphism.

What is the class of objects that present functors like that?

In what category is it natural to consider automorphisms of nonhomogeneous objects? (Such as a nonhomogeneous nondegenerate bilinear form. What is its canonical form?)

(1b) Similarly to the above, we only consider homogeneous subalgebras and ideals of superalgebras, whereas the manifolds could be well ringed with sheaves of arbitrary subalgebras and quotients of commutative superalgebras. In 1975 V. Kac conjectured that the PI-category containing commutative superalgebras and their arbitrary subquotients is described by the identity $[x,[y, z]] = 0$. Even if it were true, it was not clear how to work with such algebras. Recently I. Volichenko proved Kac's conjecture and introduced a basis suitable to writing Taylor series expansions in the free algebras of this category.

Is it possible to rewrite the differential geometry on the spaces ringed with such algebras, and to what extent?

What are simple superalgebras for the corresponding supergroups?

(c) Everybody knows the importance of orbits of group actions, e.g., those in the coadjoint representation host all the classical mechanics. Now the category of supermanifolds is not closed with respect to supergroup actions. Consider for instance $GL(n)$ acting in the space of the identity representation. There are two orbits: the origin and the rest. If we now look at the space as an $(0, n)$-dimensional supermanifold we see that the complement to the origin is just a kind of halo, indescribable except in the language of the point functor. The functor corresponding to the complement to the origin is not presented by a supermanifold.

Functors on the category of commutative (super)algebras represented by manifolds or supermanifolds are good because you can construct differential or "at least" algebraic geometries on them. *How to distinguish subfunctors corre-*

sponding to the orbits of supergroup action? (A similar problem takes place for groups and their orbits in prime characteristics.)

Is it possible to construct mechanics on such orbits, i.e., integral and differential calculus?

(d) At first I believed that colored algebras (graded by a "color" group, greater than $\mathbb{Z}_2$) reduce to (super)algebras. (For instance, $(\mathbb{Z} \times \mathbb{Z})$-graded-commutativity turns into supercommutativity if we simultaneously replace multiplication by a twisted one; $(ab)_{\text{new}} = (-1)^{\deg_1(a)\deg_2(b)}ab$ where $\deg_i$ is the degree with respect to the i-th copy of $\mathbb{Z}$.) Theorems by Neklyudova and Mosolova prove this for commutative and Lie algebras respectively (but only for finite gradation groups; for infinite groups this is false and their result lacks precision, since it is only stated for discrete groups). But you cannot modify at the same time both the multiplication and the action on a module. I thought that there were too many structures like that, but the recent results "around the Yang-Baxter equation," especially on quantum groups, show that among the set of such structures there are especially nice ones. *How to distinguish them?* Drinfeld's quantum groups, algebras of E. Sklyunin and A. Odessky are examples where ± 1 in the identities describing the algebras is replaced by an *arbitrary* number. V. Drinfeld, D. Gurevich, V. Lyubashenko and A. Rosenberg constructed quite a number of theories-noncommutative geometries (see "Seminar"). *Is it possible to develop differential geometry there?*

(2) *Classical Lie superalgebras.* Around 1966 V. Kac distinguished an important class of Lie algebras: simple $\mathbb{Z}$-graded of finite growth. Among them the class of contragredient is most often encountered in applications because of their "symmetricity" (recall that if $T \subset G$ is a maximal torus, then G is *contragredient* if $G = \bigoplus_{a \in T^*} G_a$ and $\dim G_a = \dim G_{-a}$).

Superization of this class shows new phenomena: some series of Lie superalgebras are only contragredient for small values of parameter (in string theories); twisted loops with values in a contragredient Lie superalgebras with Cartan matrix may be indescribable by Cartan matrices or the other way around.

But even more interesting is the generalization to $\mathbb{Z}^n$-gradation. Examples of such algebras have begun to appear in recent works on integrable systems. In order to describe them we use the key notion of finiteness of growth due to I. Kantor: Let $G = \bigoplus_{I \in \mathbb{Z}^n} G_I$ be a $\mathbb{Z}^n$-graded Lie algebra (or superalgebra, in which case the parity or the $\mathbb{Z}_2$-grading is in no way connected with the $\mathbb{Z}^n$-grading) such that $\dim G_i < \infty$ for all I. Then the *k-growth* of G at $I_0 = (i_0^1, \ldots, i_0^{k-1}, i_0^{k+1}, \ldots, i_0^n) \in \mathbb{Z}^{n-1}$ is

$$\varlimsup_{n \to \infty} \left(\log\left(\dim \bigoplus_{I = (i_0^1, \ldots, i_0^{k-1}, i, i_0^{k+1}, \ldots, i_0^n), |i| \leqslant m} G_i \right) \right) / \log m.$$

If all the k-growths are finite for all I_0 and k, $1 \leqslant k \leqslant n$, then G is said to have a *finite growth*.

Even for $n = 1$, $\mathbb{Z}$-graded simple Lie algebras of finite growth are only classified under a technical assumption that they are generated by elements of degree ± 1. Nevertheless, it seems plausible to describe all such algebras for $n = 2$ (there are few two-dimensional manifolds) or modulo the classification of manifolds. But perhaps we should not do that at all. Since any manifold can be endowed with a metric, we may consider only Riemannian manifolds and embed them into their jacobians. Therefore, the most important examples are loops and vector fields on tori, and their twisted versions.

The simple algebras of the class above is the necessary first step, the skeleton, but the body to be explored has central extensions, algebras of derivations and their iterations and the real forms of them. Recent results by V. Kac and I. Todorov suggest also that tori should be replaced by supertori.

Examples. (a) $G^T = G \otimes \mathbb{C}[t_1, \ldots, t_n, t_1^{-1}, \ldots, t_n^{-1}]$. Here G is a simple $\mathbb{Z}$-graded Lie superalgebra of finite growth (not necessarily finite-dimensional), $T = T^n$ a torus. If G is contragredient, the residue ensures the existence of central extension. If we want to construct a Verma-like module, we should set $\deg t_i = a_i$, where a_i's are incommensurable; otherwise the space of the highest weight vectors depends on infinitely many parameters (which is still another possibility to consider). Do the constructions corresponding to different sets (a) *differ essentially?* It would be a pity if they do.

(b) Consider polynomials in $q_1, \ldots, q_n, p_1, \ldots, p_n$ and some of their inverses. The Poisson bracket makes the space of these polynomials into a Lie algebra. As a category of representation we can take modules constructed as in Example (a), but the natural grading $\deg p_i = \deg q_i = 1$ is more meaningful here, since the algebra of fields of degree 0 is just the algebra of contact vector fields on $(2n - 1)$-dimensional space. For $n = 1$ it seems plausible to reduce the description of irreducible representations to that of Virasoro algebra.

(3) *Singularities and Lie algebras in characteristic* 2. Recent investigations of singularities show their mysterious relation to root systems over $\mathbb{Z}_2$. An important subclass of root systems is that of simple Lie algebras.

In characteristic 2 contragredient Lie algebras (finite-dimensional over an algebraically closed field) were listed by B. Weisfeiler and V. Kac, and besides Lie algebras of vector fields of series w, s, h and k some odd examples were known. My conjecture is that

(a) *the known simple $\mathbb{Z}$-graded Lie superalgebras of finite growth exhaust the list* (for algebras of the form $\bigoplus_{i \geqslant -l} G_i$ see the scheme of the proof with clearly outlined and, as the authors believe, reparable gaps in "Seminar");

(b) *all simple finite-dimensional Lie superalgebras over (an algebraically closed) field k of* char 2 *are obtained from simple $\mathbb{Z}$-graded Lie superalgebras of finite growth choosing a Chevalley basis (in which structure constants are integers) tensoring by* k *over $\mathbb{Z}$ and taking the simple piece of the subalgebra of elements with zero square; sometimes these algebras may be deformed.*

With this in mind I investigated the contragredient case and Yu. Kochetkov the noncontragredient one. I just rediscovered the Weisfeiler-Kac list giving interpretation to some of their algebras (known to Kac since 1975), while Yu. Kochetkov discovered new series and exceptional algebras. It now remains to describe their deformations.

What are these algebras to singularities?

(4) *Nonstandard realizations of Lie superalgebras of vector fields.* Unlike simple Lie algebras of vector fields L that preserve some structure on the manifold $(L/L_>)^*$, where $L_>$ is the unique maximal subalgebra of finite codimension, there are several such subalgebras in Lie superalgebras. For the general series $w(n, m)$, say, these subalgebras are connected with the gradings when some of odd variables are of degree 0, whereas all the others are of degree 1. *Not for all gradings is it clear what structure the algebra preserves.*

For the standard realizations all the unary invariant differential operators are described. The answer is boring: this is essentially the exterior differential only. The invariant differential operators for nonstandard realizations are quite new. *List them and interpret.* Is it possible to reduce the problem to computing Massey products of the Lie superalgebra of linear fields in the "maximal" realization?

Physical applications of these structures have never been investigated.

(5) *G-structures on classical superdomains. Einstein-like equations.* I. Kantor was the first to notice the following one-to-one correspondence:

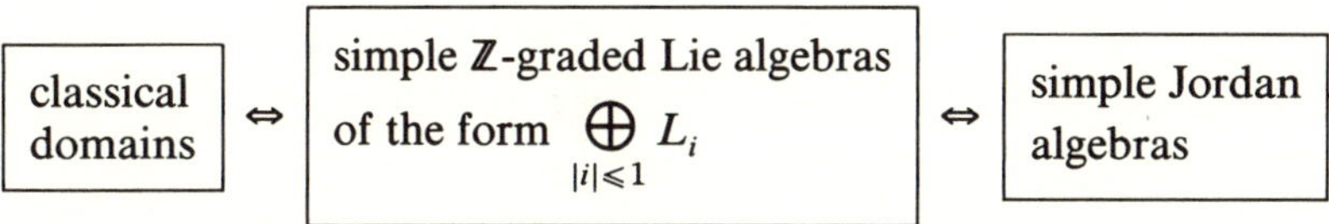

He also generalized these correspondences to the algebras of the form $\bigoplus_{i\geqslant -l} L_i$ and in particular $\bigoplus_{|i|\leqslant 2} L_i$, $\bigoplus_{i\geqslant -1} L_i$ and $\bigoplus_{i\geqslant -2} L_i$. In "Seminar" all these gradings for the known simple Lie superalgebras are enumerated. The most important case is clearly the one corresponding to Jordan superalgebras and its generalization of the form $\bigoplus_{i\geqslant -1} L_i$.

Let G/P be a classical superdomain, where $L = \mathrm{Lie}\,(G)$ and $\mathrm{Lie}\,(P)$ is either $L_0 \oplus L_{-1}$ or $L_0 \oplus L_1$ (perhaps $\oplus L_2 \oplus \ldots$). It is interesting to calculate (a) $H^*(L_{-1}; L)$ and (b) $H^*(L_{-1}; L_{-1} \oplus L_0^s)$, where L^s is the semisimple part of L. The sections of the bundle over G/P with these cohomology groups as a fiber, L_{-1} being identified with the tangent space, are just structure functions of the corresponding generalized (a) conformal G_0-structure ($L_0 = \mathrm{Lie}\,G_0$) or (b) Riemannian G_0^s-structure ($L_0^s = \mathrm{Lie}\,G_0^s$).

New phenomena are *the odd metric* and *the odd complex structure.* But even more exciting are the structures on G/P, when $\mathrm{Lie}\,G = sh$ (special Hamiltonian algebra) with the grading (discovered by V. Serganova) $\deg \xi = - \deg \eta = 1$ for a pair of conjugate variables (with respect to the symplectic form), deg of the other variables being 0.

Since there are deformations ("quantizations")

$$sh(2n) \to psl(2^{n-1}/2^{n-1}) \quad \text{and} \quad sh(2n-1) \to psq(2^{n-1})$$

(which should interest the experts in Lie algebras of prime characteristics) compatible with the grading above, then there are deformations of G/P into Grassmannians $GR_{a,a}^{2a,2a}$ and QGR_a^{2a}, where $a = 2^{n-2}$.

For symmetric affine connections the notion of compatibility with metric leads to two versions of Levi-Civita connections. The corresponding conformal structure functions are only calculated for the even metric: the Riemannian tensor has the same components as on manifolds. The odd case is in progress (E. Poletaeva).

The space $H^*(L_{-1}; L)$ contains several irreducible L_0-modules (not necessarily completely reducible for superalgebras); denote R the trivial component. It is one-dimensional, at least for $L = osp$, $s\pi$ and sh. Let r be a section of the corresponding bundle, g a section of the trivial submodule in $S^2(L_{-1})$, a metric in the given conformal class, ω the symplectic form preserved by sh.

By *Einstein equation* we will mean

$$g = ar \quad \text{or} \quad \omega = ar \quad \text{for a constant } a.$$

Physicists do not like this equation for *osp* except when *osp* turns into $o(4)$, since this equation contains too many components that do not correspond to known fields; besides, they believe nowadays that the supergroup of motions of the superspace we live in is a central extension of a contraction of *osp*, not *osp* itself. Other cases are not even considered, especially infinite-dimensional purely even or mixed versions of symplectic structure. I can't believe that such beautiful equations have no meaning. *What is their meaning?*

(6) *Quantization* is deformation of the Lie superalgebra structure (at least locally). (The commutative associative algebra structure on the same space is an opium for people.) The quantization of the Poisson bracket is the deformation

$$Po(2n/2m) \to \text{Diff}(n/m),$$

the space of differential operators with polynomial coefficients,

$$Po(2n/2m - 1) \to Q\,\text{Diff}(n/m) = \{ D \in \text{Diff}(n/m) | [D, D_0] = 0, \text{ where } D_0$$
$$\text{is odd and such that } D_0^2 = -\text{id} \}.$$

(We may set $D_0 = \sqrt{-1}\,(\xi_1 + \partial/\partial\xi_1)$.)

Except for small n, m there is only this deformation of *Po*; the other deformations are "silly" and hence there is the unique quantization.

The same is true of the antibracket or $B(n)$, Buttin superalgebra, which is quantized with the help of an odd (nilpotent) Planck's constant, except for the presence of a global deformation with even parameter, which, though not quantization, is by no means silly. Yu. Kochetkov showed that the deformed algebras (they multiply the odd contact structure by a certain factor of divergence; this factor is the parameter) are deformed at the values of parameter $-n/(n \pm 2)$ and -1. *Interpret the deformed algebras and describe their representations.* Yu. Kochetkov also showed that SB can be deformed with odd parameter, the deformation can be deformed with odd parameter, ad infinitum. What is *their* interpretation?

(7) *Weyl supergroups and Hecke algebras.*   V. Serganova listed all the systems of simple roots of finite-dimensional and (twisted) loop simple Lie superalgebras with Cartan matrices. The neighboring systems are related by "reflections." It was only natural to see what group these "reflections" generate. It turns out that there are at least two ways to construct the group. One acts linearly on the Cartan subalgebra but does not always preserve the roots (only the root lattice); the other is defined as a subgroup permuting roots, and from it a universal infinite group that permutes the weights of any irreducible representation is constructed. These groups enabled Serganova to get the restrictions on the labels of the highest weights of irreducible finite-dimensional representation with respect to *any* Borel subalgebra, list all systems of simple roots, etc.

CONJECTURE.   *These groups are Coxeter, not just quotients of Coxeter groups.*

These are two versions of Weyl group for superalgebras. But in the "even" case the Weyl group often appears as the index set; its group structure is not usually needed (Borel-Weyl-Bott-... theorem, character formula, etc.: in "Seminar" are listed about a dozen instances). All are different, essentially. One of them is connected by Yu. Manin and A. Voronov with the following question:

(8) *How to label cohomology of supermanifolds by two numbers* (so as to make it dual to homology, which is labelled by the dimension and which is not even a pair of numbers but an element of $\mathbb{Z}[x]/(x^2 - 1)$)?

(Since differential forms on a supermanifold of dimension $(n, 1)$ may be of any complex order, the same as pseudodifferential or pseudointegrable forms, it is quite a question: *What is the index set for (co)homology?*)

Yu. Manin and A. Voronov described Schubert supercells for Grassmannians related with matrix supergroups of motion. Their construction can be generalized to supergroups with Lie superalgebras $w(0, n)$, $s(0, n)$, $s'(n)$ and $sh(n)$ by the same lines ($w(0, n)$ should be realized as a subalgebra in $gl(\mathbb{C}[\xi])$ preserving the tensor of type $(2, 1)$ defining the multiplication in the Grassmann superalgebra $\mathbb{C}[\xi]$, etc.), and this makes one wonder what the infinite-dimensional version of superWeyl group that indexes the supercells for $w(n, m)$ might mean?

Another attempt to index cohomology by two numbers was made by A. Zorich and F. Voronov. Its elucidation, due to V. Shander, is found in "Seminar."

(9) *To get Bruhat decomposition* the Weyl group should be contained in the group itself. This is not the case for Lie supergroups; only the Weyl group of the underlying group belongs. We should generate a bigger group containing the supergroup and its superWeyl group, but how do we do this?

(10) *What is time?* V. Shander proved the straightening of vector field theorem (the main theorem of ordinary differential equation theory) for supermanifolds and found out that the dynamical parameter (time) is $(1,1)$-dimensional; the square root of the generator of shifts exists along the even time. The additional solutions this provides for classical equations (e.g., Schrödinger's equations) have not been investigated.

(11) *The Stokes formula* on supermanifolds is connected with the notion of integrable forms, which in turn are related to the Beresinian (superdeterminant for $GL(n, m)$). But the GL group has at least two analogues on supermanifolds: GQ (queer series) and the one connected with the Poisson superalgebras (see deformations in (5) and (6) above), to say nothing of the Buttin pseudogroup. For all of them there exists an analogue of determinant and, I believe, an integration theory.

(12) In *representation theory* there are too many interesting questions, so I list only the most interesting to me at present.

Models of representations constructed by I. Gelfand and A. Zelevinsky provide a natural class of representations. What are they in the case of Lie superalgebras (J. Bernstein, I. Frenkel, 1976)?

The results of Berezin-Kac-Sergeev on *invariant* polynomials on simple Lie superalgebras show that *rational functions* are more natural than the polynomial ones (the latter do not even constitute a Noetherian ring). But no *invariant theory* exists so far; even the classical one is not superized yet.

What are Kazhdan-Lustig polynomials for superalgebras? Attempts to take the D-module approach to representations of superalgebras are reviewed by Yu. Manin, I. Penkov, I. Skornyakov and A. Voronov in a book to be published by Asterisque.

References

1. V. Kac, *Lie superalgebras*, Adv. Math, **26** (1977) 8–96.
2. Yu. Manin, *New dimension in geometry*, Russian Math. Surveys **40** (1985).
3. ______, *Gauge fields and complex geometry*, Nauka, Moscow (1985) (in Russian).
4. F. Berezin, *Algebra and Analysis with Anticommuting Parameters* (1983) (in Russian), Moscow Univ. Press, Moscow (enlarged and edited version in English to be published by D. Reidel).
5. D. Leites, *Introduction to supermanifold theory*, Russian Math. Surveys, **35** (1980) n1, 3–50.
6. ______, *Lie superalgebras*, JOSMAR, **31** (1980).

Stockholm University, Mathematics Institute, Box 6701, S-11385 Stockholm, Sweden.

CUBIC HYPERSURFACES AND A RESULT OF HERMITE

DANIEL F. CORAY

Dedicated to Yu. I. Manin

Let k be an infinite field, and K/k a separable extension of degree 5. The following result is due to Hermite [5]: *there exists an element η in K whose minimal polynomial (over k) is of the form $x^5 + ax^3 + bx + c$.*

In his proof Hermite used a rather complicated invariant. Recently, Serre asked whether there is a more natural approach in terms of cubic surfaces. Indeed it is easy to reformulate Hermite's result as asserting the existence of a k-point on some specific cubic surface (1.8), which depends on the extension K/k.

It seems that this question originated from a letter of Joe Buhler to Serre (July 1982), in which Buhler mentioned Hermite's result and asked about possible generalizations to higher degrees.

In this paper we give a new proof of Hermite's result by studying the lines lying on the corresponding cubic surface. We also prove a generalization to degree 6 by constructing some curves in $\mathbb{P}^4$. Both techniques were already used in [2], though in simpler situations. The results of [2] also imply generalizations to degree 7 or 8, provided k belongs to some special class of fields, which includes all local fields.

I am grateful to J-P. Serre for suggesting the problem and to J-F. Mestre for communicating the information he had on the question. I also thank J. Buhler and A. Brumer for their comments on the manuscript.

§1. Reformulation of Hermite's result and generalization. The main purpose of this section is to introduce some notation. The equivalence of (H) and (H′) is almost evident, but the argument supplied below (Proposition 1.1) gives us the opportunity to prove some auxiliary results, which will be needed later on.

Notation. If $K = k(\theta)/k$ is a separable extension of degree d, we denote by $\tau_i (i = 1, \ldots, d)$ the various k-embeddings of K in a fixed algebraic closure $\bar{k}$ of k. We consider the formal expression

$$(1.1) \qquad\qquad x = x_0 + x_1\theta + \cdots + x_{d-1}\theta^{d-1},$$

Received November 29, 1986.

and set

(1.2)
$$y_i = \tau_i(x),$$

where the τ_i act trivially on the x_j.

Note that the row vector $\mathbf{y} = (y_1, \ldots, y_d)$ is obtained from $\mathbf{x} = (x_0, \ldots, x_{d-1})$ by the nondegenerate linear transformation

(1.3)
$$\mathbf{y} = \mathbf{x}\Theta,$$

where Θ is the van der Monde matrix

(1.4)
$$\Theta = \begin{pmatrix} 1 & \cdots & 1 \\ \theta_1 & \cdots & \theta_d \\ \vdots & & \vdots \\ \theta_1^{d-1} & \cdots & \theta_d^{d-1} \end{pmatrix},$$

with $\theta_i = \tau_i(\theta)$.

For $\nu \in \mathbb{N}$, we shall consider the symmetric function

(1.5)
$$s_\nu(x) = \mathrm{Tr}(x^\nu) = \sum_{i=1}^{d} y_i^\nu.$$

Note that $s_\nu(x)$ is a form of degree ν in the variables $x_0, \ldots, x_{d-1}$, with coefficients in k. The same remark applies to the elementary symmetric functions $\sigma_\nu(x)$, which are defined in a similar way. For instance,

(1.6)
$$\sigma_3(x) = y_1 y_2 y_3 + y_1 y_2 y_4 + \cdots + y_{d-2} y_{d-1} y_d.$$

We shall need the following special cases of Newton's formula ([12], §33, Exercise 2):

(1.7)
$$\begin{cases} s_1 = \sigma_1 \\ s_2 = \sigma_1^2 - 2\sigma_2 \\ s_3 = \sigma_1^3 - 3\sigma_1\sigma_2 + 3\sigma_3. \end{cases}$$

(H). *There exists $\eta \in K$ such that $K = k(\eta)$ and $\sigma_1(\eta) = \sigma_3(\eta) = 0$.*

Clearly, this assertion does not hold in general for $d \leq 4$. In this paper we shall prove it for $d = 5$ or 6, and discuss it for higher degrees. Interestingly, for any d, it is usually impossible to find an element $\eta \in K^*$ such that $\sigma_1(\eta) = \sigma_2(\eta) = 0$. Indeed, by (1.7), such an element would satisfy $s_2(\eta) = \Sigma\tau_i(\eta)^2 = 0$, which is clearly impossible if, for instance, $k = \mathbb{Q}$ and K is totally real.

With the notation introduced above, it is easy to reformulate (H) in geometric terms. Indeed, with x as in (1.1), the equations

$$(1.8) \qquad\qquad \sigma_1(x) = \sigma_3(x) = 0$$

define a subvariety V of projective space $\mathbb{P}_k^{d-1}$ with coordinates $(x_0, \ldots, x_{d-1})$. This variety is the intersection of a hyperplane and a cubic hypersurface, i.e., it is a cubic hypersurface in some $\mathbb{P}_k^{d-2}$.

Definition.   On going over to the algebraic closure, we shall consider two systems of coordinates in $\mathbb{P}^{d-1}$: the **x**-*coordinates* $(x_0, \ldots, x_{d-1})$, and the **y**-*coordinates*, which are derived from them by the transformation (1.3). Given an algebraic extension L of k, $V(L)$ denotes the set of *L-points* of V, i.e., the set of points on V with **x**-coordinates in L.

If $\mathrm{char}(k) \neq 3$, it follows from (1.7) that V can also be defined by the equations

$$(1.9) \qquad\qquad s_1(x) = s_3(x) = 0.$$

With the notation of (1.2) and (1.5), we are therefore led to the following statement:

(H'). *The projective variety* $V \subset \mathbb{P}_k^{d-1}$ *defined by*

$$(1.9) \qquad \begin{cases} \mathrm{Tr}(x) = \displaystyle\sum_{i=1}^{d} y_i = 0 \\[2ex] \mathrm{Tr}(x^3) = \displaystyle\sum_{i=1}^{d} y_i^3 = 0 \end{cases}$$

has a k-point, i.e., a point with **x**-*coordinates in k.*

In fact we have the following

PROPOSITION 1.1.   *If* $d \geq 5$ *then assertions* (H) *and* (H') *are equivalent, provided* $\mathrm{char}(k) \neq 3$ *and, if d is even,* $\mathrm{char}(k) \neq 2$.

Proof.   If (H) holds, write $\eta \in K$ in the form

$$(1.10) \qquad\qquad \eta = a_0 + a_1\theta + \cdots + a_{d-1}\theta^{d-1},$$

with $a_j \in k$. Since the action of τ_i is trivial on the a_j as well as on the x_j, τ_i commutes with specialization, i.e., $\tau_i(\eta) = y_i|_{\mathbf{x}=(a_0, \ldots, a_{d-1})}$. Hence $(a_0, \ldots, a_{d-1})$ are the **x**-coordinates of a point lying on the variety defined by (1.8), and hence also a solution of (1.9).

Conversely, if $(a_0, \ldots, a_{d-1}) \in V(k)$, define η by (1.10). Since $\mathrm{char}(k) \neq 3$, (1.9) implies (1.8), whence $\sigma_1(\eta) = \sigma_3(\eta) = 0$. We still need to show that $K = k(\eta)$. For $d = 5$ this is almost trivial: η fails to generate K only if $\eta \in k$; then $\sigma_1(\eta) = 5\eta \neq 0$, unless $\mathrm{char}(k) = 5$. The case of characteristic 5 must be handled separately: The point with x-coordinates $(1, 0, 0, 0, 0)$, which corresponds to $\eta = 1$, does belong to the cubic surface V. In fact one can check that it is a double point (see Lemma 1.2 below). It follows that V is k-rational (cf. [4], Lemma 1.1). Hence $V(k)$ contains also some other points.

For arbitrary d, the proof is a little more subtle, because $k(\eta)$ might be a proper subfield of K. But V is a cubic hypersurface in some $\mathbb{P}_k^{d-2}$, with only finitely many singular points. More precisely:

LEMMA 1.2.   *If* $\mathrm{char}(k) \neq 3$, *the variety* V *defined by* (1.9) *has only finitely many singular points. They are only double, provided* $\mathrm{char}(k) \neq 2$. *In characteristic* 2, V *is smooth if* d *is odd, and is a cone with vertex given by* $y_1 = \cdots = y_d = 1$ *if* d *is even. In characteristic* 0, V *is smooth if* d *is odd, and has precisely* $\frac{1}{2}\binom{d}{d/2}$ *double points if* d *is even.*

Proof.   To prove these assertions, we are free to go over to the algebraic closure of k and, via (1.3), to examine V in the y-coordinates. Then a point of V is singular if and only if the matrix

$$\begin{pmatrix} 1 & \cdots & 1 \\ y_1^2 & \cdots & y_d^2 \end{pmatrix}$$

is not of maximal rank, i.e., if and only if $y_1^2 = \cdots = y_d^2 = 1$.

In characteristic zero, no such point satisfies (1.9) if d is odd. For d even, the point with y-coordinates $y_1 = \cdots = y_{d/2} = 1$, $y_{d/2+1} = \cdots = y_d = -1$, and those obtained from it by permuting the y_i, are singular. A simple consideration of symmetry shows that they cannot be more than double. Indeed, either none of them is triple, or all of them are; but a cubic with finitely many singular points cannot have more than one triple point! For the count of these singularities, remember that y and $-$y define one and the same point.

Small characteristics require a little more work, since there may be only one singular point (for instance if $d = \mathrm{char}(k) = 5$). To show that such a point P, with y-coordinates $(y_1, \ldots, y_d)$, is not triple, under the assumption that $\mathrm{char}(k) \neq 2$, one may assume that $y_1 = y_2$ and consider the line through P and the point $\mathbf{y} = (1, -1, 0, \ldots, 0)$. This line is not entirely contained in V; hence V is not a cone and P is not triple. We leave the discussion of characteristic 2 to the reader. $\qquad\square$

As $d \geq 5$, $\dim V \geq 2$. Hence Lemma 1.2 has the following consequence:

COROLLARY 1.3.   *Under the assumptions of Proposition* 1.1, *if* $V(k) \neq \varnothing$ *then* V *is* k-*unirational.*

Proof. Either $V(k)$ contains a double point, and then V is k-rational, or V contains a smooth k-point. In the latter case, the result is well-known in dimension 2 (cf. [8], chap. IV, 7.8 and 8.1), since we may assume that $\mathrm{char}(k) \neq 5$ and then V is smooth (see the proof of Lemma 1.2). For $\dim V \geq 3$, one can apply the argument of [1], Remark 7.4, since V has only finitely many singular points (and is smooth in characteristic 2). $\square$

Thus, in order to prove that η can be found such that $k(\eta) = K$, it suffices to establish the following:

LEMMA 1.4.   *The points* $(a_0, \dots, a_{d-1}) \in V(k)$ *such that* η, *defined by* (1.10), *has degree smaller than d lie in some proper subvariety of V.*

Proof.   In fact, suppose η is of degree $r < d$. Then there is a non-trivial linear combination $c_0 + c_1\eta + \cdots + c_r\eta^r = 0$ with $c_j \in k$. Hence the y-coordinates of our point, which are just the conjugates of η, also satisfy $c_0 + c_1 y_i + \cdots + c_r y_i^r = 0$. In other words, the van der Monde matrix

$$\left(y_i^j \right)^{j = 0, \dots, r}_{i = 1, \dots, d}$$

is not of maximal rank. In particular we have:

$$(1.11) \qquad\qquad \prod_{i < j \leq r+1} (y_i - y_j) = 0.$$

Now this relation defines a union of proper subvarieties of V. Indeed, the subvariety defined by (1.9) and (say) $y_1 = y_3$ is proper, since it does not contain the point with y-coordinates $(1, -1, 0, \dots, 0)$, which lies on V. $\square$

§2. The case of degree 5.   In this section we prove Hermite's result, i.e., assertion (H) for $d = 5$, under the assumption that $\mathrm{char}(k) \neq 3$. By Proposition 1.1, this enables us to reformulate it in the form (H$'$). The case of characteristic 3 will be discussed at the end (Remark (b)).

We begin with a general remark on how to find a k-point on the variety $V \subset \mathbb{P}^{d-1}_k$. Let N be the normal closure of K/k; then $G = \mathrm{Gal}(N/k)$ is naturally identified with a subgroup of the symmetric group $\mathscr{S}_d$. In fact, G acts on the y_i, defined as in (1.2), as a set of permutations. In given instances, it can even be the full group $\mathscr{S}_d$. Now, suppose we are in the special situation in which $\mathrm{Tr}_{K/k}(\theta) = \mathrm{Tr}_{K/k}(\theta^3) = 0$. Then V contains the point with x-coordinates $(0, 1, 0, \dots, 0)$. This is a k-point of V, but its y-coordinates, viz., $(\theta_1, \dots, \theta_d)$, are not invariant under the action of $\mathscr{S}_d$. This example shows that usually there is no easy way to detect a k-point from the action of G on its y-coordinates. However, if the y-coordinates are in k, we are slightly better off:

LEMMA 2.1.   *Let Q be a point whose y-coordinates* $(y_1, \dots, y_d)$ *are in k. Then Q is defined over the normal closure of K/k, i.e., its x-coordinates are in N. If*

$\sigma \in \mathrm{Gal}(N/k)$ *then the* y-*coordinates of* Q^{σ} *are of the form* $\mathbf{y}P_{\sigma}$, *where* P_{σ} *is a permutation matrix. Hence the orbit of* Q *under the action of the full group* $\mathscr{S}_d$ *on the* y_i, *is a* k-*rational* 0-*cycle.*

Proof. From (1.3), we get $\mathbf{x} = \mathbf{y}\Theta^{-1}$. Hence the x-coordinates of Q are certainly in N. For $\sigma \in G = \mathrm{Gal}(N/k)$, let $\mathbf{y}_{\sigma}$ be the y-coordinates of Q^{σ}. Then, by (1.3) again, $\mathbf{y}_{\sigma} = \mathbf{x}^{\sigma}\Theta$. Now $\Theta = \Theta^{\sigma}P_{\sigma}$, where P_{σ} is a permutation matrix. Hence $\mathbf{y}_{\sigma} = (\mathbf{x}\Theta)^{\sigma}P_{\sigma} = \mathbf{y}^{\sigma}P_{\sigma} = \mathbf{y}P_{\sigma}$ since, by assumption, the y-coordinates of Q are in k.

The final assertion is clear, since N/k is separable. It suffices to notice that, if a point R belongs to the orbit of Q for the action of $\mathscr{S}_d$ on the y_i, then so does R^{σ}, for any $\sigma \in G$. $\square$

THEOREM 2.2.   *If* $\mathrm{char}(k) \neq 3$ *then assertion* (ℍ) *holds for* $d = 5$.

Proof. In view of Proposition 1.1, we prove assertion (ℍ′). $V \subset \mathbb{P}_k^4$ is a cubic surface. If $\mathrm{char}(k) = 5$ we saw in the proof of Proposition 1.1 that V is k-rational. For other characteristics, V is smooth (cf. Lemma 1.2 and its proof). Thus we shall assume that $\mathrm{char}(k) \neq 5$, and then V will be a smooth cubic surface contained in the hyperplane defined by $\mathrm{Tr}(x) = 0$. To study this surface, we begin by determining its 27 lines.

Working in the y-coordinates, one sees immediately that V contains the line

$$(2.1) \qquad\qquad d_1 = \{\, y_1 + y_3 = y_2 + y_4 = y_5 = 0 \,\},$$

which is left invariant by the action of the dihedral group

$$D_4 = \langle (13); (1234) \rangle$$

on its equations. It follows that the orbit of d_1 under the action of $\mathscr{S}_5$ consists of at most $5!/8 = 15$ distinct lines. This is in fact the exact number, as an easy verification shows. It is important to observe that this set of fifteen lines is defined over k. Indeed, if we refer to (1.2), we see that $\sigma \in \mathrm{Gal}(\overline{k}/k)$ acts on d_1 by permuting the y_i in its equations. Hence the set of fifteen lines, which is the orbit of d_1 under the action of the whole of $\mathscr{S}_5$, is left invariant by $\mathrm{Gal}(\overline{k}/k)$. (It may, of course, split into several orbits for this latter action.)

Thus V contains 12 other lines, which form a k-rational set. We shall determine these lines and show that they form a double-six (cf. [6], §25, and [10], chap. I). This is easy to do because, once a line d_1 has been determined on a smooth cubic surface, it is possible to find all the others by a systematic procedure (cf. [13], §35). It suffices to consider the pencil of planes through d_1, which defines a pencil of conics on V. By evaluating a determinant, one discovers the five planes in which the conic degenerates into a pair of lines. We omit this computation, which reveals the following line lying on V:

$$(2.2) \qquad d_2 = \{\, y_1 + \rho y_2 + y_3 = \rho y_1 + \rho y_2 - y_4 = \rho y_1 + y_2 + y_5 = 0 \,\},$$

where ρ is the "golden number", i.e., one of the two roots of

$$(2.3) \qquad\qquad \rho^2 - \rho - 1 = 0,$$

which we *fix*. The reader will find it interesting to check that d_2 lies on V, and also that it is left invariant by the action of the dihedral group

$$D_5 = \langle (12)(35); (12345) \rangle$$

on its equations. As expected, the orbit of d_2 under the action of $\mathscr{S}_5$ consists of $5!/10 = 12$ lines. As $D_5 \subset \mathscr{A}_5$, the easiest way to investigate the structure of this set of lines is to write out explicitly the orbit of d_2 under the action of the alternating group $\mathscr{A}_5$. This consists of the following six lines:

$$(2.4) \qquad \textbf{1.} \ \ y_1 + \rho y_2 + y_3 = \rho y_1 + \rho y_2 - y_4 = \rho y_1 + y_2 + y_5 = 0$$

$$\textbf{2.} \ \ y_1 + \rho y_2 + y_4 = \rho y_1 + \rho y_2 - y_5 = \rho y_1 + y_2 + y_3 = 0$$

$$\textbf{3.} \ \ y_1 + \rho y_2 + y_5 = \rho y_1 + \rho y_2 - y_3 = \rho y_1 + y_2 + y_4 = 0$$

$$\textbf{4.} \ \ y_1 + \rho y_3 + y_4 = \rho y_1 + \rho y_3 - y_2 = \rho y_1 + y_3 + y_5 = 0$$

$$\textbf{5.} \ \ y_1 + \rho y_4 + y_5 = \rho y_1 + \rho y_4 - y_2 = \rho y_1 + y_4 + y_3 = 0$$

$$\textbf{6.} \ \ y_1 + \rho y_5 + y_3 = \rho y_1 + \rho y_5 - y_2 = \rho y_1 + y_5 + y_4 = 0.$$

These systems of equations are obtained by subjecting d_2 to the transformations (345), $(345)^2$, (234), $(234)(345)$, $(234)(345)^2$, respectively. Because of the group action, it is not hard to verify that these lines are pairwise skew, so that they form a sextuplet; call it Σ.

Like every sextuplet on a smooth cubic surface, Σ has a unique "complementary sextuplet". This can be obtained in either of two (necessarily equivalent) ways:

(i) apply any odd permutation to the equations (2.4) of the lines composing Σ; or:

(ii) replace ρ in (2.4) by the other root, $\bar{\rho}$, of (2.3).

It is a straightforward, but tedious, exercise to check that the twelve lines thus described do form a double-six. Fortunately, this will not be needed, since only the first sextuplet, Σ, is used in what follows.

In general Σ is not defined over k, but over some quadratic extension (because a double-six consists of exactly two sextuplets). For simplicity we extend the ground field a bit more by adjoining both ρ and the square root of the discriminant of θ:

$$(2.5) \qquad\qquad k' = k\left(\rho, \sqrt{\mathrm{disc}(\theta)}\right).$$

(Replace by the appropriate thing in characteristic 2.) Then every $\sigma \in \mathrm{Gal}(\bar{k}/k')$ acts on the y_i as an even permutation. Since, moreover, $\rho \in k'$, the set of equations (2.4) is left invariant by this action. It follows that the sextuplet Σ is k'-rational.

Now let $V' = V \times_k k'$, i.e., V viewed as a cubic surface over the field k'. The sextuplet Σ, being defined over k', can be blown down on V'. This yields a birational k'-morphism

$$(2.6) \qquad\qquad\qquad g : V' \to S,$$

where S is a Severi-Brauer surface over k'. We will show that $S(k') \neq \varnothing$, i.e., that S is isomorphic to $\mathbb{P}^2_{k'}$ (cf. [11], chap. X, §6; in characteristic 2, note that k'/k, as a subextension of $N(\rho)/k$, is separable!). By a classical result (cf. [3], §2), it suffices to show that $S(L) \neq \varnothing$ for some separable extension L/k' of degree prime to 3. Now, there lies on d_1 the point Q with y-coordinates $(1, 0, -1, 0, 0)$, which is left invariant by the action of the group

$$\mathscr{S}_2 \times \mathscr{S}_3 = \langle (13); (24); (245) \rangle$$

on the y_i. By Lemma 2.1, the orbit of Q under the action of the full group $\mathscr{S}_5$ is a k-rational 0-cycle of degree $5!/12 = 10$. This cycle may split into various orbits for the action of $\mathrm{Gal}(\bar{k}/k')$, but certainly one of them is of order prime to 3. A point picked in that orbit is defined over a separable extension L/k' of degree prime to 3. This field L does what is required, since $V'(L) \neq \varnothing$ implies $S(L) \neq \varnothing$. Thus we have shown that S is isomorphic to $\mathbb{P}^2_{k'}$. It follows, in view of (2.6), that $V(k') \neq \varnothing$.

Finally, we observe that k'/k is a succession of quadratic extensions. Now, by a classical argument, which consists of drawing the line through a quadratic point and its conjugate, and looking at the residual intersection with the cubic surface (see [2], Proposition 2.2), one proves:

LEMMA 2.3.   *Let ℓ/k be a quadratic extension. Then if a cubic form f, defined over k, represents zero in ℓ, it also represents zero in k.* $\square$

Thus from $V(k') \neq \varnothing$ one derives that $V(k(\rho)) \neq \varnothing$, and then $V(k) \neq \varnothing$. This completes the proof of the Theorem. $\square$

Remarks.

(a) The point Q, which appeared near the end of the proof, has a nice geometric interpretation. Namely, it is an *Eckardt point*: this means that Q lies on three (coplanar) lines of the surface. The orbit of Q under the action of $\mathscr{S}_5$ consists of the ten points with y-coordinates satisfying $y_i = y_j = y_k = 0$ (with i, j, k all distinct). It can be shown that these are precisely all the Eckardt points on V. This is the geometric origin of a k-rational zero-cycle of degree 10 on V. It is worth remarking that a general cubic surface does not contain any Eckardt point. This means that the cubics occurring in our problem are very special indeed.

(b) In characteristic 3, the equivalence between (H) and (H′) does not hold. Therefore, to prove assertion (H), one must work with the equations (1.8), instead of (1.9). By a slightly tedious computation, one shows that they define a smooth cubic surface. Then the reader will check that the argument of Theorem 2.2 carries through, with virtually no change. However, the verification becomes laborious at places, so we chose to forget about characteristic 3 in the statement. Another reason is that there is nothing to prove in case the field k is perfect. Indeed one has the following result, which was communicated to me by Andreas Dress:

LEMMA 2.4 (A. Dress). *Let k be a perfect field of characteristic* 3, *and* $f(x, y, z, t)$ *a cubic form with coefficients in k. Then f represents zero in k.*

Proof. Write $f(x, y, z, t) = at^3 + \ell(x, y, z)t^2 + q(x, y, z)t + c(x, y, z)$. There exists an extension k'/k, of degree 1 or 2, in which ℓ and q have a common solution (ξ, η, ζ). (If ℓ and q are non-trivial, consider the intersection of the line $V(\ell)$ with the conic $V(q)$ in $\mathbb{P}_k^2$.) Hence $f(\xi, \eta, \zeta, t) = at^3 + c(\xi, \eta, \zeta)$; without loss of generality, $a \neq 0$. Since k' is perfect of characteristic 3, one can find $\tau \in k'$ such that $a\tau^3 + c(\xi, \eta, \zeta) = 0$. Thus f represents zero in k', and hence also in k (by Lemma 2.3). $\qquad\square$

§3. The case of degree 6. In this section we establish:

THEOREM 3.1.   *If* $\mathrm{char}(k) \neq 2, 3$ *then assertion* (H) *holds for $d = 6$.*

Proof.   In view of Proposition 1.1, we prove assertion (H′). $V \subset \mathbb{P}_k^5$ is a cubic threefold, contained in the hyperplane $\mathbb{P}_k^4$ defined by $\mathrm{Tr}(x) = 0$. By Lemma 1.2 (and its proof), we know that V has precisely 10 double points, whose y-coordinates are $(1, 1, 1, -1, -1, -1)$ and permutations thereof. This set of 10 points is of course k-rational. Even if k is imperfect, the corresponding 0-cycle of degree 10 is k-rational (i.e., there is no inseparability), by virtue of Lemma 2.1. Thus it suffices to prove the following result:

PROPOSITION 3.2.   *Every cubic threefold $V \subset \mathbb{P}_k^4$ with precisely* 10 *double points* (*forming a k-rational* 0-*cycle of degree* 10) *has a k-point.*

Preliminary discussion.   We first quote a geometric result, which is not entirely satisfactory for our purpose, but gives some insight into what we are doing. In [9], Marletta described a large number of Cremona transformations of $\mathbb{P}^n$, where n is arbitrary. On specializing one of his results ([9], §2, no. 7, with $r = s + 1$ and $\ell = 0$), we obtain the following

LEMMA 3.3 (Marletta).   *Given a general $\mathbb{P}^{n-2}$ and ns general points in $\mathbb{P}^n$, there is a one-to-one correspondence between the lines of $\mathbb{P}^n$ and the rational curves of* *degree $(n - 1)s + 1$ passing through the ns points and meeting the assigned $\mathbb{P}^{n-2}$ in $(n - 1)s$ points.*

Since there is a unique line through two given points of $\mathbb{P}^n$, we also have the following

COROLLARY 3.4 (Marletta).   *Given a general $\mathbb{P}^{n-2}$ and $ns + 2$ general points in $\mathbb{P}^n$, there is one and only one rational curve of degree $(n - 1)s + 1$ that passes through the $ns + 2$ points and meets the assigned $\mathbb{P}^{n-2}$ in $(n - 1)s$ points.*

The relationship between this result and our problem is quite simple. For $n = 4$ and $s = 2$, Corollary 3.4 asserts the existence of a unique rational curve Γ of degree 7 which passes through a set of 10 points in general position in $\mathbb{P}^4$ and in addition meets an assigned $\mathbb{P}^2$ with multiplicity 6. For k perfect, if the set of 10 points is k-rational and the assigned $\mathbb{P}^2$ is defined over k, then Γ is also defined over k (because it coincides with its conjugates). This has the following application to Proposition 3.2: If the 10 double points are in sufficiently general position, there is a k-rational curve Γ of degree 7 and genus 0 through them. If $\Gamma \subset V$ then $V(k) \neq \varnothing$, since $\Gamma(k) \neq \varnothing$ (because Γ has genus 0 and odd degree). Otherwise, by the Bézout theorem, Γ meets V in $7.3 = 21$ points, counted with the relevant multiplicities. Now each double point of V counts for 2 intersections. Hence there is a residual cycle of $21 - 10.2 = 1$ point. This is a k-point of V. (We have used implicitly some of the lemmas in [2], §1.)

Although this argument is reasonably nice and intuitive, we cannot consider it to be a proof of Proposition 3.2. Indeed, first of all one would need to check the proof of Lemma 3.3. (In our special case it might be possible to give a direct construction for the curve Γ, which is the residual intersection of three cubic threefolds passing through the 10 given points and having the assigned $\mathbb{P}^2$ as a double plane.) More importantly, the 10 double points on V need not be in general position. (Even those corresponding to Theorem 3.1 are not! Indeed, the plane defined by $y_1 + y_2 = y_3 + y_4 = y_5 + y_6 = 0$ contains 4 double points of V.) In particular, Γ need not be irreducible, and the various degenerate cases would have to be handled by some specialization argument.

For all these reasons, we shall replace Corollary 3.4 by a purely algebraic argument, which generalizes a result of Cassels ([2], Proposition 3.1). However, in some sense this is just a rigorous way of constructing Marletta's curve in our context and of coping with the degenerate situations. (The corresponding $\mathbb{P}^2$ is defined by the vanishing of the first two coordinates.)

Proof of Proposition 3.2.   Let P be a double point of V. We denote by $k(P)$ its least field of definition over k. By assumption, $k(P)/k$ is separable, of degree $\delta \leq 10$. In what follows we consider each Galois orbit individually. This is because Proposition 3.5 below always yields an irreducible curve. Now, if eight of the double points lie in, and span, a hyperplane which does not contain the remaining two double points, then it is clearly impossible to find an irreducible curve of degree 7 through all ten points.

PROPOSITION 3.5. *Given a point $P \in \mathbb{P}^n$ such that $k(P)/k$ is a separable extension of degree δ, suppose μ is an integer satisfying:*

$$(3.1) \qquad \frac{n-1}{n}(\delta - 1) < \mu \leq \delta - 1.$$

Then there exists an absolutely irreducible k-rational curve $\Gamma_m \subset \mathbb{P}^n_k$, of degree $m \leq \mu$ and genus 0, that passes through P and all its conjugates. Moreover, $\Gamma_m(k) \neq \varnothing$.

Proof (argument of Cassels). We can choose coordinates so that P is of the form

$$(3.2) \qquad P = (1, \eta, \alpha_1, \ldots, \alpha_{n-1}), \quad \text{with } k(P) = k(\eta).$$

Indeed we may clearly assume that the first coordinate is equal to 1. Then, if $k(\eta) \neq k(P)$, it follows from the primitive element theorem that $k(P)$ is generated by an element η' of the form $\eta + c_1\alpha_1 + \cdots + c_{n-1}\alpha_{n-1}$, with $c_i \in k$. This defines a projective change of coordinates (over k) which replaces η by η'.

Let us write

$$(3.3) \qquad \alpha_i = \sum_{j=0}^{\delta-1} a_{ij}\eta^j,$$

with $a_{ij} \in k$. We now multiply all the coordinates of P by a polynomial

$$(3.4) \qquad A_0(\eta) = c_0 + c_1\eta + \cdots + c_{\mu-1}\eta^{\mu-1},$$

whose coefficients $c_i \in k$ will be determined later. Then, provided $A_0(\eta) \neq 0$, we have:

$$(3.5) \qquad P = \big(A_0(\eta), \eta A_0(\eta), A_1(\eta), \ldots, A_{n-1}(\eta)\big),$$

where

$$(3.6) \qquad A_i(\eta) = \alpha_i A_0(\eta) = \sum_{j=0}^{\delta-1} a'_{ij}\eta^j,$$

with $a'_{ij} \in k$. This relation defines $n-1$ new polynomials $A_i(t)$, whose coefficients, a'_{ij}, depend k-linearly and homogeneously on those of $A_0(t)$. In view of (3.1), we can therefore determine the coefficients c_i in such a way that the degree of each $A_i(t)$ does not exceed μ, since this amounts to solving (non-trivially) a linear system of $(n-1)(\delta - 1 - \mu) < \mu$ homogeneous equations in μ unknowns.

The image of the map $\varphi : \mathbb{P}^1_k \to \mathbb{P}^n_k$ defined (on an open subset of $\mathbb{P}^1_k$) by

$$(3.7) \qquad \varphi : t \mapsto \left(A_0(t), tA_0(t), A_1(t), \ldots, A_{n-1}(t) \right)$$

is an absolutely irreducible k-rational curve $\Gamma_m \subset \mathbb{P}^n_k$, of degree $m \leq \mu$ and genus 0. As $\mu < \delta$, we see from (3.4) that $A_0(\eta) \neq 0$. Hence P is given by (3.5), i.e., $P = \varphi(\eta)$. It follows that Γ_m passes through P and all its conjugates. Moreover, $\Gamma_m(k) \neq \varnothing$, since it contains points of the form $\varphi(c)$ for $c \in k$. $\qquad\square$

To prove Proposition 3.2, we shall apply this result to the double points of $V \subset \mathbb{P}^4_k$. Thus $n = 4$, $\delta \leq 10$, and μ will be the least integer satisfying

$$(3.8) \qquad\qquad 3(\delta - 1) < 4\mu.$$

There are various cases, which correspond to the various decompositions of the set of double points into orbits for the Galois action. If, for instance, there is only one orbit, we apply Proposition 3.5 with $\delta = 10$ and $\mu = 7$: there exists an absolutely irreducible curve Γ_m, of degree $m \leq 7$, that passes through all 10 points. If $m < 7$ then $\Gamma_m \subset V$, by the Bézout theorem, and since $\Gamma_m(k) \neq \varnothing$ we are done. Thus we can assume that $m = 7$ and $\Gamma_7 \not\subset V$. Then the Bézout theorem, applied as explained after Corollary 3.4, shows that V contains a residual k-rational 0-cycle of degree $7.3 - 10.2 = 1$, i.e., $V(k) \neq \varnothing$. (It could be argued that the double points need not count for 2 in the intersection $\Gamma_7.V$. However, if one of them counted for more than two intersections, then all of them would. Hence the degree of the residual 0-cycle would be negative, which is absurd, since we have assumed that $\Gamma_7 \not\subset V$.)

If, on the other hand, the Galois action is not transitive on the double points of V, it is easy to see that there is an orbit of order $\delta = 1, 2, 4, 5$ or 7. (For instance, if there is an orbit of order 6 then there is also one of order $\delta = 1, 2$ or 4.) Hence we may assume that δ takes on one of these values and apply Proposition 3.5 as above with μ chosen as in (3.8). The method being identical, we summarize all these cases (including $\delta = 10$) in the form of a table. The last column contains the degree $\rho = 3\mu - 2\delta$ of the residual cycle, provided $\Gamma \not\subset V$.

δ	μ	$\rho = 3\mu - 2\delta$
10	7	1
7	5	1
5	4	2
4	3	1
2	1	-1
1	—	—

For $\delta = 1$, the double point is already k-rational. For $\delta = 2$, one finds $\rho = -1$, which reflects the well-known fact that the line joining two double points lies on the cubic hypersurface. For $\delta = 5$, one finds a residual 0-cycle of degree 2, whence $V(k) \neq \varnothing$ by virtue of Lemma 2.3. In all other cases, the residual 0-cycle is of degree 1. This completes the proof of Proposition 3.2 and of Theorem 3.1. $\square$

Remark. With the same method one can prove:

PROPOSITION 3.6. *If a cubic threefold $V \subset \mathbb{P}_k^4$ contains a set of δ double points, forming a k-rational 0-cycle of degree δ prime to 3, then $V(k) \neq \varnothing$, provided $\delta \leq 11$.* $\square$

By considering the degree of the dual variety, one shows easily that $\delta \leq 12$ if V has only isolated singularities. In fact, it is asserted in the classical literature that $\delta \leq 10$, so that the maximum is attained on the Segre threefold (1.9).

§4. An incursion into higher degrees. Looking back on our proof of assertion (ℍ) for degree 5 or 6, we see that a common feature of the two cases is that we first discovered a point of degree prime to 3 and then derived from it the existence of a k-point on V. Therefore the proper perspective seems to be provided by the following *conjecture of Cassels and Swinnerton-Dyer*, which was analyzed in [2]:

(ℂ𝕊) *Let $V \subset \mathbb{P}_k^n$ be a cubic hypersurface defined over k. Suppose V contains a k-rational zero-cycle of degree δ prime to 3. Then $V(k) \neq \varnothing$.*

For $\delta = 2$, this is Lemma 2.3. (ℂ𝕊) is also known to hold in general for $n \leq 2$. For $n \geq 3$, very little is known about this conjecture, except in some special cases (cf. [2]). There is also a conditional result of Kanevsky ([7], Theorem 2) for the case $n = 3$ over $k = \mathbb{Q}$.

In §2 we succeeded in proving (ℂ𝕊) for the relevant smooth cubic surface because V contained a sextuplet defined over a quadratic extension of k. This situation had already been considered in [2], Proposition 8.1, with a different, perhaps more constructive, proof.

The case of singular cubic surfaces was also settled there ([2], Proposition 8.2). In §3 an adaptation of that argument enabled us to handle the case of a cubic threefold with 10 double points.

It is possible to go further by restricting the class of fields we consider. Indeed the following fact is established in [2], §4:

PROPOSITION 4.1. (ℂ𝕊) *holds (for arbitrary n) if k is a quasi-local field.*

A field is called *quasi-local* if it is complete with respect to a discrete valuation and has the property that (ℂ𝕊) holds over its residue field. As (ℂ𝕊) holds for all C_1 fields ([2], Corollary 2.4), it follows from Proposition 4.1 that the class of quasi-local fields includes all local fields, and also fields like $\mathbb{C}(t)((u_1))\ldots((u_s))$, $\mathbb{Q}_p((u_1))\ldots((u_s))$, $\mathbb{F}_p((u_1))\ldots((u_s))$, etc.

Note that Proposition 4.1 provides us with a new proof of assertion (ℍ) for $d = 5$ or 6 if k is quasi-local. It suffices to notice that V contains precisely 10 Eckardt points for $d = 5$, and 10 double points for $d = 6$. The rest follows from property ($C\mathbb{S}$). But, of course, the proofs we gave in §2 and §3 are valid for an arbitrary field. To conclude, we establish the following

THEOREM 4.2. *Assertion* ($\mathbb{H}$) *holds for $d = 7$ or 8 if the field k is quasi-local of characteristic not equal to 2 or 3.*

Proof. In view of Proposition 1.1, we prove assertion ($\mathbb{H}'$). The cubic $V \subset \mathbb{P}_k^{d-1}$ contains the point Q with y-coordinates $(1, 1, 1, -1, -1, -1, 0)$ if $d = 7$, respectively $(1, 1, 1, -1, -1, -1, 0, 0)$ if $d = 8$. This point is left invariant by the action of the subgroup of $\mathscr{S}_d$ generated by (123) and (456). Hence the orbit of Q under the action of the full group $\mathscr{S}_d$ has order prime to 3 (for it is a divisor of $8!/3^2$). Now, by Lemma 2.1, this is a k-rational 0-cycle. It follows from Proposition 4.1 that $V(k) \neq \varnothing$. $\qquad\square$

For $d = 8$ one may also observe that the double points of V form a k-rational 0-cycle of degree 35 (by Lemma 1.2). The conclusion follows again from Proposition 4.1. For the same reason, Theorem 4.2 holds also for $d = 18, 20, 24, 26, 54, 56$, etc. Indeed, by Lemma 1.2, the number of double points in these exotic cases is also prime to 3.

Added in proof: A. Brumer points out that Theorem 3.1 was also known to the nineteenth-century invariant theorists. In fact, it appears due to (Father) Joubert, "Sur l'équation du sixième degré," *C-R. Acad. Sc. Paris*, **64** (1867), 1025–1029, 1081–1085. See also H. W. Richmond, "Note on the invariants of a binary sextic"; *Quarterly J. of Maths*, **31** (1900), 57–59. Interestingly, Richmond devoted several papers to a geometric study of Segre's cubic threefold (1.9): *ibidem*, **31** (1900), 125–160, and **34** (1903), 117–154.

REFERENCES

1. J.-L. COLLIOT-THÉLÈNE, D. CORAY AND J.-J. SANSUC, *Descente et principe de Hasse pour certaines variétés rationnelles*, J. Crelle **320** (1980), 150–191.
2. D. F. CORAY, *Algebraic points on cubic hypersurfaces*, Acta Arithmetica **30** (1976), 267–296.
3. ______, *Points algébriques sur les surfaces de del Pezzo*, C. R. Acad. Sc. Paris **284** (1977), 1531–1534.
4. D. F. CORAY AND M. A. TSFASMAN, *Arithmetic on singular Del Pezzo surfaces*, to appear in Proc. London Math. Soc.
5. CH. HERMITE, *Sur l'invariant du dix-huitième ordre des formes du cinquième degré* (extrait de deux lettres à M. Borchardt), J. Crelle **59** (1861), 304–305 (= *œuvres*, **II**, 107–108).
6. D. HILBERT AND S. COHN-VOSSEN, *Anschauliche Geometrie*, Springer, Berlin, 1932.
7. D. KANEVSKY, *Application of the conjecture on the Manin obstruction to various diophantine problems*, in: Journées Arithmétiques *1985* (Besançon); to appear.
8. YU. I. MANIN, *Cubic Forms*, Nauka, Moscow, 1972; English translation: North-Holland, Amsterdam, 1974.
9. G. MARLETTA, *Alcuni sistemi omaloidici nell'S_n*, Rendiconti Circ. Mat. Palermo **49** (1925), 252–262.
10. B. SEGRE, *The Non-Singular Cubic Surfaces*, Clarendon Press, Oxford, 1942.
11. J.-P. SERRE, *Corps Locaux*, 2nd ed. Hermann, Paris, 1968.
12. B. L. VAN DER WAERDEN, *Algebra*, 7th ed. Springer, Berlin, Heidelberg, 1966.
13. ______, *Einführung in die algebraische Geometrie*, Springer, Berlin, 1939.

SECTION DE MATHÉMATIQUES, UNIVERSITÉ DE GENÈVE, 2-4, RUE DU LIÈVRE, CH-1211 GENÈVE 24, SWITZERLAND.

R-EQUIVALENCE ON CONIC BUNDLES OF DEGREE 4

J.-L. COLLIOT-THÉLÈNE and A. N. SKOROBOGATOV

To Yu. I. Manin on his fiftieth birthday

0. Introduction. Let k be a field of characteristic zero, $\bar{k}$ an algebraic closure of k, and $\mathfrak{g} = \mathrm{Gal}(\bar{k}/k)$. Let X/k be a smooth projective surface which admits a dominant k-morphism $p\colon X \to \mathbb{P}^1_k$ with smooth generic fibre of genus zero and with 4 geometric degenerate fibres, and assume that p is relatively minimal. Then any geometric degenerate fibre is the transverse intersection of two exceptional curves of the first kind ([10], 1.6) and the degree $(\omega_X \cdot \omega_X)$ of X is 4. According to Iskovskikh [7], [8] (see also [9], §3), two cases may occur: either the anticanonical bundle ω_X^{-1} is not ample, in which case X is a generalized Châtelet surface as soon as $X(k) \neq \varnothing$ (see [9]), or ω_X^{-1} is ample, in which case X is a k-minimal Del Pezzo surface of degree 4 with Pic X free of rank 2 (see [8] and [9]).

These last surfaces, as soon as the set $X(k)$ of their rational points is not empty, may also be characterized [9] as those smooth complete intersections of two quadrics $X \subset \mathbb{P}^4_k$ which are given in homogeneous coordinates by a system of two quadratic forms with coefficients in k of the following type:

$$\begin{cases} f(x_0, \ldots, x_4) = 0, \\ x_1 x_2 + x_3 x_4 = 0. \end{cases}$$

In Manin's description of the possible actions of $\mathfrak{g}$ on the 16 lines of a Del Pezzo surface of degree 4 ([10]; [12], 2nd edition, p. 178), they correspond to those actions such that no orbit crosses the middle vertical line in Table 2 (see [8] and [9]).

In descent theory over a given smooth proper rational k-surface X [3], [4], which was developed after work of F. Châtelet [1] and Yu. I. Manin ([12], chap. VI), J.-J. Sansuc and the first named author have raised two basic questions, an affirmative answer to which would solve most natural arithmetico-geometric problems concerning X:

(Q1) *If $\mathcal{T}$ is a universal torsor over X with $\mathcal{T}(k) \neq \varnothing$, is $\mathcal{T}$ a k-rational variety?*

(Q2) *If k is a number field and $\mathcal{T}$ a universal torsor over X, does $\mathcal{T}$ satisfy the Hasse principle?*

In [6], an affirmative answer was given to both questions when X is (a suitable model of) a generalized Châtelet surface, i.e. when X belongs to the first

Received January 22, 1987.

nontrivial class of rational surfaces from the point of view of the Enriques-Manin-Iskovskih classification ([10], [11], [8]). It is the purpose of this note to show that for the next class of nontrivial rational surfaces, question **(Q1)** still has a positive answer:

THEOREM 1. *Any universal torsor $\mathcal{T}$ with $\mathcal{T}(k) \neq \varnothing$ over a relatively minimal conic bundle $X/\mathbb{P}^1_k$ of degree 4 is a k-rational variety.*

For simplicity, we give the proof when $\mathrm{char}(k) = 0$, but the proof given works for any perfect field k with $\mathrm{char}(k) \neq 2$ and which is not too small. We refer the reader to [3], §5, and [4], §2.8 and §3.8, for the list of consequences which such a theorem has—and which can be neatly formulated when $\mathrm{char}(k) = 0$. We shall only repeat two of these consequences.

THEOREM 2. *For X over k as above with a k-rational point O, the natural map $X(k)/R \to A_0(X)$ from the set of R-equivalence classes to the Chow group of 0-cycles of degree 0 on X modulo rational equivalence, induced by $P \to (P - O)$, is a bijection.*

That the map is surjective was proved in [2]: it depends on the fact that p has less than 5 geometric degenerate fibres. The injectivity is a general consequence of Theorem 1.

THEOREM 3. *For X over k as above, each R-equivalence class is parametrized by the k-points of one smooth k-rational variety, and if k is finitely generated (as a field) over $\mathbf{Q}$, there are only finitely many R-equivalence classes on $X(k)$.*

When k is a number field, the finiteness of $X(k)/R$ in the last theorem was already obtained in [5], using some results of [6].

Let us point out that the paper of Salberger [14], which builds upon [6], may be interpreted as an affirmative answer to question **(Q2)** for all conic bundles X over $\mathbb{P}^1_k$ of degree 4 which are Del Pezzo surfaces of degree 4 and are not of type (XV) in Manin's classification ([12], p. 178).

The reader is referred to [3] and [4] for the basic concepts and results of descent theory on rational varieties which we shall freely use, in particular for the definition of universal torsors and the "local description" of these torsors. He is advised to consult the survey [13] for a comprehensive overview.

1. Complement to a previous paper. Let $p\colon X \to \mathbb{P}^1_k$ be a relatively minimal conic bundle. In [5] and [4] §2.6, down-to-earth equations for the universal torsors over X were given. In this section, we recall a few of the notations and results of [5], and we complement them in a few places, which are essential for the argument given in §2.

We shall assume that the fibre at infinity of $X/\mathbb{P}^1_k$ is smooth, and that it contains a k-point (hence is k-isomorphic to $\mathbb{P}^1_k$). The degeneracy locus of the fibration can be described as $Y = \mathrm{spec}\, A \subset \mathbf{A}^1_k = \mathrm{Spec}\, k[x]$ for A a finite étale k-algebra $A = k[x]/P(x) = k[\theta]$ of rank r. There is a finite free A-algebra A'

of rank 2, which may be described by $A' = A[t]/(t^2 - a)$ for a suitable unit $a(\theta) \in A^*$, which describes the local quadratic extensions which prevent the fibration $X/\mathbb{P}^1_k$ from being smooth at the points of Y. Let U_0 be the complement of Y in $\mathbb{A}^1_k$ and let $U = p^{-1}(U_0)$. Let $\hat{S}$ be the kernel of the restriction map $\hat{S}_0 = \operatorname{Pic} \overline{X} \to \operatorname{Pic} \overline{U} = \mathbb{Z}$, and let λ denote the inclusion $\hat{S} \subset \hat{S}_0$.

Let $W \subset \mathbb{A}^2_k \times_k R_{A/k} \mathbb{A}^2_k \ (\simeq \mathbb{A}^{2r+2}_k)$ be the k-variety defined by the equation

$$
(1) \qquad u - \theta v = \left(u_0 + u_1\theta + \cdots + u_{r-1}\theta^{r-1}\right)^2
$$

$$
- a(\theta)\left(v_0 + v_1\theta + \cdots + v_{r-1}\theta^{r-1}\right)^2.
$$

Note that this variety is a geometrically integral complete intersection of $(r - 2)$ quadratic cones in affine space $\mathbb{A}^{2r}_k$ (with coordinates u_i, v_i), which is the affine cone over a geometrically integral nonconical complete intersection of $(r - 2)$ quadrics $V \subset \mathbb{P}^{2r-1}_k$.

PROPOSITION 1. *The universal torsor over X whose fibre at the k-points at infinity is trivial is k-birational to $\mathbb{A}^2_k \times_k W$, hence to $\mathbb{A}^3_k \times_k V$.*

Proof. Let $\mathscr{T}$ be the universal torsor on X with trivial fibre at any of the k-points of X at infinity. Applying the change of structural group $S_0 \to S$ defined by λ we get a torsor $\mathscr{T}_1$ over X under S. This torsor is of type λ, has a trivial fibre at the k-points of X at infinity and $\mathscr{T}$ is a torsor over $\mathscr{T}_1$ under $\mathbf{G}_m$ hence $\mathscr{T}$ is k-birational to $\mathscr{T}_1 \times_k \mathbf{G}_m$. Given $\alpha(\theta) \in A^*$, let W_α be the k-variety defined in $\mathbb{A}^2_k \times_k R_{A/k} \mathbb{A}^2_k \ (\simeq \mathbb{A}^{2r+2}_k)$ by the equation

$$
(2) \qquad \alpha(\theta)(u - \theta v) = \left(u_0 + u_1\theta + \cdots + u_{r-1}\theta^{r-1}\right)^2
$$

$$
- a(\theta)\left(v_0 + v_1\theta + \cdots + v_{r-1}\theta^{r-1}\right)^2.
$$

As explained in [5] and [4], just as for any torsor of type $\lambda: S \to S_0$, the restriction of $\mathscr{T}_1$ to the open set U of X is given by a diagram of the following type:

$$
\begin{array}{ccc}
\mathscr{T}_{1,U} \longrightarrow & W_0 \longrightarrow & M \\
\downarrow & \downarrow & \downarrow \\
U \ \xrightarrow{\ p\ } & U_0 \longrightarrow & R,
\end{array}
$$

where the two squares are fibre products, where $M \to R$ is the map

$$
\mathbf{G}_{m,k} \times_k R_{A'/k}\mathbf{G}_m \to R_{A/k}\mathbf{G}_m
$$

$$
\left(t, X + \sqrt{a(\theta)}\,Y\right) \to t^{-1} \cdot \left(X^2 - a(\theta)Y^2\right),
$$

(X and Y being "variables" in A), whose kernel is the k-torus S, and where the map $U_0 \to R$ is given by the function $\alpha(\theta) \cdot (x - \theta)$, for some $\alpha(\theta) \in A^*$ whose class in $A^*/k^* \cdot N_{A'/A} A'^*$ only depends on the choice of $\mathscr{T}_1$ among torsors of type λ.

Now W_0 is simply a nonempty open set of the variety W_α. In order to prove the proposition, it is enough to show that the hypothesis that $\mathscr{T}_1$ has trivial fibre at infinity implies the two facts:

(a) The conic fibration $\mathscr{T}_{1,U} \to W_0$ admits a section over an open set of W_0.

(b) In the above description, the class of $\alpha \in A^*/k^* \cdot N_{A'/A} A'^*$ is 1.

Let $U_1 \subset \mathbb{P}^1_k$ be the open set $\mathbb{P}^1_k - \{Y \cup 0\}$, let $y = 1/x$ be the usual parameter for $\mathbb{A}^1_k = \mathbb{P}^1_k - \{0\}$, and consider the torsor $\mathscr{T}_2$ under S over $V = p^{-1}(U_1)$ given by the diagram of fibre products:

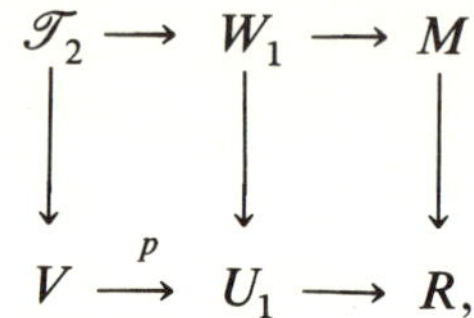

where the map $U_1 \to R$ is given by the function $\alpha(\theta) \cdot (1 - \theta y)$. An obvious change of variables reveals that the restrictions of $\mathscr{T}_1$ and $\mathscr{T}_2$ to $U \cap V$ are isomorphic torsors under S. But two torsors under a k-torus S over a smooth k-variety Z which are isomorphic when restricted to some open set of Z have isomorphic restriction to any local ring of this variety hence have equal fibres in $H^1(k, S)$ over any k-point of Z (see [4], 2.7.5). By definition, the fibre of $\mathscr{T}_1$ over a k-point of X at infinity is equal to 1. But the fibre of $\mathscr{T}_2$ over any k-point of V with $y = 0$ is clearly equal to the class of $\alpha(\theta)$ in $A^*/k^* \cdot N_{A'/A} A'^* = H^1(k, S)$. Thus $\alpha(\theta)$ in (2) may be taken equal to 1, which proves (b).

The same computation also gives (a). Indeed, [5] and [4] (proof of Theorem 2.6.4 and Remark 2.6.8) show that the class say ξ of the generic fibre of the conic fibration $\mathscr{T}_{1,U} \to W_0$ in the Brauer group $\operatorname{Br} k(W_0)$ of the function field $k(W_0)$ of W_0 comes from the Brauer group of k. Hence so does the generic fibre of the smooth conic fibration $\mathscr{T}_2 \to W_1$. But we have just seen that the fibres of this fibration over the k-points of W_1 which project down to the point at infinity of $\mathbb{P}^1_k$ are trivial. A specialization argument then shows $\xi = 0 \in \operatorname{Br} k(W_0)$, which implies that the conic fibration admits a section over an open set.

Remark. When $X/\mathbb{P}^1_k$ is a generalized Châtelet surface, as described in §7 of [6], a universal torsor $\mathscr{T}$ over X with $\mathscr{T}(k) \neq \varnothing$ is a k-rational variety (loc. cit., theorem 8.1). A slightly simpler proof has since been shown to us by P. Salberger. With the same notations as above, we here have $r = 4$ and $a(\theta) = a \in k^*$. The main theorem of [5] accounts for the change of variables in [6] which shows that $\mathscr{T}$ is k-birational to $\mathbb{P}^3_k \times_k V$, where V is a geometrically integral nonconical complete intersection of two quadrics in $\mathbb{P}^7_k$ which contains two skew linear spaces of dimension 3, Π_1 and Π_2, each defined over $k(\sqrt{a})$ and conjugate (Π_1 is defined by the equations $u_i = \sqrt{a}\,v_i$, $i = 0,\ldots,3$, and Π_2 by $u_i = -\sqrt{a}\,v_i$, $i = 0,\ldots,3$). Also, $\overline{V} = V \times_k \overline{k}$ contains exactly 8 singular points. For all this, see [6], Theorem 7.1. The hypothesis $\mathcal{T}(k) \neq \varnothing$ and a simple lemma imply $V(k) \neq \varnothing$. Let M be in $V(k)$.

Salberger's nice remark is that the line L which is the intersection of the 4-dimensional linear spans $\{M, \Pi_1\}$ and $\{M, \Pi_2\}$, which is clearly defined over k, lies on V. It is enough to show that L lies on any quadric Q which contains V, and this follows from the fact that Q contains at least three points of L: the point $\{M, \Pi_1\} \cap \Pi_2$, the point $\{M, \Pi_2\} \cap \Pi_1$ and the point M itself. Now V contains the k-rational line L, which does not belong entirely to the singular locus, hence V is a k-rational variety (cf. [6], Prop. 2.2).

2. The proof of Theorem 1.

Let $p\colon X \to \mathbb{P}^1_k$ be a relatively minimal conic bundle of degree 4 with $X(k) \neq \varnothing$. In order to prove Theorem 1, we shall only consider the case where X is also a Del Pezzo surface of degree 4, since the case of Châtelet surfaces is handled in [6] and in the above remark. According to Iskovskih [8], X may be given a second conic fibration structure $q\colon X \to \mathbb{P}^1_k$, which is also k-minimal (actually, X itself is a k-minimal surface) and if F_1, resp. F_2 are geometric fibres of p, resp. q, their intersection number $(F_1 \cdot F_2)$ equals 2. Also, the 16 lines on the Del Pezzo surface of degree 4 $\overline{X} = X \times_k \overline{k}$ are none but the components of the 4 degenerate geometric fibres of each of the two fibrations.

Let now $\mathcal{T}$ be a universal torsor over X with $\mathcal{T}(k) \neq \varnothing$, and let M be a k-point in $X(k)$ which belongs to the projection of $\mathcal{T}(k)$ under the structural map.

The fibres $p^{-1}(p(M))$ and $q^{-1}(q(M))$ cannot both be degenerate: since both fibrations p and q are k-minimal, M would lie simultaneously on 4 lines of $\overline{X}$, and this is impossible for a Del Pezzo surface of degree 4. Changing p and q if need be, we may assume that the fibre $F_1 = p^{-1}(p(M))$ is a smooth curve, hence isomorphic to $\mathbb{P}^1_k$. Because k is infinite (actually, only small finite fields would be a problem), we may replace M by another k-point N on F_1 so that the fibre $F_2 = q^{-1}(q(N))$ is smooth, does not pass through the 4 geometric singular points of the degenerate fibres of the fibration p, and intersects F_1 in N and in a different k-point.

Performing a suitable k-automorphism of the basis of the first fibration, we find that there lies on X the smooth proper curve $C = F_2$ k-isomorphic to $\mathbb{P}^1_k$, which the map p makes into a double cover of $\mathbb{P}^1_k$, and which does not pass through the singular points of the degenerate fibres of p, and such that the (smooth) fibre of p at infinity intersects C in two distinct k-rational points. Note that the universal torsor $\mathcal{T}$ on X, which has trivial fibre at M, also has trivial fibre at N, since these two k-points are R-equivalent ([4], 2.7.2).

After another suitable change of variables, this implies that the restriction of X over $\mathbf{A}^1_k = \operatorname{Spec} k[x]$ splits when going over to the cover

$$\operatorname{Spec} k[x][y]/(y^2 - x^2 - ax - b)$$

for suitable $a, b \in k$. That the coefficient of x^2 may be chosen to be 1 follows from the fact that $C \cap p^{-1}(\infty)$ consists of two k-rational points.

The above choices now imply that the element $a(\theta)$ which defines the extension A'/A may be taken equal to $\theta^2 + a\theta + b$. Thus the equation (1) of the k-variety $W \subset \mathbf{A}^{10}_k$ here reads:

$$(3) \qquad u - \theta v = \left(u_0 + u_1\theta + u_2\theta^2 + u_3\theta^3\right)^2$$

$$- (\theta^2 + a\theta + b)\left(v_0 + v_1\theta + v_2\theta^2 + v_3\theta^3\right)^2.$$

This k-variety contains the k-variety defined by:

$$(4) \qquad u_2 = u_3 = v_1 = v_2 = v_3 = 0, \qquad u_1 = v_0,$$

$$u = u_0^2 - bv_0^2, \qquad v = av_0^2 - 2u_0u_1.$$

Now the projection of $\mathbf{A}^{10}_k$ onto the affine space $\mathbf{A}^8_k$ which forgets the coordinates u and v induces a k-isomorphism of W with an intersection of two quadrics in $\mathbf{A}^8_k$, which is itself a cone over a complete intersection Z of two quadrics in $\mathbf{P}^7_k$, whose equations are immediately deduced from (3). It now follows from (4) that Z contains the k-line given by

$$(5) \qquad u_2 = u_3 = v_1 = v_2 = v_3 = 0, \qquad u_1 = v_0,$$

and one checks that this line is not contained in the singular locus of Z. It now follows from [6], Prop. 2.2. that Z is a k-rational variety, hence also W, hence finally the universal torsor $\mathscr{T}$ over X with trivial fibre at the k-points of X at infinity.

The work for this paper was done while the first author was staying in Moscow on an exchange between C.N.R.S. (France) and the Academy of Sciences of the U.S.S.R. Both authors warmly thank M. A. Tsfasman for many discussions. In the first version of this paper, we only proved k-rationality for a specific universal torsor $\mathscr{T}$ on X, and then deduced stable rationality for the other ones by using the behaviour of universal torsors under k-birational transformations ([4], §2.9) and the many k-birational automorphisms of a Del Pezzo surface of degree 4. We are grateful to P. Salberger for pointing out that our arguments actually give k-rationality for any universal torsor $\mathscr{T}$ with $\mathscr{T}(k) \neq \varnothing$.

REFERENCES

1. F. CHÂTELET, *Points rationnels sur certaines courbes et surfaces cubiques*, Enseignement Math. (2) **5** (1959), 153–170.

2. J.-L. COLLIOT-THÉLÈNE ET D. CORAY, *L'équivalence rationnelle sur les points fermés des surfaces rationnelles fibrées en coniques*, Compositio Math. **39** (1979) 301–332.

3. J.-L. COLLIOT-THÉLÈNE ET J.-J. SANSUC, "La descente sur les variétés rationnelles," in: *Journées de géométrie algébrique d'Angers* 1979, éd. A. Beauville, Sijthoff and Noordhoff, Alphen aan den Rijn 1980, 223–237.

4. ______, *La descente sur les variétés rationnelles*, II, this volume, 375–492.

5. ______, *La descente sur les surfaces rationnelles fibrées en coniques*, C. R. Acad. Sci. Paris **303** (1986), Série 1, n∘7, 303–306.

6. J.-L. COLLIOT-THÉLÈNE, J.-J. SANSUC AND SIR PETER SWINNERTON-DYER, *Intersections of two quadrics and Châtelet surfaces*, I, J. für die reine und ang. Math. **373** (1987) 37–107; II, ibid. **374** (1987), 72–168.

7. V. A. ISKOVSKIH, *Birational properties of surfaces of degree* 4 *in* $\mathbb{P}_k^4$, Mat. Sb. (N.S.), **88** (1972), 31–37 = Math. USSR-Sb. **17** (1972), 575–577.

8. ______, *Minimal models of rational surfaces over arbitrary fields*, Izv. Akad. Nauk SSSR Ser. Mat. **43** (1979), 19–43 = Math. USSR-Izv. **14** (1980), 17–39.

9. B. È. KUNYAVSKIĬ, A. N. SKOROBOGATOV AND M. A. TSFASMAN, *Del Pezzo surfaces of degree* 4. Preprint deposited at VINITI, 24.09.86, N 6840-86.

10. YU. I. MANIN, *Rational surfaces over perfect fields*, Publ. Math. I.H.E.S. **30** (1966), 55–113 = Transl. Amer. Math. Soc. (2) **84** (1969), 137–186.

11. ______, *Rational surfaces over perfect fields*, II. Mat. Sb. (N.S.), **72** (114) 1967), 161–192 = Math. USSR-Sb. **1** (1967), 141–168.

12. ______, *Cubic forms*: *algebra, geometry, arithmetic*, Nauka, Moscou 1972, engl. transl. North-Holland, Amsterdam-London, 1974, 2nd ed. 1986.

13. YU. I. MANIN AND M. A. TSFASMAN, *Rational varieties*: *algebra, geometry, arithmetic*, Uspekhi Mat. Nauk **41** (1986), 43–94 = Russian Math. Surveys **41** (1986), 51–116.

14. P. SALBERGER, "On the arithmetic of conic bundle surfaces," in *Séminaire de théorie des nombres de Paris*, 1985–1986 (see also C. R. Acad. Sci. Paris Série I, **303** (1986) 273–276).

J.-L. COLLIOT-THÉLÈNE, C.N.R.S., MATHÉMATIQUES, BÂT. 425, UNIVERSITÉ DE PARIS-SUD, F-91405 ORSAY, FRANCE

A. N. SKOROBOGATOV, INSTITUTE FOR PROBLEMS OF INFORMATION TRANSMISSION, ACADEMY OF SCIENCES OF THE USSR, 19, ERMOLOVOY UL., MOSCOW 101447, USSR

Vol. 54, No. 2   DUKE MATHEMATICAL JOURNAL ©   1987

NOTES ON MOTIVIC COHOMOLOGY

A. BEILINSON, R. MACPHERSON, AND V. SCHECHTMAN

Dedicated to Yu. I. Manin on the occasion of his 50th birthday

Introduction. Imagine a world in which the K-theory $K^*(X)$ of a topological space X had been defined, but the ordinary cohomology groups $H^*(X)$ had not yet been discovered. Then the rational cohomology groups $H^*(X, \mathbf{Q})$ would be known at least as functors, since they could be defined as the associated graded of the Atiyah–Hirzebruch filtration of $K^*(X) \otimes \mathbf{Q}$. However, there would be at least three difficulties in the situation:

1. This world would not have our explicit cocycles representing cohomology classes. These are often interesting geometric objects like differential forms or Cech cocycles which relate cohomology to other mathematical ideas.

2. The integral cohomology groups would not be defined.

3. The powerful computational techniques of cohomology theory would not be available.

In such circumstances, one might well expect to find a quest going on for a geometrically defined cohomology theory which, when tensored with the rationals, would admit a Chern character isomorphism from $K^*(X) \otimes \mathbf{Q}$.

Present day algebraic geometry is such a world. The algebraic K groups $K_i(X)$ of an algebraic variety X have been known since 1973 [Q], but a cohomology theory which is appropriately related to these algebraic K-groups has not been found. Precise conjectures as to what properties this hoped-for cohomology theory should have were given in [B2] for the Zariski topology, and then in [L1] for the étale topology. In a sense the quest for the hoped-for cohomology theory dates back to Grothendieck who, working before algebraic K-theory was defined, was looking for a cohomology theory of algebraic varieties with certain universal properties. (We know about Grothendieck's ideas through the work of Manin [M].) Following Grothendieck, we call the hoped-for cohomology theory *motivic cohomology*. (Lichtenbaum calls it arithmetic cohomology in honor of its conjectured relations with number theory [B1], [B2], [L1], [L2], which constitute one of the main reasons for seeking it.)

In this introduction, we give some of the most basic expected properties of motivic cohomology, based on its conjectured connection with algebraic K-theory. For a more detailed account, see [B2].

0.1 *The γ-filtration.* The analogue for algebraic K-theory of the Atiyah-Hirzebruch filtration on topological K-theory is the γ-filtration [So]. Consider the

Received October 27, 1986.

associated graded groups $\mathrm{gr}_{\gamma}^{p}K_{j}(X)$. We abbreviate the rational version $\mathrm{gr}^{p}K_{j}(X) \otimes \mathbb{Q}$ to $\mathrm{gr}^{p}K_{j}$ and display these groups in an array:

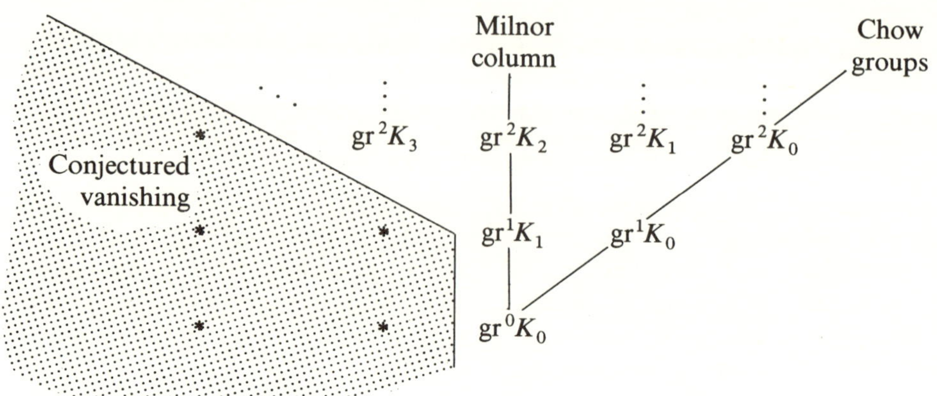

We recall some properties of the γ-filtration in terms of this array:

1. The groups on diagonal on the right are the rational Chow groups of codimension p algebraic cycles modulo rational equivalence—$\mathrm{gr}^{p}K_{0}(X) \otimes \mathbb{Q} = \mathrm{CH}^{p}(X) \otimes \mathbb{Q}$.

2. The groups in the shaded region are conjectured to be zero [B1] and independently [So].

3. If X is the spectrum of a local ring, then the groups to the right of the central vertical column vanish and the groups on the central vertical column are the Milnor K-groups [M]—$\mathrm{gr}^{p}K_{p}(X) \otimes \mathbb{Q} = K_{p}^{M}(X)$. So groups in this Milnor column have clear descriptions in terms of symbols.

4. The groups to the left of the Milnor column are more mysterious; it is difficult even to find invariants or "higher regulators" that do not vanish on them. Higher regulators for $\mathrm{gr}^{2}K_{3}$ involve the dilogarithm function. It is widely conjectured that regulators for the column p units to the left of the Milnor column will involve the p-logarithm function.

0.2 *Motivic cohomology.* Following the motivation from topology which we began with, we expect the motivic cohomology of X tensored with the rationals to be the groups $\mathrm{gr}^{p}K_{j}(X) \otimes \mathbb{Q}$. Since these groups are indexed by two integers (p and j), the same must hold for the motivic cohomology groups. (In this way, the situation in algebraic geometry is more complicated than that in topology, where cohomology is graded by a single integer. The reason is that topological K-theory is only $\mathbb{Z}/2\mathbb{Z}$ graded while algebraic K-theory is $\mathbb{Z}$-graded.) We symbolize the hoped-for motivic cohomology groups by the notation $H_{\mathcal{M}}^{i}(X, \mathbb{Z}(p))$ and its rational version by $H_{\mathcal{M}}^{i}(X, \mathbb{Q}(p)) = H_{\mathcal{M}}^{i}(X, \mathbb{Z}(p)) \otimes \mathbb{Q}$.

The indices i and p are chosen to satisfy $H^i_{\mathcal{M}}(X, \mathbb{Q}(p)) \cong \mathrm{gr}^p_\gamma K_{2p-i}(X) \otimes \mathbb{Q}$ so the rational motivic cohomology groups fit into an array isomorphic to the one above:

$$
\begin{array}{cccc}
H^1_{\mathcal{M}}(X, \mathbb{Q}(2)) & H^2_{\mathcal{M}}(X, \mathbb{Q}(2)) & H^3_{\mathcal{M}}(X, \mathbb{Q}(2)) & H^4_{\mathcal{M}}(X, \mathbb{Q}(2)) \\[2mm]
H^1_{\mathcal{M}}(X, \mathbb{Q}(1)) & H^2_{\mathcal{M}}(X, \mathbb{Q}(1)) & & \\[2mm]
H^0_{\mathcal{M}}(X, \mathbb{Q}(0)) & & &
\end{array}
$$

For fixed p, the groups $H^i_{\mathcal{M}}(X, \mathbb{Z}(p))$ should be receivers of a generalized pth Chern class. More specifically, there should be Chern classes $c^p_j : K_j(X) \to H^{2p-j}_{\mathcal{M}}(X, \mathbb{Z}(p))$. The usual formulas, using a conjectured multiplication $H^i_{\mathcal{M}}(X, \mathbb{Z}(p)) \otimes H^{i'}_{\mathcal{M}}(X, \mathbb{Z}(p')) \to H^{i+i'}_{\mathcal{M}}(X, \mathbb{Z}(p + p'))$, give a Chern character

$$
\mathrm{Ch}\colon K_j(X) \to \bigoplus_p H^{2p-j}_{\mathcal{M}}(X, \mathbb{Q}(p))
$$

This Chern character should induce the isomorphisms $H^i_{\mathcal{M}}(X, \mathbb{Q}(p)) \cong \mathrm{gr}^p_\gamma K_{2p-i}(X) \otimes \mathbb{Q}$.

For fixed i, the motivic cohomology groups $H^i_{\mathcal{M}}(X, \mathbb{Z}(p))$ should be refinements of the usual cohomology groups $H^i(X)$—there should be cycle maps $H^i_{\mathcal{M}}(X, \mathbb{Z}(p)) \to H^i(X)$. (Here $H^i(X)$ can be taken to be any of the usual cohomology theories. For example, if X is a variety over the complex numbers, $H^i(X)$ could mean singular cohomology.) The direct sum $\bigoplus_p H^i_{\mathcal{M}}(X, \mathbb{Z}(p))$ should be thought of as universal with respect to refinements of $H^i(X)$: it should map to other ones. For example the Deligne cohomology groups are one such refinement, and the cycle map should factor

$$
H^i_{\mathcal{M}}(X, \mathbb{Z}(p)) \to H^i_{\mathcal{D}}(X, \mathbb{Z}(p)) \to H^i(X)
$$

where $H^i_{\mathcal{D}}(X, \mathbb{Z}(p))$ represents the Deligne cohomology of X [B1].

0.3 *Motivic sheaves.* Conjecturally, for fixed p the motivic cohomology groups $H^i_{\mathcal{M}}(X, \mathbb{Z}(p))$ should be the hypercohomology groups of X with coefficients in a complex of sheaves $\mathbb{Z}(p)^{\cdot}_{\mathcal{M}}$. We can view these complexes on the same array as before:

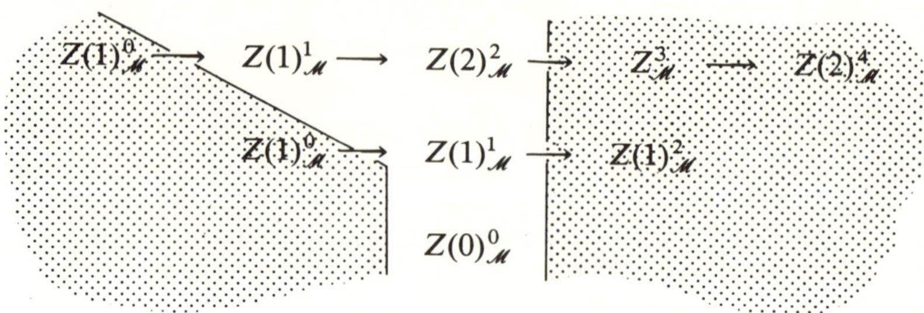

The stalk cohomology of these complexes should be zero in the shaded region. (The stalk cohomology at a point $x \notin X$, when tensored with $\mathbf{Q}$, should be the same as the similarly placed $\mathrm{gr}^p K_j$ of X localized at x. So this region is consistent with properties 2 and 3 of the γ-filtration above.) Therefore, up to quasi-isomorphism, the complex $\mathbf{Z}(p)^{\cdot}_{\mathcal{M}}$ can be chosen to be zero in the shaded region. The first three complexes should be as follows:

$\mathbf{Z}(0)_{\mathcal{M}} \cong \mathbf{Z}$, the constant sheaf, placed in degree 0.

$\mathbf{Z}(1)_{\mathcal{M}} \cong \mathcal{O}^*$, placed in degree 1, so that $\mathbf{Z}(1)_{\mathcal{M}}$ is quasi-isomorphic to the Deligne cohomology sheaf $\mathbf{Z}(1)_{\mathcal{D}}$ [B1].

$\mathbf{Z}(2)_{\mathcal{M}}$ is the Bloch-Suslin complex [Bl1], [S1]: $\mathbf{Z}\{\mathcal{O}^{**}\}/R \xrightarrow{d} \wedge^2 \mathcal{O}^*$ placed in degrees 1 and 2, where $\wedge^2 \mathcal{O}^*$ is the exterior power considering $\mathcal{O}^*$ as a $\mathbf{Z}$-module with multiplication instead of addition, $\mathcal{O}^{**}$ is the elements of $\mathcal{O}^*$ which do not map to 1 in the residue field, d takes t to $t \wedge (1 - t)$, and R is the subgroup generated by the expressions

$$\left(x - y + \frac{y}{x} - \frac{y-1}{x-1} + \frac{x(y-1)}{y(x-1)} \right)$$

occurring in the functional equation of the dilogarithm. It is known that this complex calculates K_2 and the non-Milnor part of K_3 for a local ring.

0.4 *Contents of this paper.* In this paper, we give five different but related geometric constructions, each of which we conjecture is a description of motivic cohomology (§1.1.5, §1.1.7, §1.4.4, §2.2.4, §2.3.1). These fall into two basic approaches, corresponding to the two parts of the paper. In chapter 1, we construct candidates for the complexes $\mathbf{Z}(p)_{\mathcal{M}}$ by representing elements of the complexes as maps into Grassmann manifolds. This has the merit that it explains the relation to the polylogarithm functions. In chapter 2, we construct a candidate for $H^i_{\mathcal{M}}(X, \mathbf{Z}(p))$ as an ext group in a category of Tate motives. This has the merit that the vanishing corresponding to property 2 of the γ-filtration is naturally evident. The reason for incorporating both chapters in one paper is that they both use elementary geometry of linear subspaces of projective space as the primary tool.

There is also another construction: the beautiful higher Chow groups of Bloch and Landsburg [Bl2], [La]. They conjecture, and we believe, that the appropriately sheafified higher Chow complexes are the motivic complexes $\mathbb{Z}(p)_{\mathcal{M}}$. The cochains for higher Chow groups are of the complexity of all algebraic subvarieties of projective space, rather than all linear ones.

We assume that all algebraic varieties considered are nonsingular. We have no idea how to handle the singular case. Complete proofs of the theorems announced in this paper will appear elsewhere.

We would like to thank J.-L. Brylinski, C. DeConcini, P. Deligne, B. Feigin, W. Fulton, I. M. Gelfand, W. Gerdes, M. Goresky, R. Hain, P. Loday, M. Hanamura, A. Nicas, C. Procesi, D. Ramakrishnan, C. Soulé, A. Suslin, and B. Yusin for useful and stimulating conversations on material related to this paper.

CHAPTER 1. GRASSMANNIAN HOMOLOGY

In the projective space $\mathbb{P}^{p+q}$ there is a "simplex" consisting of the $p + q + 1$ coordinate hyperplanes defined by the vanishing of one of the homogeneous coordinates. Let $\hat{G}_q^p$ denote the open subset of the Grassmannian of q-dimensional subspaces of $\mathbb{P}^{p+q}$ which are in transverse to this coordinate simplex. We call this the *generic part* of the Grassmannian. There are $p + q + 1$ maps $A_i: \hat{G}_q^p \to \hat{G}_{q-1}^p$, $i = 0, \ldots, p + q$, defined by intersecting with the ith coordinate hyperplane. These maps fit into an array called the *cosimplicial Grassmannian* $\hat{G}^p$

$$
\cdots \hat{G}_2^p \underset{A_{p+2}}{\overset{A_0}{\rightrightarrows}} \hat{G}_1^p \underset{A_{p+1}}{\overset{A_0}{\rightrightarrows}} \hat{G}_0^p
$$

If X is an algebraic variety, a complex of sheaves on X called the *Grassmannian complex* $^0\mathbf{GC}^p$ is constructed as follows: For a point $x \in X$, the stalk of $^0\mathbf{GC}_q^p$ at x is the formal linear combinations of germs at x of maps from X to $\hat{G}_q^p$. The differential from $^0\mathbf{GC}_q^p$ to $^0\mathbf{GC}_{q-1}^p$ is the alternating sum of the maps induced by composition with A_i. The hyperhomology of the Grassmannian complex is the *Grassmannian homology* of X. After appropriate truncation, $^0\mathbf{GC}^p/^0\mathbf{GC}^{p-2}$ is our first candidate for $\mathbb{Z}(p)_{\mathcal{M}}$.

Section 1.1 is devoted to the construction of $^0\mathbf{GC}^p$, its relation with the higher Chow groups, and its relation to Suslin's theorem about Milnor K-theory. Section 1.2 gives a geometric way of representing elements of algebraic K-groups as vector bundles over "mixed varieties". Section 1.3 uses this to construct highest Chern classes of vector bundles over mixed varieties with values in Grassmannian homology. In section 1.4, a generalized Grassmannian complex is constructed, and Chern classes from algebraic K-theory to its cohomology are constructed.

In the rest of this introduction, we will describe what the cosimplicial Grassmannian $\hat{G}^p_\bullet$ has to do with polylogarithms and how this gives a map from Grassmannian homology to Deligne homology, as predicted in §0.4. The material about polylogarithms is not needed for the rest of the paper.

1.0.1 *Polylogarithms.*   We wish to sketch the relation between the cosimplicial Grassmannian $\hat{G}^p_\bullet$ and p-log functions. In this section, we work over the complex numbers. The material in this section will be treated more fully in [HM].

First, we observe that $\hat{G}^p_0$ is just a torus $(\mathbb{C}^*)^p$. On it, there is a holomorphic p-form called the "volume form" vol, defined by

$$\text{vol} = \frac{dz_1}{z_1} \wedge \frac{dz_2}{z_2} \wedge \cdots \wedge \frac{dz_p}{z_p}$$

Consider the multivalued Deligne complex $\tilde{\mathbb{Q}}(p)$ on a variety Y.

$$\mathbb{Q} \xrightarrow{(2\pi i)^p} \tilde{\Omega}^0(Y) \xrightarrow{d} \tilde{\Omega}^1(Y) \xrightarrow{d} \cdots \xrightarrow{d} \tilde{\Omega}^{p-1}(Y) \to 0$$

Here $\tilde{\Omega}^i$ represents *multivalued* holomorphic differential forms, i.e. holomorphic differential forms defined on the simply connected covering space $\tilde{Y}$ of Y. We wish to consider a double complex $\mathbb{D}$ which is the multivalued Deligne complex $\tilde{\mathbb{Q}}(p)$ in one direction and is a complex constructed from the cosimplicial Grassmannian in the other direction.

We fix a positive integer p, and we define the symbol $\tilde{\Omega}^k_q$ to represent $\tilde{\Omega}^k(\hat{G}^p_q)$, multivalued holomorphic k-forms on the generic part of the Grassmannian of projective planes in $\mathbb{C}\mathbb{P}^{p+q}$. The double complex $\mathbb{D}$ is:

$$
\begin{array}{ccccccccccc}
[\mathbb{Q}] & \hookrightarrow & \tilde{\Omega}^0_p & & & & & & & & \\
\uparrow & & {\scriptstyle A^*}\uparrow & & & & & & & & \\
\mathbb{Q} & \hookrightarrow [\tilde{\Omega}^0_{p-1}] & \xrightarrow{d} & \tilde{\Omega}^1_{p-1} & & & & & & & \\
\uparrow & & {\scriptstyle A^*}\uparrow & & {\scriptstyle A^*}\uparrow & & & & & & \\
\mathbb{Q} & \hookrightarrow & \tilde{\Omega}^0_{p-2} & \xrightarrow{d} & [\tilde{\Omega}^1_{p-2}] & \xrightarrow{d} & & & & & \\
\vdots & & \vdots & & \vdots & & & & & & \\
\mathbb{Q} & \hookrightarrow & \tilde{\Omega}^0_2 & \xrightarrow{d} & \tilde{\Omega}^1_2 & \xrightarrow{d} & \cdots & \tilde{\Omega}^{p-2}_2 & & & \\
\uparrow & & {\scriptstyle A^*}\uparrow & & {\scriptstyle A^*}\uparrow & & & {\scriptstyle A^*}\uparrow & & & \\
\mathbb{Q} & \hookrightarrow & \tilde{\Omega}^0_1 & \xrightarrow{d} & \tilde{\Omega}^1_1 & \xrightarrow{d} & \cdots & [\tilde{\Omega}^{p-2}_1] & \xrightarrow{d} & \tilde{\Omega}^{p-1}_1 & \to 0 \\
\uparrow & & {\scriptstyle A^*}\uparrow & & {\scriptstyle A^*}\uparrow & & & {\scriptstyle A^*}\uparrow & & {\scriptstyle A^*}\uparrow & \\
\mathbb{Q} & \hookrightarrow & \tilde{\Omega}^0_0 & \xrightarrow{d} & \tilde{\Omega}^1_0 & \xrightarrow{d} & \cdots & \tilde{\Omega}^{p-2}_0 & \xrightarrow{d} & [\tilde{\Omega}^{p-1}_0] & \to 0
\end{array}
$$

Here A^* is $\Sigma(-1)^i A_i^*$ (which makes sense and forms a complex with proper attention to lifting the maps A_i to maps of the simply connected covering spaces).

We are interested in total cocycles Z in this double complex of degree p. The groups in which the components of Z live are indicated by brackets in the diagram. To such a cocycle, we can associate a p-form on $\hat{G}_0^p$, namely $d(Z|\tilde{\Omega}_0^{p-1})$ which we call δZ.

CONJECTURE. *There exists a p-cocycle Z in the double complex $\mathbb{D}$ such that* $\delta Z = \mathrm{vol}$.

Definition. A polylogarithm p-log is the component $Z|\tilde{\Omega}_{p-1}^0$ for any p-cocycle Z in $\mathbb{D}$ such that $\delta Z = 0$.

So the p-log is a multivalued analytic function on $\hat{G}_{p-1}^p$. It will necessarily satisfy a functional equation with $2p + 1$ terms. This is just the expression of the fact that $A^* p$-log is $(2\pi i)^p$ times a rational number. Recalling the formula for A^*, we get:

COROLLARY. *The p-log function satisfies the functional equation*

$$\sum_{i=0}^{2p} (-1)^i A_i^* p\text{-log} = \text{constant}.$$

The idea that p-logarithms should naturally be considered on open subsets of the Grassmannian first occurred in [GM]. The work we are describing here involves a generalization of that paper from the real numbers to the complexes.

1.0.2 Polylogarithms and Deligne cohomology. The polylogarithms of the last section give an explicit construction of a map form the Grassmannian cohomology of a space to its Deligne cohomology.

The fundamental group $\pi_1(Y)$ acts on the multivalued Deligne complex $\tilde{\mathbb{Q}}(p)$ of Y by deck transformations of the simply connected covering space $\tilde{Y}$ of Y. Then we define the multivalued Deligne cohomology groups $MH_{\mathscr{D}}^*(Y, \mathbb{Q}(p))$ to be the hypercohomology of $\pi_1(Y)$ with coefficients in $\tilde{\mathbb{Q}}(p)$, i.e. the total cohomology of the double complex $C^s(\pi_1(Y), \tilde{\mathbb{Q}}^r(p))$ where C^s is the complex that computes group cohomology. (More precisely, one considers only nilpotent quotients of $\pi_1(Y)$, and only differential forms on $\tilde{Y}$ which are iterated integrals on Y. This allows treatment of more general algebraic varieties than complex ones.)

Multivalued Deligne cohomology $MH_{\mathscr{D}}^*(Y, \mathbb{Q}(p))$ maps to the usual Deligne cohomology $H_{\mathscr{D}}^{\cdot}(Y, \mathbb{Z}(p))$. This may be seen by calculating a Cech approximation to $H_{\mathscr{D}}^{\cdot}(Y, \mathbb{Z}(p))$ using the Grothendieck open cover of Y consisting of just one open set $\tilde{Y}$ projecting to Y. This Cech approximation calculates $MH_{\mathscr{D}}^*(Y, \mathbb{Q}(p))$.

Since the Grassmannian complex $^0\mathbf{GC}_{\cdot}^p$ is representable by the cosimplicial Grassmannian $\hat{G}_{\cdot}^p$, in order to map $^0\mathbf{GC}_{\cdot}^p$ to the Deligne complex $\mathbb{Q}(p)$ for all X, it suffices to find an element of the group $MH^p_{\mathscr{D}}(\hat{G}^p_{\cdot}, \mathbb{Q}(p))$. This group in turn is calculated by the triple complex $\mathscr{K}^{q,r,s} = C^s(\pi_1(\hat{G}_q^p), \tilde{\mathbb{Q}}^r(p))$ where the q differential is induced from the maps A_i of generalized Grassmannians, the r differential is d on multivalued differential forms, and the s differential computes group cohomology of the fundamental group. Note that for the usual choice of a complex $C^{\cdot}$ to compute group cohomology, $\mathscr{K}^{q,r,0} = \tilde{\mathbb{Q}}^r(p)(\hat{G}_q^p)$ so this triple complex includes the double complex $\mathbb{D}$ of §1.0.1 as one of its faces. The conjecture of §1.0.1 is strengthened to the following.

CONJECTURE. *Up to homology class, there is a unique total p-cocycle $\{Z^{q,r,s}\}$ in the triple complex $\mathscr{K}^{\cdot,\cdot,\cdot}$ computing $MH^p_{\mathscr{D}}(\hat{G}^p_{\cdot}, \mathbb{Q}(p))$ such that $\delta(Z^{0,p,0}) = $ vol.*

The complex valued function $z^{p-1,1,0}$ on $\hat{G}^p_{p-1}$ is the p-log function of §1.0.1.

At the moment the conjecture is proved for $p = 1, 2,$ and 3 [HM].

§1.1 The Grassmannian complex

1.1.1 *The generic part $\hat{G}_q^p$ of the Grassmannian.* In §1.1.1–§1.1.3, we recall some definitions from [GM] (which we modify by using projective dimensions in numbering Grassmannians instead of affine dimensions). We work over the complex numbers, but an arbitrary field would do as well.

Denote by $G_q^p = G^p(\mathbb{CP}^{p+q})$ the Grassmann manifold of all q dimensional projective subspaces of $\mathbb{CP}^{p+q}$, complex projective $p + q$ space. (In general, in our notation for Grassmannians, we denote the dimension of the plane by a subscript on the G and the codimension by a superscript.) We consider $\mathbb{CP}^{p+q}$ to be endowed with a simplex of coordinate hyperplanes obtained by setting one of the homogeneous coordinates equal to zero. A q-plane ξ is said to be *generic* if it is transverse to the stratification of $\mathbb{CP}^{p+q}$ by coordinate hyperplanes and their intersections. (It is easy to see that for ξ to be generic it suffices that it miss all the $p + 1$ fold intersections of distinct coordinate hyperplanes.) The (Zariski open) subspace of G_q^p consisting of generic q-planes is denoted by $\hat{G}_q^p$ or $\hat{G}^p(\mathbb{CP}^{p+q})$.

1.1.2 *The intersection and projection maps.* For each integer i such that $0 \leqslant i \leqslant p + q$, there are intersection maps A_i and projection maps B_i

$$\hat{G}_q^{p-1} \overset{B_i}{\leftarrow} \hat{G}_q^p \overset{A_i}{\rightarrow} \hat{G}_{q-1}^p$$

Here the subspace $A_i(\xi)$ is the intersection of ξ with the ith coordinate hyperplane. More specifically, let z denote the point with homogeneous coordinates $(z_0 : \cdots : z_{p+q-1})$. Let $J_i : \mathbb{CP}^{p+q-1} \to \mathbb{CP}^{p+q}$ be the inclusion of the ith coordinate hyperplane by sending

$$z = (z_0 : \cdots : z_{p+q-1}) \quad \text{to} \quad (z_0 : \cdots : z_{i-1} : 0 : z_i : \cdots : z_{p+q-1}).$$

Then $A_i(\xi) = J_i^{-1}\xi$.

The subspace $B_i(\xi)$ is the projection of ξ on the ith hyperplane by the projection that sends $(z_0 : \cdots : z_{p+q})$ to $(z_0 : \cdots : z_{i-1} : z_{i+1} : z_{p+q})$.

1.1.3 *The torus action.* The torus $\tilde{H} = (\mathbb{C}^*)^{p+q+1}$ acts on G_q^p as follows: the element $h = (h_0, \ldots, h_{p+q}) \in (\mathbb{C}^*)^{p+q+1}$ acts on the point $z = (z_0 : \cdots : z_{p+q})$ by multiplying the homogeneous coordinate z_i by h_i. The diagonal acts trivially. Let H denote the quotient of $\tilde{H}$ by the diagonal. ($\tilde{H}$ is the maximal torus in $\mathrm{GL}(p+q+1)$ and H is the maximal torus in $\mathrm{PGL}(p+q+1)$.)

PROPOSITION. *H preserves $\hat{G}_q^p$ and acts freely on it.*

This is proposition 1.1.2 of [GM].

1.1.4 *Remark.* For fixed p, the spaces $\mathbb{CP}^{p+q}$ together with the inclusions J_i form a strict simplicial algebraic variety or Δ-variety [RS], i.e. a simplicial variety without the degeneracy maps. The spaces $\hat{G}_q^p$ together with the mappings A_i form a strict cosimplicial variety except that the first p cosimplices are missing.

1.1.5 *Grassmannian homology.* For an algebraic variety X, let ${}^0\mathbf{GC}^p$ denote the following complex of sheaves: For a Zariski open set $U \subset X$, the sheaf ${}^0\mathbf{GC}_q^p$ evaluated on U is the set of all formal linear combinations $\Sigma n_\ell f_\ell$ where the n_ℓ are integers and the f_ℓ are maps from U into $\hat{G}_q^p$. The differential $A : {}^0\mathbf{GC}_q^p \to {}^0\mathbf{GC}_{q-1}^p$ sends $f : U \to \hat{G}_q^p$ to $\Sigma(-1)^j A_j \circ f$. Finally, the Grassmann homology groups ${}^0GH_q^p(X)$ are the hyperhomology groups of the complex ${}^0\mathbf{GC}^p$. If $\hat{C}$ stands for Cech cochains, then ${}^0GH_q^p(X)$ is the homology of the complex $\hat{C}^k({}^0\mathbf{GC}_\ell^p)$ at the degree $q = \ell - k$.

CONJECTURE. *The truncated complex $\tau_{<p}{}^0\mathbf{GC}^p$ is $\mathbb{Z}(p)_{\mathcal{M}} \oplus \mathbb{Z}(p-2) \oplus \mathbb{Z}(p-4)_{\mathcal{M}} \oplus \ldots$. Specifically, $\tau_{<p}{}^0\mathbf{GC}_q^p \cong \mathbb{Z}(p)_{\mathcal{M}}^{p-q} \oplus \mathbb{Z}(p-2)_{\mathcal{M}}^{p-q} \oplus \ldots$.*

This conjecture would imply that the Grassmannian cohomology group ${}^0GH_q^p(X)$ naturally maps to the Motivic cohomology group $H_{\mathcal{M}}^{p-q}(X, \mathbb{Z}(p))$. Note that the truncation has been done so as to agree with the vanishing conjecture, property 2 of §0.1.

The complexes $\tau_{<p}{}^0\mathbf{GC}^p$ admit natural maps to $\mathbb{Z}(p)_{\mathcal{M}}$ for $p = 0, 1,$ or 2 as described in §0.3 of the introduction. For $p = 1$ this is because $\hat{G}_0^1$ is $\mathbb{C}^*$, and for $p = 2$ it is because $\hat{G}_0^2$ is $\mathbb{C}^* \times \mathbb{C}^*$ and $\hat{G}_1^2/H$ is $\mathbb{C}^* - \{1\}$ ([GM], §2.2).

§1.1.6 *The map to the higher Chow groups.* There is an obvious map from the sheaf ${}^0\mathbf{GC}_q^p$ to the sheaf $U \to z^p(U, p+q)$ of the chain complex in [Bl1] defining higher Chow groups. Recall that $z^p(U, p+q)$ is linear combinations $\Sigma n_\ell V_\ell$ where the n_ℓ are integers and the V_ℓ are suitably transverse irreducible subvarieties of $U \times A^{p+q}$ of codimension p. Here A^{p+q} is affine space and "suitably transverse" means transverse to a simplex of affine subspaces. We embed A^{p+q} in projective space $\mathbb{P}^{p+q}$ so that its complement hyperplane is not one of the coordinate hyperplanes. Then a map f_ℓ from U to $\hat{G}_q^p$ gives a subvariety of $U \times \mathbb{P}^{p+q}$ which is linear in each fiber, which restricts to a subvariety of $U \times A^{p+q}$ which will automatically have the required transversality property. This map from ${}^0GC_q^p(U)$ to $z^p(U, p + q)$ is compatible with the differentials of the two complexes, and hence induces a map $GH_q^p(X) \to CH^p(X, p + q)$ from the Grassmannian homology groups to the higher Chow groups.

1.1.7 *An equivalent conjecture.* In this section we give another conjecture as to the identity of $\mathbb{Z}(p)_{\mathcal{M}}$ which is inspired by Suslin's beautiful result that for a local ring R, the Milnor K-group $K_p^M(R) \cong H_p\mathrm{GL}_p(R)/H_p\mathrm{GL}_{p-1}(R)$ [S1].

CONJECTURE. *The stalk of $\mathbb{Z}(p)_{\mathcal{M}}$ at x is $\tau_{<2p}\mathrm{Cone}(C.\mathrm{GL}_{p-1}(\mathcal{O}_x) \to C.\mathrm{GL}_p(\mathcal{O}_x))$.*

(The degrees are placed so that an i-cycle of the cone is a $2p - i$ cocycle of $\mathbb{Z}(P)_{\mathcal{M}}$. Once again, the truncation guarantees §0.1 property 2.)

THEOREM. *The complex $\oplus_{p \leqslant r, \, p \equiv r(\mathrm{mod}\, 2)}\tau_{<2p}\mathrm{Cone}(C.\mathrm{GL}_{p-1}(R) \to C.\mathrm{GL}_p(R))$ is quasi-isomorphic to the complex $\tau_{<r}{}^0\mathbf{GC}^r.$ of §1.1.5.*

The proof is an application of the Suslin spectral sequence [S1].

§1.2 Mixed varieties and vector bundles over them.

In order to deal geometrically with algebraic K-theory, we introduce the idea of mixed varieties and vector bundles over them. Just as an element of $K_0(X)$ can be represented by an algebraic vector bundle over X, an element of a (rational) higher algebraic K-group can be represented by a vector bundle over a mixed variety. In order to construct the whole family of pth Chern classes c_j^p mentioned in the introduction, it suffices to construct a single pth Chern class c_p for vector bundles over mixed varieties.

The idea of a mixed variety is essentially a special case of the idea of a simplicial variety, in a language influenced by Karoubi's construction of K-theory. Our use of mixed varieties apparently restricts the validity of this chapter to algebraic varieties over the complex numbers. However this restriction is illusory since the results could be translated to a purely algebraic context by use of simplicial varieties.

§1.2.1 *Vector bundles over mixed varieties.* A *mixed variety* $X \times P$ is just a topological space which is product of a complex algebraic variety and a compact polyhedron. We keep track of the product structure and the algebraic and polyhedral structures of the respective factors. A *vector bundle over a mixed variety* is a topological vector bundle E over $X \times P$ whose restriction to every algebraic slice $X \times \{p\}$, $p \in P$ has the structure of an algebraic vector bundle and whose restriction to every polyhedral slice $\{x\} \times P$, $x \in X$ has the structure of a flat vector bundle. These structures are subject to the condition below. Given any path from p to p' in P, there is a *monodromy map* from the restriction of E to $X \times \{p\}$ to the restriction of E to $X \times \{p'\}$. For each $x \in X$, the monodromy map restricts to the map from the fiber over $x \times p$ to the fiber over $x \times p'$ obtained from the flat structure on $\{x\} \times P$. The condition relating the two structures on E is that all of these monodromy maps must be isomorphisms of algebraic vector bundles.

§1.2.2 *Circuits.* An *m-circuit with boundary* is a compact polyhedron P with a regular cellulation whose m-cells are oriented, such that (1) P is the closure of its m-cells and (2) if $[P]$ is the cellular chain which is the sum of all the oriented m-cells of P, then its homological boundary $\partial[P]$ has multiplicity 0, -1, or $+1$ on each $m - 1$ cell. The boundary ∂P of P is the union of the $(m - 1)$-cells on which $\partial[P]$ is nonzero, together with the induced orientation. So $\partial[P] = [\partial P]$. An *m-circuit* is an m-circuit with boundary whose boundary is empty, so $[P]$ is a cycle. If P is an m-circuit with boundary, then ∂P is an m-circuit. The notions of an m-circuit and an m-circuit with boundary are independent of the cellulations chosen. The choice of compatible orientations on the cells is equivalent to choosing an orientation class $[P]$ in $H_m(P, \partial P; \mathbb{Z})$. If P is an m-circuit, then $-P$ denotes the same m-circuit with the orientation reversed.

§1.2.3 *Mixed varieties and K-theory.* A *mixed m-circuit variety* $X \times P$ is a mixed variety where P is an m-circuit. There are three natural equivalence relations on the set of vector bundles E over mixed m-circuit varieties $X \times P$ for fixed X. If E and F are vector bundles over mixed m-circuit varieties $X \times P$ and $X \times P'$, then $E + F$ is the union bundle over the mixed variety $X \times P \cup X \times P'$. *Cobordism* is the equivalence relation $E \simeq F$ if $E + (-F)$ is the restriction to $X \times \partial P'$ of a bundle over $X \times P'$ for some $(m + 1)$-circuit with boundary P'. *Stable equivalence* is the equivalence relation generated by $E \simeq E \oplus 1$ where 1 is the trivial line bundle over $X \times P$. Finally, *equivalence* is the equivalence relation generated by cobordism and stable equivalence.

PROPOSITION. *Suppose X is a complex affine variety and T is the torsion subgroup of $K_m(X)$. Then there is a natural inclusion g of $K_m(X)/T$ into the set of equivalence classes of vector bundles over mixed m-circuit varieties $X \times P$.*

Proof. For $m = 0$, a mixed variety is just a union of copies of X with formal signs, $+X$ or $-X$. If $[E]$ is the class in $K_0(X)$ represented by the vector bundle E over X, define $g(E)$ to be E considered as a vector bundle over $+X$. (For $m = 0$, there is no need to divide by T.)

A vector bundle over a mixed variety is *algebraically trivial* if it is trivial on the algebraic slices $X \times \{p\}$. For fixed X, the set S of cobordism equivalence classes of algebraically trivial n-dimensional vector bundles E over a mixed m-circuit $X \times P$ can be identified with $H_m(\mathrm{GL}(n, \Gamma(X)))$ where $\Gamma(X)$ is the coordinate ring of X. The map from S to homology is constructed as follows: Pick a representative of an equivalence class in S, say E over $X \times P$, and assume that P is connected. There is a map from the fundamental group $\pi_1(P)$ to $\mathrm{Aut}(E|X \times \{p\}) = \mathrm{GL}(n, \Gamma(X))$ given by the flat structure. This gives a map from P to $K(\mathrm{GL}(n, \Gamma(X)), 1)$ defined up to homotopy, and the homology class in question is the image of the orientation class $[P]$ of P. Cobordism of vector bundles over mixed varieties is equivalent to homology of the corresponding cycles in

$K(\mathrm{GL}(n, \Gamma(X)), 1)$. Finally, $K_m(X)$ constructed through the $+$ construction maps to $\lim_{n \to \infty} H_m(\mathrm{GL}(n, \Gamma(X)))$ by the Hurewicz map, and this map is an embedding by [MM].

§1.2.4 *A Gysin map.* Let $\mathbb{Z}(p)$ be a complex of Zariski sheaves on X and let $H^*(X, \mathbb{Z}(p))$ denote the Cech hypercohomology. In this section, we construct a Gysin map $(\mathrm{Pr}_X)_* : H^{2p}(X \times P, \mathrm{Pr}_X^* \mathbb{Z}(p)) \to H^{2p-j}(X, \mathbb{Z}(p))$ for every mixed j-circuit variety $X \times P$, where Pr_X is the projection onto the X factor. The moral reason for the existence of this Gysin map is that Pr_X has a "nearly nonsingular" fiber (see [FM], p. 83).

We carry out the construction with respect to a particular Cech covering $\{U_a\}$ of X. Given such a covering, let N be the nerve of $\{U_a\}$ and consider the polyhedron $N \times P$. Choose a regular cell D division of $N \times P$ such that the projection $\mathrm{Pr}_N : N \times P \to N$ is cellular. Then we regard $\mathrm{Pr}_X^* \mathbb{Z}(p)$ as a cellular sheaf on D which to each cell σ associates the sheaf $\mathbb{Z}(p)$ evaluated on the intersection of open sets U_a in X represented by $\mathrm{Pr}_N(\sigma)$.

Now, we suppose that c is a cochain for $H^{2p}(X \times P, \mathrm{Pr}_X^* \mathbb{Z}(p))$ calculated with respect to this cellular sheaf. Then under the additional assumption that P is an $(p + q)$-circuit, we want to produce a Cech cochain $(\mathrm{Pr}_X)_* c$ for $H^{2p-j}(X, \mathbb{Z}(p))$ calculated with respect to $\{U_a\}$. Given a collection of $2p - j + 1$ open sets U_a, ordered up to sign, we assign to its intersection the element $\Sigma_\sigma c(\sigma)$ where the sum runs over all simplices σ of (maximal) dimension $2p$ such that $\mathrm{Pr}_N(\sigma)$ is the cell in N associated to that intersection. Note that to evaluate $c(\sigma)$, one must orient σ by the product of the orientation of $\mathrm{Pr}_N(\sigma)$ from the ordering of the open sets U_a and the orientation of the fiber of $\mathrm{Pr}_N|\sigma$ obtained from the orientation of P as a j-circuit.

§1.2.5 *Chern classes of mixed bundles.* Suppose that we have a pth Chern class $c_p(E)$ for any vector bundle E over a mixed variety $X \times P$ with values in $H^{2p}(X \times P, \mathrm{Pr}_X^* \mathbb{Z}(p))$. Suppose further that this Chern class has two properties: it is natural with respect to inclusions $X \times P \subset X \times P'$ and it depends only on the stability class of E.

PROPOSITION. *The class* $(\mathrm{Pr}_X)_* c_p(E)$ *in* $H^{2p-j}(X, \mathbb{Z}(p))$ *depends only on the equivalence class of* E.

It must be shown that $(\mathrm{Pr}_X)_* c_p(E)$ is cobordism invariant. This is a standard argument using the naturality property on the two ends of the cobordism.

Given such a pth Chern class c_p, this proposition shows that we can construct a whole family of Chern classes $c_j^p : K_j(X) \to H^{2p-j}(X, \mathbb{Z}(p))$ as in §0.3. By applying the inclusion g of §1.2.3 to an element k of $K_j(X)$, we get an equivalence class containing the vector bundle E over $X \times P$. Then $c_j^p(k) = (\mathrm{Pr}_X)_* c_p(E)$.

§1.3 Geometric construction of the highest Chern class. In this section, we construct a highest Chern class c^p for a p-dimensional vector bundle over a mixed variety $X \times P$ with values in $H^{2p}(X \times P, \mathrm{Pr}_X^{*0}\mathbf{GH}^p)$, which is natural with respect to inclusions $X \times P \subset X \times P'$. This does not suffice to define Chern classes c_j^p on $K_j(X)$ with values in the Grassmannian homology of X. For that, by §1.2.5, we would need c^p to be defined on arbitrary vector bundles over a mixed variety and to be invariant under stable equivalence. However, it does give an unstable version that fits into the following diagram for if X is spec R:

$$
\begin{array}{ccc}
H_j(\mathrm{GL}_p(R)) & \xrightarrow{\;\;c_j^p\;\;} & {}^0\mathrm{GH}^p_{j-p}(X) \\
\downarrow & & \downarrow \\
K_j(X) \longrightarrow H_j(\mathrm{GL}_\infty(R)) & \to & \mathrm{CH}^p(X, j)
\end{array}
$$

1.3.1 *Open coverings.* For any polyhedron P with a regular cell division, the *star* of a vertex v, symbolized $\mathrm{st}(v)$, is the union of all the open cells that contain that vertex in their closure. The star of any vertex is a contractible open set in P. The set of stars of all the vertices forms an open cover of P. If the regular cell division was a triangulation, then the nerve of that open cover is homeomorphic to P in a cell-preserving way.

If $X \times P$ is a mixed variety, P has a regular cell division, and X has an open cover $\{U_a\}$ then $X \times P$ has an open cover by products $U_a \times \mathrm{st}(v)$. We call this the *product covering*. Its nerve maps to the nerve of $\{U_a\}$ in a canonical way.

1.3.2 *Genericity of sections.* Suppose that E is a vector bundle of fiber dimension d over a mixed variety $X \times P$ and that U is an open set in $X \times P$. By a *section* of E over U we mean a section $s : U \to E$ which is algebraic on the slices $X \times p$ and is locally constant with respect to the flat structure on the slices $x \times P$. A set of sections $\{s_j : U \to E\}$ over U is called *generic* if no subset of the sections is linearly dependent in the fiber over any point in U unless that subset has at least $d + 1$ elements.

TRANSVERSALITY LEMMA. *Given any mixed variety $X \times P$ with a fixed regular cell decomposition of P, there exists an open cover $\{U_a\}$ of X and a choice of a section $s_{a,v}$ over each open set $U_a \times \mathrm{st}(v)$ in the product open cover, such that over any intersection U of open sets in the open cover, the set of all sections defined over U by restriction is generic.*

1.3.3 *Construction of the Chern classes.* We now construct a pth Chern class $c^p(E)$ for a p-dimensional vector bundle over a mixed variety $X \times P$ with values in $H^{2p}(X \times P, \mathrm{Pr}_X^{*0}\mathbf{GH}^p)$.

First choose a triangulation T of P. Next, apply the transversality lemma of §1.3.2 to find an open cover for the regular cell decomposition T and a generic set of sections $s_{a,v}$. Next, choose a cell decomposition D of $N \times P$ as in §1.2.4 by finding a triangulation with no new vertices that is subordinate to the product cellulation arising from T. This can be done by a standard lemma from simplicial geometry. Since D has no new vertices, its vertices are indexed by pairs a, v.

Now, given a simplex σ of dimension k in D, $c^p(E)\sigma$ should be a section of $^0GC_\ell^p$ over the intersection U in X corresponding to $\mathrm{Pr}_N(\sigma)$ in N, where $\ell - k = -p$ or $\ell = k - p$. A section of $^0GC_\ell^p$ over U is just a map from U to $\hat{G}_\ell^p \subset G_\ell(\mathbb{C}\mathbb{P}^{p+\ell}) = G_{\ell+1}(\mathbb{C}^{p+\ell+1}) = G_{k+1-p}(\mathbb{C}^{k+1})$.

At this point we *have* the $k + 1$ sections $s_{a,v}$ of the p dimensional bundle E over U corresponding to the vertices of σ. The $k + 1$ sections give a linear map from $\mathbb{C}^{k+1}$ to the fiber of E parameterized by the point in U. We *want* a map from U to the Grassmann manifold $G_{k+1-p}(\mathbb{C}^{k+1})$. The key idea is to obtain the plane in the Grassmannian by taking the kernel of the map to the fiber. (This is a $(k + 1 - p)$-plane since the map is always surjective over U by the transversality lemma.)

That the chain for $^0GH^p_{-p}(X \times P)$ obtained in this way is a cycle is easy to check. The essential point is that the intersection maps A_j correspond by taking kernels to the operation of forgetting one of the sections.

The idea of obtaining a point in a Grassmannian from a configuration of vectors in $\mathbb{C}^p$ was exploited in [Mac] and in [GM].

§1.4 The generalized Grassmannian complex.   In the last section, we found a top Chern class for a vector bundle over a mixed variety with values in the hypercohomology of the Grassmannian complex $^0\mathbf{GC}^p$. In this section, we construct a complex of sheaves $\mathbf{GC}^p$ called the generalized Grassmann complex whose hypercohomology admits all Chern classes for a vector bundle over a mixed variety. This results in Chern classes from algebraic K-theory.

The construction of $\mathbf{GC}^p$ is similar to that of $^0\mathbf{GC}^p$ with the added complication that some new varieties called generalized Grassmannians (see §1.4.5) replace ordinary Grassmannians.

We conjecture that, after a truncation similar to the previous one, $\mathbf{GC}^p/\mathbf{GC}^{p-2}$ is the motivic sheaf $\mathbb{Z}(p)_\mathcal{M}$.

1.4.1 *Multi-indices.*   Let $\mathcal{N}$ denote a *multi-index*, by which we mean a sequence of non-negative integers $\mathcal{N} = \{n_0, n_1, \ldots, n_i, \ldots, n_e\}$. We call the multi-index $\mathcal{N}$ *positive* if each integer n_i is positive. We call the number $e = e(\mathcal{N})$ the excess of $\mathcal{N}$. We denote the sum $n_0 + \cdots + n_e$ by $|\mathcal{N}|$ and we define $t = t(\mathcal{N}) \equiv |\mathcal{N}| + e$.

To any multi-index $\mathcal{N}$, we associate its *standard indexing set*, symbolized $I(\mathcal{N})$. This is a set of pairs of integers

$$
\begin{array}{llcl}
(0,0), & (0,1), & \cdots & (0, n_0) \\
(1,0), & (1,1), & \cdots & (1, n_1) \\
\vdots & \vdots & & \vdots \\
(e,0), & (e,1), & \cdots & (e, n_e)
\end{array}
$$

i.e. the standard indexing set consists of the pairs (i, j) such that $0 \leqslant i \leqslant e$ and

$0 \leqslant j \leqslant n_i$. The standard indexing set has $t + 1$ indices divided into $e + 1$ subsets (the rows), the ith one of which has $n_i + 1$ elements.

For $\mathcal{N}$ positive, we define the symbol $\mathcal{N} - 1_i$ to be the positive multi-index

$$\mathcal{N} - 1_i = \begin{cases} \{n_0, \ldots, n_i - 1, \ldots, n_e\} & \text{if } n_i > 1 \\ \{n_0, \ldots, n_{i-1}, n_{i+1}, \ldots, n_e\} & \text{if } n_i = 1 \end{cases}$$

Note that $|\mathcal{N} - 1_i| = |\mathcal{N}| - 1$ and $e(\mathcal{N} - 1_i) = e(\mathcal{N})$ or $e(\mathcal{N}) - 1$.

1.4.2 *The varieties $\hat{G}^p(\mathcal{N})$.* Fix a positive multi-index $\mathcal{N}$. The variety $\hat{G}^p(\mathcal{N})$ is a quotient by a particular torus action of the generic part of the Grassmannian $\hat{G}^{p+e}(\mathbb{CP}^t)$ defined in §1.1.1. Think of $\mathbb{CP}^t$ as $\mathbb{P}(\mathbb{C}^{n_0+1} \oplus \cdots \oplus \mathbb{C}^{n_e+1})$ and choose homogeneous coordinates for it indexed by the standard indexing set:

$$(z_{0,0} : z_{0,1} : \cdots : z_{0,n_0} : z_{1,0} : z_{1,1} : \cdots z_{1,n_1} : \cdots : z_{e,0} : z_{e,1} : \cdots : z_{e,n_e})$$

Now we have a projective space $\mathbb{CP}^t$ with a choice of homogeneous coordinates and therefore distinguished coordinate hyperplanes indexed by the standard indexing set $I(\mathcal{N})$. Therefore all of constructions of §1.1.1–§1.1.3 apply. So we have varieties $\hat{G}^{p+e}(\mathbb{CP}^t)$ and the action of the torus $\tilde{H}$ on them. Define a subgroup $\tilde{H}(\mathcal{N}) \subset \tilde{H}$ by the condition that for fixed i, each homogeneous coordinate $z_{i,j}$ $(0 \leqslant j \leqslant n_i)$ is multiplied by the same number. In other words $\tilde{H}(\mathcal{N}) \subset \tilde{H}$ consists of elements $(h_{0,0}, h_{0,1}, \ldots, h_{0,n_0}, h_{1,0}, h_{1,1}, \ldots, h_{1,n_1}, \ldots, h_{e,0}, h_{e,1}, \ldots, h_{e,n_e})$ such that $h_{i,0} = h_{i,1} = \cdots = h_{i,n_i}$ for all i. Let $H(\mathcal{N})$ be $\tilde{H}(\mathcal{N})$ divided by the diagonal. Then $H(\mathcal{N})$ is a torus of dimension $e(\mathcal{N})$.

Definition. The variety $\hat{G}^p(\mathcal{N})$ is the quotient $\hat{G}^{p+e}(\mathbb{CP}^t)$ by $H(\mathcal{N})$.

The variety $\hat{G}^p(\mathcal{N})$ is irreducible and smooth by proposition 1.1.3. It has dimension $(p + e)(|\mathcal{N}| - p + 1) - e$. In §1.4.5 it will be given a geometric interpretation as the generic part of a generalized Grassmann manifold.

1.4.3 *The mappings $\bar{A}_{ij}$.* Fix a positive multi-index $\mathcal{N}$. For each pair of integers i, j of the standard indexing set $I(\mathcal{N})$, we will define mappings $\bar{A}_{ij} : \hat{G}^p(\mathcal{N}) \to \hat{G}^p(\mathcal{N} - 1_i)$. Let $t = t(\mathcal{N})$, $t' = t(\mathcal{N} - 1_i)$, $e = e(\mathcal{N})$, and $e' = e(\mathcal{N} - 1_i)$. Choose and label homogeneous coordinates in $\mathbb{CP}^t$ and $\mathbb{CP}^{t'}$ indexed by $I(\mathcal{N})$ and $I(\mathcal{N} - 1_i)$ as in §1.4.2. We consider two cases.

Case I when $n_i > 1$ so $t' = t - 1$ and $e' = e$. There is an intersection mapping $A_{ij} : \hat{G}^{p+e}(\mathbb{CP}^t) \to \hat{G}^{p+e}(\mathbb{CP}^{t'})$ defined as in §1.1.2. This is compatible with the actions of $H(\mathcal{N})$ and $H(\mathcal{N} - 1_i)$ and so defines $\bar{A}_{ij}$ by passage to the quotient.

Case II when $n_i = 1$ so $t' = t - 2$ and $e' = e - 1$. Let the (non-positive) multi-index $\bar{\mathcal{N}}$ be $\{n_0, \ldots, n_i - 1, \ldots, n_e\}$ so that $\bar{t} = t(\bar{\mathcal{N}}) = t - 1$ and $\bar{e} = e(\bar{\mathcal{N}}) = e$. Then the composition

$$\hat{G}^{p+e}(\mathbb{CP}^t) \overset{A_{ij}}{\to} \hat{G}^{p+\bar{e}}(\mathbb{CP}^{\bar{t}}) \overset{B_{i,0}}{\to} \hat{G}^{p+e'}(\mathbb{CP}^{t'})$$

is likewise compatible with the actions of $H(\mathcal{N})$ and $H(\mathcal{N}-1_i)$ and so it defines $\bar{A}_{ij}$ by passage to the quotient.

1.4.4 *The generalized Grassmannian complex.* For an algebraic variety X, let $\mathbf{GC}_{\cdot}^{p}(\mathcal{N})$ denote the following sheaf: For a Zariski open set $U \subset X$, the sheaf $\mathbf{GC}_{\cdot}^{p}(\mathcal{N})$ evaluated on U is the set of all formal linear combinations $\Sigma n_{\ell} f_{\ell}$ where the n_{ℓ} are integers and the f_{ℓ} are maps from U into $\hat{G}^{p}(\mathcal{N})$. The generalized Grassmannian complex $\mathbf{GC}_{\cdot}^{p}$ is defined to be the direct sum

$$\mathbf{GC}_{\cdot}^{p} = \bigoplus_{\substack{|\mathcal{N}|=p+q \\ \mathcal{N} \text{ positive}}} \mathbf{GC}_{\cdot}^{p}(\mathcal{N})$$

The sum is finite since the multi-indices are required to be positive. The differential $A : \mathbf{GC}_{q}^{p} \to \mathbf{GC}_{q-1}^{p}$ sends $f : U \to \hat{G}^{p}(\mathcal{N})$ to $\Sigma(-1)^{i+j}\bar{A}_{i,j} \circ f$. One may verify that the square of A is zero. The generalized Grassmannian homology groups $GH_{\cdot}^{p}(X)$ are the hyperhomology groups of the complex $\mathbf{GC}_{\cdot}^{p}$:

CONJECTURE. *The complex $\mathbf{GC}_{\cdot}^{p}$ is quasi-isomorphic to the complex $^{0}\mathbf{GC}_{\cdot}^{p}$*

1.4.5 *Generalized Grassmannians.*   In this section, we give an interpretation of the variety $\hat{G}^{p}(\mathcal{N})$ as the generic part of a generalized Grassmannian. This interpretation is needed to define the map of generalized Grassmannian cohomology to the higher Chow groups, but it is not necessary for the other parts of this paper.

A Grassmann manifold G_{q}^{p} is the space of all subvarieties of $\mathbb{CP}^{p+q}$ which are as simple as possible, i.e. which are linear. A generalized Grassmannian is a space of similarly simple or "straight" subvarieties of a product of projective spaces.

Let $\mathcal{N} = \{n_0, \ldots, n_e\}$ be a positive multi-index. We denote by $\mathbb{CP}^{\mathcal{N}}$ the product of projective spaces $\mathbb{CP}^{\mathcal{N}} \equiv \mathbb{CP}^{n_0} \times \mathbb{CP}^{n_1} \times \cdots \times \mathbb{CP}^{n_e}$. We want to view it as a quotient of a Zariski open subset $S_{\mathcal{N}}$ of $\mathbb{CP}^{t}$. This can be done since $\mathbb{CP}^{t}$ is the join of the same set of projective spaces. As in §1.4.2, think of $\mathbb{CP}^{t}$ as $\mathbb{P}(\mathbb{C}^{n_0+1} \oplus \cdots \oplus \mathbb{C}^{n_e+1})$ and choose the homogeneous coordinates z_{ij} indexed by $I(\mathcal{N})$. Let F_i be the subset of $\mathbb{CP}^{t}$ defined by the condition that all the coordinates $z_{i,0}, \ldots, z_{i,n_i}$ vanish. Let $F_{\mathcal{N}}$ be the union of the F_i. Then there is clearly a map $\mathrm{Pr}: \mathbb{CP}^{t} \backslash F_{\mathcal{N}} \to \mathbb{CP}^{\mathcal{N}}$ that sends point with the above homogeneous coordinates to the point $(z_{0,0} : z_{0,1} : \cdots : z_{0,n_0}) \times (z_{1,0} : z_{1,1} : \cdots z_{1,n_1}) \times \cdots \times (z_{e,0} : z_{e,1} : \cdots : z_{e,n_e})$. In fact, $\mathbb{CP}^{\mathcal{N}}$ is the quotient of $\mathbb{CP}^{t} \backslash F_{\mathcal{N}}$ by the torus $H(\mathcal{N})$.

A *straight subvariety* of the product of projective spaces $\mathbb{CP}^{\mathcal{N}}$ is the closure of the projection by Pr defined above of a linear subspace of $\mathbb{CP}^{\mathcal{N}}$ which is transverse to the stratification by the subvarieties F_i and their intersections. We denote the space of all straight subvarieties of $\mathbb{CP}^{\mathcal{N}}$ of codimension p by $G^{p}(\mathbb{CP}^{\mathcal{N}})$ and we call it a *generalized Grassmannian*. It forms a smooth algebraic variety.

Examples. A straight subvariety of a single projective space is a linear subvariety, so an ordinary Grassmannian is an example of a generalized Grassmannian. A straight subvariety of dimension n of $\mathbb{CP}^n \times \mathbb{CP}^n$ is the graph of a projectivity. A straight subspace of dimension n of $\mathbb{CP}^n \times \mathbb{CP}^1$ is a pencil of hyperplanes in $\mathbb{CP}^n$ through a subspace of codimension two.

Remark. Generalized Grassmannians are in general not compact. There are a number of equally canonical ways to compactify $G^p(\mathbb{CP}^{\mathcal{N}})$. In fact, for every interior point of the image of the moment map for $H(\mathcal{N})$ acting on $\mathbb{CP}^t$, the symplectic quotient is such a compactification [K]. The points added at ∞ in these compactifications correspond to reducible subvarieties of $\mathbb{CP}^{\mathcal{N}}$.

For each pair of integers i, j in the standard indexing set $I(\mathcal{N})$ for $\mathcal{N}$, we define the "coordinate hyperplane" $\mathbb{H}_{i,j}$ of $\mathbb{CP}^{\mathcal{N}}$ to be the set of points whose projection in $\mathbb{CP}^{n_i}$ lies in the jth hyperplane of $\mathbb{CP}^{n_i}$ defined by $z_{ij} = 0$. The subvariety $\mathbb{H}_{ij}$ is canonically isomorphic to $\mathbb{CP}^{\mathcal{N}-1_i}$; we can choose an isomorphism $J_{i,j} : \mathbb{CP}^{\mathcal{N}-1_i} \to \mathbb{H}_{i,j}$ as in §1.1.2.

A straight subvariety of $\mathbb{CP}^{\mathcal{N}}$ is *generic* if it is transverse to the stratification of $\mathbb{CP}^{\mathcal{N}}$ by coordinate hyperplanes $\mathbb{H}_{ij}$ and their intersections. The variety $\hat{G}^p(\mathcal{N})$ has the interpretation as the subset of the generalized Grassmannian $G^p(\mathbb{CP}^{\mathcal{N}})$ consisting of generic straight subvarieties. The maps $\overline{A}_{i,j} : \hat{G}^p(\mathcal{N}) \to \hat{G}^p(\mathcal{N} - 1_i)$ have the interpretation that $\overline{A}_{ij}(\xi)$ is the intersection of the straight subvariety ξ with $\mathbb{H}_{i,j}$. Specifically, $\overline{A}_{ij}(\xi) = J_{ij}^{-1}(\xi)$.

The interpretation of $\hat{G}^p(\mathcal{N})$ as a generalized grassmannian provides a map of the sheaf $\mathbf{GC}.^p$ to the sheafified version of the higher Chow groups. The higher Chow groups $CH^p(X, p + q)$ are defined using appropriately transverse cycles in $X \times A^{p+q}$ where A is affine space, but cycles in $X \times \mathbb{CP}^{\mathcal{N}}$ would do as well. A map of X to $\hat{G}^p(\mathcal{N})$ creates such an appropriately transverse cycle all of whose fibers are straight subvarieties.

1.4.6 *Chern classes.* The original constructions of Chern and Whitney obtained the pth Chern class c^p of a vector bundle E of fiber dimension $p + e$ as the obstruction to finding $e + 1$ independent sections of E. We follow this original idea, rather than use the projective bundle trick of Grothendieck which has become more usual in algebraic geometry.

The plan is the same as that for constructing a top Chern class with values in Grassmannian cohomology described in §1.2 and §1.3. In order to construct Chern classes $c_j^p : K_j^p(X) \to GH_{j-p}^p(X)$ for affine varieties X, it suffices to construct a pth Chern class $c^p(E)$ for a $p + e$ dimensional vector bundle over mixed variety $X \times P$ with values in $H^{2p}(X \times P, \mathrm{Pr}_X^* \mathbf{GH}.^p)$. Only §1.3.3 needs to be modified. The procedure in that section must be replaced by one which is very similar except that the construction of the regular cell decomposition D is more complicated. In §1.3.3, D was a cellulation by simplices Δ^n, each vertex of which corresponded to a section of E over some open set in X. Here, D will be a cellulation by products of simplices $\Delta^{\mathcal{N}}$ each vertex of which correspond to $e + 1$ sections of E over an open set in X.

First choose a triangulation T of P. Now, multiply the mixed variety by Δ^e, the standard e-simplex with vertices labeled $\{w_i\}$, $0 \leqslant i \leqslant e$, to obtain a new mixed variety $X \times P \times \Delta^e$. Next, apply the transversality lemma to $X \times P \times \Delta^e$ to find $\{U_a\}$ and a generic set of sections $s_{a,v,i}$. Choose a cell decomposition $\overline{D}$ of $N \times P \times \Delta^e$ by finding a triangulation with no new vertices that is subordinate to the product cellulation. Finally, we arrive at the cellulation D of $X \times P$ by intersecting $\overline{D}$ with the fiber over the barycenter of Δ^e.

We want to analyze the nature of a cell σ of D. It is the intersection with the fiber over the barycenter of a cell $\bar{\sigma}$ of $\overline{D}$ which projects to all of Δ^e. This, in turn, is a join of $\Delta^{n_0}, \Delta^{n_1}, \ldots$ and Δ^{n_e} where Δ^{n_i} is the n_i simplex of $\bar{\sigma}$ lying over w_i. The cell σ itself is the product $\Delta^{n_0} \times \Delta^{n_1} \times \cdots \times \Delta^{n_e}$. Note that if $\mathcal{N} = \{n_0, \ldots, n_e\}$, then the dimension k of σ is $|\mathcal{N}|$. The vertices of $\bar{\sigma}$ are indexed by the standard indexing set for $\mathcal{N}$. So over the intersection U in X corresponding to $\mathrm{Pr}_N(\sigma)$, there is a set of $t+1$ sections $s_{i,j}$, $0 \leqslant i \leqslant e$, $0 \leqslant j \leqslant n_i$ indexed by the standard indexing set. The vertices of σ itself correspond to e-simplices of $\bar{\sigma}$ which project homeomorphically to Δ^e. These are indexed by all $e+1$ element subsets of the standard indexing set which contain exactly one $s_{i,j}$ for each i.

Now, $c^p(E)\sigma$ should be a section of $\mathbf{GC}_\ell^p$ over the intersection U where $\ell - |\mathcal{N}| = -p$ or $\ell + p = |\mathcal{N}|$. By §1.4.4, a map from U to $\hat{G}^p(\mathcal{N})$ is such a section. By the definition of $\hat{G}^p(\mathcal{N})$, §1.4.2, such a map is produced by a map to $\hat{G}^{p+e}(\mathbf{CP}^t) \subset G^{p+e}(\mathbf{C}^{t+1})$. So just as in §1.3.3, we use the $t+1$ sections indexed by the standard indexing set for $\mathcal{N}$ to produce the required map. The chain for $H^{2p}(X \times P, \mathrm{Pr}_X^* \mathbf{GH}_\ell^p)$ obtained in this way is a cycle.

CHAPTER 2. MOTIVIC SHEAVES

In this chapter, we present the second class of conjectural constructions of the hoped-for motivic cohomology of algebraic varieties. Its form is a realization of the conjectures set forth in [B2], which could be read as a technical introduction to this section. The following is a more naive introduction.

Recall from §0.2 that for fixed i, the motivic cohomology groups $H_{\mathcal{M}}^\ell(X, \mathbb{Z}(p))$ of a complex algebraic variety X should be refinements of the usual singular cohomology groups $H^\ell(X, \mathbf{Q})$. How should one construct such refinements of $H^\ell(X, \mathbf{Q})$? Recall that $H^\ell(X, \mathbf{Q})$ coincides with the ext group $\mathrm{Ext}_{sheaves}^\ell(\mathbf{Q}, \mathbf{Q})$ in the category of sheaves of the constant sheaf $\mathbf{Q}$ (whose stalk is $\mathbf{Q}$) with itself. The first basic idea of chapter 2 is to construct a refinement of the singular homology groups by constructing a more refined category $\mathscr{C}$ in which the ext groups are taken. If $\mathscr{C}$ is a category equipped with a functor F to *sheaves* and $\tilde{\mathbf{Q}}_1$ and $\tilde{\mathbf{Q}}_2$ are objects in $\mathscr{C}$ such that $F\tilde{\mathbf{Q}}_1 = F\tilde{\mathbf{Q}}_2 = \mathbf{Q}$, then $\mathrm{Ext}_{\mathscr{C}}^\ell(\tilde{\mathbf{Q}}_1, \tilde{\mathbf{Q}}_2)$ maps to $\mathrm{Ext}_{sheaves}^\ell(\mathbf{Q}, \mathbf{Q})$. The categories $\mathscr{C}$ considered will be categories of sheaves with extra structure.

Which extra structures should one put on the sheaves? Recall that an important class of examples of sheaves over X is the cohomology of an algebraic family of varieties parameterized by X. We will call sheaves arising in this way *fiber cohomology sheaves* for short. The second idea is to endow the sheaves in $\mathscr{C}$

with as many as possible of the structures possessed by fiber cohomology sheaves. There is a set of fiber cohomology sheaves which are lifts of $\mathbf{Q}$, namely the Tate sheaves $\mathbf{Q}(p)$ of Hodge type $(-p, -p)$. We may now state the main principle: as $\mathscr{C}$ acquires more and more of the structures possessed by fiber cohomology sheaves, $\mathrm{Ext}^{\ell}_{\mathscr{C}}(\mathbf{Q}(0), \mathbf{Q}(p))$ becomes closer to the motivic cohomology group $H^{\ell}_{\mathscr{M}}(X, \mathbf{Q}(p))$.

The best known structure possessed by the cohomology of an algebraic variety is a mixed Hodge structure. If $\mathscr{C}$ is an appropriately defined category of Hodge structures over X, then the groups $\mathrm{Ext}^{\ell}_{\mathscr{C}}(\mathbf{Q}(0), \mathbf{Q}(p))$ are the Deligne cohomology groups $H^{\ell}_{\mathscr{D}}(X, \mathbf{Q}(p)))$ of X, as shown in [B3]. The first illustration of the main principle stated above is that these are already a large step from singular cohomology in the direction of motivic cohomology.

Pursuing this main principle to its extreme, one would expect $H^{\ell}_{\mathscr{M}}(X, \mathbf{Q}(p))$ itself to be $\mathrm{Ext}^{\ell}_{\mathscr{M}}(\mathbf{Q}(0), \mathbf{Q}(p))$ where $\mathscr{M}$ is the category of sheaves possessing *all* of the structures of fiber cohomology sheaves. The dream of describing all of the structures possessed by the cohomology of an algebraic variety was due to Grothendieck. He called a $\mathbf{Q}$-vector space endowed with all of these structures a motive. We have no idea how to define motives precisely. However, we do give a definition for the category $\mathscr{TM}$ of Tate motives, i.e. motives which are successive extensions of $\mathbf{Q}(p)$ for various p.

In this chapter, the program of defining the motivic cohomology group $H^{\ell}_{\mathscr{M}}(X, \mathbf{Q}(p))$ as $\mathrm{Ext}^{\ell}_{\mathscr{TM}}(\mathbf{Q}(0), \mathbf{Q}(p))$ is carried out for the generic point of a variety—Spec k where k is a field. Hopefully, it should work without modification for a local ring. In §2.1, there is a general procedure for constructing categories by displaying them as categories of modules over certain algebras. In §2.2, the category $\mathscr{TM}$ of Tate motives is defined, using the procedure of §2.1, by a universal property. In §2.3, the algebra which should give rise to $\mathscr{TM}$ is constructed directly by generators and relations.

§2.1 Mixed categories. In this section, all additive categories will be supposed to be $\mathbf{Q}$-*categories*, i.e., categories in which the homomorphism groups $\mathrm{Hom}(x, y)$ are vector spaces over the rationals, $\mathbf{Q}$. We will denote by $\mathscr{V}$ the category of finite dimensional vector spaces over $\mathbf{Q}$. As usual V^* is the dual vector space, i.e. for $V \in \mathrm{Ob}\,\mathscr{V}$, $V^* := \mathrm{Hom}(V, \mathbf{Q})$.

Recall that if C and D are categories and Φ and Ψ are functors from C to D, then a natural transformation T from Φ to Ψ is a collection of morphisms $T(x)$: $\Phi(x) \to \Psi(x)$ in D, one for each $x \in \mathrm{Ob}\,C$, such that for all morphisms f: $x \to y$ in C, we have $\Psi(f) \circ T(x) = T(y) \circ \Phi(f)$. We denote by $\mathrm{Hom}(\Phi, \psi)$ the set of all natural transformations from Φ to Ψ. If Φ and Ψ are additive, then $\mathrm{Hom}(\Phi, \Psi)$ is a rational vector space.

§2.1.1 *Pure categories.*

Definition. A *pure category* C is an Abelian semisimple category such that all homomorphism groups $\mathrm{Hom}(x, y)$ are finite dimensional rational vector spaces.

(Here semisimple means that every object is semisimple.) It follows from the definition that every object has finite length.

A *fiber functor* on C is an additive functor $\Phi\colon C \to \mathscr{V}$ such that for all nonzero objects x of C, $\Phi(x) \neq 0$.

A *pure algebra* A is a $\mathbf{Q}$-algebra isomorphic to a product of semisimple finite dimensional $\mathbf{Q}$-algebras. We will consider a pure algebra as a topological algebra, with the topology it receives as an infinite direct product.

Denote by $\mathscr{M}(A)$ the category of left A-modules M which are finite dimensional as $\mathbf{Q}$ vector spaces, and such that the action of A on M is continuous (with respect to the discrete topology on M. Continuity means that all but finitely many of the simple constituents of A act trivially on the module). There is an obvious fiber functor $\Phi(A)$ on $\mathscr{M}(A)$ which associates to an A-module its underlying $\mathbf{Q}$ vector space.

If Φ is a fiber functor on a pure category C, we denote by A^{Φ} the space $\mathrm{Hom}(\Phi, \Phi)$ of all natural transformations from Φ to itself. Composition of transformations gives it an algebra structure. It is a pure algebra whose simple constituents are algebras of $n_i \times n_i$ matrices over $\mathrm{Hom}(x_i, x_i)$ where x_i is a simple object of C and n_i is the dimension of $\Phi(x_i)$, over $\mathrm{Hom}(x_i, x_i)$.

PROPOSITION. *The pair of inverse correspondences* $(C, \Phi) \to A^{\Phi}$ *and* $((\mathscr{M}(A), \Phi(A)) \leftarrow A$ *induce an equivalence of categories*

$$\{\text{Pairs (A pure category } C, \text{ a fiber functor on } C)\} \leftrightarrow \{\text{Pure algebras}\}^{\mathrm{opp}}$$

The morphisms $(C, \Phi) \to (D, \Psi)$ of the category of pairs are additive functors $F\colon C \to D$ such that $\Psi \circ F = \Phi$. Morphisms of pure algebras are required to be continuous.

§2.1.2 *Mixed categories.* A *mixed category* $\mathscr{M}$ is an abelian category in which every object is supplied with an increasing filtration $W.$, called the weight filtration, such that

(0) $W.$ is finite, i.e., for some n, $W_{-n}(M) = 0$ and $W_n(M) = M$.

(1) Morphisms in $\mathscr{M}$ are strictly compatible with $W.$, i.e. the functors $M \to W_i(M)$ for $i \in \mathbb{Z}$ are all exact.

(2) For each object M of $\mathscr{M}$, and each integer i, the object $\mathrm{gr}_i^W M = W_i(M)/W_{i-1}(M)$ is semisimple.

(4) For all objects M and N, the $\mathbf{Q}$-vector space $\mathrm{Hom}(M, N)$ is finite dimensional. (It follows that every object has finite length.)

We denote by $\mathscr{M}_{\leqslant i}$ the full subcategory of $\mathscr{M}$ whose objects satisfy $W_i(M) = M$. Similarly, $\mathscr{M}_{\geqslant i}$ is the full subcategory whose objects satisfy $W_{i-1}(M) = 0$. Also $\mathscr{M}_{[i, j]} = \mathscr{M}_{\geqslant i} \cap \mathscr{M}_{\leqslant j}$ and $\mathscr{M}_i = \mathscr{M}_{[i, i]}$.

It follows from the definitions that all of the categories $\mathscr{M}_i$ are pure, and that gr_i^W is a functor from $\mathscr{M}$ to $\mathscr{M}_i$. In this paper, since it suffices for the applications we have in mind, we assume that $\mathrm{Hom}(x, x) = \mathbf{Q}$ for all simple objects in $\mathscr{M}_i$.

A functor $F\colon \mathscr{M} \to \mathscr{N}$ between two mixed categories is called *pure* if it is additive and it maps $\mathscr{M}_i$ to $\mathscr{N}_i$.

LEMMA. *A pure functor is exact.*

A mixed category is called *split* if it is semisimple. For any mixed category $\mathscr{M}$, there is a canonical split mixed category $\mathscr{M}^S := \bigoplus M_i$. There are pure functors

$$\mathscr{M} \underset{i}{\overset{\mathrm{gr}_{\cdot}^W}{\rightleftarrows}} \mathscr{M}^S$$

defined by $\mathrm{gr}_{\cdot}^W(M) = \bigoplus \mathrm{gr}_i^W M$ and $i(\bigoplus M_i) = \bigoplus M_i$. Clearly, $\mathscr{M}$ is split if and only if these are equivalences.

A *fiber functor* Φ on a mixed category $\mathscr{M}$ is a set of fiber functors $\Phi_i\colon \mathscr{M}_i \to \mathscr{V}$, one for each integer i. If Φ is a fiber functor, we will also denote by Φ_i the composition

$$\mathscr{M} \overset{\mathrm{gr}_i^W}{\longrightarrow} \mathscr{M}_i \overset{\Phi_i}{\longrightarrow} \mathscr{V}$$

A *mixed algebra* A is a rational vector space (A_{ij}) for every pair of integers i and j, topologized as an inverse limit of finite dimensional vector spaces, together with continuous multiplication mappings $A_{ij} \otimes A_{jk} \to A_{ik}$ satisfying the following properties:
   (1) The multiplication mappings are associative.
   (2) All of the subalgebras A_{ii} are pure algebras.
   (3) $A_{ij} = 0$ for $j < i$.
If Φ is a fiber functor on a mixed category $\mathscr{M}$, then put $A_{ij}^\Phi = \mathrm{Hom}(\Phi_j, \Phi_i)$. These $\mathbf{Q}$-vector spaces, together with the multiplication given by composition, form a mixed algebra, which we denote by A^Φ.

   Dually, a *mixed coalgebra* is a rational vector space A_{ij} for each pair of integers together with comultiplication mappings $A_{ik} \to A_{ij} \otimes A_{ik}$ such that $\{A_{ij}^*\}$ is a mixed algebra.

   Let A be a mixed algebra. A *mixed (left) A-module* M is a finite dimensional rational vector space M_i for each integer i, together with an associative collection of continuous multiplication mappings $A_{ij} \otimes M_j \to M_i$. Denote by $\mathscr{M}(A)$ the category of mixed A-modules. There is an obvious fiber functor $\Phi(A)$ on $\mathscr{M}(A)$ which, for each i, associates to an A-module M the underlying $\mathbf{Q}$ vector space of M_i.

THEOREM. *The pair of inverse correspondences* $(\mathscr{M}, \Phi) \to A^\Phi$ *and* $((\mathscr{M}(A), \Phi(A)) \leftarrow A$ *induce an equivalence of categories*

$$\{\text{Pairs (A mixed category } \mathscr{M}, \text{ a fiber functor on } \mathscr{M})\} \leftrightarrow \{\text{Mixed algebras}\}^{\mathrm{opp}}.$$

The morphisms $(\mathcal{M}, \Phi) \to (\mathcal{N}, \Psi)$ of the category of pairs are pure functors $F: \mathcal{M} \to \mathcal{N}$ such that $\Psi_i \circ F = \Phi_i$.

Under this correspondence, split mixed categories correspond to mixed algebras $A = (A_{ij})$ such that $A_{ij} = 0$ for $i \neq j$.

§2.1.3 *Tate twists.* Let $\mathcal{M}$ be a mixed category. Then $\mathcal{M}(1)$ denotes the same category with the indexing of the weight filtration shifted by 2: $W_i(M)$ considered as an object of $\mathcal{M}(1)$ is $W_{i+2}(M)$ considered as an object of $\mathcal{M}$. A *Tate twist* on $\mathcal{M}$ is an isomorphism $Tw: \mathcal{M} \to \mathcal{M}(1)$ of mixed categories. A mixed category with a tate twist is called a *T-mixed category*. A *T-fiber functor* on a T-mixed category $\mathcal{M}$ is a fiber functor $\Phi = \{\Phi_i\}$ such that $\Phi_{i+2} = \Phi_i(1)$.

A *T-mixed algebra* is a mixed algebra $A = (A_{ij})$ together with isomorphisms $A_{ij} \to A_{i+2, j+2}$ which are compatible with the multiplications.

PROPOSITION. *The correspondences of Theorem 2.2.2 induce an equivalence of categories*

$$\{\text{Pairs (A } T\text{-mixed category } \mathcal{M}, \text{ a } T\text{-fiber functor on } \mathcal{M})\}$$

$$\leftrightarrow \{T\text{-mixed algebras}\}^{\mathrm{opp}}.$$

Example. The *standard split Tate category* $\mathcal{V}^{ev}$ is the T-mixed category of finite dimensional vector spaces graded by the even integers

$$\mathcal{V}^{ev} = \bigoplus_{i \in \mathbf{Z}} \mathcal{V}^{ev}_{2i}, \qquad \mathcal{V}^{ev}_{2i} = \mathcal{V}$$

If $v = \bigoplus v_{2i}$ then $(Tw\, v)_j = v_{j-2}$.

§2.1.4 *Mixed categories over a fixed split one.* Let $\overline{\mathcal{M}}$ be a fixed split mixed category. A *category over $\overline{\mathcal{M}}$* is a mixed category $\mathcal{M}$ with an identification of $\mathcal{M}^S$ with $\overline{\mathcal{M}}$. In other words, a category over $\overline{\mathcal{M}}$ is a mixed category together with a pure functor $\pi: \mathcal{M} \to \overline{\mathcal{M}}$ which induces an equivalence $\mathcal{M}^S \overset{\sim}{\to} \overline{\mathcal{M}}$. The collection of all categories over $\overline{\mathcal{M}}$ itself forms a category which we denote by $\mathrm{Ov}(\overline{\mathcal{M}})$. The morphisms of $\mathrm{Ov}(\overline{\mathcal{M}})$ are functors $F: \mathcal{M}_1 \to \mathcal{M}_2$ which commute with the projections to $\overline{\mathcal{M}}$

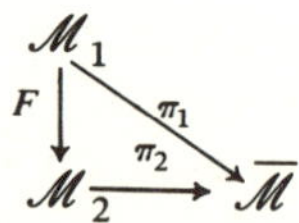

We want to understand the structure of the category $\mathrm{Ov}(\overline{\mathcal{M}})$ of mixed categories over a fixed split category $\overline{\mathcal{M}}$. In the case that $\overline{\mathcal{M}}$ has a fiber functor Φ, this can be done in terms of Φ. Note that a fiber functor Φ on $\overline{\mathcal{M}}$ is simultaneously a fiber functor on every object $\mathcal{M}$ of $\mathrm{Ov}(\overline{\mathcal{M}})$, since $\mathcal{M}_i$ is canonically equivalent to $\overline{\mathcal{M}}_i$ for all i.

PROPOSITION. *For a fixed split mixed category $\overline{\mathcal{M}}$ and a fixed fiber functor Φ on $\overline{\mathcal{M}}$, the correspondences of Theorem 2.1.2 induce an equivalence of categories between $\mathrm{Ov}(\overline{\mathcal{M}})$ and the opposite of the category of mixed algebras $A = (A_{ij})$ with an identification of A_{ii} with $A_{ii}^{\Phi}(\overline{\mathcal{M}})$.*

(Recall that $A_{ij}^{\Phi}(\overline{\mathcal{M}}) = 0$ for $i \neq j$.) Of course, this theorem has a T-mixed analogue: If $\overline{\mathcal{M}}$ is a T-mixed category, then the equivalence identifies the T-mixed versions of the two sides.

§2.1.5 *Quotients by a group action.* Suppose that we are given an action of a group G on a mixed category $\mathcal{M}$, and that the elements of G act by pure automorphisms of $\mathcal{M}$. Suppose further that G acts trivially on the $\mathcal{M}^S$. Then there exists a universal quotient category $\mathcal{M}/G$ together with a pure functor p: $\mathcal{M} \to \mathcal{M}/G$ such that p is G invariant and p induces an equivalence between $\mathcal{M}^S$ and $(\mathcal{M}/G)^S$.

To construct $\mathcal{M}/G$, consider $\mathcal{M}$ as a category over $\mathcal{M}^S$ and choose a fiber functor for $\mathcal{M}^S$ (and hence for $\mathcal{M}$). The group G acts on the corresponding algebra (A_{ij}) and is the identity on A_{ii}. Then $\mathcal{M}/G$ corresponds to the subalgebra of invariants A^G.

§2.1.6 *A universal construction.* The category $\mathrm{Ov}(\overline{\mathcal{M}})$ has an obvious final object but no initial object. We will consider a refinement of it in which we can construct an initial object.

Let $\mathcal{N}$ and $\mathcal{M}$ be mixed categories. An *exact ∂-functor H^*: $\mathcal{N} \to \mathcal{M}$* is a set of functors $\{H^i\}$, one for each integer i, such that a short exact sequences $0 \to M_1 \to M_2 \to M_3 \to 0$ in $\mathcal{N}$ gives rise to a long exact sequence

$$\cdots \to H^i(M_1) \to H^i(M_2) \to H^i(M_3) \xrightarrow{\partial} H^{i+1}(M_i) \to \cdots$$

in $\mathcal{M}$, and a morphism of short exact sequence gives rise to a morphism of long exact sequences. The exact ∂-functor H^* is called *pure* if for each object M in $\mathcal{N}_i$, $H^a(M)$ lies in $\mathcal{M}_{i+a}$. A pure ∂-functor H^*: $\mathcal{N} \to \mathcal{M}$ gives rise to a pure ∂-functor $\mathcal{H}^*$: $\mathcal{N} \to \mathcal{M}^S$ by composing with $\pi = \mathrm{gr}^W_\cdot$.

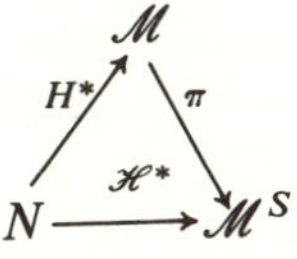

We express the commutativity of this diagram by saying that H^* is an *improvement* of $\mathcal{H}^*$.

Now fix a pure exact ∂-functor $\mathcal{H}^*$: $\mathcal{N} \to \overline{\mathcal{M}}$ from a mixed category $\mathcal{M}$ to a split mixed category $\overline{\mathcal{M}}$. We denote by $\mathrm{Ov}(\mathcal{H}^*)$ the category whose objects are improvements of $\mathcal{H}^*$ and whose morphisms are pure functors F: $\mathcal{M}_1 \to \mathcal{M}_2$

which fit into a commutative diagram

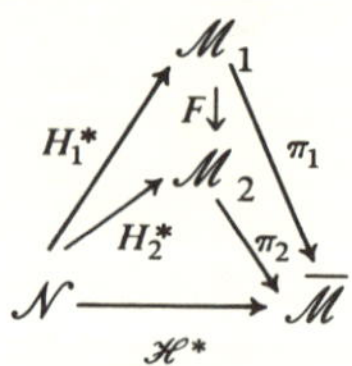

We call an initial object in the category $\mathrm{Ov}(\mathcal{H}^*)$ a *universal improvement* of $\mathcal{H}^*$. In other words, a universal improvement of $\mathcal{H}^*$ is an improvement which has a unique functor in $\mathrm{Ov}(\mathcal{H}^*)$ to any other improvement of $\mathcal{H}^*$. If a universal improvement exists, then it is unique.

PROPOSITION. *If $\overline{\mathcal{M}}$ has a fiber functor Φ, then $\mathcal{H}^*\colon \mathcal{N} \to \overline{\mathcal{M}}$ has a universal improvement.*

The proof is to construct a mixed algebra $A^{\mathcal{H}}$ which corresponds to the universal improvement by Theorem 2.1.2. The algebra is given by $A_{ij}^{\mathcal{H}} = \mathrm{Hom}(\Phi_j \mathcal{H}^*, \Phi_i \mathcal{H}^*)$, where Hom means natural transformations of ∂-functors. (That elements of $A_{ij}^{\mathcal{H}}$ are natural transformations of ∂-functors means that they are natural transformations which, for every short exact sequence $0 \to M_1 \to M_2 \to M_3 \to 0$ in $\mathcal{N}$ the following diagram in $\mathcal{V}$ commutes:

$$\begin{array}{ccc}
\Phi_j \mathcal{H}^k(M_3) & \xrightarrow{\ \partial\ } & \Phi_j \mathcal{H}^{k+1}(M_1) \\
\downarrow & & \downarrow \\
\Phi_i \mathcal{H}^k(M_3) & \xrightarrow{\ \partial\ } & \Phi_i \mathcal{H}^{k+1}(M_1)
\end{array}$$

§2.1.7 *A Yoneda-style description of mixed coalgebras.* Let $\mathcal{M}$ be a mixed category and let x and y be two pure objects in $\mathcal{M}$, $x \in \mathcal{M}_i$ and $y \in \mathcal{M}_j$, $i \leqslant j$. Denote by $\mathcal{M}(x, y)$ the category whose objects are $z \in \mathrm{Ob}\,\mathcal{M}_{[i, j]}$ such that $\mathrm{gr}_i^W z = x$ and $\mathrm{gr}_j^W z = y$, and whose morphisms are morphisms in $\mathcal{M}$ which are the identity on gr_i^W and gr_j^W. We define $A(x, y)$ to be $\pi_0 \mathcal{M}(x, y)$ (the set of connected components of the classifying space of the category). For $i > j$, we put $A(x, y) = *$.

An object z in $\mathcal{M}(x, y)$ is called *minimal* if it has no nontrivial subobject or quotient object in $\mathcal{M}(x, y)$.

LEMMA. (1) *Every object in $\mathcal{M}(x, y)$ has a unique minimal subquotient in $\mathcal{M}(x, y)$.*

(2) *Every morphism between minimal objects in $\mathcal{M}(x, y)$ is an isomorphism.*

COROLLARY. *The set $A(x, y)$ is equivalent to the set of isomorphism classes of minimal objects in $\mathcal{M}(x, y)$.*

Exercise. Construct an abelian group structure on $A(x, y)$ à la Yoneda [Yo]. For every triple of simple objects x, y, z construct a comultiplication $A(x, z) \to A(x, y) \otimes A(y, z)$.

Let us fix in $\mathcal{M}$ a set S of representatives of isomorphism classes of simple objects. Let S_i denote $S \cap \mathrm{Ob}\,\mathcal{M}_i$. This choice fixes a fiber functor Φ on $\mathcal{M}$, namely

$$\Phi_i(x) = \bigoplus_{x_\alpha \in S_i} \mathrm{Hom}(x_\alpha, x)$$

for x in $\mathcal{M}_i$. One has evident maps

$$\varphi_{ij} \colon \bigoplus A(x, y) \to \left(A_{ij}^\Phi\right)^*$$

THEOREM. *The maps φ_{ij} are isomorphisms of coalgebras.*

§2.1.8 *Ext groups in mixed categories.* We construct an explicit cochain complex for computing ext groups. We maintain the choice of S as in the last section.

Suppose $x \in \mathcal{M}_i$, $y \in \mathcal{M}_{i+n}$, $n > 0$, and $1 \leqslant k \leqslant n$. Then we put

$$A^k(x, y) = \otimes \left(A(x_0, x_1) \otimes A(x_1, x_2) \otimes \cdots \otimes A(x_{k-1}, x_k) \right)$$

where the sum is taken over all sequences $(x_0, x_1, \ldots, x_k)$ with $x_0 = x$, $x_k = y$ and $x_p \in S_{i_p}$ for $i = i_0 < i_1 < \cdots < i_k = i + n$. Comultiplication in A induces evident maps $\delta_p \colon A^k(x, y) \to A^{k+1}(x, y)$ for $p = 1, 2, \ldots, k - 1$. Put $d^k = \sum_{p=1}^k (-1)^p \delta_p$. We get a complex

$$A^{\cdot}(x, y) \colon 0 \to A^1(x, y) \to A^2(x, y) \to \cdots \to A^n(x, y) \to 0$$

THEOREM. *There is a natural isomorphism $\mathrm{Ext}^i(y, x) \cong H^i A^{\cdot}(x, y)$. For $i < 1$ and $i > n$, $\mathrm{Ext}^i(y, x) = 0$.*

This theorem follows from the equivalence of categories of §2.1.2 and the description of Ext groups using the bar construction, [M].

§2.2 Motives through perverse sheaves. In order to construct the category $\mathcal{TM}$ of mixed Tate motives, we use a supply of algebraic objects with cohomology groups that is rich enough that every mixed Tate motive occurs as a subquotient of the cohomology of one of these objects. These objects will be certain perverse sheaves on projective space with some extra structures. Perverse sheaves are convenient for categorical constructions since they already form an Abelian category.

§2.2.1 *Motivic perverse sheaves without monodromy.* In this section, we construct a particular T-mixed category called *linearly constructible motivic perverse sheaves without monodromy*, or l.c.m.p.s.w.m. for short.

Let k be a fixed field and $\mathbb{P}_k^n$ be the n-dimensional projective space over k. An l.c.m.p.s.w.m. on $\mathbb{P}_k^n$ is a two dimensional diagram, almost all of whose groups are zero:

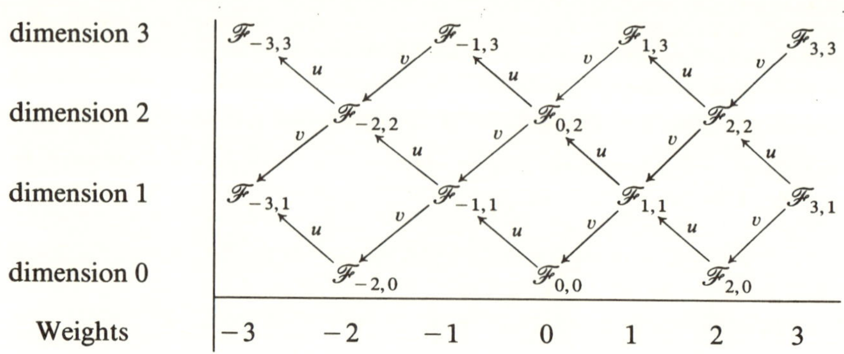

Here $\mathscr{F}_{ij}$ is a direct sum $\bigoplus_L \mathscr{F}_i(L)$ where L ranges over a finite collection of j-dimensional projective subspaces of $\mathbb{P}_k^n$ and $\mathscr{F}_i(L)$ is a $\mathbb{Q}$-vector space in $\mathscr{V}$. The u's and v's are linear maps which are subject to axioms 0), 1), and 2) as follows:

(0) The restriction of u to $\mathscr{F}_i(L) \subset \mathscr{F}_{ij}$ has its image in $\bigoplus_{L \subset L'} \mathscr{F}_{i-1}(L')$. The restriction of v to $\mathscr{F}_i(L) \subset \mathscr{F}_{ij}$ has its image in $\bigoplus_{L' \subset L} \mathscr{F}_{i-1}(L')$. In light of this condition, the maps u are determined by a collection of maps $\mathscr{F}_i(L) \to \mathscr{F}_{i-1}(L')$ whenever $L \subset L'$, which we denote by $u(L, L')$. Similarly the maps v are determined by maps $\mathscr{F}_i(L') \to \mathscr{F}_{i-1}(L)$ which we denote by $v(L', L)$.

(1) The monodromy-free condition: Whenever $L \subset L'$, 1a) the composition

$$\mathscr{F}_i(L) \xrightarrow{u(L, L')} \mathscr{F}_{i-1}(L') \xrightarrow{v(L', L)} \mathscr{F}_{i-2}(L)$$

is zero and 1b) the composition

$$\mathscr{F}_i(L') \xrightarrow{v(L', L)} \mathscr{F}_{i-1}(L) \xrightarrow{u(L, L')} \mathscr{F}_{i-2}(L')$$

is zero.

(2) The diagram is almost a double complex: 2a) $v \circ v: \mathscr{F}_{ij} \to \mathscr{F}_{i-2, j-2}$ is zero; 2b) $u \circ u: \mathscr{F}_{ij} \to \mathscr{F}_{i-2, j+2}$ is zero; and 2c) if $j > 0$, $u \circ v + v \circ u: \mathscr{F}_{ij} \to \mathscr{F}_{i-2, j}$ is zero.

Since the diagram is determined by the set of mappings $u(L, L')$ and $v(L, L')$, a l.c.m.p.s.w.m. may also be defined by the following data: For each linear subspace L of $\mathbb{P}_k^n$ we give a graded vector space $\mathscr{F}(L) = \oplus \mathscr{F}_i(L)$, with only finitely many $\mathscr{F}_i(L)$ nonzero and $\mathscr{F}_i(L) = 0$ unless $i \equiv \dim L \pmod 2$. For each pair L and L' such that L is a codimension one hypersurface in L', we give maps

$$\mathscr{F}(L) \underset{v(L', L)}{\overset{u(L, L')}{\rightleftarrows}} \mathscr{F}(L')$$

such that $u(L, L')$ takes $\mathcal{F}_i(L)$ to $\mathcal{F}_{i-1}(L')$ and $v(L', L)$ takes $\mathcal{F}_i(L')$ to $\mathcal{F}_{i-1}(L)$. We demand the following conditions, whenever the compositions named make sense:

(1a) $v(L', L) \circ u(L, L') = 0$

(1b) $u(L, L') \circ v(L', L) = 0$

(2a) For fixed L and L'', $\sum_{L \subset L' \subset L''} v(L', L) \circ v(L'', L') = 0$

(2b) For fixed L and L'', $\sum_{L \subset L' \subset L''} u(L', L'') \circ u(L, L') = 0$

(2c) $v(L_2', L'') \circ u(L'', L_1') + u(L_2', L) \circ v(L, L_1') = 0$

The morphisms of the category of l.c.m.p.s.w.m.'s are just mappings of all the groups $\mathcal{F}(L)$ which preserve the grading and commute with the maps u and v. It is a T-mixed category. To give the weight filtration, we set W_i of a diagram to be that part of it lying in or to the left of the ith column. The Tate twist just shifts the whole diagram to the right by two.

§2.2.2 *Perverse sheaves without monodromy.* In this section, we give the motivation for the definition of l.c.m.p.s.w.m.'s. The main theorem of this section is of independent interest since it amounts to an elementary construction of a certain category of perverse sheaves (see [MV] for a description of this problem).

A perverse sheaf on a smooth complex algebraic variety X is a complex of sheaves, constructible with respect to some algebraic stratification $X = \bigcup S_\alpha$, satisfying the perversity conditions of [BBD]. A perverse sheaf is said to be *without monodromy* if the local system of vanishing cycles on the regular part of the conormal bundle to each stratum S_α (see [MV], p. 422) has trivial monodromy action of the fundamental group on the fiber. A perverse sheaf on a complex projective space $\mathbb{P}_\mathbb{C}^n$ is *linearly constructible* if it is constructible with respect to a stratification $\mathbb{P}_\mathbb{C}^n = \bigcup S_\alpha$ whose strata S_α have closures which are linear subspaces. The category of linearly constructible perverse sheaves without monodromy on $\mathbb{P}_\mathbb{C}^n$ is by definition the full subcategory of the category of perverse sheaves on $\mathbb{P}_\mathbb{C}^n$ whose objects are linearly constructible and without monodromy.

We can construct a category $\mathscr{C}(\mathbb{P}_\mathbb{C}^n)$ combinatorially by forgetting the grading by i in the definition of l.c.m.p.s.w.m.'s in the last section. Specifically, the objects of $\mathscr{C}(\mathbb{P}_\mathbb{C}^n)$ are rules $\mathcal{F}$ which assign a $\mathbb{Q}$-vector space $\mathcal{F}(L)$ to each linear subspace L of $\mathbb{P}^n$ (with $\mathcal{F}(L) = 0$ for all but finitely many L) together with mappings

$$\mathcal{F}(L) \underset{v(L', L)}{\overset{u(L, L')}{\rightleftarrows}} \mathcal{F}(L')$$

whenever L is contained in L' as a codimension one hyperplane, satisfying the same list of conditions listed in the previous section.

THEOREM. *The category of linearly constructible perverse sheaves without monodromy is equivalent to the category* $\mathscr{C}(\mathbb{P}_\mathbb{C}^n)$.

This theorem will not be proved here, but we remark that for a perverse sheaf $\mathscr{S}$, $\mathscr{F}(L)$ has the interpretation as the vanishing cycles of $\mathscr{S}$ in the conormal bundle of the stratum whose closure is L. So, we see that a l.c.m.p.s.w.m. is essentially a linearly constructible perverse sheaf without monodromy with compatible gradings of all of the spaces of vanishing cycles.

There is an obvious functor from the category of l.c.m.p.s.w.m.'s on $\mathbb{P}_{\mathbb{C}}^n$ to the category $\mathscr{C}(\mathbb{P}_{\mathbb{C}}^n)$ which forgets the grading by i. We now define a functor called *motivization* from $\mathscr{C}(\mathbb{P}_{\mathbb{C}}^n)$ to the category of l.c.m.p.s.w.m.'s on $\mathbb{P}_{\mathbb{C}}^n$. Given an object $\mathscr{F}$ in $\mathscr{C}(\mathbb{P}_{\mathbb{C}}^n)$, there is a canonical weight filtration $F.$ on each vector space $\mathscr{F}(L)$ which is defined inductively, in order of descending dimension of L, starting with $L = \mathbb{P}_{\mathbb{C}}^n$ itself. We define $F_{n-1}\mathscr{F}(\mathbb{P}_{\mathbb{C}}^n) = 0$ and $F_n\mathscr{F}(\mathbb{P}_{\mathbb{C}}^n) = \mathscr{F}\mathbb{P}_{\mathbb{C}}^n$. Now if the filtrations have been defined for all linear spaces of dimension greater than L, we define $F_k\mathscr{F}(L)$ to be $\bigcap_{L \subset L'} u(L, L')^{-1}F_{k-1}\mathscr{F}(L')$. Now we define the graded vector space $\oplus \mathscr{F}_i(L)$ to be the associated graded of $F.\mathscr{F}(L)$. Clearly each $u(L, L')$ maps $F_k\mathscr{F}(L)$ to $F_{k-1}\mathscr{F}(L')$. Also it follows from property 2c) that $v(L', L)$ maps $F_k\mathscr{F}(L')$ to $F_{k-1}\mathscr{F}(L)$. So $u(L, L')$ and $v(L', L)$ induce maps between the associated gradeds.

§2.2.3 *The cohomology of a perverse sheaf.* Any perverse sheaf $\mathscr{S}$ has cohomology groups $\mathbb{H}^i(\mathscr{S})$, (the hypercohomology groups of the complex of sheaves). If $\mathscr{S}$ is linearly constructible and without monodromy, by the result of the last section it is determined by geometric data of the positions of the linear subspaces $L \subset \mathbb{P}_{\mathbb{C}}^n$ and by the combinatorial data of the vector spaces $\mathscr{F}(L)$ and the maps u and v. One might ask whether the cohomology groups $\mathbb{H}^\ell(\mathscr{S})$ can be determined functorially in terms of the combinatorial data alone. The answer is no. In fact there are algebraic families of such perverse sheaves in which the combinatorial data remains constant but the cohomology has monodromy. However if $F.$ is the weight filtration on $\mathbb{H}^*(\mathscr{S})$, the associated graded can be computed in terms of the combinatorial data alone.

Given any l.c.m.p.s.w.m. $\mathscr{F}$ and any even integer i, we may construct a cochain complex $C^*(\mathscr{F})_i$ as follows: In the two dimensional diagram which gives $\mathscr{F}$, we extract the subdiagram consisting of those $\mathscr{F}_{i'j'}$ so that $|(i' - i)| \leqslant j'$. It lies in a V-shaped region with lower vertex at $\mathscr{F}_{i,0}$. This is a double complex by condition 2) of the definition of a l.c.m.p.s.w.m. Then $C^*(\mathscr{F})_i$ is the associated single complex, numbered so that the groups $\mathscr{F}_{ij}$ in the column over the vertex lie in degree zero. The following diagram shows $C^*(\mathscr{F})_2$.

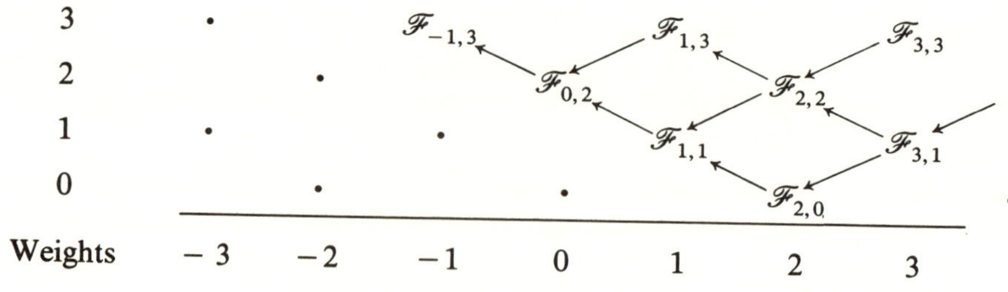

Finally, we denote by $\mathscr{H}^*(\mathscr{F})_i$ the cohomology groups $\mathscr{H}^\ell(C^*(\mathscr{F})_i)$ of this complex.

THEOREM. *The ith graded weight factor of the cohomology group $\mathbb{H}^\ell(\mathscr{S})$ of a linearly constructible perverse sheaf $\mathscr{S}$ without monodromy on $\mathbb{P}^n_{\mathbb{C}}$ is $\mathscr{H}^\ell(\mathscr{F})_i$ where $\mathscr{F}$ is the motivization of $\mathscr{S}$.*

§2.2.4 *Mixed Tate motives.* Returning to a general field k, the δ-functor $\mathscr{H}^*(\mathscr{F})_i$ defined as in the last section still makes sense. Denote $\mathbb{P}^\infty_k = \bigcup_n \mathbb{P}^n_k$ by $\mathbb{P}_k$. We define the category $\mathscr{M}_{wm}(\mathbb{P}_k)$ to be the union over n of the categories of l.c.m.p.s.w.m.'s on $\mathbb{P}^n_k$. Putting together the functors $\mathscr{H}^*(\mathscr{F})_i$ for all i gives us a ∂-functor $\mathscr{H}^*$ from $\mathscr{M}_{wm}(P_k)$ to the category $\mathscr{V}^{\mathrm{ev}}$. Let the functor H^* from $\mathscr{M}_{wm}(\mathbb{P})$ to $\mathscr{T}\tilde{\mathscr{M}}(k)$ be the universal improvement of $\mathscr{H}^*$ (see §2.1.6)

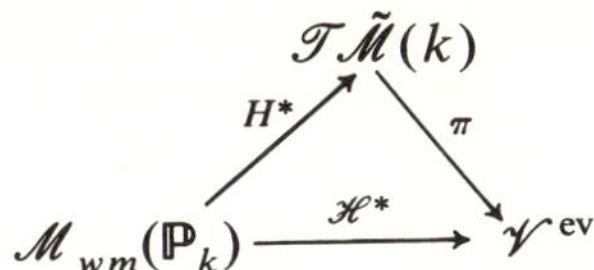

Then $\mathscr{T}\tilde{\mathscr{M}}(k)$ is a T-mixed category. The obvious action of $\mathrm{PGL}_\infty(k)$ on $\mathbb{P}_k$ induces an action on $\mathscr{M}_{wm}(\mathbb{P}_k)$ for which $\mathscr{H}^*$ is an invariant functor. So we get an action of $\mathrm{PGL}_\infty(k)$ on $\mathscr{T}\tilde{\mathscr{M}}(k)$. Put $\mathscr{T}\mathscr{M}(k) := \mathscr{T}\tilde{\mathscr{M}}(k)/\mathrm{PGL}_\infty(k)$ (see §2.1.5 for the construction of the quotient category). This is also a T-mixed category.

Definition. The category $\mathscr{T}\mathscr{M}(k)$ is the category of *mixed Tate motives* on Spec k.

CONJECTURE. *The hoped-for motivic cohomology group $H^\ell_{\mathscr{M}}(\mathrm{Spec}\,k, \mathbb{Q}(p))$ is $\mathrm{Ext}^\ell_{\mathscr{T}\mathscr{M}}(\mathbb{Q}(0), \mathbb{Q}(p))$.*

Remark. For any T-mixed category $\mathscr{M}$ if A is in $\mathscr{M}_0$ and B is in $\mathscr{M}_{-p}$ then $\mathrm{Ext}^1_{\mathscr{M}}(A, B) = 0$ for $\ell > p$ and for $\ell \leqslant 0$ if $p \neq 0$, so this conjecture is compatible with the expected bounds on the stalk cohomology of $\mathbb{Z}(p)_{\mathscr{M}}$ explained in §0.3.

§2.3 Motivic sheaves through coalgebras.

Let k be a field and $X = \mathrm{Spec}\,k$. In this section we give an explicit description of a T-mixed coalgebra A which presumably corresponds to the category of motivic sheaves over X, $\mathscr{M}(X)$. In other words, conjecturally $\mathscr{M}(X) = \mathscr{M}(A^*)$.

All of the projective spaces that we consider in this section will be defined over k.

§2.3.1 *The definition of the coalgebra.* We say that a collection of hyperplanes $H_1, H_2, \ldots, H_m$ in the projective space $\mathbb{P}^n$ are in *general position* if they intersect mutually transversely, that is if every subset of p distinct hyperplanes intersects in a set of dimension $\leqslant n - p$.

First, we define a group B_n for nonnegative integers n. The group B_0 is $\mathbb{Z}$. For $n > 0$, B_n is presented by generators and relations:

Generators. $(2n + 2)$-tuples $(L, M) = (L_0, \ldots, L_n; M_0, \ldots, M_n)$ of hyperplanes in $\mathbb{P}^n$ in general position.

Relations

(R1) If the $(2n + 3)$-tuple of hyperplanes $(L_0, \ldots, L_{n+1}; M_0, \ldots, M_n)$ is in general position, then $\Sigma_0^{n+1}(-1)^i(L_0, \ldots, \hat{L}_i, \ldots, L_{n+1}; M_0, \ldots, M_n) = 0$.

(R2) $(L, M) = -(M, L)$.

(R3) For every automorphism of projective space $g \in \mathrm{PGL}(n + 1)$, $(L, M) = (gL, gM)$.

(R4) For every permutation $\sigma \in \Sigma_{n+1}$, $(L_0, \ldots, L_n; M_0, \ldots, M_n) = (L_{\sigma(0)}, \ldots, L_{\sigma(n)}; M_0, \ldots, M_n)$

Next, we define a comultiplication $\mu = \mu_{k, n-k} \colon B_n \to B_k \otimes B_{n-k}$. Let $[n]$ denote the set of integers $\{1, \ldots, n\}$. For a subset $I = \{i_1, \ldots, i_k\}$ of $[n]$, where $i_1 < \cdots < i_k$, we denote by $\bar{I}$ its complement $\bar{I} = [n] - I = \{\bar{i}_1, \ldots, \bar{i}_{n-k}\}$ where $\bar{i}_1 < \cdots < \bar{i}_{n-k}$. Similarly, for $J = \{j_1, \ldots, j_k\} \subset [n]$, where $j_1 < \cdots < j_k$, we write $\bar{J} = [n] - J = \{\bar{j}_1, \ldots, \bar{j}_{n-k}\}$ where $\bar{j}_1 < \cdots < \bar{j}_{n-k}$. Given choices of I and J, we define projective subspaces P^L and P^R of $\mathbb{P}^n$ by

$$P^L = \bigcap_{i \in \bar{I}} L_i \quad \text{and} \quad P^R = \bigcap_{j \in \bar{J}} M_j$$

Now for the element $x = (L_0, \ldots, L_n; M_0, \ldots, M_n)$ of B_n, we set

$$\mu^{I, J}(x) = \Big(L_0 \cap P^L, L_{i_1} \cap P^L, \ldots, L_{i_k} \cap P^L; M_0 \cap P^L,$$

$$M_{\bar{j}_1} \cap P^L, \ldots, M_{\bar{j}_k} \cap P^L \Big)$$

$$\otimes \Big(L_0 \cap P^R, L_{\bar{i}_1} \cap P^R, \ldots, L_{\bar{i}_{n-k}} \cap P^R; M_0 \cap P^R,$$

$$M_{j_1} \cap P^R, \ldots, M_{j_{n-k}} \cap P^R \Big)$$

Finally, define the comultiplication μ by

$$\mu(x) = \bigoplus_{I, J} \big((-1)^{\Sigma(I, J)} \big) \mu^{I, J}(x)$$

where $\Sigma(I, J)$ is the sign of the permutation $(I, J, \bar{I}, \bar{J})$.

Definition. The T-mixed coalgebra $A = (A_{ij})$ is defined by $A_{0, 2i} = B_i \otimes \mathbb{Q}$ and $A_{0, 2i+1} = 0$ for nonnegative integers i.

§2.3.2 *Some computations.*

1. The cross ratio induces an isomorphism $B_1 \to k^* = K_1(k) = K_1(X)$.

2. Generators for the group B_2 are pairs of triangles (L_0, L_1, L_2) and (M_0, M_1, M_2) in general position. Consider $\mathbb{P}^2$ as a compactification of the affine

plane A^2 with coordinates (x, y). Using PGL(2) invariance (relation R3) the pair of triangles (L, M) may be uniquely reduced to the following:

L_0 is the line at ∞ L_1 is $\{y = 0\}$ L_2 is $\{x = 0\}$

M_0 is $\{y = x + 1\}$ M_1 is $\{y = a_1 x + b_1\}$ M_2 is $\{y = a_2 x + b_2\}$

The general position conditions reduce to the statement that $a_i \neq 0, 1$; $b_i \neq 0, 1$; $a_1 \neq a_2$; $b_1 \neq b_2$; $a_i \neq b_i$; and

$$\Delta := \begin{vmatrix} a_1 & a_2 \\ b_1 & b_2 \end{vmatrix} \neq 0.$$

Let us denote this generator by $\langle a_1, a_2 : b_1, b_2 \rangle$.

3. The comultiplication $\mu_2 : B_2 \to B_1 \otimes B_1$ is given by the following formula:

$$\mu_2 \langle a_1, a_2 : b_1, b_2 \rangle = -a_1 \otimes (1 - a_1) + a_2 \otimes (1 - a_2) + b_1 \otimes (1 - b_1)$$

$$-b_2 \otimes (1 - b_2)$$

$$+ a_1 a_2^{-1} \otimes \left(1 - a_1 a_2^{-1}\right) - b_1 b_2^{-1} \otimes \left(1 - b_1 b_2^{-1}\right)$$

$$- a_1 b_1^{-1} \otimes \left(1 - a_1 b_1^{-1}\right) - a_2 b_2^{-1} \otimes \left(1 - a_2 b_2^{-1}\right)$$

$$- a_1 a_2^{-1} b_1^{-1} b_2 \otimes \left(1 - a_1 a_2^{-1} b_1^{-1} b_2\right) + 2(b_2 \otimes b_2)$$

$$- b_1 \otimes b_2 - b_2 \otimes b_1 - a_2 \otimes b_2 - b_2 \otimes a_2$$

$$+ a_2 \otimes b_1 + b_1 \otimes a_2.$$

THEOREM. *Put $R = \operatorname{Im} \mu_2$. Then $(B_1 \otimes B_1/R) \otimes \mathbf{Q} \cong K_2(k) \otimes \mathbf{Q}$ by a map which sends $p \otimes q$ to the symbol $\{p, q\}$.*

COROLLARY. *In the category $\mathcal{M}(A^*)$, we have that $\operatorname{Ext}^p(\mathbf{Q}(0), \mathbf{Q}(p)) \cong K_p^M(k) \otimes \mathbf{Q}$.*

This follows from the theorem and §2.1.8.

REFERENCES

[BBD]   A. BEILINSON, J. BERNSTEIN, AND P. DELIGNE, *Faisceaux Pervers*, *Analyse et topologie sur les espaces singuliers*, Astérisque **100** (1982), 7–171.

[B1]   A. BEILINSON, *Higher regulators and values of L-functions*, Itogi Nauki i Tekhniki, Seriya Sovremennye Problemy Matematiki (Noveishie Dostizheniya) **24** (1984), 181–238.

[B2]   ______, *Height pairings between algebraic cycles*, Preprint, 1984.

[B3]   ______, *Notes on absolute Hodge cohomology*, Contemporary Mathematics **55**, Part I (1986), 35–68.

[Bl1]   S. BLOCH, *Higher regulators*, preprint.

[Bl2]   ———, *Algebraic cycles and higher K-theory*, Advances in Math. **61** (1986), 267–304.

[C]   P. CARTIER, *Decomposition des polyedres: le point sur le troisieme probleme de Hilbert*, Seminaire Bourbaki No. **646**, Juin 1985.

[FM]   W. FULTON AND R. MACPHERSON, *Categorical framework for the study of singular spaces*, Memoirs of the Amer. Math. Soc. **31**, No. 243 (1981).

[GGL]   A. M. GABRIELOV, I. M. GELFAND, AND M. V. LOSIK, *Combinatorial computation of characteristic classes*, Functional Analysis and its Applications **9** No. 2 (1975), 103–115 and **9** No. 3 (1975) 5–26 (in Russian).

[GM]   I. M. GELFAND AND R. MACPHERSON, *Geometry in Grassmannians and a generalization of the dilogarithm*, Advances in Mathematics **44** (1982), 279–312.

[G]   H. GILLET, *Riemann-Roch Theorem in Higher K-Theory*, Advances in Math. **19**.

[HM]   R. HAIN AND R. MACPHERSON, *Higher logarithms*, to appear.

[K]   F. KIRWAN, *Cohomology of quotients in symplectic and algebraic geometry*, Princeton Univ. Press Math. Notes No. 31 (1984).

[La]   S. E. LANDSBURG, *Relative cycles and algebraic K-theory*, preprint, 1985.

[L1]   S. LICHTENBAUM, *Values of zeta-functions at non-negative integers*, Journées Arithmetiques, Noordwykinhoot, Netherlands, Springer Verlag, 1983.

[L2]   ———, *The construction of weight two arithmetic cohomology*, to appear in Inventiones Math.

[Ma]   S. MACLANE, *Homology*, Academic Press, New York, 1963.

[Mac]   R. MACPHERSON, *The combinatorial formula of Gabrielov, Gelfand, and Losik for the first Pontrjagin class*, Seminaire Bourbaki 497, Fev. 1977.

[MV]   R. MACPHERSON AND K. VILONEN, *Elementary construction of perverse sheaves*, Inventiones Math. **84** (1986), 403–435.

[M]   YU. I. MANIN, *Correspondences, motifs, and monoidal transformations*, Mat. Sb. (N.S.) **77**, **119** (1968), 475–507.

[MM]   J. MILNOR AND J. MOORE, *On the structure of Hopf algebras*, Ann. of Math. (2) **81** (1965), 211–264.

[Q]   D. QUILLEN, *Higher algebraic K-theory* I, Proc. conf. on higher algebraic K-theory, Springer Lect. Notes in Math. **341** (1973), 85–147.

[RS]   C. ROURKE AND B. SAUNDERSON, *Δ-sets* I: *Homotopy theory*, Quart Jour. Math. Oxford (ser2) **22** (1971), 321–338.

[So]   C. SOULÉ, *Operations en K-theorie Algebrique*, Can. J. Math. **27** (1985), 488–550.

[S1]   A. A. SUSLIN, *Homology of* GL_n, *characteristic classes, and Milnor's K-theory*, Proceedings of the Steklov Institute of Mathematics 1985, Issue 3, 207–226 and Springer Lect. Notes in Math. **1046** (1984), 357–375.

[S2]   ———, lecture, Moscow, 1983.

[Yo]   YONEDA, *On* Ext *and exact sequences*, J. Fac. Sci. Univ. Tokyo, Sect. I, **8** (1960), 507–576.

[Y]   B. V. YUSIN, *Sur les formes* $S^{p,q}$ *apparaissant dans le calcul combinatoire de la deuxieme classe de Pontrjaguine par la methode de Gabrielov, Gelfand, et Losik*, C. R. Acad. Sci. Paris, Ser I Math. **292** (1981), 641–644.

MACPHERSON: DEPARTMENT OF MATHEMATICS, BROWN UNIVERSITY, PROVIDENCE, RHODE ISLAND 02912

Vol. 54, No. 2   DUKE MATHEMATICAL JOURNAL ©   1987

REFINED CONJECTURES OF THE "BIRCH AND SWINNERTON-DYER TYPE"

B. MAZUR and J. TATE

To Yuri Manin on the occasion of his fiftieth birthday

The idea behind the present article is that, in certain instances, arithmetic conjectures concerning the special values of derivatives of p-adic L functions can be "refined" to obtain formulations of *stronger* conjectures. These stronger conjectures *avoid any mention of p-adic L functions* and therefore obviate the necessity of constructing the p-adic L functions for the statement of the conjectures. Moreover, they avoid any reference to a prime number p, and require no p-adic limiting process; they should ultimately be phrased, perhaps, in adelic language. In this paper, however, we state our conjectures "at a finite layer M", where M is a possibly composite number (somewhat restricted). Even when $M = p$, however, our conjecture "at layer p" is not implied by the analogous conjecture for the p-adic L function. Indeed, our conjecture predicts congruence formulas modulo divisors of $p - 1$, in this case. When M is a product of distinct primes, our conjectured congruence formulas involve what seems to us to be a thoroughgoing mixture of phenomena related to those prime divisors.

In 1974 ([Man]) Yuri Manin expressed the hope that there should exist "functions with adelic type domains of definition and ranges of values" for which p-adic L functions are only a component. We are pleased, on the occasion of Manin's fiftieth birthday, to be able to offer, at least, a very fragmentary piece of "experimental mathematics" which is resonant with Manin's hope.

Here is the specific context in which we work: Let $A_{/\mathbf{Q}}$ be an elliptic curve admitting a modular parametrization,

$$f: X_0(N)_{/\mathbf{Q}} \to A_{/\mathbf{Q}}.$$

In [M-T-T], p-adic analogues of the conjectures of Birch and Swinnerton-Dyer for $A_{/\mathbf{Q}}$ were formulated, in the case when p has ordinary reduction for A. Numerical evidence was gathered in support of these conjectures. The conjectures of [M-T-T] are formulas, whose *left hand sides* are special values of derivatives of the p-adic L function $L_p(A, s)$ and whose *right hand sides* are expressions involving arithmetic invariants of $A_{/\mathbf{Q}}$ among which is a "regulator term", or, as it is referred to in [M-T-T], the "*sparsity*". In the case where p is a prime of split multiplicative reduction for A, the "regulator term" involves the p-adic logarithm

Received February 4, 1987.

711

of $q_p(A)$, *the p-adic multiplicative period of A*, as well as the canonical p-adic height pairing on the Mordell-Weil group $A(\mathbf{Q})$.

As for the *left hand side*, the p-adic L function is obtained as the p-adic Mellin transform of a p-adic measure obtained from the modular symbols

$$[\ ]_A : \mathbf{Q} \to \mathbf{Q}$$

attached to the curve A. The p-adic measure is obtained from the above modular symbols by a well-known "smoothing process".

To indicate how we came upon the notion that there might indeed be "refined Birch and Swinnerton-Dyer conjectures", consider the following conjectural formula, which is essentially a special case of Conjecture 1 [M-T-T], Chap. 11 §13. Let p be of split multiplicative reduction for A, and let $m_p = \mathrm{ord}_p(q_p(A))$. Put

$$\tilde{q}_p = q_p(A)/p^{m_p} \in \mathbf{Z}_p^*.$$

For any integer N, (if the relevant modular symbols are p-integral) we have the conjectural congruence mod p^N:

"Old conjecture":

$$m_p \cdot \sum_{\substack{a \bmod p^N \\ (a,p)=1}} [a/p^N]_A \cdot \log_p(a) \stackrel{?}{\equiv} [0]_A \cdot \log_p(\tilde{q}_p).$$

Last Spring, Oesterlé communicated to us some of his conjectures concerning the reduction mod p of $\tilde{q}_p$. Our first impulse, upon hearing Oesterlé's conjectures was to try to amalgamate the "old conjecture" above with Oesterlé's conjecture, so as to obtain a conjectural formula of $\tilde{q}_p$ itself (in cases where $[0]_A$ is not a bothersome factor).

But then, a more naive—but at that point, utterly unjustified—strategy for obtaining a conjectural formula for $\tilde{q}_p$ occurred to us: "Exponentiate the above conjecture!" Specifically,

Refined conjecture:

$$\prod_{\substack{a \bmod p^N \\ (a,p=1)}} a^{m_p \cdot [a/p^N]_A} \stackrel{?}{\equiv} \tilde{q}^{[0]_A} \quad \text{in } (\mathbf{Z}/p^N\mathbf{Z})^*.$$

Visibly, the "old conjecture" for general N, and the refined conjecture for $N = 1$ imply the refined conjecture for general N. Thus our "refinement" amounts to one extra equality in $(\mathbf{Z}/p\mathbf{Z})^*$. We have computed enough special cases to convince ourselves that this "refinement" is justified.

With some reflection, one sees that this naive process of refinement is something that one might try out on many conjectures involving special values of derivatives of p-adic L functions. At a seminar at Harvard, we presented various such "refined conjectures," Gross and Hayes framed "refined" versions of the p-adic Gross-Stark conjectures and provided numerical evidence in their support, and Glenn Stevens contributed some key ideas and important numerical data. What emerged in the course of our seminar is the following "sophisticated" format of the naive refinement above.

We define, for any integer $M \geq 1$, the *modular element*

$$\theta_{A, M} = \sum_a [a/M]_A \sigma_a \in \mathbb{Q}[(\mathbb{Z}/M\mathbb{Z})^*/(\pm 1)],$$

where a runs through a system of integers mod M, representing elements σ_a in $(\mathbb{Z}/M\mathbb{Z})^*/(\pm 1)$. We view $\theta_{A, M}$ as our analogue "at layer M" of the L function of A. Let R be a subring of $\mathbb{Q}$ containing the coefficients of $\theta_{A, M}$. Let $I \subset R[(\mathbb{Z}/M\mathbb{Z})^*/(\pm 1)]$ be the augmentation ideal. The analogue of saying that the "L function vanishes to order $\geq r$ at $s = 1$" is simply to say that $\theta_{A, M}$ is contained in the r-th power of the augmentation ideal I.

If $\theta_{A, M}$ lies in I^r (i.e., "vanishes to order $\geq r$ at $s = 1$"), the analogue of the "r-th coefficient of the Taylor expansion of the L function at $s = 1$" is simply the *image $\tilde{\theta}_{A, M} \in I^r/I^{r+1}$* of $\theta_{A, M}$.

Our "refined Birch and Swinnerton-Dyer conjectures" *at layer M* will then be statements about (a) the "order of vanishing" of the element $\theta_{A, M}$ and (b) the image of its "leading coefficient" in I^r/I^{r+1}, if the order of the torsion subgroup of $A(\mathbb{Q})$ is invertible in R.

As for "order of vanishing", let s denote the number of primes of split multiplicative reduction for A which divide M. Let $r = \operatorname{rank} A(\mathbb{Q}) + s$. We conjecture that $\theta_{A, M}$ lies in I^r.

To produce a conjectural formula for the "r-th Taylor coefficient" of $\theta_{A, M}$ is significantly more difficult. In this paper we do this only under the hypothesis that if p is not split multiplicative for A then p^2 does not divide M. This is not an obligatory hypothesis when p is a prime of ordinary reduction for A (for the techniques of [M-T-T] could be used even if p^2 divides M) but in the case when p is of supersingular reduction for A, the limitations of our present technique force such a hypothesis upon us.

We define a "regulator term" by a circuitous process using "tame height pairings." See Chapter 2 below in which we manage to define an element in $R \otimes \operatorname{Sym}_r((\mathbb{Z}/M\mathbb{Z})^*/(\pm 1))$ called the "corrected discriminant", where R is any commutative ring in which the order of the torsion subgroup in $A(\mathbb{Q})$ is invertible. We then project this "corrected discriminant" to $R \otimes I^r/I^{r+1}$ via the natural mapping. The *right hand side* of our conjectural formula is obtained by multiplying this projected "corrected discriminant" by a scalar term (which involves, among other things, the order of the Shafarevich group III; see Conjecture 4 of Chapter 3).

Since our conjectural formula takes place in the R-module $R \otimes (I^r/I^{r+1})$, it is fortunate that much is known about the structure of I^j/I^{j+1} ($j \geqslant 0$) where I is the augmentation ideal in the group ring $\mathbb{Z}[G]$, for G a finite abelian group ([Ha], [Pa]).

Why do we call the "corrected discriminant" *corrected*? The answer is that it corresponds (in a fairly precise way; see the heuristical explanations in (2.7)) to the regulator term of the classical Birch and Swinnerton-Dyer formula multiplied by some of the "fudge factors" that also occur on the right hand side of that conjectured formula. It seems unlikely to us that, in general, an analogue of a "discriminant" (which does not include some of these auxiliary factors) can be defined, in our context.

We have referred to the process of its definition as "circuitous" because the "corrected discriminant" is not defined as the determinant of a matrix constructed from a *single* self-pairing. Rather, our "corrected discriminant" is given as a weighted alternating sum of discriminants coming from tame height pairings at various different layers. Is there an elegant algebraic format to handle such "corrected discriminants"? Although most of our paper is in the context of elliptic curves over $\mathbb{Q}$ the definition of "corrected discriminants" is given in Chapter 2 for an arbitrary abelian variety over an arbitrary global field.

As we have already mentioned, our "Birch and Swinnerton-Dyer-type conjecture" at layer M involves a serious mixing of information relative to the various primes dividing M. For a striking example of this, consider conjecture 5 of Chapter 3 below, where, in the case when $A(\mathbb{Q})$ is finite, and all primes dividing M are of split multiplicative reduction for A, our conjectured formula connects the "leading coefficient" $\tilde{\theta}_{A,M}$ to a determinant built out of the various p-adic multiplicative periods $q_p(A)$ for *all* primes p dividing M.

At the end of Chapter 3 we provide a certain amount of numerical evidence in support of these conjectures. We have checked many instances of our conjecture, including cases where M is a product of two primes of split multiplicative reduction for A, and cases where M is divisible by a prime of supersingular reduction for A, or of non-split multiplicative reduction. However we have not yet tested any case in which M is divisible by a prime of additive reduction for A.

Table of Contents

Chapter 1. Modular elements (*The left hand side*).

Chapter 1. Modular elements (*The left hand side*).

(1.1) *Modular symbols.* Let $A_{/\mathbf{Q}}$ be an elliptic curve and ω a Néron differential for $A_{/\mathbf{Q}}$, i.e., a regular differential on $A_{/\mathbf{Q}}$ which extends to a regular differential on the Néron model of A over the base $\mathbf{Z}$. Such an ω is unique up to sign. By the *Néron lattice* $\Lambda_A \subset \mathbf{C}$ we mean the set of "periods" $\int_\gamma \omega \in \mathbf{C}$, where γ runs through all closed oriented (piece-wise C^∞) 1-cycles on the Riemann surface $A(\mathbf{C})$.

Let Ω_A^+, Ω_A^- denote the unique pair of positive real numbers such that one of these two conditions holds:

(I) ("rectangular case")

$$\Lambda_A = \mathbf{Z} \cdot \Omega_A^+ + \mathbf{Z} \cdot \sqrt{-1} \cdot \Omega_A^-$$

(II) ("nonrectangular case")

$$\Lambda_A \subset \mathbf{Z} \cdot \Omega_A^+ + \mathbf{Z} \cdot \sqrt{-1} \cdot \Omega_A^-$$

is the sublattice consisting of the complex numbers $a \cdot \Omega_A^+ + b \cdot \sqrt{(-1)}\,\Omega_A^-$ such that $a \equiv b \bmod 2$. Note that $\Omega_A^+ = \frac{1}{2}\int_{A(\mathbf{R})}|\omega|$ so that in the "nonrectangular case" Ω_A^+ is a "half-period" rather than a period.

By a celebrated conjecture to which the names Shimura, Taniyama, and Weil have been attached, any elliptic curve $A_{/\mathbf{Q}}$ can be parametrized by a $\mathbf{Q}$-rational mapping from the modular curve $X_0(N)$, where N is the arithmetic conductor of A. This, together with the theorem of Manin-Drinfeld (cf. [La], Chap. IV, Thm. 2.1) implies that there is a unique cuspidal newform ν of weight two on $\Gamma_0(N)$, $\nu(z) = 1 \cdot q + \lambda_2 q^2 + \cdots + \lambda_n q^n + \cdots$. ($q = e^{2\pi i z}$) such that $\sum_{n \geqslant 1}\lambda_n n^{-s} = L_A(s)$ is the Hasse-Weil L function of A, and such that for any rational number $a/b \in \mathbb{Q}$, we can express the integral of v along the vertical line $z = a/b + iy$ $(0 \leqslant y \leqslant \infty)$ as follows:

$$+2\pi \int_0^\infty v\left(\frac{a}{b} + iy\right) dy = \left[\frac{a}{b}\right]_A^+ \cdot \Omega_A^+ + \left[\frac{a}{b}\right]_A^- \cdot \sqrt{-1}\,\Omega_A^-$$

where

$$\left[\frac{a}{b}\right]_A^+, \left[\frac{a}{b}\right]_A^-$$

are rational numbers. The rational numbers

$$\left[\frac{a}{b}\right]_A^+, \left[\frac{a}{b}\right]_A^-$$

are called the "$+$" and "$-$" *modular symbols* for A, evaluated at a/b, and are clearly dependent only upon $a/b \bmod 1$.

We are primarily interested in the "$+$" modular symbol, so we put

$$\left[\frac{a}{b}\right]_A := \left[\frac{a}{b}\right]_A^+.$$

(1.2) *Modular elements.* Fix an integer $M > 0$ and put $G_M :=$ $(\mathbb{Z}/M\mathbb{Z})^*/(\pm 1)$. We may view G_M alternatively as the Galois group of an abelian extension of $\mathbb{Q}$ or as a quotient of the idele class group of $\mathbb{Q}$ via the standard isomorphisms:

$$G_M \xrightarrow[\cong]{} \mathrm{Gal}\big(\mathbb{Q}(\mu_M)^+/\mathbb{Q}\big) \xleftarrow[\cong]{} \mathbb{Q}^* \backslash A_\mathbb{Q}^* / U_M$$

where the isomorphism on the left comes from letting, for $a \in (\mathbb{Z}/M\mathbb{Z})^*$, σ_a be the automorphism of $\mathbb{Q}(\mu_M)^+$ induced by "raising to the a-th power" in μ_M, and we choose the sign of the isomorphism on the right so that σ_a is associated to the idele with components a at primes dividing m and 1 at the other places (here U_M is the subgroup of ideles given by the condition that their p-th component, for any finite prime p, be congruent to 1 mod $M \cdot \mathbb{Z}_p$). If a is an integer prime to M, then we sometimes let σ_a denote the associated element in G_M.

We define the *modular element* $\theta_{A,M} \in \mathbb{Q}[G_M]$ by the formula:

$$\theta_{A,M} = \frac{1}{2} \sum_{a \bmod M} \left[\frac{a}{M}\right]_A \cdot \sigma_a.$$

Remark. Our conjectures in this article involve S-pairings where S is a finite set of prime numbers, i.e., nonarchimedean places of $\mathbb{Q}$. If we were to include ∞, our first guess would be to use modular elements defined by

$$\theta_{A,\,M\cdot\infty} = (-1)^{2/n_\infty} \cdot \frac{n_\infty}{2} \sum_a \left(\left[\frac{a}{M}\right]_A^+ + \left[\frac{a}{M}\right]_A^- \right) \cdot \sigma_a$$

in $\mathbb{Q}[(\mathbb{Z}/M\mathbb{Z})^*]$, where the summation is over all a mod M, relatively prime to M, and where σ_a denotes the element in the group ring corresponding to the congruence class of a mod M Here n_∞ denotes the number of components of $A(\mathbb{R})$, i.e., is 2 if Λ_A is "rectangular" and is 1 if not.

(1.3) *Compatibility formulas for modular elements.* Let M' be a multiple of M and let

$$z_M \colon \mathbb{Q}[G_{M'}] \to \mathbb{Q}[G_M]$$

be the natural projection. Let

$$v_{M'} \colon \mathbb{Q}[G_M] \to \mathbb{Q}[G_{M'}]$$

be the trace mapping which for $x \in G_M$, sends $[x] \in \mathbb{Q}[G_M]$ to $\sum_{x'}[x'] \in \mathbb{Q}[G_{M'}]$, where x' ranges through all elements of $G_{M'}$ projecting to x in G_M. Using the classical formulas for the correspondences $T_\ell(\ell \nmid N)$ and $U_\ell(\ell | N)$ on the upper half plane, a straightforward computation yields:

Let ℓ be a prime number.
(1) If $\ell \nmid M$, $\ell \nmid N$, then:

$$z_M(\theta_{A,\,M\ell}) = \left(\lambda_\ell - \sigma_\ell - \sigma_\ell^{-1}\right) \cdot \theta_{A,\,M}.$$

(2) If $\ell \nmid M$, $\ell | N$, then:

$$z_M(\theta_{A,\,M\ell}) = \left(\lambda_\ell - \sigma_\ell^{-1}\right) \cdot \theta_{A,\,M}.$$

(3) If $\ell | M$, $\ell | N$, then:

$$z_M(\theta_{A,\,M\ell}) = \lambda_\ell \cdot \theta_{A,\,M}.$$

(4) If $\ell | M$, $\ell \nmid N$, then:

$$z_M(\theta_{A,\,M\ell}) - \lambda_\ell \cdot \theta_{A,\,M} + v_M(\theta_{A,\,M/\ell}) = 0.$$

(1.4) *Special values of L-functions and modular elements.* We state the connection between $\theta_{A,\,M}$ and special values of L functions only in the case where $(M, N) = 1$ and we are given an even Dirichlet character $\chi \colon G_M \to \mathbb{C}^*$ which is primitive, of conductor M. Denote by the same letter $\chi \colon \mathbb{Q}[G_M] \to \mathbb{C}$ the induced ring homomorphism. Let $\tau(\chi) = \Sigma\chi(a) \cdot \exp(2\pi i a/M)$ denote the Gauss sum.

Let

$$L(A, \chi, s) = \sum \lambda_n \chi(n) n^{-s}$$

be the L function on the elliptic curve A, twisted by χ. This complex Dirichlet series extends, cf. [Sh], to an entire function, since A is endowed with a modular parametrization.

We have the formula (cf., e.g., [M-T-T])

$$(1) \qquad \chi(\theta_{A, M}) = \frac{\tau(\chi) \cdot L(A, \bar{\chi}, 1)}{2 \cdot \Omega_A^+} .$$

(1.5) *Weak vanishing conjectures.*   Let A, M be as above, and let

$$\chi \colon G_M \to \mathbb{C}^*$$

be a homomorphism, i.e, χ is a *not necessarily primitive*, even, Dirichlet character (of conductor M). Let $R \subset \overline{\mathbb{Q}}$ be a ring containing the coefficients of $\theta_{A, M}$. Let $R[\chi] \subseteq \overline{\mathbb{Q}}$ be the R-algebra generated by the values of χ. Let

$$\tilde{\chi} \colon R[\chi][G_M] \to R[\chi]$$

be the ring homomorphism obtained by sending $x \in G_M$ to $\chi(x) \in R[\chi]^*$. Define I_χ, the *augmentation ideal at* χ, to be the kernel of $\tilde{\chi}$. Given an element $\vartheta \in R[\chi][G_M]$, say that ϑ *vanishes to* ∞ *order at* χ if ϑ is contained in all powers of I_χ. Otherwise, say that it *vanishes to order* r (*at* χ) if ϑ is contained in I_χ^r but not in I_χ^{r+1}. If ϑ vanishes to finite order r, define the *leading term of* ϑ (*at* χ) to be the image $\tilde{\vartheta}$ of ϑ in I_χ^r / I_χ^{r+1}.

By the χ-*part of Mordell-Weil* of $A_{/\mathbb{Q}}$ we mean the complex vector space V_χ given as the subspace of $A(\mathbb{Q}(\mu_M)^+) \otimes_{\mathbb{Q}} \mathbb{C}$ consisting of all vectors v such that

$$\sigma_A \cdot v = \chi(a) \cdot v$$

for all a relatively prime to M.

CONJECTURE 1 ("Weak vanishing").   *The order of vanishing of $\theta_{A, M}$ at χ is greater than or equal to the dimension of the χ-part of the Mordell-Weil group of $A_{/\mathbb{Q}}$.*

Remark.   Note that there are cases where the order of vanishing is strictly greater than the dimension of the χ-part of Mordell-Weil. A banal instance of this is, e.g., when $R[\chi] = \overline{\mathbb{Q}}$ in which case $I_\chi^r = I_\chi$ for all $r \geqslant 1$. But there are more interesting instances, e.g., connected to the "exceptional cases" of [M-T-T], and (3.3) below.

As for the connection between the "order of vanishing" of $\theta_{A,M}$ at χ, and the order of vanishing of the classical L function $L(A, \chi, s)$ at $s = 1$ we can prove nothing along these lines that is not a trivial consequence of formula (1).

Nevertheless, combining the above conjecture with the "classical" conjecture of Birch and Swinnerton-Dyer, we are led to formulate

CONJECTURE 2. *The order of vanishing of $\theta_{A,M}$ at χ is greater than or equal to the order of vanishing of $L(A, x, s)$ at $s = 1$.*

(1.6) *Parity of vanishing and the functional equation.* Let N and M be as above, and assume that N can be decomposed into the product $Q \cdot Q'$ of two relatively prime factors such that $(Q, M) = 1$ and Q' divides M. Denote by ν the newform (of weight two on $\Gamma_0(N)$) parametrizing A. Let w_Q denote the standard involution, attached to Q, of cuspforms of weight 2 on $\Gamma_0(N)$ (cf. [A-L], [M-T-T] Chap. 1 §5). Then ν is an eigenvector for the action of w_Q. Let ε denote the *negative* of the sign of its eigenvalue under the action of w_Q. We have the functional equation for modular symbols:

$$(1.6.1) \qquad\qquad [a/M]_A = \varepsilon[a'/M]_A,$$

where a' is an integer such that $a'aQ \equiv -1 \bmod M$ ([M-T-T], Chap. 1 §6).

This translates into the following functional equation for the modular elements:

$$(1.6.2) \qquad\qquad \theta_{A,M} = \varepsilon \cdot \sigma_Q^{-1} \cdot \theta_{A,M}^*,$$

where $x \to x^*$ denotes the canonical involution of $\mathbb{Q}[G_M]$ induced by sending σ_a to σ_a^{-1} for all a relatively prime to M.

The involution $*$ induces the automorphism multiplication by (-1) on I/I^2 and hence multiplication by $(-1)^r$ on I^r/I^{r+1}.

Let $R \subset \mathbb{Q}$ be a subring containing the coefficients of $\theta_{A,M}$, and let I denote the augmentation ideal in the ring $R[G_M]$.

Suppose that $\theta_{A,M}$ vanishes ("exactly") to order $r(< \infty)$. Then, by (1.6.2), we have:

$$(1.6.3) \qquad \tilde{\theta}_{A,M} = \varepsilon\sigma_Q^{-1} \cdot \tilde{\theta}_{A,M}^* = (-1)^r \cdot \varepsilon \cdot \tilde{\theta}_{A,M} \in I^r/I^{r+1}$$

where $\tilde{\ }$ denotes image in I^r/I^{r+1}. It follows that if multiplication by 2 is an automorphism of I^r/I^{r+1} (e.g., if 2 is invertible in R) then

$$(1.6.4) \qquad\qquad (-1)^r = \varepsilon,$$

i.e., we can predict the parity of the order of vanishing of $\theta_{A,M}$ purely in terms of the sign of the functional equation for the newform ν.

(1.7) *A weak "main conjecture."* By the (*pro-finite*) *Selmer group* of the elliptic curve A over a number field K, $\hat{S}(A/K)$, we mean the group which fits into the following cocartesian diagram:

$$(1.7.1)$$

$$
\begin{array}{ccc}
H^1(\mathrm{Gal}(\overline{K}/K), A(\overline{K}))^* & \overset{j^*}{\hookrightarrow} & H^1(\mathrm{Gal}(\overline{K}/K), A(\overline{K})_{\mathrm{tors}})^* \\
\downarrow{\scriptstyle i^*} & & \downarrow \\
\mathrm{III}(A/K)^* & \rightarrow & \hat{S}(A/K)
\end{array}
$$

where III denotes the Shafarevich group, $*$ means Pontrjagin duality, and i, j are the natural mappings.

The (profinite) Selmer group is naturally a $\hat{\mathbb{Z}}$-module, and also fits into the exact sequence which is the first horizontal line of the diagram below:

$$
\begin{array}{ccccccccc}
0 & \rightarrow & \mathrm{III}(A/K)^* & \rightarrow & \hat{S}(A/K) & \rightarrow & \mathrm{Hom}(A(K), \hat{\mathbb{Z}}) & \rightarrow & 0 \\
 & & \uparrow{\scriptstyle =} & & \uparrow & & \uparrow & & \\
0 & \rightarrow & \mathrm{III}(A/K)^* & \rightarrow & S(A/K) & \rightarrow & \mathrm{Hom}(A(K), \mathbb{Z}) & \rightarrow & 0.
\end{array}
$$

Define the *integral Selmer group* $S(A/K)$ to be the full inverse image in $\hat{S}(A/K)$ of $\mathrm{Hom}(A(K), \mathbb{Z}) \subset \mathrm{Hom}(A(K), \hat{\mathbb{Z}})$ as indicated in the diagram. If $\mathrm{III}(A/K)$ is finite, then $S(A/K)$ is a finitely generated abelian group—we assume this.

When $K/\mathbb{Q}$ is Galois, with Galois group G, then G acts on $\hat{S}(A/K)$ and $S(A/K)$ in a natural way, via its action on the Galois cohomology groups in (1.7.1). Thus we may view $S(A/K)$ as a $\mathbb{Z}[G]$-module. If G is abelian, we let $\Phi(A/K) \subset \mathbb{Z}[G]$ denote the (initial) Fitting ideal (cf. [N]) of the $\mathbb{Z}[G]$-module $S(A/K)$.

CONJECTURE 3. ("Weak main conjecture"). *Let M be a positive integer, and let $K = \mathbb{Q}(\mu_M)^+$ so that $K/\mathbb{Q}$ is Galois, with Galois group $G = G_M$. Let R be a subring of $\mathbb{Q}$ containing the coefficients of $\theta_{A,M}$. Then*

$$
\theta_{A,M} \in \Phi(A/K) \bigotimes_{\mathbb{Z}} R \subseteq R[G_M].
$$

Remark. We do not expect that $\theta_{A,M}$ be a generator (in general) for the ideal $\Phi(A/K) \otimes_{\mathbb{Z}} R$, especially in the context of "extra vanishing". It would be very instructive to formulate a convincing "strong main conjecture", i.e., to describe explicitly a natural module containing $S(A/K)$, whose Fitting ideal can be described concretely (say, in terms of modular symbols).

PROPOSITION 3. *Conjecture 3 implies Conjecture 1.*

Proof. Let χ be as in Conjecture 1. Suppose conjecture 3 to be true.

We shall show that $\Phi(A/K)$ is contained in I_χ^r, where r is the rank of the χ-part of the Mordell-Weil group of A. Let R be as in conjecture 3.

Let H denote the $R[\chi]$-module, $\operatorname{Hom}(A(K), Z) \otimes_{\mathbf{Z}[G_M]} R[\chi]$, where the tensor product is via $\tilde{\chi}: \mathbf{Z}[G_M] \to R[\chi]$. We have that H maps surjectively to a locally free $R[\chi]$-module of rank r. Hence $S(A/K)$ maps surjectively to a locally free $R[\chi][G_M]/I_\chi$ module of rank r (the surjection being compatible with $\mathbf{Z}[G_M]$-module structure). It follows that $\Phi(A/K)$ is contained in the $R[\chi][G_M]$-Fitting ideal of a locally free $R[\chi][G_M]/I_\chi$ module of rank r, i.e., it is contained in I_χ^r as was to be proved.

Chapter 2. Canonical heights and "corrected" discriminants (*The right hand side*)

(2.1) *Local modifications and trivializations.* The first part of this section can be viewed as a commentary on our paper [M-T]. We shall make some minor improvements on results proved there. Keep the notation of [M-T] §1 until we signal otherwise. Thus, K is a field complete with respect to a place v which is archimedean or discrete. If v is discrete, $\mathcal{O} \subset K$ denotes its ring of integers and $k = \mathcal{O}/\pi\mathcal{O}$ its residue field.

Let $A_{/K}$ be an abelian variety and $B_{/K}$ its dual. Let $E_{/K}$ be the (canonical) biextension of $(A_{/K}, B_{/K})$ by $\mathbf{G}_m$ expressing the duality. Taking K-rational points, we obtain $E(K)$, a set-theoretic biextension of $(A(K), B(K))$ by $\mathbf{G}_m(K) = K^*$.

By a *modification* $(E', \alpha, \beta, \rho)$ of $E(K)$ we mean the biextension E' obtained from $E(K)$ by pullback via two homomorphisms

$$\alpha: P \to A(K), \qquad \beta: Q \to B(K)$$

and pushout via a homomorphism

$$\rho: K^* \to G,$$

together with the triple (α, β, ρ) of homomorphisms relating E' to $E(K)$. By a *trivialization* $(\alpha, \beta, \rho, \psi)$ of $E(K)$ we mean a modification as above and a splitting ψ of E'. Such a splitting is the same thing as a ρ-splitting ψ (cf. [M-T]) of the pullback $(\alpha^*, \beta^*)(E(K))$, i.e., a map

$$(\alpha^*, \beta^*)(E(K)) \overset{\psi}{\to} G$$

with certain properties, and we shall usually view ψ in that way. In §2.2 we will give several examples of natural trivializations of $E(K)$.

A useful lemma is

LEMMA. *Let P, Q and G be abelian groups, and X a biextension of (P, Q) by G. Let P_0 be a subgroup of P, let X_0 be the part of X above $P_0 \times Q$, and suppose $\psi_0: X \to G$ is a splitting of X_0. Let M be an integer $\geqslant 1$. If either of the following* *two conditions holds, then ψ_0 has a unique extension to a splitting ψ of X. (In particular, if either condition holds with $P_0 = (0)$, then X has a unique splitting.)*

(a) One of the three groups $P/P_0, Q, G$ is killed by M, and another is uniquely divisible by M.

(b) P/P_0 and Q are divisible by M, and G is killed by M.

Proof. That X splits follows easily using the commutative diagram made with the exact sequences (cf. [SGA7], exp VIII, 1.1.4)

$$0 \to \mathrm{Ext}^1(\tilde{P}, \mathrm{Hom}(Q, G)) \to \mathrm{Biext}(\tilde{P}, Q; G) \to \mathrm{Hom}(\tilde{P}, \mathrm{Ext}(Q, G)),$$

for $\tilde{P} = P_0$, P, and P/P_0, together with the fact that the groups $\mathrm{Ext}^1(P/P_0, \mathrm{Hom}(Q, G))$ and $\mathrm{Hom}(P/P_0, \mathrm{Ext}^1(Q, G))$ are zero under either of the two conditions (a) or (b). That ψ_0 has a *unique* extension follows from the fact that the group

$$\mathrm{Hom}(P/P_0, \mathrm{Hom}(Q, G))$$

is zero under either of the two conditions.

(2.2) *Examples of canonical trivializations of $E(K)$.* In each of the following examples, the "pushout" map ρ will be the canonical map from K^* to some quotient K^*/Y of itself, and in all but the last example, the "pullback" maps α and β will be inclusions of some subgroups $P \subset A(K)$ and $Q \subset B(K)$. Thus, in these cases, the modification is determined by the triple of groups $(P, Q, K^*/Y)$, and its splitting ψ is a map from the part of $E(K)$ above $P \times Q$ to K^*/Y. Only examples (3), (4) and (6) will be used in the sequel. For those, and also for the new part of example (2) we will give a down-to-earth description of the splitting in terms of zero-cycles and divisors, at least in the elliptic curve case. For this, recall (cf. [M-T] (2.2)) that a ρ-splitting ψ of $E(K)$ determines and is determined by a function $\tilde{\psi}$ of pairs $(\mathfrak{a}, D)$ consisting of a zero cycle $\mathfrak{a}$ made up of K-rational points of A and a divisor D on $A_{/K}$ such that $\mathfrak{a}$ is of degree 0, D is algebraically equivalent to 0, and $\mathfrak{a}$ and D have disjoint supports. Such a function $\tilde{\psi} = \tilde{\psi}(\mathfrak{a}, D)$ corresponds to a splitting if and only if it is biadditive, satisfies $\tilde{\psi}(\mathfrak{a}, (f)) = \rho(f(\mathfrak{a}))$ for rational functions f on A/K, and is invariant under simultaneous translation of the arguments $\mathfrak{a}$ and D. By slight variants of this principle, adapted to the modifications of $E(K)$ in question, it is easy to check that our explicit description of $\tilde{\psi}$ for elliptic curves is the correct one in each case. We carry out the details in the last example, and leave the others to the reader. Here are the examples.

(1) *The archimedean Néron unramified trivialization.* For v archimedean and $U = \{c \in K^*: |c| = 1\}$, the modification of $E(K)$ determined by the triple $(A(K), B(K), K^*/U)$ has a unique continuous splitting

$$\psi: E(K) \to K^*/U \xrightarrow{\sim} \mathbb{R}; \text{ cf. [M-T], (1.8.1).}$$

(2) *The real ramified trivialization.* For v real, let A^0 and B^0 denote the connected components of $A(K)$ and $B(K)$, respectively. The modification of $E(K)$ determined by the triple (A^0, B^0, K^*) has a unique continuous splitting. Indeed, in view of the direct product decomposition $K^* = \mathbb{R}^* = (\pm 1) \times \mathbb{R}_+^*$, it suffices to treat separately the triples $(A^0, B^0, \mathbb{R}^*/\mathbb{R}_+^*)$ and $(A^0, B^0, \mathbb{R}^*/(\pm 1))$. For the first we can use part (b) of the lemma in §2.1, since A^0 and B^0 are 2-divisible. For the second, use the restriction of the unramified splitting (1).

For an elliptic curve A, the connected component A^0 is a "circle group." If

$$\mathfrak{a} = \sum m_i(P_i) \quad \text{and} \quad D = \sum n_j(Q_j)$$

are zero cycles of degree 0 made up of points P_i and Q_j of A^0, then the splitting for the first of the two triples above corresponds to the function $\tilde{\psi}$ with values in $\mathbb{R}^*/\mathbb{R}_+^* = (\pm 1)$ given by

$$\tilde{\psi}(\mathfrak{a}, D) = \text{sign}(f_D(\mathfrak{a})) \overset{\text{defn}}{=} \text{sign}\left(\prod_i f_D(P_i)^{m_i} \right),$$

where f_D is any analytic function on A^0 with divisor D. In particular, for four distinct points P_1, P_2, Q_1, Q_2 on A^0,

$$\tilde{\psi}((P_1) - (P_2), (Q_1) - (Q_2)) = \begin{cases} +1, & \text{if the } P\text{'s do not "separate" the } Q\text{'s;} \\ -1, & \text{if the } P\text{'s "separate" the } Q\text{'s.} \end{cases}$$

In the remaining examples the valuation v is discrete. Let $A = A_{/\mathcal{O}}$ and $B = B_{/\mathcal{O}}$ denote the Néron models of A and B over $\mathcal{O}$, and let $E = E_{/\mathcal{O}}$ denote the canonical biextension of (A, B^0) by $\mathbb{G}_{m/\mathcal{O}}$, where B^0 denotes the connected component of 0 in $B_{/\mathcal{O}}$ (cf. [SGA7, exp VIII Thm. 7.1]). The general fiber of E is the biextension $E_{/K}$ we have been considering, and $E^0(\mathcal{O})$ is a certain subset of the part of $E(K)$ above $A(K) \times B^0(\mathcal{O})$ whose fibers are acted on transitively by $\mathcal{O}^*$. Let

$$B(K) \supset B^0(K) \supset B^1(K) \supset \cdots$$

be the filtration of $B(K) = B(\mathcal{O})$ defined by $B^0(K) =$ group of points in $B(\mathcal{O})$ whose reduction mod v is in the connected component of 0 in the fiber $B_{/k}$, and

$$B^n(K) = \text{Ker}(B(\mathcal{O}) \to B(\mathcal{O}/\pi^n\mathcal{O})) \quad \text{for } n \geqslant 1.$$

Similarly, let

$$K^* \supset U^0 \supset U^1 \supset \cdots$$

be the filtration of K^* defined by $U^0 = \mathcal{O}^*$ and $U^n = \text{Ker}(\mathcal{O}^* \to (\mathcal{O}/\pi^n\mathcal{O})^*)$ for $n \geqslant 1$. Our first example of a non-archimedean canonical trivialization is

(3) *The Néron unramified trivialization.* For v discrete the modification of $E(K)$ given by the triple $(A(K), B^0(K), K^*/\mathcal{O}^* = \mathbb{Z})$ has a canonical splitting ψ which is characterized by the fact that $\psi(E(\mathcal{O})) = 0$ (cf. [M-T] §1.9). For an elliptic curve $A_{/K}$, we can view $A = A_{/\mathcal{O}}$ as a regular "surface", which is mapped to the "bit of curve" Spec $\mathcal{O}$, with fibers of dimension 1, and the splitting ψ is given in terms of zero-cycles by an intersection multiplicity:

$$\tilde{\psi}(\mathfrak{a}, D) = \mathrm{degree}(\mathfrak{a}' \cdot D'),$$

where $\mathfrak{a}'$ (resp. D') is a divisor on A whose intersection with the general fiber A_K is $\mathfrak{a}$ (resp. D), and such that, for each component F_i of the special fiber A_k the degree of $D' \cdot F_i$ is zero. (Such a D' exists because the class of D corresponds to a point in $B^0(\mathcal{O})$; we do not need to complete A to the $\bar{A}$ such that $\bar{A} \to$ Spec $\mathcal{O}$ is proper, because we consider only zero cycles whose points are K-rational.) In particular, if P_1, P_2, Q_1, Q_2 are four distinct points of $A(K)$ such that Q_1 and Q_2 reduce to the same connected component of A_k, then

$$\tilde{\psi}((P_1) - (P_2), (Q_1) - (Q_2))$$

$$= \mathrm{degree}(P_1' \cdot Q_1' - P_1' \cdot Q_2' - P_2' \cdot Q_1' + P_2' \cdot Q_2')$$

$$= \lambda(P_1 - Q_1) - \lambda(P_1 - Q_2) - \lambda(P_2 - Q_1) + \lambda(P_2 - Q_2),$$

where

$$\lambda(P) = \deg(P' \cdot Q') = \begin{cases} 0, & \text{if } x(P) \in \mathcal{O} \\ -\tfrac{1}{2}v(x(P)), & \text{if } x(P) \notin \mathcal{O}, \end{cases}$$

where x is the "x-coordinate" in some minimal Weierstrass model for A at v. (For a slightly different point of view, and the generalization to abelian varieties, see Lang [La], Ch. 11, §5.)

(4) *The tamely ramified trivialization.* For v discrete, the modification of $E(K)$ given by the triple $(A(K), B^1(K), K^*/U^1)$ has a canonical splitting ψ. To see this, let $E^1(\mathcal{O})$ denote the part of $E(\mathcal{O})$ above $A(K) \times B^1(K)$. For $x \in E^1(\mathcal{O})$, the reduction $\tilde{x}$ of $x \bmod v$ lies in $E(k)_0$, the part of $E(k)$ lying above $A(k) \times (0)$. Let $s\colon E(k)_0 \to k^* = \mathcal{O}^*/U^1$ be the canonical splitting of $E(k)_0$. Then $x \mapsto s(\tilde{x})$ is an $(\mathcal{O}^* \to \mathcal{O}^*/U^1)$-splitting of $E^1(\mathcal{O})$, and extends uniquely to the splitting sought, because our modification of $E(K)$ is obtained from $E^1(\mathcal{O})$ by pushout via the inclusion $\mathcal{O}^* \to K^*$.

For an elliptic curve A we can describe the function $\tilde{\psi}$ in a "geometric" way as in the preceding example (3). With $\mathfrak{a}'$, D', $\mathfrak{a}$ and D as in that example, it suffices to treat the case in which $\mathfrak{a}'$ and D' are "horizontal" (i.e., contain no fibral irreducible components), and are disjoint on the surface. Then we have

$$\tilde{\psi}(\mathfrak{a}, D) = f_{D'}(\mathfrak{a} \cdot A_{/k}) \in k^* = \mathcal{O}^*/U' \subset K^*/U',$$

where $f_{D'}$ is a rational function on the special fiber $A_{/k}$ whose divisor is $D' \cdot A_{/k}$, which exists because the class of D corresponds to a point of $B^1(K)$, i.e., has trivial reduction mod r. In practice, for computing the tame global height pairing in special cases (see (3.2) below), it suffices to consider the case in which $D' \cdot A_k = (0)$, and hence $\tilde{\psi}(\mathfrak{a}, D) = 1$. For example, suppose P_1, P_2, Q_1, Q_2 are four distinct points of $A(K)$ such that Q_1 and Q_2 are equal mod v but are unequal to P_1 and P_2 mod v. Then

$$\tilde{\psi}(P_1) - (P_2), (Q_1) - (Q_2) = 1 \quad \text{because } ((Q_1) - (Q_2)) = 0 \ (\mathrm{mod}\ v).$$

The trivializations (3) and (4) which we have just discussed can be viewed as the cases $n = 0$ and $n = 1$ of a canonical trivialization involving the modification of $E(K)$ given by the triple $(A(K), B^n(K), K^*/U^n)$ for any n, for which a splitting is constructed exactly as above with k replaced by $\mathcal{O}/\pi^n\mathcal{O}$. However these trivializations seem less useful for $n \geqslant 2$ than for $n = 0$ and 1; passage to the subgroup $B^n(K)$ seems too high a price to pay for a splitting mod U^n. More useful splittings involving "wild" quotients of K^* (in fact, K^* itself) are the ones considered in [M-T], which require that A have ordinary reduction at v.

(5) Suppose the residue field k has characteristic $p > 0$ and that A has ordinary reduction at v. Consider the triples

(a) $(A^t(K), B^t(K), K^*)$

(b) $(A'(K), B^1(K), K^*)$,

where $A^t(K)$ denotes the subgroup of $A(K)$ whose reduction mod v is the toric part of $A(K)$, and where $A'(K)$ is the subgroup of points of $A(K)$ which are of finite order prime to p modulo $A^t(K)$. A canonical splitting of the modification for case (a) is produced in [M-T] §5. To take care of case (b), notice that the restriction to $A^t(K) \times A'(K)$ of the splitting for (a) extends uniquely to $A'(K) \times B^1(K)$ by part (a) of the lemma in §2.1, since $B^1(K)$ is uniquely divisible by every integer prime to p.

We will not use the trivialization (5) in the sequel. However we will use

(6) *The split multiplicative trivialization.* Suppose v is discrete and A has "split multiplicative reduction" at v, i.e., that the connected component of the Néron fiber $A_{/k}$ at v is a split torus. In this case, $A(K)$ and $B(K)$ are rigid analytic quotients of split tori $T_X(K)$ and $T_Y(K)$ by discrete subgroups of periods Y and X respectively (see immediately below) and the modification of $E(K)$ in question is its pullback E' to $T_X(K) \times T_Y(K)$.

Let X (resp. Y) be the character group of the split torus $A^0_{/k}$ (resp. $B^0_{/k}$). Let T_X (resp. T_Y) denote the torus defined over K with character group X (resp. Y). There is a canonical pairing $\langle \ , \ \rangle \colon X \times Y \to K^*$ and exact sequences

$$0 \to Y \xrightarrow{i} T_X(K) \xrightarrow{\alpha} A(K) \to 0$$

$$0 \to X \xrightarrow{j} T_Y(K) \xrightarrow{\beta} B(K) \to 0,$$

where the maps i and j are those induced by the pairing $\langle\ ,\ \rangle$, and α and β are (induced by) rigid analytic homomorphisms. The maps i and j are injective and we will view them as inclusions from now on. Then Y and X are discrete subgroups of $T_X(K)$ and $T_Y(K)$ respectively; more precisely, there exists a homomorphism $\omega\colon Y \to X$ such that $y \mapsto v\langle\omega(y),\, y\rangle$ is the restriction to Y of a positive definite quadratic form on $Y \otimes \mathbb{R}$, and such an ω induces a polarization $A \to B$.

To obtain our trivialization we prove that *the modification $E' = (\alpha^*, \beta^*)(E(K))$ has a unique rigid analytic splitting.* The uniqueness comes from the fact that, in the rigid analytic category,

$$\mathrm{Hom}(\mathbb{G}_m, \mathbb{G}_m) = \mathbb{Z} \quad\text{and}\quad \mathrm{Hom}(\mathbb{G}_m, \mathbb{Z}) = 0,$$

so there are no non-zero pairings of tori to $\mathbb{G}_m$.

To prove the existence of the splitting we use theta-functions. This method has the advantage that it yields convenient explicit formulas. We suspect that in fact *every* biextension of $(\mathbb{G}_m, \mathbb{G}_m)$ by $\mathbb{G}_m$ in the rigid analytic category splits so that the existence could also be obtained from that principle.

The following diagram is cartesian:

$$
\begin{array}{ccc}
E' & \xrightarrow{\ \ \gamma\ \ } & E(K) \\
\big\downarrow{\scriptstyle\pi'} & & \big\downarrow{\scriptstyle\pi} \\
T_X(K) \times T_Y(K) & \xrightarrow{\ \alpha\times\beta\ } & A(K) \times B(K).
\end{array}
$$

As explained in [M-T, §2] points of $E(K)$ can be represented by symbols $[\mathfrak{a}, D, c]$. Analogously, we will now explain how points of E' are determined by triples $[\mathfrak{a}', \theta, c]$, where

$\mathfrak{a}'$ is a zero cycle of degree 0 made of points $t \in T_X(K)$;

θ is a "meromorphic theta function" on $T_X(K)$ such that, for each $y \in Y$ the ratio $\theta(t + y)/\theta(t)$ is independent of t, and such that the divisor of θ is disjoint from $\mathfrak{a}'$;

$c \in K^*$.

Given such a triple, let $s(\mathfrak{a}') \in T_X(K)$ denote the "sum" of $\mathfrak{a}'$ (if $\mathfrak{a}' = \Sigma m_i(t_i)$, then $s(\mathfrak{a}') = \Sigma m_i t_i$), and let $u_\theta \in T_Y(K) = \mathrm{Hom}(Y, K^*)$ be the element defined by

$$u_\theta(y) = \frac{\theta(t + y)}{\theta(y)}, \quad\text{for all } t \in T_X(K) \text{ and } y \in Y.$$

Then we let the symbol $[\mathfrak{a}', \theta, c]$ stand for the element of E' such that (cf. the diagram above)

$$\pi'([a', \theta, c]) = (s(a'), u_\theta) \in T_X(K) \times T_Y(K)$$

and

$$\gamma([a', \theta, c] = [\alpha(a'), D_\theta, c] \in E(K),$$

where the last symbol [,,] is that discussed in [M-T] §2.1, and D_θ is the divisor on $A_{/K}$ such that $\alpha^*(D_\theta) = (\theta)$. This makes sense, because $s(\alpha(a')) = \alpha(s(a'))$, and the class of D_θ is $\beta(u_\theta) \in B(K)$, according to the analytic description of the isomorphism $\mathrm{Pic}^0(A_{/K}) = B(K)$.

Every point of E' is of the form $[a', \theta, c]$, because, for suitable choices of a' and θ, $f(a')$ can be an arbitrary zero cycle of degree 0 on $A(K)$, and D_θ an arbitrary divisor algebraically equivalent to 0 on $A_{/K}$ whose support is disjoint from that of $f(a')$.

The splitting ψ of E' is given simply by

$$\psi([a', \theta, c]) = c\theta(a'),$$

where $\theta(a)$ is defined by $\theta(\sum m_i(t_i)) = \prod \theta(t_i)^{m_i}$. If we can show that this ψ is well defined, i.e., is independent of the triple (a', θ, c) chosen to represent a given point of E', then it is obviously a splitting, for it is bilinear in a and θ and is compatible with the operation of $c \in K^*$. To show that ψ is independent of the choice of θ we use the rule

$$[a', \theta_1, c] = \left[a', \theta, c\frac{\theta_1(a')}{\theta(a')}\right] \quad \text{for } u_{\theta_1} = u_\theta.$$

This follows from the rule

$$[a, D + (\phi), c] = [a, D, c\phi(a)]$$

for the symbol on A. To show that ψ is independent of the choice of a' we must show that if $s(a'_1) = s(a')$, then

$$[a'_1, \theta, c] = \left[a', \theta, c\frac{\theta(a'_1)}{\theta(a')}\right].$$

To prove this, it suffices to treat the case $a'_1 = a'_t$ is the translation of a' by an element $t \in T_X(K)$, because any zero cycle of degree 0 of sum 0 is a sum of cycles of the form $a_t - a$. Since the symbol [,,] on $A(K)$ is invariant under simultaneous translation of its first two arguments, we have

$$[a'_t, \theta, c] = [a', \theta_{-t}, c] = \left[a', \theta, c\frac{\theta_{-t}(a')}{\theta(a')}\right] = \left[a', \theta, c\frac{\theta(a'_t)}{\theta(a')}\right],$$

as claimed.

Suppose A is an elliptic curve. Then we can take $T_X = T_Y = \mathbf{G}_m$ and there is an element $q \in K^* = \mathbf{G}_m(K)$ such that $v(q) > 0$ and $Y = X = q^{\mathbf{Z}} = \{q^n\}_{n \in \mathbf{Z}}$. There is a basic theta function

$$\theta(t) = (1 - t) \prod_{n=1}^{\infty} (1 - q^n t)(1 - q^n t^{-1})$$

for $t \in K^*$. By construction, θ has simple zeros at the points q^n, $n \in \mathbf{Z}$, and no other zeros or poles, i.e., its divisor is $\alpha^*((0))$. It satisfies the identities

$$\theta(qt) = -t^{-1}\theta(t) = \theta(t^{-1}).$$

Let $\mathfrak{a}' = \Sigma m_i(a_i)$ and $\mathfrak{b}' = \Sigma n_j(b_j)$ be zero cycles of degree 0 on K^* such that their images $\mathfrak{a} = \alpha(\mathfrak{a}')$ and $\mathfrak{b} = (\beta(\mathfrak{b}')$ on $A(K) = B(K)$ have disjoint supports (i.e., such that none of the ratios a_i/b_j is a power of q). Put

$$\theta_{\mathfrak{b}'}(t) = \prod_j \theta\left(\frac{t}{b_j}\right)^{n_j}.$$

Then the divisor of $\theta_{\mathfrak{b}'}$ is $\beta^*(\mathfrak{b})$, and

$$\theta_{\mathfrak{b}'}(qt) = s(\mathfrak{b}')\theta_{\mathfrak{b}'}(t), \quad \text{where } s(\mathfrak{b}') = \prod_j b_j^{n_j}.$$

Consequently, the triple $[\mathfrak{a}', \theta_{\mathfrak{b}'}, 1]$ represents the point $e' \in E'$ such that

$$\gamma(e') = [\mathfrak{a}, \mathfrak{b}, 1] \in E(K)$$

and

$$\pi'(e') = s(\mathfrak{a}') \times s(\mathfrak{b}') \in K^* \times K^*.$$

Therefore the splitting of E' is given in these terms by

$$\psi(e') = \prod_{i,j} \theta\left(\frac{a_i}{b_j}\right)^{m_i n_j}.$$

Suppose for simplicity $\mathfrak{a}' = (a_1) - (a_2)$ and $\mathfrak{b}' = (b_1) - (b_2)$. Then

$$\psi(e') = \frac{\theta\left(\dfrac{a_1}{b_1}\right)\theta\left(\dfrac{a_2}{b_2}\right)}{\theta\left(\dfrac{a_1}{b_2}\right)\theta\left(\dfrac{a_2}{b_1}\right)}.$$

Special cases: (i) If $b_1 = 1$ and $b_2 = q^{-1}$ (so $s(\mathfrak{b}') = q$), then

$$\psi(e') = \frac{a_1}{a_2} = s(\mathfrak{a}').$$

(ii) If $a_1, a_2, b_1, b_2 \in \mathcal{O}^*$, then, using the fact that $\theta(t) \equiv 1 - t \pmod{U^1}$ for t a unit, $t \neq 1$, we find

$$\psi(e') \equiv \frac{\left(1 - \dfrac{a_1}{b_1}\right)\left(1 - \dfrac{a_2}{b_2}\right)}{\left(1 - \dfrac{a_1}{b_2}\right)\left(1 - \dfrac{a_2}{b_1}\right)} = \frac{(b_1 - a_1)(b_2 - a_2)}{(b_2 - a_1)(b_1 - a_2)} \pmod{U^1}.$$

This formula will be useful for computing tame pairings (cf. 2.4) involving split multiplicative places. It is especially useful when a_1, a_2, b_1, b_2 are in distinct residue classes; then knowing them mod π is enough to compute $\psi(e') \bmod \pi$.

(2.3) *The global pairing defined by a family of local trivializations.* Let K be a field, and let $\mathscr{V}$ be a set of places of K, as in §3 of [M-T], i.e., such that each $v \in \mathscr{V}$ is either archimedean or discrete, and such that, for each $c \in K^*$, we have $|c|_v = 1$ for all but a finite number of $v \in \mathscr{V}$. We keep to the notation of §3 of [M-T]. In particular, for each $v \in \mathscr{V}$, we let K_v denote the completion of K at v. Let $A_{/K}, B_{/K}$ be dual abelian varieties and $E_{/K}$ the biextension of (A, B) by $\mathbf{G}_m$ which expresses the duality. By *global pairing data* δ we mean a collection of local trivializations $\delta_v = (\alpha_v, \beta_v, \rho_v, \psi_v)$, one for each $v \in \mathscr{V}$, such that, for all but a finite number of places v, the local trivialization (δ_v) is given by example (3) in (2.2), i.e., is the "Néron unramified local trivialization".

Given a global pairing data, δ, we define three abelian groups $A_\delta, B_\delta, C_\delta$ together with homomorphisms

$$A_\delta \xrightarrow{\alpha} A(K) \qquad B_\delta \xrightarrow{\beta} B(K)$$

$$I/K^* \xrightarrow{\rho} C_\delta,$$

where

$$I = \prod_{v \in \mathscr{V}} K_v^*$$

is the "idele group", i.e., the restricted direct product of the groups K_v^* relative to the subgroups $\mathcal{O}_v^*$, and we define a "height pairing"

$$\langle\ ,\ \rangle_\delta \colon A_\delta \times B_\delta \to C_\delta.$$

We call A_δ and B_δ the extended Mordell-Weil groups. In applications they will usually be extensions of subgroups of finite index in $A(K)$ and $B(K)$ by free abelian groups of finite rank, and C_δ will be a quotient of the "idele class group" I/K^*.

To define these things, let $\delta = (\delta_v) = (\alpha_v, \beta_v, \rho_v, \psi_v)$ be global pairing data. Changing notational letters from those in (2.1), let us denote the modifying triple at the place v by (A_v, B_v, C_v). Thus we are given, for each $v \in \mathcal{V}$, abelian groups and homomorphisms

$$A_v \stackrel{\alpha_v}{\to} A(K_v) \qquad B_v \stackrel{\beta_v}{\to} B(K_v)$$

$$K_v^* \stackrel{\rho_v}{\to} C_v,$$

such that, for almost all v, they are simply

$$A_v = A(K_v) \qquad B_v = B(K_v)$$

$$K_v^* \to K_v^*/\mathcal{O}_v^* = \mathbb{Z}.$$

With the A_v and α_v we define A_δ and α to be the abelian group and homomorphism such that the following diagram is cartesian,

$$(2.3.1) \qquad \begin{array}{ccc} A_\delta & \stackrel{\alpha}{\longrightarrow} & A(K) \\ \downarrow & & \downarrow \\ \prod_v A_v & \stackrel{\Pi \alpha_v}{\longrightarrow} & \prod_v A(K_v), \end{array} \qquad i = (i_v)$$

where for each v, $i_v\colon K \to K_v$ is the canonical map. Thus, A_δ fits in an exact sequence

$$(2.3.2) \qquad 0 \to \prod_v \operatorname{Ker} \alpha_v \to A_\delta \stackrel{\alpha}{\to} A(K) \to \prod_v \operatorname{Coker} \alpha_v.$$

A point $a \in A_\delta$ is described by a point $P \in A(K)$ together with, for each v, an element $a_v \in A_v$ such that $\alpha_v(a_v) = i_v(P)$; we write $a = (P, (a_v))$. Define similarly the group B_δ, fitting in an exact sequence.

$$(2.3.3) \qquad 0 \to \prod_v \operatorname{Ker} \beta_v \to B_\delta \stackrel{\beta}{\to} B(K) \to \prod_v \operatorname{Coker} \beta_v.$$

Define C_δ and ρ to be the group and homomorphism such that the diagram

$$\begin{array}{ccc} I/K^* & \stackrel{\rho}{\longrightarrow} & C_\delta \\ \theta \uparrow & & \uparrow \theta \\ I = \prod_v K_v^* & \stackrel{\Sigma_v \rho_v}{\longrightarrow} & \bigoplus_{v \in \mathcal{V}} C_v. \end{array}$$

is co-cartesian. Thus C_δ fits in the exact sequence

$$(2.3.4) \qquad K^* \xrightarrow{\Sigma_v \rho_v \circ i_v} \bigoplus_{v \in \mathscr{V}} C_v \xrightarrow{\theta} C_\delta \to 0,$$

and if C_v is a quotient K_v^*/U_v of K_v^* for each v, then

$$C_\delta = I \Big/ \Big(K^* \prod_{v \in \mathscr{V}} U_v \Big)$$

is a quotient of the "idele class group".

We have then a pairing

$$(2.3.5) \qquad \langle \ , \ \rangle_\delta \colon A_\delta \times B_\delta \to C_\delta$$

defined as follows. Let $a = (P, (a_v)) \in A_\delta$ and $b = (Q, (b_v)) \in B_\delta$. To obtain $\langle a, b \rangle_\delta$, choose a point $x \in E(K)$ lying over $P \times Q$, such that $i_v(x) \in E(\mathcal{O}_v)$ for all but a finite number of v, and let, for each v, $x_v \in (\alpha_v \times \beta_v)^* E(K_v)$ be the (unique) point lying over $a_v \times b_v$ in $A_v \times B_v$ and over $i_v(x)$ in $E(K_v)$. Then

$$(2.3.6) \qquad \langle a, b \rangle_\delta \stackrel{\mathrm{defn}}{=} \theta\Big(\sum_v \psi_v(x_v) \Big) \in C_\delta,$$

with θ as in (2.3.4). Changing the choice of x by $c \in K^*$ changes each x_v by $i_v(c) \in K_v^*$, so changes the right side of (2.3.6) by $\theta(\Sigma_v \rho_v i_v(c)) = 0$.

(2.4) *The S-pairings.* We keep to the notation of the preceding paragraph. In particular, $A_{/K}$ and $B_{/K}$ are our dual abelian varieties. Let $S \subset \mathscr{V}$ be a finite set of nonarchimedean places of K, and let $S_m \subset S$ denote the subset of S consisting of those v, such that $A^0_{/k_v}$ is isomorphic to a split torus over k_v, i.e., such that v is a place of split multiplicative reduction for A. We make the hypothesis that for $v \in S - S_m$, the toric part of $A^0_{/k_v}$ possesses no k_v-rational characters. (This is no restriction in the case where A is of dimension one).

Such a set S determines global pairing data δ_S, as follows:

(i) For v archimedean, α_v, β_v are the identity maps, and $C_v = 0$ (i.e., we are "suppressing" the archimedean primes).

(ii) For v discrete, $v \notin S$, δ_v is the unramified Néron trivialization of example (3) in (2.2).

(iii) For $v \in S - S_m$, δ_v is the tamely ramified trivialization of example (4) in (2.2).

(iv) For $v \in S_m$, δ_v is the split multiplicative trivialization of example (6) in (2.2).

We will write simply $\langle \ , \ \rangle_S, A_S$, etc., for $\langle \ , \ \rangle_{\delta_S}, A_{\delta_S}$, etc., and will refer to $\langle \ , \ \rangle_S$ as the (*canonical*) S-pairing:

$$\langle \ , \ \rangle_S \colon A_S \times B_S \to C_S.$$

However, we warn the reader that our notation A_S, B_S is *not* symmetric (we have chosen an *ordering* of our abelian varieties A, B and if we switch this ordering the groups A_S, B_S may change; see (2.4.1), (2.4.2) below).

The target group C_S is a quotient of the idele class group;

$$C_S = I \Big/ \Big(K^* \prod_v U_v \Big),$$

where

$$U_v = \begin{cases} K_v^*, & \text{if } v \text{ is archimedean,} \\ \mathcal{O}_v^*, & \text{if } v \text{ is non-archimedean, } v \notin S, \\ 1 + \mathfrak{m}_v, & \text{if } v \in S - S_m, \\ (1), & \text{if } v \in S_m. \end{cases}$$

For example, if $K = \mathbf{Q}$ and $\mathscr{V}$ consists of all places of $\mathbf{Q}$, then

$$C_S = \Big(\prod_{p \in S - S_m} \mathbb{F}_p^* \times \prod_{p \in S_m} \mathbb{Z}_p^* \Big) / (\pm 1).$$

As for A_S and B_S, the exact sequences (2.3.2), (2.3.3) become:

$$(2.4.1) \qquad 0 \to \prod_{v \in S_m} Y_v \to A_S \to A(K) \to 0$$

$$(2.4.2) \quad 0 \to \prod_{v \in S_m} X_v \to B_S \to B(K) \to \prod_{v \in S - S_m} B(k_v) \times \prod_{v \notin S} (B/B^0)(k_v).$$

The last group on the right is finite because the Néron fibers of B are connected for all v of good reduction for B. Note that A_S depends only upon S_m.

If $A(K)$ is finitely generated, (e.g., if K is finitely generated over the prime field), then A_S and B_S are finitely generated groups of the same rank

$$(2.4.3) \qquad r = \text{rank}(A(K)) + \#(S_m) \cdot \dim(A).$$

For subsets $T \subset S$ we have the natural mappings:

$$A_S \xrightarrow{x} A_T, \qquad B_S^T \xrightarrow{y} B_T, \qquad C_S \xrightarrow{z = z_{T,S}} C_T$$

where

$$B_S^T = \text{Ker}\Big(B_S \to \prod_{v \in S_m - T_m} (B/B^0)(k_v) \Big) \quad \text{and} \quad T_m = T \cap S_m.$$

The S-pairing and the T-pairing are compatible with respect to the homomorphisms x, y, z in the sense that

$$(2.4.4) \qquad \langle x(a), y(b)\rangle_T = z \cdot \langle a, b\rangle_S.$$

Explicit formulas for the S-pairing on an elliptic curve.   Suppose A is an elliptic curve. Note that although $A = B$ in this case, we do not have $A_S = B_S$ in general; our notation is not symmetric. Therefore we keep using the two letters A and B. Let $a = (P,(a_v))$ and $b = (Q,(b_v))$ be points of A_S and B_S (notation as in §2.3). In other words, let $P \in A(K)$ and for each $v \in S_m$ let $a_v \in K_v^*$ such that $\alpha(a_v) = i_v(P)$ and let $Q \in B(K)$ such that

$$i_v(Q) \in B^0(K_v) \quad \text{for } v \text{ non-archimedean, } v \notin S,$$

and

$$i_v(Q) \in B^1(K_v) \quad \text{for } v \in S - S_m,$$

and for $v \in S_m$, let $b_v \in K_v^*$ such that $\beta(b_v) = i_v(Q)$. (Note that we do not need to give a_v and b_v for $v \notin S_m$ because α_v and β_v are injective for such v.)

We want to give an explicit prescription for computing the value of $\langle a, b\rangle_S$. For simplicity we do this only under the following assumption, which is usually satisfied in practice. We suppose *either* $P = 0$ *or* $Q = 0$ *or* there is a point $P' \in A(K)$ such that none of the four points

$$P', P + P', Q + P', P + Q + P'$$

reduces to 0 on any fiber above $S - S_m$, i.e., such that for each $v \in S - S_m$, no one of those four points is $\equiv 0 \pmod{v}$.

Suppose first that such a P' exists. Choose one, and for each $v \in S_m$, choose a point $a_v' \in K_v^*$ such that $\alpha_v(a_v') = i_v(P')$. To compute $\langle a, b\rangle_S$, simply follow the definition in §2.3, taking

$$x = [(P + P') - (P'), (0) - (-Q), 1] \in E(K)$$

and for $v \in S_m$

$$x_v = \left[(a_v a_v') - (a_v'), \theta_{(1) - (b_v^{-1})}, 1\right] \in (\alpha_v \times \beta_v)^* E(K_v).$$

Then, using the explicit descriptions of the splittings ψ_v in terms of zero-cycles and divisors given in (2.2) one finds that an idele $c = (c_v)$ representing the idele class $\langle a, b\rangle_S$ is given as follows:

$$c_v = \text{arbitrary in } K_v^*, \quad \text{for } v \text{ archimedean};$$

$$c_v = \frac{t_v(P + P')t_v(Q + P')}{t_v(P')t_v(P + Q + P')} \bmod \mathcal{O}_v^*, \quad \text{for } v \text{ discrete, } v \notin S,$$

where $t_v(P)$ denotes an element of K_v^* such that $t_v(P)^2$ is a denominator for the x-coordinate of P in a local minimal Weierstrass model for A at v;

$$c_v = 1 \bmod (1 + \mathfrak{m}_v), \quad \text{for } v \in S - S_m,$$

and

$$c_v = \frac{\theta_v(a_v a_v')\theta_v(b_v a_v')}{\theta_v(a_v')\theta_v(a_v b_v a_v')}, \quad \text{for } v \in S_m.$$

Here θ_v is the basic theta function for A at v, namely,

$$\theta_v(w) = (1 - w)\prod_{n=1}^{\infty}(1 - q_v^n w)(1 - q_v^n w^{-1}),$$

where q_v is the multiplicative period of A at v.

If $P = 0$ or $Q = 0$, then the situation is even simpler. We do not need a P'. Suppose $Q = 0$. Then the "v-th coordinate", b_v, of the point $b = (0, (b_v))$ is a power of the multiplicative period q_v. For each $v \in S_m$, let $q(v)$ denote the point of this type whose v-th coordinate is q_v and whose other coordinates are 1. Then (taking $x = [(P) - (0), 0, 1]$ and $x_v = [(a_v q_v^m) - (q_v^m), (1) - (q_v^{-1}), 1]$ for suitable $m \in \mathbb{Z}$, and $x_{v'} = [(a_{v'}) - (1), 0, 1]$ for $v' \in S_m - \{v\}$) we find that $\langle a, q(v)\rangle$ is represented by the idele whose entry at v is a_v and whose entry at all other places is 1. In particular, in case $P = 0$ and $Q = 0$, we have, for $v, v' \in S_m$,

$$\langle q(v'), q(v)\rangle_S = \begin{cases} 1, & \text{if } v' \neq v, \\ \gamma_v, & \text{if } v' = v, \end{cases}$$

where γ_v is the class of the idele whose v-th entry is q_v and whose entry at all other places is 1.

For abelian varieties of arbitrary dimension this last formula generalizes as follows. Let v, v' be places in S_m; let $y(v') \in Y_{v'}$ and $x(v) \in X_v$ be viewed as elements of A_S and B_S respectively, via the mappings of (2.4.1) and (2.4.2).

Then $\langle y(v'), x(v)\rangle_S$ is 1 if $v' \neq v$ and if $v' = v$ it is the class of the idele whose v-th component is $\langle x(v), y(v)\rangle_v$ and whose other components are 1.

(2.5) *Discriminants and compatible orientations.* Let M and N be free $\mathbb{Z}$-modules of the same rank, r, let $\mathscr{S}$ be a commutative ring, and let $h: M \times N \to \mathscr{S}$ be a biadditive map. Choosing bases (P_i) and (Q_j) for M and N we obtain a *discriminant*:

$$(2.5.1) \qquad \operatorname{disc}(h) = \det_{1 \leqslant i, j \leqslant r}\big(h(P_i, Q_j)\big) \in \mathscr{S}.$$

This is defined only up to sign, but if we have a notion of "compatible orientations" of $M \otimes \mathbb{R}$ and $N \otimes \mathbb{R}$, then, choosing bases with compatible orientations removes the ambiguity. Assuming this, if M_1 and N_1 are submodules of finite index in M and N, and h_1 is the restriction of h to $M_1 \times N_1$, then

$$(2.5.2) \qquad \mathrm{disc}(h_1) = (M : M_1)(N : N_1)\mathrm{disc}(h).$$

If the indices $(M : M_1)$ and $(N : N_1)$ are invertible in $\mathscr{S}$, then

$$(2.5.3) \qquad \mathrm{disc}(h) = \frac{\mathrm{disc}(h_1)}{(M : M_1)(N : N_1)}.$$

If M and N are not necessarily free, but are finitely generated *with orders of their torsion subgroups invertible in $\mathscr{S}$*, then we can still define the discriminant of a pairing $h\colon M \times N \to \mathscr{S}$, by the formula (2.5.3), taking M_1 and N_1 to be complementary subgroups to the torsion subgroups of M and N. (Thus $\mathrm{disc}(h_1)$ $= \det h(P_i, Q_j)$ where (P_i) and (Q_j) are "bases mod torsion", compatibly oriented. With this extension of the notion of discriminant, formula (2.5.2) holds for all pairs M_1, N_1.

Remarks on orientations. An *orientation* on a real, finite-dimensional, vector space V is given by a class of bases $(P_i)_i$ for V such that any two bases in the class can be brought one into another by a linear transformation of positive determinant, and conversely, if $(P_i)_i$ is in the class, any basis obtained by applying a linear transformation of positive determinant to $(P_i)_i$ is also in the class. To give an orientation to V is the same as giving an orientation to the one-dimensional vector space $\det V$, the d-fold exterior power of V, where $d = \dim V$. An orientation on the dual V^* of V is determined from an orientation of V by the rule that the determinant of the matrix $\langle P_i, P_j^* \rangle$ be positive, where $(P_i)_i$ is an oriented basis of V and $(P_j^*)_j$ is an oriented basis of V^*. An orientation on any two of the three vector spaces V_i $(i = 1, 2, 3)$ in an exact sequence $0 \to V_1 \to V_2 \to V_3 \to 0$ determines an orientation on the third by the standard rule.

Let $A_{/K}$ and $B_{/K}$ be dual abelian varieties. An orientation on the vector space $A(K) \otimes \mathbb{R}$ determines an orientation on $B(K) \otimes \mathbb{R}$ by taking any polarization defined over K of A, $f\colon A \to B$, and transporting the orientation on $A(K) \otimes \mathbb{R}$ to $B(K) \otimes \mathbb{R}$ via the isomorphism induced by f. For this to be well-defined, we must show:

LEMMA. *Let $A(K)$ be finitely generated. Let $\phi, \phi'\colon A \to B$ be two polarizations defined over K (cf. [Mu-F] Chap. 6, §2). Then the induced isomorphisms*

$$A(K) \otimes \mathbb{R} \underset{\phi'}{\overset{\phi}{\rightrightarrows}} B(K) \otimes \mathbb{R}$$

have the property that $\det(\phi^{-1} \cdot \phi')$ is positive.

Proof. Let $f = \phi^{-1}\phi' \in \mathbf{Q} \otimes \text{End } A$. It suffices to show that there is a polynomial $Q(t)$ all of whose roots are real and > 0, such that $Q(f) = 0$, for then the action of f on any $(\mathbf{R} \otimes \text{End } A)$-module V will have positive eigenvalues and preserve orientation. By Mumford [Mu], there is a polynomial $P(t)$ of degree $\dim A = g$, with all its roots real and < 0 such that for every integer $n > 0$, the degree of the homomorphism $n\phi + \phi'$ is $P(n)^2$. Thus the degree of $n + \phi^{-1}\phi'$ is of the form $\prod_{j=1}^{g}(n + \alpha_j)^2$, with $\alpha_j > 0$ for each j. By [Mu] §19, it follows that $f = \phi^{-1}\phi'$ satisfies $Q(f) = 0$, where $Q(t) = \prod_{j=1}^{g}(t - \alpha_j)^2$.

Now tensor (2.4.1 and 2.4.2) with $\mathbf{R}$ to get:

$$0 \to \prod_{v \in S_m} Y_v \otimes \mathbf{R} \to A_S \otimes \mathbf{R} \to A(K) \otimes \mathbf{R} \to 0$$

$$0 \to \sum_{v \in S_m} X_v \otimes \mathbf{R} \to B_S \otimes \mathbf{R} \to B(K) \otimes \mathbf{R} \to 0.$$

Note that by composition of the natural pairing

$$X_v \times Y_v \to K_v^*$$

with ord_v we may identify $X_v \otimes \mathbf{R}$ with the dual vector space to $Y_v \otimes \mathbf{R}$. It follows from the above discussion that if we put an orientation on $Y_v \otimes \mathbf{R}$ for $v \in S_m$ we have a natural orientation on:

$$\prod_{v \in S_m} Y_v \otimes \mathbf{R} \quad \text{(product)}$$

and hence on:

$$\prod_{v \in S_m} X_v \otimes \mathbf{R} \quad \text{(dual of the above).}$$

Putting an orientation on $A(K) \otimes \mathbf{R}$ determines a compatible orientation on $B(K) \otimes \mathbf{R}$, by the above discussion. Then, using the two exact sequences above, we have orientations on $A_S \otimes \mathbf{R}$ and on $B_S \otimes \mathbf{R}$.

Orientations on $A_S \otimes \mathbf{R}$ and $B_S \otimes \mathbf{R}$ obtained from the above construction will be called *compatible*.

Let R be a ring in which the orders of the torsion subgroups of $A(K)$ and $B(K)$ are invertible. The sequences (2.4.1) and (2.4.2) show then that the same is true for the torsion in A_S and B_S, because the X_v and Y_v are torsion-free. Taking

$$(2.5.4) \qquad \mathscr{S} = R \bigotimes_{\mathbf{Z}} \text{Sym}_{\mathbf{Z}}(C_S) = R \otimes (\mathbf{Z} \oplus C_S \oplus \text{Sym}_2 C_S \oplus \cdots)$$

we obtain an element

$$(2.5.5) \qquad\qquad d_S = 1 \otimes \text{disc}(\langle \ , \ \rangle_S) \in R \otimes \text{Sym}_r(C_S)$$

where, in the definition of $\mathrm{disc}(\langle\ ,\ \rangle_S)$ we make use of the compatible orientations on $A_S \otimes \mathbb{R}$ and $B_S \otimes \mathbb{R}$ described above.

(2.6) *Corrected discriminants.* If $T \subset S$ is a smaller set of places containing S_m, we have an exact sequence

$$(2.6.1) \qquad \prod_{v \in S-T} \left(\mathcal{O}_v^* / (1 + \mathfrak{m}_v) \right) \to C_S \xrightarrow{z_{T,S}} C_T \to 0.$$

The order of the lefthand group in (2.6.1) is $\prod_{v \in S-T}(q_v - 1)$ where q_v is $\#(k_v)$, and consequently there is a unique mapping

$$(2.6.2) \qquad \mu_{S,T} \colon C_T \to C_S$$

such that

$$\mu_{S,T} \cdot z_{T,S}(c) = \prod_{v \in S-T} (q_v - 1) \cdot c$$

for all $c \in C_S$.

The sequence

$$(2.6.3) \qquad 0 \to B_S \to B_{S_m} \xrightarrow{b_S} \prod_{v \in S-S_m} B^0(k_v)$$

is exact, as is seen from (2.4.2). Let j_S be the order of the cokernel of b_S. Then

$$(2.6.4) \qquad \left(B_{S_m} \colon B_S \right) \cdot j_S = \prod_{v \in S-S_m} n_v.$$

where $n_v = \#(B^0(k_v))$.

We can now define the *corrected discriminant of the S-pairing*

$$\mathrm{Disc}_S = \mathrm{Disc}_S(A) \in R \otimes \mathrm{Sym}_r(C_S)$$

as follows:

$$(2.6.5) \quad \mathrm{Disc}_S(A) = \sum_{S_m \subset T \subset S} (-1)^{\#(T-S_m)} \cdot \mu_{S,T}(j_T \cdot d_T) \in R \otimes \mathrm{Sym}_r(C_S)$$

where d_T is the discriminant (attached to T) of (2.5.5).

(2.7) *Heuristic explanation of the "corrected discriminant."* It may seem strange to make use of *all* the T-pairings, for $T \subset S$ containing S_m in order to come up with a term which will play the role of the "regulator" in our conjectured formula. The essential problem is that if we consider only the S-pairing, it is defined on $A_{S_m} \times B_S$, and B_S is a subgroup of finite index in B_{S_m}.

The "desired term" would then be obtained by taking the discriminant d_S (2.5.5) of the S-pairing, and "dividing it" by the index $(B_{S_m} : B_S)$. But, of course, in the value group $R \otimes \mathrm{Sym}_r(C_S)$, it may not be possible to do this. To get a feel for the formula (2.6.5), then, let us make the hypothesis that $A(K)$ and $B(K)$ *have no torsion*, and that the index $(B_{S_m} : B_S)$ is *invertible* in R. In this case we do have a viable candidate for the "naive discriminant", call it D_S:

$$D_S = d_S / (B_{S_m} : B_S) \in R \otimes \mathrm{Sym}_r(C_S).$$

Under this hypothesis, the naive discriminants D_T (for $S_m \subset T \subset S$) exist also, and the image of D_S in $R \otimes \mathrm{Sym}_r(C_T)$ is D_T. Moreover, the image of D_T under $\mu_{S,T}$ is just $\prod_{v \in S-T}(q_v \cdot - 1) \cdot D_T$.

It follows that, under the same hypotheses, formula (2.6.5) gives:

$$(2.7.1) \qquad \mathrm{Disc}_S(A) = \prod_{v \in S - S_m} (q_v - 1 - n_v) \cdot D_S.$$

In this case, the "correcting term" is just $\prod_{v \in S - S_m}(q_v - 1 - n_v)$, which is happily consistent with the "compatibility term" encountered in Chapter 1 (1.3), which will occur on the left-hand side of our conjectured formula.

(2.8) *How Disc_S changes with S.* Let v be a discrete place not in S which is either split multiplicative for A, or for which the toric part of $A_{/k_v}$ has no k_v-rational character. Put $S' = S \cup \{v\}$. If r denotes, as above, the rank of A_S and r' the rank of $A_{S'}$ then we have:

$$(2.8.1) \qquad r' = \begin{cases} r, & \text{if } v \text{ is not split multiplicative for } A; \\ r + \dim A, & \text{if } v \text{ is split multiplicative for } A. \end{cases}$$

Let $z_{S,S'} \colon R \otimes \mathrm{Sym}_{r'}(C_{S'}) \to R \otimes \mathrm{Sym}_{r'}(C_S)$ be the mapping induced from the natural mapping $C_{S'} \to C_S$.

We wish to compare the two elements

$$(2.8.2) \quad \mathrm{Disc}_S \in R \otimes \mathrm{Sym}_r(C_S), \quad \text{and} \quad z_{S,S'}(\mathrm{Disc}_{S'}) \in R \otimes \mathrm{Sym}_{r'}(C_S).$$

PROPOSITION 1. *If v is not split multiplicative for A, then:*

$$(2.8.3) \qquad z_{S,S'}(\mathrm{Disc}_{S'}) = (q_v - 1 - n_v) \cdot \mathrm{Disc}_S.$$

Proof. By (2.4.1), $A_{S'} = A_S$; by (2.6.4), $(B_S : B_{S'}) = (j_S / j_{S'}) \cdot n_v$.

Hence, by the compatibility of S-pairing and S'-pairing (2.4.6) and by the general rule (2.5.2) for change of discriminant, we have

$$(2.8.4) \qquad z_{S,S'}(j_{S'} d_{S'}) = j_{S'} \cdot (B_S : B_{S'}) \cdot d_S = n_v \cdot j_S \cdot d_S.$$

Our proposition then follows from a straightforward computation using formula (2.6.5), the above relation (2.8.4) applied to all pairs of subsets (T, T') where $T \subset S$, $T \supset S_{m'}$ and $T' = T \cup \{v\}$, and the commutation relations:

$$z_{S, S'}\mu_{S', T} = (q_v - 1)z_{S, T}$$

$$\mu_{S, T}z_{T, T'} = z_{S, S'}\mu_{S', T'}.$$

To describe the comparison of the elements (2.8.2) when v is split multiplicative, let $d = \dim A$, $\theta \colon \bigoplus_v C_v \to C_S$ the natural mapping (2.3.4), and let $(x_i)_i$ be a basis of X_v and $(y_j)_j$ a basis of Y_v with compatible orientation. We may form the element

$$\det\big(\theta\big(\rho_v\langle x_i, y_j\rangle\big)\big) \in \operatorname{Sym}_d(C_S).$$

We have a natural mapping

$$\operatorname{Sym}_d(C_S) \otimes R \otimes \operatorname{Sym}_r(C_S) \to R \otimes \operatorname{Sym}_{r'}(C_S)$$
$$\alpha \quad \otimes \quad \beta \longmapsto \alpha \cdot \beta.$$

Define $c = c(S_m, v)$ to be the order of the image of the natural mapping,

$$B_{(S_m \cup \{v\})} \to (B/B^0)(k_v).$$

PROPOSITION 2.   *If v is split multiplicative*

$$c \cdot z_{S, S'}(\operatorname{Disc}_{S'}) = \det\big(\theta\big(\rho_v\langle x_i, y_j\rangle_v\big)\big) \cdot \operatorname{Disc}_S \in R \otimes \operatorname{Sym}_{r'}(C_S).$$

Proof.   This is, in effect, a matrix calculation. To set up the matrices, first note:

LEMMA 1.   *There are natural exact sequences,*

(2.8.5)

$$\text{(i) } 0 \to X_v \times B_S \to B_{S'} \xrightarrow{\phi} (B/B^0)(k_v)$$

$$\text{(ii) } 0 \to Y_v \to A_{S'} \to A_S \to 0.$$

Proof of Lemma 1.   We let ϕ above be the natural mapping, and put $\tilde{B}_{S'} = \ker(\phi)$. An appeal to definitions (2.4.2) yields the exact sequence

(2.8.6)
$$0 \to X_v \to \tilde{B}_{S'} \to B_S \to 0,$$

fitting into the following diagram, whose squares are Cartesian and whose columns are exact:

$$
\begin{array}{ccccc}
0 & & 0 & & 0 \\
\downarrow & & \downarrow & & \downarrow \\
X_v & \overset{\sim}{\leftarrow} & X_v & \overset{\sim}{\leftarrow} & X_v \\
\downarrow & & \downarrow & & \downarrow \\
T_{X_v}(K_v) & \leftarrow & T^0_{X_v}(K_v) & \leftarrow & \tilde{B}_{S'} \\
\downarrow & & \downarrow & & \downarrow \\
B(K_v) & \leftarrow & B^0(K_v) & \leftarrow & B_S \\
\downarrow & & \downarrow & & \downarrow \\
0 & & 0 & & 0
\end{array}
$$

The middle column, however, has a natural splitting, coming from the diagram

$$
\begin{array}{c}
T_{X_v}(\mathcal{O}_v) \to T^0_{X_v}(K_v) \\
\searrow \qquad \downarrow \\
B^0(K_v)
\end{array}
$$

which provides a canonical splitting of the exact sequence (2.8.6). This establishes (2.8.5)(i). The exact sequence (2.8.5)(ii) comes in an evident way from the definition of A_S, $A_{S'}$ (2.4.1).

LEMMA 2. *Let $x \in X_v \times \{0\} \subset B_{S'}$ and $y \in Y_v \subset A_{S'}$. Then*

$$
z_{S, S'}\langle y, x \rangle_{S'} = \theta\big(\rho_v(\langle y, x \rangle_v)\big).
$$

Proof. This is the formula described at the end of (2.4), together with the compatibility (2.4.6).

LEMMA 3. *Let $b \in B_S = \{0\} \times B_S \subset B_{S'}$. Then if $a' \in A_{S'}$ maps to $a \in A_S$ under (2.8.5)(ii), we have $z_{S, S'}\langle a', b \rangle_{S'} = \langle a, b \rangle_S$, and in particular if $y \in Y_v \subset A_{S'}$ we have $z_{S, S'}\langle y, b \rangle_{S'} = 0$.*

Proof. This is just the compatibility (2.4.6).

Now let i_T (for a subset $T \subset S$ containing S_m) denote the order of the cokernel of the injection

$$
X_v \times B_T \to B_{T'}
$$

of (2.8.5)(i), applied to T and $T' := T \cup \{v\}$. Lemmas 2 and 3 together imply

LEMMA 4. *We have*

$$i_T \cdot z_{T,T'}(d_{T'}) = \det\big(\theta(\rho_v\langle x_i, y_j\rangle_v)\big) \cdot d_T \in R \otimes \mathrm{Sym}_r(C_T).$$

Recall that $c = c(S_m, v)$ is the order of the image of the natural mapping

$$B_{S_m \cup \{v\}} \to (B/B^0)(k_v).$$

LEMMA 5. $i_T \cdot j_T = c \cdot j_{T'}.$

This formula comes from a straightforward diagram-chase using the definitions. The diagram that one should chase is the following, where the groups J_T,
$J_{T'}$, C, and I_T have as orders, their lower case counterparts:

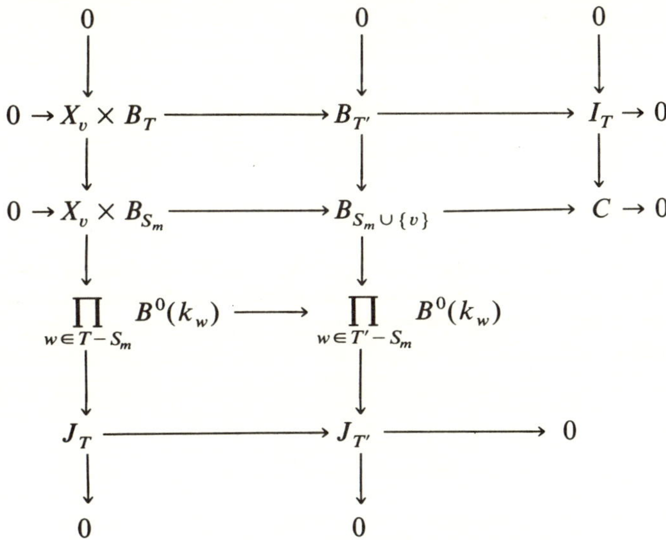

Proof of Proposition 2.   This is now a straightforward computation, given the
above discussion:

$$c \cdot z_{S,S'}(\mathrm{Disc}_{S'}) = \sum_{S_m \cup \{v\} \subset T' \subset S'} (-1)^{\#(T - S_m)} c \cdot z_{S,S'}\mu_{S',T'}(j_{T'} \cdot d_{T'})$$

$$= \sum_{S_m \subset T \subset S} (-1)^{\#(T - S_m)} \mu_{S,T} z_{T,T'}(c \cdot j_{T'} \cdot d_{T'})$$

$$= \sum_{S_m \subset T \subset S} (-1)^{\#(T - S_m)} j_T \cdot \mu_{S,T} z_{T,T'}(i_T \cdot d_{T'})$$

$$= \det\big(\theta(\rho_v\langle x_i, y_j\rangle_v)\big) \cdot \mathrm{Disc}_S(A).$$

(2.9) *Example.* Let A be an elliptic curve. Suppose that the Mordell-Weil group $A(K)$ is $\{0\}$. Let the ring R be $\mathbb{Z}$. Let $S = S_m$ be a finite set of places all of which are split multiplicative for A. For each $v \in S$, let $q_v = q_v(A) \in K_v^*$ be the v-adic multiplicative period for A. Let $\underline{q}_v \in I$ denote the K-idele whose entry in the v-th component is q_v and whose entry in all other components is 1. Let $\gamma_v \in C_S$ denote the image of the K-idele $\underline{q}_v$ under the natural projection $I \to C_S = K^* \backslash I / \prod_{v \notin S} U_v$ (here $U_v = K_v^*$ for v archimedean and $U_v = \mathcal{O}_v^*$ for v nonarchimedean, $v \notin S$).

Viewing $\gamma_v \in C_S = \mathrm{Sym}_1(C_S)$ we have then

PROPOSITION. $\mathrm{Disc}_S(A) = \prod_{v \in S} \gamma_v \in \mathrm{Sym}_r(C_S)$, where $r = \#S$.

Proof. Straightforward. Indeed, $A_S = B_S$ is a free $\mathbb{Z}$-module of rank r with basis consisting of the elements $q(v)$, $v \in S_m$, defined near the end of 2.4, and, as discussed there,

$$\langle q(v), q(v') \rangle_S = \begin{cases} 0, & \text{if } v \neq v'; \\ \gamma_v, & \text{if } v = v'. \end{cases}$$

Chapter 3. The Conjectures (*the two sides related*).

(3.1) *Statements of the conjectures.* Let $A_{/\mathbb{Q}}$ be an elliptic curve, which we identify with its dual $B_{/\mathbb{Q}}$. Let S be a finite set of rational primes, and let $S_m \subset S$ be the subset of primes with respect to which A has split multiplicative reduction. Thus (cf. (2.4)),

$$C_S = \left(\prod_{p \in S - S_m} \mathbb{F}_p^* \times \prod_{p \in S_m} \mathbb{Z}_p^* \right) / (\pm 1).$$

For each $p \in S_m$ fix an integer $e_p \geq 0$, and put:

$$M := \prod_{p \in S - S_m} p \prod_{p \in S_m} p^{e_p},$$

$$G_M := (\mathbb{Z}/M\mathbb{Z})^* / (\pm 1).$$

We have a natural mapping $C_S \twoheadrightarrow G_M$.

Let R be a subring of $\mathbb{Q}$ in which the order of the torsion subgroup of the Mordell-Weil group $A(\mathbb{Q})$ is invertible. If $I \subset R[G_M]$ is the augmentation ideal, and m an integer, let

$$\eta_m : R \otimes \mathrm{Sym}_m(C_S) \to I^m / I^{m+1}$$

be the homomorphism induced from the natural mapping $C_S \to G_M \to I/I^2$.

From Chapter II, recall the "corrected discriminant"

$$\mathrm{Disc}_S(A) \in R \otimes \mathrm{Sym}_r(C_S),$$

where $r = \mathrm{rank}\ A_S = \mathrm{rank}\ A(\mathbf{Q}) + \#(S_m)$. Define also

$$\phi_{S_m} = \#\ \mathrm{coker}\!\left(B(\mathbf{Q}) \to \prod_{p \notin S_m} (B/B^0)(\mathbb{F}_p) \right),$$

and suppose that R contains the coefficients of the modular element,

$$\theta_{A,M} \in R[G_M].$$

CONJECTURE 4. (A "Birch-Swinnerton-Dyer type" conjecture, relative to G_M).
*Let $r = \mathrm{rk}(A(K)) + \#S_m$. The modular element $\theta_{A,M} \in R[G_M]$ lies in the r-th
power of the augmentation ideal, $I^r \subset R[G_M]$, and if $\tilde{\theta}_{A,M}$ denotes its image in
I^r/I^{r+1}:*

$$(3.1.1) \qquad \tilde{\theta}_{A,M} \overset{?}{=} \#(\text{III}) \cdot \phi_{S_m} \cdot \eta_r\!\left(\mathrm{Disc}_S(A)\right) \in I^r/I^{r+1}.$$

Here III is the Shafarevich group, which we assume to be finite.

Remark. To see the analogy between the conjectured formula (3.1.1) in the
case where S_m is empty and the classical Birch and Swinnerton-Dyer conjecture,
for any r, the reader might return to (2.7) where a heuristic description of
$\mathrm{Disc}_S(A)$ was given.

Note that if $S_m = \varnothing$, then $A_{S_m} = A(\mathbf{Q})$, but B_{S_m} is of index

$$\left(\prod_p m_p\right)\!\Big/\phi_{S_m}$$

in $B(\mathbf{Q})$, were $m_p = \#(B/B^0)(\mathbb{F}_p)$. If this index is invertible in R then $\phi_{S_m} \cdot
\mathrm{Disc}_S(A)$ is equal to

$$\prod_{p \in S} (p - 1 - n_p) \cdot \prod_p m_p$$

times the discriminant of a single pairing $A(\mathbf{Q}) \times B(\mathbf{Q}) \to R \otimes G_M$. For example, if $r = 0$, this discriminant is necessarily $1/\tau^2$, where $\tau = \#A(\mathbf{Q})$ and
conjecture 4 reads:

$$\bar{\theta}_{A,M} \overset{?}{=} \#\text{III} \cdot \prod_{p \in S} (p - 1 - n_p) \cdot \prod_p m_p/\tau^2$$

in $I^0/I^1 = R$. By the compatibility proposition 1 just below, this follows for general S (with $S_m = \varnothing$) from the special case where $S = \varnothing$. In this special case (cf. remark (1) in (1.2)) the formula above reads as follows:

$$\frac{1}{2}[0] = \frac{\#(\text{III})}{\tau^2} \prod_p m_p.$$

But this is the classical conjecture, since (cf. [Ta])

$$\frac{1}{2}[0] = \frac{L_f(1)}{\int_{A(R)}|\omega|}.$$

Thus, for $r = 0$, conjecture 4 is implied by the classical conjecture of Birch and Swinnerton-Dyer and is in fact equivalent to it if $R = \mathbb{Z}[1/\tau]$, *and if $n_p \neq p - 1$ for all $p \in S$.*

Remarks on compatibility. Suppose that we are given a situation $(A_{/\mathbb{Q}}, R, S, M)$ as above. Let p be a prime number not in S, and let $e_p > 0$ be an integer, supposed equal to 1 unless p is of split multiplicative reduction for A. Let $S' = S \cup \{p\}$ and $M' = M \cdot p^{e_p}$.

For short, let $\text{LHS}_S, \text{RHS}_S$ denote the left hand side and right hand side, respectively, of the conjectured formula (3.1.1) for $(A_{/\mathbb{Q}}, R, S, M)$ and let $\text{LHS}_{S'}, \text{RHS}_{S'}$ be the corresponding quantities for $(A_{/\mathbb{Q}}, R, S', M')$. Let $z_{S, S'}$ denote maps induced by the canonical map $G_{M'} \to G_M$.

PROPOSITION 1. *If p is not of split multiplicative reduction for A, then*

$$z_{S, S'}(\text{LHS}_{S'}) = (p - 1 - n_p) \cdot \text{LHS}_S$$

$$z_{S, S'}(\text{RHS}_{S'}) = (p - 1 - n_p) \cdot \text{RHS}_S.$$

Proof. This is just (1.3), (1) and (2), and (2.8) Proposition 1.

Now let $f_p \in G_M$ be the image of Frobenius at p (i.e., the image of the idele $\underline{p}$ whose entries are: p in the p-th component and 1 in all other components). Let $m_p = \#(B/B^0)(\mathbb{F}_p)$.

PROPOSITION 2. *If p is of split multiplicative reduction for A, then*

$$z_{S, S'}(\text{LHS}_{S'}) = ([f_p \cdot] - [1]) \cdot \text{LHS}_S$$

$$m_p \cdot z_{S, S'}(\text{RHS}_{S'}) = m_p \cdot ([f_p] - [1]) \cdot \text{RHS}_S.$$

Proof. The first equality is just 1.4, (2). For the second equality, first note that $c \cdot \phi_{S_m} = m_p \phi_{S_m \cup \{p\}}$, where the constant c as in Proposition 2 of (2.8). Also, the matrix $\theta(\rho_p \langle x_i, y_j \rangle_p)$ is just a 1×1 matrix, i.e., a scalar, whose image in I/I^2 is:

$$[f_p^{m_p}] - [1] = m_p \cdot ([f_p] - [1]).$$

The second equality then follows from the above discussion, and from Proposition 2 of (2.8). We do not know whether one can strike the m_p's from the second equality.

A special case of conjecture 4 is:

CONJECTURE 5. *Let $A(\mathbf{Q})$ be finite, of order τ, supposed invertible in R. Let M, S as above, and suppose that $S = S_m$ is composed of split multiplicative primes for A. Let $r = \#(S)$. Then the modular element $\theta_{A,M}$ is contained in the r-th power of the augmentation ideal, $I^r \subset R[G_M]$, and if $\tilde{\theta}_{A,M}$ denotes its image in I^r/I^{r+1}, we have the formula:*

$$(3.1.2) \qquad \tilde{\theta}_{A,M} \overset{?}{=} \#(\text{III}) \cdot \prod_{p \notin S_m} m_p \cdot (1/\tau^2) \cdot \eta_r\left(\bigotimes_{p \in S_m} \gamma_p \right)$$

where $m_p = \#(B/B^0)(\mathbb{F}_p)$, and the term $\bigotimes \gamma_p$ is, as in (2.9), the symmetric tensor product of the images of the ideles $\underline{q}_p$, for $p \in S_m$, where $\underline{q}_p$ has entry 1 in all components except for the p-th, at which its entry is q_p, the p-adic multiplicative period for A.

A surprising aspect of conjecture 5 is that the various p-adic multiplicative parameters are all combined in one (conjectural) formula, the left-hand side of which is computed by modular symbols. One can give conjecture 5 a slightly simpler shape if one notes that, taking $S = \varnothing$, conjecture 3 for $\theta_{A,1}$ becomes, in the case where $A(\mathbf{Q})$ is finite, the classical Birch-Swinnerton-Dyer conjecture:

$$(3.1.3) \qquad \frac{[0]_A}{2} \overset{?}{=} \#(\text{III}) \cdot \prod_p m_p \cdot (1/\tau^2).$$

Since τ is invertible in R, conjecture (3.1.3) implies

$$(3.1.4)(?) \qquad [0]_f \Big/ \left(2 \prod_{p \in S_m} m_p \right) \in R$$

Combining (3.1.2) and (3.1.3) we are led to

CONJECTURE 6. *Suppose R contains the coefficients of $\theta_{A,M}$, and contains $[0]_A/(2\prod_{p \in S_m} m_p)$ (but not necessarily $1/\tau$). Suppose $S = S_m$. Let $s = \#S$. The modular element $\theta_{A,M}$ lies in I^s and:*

$$\tilde{\theta}_{A,M} \overset{?}{=} \left([0]_A \Big/ 2 \prod_{p \in S_m} m_p \right) \cdot \eta_s\left(\bigotimes_{p \in S_m} \gamma_p \right).$$

If $A(\mathbf{Q})$ is not finite, then both sides are conjecturally 0.

Remark on compatibility. Let A, A' be isogenous elliptic curves over $\mathbf{Q}$ such that the Néron lattice Λ_A is contained in the Néron lattice $\Lambda_{A'}$. Then, using the known 'compatibility under isogenies' of the classical conjecture of Birch and Swinnerton-Dyer, the reader can check that conjectures 5 and 6 for (A, R, S) imply the same conjectures for (A', R, S). Applying conjecture 6 to S and to disjoint subsets $S_1, S_2 \subset S$ such that $S_1 \cup S_2 = S$, one may deduce (conjectural, of course) quadratic congruence relations satisfied by modular symbols. Here is an example:

Let $S = (p_1, p_2)$, put $q_i = q_{p_i}$, $m_i = m_{p_i}$ $(i = 1, 2)$. Let $\tilde{p}_1 \in (\mathbf{Z}/p_2\mathbf{Z})^*$ and $\tilde{p}_2 \in (\mathbf{Z}/p_1\mathbf{Z})^*$ be the images of p_1 and p_2, respectively.

Let W be a quotient ring of R in which 2 is invertible. Let $\lambda_i: (\mathbf{Z}/p_i\mathbf{Z})^* \to W^+$ be logarithmic characters, i.e., homomorphisms to the additive group of W.

CONJECTURE 7. *Suppose that p_1, p_2 are of split multiplicative reduction for A. Then we have the following formula in W:*

$$(3.1.5) \qquad \sum_{\substack{a \bmod p_1 p_2 \\ (a, p_1 p_2) = 1}} ([0]_A \cdot [a/p_1 p_2]_A - [a/p_1]_A \cdot [a/p_2]_A) \lambda_1(a) \cdot \lambda_2(a))$$

$$\overset{?}{=} [0]_A^2 \cdot \lambda_1(\tilde{p}_1) \lambda_2(\tilde{p}_1).$$

(3.2) *Experimental evidence for the conjectures.* We have checked that conjectures 4 and 6 are true in many special cases, using programs written in the Pascal language and run on a Macintosh 512K computer. Glenn Stevens was of great help to us in this project. He had made for his own researches a floppy disk which computes the modular symbols for the elliptic curves of square-free conductor $N < 50$ listed in the Antwerp tables ([A IV], table 1). We heartily thank him for giving us a copy of this disk to use in computing the modular elements in our experiments. We thank him equally heartily for explaining to us how to interpret satisfactorily our computational results involving the twists by quadratic characters of the symbols supplied by the disk.

Our first large-scale experiment was to try to check a conjecture like conjecture 6, with $S = \{p\}$ consisting of one-split-multiplicative prime, and $M = p$, in a large number of cases. We did this quite early and later realized that we had checked only conjecture 6 multiplied through by 2, rather than conjecture 6 itself. This weaker version can be expressed as the equality

$$(*) \qquad \prod_{a=1}^{p-1} a^{[a/p]_A^+} = \pm(\tilde{q}_p)^{[0]_A}, \quad \text{in } \mathbf{F}_p^*/(\pm 1),$$

where $\tilde{q}_p$ is the residue mod p of the "unit part" of the multiplicative parameter q_p of the curve at p. Using the remark after conjecture 6, we can assert that we have checked that $(*)$ holds for any twist by a quadratic character χ_d with odd conductor $|d|$ such that $I < |d| < 100$ and $(|d|, N) = 1$, of any curve in the Antwerp tables of conductor $N = 11, 14, 15, 17, 19, 26, 38, 39, 43$, or 46, and for each prime p which the split-multiplicative for the twist in question. In every case we could (and did) take $R = \mathbb{Z}$. Incidentally, in every case, the quantity $[0]_A$ was even, so that conjecture 6 itself, which is obtained by replacing $p - 1$ by $(p - 1)/2$ and $[0]_A$ by $1/2[0]_A$ in $(*)$, makes sense with $R = \mathbb{Z}$. For $p \equiv 1 \pmod 4$ this is a real strengthening and should be tested.

Our next experiment was to test conjecture 4 for some cases $r = (\mathrm{rk}\ A(\mathbb{Q})) = 1$ and $S = \{ p \}$ a set consisting of one prime p, at which the reduction of A is *not* split multiplicative. Here we are free to vary p, so we worked with only two different curves A, but with many primes p. The curves are those designated by 37A and 43A in the Antwerp tables, whose global minimal equations are

$$(37A) \qquad y^2 + y = x^3 - x \quad (\Delta = 37),$$

and

$$(43A) \qquad y^2 + y = x^3 + x^2 \quad (\Delta = -43).$$

Let $A_{/\mathbb{Q}}$ be one of these. Then A is the only curve in its $\mathbb{Q}$-isogeny class, and $A(\mathbb{Q})$ is infinite cyclic, generated by the point $P = (0, 0)$. At the one prime of bad reduction (37 or 43 as the case may be) A has *non-split multiplicative reduction*, with connected fiber.

Let p be a prime number and $n_p = \#(A(\mathbb{F}_p))$. For $S = \{ p \}$, Conjecture 4 with $R = \mathbb{Z}$ states, when unravelled (assuming $\mathrm{III} = (0)$) that the following equality holds in $\mathbb{F}_p^* / (\pm 1)$:

$$\prod_{a=1}^{(p-1)/2} a^{[a/p]_A} = \pm g\big(\langle P, n_p P \rangle\big)^{-1},$$

where $g\colon C_S \to \mathbb{F}_p^* / (\pm 1)$ is the canonical map. We have checked that this is true for all primes p in the interval $30 \leqslant p \leqslant 1171$, for the curve 37A, and in the interval $30 \leqslant p \leqslant 863$, for the curve 43A.

Glenn Stevens' disk furnished directly the integer exponents $[a/p]_A$ needed to compute the left hand side. For the right side, let d_ν denote the square-root of the denominator of the rational number $x(\nu P)$. Then $\langle P, n_p P \rangle_S$ is represented by the idele whose p-component is 1 and whose other components are, for suitable j,

$$\frac{d_{1-j}\, d_{n_p - j}}{d_{-j}\, d_{1 + n_p - j}}.$$

This follows from the prescription at the end of §2.4, taking

$$P = P, \quad Q = n_p P, \text{ and } P' = -jP.$$

Thus j must be chosen so that none of the four d's is $\equiv 0 \pmod{p}$, i.e., such that none of the four points is zero modulo p. Dividing the idele in question by the appropriate principal idele we find its image in $\mathbb{F}_p^*/(\pm 1)$ is

$$g(\langle P, n_p P \rangle_S) = \pm \frac{d_j\, d_{n_p+1-j}}{d_{j-1}\, d_{n_p-j}} \pmod{p}.$$

How can one compute $d_\nu \pmod{p}$ for large ν? For example, in case of the 37-curve, for $p = 1171$ we find $n_p = 1194$; how do we compute the residue class modulo p of the *denominators* of the rational numbers

$$x(1192P) \quad \text{and} \quad x(1193P)?$$

For the numbers $x(\nu P)$ themselves we have (recall $x(P) = 0$) the formula

$$x(\nu P) = -\frac{f_{\nu+1}f_{\nu-1}}{f_\nu^2},$$

where the integers f_ν (which are values of "division polynomials", cf. [La3], Ch. 11) can be computed recursively. Doing the recursions $\pmod{p}$, it is easy to get the f_ν's $\pmod{p}$ very quickly with a small computer. If f_ν and $f_{\nu+1}$ are relatively prime for all ν, then $d_\nu = \pm f_\nu$, and we are done. The necessary relative primeness is in fact true in the cases at hand. Indeed, it holds whenever A has semi-stable reduction everywhere and P is a point which is in $A^0(\mathbb{Q}_p)$ for all p, but in $A^1(\mathbb{Q}_p)$ for no p.

We tested one curve A with rk $A(\mathbb{Q}) = 2$, namely $A = X_0(11)$ twisted by χ_{-47}. This curve has minimal equation

$$y^2 + (47)^2 y = (x - 47)(x - 94)(x + 1888) \quad (\Delta = -11^5 47^6).$$

The group $A(\mathbb{Q})$ is almost certainly $\approx \mathbb{Z} \times \mathbb{Z}$, generated by any two of the three points on the line $y = 0$. Assuming this, and that $\text{III} = (0)$, we checked conjecture 4 for this curve for $S = \{p\}$, $13 \leqslant p \leqslant 97$, and $R = \mathbb{Z}[1/6]$. (Except that we only got $R = \mathbb{Z}[1/30]$ for the case $p = 61$; the reason for the $R = \mathbb{Z}[1/m]$ is that we computed only

$$\det\begin{pmatrix} \langle P_1, 2n_p P_1 \rangle_S & \langle P_1, 2n_p P_2 \rangle \\ \langle P_2, 2n_p P_1 \rangle_S & \langle P_2, 2n_p P_2 \rangle \end{pmatrix} \in \text{Sym}_2\big(\mathbb{F}_p^*/(\pm 1)\big)$$

which is a certain multiple of the corrected discriminant. Thus our computation checked the conjecture only if the gcd of that multiple and $p - 1$ is invertible in R.)

More interesting perhaps are tests of cases in which $r = 2$ and S consists of two primes p and p', and $M = pp'$. For any finite abelian group G, the map

$$\mathrm{Sym}_2 G \to I_G^2/I_G^3$$

is an isomorphism. Thus we can view the conjecture in case $M = pp'$ as asserting an equality in $\mathrm{Sym}_2(G_{pp'})$. This group maps canonically onto

$$\mathrm{Sym}_2(G_p \times G_{p'}) = G_p^{\otimes 2} \times (G_p \otimes G_{p'}) \times G_{p'}^{\otimes 2},$$

with a kernel of order 2. Ignoring that kernel, the difference between the conjecture for $S = \{p, p'\}$ and the union of the conjectures for $S = \{p\}$ and $S = \{p'\}$ separately, is the conjectured equality in the "cross-term", $G_p \otimes G_{p'}$. This is a cyclic group of order

$$d = \gcd\left(\frac{p-1}{2}, \frac{p'-1}{2}\right).$$

Thus we choose pairs of primes p and p' for which d is large, in order to have a good test of the new feature, i.e., the "interaction" between p and p'. But we did not confine our tests to the cross-term; we tested equality in $\mathrm{Sym}_2 G_{pp'}$, not even ignoring the kernel of order 2 alluded to above, in the following cases where $r = \#(S) = 2$.

Cases with $\mathrm{rk}(A) = 1$, $S = \{p, p'\}$, $S_\infty = p$.   The curve 43A twisted by $\sqrt{-15}$ has equation

$$y^2 + 225y = x^3 + 75x^2 + 1800x$$

and the prime 43 is split-multiplicative. Assuming $A(\mathbb{Q})$ generated by $P = (0,0)$ and $\mathrm{III} = (0)$, we checked our conjecture for the four sets $S = \{43, p'\}$ with $p' = 29, 71, 113,$ and 127.

The curve 46A, twisted by $\sqrt{-15}$ has the equation

$$y^2 + xy + 225y = x^3 + 56x^2 - 1350x$$

and has split multiplicative reduction at 23. Assuming the same things, we checked Conjecture 4 for three sets $S = \{23, p'\}$ with $p' = 67, 89,$ and 199.

Cases with $\mathrm{rk}(A) = 0$, $S = S_\infty = \{p, p'\}$.   Glenn Stevens obligingly supplied us with a supplement to his disk, adding to it the curve 91A, with equation

$$y^2 + y = x^3 + x.$$

This curve is alone in its $\mathbb{Q}$-isogeny class. The smallest d's for which its twist by

χ_d has both 7 and 13 as split multiplicative primes and also has $[0]_A \neq 0$ are $d = -8, -11, -15, -67, -71, -148, -151,$ and $-163,$ for which $[0]_A = 4, 4, 4, 64, 4, 36, 4,$ and 64, respectively. We have checked conjecture 6 with $S = S_\infty = \{7, 13\}$ for each of these twisted curves.

We have not yet computed a case with $\mathrm{rk}(A(\mathbb{Q})) = 2$ and S consisting of two non-split primes.

References

[A IV] *Modular functions of One Variable IV*, Proceedings of the International Summer School, U. of Antwerp, RUCA, 1972. Lecture Notes in Math. **476**, Springer, 1975.

[A-L] O. Atkin and W. Li, *Twists of newforms and pseudo-eigenvalues of w-operators*, Invent. Math. **48** (1978), 221–243.

[Ha] A. Hales, *Stable augmentation quotients of abelian groups*, Pac. Journ. of Math. **118**, No. 2, (1985), 401–410.

[La 1] S. Lang, *Introduction to Modular Forms*, Springer-Verlag, Berlin-Heidelberg-New York, 1976.

[La 2] ______, *Fundamentals of Diophantine Geometry*, Springer-Verlag, New York-Berlin-Heidelberg-Tokyo, 1983.

[LA 3] ______, *Elliptic Curves. Diophantine Analysis*, Springer-Verlag, Berlin-Heidelberg-New York, 1978.

[Man] Y. Manin, *Values of p-adic Hecke Series at Integral Points of the Critical Strip*, Matematiceskii Sbornik. **93**, 1974 (135).

[M-T] B. Mazur, and J. Tate, "Canonical Height Pairings via Biextensions," in *Arithmetic and Geometry*, Progr. Math., vol. **35**, (195–237) Boston-Basel-Stuttgart: Birkhäuser 1983.

[M-T-T] B. Mazur, J. Tate, and J. Teitelbaum, *On p-adic analogues of the conjectures of Birch and Swinnerton-Dyer*, Invent. Math. **84** (1–48), 1986.

[Mu] D. Mumford, *Abelian Varieties*, Oxford University Press, 1970.

[M-F] D. Mumford, and J. Fogarty, *Geometric Invariant Theory*, (Second Enlarged Edition) Springer Verlag. Berlin-Heidelberg-New York 1982.

[N] D. G. Northcott, *Finite Free Resolutions*, Cambridge Univ. Press, Cambridge-New York, 1976.

[Pa] I. B. S. Passi, *Group Rings and Their Augmentation Ideals*, Lecture notes in Math. **715**, Springer, 1979.

[Sh] G. Shimura, *Introduction to the Theory of Automorphic Forms*. Publ. Math. Soc. Japan **11**, Tokyo-Princeton, 1971.

[SGA7] *Groupes de Monodromie en Géométrie Algébrique*, Séminaire de Géométrie Algébrique du Bois-Marie 1967–1969. Directed by A. Grothendieck with the collaboration of M. Raynaud and D. S. Rim. Lecture Notes in Math. **288** Springer-Verlag, Berlin-Heidelberg-New York, 1972.

[Ta] J. Tate, *The Arithmetic of Elliptic Curves*, Invent. Math. **23** (1974), 179–206.

Department of Mathematics, Harvard University, Cambridge, Massachusetts 02138